AF579723

Coherence and Time Delay Estimation

IEEE PRESS
445 Hoes Lane, PO Box 1331
Piscataway, NJ 08855-1331

Underwater Acoustics Signal Processing
Technical Committee of the IEEE, *Sponsor*

Chair

John Fay

Technical Reviewers
José Moura
Carnegie Mellon University

Hua Lee
University of California

Kent Scarbrough
Tracor Applied Sciences

Coherence and Time Delay Estimation

An Applied Tutorial for Research, Development, Test, and Evaluation Engineers

Edited by

G. Clifford Carter
NUWC
and
University of Connecticut

A Selected Reprint Volume
Underwater Acoustics Signal Processing Technical Committee of the IEEE, *Sponsor*

The Institute of Electrical and Electronics Engineers, Inc., New York

This book may be purchased at a discount from the publisher when ordered in bulk quantities. For more information contact:

IEEE PRESS Marketing
Attn: Special Sales
P.O. Box 1331
445 Hoes Lane
Piscataway, NJ 08855–1331

Printed in the United States of America

10 9 8 7 6 5 4 3 2 1

ISBN 0–7803–1006–3

IEEE Order Number: PC0355–8

Library of Congress Cataloging-in-Publication Data

Coherence and time delay estimation: an applied tutorial for research, development, test, and evaluation engineers/edited by G. Clifford Carter.
p. cm.
Includes bibliographical references and index.
ISBN 0–7803–1006–3
1. Signal processing—Data processing. 2. Spectral theory (Mathematics)—Data processing. 3. Estimation theory—Data processing. 4. Delay differential equations. I. Carter, G. Clifford.
TK5102.5.C58 1993
621.382′2—dc20 92–46209
CIP

Contents

Preface

Intended Audience

This reprint book is compiled for beginning Research & Development (R&D) engineers and graduate students and for working Test and Evaluation (T&E) Electronics Engineers (EEs), all of whom would benefit from a tutorial explanation of time delay and coherence estimation. It can be used as a supplement to either a senior-level undergraduate or a first-year graduate-level signal processing electrical engineering course. The academic preparation expected is at least that of a senior-level undergraduate in electrical engineering. It is expected that the reader reads the *IEEE Transactions* (e.g., *Transactions on Signal Processing*) or is working in the spectral estimation, coherence estimation, time delay, or systems identification field. The position titles of readers include (1) R&D engineer, (2) T&E engineer, (3) sonar or radar engineer, (4) acoustician, and (5) geophysical engineer/physicist/geophysicist.

Background

This book of reprints was introduced as an idea more than 10 years ago. At that time, the IEEE Underwater Acoustic Signal Processing Technical Committee of the IEEE Acoustics, Speech, and Signal Processing Society (now the IEEE Signal Processing Society) began work on a sonar signal processing book that would emphasize passive sonar. The project soon expanded to include active sonar and then radar; with this expansion, the number of contributing authors became so large that it was impossible to select the benchmark papers.

In 1990, following my 1988 election to Fellow of the IEEE for contributions to the theory of coherence and time delay estimation, I was asked by the IEEE Educational Activities Board (EAB), through Dr. John Woods and Dr. John Fay of the IEEE Signal Processing Society training committee (in concert with a recommendation by Dr. Kent Scarbrough of the Underwater Acoustic Signal Processing Committee), to prepare a 3-hour educational course on coherence and time delay estimation. Because the topic was so focused, I was able to review all the contributions that have had a significant impact in this important technical area.

This book is a collection of the best papers on coherence estimation and time delay estimation. It can be used as a supplement to the IEEE EAB tutorial (Carter [1991]) given at ICASSP 91 in Toronto, which has been videotaped and marketed through and for the IEEE (product #HV0230-3). The tutorial focused on signal processing methods for the nonparametric estimation of auto and cross power spectral density functions and the use of these methods in extracting fundamental characteristics of interest for the applied signal processing community, including coherence and time delay. The tutorial also reviewed the generalized framework for nonparametric spectral estimation and the special cases of the weighted overlapped segment averaging (WOSA) and lag reshaping methods for coherence and time delay estimation. Application of these methods to underwater acoustic problems of passive detection and localization are discussed in Part 1.

Because current concepts in time delay estimation and coherence estimation are being inserted into computer hardware and software so rapidly, working engineers often need assistance in learning to use such equipment and interpret these new functions. This book will help users of the new technology to understand the fundamentals of coherence and time delay estimation.

Technical and Personal Note

For some time I have been concerned about the lack of attention that coherence and time delay estimation receive at the undergraduate educational level. I believe this instructional bias is due both to the mathematical immaturity of most undergraduates and to the computational complexities and subtleties of estimation. For example, although a system consisting of a time delay is linear, the input and output cannot be characterized by ordinary or partial differential equations of the type that the student is exposed to early in a classical undergraduate engineering program. Moreover, as a fourth-year student, one is typically only just becoming exposed to the concepts of correlation coefficient and second-order statistics and the concept of a correlation coefficient in the frequency domain. In addition, because of the widespread application of modern computers to the problem of estimating coherence and time delay in recent years, it is now commonplace for T&E engineers and even highly trained technicians to employ these technologies; however, we find that the estimation process is often computationally complex. Many of these complexities are discussed in the tutorial and reprint sections of this book.

As a personal note, after I had accepted the invitation (in the early fall of 1990) to give a 3-hour training course on coherence and time delay estimation, but before the lecture actually occurred in May of 1991, I developed a medical problem that was diagnosed with time delay estimation techniques. The problem was subsequently resolved. It gives me immense satisfaction to know that the technologies that I helped to develop for defense have application in the medical field.

I would also emphasize how rapidly technology changes. The half-life of an engineer (that is, the time by which half our learned knowledge is obsolete) is very brief. This condition requires that students master the art of learning. To remain technically well informed, an engineer must be active in professional societies such as the IEEE, read the *IEEE Transactions,* and attend professional conferences. Even if you do not plan to become an expert in coherence and time delay estimation, the discipline of reading and extracting information from these reprints will provide important training in keeping well informed.

Acknowledgments

I gratefully acknowledge the support and encouragement of Mr. Dudley Kay of IEEE, and also the assistance of IEEE staff members Valerie Zaborski and Marybeth Hunter.

I also am very appreciative of the help and support of Dr. Neil Gerr, Prof. José Moura, Dr. John Ianniello, Dr. Kent Scarbrough, Dr. John Fay, Prof. John Woods, Prof. Stuart Schwartz, Prof. Don Tufts, Mr. John Sikorski, and to the many individuals who have coauthored work with me in this technical area.

In addition, I am especially grateful for the editorial assistance of Ms. Karen Holt and for the hard work of Ms. Joyce Manzella and Mr. Paul Christensen.

Special thanks to past and present members of the IEEE Underwater Acoustic Signal Processing Committee for their support and advice for this work. The members of this committee, as of June 5, 1992, were:

Dr. John Fay, chair
Dr. Michel Bouvet
Dr. G. C. Carter
Prof. Y. T. Chan
Dr. M. L. Cohen
Dr. Neil Gerr
Dr. Don Gingras
Dr. J. P. Ianniello
Dr. Ivars P. Kirsteins
Dr. Emil Modugno
Prof. José M. F. Moura
Dr. Dan E. Ohlms
Dr. Norman Owsley
Dr. Kent Scarbrough
Prof. Peter M. Schultheiss
Prof. Stuart Schwartz
Dr. Hugh M. South
Prof. Don Tufts
Prof. Richard Vaccaro
Dr. James Zeidler

Table of Abbreviations

AAMI	average amount of mutual information
ABCTDE	adaptive basic cross-correlation time delay estimation
ABR	auditory brain stem response
ACV	asymptotic covariance
ADDLD	adaptive digital delay-lock discriminator
AEP	auditory evoked potential
AFFDL	Air Force Flight Dynamics Laboratory
AM	amplitude modulation
AMDF	average magnitude difference function
AMFR	amplitude-modulated following response
AML	approximate maximum likelihood
ANSI	American National Standards Institute
a.p.	action potential
APSD	auto power spectral density
AR	autoregressive
ARMA	autoregressive moving average
AS	amplitude square
ASA	Acoustical Society of America
ASI	Advanced Study Institute
ASJ	Acoustical Society of Japan
ASL	Atmospheric Sciences Laboratory
ASSP	Acoustics, Speech, and Signal Processing
ASW	antisubmarine warfare
ATDC	adaptive time delay estimation based on cumulants
AVCN	anteroventral portion of the cochlear nucleus
avg.	average
BB	Barankin bound
BCC	basic cross correlation
BF	best frequency
BM	basilar membrane
BPSK	biphase-shift keying
BSD	bearing standard deviation
BT	bandwidth-time product
BW	bandwidth
BWR	boiling water reactor
CC	cross correlation
CCBG	cross-correlation-based gradient
CDF	cumulative distribution function
CDSPLOT	constant delay, stationary processes, and long observation interval
CFAR	constant false-alarm rate
CON	complete and orthonormal
CPE	correlator performance estimate
CPSD	cross-power spectral density
CPU	central processing unit
CRLB	Cramér-Rao Lower Bound
CRP	Chan-Riley-Plant
CSM	cross-spectral matrix
dB	decibel
DC	direct correlation
deg.	degrees
DFT	discrete Fourier transform
diam.	diameter
DSTC	deskewed short-time correlator
DTD	differential time delay
EAB	Educational Activities Board
e.a.p.	extracellular action potential
ed.	editor
eds.	editors
EE	electronics engineer
EEG	electroencephalographic
EK	Eckart
eqn.	equation
EUSIPCO	European Signal Processing Conference
fig.	figure
FIR	finite impulse response
FFR	frequency-following response
FFT	fast Fourier transform
FM	frequency modulation
FR	fixed receiver
FS	fixed source
ft.	feet
GAF	generalized ambiguity function
GCC	generalized cross correlation
GKBF	generalized Kalman-Bucy filter
GPS	generalized phase spectrum
GSM	generic sonar model
GWF	generalized Wiener filter
HB	Hassab-Boucher
HH	Hamon-Hannan
HOS	higher-order spectra
HR	high resolution
HRM	high-resolution method
HRTDE	high-resolution time delay estimation
HS	hybrid sign
HT	Hannan-Thomson
Hz	hertz
ICASSP	International Conference on Acoustics, Speech, and Signal Processing
IEEE Trans.	*IEEE Transactions*
IFT	inverse Fourier transform
IGKBF	time-invariant generalized Kalman-Bucy filter
IIR	infinite impulse response
IPFM	integral pulse frequency modulator
ISR	interference-to-signal ratio
JASA	*Journal of the Acoustical Society of America*
kHz	kilohertz
km	kilometers
LHS	left-hand side
LLF	log-likelihood function
LMMSE	linear minimun mean square error
LMS	least mean square
LMSTDE	least-mean-square time delay estimation
LPRM	local power range monitor
LRT	likelihood ratio test
m	meters
MAP	maximum a posteriori
m.a.p.	membrane action potential
MC	magnitude coherence
MELECON	Mediterranean Electrotechnical Conference
ML	maximum likelihood
MLBS	maximal length binary sequence
MLE	maximum likelihood estimator

MMPCP	multitarget multisensor postcorrelation
MMSE	minimum mean square error
MR	moving receiver
MS	Meyr-Spies; moving source
ms	milliseconds
m/s	meters per second
MSC	magnitude-squared coherence
MSE	mean square error
msec.	milliseconds
MUSIC	multiple signal classification
MYW	modified Yule-Walker
NATO	North Atlantic Treaty Organization
NB	narrowband
NDE	nondestructive evaluation
NLLF	negative log-likelihood function
NMNL	no-memory nonlinear
no.	number
NUSC	Naval Underwater Systems Center
NUWC	Naval Underwater Warfare Center
OMP	optimum multitarget processor
ONR	Office of Naval Research
ORIV	overdetermined recursive instrumental
PAR	Princeton Applied Research Laboratory
PC	polarity coincidence
PD	phase data
PDF	probability density function
PHAT	phase transform
PN	pseudonoise
PPM	pulse position modulation
Proc. IEEE	*Proceedings of the IEEE*
PSD	power spectral density
PSK	phase-shift keying
pt.	part
pWD	pseudo-Wigner distribution
PWR	pressurized water reactor
R&D	research and development
r.a.p.	recorded action potential
RDT&E	research, development, test and evaluation
ref.	reference
RGN	recursive Gauss-Newton
RHS	right-hand side
RMS	root mean squared
ROC	receiver operating characteristics
RPE	recursive prediction error
RSD	range standard deviation
RTC	relative time companding
RWP	robust Wiener processor
s	seconds
SCC	standard cross correlation
SCOT	smoothed coherence transform
SDC	Smith delay compensator
sec.	seconds
SGN	algebraic sign function {SGN(A) = 1 for A > 0, SGN(A) = −1 for A < 0}
SIO	Scripps Institute of Oceanography
SIR	signal-to-interference ratio
SL	sensation level
SLOT	stationary long observation time
SM	sample mean
SNR	signal-to-noise ratio
SNR_{av}	average signal-to-noise ratio
SNR_{th}	threshold signal-to-noise ratio
SOSS	synthetic observer/stationary source
SP	Signal Processing
SPL	sound pressure level
T&E	test and evaluation
TCC	time-companding correlator
TD	time delay variable
TDE	time delay estimation
TDOA	time difference of arrival
TDPE	time delay parameter estimator
TOAD	time of arrival difference
Trans. ASME	*Transactions of the American Society* of Mechanical Engineers
vol.	volume
vs.	versus
WD	Wigner distribution
WOSA	weighted overlapped segment averaging
WP	Wiener processor
ZZLB	Ziv-Zakai Lower Bound

Part 1
Tutorial Overview of Coherence and Time Delay Estimation

1.1 Introduction

This chapter is a tutorial based upon the IEEE videotape described in the preface. Rigorous derivations can be found in the references.

This part is divided into two major sections: One treats coherence and the other, time delay. All references to coherence estimation address nonparametric estimation. Parametric estimation (of model parameters) is outside the scope of this book. In the first section of this part the emphasis is on coherence, both as a theoretical quantity and as a quantity to be estimated if it is to be used. Coherence and its estimation turn out to be intimately related to time delay and time delay estimation, which is the focus of the second major section of this part. Thus, this part fully treats coherence, time delay, and their estimation. First we examine what coherence is, how it is estimated (including example estimates), and what is known about the statistics of the estimates.

1.2 Coherence

Coherence, $\gamma_{xy}(f)$, is a complex quantity; it is the cross-power spectral density between two random processes, x and y, divided by the square root of the product of their auto power spectral densities:

$$\gamma_{xy}(f) = \frac{G_{xy}(f)}{\sqrt{G_{xx}(f)G_{yy}(f)}}. \tag{1}$$

The magnitude-squared coherence (MSC), denoted by $C_{xy}(f)$, is a quantity bounded by zero and unity; that is,

$$C_{xy}(f) \equiv |\gamma_{xy}(f)|^2 \quad \text{and} \tag{2}$$

$$0 \leq C_{xy}(f) \leq 1, \qquad \forall f. \tag{3}$$

Consider the system shown in Figure 1. Here, $x(t)$ and $y(t)$ are sample or member functions of two random processes. The function $x(t)$ is the input to the filter, with transfer function H. The output of this filter, $y_0(t)$, is differenced with $y(t)$, and the result is the error signal output, $e(t)$. The auto power spectral density of the error signal, $G_{ee}(f)$, is given next:

$$G_{ee}(f) = G_{yy}(f) + G_{xx}(f)|H(f)|^2 - H(f)G_{xy}^*(f) - H^*(f)G_{xy}(f). \tag{4}$$

If we select the minimum mean square error (MMSE) filter (i.e., if we minimize the power in the error signal $e(t)$), the optimum filter is given by

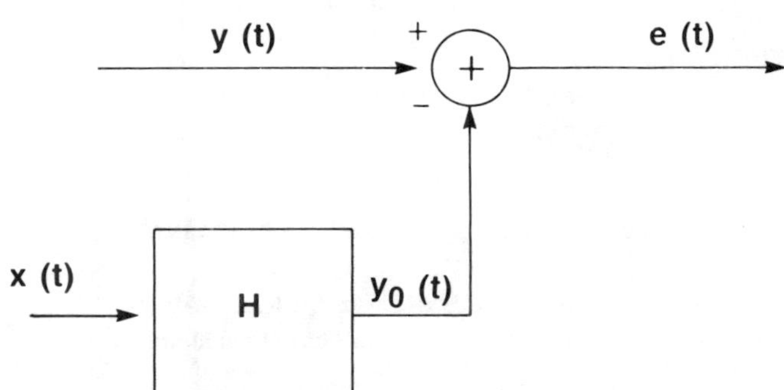

Fig. 1. Error resulting from linearly filtering $x(t)$ to match any desired signal $y(t)$ (Carter and Knapp 1975).

$$H_0(f) = \frac{G_{xy}(f)}{G_{xx}(f)}. \tag{5}$$

The complex coherence is this optimal filter normalized by a square root ratio of the two autopower spectral densities of x and y; it can be interpreted as:

$$\gamma_{xy}(f) = H_0(f)\sqrt{\frac{G_{xx}(f)}{G_{yy}(f)}}. \tag{6}$$

Hence, the complex coherence can be interpreted as a normalized optimal filter, with the autopower spectral density of the error process given by

$$G_{ee}(f) = G_{yy}(f)[1 - C_{xy}(f)]. \tag{7}$$

Thus, if the MSC, $C_{xy}(f)$ (defined in equation (2)), is equal to zero (indicating that x and y are uncorrelated), then $G_{ee}(f)$ is equal to $G_{yy}(f)$. On the other hand, if $C_{xy}(f)$ is equal to unity (indicating full coherence between x and y), $G_{ee}(f)$ is equal to zero; that is, there is no power in the error process.

Let us again consider the coherence between $x(t)$ and $y(t)$ in the model shown in Figure 1. We want to adjust H so that we have an optimum output, y_0. Using elementary algebra, G_{yy} can be written as

$$G_{yy} = CG_{yy} + [1 - C]G_{yy}, \tag{8}$$

where C is the magnitude-squared coherence and the dependence upon frequency is dropped for convenience. It can be seen that G_{yy} comprises an optimum part and an error part:

$$G_{yy} = G_{y_0y_0} + G_{ee}. \tag{9}$$

Hence, C is the proportion of the linear component of the random process y, and the quantity $1 - C$ is the error or the

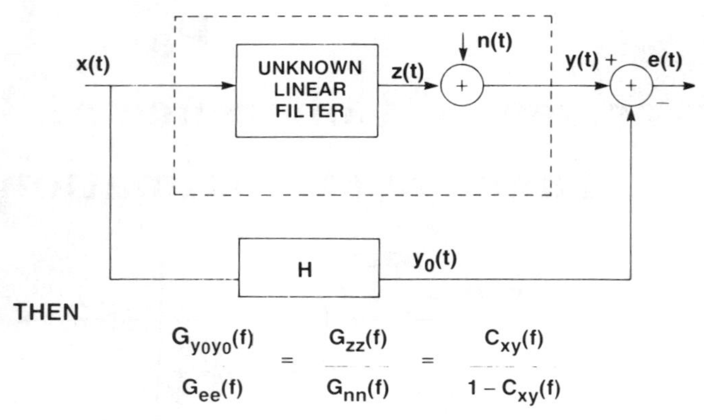

C = 0 G_{zz}/G_{nn} = 0 ie NOISE DOMINATES

C = 1 INFINITE SNR ie SIGNAL DOMINATES

Fig. 2. Error resulting from linear approximation of an unidentified linear system (Carter and Knapp 1975).

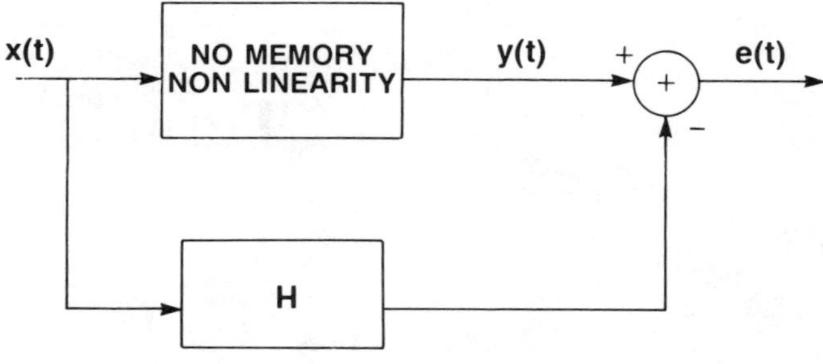

x(t) ZERO MEAN GAUSSIAN RANDOM PROCESS

Fig. 3. Error resulting from linear approximation of an unidentified NMNL system (Carter and Knapp 1975).

nonlinear portion of y. These results are discussed in greater detail in Carter and Knapp (1975).

Another system model is shown in Figure 2. Here, we observe an input, $x(t)$, and an output, $y(t)$, of an unknown system. This system is a linear filter with response $z(t)$; uncorrelated noise, $n(t)$, is added to $z(t)$ to give the output, $y(t)$. If we develop a linear filter that will model this network, we minimize the error power in this system, so that the power spectrum of the optimum output relative to the error power spectrum is as shown in Figure 2. The $C/(1 - C)$ term is a fundamental quantity that will appear repeatedly throughout this work; as before, we drop the dependence on f and x, y for notational simplicity. If the coherence between x and y is equal to zero, noise dominates, meaning there is so much attenuation in the linear system that only noise is present. If, on the other hand, the coherence between the input and the output is equal to unity, only the signal is present, and noise is absent.

Let us now consider a no-memory, nonlinear (NMNL) model (see Figure 3). Since this system is not linear, the results here are not as obvious as before. For example, let the output, $y(t)$, equal the cube of the input, $x(t)$, plus a constant, b, times the input:

$$y(t) = x^3(t) + bx(t). \tag{10}$$

Hence, $y(t)$ has a nonlinear component, $(x^3(t))$, and a linear component, $(bx(t))$. However, if the filter, H, is constructed in an optimum way (such that the complex cross-power spectral density is divided by the input autopower spectral density), then the optimum linear part is *not* $bx(t)$, as might be expected.

There is a class of random processes called separable processes, of which Gaussian processes are a subset. If the cross-correlation function between separable random processes x and y (the inverse Fourier transform of their power spectral density) is used, then the cross-correlation function between the input and the output of the NMNL network is simply a constant, K, times the autocorrelation of the input; that is,

$$R_{xy}(\tau) \equiv \text{FFT}^{-1}\,[G_{xy}(f)], \tag{11}$$

$$R_{xy}(\tau) = KR_{xx}(\tau). \tag{12}$$

It is interesting to note that for separable random processes and an NMNL system, the cross-correlation function between the input and the output is determined only by the autocorrelation of the input. The expression for the constant K is

$$K = \frac{1}{\sigma^2}\int f(x)(x - \mu)p(x)\,dx, \tag{13}$$

where σ^2 is the variance of $x(t)$, μ is the mean of $x(t)$, $y = f(x)$ is the function describing the output of the NMNL system, and $p(x)$ is the first-order probability density function (PDF). For this particular problem,

$$K = 3\sigma^2 + b; \tag{14}$$

consequently, the optimum part of the output is

$$y_0(t) = (b + 3\sigma^2)x(t) \tag{15}$$

and *not* $bx(t)$. (See also Carter and Knapp (1975).)

So far, we have considered two types of models—a linear system corrupted by additive noise and an NMNL system. Now, let us consider a new time delay model (Figure 4), where an input (source) signal, $s(t)$, travels to two different receivers, $r_1(t)$ and $r_2(t)$. The $r_1(t)$ receiver observes $s(t)$ delayed by an amount A plus an additive noise $n_1(t)$, whereas $r(t)$ observes $s(t)$ plus an additive noise $n_2(t)$. For this system, the power signal-to-noise ratio (SNR) is related as shown in Note 1 in Figure 4, which is similar to (but not the same as) the quantity $C/(1 - C)$ obtained previously. Fundamentally, different models give different results. Thus, we must consider the under-

CONSIDER NEW TIME DELAY MODEL

NOTE 1

$$\frac{G_{ss}(f)}{G_{nn}(f)} = \frac{\sqrt{C}}{1-\sqrt{C}} \quad \text{VICE} \quad \frac{C}{1-C}$$

NOTE 2

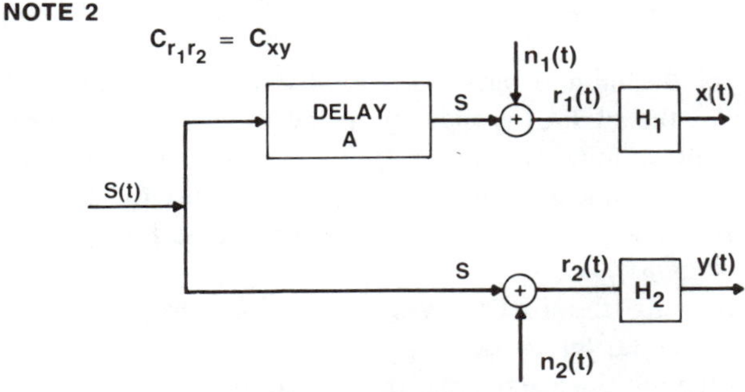

MAG.SQ.COH. INVARIANT TO LINEAR FILTERING

Fig. 4. Model of a directional signal corrupted with additive noise and processed (Carter and Knapp 1975).

lying physical and mathematical models before blindly "plugging in" formulas for SNRs, coherences, or magnitude-squared coherences.

There is another characteristic associated with the model shown in Figure 4. (As before, capital letters indicate frequency functions and the f dependence is left out for simplicity.) If $r_1(t)$ enters filter H_1 with output $x(t)$ and $r_2(t)$ enters filter H_2 with output $y(t)$, then the magnitude-squared coherence, C, between $r_1(t)$ and $r_2(t)$ is exactly the same as C between x and y (Note 2). Hence, if there is an input to two different linear filters, then C between the two inputs is the same as C between the outputs, which is a theoretical result. Care must be taken not to misinterpret forms such as 0/0, which are undefined. Care must also be taken when these results are applied in practice; a particular concern is whether it is the true MSC or estimates of the MSC that are being dealt with; for example, an estimate of zero divided by an estimate of zero will give a misleading result.

To summarize up to this point, the magnitude-squared coherence, C, is bounded by zero and unity for all frequencies. It is a measure of the SNR or a "correlatedness" between two random processes. Now that we have a theoretical description of coherence, the next logical question is, "How can coherence be estimated?"

1.3 Coherence Estimation

There are two fundamental types of spectral estimation, parametric and nonparametric. In nonparametric estimation we use prespecified processing, such as the fast Fourier transform (FFT), which gives fixed bandwidths for analyzing received signals. These fixed bandwidths are predetermined, independent of the signal, providing a robust detection and estimation

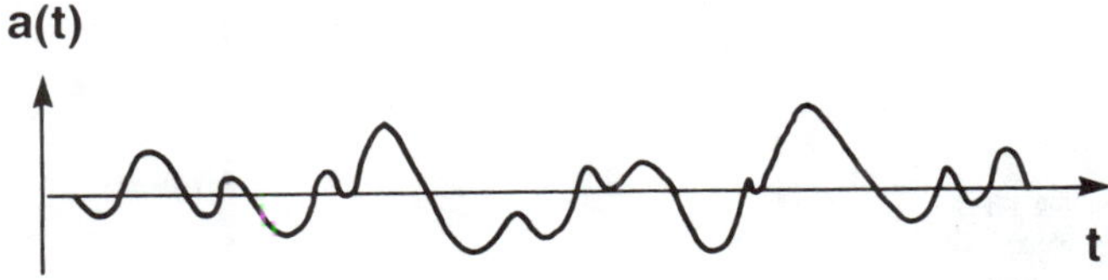

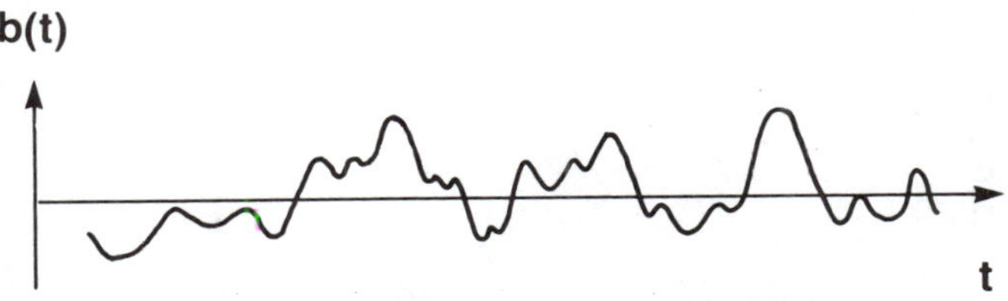

Fig. 5. Two time series from processes $a(t)$ and $b(t)$ (Carter 1972).

capability but without intrinsic ability to adapt. In parametric estimation, however, we can effectively adapt the position of model parameters (poles and zeros) to where the energy is located. This text will consider only the robust nonparametric spectral estimation techniques. These techniques have been unified by a generalized framework (Carter and Nuttall 1980a). One subset of the generalized framework is "weighted overlapped segment averaging" (WOSA), which is very popular and widely used (Carter and Nuttall 1980b). Another subset is a relatively new (approximately 10 years old) method called *lag reshaping* (Carter and Nuttall 1983). Although most of the examples in the following discussion use the WOSA method, the newer lag reshaping method is a computationally efficient technique that should be investigated more thoroughly.

We start with the two waveforms $a(t)$ and $b(t)$ shown in Figure 5. These waveforms could be sample or member functions from any physical random process, such as voltage versus time or pressure versus time. We wish to process the waveforms and estimate their coherence. In general, the limited amount of time available will influence how the data are to be processed. In the WOSA method, the waveform is divided into N segments, and each segment is multiplied by a smooth weighting function, such as a Hanning weighting function (Figure 6a). Next, the data are processed, which generally involves using an FFT, squaring (for autospectral estimation) or multiplying by the FFT from the other process (for cross-spectral

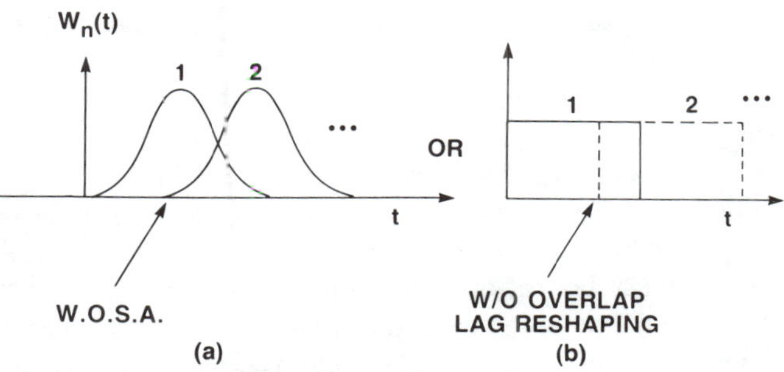

Fig. 6. Overlapped time-weighting functions. (a) Smooth. (b) Rectangular.

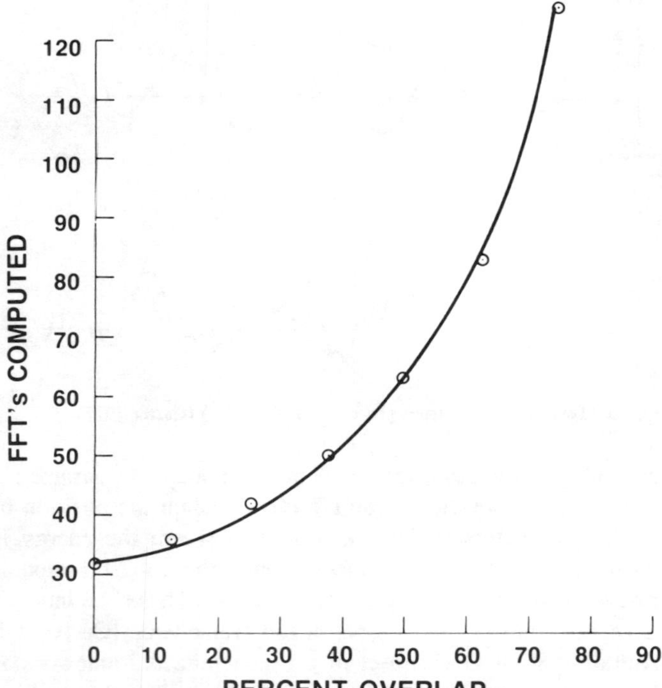

Fig. 7. Effect of overlap on amount of required FFTs (using 32 segments) (Carter, Knapp, and Nuttall 1973a).

estimation) and then time averaging over each fixed frequency bin. Intuitively, we can see that multiplying the time segments by the smooth weighting function "wastes" the data at the segment edges. We compensate for that waste by overlapping the time segments. Because the second segment overlaps the first, the effective time weight has about the same amount of weight for all the data. (This method, while not originally called WOSA, is described in Carter, Knapp, and Nuttall (1973a), Welch (1967), and Carter and Nuttall (1980b).)

In contrast to the smooth time-weighting function used in the WOSA method, in the lag reshaping method we use a rectangular time-weighting function (see Carter and Nuttall (1983)). Since no multiplication is involved before processing the data, rectangular weighting is sometimes referred to as "no" weighting.

As shown in Figure 6b the overlap is exaggerated; however, no overlap is needed, since each segment with rectangular weighting should abut the next. Because no overlap is needed, there is no waste of the data. With abrupt rectangular time weighting, we do not taper down the ends of the data segments.

When overlapped-segment averaging is used, the data are overlapped usually by 50 percent, which requires twice as many FFTs (assuming FFTs of the same size). Figure 7 shows the number of FFTs required as a function of the amount of overlap. For the case of 50 percent overlap and 32 segments, 64 FFTs must be performed. More overlap requires more FFTs. Note the rapid increase in computational cost with each attempt to extract more from the data. The results will determine if doing this is worthwhile. Note also that multiplying time segments is equivalent to a frequency domain convolution that broadens the frequency resolution; hence, if the frequency resolution is to remain constant, the FFT size must become larger. When the FFT size becomes larger for a fixed duration of data, the number of FFTs is reduced.

The computation of the FFTs is performed as follows:

$$A_n(\ ) = \text{FFT}\,[a(t)W_n(t)], \tag{16a}$$

$$B_n(\ ) = \text{FFT}\,[b(t)W_n(t)]. \tag{16b}$$

Some of the indices have been removed for simplicity—for example, the FFT of a quantity in the time domain yields a result with an index that refers to the frequency domain. Thus, in the empty brackets, there would be a k index for each discrete frequency. This processing must be done for each segment of data.

Performing the FFTs gives a collection of vectors, A_n and B_n (capital letters denoting the frequency domain, n denoting the nth time segment). In order to obtain some stability in these estimates, averaging over the N time segments is performed:

$$\hat{G}_{ab}(\ ) = \frac{1}{N}\sum_{n=1}^{N} \mathbf{A}_n(\ )\mathbf{B}_n^*(\ ). \tag{17}$$

To obtain the cross spectrum, at each individual frequency A( ) is multiplied by B*( ) and then averaged over time. Similarly, to obtain the autospectrum, A( ) is multiplied by A*( ) or B( ) by B*( ). This procedure provides an estimate of the auto- and cross-power spectral densities. For some processing techniques, this information would be adequate. In the generalized framework, however, we obtain an estimate of the cross-correlation function by taking the inverse FFT of the estimate of the cross-power spectral density:

$$\hat{R}_{ab}(\tau) = \text{FFT}^{-1}[\hat{G}_{ab}(\ )]. \tag{18}$$

We compute an estimate of the autocorrelation function by following a similar procedure. We next modify the estimate of the cross- or autocorrelation function before transforming back to the frequency domain; this correlation function modification is to improve the spectral estimator.

In the 1950s, a spectral estimation technique was developed called the Blackman-Tukey method (see Blackman and Tukey (1958) or Jenkins and Watts (1968)), where an estimate of auto- and cross-correlation functions was made and then weighting or taper functions were applied, as illustrated in Figure 8 by $W_2(\tau)$. For the lag reshaping method, $W_2(\tau)$ is divided by the triangular $\phi(\tau)$ waveform.

In this study, we take the estimated cross-correlation function and multiply it by a weighting function. But now we are also going to divide by a triangular function, $\phi(\tau)$. In the lag reshaping method (within the generalized framework), we use the knowledge that the autocorrelation of the rectangular weighting function is a triangular function. By using rectangular time weighting as opposed to smooth time weighting, shown in Figure 9a, we effectively impose a triangular autocorrelation function, shown in Figure 9b. Thus the approach in lag reshaping is to divide out the undesirable effects of the

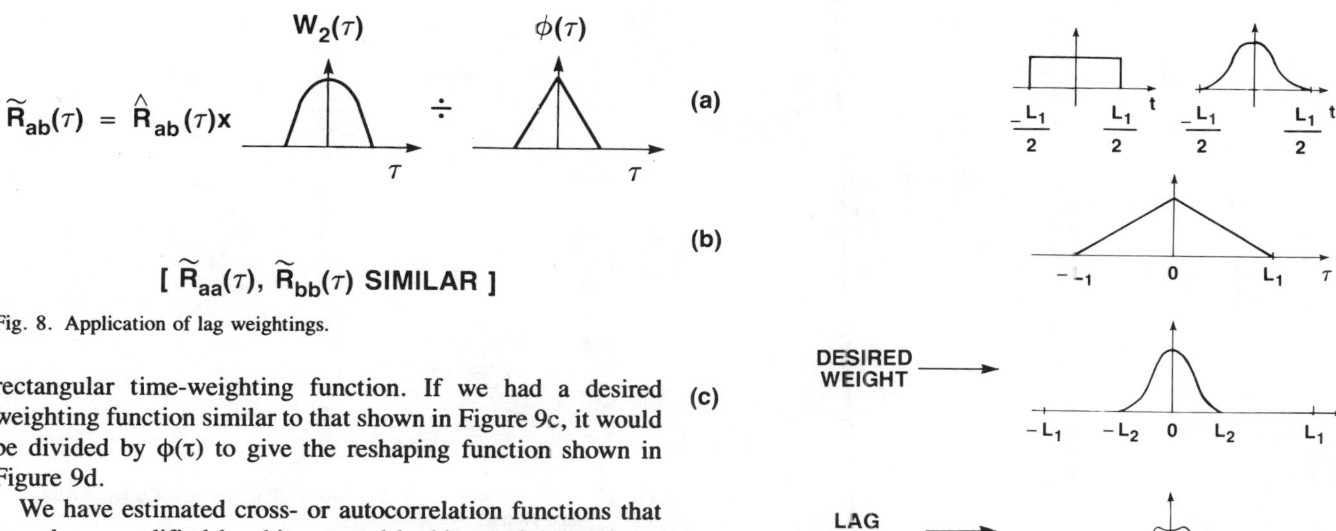

Fig. 8. Application of lag weightings.

Fig. 9. Time-weighting and lag reshaping weighting functions (Nuttall and Carter 1982). (a) Sketch of two examples of time weighting (rectangular on left, cosine on right). (b) Sketch of autocorrelation of rectangular time weighting. (c) Sketch of typical desired lag weighting. (d) Sketch of lag reshaping to compensate rectangular time weighting and achieve desired effective weighting ((c) + (b)).

rectangular time-weighting function. If we had a desired weighting function similar to that shown in Figure 9c, it would be divided by $\phi(\tau)$ to give the reshaping function shown in Figure 9d.

We have estimated cross- or autocorrelation functions that have been modified by this unusual-looking weighting function. We next take the FFT in order to return to the frequency domain:

$$\tilde{G}_{ab}(\ ) = \mathrm{FFT}[\tilde{R}_{ab}(\tau)]. \tag{19}$$

We then obtain the cross- and autopower spectral density estimates.

Since our primary intent has been to obtain C (the MSC), we take the magnitude squared of the cross-power spectral density and divide by the product of the two autopower spectral densities of the input and output processes to form magnitude-squared coherence:

$$|\tilde{\gamma}_{ab}(f)|^2 = \frac{|\tilde{G}_{ab}(f)|^2}{\tilde{G}_{aa}(f)\tilde{G}_{bb}(f)}. \tag{20}$$

Forming the complex coherence requires a complex quantity in the numerator and a square root in the denominator.

There was a debate in the literature about whether or not the WOSA was an effective technique (see, for example, Carter and Nuttall (1980b)). Those who have used the WOSA method correctly know that it is excellent, particularly so for signals with large dynamic range (that is, when the signal power level in one frequency differs by tens of decibels from the signal power level in different frequency bands). When research was first done on the WOSA method, we felt it necessary to use a smooth weighting function in the time domain to avoid irreconcilably distorting spectral estimates of the data. However, we later learned this was not necessary because the lag reshaping method can, in fact, undo (in terms of the mean value) the negative effects (spectral leakage) of the rectangular weighting function. Moreover, in terms of the computation time required by the lag reshaping method, which does not require overlapping, the number of necessary computations can be reduced by a factor of 2; also, in the lag reshaping method there is no multiplication required to apply the smooth time-weighting function.

In summary, the emphasis to this point has been on the theory of coherence and coherence estimation. Now we will treat an important aspect of any estimation method. Imagine that we are going to take a measurement (a person's height, for example) with a tape measure. It is necessary to know whether or not the measuring device has a bias (e.g., if 6 in. were cut off the end of the tape measure, on average every measurement would then be shortened by 6 in., which is a significant bias). Even though excellent spectral estimation techniques are available, if there is no sense of their underlying statistics [either the bias, the variance, or the probabilistic density function (PDF)], these techniques are not very valuable. For the segment-averaging techniques, we understand—either through theory or through experiments—how well they work. In particular, we understand the WOSA spectral estimation method.

Figure 10 shows a plot of four different statistical functions. Figure 10a shows the first-order PDF of the MSC estimate for different amounts of independent segment averaging. So, for example, if we increase the number of averaged segments from 32 to 64, the PDF "tightens up"—in other words, we have less variability in the estimate and a taller PDF. Conversely, reducing the number of averaged segments increases the amount of variability in the estimate. We show shortly that variability is also a function of the true underlying coherence. Plotting the cumulative distribution function (CDF) from the PDF for segments that are independent of one another (Figure 10b), we obtain CDFs that are sums of hypergeometric functions, which are themselves hypergeometric functions. (Hypergeometric functions are easily evaluated by computer.)

Normalized PDFs are shown in Figure 11. (The integral of any PDF from $-\infty$ to $+\infty$ is 1, so that sharp PDFs must be taller and broad ones shorter.) The PDFs have different

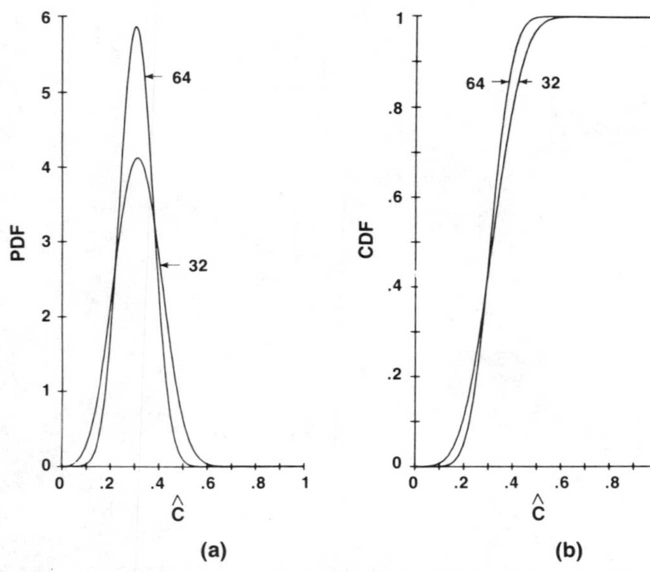

Fig. 10. PDF and CDF of the MSC estimate for MSC = 0.3, using 32 and 64 segments (Carter, Knapp, and Nuttall 1973a). (a) PDF. (b) CDF.

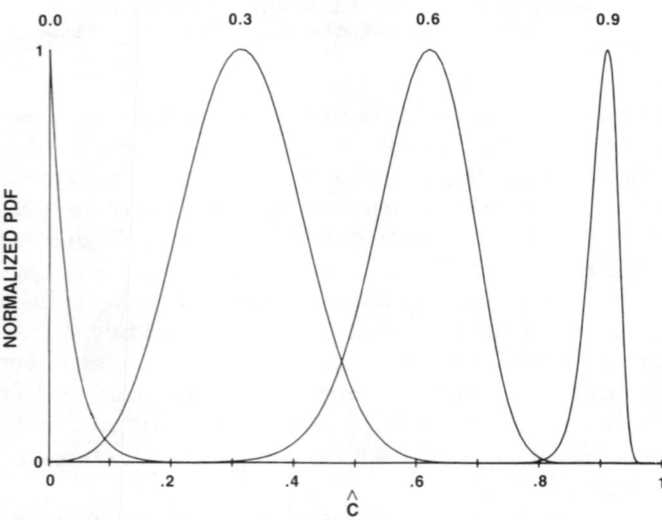

Fig. 11. Normalized PDFs for different MSCs (using 32 segments) (Carter, Knapp, and Nuttall 1973a).

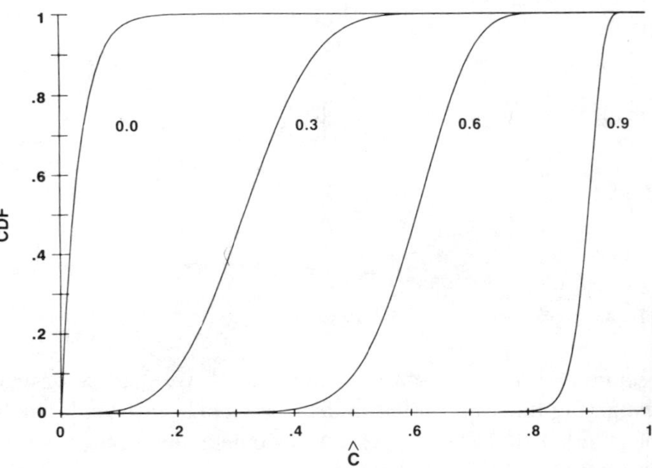

Fig. 12. CDFs for different MSCs (using 32 segments) (Carter, Knapp, and Nuttall 1973a).

behavior, depending on the true MSC. Figure 11 shows that as the true MSC varies from 0 to 1, the shape is Gaussian-like only in the middle (such as for the PDFs corresponding to MSCs of 0.3 and 0.6). In these estimation techniques there are no numbers less than zero or greater than unity, so the PDF encounters a hard constraint at both the left and right ends. The results (presented here from Carter, Knapp, and Nuttall (1973a)) are derived based on Gaussian assumptions for stationary signals. For the case of zero coherence, however, there has been important work done indicating that the results are very robust with regard to the Gaussian assumption (Gish and Cochran 1987). If one of the processes is not Gaussian, the statistics of the coherence estimate still hold. This is a very important point for setting constant false-alarm-rate (CFAR) detectors in the signal-absent case.

We integrate the PDFs of Figure 11 to get the CDFs shown in Figure 12. From the CDFs, confidence bounds can be obtained (see, e.g., Cramér (1946)).

Figure 13a shows confidence bounds for the confidence interval of 80 percent and for certain amounts of averaging (N = 8, 16, 32, . . . independent segments averaged). The widest confidence curves represent the least amount of averaging. These curves are constructed from CDFs (like those of Figure 12) by obtaining the numbers from the CDFs, plotting them on the confidence limits graph, and then drawing the curves by hand. (A challenge to the reader is to develop a closed-form expression for confidence bounds.)

The curves are used as follows. Suppose we conduct an experiment and obtain an estimate of the MSC, having used a fixed number of segments. We then refer to the confidence bounds and draw a horizontal line from the y-axis of the experimentally estimated value of the MSC. The two x-axis values where the horizontal line intersects with the pair of curves (for fixed N) give the range of the true MSC (with 80 percent confidence). If the experiment is conducted repeatedly, then the true MSC will fall within the range determined by this procedure 80 percent of the time (Cramér 1946). Note that for estimated MSC less than about 0.3, the curves drop abruptly to zero. The range of true coherence given by the curves at this end of the scale is still valid for the given 80 percent confidence interval. The original pioneering work in this area was done by Benignus (1969b). (Without detracting from the contributions of that work, we note here that, with the exception of the abrupt drop to zero, the curves in Figure 13 do not possess the unusual discontinuities of the Benignus (1969b) paper.)

Another set of curves is shown in Figure 13b for 95 percent confidence intervals. If we wish to be 95 percent confident that results will fall within a certain range, that range will be much larger (note the wider spread of the pairs of curves for a confidence level of 95 percent versus 80 percent).

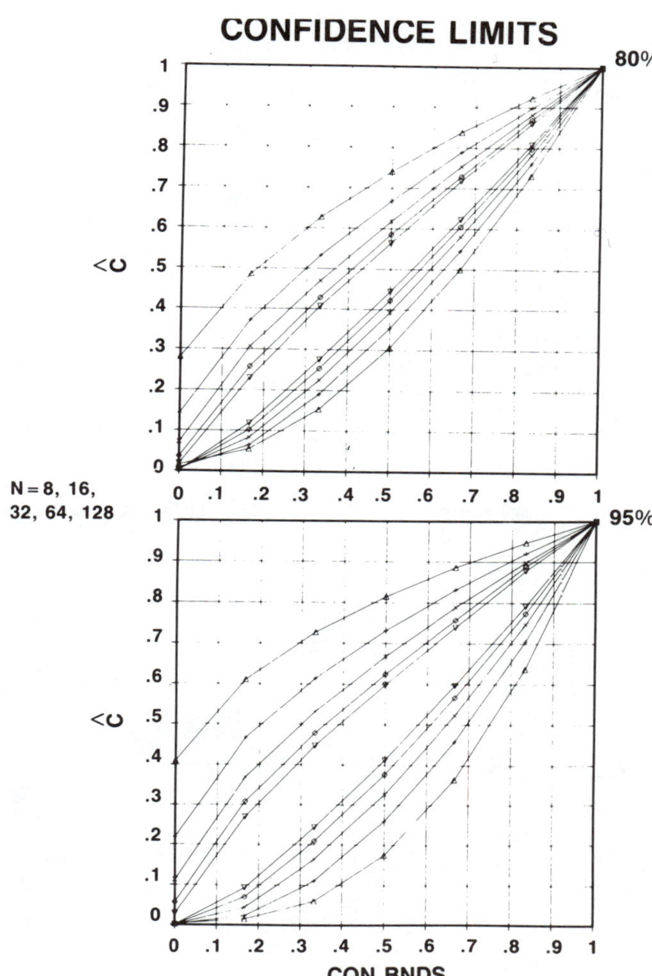

Fig. 13. Confidence bounds for MSC estimates for various numbers of segments averaged (Scannell and Carter 1978b). (a) 80 percent confidence. (b) 95 percent confidence.

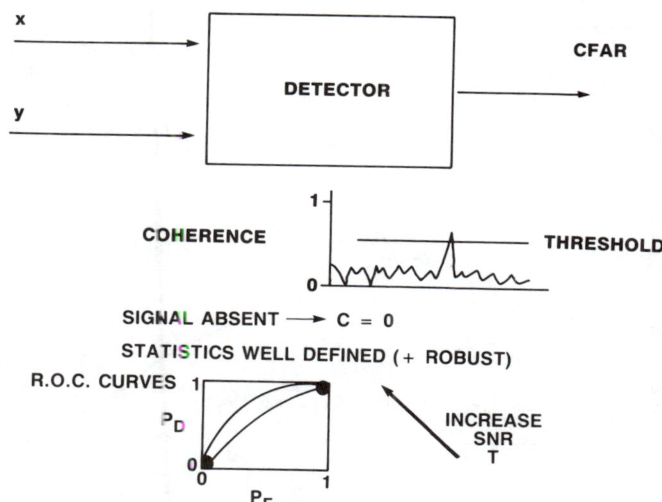

Fig. 14. Constant false-alarm-rate (CFAR) receiver.

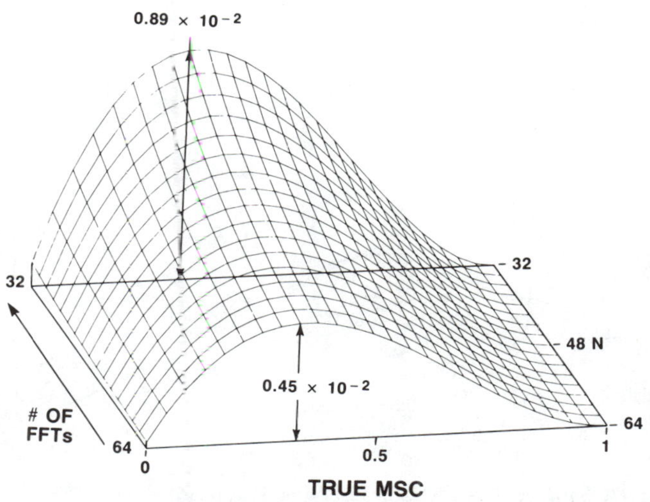

Fig. 15. Variance of the MSC estimate (Carter, Knapp, and Nuttall 1973a).

The estimated MSC can be used as a detection statistic. Figure 14 shows a particularly robust detection method called a CFAR receiver. When the signal is absent, the MSC is zero. A simple formula for the PDF (which is a cusp-shaped function) is given in Carter, Knapp, and Nuttall (1973a) based upon Gaussian assumptions. However, the MSC detector is not very sensitive to the Gaussian assumption. For example, if one of the processes is Gaussian and the other is not, the results still hold (Gish and Cochran 1987).

Empirically speaking, MSC detectors are very robust. The performance of any detector can be analyzed in terms of the receiver operating characteristics (ROC). Typical ROC curves, plotted as the probability of detection, P_D, versus the probability of false alarm, P_F, are shown at the bottom of Figure 14. The ROC curve always contains the point (0, 0) and the point (1, 1). Setting the detection threshold arbitrarily low will ensure detection but will also increase the number of false alarms. Moving the threshold higher will reduce the number of false alarms but will also reduce the number of detections. If the threshold is set arbitrarily high, there will be neither false alarms nor detections. Thus the ROC curves predict how well the detector performs. Performance will also improve and the ROC curves will change with higher SNR and increased processing time.

From the PDFs, we can extract the bias and variance. These statistics will differ, depending on whether MSC or magnitude coherence (MC) is estimated. The MSC variance, plotted in Figure 15, is a function of how many FFT segments have been averaged. The maximum variance of the MSC estimate occurs for a true MSC of approximately one-third, which is the most difficult MSC to estimate accurately. As unity coherence is approached, the PDF becomes narrower; consequently, the variance approaches zero. When unity is reached, there is no variance in estimating the MSC; at the other extreme, there is a finite variance with zero coherence.

Since expressions for the variance are complicated, the literature offers some guidelines for its approximation. Figure 16

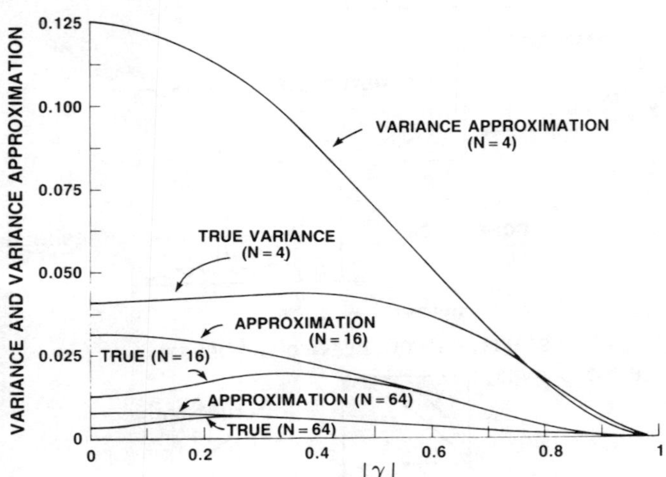

Fig. 16. Variance approximation of the MC estimate (Carter, Knapp, and Nuttall 1973b).

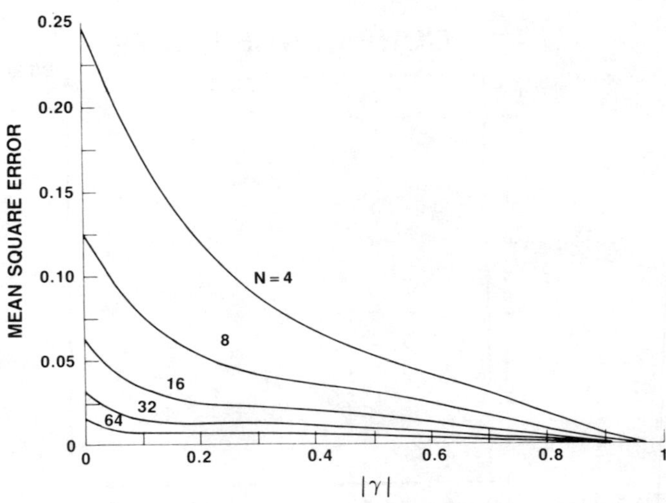

Fig. 18. Mean square error of the MC estimate (Carter, Knapp, and Nuttall 1973b).

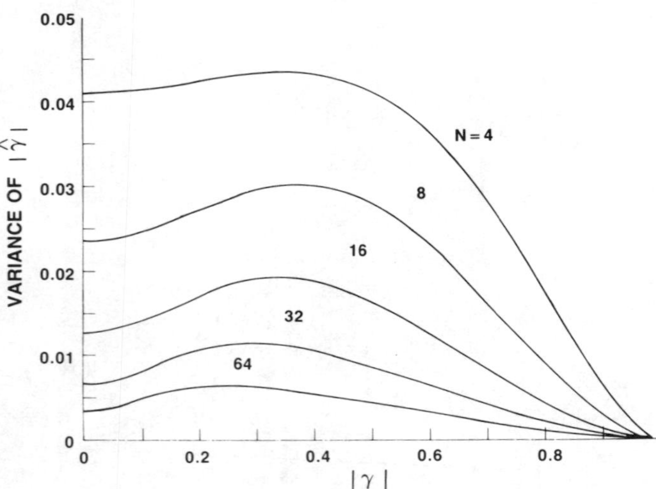

Fig. 17. Variance of the MC estimate (Carter, Knapp, and Nuttall 1973b).

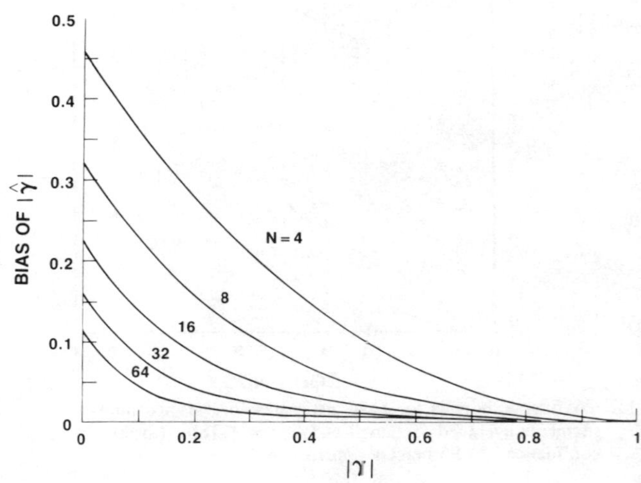

Fig. 19. Bias of the MC estimate (Carter, Knapp, and Nuttall 1973b).

shows that the approximation becomes more accurate as a greater number of FFTs is used. When based on four FFTs ($N = 4$), there is a large deviation from the true variance. For $N = 16$, however, the approximation is closer to the true variance, and it is even closer for $N = 64$. Many approximations behave in this same manner, doing well as long as the number of FFTs used is not too small.

Figure 17 also shows the variance of the MC estimate. Note that the statistics are different for the MC and the MSC estimates. The mean square error (MSE) of the MC estimates can be computed and is shown in Figure 18. The bias of the MC estimates (Figure 19) or of the MSC estimates (Figure 20) can also be computed. The bias depends on the true MSC and on the number of FFT segments (plotted here from $N = 32$ to $N = 64$). The maximum bias of the MSC estimate is equal to one over the number of FFT segments—that is, $1/N$—which occurs when $C = 0$.

It should be noted, however, that the bias expressions de-

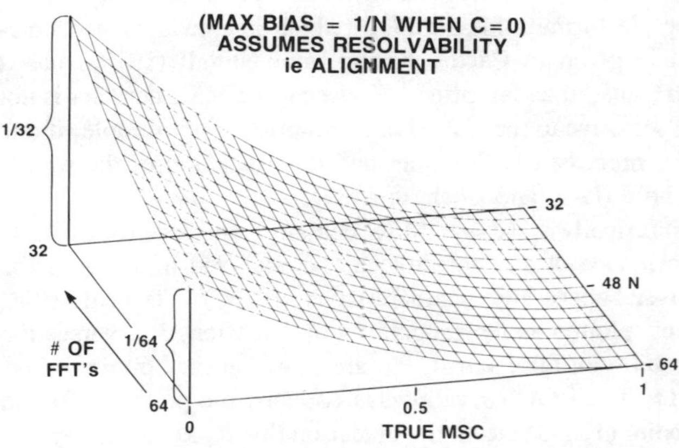

Fig. 20. Bias of the MSC estimate (Carter, Knapp, and Nuttall 1973a).

picted in Figures 19 and 20 assume frequency resolvability. In other words, there is an implicit assumption that the FFT size

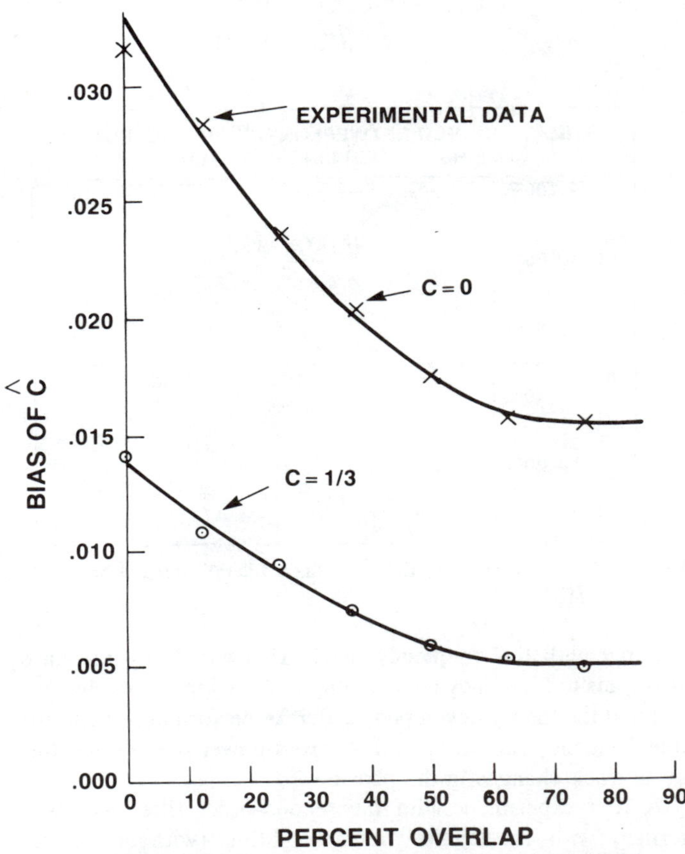

Fig. 21. Bias of the MSC estimate vs. the amount of overlap (using 32 segments) (Carter, Knapp, and Nuttall 1973a).

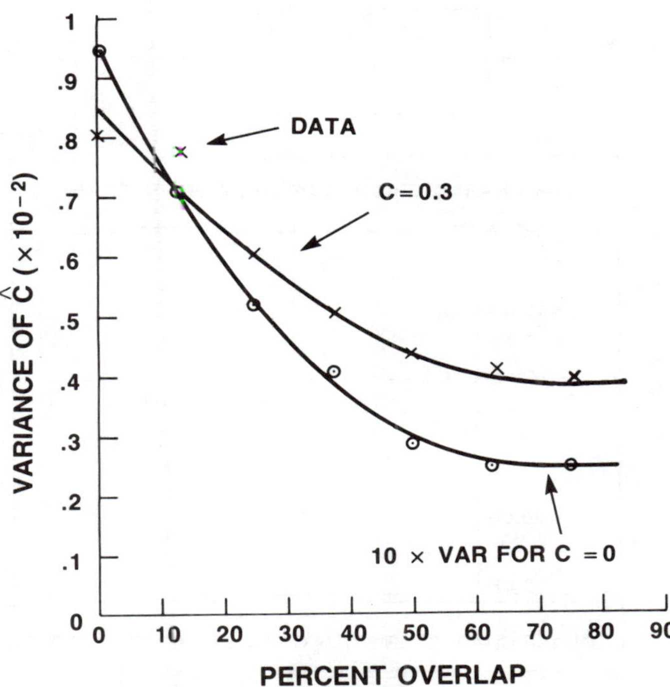

Fig. 22. Variance of the MSC estimate vs. the amount of overlap (using 32 segments) (Carter, Knapp, and Nuttall 1973a).

is large enough to resolve the detail in the true power spectrum. In dealing with autopower spectral estimation, a frequency "window" that is narrower than the finest detail of the autopower spectral density must be used. Estimation of the complex cross-power spectral density also requires the use of a window that will track the detail of both the magnitude and the phase of the cross-power spectral density. Then, too, for a fixed amount of data, there is a trade-off between the size of each FFT segment and the number of independent segments available to average. When the assumption of resolvability is incorrect, bias and variance predictions can be in error.

Some experimental estimates of bias versus overlap for WOSA data are shown in Figure 21. Since multiplication by a Hanning time-weighting function causes the data to taper smoothly to zero at both ends of each time segment, the weighted segments must be overlapped to avoid wasting data. This amount of overlap can be varied. As the amount of overlap is increased, the number of dependent segments increases, and the bias and the variance of the MSC estimate are reduced. Conducting a series of independent experiments provides a better understanding of the confidence bounds about these experimental data points. What is seen for Hanning time weighting as a function of overlap is a knee developing in the curve at about 50 percent or 62.5 percent overlap. In other words, as overlap is increased, a point of diminishing returns (in terms of bias reduction) is reached. This holds whether the true MSC is one-third, as in the bottom curve, or whether the true MSC is zero. The computational aspects are not addressed here, because we are concerned only with obtaining the maximum amount from the data. After about 50 or 62.5 percent overlap, with the use of Hanning time weighting, it seems that everything (that is practical) has been extracted from the data. A time-weighting function other than a Hanning function would require a different amount of overlap to extract the most from the data; for example, a time-weighting function that was very peaked in the time domain would require increased overlap.

Figure 22 depicts the experimentally determined variance of the MSC estimate versus overlap. Here, the $C = 0$ curve has been scaled by a factor of 10, so that the two curves appear on the same plot. Again, at around 50 or 62.5 percent overlap, the knee in the curve, or the point of diminishing returns, is reached. At this point we have extracted all that is possible from the data with Hanning time weighting.

Now let us consider experiments with linear systems. It was earlier stated that in a linear system, the true MSC between the input and the output is unity for all frequencies. This is shown in the results of experiments performed by Carter and Arnold (1971).

A first experiment used the first-order linear filter shown in Figure 23, where H is the filter, x is the input, and y is the output. The magnitude of the transfer function is shown in the top figure, and the phase response is shown in the bottom one. Since a digital filter was used, the x-axis could have been numbered from 0 to π. An arbitrary Nyquist frequency of 1024 Hz was chosen for the labeling process. Note the smooth

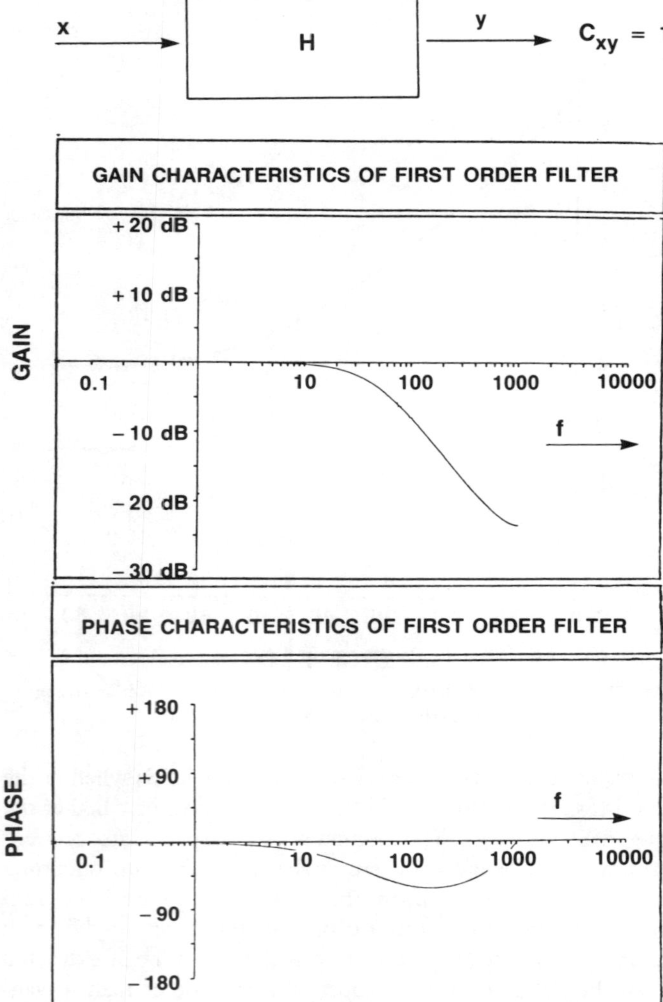

Fig. 23. Transfer function of a first-order linear filter (Carter and Arnold 1971).

$$C_{xy}(f) = 1 \;\forall f, \quad \text{AND} \quad \hat{C}_{xy}(f) = 1 \;\forall f$$

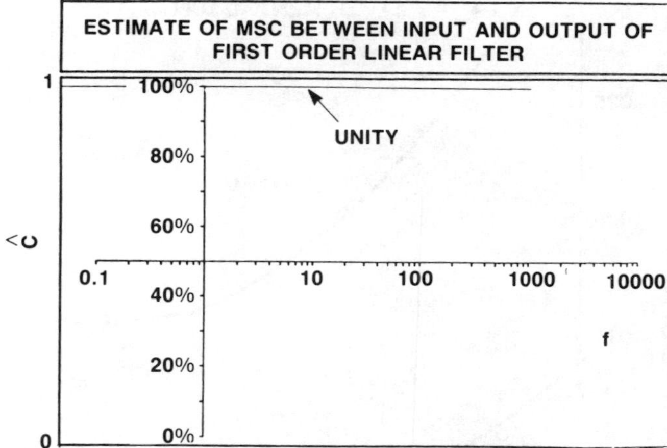

Fig. 24. MSC estimate for the first-order linear filter (Carter and Arnold 1971).

roll-off behavior and the smooth phase. The true MSC for all frequencies is unity, because the true MSC between the input and output of any linear system is unity for all frequencies. Since there is no variance in the MSC estimate when the true MSC is unity (the PDF is a delta function), the estimates of the MSC should be unity for all frequencies.

An experiment with the amount of averaging shown here (approximately 32 segments of averaging) was conducted, and the expected estimates of unity for all frequencies were achieved, as shown in Figure 24. A second experiment used the second-order linear filter shown in Figure 25. Again, since the true MSC is unity for all frequencies, the estimated values for the MSC should be unity for all frequencies.

Note that at approximately 30 Hz, this linear system possesses a sharp resonance where the phase changes rapidly. The units of the slope of the phase are radians divided by Hertz; Hertz are cycles per second. Hence, the slope of the phase has units of seconds, or time delay. It can be seen by the steep slope of the phase response that there is a large time delay associated with this frequency band. This variability in phase also leads to frequency resolvability and tracking problems; recall that the theory developed earlier for performance assumed fine frequency resolution and for cross-power spectra the ability to track changes in the phase.

A first experiment with the second-order filter was performed with rectangular time weighting (without lag reshaping). Because a rectangular weighting function was used instead of a Hanning weighting function, the MSC estimates differed from unity (see Figure 26). This initially surprising result led to a reprocessing of the data with a smoother time-weighting function. Shown in Figure 27 is the MSC estimate reprocessed with Hanning weighting. The Hanning weighting has sidelobes that decay more rapidly than the rectangular time weighting; hence, the Hanning weighting can better resolve behavior in one frequency band without leakage from another band. One concludes that, in the WOSA method, one needs smooth-time weighting like Hanning weighting.

In the early 1970s, using an FFT size of 1024 strained the capability of the digital computer, but even so, the MSC estimates were good, except for a small dip in the region of rapidly changing phase. However, if a larger FFT size, such as 4096, is used, that dip virtually disappears. Although there are other ways of eliminating the dip (such as prefiltering to whiten the complex cross spectrum), resolution of the phase slope by use of a large FFT size will also remove the biases. These biases are analogous to those connected with misalign ment because there is a delay involved with the frequency domain that manifests itself as a time delay type of bias. One of the reasons coherence has not historically been used as a tool is that it is difficult to estimate. In other words, there are many subtleties to consider, even in the case of stationary Gaussian random processes, without taking into account "real-world" problems. It is only with the advent: (1) of the FFT, (2) com-

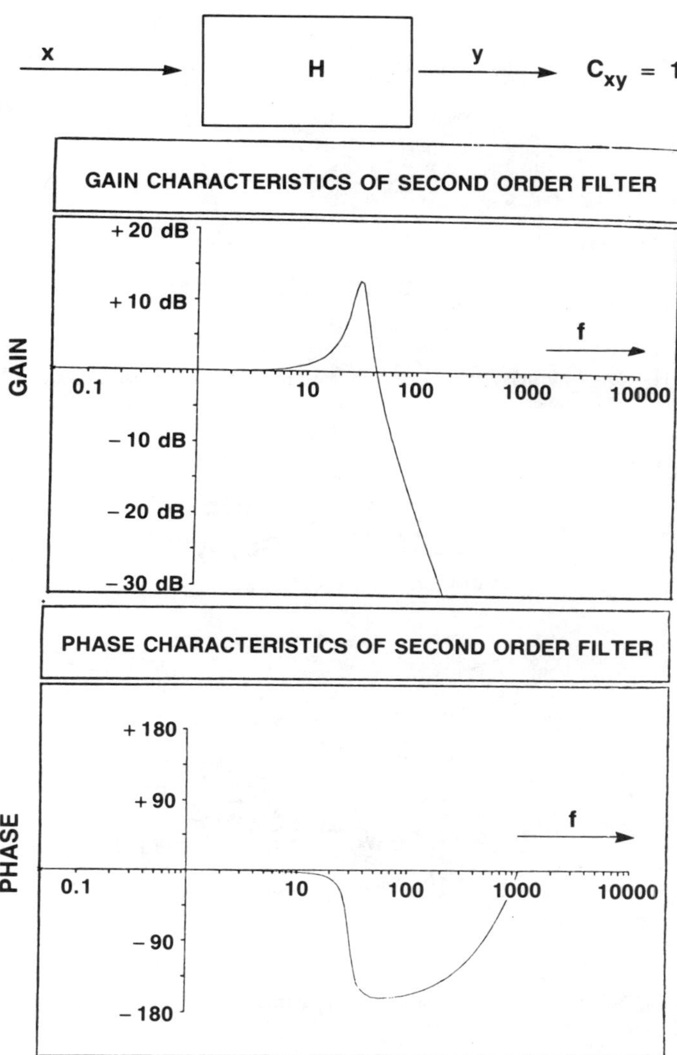

Fig. 25. Transfer function of a second-order linear filter (Carter and Arnold 1971).

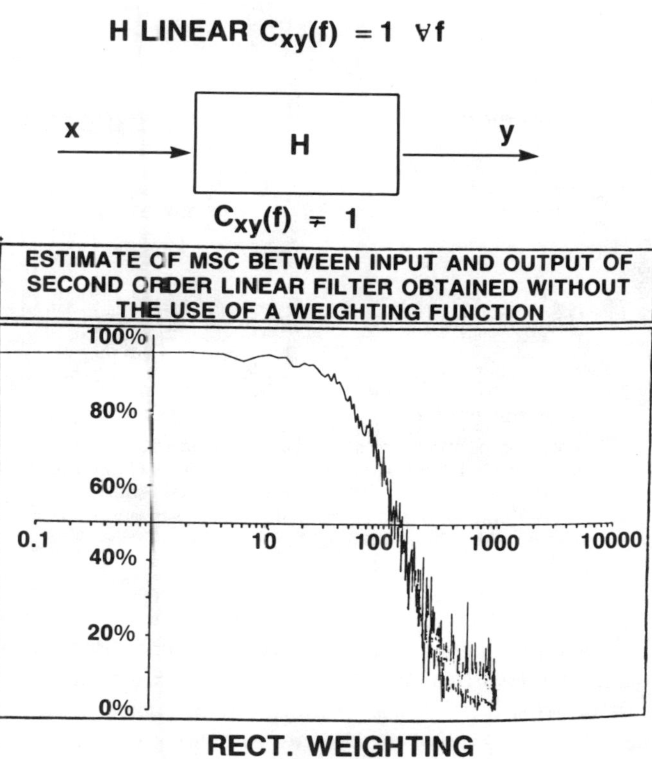

Fig. 26. MSC estimate for the second-order linear filter (using rectangular weighting) (Carter and Knapp 1975).

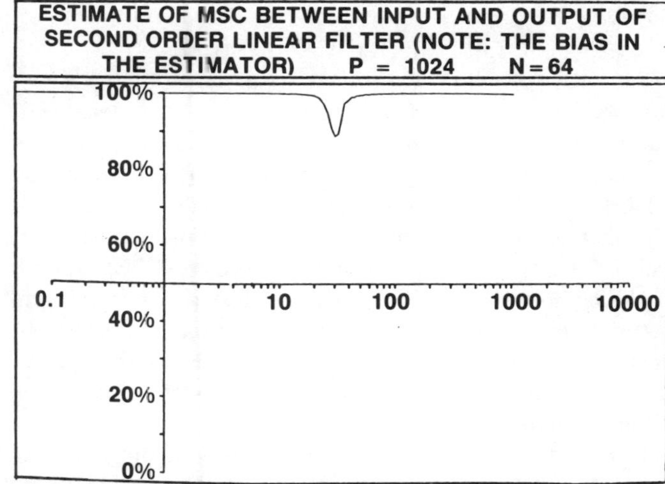

Fig. 27. MSC estimate for the second-order linear filter (using smooth weighting) (Carter and Knapp 1975).

puters that can do many large-size FFTs, and (3) increased understanding of the estimation pitfalls that coherence can be used to solve critical engineering problems.

Figure 28 shows the results of an experiment conducted a number of years ago in which we used an underwater sound source and a receiver located a certain distance from the source. The signals at the source and at the receiver were recorded, and the MSC was estimated between them. There was significant misalignment due to the travel time (some number of seconds) between the two. A process was devised in the laboratory to line up the segments and to estimate coherence. At this time it was also noticed that the estimate of the coherence was very sensitive to small differences in how accurately the segments were lined up. This observation led to a prediction of bias due to misalignment (Carter 1980). (The smoothed coherence transform (SCOT) is a member of a family of generalized cross-correlation (GCC) functions and is covered in more detail later.)

Coherence estimates can change significantly, depending on the amount of misalignment. If there is a misalignment between the two time series (which is analogous to not being able to resolve or track the detail of the complex coherence or the phase of the cross-power spectral density), a very different estimate of the MSC is obtained. This bias is documented in the literature with a simple formula (Carter 1980).

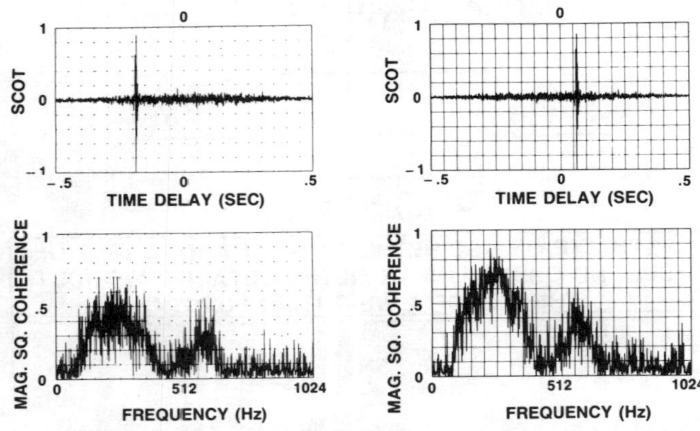

Fig. 28. Bias due to misalignment (Carter 1980). (a) SCOT estimate showing a −180-ms delay and corresponding MSC estimate of 0.45 at 250 Hz with a −180-ms delay. (b) SCOT estimate showing a 70-ms delay and corresponding MSC estimate of 0.7 at 250 Hz with a 70-ms delay.

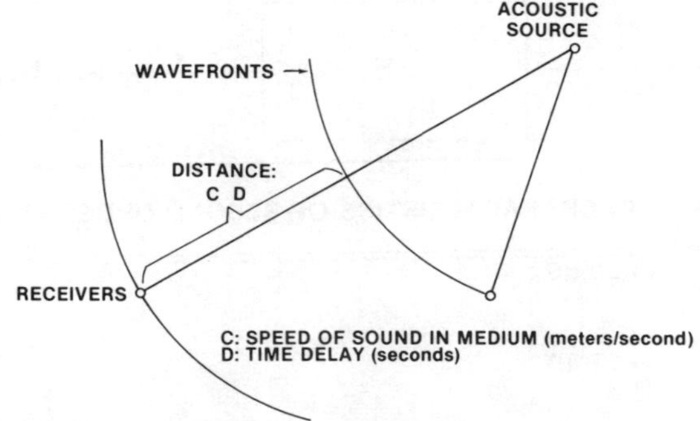

Fig. 29. Time delay associated with wave fronts emitted by an acoustic source.

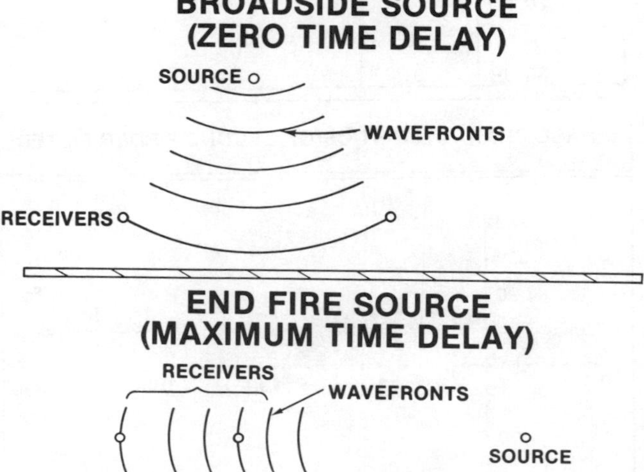

Fig. 30. Time delays associated with wave fronts from a broadside source and an endfire source.

To briefly summarize this section, coherence is a useful quantity but is difficult to estimate. The MSC is a normalized, cross-power spectral density between zero and unity for all frequencies. The two best methods for estimating coherence are the WOSA method and the lag reshaping method. Both are described in the literature and the references. It was emphasized that understanding the statistics of an estimator is critical to being able to use it wisely as a tool. Subsequent sections examine how coherence plays a role in time delay estimation.

1.4 Time Delay Estimation

This section provides a brief introduction to time delay estimation (TDE). A maximum likelihood estimator is derived and its interpretation is given. The Cramér-Rao lower bound (CRLB) is shown, and some simulation results are compared with that bound. Other methods of predicting performance, such as the correlator performance estimate (CPE) and the Ziv-Zakai lower bound (ZZLB), are also described. These methods are discussed in terms of the motion of acoustic objects, which are being tracked with variable length observation time. Finally, more complex techniques of modeling the ocean medium will be presented.

TDE is often referred to by other names, including time delay (TD), time difference of arrival (TDOA), group delay, time-of-arrival difference (TOAD), delay, delay time, and phase delay. For our purposes here these terms have the same meaning.

The physical problem in two dimensions is shown in Figure 29. Here, a source radiates acoustic energy by means of propagating wave fronts, but, more generally, the source in Figure 29 could be any other wave-propagating source, such as electromagnetic. In this example there are two receivers. The distance between the wave fronts impinging upon one receiver and the next—that is, between when the wave front reaches the first receiver and when it reaches the second receiver—is equal to the propagation velocity in the medium times the time delay.

Figure 30 (top) shows that when an acoustic source is broadside to an array, the wave fronts will reach both sensors at the same time. This is true for near-field and far-field sources. If the source is endfire to the array, as shown in Figure 30 (bottom), the acoustic wave fronts will reach one sensor before the other, with maximum delay. The maximum time delay associated with endfire reception is a prior physical constraint. The time delay cannot be greater than this physical constraint or the signal will be assumed to be propagating slower than it can in the medium. If a physical constraint is known a priori, that information can be built into the estimator and the system will perform better than an estimator without prior knowledge.

If bearing is estimated and the source is in the far feild, as shown in Figure 31, the arriving curved wave fronts look similar to planar waves, and the cosine of the bearing is equal to C times the time delay divided by the length of the array. (Note that C as used here is the speed of sound, not the MSC,

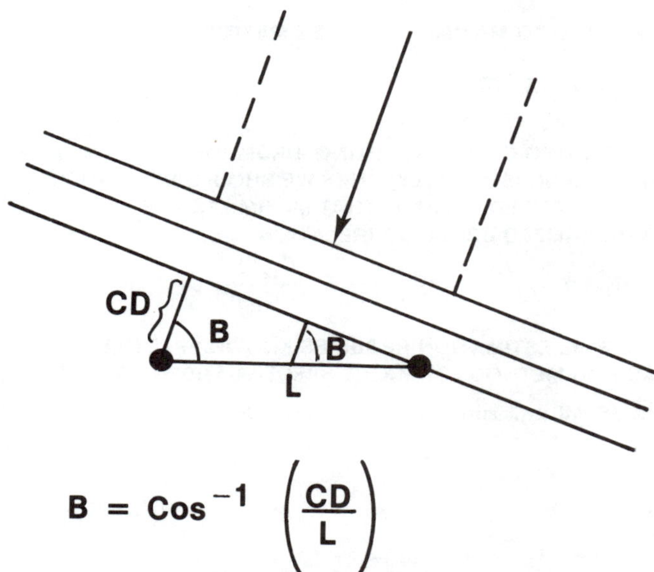

Fig. 31. Geometry used to estimate the bearing of an acoustic source passively.

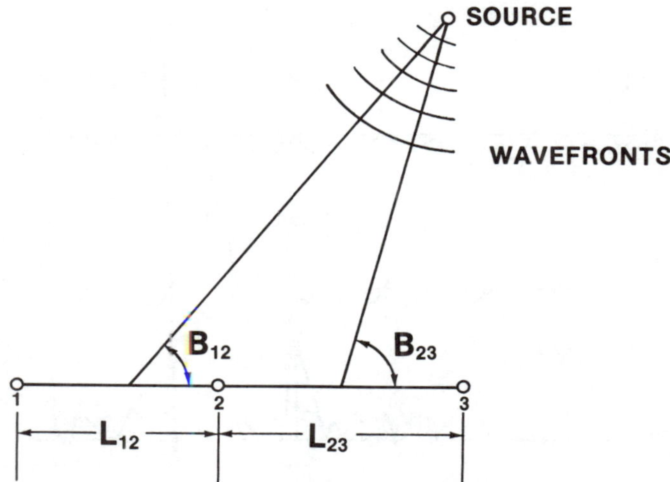

Fig. 32. Geometry used to estimate the range and bearing of an acoustic source passively.

and D is the time delay.) The expression for the bearing, B, is given in Figure 31 (bottom). One of the pioneering works in bearing estimation and performance of a linear array was MacDonald and Schultheiss (1969).

The fundamental physical problem can be described as follows. There are two receiving sensors, which receive waveforms $r_1(t)$ and $r_2(t)$. An attempt is made to determine the bearing of the source by estimating the time delay between the received signals. Essentially there is a signal plus a noise and a delayed signal plus a noise:

$$r_1(t) = s(t) + n_1(t), \tag{21a}$$

$$r_2(t) = s(t - D) + n_2(t). \tag{21b}$$

In the ideal case, or at least to make the problem mathematically tractable, we assume that the noises are uncorrelated with each other and also with the source signal. We also assume that the signals and noises are stationary Gaussian random processes and that the observation time is large compared to the delay D. The problem becomes more complicated if this is not the case.

If there were three sensors instead of two, the first two sensors could provide one (time delay) bearing estimate, and the other two could provide another (time delay) bearing estimate. The point at which these bearings intersected would yield an estimate of the source location in both range and bearing. This process is called localization (see Figure 32).

Note that a small error in making bearing estimates or a small error in knowing the location of the sensors can make a significant difference in the estimated range of the acoustic source. Still, in localization, the key question is, How should we estimate time delay?

To paraphrase the literature approximately (rigorous derivations can be found in standard texts such as Van Trees (1968)), the CRLB provides a lower bound on the best performance because it gives the minimum variance. It can be shown that the maximum likelihood (ML) estimator will achieve the CRLB with sufficient observation time. However, how much observation time is required is not always known, nor is it always practical or possible to achieve the necessary observation time. Therefore, this approach will be used in the following pages: (1) derive the ML estimator, (2) compute the CRLB, and (3) run a simulation experiment to see if the estimator works as predicted—that is, achieves the bound.

The concept of variance of a time delay estimate is illustrated in Figure 33, where a number of trial GCC functions are sketched. For each GCC function, the peak is identified and its x-axis location is noted. By averaging over these trials, the variance and bias in the time delay estimate are determined. One sketch, trial 5, is intentionally drawn to have some large spurious peaks away from the true peak to show how, with additional noise, an anomaly could suddenly appear and tracking of the correct peak could be "lost." There are also other problems to be addressed.

Figure 34 shows the equations used for the derivation of the ML estimator.

Essentially, the power spectral density (PSD) matrix, **Q**, is obtained from a time series and its Fourier transform. This matrix has components of autopower spectral density on the diagonal and cross-power spectral density on the off-diagonal. The Fourier transform of these received data conditioned on **Q** is the probability of the vector **X** conditioned on the signal PSD. Note that **Q** is also conditioned on the delay. This derivation can be found in the literature (Knapp and Carter 1976). If quantities that weakly depend only on the delay are neglected, we obtain a function that has to be maximized (see Figure 35).

In order to estimate the delay, then, the resulting function to maximize (in Figure 35) is computed by multiplying the cross PSD function by a weighting function proportional to $C/(1 - C)$, where C is the MSC. This result is normalized by the magnitude of the cross-spectral density; then the Fourier

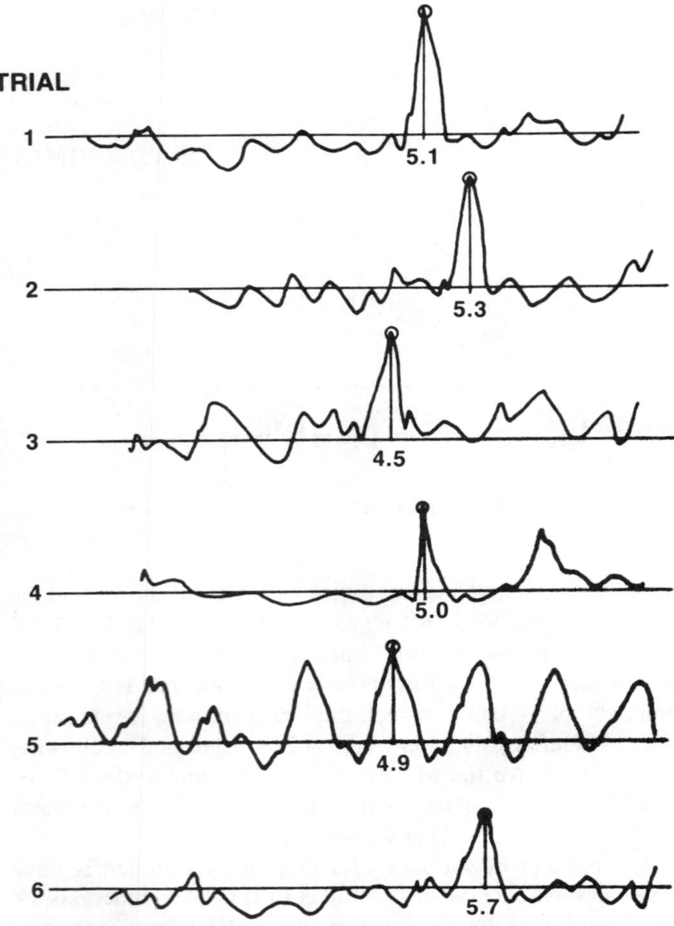

Fig. 33. Sketch of six hypothetical generalized cross-correlation (GCC) functions (Carter 1976a).

FOURIER RELATIONSHIP:

$$X_i(k) = \frac{1}{T}\int_{-T/2}^{T/2} X_i(t)e^{-jkt\omega_\Delta}\,dt$$

SPECTRAL DENSITY MATRIX:

$$Q = TE[X(k)X^{*'}(k)] = \begin{bmatrix} G_{11} & G_{12} \\ G_{12}^{*} & G_{22} \end{bmatrix}$$

PROBABILITY DENSITY FUNCTION:

$$P(X \mid Q) = P(X \mid G_{ss}, G_{n_1n_1}, G_{n_2n_2}, G_{n_1n_2}, D)$$

$$= C_P \, \mathrm{EXP}(-\tfrac{1}{2} J_1)$$

WHERE (C_P ONLY WEAKLY DEPENDS ON D) AND

$$J_1 = T\,\Sigma\, X^{*'}\,Q^{-1}X$$

Fig. 34. Derivation of the ML estimator (Knapp and Carter 1976).

transform of that whole quantity is taken. The Fourier transform of a cross-power spectral density times this whole expression (given in Figure 35) is called the generalized cross-correlation (GCC) function. Finally, the peak in the GCC function is located. It is important to note that ML estimation of time delay requires knowledge of the signal and noise spec-

SELECT D TO MAXIMIZE P(X|Q) BY MAXIMIZING

$$J_1 = \int_{-\infty}^{\infty} X^{*'}\,Q^{-1}X$$

COMPUTING Q^{-1}, SUBSTITUTING, DROPPING TERMS NOT DEPENDING (STRONGLY) ON D, WE SHOULD SELECT THE HYPOTHESIZED DELAY τ THAT MAXIMIZES THE GENERALIZED CROSS CORRELATION.

$$\hat{R}_{12}(\tau) = \int_{-\infty}^{\infty} \hat{G}_{12}(f)\,\frac{1}{|G_{12}(f)|}\cdot\frac{C_{12}(f)}{[1-C_{12}(f)]}\,e^{j2\pi f\tau}\,df$$

NOTE ML ESTIMATION REQUIRES KNOWN SPECTRA
AD HOC METHOD ESTIMATES SPECTRA AND SUBSTITUTES

Fig. 35. ML estimation (Knapp and Carter 1976).

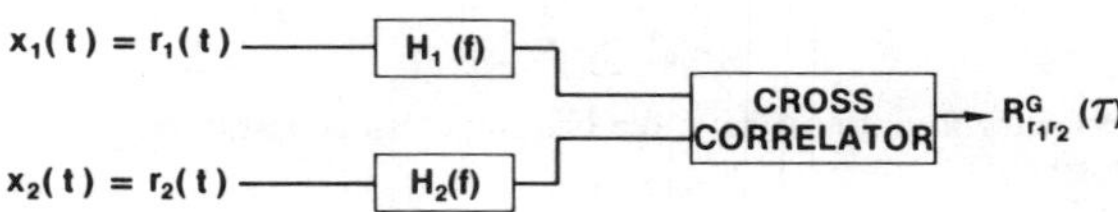

GCC FUNCTION

$$R^G_{x_1x_2}(\tau) = R^G_{r_1r_2}(\tau) = \int_{-\infty}^{\infty} W(f)\,G_{r_1r_2}(f)\,e^{j2\pi f\tau}\,df = \int_{-\infty}^{\infty} W_\phi(f)\,e^{j\phi(f)}\,e^{j2\pi f\tau}\,df$$

WEIGHTING FUNCTION

$$W(f) = H_1(f)\,H_2^{*}(f), \quad W_\phi(f) = |G_{r_1r_2}(f)|\,W(f)$$

Fig. 36. The GCC method for time delay estimation (Scarbrough, Ahmed, and Carter 1981b).

tra and their characteristics. When these quantities are not known, they must be estimated. Friedlander (1983, 1984) and Friedlander and Porat (1984) investigate the joint estimation of the spectra and the time delay.

The processing is done as follows: We estimate the complex cross spectrum, multiply by a weighting function, and then compute the Fourier transform. This procedure gives a GCC function. A GCC function is implemented by filtering with H_1 and H_2, as shown in Figure 36, and cross correlating. Because processor interpretation is compiled from several different sources, the notations sometimes change (e.g., $x_1(t)$ and $r_1(t)$ are used interchangeably in Figure 36).

The only difference between the GCC function and a standard cross-correlation function is the introduction of two prefilters, which, in general, enhance the frequency bands where the signal is strong and attenuate the bands where the noise is strong. Looking at this in another way, the GCC function can be thought of as the Fourier transform of the cross spectrum scaled by a weighting function, $W(f)$. Or, since the slope of the phase has units of time delay, another interpretation is that the GCC function is the Fourier transform of the phase of the cross spectrum scaled by a phase weighting function, $W_\phi(f)$. (It turns out that the phase weighting is related to the variance of the phase.) One can choose either interpretation or use both.

Figure 37 shows some weighting functions that are commonly used in time delay estimation, including the weighting

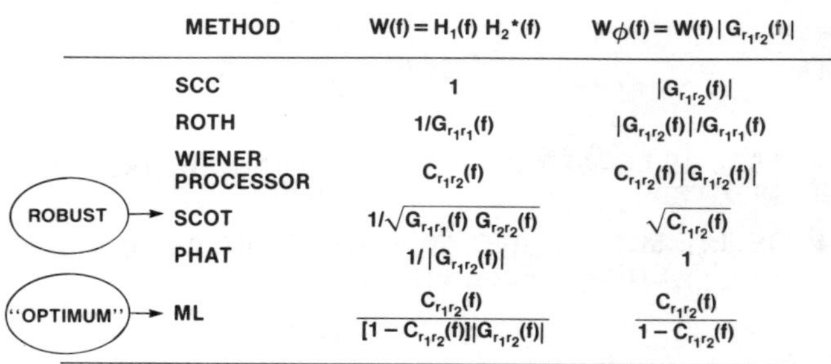

	METHOD	$W(f) = H_1(f)\, H_2^*(f)$	$W_\phi(f) = W(f)\,\lvert G_{r_1r_2}(f)\rvert$
	SCC	1	$\lvert G_{r_1r_2}(f)\rvert$
	ROTH	$1/G_{r_1r_1}(f)$	$\lvert G_{r_1r_2}(f)\rvert / G_{r_1r_1}(f)$
	WIENER PROCESSOR	$C_{r_1r_2}(f)$	$C_{r_1r_2}(f)\,\lvert G_{r_1r_2}(f)\rvert$
ROBUST →	SCOT	$1/\sqrt{G_{r_1r_1}(f)\, G_{r_2r_2}(f)}$	$\sqrt{C_{r_1r_2}(f)}$
	PHAT	$1/\lvert G_{r_1r_2}(f)\rvert$	1
"OPTIMUM" →	ML	$\dfrac{C_{r_1r_2}(f)}{[1 - C_{r_1r_2}(f)]\lvert G_{r_1r_2}(f)\rvert}$	$\dfrac{C_{r_1r_2}(f)}{1 - C_{r_1r_2}(f)}$

$$C_{r_1r_2}(f) = \frac{\lvert G_{r_1r_2}(f)\rvert^2}{G_{r_1r_1}(f)\, G_{r_2r_2}(f)}$$

Fig. 37. Various GCC functions (Carter 1987).

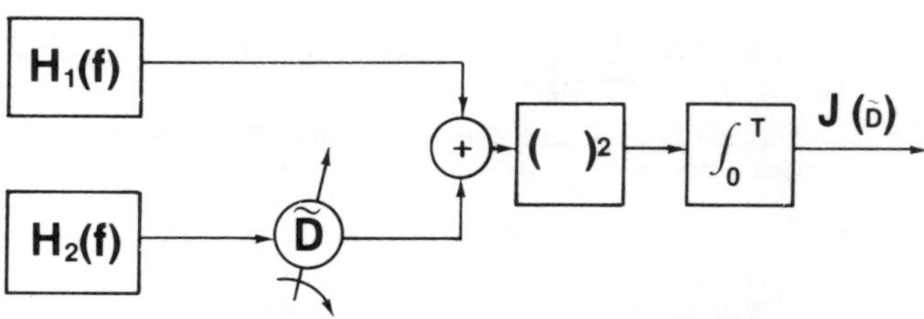

Fig. 38. Filter and sum realization of the ML time delay estimator (Carter 1976a).

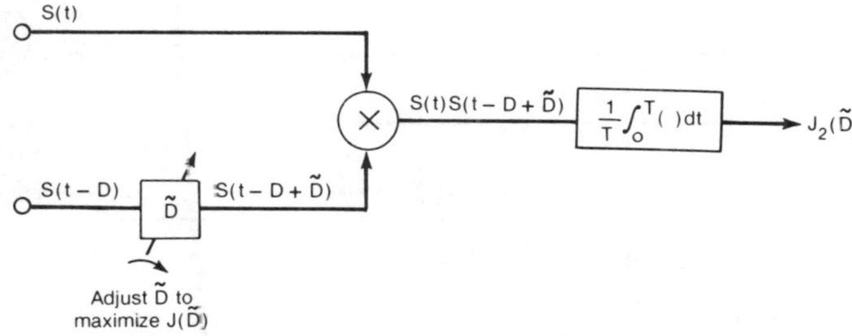

Fig. 39. Conceptual cross correlator configuration (Carter 1981b).

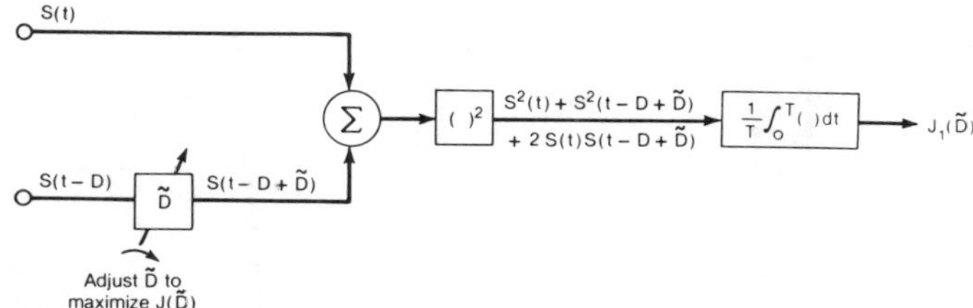

Fig. 40. Conceptual delay, sum, square, and integrate configuration (Carter 1981b).

functions for the ML estimator just derived. The ML, or optimum weighting function, contains the fundamental term $C/(1 - C)$. Thus, greater weight is placed on frequency bands that give near-unity coherence. In contrast, frequency bands in which the coherence is near zero are deemphasized. It should be emphasized, however, that the ML estimator achieves the minimum variance for time delay estimation only while "in track," or "in lock." It does *not* take into account large spurious anomalies that can occur. There are several other processing techniques, such as the smoothed coherence transform (SCOT) mentioned earlier (Carter, Nuttall, and Cable 1973). The SCOT can be seen as the Fourier transform of the complex coherence and is a very robust processor. It has a 1/noise kind of behavior at low SNRs and thus attenuates bands of noise. But, it also has a signal spectrum term that tends to deemphasize areas where the signal spectrum is very strong. This has a fortuitous effect on many practical problems. There does yet not seem to be a mathematical justification for why the SCOT is so robust. A challenge for the reader is to determine why the SCOT is such a good practical estimator. There are a number of other estimators, including the Roth (Roth 1971) and Wiener processors and the phase transform (PHAT), which, in practice, perform very well. The weighting functions for these are also given in Figure 37.

Another interpretation of ML time delay estimation is depicted in Figure 38, where instead of multiplication, summation is used. This system can be thought of as a two-sensor beamformer with prefilters.

As with the GCC processing, the linear filters, in general, will accentuate the frequency bands where the signal is strong and attenuate the bands where the noise is strong. For example, in signal detection theory at low SNRs, Eckart filters (square root of signal spectrum over noise spectrum) are used. Time delay estimation also results in the same Eckart filters at low SNRs. Instead of cross correlating (multiplying), the signals can be combined by squaring and integrating and then evaluating the "cost" function as the hypothesized time delay is varied. When the peak "cost" output is achieved, this is also the ML estimator for delay.

Considering the noise-absent case in terms of multiplication (Figure 39), the delayed signal is the input to a hypothesized variable delay. The signal minus a delay plus a hypothesized delay must be made to match the signal without delay as closely as possible. Multiplication then produces a signal-squared term, and integration is performed to obtain the maximum.

If a summer (that is, an adder) instead of a multiplier is used, as shown in Figure 40, we again obtain a signal-squared term in the region of the correct delay, where the hypothesized delay is approximately equal to the true delay. Another signal-squared term is then acquired together with two more signal-squared terms, resulting in four signal-squared terms. This cost function, in the neighborhood of the correct time delay, again behaves as signal-squared. Consequently, a summer and a multiplier both produce similar signal-squared results.

Figure 41 conceptually illustrates how TDE can be implemented in hardware for three discrete units of hypothesized delay. Both a signal and a delayed signal are received. Three delays are inserted; then the signals are added, squared, and integrated. A memory bank of as many of these delayed versions (sort of a tapped delay line) as are needed for the search is kept.

Figure 42 shows a simple example of the time delay estimation process. Two signals, $s(t)$ and $s(t - 2)$, are present. Time, t, is given in the first column. The second column describes the waveform. The third column shows the signal

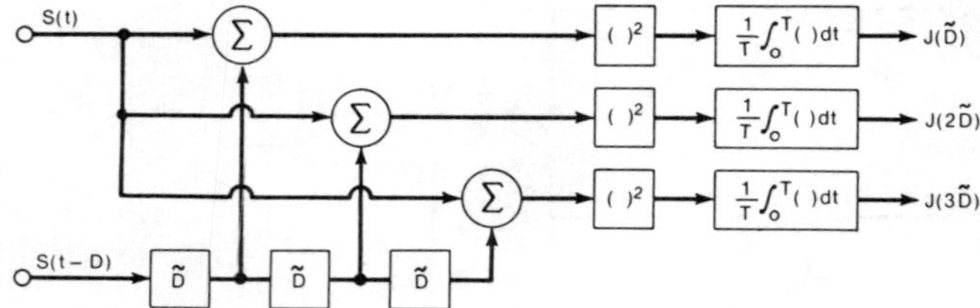

Fig. 41. Time domain beamformer instrumentation for three hypothesized delays (Carter 1981b).

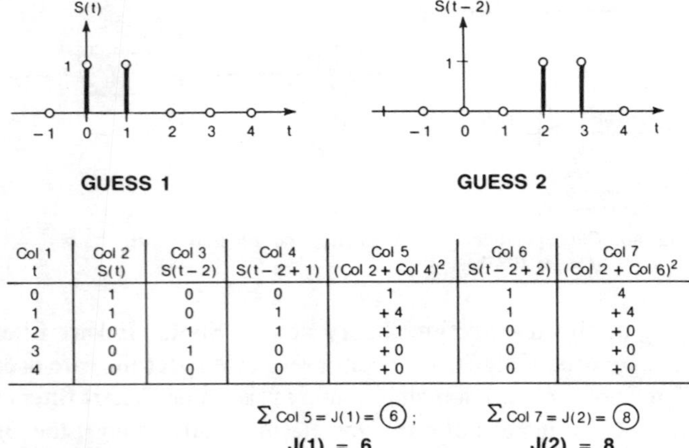

Col 1 t	Col 2 S(t)	Col 3 S(t − 2)	Col 4 S(t − 2 + 1)	Col 5 (Col 2 + Col 4)²	Col 6 S(t − 2 + 2)	Col 7 (Col 2 + Col 6)²
0	1	0	0	1	1	4
1	1	0	1	+ 4	1	+ 4
2	0	1	1	+ 1	0	+ 0
3	0	1	0	+ 0	0	+ 0
4	0	0	0	+ 0	0	+ 0

Σ Col 5 = J(1) = (6) ; Σ Col 7 = J(2) = (8)

J(1) = 6 J(2) = 8

Fig. 42. Example of pulsed arrival for one instrumentation (Carter 1981b).

delayed by 2. Column four is guess 1 of one unit of delay, and column 6 is guess 2 for two units of delay. In columns 5 and 7 the cost function is computed. The entries in column 5 add up to 6, and the entries in column 7 add up to 8. In other words, with a hypothesis or guess of two units of delay, the maximum is achieved, which becomes the estimate of the delay.

In the *IEEE Transactions on Acoustics, Speech, and Signal Processing* special issue on time delay estimation (June 1981) edited by Carter, Quazi (1981b) evaluated the CRLB based on the theory that Knapp and Carter (1976) developed; these evaluations are very helpful because they provide actual performance curves. The CRLB is dependent on the $C/(1 - C)$ term seen earlier in the theory of coherence as well as on the amount of observation or integration time. Making assumptions about limited bandwidth and low-pass characteristics results in an SNR-type term, $1/(B^3T)$, where B is the bandwidth and T is the observation time (see Figure 43).

Up to this point, it has been assumed that tracking has "locked" on the right delay. The CRLB does not take into account large (or spurious) estimation errors, such as those possible with GCC functions like the one shown in trace 5 of Figure 33. If we consider a process that occurs at many different frequencies or the center (RMS) frequency of a broadband process, we obtain the expression given for the variance (σ^2) in Figure 43; later we will evaluate the performance due to spurious peaks.

$$\sigma^2_{CRLB} = \left| T\int_{-\infty}^{\infty} (2\pi f^2) \frac{C_{r_1r_2}(f)}{1-C_{r_1r_2}(f)}\, df \right|^{-1}$$

- **ASSUME LOW-PASS, FLAT SIGNAL AND NOISE POWER SPECTRA**
- **DEFINE** $SNR = G_{ss}(f)/G_{nn}(f) = CONSTANT,\ 0 \leq |f| \leq B$
  $SNR = 0$, **ELSEWHERE**

$$\sigma^2_{CRLB} = \left| T\int_{-B}^{B} (2\pi f)^2 \frac{SNR^2}{2\,SNR+1}\, df \right|^{-1}$$

$$= \frac{3}{8\pi^2} \cdot \frac{2\,SNR+1}{SNR^2} \bullet \frac{1}{B^3T}$$

B = BANDWIDTH
T = OBSERVATION TIME

Fig. 43. Overview of the Cramér-Rao lower bound (CRLB) for TDE (Quazi 1981b).

$$\sigma_D = \frac{1}{2\pi}\ \frac{1}{\sqrt{2TW}}\ \frac{1}{f rms}\ \sqrt{\frac{1+SNR_1+SNR_2}{SNR_1 \cdot SNR_2}}$$

$$\simeq \frac{1}{2\pi}\ \frac{1}{\sqrt{2TW}}\ \frac{1}{f rms}\ \frac{1}{SNR};\quad SNR_1 = SNR_2 = SNR << 1$$

$$\simeq \frac{1}{2\pi}\ \frac{1}{\sqrt{2TW}}\ \frac{1}{f rms}\ \frac{1}{\sqrt{SNR/2}};\quad SNR >> 1$$

Fig. 44. Equations for the standard deviation of time delay errors (σ_D) (Robinson and Quazi 1985).

Figure 44 gives equations for the standard deviation of time delay errors, with approximations for low and high SNR. (In Figure 44, W is the bandwith.) Figure 45 shows a plot of the standard deviation of time delay errors versus average SNR in the band for integration times of 20 s and 60 s.

Other errors besides time delay estimation errors, such as uncertainties in knowing the exact sensor positions (Carter 1979a), can influence bearing estimation. In the ideal case (including known sensor position), for larger and larger SNRs, arbitrarily good performance can be achieved (Figure 46). However, if there is uncertainty about the array's position, a limiting performance based on the size of the uncertainty occurs. From a performance point of view, these errors may dominate and become much more critical than the SNR.

Earlier, we stated that we would derive the ML estimator, evaluate the CRLB, and then run a simulation to ensure that the derivation was correct. In our study so far, an examination of the CRLB theory at high SNR after the simulation was run

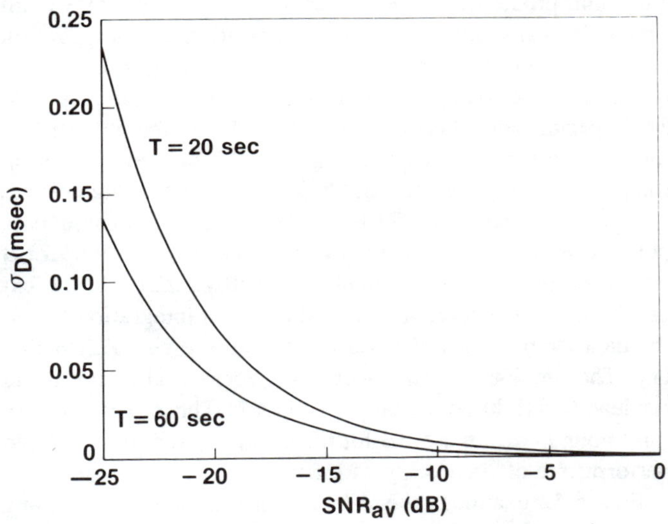

Fig. 45. Standard deviation of time delay errors as a function of SNR_{av} (W = 500 Hz, f_0 = 1500 Hz) (Quazi and Lerro 1985).

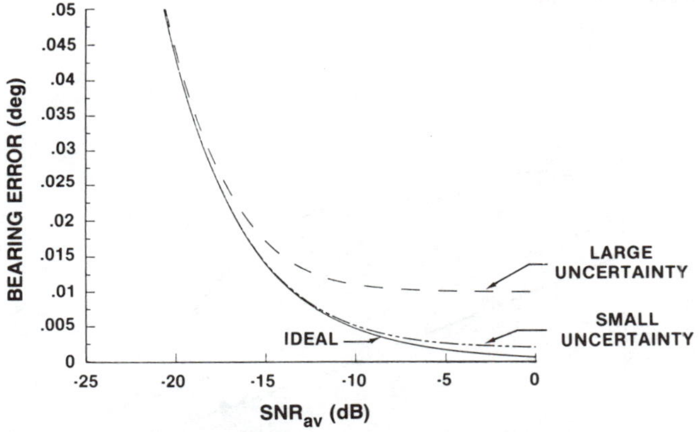

Fig. 46. Standard deviation of bearing error with positional uncertainty as a function of SNR_{av} (W = 500 Hz, f_0 = 1500 Hz, L = 500 ft, T = 20 s) (Quazi and Lerro 1985).

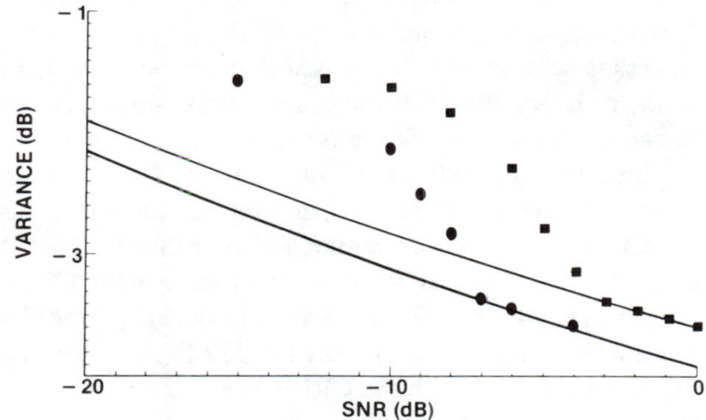

Fig. 47. Comparison of simulation results and the CRLB (T = 2, 8 s) (Scarbrough, Carter, and Tremblay 1984).

1) NARROW-BAND SIGNALS (AMBIGUITIES)

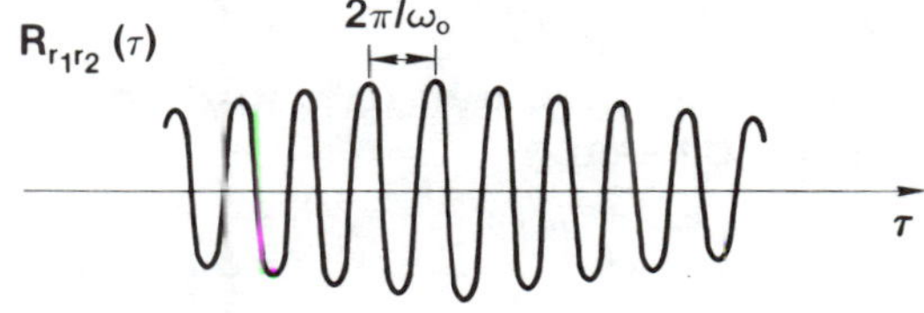

2) LOW-BAND SIGNALS (ANOMALOUS ESTIMATES)

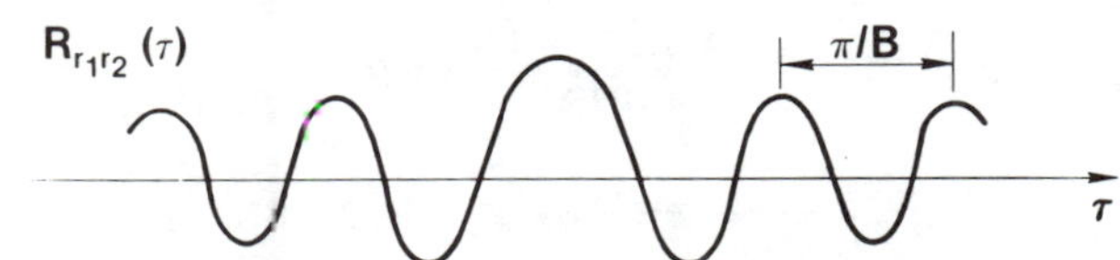

Fig. 48. Typical narrowband and low-band signal cross correlation (Weinstein and Weiss 1984). (a) Narrowband signals. (b) Low-band signals.

showed that the theory and data agreed. To verify the data further, additional experiments were conducted. The results of two of these tests, based on two different processing times, T, are shown in Figure 47.

There is some agreement between the theory and the experiment but there is also disagreement. Recall that the theory stated that a large-enough integration time would achieve the CRLB, but it did not specify how large that time was required to be. The bound can be achieved, provided that the integration time is long enough. As the SNR decreases, the integration time must increase in order to achieve the CRLB.

From a different perspective, Fourier transform pairs are sharp in one domain and broad in the other. Therefore, a waveform with deltalike behavior in one domain yields a sinelike wave in the other domain. If there were one or two strong narrowband components in the frequency domain, then when the Fourier transform was taken, we would obtain a cross-correlation function with significant oscillation or ringing (see Figure 48).

Indeed, one of the reasons the SCOT (a non-ML estimator) may be a good technique is that it prewhitens the signal and tones down those strong sinusoids. The CRLB theory developed (for the ML estimation) applies to the performance only as long as the local peak is tracked but does not address the fact that we may lose track and follow another, incorrect peak. These concerns must be addressed in order to modify the theory. There is some important work by Ianniello, for instance, on the correlator performance estimate (CPE) for time delay estimation during lost track, which has been presented at International Conferences on Acoustics, Speech, and Signal Processing. There has also been fundamental work by Weinstein and Weiss (1984) and statistical work by Chazan, Zakai, and Ziv (1975).

Figure 49 is an overview of some mathematics relating to a new bound, based on Ziv and Zakai's work, called the ZZLB. If a process is set up where it is possible to arrive at an incorrect peak, a theory that will more accurately predict performance can be developed. Figure 50 shows several subtleties. First, at high-enough SNRs, the ZZLB and the CRLB are the same. Second, the ZZLB is a larger (that is, tighter) bound than the CRLB. Although we cannot do better than the CRLB, we can do worse. At low SNR, we will not achieve the CRLB; we can achieve only the ZZLB. (See Scarborough, 1984)

Also, a source of confusion is why the ZZLB curve looks as if it will intersect again with the CRLB at a low-enough SNR. The reason is that the ZZLB takes into account prior physical information about the maximum time delay (as applied to this particular problem). Recall that there is a (physical) maximum time delay caused by fixed sensor separation and propagation velocity. There exists a transition region having to do with a threshold SNR analogous to losing track. Below the threshold SNR, performance becomes very poor due to the process losing track. Once this happens, great difficulty in estimating time delay or equivalently in tracking source bearing occurs.

Figure 51 shows a CRLB, a ZZLB, and simulated data points from two experiments. Most critical now are the ZZLB and those data points associated with that ZZLB. In the experiments shown here, which used different integration times, the data from the simulations agree with the new ZZLB theory. The previous CRLB theory was correct, although it did not lead to a tight bound on performance. The ZZLB is a very tight bound (within 0.5 dB for the cases studied) on predicted performance of time delay estimators.

Figure 52 examines the ZZLB as a function of increasing integration time. The threshold SNR point at which the process loses track can be reduced to progressively lower and lower SNRs by increasing the coherent integration time. This, in turn, can give significant performance improvement because track is not lost until a much lower SNR.

H_0: TRUE DELAY $= \tau + \Delta,\ \Delta \geq 0$
H_1: TRUE DELAY $= \tau,\ -D_o \leq \tau \leq D_o - \Delta$

- DEFINE $P_e(\tau, \tau+\Delta)$ = MINIMUM ATTAINABLE PROBABILITY OF ERROR

$$\sigma^2_{ZZLB} = \frac{1}{2D_o}\int_o^{D_o} \Delta\, G\left[\int_o^{D_o-\Delta} P_e(\tau, \tau+\Delta)\, d\tau\right] d\Delta$$

$$= \sigma^2_{CRLB}\left[1 + 2(a^2-1)\,Q(a) - \frac{2a}{\sqrt{2\pi}} e^{-a^2/2}\right] + 2P_e D_o^2/3$$

WHERE

$$a^2 = 2BT\,\frac{\sqrt{1+SNR'}-1}{\sqrt{1+SNR'}},\quad SNR' = \frac{SNR^2}{2SNR+1},$$

$$Q(a) = \int_a^\infty \frac{1}{\sqrt{2\pi}} e^{-x^2/2}\,dx,\quad P_e = F(B, T, SNR')$$

- AS $SNR \to \infty$, $P_e \to 0$, $\sigma^2_{ZZLB} \to \sigma^2_{CRLB}$
- AS $SNR \to 0$, $P_e \to 1/2$, $\sigma^2_{ZZLB} \to D_o^2/3$

Fig. 49. Overview of the Ziv-Zakai lower bound (ZZLB) for TDE (Scarbrough 1984).

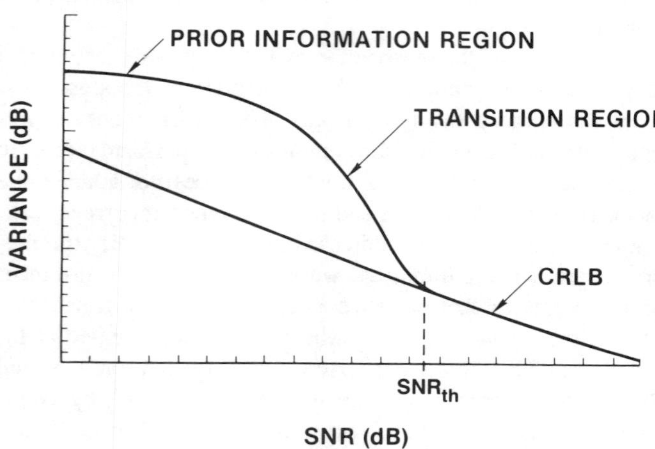

Fig. 50. Comparison of the ZZLB and the CRLB (Scarbrough, Carter, and Tremblay 1984).

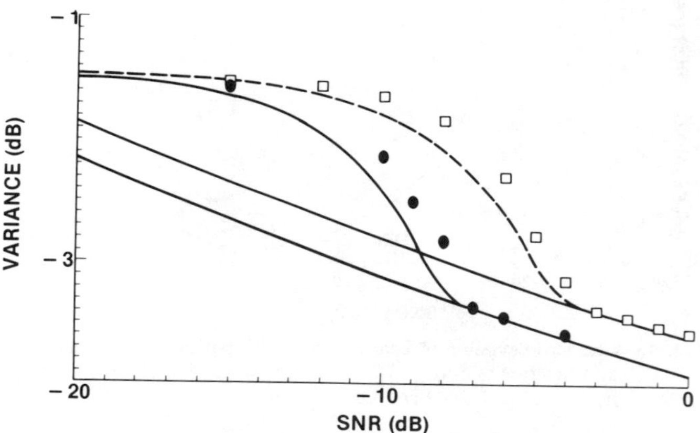

Fig. 51. Comparison of simulation results and the ZZLB ($T = 2, 8$ s) (Scarbrough, Carter, and Tremblay 1984).

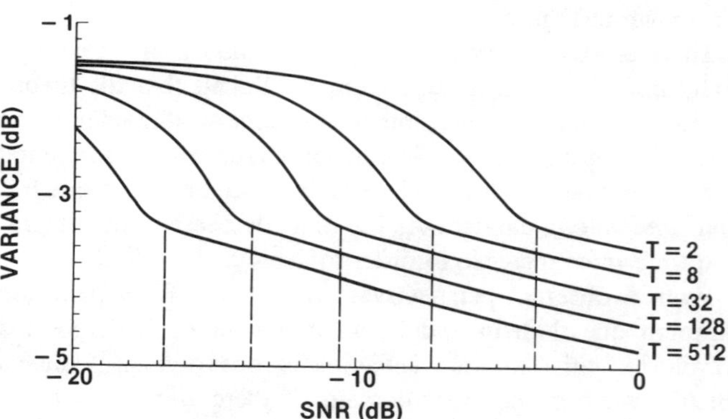

Fig. 52. Effect of increasing the integration time (T) on the ZZLB. (Scarbrough, Carter, and Tremblay 1984).

CPE

$$SNR_{th} = \frac{X}{\sqrt{1 - X^2/2} - X}, \quad X = \frac{1}{\sqrt{BT}} \, Q^{-1}\left(\frac{P[A]}{4D_oB - 1}\right)$$

- FOR LARGE BT, X IS SMALL

$$SNR_{th} \simeq \frac{1}{\sqrt{BT}} \, Q^{-1}\left(\frac{P[A]}{4D_oB-1}\right)$$

- EMPIRICALLY, $P[A] \simeq 5 \times 10^{-6}$

ZZLB

$$SNR'_{th} \simeq \frac{1}{BT}\left[F_{+}^{-1}\left[\left(\frac{3}{4\pi BD_o}\right)^2\right]\right]^2, \quad SNR' = \frac{SNR^2}{2\,SNR + 1}$$

$$F(X) = X^2 \, Q(X)$$

- FOR LARGE BT, $SNR' \simeq SNR^2$

$$SNR_{th} \simeq \frac{1}{\sqrt{BT}} \, F_{+}^{-1}\left[\left(\frac{3}{4\pi BD_o}\right)^2\right]$$

Fig. 53. Threshold SNR behavior (Scarbrough 1984).

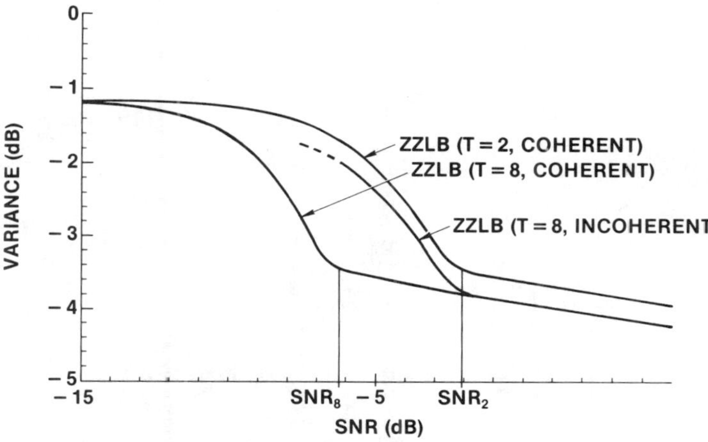

Fig. 54. Impact of coherent and incoherent processor performance on the ZZLB (Scarbrough, Carter, and Tremblay 1984).

Figure 53 describes two approaches to developing the variance of time delay estimates. One, called the correlator performance estimate (CPE), was developed by Ianniello (1982B). Unlike the ZZLB, the CPE is not a bound but a prediction. It recognizes that usually we are going to implement these time delay estimators with a cross correlator or GCC function and predicts the performance of these correlators (see Ianniello (1982b), Weiss and Weinstein (1982), or Scarbrough (1984)). From these results, the threshold SNR where track is lost can be computed. Global predictions of performance for bearing, range, and Doppler have been provided by Moura and Baggeroer (1978) and Moura (1979a, 1979b), including corroboration of results with simulations.

The point at which we lose track affects how we do our processing. For example, a certain amount of data—say eight units of time—could be processed that in a way that was partially coherent and partially incoherent. For example, if we add eight complex numbers coherently, we add eight real parts and eight imaginary parts. If we add them partially coherently, we might add subsets coherently and then compute and add magnitudes. (See Scarbrough, Carter, and Tremblay (1984) for a fuller explanation.) This concept is illustrated in Figure 54.

If the data were processed as partially incoherent, we might move from one of these curves to the other (incrementing the gain slightly), but if the data were processed as fully coherent—say eight units of time—threshold performance would be improved. Because there is loss of track, there is also a tremendous opportunity for performance improvement—that is, to regain lost performance (see Scarbrough, Carter, and Tremblay (1984)).

It has been established that time delay is directly related to bearing. The ML estimator for time delay has been derived and performance has been evaluated. It has been shown that the bounds, the CRLB and ZZLB, experience a significant decrease in the threshold SNR before losing track when coherent integration time is increased, thereby improving performance. Now we extend bearing estimation to range estimation and examine some fundamental performance limits. Finally, we investigate the moving time delay case, because as the processing time for these coherent processors is increased, longer integration is performed, and thus a time-varying time delay must be addressed.

1.5 Passive Sonar Localization and Fundamental Limits of Performance

In an article on *focused beamforming,* Carter (1977b) shows that estimating range and bearing with a number of hydrophone receivers (in this case M sensors) involves running the sensor outputs through M filters and then delaying, summing, squaring, and integrating the signals. Without the hypothesized range and bearing constraints on the lower part of Figure 55, the cost function J would be a function of a time delay (TD) vector (for example, see Hahn and Tretter (1973)). If there were no focusing—that is, no slaving or ganging of delays—the cost J would be a function of the $M - 1$ delays. All these delays could be adjusted to achieve a maximum in this M-dimensional space. Because there is a physical problem (acoustic energy coming from one point in space), however, that is causing the signal to propagate to the M sensors, hypothesized range and bearing dials could be adjusted instead, focusing all the time delays to some point in space. As hypothesized range and bearing are adjusted, all the delays are "ganged" together so that only certain combinations of the delays (those that make physical sense) are inserted into the network. This is the ML instrumentation for range (R) and bearing (B) and is called the focused beamformer (see Carter (1977b)); it is also called focalization.

Figure 56 shows an array of length L with M receiving sensors, a source at bearing B, and range R. For a particular integration time and SNR, there is an uncertainty region due to our inability to estimate the source location. This uncertainty region is extremely elongated in the range direction; it is very difficult to estimate passively the range (distance) of a source. Pioneering ranging work by Bangs (1971) quantified ranging

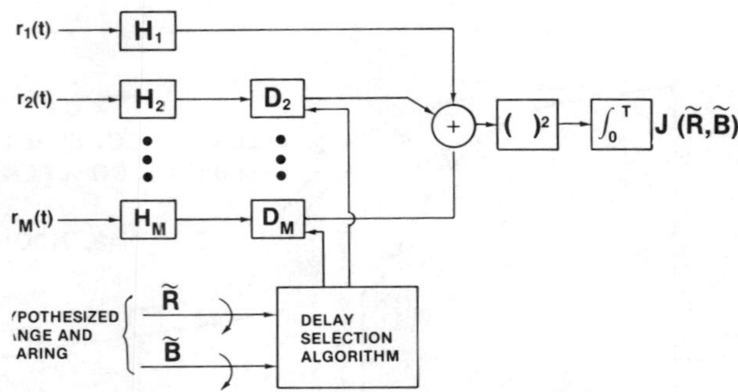

Fig. 55. Focused beamforming by maximum likelihood delay selection (Carter 1977b).

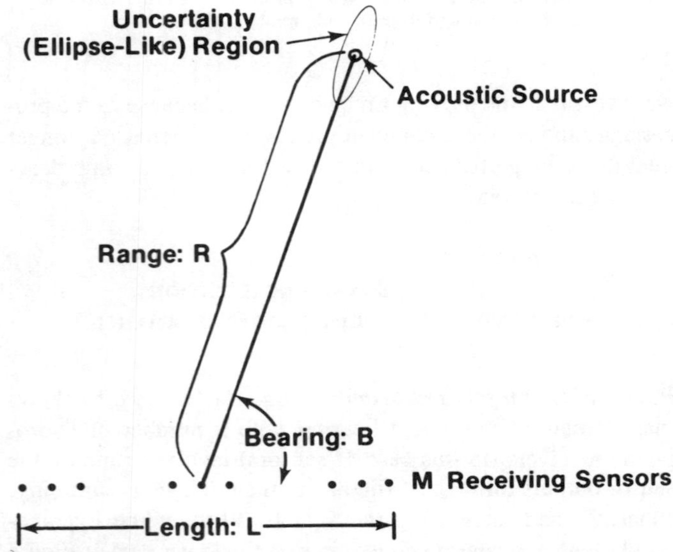

Fig. 56. Array geometry used to estimate source position passively (Carter 1977b).

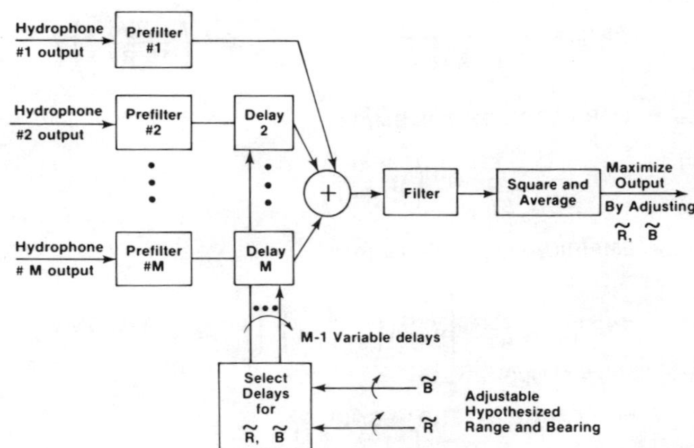

Fig. 57. Maximum likelihood estimator for range and bearing (Carter 1977b).

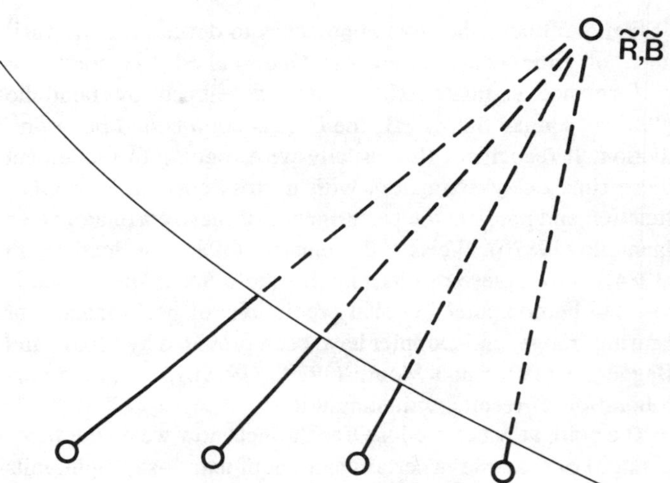

Fig. 58. Focusing on a hypothesized range and bearing by adjusting individual delays.

methods and described the extent of the difficulty. Bangs used a coordinate system, which, in retrospect, makes the problem mathematically difficult (for some cases). If a coordinate system is chosen for a symmetric line array that is situated in the middle of the array, the mathematics become simpler, and the errors in range and bearing (or cross range) become uncorrelated. (See Carter (1977b) for a discussion of the assumptions and approximations leading to this result.)

The ML estimator for range and bearing is depicted in Figure 57. For low SNR, classical Eckart filters are used for the prefilters. For estimates of time delay, for detection, or for focused beamforming, the transfer function of the prefilters at low SNR is given by

$$H(f) = \frac{\sqrt{G_{ss}(f)}}{G_{nn}(f)}. \tag{22}$$

For our simplified purposes, the postfilter (after the "summer") is unity or a low-pass smoothing filter.

Figure 58 illustrates the process of hypothesizing a range and bearing, inserting the delays needed to focus on the location, adjusting the hypothesized position by indirectly adjusting the delays slaved to the hypothesized range and bearing adjustments, and maximizing the output J of the system.

An expression for the CRLB for estimating bearing and for estimating range is

$$\sigma^2(\hat{\xi}) = \frac{2\pi C^2}{TU_\xi \int_0^\Omega V(\omega)\omega^2\, d\omega}. \tag{23}$$

In this expression, the symbol ξ is to be replaced by R or B (i.e., $\xi = R$ or $\xi = B$). In this CRLB, there are three terms in the denominator (T, U, V). (Note the change in notation here; C is not the MSC but is, instead, the speed of sound in the medium.) The terms T, U, and V should be made large to reduce the variance of the estimated range or bearing. V is a function of the output SNR, U is a function that depends on whether the estimates are of bearing or range, and T is the observation time. Since V is in the denominator, it should be made large. Expanding V yields two different forms, one for low-output SNR and one for high-output SNR. For low-output SNR, V is described by

$$V(\omega) \cong M^2 \frac{S^2(\omega)}{N^2(\omega)} \quad \text{(low SNR)}. \tag{24a}$$

[Note that S is now the signal power spectrum G_{ss}; that is, these terms are power spectral densities (PSDs) and not Fourier transforms.] When operating at low-output-power SNR, it helps greatly to increase the number of sensors (M). If the number of sensors is doubled at low-output SNR, a factor-of-four improvement is observed in variance. For high-output SNR, however, the expression for V changes:

$$V(\omega) \cong M \frac{S(\omega)}{N(\omega)} \quad \text{(high SNR)}. \tag{24b}$$

Therefore, until high-output SNR is reached, the addition of new receiving sensors (hydrophones) in the array contributes greatly to improved performance. Once high-output SNR is achieved, though, doubling M only doubles the array performance. Therefore, changes in time, T, and in the number of sensors, M, in the denominator have the same effect at high-output SNR. Recall that U is different whether we estimate bearing (U_B) or range (U_R); therefore, U_B should be made large in Figure 59.

Also, from Figure 59, the effective length of the array (L_e) should be made large—that is, the length of the array times the sine (or cosine, depending on the definition of bearing angle) of the array projected toward the source. To make L_e large, (1) the array aperture must be made physically large (a long baseline), and (2) the array must be physically steered broadside to the acoustic source. For quantifying bearing estimation performance, there is a constant that depends on the location of the M receiving sensors on a line. There is a theoretical limit on what occurs if half the elements are placed at one end of the array and half at the other end (see MacDonald and Schultheiss (1969). This constant is 2, and since U_B should be large, K_B must be made small. The smallest K_B in this table is 2. Note, though, that the values shown are bounds. No group of hydrophone elements could ever be physically located at one spot on the array; in practice, the sensors would have to be placed at half-wavelength spacing. However, with a large aperture and a limited number of hydrophones, bearing estimation is best accomplished by putting half the hydrophones toward each end of the array.

$$U_B = \frac{L_e^2}{K_B}$$

Array Type	K_B	
Equispaced Line	6	
M/2, 0, M/2	2	⬅Bound
M/3, M/3, M/3	3	
M/4, M/2, M/4	4	

Fig. 59. Comparison of various array configurations for bearing estimation (Carter 1987).

$$U_R = \frac{L_e^4}{K_R\, R^4}$$

Array Type	K_R	
Equispaced Line	360	
M/2, 0, M/2	∞	
M/3, M/3, M/3	144	
M/4, M/2, M/4	128	⬅Bound

Fig. 60. Comparison of various array configurations for range estimation (Carter 1987).

For range estimation, the expression for U_R shown in Figure 60 is needed to compute the variance of the range estimate. Again, with T, U, and V in the denominator of Equation 23, U_R should be made large. Here, the expression for U_R is different than for bearing, U_B. Also, the variance expression now has an $(L_e/R)^4$ effect. Fourth-power effects have great significance. Passive ranging, for instance, is extremely difficult (Carter 1978a). In this case, half the elements should be placed in the middle of the array and a quarter toward each end. The constant, K_R, of 128 in the performance equation serves as the bound on performance. Note, though, that the constants (namely, 144 and 128) are approximately equal.

In Figure 61, the passive ranging problem is examined in terms of relative range error. Relative range error is proportional to how well bearing estimation is accomplished. The relative range errors are related to the ratio of the range to the length of the array. As the true range doubles, the standard deviation quadruples, reflecting a critical inability to estimate range passively. (For different array configurations, the performance constants remain about the same.)

To minimize the location uncertainty region, one-third of the elements should be placed toward each end and one-third in the middle of the aperture (see Figure 62).

Another complexity in time passive ranging is illustrated in Figure 63, which shows an array of three sensors with assumed positions, but the actual sensors are not in these positions. For this case, bearing lines are drawn from where the sensors are thought to be and from where they actually are, showing a mismatch between the actual source position and its assumed position.

The relative range error is a function of the standard deviation of the perturbations of the array:

$$\frac{\sigma(\hat{R})}{R} \cong \frac{2R}{L_e} \circ \frac{\sigma(\hat{p})}{L}. \tag{25}$$

NOTE: AS R DOUBLES σ(R) QUADRUPLES

$$\frac{\sigma(\hat{R})}{R} = \sqrt{\frac{K_R}{K_B}} \left[\frac{R}{L_e}\right] \sigma(\hat{B})$$

Array Type	$\sqrt{\frac{K_R}{K_B}}$
Equispaced Line	7.75
M/2, 0, M/2	∞
M/3, M/3, M/3	6.9
M/4, M/2, M/4	5.7 ← BEST

NOTE 5.7 ≅ 6.9 ≅ 7.75

Fig. 61. Comparison of the relative range errors for various array configurations (Carter 1987).

FOR ESTIMATING	BEST ARRAY CONFIGURATION
BEARING	• • • • • • M/2 • • • • • • M/2
RANGE	• • • M/4 • • • • • • M/2 • • • M/4
POSITION	• • • • M/3 • • • • M/3 • • • • M/3

Fig. 62. Optimum sensor configurations for a linear array (Carter 1977b).

Note that by multiplying by R and squaring, the variance again shows a fourth-power dependence on range. If the true range doubles, the standard deviation of the range error quadruples, whereas the variance increases by a factor of 16. Moreover, with the addition of other real-world complexities, it becomes even more difficult to estimate range passively.

Another difficulty involves source motion during long processing times. A benefit to be derived by increasing the integration time is the reduction of the threshold SNR at which track is lost. Longer processing time works well for a stationary acoustic source, but with an acoustic source that changes bearing, complexity increases, and the question becomes, How do we estimate time delay with a moving source?

1.6 Time-Varying Time Delay Estimation

Knapp and Carter (1977) describe the modeling of motion as a time compression (either a Doppler or a generalized Doppler). The signal gets compressed so that different frequencies

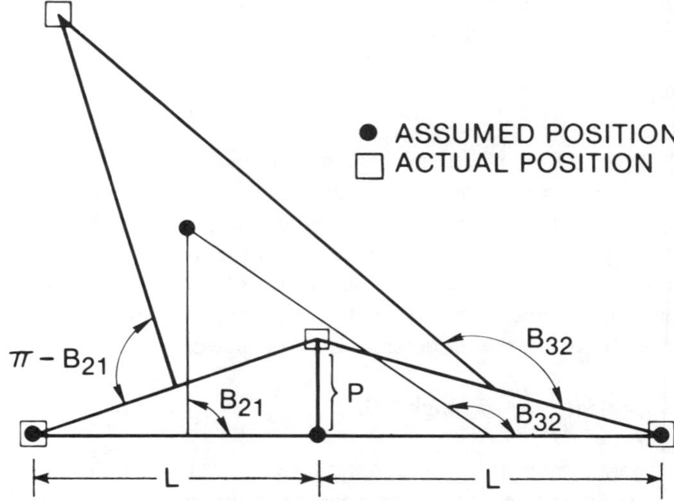

Fig. 63. Effect of sensor positional uncertainty on localization (Carter 1979a).

shift by varying amounts. The modeling can be done as a time compression or a time-scaling factor, β:

$$r_1(t) = s(t) + n_1(t), \tag{26a}$$

$$r_2(t) = s(\beta\,[t + D]) + n_2(t). \tag{26b}$$

Substituting

$$d_0 = \beta D, \qquad d' = \beta - 1, \tag{26c}$$

yields

$$r_2(t) \cong s(t + d_0 + d't) + n_2(t). \tag{26d}$$

The paper by Knapp and Carter (1977) refers to events that occur when the wide-sense stationarity of these received processes $r_1(t)$ and $r_2(t)$ is examined. The terms $r_1(t)$ and $r_2(t)$ are not jointly wide-sense stationary, so the usual graduate-level mathematics and typical tools do not apply to this particular problem. Not surprisingly, developing an estimator involves a fair amount of educated guesswork for determining the proper amount of compression and then for trying to locate the estimator in a region that allows quasi-stationarity to be achieved.

A Taylor series expansion of time compression yields just what we expect. Thus, instead of only an estimate of the time delay, d_0, an estimate of the time delay rate, d', must also be made. A coherent processor must compensate for d'. Physical models where this would apply are a stationary case with a time delay and a physical case where there is time delay rate. Consideration must also be given to time delay acceleration in cases where the distance between the acoustic source and the sensors is small. Such a condition contributes an added level of complexity to the signal processing. Depending on the geometries involved, significant gains can be made or significant losses may have to be tolerated.

The ML estimator, derived by Knapp and Carter (1977) and shown in Figure 64, depicts an interesting result for moving time delay cases. This figure explores a very difficult problem with some complicated mathematics (note the change in

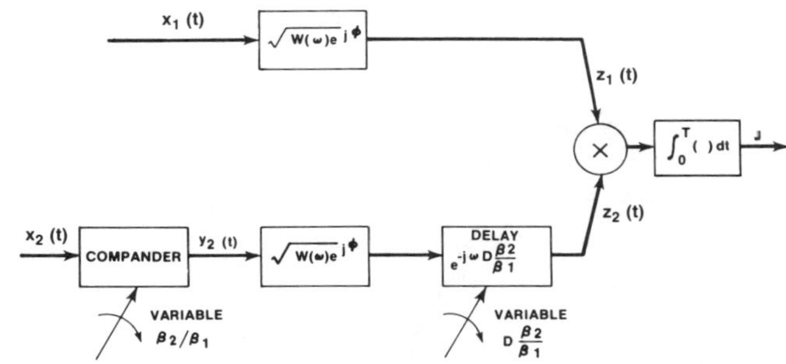

Fig. 64. ML estimator including delay and delay rate (Knapp and Carter 1977).

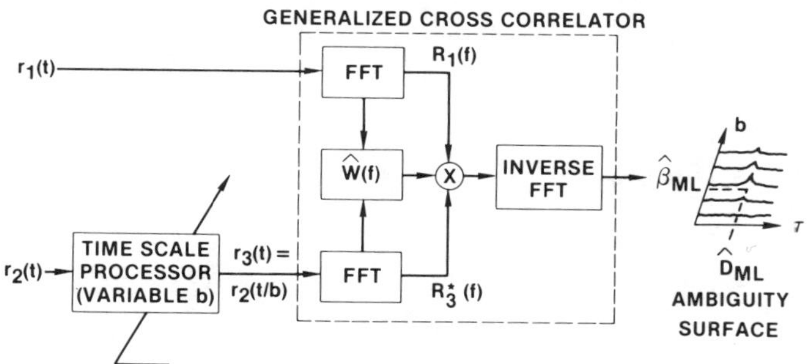

Fig. 65. The ML estimator (Scarbrough 1984).

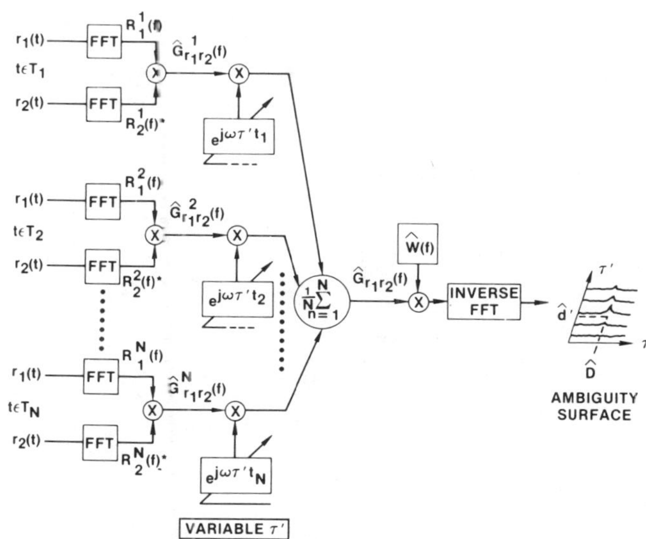

Fig. 66. Frequency domain compensator (Scarbrough 1984).

notation: $x_1 = r_1$). In this situation, the signal is received, run through an expander or contractor (compander), filtered according to the SNR, and delayed in time. This signal is then multiplied with the other signal, yielding a new function, which is a function of how much the Doppler shifted, expanded, or contracted, along with another parameter, the time delay. The result is a two-dimensional ambiguity surface that is a function of delay and delay rate.

Another way to treat the ML estimator is shown in Figure 65 (see Scarbrough (1984)). Here the received signal $r_2(t)$ must be either variably sampled or variably expanded and contracted. This is difficult to do digitally, once a waveform has been sampled, since it is very difficult to sample again at a different rate. (It could be done, however, with a bank of different sampling rates on the original analog waveform or with great difficulty with a very sophisticated digital processing box.) Next, the weighting function is estimated and cross PSDs are formed (taking into account the estimated weighting function). The inverse FFT is computed, resulting in an ambiguity surface that is a function of delay and delay rate. The (x, y) location of the peak of the ambiguity surface provides the estimate of delay and delay rate.

The frequency domain compensator developed by Scarbrough (1984) is shown in Figure 66. Here, an FFT is computed for each pair of $r_1(t)$ and $r_2(t)$ from the N time segments, then the signals are multiplied to give the cross-power spectrum at a certain frequency for each time segment. Without the Doppler compensation, the processing would be similar to the earlier WOSA spectral estimation method, where summing was performed (Equation 17). This sum over the available time segments, for $n = 1$ to N, would then just be averaging over all the time segments. Because Doppler compensation is being performed, however, each of these must be multiplied by a complex exponential, depending on the hypothesized time delay rate. The result is an ambiguity surface of delay rate and delay; the location of the peaks in that surface are the delay and delay rate estimates.

As before, a simulation was performed to determine the effects at a particular SNR and whether the estimator would work for a well-defined delay and delay rate problem. (Note the two incorrect data points in Figure 67.) With the exception of an occasional incorrect point, the method works.

Another simulation with a lower SNR is shown in Figure 68 (see also Scarbrough (1984)). Although lower SNR reduces performance, the simulations demonstrate that these estimators can be built and implemented (with 1984 technology) successfully.

1.7 Complexity of the Ocean Environment

In the ocean environment, the passive sonar problem consists of listening devices (hydrophones) that receive radiating energy from an acoustic source (Figure 69).

The energy propagates along ray paths through the medium (possibly reflecting off the layered bottom or off the ocean surface) and perhaps travels through ducted channels to the receivers. There are many variables: The receivers may be stationary or moving, the source may be stationary or moving, there may be more than one source, or there may be multipath. The problem is to detect the presence of the acoustic source, estimate its state vector, analyze performance, and compare that analysis with the performance bounds.

In a multipath ocean environment, the acoustic waves radiating from the source are considered and then perpendiculars

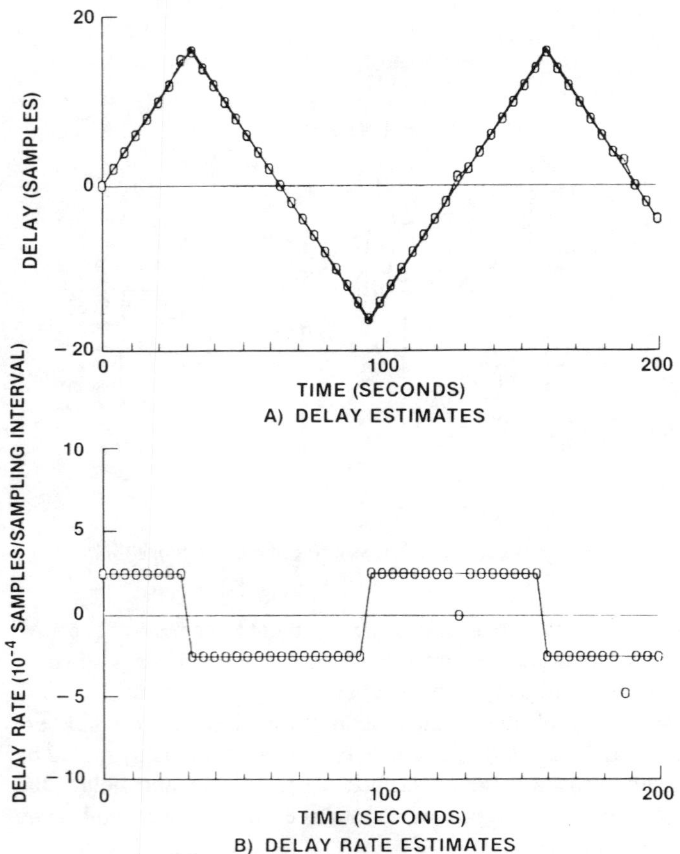

Fig. 67. Simulation of delay and delay rate estimation (1) (Scarbrough, Carter, and Tremblay 1984). (a) Delay estimates. (b) Delay rate estimates.

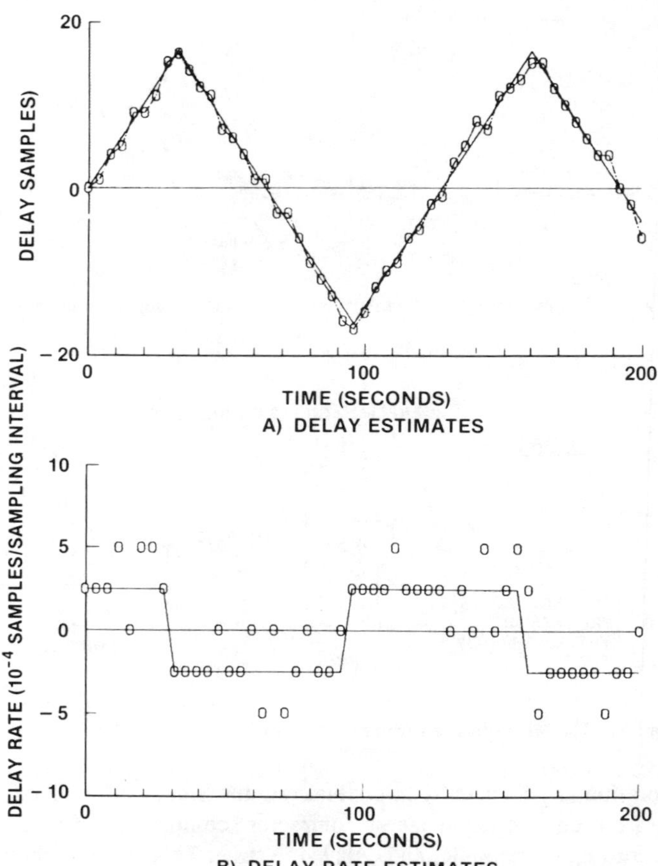

Fig. 68. Simulation of delay and delay rate estimation (2) (Scarbrough 1984). (a) Delay estimates. (b) Delay rate estimates.

to the wave front are drawn. These are acoustic rays; a ray from the source to the receiver and a ray from the source to a reflecting surface and then to the receiver form a very simple receiver model of a signal plus a delayed signal (Figure 70).

Typically, there is an attenuation constant, shown in the figure as α, associated with reflection. When energy impinges on a surface such as the ocean surface, which is a pressure-release surface, there is a 180° phase shift or a negative sign associated with the attenuation constant. Also, there is usually additive noise present at the receiver, shown in Figure 70 as $n(t)$, which in the simplest case is modeled as being uncorrelated from sensor to sensor.

The PSD of a two-path signal is $s(t) + \alpha[s(t + D)]$, such as that given in Figure 70, has a Fourier transform with a form of $\alpha e^{j2\pi ft}$; the complex exponential in the frequency domain will look like a sine wave (see Figure 71). In the time domain, it will look like a delta function.

Hence, the Fourier transform of the PSD function appears to be an autocorrelation function that has a pattern with a secondary peak besides the zero-delay peak. Indeed, as more paths are added, more complicated autocorrelation functions are present (Chan (1989)). For this simple case, the result is a displaced peak at the delay of this modulation.

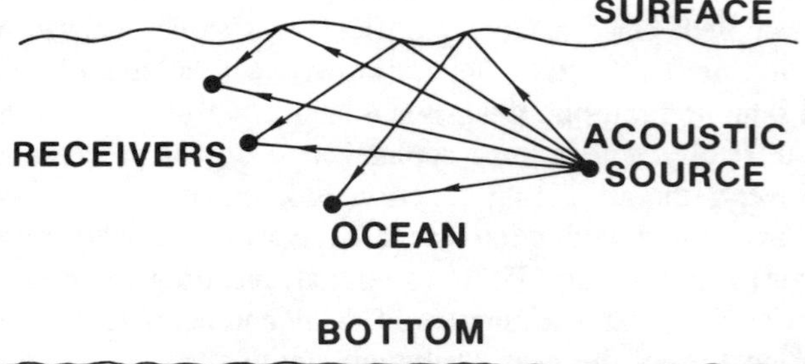

Fig. 69. Direct and surface-reflected sound ray paths received at three sensors (Carter 1981b).

Now, let us consider the case shown in Figure 72, where there are two receivers. There is a direct path and a surface-reflected path from the source to each of the two receivers.

For now, let us neglect the details of ocean propagation and consider only a single pulse leaving the source (Figure 73a).

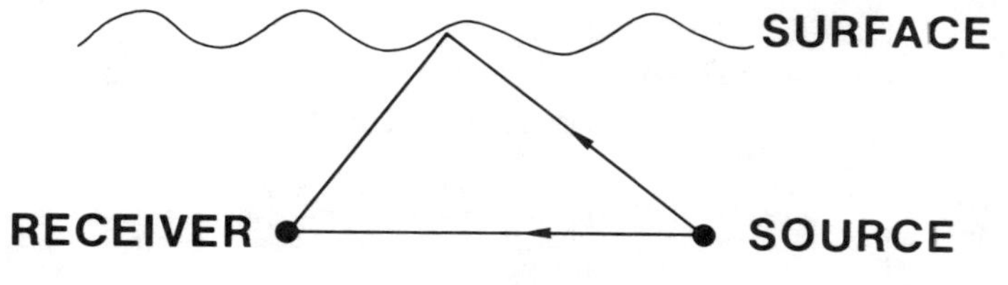

$$r(t) = s(t) + \alpha s(t + D) + n(t)$$

Fig. 70. Two-dimensional multipath ocean environment.

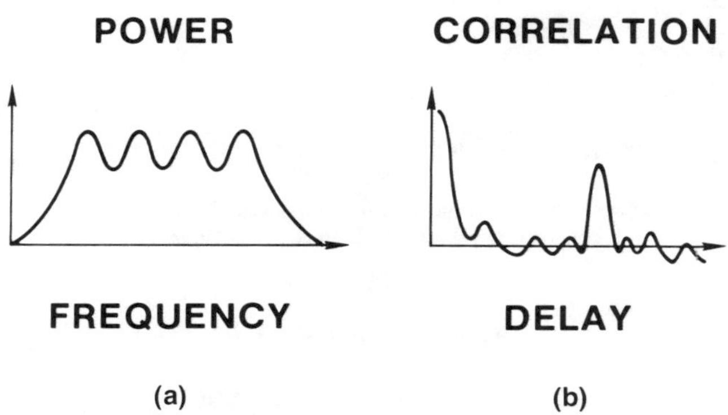

Fig. 71. Power spectral density (PSD) of a two-path signal. (a) Frequency domain. (b) Time domain.

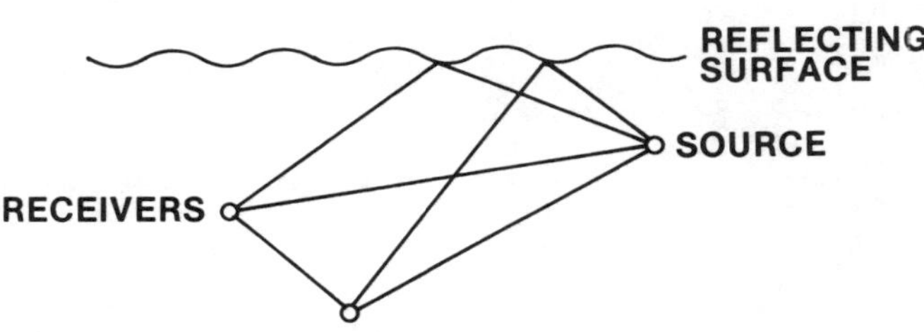

Fig. 72. Two-receiver three-dimensional ocean multipath model.

At one of the receivers, the pulse and the surface-reflected version of that pulse are received (Figure 73b). At the other receiver, also, the pulse and then the surface-reflected pulse are received (Figure 73c). At some time, there are two pulses at each receiver, which are cross correlated. As they slide by each other, they first line up to produce one peak; then, as they continue to slide by, they line up again to produce a stronger peak, and as they slide by and line up once again, another weaker peak appears. Thus the progression is from a weak peak to a strong peak and then to a weak peak again. A simple explanation (whether there is a broadband pulse or other broadband acoustic energy) is that the cross-correlation functions have a type of triplet (shown in Figure 73d).

Robinson and Quazi (1985) have extended these results with an ocean model, called the generic sonar model (GSM). The GSM is a particular model developed by Henry Weinberg. This ocean model is widely used and is the subject of a short

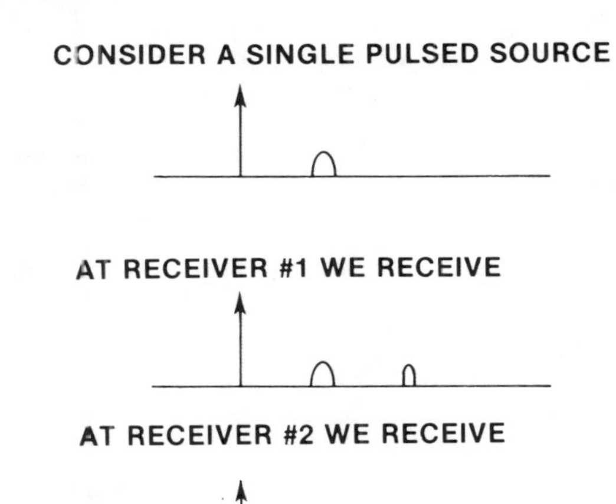

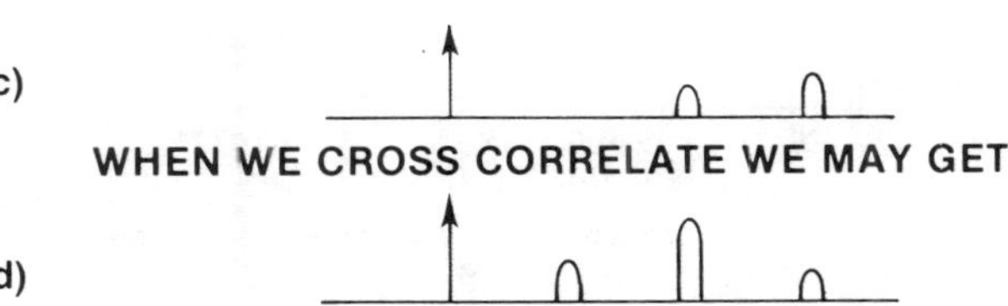

Fig. 73. Cross correlation for a single pulsed source and two receivers. (a) Sketch of a hypothetical pulse signal from one source. (b) Sketch of the signal detected at receiver 1. (c) Sketch of the signal detected at receiver 2. (d) Sketch of the cross-correlation function for the two received signals.

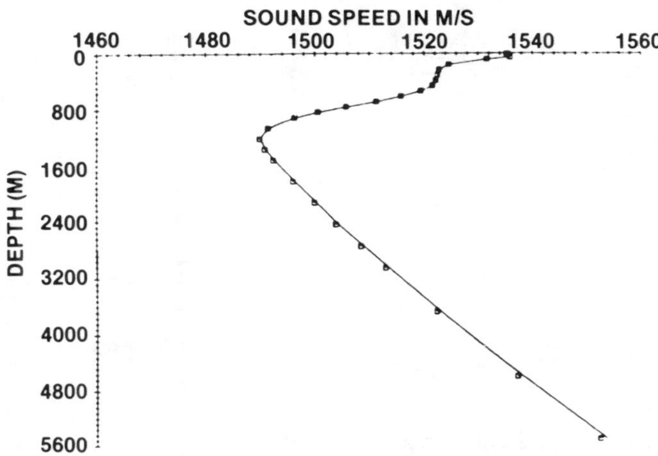

Fig. 74. Ocean sound speed versus depth (Robinson and Quazi 1985).

course on ocean acoustics. Physical properties in the ocean, such as temperature, salinity, and the speed of sound, vary as a function of depth (see Figure 74).

This variable speed acts as an acoustic lens to curve the rays of sound as they propagate through the ocean. Accounting for this particular sound velocity profile, the sophisticated GSM produces (much as expected for a single pulse) the cross-correlation function that forms the triplet given in Figure 75.

With the cross correlation of signals off broadside by 30° (see Figure 76), as expected, the peak no longer occurs at zero but at a delay—corresponding to the correct delay or bearing.

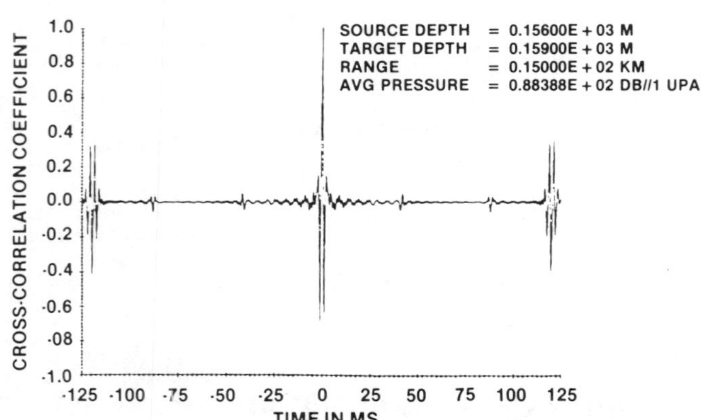

Fig. 75. Cross-correlation function for a broadside source with a range of 15 km (Robinson and Quazi 1985).

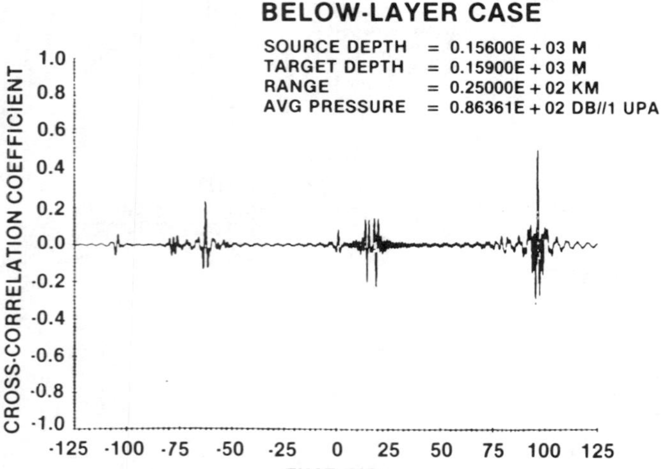

Fig. 76. Cross-correlation function for a source at 30° with a range of 25 km (Robinson and Quazi 1985).

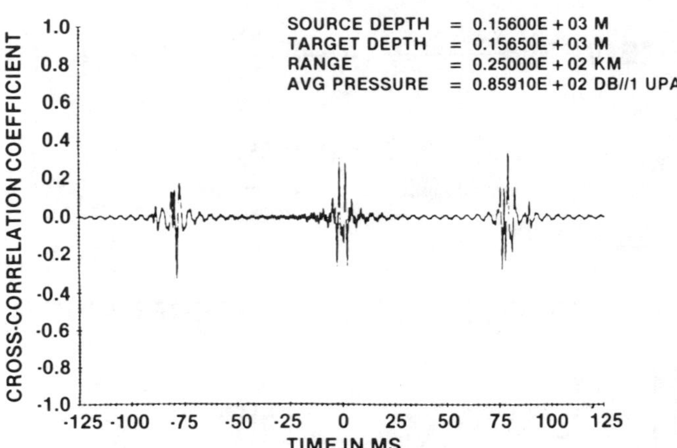

Fig. 77. Cross-correlation function for a broadside source with a range of 25 km, with sensor droop (Robinson and Quazi 1985).

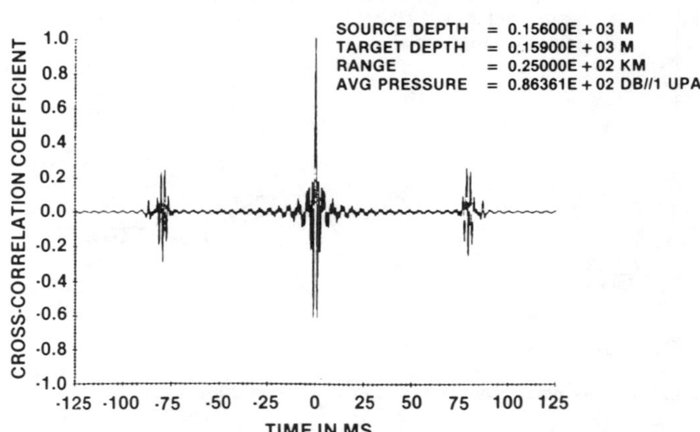

Fig. 78. Cross-correlation function for a broadside source with a range of 25 km (Robinson and Quazi 1985).

Two other peaks also appear with sidelobes that are more complex. These peaks look similar neither to each other nor to the ones before. The GSM predicts this complicated cross-correlation function. Nonetheless, the value of the largest peak is selected and the time delay for that peak is estimated.

Suppose, however, that one of the two receivers sags (droops). Figure 77 shows the cross-correlation function resulting from a droop of 5 m, although the absolute numbers here are not important. Note that even for a broadside source, if one receiver sags, an automatic algorithm to select the peak would be unable to locate the correct peak at the right time delay. The ocean environment has suddenly complicated the selection simply because a sensor is drooping.

Figure 78 shows the cross-correlation function at broadside for a range of 25 km. A comparison of this figure with Figure 75 shows that when the target range increases from 15 km to 25 km, the sidelobes in the cross-correlation function move in. This movement implies that encoded in the GCC function is information about range. Recall that there were only two sensors in this case. Earlier a simpler theory for the planar case was discussed; in that simple planar theory development, it was determined that range could not be estimated with only two sensors. Therefore, this example demonstrates that, with only two sensors in a three-dimensional ocean, information can be obtained about range from the GCC function (see Robinson and Quazi (1985) and Rosenberger (1989)).

In conclusion, the results presented in this tutorial chapter suggest that time delay estimation is of fundamental importance. Also, it should be noted that some real-world complexities have not been included in the models presented in this chapter, such as the nonlinear effects upon a reflected signal from the up-and-down modulation of the ocean surface. Fundamentally, there is a trade-off between a model's simplicity and ease of use and the model's accuracy. The approach used herein has been to utilize the simplest model that captures the essence of the problem—that is, to begin initially at a simplistic level and work up from there, introducing additional complexity only when needed. Sometimes, when the analytical work would be too difficult, a simulation is performed in order to include complex effects.

In any case, the papers reprinted in this book as well as those listed in the bibliography present the many different viewpoints and approaches to the problems inherent in coherence and time delay estimation. Even though the importance of coherence and time delay estimation in the sonar context has been emphasized in this tutorial chapter, this field has many applications in other areas, as will be evident from the selected papers. Indeed, the papers have been selected to provide a broad sense of the applicability of coherence and time delay estimation to practical engineering problems. Often this has required omitting outstanding papers to make room for other samples of outstanding work. Although some of the questions in coherence and time delay estimation have been addressed, there are many remaining questions with answers yet to be found.

Part 2
Selected Papers on Coherence

Estimation of the Coherence Spectrum and Its Confidence Interval Using the Fast Fourier Transform

VERNON A. BENIGNUS

Abstract

Correction factors were devised for correction of bias and computation of confidence intervals around estimates of coherence spectra computed upon Gaussian random time series. Using Monte Carlo methods to compute approximate sampling distributions for coherence, statistical descriptions of the bias and standard deviations were obtained. Correction factors were devised which extend the utility of coherence estimates to a value of coherence=0. Previously coherence=0.4 was a lower limit.

Introduction

With the rediscovery of the fast Fourier transform (FFT) [1], [2] it has become economically feasible to compute power spectrum estimates on a wide range of signals. Methods for computing the cross power spectral estimates from the two power spectral estimates are available [3], [4], and may be used with smoothing to compute estimates of coherence functions [5]–[7]. This paper follows the convention of Tukey [8] and Enochson and Goodman [9] in defining estimated coherence as

$$\hat{\gamma}^2(F) = \frac{\hat{G}_{xy}{}^2(F)}{\hat{G}_{xx}(F)\cdot\hat{G}_{yy}(F)} \tag{1}$$

where $\hat{G}_{xy}(F)$ is the cross power spectrum estimate, $\hat{G}_{xx}(F)$ is the power spectrum estimate of time series x, $\hat{G}_{yy}(F)$ is the power spectrum estimate of time series y, and F is the frequency index. Notice that $\gamma^2(F)$ is called coherence and is the square of the quantity defined as coherence in other references [5], but agrees with Tukey [8] and others [9], [10]. The quantities with a caret in the coherence ratio imply smoothed estimates, where smoothing is accomplished by averaging over either adjacent frequency bands or over spectra of segments into which the original time series is divided. Such smoothing procedures are well documented [5]–[7].

It is known that coherence estimates are approximately normally distributed if they are transformed by the Fisher "z transform" [9]. Estimates of coherence are positively biased, however (are overestimates), and it has been shown by Enochson and Goodman [9] that this bias is both a function of "effective degrees of freedom" and of true coherence $\gamma^2(F)$. Enochson and Goodman [9] describe a bias correction factor which works well for coherence above about 0.4, but values less than this are still overestimated. Computation of confidence intervals is further complicated according to Enochson and Goodman [9] by the fact that variance computation becomes "less accurate" for coherence below about 0.4. A coherence value of 0.4 represents an amplitude signal-to-noise ratio of about 1.71:1 [12] which is not an uncommon case in such applications as biological data processing. It therefore seems desirable to improve the procedure for coherence and confidence interval computation so that lower values of coherence can be meaningfully utilized.

Monte Carlo Methods

In order to arrive at improved estimates for coherence it was necessary to have adequate descriptions of estimate bias and sampling distribution variances. Enochson and Goodman [9] show curves for the bias of the transformed coherence estimates as a function of effective degrees of freedom and true coherence. No such description was available, however, for variance of the sampling distribution of coherence. It was therefore decided to use Monte Carlo methods to explore these factors empirically in a way similar to Foster and Guinzey [12] and Haubrich [11].

Using the FFT on an IBM 1800, a large number of coherences were computed at various true coherence values and levels of effective degrees of freedom in order to provide approximate sampling distributions for coherence. The coherences were computed upon time series constructed by sampling from large Gaussian random number tables [13]. Sampling from separate parts of the random number table produced two independent Gaussian random time series, x and y. By adding a proportion of x and y the true coherence between x and y can be adjusted to any desired value [12]. Coherence estimates were computed by smoothing over spectra of segments of the original time series. The number of segments N over which smoothing was carried out determined the variance of the sampling distributions [5]–[7]. Coherence sampling distributions were approximated at values shown in Table I. The entries in the table are the number of coherences n in each approximate sampling distribution. Coherences were computed for 64 frequency

Manuscript received February 23, 1969.

[1] The Research Computation Center is supported by the Department of Health, Education and Welfare, Special Research Resources Branch, Grant 5 PO7-FR00024 (05).

Reprinted from *IEEE Trans. Audio and Electroacoustics*, vol. 17, no. 2, pp. 145–150, June 1969.

TABLE I

Size of Approximate Sampling Distributions n Computed at Each Value of True Coherence γ^2 (F) and Extent of Smoothing N

$N \downarrow$ / $\gamma^2 \rightarrow$	0	.1	.2	.3	.4	.5	.6	.7	.8	.9
4	896	-	-	840	-	896	-	896	-	840
8	840	840	840	840	840	840	840	840	840	840
16	672	560	-	-	560	-	616	-	-	672
32	448	-	-	336	-	448	-	448	-	448
64	280	-	280	-	-	280	-	280	-	280

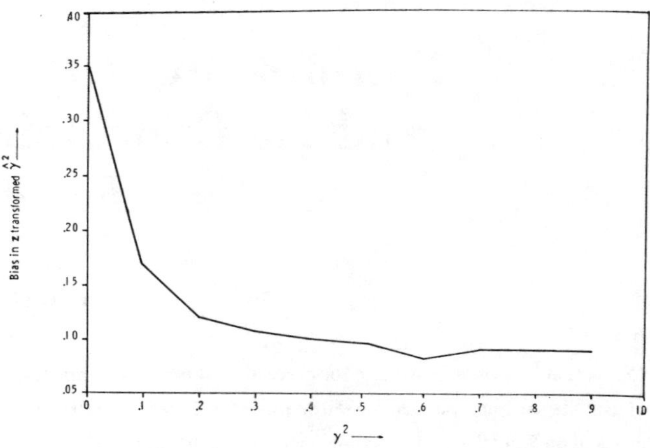

Fig. 1. Empirical curve of bias in z transformed $\hat{\gamma}^2$ as a function of γ^2 for $N=8$.

values in each sample from the Gaussian random number table and these estimates were then pooled with similar estimates from other independently drawn samples to construct approximations to sampling distributions. The approximate sampling distributions were studied for bias and variance, and then correction factors were devised.

Bias

After performing a Fisher z transform [9] on all coherences, the biases of the means of the approximate transformed sampling distributions were computed and these bias estimates plotted against true coherence. Fig. 1 shows such a plot for $N=8$. These bias curves agreed very well with those published by Enochson and Goodman [9]. Another set of plots was made of bias estimates transformed back to coherence by a hyperbolic tangent transform [9]. This was done rather than plotting estimates of bias in z transformed coherence as was done above. Such a family of curves is shown for various levels of N in Fig. 2. Apparently if true coherence were known, an expression which is a linear function of coherence could be used to subtract bias from an estimate and its confidence limits. In practice an estimate of true coherence must be used to estimate bias which in turn can be subtracted from the coherence estimate and its confidence limits. This procedure was proposed by Enochson and Goodman [9] but there the bias was subtracted from the z transformed coherence and its confidence limits, a procedure which optimally necessitates a nonlinear form for the expression of bias as a function of coherence.

A least-squares procedure was used to fit an expression for bias to the family of curves in Fig. 2 and the result was

$$\hat{B}(\hat{\gamma}^2) = \frac{1}{N}(1 - \hat{\gamma}^2). \tag{2}$$

Here the frequency index is omitted for simplicity and because the expression is true for any frequency; this practice will be continued for the remainder of this paper.

Fig. 2. Empirical curves of bias in $\hat{\gamma}^2$ as a function of γ^2 and N.

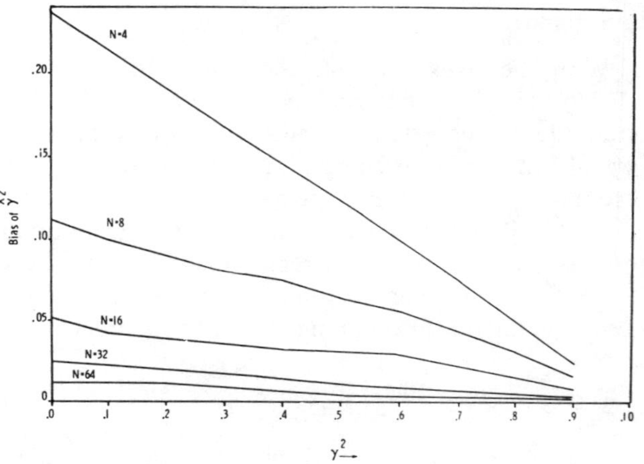

The estimate of bias, $\hat{B}(\hat{\gamma}^2)$, can be subtracted from $\hat{\gamma}^2$ to produce $\tilde{\gamma}^2$ an improved estimate of γ^2:

$$\tilde{\gamma}^2 = \hat{\gamma}^2 - \hat{B}(\hat{\gamma}^2). \tag{3}$$

By computing $\hat{B}(\hat{\gamma}^2)$ for the various sampling distribution means and subtracting this bias estimate from the means, an estimate of remaining error was obtained. Table II shows these errors. The noticeable fact about these data is that the row for $N=4$ has a slightly greater mean error and variability than other rows. If the row for $N=4$ is excluded there does not seem to be a trend in mean error as N increases. In view of the size of the errors for all degrees of smoothing at various coherence values, the question of trends either across N or across coherence seems irrelevant. Elimination of such small remaining error would necessitate a more complex expression for estimated bias, and this seems unwarranted. In view of the possibly larger average error in the case of $N=4$, es-

TABLE II

Residual Errors in $\tilde{\gamma}^2$; Means Computed for Columns by Omitting All Row 1 Entries

N ↑ \ γ^2 →	0	.1	.2	.3	.4	.5	.6	.7	.8	.9	M
4	.046			.034		.029		.021		.009	.0278
8	.004	.000	.001	.006	.010	.013	.006	.012	.009	.005	.0066
16	-.007	-.007			-.009		-.002			.001	-.0048
32	-.006			-.003		-.002		-.001		.000	-.0024
64	-.004		-.009			-.007		-.004		-.001	-.0050
M	-.0033	-.0035	-.0040	.0015	.0005	.0013	.0020	.0023	.0090	.0013	-.0002

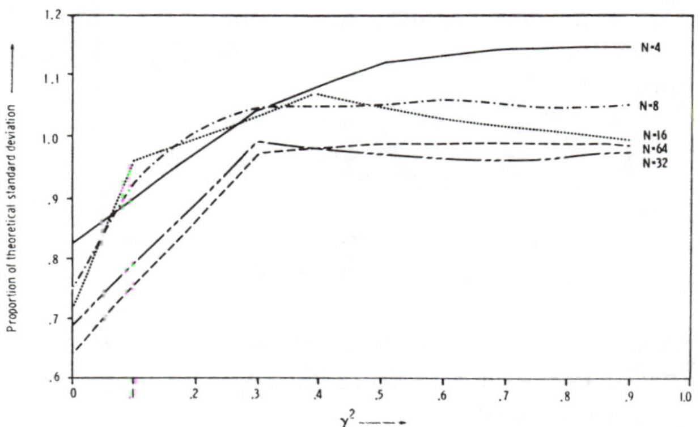

Fig. 3. Errors in theoretical standard deviation plotted as a function of γ^2.

timates made with this degree of smoothing should be interpreted with more caution.

Confidence Intervals

The theoretical variance of the z transformed coherence sampling distributions according to Hinich and Clay [5] can be expressed [6] as

$$\text{Var}\ (\hat{z}) = \frac{1}{2N} \tag{4}$$

where $\hat{z}$ is the z transformed coherence. There is some indication that the theoretical variance is not correct [9] according to empirical findings. For this reason estimates of population standard deviations of all available z transformed coherence sampling distributions were computed using

$$\hat{\sigma}_{\hat{z}} = \sqrt{\frac{\Sigma(\hat{z} - M_{\hat{z}})^2}{N - 1}}\,. \tag{5}$$

Here $M_{\hat{z}}$ is the mean of the biased, z transformed sampling distribution, $\hat{z}$ is the z transformed, biased estimate of coherence, and $\hat{\sigma}_{\hat{z}}$ is the best estimate of biased population standard deviation [14]. The standard deviation of the biased, z transformed distribution was used because subtraction of $\hat{B}(\hat{\gamma}^2)$ from all estimates constitutes subtraction of a variable from the distribution of coherences and will skew the distribution. Therefore confidence intervals must be computed first and bias subtracted later. Each standard deviation was then converted to a proportion of theoretical standard deviation and plotted as a function of true coherence as shown in Fig. 3.

By examination of Fig. 3 it may be seen that as N decreases, the empirical standard deviation is increasingly underestimated by (2). Another feature of the curves in Fig. 3 is that the empirical standard deviation decreases as coherence approaches zero. Neither of the above discrepancies from the theoretical standard deviation is readily explainable. The fact that Enochson and Goodman [9] found similar data for the correlogram method and the fact that very large samples were utilized here forces the conclusion that these discrepancies are not fortuitous sampling errors.

It was undertaken to devise correction procedures for estimating standard deviations and confidence intervals. Correction for underestimates of standard deviations were approached in a manner similar to Enochson and Goodman [9] who computed "effective degrees of freedom" based on sampling from an extensive coherence probability table.

In this study an estimate of "effective degrees of freedom" was made by fitting the following equation to the empirical standard deviations for coherences greater than 0.6:

$$\bar{\sigma}_{\hat{z}} = \sqrt{\frac{1}{2(N - k)}}\,. \tag{6}$$

The optimum value of k was 1 so that the specific best fit case for (6) is

$$\bar{\sigma}_{\hat{z}} = \sqrt{\frac{1}{2(N - 1)}}\,. \tag{6a}$$

This solution is similar to the one published in [9].

Expression (6a) would still yield a substantial overestimate of $\sigma_{\hat{z}}$ for $\gamma^2 < 0.3$. In order to study this problem a graph was constructed by averaging the proportional standard deviations using (6a) as theoretical standard deviation. A function of the form

$$E(\sigma_{\hat{z}}) = 1 - c^{(a\bar{\gamma}^2 + b)} \tag{7}$$

was fitted to the empirical data using a least squares method. Here $E(\sigma_{\hat{z}})$ is an estimate of proportional error and a, b, and c are empirical constants. The specific case finally adapted is

$$E(\sigma_{\hat{z}}) = 1 - 0.004^{(1.6\bar{\gamma}^2 + 0.22)}. \tag{7a}$$

The following was used to compute estimated standard deviation to be used in the computation of confidence intervals:

$$\tilde{\sigma}_{\hat{z}} = \left[\sqrt{\frac{1}{2(N - 1.0)}}\right] [E(\sigma_{\hat{z}})]. \tag{8}$$

Fig. 4 shows a graph of confidence intervals as computed by the above method. The straight center line represents the best estimate of γ^2. Examination of Fig. 4 reveals a problem of interpretation where $\gamma^2 \rightarrow 0$. In this region the lower confidence limit becomes negative, which is a nonsensical value for $\tilde{\gamma}^2$. The solution to this problem was to set all negative lower limit coherence values to zero. This folds the lower tail of the distribution into the confidence interval and produces an error of $\alpha/2$ where α is the area outside the confidence interval before truncation. If the same confidence interval is to be maintained, a one-tailed estimate must be used with all of the excluded area in the upper tail of the distribution. When the lower confidence limit becomes negative then a discontinuity occurs in the upper limit when a one-tailed rather than a two-tailed estimate is used.

A program was written to compare the theoretical confidence to a measure of empirical confidence. To perform this the following procedure was followed for each approximate sampling distribution. All sample coherences $\hat{\gamma}^2$ were z transformed and the mean of this biased distribution was converted back to coherence. A bias correction was made to the biased mean using (3) and then (8) was used to compute the expected standard deviation and confidence limits. Notice that what is estimated by (8) is the standard deviation of the biased z transformed distribution. After finding the confidence limits in z transformed form, these limits were converted back to coherences and (2) was used to estimate bias for each limit after which these bias estimates were subtracted from the biased limits as in (3). Each value of $\hat{\gamma}^2$ was then corrected for bias, estimating bias with (2) for each specific value. The bias-corrected estimates were then tested for inclusion in the computed confidence interval. The empirical confidence is defined as the percent of corrected estimates which were in fact found to be within the interval. If a negative or zero lower limit occurred, its value was set to zero and a one-tailed confidence interval was used. If a bias-corrected coherence was found to be negative its value was set to zero. Tables III–V show the percent errors in empirical versus computed confidence for the 90, 95, and 99 percent confidence levels. A negative error implies that the empirical confidence was larger than expected; i.e., the expected confidence was too low.

Examination of Tables III–V reveals that overall error tends to be positive for the 99 percent level and slightly negative for the 90 and 95 percent levels. All errors tend to be slightly negative as $\gamma^2 \rightarrow 1$. The difference in errors across levels of N is not systematic. Since the general level of errors is relatively low and certainly an improvement over previous solutions it seems justified to utilize (7a) until either a better empirical solution or a general theoretical solution becomes available.

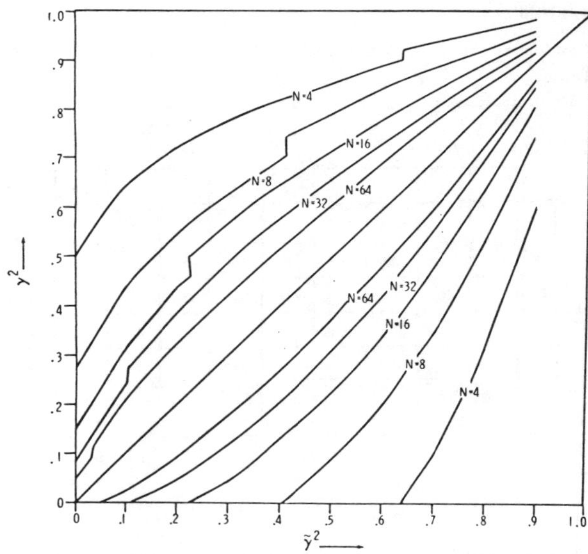

Fig. 4. Graph of 95 percent confidence intervals for $N = 4$, 8, 16, 32, and 64. Discontinuities in upper bounds result from folding of lower confidence limits into the first quadrant and the use of one-tailed intervals in the folded region.

Discussion and Conclusions

A method has been devised, using Monte Carlo methods and empirical corrections, for estimating the coefficient of coherence and confidence intervals for true coherence values over the entire range. The procedure for estimation of coherence is as follows: compute $\hat{\gamma}^2$ by (1), compute a bias estimate using (2), and subtract bias as shown in (3) to arrive at $\tilde{\gamma}^2$, the bias-corrected coherence estimate. The estimate $\hat{\gamma}^2$ is then z transformed and the standard deviation estimated using (8). Compute confidence intervals for the z transformed, biased coherence and then convert these limits back to coherence and subtract a bias estimate from each, based upon itself. If the lower limit is found to be negative use a one-tailed upper limit and set the lower limit equal to zero. If $\tilde{\gamma}^2$ is negative set it equal to zero and proceed appropriately.

The results of this study are based upon the assumptions that the time series x and y have a Gaussian distribution and a relatively flat spectrum [9], [12]. Foster and Guinzey [12] showed that for relatively large values of γ^2 the Gaussian assumption may be to a great extent violated with impunity. No data is yet available on the effect of such violation for small values of γ^2. There are also no data available on the effect of violation of the Gaussian assumption on confidence intervals since such a violation could conceivably leave the sampling distribu-

TABLE III

Percent Errors for 90 Percent Confidence Interval Estimates

N \ γ^2	0	.1	.2	.3	.4	.5	.6	.7	.8	.9	M
4	-1.74		-0.95		-0.74		-0.51			-0.60	-0.91
8	-0.60	0.36	0.60	0.00	-1.31	-1.31	-0.71	-0.95	-0.95	-0.95	-0.58
16	0.42	1.96			0.71		-0.75			-1.52	-0.16
32	2.05			-1.37		-1.07		-1.30		-0.85	-0.51
64	-1.42		-0.36			-0.36		-2.14		-1.43	-1.14
M	-0.26	1.16	-0.24	-0.69	-0.45	-0.91	-0.66	-1.46	-0.95	-1.07	-0.59

TABLE IV

Percent Errors for 95 Percent Confidence Interval Estimates

N \ γ^2	0	.1	.2	.3	.4	.5	.6	.7	.8	.9	M
4	0.25			-0.48		-0.20		-0.20		-0.60	-0.25
8	0.36	0.48	-0.12	0.00	-0.48	-0.12	-0.24	-0.36	-1.07	-0.12	-0.17
16	1.40	0.54			-0.18		-0.94			-1.13	-0.06
32	3.26			0.06		-1.21		-1.43		-0.98	-0.06
64	1.07		0.36			-1.43		-0.71		-0.71	-0.28
M	1.27	0.51	0.12	-0.21	-0.33	-0.74	-0.59	-0.68	-1.07	-0.71	-0.16

TABLE V

Percent Errors for 99 Percent Confidence Interval Estimates

N \ γ^2	0	.1	.2	.3	.4	.5	.6	.7	.8	.9	M
4	0.67			-0.05		-0.11		-0.11		0.31	0.142
8	1.02	0.42	0.19	0.54	0.31	0.07	-0.29	0.07	-0.17	-0.05	0.197
16	1.38	0.43			0.43		0.14			-0.26	0.424
32	1.90			0.19		-0.11		-0.12		-0.33	0.354
64	.79		0.07			0.43		0.43		0.07	0.358
M	1.152	0.425	0.23	0.227	0.370	0.070	0.075	0.51	-0.17	-0.052	0.283

tion in Gaussian form, but affect its standard deviation, especially in the tails of the curves.

In view of the empirical nature of the results, the nonuniqueness of some of the solutions and the possibly nonzero average errors, users of the above methods might be well advised to proceed with caution. Confidence estimates ought to be made either type I or type II conservative depending upon the least abhorrent error. Significance tests which are near rejection limits should be accepted guardedly. Other methods of analysis and replication of measurements should be used to validate results. While the above methods of estimating γ^2 and confidence intervals are far more reliable, especially for small values of γ^2, than previous methods [9], careless use of the statistic can lead to severe inferential errors.

References

[1] J. W. Cooley and J. W. Tukey, "An algorithm for the machine calculation of complex Fourier series," *Math. Comp.*, vol. 19, pp. 297–301, April 1965.

[2] G-AE Subcommittee on Measurement Concepts, "What is the fast Fourier transform?," *IEEE Trans. Audio and Electroacoustics*, vol. AU-15, pp. 45–55, June 1967.

[3] G. M. Jenkins and D. G. Watts, *Spectral Analysis and Its Applications.* San Francisco, Calif.: Holden-Day, 1968.

[4] A. J. Villasenor, "Digital spectral analysis," NASA, Washington, D. C., Tech. Note D-4510, 1968.

[5] M. J. Hinich and C. S. Clay, "The application of the discrete Fourier transform in estimation of power spectra coherence and bispectra of geophysical data," *Rev. Geophys.*, vol. 6, pp. 347–363, August 1968.

[6] V. A. Benignus, "Computation of coherence and regression spectra using the FFT," *COMMON, Proc. Houston Meeting* (December 9–11, 1968), April 1969.

[7] L. D. Enochson and A. G. Piersol, "Application of fast Fourier transform procedures to shock and vibration data analysis," presented at the Aeronautics and Space Engineering and Manufacturing Meeting, Los Angeles, Calif., 1967 (Soc. of Automotive Engrs. Reprint 670874).

[8] J. W. Tukey, ch. 2 in *Spectral Analysis of Time Series*, B. Harris, Ed. New York: Wiley, 1966.

[9] L. D. Enochson and N. R. Goodman, "Gaussian approximations to the distribution of sample coherence," Air Force Flight Dynamics Lab., Research and Tech. Div., AF Systems Command, Wright-Patterson AFB, Ohio, Bull. AFFDL-TR-65-67, 1965.

[10] J. S. Bendat and A. G. Piersol, *Measurement and Analysis of Random Data.* New York: Wiley, 1966.

[11] R. A. Haubrich, "Earth noises, 5 to 500 millicycles per second, 1" *J. Geophys. Research*, vol. 70, pp. 1415–1427, March 1965.

[12] M. R. Foster and N. J. Guinzy, "The coefficient of coherence: its estimation and use in geophysical data processing," *Geophys.*, vol. 32, pp. 602–616, August 1967.

[13] RAND Corp., *A Million Random Digits with 100,000 Normal Deviates.* Glencoe, Ill.: Glencoe Free Press, 1955.

[14] W. L. Hays, *Statistics for Psychologists.* New York: Holt, Rinehart and Winston, 1963.

The Key Role of Tapering in Spectrum Estimation

DAVID R. BRILLINGER

Abstract—A number of advantages of employing a taper (or data window) prior to computing the Fourier transform of an observed stretch of time series are reviewed.

In a recent correspondence, "Comments on Modern Methods for Spectrum Estimation" [1], Yuen argues that the operation of tapering "is statistically quite unsound" and that there is "mathematical difficulty of comparing different procedures in theory." In fact, tapering may be dealt with in a direct analytic fashion and there are clear statistical advantages in employing the procedure.

We will adopt the notation of [2]. The time series data $X(t)$, $t = 0, 1, \cdots, T - 1$ will be assumed available. The taper $\phi^T(t)$, vanishing for $t < 0$, $t > T - 1$, will be employed. The finite Fourier transform of the tapered values is then given by

$$d^T(\lambda) = \sum_t \phi^T(t) \exp\{-i\lambda t\} X(t). \tag{1}$$

(Not tapering corresponds to defining $\phi^T(t) = 1$ for $0 \leqslant t \leqslant T - 1$ and $= 0$ otherwise.) Now suppose the series $X(\cdot)$ is wide-sense stationary with mean 0 and power spectrum $f(\lambda)$. Then it has (see [2]) Cramér representation.

$$X(t) = \int_{-\pi}^{\pi} \exp\{i\alpha t\} dZ(\alpha) \tag{2}$$

with $E\{dZ(\alpha)\} = 0$, $\operatorname{cov}\{dZ(\alpha), dZ(\beta)\} = \delta(\alpha - \beta) f(\alpha)\, d\alpha\, d\beta$. (Here $\delta(\cdot)$ is the Dirac delta function.)

Defining

$$\Phi^T(\lambda) = \sum_t \phi^T(t) \exp\{-i\lambda t\} \tag{3}$$

one sees directly from (1) and (2) that

$$d^T(\lambda) = \int \Phi^T(\lambda - \alpha)\, dZ(\alpha). \tag{4}$$

At this point, if nothing else, the results of classical Fourier analysis (see, for example, [3]) suggest that substantial advantages may result from the insertion of nonconstant convergence factors in the finite Fourier transform (1). The book [5] indicates a broad array of the effects of employing different specific ϕ^T. In particular, the nearness of expression (1), or equivalently (4), to $dZ(\lambda)$ is investigated.

It is immediate from expression (4) that the expected value of the periodogram $I^T(\lambda) = (2\pi T)^{-1} |d^T(\lambda)|^2$, on which many spectrum estimates are based, is given by

$$E\{I^T(\lambda)\} = (2\pi T)^{-1} \int |\Phi^T(\lambda - \alpha)|^2 f(\alpha)\, d\alpha. \tag{5}$$

The tapering procedure is thus seen to affect the window through which the power spectrum is seen in the estimation process. Not tapering corresponds to

$$|\Phi^T(\alpha)|^2 = \left(\frac{\sin T\alpha/2}{\sin \alpha/2}\right)^2, \tag{6}$$

a window with rippling sidelobes and dying off only as α^2 (approximately). By employing a taper one can reduce the rippling and increase the speed of dying off substantially. A first advantage of tapering is thus seen to be an improvement in the resolution (reduction in the bias) of the estimate in the case that the power spectrum has a split peak or a very large peak at some location.

As Yuen [1] himself argues (in the case of quadratic (spectrum) windows), one has to consider the bias/variance tradeoff of an estimate. It is shown in [5] that the effect of tapering is to multiply the (asymptotic) variance of the spectrum estimate by the factor $T^{-1} \Sigma \phi^T(t)^2$. In the case of the cosine taper over the first and last 10 percent of the data, this factor is 1.116, for large T. If the power spectrum is near constant, then the little reduction in bias resulting from tapering will not be worth having in the light of the increased variance. If, however, the spectrum has substantial structure, then the gain may prove very worthwhile.

A striking example of the beneficial effect of employing a taper is presented in [6]. That work was concerned with the estimation of the curvature spectrum of a prototype waveguide. The beneficial effect results from the fact that such spectra may fall off 10 to 16 decades. The taper employed in [6] is a prolate spheroidal function (which minimizes the integral of $|\Phi^T(\cdot)|^2$ outside the neighborhoods of 0). The spectrum estimates are shown to be substantially misleading in an important frequency range if no tapering is employed. Another compelling example of the use of a taper is provided in [7, ch. 5].

The importance of the operation of tapering with respect to bias reduction reduces as the spectrum at issue becomes flatter. So too does the effect of employing different tapering functions. Through the use of sensible prewhitening filters, the necessity for tapering may be much reduced in the formation of the final spectrum estimate; however, it remains essential in the pilot estimation to determine the prewhitening filter.

Tapering has a second useful effect on the statistics of spectrum analysis. From expression (4) one can further see that

$$\operatorname{cov}\{d^T(\lambda), d^T(\mu)\} = \int \Phi^T(\lambda - \alpha)\, \overline{\Phi^T(\mu - \alpha)}\, f(\alpha)\, d\alpha. \tag{7}$$

By employing a taper whose Fourier transform (3) dies off more rapidly than (6), one can arrange that the finite Fourier transform values at distinct frequencies be more nearly uncorrelated and, indeed, as an analysis of higher order moments

Manuscript received October 20, 1980; revised April 7, 1981. This work was supported by the National Science Foundation under Grant PFR-7901642.

The author is with the Department of Statistics, University of California, Berkeley, CA 94720.

Reprinted from *IEEE Trans. Acoust., Speech, Signal Processing*, vol. 29, no. 5, pp. 1075–1076, October 1981.

shows, more nearly statistically independent. This result can prove useful in developing approximate distributions for test statistics and for estimates of quantities of interest.

A further advantage of tapering is that for some stationary processes, for example, stationary random Schwarz distributions (see [2, sect. XV]), the Fourier transform may not even be able to be evaluated via indicator functions $\phi^T(\cdot)$; it may only be defined for sufficiently regular functions.

A final remark applies to the advantage of employing the definition (1) even in the case that $\phi^T(\cdot)$ takes on only the values 1 or 0. The region where $\phi^T(\cdot)$ is not zero gives the domain of observation of the process. By employing the expression (1) irregular domains of observation are handled just as easily as simple domains. This is useful with discrete time series missing stretches of values, or for spatial series observed over irregular regions.

The comments of this note apply *mutatis mutandis* to the problems of the estimation of the joint spectra of multivariate processes, of spatial processes, and of higher order spectra.

A rebuttal to the charges of Yuen was provided earlier in [8].

REFERENCES

[1] C. K. Yuen, "Comments on modern methods for spectrum estimation," *IEEE Trans. Acoust., Speech, Signal Processing*, vol. ASSP-27, pp. 298–299, June 1979.

[2] D. R. Brillinger, "Fourier analysis of stationary processes," *Proc. IEEE*, vol. 62, pp. 1628–1643, Dec. 1974.

[3] A. Zygmund, *Trigonometric Series*. Cambridge, England: Cambridge Univ. Press, 1968.

[4] P. L. Butzer and R. J. Nessel, *Fourier Analysis and Approximation*, vol. I. New York, Academic, 1971.

[5] D. R. Brillinger, *Time Series: Data Analysis and Theory*. New York: Holt, Rinehart and Winston, 1975 and San Francisco, CA: Holden-Day, 1981 (expanded edition).

[6] D. J. Thomson, "Spectrum estimation techniques for characterization and development of WT4 waveguide–II," *Bell Syst. Tech. J.*, vol. 56, pp. 1983–2005, Dec. 1977.

[7] P. Bloomfield, *The Fourier Analysis of Time Series: An Introduction*. New York: Wiley, 1976.

[8] A. H. Nuttall and G. C. Carter, "A generalized framework for power spectral estimation," *IEEE Trans. Acoust., Speech, Signal Processing*, vol. ASSP-28, pp. 334–335, June 1980.

Linear Modeling and the Coherence Function

JAMES A. CADZOW, FELLOW, IEEE, AND OTIS M. SOLOMON, JR., MEMBER, IEEE

***Abstract*—In this paper, a tutorial introduction to the magnitude squared (MS) coherence function, and its relation to linear modeling, is interleaved with new methods for characterizing bidirectional and unidirectional linear association between two time series. A new method for characterizing bidirectional linear association is based on the new concept of an MS coherence sequence which is the inverse *z*-transform of the MS coherence function. The MS coherence function is approximated as a single real-valued ratio of polynominals in *z*. The ratio is determined by a matrix equation whose entries are convolutions of correlation functions. Solutions for the corresponding MS coherence parameters employ an eigenvalue–eigenvector decomposition. The algorithm is compared to a traditional periodogram-based method using computer simulations. The computer simulations also demonstrate the effects of parameter overdetermination and overordering. The sensitivity of the matrix equations for the computer simulations is computed via the numerical linear algebra concept of condition number.**

I. Introduction

In a variety of signal processing applications, the linearity between two time series is hypothesized, but not verified. All too frequently, a linear association is stipulated even in situations where such an assumption is not warranted. Linearity is an important underlying assumption in such applications as system identification and source bearing estimation, where linear models play a prominent role. Although in principle one can always construct the linear model which *best* represents a given pair of time series, the worthiness of the model must be carefully assessed. If it can somehow be ascertained that a linear data fit is appropriate, then the corresponding "optimum" linear model can provide valuable insight into the phenomenon being characterized. On the other hand, if a linear model assumption is inappropriate, then it would be imprudent to make inferences based on such a model.

The *coherence function* provides a mechanism for measuring the linear association existent between two time series. It has been used by statisticians and scientists alike for this purpose, but its utilization by the signal processing community is somewhat limited (e.g., see [2]–[9]). With this in mind, a brief review of the coherence function is made in the next section. This review is followed by the development of a new effective procedure for estimating the coherence from raw time-series observations.

Manuscript received June 22, 1985; revised May 17, 1986. This work was performed in part at Sandia National Laboratories under U.S. Department of Energy Contract DE-AC04-76DP00789 and at Arizona State University under Office of Naval Research Contract N00014-86-K-0540 and Navy Contract N00039-86-C-0429.

J. A. Cadzow is with the Department of Electrical and Computer Engineering, Arizona State University, Tempe, AZ 85287.

O. M. Solomon, Jr. is with Sandia National Laboratories, Albuquerque, NM 87185.

IEEE Log Number 8611232.

II. Optimum Linear Modeling

The coherence function provides a frequency domain means for quantifying the bidirectional linear association existent between any pair of wide-sense stationary time series. A *perfect bidirectional linear association* requires that each of the time series be related to one another by means of a time-invariant linear operation (not necessarily causal). For instance, a white noise input to an *ideal* low-pass filter does not share a perfect bidirectional linear association with the filter's output outside of the filter's passband. No linear operator can resurrect the input data from the response data in the rejection band of an ideal low-pass filter. However, the output of an ideal low-pass filter driven by a sine wave in the passband is perfectly bidirectionally linearly associated with the sine wave input.

Linear operators whose frequency responses are square integrable generate stationary outputs from stationary inputs. When the frequency response of a linear operator is bounded away from zero and infinity, both it and its inverse are square integrable (see the matching condition in [5]). Consequently, all inputs and corresponding outputs of linear operators whose frequency response satisfies the above conditions will be perfectly bidirectionally linearly associated. For an autoregressive moving-average (ARMA) linear operator, its frequency response is bounded away from zero and infinity provided that it has no poles or zeros on the unit circle.

With the above definition of bidirectional linear association in mind, let us consider the two real-valued *wide-sense stationary* time series $\{x(n)\}$ and $\{y(n)\}$. To measure how closely these two time series are linearly related, one time series is designated as the linear operator's *excitation* and the other is designated as the *target response*. In certain situations, this assignment is natural, while in others it is completely arbitrary. To begin, we first select $\{x(n)\}$ as the excitation time series, and then synthesize the linear model whose response to $\{x(n)\}$ best matches the target response $\{y(n)\}$. Then the roles of the two time series are reversed to demonstrate a symmetry in the resultant linear modeling characterization. These two configurations are shown in Fig. 1. In the event that the two time series are actually bidirectionally linearly associated, both of these modeling hypotheses will be perfectly fulfilled by a pair of linear time-invariant operators $H(z)$ and $1/H(z)$.

Reprinted from *IEEE Trans. Acoust., Speech, Signal Processing*, vol. 35, no. 1, pp. 19–28, January 1987.

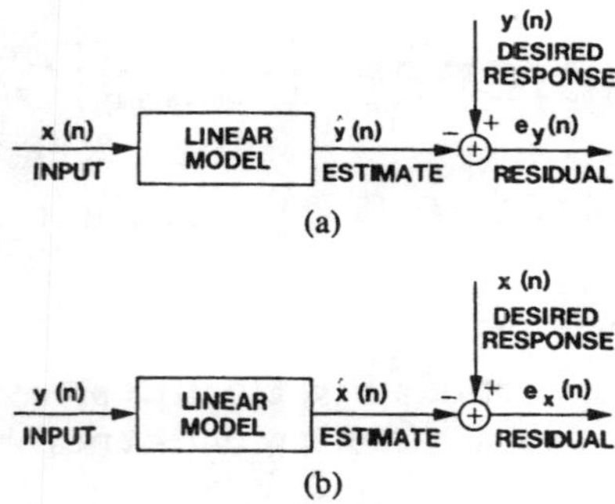

Fig. 1. The approximation of one time series by a linear operation on another time series.

Let the time series $\{x(n)\}$ be applied to a linear operator whose weighting sequence (i.e., its unit-impulse response) is given by $\{h_{xy}(n)\}$. The subscript "xy" on the weighting sequence is used to explicitly call out the fact that the linear operator's excitation is chosen as $\{x(n)\}$ while the target response is $\{y(n)\}$. The linear operator's response is then specified by the convolution summation

$$\hat{y}(n) = \sum_{k=-\infty}^{\infty} h_{xy}(k)\, x(n-k) \tag{1}$$

where it is to be noted that no causality constraints have been imposed on this linear operator (i.e., $h_{xy}(k)$ is not necessarily zero for $k < 0$). To quantify the goodness of modeling, the model error is introduced

$$e_y(n) = y(n) - \hat{y}(n) \tag{2}$$

and is generated in accordance with Fig. 1(a). In the *classical linear modeling* problem, the weighting sequence $\{h_{xy}(n)\}$ is chosen to *minimize* the mean squared error (MSE) criterion $E\{e_y^2(n)\}$, where E denotes the expected value operator [5], [6].

It is well known that the minimum MSE solution is specified by the *Wiener–Hopf* (or *normal*) equations [5], [6]

$$\sum_{k=-\infty}^{\infty} h_{xy}^o(k)\, r_{xx}(n-k) = r_{yx}(n) \quad \text{for all } n, \tag{3}$$

where the nth cross-correlation value $r_{yx}(n)$ is defined by $E[y(n+m)\,x(m)]$. The superscript "o" on the sequence $\{h_{xy}^o(n)\}$ denotes that this weighting sequence is optimum. Solving this infinite set of Wiener–Hopf equations appears rather formidable. However, upon taking the z-transform of relationship (3) and solving for the corresponding *optimal transfer function*, we obtain [5], [6]

$$H_{xy}^o(z) = S_{yx}(z)/S_{xx}(z) \tag{4}$$

where $S_{yx}(z) = Z\{r_{yx}(n)\}$ is the cross-spectrum function associated with the two time series $\{x(n)\}$ and $\{y(n)\}$. Thus, given the spectral functions $S_{xx}(z)$ and $S_{yx}(z)$, relationship (4) provides the optimal linear model in the sense of minimizing the MSE criterion $E\{e_y^2(n)\}$.

The quality of the linear association existent between the two time series is revealed through the behavior of the *optimum model error autocorrelation sequence* as specified by

$$\begin{aligned} r_{e_y^o e_y^o}(n) &= E\{e_y^o(n+m)\, e_y^o(m)\} \\ &= E\left\{\left[y(n+m) - \sum_{k=-\infty}^{\infty} h_{xy}^o(k)\, x(n+m-k)\right] \cdot \left[y(m) - \sum_{k=-\infty}^{\infty} h_{xy}^o(k)\, x(m-k)\right]\right\} \\ &= r_{yy}(n) - \sum_{k=-\infty}^{\infty} h_{xy}^o(k)\, r_{yx}(n-k), \end{aligned} \tag{5}$$

where (3) was used to obtain the final result. A convenient scalar measure of the effectiveness of the optimum linear model is given by the *normalized MSE index*

$$\rho_{xy} = \frac{E\{e_y^o(n)^2\}}{E\{y^2(n)\}} = \frac{r_{yy}(0) - \sum_{k=-\infty}^{\infty} h_{xy}^o(k)\, r_{yx}(k)}{r_{yy}(0)}. \tag{6}$$

Because the optimum MSE can never be greater than the variance of the target response (i.e., $E\{y^2(n)\}$), the normalized MSE index will always take on a value between zero and one. A high degree of unidirectional linear association between $\{x(n)\}$ and $\{y(n)\}$ is indicated whenever this index is close to zero. On the other hand, little linear association exists when this index is close to one.

When $\{y(n)\}$ is chosen as the excitation and $\{x(n)\}$ the target response, the modeling configuration depicted in Fig. 1(b) arises. Using the procedure taken above, it immediately follows that the optimum transfer function in this case is specified by

$$H_{yx}^o(z) = S_{xy}(z)/S_{yy}(z) \tag{7}$$

while the corresponding *optimum model error autocorrelation* is given by

$$r_{e_x^o e_x^o}(n) = r_{xx}(n) - \sum_{k=-\infty}^{\infty} h_{yx}^o(k)\, r_{yx}(n-k), \tag{8}$$

and the normalized optimum MSE index is specified by

$$\rho_{yx} = \frac{E\{e_x^{o2}(n)\}}{E\{x^2(n)\}} = \frac{r_{xx}(0) - \sum_{k=-\infty}^{\infty} h_{yx}^o(k)\, r_{xy}(k)}{r_{xx}(0)}. \tag{9}$$

It should be noted that the values assumed by the two normalized MSE indexes ρ_{xy} and ρ_{yx} are generally different. When ρ_{xy} and ρ_{yx} are both zero, the two time series $\{x(n)\}$ and $\{y(n)\}$ are perfectly bidirectionally linearly associated.

In terms of determining the level of linear association between two time series, it is important to keep in mind that the optimum model error autocorrelations (5) and (8) hold only for the optimum linear model selections. A suboptimal linear model choice is characterized by different autocorrelation expressions. We emphasize this point since we are hereafter exclusively concerned with analyzing the optimal linear model case.

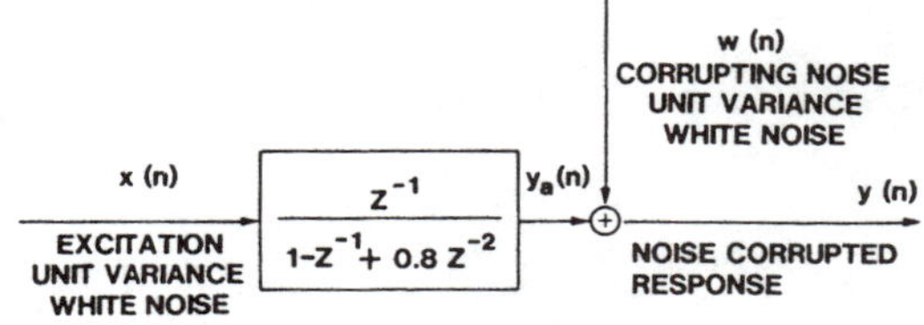

Fig. 2. A numerical example with white corrupting noise.

Example 1: To illustrate these concepts, let us consider the two time series as generated in accordance with Fig. 2. The minimum MSE $E[e_y^{o2}(n)]$ for the linear association using configuration in Fig. 1(a) is found to be 1.0 while the minimum MSE $E[e_x^{o2}(n)]$ for the configuration of Fig. 1(b) is found to be approximately 0.515 from (8). Because the minimum MSE's are nonzero, the two time series $\{x(n)\}$ and $\{y(n)\}$ are not perfectly linearly associated. Since $\sigma_y^2 = 5.018$ and $\sigma_x^2 = 1.0$, the normalized minimum MSE's ρ_{xy} and ρ_{yx} per (6) and (9) are

$$\rho_{xy} = 0.199$$

and

$$\rho_{yx} = 0.515.$$

To choose the best unidirectional modeling configuration, it is necessary to first select a criterion for measuring the degree of unidirectional linear association. Using the normalized MSE as the measure of unidirectional linear association leads to the conclusion that the best linear modeling configuration interprets $\{x(n)\}$ as the excitation and $\{y(n)\}$ as the response. Note that using the MSE as the measure of unidirectional linear association reverses the roles of $\{x(n)\}$ and $\{y(n)\}$.

III. Coherence Function

Although a meaningful basis for measuring the linear association between two time series is given by the normalized MSE indexes, the coherence function to be now described can yield significantly more information. This function is obtained by taking the z-transforms of the optimum model error autocorrelation relationships (5) and (8), and then substituting in the optimum transfer functions (4) and (7). After rearrangement of terms, the optimum model error spectra associated with the optimum model error autocorrelations are specified by [5], [6]

$$S_{e_y^o e_y^o}(z) = [1 - \Gamma_{xy}(z)]\, S_{yy}(z), \tag{10}$$

$$S_{e_x^o e_x^o}(z) = [1 - \Gamma_{xy}(z)]\, S_{xx}(z). \tag{11}$$

The magnitude squared (MS) coherence function $\Gamma_{xy}(z)$ which appears in this expression is given by [5]

$$\Gamma_{xy}(z) = \begin{cases} \dfrac{S_{xy}(z)\, S_{yx}(z)}{S_{xx}(z)\, S_{yy}(z)}, & \text{for } S_{xx}(z)\, S_{yy}(z) \neq 0 \\ 0, & \text{for } S_{xx}(z)\, S_{yy}(z) = 0. \end{cases} \tag{12}$$

The terminology *magnitude squared* coherence function arises from the fact that the frequency domain equivalent

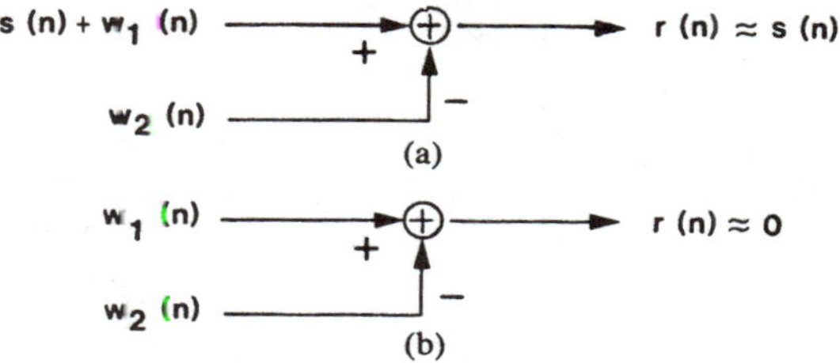

Fig. 3. Noise subtraction scheme. (a) Desired result. (b) Test.

of this function can be expressed as $\Gamma_{xy}(e^{j\omega}) = |C_{xy}(e^{j\omega})|^2$ where

$$C_{xy}(e^{j\omega}) = S_{xy}(e^{j\omega})/(S_{xx}(e^{j\omega})\, S_{yy}(e^{j\omega}))^{1/2}$$

is the *complex coherence function*. An important property of the MS coherence function is revealed by evaluating either (10) or (11) on the unit circle (i.e., $z = e^{j\omega}$) to determine that

$$0 \leq \Gamma_{xy}(e^{j\omega}) \leq 1 \tag{13}$$

since the optimum error autospectra may never be nonnegative.

To determine the degree of bidirectional linear association between the time series $\{x(n)\}$ and $\{y(n)\}$, a plot of the associated MS coherence function $\Gamma_{xy}(e^{j\omega})$ on the interval $[0, \pi]$ is helpful. This is made apparent upon examining the optimum model error spectra (10) and (11) with the substitution $z = e^{j\omega}$. Clearly, whenever the MS coherence is one for all frequencies, both optimum model error power spectral functions are identically zero. In such situations, the corresponding normalized MSE indexes ρ_{xy} and ρ_{yx} are zero, thereby indicating the existence of a perfect bidirectional linear relationship between the two time series $\{x(n)\}$ and $\{y(n)\}$. It often happens, however, that the coherence function has values chosen to one for certain ranges of frequencies and values significantly smaller than one for other frequencies. This behavior in and of itself does not indicate the inappropriateness of linear modeling. It does suggest, however, that linear modeling is more applicable over certain frequency intervals than others.

Example 2: To demonstrate the use of the MS coherence function, we consider the noise subtraction scheme shown in Fig. 3. The sequences $\{w_1(n)\}$ and $\{w_2(n)\}$ are two physically independent measurements of the corrupting noise. When the method works, the noise sequences $\{w_1(n)\}$ and $\{w_2(n)\}$ are nearly equal, and the result of the subtraction $\{r(n)\}$ is approximately equal to the signal $\{s(n)\}$. If the signal $\{s(n)\}$ can be set to zero, the propriety of the method can be investigated by plotting the result $\{r(n)\}$. Any significant deviation of $\{r(n)\}$ from zero indicates that the method is not appropriate for this particular problem. However, what counts as significant is not clear.

In [8], a formal derivation shows that this simple noise subtraction method is appropriate for correcting errors induced by tape speed variation in frequency modulated telemetry data recorded on analog tape machines. Here

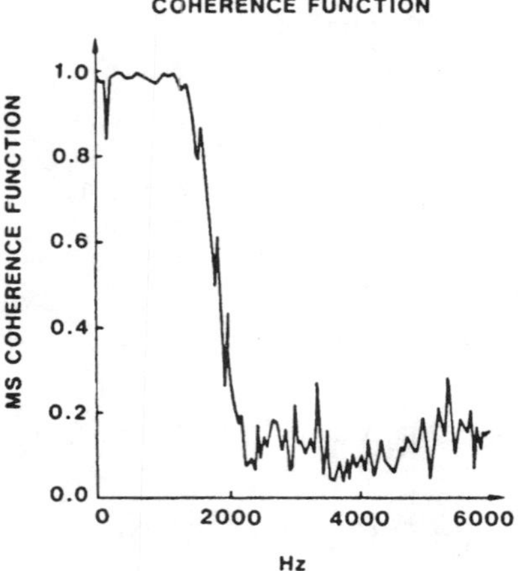

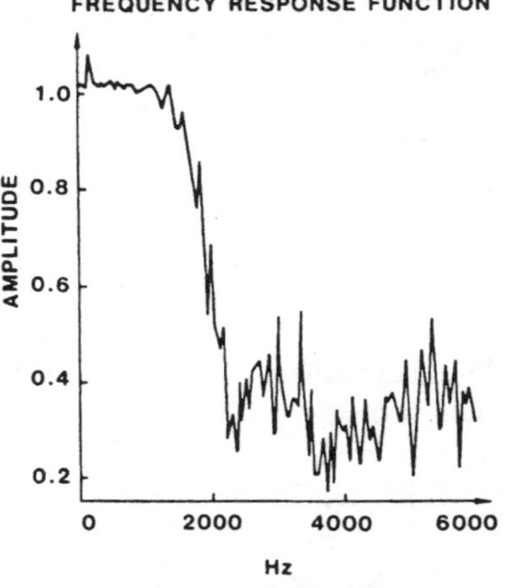

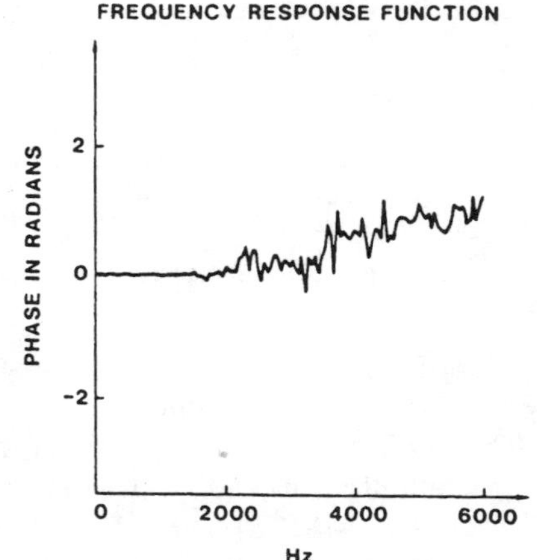

Fig. 4. MS coherence along with magnitude and phase of the frequency response function between $\{w_1(n)\}$ and $\{w_2(n)\}$ for Fig. 3(b).

we illustrate an experimental method for determining when noise subtraction schemes are appropriate. Two sequences $\{w_1(n)\}$ and $\{w_2(n)\}$, each of length 8000, were digitized from an analog tape machine (see [8] for a full description of the experimental setup). The signal $\{s(n)\}$ was set to zero and, consequently, $\{r(n)\}$ should have been zero. Fig. 4 shows the MS coherence function along with the amplitude and phase of the frequency response $H_{w_2w_1}(e^{j2\pi f})$ between $\{w_2(n)\}$ and $\{w_1(n)\}$. For frequencies where the MS coherence is near one, the amplitude is near one and the phase is near zero. The MS coherence drops to zero for frequencies where the amplitude is not one and the phase is not zero. An unshown plot of the autospectrum of $\{w_2(n)\}$ disclosed that $\{w_2(n)\}$ has little power where the MS coherence is zero. These observations demonstrate that noise subtraction is an appropriate signal processing method for these data sequences. Very little improvement would result from passing $\{w_2(n)\}$ through $H_{w_2w_1}(e^{j2\pi f})$ and then subtracting the result from $\{w_1(n)\}$. For problems in which $\{w_2(n)\}$ has significant power for frequencies where the MS coherence is near zero, then the additional filtering is justified.

Transfer Function Invertibility

Upon inserting the optimal transfer functions (4) and (7) into the MS coherence function's definition, we see that

$$\Gamma_{xy}(z) = H_{xy}^o(z)\, H_{yx}^o(z). \tag{14}$$

This latter relationship suggests that when linear modeling is appropriate (i.e., $\Gamma_{xy}(e^{j\omega}) = 1$ for all ω) then the *optimal frequency responses* $H_{xy}^o(e^{j\omega})$ and $H_{xy}^o(e^{j\omega})$ will tend to be the inverses of one another. In fact, they are exact inverses when $\Gamma_{xy}(e^{j\omega}) \equiv 1$.

MS Coherence Sequence

Due to the readily established property that $\Gamma_{xy}(z) = \Gamma_{xy}(z^{-1})$, it follows that the sequence associated with the MS coherence function must be even. This MS *coherence sequence* is formally given by the inverse z-transform relationship

$$\gamma_{xy}(n) = \frac{1}{2\pi j} \oint \Gamma_{xy}(z) z^{n-1}\, dz \quad \text{for } n = 0, \pm 1, \pm 2, \cdots,. \tag{15}$$

The MS coherence sequence can be given an interesting interpretation in relationship to linear modeling. Namely, upon taking the inverse z-transforms of relationships (10) and (11), we obtain

$$r_{e_y^o e_y^o}(n) = r_{yy}(n) - \sum_{k=-\infty}^{\infty} \gamma_{xy}(k)\, r_{yy}(n-k) \tag{16}$$

and

$$r_{e_x^o e_x^o}(n) = r_{xx}(n) - \sum_{k=-\infty}^{\infty} \gamma_{xy}(k)\, r_{xx}(n-k), \tag{17}$$

respectively. It should be noted that $\Gamma_{xy}(e^{j\omega}) = 1$ for all ω if and only if $\{\gamma(n)\} = \{\delta(n)\}$. If the MS coherence sequence $\{\gamma_{xy}(n)\}$ equals the unit Kronecker delta sequence [i.e., $\gamma(n) = \delta(n)$], then each of these model error autocorrelations will be identically zero, thereby indicating a bidirectional linear relationship between $\{x(n)\}$ and $\{y(n)\}$. Thus, the degree to which the MS coherence sequence approximates the unit Kronecker delta sequence provides yet another measure of linear association.

IV. Rational Magnitude Squared Coherence Functions

Whenever the autospectra and cross-spectrum characterizing two time series are rational (i.e., ratio of polynomials) in nature, their associated MS coherence function is also rational. Since the assumption of rational spectra is often invoked in practical applications, the study of rational MS coherence functions is important [4]. With this in mind, we now examine the case in which the underlying spectra cause the associated MS coherence function to have the rational form of order $(2p, 2q)$ as specified by

$$\Gamma_{xy}(z) = \frac{f_q z^q + \cdots + f_1 z + 2f_o + f_1 z^{-1} + \cdots + f_q z^{-q}}{g_p z^p + \cdots + g_1 z + 2 + g_1 z^{-1} + \cdots + g_p z^{-p}} = \frac{F_{2q}(z)}{G_{2p}(z)}. \quad (18)$$

The middle coefficients of these symmetric polynomials have been arbitrarily taken as $2f_o$ and 2 to simplify a matrix notation to be subsequently employed. It is to be noted that the real-valued coefficients characterizing the constituent polynomials $F_{2q}(z)$ and $G_{2p}(z)$ are symmetric. This is a direct consequence of the property $\Gamma_{xy}(z) = \Gamma_{xy}(z^{-1})$ possessed by the MS coherence function (12).

A convenient procedure for determining the f_k and g_k coefficients of representation (18) from the true auto- and cross-correlation functions is obtained by first introducing the auxiliary functions

$$A(z) = S_{xy}(z)\, S_{yx}(z) \quad (19)$$

and

$$B(z) = S_{xx}(z)\, S_{yy}(z), \quad (20)$$

which constitute the numerator and denominator, respectively, of the MS coherence function. The deterministic time series associated with these auxiliary functions are readily computed using the corresponding time domain convolution summations

$$a(m) = \sum_{k=-\infty}^{\infty} r_{xy}(k)\, r_{yx}(m-k) \quad (21)$$

and

$$b(m) = \sum_{k=-\infty}^{\infty} r_{xx}(k)\, r_{yy}(m-k) \quad (22)$$

obtained by taking the inverse z-transform of (19) and (20). From the defining z-transform expressions (19) and (20), it immediately follows that both of these time series are even [i.e., $a(-m) = a(m)$ and $b(m) = b(-m)$]. Upon using the equality $\Gamma_{xy}(z) = A(z)/B(z)$ and taking into account the rational representation (18), the following time domain recursive expression

$$\sum_{n=0}^{p} g_n[a(m-n) + a(m+n)] - \sum_{n=0}^{q} f_n[b(m-n) + b(m+n)] = 0 \quad \text{for all } m \quad (23)$$

is seen to arise. The symmetry of the polynomial coefficients f_n and g_n has been used in arriving at this result.

To determine the values associated with the underlying f_n and g_n coefficients, we need only evaluate relationship (23) over an appropriate set of indexes "m" and then solve the resultant system of linear equations for these coefficients. Normally, an appropriate set of indexes is 0 to $M-1$ for $M = p + q + 1$. This choice of M provides M equations from which to calculate $M = p + q + 2$ unknown parameters (see theorem 1). In this procedure, the computed values for the $\{a(m)\}$ and $\{b(m)\}$ time series, as given by expressions (21) and (22), are used. To illustrate this procedure, let us evaluate recursion (23) over the set of indexes $0 \le m \le M-1$ to arrive at the homogeneous system of partitioned equations

$$[A_p \;\vdots\; B_q]\boldsymbol{\theta} = \mathbf{0} \quad (24)$$

where $\boldsymbol{\theta}$ is the $(p + q + 2) \times 1$ parameter vector as specified by

$$\boldsymbol{\theta} = [1, g_1, g_2, \cdots, g_p, f_o, f_1, \cdots, f_q]' \quad (25)$$

while A_p and B_q are $M \times (p+1)$ and $M \times (q+1)$ matrices, respectively, whose elements are specified by

$$A_p(m, n) = a(m-n) + a(m+n-2) \qquad 1 \le m \le M, \quad 1 \le n \le p+1 \quad (26)$$

$$B_q(m, n) = -(b(m-n) + b(m+n-2)) \qquad 1 \le m \le M, \quad 1 \le n \le q+1. \quad (27)$$

The prime symbol "$'$" appearing in (25) denotes the operation of vector transposition. The subscripts p and q have been appended to the matrices in order to explicitly recognize the order of the MS coherence function as specified in expression (18). Since the entries of $[A_p \;\vdots\; B_q]$ are convolutions of correlation functions, we will call $[A_p \;\vdots\; B_q]$ the convolved correlation matrix. Given entries for the A_p and B_q matrices, a solution for the MS coherence function coefficients is then obtained by finding a normalized vector $\boldsymbol{\theta}$ (i.e., its first component is one) which satisfies the homogeneous system (24). Such a vector lies in the null space of the convolved correlation matrix. The nullity of $[A_p \;\vdots\; B_q]$ is defined to be the number of linearly independent solutions to (24) (i.e., the dimension of the null space of $[A_p \;\vdots\; B_q]$). Clearly, the existence of a *unique* solution implies that

$$\text{nullity } [A_p \;\vdots\; B_q] = 1. \quad (28)$$

In many applications, the precise order of the MS coherence function is unknown. This fact is not a problem for the approach just described because choosing p and q larger than necessary (i.e., overordering) does not impair our determination of the MS coherence. Furthermore, to achieve a noise desensitization in the practical case in which only time series observations (and not correlation lags) are available for estimating the MS coherence function, it has been found beneficial to use an overordered rational model for spectrum estimation problems [1]. With this in mind, the following theorem describing the ideal situation being considered here is offered.

Theorem 1: Let the autocorrelation and cross-correlation lags associated with the two wide-sense stationary time series $\{x(n)\}$ and $\{y(n)\}$ be such that their MS coherence function is rational with reduced order $(2\bar{p}, 2\bar{q})$. Let this MS coherence function be modeled by a rational model of extended order $(2p, 2q)$ in which $p \geq \bar{p}$ and $q \geq \bar{q}$. A selection for the coefficients characterizing this model must then satisfy the consistent system of linear equations

$$[A_p \,\vdots\, B_q]\,\theta = \mathbf{0} \tag{29}$$

where the entries of the constituent matrices are given by (25)–(27). If the full rank conditions

$$\text{rank}\,[A_p] = p + 1, \quad \text{rank}\,[B_q] = q + 1 \tag{30}$$

are satisfied, it then follows that

$$s = \text{nullity}\,[A_p \,\vdots\, B_q] = 1 + \min\,(p - \bar{p}, q - \bar{q}). \tag{31}$$

Thus, the solutions to the modeling equations (29) lie in a space of dimension s. Furthermore, the MS coherence function associated with any solution to relationship (29) simplifies to the underlying reduced order $(2\bar{p}, 2\bar{q})$ rational model after pole-zero cancellations; that is,

$$\Gamma_{xy}(z) = \frac{F_{2q}(z)}{G_{2p}(z)} = \frac{F_{2\bar{q}}(z)}{G_{2\bar{p}}(z)}. \tag{32}$$

This theorem provides us with the comforting knowledge that a rational model overordering does not impair our determination of the underlying coherence function. This theorem is proven in [7].

Eigenvector Generated Solution

To determine a solution to the consistent system of linear equations (29), it is fruitful to employ an eigenspace approach [1], [7], [11]. In particular, let the $M \times (p + g + 2)$ matrix characterizing this system of equations be denoted by

$$C_{p,q} = [A_p \,\vdots\, B_q]. \tag{33}$$

We next compute the eigenvalue–eigenvector pairs associated with the matrix product $C'_{p,q}C_{p,q}$; that is,

$$C'_{p,q}C_{p,q}\mathbf{v}_k = \lambda_k \mathbf{v}_k \quad 1 \leq k \leq p + q + 2. \tag{34}$$

Since this matrix product is a positive-semidefinite symmetric matrix, it then follows that its eigenvalues are all nonnegative real. They shall be here ordered in the nondecreasing fashion

$$0 \leq \lambda_1 \leq \lambda_2 \leq \cdots \leq \lambda_{p+q+2}. \tag{35}$$

Moreover, the symmetrical structure of the matrix product also indicates that a full set of orthonormal eigenvectors $\{\mathbf{v}_k\}$ for $1 \leq k \leq p + q + 2$ exist where

$$\mathbf{v}'_k \mathbf{v}_m = \begin{cases} 1, & \text{for } k = m \\ 0, & \text{for } k \neq m. \end{cases} \tag{36}$$

For zero eigenvalues, (34) reduces to

$$C'_{p,q}C_{p,q}\mathbf{v}_k = \mathbf{0}.$$

Multiplying this latter equation on the left by $\mathbf{v}'_k$ shows that the norm squared of $C_{p,q}\mathbf{v}_k$ is zero and, consequently, that $C_{p,q}\mathbf{v}_k = 0$. Thus, eigenvectors corresponding to zero eigenvalues lie in the null space of $C_{p,q}$.

If the conditions as spelled out in theorem 1 are satisfied, it follows that "s" of the eigenvalues (35) must be zero. This then implies that the eigenvectors associated with these zero eigenvalues

$$\{\mathbf{v}_1, \mathbf{v}_2, \cdots, \mathbf{v}_s\} \tag{37}$$

form a basis for the solutions to the homogeneous modeling relationship (29). Namely, any solution to this system of homogeneous equations may be expressed as

$$\theta = \sum_{k=1}^{s} \alpha_k \mathbf{v}_k \tag{38}$$

where the scalars α_k are selected so as to ensure that the first component of θ is one as required, but are otherwise arbitrary. When the parameter $s > 1$, there will exist an infinity of solutions of the form (38). From this infinity of solutions, the minimum norm solution as specified by

$$\theta = \left\{\sum_{k=1}^{s} v_k^2(1)\right\}^{-1} \sum_{k=1}^{s} v_k(1)\mathbf{v}_k \tag{39}$$

is here used [1].

V. Coherence Function Estimation

In practical applications, the autocorrelation and cross-correlation values for coherence function identification are not usually known. More typically, there is given a set of empirically obtained observations

$$x(1), x(2), \cdots, x(N) \quad \text{and}\; y(1), y(2), \cdots, y(N) \tag{40}$$

from which a coherence function estimate is to be made. It is possible to straightforwardly adapt the procedure described in the last section to this case so as to achieve effective estimates. In particular, these observations are used to first generate estimates for the autocorrelation and cross-correlation functions needed for computing the entries of matrices A_p and B_q as specified by (21), (22), (26), and (27). Estimates will be distinguished from true values by placing a caret (ˆ) over the corresponding true value. Due to errors inherent in estimating correlation functions, and the truncated summations necessary to estimate the convolutions for the $\hat{a}(n)$ and $\hat{b}(n)$ in (21) and (22), it is generally found that the matrix estimate $\hat{C}_{p,q}$ is of full rank. This, in turn, indicates that the associated model residual vector as specified by

$$\varepsilon = [\hat{A}_p \,\vdots\, \hat{B}_q]\,\theta = \hat{C}_{p,q}\theta \tag{41}$$

cannot be made equal to the zero vector for any feasible choice of the parameter vector θ (recall that $\theta(1) = 1$). This being the case, acceptable choices for the coefficients vectors intuitively would be those which cause the squared residual criterion

$$\varepsilon'\varepsilon = \theta' \hat{C}'_{p,q} \hat{C}_{p,q} \theta \tag{42}$$

to take on a value suitably close to zero.

Least Squares Solution

The most obvious choice for the coefficient vectors are those which render the squared residual criterion (42) a minimum subject to the constraint that the first component of θ is one. It is readily shown that this *least squares solution* is given by

$$\theta_{LS} = \lambda[\hat{C}'_{p,q}\hat{C}_{p,q}]^{-1} e_1, \tag{43}$$

where e_1 is the $(p + q + 2) \times 1$ vector whose components are all zero except for its first component which is one. The normalizing scalar $\hat{\lambda}_k$ appearing here is selected so that the first component of θ_{LS} is one as required. Unfortunately, this least squares solution does not always result in suitably good MS coherence function estimates. Earlier it was shown that the required parameter vector θ is determined by the eigenvectors associated with the zero eigenvalues of the true matrix product $C'_{p,q}C_{p,q}$. The problem with the least squares estimate is that, for finite data lengths, the eigenvectors corresponding to the eigenvalues which are theoretically zero are not given sufficient weight relative to the other eigenvectors.

Near Null Space Approach

The procedure to be presented now typically yields better estimates than the above least squares method by not using the eigenvectors corresponding to the larger eigenvalues at all. With this in mind, let us now generate the eigencharacterization for the estimate of this matrix product; that is,

$$\hat{C}'_{p,q}\hat{C}_{p,q}\hat{v}_k = \hat{\lambda}_k \hat{v}_k \quad 1 \le k \le p + q + 2 \tag{44}$$

where the eigenvalue ordering $\hat{\lambda}_k \le \hat{\lambda}_{k+1}$ has been stipulated. Due to the aforementioned error sources, it is generally found that all of the $\hat{\lambda}_k$ eigenvalues are positive. These errors have therefore perturbed the underlying "s" zero eigenvalues to positive values. Fortunately, it often happens that the "s" smallest eigenvalues arising from (44) are still smaller than their $p + q + 2 - s$ larger valued counterparts. In such cases, the associated eigenvectors $\{\hat{v}_1, \hat{v}_2, \cdots, \hat{v}_s\}$ form a reasonably good approximation to the underlying basis (37). As such, the minimum norm solution approximation as given by

$$\hat{\theta} = \left(\sum_{k=1}^{s} \hat{v}_k^2(1)\right)^{-1} \sum_{k=1}^{s} \hat{v}_k(1)\hat{v}_k \tag{45}$$

appears as a logical choice for the coefficient vectors. It has been empirically determined that the minimum norm solution approximation has provided good coherence function estimates on a variety of examples. The next section provides one such example. The near null space procedure is summarized in Table I.

TABLE I
NEAR NULL SPACE PROCEDURE

1) Estimate the correlation functions from the raw data.
2) Estimate $\hat{a}(n)$ and $\hat{b}(n)$ by evaluating (21) and (22) with the true correlation functions replaced by their estimates.
3) Choose p, q and M; and then form $\hat{C}_{p,q} = [\hat{A}_p \vdots \hat{B}_q]$ according to (26) and (27).
4) Compute the eigendecomposition of $\hat{C}'_{p,q}\hat{C}_{p,q}$ as in (34).
5) Select the s smallest eigenvalues, and construct the minimum norm solution θ as in (45).
6) Form the rational MS coherence estimate according to (18) and (25).

VI. Numerical Example

To demonstrate the effectiveness of the proposed estimation method, we shall now examine the MS coherence between a white noise excitation and the corresponding response corrupted by a colored noise process as depicted in Fig. 5. Namely, the noise-free excitation-response data are governed by the ARMA relationship

$$y_a(n) = y_a(n-1) - 0.8\, y_a(n-2) + x(n-1). \tag{46a}$$

The data sequences were generated by exciting this system with a white Gaussian noise process of variance $\sigma_x^2 = 1$. After transient response effects have disappeared, a set of steady-state excitation-response measurements $x(n)$ and $y_a(n)$ for $1 \le n \le 2500$ was recorded. The corrupted response measurements were generated in accordance with the equations

$$y(n) = y_a(n) + v(n), \tag{46b}$$

$$v(n) = 1.212\, v(n-1) - 0.49\, v(n-2) + w(n), \quad 1 \le n \le 2500, \tag{46c}$$

where $\{w(n)\}$ is a white Gaussian process of variance $\sigma_w^2 = 1$. The actual MS coherence function between $\{x(n)\}$ and $\{y(n)\}$ is given by

$$\Gamma_{xy}(e^{j\omega}) = \frac{0.5066 - 0.6754\cos(\omega) + 0.1832\cos(2\omega)}{1.0 - 1.3482\cos(\omega) + 0.4811\cos(2\omega)}. \tag{47}$$

In order to provide a statistical basis for comparing the different estimator's performance, 100 independent repetitions of this data generation procedure were realized. This resulted in the following observation records:

$$(x^{(k)}(n), y^{(k)}(n)) \quad \text{for } 1 \le n \le 2500 \quad \text{and } 1 \le k \le 100.$$

Using these data records, estimates of the coherence function were computed by three versions of this paper's near null space approach and a traditional periodogram-based method [3]. The different versions of this paper's method demonstrate the effects of parameter overdetermination and overordering on the MS coherence esti-

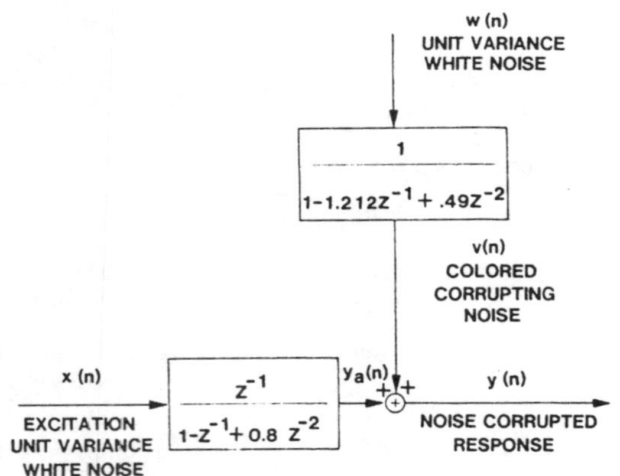

Fig. 5. A numerical example with colored corrupting noise.

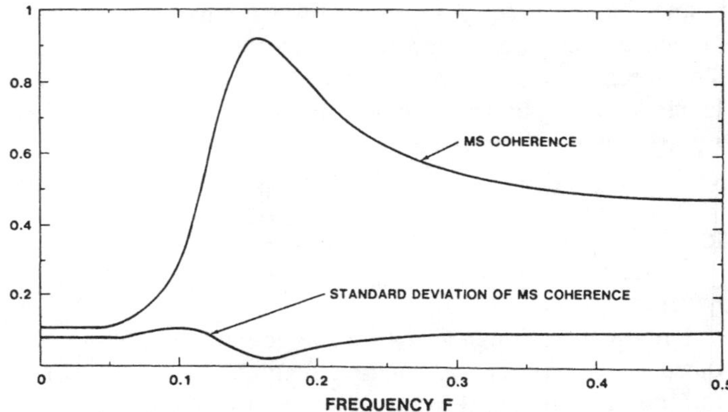

Fig. 6. The true MS coherence along with its approximate standard deviation for a large number of degrees of freedom.

mates. Version I uses $p = q = 2$, $s = 1$, and $M = 2 \times (p + q + 2)$ which corresponds to the true MS coherence order with a factor of two overdetermination. Version II uses $p = q = 4$, $s = 3$, and $M = 10$ which overorders the numerator and denominator of the MS coherence function by two independent parameters each. This choice of M causes $C_{4,4}$ to be square. Version III uses $p = q = 2$, $s = 1$, and $M = p + q + 2$ which corresponds to the true MS coherence order with no overdetermination. In all three versions, the range of auto- and cross-correlation lags computed was -30 to 30. These functions were assumed to be zero for lags outside of this range.

A very popular periodogram-based method estimates the autospectra and cross spectrum via fast Fourier transforms (FFT) of windowed overlapping segments [3]. Before segmenting for the FFT method, the sequences $\{x^{(k)}(n)\}$ and $\{y^{(k)}(n)\}$ are aligned to make their peak correlation occur at lag 0. This alignment eliminates one possible source of bias [10]. Next, the two sequences are partitioned into K overlapping segments of equal length L, and a Hanning time domain window is applied to each segment. The amount of overlap is 62.5 percent which produces maximum reduction in bias and variance of the MS coherence [3]. Estimates of the autospectra and cross spectrum are calculated by averaging FFT's of the windowed overlapping segments. An estimate of the MS coherence is formed by substituting these power spectral density estimates into (12). The FFT segment length was chosen to be 256, which is the minimum length that provides sufficient resolution to avoid significant bias at the peak value of the MS coherence function.

The statistics of this estimator are well known [3], [6]. Fig. 6 shows the true MS coherence along with its approximate standard deviation (SD) calculated from equation (IIg) in [3]:

$$SD = \{2\Gamma_{xy}(e^{j2\pi f}) [1 - \Gamma_{xy}(e^{j2\pi f})]/n_d\}^{0.5}$$

for $n_d = 25$, which is the number of overlapping segments for 2500 points with a 62.5 percent overlap factor.

For each of the four estimation methods described in the preceding paragraph, a set of 100 (one for each data set) MS coherence estimates $\hat{\Gamma}_k(e^{j\omega})$ was generated. The "sample mean" of these MS coherence estimates as defined by

$$\overline{\Gamma}_{xy}(e^{j\omega}) = \frac{1}{100} \sum_{k=1}^{100} \hat{\Gamma}_k(e^{j\omega}) \tag{48}$$

was then computed for each of the three methods. In a similar fashion, the "sample root mean squared error" ($\overline{\text{rmse}}$) associated with these MS coherence estimates was computed according to

$$\overline{\text{rmse}}(e^{j\omega}) = \left\{\frac{1}{100} \sum_{k=1}^{100} [\hat{\Gamma}_k(e^{j\omega}) - \Gamma(e^{j\omega})]^2\right\}^{1/2}. \tag{49}$$

The sample means and $\overline{\text{rmse}}$'s associated with each of the modeling methods being compared were then plotted as a function of f along with a plot of the true MS coherence function (47). These plots are shown in Fig. 7. Fig. 8 shows the true MS coherence function along with the first 5 estimates of the MS coherence function from the 100 independent data records. Figs. 7 and 8 clearly show that this paper's version I (overdetermination) outperforms this paper's versions II (overordering) and III (neither overdetermination nor overordering) as well as the periodogram-based method for this particular example. This computer experiment should not be construed as showing that overdetermination is always better than overordering. However, overordering did not significantly outperform overdetermination for any of our computer simulated examples. More research is needed to definitely describe when to and when not to use overdetermination and overordering.

It is appropriate to ask about the computational time of the methods. In order to approximately answer this question, the CPU time in seconds for version I and the periodogram method was measured. There was no attempt to streamline either computer code. Version I averaged 1.8 CPU s per run and the periodogram method averaged 4.6 CPU s per run. These execution times include data generation as well as the MS coherence computation.

In [7], the sensitivity of the matrix equation

$$C_{p,q}\boldsymbol{\theta} = \mathbf{0}$$

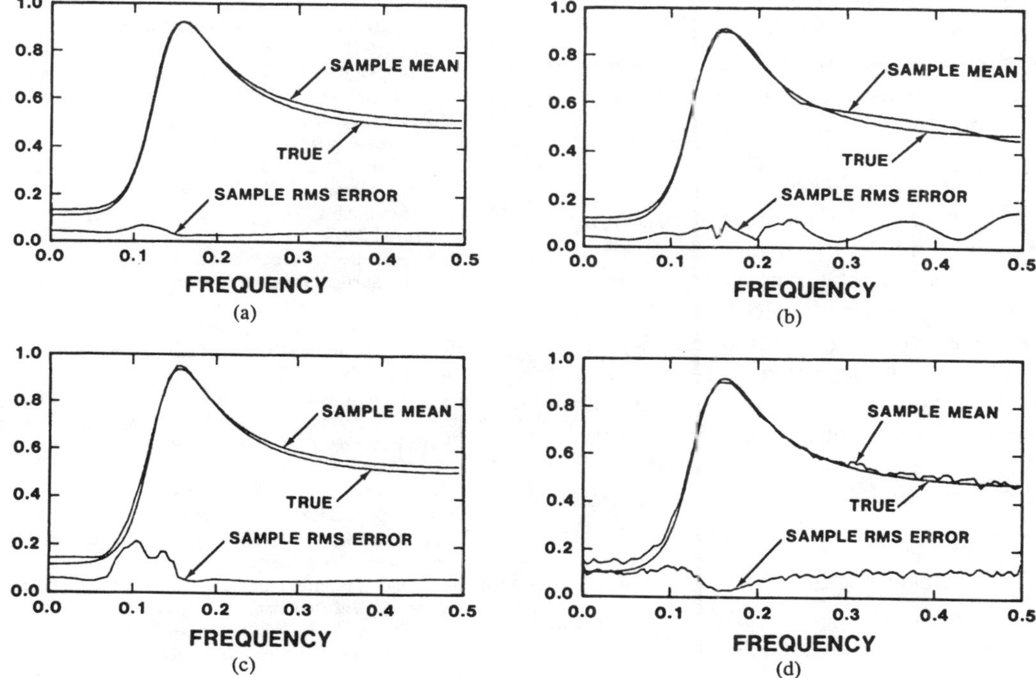

Fig. 7. True MS coherence function along with the sample mean and sample RMS error from 100 runs of 4 different estimators. (a) Version I—overdetermination. (b) Version II—overordering. (c) Version III—neither overdetermination nor overordering. (d) A periodogram method.

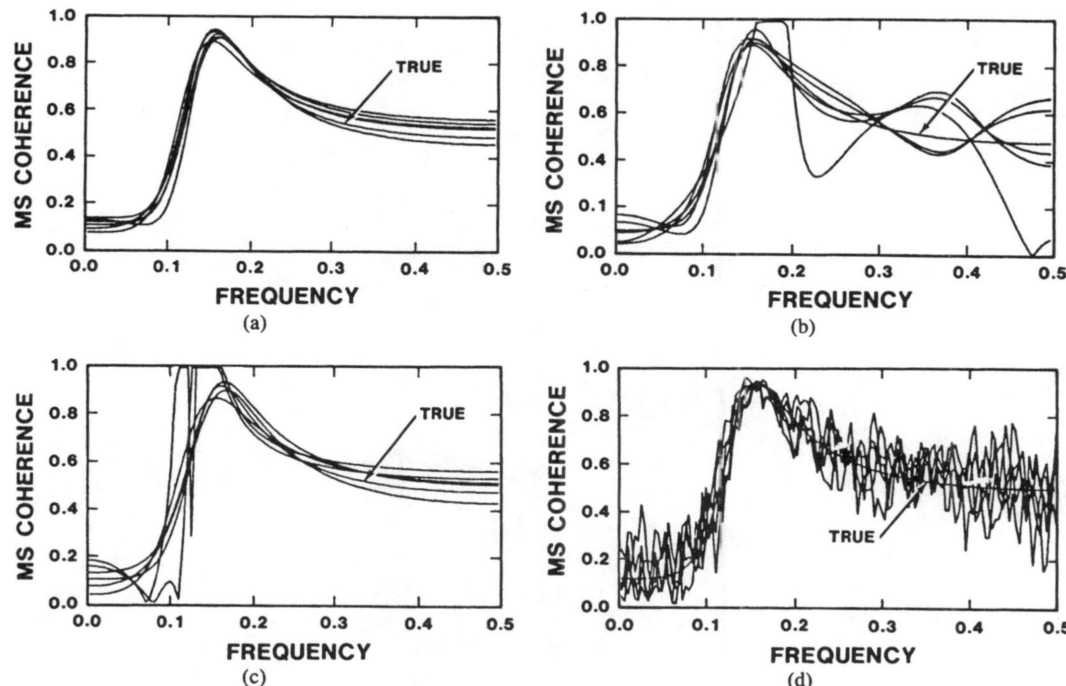

Fig. 8. True MS coherence function along with 5 estimates from 4 different estimators. (a) Version I—overdetermination. (b) Version II—overordering. (c) Version III—neither overdetermination nor overordering. (d) A periodogram method.

is investigated via the numerical linear algebra concepts of numerical stability and condition number. The condition number is a measure of the sensitivity of $\boldsymbol{\theta}$ to perturbations in $C_{p,q}$. When estimating the MS coherence function from sequences $\{x(n)\}$ and $\{y(n)\}$, we work with a perturbed $\hat{C}_{p,q}$ rather than the true $C_{p,q}$. The perturbations in $\hat{C}_{p,q}$ are introduced in steps 1) and 2) of the near null space procedure described in Table I. The condition numbers for versions I, II, and III of the near null space procedure are 54, 86, and 97, respectively (see [7] for details). Thus, version I (overdetermination) is almost a factor of 2 less sensitive to perturbations than version III (neither overdetermination nor overordering).

VI. Conclusion

Several properties of the MS coherence function were established in conjunction with linear modeling. An estimation procedure for the MS coherence function has been proposed, which is based on a null space characterization of a convolved correlations matrix. Some useful properties of the convolved correlations matrix were established for the case of ideal correlation information. These properties formed the basis for MS coherence estimation in the nonideal case of finite length observation records. The procedure was compared to a periodogram-based estimation method.

References

[1] J. A. Cadzow, "Spectral estimation: An overdetermined rational model equation approach," *Proc. IEEE*, vol. 70, pp. 907–939, Sept. 1982.

[2] G. C. Carter and C. H. Knapp, "Coherence and its estimation via the partitioned modified chirp-Z transform," *IEEE Trans. Acoust., Speech, Signal Processing*, vol. ASSP-23, June 1975.

[3] G. C. Carter, C. H. Knapp, and A. H. Nuttall, "Estimation of the magnitude-squared coherence function via overlapped fast Fourier transform processing," *IEEE Trans. Audio Electroacoust.*, vol. AU-21, Aug. 1973.

[4] Y. T. Chan and R. K. Miskowisz, "Estimation of coherence and time delay with ARMA models," *IEEE Trans. Acoust., Speech, Signal Processing*, vol. ASSP-32, June 1984.

[5] L. H. Koopmans, *The Spectral Analysis of Time Series.* New York: Academic, 1974.

[6] M. B. Priestley, *Spectral Analysis and Time Series, Volume 2: Multivariate Series, Prediction and Control.* New York: Academic, 1981.

[7] O. M. Solomon, Jr., "Linear modeling and the coherence function: An algebraic approach," Ph.D. dissertation, Univ. New Mexico, Albuquerque, Dec. 1985.

[8] —, "A nonmystical treatment of tape speed compensation for frequency modulated signals," in *Proc. Int. Telemeter. Conf.*, vol. XVIII, San Diego, CA, 1982, pp. 565–576.

[9] S. D. Stearns, "Applications of the coherence function in comparing seismometers," Sandia National Laboratories, Albuquerque, NM, Tech. Rep. SAND79-1633, Dec. 1979; available from National Tech. Information Service, U.S. Dept. of Commerce, 5285 Port Royal Rd., Springfield, VA 22161.

[10] —, "Tests of coherence unbiasing methods," *IEEE Trans. Acoust., Speech, Signal Processing*, vol. ASSP-29, Apr. 1981.

[11] D. H. Youn, Nasir Ahmed, and G. C. Carter, "Magnitude-squared coherence function estimation: An adaptive approach," *IEEE Trans. Acoust., Speech, Signal Processing*, vol. ASSP-31, Feb. 1983.

Estimation of the Magnitude-Squared Coherence Function Via Overlapped Fast Fourier Transform Processing

G. CLIFFORD CARTER, CHARLES H. KNAPP, AND ALBERT H. NUTTALL

Abstract—A method for estimating the magnitude-squared coherence function for two zero-mean wide-sense-stationary random processes is presented. The estimation technique utilizes the weighted overlapped segmentation fast Fourier transform approach. Analytical and empirical results for statistics of the estimator are presented. The analytical expressions are limited to the nonoverlapped case; empirical results show a decrease in bias and variance of the estimator with increasing overlap and suggest a 50-percent overlap as being highly desirable when cosine (Hanning) weighting is used.

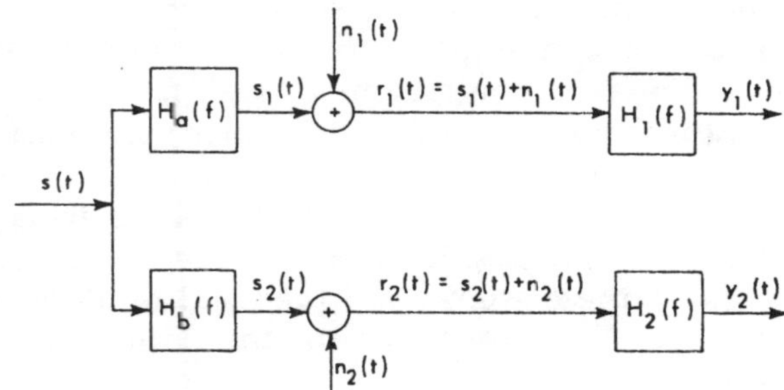

Fig. 1. Configuration for examining properties of MSC.

I. Introduction

The complex coherence function is the normalized cross spectral density,

$$\gamma_{xy}(f) = \frac{\Phi_{xy}(f)}{\sqrt{\Phi_{xx}(f)\Phi_{yy}(f)}} \tag{1}$$

where $\Phi_{xy}(f)$ is the cross spectral density at frequency f between two zero-mean stationary random processes $x(t)$ and $y(t)$ with auto spectra $\Phi_{xx}(f)$ and $\Phi_{yy}(f)$. In particular, the magnitude-squared coherence (MSC)

$$|\gamma_{xy}(f)|^2 = \frac{|\Phi_{xy}(f)|^2}{\Phi_{xx}(f)\Phi_{yy}(f)} \tag{2}$$

has useful properties, which has led investigators to examine the errors arising from estimating this function [1]-[18].

Current techniques for estimating the MSC generally use the fast Fourier transform (FFT) [19]-[21]. Briefly, the method consists of obtaining two finite-time series from the random processes being investigated. Each time series is partitioned into n equal length segments and sampled at P equally spaced data points. The segments may be overlapped or disjoint. Samples from each segment are multiplied by a weighting function, and the FFT of the weighted P-point sequence is performed. Then the Fourier coefficients for each weighted segment are used to estimate the auto- and cross-power spectral densities. The spectral density estimates thus obtained are used in (2) to form the MSC estimate.

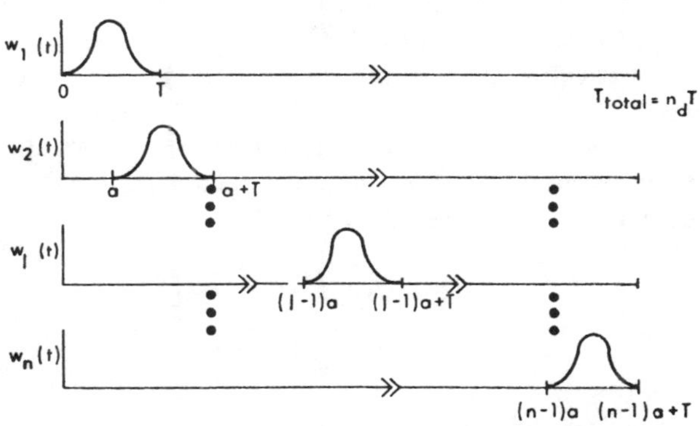

Fig. 2. Overlapped weighting functions.

Spectral resolution of the estimates varies inversely with the segment length T. Proper weighting or "windowing" of the T-seconds segment is also helpful in achieving good sidelobe reduction. On the other hand, for independent segments with ideal windowing, the bias and the variance of the MSC estimate vary inversely with the number of segments n. Therefore, to generate a good estimate with limited data, one may be faced with conflicting requirements on n and T. Segment overlapping can be used to increase both n and T. As the percentage of overlap increases, however, the computational requirements increase rapidly, while the improvement stabilizes owing to

Manuscript received July 18, 1972.
G. C. Carter and A. H. Nuttall are with the Naval Underwater Systems Center, New London Laboratory, New London, Conn. 06320.
C. H. Knapp is with the Department of Electrical Engineering, University of Connecticut, Storrs, Conn. 06268.

Reprinted from *IEEE Trans. Audio and Electroacoustics,* vol. 21, no. 4, pp. 337–344, August 1983.

the greater correlation between data segments [16].

This paper presents the results of two studies that provide helpful insight into the choice of n and the amount of overlap for MSC estimation for a given frequency resolution T^{-1}. In the first study, results of Fisher [1] and Goodman [2] were utilized to obtain analytical expressions for the probability distribution and density functions of the MSC estimate and, thence, for the bias and variance of the estimate. The results are valid for Gaussian processes with independent data segments and ideal windowing.

In the second study, the effect of overlap on MSC estimation is examined through a computational study that uses cosine (Hanning) weighting, a finite time history, and a specified frequency resolution. Under these conditions, estimates of the bias and variance of the MSC estimate have been calculated as a function of overlap for two values of coherence.

Results of these studies are preceded by brief discussions of the properties of the MSC and its estimation.

II. Properties of the MSC Function

In order to determine the basic properties and certain applications of the MSC function, special cases of the configuration depicted in Fig. 1 will be considered. It is assumed that $s(t)$, $n_1(t)$, and $n_2(t)$ in Fig. 1 are sample functions of stationary uncorrelated random processes. By the application of basic relationships, it may be verified that [18]

$$|\gamma_{y_1 y_2}(f)|^2 = |\gamma_{r_1 r_2}(f)|^2 \tag{3}$$

$$= \frac{\Phi_{ss}^2(f)\,|H_a(f)\,H_b^*(f)|^2}{[\Phi_{ss}(f)\,|H_a(f)|^2 + \Phi_{n_1 n_1}(f)]\,[\Phi_{ss}(f)\,|H_b(f)|^2 + \Phi_{n_2 n_2}(f)]}. \tag{4}$$

Examination of (2) and (4) and Fig. 1 yields the following conclusions.

1) The MSC is a real number, conveniently normalized to lie between zero and unity.

2) Since $|\gamma_{y_1 y_2}(f)|^2$ is independent of $H_1(f)$ and $H_2(f)$ in (4), the MSC between two signals is not affected by the arbitrary linear processing of these signals.

3) With $n_1(t) = n_2(t) = 0$ in Fig. 1, $|\gamma_{y_1 y_2}(f)|^2 = 1$. Consequently, the MSC of two signals derived from linear operations on the same signal is unity. Equivlently, the MSC of two linearly related signals is unity.

4) If $\Phi_{n_1 n_1}(f) = \Phi_{n_2 n_2}(f) = \Phi_{nn}(f)$ and $|H_a(f)|^2 = |H_b(f)|^2 = 1$, then

$$|\gamma_{y_1 y_2}(f)|^2 = \frac{\Phi_{ss}^2(f)}{[\Phi_{ss}(f) + \Phi_{nn}(f)]^2} \tag{5}$$

and the signal-to-noise power ratio (S/N) is

$$\frac{\Phi_{ss}(f)}{\Phi_{nn}(f)} = \frac{|\gamma_{y_1 y_2}(f)|}{1 - |\gamma_{y_1 y_2}(f)|}. \tag{6}$$

5) If $n_2(t) = 0$ and $|H_a(f)|^2 = 1$, then

$$|\gamma_{y_1 y_2}(f)|^2 = \frac{\Phi_{ss}^2(f)}{[\Phi_{ss}(f) + \Phi_{nn}(f)]\,\Phi_{ss}(f)} \tag{7}$$

and the S/N is

$$\frac{\Phi_{ss}(f)}{\Phi_{nn}(f)} = \frac{|\gamma_{y_1 y_2}(f)|^2}{1 - |\gamma_{y_1 y_2}(f)|^2}. \tag{8}$$

Conclusions 4) and 5) indicate ways in which the MSC can be used to estimate the S/N. Conclusion 3) suggests that the MSC can be used as a measure of linearity.

III. Estimation of the MSC

We assume throughout the paper that $x(t)$ and $y(t)$ are sample functions of random processes that are zero-mean, wide-sense stationary, and ergodic. The basic MSC estimation procedure is to obtain estimates of the spectral densities $\Phi_{xx}(f)$, $\Phi_{xy}(f)$, $\Phi_{yy}(f)$ and use these estimates in (2) in place of their exact counterparts. The method to be described for generating power spectral density estimates is the direct method discussed in [8]–[18]. It employs (cosine) weighting, overlapped processing, and the FFT.

Let the time-limited sample functions $x(t)$ and $y(t)$ be sampled (at a frequency f_s Hz that is greater than twice the larger bandwidth of the two processes) and converted to two (quantized) sequences of numbers or time series. Each time series is effectively divided into n segments through the multiplication of a series of real weighting functions, as illustrated in Fig. 2. If T is the length of each segment, n_d is the possible number of disjoint segments, and $n_d\,T$ seconds is the total record length, then the total number of (overlapped) segments is

$$n = \frac{n_d - O_f}{1 - O_f} \tag{9}$$

where

$$O_f = \frac{T - a}{T} = 1 - \frac{a}{T} \tag{10}$$

is the overlap fraction. When O_f is zero, $n = n_d$.

A P-point ($P = f_s T$) FFT is obtained for the P-point sequence representing each of the weighted data segments. The frequency domain equivalent of multiplying each segment by a weighting function is a convolution of the spectrum with the Fourier transform of the weighting function. Ideally, there-

TABLE I
Probability Density and Distribution Functions

Density Function
$p(\lvert\hat{\gamma}\rvert^2 \mid n_d, \lvert\gamma\rvert^2) = (n_d - 1)(1 - \lvert\gamma\rvert^2)^{n_d}(1 - \lvert\hat{\gamma}\rvert^2)^{n_d-2} \cdot {}_2F_1(n_d, n_d; 1; \lvert\gamma\rvert^2 \lvert\hat{\gamma}\rvert^2), \quad 0 \leq \lvert\gamma\rvert^2 \lvert\hat{\gamma}\rvert^2 < 1$ (Ia)
$= (n_d - 1)(1 - \lvert\gamma\rvert^2)^{n_d}(1 - \lvert\hat{\gamma}\rvert^2)^{n_d-2} \cdot (1 - \lvert\gamma\rvert^2 \lvert\hat{\gamma}\rvert^2)^{1-2n_d}\, {}_2F_1(1 - n_d, 1 - n_d; 1; \lvert\gamma\rvert^2 \lvert\hat{\gamma}\rvert^2)$ (Ib)
$= (n_d - 1)\left[\frac{(1 - \lvert\gamma\rvert^2)(1 - \lvert\hat{\gamma}\rvert^2)}{(1 - \lvert\gamma\rvert^2 \lvert\hat{\gamma}\rvert^2)^2}\right]^{n_d} \cdot \frac{(1 - \lvert\gamma\rvert^2 \lvert\hat{\gamma}\rvert^2)}{(1 - \lvert\hat{\gamma}\rvert^2)^2}\, {}_2F_1(1 - n_d, 1 - n_d; 1; \lvert\gamma\rvert^2 \lvert\hat{\gamma}\rvert^2).$ (Ic)
Distribution Function
$P(\lvert\hat{\gamma}\rvert^2 \mid n_d, \lvert\gamma\rvert^2) = \lvert\hat{\gamma}\rvert^2 \left(\frac{1 - \lvert\gamma\rvert^2}{1 - \lvert\gamma\rvert^2 \lvert\hat{\gamma}\rvert^2}\right)^{n_d} \sum_{k=0}^{n_d-2} \left(\frac{1 - \lvert\hat{\gamma}\rvert^2}{1 - \lvert\gamma\rvert^2 \lvert\hat{\gamma}\rvert^2}\right)^k \cdot {}_2F_1(-k, 1 - n_d; 1; \lvert\gamma\rvert^2 \lvert\hat{\gamma}\rvert^2).$ (Id)

fore, the weighting function transform should be an impulse, which would require a unit weighting for all t. Equivalently, T should be large. Practical factors affecting selection of the segment length, T and window shape include the following:

1) $n_d = T_{total}/T$ should be large in order to reduce the bias and variance of the spectral estimates.

2) The Fourier transform of the weighting function $w_l(t)$ should have a main lobe that is narrower than the finest detail of the true spectra of processes $x(t)$ and $y(t)$. This lobe width is inversely proportional to T.

3) In order to minimize the sidelobes of the Fourier transform of $w_l(t)$, higher order derivatives of $w_l(t)$ should be continuous, if possible.

Thus good spectral resolution (large T) and small bias and variance (large n) provide conflicting requirements on n and T for a specified record length T_{total}.

If $X_l(f_k)$ denotes the FFT of the lth weighted segment of $x(t)$ at frequency f_k, then the spectral density estimates are given by

$$\hat{\Phi}_{xx}(f_k) = \alpha \sum_{l=1}^{n} \left| X_l(f_k) \right|^2,$$

$$\hat{\Phi}_{xy}(f_k) = \alpha \sum_{l=1}^{n} X_l(f_k) Y_l^*(f_k),$$

$$\hat{\Phi}_{yy}(f_k) = \alpha \sum_{l=1}^{n} \left| Y_l(f_k) \right|^2, \tag{11}$$

where $\alpha = 1/(nPf_s)$. Finally, the MSC estimate is

$$\lvert\hat{\gamma}_{xy}(f_k)\rvert^2 \triangleq \frac{\lvert\hat{\Phi}_{xy}(f_k)\rvert^2}{\hat{\Phi}_{xx}(f_k)\hat{\Phi}_{yy}(f_k)}. \tag{12}$$

Observe that when $n = 1$, $\lvert\hat{\gamma}_{xy}(f_k)\rvert^2 = 1$, regardless of the true value of MSC. Consequently, the estimate is biased, in general, the actual bias being a function of n and true MSC, $\lvert\gamma_{xy}(f)\rvert^2$.

Note that if $O_f = 0$, there will be no overlap, and each segment is virtually independent of the previous one (except for correlated edge effects). All theoretical results presented in the next section are valid for the case of independent segments, that is, no overlap. A detailed analysis of the effect of overlapped weighted segmentation for estimating auto spectral densities is given in [16].

IV. Statistics for the Estimate of the MSC

Goodman [2] derived an analytical expression for the probability density function of the magnitude-coherence estimate $\lvert\hat{\gamma}\rvert$ based on (12).[1] His results are valid for two Gaussian random processes that have been partitioned into n_d independent segments. Each segment was assumed large enough to ensure adequate spectral resolution. Furthermore, each segment was assumed perfectly weighted in the sense that the Fourier coefficient at the kth frequency received no power from other bins. (However, Hannan [13] points out that the statistics do not hold at the zeroeth or folding frequencies.)

The first-order probability density and distribution functions for the estimate of MSC, given the true value of MSC, $\lvert\gamma\rvert^2$, and the number of independent segments n_d are given in Table I. Equation (Ia) in Table I is obtained directly by using the relation

$$p(\lvert\hat{\gamma}\rvert^2 \mid n_d, \lvert\gamma\rvert^2) = \frac{1}{2\lvert\hat{\gamma}\rvert}\, p(\lvert\hat{\gamma}\rvert \mid n_d, \lvert\gamma\rvert^2). \tag{13}$$

[1] The f dependence and subscripts are suppressed for notational simplicity.

Alternate expressions (Ib) and (Ic) are obtained with the help of [22, eq. (15.35)]. Equations (Ib) and (Ic) are useful because ${}_2F_1(1 - n_d, 1 - n_d; 1; |\gamma|^2 |\hat{\gamma}|^2)$ is an $(n_d - 1)$st-order polynomial in $|\gamma|^2 |\hat{\gamma}|^2$.

Fisher [1], working on statistics of the estimate of the squared correlation coefficient, derived the probability density for that random variable. He integrated the result and achieved a closed-form solution for the distribution function; specifically, the solution was a finite sum of ${}_2F_1$ functions, each one a polynomial. Although these statistics are for a different problem, proper identification of the variables yields exactly the integration formula needed to find the probability distribution of the estimate of MSC, namely, (Id) in Table I.

Figs. 3 and 4 illustrate the probability density and distribution functions for several cases, as computed using (Ib) and (Id) from Table I. It is evident from Fig. 4 that the bias and variance of the MSC estimate decrease when n_d is increased.

The bias and variance of the MSC estimate can be evaluated using a general expression for the mth moment of the MSC estimate. This expression, obtained from (Ia) in Table I and [23, eq. (7.512 12)], is

$$E[(|\hat{\gamma}|^2)^m | n_d, |\gamma|^2] = \int_{-\infty}^{+\infty} p(|\hat{\gamma}|^2 | n_d, |\gamma|^2)(|\hat{\gamma}|^2)^m d|\hat{\gamma}|^2$$

$$= (1 - |\gamma|^2)^{n_d} \frac{\Gamma(n_d)\,\Gamma(m+1)}{\Gamma(n_d + m)} \cdot {}_3F_2(m+1, n_d, n_d; m + n_d, 1; |\gamma|^2) \tag{14}$$

where

$${}_3F_2(a, b, c; d, e; z) = \sum_{k=0}^{\infty} \frac{(a)_k (b)_k (c)_k}{(d)_k (e)_k} \frac{z^k}{k!} \tag{15}$$

$$(a)_k \triangleq \frac{\Gamma(a+k)}{\Gamma(a)}. \tag{16}$$

These results can be verified through the proper identification of the variables in the work of Anderson [3], who extended Fisher's original work on the squared correlation coefficient [1].

Bias and variance expressions obtained using (14) are summarized in Table II. Approximations (IIc) and (IId) are the result of truncating the series (IIa) and (IIb). Equations (IIe) through (IIg) then follow for large n_d; they indicate that the MSC estimate is asympototically unbiased, and that for large n_d the following is true.

1) The bias is greatest, $1/n_d$, at $|\gamma|^2 = 0$, and smallest, 0, at $|\gamma|^2 = 1$.

2) The variance is zero for $|\gamma|^2 = 1$ and greatest, $(2/3)^3/n_d$, at $|\gamma|^2 = 1/3$.

3) The mean-square error from the true value is equal to the variance, provided $|\gamma|^2 \neq 0$.

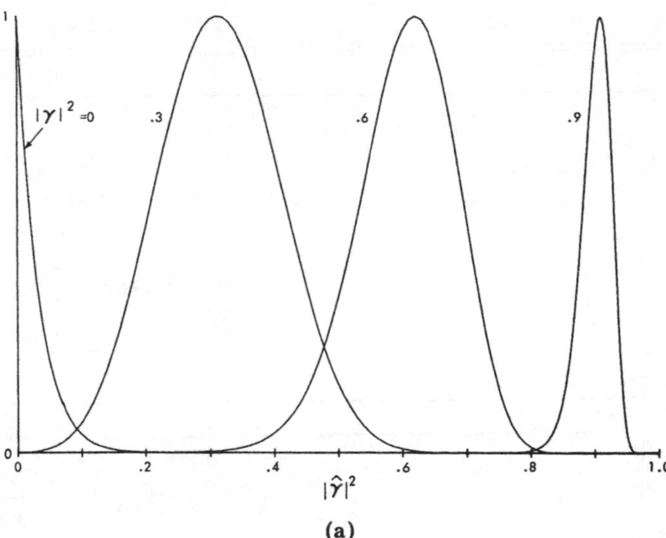

(a)

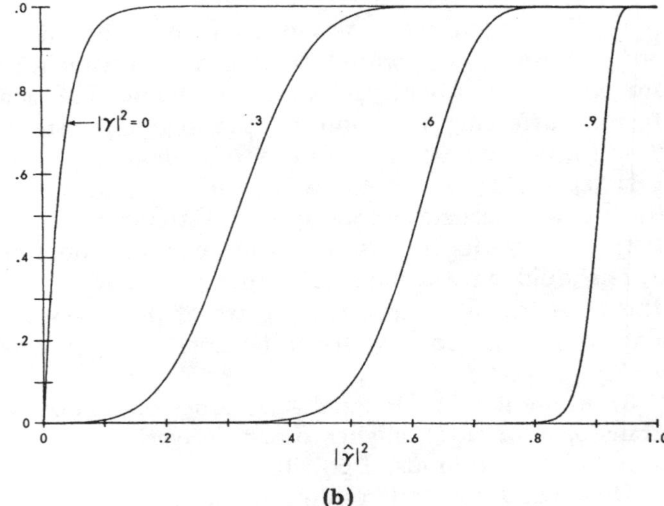

(b)

Fig. 3. (a) Probability density functions. (Functions have been normalized by maximum values, which are 31.0, 4.13, 5.23, and 17.5.) (b) Distribution functions.

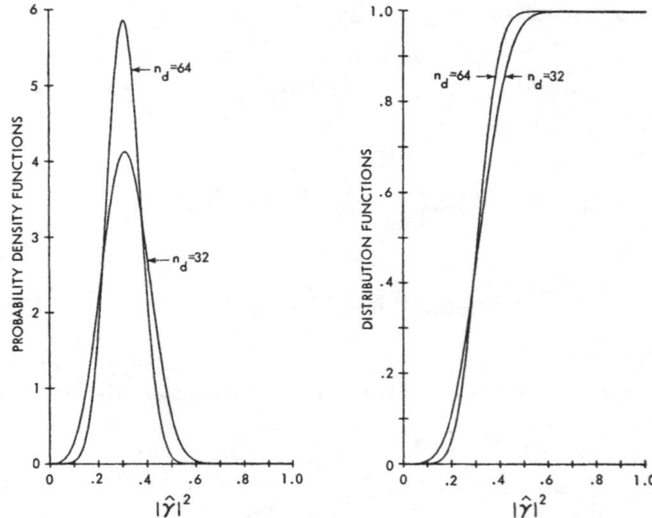

Fig. 4. Probability density and distribution functions of $|\hat{\gamma}|^2$ for $|\gamma|^2 = 0.3$.

TABLE II
Bias and Variance of the Estimation Error

Exact

Bias

$$B = E[|\hat{\gamma}|^2 \,|\, n_d, |\hat{\gamma}|^2] - |\gamma|^2 = \frac{(1-|\gamma|^2)^{n_d}}{n_d}\, {}_3F_2(2, n_d, n_d; n_d+1, 1; |\gamma|^2) - |\gamma|^2$$
$$= \frac{1}{n_d} + \frac{n_d - 1}{n_d + 1} |\gamma|^2 \, {}_2F_1(1, 1; n_d + 2; |\gamma|^2) - |\gamma|^2. \qquad \text{(IIa)}$$

Variance

$$V = E(|\hat{\gamma}|^4 \,|\, n_d, |\gamma|^2) - E^2(|\hat{\gamma}|^2 \,|\, n_d, |\gamma|^2)$$
$$= \frac{2(1-|\gamma|^2)^{n_d}}{n_d(n_d+1)}\, {}_3F_2(3, n_d, n_d; n_d+2, 1; |\gamma|^2)$$
$$- \left[\frac{(1-|\gamma|^2)^{n_d}}{n_d}\, {}_3F_2(2, n_d, n_d; n_d+1, 1; |\gamma|^2)\right]^2. \qquad \text{(IIb)}$$

Approximate

$$B_0 \cong \frac{1}{n_d} - \frac{2}{n_d+1}|\gamma|^2 + \frac{1!(n_d-1)}{(n_d+1)(n_d+2)}(|\gamma|^2)^2 + \frac{(n_d-1)\,2!}{(n_d+1)(n_d+2)(n_d+3)}(|\gamma|^2)^3; \quad B \cong \begin{cases} B_0, & B_0 \geq 0 \\ 0, & B_0 < 0 \end{cases}. \qquad \text{(IIc)}$$

$$V_0 \cong \frac{(n_d-1)}{n_d(n_d+1)}\left[\frac{1}{n_d} + 2\frac{n_d-2}{n_d+2}|\gamma|^2 - 2\frac{2n_d^3 - n_d^2 - 2n_d + 3}{(n_d+1)(n_d+2)(n_d+3)}(|\gamma|^2)^2\right.$$
$$+ 2\frac{n_d^4 - 6n_d^3 - n_d^2 + 10n_d - 8}{(n_d+1)(n_d+2)(n_d+3)(n_d+4)}(|\gamma|^2)^3$$
$$\left. + \frac{13n_d^5 - 15n_d^4 - 113n_d^3 + 27n_d^2 + 136n_d - 120}{(n_d+1)(n_d+2)^2(n_d+3)(n_d+4)(n_d+5)}(|\gamma|^2)^4\right]; \quad V \cong \begin{cases} V_0, & V_0 \geq 0 \\ 0, & V_0 < 0 \end{cases}. \qquad \text{(IId)}$$

Approximation for Large n_d

$$B \cong \frac{1}{n_d}[1 - |\gamma|^2]^2 \qquad \text{(IIe)}$$

$$\leq \frac{1}{n_d}[1 - |\gamma|^2] \qquad \text{(IIf)}$$

$$V \cong \begin{cases} \dfrac{1}{n_d^2}, & |\gamma|^2 = 0 \\[2mm] \dfrac{2|\gamma|^2}{n_d}(1 - |\gamma|^2)^2, & 0 < |\gamma|^2 \leq 1 \end{cases}. \qquad \text{(IIg)}$$

Equation (IIe) is a more accurate result than the empirically determined bias, $(1 - |\gamma|^2)/n_d$, given in [12]. However, Benignus' result, (IIf), is an upper bound on our approximation (IIe).

Figs. 5 and 6, respectively, show the bias and variance as functions of n_d and $|\gamma|^2$. For values of n_d in the range from 32 to 64, expressions (IIe) through (IIg) are good approximations; however, the curves in Figs. 5 and 6 were obtained using the exact formulas, (IIa) and (IIb). Peaks in the variance curves at $|\gamma|^2 \cong 1/3$ are evident.

V. Experimental Investigation of Overlap Effects

An experimental study has been made of the effect of overlap of data on the MSC estimate, $|\hat{\gamma}|^2$. The analytical results presented earlier relate only to the case of independent segments, that is, the case of zero overlap. Intuitively, the application of nonoverlapped weighting functions does not make the most efficient use of the data when forming the estimate, $|\hat{\gamma}|^2$. The experiment described herein examines this inefficiency, in terms of bias and variance of $|\hat{\gamma}|^2$, as a function of different amounts of overlap.

The method of evaluating overlap is straightforward in concept. Data are generated with an accurately prespecified value of MSC that is independent of frequency. Then the sample mean and variance of $|\hat{\gamma}|^2$ can be measured for the given overlap by averaging over frequency. In order to generate two processes with an MSC independent of frequency, let

$$x(t) = n_1(t) + Gn_2(t)$$
$$y(t) = n_2(t) + Gn_1(t) \qquad (17)$$

where

$$\Phi_{n_1 n_2}(f) = 0, \; \Phi_{n_1 n_1}(f) = \Phi_{n_2 n_2}(f) = \Phi_{nn}(f).$$

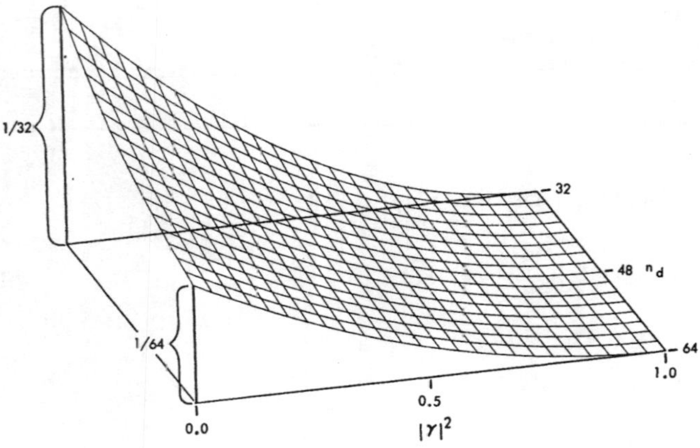

Fig. 5. Bias of $|\hat{\gamma}|^2$ versus $|\gamma|^2$ and n_d.

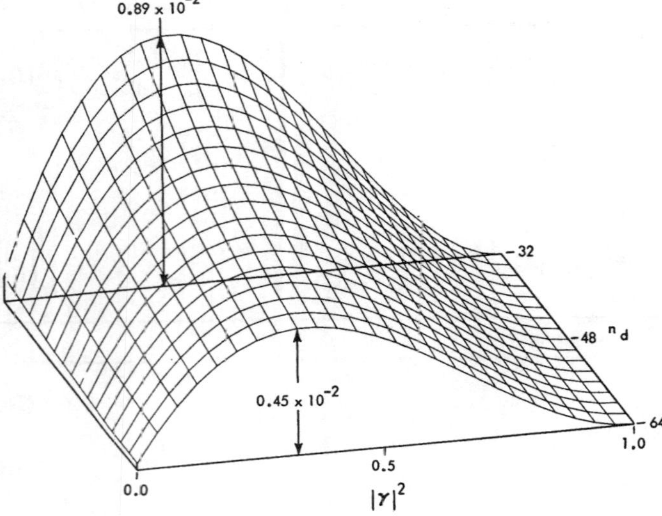

Fig. 6. Variance of $|\hat{\gamma}|^2$ versus $|\gamma|^2$ and n_d.

It follows from (2) and (3) that (using more explicit notation again)

$$|\gamma_{xy}(f)|^2 = \frac{4G^2\ \Phi_{nn}^2(f)}{(1+G^2)^2\ \Phi_{nn}^2(f)} = \frac{4G^2}{(1+G^2)^2} \tag{18}$$

an expression that is independent of frequency. The gain G can be selected according to (18) to achieve a desired value of $|\gamma_{xy}(f)|^2$. The processes generated according to (17) can be shown to be maximally insensitive to minor differences in the power spectral density functions of the original uncorrelated waveforms, $n_1(t)$ and $n_2(t)$.

In order to generate variable-coherence time series, one Gaussian noise source was used to generate a time-limited (32-s) sample function of $n_1(t)$ and, later, of $n_2(t)$. (This method eliminates the need for two identical filters.) The waveforms were first band limited to 1200 Hz using a low-pass filter and then digitized. Next, the digital data were stored on magnetic tape in a format compatible with overlapped processing. Digital versions of $x(t)$ and $y(t)$ were

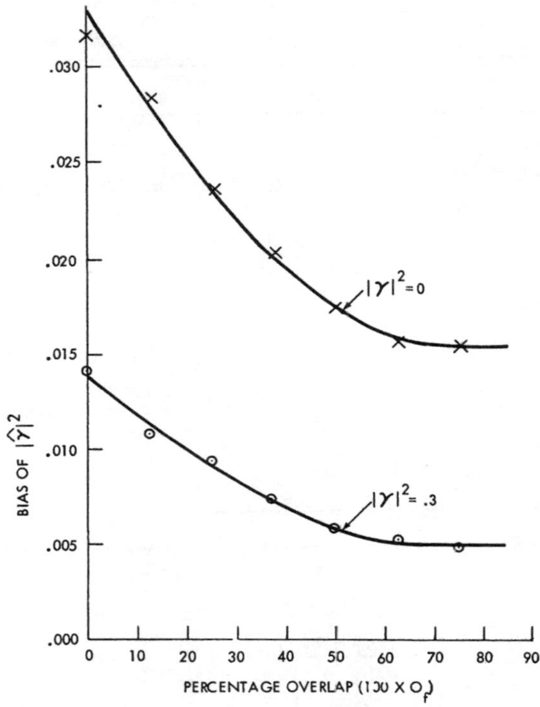

Fig. 7. Bias of $|\hat{\gamma}|^2$ when $n_d = 32$.

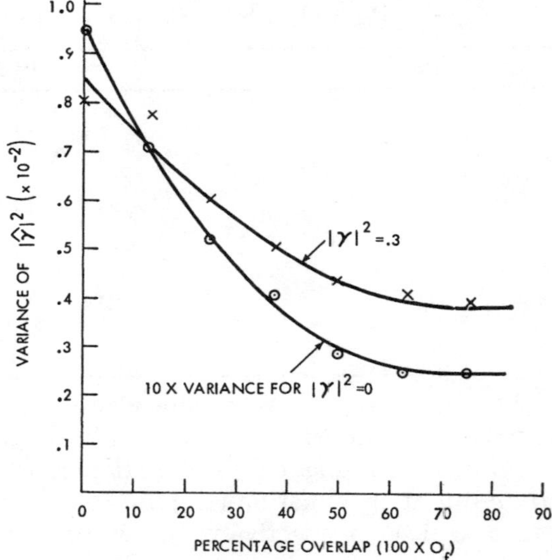

Fig. 8. Variance of $|\hat{\gamma}|^2$ when $n_d = 32$.

generated from digital versions of $n_1(t)$ and $n_2(t)$ for two values of $|\gamma_{xy}(f)|^2$.

Pertinent information regarding the data includes

1) bandwidth of $\Phi_{nn}(f) = 1200$ Hz
2) record length of $x(t)$ and $y(t) = 32$ s
3) sampling rate $f_s = 4096$ Hz
4) FFT interval $T = 1$ s (4096 samples).

Estimates of the bias and variance were performed according to

$$\widehat{\text{bias}} = \left[\frac{1}{1000}\sum_{k=1}^{1000} |\hat{\gamma}_{xy}(f_k)|^2\right] - |\gamma_{xy}(f)|^2 \tag{19}$$

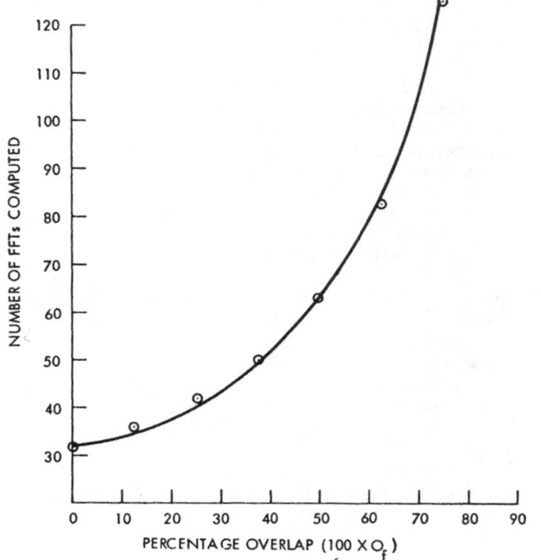

Fig. 9. Number of FFT's required for overlapped processing when n_d = 32.

and

$$\hat{\text{var}} = \frac{1}{999} \sum_{k=1}^{1000} [|\hat{\gamma}_{xy}(f_k)|^2 - \hat{\text{bias}}]^2 . \qquad (20)$$

Results of the experiment are summarized in Figs. 7 and 8. It is apparent from these results that the bias and variance of $|\hat{\gamma}_{xy}(f_k)|^2$ can be reduced through overlapped processing. For example, when $|\gamma_{xy}(f_k)|^2 =$ 0.0, the variance of the estimator with a 50-percent overlap equals 31 percent of the variance of the estimator with no overlap. With a 50-percent overlap, the bias is 55 percent as large as with no overlap. Similarly, when $|\gamma_{xy}(f)|^2 = 0.3$ and the overlap is 50 percent, the variance is 55 percent of the nonoverlapped estimator, and the bias is 50 percent as large. Observe that the bias and variance for zero overlap agree very well with (IIe) through (IIg) in Table II. With a 62.5-percent overlap, or greater, the bias and variance achieve values corresponding to an effective n_d of about 64 with nonoverlapped processing.

Clearly, as the overlap increases, the computational cost must also increase. Fig. 9 [a plot of (9) with n_d = 32] illustrates the number of FFT's per sample function required as a function of overlap. Increasing the overlap from 50 to 62.5 percent requires 32 percent more FFT's, but the variance of the MSC estimator decreases only from 80 to 95 percent of its value at a 50-percent overlap. It is doubtful, therefore, that the improvement derived from using a 62.5-percent overlap, as opposed to a 50-percent overlap, will warrant the increased computational costs, except in unusual circumstances. Overlap percentages of 25 to 50 percent are quite reasonable, however.

VI. Conclusions

A detailed analytical analysis of the statistics for estimating the MSC function has been made. When such estimates are made, time-limited sample functions of long duration, which are stationary (in the wide sense) over the period of observation, must be available. Expressions for the probability density, the cumulative distribution, and the bias and variance of the MSC estimate have been presented for Gaussian processes and for the case where nonoverlapped processing is used. Evaluation of these expressions, which are dependent on both the true value of the MSC and the number of disjoint segments n_d, dramatically portrays the requirement that n_d be large.

The application of a cosine-weighting function, in order to reduce errors from sidelobe leakage, makes inefficient use of the available data. As shown empirically, much better use of these data can be achieved through overlapped processing. An overlap of about 62.5 percent provides the limit of improvement achievable. This limit is roughly equivalent to having twice as much data as with nonoverlapped processing. The reduced bias and variance of the estimator achieved through 62.5-percent overlapped processing can be realized almost entirely through a 50-percent overlap. The computational cost associated with 50-percent overlap is not unreasonable.

Acknowledgment

The authors wish to acknowledge the assistance of Dr. H. M. Lucal and Dr. R. J. Kochenburger of the University of Connecticut, for reviewing portions of the manuscript; Dr. P. G. Cable and Dr. F. J. Kingsbury of the Naval Underwater Systems Center, for their support; R. M. Kennedy, R. R. Kneipfer, C. R. Arnold, and J. F. Ferrie, of the Naval Underwater Systems Center, for their recommendations; and J. C. Sikorski of the Naval Underwater Systems Center, for digitizing the data.

References

[1] R. A. Fisher, *Contributions to Mathematical Statistics.* New York: Wiley, 1950. (Chapter 14 was originally published as "The general sampling distribution of the multiple correlation coefficient," in *Proc. Roy. Soc.*, ser. A, vol. 121, pp. 654-673, 1928.)

[2] N. R. Goodman, "On the joint estimation of the spectra, cospectrum, and quadrature spectrum of a two-dimensional stationary Gaussian process," New York Univ., New York, Sci. Paper 10 (AD 134 919), Mar. 1957.

[3] T. W. Anderson, *An Introduction to Multivariate Statistical Analysis.* New York: Wiley, 1958.

[4] D. E. Amos and L. H. Koopmans, "Tables of the distribution of the coefficient of coherence for stationary bivariate Gaussian processes," Sandia Corp., Mono. SCR-483, 1963.

[5] R. A. Haubrich, "Earth noise, 5 to 500 millicycles per second. I. Spectral stationarity, normality, and nonlinearity," *J. Geophys. Res.*, vol. 70, pp. 1415-1427, Mar. 1965.

[6] L. D. Enochson and N. R. Goodman, "Gaussian approximations to the distribution of sample coherence," Air Force Flight Dynamics Lab., Res. Tech. Div., AF Systems Command, Wright-Patterson AFB, Ohio, Bull. AFFDL-TR-65-67, June 1965.

[7] L. J. Tick, "Estimation of Coherency," in *Spectral Analysis of Time Series*, B. Harris, Ed. New York: Wiley, pp. 133-152, 1967.

[8] C. Bingham, M. D. Godfrey, and J. W. Tukey, "Modern techniques of power spectrum estimation," *IEEE Trans. Audio Electroacoust.*, vol. AU-15, pp. 56-66, June 1967.

[9] P. D. Welch, "The use of fast Fourier transform for the estimation of power spectra: A method based on time averaging over short, modified periodograms," *IEEE Trans. Audio Electroacoust.*, vol. AU-15, pp. 70-73, June 1967.

[10] G. M. Jenkins and D. G. Watts, *Spectral Analysis and Its Applications.* San Francisco, Calif.: Holden-Day, 1968.

[11] C. H. Knapp, "An algorithm for estimation of the inverse spectral matrix," General Dynamics Electric Boat Res. Proj., SUBIC Rep. 1970-U417-70-010, Apr. 1969.

[12] V. A. Benignus, "Estimation of the coherence spectrum and its confidence interval using the fast Fourier transform," *IEEE Trans. Audio Electroacoust.*, vol. AU-17, pp. 145-150, June 1969.

[13] E. S. Hannan, *Multiple Time Series.* New York: Wiley, 1970.

[14] P. R. Roth, "Effective measurements using digital signal analysis," *IEEE Spectrum*, vol. 8, pp. 62-70, Apr. 1971.

[15] G. C. Carter and A. H. Nuttall, "Statistics of the estimate of coherence," *Proc. IEEE* (Lett.), vol. 60, pp. 465-466, Apr. 1972.

[16] A. H. Nuttall, "Spectral estimation by means of overlapped FFT processing of windowed data," Naval Underwater Systems Cent., New London, Conn., Rep. 4169, Oct. 1971.

[17] J. S. Bendat and A. G. Piersol, *Random Data: Analysis and Measurement Procedures.* New York: Wiley, 1971.

[18] G. C. Carter, "Estimation of the magnitude-squared coherence function (spectrum)," Naval Underwater Systems Cent., New London, Conn., Tech. Rep. TR 4343 (AD 743 945), May 19, 1972.

[19] J. W. Cooley and J. W. Tukey, "An algorithm for the machine calculation of complex Fourier series," *Math. Comput.*, vol. 19, pp. 297-301, Apr. 1965.

[20] R. C. Singleton, "An algorithm for computing the mixed radix fast Fourier transform," *IEEE Trans. Audio Electroacoust.*, vol. AU-17, pp. 93-102, June 1969.

[21] B. Gold and C. M. Rader, *Digital Processing of Signals.* New York: McGraw-Hill, 1969.

[22] M. Arbramowitz and I. A. Stegun, Ed., *Handbook of Mathematical Functions With Formulas, Graphs, and Mathematical Tables.* Washington, D.C.: U.S. Gov. Printing Office, 1964.

[23] I. S. Gradshteyn and I. M. Ryshik, *Table of Integrals, Series, and Products.* New York: Academic Press, 1965.

Part 3
Selected Papers on Time Delay Estimation

Robust Eckart Filters for Time Delay Estimation

EMAD K. AL-HUSSAINI, MEMBER, IEEE, AND SALEEM A. KASSAM, SENIOR MEMBER, IEEE

***Abstract*—Minimax or robust filters which optimize worst-case performance have been considered previously for matched and Wiener filtering. In this paper we show how similar ideas can be applied to obtain robust versions of one particular filter, the Eckart filter, for time delay estimation. The use of a robust filter can eliminate the requirement of estimation of the signal and noise power spectral densities (PSD's) and cross PSD's, robust filters being less sensitive in their performance to variations of the input PSD's within specified classes of PSD's. Explicit solutions for specific classes of PSD's are derived, and several numerical examples and simulation results are given. Similar formulations for robust filters should also be possible for other types of filters used in time delay estimation.**

I. Introduction

The problem of estimating the time delay between signals received at two spatially separated sensors in the presence of noise has been of interest in various applications, such as in seismology, acoustics, radar, and sonar problems [1]. The mathematical model is often given by

$$r_1(t) = s(t) + n_1(t), \qquad 0 \leq t \leq T \qquad (1)$$
$$r_2(t) = s(t - D) + n_2(t),$$

where $r_1(t)$ and $r_2(t)$ are the outputs of two spatially separated sensors, $s(t)$ is the source signal, and $n_1(t)$ and $n_2(t)$ represent additive noises. The signal and additive noises are assumed to be uncorrelated, zero-mean, stationary, Gaussian random processes. Basically, the solution has consisted of cross-correlating the sensor outputs and using the time argument that corresponds to the maximum peak in the output as the time delay estimate. To improve the detection and estimation processes, various filters, weighting functions, or windows have been suggested for use after the cross correlation [2]-[7].

In order to implement such filters, the PSD's of the input processes and cross PSD's should be known. In practice, estimates can be obtained from the finite-length observations of $r_1(t)$ and $r_2(t)$. This approach, besides increasing the complexity of the system, may also lead to performance degradation due to the errors inherent in spectral estimates obtained from finite-length data. It would therefore be useful if fixed filters were specifically designed to have good performance over entire classes of spectral densities. One such design approach has been successfully applied for Wiener and matched filtering under spectral uncertainty. In this approach, minimax robust filters are used which optimize worst-case performance over reasonable classes of signal and noise PSD's [8]-[11]. These robust filters are distinguished by their guaranteed minimum level of performance for the assumed classes of signal and noise PSD's, and by their fixed characteristics so that spectral estimation is not required.

Manuscript received August 24, 1982; revised August 29, 1983. This work was supported by the Office of Naval Research under Contract N00014-80-K-0945.

E. K. Al-Hussaini is with the Department of Electronics and Communications, Cairo University, Giza, Egypt.

S. A. Kassam is with the Moore School of Electrical Engineering, University of Pennsylvania, Philadelphia, PA 19104.

In this paper we obtain useful robust versions of the simple Eckart filter for time delay estimation. The Eckart filter is defined for given signal and noise PSD shapes to have a specific low-SNR optimality property. In obtaining the robust Eckart filter, we will draw upon several previously established results on robust filtering for detection and estimation, synthesizing and extending them for this application in time delay estimation. Section II contains the theoretical developments. Here we introduce the class of piecewise constant Eckart filters, and then show that they are minimax robust for a specific type of signal and noise PSD classes. After this we establish some results for the robust Eckart filter for more general classes of PSD's. In Section III we consider numerical examples and present simulation results for the performances of some of the robust filters of Section II.

We will see that in many cases the robust Eckart filter does much better than the filter which is optimum for an assumed pair of signal and noise PSD's, for deviations of the actual spectra from the nominal assumptions. When the nominal assumptions hold, a robust Eckart filter need not suffer a significant performance degradation. The good results which have been obtained in making the Eckart filter robust suggest that other types of filters used in time delay estimation can also be expected to have useful robust versions.

II. Robust Eckart Filters

Many of the methods devised to estimate the time delay are related to the generalized cross-correlation (GCC) method [4]. In this approach the cross-power spectral estimate $G_{r_1 r_2}(\omega)$ is multiplied by a weighting function or filter $W(\omega)$, and then the inverse Fourier transform is taken to obtain the GCC function

$$R^G_{r_1 r_2}(\tau) = \frac{1}{2\pi} \int_{-\infty}^{\infty} W(\omega)\, G_{r_1 r_2}(\omega) e^{j\omega\tau}\, d\omega. \qquad (2)$$

The argument τ that maximizes $R^G_{r_1 r_2}(\tau)$ is the estimate $\hat{D}$ of the true time delay D. The Eckart filter [4], [12], [13] is designed to maximize the output signal-to-noise ratio (SNR) at $\tau = D$ in the limit of low input SNR. Denoting this output SNR by d, for long averaging time we obtain [13]

Reprinted from *IEEE Trans. Acoust., Speech, Signal Processing,* vol. 32, no. 5, pp. 1052–1063, October 1984.

$$d(W; S, N_1 N_2) = \frac{\left[\int_{-\infty}^{\infty} W(\omega) S(\omega)\, d\omega\right]^2}{\int_{-\infty}^{\infty} W^2(\omega) N_1(\omega) N_2(\omega)\, d\omega}. \quad (3)$$

where $S(\omega)$, $N_1(\omega)$, and $N_2(\omega)$ are the PSD's of $s(t)$, $n_1(t)$, and $n_2(t)$, respectively. The Eckart filter $W_{EO}(\omega)$ maximizing the distance or SNR $d(W; S, N_1 N_2)$ is given by

$$W_{EO}(\omega) = \frac{S(\omega)}{Q(\omega)}, \quad \omega \in \Omega \quad (4)$$

where $Q(\omega) = N_1(\omega) N_2(\omega)$ and Ω is the region of interest in the frequency domain. The optimum distance is now given by

$$d_0(S, Q) = \int_{\Omega} \frac{S^2(\omega)}{Q(\omega)}\, d\omega. \quad (5)$$

From (4) it is clear that in order to implement the Eckart filter, the signal and noise PSD's must be known exactly or be estimated accurately. Small deviations from these assumed or estimated PSD's may result in large degradation in performance. Here we will develop filters which are less sensitive to variations in the input spectra and which do not require spectral estimation. In Section II-A we will consider Eckart filters with optimum *piecewise-constant* frequency responses. In Section II-B we will show that these filters are robust in a minimax sense for a specific class of possible spectral densities, so that they perform well despite deviations from nominal assumed spectra in the class. In Section II-C robust filters for more general spectral classes will be considered.

A. Optimum Piecewise-Constant Eckart Filters

Let us now assume a piecewise-constant filter of the form

$$W_{PC}(\omega) = c_j \quad \omega \in R_j,\ j = 1, 2, \cdots, m \quad (6)$$

where the R_j form a partition of Ω. Due to symmetry of the spectra, the regions R_j include both the negative and positive complementary parts of the frequency domain Ω. With $W_{PC}(\omega)$ in place of $W(\omega)$ in (3), we can maximize the distance $d(W_{PC}; S, Q)$ with respect to the c_j for a fixed partition $\{R_1, R_2, \cdots, R_m\}$. The result for the optimum levels c_j^* is

$$c_j^* = \frac{\int_{R_j} S(\omega)\, d\omega}{\int_{R_j} Q(\omega)\, d\omega}, \quad j = 1, 2, \cdots, m. \quad (7)$$

The c_j^* can be multiplied by any constant $K > 0$ without changing their optimality for given regions.

The next step is to consider further optimization with respect to the choice of the regions $\{R_j\}_{j=1}^m$. The following theorem and corollary give the structure of the optimum piecewise-constant Eckart filter partitioning. The proof is similar to that given in [14] for a data-quantization problem in signal detection, and is therefore omitted.

Theorem 1: The set of regions $\{R_j^*\}_{j=1}^m$ maximizing the distance $d(W_{PC}; S, Q)$ with optimum levels for $W_{PC}(\omega)$ satisfies

$$R_j^* \subset \left\{\omega \in \Omega \,|\, h_{j-1} < \frac{S(\omega)}{Q(\omega)} < h_j\right\}, \quad j = 1, 2, \cdots, m \quad (8)$$

for some set of constants $\{h_j\}_{j=0}^m$ such that $h_0 = 0$ and $h_m = \infty$.

Corollary 1: The set of constants $\{h_j\}_{j=1}^{m-1}$ maximizing the distance $d(W_{PC}; S, Q)$ for optimum levels for an m-level Eckart filter satisfy

$$h_j = \frac{c_j^* + c_{j+1}^*}{2}, \quad j = 1, 2, \cdots, m-1 \quad (9)$$

where c_j^* is given by (7).

The optimum m-level Eckart filter can thus be obtained by solving (7) and (9) simultaneously, using condition (8). When $W_{EO}(\omega)$ is monotonically increasing or decreasing, the R_j^* are complementary intervals, and numerical solutions can be obtained easily using iterative methods.

The optimum m-level piecewise-constant Eckart filter can be expected to be more robust than the optimum Eckart filter $W_{EO}(\omega)$, especially for a given set of regions $\{R_j\}_{j=1}^m$. This is because its levels are insensitive to variations in the fine details of $S(\omega)$ and $Q(\omega)$. A similar observation for piecewise-constant Wiener filters was made in [15].

B. Robust Eckart Filter for p-Point Spectral Classes

Let us assume that the signal PSD, $S(\omega)$, and the product $Q(\omega)$ of noise PSD's are known only to be members of some general classes Φ and Π, respectively. That is, we know that $S \in \Phi$ and $Q \in \Pi$, but we do not know which members of Φ and Π are the true signal and the product of noise spectra. In practical situations the class Φ may be determined by measurements at the output of the correlator when a signal is present and Π may be determined from the powers in different frequency bands at the correlator output when noise alone is present.

Formally, a robust Eckart filter $W_{ER}(\omega)$ is defined to be a minimax filter such that

$$\max_{W \in \Lambda} \min_{(S, Q) \in \Phi \times \Pi} d(W; S, Q) = \min_{(S, Q) \in \Phi \times \Pi} d(W_{ER}; S, Q) \quad (10)$$

where Λ is some (large) class of allowable weighting functions. A stronger definition of a robust Eckart filter $W_{ER}(\omega)$ is that it is optimum for some pair $(S_R, Q_R) \in \Phi \times \Pi$, such that $[W_{ER}, (S_R, Q_R)]$ forms a *saddlepoint* for $d(W; S, Q)$:

$$\min_{(S, Q) \in \Phi \times \Pi} d(W_{ER}; S, Q) = d(W_{ER}; S_R, Q_R) = \max_{W \in \Lambda} d(W; S_R, Q_R). \quad (11)$$

This says that for given classes Φ and Π, the robust Eckart filter $W_{ER}(\omega)$ has performance lower bounded by its performance for the pair $[S_R(\omega), Q_R(\omega)]$. This is the best lower bound possible because $W_{ER}(\omega)$ is the optimum filter for $[S_R(\omega), Q_R(\omega)]$.

With $d_0(S, Q)$ the optimum distance (5) for $(S, Q) \in \Phi \times \Pi$, obtained using the optimum Eckart filter, (11) implies that

$$d_0(S, Q) \geqslant d_0(S_R, Q_R), \quad \text{all} \quad (S, Q) \in \Phi \times \Pi. \tag{12}$$

In general, any pair $(S_R, Q_R) \in \Phi \times \Pi$ satisfying (12) is called the *least favorable* pair. It is to be noted that (11) implies that (12) is true, but the converse is not always valid.

The p-point classes of PSD's which we will consider are defined explicitly in terms of a *given* vector $(a_1, a_2, \cdots, a_{m-1})$, *given* fractions u_j and v_j, $j = 1, 2, \cdots, m$, and *given* total powers σ_s^2, σ_q^2, by

$$\Phi \triangleq \left\{ S \mid \frac{1}{\pi} \int_{a_{j-1}}^{a_j} S(\omega)\, d\omega = u_j \sigma_s^2, \quad j = 1, 2, \cdots, m \right\} \tag{13}$$

and

$$\Pi \triangleq \left\{ Q \mid \frac{1}{\pi} \int_{a_{j-1}}^{a_j} Q(\omega)\, d\omega = v_j \sigma_q^2, \quad j = 1, 2, \cdots, m \right\} \tag{14}$$

where $a_0 \triangleq 0$, $a_m \triangleq \infty$ and $\Sigma_{j=1}^m u_j = \Sigma_{j=1}^m v_j = 1$. The u_j's represent the fractions of the correlator output in the band $a_{j-1} < |\omega| \leqslant a_j$ when a signal is present. The v_j's are the fractions of the correlator output power in the same band when noise alone is present. The saddlepoint solution for these simple classes can be derived easily.

Specifically, it can be easily shown that the piecewise-constant Eckart filter defined by

$$\begin{aligned} W_{ER}(\omega) &= \frac{u_j \sigma_s^2}{v_j \sigma_q^2} \\ &\triangleq K_j, \quad a_{j-1} < |\omega| \leqslant a_j \end{aligned} \tag{15}$$

is the robust Eckart filter satisfying the saddlepoint condition (11) for the p-point classes Φ and Π defined above, for any pair of least favorable spectral functions S_R and Q_R constrained by

$$\frac{S_R(\omega)}{Q_R(\omega)} = K_j, \quad a_{j-1} < |\omega| \leqslant a_j. \tag{16}$$

The least-favorable pair is not unique. The left-hand side equality in (11) happens to be satisfied for *all* $(S, Q) \in \Phi \times \Pi$, and (12) is true with equality. Thus, $W_{ER}(\omega)$ gives the *same* value for $d(W_{ER}; S, Q)$, for $S(\omega)$, and for $Q(\omega)$ in the p-point classes.

The result of Section II-A allows us to find the optimum m-level piecewise-constant Eckart filter for given $S(\omega)$ and $Q(\omega)$. We have just seen that the resulting filter is minimax robust for p-point classes of signal and noise spectra, generated from the given $S(\omega)$ and $Q(\omega)$ so that the fractional powers in the intervals defining the optimum piecewise-constant filter are the same. Although we defined the p-point classes above in terms of *intervals* on the frequency axis, clearly we could just as well have used a more general partitioning into m subsets.

It is also interesting to note that although we assumed that σ_s^2 and σ_q^2 are known, from (15) we find that the robust filter does not depend in an essential way on σ_s^2 and σ_q^2. This is because the output SNR or distance d is invariant to amplitude scaling of the filter. Thus, we could have required only that the classes Φ and Π contain densities with *identical* total powers for members of the same class.

C. Robust Eckart Filters for General Spectral Classes

We now consider robust Eckart filters for general convex classes of signal and noise spectra which include as a special case the p-point classes. A result is established (Lemma 1) which shows that the saddlepoint robust filter can be obtained as the optimum filter for some least favorable pair $(S_R, Q_R) \in \Phi \times \Pi$. That is, the problem is finally resolved to finding least favorable pairs $(S_R, Q_R) \in \Phi \times \Pi$. The proof of Lemma 1 is traced in the Appendix.

Lemma 1: Suppose Φ and Π are convex sets, $(S_R, Q_R) \in \Phi \times \Pi$ and $W_{ER}(\omega) = S_R(\omega)/Q_R(\omega)$. Then the following statements are *equivalent.*

1) $(S_R, Q_R) \in \Phi \times \Pi$ is *least favorable* for Eckart filtering for Φ and Π.

2) $[W_{ER}, (S_R, Q_R)]$ is a *saddlepoint* for Eckart filtering for Λ and $\Phi \times \Pi$.

3)

$$2\int_{-\infty}^{\infty} W_{ER}(\omega) S(\omega)\, d\omega \geqslant \int_{-\infty}^{\infty} W_{ER}(\omega) S_R(\omega)\, d\omega + \int_{-\infty}^{\infty} W_{ER}^2(\omega) Q(\omega)\, d\omega, \quad \text{any } (S, Q) \in \Phi \times \Pi. \tag{17}$$

This lemma is similar to the two conjugate lemmas given in [16] for the matched filtering case. This result is significant because a least-favorable pair minimizing $d_0(S, Q)$ is generally easier to obtain than the robust filter directly.

In what follows we will consider some particular spectral classes using the analogy between robust *hypothesis testing* and robust filtering problems. In robust binary hypothesis testing it is required to decide which of two hypotheses is true, based on a set of observations known to be governed by one of the two hypotheses. There, the performance is measured by the *risk* function, and the problem is to find the test statistic which minimizes the maximum risk, when the probability density functions of the observations under each hypothesis are members of some general classes of probability densities. In most cases the problem can be reduced to that of finding least-favorable pairs of probability density functions. Now consider for the robust Eckart filtering problem convex classes Φ and Π which are constrained to contain densities with fixed total powers. Thus, in addition to any other conditions describing the classes, we will assume that

$$\int_{\Omega} S(\omega)\, d\omega = 2\pi\sigma_s^2, \quad \text{all} \quad S \in \Phi \tag{18}$$

$$\int_{\Omega} Q(\omega)\, d\omega = 2\pi\sigma_q^2, \quad \text{all } Q \in \Pi \tag{19}$$

where σ_s^2 and σ_q^2 are fixed and known. This makes Φ and Π classes of nonnormalized probability density functions. Some general results on the equivalence between robust hypothesis testing and a class of robust filtering problems for power-constrained spectral density classes have been given in [10]. From these it follows that our problem of maximizing the minimum SNR for robust Eckart filtering is equivalent to that of finding the least-favorable pair of probability densities from the corresponding normalized versions of Φ and Π. This allows known results for robust hypothesis testing to be used to give solutions for the robust filtering problem. Let us now consider some specific classes of interest, where we will assume that (18) and (19) hold.

1) Bounded Spectral Classes: Let the convex classes Φ and Π be defined by

$$\Phi = \{S \mid S_L(\omega) \leqslant S(\omega) \leqslant S_U(\omega)\} \tag{20}$$

$$\Pi = \{Q \mid Q_L(\omega) \leqslant Q(\omega) \leqslant Q_U(\omega)\} \tag{21}$$

where the upper and lower bounds $S_U(\omega)$, $Q_U(\omega)$, $S_L(\omega)$, and $Q_L(\omega)$ are assumed known, with powers larger than and less than the constraints, respectively. This is a reasonable model when the signal and noise PSD's are estimated from data as in the time delay estimation problem. The solution for this case can be obtained directly from the results obtained in [17] for the corresponding robust hypothesis testing problem.

A special case of this model is the *ϵ-contamination* model which defines possible signal and product of noise PSD's by

$$S(\omega) = (1 - \epsilon_s) S_0(\omega) + \epsilon_s S_c(\omega) \tag{22}$$

$$Q(\omega) = (1 - \epsilon_q) Q_0(\omega) + \epsilon_q Q_c(\omega) \tag{23}$$

where ϵ_s and ϵ_q are assumed degrees of contamination of the *nominal* PSD's $S_0(\omega)$, $Q_0(\omega)$, and $S_c(\omega)$, $Q_c(\omega)$ are any *contaminating* PSD's with total powers $2\pi\sigma_s^2$ and $2\pi\sigma_q^2$. Thus, the total power constraints of (18) and (19) are always satisfied. The ϵ-contamination model can be viewed as a special case of the *band model* of (20) and (21) with $S_L(\omega) \triangleq (1 - \epsilon_s) S_0(\omega)$ and $Q_L(\omega) \triangleq (1 - \epsilon_q) Q_0(\omega)$ and with $S_U(\omega)$ and $Q_U(\omega)$ equal to ∞ for all ω.

Because we will use it in the next section, we give here the explicit solution for the robust Eckart filter with respect to the ϵ-contamination model. To state the solution we have to define two subsets as functions of two finite, positive parameters k_s and k_q:

$$\gamma(k_s) \triangleq \{\omega \mid S_L(\omega) < k_s Q_L(\omega)\} \tag{24}$$

$$\beta(k_q) \triangleq \{\omega \mid k_q Q_L(\omega) \leqslant S_L(\omega)\}. \tag{25}$$

The robust filter can now be obtained through the following theorem. A proof can also be given paralleling the direct proof for a Wiener filtering problem in [8], and is omitted.

Theorem 2: For any signal and product of noise PSD's $S(\omega)$ and $Q(\omega)$ which are members of the ϵ-contamination classes defined by (22) and (23), respectively, the robust filter $W_{ER}(\omega)$ satisfying the saddlepoint condition (11) is given by $W_{ER}(\omega) = S_R(\omega)/Q_R(\omega)$ where the least favorable PSD's $S_R(\omega)$ and $Q_R(\omega)$ are obtained according to the following.

1) If $k_s \leqslant k_q$ exist satisfying

$$k_s \int_{\gamma(k_s)} (1 - \epsilon_q) Q_0(\omega)\, d\omega + \int_{\bar{\gamma}(k_s)} (1 - \epsilon_s) S_0(\omega)\, d\omega = 2\pi\sigma_s^2 \tag{26}$$

$$\frac{1}{k_q} \int_{\beta(k_q)} (1 - \epsilon_s) S_0(\omega)\, d\omega + \int_{\bar{\beta}(k_q)} (1 - \epsilon_q) Q_0(\omega)\, d\omega = 2\pi\sigma_q^2 \tag{27}$$

then

$$S_R(\omega) = \begin{cases} k_s(1 - \epsilon_q) Q_0(\omega), & \omega \in \gamma(k_s) \\ (1 - \epsilon_s) S_0(\omega), & \omega \in \bar{\gamma}(k_s) \end{cases} \tag{28}$$

$$Q_R(\omega) = \begin{cases} \dfrac{1}{k_q}(1 - \epsilon_s) S_0(\omega), & \omega \in \beta(k_q) \\ (1 - \epsilon_q) Q_0(\omega), & \omega \in \bar{\beta}(k_q). \end{cases} \tag{29}$$

2) Otherwise, with

$$k = \frac{\sigma_s^2}{\sigma_q^2}, \tag{30}$$

$$S_R(\omega) = \begin{cases} k(1 - \epsilon_q) Q_0(\omega) + S_e(\omega), & \omega \in \gamma(k) \\ (1 - \epsilon_s) S_0(\omega) + S_e(\omega), & \omega \in \bar{\gamma}(k) \end{cases} \tag{31}$$

$$Q_R(\omega) = \begin{cases} \dfrac{1}{k}(1 - \epsilon_s) S_0(\omega) + Q_e(\omega), & \omega \in \beta(k) \\ (1 - \epsilon_q) Q_0(\omega) + Q_e(\omega), & \omega \in \bar{\beta}(k) \end{cases} \tag{32}$$

where $S_e(\omega)$, $Q_e(\omega)$ are arbitrary spectral density functions with

$$S_e(\omega) = k Q_e(\omega) \tag{33}$$

such that $S_R(\omega)$, $Q_R(\omega)$ satisfy the power constraints.

Looking at the form of $S_R(\omega)$ and $Q_R(\omega)$ given by (28) and (29), we see that the ratio $W_{ER}(\omega) = S_R(\omega)/Q_R(\omega)$ is $(1 - \epsilon_s) S_0(\omega)/(1 - \epsilon_q) Q_0(\omega)$ *limited* at the maximum and minimum values k_q and k_s, respectively. As ϵ_s and ϵ_q become smaller k_s approaches zero and k_q approaches ∞, and the asymptotic case is the optimum Eckart filter for $S_0(\omega)$ and $Q_0(\omega)$. The second part of the theorem completes the solution by giving the solution for large uncertainties (large ϵ_s and ϵ_q).

2) Bounded p-Point Spectral Classes: The p-point spectral classes we considered in Section II-B above were power-constrained convex classes specified by *known* fractions of total powers in subsets of a given partition. These classes can be generalized to those where the fractions u_j and v_j of total powers are not known precisely but are *bounded* by known upper and lower bounds u_{jU}, u_{jL} and v_{jU}, v_{jL}, respectively, $j = 1, 2, \cdots, m$. Again, the least-favorable pair of spectra for these classes for robust Eckart filtering can be obtained from recent results on robust hypothesis testing for such classes of probability density functions [18].

III. Numerical Examples and Simulation Results

In this section we present specific results which illustrate the applicability of our results of Section II in practical situations. The first three examples forming Section III-A below are concerned with robust piecewise-constant Eckart filters. Example 4 in Section III-B is for the ϵ-contamination classes of spectral densities.

A. Piecewise-Constant Eckart Filters

Example 1: We take nominal signal and noise spectra to be given by

$$S_0(\omega) = \begin{cases} \dfrac{A\alpha^2}{(\alpha^2 + \omega^2)^2}, & |\omega| \leq \pi \\ 0, & |\omega| > \pi \end{cases} \tag{34}$$

and

$$N_{10}(\omega) = N_{20}(\omega)$$

$$= \begin{cases} 1, & |\omega| \leq \pi \\ 0, & |\omega| > \pi. \end{cases}$$

The constant A can be chosen to get different input SNR's. The optimum Eckart filter is, from (4),

$$W_{EO}(\omega) = \frac{A\alpha^2}{(\alpha^2 + \omega^2)^2}, \quad |\omega| \leq \pi.$$

Let us now use the values $\alpha = \frac{1}{3}$ and input SNR = 0 dB ($A = 1.33$). From (5), the optimum value for the output SNR or distance is

$$d_0(S_0, Q_0) = 47.17.$$

Consider now the optimum piecewise-constant Eckart filter $W_{EPC}(\omega)$ for this case. For two-level filtering we obtain, by solving (7) and (9) with (8) defining the regions, with $\alpha = \frac{1}{3}$ and without loss of generality $A = 1$,

$$W_{EPC}(\omega) = \begin{cases} 6.65, & |\omega| \leq 0.26 \\ 0.21, & 0.26 < |\omega| \leq \pi. \end{cases}$$

The output SNR for an input SNR of 0 dB is now

$$d(W_{EPC}; S_0, Q_0) = 41.05.$$

For three-level filtering we obtain for $\alpha = \frac{1}{3}$ and $A = 1$ the result

$$W_{EPC}(\omega) = \begin{cases} 7.61, & |\omega| \leq 0.18 \\ 3.21, & 0.18 < |\omega| \leq 0.38 \\ 0.12, & 0.38 < |\omega| \leq \pi \end{cases}$$

and for 0 dB input SNR, an output SNR of

$$d(W_{EPC}; S_0, Q_0) = 44.76.$$

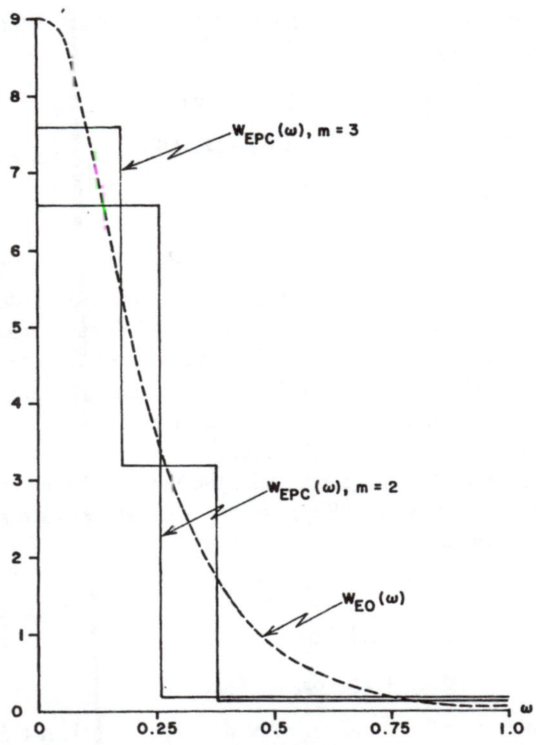

Fig. 1. Optimum Eckart filter $W_{EO}(\omega)$ and optimum piecewise-constant Eckart filters $W_{EPC}(\omega)$ for m=2 and m=3, for spectra of Example 1 [A = 1 in (34)].

As per our remark at the end of Section II-B, these piecewise-constant filters are the robust Eckart filters $W_{ER}(\omega)$ for the p-point classes generated by the $S_0(\omega)$ and $Q_0(\omega) = N_{10}(\omega)\, N_{20}(\omega)$ and the optimum partitions $\{R_j^*\}_{j=1}^m$ of the piecewise-constant filters. Fig. 1 shows the characteristic $W_{EO}(\omega)$ as well as $W_{EPC}(\omega)$ for $m = 2$ and $m = 3$ for this example.

Example 2: Let us now take the nominal signal to have the spectrum $S_0(\omega)$ defined by (34), as in the previous example, and assume the *same* form for the noise nominal spectra,

$$N_{10}(\omega) = N_{20}(\omega)$$

$$= \begin{cases} \dfrac{\alpha^2}{(\alpha^2 + \omega^2)^2}, & |\omega| \leq \pi \\ 0, & |\omega| > \pi. \end{cases}$$

Then the parameter A in (34) may be adjusted to get different input SNR's. When the signal and noise spectra have the same shape, as here, filters defined to be optimum under commonly used performance criteria for time-delay estimation all turn out to be identical to the Eckart filter. This is now given by

$$W_{EO}(\omega) = \frac{(\alpha^2 + \omega^2)^2}{A\alpha^2}, \quad |\omega| \leq \pi.$$

For an input SNR of 0 dB we take $A = 1$ in (34). Clearly, the performance will be invariant to any amplitude scaling of $W_{EO}(\omega)$ in any case. The optimum filter gives a distance or output SNR of

$$d_0(S_0, Q_0) = 2\pi$$

for all values of α as long as the input SNR is 0 dB.

An optimum two-level piecewise-constant Eckart filter is, for $\alpha = \frac{1}{3}$ and $A = 1$,

$$W_{EPC}(\omega) = \begin{cases} 0.18, & |\omega| \leq 1.61 \\ 132.67, & 1.61 < |\omega| \leq \pi \end{cases} \tag{35}$$

giving, for input SNR of 0 dB,

$$d(W_{EPC}; S_0, Q_0) = 2.76.$$

For optimum three-level filtering the distance can be increased to 3.88. We note from (35) that $W_{EPC}(\omega)$ has a maximum level of 132.67, whereas $W_{EO}(\omega)$ for this problem reaches a value of $[(\alpha^2 + \pi^2)/\alpha]^2 \approx 900$ at $\omega = \pi$. The flattening out of this optimum characteristic is what gives the piecewise-constant filter its robustness property.

Example 3: Once again we take the signal to have the nominal spectrum of (34). Suppose the noise spectra are now of the ideal low-pass type

$$N_1(\omega) = N_2(\omega) = \begin{cases} \eta, & |\omega| \leq \omega_c \\ 0, & |\omega| > \omega_c. \end{cases}$$

If we pick $\eta \approx 2.87$ and ω_c only slightly larger than 1.61 [the optimum breakpoint of the two-level $W_{EPC}(\omega)$ characteristic (35)], we find that $Q = N_1 N_2$ above and $Q_0 = N_{10}N_{20}$ in the previous example ($\alpha = \frac{1}{3}$) both fall in the p-point class defined by the single p-point $a_1 = 1.61$. The filter $W_{EPC}(\omega)$ of (35) is the *robust* filter $W_{ER}(\omega)$ for this class, and maintains the value $d(W_{ER}; S_0, Q) = 2.76$ for all Q in it. Note that the input SNR changes because it is Q and not N_1, N_2 for which total power is fixed; in the above case the input SNR is -3 dB.

If the filter $W_{EO}(\omega)$, which is optimum for the spectra of Example 2, were used here, the distance or output SNR would fall to the very low value of 0.01. We see, therefore, that the two-level robust (piecewise-constant) filter is able to maintain performance at a reasonable level, whereas the filter optimum for the nominal spectra of Example 2 undergoes a major performance degradation.

Variance of the Time-Delay Estimates: An expression for the theoretical variance of a time-delay estimate from any generalized cross-correlation scheme has been given previously [4, eq. (51)]. This is a function of the filter characteristic as well as the signal and noise spectra. Thus, it is possible to further compare the predicted performances of the optimum Eckart and piecewise-constant and robust Eckart filters on the basis of the variance. We will give some numerical values for the variances also in describing the simulation results next.

Simulation Results for Examples 1-3: To simulate the behavior of the filters, ten independent trials were run for each case that we considered. For each trial, three independent sequences of uncorrelated Gaussian variates were generated. Different spectral characteristics were generated by linear filtering of these uncorrelated sequences. Two separate signal-plus-noise sequences were formed by adding relatively delayed versions of one sequence to each of the remaining two, using a delay of 20 units. The total length of each sequence was 512 units. The optimum and piecewise-constant filters were implemented using a 512-point FFT, and the dominant peak was located at the processor output using parabolic fitting [19]. For each different case the same set of ten pairs of signal-plus-noise sequences were used, to get a fair comparison.

Fig. 2(a)-(c) shows the results for the filters (and spectra) of Example 1. These figures show the averages of the outputs on the ten trials for the optimum, the three-level and the two-level Eckart filters, respectively, for an input SNR of -3 dB. Table I summarizes the results for Example 1, and includes those for an input SNR of 0 dB. Table II summarizes the results obtained for the filters and spectra of Example 2. These results show that the piecewise-constant filters do perform very well, giving results which are quite close to those obtained with the optimum Eckart filter.

Fig. 3(a) and (b) shows the simulation results obtained for the signal and noise spectra of Example 3; Fig. 3(a) is for the case where the nominally optimum Eckart filter of Example 2 is used, and Fig. 3(b) is for the case where the robust two-level piecewise-constant filter $W_{ER}(\omega) = W_{EPC}(\omega)$ of (35) is used. Once again, each of these figures shows the average over the ten independent trials of the filter outputs. The input SNR in both cases was -3 dB.

We see from Fig. 3(a) that the filter $W_{EO}(\omega)$, which was optimum for the nominal spectra of Example 2, deteriorates almost completely in its performance for the noise spectra of Example 3. Of the ten trials for which the average output is shown in Fig. 3(a), this filter succeeded only three times in estimating the time delay as lying in the range (19, 21). The other seven times the peak value at the processor output occurred at argument values larger than 40, with an average value over the ten trials of approximately 185. The noise spectra in Example 3 have a product $Q(\omega)$ which lies in the same p-point class for which the two-level piecewise-constant filter of Example 2 is the robust filter. Indeed, we see from Fig. 3(b) that the robust filter does give very good performance. The average value $\hat{D}$ of the time-delay estimates for the ten trials was 20.004 for this case, with a sample variance of $\widehat{\sigma^2} = 8.0 \times 10^{-4}$. The theoretical variance σ_{th}^2 for this case is 8.1×10^{-4}, whereas for the nominally optimum filter it is 2.2×10^{-1}, for the spectra of Example 3. For the nominal spectra of Example 2 the two-level piecewise-constant filter has a theoretical variance of 3.5×10^{-3} and the optimum Eckart filter has a theoretical variance of 1.8×10^{-3}, as indicated (for input SNR = 0 dB) in Table II. We will observe a similar behavior for the variance in the next example, and will make some further comments on this behavior there.

B. Robust Eckart Filter for ϵ-Contaminated Spectral Classes

Example 4: We will once again assume that the nominal signal and noise spectral densities $S_0(\omega)$ and $N_{10}(\omega)$, $N_{20}(\omega)$ are the same as those defined for Example 1 above. Now, however, we will assume that the true signal and product of noise spectral densities $S(\omega)$ and $Q(\omega)$ are in the classes described by (22) and (23), respectively, with fixed total powers.

For a numerical solution let us take $\epsilon_s = \epsilon_q = 0.2$, or a 20 percent possible contamination. Solving (26) and (27), we get for $\alpha = \frac{1}{3}$ and $A = 1$ in (34) the result $k_q = 1.734$ and $k_s = 0.275$. The least-favorable pair $[S_R(\omega), Q_R(\omega)]$ is therefore given by

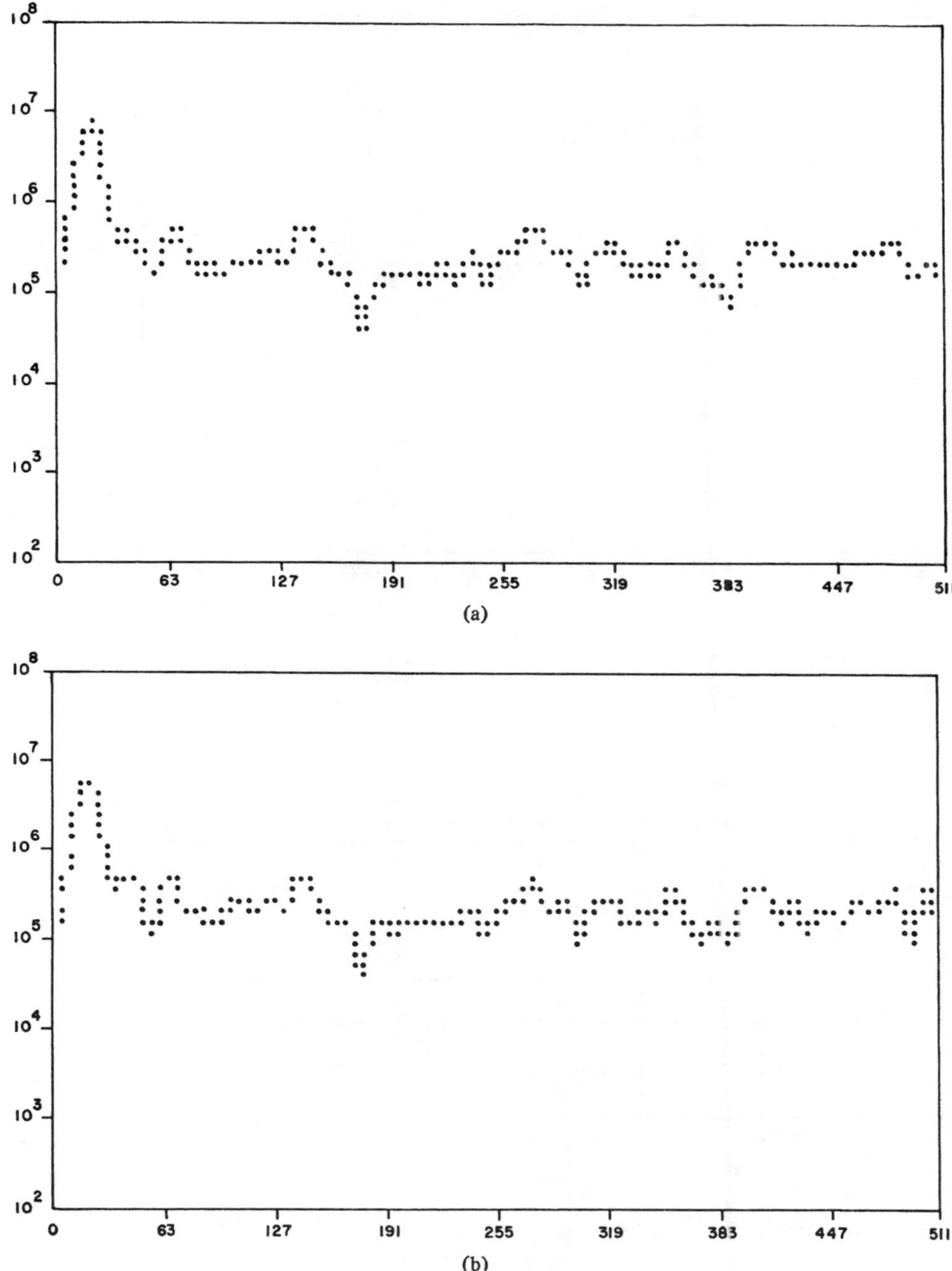

Fig. 2. Output (averaged for 10 simulation runs) for signal and noise spectra of Example 1, SNR = −3 dB. (a) Using optimum Eckart filter $W_{EO}(\omega)$. (b) Using optimum three-level piecewise constant Eckart filter $W_{EPC}(\omega)$. (c) Using optimum two-level piecewise constant Eckart filter $W_{EPC}(\omega)$.

$$S_R(\omega) = \begin{cases} 0.8 \dfrac{0.111}{(0.111 + \omega^2)^2}, & |\omega| \leq 0.723 \\ 0.8\, k_s, & 0.723 < |\omega| \leq \pi \end{cases}$$

and

$$Q_R(\omega) = \begin{cases} \dfrac{0.8}{k_q} \dfrac{0.111}{(0.111 + \omega^2)^2}, & |\omega| \leq 0.377 \\ 0.8, & 0.377 < |\omega| \leq \pi. \end{cases}$$

The robust Eckart filter for this problem is therefore specified as

$$W_{ER}(\omega) = \begin{cases} 1.734, & |\omega| \leq 0.377 \\ \dfrac{0.111}{(0.111 + \omega^2)^2}, & 0.377 < |\omega| \leq 0.723 \\ 0.275, & 0.723 < |\omega| \leq \pi. \end{cases}$$

Fig. 4(a) shows the characteristics $S_R(\omega)$ and $Q_R(\omega)$, and $W_{ER}(\omega)$ is shown in Fig. 4(b). Now from (26) and (27) it is clear that once k_s and k_q have been obtained for a given $S_0(\omega)$ and $Q_0(\omega)$, for an amplitude-scaled version of $S_0(\omega)$ we simply scale by the same factor the original solutions for

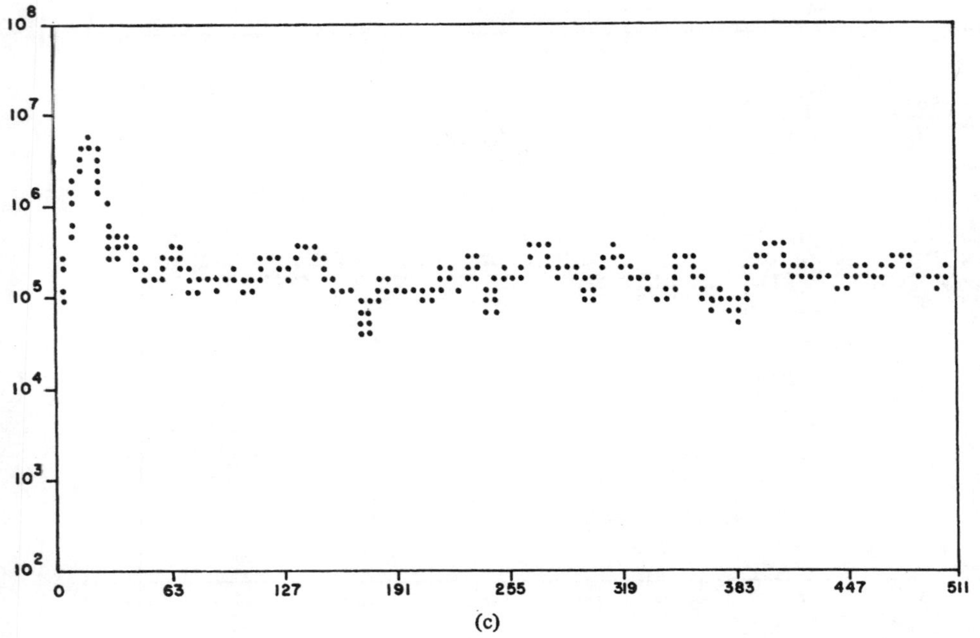

(c)

Fig. 2. *(Continued)*

TABLE I
SUMMARY OF SIMULATION RESULTS FOR EXAMPLE 1, AVERAGE OF 10 RUNS

SNR dB	W_{EO} $\hat{D}$	W_{EO} $\hat{\sigma^2}$	W_{EO} σ^2_{th}	W_{EPC} (m = 3) $\hat{D}$	W_{EPC} (m = 3) $\hat{\sigma^2}$	W_{EPC} (m = 3) σ^2_{th}	W_{EPC} (m = 2) $\hat{D}$	W_{EPC} (m = 2) $\hat{\sigma^2}$	W_{EPC} (m = 2) σ^2_{th}
0	19.79	0.165	0.135	19.72	0.260	0.177	19.92	0.390	0.300
-3	19.71	0.327	0.289	19.61	0.540	0.400	19.95	0.877	0.644

$$S(\omega) = \frac{0.111A}{(0.111 + \omega^2)^2}, \qquad |\omega| \leq \pi$$

$$N_1(\omega) = N_2(\omega) = 1, \qquad |\omega| \leq \pi$$

TABLE II
SUMMARY OF SIMULATION RESULTS FOR EXAMPLE 2, AVERAGE OF 10 RUNS

SNR (dB)	W_{EO} $\hat{D}$	W_{EO} $\hat{\sigma^2}$	W_{EO} σ^2_{th}	W_{EPC} (p = 3) $\hat{D}$	W_{EPC} (p = 3) $\hat{\sigma^2}$	W_{EPC} (p = 3) σ^2_{th}	W_{EPC} (p = 2) $\hat{D}$	W_{EPC} (p = 2) $\hat{\sigma^2}$	W_{EPC} (p = 2) σ^2_{th}
0	20.00	0.0003	0.0018	19.99	0.0012	0.0025	20.00	0.0046	0.0035
-3	19.99	0.0007	0.0048	19.98	0.0038	0.0067	19.78	0.16	0.0093

$$S(\omega) = AN_1(\omega),$$

$$N_1(\omega) = N_2(\omega) = \frac{0.111}{(0.111 + \omega^2)^2}, \qquad |\omega| \leq \pi$$

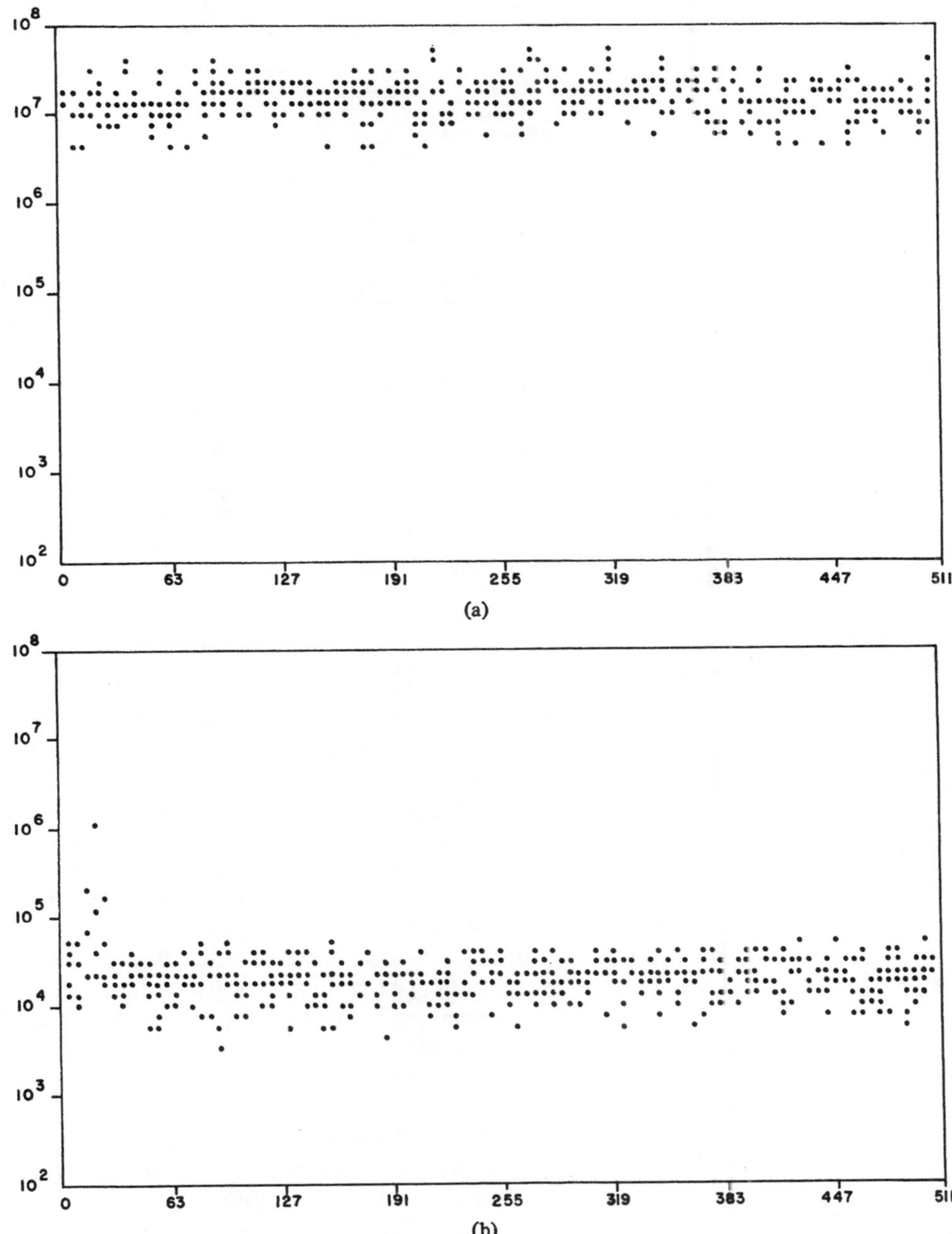

Fig. 3. Output (averaged for 10 simulation runs) for nonnominal signal and noise spectra of Example 3, SNR = -3 dB. (a) Using nominally optimum Eckart filter $E_{WO}(\omega)$ of Example 2. (b) Using robust two-level piecewise constant Eckart filter of Example 2 [see (35)].

k_s and k_q. From (28) and (29) the least-favorable signal PSD $S_R(\omega)$ will also be scaled by the same factor, $Q_R(\omega)$ remaining unchanged. Thus, the robust Eckart filter is essentially invariant to amplitude scaling of $S_0(\omega)$, and remains the same for different input SNR's obtained for a fixed $Q_0(\omega)$ and scaled versions of $S_0(\omega)$. In a similar way we can conclude that the robust Eckart filter is essentially invariant to amplitude scaling of $Q_0(\omega)$.

Simulations were run for the performance of the robust Eckart filter in the same way that they were performed for the previous examples. The robust Eckart filter was used both for nominal conditions $[S_0(\omega), Q_0(\omega)]$ and for least-favorable spectra $[S_R(\omega), Q_R(\omega)]$. The filter W_{EO}, which is optimum for $[S_0(\omega), Q_0(\omega)]$, was also simulated for performance under both conditions. A summary of the theoretical and simulation results is given in Table III.

The results in Table III indicate that the robust filter W_{ER} does have a higher minimum level of performance (as measured by output SNR, the quantity d). Its performance is somewhat degraded relative to the optimum performance under nominal conditions. It should be noted, however, that $[S_R(\omega), Q_R(\omega)]$ is *not* least-favorable for $W_{EO}(\omega)$, as it is for $W_{ER}(\omega)$. In fact, it can be shown that within the ϵ-contamination classes the worst-case performance of $W_{EO}(\omega)$ gives $d = 6.5$ for nominal input SNR of 0 dB and $d = 1.6$ at nominal input SNR of -3 dB. This means that in terms of worst-case performance, the output SNR is roughly twice as high for the robust filter.

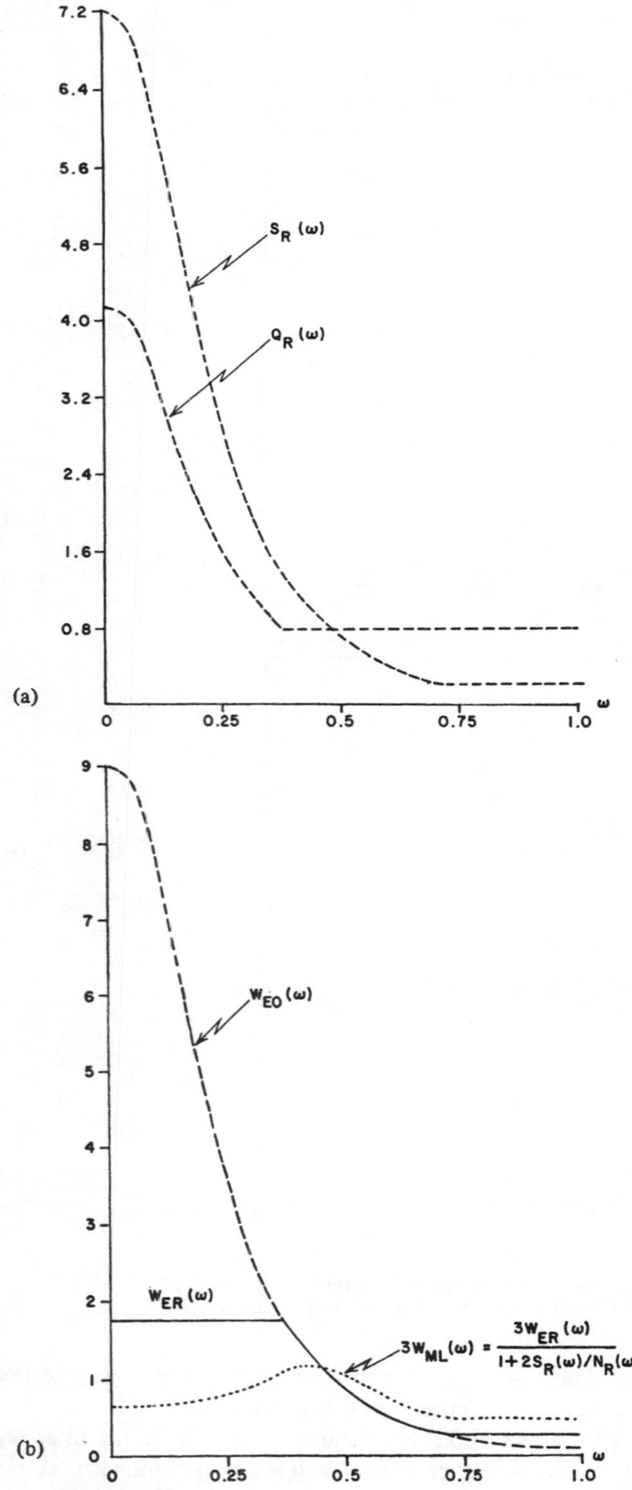

Fig. 4. (a) Least favorable signal spectrum $S_R(\omega)$ and product of noise spectra $Q_R(\omega)$ in ϵ-contaminated classes for Example 4. (b) Comparison of optimum Eckart filter $W_{EO}(\omega)$, robust Eckart filter $W_{ER}(\omega)$, and maximum likelihood filter $W_{ML}(\omega)$ for least favorable spectra, for Example 4.

Table III also indicates some very interesting behavior for the variance of the time-delay estimate. We find that the theoretical and estimated variances of the time-delay estimates are in reasonable agreement, with the estimated variances generally being higher. The results for the nominally optimum and robust filters are about the same under nominal conditions. However, the behavior for nonnominal conditions is quite different. In particular, the robust filter *improves* markedly in its variance at an input SNR of 0 dB, whereas the nominally optimum filter variance is increased at this SNR, for $[S_R(\omega), Q_R(\omega)]$. At the lower SNR of -3 dB the theoretical values of the variance indicate a similar behavior. In the simulations the marked theoretical improvement for W_{ER} is not observed, although it does perform better than does W_{EO} for $[S_R(\omega), Q_R(\omega)]$. At the lower SNR the occurrence of grossly erroneous peaks renders the theoretical variance expression inapplicable.

The reason why W_{ER} has very good performance for $[S_R(\omega), Q_R(\omega)]$ as far as variance is concerned, at least at input SNR of 0 dB and higher, is that $W_{ER}(\omega)$ has a characteristic resembling the frequency response $W_{ML}(\omega)$ of the *maximum-likelihood* filter [4] which minimizes the estimation variance, for $[S_R(\omega), Q_R(\omega)]$. Fig. 4(b) compares $W_{EO}(\omega)$, $W_{ER}(\omega)$, and $W_{ML}(\omega)$ for our problem, from which the above conclusion may be drawn.

TABLE III

SUMMARY OF THEORETICAL AND SIMULATION RESULTS (AVERAGE OF 10 RUNS) FOR EXAMPLE 4

SNR (dB)	Filter-Spectra Combination	Output SNR, d	$\hat{D}$	$\hat{\sigma}^2$	σ^2_{th}
0	W_{EO}; S_0, Q_0	47.17	19.79	0.165	0.135
	W_{ER}; S_0, Q_0	32.17	19.84	0.169	0.154
	W_{ER}; S_R, Q_R	11.20	19.94	0.028	0.010
	W_{EO}; S_R, Q_R	9.20	19.68	0.457	0.238
-3	W_{EO}; S_0, Q_0	11.79	19.71	0.327	0.289
	W_{ER}; S_0, Q_0	8.04	19.87	0.384	0.480
	W_{ER}; S_R, Q_R	2.80	19.75	0.576	0.031
	W_{EO}; S_R, Q_R	2.30	19.59	1.138	0.530

$$S_0(\omega) = \frac{0.111}{(0.111 + \omega^2)^2}, \quad |\omega| \le \pi$$

$$N_{10}(\omega) = N_{20}(\omega) = 1, \quad |\omega| \le \pi$$

ϵ-model, $\epsilon_s = \epsilon_q = 0.2$

IV. CONCLUSION

We have studied robust versions of the Eckart filter in time delay estimation. A robust Eckart filter is a minimax filter optimizing worst-case performance (measured as an output SNR for low input SNR) over classes of signal and product of noise power spectral densities. Simple piecewise-constant frequency responses were shown to result in good performance and lead to systems which were robust for *p*-point classes of spectra. The parameters defining such classes are quantities which can be easily estimated in practice. A similarity between robust filtering problems and robust hypothesis testing problems allows other general solutions to be obtained. One such solution for the robust filter was obtained for the class of ϵ-contaminated spectra.

The numerical and simulation results we discussed show that the optimum Eckart filter for assumed signal and noise spectral characteristics can degrade quite considerably in its performance when the actual spectra are different. Robust Eckart filters generally suffer minor performance losses under nominal conditions, and can maintain a useful minimum level of performance over entire classes of spectral characteristics for which they are designed. In addition, we have noted that robust Eckart filters can have better overall estimation performance, as measured by the variance, relative to nominally optimum filters.

In this paper we used as a criterion of performance the output SNR at the correct time delay for weak input SNR conditions. It should be possible to consider in the same way the variance of the time delay estimate and obtain robust filters minimizing the worst-case variance. Additionally, it would be interesting to consider the case of more than two sensors and allow correlation between noise processes in different sensors.

Appendix

Proof of Lemma 1

The following will be used in our proof.

Lemma: If x_1 and x_2 are positive numbers and $0 \leqslant \delta \leqslant 1$, then

$$\frac{[\delta y_1 + (1-\delta) y_2]^2}{\delta x_1 + (1-\delta) x_2} \leqslant \delta \frac{y_1^2}{x_1} + (1-\delta) \frac{y_2^2}{x_2}. \qquad \text{(A-1)}$$

Now define a continuous function of δ by

$$D_0[\delta \mid S, Q] = d_0[(1-\delta)S_R + \delta S, (1-\delta)Q_R + \delta Q]$$

$$= \int_{-\infty}^{\infty} \frac{[(1-\delta)S_R + \delta S]^2}{(1-\delta)Q_R + \delta Q} \, d\omega. \qquad \text{(A-2)}$$

Inequality (A-1) implies this is convex. Furthermore, since the general classes Φ, Π are convex, the pair (S_R, Q_R) is least favorable if and only if

$$D_0[\delta \mid S, Q] \geqslant D_0[0 \mid S, Q] \qquad \text{for all } (S, Q) \in \Phi \times \Pi. \qquad \text{(A-3)}$$

This is true if and only if

$$\left. \frac{\partial D_0[\delta \mid S, Q]}{\partial \delta} \right|_{\delta=0} \geqslant 0. \qquad \text{(A-4)}$$

Performing the differentiation on the left-hand side of (A-4), inequality (17) is obtained and the proof of the first part of the lemma is complete.

For the second part, similarly define a continuous function D by

$$D[\delta \mid W_{ER}; S, Q] = d[W_{ER}; (1-\delta)S_R + \delta S, (1-\delta)Q_R + \delta Q]$$

$$= \frac{\left[\int_{-\infty}^{\infty} [(1-\delta)S_R + \delta S] \, W_{ER} \, d\omega\right]^2}{\int_{-\infty}^{\infty} [(1-\delta)Q_R + \delta Q] \, W_{ER}^2 \, d\omega}. \qquad \text{(A-5)}$$

The saddle point solution of (11) is now true if and only if

$$D[\delta \mid W_{ER}; S, Q] \geqslant D[0 \mid W_{ER}; S, Q] \qquad \text{(A-6)}$$

for all $\delta \in [0, 1]$ and all $(S, Q) \in \Phi \times \Pi$. Since $D[\delta \mid W_{ER}; S, Q]$ is convex in δ for each $(S, Q) \in \Phi \times \Pi$, (A-6) holds if and only if

$$\left. \frac{\partial D[\delta \mid W_{ER}; S, Q]}{\partial \delta} \right|_{\delta=0} \geqslant 0. \qquad \text{(A-7)}$$

Again (A-7) results in the same inequality (17) and the entire proof is now complete. That is, the saddle point solution for the Eckart filtering problem is obtained by finding the least favorable pair (S_R, Q_R).

References

[1] G. C. Carter, "Time delay estimation for passive sonar signal processing," *IEEE Trans. Acoust. Speech, Signal Processing*, vol. ASSP-29, pp. 463–470, June 1981 (Special Issue on Time Delay Estimation).

[2] J. P. Ianniello, "Time delay estimation via cross-correlation in the presence of large estimation error," *IEEE Trans. Acoust., Speech, Signal Processing*, vol. ASSP-30, pp. 998–1003, Dec. 1982.

[3] K. Scarbrough, N. Ahmed, and G. C. Carter, "On the simulation of a class of time delay estimation algorithms," *IEEE Trans. Acoust., Speech, Signal Processing*, vol. ASSP-29, pp. 534–540, June 1981.

[4] C. H. Knapp and G. C. Carter, "The generalized correlation method for estimation of time delay," *IEEE Trans. Acoust., Speech, Signal Processing*, vol. ASSP-24, pp. 320–327, Aug. 1976.

[5] R. L. Kirlin and J. N. Bradley, "Delay estimation simulations and a normalized comparison of published results," *IEEE Trans. Acoust., Speech, Signal Processing*, vol. ASSP-30, pp. 508–511, June 1982.

[6] J. C. Hassab and R. E. Boucher, "Optimum estimation of time delay by a generalized correlator," *IEEE Trans. Acoust., Speech, Signal Processing*, vol. ASSP-27, pp. 373–380, Aug. 1979.

[7] —, "An experimental comparison of optimum and sub-optimum filters effectiveness in the generalized correlator," *J. Sound Vibrations*, vol. 76, pp. 117–128, 1981.

[8] S. A. Kassam and T. L. Lim, "Robust Wiener filters," *J. Franklin Inst.*, pp. 171–185, Oct.–Nov. 1977.

[9] S. A. Kassam, T. L. Lim, and L. J. Cimini, "Two-dimensional filters for signal processing under modeling uncertainties," *IEEE Trans. Geosci. Remote Sensing*, vol. GRS-18, pp. 331–336, Oct. 1980.

[10] H. V. Poor, "On robust Wiener filtering," *IEEE Trans. Automat. Contr.*, vol. AC-25, pp. 531–536, June 1980.

[11] S. A. Kassam and H. V. Poor, "Robust signal processing for communication systems," *IEEE Commun.*, vol. C-21, pp. 20–28, Jan. 1983.

[12] A. H. Nuttal and D. W. Hyde, "A unified approach to optimum and suboptimum processing for arrays," Naval Underwater Syst. Cent., New London Lab., New London, CT, Rep. 992, Apr. 1969.

[13] V. H. MacDonald and P. M. Schultheiss, "Optimum passive bearing estimation," *J. Acoust. Soc. Amer.*, vol. 46, pp. 37–43, July 1969.

[14] J. G. Shin and S. A. Kassam, "Multilevel coincidence correlators for random signal detection," *IEEE Trans. Inform. Theory*, vol. IT-17, pp. 47–53, Jan. 1979.

[15] L. J. Cimini and S. A. Kassam, "Optimum piecewise-constant Wiener filters," *J. Opt. Soc. Amer.*, vol. 77, pp. 1162–1171, Oct. 1981.

[16] H. V. Poor, "Robust matched filters," *IEEE Trans. Inform. Theory*, vol. IT-29, pp. 677–687, Sept. 1983.

[17] S. A. Kassam, "Robust hypothesis testing for bounded classes of probability densities," *IEEE Trans. Inform. Theory*, vol. IT-27, pp. 242–247, Mar. 1981.

[18] —, "The bounded *p*-point classes in robust hypothesis testing and filtering," in *Proc. 20th Annu. Allerton Conf. Commun., Contr., Comput.*, Oct. 1982, pp. 526–534.

[19] R. E. Boucher and J. C. Hassab, "Analysis of discrete implementation of generalized cross correlator," *IEEE Trans. Acoust., Speech, Signal Processing*, vol. ASSP-29, pp. 609–611, June 1981.

Delay Estimation Using Narrow-Band Processes

SIU-KAY CHOW, MEMBER, IEEE, AND PETER M. SCHULTHEISS

Abstract—Array processing of narrow-band Gaussian signals is studied with emphasis on delay estimation. The Barankin bound is used to examine the effect of ambiguity on mean-square measurement error. When the bound is plotted as a function of signal-to-noise ratio one observes a distinct threshold. Above the critical signal-to-noise ratio the lower bound on mean-square error is given by the Cramér-Rao inequality, which is approached by the Barankin inequality under these conditions. Below the threshold the Barankin bound can exceed the Cramér-Rao bound by large factors. The relative magnitude of the bounds in that region depends critically on the ratio of signal center frequency to signal bandwidth.

I. Introduction

THE purpose of this study is to investigate the problem of *parameter estimation* of *narrow-band processes* with arrays. The parameters of primary interest are the *relative delays* between sensors which determine bearing and range to the source. The signal is modeled as a sample function of a random process, for our purposes a *Gaussian process*. We assume that the transmission medium is homogeneous so that the signal wavefront is perfectly coherent over the entire array, i.e., the signal components received by various sensors are delayed replicas of each other. The noise is considered to be additive, white, Gaussian, and incoherent from sensor to sensor.

Array processing of wide-band signals has attracted widespread interest [1]-[6]. Various estimation techniques have been studied, and special estimators such as the split-beam tracker have been developed. In most cases the *Cramér-Rao* (CR) *bound* has been used as the standard of reference against which specific instrumentations were compared. This procedure is usually justified by invoking an asymptotic theorem which asserts that the CR bound can be approached by physically realizable instrumentations if the observation time is sufficiently large. Unfortunately, there are important practical problems in which a "sufficiently large" observation time is in fact unreasonably large. This paper is concerned with that class of problems.

Probably the most common setting in which this difficulty occurs is in delay measurement using very narrow-band signals. The preferred—and for the case of only two sensors optimum—instrumentation cross correlates the sensor outputs and obtains the desired delay from the peak of the cross-correlation function. Since a narrow-band signal has a highly oscillatory correlation function, there will be many peaks of only slightly different magnitude. The resulting ambiguities can be resolved only for very large observation times or for extremely high *signal-to-noise ratios* (SNR's), and only when they have been resolved does the CR bound furnish a realistic indication of the attainable mean square error.

A procedure that has been used with some success to address the ambiguity problem is based on the *Barankin bound* [7]. Its most general version is actually a greatest lower bound and hence always approximately realizable. Unfortunately, it prescribes a computational procedure of rather overwhelming complexity. Therefore, one usually turns to simplified versions of the Barankin bound which preserve the sensitivity to ambiguities but are no longer greatest lower bounds. This approach was first used by Swerling [8] who studied parameter estimation of deterministic signals. Later McAulay [9], [10] studied the pulse-position modulation problem and the problem of joint estimation of delay and phase of a radar pulse.

The first application of the Barankin bound to array processing was made by Baggeroer [11] who used a simplified version of the bound derived by Chapman and Robbins [12]. Most recently Becker [13] applied Chapman-Robbins' version to the estimation of relative delay between two sensors when the signal is a narrow-band Gaussian process. His calculation shows that the mean-square error can exceed the CR bound by a factor of the order $(\omega_c/W)^2$, where ω_c is the center frequency and W the bandwidth of the signal. However, computational approximations in his analysis prevent it from dealing with the threshold phenomenon which is a primary concern of the present paper.

The present study is also based on the Chapman-Robbins version of the Barankin bound and deals with the estimation of delay using narrow-band signals. In line with Becker's result, we find that the mean-square error at moderate SNR can now be several orders of magnitude above the CR bound. As the SNR increases, we discover a *threshold* at which the Barankin bound quickly approaches the CR bound. This suggests that the *ambiguity error* is, in a sense, additive and can be eliminated when the SNR exceeds a certain level. The point at which the threshold occurs, as well as its magnitude, are factors of obvious practical importance. After discussing the threshold problem we deal briefly with an ad hoc estimator whose performance comes relatively close to the Barankin bound, thus showing that the bound is quite tight and at least approximately attainable.

II. The Barankin Bound

The basic system of interest consists of a stationary signal source generating quasi-sinusoidal signals and an array of M observing sensors. The waveform received by each sensor contains a signal portion $s(t)$ and a noise portion $n(t)$. The wave-

Manuscript received June 30, 1980; revised November 5, 1980. This work was supported in part by the Naval Ocean Systems Center under Contract N6601-77-0396.

S. K. Chow is with Ocean Systems Studies Center, Bell Laboratories, Whippany, NJ 07981.

P. M. Schultheiss is with the Department of Engineering and Applied Science, Yale University, New Haven, CT 06520.

Reprinted from IEEE Trans. Acoust., Speech, Signal Processing, vol. 29, no. 3, pt. 2, pp. 478–484, June 1981.

forms $r_1(t), r_2(t), \cdots, r_M(t)$, observed at the M sensors are

$$r_i(t) = s_i(t) + n_i(t) \quad i = 1, 2, \cdots, M, \quad t \in [0, T] \tag{1}$$

where $[0, T]$ is the observation interval.

The transmission medium will be regarded as homogeneous. Under the assumption that the distance between the signal source and the receiving array is always very large compared with the array length, the wavefront is essentially planar over the dimensions of the array. For analytical simplicity we shall confine ourselves to linear arrays with equally spaced sensors. In that case the received waveforms can be written as

$$r_i(t) = s[t + \tau_1 + (i - 1) D] + n_i(t)$$
$$i = 1, 2, \cdots, M, \quad t \in [0, T] \tag{2}$$

where τ_1 is the delay between the signal source and the first sensor. D denotes the relative delay between adjacent sensors. Clearly, D is a function of the bearing B:

$$D(B) = \frac{d}{c} \sin B. \tag{3}$$

Here d is the spacing between adjacent sensors and c is the velocity of wave propagation. The relevant geometry is illustrated in Fig. 1. Because of the simple but nonlinear relation between B and D we avoid the added analytical complexity of dealing with B and confine our attention to the estimation of the delay D.

The signal $s(t)$ is assumed to be a sample function of a zero mean narrow-band Gaussian process. The noises $n_1(t)$, $n_2(t)$, $\cdots$, $n_M(t)$ are considered to be uncorrelated white Gaussian noises with zero mean and spectral height $N_0/2$. We shall consider the waveform received by each sensor to be represented by its Fourier coefficients

$$r_{ik} \triangleq \frac{1}{\sqrt{T}} \int_0^T r_i(t)\, e^{j\omega_k t}\, dt. \tag{4}$$

The data vector $\boldsymbol{r}_k$ is formed from the r_{ik} and the complete data vector $\boldsymbol{r}$ is the concatenation of the $\boldsymbol{r}_k$. A well-known theorem of statistics [14] asserts that Fourier coefficients associated with different frequencies are uncorrelated if the observation time is large compared with the correlation time of signal and noise. We shall assume that this condition is satisfied and can then write the elements of the data covariance matrix as follows:

$$E[\boldsymbol{r}_k \boldsymbol{r}_1^+] = \left[S(\omega_k)\, \boldsymbol{v}_k \boldsymbol{v}_k^+ + \frac{N_0}{2} I\right] \delta_{k1}$$
$$\triangleq \boldsymbol{K}_k \delta_{k1}. \tag{5}$$

Here + denotes the conjugate transpose, $S(\omega)$ is the signal spectrum, and δ_{k1} is the Kronecker delta function. $\boldsymbol{K}_k$ is the data covariance matrix for frequency ω_k and $\boldsymbol{v}_k$ is the "steering vector" characterizing the signal delays from sensor to sensor.

$$\boldsymbol{v}_k^+ \triangleq [e^{j\omega_k D}, e^{j\omega_k 2D}, \cdots, e^{j\omega_k (M-1) D}].$$

It is now a simple matter to write down the probability density of the complete data vector $\boldsymbol{r}$ for fixed differential

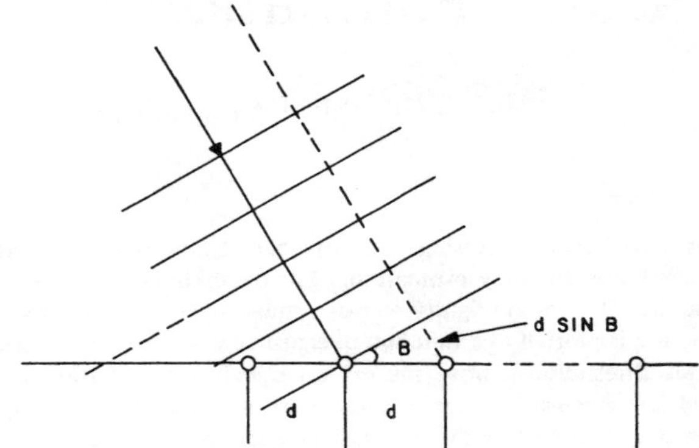

Fig. 1. Far-field source model with linear array.

delay D,

$$p(\boldsymbol{r}|D) = \prod_{k=C-P}^{C+P} \frac{1}{\det(\pi \boldsymbol{K}_k)} \exp\{-\boldsymbol{r}_k^+ \boldsymbol{K}_k^{-1} \boldsymbol{r}_k\} \tag{7}$$

where C and P are integers corresponding to the center frequency ω_c and signal bandwidth W, respectively.

We are interested in setting a lower bound on the mean-square error of the estimate $\hat{D}$ of D obtainable from the data vector $\boldsymbol{r}$. The Chapman-Robbins version [12] of the Barankin bound asserts

$$\operatorname{var}(\hat{D}) \geq \left\{\frac{1}{H^2}\left[\int \frac{p^2(\boldsymbol{r}|D+H)}{p(\boldsymbol{r}|D)}\, d\boldsymbol{r} - 1\right]\right\}^{-1}$$
$$\triangleq [J_B(\hat{D})]^{-1}. \tag{8}$$

Equation (8) holds for arbitrary H; hence, it holds in particular when H is chosen to maximize the right side. This is the choice made by Chapman and Robbins. Notice that (8) is not the greatest lower bound; however, it is sufficiently general to include the CR bound as a special case. In particular, if $p(\boldsymbol{r}|D)$ is twice differentiable with respect to D, then $J_B(\hat{D}) \rightarrow J_{CR}(\hat{D})$, the inverse of the CR bound, as $H \rightarrow 0$. It is easy to see that the difference of the two bounds can be very significant. When $H \rightarrow 0$ the right side of (8) is a ratio of two small numbers. If $r_i(t)$ possesses a quasi-sinusoidal correlation function, $p(\boldsymbol{r}|D)$ will be quasi-periodic. Then there will be nonzero values of H for which the numerator of $J_B(\hat{D})$ is near zero while the denominator H^2 is quite finite. In such a case the Barankin bound can easily exceed the CR bound by orders of magnitude.

The next step is to substitute (7) into (8). Using the symbol $\boldsymbol{K}_{kH}$ to designate the data covariance matrix of frequency ω_k when D is replaced by $D + H$, one quickly obtains

$$\int \frac{p^2(\boldsymbol{r}_k|D+H)}{p(\boldsymbol{r}_k|D)}\, d\boldsymbol{r}_k = \frac{\det(\boldsymbol{K}_k)}{[\det(\boldsymbol{K}_{kH})]^2} \det[(2\boldsymbol{K}_{kH}^{-1} - \boldsymbol{K}_k^{-1})^{-1}]$$
$$\cdot \int \frac{\exp[-\boldsymbol{r}_k^+(2\boldsymbol{K}_{kH}^{-1} - \boldsymbol{K}_k^{-1})\boldsymbol{r}_k]}{\det[\pi(2\boldsymbol{K}_{kH}^{-1} - \boldsymbol{K}_k^{-1})^{-1}]}\, d\boldsymbol{r}_k. \tag{9}$$

If $2\boldsymbol{K}_{kH}^{-1} - \boldsymbol{K}_k^{-1}$ is positive definite, the integration in (9) can be performed without difficulty, yielding

$$\int \frac{p^2(r_k|D+H)}{p(r_k|D)} dr_k$$
$$= [(\det K_k)^{-1} (\det K_{kH})^2 \det (2K_{kH}^{-1} - K_k^{-1})]^{-1}. \quad (10)$$

To proceed further, we introduce without proof two theorems of matrix theory.

Theorem 1 [15]: Let A be an $M \times M$ nonsingular matrix, b an $M \times 1$ column vector, and S, N nonzero constants. Then

$$(Sbb^+ + NA)^{-1} = \frac{1}{N}A^{-1} - \frac{S/N^2}{1+GS/N} A^{-1} bb^+ A^{-1} \quad (11)$$

where $G \triangleq b^+A^{-1}b$.

Theorem 2 [16]: Let A be an $M \times M$ diagonal matrix whose ith diagonal element is a_i and b an $M \times 1$ column vector with elements b_i. Then

$$\det (A + bb^+) = \left(1 + \sum_{i=1}^{m} \frac{b_i b_i^*}{a_i}\right) \det A. \quad (12)$$

Using these two theorems and (5), the determinants required in (10) can be evaluated by a straightforward though tedious computation [17]. The result is

$$(\det K_k)^{-1} (\det K_{kH})^2 \det (2K_{kH}^{-1} - K_k^{-1})$$
$$= 1 - \frac{8S^2(\omega_k)/N_0^2}{1+2MS(\omega_k)/N_0} \left(M^2 - \frac{\sin^2 \frac{M}{2}\omega_k H}{\sin^2 \frac{1}{2}\omega_k H} \right). \quad (13)$$

Then the inverse of the Barankin bound (8) is

$$J_B(\hat{D}) = \frac{1}{H^2} \left\{ \prod_{k=C-P}^{C+P} \left[1 - \frac{8S^2(\omega_k)/N_0^2}{1+2MS(\omega_k)/N_0} \cdot \left(M^2 - \frac{\sin^2 \frac{M}{2}\omega_k H}{\sin^2 \frac{1}{2}\omega_k H} \right) \right]^{-1} - 1 \right\}. \quad (14)$$

In the light of our earlier comment, this expression is valid only if $2K_{kH}^{-1} - K_k^{-1}$ is positive definite. A sufficient condition for an hermitian matrix to be positive definite is that its principle minors are all positive. This condition is satisfied if

$$1 - \frac{8S^2(\omega_k)/N_0^2}{1+2NS(\omega_k)/N_0} \left(N^2 - \frac{\sin^2 \frac{N}{2}\omega_k H}{\sin^2 \frac{1}{2}\omega_k H} \right) > 0 \quad (15)$$

for all k and $N = 1, 2, \cdots, M$. If $2K_{kH}^{-1} - K_k^{-1}$ has negative eigenvalues, a diagonalizing orthogonal transformation converts (9) into a product of integrals, some of which diverge. A value of H for which this is true yields a bound of zero so that a better bound is given by the CR inequality. A striking feature of (15) is that its left side is a monotone decreasing function of SNR. For any given $H \neq 0$ there exists an SNR at which the expression becomes negative so that one is ultimately forced back to the CR bound (because one can always obtain the CR bound by choosing $H \to 0$). This leads to the threshold phenomenon discussed earlier.

The next important question concerns the choice of H. We are interested in setting the largest possible lower bound on the mean-square error, which means minimizing (14). The nonnegative factor

$$M^2 - \frac{\sin^2 \frac{M}{2}\omega_k H}{\sin^2 \frac{1}{2}\omega_k H}, \quad (16)$$

considered as a function of H, is periodic with period $2\pi/\omega_k$ and assumes a value of zero for $H = 2n\pi/\omega_k$, $n = 0, 1, 2, \cdots$. By choosing $H = 2n\pi/\omega_c$ we therefore ensure that the term in square brackets in (14) is precisely equal to unity for $\omega_k = \omega_c$. If the signal is confined to a narrow band centered at ω_c, the corresponding terms for other ω_k differ only slightly from unity so that the term in braces becomes small, while the denominator H^2 remains relatively large.

This choice of H also has a good deal of intuitive appeal. H is the difference between the nominal sensor to sensor differential delay $D + H$ and the actual sensor to sensor delay D. A narrow-band signal with center frequency ω_c has a quasi-periodic autocorrelation function $R(\tau)$ and hence, in the Gaussian case, a quasi-periodic probability density function. An increment of $2\pi/\omega_c$ represents a displacement from the true peak of the autocorrelation function to a secondary peak (see Fig. 2). When the signal bandwidth is small the envelope of the correlation function decreases slowly so that it becomes difficult to select the global maximum from a number of nearly equal maxima. This is, of course, the physical cause of the ambiguity problem we are addressing.

When $H = 2n\pi/\omega_c$, (14) assumes the form

$$J_B(\hat{D}) = \frac{1}{\left(\frac{2n\pi}{\omega_c}\right)^2} \left\{ \prod_{k=C-P}^{C+P} \cdot \left[1 - \frac{8S^2(\omega_k)/N_0^2}{1+2MS(\omega_k)/N_0} \left(M^2 - \frac{\sin^2 Mn\pi \frac{\omega_k}{\omega_c}}{\sin^2 n\pi \frac{\omega_k}{\omega_c}} \right) \right]^{-1} - 1 \right\}. \quad (17)$$

Equation (17) sheds further light on the threshold problem. The smaller the factor

$$M^2 - \frac{\sin^2 Mn\pi \frac{\omega_k}{\omega_c}}{\sin^2 n\pi \frac{\omega_k}{\omega_c}},$$

the larger the SNR can become before (15) is violated. For very narrow-band signals the above factor will be largest near the edge of the signal band, where $|\omega_k - \omega_c|$ is at a maximum.

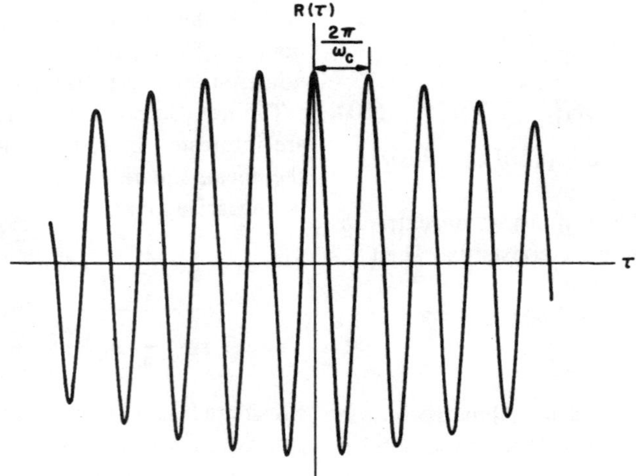

Fig. 2. A quasi-periodic autocorrelation function.

One would therefore expect the threshold SNR to increase as the signal bandwidth diminishes.

Aside from establishing the threshold phenomenon we are particularly interested in behavior below the threshold, in the ambiguity dominated region. Since the signal bandwidth is small, we can approximate the trigonometric functions in (17) by the first few terms of a Taylor expansion about the center frequency ω_c

$$J_B(\hat{D}) \cong \frac{1}{\left(\frac{2n\pi}{\omega_c}\right)^2} \left\{ \prod_{k=C-P}^{C+P} \left[1 - \frac{8S^2(\omega_k)/N_0}{1 + 2MS(\omega_k)/N_0} \cdot \frac{M^2(M^2-1)}{3} \left(\frac{n\pi}{\omega_c}\right)^2 (\omega_k - \omega_c)^2 \right]^{-1} - 1 \right\}. \quad (18)$$

With the same approximation (15) becomes

$$1 - \frac{8S^2(\omega_k)/N_0^2}{1 + 2NS(\omega_k)/N_0} \frac{N^2(N^2-1)}{3} \left(\frac{n\pi}{\omega_c}\right)^2 (\omega_k - \omega_c)^2 > 0 \quad (19)$$

for all k, n, and $N = 1, 2, \cdots, M$. Under this condition expansion of the production in (18) leads to a positive term power series in n. Hence the choice $n = 1$ minimizes $J_B(\hat{D})$.

$$J_B(\hat{D}) = \left(\frac{\omega_c}{2\pi}\right)^2 \left\{ \exp\left[- \sum_{k=C-P}^{C+P} \ln \left(1 - \frac{(2MS(\omega_k)/N_0)^2}{1 + 2MS(\omega_k)/N_0} \cdot \frac{(M^2-1)}{6} \left(\frac{2\pi}{\omega_c}\right)^2 (\omega_k - \omega_c)^2 \right) \right] - 1 \right\}. \quad (20)$$

Since the observation time is much larger than the correlation time, we may assume that the signal and noise spectra will not change significantly over the frequency increment $2\pi/T$. The summation of (20) can now be written in integral form. In order to proceed further we postulate a particular form of $S(\omega)$. Exact analytical results can be obtained for the simple shape

$$S(\omega) = \begin{cases} S & \omega \in \left[\omega_c - \frac{W}{2},\ \omega_c + \frac{W}{2}\right] \\ 0 & \text{elsewhere.} \end{cases} \quad (21)$$

If we define a constant K as

$$K^2 \triangleq \frac{(2MS/N_0)^2}{1 + 2MS/N_0} \frac{(M^2-1)}{6} \left(\frac{2\pi}{\omega_c}\right)^2 \quad (22)$$

(20) assumes the form

$$J_B(\hat{D}) = \left(\frac{\omega_c}{2\pi}\right)^2 \left\{ \exp\left[-\frac{T}{\pi} \int_0^{W/2} \ln(1 - K^2 z^2)\, dz \right] - 1 \right\}$$
$$= \left(\frac{\omega_c}{2\pi}\right)^2 \left\{ \left[\frac{(1 - KW/2)^{(2/KW)-1} \cdot e^2}{(1 + KW/2)^{(2/KW)+1}} \right]^{TW/2\pi} + 1 \right\}, \quad (23)$$

with $KW/2 < 1$. For $KW/2 > 1$, $J_B(\hat{D}) = 0$ because the integral in (9) diverges. $KW/2 = 1$ is therefore the threshold value of our earlier, more qualitative discussion. For $KW/2 > 1$ the integral in (23) is clearly meaningless because the argument of the logarithmic function becomes negative over part of the range of integration. Equation (23) must be treated with caution even when $KW/2$ comes close to unity: as the integrand tends to infinity, the assumption of small variation over the interval $2\pi/T$ is violated, so that the conversion from the sum in (20) to the integral in (23) is no longer justified. When $KW/2 = 1$, (23) converges while (20) diverges. However, for the relative large TW products considered here the discrepancy is confined to a small neighborhood of $KW/2 = 1$ and (23) will be used as the basis of further analysis.

Fig. 3 shows the Barankin bound $[J_B(\hat{D})]^{-1}$, normalized with respect to $(\omega_c/2\pi)^2$ plotted as a function of $KW/2$ for TW values of 100, 200, and 1000. A logarithmic (dB) scale is used for both variables. While each curve approaches a variance of zero at $KW/2 = 1$ the transition begins earlier for larger values of TW. $KW/2 = 1$ is the point at which the transition is complete. We shall later determine the point at which it begins.

Fig. 3 presents the information contained in (23) very compactly, but probably not in its most easily interpreted form. In practice we are most likely to define the threshold in terms

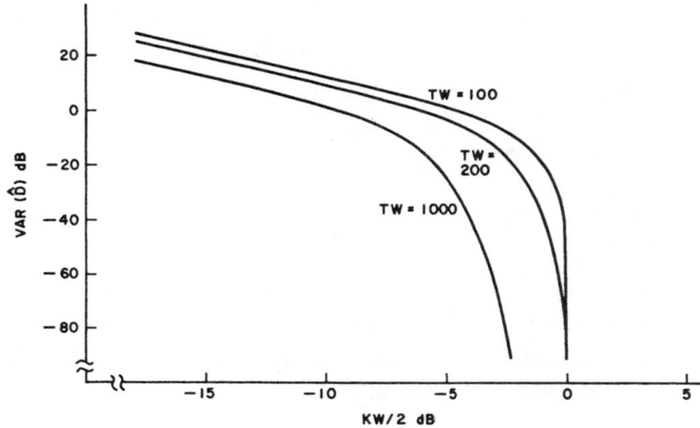

Fig. 3. Behavior of the Barankin bounds.

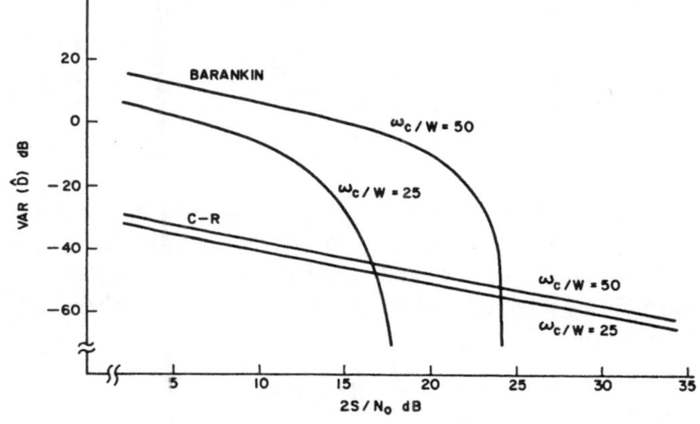

Fig. 5. Error bounds normalized with respect to $(\omega_c/2\pi)^2$; $M = 2$, $T = \frac{1}{4}\pi$ and $\omega_c = 2\pi \cdot 10^4$.

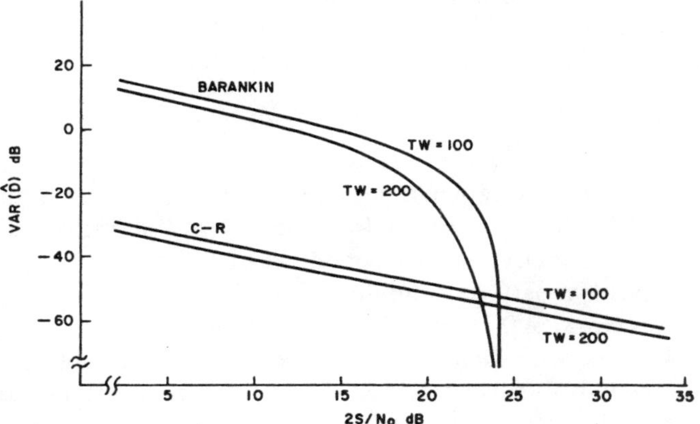

Fig. 4. Error bounds normalized with respect to $(\omega_c/2\pi)^2$; $M = 2$ and $\omega_c/W = 50$.

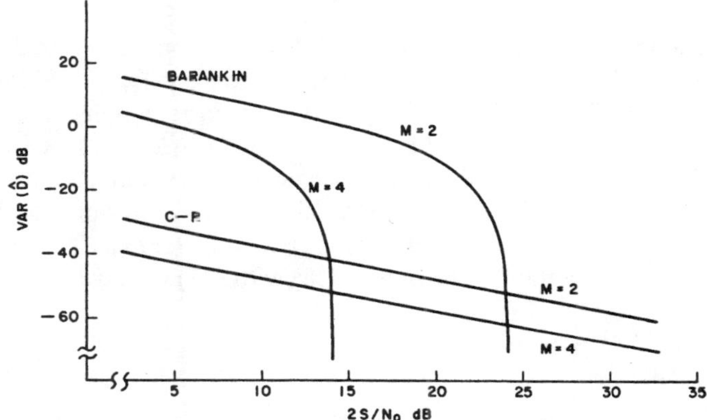

Fig. 6. Error bounds normalized with respect to $(\omega_c/2\pi)^2$; $TW = 100$ and $\omega_c/W = 50$.

of a critical SNR and are concerned with its dependence on observation time, bandwidth, and number of sensors. Figs. 4-6 examine the effect of each of these parameters in turn. Also shown for comparison is the CR bound, obtained by allowing H in (14) to approach zero. After a few steps of computation one obtains

$$J_{\text{CR}}(\hat{D}) = \frac{T}{2\pi} \int_{\omega_c - W/2}^{\omega_c + W/2} \frac{(2MS(\omega)/N_0)^2}{1 + 2MS(\omega)/N_0} \frac{M^2 - 1}{6} \omega^2 d\omega. \tag{24}$$

Thus, for the signal spectrum (21),

$$J_{\text{CR}}(\hat{D}) = \frac{TW}{2\pi} \frac{(2MS/N_0)^2}{1 + 2MS/N_0} \frac{M^2 - 1}{6} \left(\omega_c^2 + \frac{W^2}{12}\right). \tag{25}$$

Fig. 4 plots the normalized Barankin and CR bound for $M = 2$ as a function of SNR, with observation time T as a parameter. As T doubles, the CR bound shifts to the left (or downward) by 3 dB and the linear portion of the Barankin bound behaves similarly. The formal threshold value remains unchanged but the transition begins earlier, as already observed in connection with Fig. 3. The fact that the Barankin bounds dip below the corresponding CR bounds is simply evidence of the fact that the choice $H = 2\pi/\omega_c$ has become inappropriate even before the integral in (9) formally diverges.

Fig. 5 plots the normalized bounds for $M = 2$ still as a function of SNR, with bandwidth W as a parameter. Since the threshold occurs at a value of SNR well above unity, K varies as $(2S/N_0)^{1/2}$ so that an increase in bandwidth by a factor of 2 reduces the threshold by approximately 6 dB.

Fig. 6, finally, explores the effect of changes in the number of sensors M. For $MS/N_0 \gg 1$ the factor K^2 of (22) varies as $M(M^2 - 1)$. The CR bound and the linear portion (below threshold) of the Barankin bound are both proportional to K^2. For M appreciably larger than unity, doubling the number of sensors should therefore lower the level of both curves by approximately 9 dB. Under the same assumptions K is proportional to $M(MS/N_0)^{1/2}$, so that doubling M should also reduce the threshold SNR by approximately 9 dB. However, Fig. 6 uses relatively small value of M (2 and 4) and therefore the shift of threshold is slightly larger.

All of the curves in Figs. 4-6 exhibit linear behavior with respect to SNR below the threshold. This is easy to understand by looking at (18). Below the threshold the SNR dependent term is smaller than unity. If it is substantially smaller than unity, one can multiply out the products and discard all but the lowest order SNR dependent terms. With $n = 1$ the result is

$$J_B(\hat{D}) = \sum_{k=C-P}^{C+P} \frac{(2MS(\omega_k)/N_0)^2}{1 + 2MS(\omega_k)/N_0} \frac{M^2 - 1}{6} (\omega_k - \omega_c)^2. \tag{26}$$

With the usual integral approximation, this becomes

$$J_B(\hat{D}) = \frac{T}{2\pi} \int_{\omega_c - W/2}^{\omega_c + W/2} \frac{(2MS(\omega)/N_0)^2}{1 + 2MS(\omega)/N_0} \frac{M^2 - 1}{6} (\omega - \omega_c)^2 \, d\omega. \tag{27}$$

Equations (24) and (27) differ only in one component of the integrand: the factor ω^2 in (24) is replaced by $(\omega - \omega_c)^2$ in (27). For a narrow-band signal, (24) should therefore exceed (27) by a factor of the order of $(\omega_c/W)^2$. For the particular spectrum given by (21)

$$\begin{aligned} J_B(\hat{D}) &\cong \frac{TW}{2\pi} \frac{(2MS/N_0)^2}{1 + 2MS/N_0} \frac{M^2 - 1}{6} \frac{W^2}{12} \\ &= \left(\frac{\omega_c}{2\pi}\right)^2 \frac{1}{6\pi} (TW) \left(\frac{KW}{2}\right)^2. \end{aligned} \tag{28}$$

The second power variation with $KW/2$ corresponds to a straight line with -2 slope on the logarithmic scale of Fig. 3. Since $K^2 \propto 2MS/N_0$ when $2MS/N_0 \gg 1$, the corresponding slope on Figs. 4–6 is -1 in that region. For $2MS/N_0 \ll 1$, $K^2 \propto (2MS/N_0)^2$ so that the slope becomes -2. We note also that the level of (28) for any fixed value of KW is proportional to TW. Thus the indicated mean-square error varies as T^{-1}.

Comparing the first version of (28) with (25) we have

$$\frac{J_{CR}(\hat{D})}{J_B(\hat{D})} = 12 \left(\frac{\omega_c}{W}\right)^2 + 1. \tag{29}$$

Throughout the linear region the Barankin bound therefore exceeds the CR bound by a constant factor which can easily become large. For $\omega_c/W = 50$, for instance, the ratio is 44.8 dB.

To gain more insight into the extent of the linear region we examine the transition from (20) to (27) more carefully. Using the spectrum given by (21) one can write (20) in the form

$$J_B(\hat{D}) = \left(\frac{\omega_c}{2\pi}\right)^2 \exp\left\{ -\sum_{k=C-P}^{C+P} \ln\left[1 - K^2(\omega_k - \omega_c)^2\right] \right\}. \tag{30}$$

Below the threshold, $K^2(\omega_k - \omega_c)^2 < 1$ for all k so that the logarithmic function can be expanded into a power series:

$$\begin{aligned} &-\sum_{k=C-P}^{C+P} \ln\left[1 - K^2(\omega_k - \omega_c)^2\right] \\ &= K^2 \sum_{k=C-P}^{C+P} (\omega_k - \omega_c)^2 + \frac{K^2}{2} \sum_{k=C-P}^{C+P} (\omega_k - \omega_c)^4 \\ &\quad + \frac{K^6}{3} \sum_{k=C-P}^{C+P} (\omega_k - \omega_c)^6 + \cdots. \end{aligned} \tag{31}$$

Once again we approximate each of the sums by an integral and obtain for the first sum

$$K^2 \sum_{k=C-P}^{C+P} (\omega_k - \omega_c)^2 = \frac{TW}{6\pi} \left(\frac{KW}{2}\right)^2 \triangleq y. \tag{32}$$

The quantity y defined above is, of course, merely (28) divided by $(\omega_c/2\pi)^2$. The other sums are simply related to y as

$$\frac{K^4}{2} \sum_{k=C-P}^{C+P} (\omega_k - \omega_c)^4 = \frac{9}{10} \left(\frac{2\pi}{TW}\right) y^2 \tag{33}$$

$$\frac{K^6}{3} \sum_{k=C-P}^{C+P} (\omega_k - \omega_c)^6 = \frac{9}{7} \left(\frac{2\pi}{TW}\right)^2 y^3. \tag{34}$$

Successive terms of (31) contain successively larger negative powers of $(TW/2\pi)$, a quantity much larger than unity in our analysis. Hence, at least for y up to the order of unity, one has to an excellent approximation

$$J_B(\hat{D}) \cong \left(\frac{\omega_c}{2\pi}\right)^2 (e^y - 1). \tag{35}$$

The linear region of the Barankin bound corresponds to the range of y for which $e^y \cong 1 + y$, so that the transition to the CR bound should begin near $y = 1$. For $y \ll 1$

$$J_B(\hat{D}) \cong \left(\frac{\omega_c}{2\pi}\right)^2 y. \tag{36}$$

At $y = 1$ this corresponds to an rms delay error equal to the period of the signal center frequency. Fig. 3 normalizes the error with respect to that period, so that one would expect significant departures from linear behavior to begin where the curves cross the 0 dB line of the vertical scale.

We close with some remarks concerning the attainability of the bounds calculated above. Since the analysis did not use the most general version of the Barankin bound, there is no guarantee that the calculated mean-square errors can actually be reached. It therefore becomes important to bound the mean-square error from above as well as below. An error analysis for any particular instrumentation furnishes such an upper bound. A promising instrumentation is suggested by the following line of argument. The CR bound is reachable in the limit of large observation times. For the case of only two sensors a simple cross correlator performs the delay measurement with minimum mean-square error. Equation (29) says that the attainable mean-square error in the ambiguity-dominated region exceeds the CR bound by a factor of the order $(\omega_c/W)^2$. This suggests use of signal envelopes as inputs to a cross correlator. The signal envelopes do not have the ambiguity problems associated with narrow-band waveforms, and one could therefore hope to realize mean-square errors close to those predicted by a CR bound calculation based on the envelope processes. The calculation has been carried out for an instrumentation composed of square law envelope detectors at the output of each of two sensors, followed by a simple cross correlator [17]. Computational details are tedious but the result shows a mean-square error exceeding the Barankin bound of Fig. 4 by only 6 dB for SNR ranging from 0 dB up to the threshold region. For that range of SNR the mean-square error of the best realizable instrumentation is therefore known to within ±3 dB.

References

[1] D. G. Tucker and B. K. Gazey, *Applied Underwater Acoustics*. New York: Pergamon, 1966, ch. 6, sec. 3.2, pp. 184–185.

[2] E. J. Kelly, Jr. and M. J. Levin, "Signal parameter estimation for seismometer arrays," M.I.T. Lincoln Laboratory, Tech. Rep. 339, Jan. 1964.

[3] V. H. MacDonald and P. M. Schultheiss, "Optimum passive bearing estimation in a spatially incoherent noise environment," *J. Acoust. Soc. Amer.*, vol. 46, pp. 37–43, July 1969.

[4] H. L. Van Trees, "A unified theory for optimum array processing," A. D. Little, Inc., Rep. 4160866, Aug. 1966.

[5] W. J. Bangs, II, "Array processing with generalized beam-formers," Ph.D. dissertation, Yale Univ., New Haven, CT, Sept. 1971.

[6] C. H. Knapp and G. C. Carter, "The generalized correlation method for estimation of time delay," *IEEE Trans. Acoust., Speech, Signal Processing*, vol. ASSP-24, pp. 320–327, Aug. 1976.

[7] E. W. Barankin, "Locally best unbiased estimates," *Ann. Math. Statist.*, vol. 20, pp. 477–501, 1949.

[8] P. Swerling, "Parameter estimation for waveforms in additive Gaussian noise," *J. SIAM*, vol. 7, pp. 154–166, 1959.

[9] R. J. McAulay and L. P. Seidman, "A useful form of the Barankin lower bound and its application to PPM threshold analysis," *IEEE Trans. Inform. Theory*, vol. IT-15, pp. 273–279, Mar. 1969.

[10] R. J. McAulay and E. M. Hofstetter, "Barankin bounds on parameter estimation," *IEEE Trans. Inform. Theory*, vol. IT-17, pp. 669–676, Nov. 1971.

[11] A. B. Baggeroer, "Barankin bound on the variance of estimates of a Gaussian random process," M.I.T. Res. Lab. Electron., Quart. Prog. Rep. 92, pp. 324–333, Jan. 1969.

[12] D. G. Chapman and H. Robbins, "Minimum variance estimation without regularity assumptions," *Ann. Math. Statist.*, vol. 22, pp. 581–586, 1951.

[13] D. Becker, "Develop Cramér–Rao bounds on parameter estimation accuracy," Orincon Tech Rep. OC-R-77-0505-5, Jan. 1977.

[14] W. B. Davenport, II and W. L. Root, *An Introduction to the Theory of Random Signals and Noise*. New York: McGraw-Hill, 1958, ch. 6, sec. 6-4, pp. 93–96.

[15] M. S. Bartlett, "An inverse matrix adjustment arising in discriminant analysis," *Ann. Math. Statist.*, vol. 20, pp. 107–111, 1951.

[16] B. Noble, *Applied Linear Algebra*. Englewood Cliffs, NJ: Prentice-Hall, 1969, ch. 5, p. 225.

[17] S. K. Chow, "Parameter estimation of narrow-band processes with arrays," Ph.D. dissertation, Yale Univ., New Haven, CT, Sept. 1978.

FAST TECHNIQUES FOR TIME DELAY ESTIMATION

Roberto Cusani

Dipartimento di Ingegneria Elettronica, University of Rome 'Tor Vergata', Via Orazio Raimondo, I-00173 Rome (Italy)

SUMMARY

Two multiplication-free correlators are analized as fast alternatives to the direct multiply-and-accumulate technique in time delay measurements. Their accuracy limits are given for general signal and noises spectra and their dependence from signal-to-noise ratio, observation interval and signal bandwidth is analized for a reference case.

1. INTRODUCTION

Correlation techniques have been extensively studied in communication theory for a large variety of applications [1]. As a matter of fact, correlation is the classic solution for signal detection and parameter estimation whenever the incoming signals are embebbed in additive noise. The basic feature of a correlator is the multiplication of two waveforms and the integration of the resulting function. Different structures, with various levels of hardware/software complexity, have been derived from this basic scheme.

In particular, 'clipping' the amplitude of one or both input signals gives fast multiplication-free structures. Referring to discrete-time applications, the hybrid-sign (HS) correlator is based on the accumulation of data samples, in polarity coincidence (PC) technique a simple up/down counting is required [1].

In time delay estimation (TDE) problems a random signal and its delayed version are received at two spatially separated noisy sensors. The location of the maximum of their cross-correlation function (ccf) measures the delay. TDE is an important task not only in remote sensing applications (sonar, radar, radio-astronomy, seismics, etc.) but also in linear systems analysis, echo cancelation, speech analysis, etc.

In this contribution the accuracy limits of HS and PC time delay estimators are evaluated for general signal and noises spectra and compared to the direct multiply-and-accumulate method. In particular, the dependence from signal-to-noise ratio (SNR), observation interval and signal bandwidth is examined for the case of white signal and noises.

As the main result, PC is verified to be a powerful tool if a certain loss in accuracy can be tolerated, or recovered by properly enlarging the observation interval.

2. THE PROBLEM

In classical TDE problems the received signals r_1, r_2 take on the form (discrete-time modeling and unit-sampling period is assumed):

$$r_1(i) = s(i) + n_1(i) \tag{1}$$

$$r_2(i) = s(i - D) + n_2(i) \tag{2}$$

$$i = 1, 2, 3, \ldots$$

where D is the time delay, to be estimated. The source signal s(i) and the sensor noises $n_1(i)$, $n_2(i)$ are zero-mean, jointly stationary and independent random processes with given spectral power densities (psd's) $G_{ss}(\omega)$, $G_{n_1n_1}(\omega)$, $G_{n_2n_2}(\omega)$, $-\pi<\omega<\pi$.

The direct correlator (DC) computes the ccf between $r_1(i)$ and $r_2(i)$ based on N sample pairs as:

$$R_{DC}(k) = \sum_{i=0}^{N-1} r_1(i)\, r_2(i+k) \qquad k = 0, \pm 1, \pm 2, \ldots, +L \tag{3}$$

Price' theorem [2] and Van Vleck' equation [3] enable us to replace (3) by the Hybrid Sign (HS) correlator:

Reprinted from *Proc. MELECON '89*, pp. 177–180, April 1989.

$$R_{HS}(k) = \sum_{i=0}^{N-1} r_1(i)\,\mathrm{sign}[r_2(i+k)]$$

$$= \sum_{i=0}^{N-1} r_1(i)\, y_2(i+k) \qquad k = 0, \pm 1, \pm 2, \ldots, +L \quad (4)$$

or by the Polarity Coincidence correlator:

$$R_{PC}(k) = \sum_{i=0}^{N-1} \mathrm{sign}[r_1(i)]\,\mathrm{sign}[r_2(i+k)]$$

$$= \sum_{i=0}^{N-1} y_1(i)\, y_2(i+k) \qquad k = 0, \pm 1, \pm 2, \ldots, +L \quad (5)$$

where sign $[x] = 1$ if $x > 0$ and $= -1$ if $x < 0$, and the binary signals $y_1(i)$, $y_2(i)$ are the hard-limited versions of $r_1(i)$, $r_2(i)$.

The location of the maximum of (3), (4) or (5) gives the corresponding time delay measurement. Consider that D is not, in general, a multiple of the sampling interval: the theoretical solution implies the computation of a large number L of ccf lags and interpolation of (3), (4) or (5) by sinc (or cubic, spline, etc.) functions before peak detection. In practical applications, a few ccf lags are computed, with a certain loss in accuracy with respect to the theoretical case (see, for example, [4]).

In both cases, replacing (3) by (4) or (5) gives significant computational advantages, to be exploited for fast and economic processing (real-time applications, batch processing of a large amount of data) and for specialized computing architectures (dedicated hardware, VLSI circuits).

DC implies N products and N-1 sums for each lag k in eq.(3). These reduce to N-1 sums in HS technique, where the products by the bit sign can be implemented by a low-cost add-or-subtract logic. PC correlation corresponds to an ex-or logic on $y_1(i)$, $y_2(i)$, followed by an up/down binary counting. Observe that in digital applications sign detection is implemented by hard-limiting the analogic input signal without the need of analogic-to-digital conversion, or by extracting the bit sign from its available binary representation.

3. ESTIMATION ACCURACY

The performance of DC, HS and PC time delay estimators are evaluated here in terms of signal and noises psd's . From these, the auto- and cross- psd's of $r_1(i)$, $r_2(i)$ are:

$$G_{r_jr_j}(\omega) = G_{ss}(\omega) + G_{n_jn_j}(\omega) \qquad j = 1, 2 \qquad (6)$$

$$G_{r1r2}(\omega) = G_{ss}(\omega)\exp(-j\omega D) \qquad (7)$$

From [2],[3] the psd's of the hard-limited signals are computed as:

$$G_{r1y2}(\omega) = \sqrt{\frac{2}{\pi}}\,\frac{G_{ss}(\omega)\exp(-j\omega D)}{\sqrt{R_{r2r2}(0)}} \qquad (8)$$

$$G_{y_jy_j}(\omega) = \mathrm{FT}\{\frac{2}{\pi}\sin^{-1}[\frac{R_{rjrj}(k)}{R_{rjrj}(0)}]\} \qquad j = 1, 2 \quad (9)$$

$$G_{y1y2}(\omega) = \mathrm{FT}\{\frac{2}{\pi}\sin^{-1}[\frac{R_{r1r2}(k)}{\sqrt{R_{r1r1}(0)\,R_{r2r2}(0)}}]\} \qquad j = 1, 2 \qquad (10)$$

where $R_{r_ir_j}(k)$, i,j=1,2 are the cross- and auto-correlation functions of $r_1(i)$, $r_2(i)$, and FT[.] is the Fourier Transform operator.

The variances of time delay estimators (3)-(5) for large SNR can be expressed by generalizing eq.(52) of [5] in the form:

$$\sigma^2 = \frac{2\pi}{N\displaystyle\int_{-\pi}^{+\pi} \frac{\omega^2\,|G_{\alpha\beta}(\omega)|^2}{G_{\alpha\alpha}(\omega)\,G_{\beta\beta}(\omega) - |G_{\alpha\beta}(\omega)|^2}\,d\omega} \qquad (11)$$

where $\alpha = r_1$, $\beta = r_2$ for DC, $\alpha = r_1$, $\beta = y_2$ for HS and $\alpha = y_1$, $\beta = y_2$ for PC. The derivation of the above result for DC-TDE in [5] is based on the Gaussianity of $\mathrm{FT}\{r_1(i)\}$, $\mathrm{FT}\{r_2(i)\}$ as a consequence of the Gaussianity of signal and noises. The same analysis can be extended to HS- and PC-TDE by observing that the FTs of $y_1(i)$, $y_2(i)$ approach a Gaussian probability distribution as well, no regarding that they are binary random variables. This is a good approximation for large N, when central-limit theorems can be invoked (see Appendix).

Eq.(11) gives the Cramer-Rao lower bounds (CRLB) for DC-TDE. It gives lower bounds for HS- and PC-TDE also, assuming that the available observation data are r_1, y_2 or y_1, y_2 respectively. Eq.(11) is a useful reference even for sub-optimal estimators, based on the computation of a few ccf lags.

No filtering has been taken into account in (3)-(5) and in (11). So, (3)-(5) are optimal estimators only if signal and noises exhibit a white spectrum (in this case, (3) is the maximum likelihood estimate of D). In practice, eq.(11) is a good approximation of TDE variance even for non-white signals and absence of filtering.

4. RESULTS

As mentioned before, a particular but significant case is when $s(i)$, $n_1(i)$, $n_2(i)$ are white noises in the frequency band $-\pi<\omega<\pi$. As a consequence, not only $r_1(i)$, $r_2(i)$ are white noises in the same band, but also $y_1(i)$, $y_2(i)$.

For this case, the plots of TDE variances are depicted in fig.1 as a function of SNR for N=750 data samples. From this, it is observed that PC outperforms HS for high SNR. Consider that for increasing SNR $y_1(i)$ coincides with $y_2(i)$ while $r_2(i)$ does not, i.e. the spectral 'coherence' as defined in [5] does not converge to unit for HS-TDE.

Simulation results based on 100 independent trials have been added to fig.1 by generating zero-mean Gaussian distributed sequences with variances σ_s^2, and $\sigma_{n1}^2=\sigma_{n2}^2=\sigma_n^2$. The ccf's (3)-(5) have been computed for a total of 256 lags (L=128) and interpolated by expanding with zeroes their 256-points DFTs, in order to obtain 16384-points ccf's. This corresponds to an oversampling factor of 64, and limits to a maximum of $\pm 1/32$ the error due to ccf sampling in peak detection.

From fig.1 the limit of validity of (11) is given by SNR>-5 dB about. On the other hand, consider that other bounds, valid at low SNR and taking into account the ambiguity problem, involve the computation of the CRLB.

Quite similar results were obtained for non-white signal and noises spectra and for non-Gaussian input distributions.

In fig.2 the estimation variances are evaluated at SNR=20 for different N. The general N^{-1} behavior, as deduced from (11), is immediately verified.

If the received signals $r_1(i)$, $r_2(i)$ are band-limited to $[-\omega_c, \omega_c]$, with $\omega_c<\pi$, eqs.(3)-(5) represent oversampled estimates of the ccf's. This gives a B^{-3} behavior, as shown in fig.3. For small B the theoretical curves of HS- and PC-TDE slightly overestimate the simulation results.

5. CONCLUSIONS

The accuracy of fast time delay estimators (4)-(5) has been evaluated in terms of signal and noises spectra (see eq.(11)). In particular, PC variance has been verified to decrease uniformly with increasing SNR while HS variance does not.

From our analysis, the effectiveness of HS- and PC- TDE can be evaluated as a trade-off between computational speed, complexity of computing architecture and estimation accuracy.

APPENDIX - GAUSSIAN APPROXIMATION FOR FOURIER TRANSFORMS

Let us consider the central limit theorem in the form given in [6], par.7.7-3, for two sets of independent random variables (rv's). Conditions 2,3 of [6] are satisfied for both normal- and binary-distributed samples $r_1(i)$, $r_2(i)$, $y_1(i)$, $y_2(i)$.

If $\alpha(1),\ldots,\alpha(N)$; $\beta(1),\ldots,\beta(N)$ are independent samples from signal pairs $[r_1(i), r_2(i)]$, $[r_1(i), y_2(i)]$ or $[y_1(i), y_2(i)]$, we state that the real and imaginary parts of their FT's:

$$A(\omega) = \sum_{n=1}^{N} \alpha(i) \exp(-j\omega n) \tag{A1}$$

$$B(\omega) = \sum_{n=1}^{N} \beta(i) \exp(-j\omega n) \tag{A2}$$

are asymptotically (joint) complex normal rv's for every frequency pair ω_1, ω_2. This can be demonstrated by decomposing the summations in (A1), (A2) into partial sums where the exponential factor is the same and can be taken out, so that the central limit theorem applies to each partial sum. Now, (A1),(A2) are linear transformations of (uncorrelated) normal terms and Gaussianity holds ([6], par.7.5).

Consider now the case of correlated samples, but suppose that the statistical dependence is negligible for a sample distance greater of or equal to M, with M<<N. Again, we decompose the FT computation in terms of normal contributions by picking up samples at a distance M from α and β alternatively. Finally, normality of the sum of (correlated) normal variates holds.

REFERENCES

[1] J.R. Jordan, "Correlation algorithms, circuits and measurement applications", *IEE Proc.*, vol.133, Pt.G, No.1, Feb. 1986.

[2] R. Price, "A useful theorem for nonlinear devices having Gaussian inputs", *IRE Trans. on IT* vol.IT-4, June 1958.

[3] J.H. Van Vleck, D. Middleton, "The spectrum of clipped noise", *Proc. IEEE*, vol.54(1), 1966.

[4] D. Hertz, "Time Delay Estimation by Combining Efficient Algorithms and Generalized Cross-Correlation Methods", *IEEE Trans. ASSP*, vol.ASSP-34, No.1, Feb. 1986.

[5] C.H. Knapp, G.C. Carter, "The Generalized Correlation Method for Estimation of Time Delay", *IEEE Trans. ASSP*, vol.ASSP-24, No.4, pp.320-327, Aug. 1976.

[6] D. Middleton, *An Introduction to Statistical Communication Theory*, New York: McGraw-Hill, 1960.

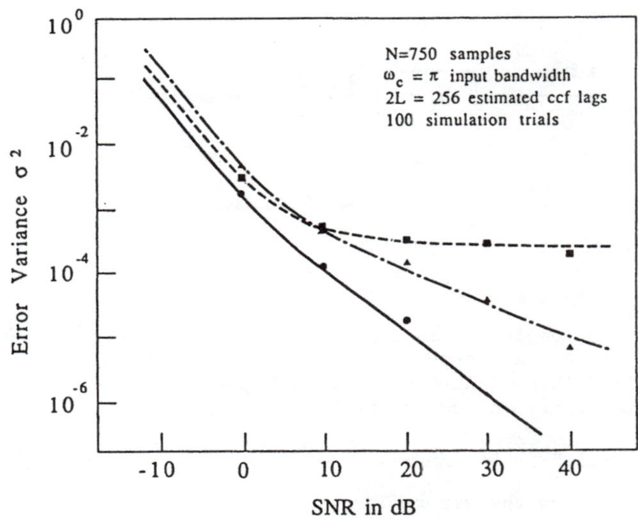

Fig.1 - Variances of DC (——), HS (- - -) and PC (—·—) TDE versus signal-to-noise ratio SNR for $G_{ss}(\omega)$ and $G_{n1n1}(\omega) = G_{n2n2}(\omega)$ constant in $-\pi<\omega<\pi$. Simulation results are also reported (● for DC, ■ for HS, ▲ for PC)

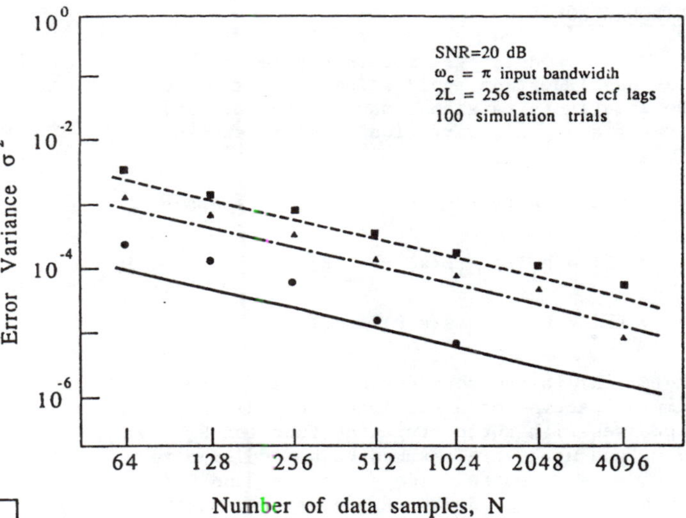

Fig.2 - Variances of DC (——), HS (- - -) and PC (—·—) TDE versus number of sample pairs N for $G_{ss}(\omega)$ and $G_{n1n1}(\omega) = G_{n2n2}(\omega)$ constant in $-\pi<\omega<\pi$. Simulation results are also reported (● for DC, ■ for HS, ▲ for PC)

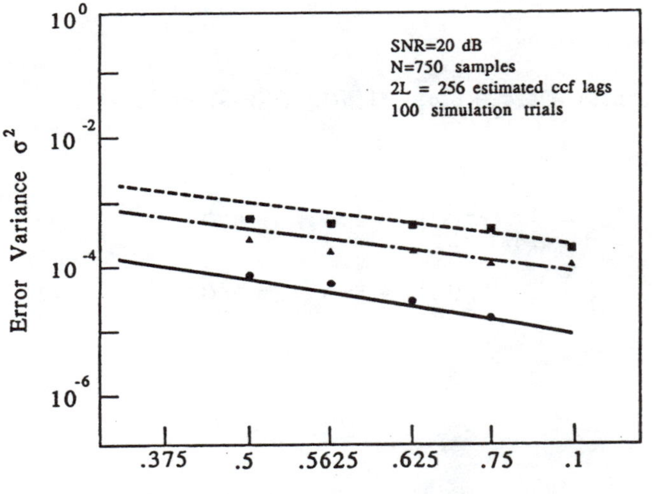

Fig.3 - Variances of DC (——), HS (- - -) and PC (—·—) TDE versus input bandwidth ω_c for $G_{ss}(\omega)$ and $G_{n1n1}(\omega) = G_{n2n2}(\omega)$ constant in $-\omega_c<\omega<\omega_c$. Simulation results are also reported (● for DC, ■ for HS, ▲ for PC)

A FINITE STEP ADAPTIVE TECHNIQUE FOR TIME-DELAY ESTIMATION

SOHEIL A. DIANAT
TAPAN K. SARKAR

Department of Electrical Engineering
Rochester Institute of Technology, Rochester, NY 14623
(716) 475-2165

ABSTRACT

The main objective of this paper is to estimate the time-delay between two-sensor outputs. The method used in this paper is based on minimization of the mean squared error, utilizing the Muller's technique. The method is an iterative procedure with a quadratic rate of convergence and the convergence is obtained in finite number of steps. The procedure is stable and easy to apply. The advantage of this technique is that it is a finite step iterative technique entirely in the time domain. The disadvantage of this method is that one has to store a vector of data instead of two data points obtained by the two-sensor outputs at each step.

INTRODUCTION

The time-delay estimation between two signals has important applications such as finding the location of a signal source, the arrival time of a signal, direction finding, and so on [1,2].

Consider a two-sensor system described by:

$$y_1(k) = h_1(k) * s(k) + n_1(k) \qquad (1)$$

$$y_2(k) = h_2(k) * s(k-D) + n_2(k) \qquad (2)$$

where s(k) is the source signal assumed to be a sample function of a stationary ergodic random processes with autocorrelation function $R_{ss}(\cdot)$. $n_1(k)$ and $n_2(k)$ are assumed to be additive zero mean Gaussian white noise processes, and D is the constant time-delay parameter to be estimated. $h_1(k)$ and $h_2(k)$ are the stable deterministic impulse responses of the transmitting media and sensors. It is also assumed that the noise signals $n_1(k)$ and $n_2(k)$ are independent of the source signal s(k). The block diagram of the model of interest is shown in Figure 1.

There are three different techniques developed so far for estimation of the time-delay parameter D, phase function data [3], cross correlation function [4,5] and time domain parameter estimation [6,7]. The first two approaches are frequency domain techniques. The parameter estimation is a time domain approach. In this paper we use a technique closely related to so-called ADDLD (Adaptive Digital Delay-Lock Discriminator) introduced by D. M. Etter [8].

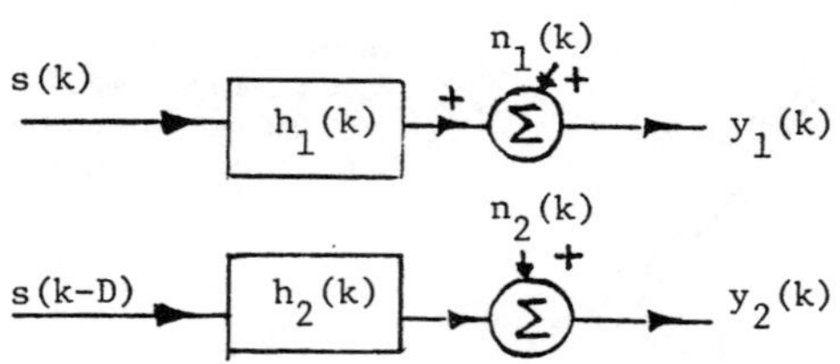

Figure 1

PROBLEM FORMULATION

Define the error signal e(k) as

$$e(k) = h_1(k) * y_2(k) - h_2(k) * y_1(k - \hat{D}) \qquad (3)$$

where $\hat{D}$ denotes the estimated delay and * stands for discrete convolution. The mean-squared error or cost function $J(D,\hat{D})$ is defined as

$$J(D, \hat{D}) = E[e^2(k)] \qquad (4)$$

By utilizing (1) and (2) in (4) we get

$$\begin{aligned} J(D,\hat{D}) &= E\{[h_1(k)*y_2(k) - h_2(k)*y_1(k-\hat{D})]^2\} \\ &= E\{[h_1(k)*h_2(k)*s(k-D) + h_1(k)*n_2(k) \\ &\quad - h_1(k)*h_2(k)*s(k-\hat{D}) - h_2(k)*n_1(k)]^2\} \end{aligned} \qquad (5)$$

which turns out to be

$$J(D,\hat{D}) = E_1\sigma_2^2 + E_2\sigma_1^2 + E\{[h(k)*(s(k-\hat{D})-s(k-D))]^2\} \qquad (6)$$

where

$$E_1 \triangleq \sum_{k=0}^{\infty} |h_1(k)|^2 = \text{Energy in } h_1(k) \qquad (7)$$

$$E_2 \triangleq \sum_{k=0}^{\infty} |h_2(k)|^2 = \text{Energy in } h_2(k) \qquad (8)$$

Reprinted from *Proc. Acoust., Speech, and Signal Processing Spectrum Estimation Workshop II*, pp. 67–69, November 1983.

σ_1^2 and σ_2^2 are the variances of $n_1(k)$ and $n_2(k)$ respectively, and

$$h(k) \triangleq h_1(k) * h_2(k) \quad (9)$$

denote the third term in equation (5) as $f(D,\hat{D})$ hence,

$$J(D,\hat{D}) = E_1\sigma_2^2 + E_2\sigma_1^2 + f(D,\hat{D}) \quad (10)$$

Therefore the problem is reduced to the following minimization problem:

$$J^* = \min_{\hat{D}} J(D, \hat{D}) = E_1\sigma_2^2 + E_2\sigma_1^2 + \min_{\hat{D}} f(D, \hat{D}) \quad (11)$$

where

$$f(D,\hat{D}) = E\{[h(k)*(s(k-D) - s(k-\hat{D}))]^2\} \quad (12)$$

Notice that $f(D,D) \geq 0$ for all values of $\hat{D}$ and $f(D,\hat{D}) = 0$. Now since $f(\cdot,\cdot)$ is a convex functional, then the solution to (11) is unique. The optimal solution is $\hat{D} = D$ and the optimal cost function J^* is

$$J^* = E_1\sigma_2^2 + E_2\sigma_1^2 \quad (13)$$

The solution is obtained by setting the gradient of J equal to zero i.e.

$$\nabla_{\hat{D}} J = \frac{\partial J(D, \hat{D})}{\partial \hat{D}} = \frac{\partial f(D, \hat{D})}{\partial \hat{D}} = 0 \quad (14)$$

The gradient $\nabla_{\hat{D}} J$ can be estimated as:

$$\hat{\nabla}_{\hat{D}} J(\hat{D}) = E\left\{2e(k)\ [h_2(k) * \frac{y_1(k+1-\hat{D})-y_1(k-\hat{D})}{1}]\right\}$$

$$= \frac{2}{M+1} \sum_{k=0}^{M} e(k)\left(h_2(k) * (y_1(k+1-\hat{D})-y_1(k-\hat{D}))\right) \quad (15)$$

where M is number of data points and

$$e(k) = h_1(k) * y_2(k) - h_2(k) * y_1(k-\hat{D}) \quad (16)$$

Equation (15) is now solved by Muller's method which is an iterative technique for finding the zero crossing of a functional. The Muller algorithm is iterative, converges quadratically and is global in the sense that one need not supply a good initial guess to obtain the solution. The sequence of steps required in Muller's technique are summarized as follows:

1) Let $\hat{D}_1$, $\hat{D}_2$ and $\hat{D}_3$ be three approximations to the zero of $\hat{\nabla}J(\hat{D})$. Compute $\hat{\nabla}J(\hat{D}_1)$, $\hat{\nabla}J(\hat{D}_2)$ and $\hat{\nabla}J(\hat{D}_3)$.

2) Compute

$$h_2 \triangleq \hat{\nabla}J(\hat{D}_3) - \hat{\nabla}J(\hat{D}_2) \quad (17)$$

$$h_1 \triangleq \hat{\nabla}J(\hat{D}_2) - \hat{\nabla}J(\hat{D}_1) \quad (18)$$

$$g(\hat{D}_3, \hat{D}_2) \triangleq \frac{1}{h_2} [\hat{\nabla}J(\hat{D}_3) - \hat{\nabla}J(\hat{D}_2)] \quad (19)$$

$$g(\hat{D}_2, \hat{D}_1) \triangleq \frac{1}{h_2} [\hat{\nabla}J(\hat{D}_2) - \hat{\nabla}J(\hat{D}_1)] \quad (20)$$

3) Compute the following expressions for $i \geq 2$

$$Z(\hat{D}_i, \hat{D}_{i-1}, \hat{D}_{i-2}) = \frac{g(\hat{D}_i,\hat{D}_{i-1}) - g(\hat{D}_{i-1},\hat{D}_{i-2})}{h_i + h_{i-1}} \quad (21)$$

and

$$C_i = h_i Z(\hat{D}_i, \hat{D}_{i-1}, \hat{D}_{i-2}) + g(\hat{D}_i, \hat{D}_{i-1}) \quad (22)$$

4) Compute

$$h_{i+1} = \frac{-2\ \hat{\nabla}J(\hat{D}_i)}{C_i \pm \sqrt{C_i^2 - 4\hat{\nabla}J(\hat{D}_i)\ Z(\hat{D}_i,\hat{D}_{i-1},\hat{D}_{i-2})}} \quad (23)$$

The sign $\pm$ is selected so that the denominator has the largest magnitude.

5) Set

$$\hat{D}_{i+1} = \hat{D}_i + h_{i+1} \quad (24)$$

6) Compute

$$g(\hat{D}_{i+1}, \hat{D}_i) = \frac{\hat{\nabla}J(\hat{D}_{i+1}) - \hat{\nabla}J(\hat{D}_i)}{h_{i+1}} \quad (25)$$

7) Set i = i+1 and repeat steps 3 through 7 until either of the following error criterion is satisfied.

a) $|\hat{D}_i - \hat{D}_{i-1}| < \varepsilon_1$ (26)

b) $|\hat{\nabla}J(\hat{D}_i)| < \varepsilon_2$ (27)

where ε_1 and ε_2 are two prescribed small positive numbers.

SIMULATION RESULTS

In this section three typical examples are presented for estimating the time-delay between two sensors outputs. The source signal s(k) is held unchanged. However, to generate different types of signals, we change $h_1(k)$ and $h_2(k)$, the three cases simulated corresponds to the following situations.

1) $H_1(j\omega) = H_2(j\omega) = 1$ (all pass signal) with time domain functions $h_1(k) = h_2(k) = \delta(k)$.

2) $H_1(j\omega) = H_2(j\omega) = u(\omega+\omega_o) - u(\omega-\omega_o)$ (low pass signal) with time domain functions

$$h_1(k) = h_2(k) = A \frac{\sin \omega_o k}{\omega_o k} \quad (28)$$

3) $H_1(j\omega) = H_2(j\omega) = u(\omega+\omega_o+\omega_1) - u(\omega+\omega_1-\omega_o)$

$$+ u(\omega-\omega_1+\omega_o) - u(\omega-\omega_1-\omega_o)$$

(Band pass signal) with time domain functions

$$h_1(k) = h_2(k) = A \frac{\sin \omega_o k}{\omega_o k} \cos \omega_1 k \quad (29)$$

It was observed that the algorithm could estimate the time-delay parameter D if the condition $D \leq D_{max}$ is satisfied where D_{max} is the first local minimum of $h_1(k)$. For a low pass signal $D_{max} = 0.75/f_o$ and for a bandpass signal $D_{max} = 0.5/f_1$. If the above condition is satisfied the estimate of D is approximately unbiased for a signal to noise ratio greater than 15 dB.

CONCLUSION

A technique based on Muller's algorithm is presented for estimating the time-delay between two received signals. The method is entirely a time domain problem similar to ADDLD (Adaptive Digital Delay-Lock Discriminator) the advantage of this technique is that it is a finite step iterative technique. The estimate is fairly unbiased if SNR is greater than 15 dB. Like ADDLD the method is useful if the autocorrelation function of the signal ($R_{ss}(\tau)$) is unimodal in the region when $\tau < D_{max}$.

REFERENCES

[1] Y. T. Chan, Richard V. Hattin and J. B. Plant, "The Least Squares Estimation of Time-Delay and its Use in Signal Detection," IEEE Trans. on Acoustics, Speech and Signal Processing, Vol. ASP-26, No. 3, pp. 217-222, 1978.

[2] G. Clifford Carter, "Time Delay Estimation for Passive Sonar Signal Processing," IEEE Trans. on Acoustics, Speech and Signal Processing, Vol. ASSP-29, No. 3, pp. 463-470, 1981.

[3] Allan G. Piersol, "Time Delay Estimation Using Phase Data," IEEE Trans. on Acoustics, Speech and Signal Processing, Vol. ASSP-29, No. 3, pp. 471-477, 1981.

[4] Charles H. Knapp and G. Clifford Carter, "The Generalized Correlation Method for Estimation of Time-Delay," IEEE Trans. on Acoustics, Speech and Signal Processing, Vol. ASP-24, No. 4, pp. 320-327, 1976.

[5] William H. Hass and Claude S. Linquist, "A Synthesis of Frequency Domain Filters for Time-Delay Estimation," IEEE Trans. on Acoustics, Speech and Signal Processing, Vol. ASP-29, No. 3, pp. 540-548, 1981.

[6] Y. T. Chan, J.M.F. Riley and J. B. Plant, "A Parameter Estimation Approach to Time-Delay Estimation and Signal Detection," IEEE Trans. on Acoustics, Speech and Signal Processing," Vol. ASP-28, No. 1 pp. 8-15, 1980.

[7] Y. T. Chan, J.M.F. Riley and J. B. Plant, "Modeling of Time-Delay and its Application to Estimation of Nonstationary Delays," IEEE Trans. on Acoustics, Speech and Signal Processing, Vol. ASP-29, No. 3, pp. 577-581, 1981.

[8] D. M. Etter and S. D. Stearns, "Adaptive Estimation of Time Delays in Sampled Data Systems," IEEE Trans. on Acoustics, Speech and Signal Processing, Vol. ASP-29, pp. 582-587, 1981.

Improved Time-Delay Estimates of Underwater Acoustic Signals Using Beamforming and Prefiltering Techniques

BRIAN G. FERGUSON

Abstract—**Passive sonar systems that localize broadband sources of acoustic energy estimate the differences in the arrival times (or time delays) of an acoustic wavefront at spatially separated hydrophones. The output amplitudes from a given pair of hydrophones are cross correlated and an estimate of the time delay is given by the time lag that maximizes the cross-correlation function. Often the time-delay estimates are corrupted by the presence of noise. By replacing each of the omnidirectional hydrophones with an array of hydrophones, and then cross correlating the beamformed outputs of the arrays, we show that the effect of noise on the time-delay estimation process is dramatically reduced. Both conventional and adaptive beamforming methods are implemented in the frequency domain, and the advantages of array beamforming (prior to cross correlation) are highlighted using both simulated and real noise-field data. Further improvement in the performance of the broadband cross-correlation processor occurs when various prefiltering algorithms are invoked. By suppressing the adjacent peaks in the observed cross-correlation function, prefiltering is shown to facilitate the identification of the peak associated with the correct differential time delay in both single path and multipath propagation environments. Each of the prefilters—Hannan-Thomson, smoothed coherence transform, and phase transform—consists of a frequency-domain weighting function which is applied to the cross-spectral density function before the cross-correlation function is computed using the inverse Fourier transform.**

I. Introduction

PASSIVE-ranging sonar systems localize broadband sources of acoustic energy by using methods that depend on the proximity of the source to the receivers. Spherical/cylindrical spreading of the acoustic energy emanating from a source leads to circular wavefronts, with the wavefront curvature being directly related to the distance from the source. With the wavefront curvature method [1], the range and bearing of a nearby source are calculated from estimates of the relative delays in the wavefront arrival times at three or more hydrophones, which are spatially distributed along a baseline. For a distant source, however, the wavefronts tend to be planar, and only the bearing of the source can be measured directly, which requires a minimum of two spatially separated hydrophones. Passive ranging by measurement of wavefront curvature is not appropriate for the far-field case, so the range of a distant source must be determined by various track-motion analysis methods.

A common method for measuring the time delay involves cross correlating the receiver outputs; an estimate of the time delay is given by the argument that maximizes the cross-correlation function. The process is often complicated by the received signal-to-noise ratio (SNR) being small, a changing time delay owing to motion of the source relative to the receivers' baseline, the received signal having a limited bandwidth, and multipath propagation which occurs whenever there is more than one propagation path between the source and the receivers.

This paper considers the use of both array beamforming and prefiltering techniques to enhance only that peak of the cross-correlation function associated with the correct differential time delay. These techniques are applied to an existing cross-correlation system with the aim of improving its time-delay estimation performance.

II. Background

A. Time-Delay Estimation [2]

Consider the signal from a remote source being received in the presence of noise at two spatially separated receivers, as illustrated in Fig. 1(a). The time histories of the receiver outputs, denoted by $r_1(t)$ and $r_2(t)$, are given by

$$r_1(t)=s(t)+n_1(t)$$
$$r_2(t)=as(t-D_{12})+n_2(t) \qquad (1)$$

where $s(t)$ is the signal waveform, $n_1(t)$ and $n_2(t)$ are the noise waveforms at the respective receivers, a is an attenuation factor, and D_{12} is the difference in the wavefront arrival times at the two receivers. For the simple case where the signal and noise terms are all mutually uncorrelated, the cross-correlation function between the received signals is given by

$$R_{12}(\tau)=E[r_1(t)r_2(t+\tau)]=aR_{ss}(\tau-D_{12}) \qquad (2)$$

where R_{ss} is the autocorrelation function of $s(t)$ transposed to have its peak value at a time delay $t = D_{12}$, and $E[\ ]$ denotes the expectation value of the quantity within the brackets. The peak of the cross-correlation function occurs at the time delay

$$D_{12}=\Delta d/c \qquad (3)$$

where $\Delta d = d_2 - d_1$ is the path length difference, and c is the speed at which the signal propagates in the medium.

The locus of points that satisfies a constant time-delay constraint is an hyperbola. As illustrated in Fig. 1(b), the time

Manuscript received June 10, 1988; revised January 3, 1989.

The author is with the Maritime Systems Division, Defence Science and Technology Organization (Sydney), P. O. Box 706, Darlinghurst, New South Wales 2010, Australia.

IEEE Log Number 8927230.

Reprinted from *IEEE J. Oceanic Eng.*, vol. 14, no. 3, pp. 238–244, July 1989.

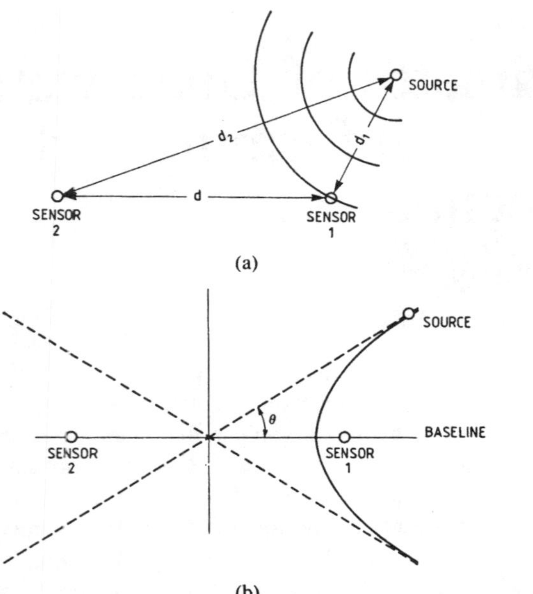

Fig. 1. (a) Acoustic source and sensors. (b) Source location and bearing-angle interpretation.

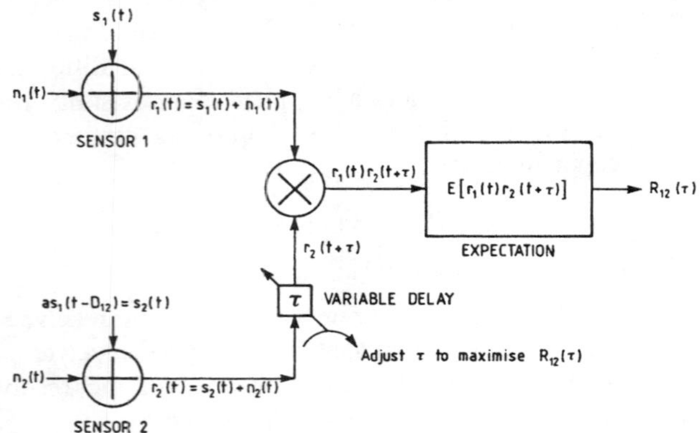

Fig. 2. Time-domain processor for estimating the differential time delay between two sensors.

delay D_{12} defines the particular hyperbola on which the source must lie. As the range of the acoustic source increases, the angle θ that the hyperbolic asymptote makes with the baseline becomes an increasingly accurate approximation to the actual bearing of the source (relative to the midpoint of the baseline). When the source is sufficiently distant for the wavefronts to be approximated by plane waves at the receiver locations, the time delay can be interpreted as a bearing angle, which is given by

$$\theta = \cos^{-1}(cD_{12}/d) \tag{4}$$

where d is the distance between the receivers.

A block diagram of a time-domain cross-correlation processor is shown in Fig. 2. Alternatively, the cross-correlation processor can be implemented in the frequency domain. The Fourier transform of the cross-correlation function is the cross-spectral density function which is given by

$$S_{12}(f) = \int_{-\infty}^{\infty} R_{12}(\tau)e^{-j2\pi f\tau}\, d\tau \tag{5}$$

where $S_{12}(f)$ is the two-sided cross-spectral function. Alternatively,

$$G_{12}(f) = 2\int_{-\infty}^{\infty} R_{12}(\tau)e^{-j2\pi f\tau}\, d\tau = aG_{ss}(f)e^{-j2\pi fD_{12}} \tag{6}$$

where $G_{12}(f)$ is the one-sided cross-spectral density function for which

$$G_{12}(f) = \begin{cases} 2S_{12}(f) & \text{for } 0<f<\infty \\ 0 & \text{otherwise} \end{cases} \tag{7}$$

and $G_{ss}(f)$ is the one-sided autospectral density function of $s(t)$. Hence, the time delay D_{12} appears in the cross spectrum as a phase function which is given by

$$\phi_{12}(f) = 2\pi f D_{12}. \tag{8}$$

Since multiplication in one domain is a convolution in the transformed domain, it follows that

$$R_{12}(\tau) = aR_{ss}(\tau) * \delta(\tau - D_{12}) \tag{9}$$

where $*$ denotes convolution and δ denotes the delta function. This equation implies that the delta function is spread by the Fourier transform of the signal spectrum. In the hypothetical case of white noise[1] for which the autospectral density function is uniform over all frequencies by definition, the Fourier transform of the signal spectrum is a delta function, and no spreading occurs. For bandwidth-limited white noise, the Fourier transform of the signal spectrum is a cosine function modulated by a sinc function, so the spreading increases as the bandwidth decreases. As the signal bandwidth becomes an increasingly smaller fraction of the center frequency, the peaks in the cross-correlation function tend to have the same magnitude, and so identification of the peak associated with the correct time delay becomes increasingly prone to ambiguity. In the limiting case of a vanishingly small bandwidth, the cross-correlation function is purely oscillatory, with a period equal to the reciprocal of the center frequency. The autospectral density functions and the associated autocorrelation functions for these three cases are given in Table I.

B. Beamforming [3]

The adverse effect of noise on the time-delay estimation process can be reduced by replacing each receiver with an array of receivers and then beamforming to improve the SNR. Beamforming involves appropriately weighting the outputs of the receivers prior to summation. The configuration for a frequency-domain beamformer is shown in Fig. 3. The output amplitude of the beamformer is given by

$$A(f, \theta) = w^H(f, \theta)R(f) \tag{10}$$

where $w(f, \theta)$ is the weighting vector for a frequency f and a beamsteer direction θ, the superscript H denotes transposition and complex conjugation, and $R(f)$ is a column vector which has the complex narrowband outputs of the receivers as its elements.

[1] White noise is a stationary random process with a constant autospectral density function.

TABLE I
SPECIAL AUTOSPECTRAL DENSITY FUNCTIONS AND THEIR FOURIER TRANSFORM PAIRS*

Type	One-Sided Autospectral Density Function	Autocorrelation Function
White Noise	$G_{ss}(f) = \begin{cases} \alpha & \text{if } f \geq 0 \\ 0 & \text{otherwise} \end{cases}$	$R_{ss}(\tau) = \alpha\delta(\tau)$
Bandwidth-Limited White Noise	$G_{ss}(f) = \begin{cases} \alpha & \text{if } 0 < f_o - (B/2) \leq f \text{ and } f \leq f_o + (B/2) \\ 0 & \text{otherwise} \end{cases}$	$R_{ss}(\tau) = \alpha B(\frac{\sin \pi B\tau}{\pi B\tau}) \cos(2\pi f_o \tau)$
Sine Wave	$G_{ss} = \beta\delta(f - f_o)$	$R_{ss}(\tau) = \beta \cos(2\pi f_o \tau)$

* α and β are constants and f_0 is the center frequency of a rectangular filter of bandwidth B.

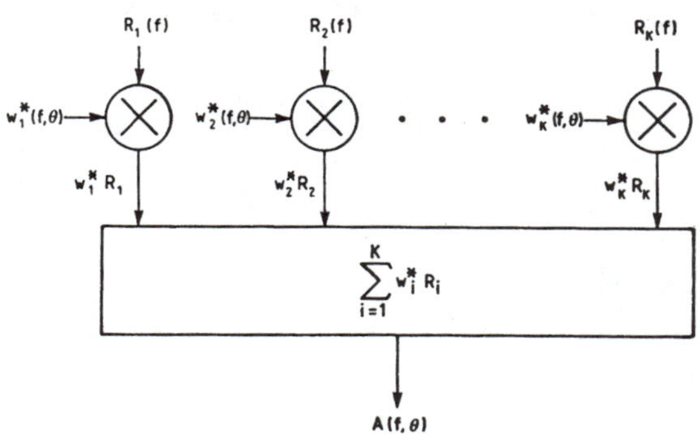

Fig. 3. Frequency-domain beamformer.

For a conventional frequency-domain beamformer, which is equivalent to the delay-and-sum beamformer in the time domain, the weight vector is given by

$$w_c(f, \theta) = v(f, \theta)/K \tag{11}$$

where $v(f, \theta)$ is the vector of phase delays required to bring all the receiver outputs of the array into phase before summing, and K is the number of receivers in the array. For a line array of equispaced receivers, the ith element of $v(f, \theta)$ is given by

$$v_i(f, \theta) = e^{-j2\pi il \cos \theta/\lambda} \tag{12}$$

where l is the distance between the sensors in the array, and $\lambda = c/f$ is the wavelength of the plane waves. For an unconstrained optimal beamformer, the set of weights that maximizes the output SNR by minimizing the beamformer's output noise power is given by

$$w_o(f, \theta) = Q_N^{-1}(f)v(f, \theta) \tag{13}$$

where $Q_N(f)$ is the cross-power spectral matrix of the noise, normalized so that all its diagonal elements are unity. The unconstrained optimal beamformer uses prior knowledge of the noise cross-power spectral matrix to suppress the noise. Unfortunately, Q_N is not usually known and needs to be estimated by measuring the array cross-power spectral matrix when the signal is absent. However, if the estimate of Q_N inadvertently contains a contribution due to the wanted signal, then the signal itself will be treated as noise and will also be suppressed. The constrained optimal (or adaptive) beamformer overcomes this difficulty be selecting the set of weights, denoted w_a, so that the output power of the beamformer is minimized while the polar response in the direction of the desired signal is constrained to unity. The adaptive weight vector is given by

$$w_a(f, \theta) = C^{-1}(f)v(f, \theta)/[v^H(f, \theta)C^{-1}(f)v(f, \theta)] \tag{14}$$

where $C(f)$ is the cross-power spectral matrix of the signal plus noise, not normalized. The ijth element of $C(f)$ is given by

$$C_{ij}(f) = E[R_i(f)R_j^*(f)]. \tag{15}$$

III. TIME-DELAY ESTIMATION PROCESSOR

Now we apply the theory (presented in the previous sections) to the problem of estimating the difference in the arrival times of a broadband signal at two spatially separated arrays of sensors. In the present case, conventional and adaptive beamforming techniques are applied to two spatially separated linear arrays, with each array having five sensors. The beamformed outputs are then cross correlated, and the cross-correlation functions associated with each beamforming technique are compared for both a simulated noise field and an actual noise field. In addition, for the real noise field case, the cross-correlation function for two single sensors located at the array midpoints is presented for comparison with the cross-correlation function for the beamformed outputs of the arrays. The separation distance between the arrays is considerably larger than the length of each array. A block diagram of the time-delay estimation processor is shown in Fig. 4. The beamforming and cross correlation are implemented in the frequency domain, with the time-domain cross-correlation function being computed from the cross-spectral density function using the inverse Fourier transform.

A. Simulated Acoustic Field

The simulated acoustic field consists of a 20-dB source of plane-wave interference at a bearing of 125°, a 0-dB signal source at a bearing of 150°, and −30 dB of uncorrelated noise. The beamsteer direction is 150°, corresponding to the signal-source bearing. Fig. 5 shows the normalized cross-correlation functions for a bandwidth-limited signal having a rectangular ("top-hat") frequency distribution whose wavelength range extends from 10 to 20 l, where l is the separation distance between adjacent sensors in an array. With conventional beamforming, the strong interference masks the weak signal source (Fig. 5(a)), and so the peak value occurs at a time displacement that corresponds to the bearing of the plane-wave interference (125°). The adaptive beamformer, however, suppresses the interference and reveals the presence of the signal source at a delay that corresponds to a bearing of 150° (Fig. 5 (b)). Conceptually, the adaptive beamformer can be thought of as using the cross terms in the cross-power spectral matrix (C_{ij}, $i \neq j$) to sense the interference direction, and then using a set of complex weights to shade the array so that a null is steered in this direction, thus suppressing the plane-wave interference [4]. Fig. 6 shows the polar response for both the conventional beamformer and the adaptive beamformer at a wavelength of about 10 l. As expected, the adaptive beamformer has steered a null in the direction of the

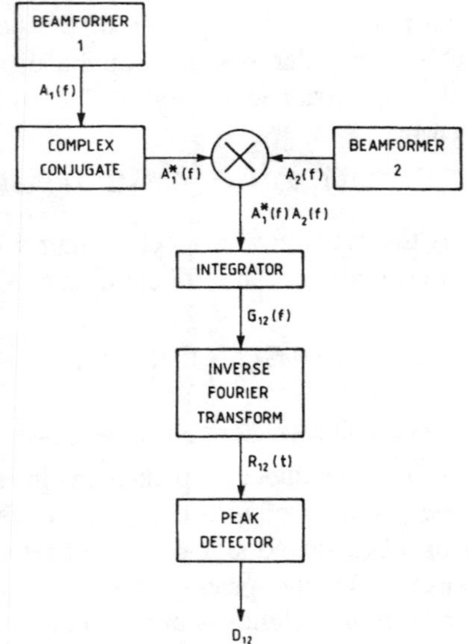

Fig. 4. Frequency-domain processor for estimating the differential time delay between two arrays.

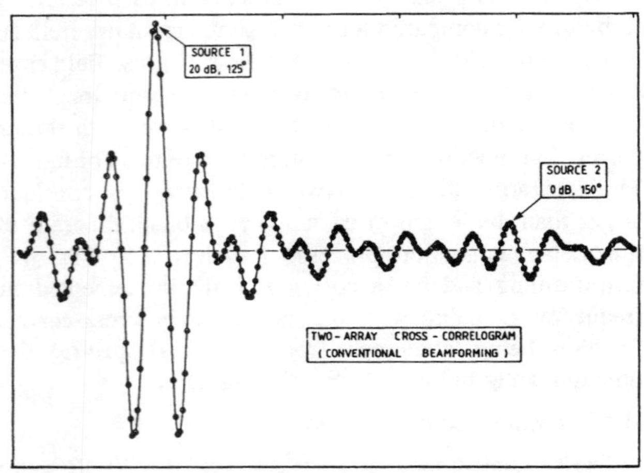

(a)

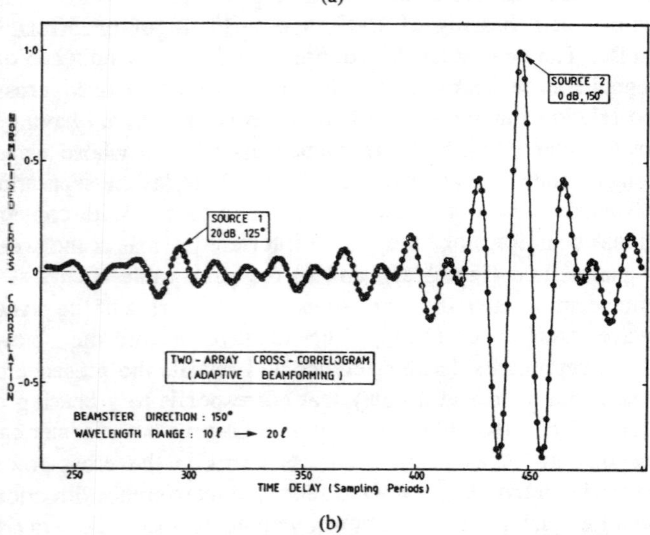

(b)

Fig. 5. Variation with time delay of the normalized cross correlation between the outputs of two conventional beamformers (a), and two adaptive beamformers (b), in a simulated acoustic field. The range of wavelengths used for the cross correlation extends from 10 to 20 l, where l is the separation distance between adjacent sensors in each of the five-element arrays.

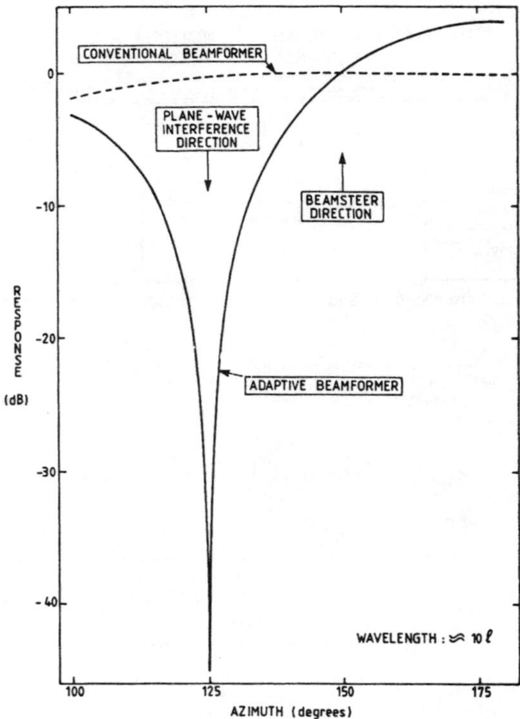

Fig. 6. Responses of a conventional beamformer and a constrained adaptive beamformer as a function of azimuth (bearing) for a five-element linear array located in a simulated acoustic field. The wavelength is about 10 l, where l is the separation distance between adjacent sensors in the array.

plane-wave interference (125°), while its response is unity in the direction of the signal (150°), which is the beamsteer direction in this case. The conventional beamformer, on the other hand, offers only marginal rejection in comparison.

B. Real Acoustic Noise Field

Next, the source of plane-wave interference is replaced by noise from the sensors of two horizontally separated arrays mounted on the hull of a submarine. The noise field has two main components: Ownship noise (or self-noise) on the sonar arrays which can be either electrical, mechanical, or hydrodynamic in origin, and ambient noise which consists of a mixture of noise generated by wind and wave action at the ocean surface, biological noise, and shipping noise. The normalized cross correlation between the outputs of two single sensors (the middle sensor of each array was selected) is shown in Fig. 7 as a function of the time delay (Fig. 7(a)—no beamforming). The effect of noise on the time-delay estimation process is dramatically reduced after the array beamforming techniques are implemented, using all five sensors of each array (Fig. 7(b)—conventional beamforming; Fig. 7(c)—adaptive beamforming).

IV. Prefiltering

The cross correlation function between $r_1(t)$ and $r_2(t)$ is related to the cross-spectral density function by the Fourier transform relationship

$$R_{12}(\tau) = \int_{-\infty}^{\infty} S_{12}(f) e^{j2\pi f t}\, df. \tag{16}$$

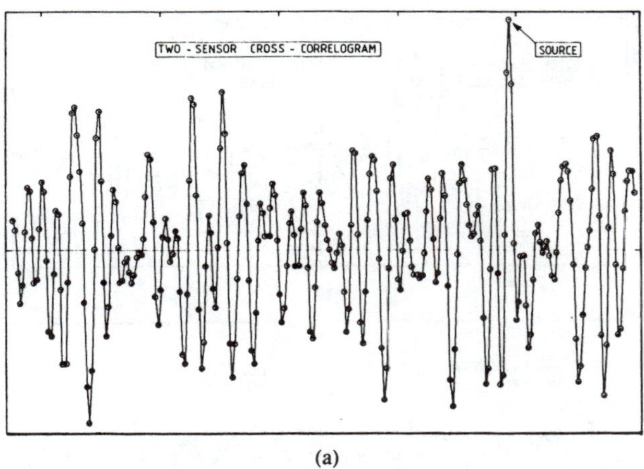

(a)

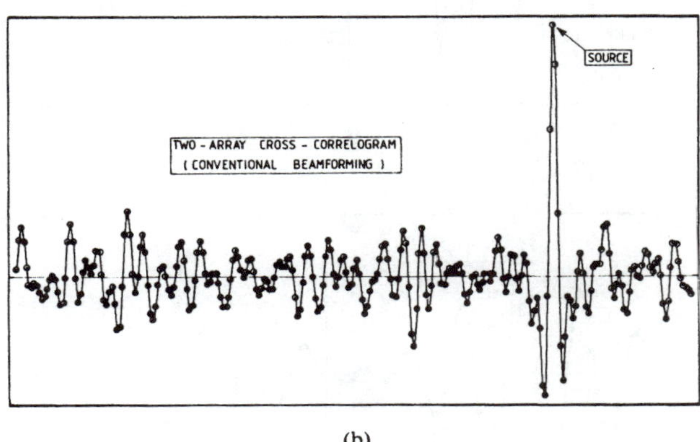

(b)

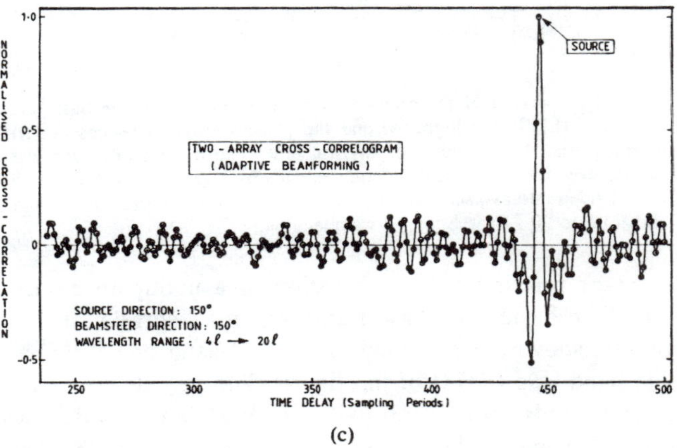

(c)

Fig. 7. Variation with time delay of the normalized cross correlation between (a) two sensors, and (b) a conventional beamforming array, and (c) a constrained adaptive beamforming array for a simulated signal source in a real noise field consisting of ownship noise (or self-noise) and ambient noise. Each array has five sensors. The range of wavelengths used for the cross correlation extends from 4 to 20 l, where l is the separation distance between adjacent sensors in an array.

For the cross-correlation function, the peak associated with the correct differential time delay can be accentuated by prefiltering the received waveforms. In the time domain, prefiltering refers to the filtering of the receiver outputs $r_1(t)$ and $r_2(t)$ prior to delay, multiplication, and integration, while in the frequency domain, prefiltering is equivalent to applying a window or weighting function to the cross-spectral density function. This leads to the generalized cross-correlation

TABLE II
CANDIDATE PREFILTERS AND THEIR FREQUENCY DOMAIN WEIGHTING FUNCTIONS*

Processor Name	Weighting Function $W(f)$
Basic Cross-Correlator	1
Smoothed Coherence Transform	$[S_{11}(f)S_{22}(f)]^{-1/2}$
Phase Transform	$\|S_{12}(f)\|^{-1}$
Hannan-Thomson	$\|\gamma_{12}(f)\|^2/[\|S_{12}(f)\|\{1 - \|\gamma_{12}(f)\|^2\}]$ where $\|\gamma_{12}(f)\|^2 = \|S_{12}(f)\|^2/[S_{11}(f)S_{22}(f)]$ is the magnitude-squared coherence

* $S_{11}(f)$ and $S_{22}(f)$ are the respective autospectral density functions, S_{12} is the cross-spectral density function, and the complex coherence function, denoted $\gamma_{12}(f)$, is the normalized cross-spectral density function, where $\gamma_{12}(f) = |\gamma_{12}(f)|e^{-j\phi_{12}(f)}$.

function, which is given by

$$R_{12}^{G}(\tau) = \int_{-\infty}^{\infty} W(f) S_{12}(f) e^{j2\pi f\tau}\, df \qquad (17)$$

where $W(f) = H_1(f)H_2^*(f)$ denotes an appropriately selected frequency weighting function, with $H_1(f)$ and $H_2(f)$ denoting the filter transfer functions which have the same phase because $W(f)$ is real-valued.

In practice, only an estimate of the cross-spectral density function can be obtained from finite observations of $r_1(t)$ and $r_2(t)$. Indeed, depending on the particular form of $W(f)$ and the *a priori* information, it may also be necessary to estimate $W(f)$. For example, when the role of the prefilters is to accentuate the signal passed to the correlator at those frequencies at which the SNR is highest, then $W(f)$ can be expected to be a function of the signal and noise spectra, which must be either known *a priori* or estimated.

Various prefilters have been proposed to optimize certain performance criteria for the cross correlator, thereby improving the time-delay estimation process [5]. Information on the candidate prefilters considered here is given in Table II. For the basic cross-correlation processor (no prefiltering), the frequency-domain weighting function is unity for all frequencies, and the cross-correlation function is a delta function convolved with the Fourier transform of the signal spectrum. The smoothed coherence transform can be interpreted as two pre-whitening filters, where $|H_1(f)| = |S_{11}(f)|^{-1/2}$ and $|H_2(f)| = |S_{22}(f)|^{-1/2}$, followed by a cross correlator. This method was developed to counteract the undesirable effects of strong tonals in broadband signals. The phase transform processor uses only the cross-spectral phase information for the time-delay estimation, and it was developed purely as an *ad hoc* technique. The Hannan–Thomson processor provides the maximum-likelihood estimate of the time delay for bandlimited random signals corrupted by white noise. This processor weights the cross-spectral phase according to the strength of the coherence, and it attaches most weight to the estimated cross-spectral phase when the variance of the estimated phase error is least.

The relative performance of the candidate prefilters is assessed by processing data from two small arrays fitted to the hull of a submarine. The beamformed outputs of the arrays are prefiltered and cross correlated, with the (conventional)

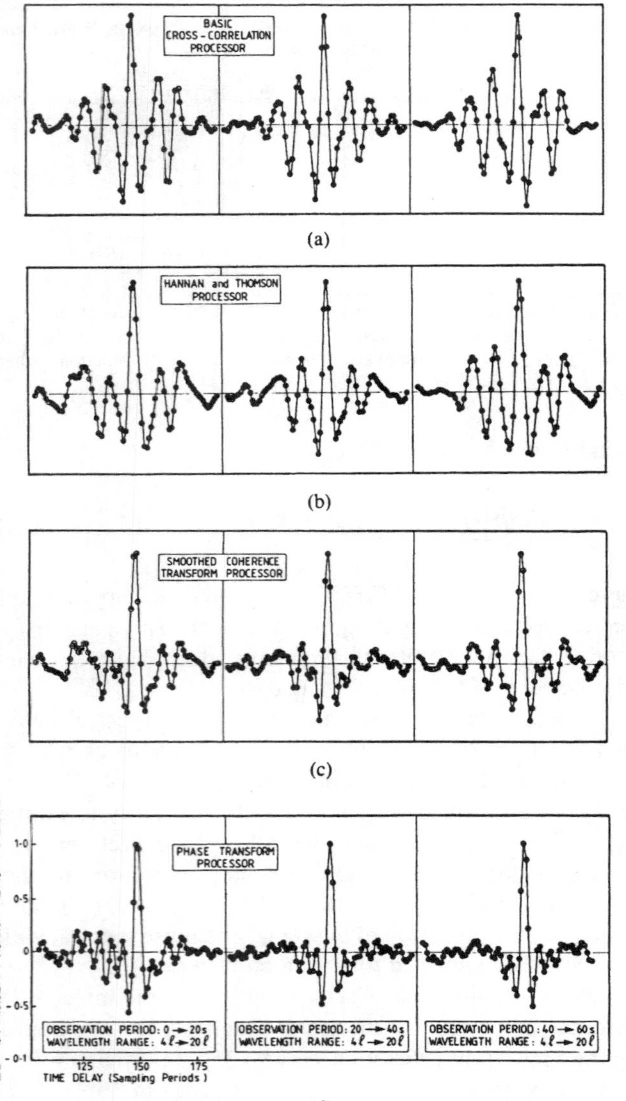

(a)

(b)

(c)

(d)

Fig. 8. Variation with time delay of the normalized cross correlation between the outputs of two arrays resulting from the processing of actual experimental data by each of the candidate processors. The range of wavelengths used for the cross correlation extended from 4 to 20 *l*, where *l* is the separation distance between adjacent sensors in an array.

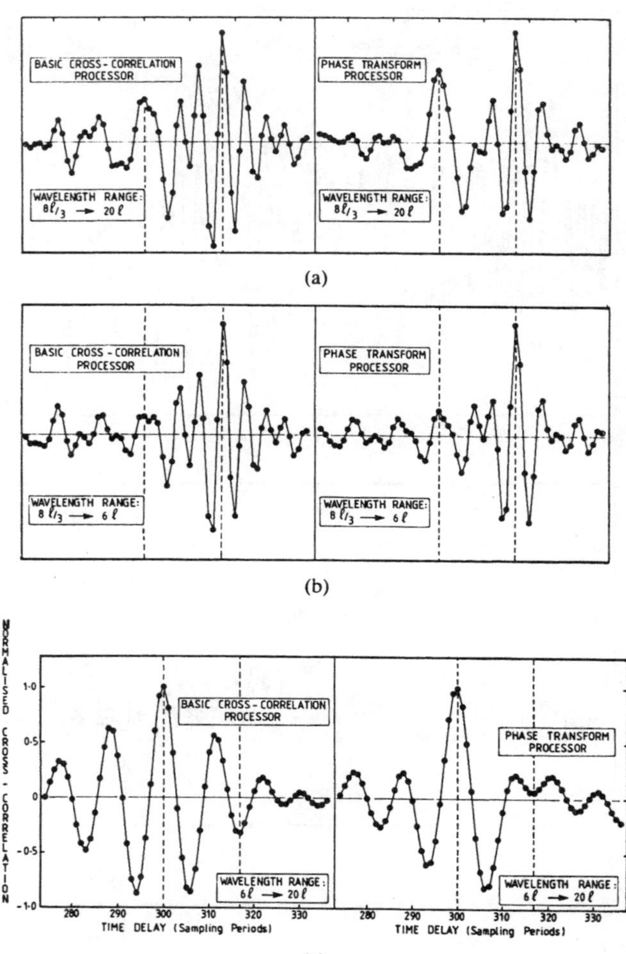

(a)

(b)

(c)

Fig. 9. Comparison of the cross-correlation functions for the basic cross correlator (LHS correlograms) and the phase-transform processor (RHS correlograms) for a multipath environment. The overall range of wavelengths used for the cross correlation extends from (a) 8*l*/3 to 20 *l*, which is then divided into two subranges: (b) 8*l*/3 to 6 *l* for the shorter wavelengths, and (c) 6 to 20 *l* for the longer wavelengths.

beamforming, prefiltering, and cross correlation all being implemented in the frequency domain. The time-domain cross-correlation functions are computed from the appropriately weighted cross-spectral density functions using an inverse Fourier transform routine. Due to the practical limitation of a finite observation time, only estimates of the autospectral density functions and the cross-spectral density function are used to compute the prefilter weighting functions. Each observation period is 20 s, which corresponds to 200 integrations. The acoustic source for the experiment was a merchant vessel located several kilometers from the submarine. Fig. 8 shows the results of processing the data with each of the prefilters. The prefilters accentuate the peak associated with the correct differential time delay, with the phase transform processor providing the best performance.

Fig. 9 demonstrates the usefulness of the phase transform processor for time-delay estimation in a multipath environment. The basic and phase transform cross-correlation functions are shown, respectively, on the left-hand side (LHS) and right-hand side (RHS) of the figure. The overall wavelength range extends from 8*l*/3 to 20 *l* (Fig. 9(a)), which was subdivided into a short wavelength range: 8*l*/3 to 6 *l* (Fig. 9(b)), and a long wavelength range: 6 to 20 *l* (Fig. 9(c)). The acoustic energy from the source arrives via two propagation paths, each propagation path being characterized by a specific set of wavelengths and a specific time delay. The energy at the longer wavelengths (lower frequencies) propagates via the path with the shorter differential time delay. The phase transform processor is preferred to the basic cross-correlation processor for distinguishing between the different propagation paths.

V. Conclusions

Time-delay estimation using the cross-correlation method is greatly improved by cross correlating the beamformed outputs of two arrays of receivers, rather than just cross correlating the outputs of two single receivers.

Prefiltering accentuates the peak of the cross-correlation function associated with the correct differential time delay, with the phase transform processor providing the best performance of the candidate processors for both single-path and multipath propagation conditions.

Acknowledgment

The author gratefully acknowledges the discussions on beamforming with Dr. A. Steele and Dr. D. Gray, Maritime Systems Division (Adelaide).

References

[1] G. C. Carter, "Time-delay estimation for passive sonar processing," *IEEE Trans. Acoust., Speech, Signal Processing,* vol. ASSP-29, pp. 463–470, June 1981.

[2] A. G. Piersol, "Time-delay estimation using phase data," *IEEE Trans. Acoust., Speech, Signal Processing,* vol. ASSP-29, pp. 471–477, June 1981.

[3] H. A. d'Assumpcao and G. E. Mountford, "An overview of signal processing for arrays of receivers," *J. EEE Aust.,* vol. 4, pp. 6–19, Mar. 1984.

[4] D. A. Gray, "Some applications of parametric and non-parametric spectral estimation techniques to passive sonar data," *J. EEE Aust.,* vol. 5, pp. 112–119, June 1985.

[5] C. H. Knapp and G. C. Carter, "The generalized correlation method for estimation of time delay," *IEEE Trans. Acoust., Speech, Signal Processing,* vol. ASSP-24, pp. 320–327, Aug. 1976.

Comparison of Various Time Delay Estimation Methods by Computer Simulation

ANTONI FERTNER AND ANDERS SJÖLUND

Abstract—**This correspondence provides qualitative estimates of the magnitude of the error in the measured delay time resulting from the error on the observed cross-correlation curve. The variances of five time delay estimators have been obtained by computer simulation to demonstrate the accuracy of all five methods. The comparison of the different correlation techniques shows that the average magnitude difference function gives results almost as accurate as direct correlation.**

I. Introduction

Correlation techniques have been used extensively in various scientific and technological fields for a number of years [1]–[7]. Time delay is a basic estimate in many applications. A common application comprises two spatially separated sensors which register the signal emanating from a remote source. The correlated signals are assumed to be bandlimited stationary Gaussian processes corrupted by noncross-correlating noise. The position of the peak in an observed cross-correlation curve is interpreted as the time delay estimate.

Because of practical interest, the implementation of different methods has been a research topic for a long time. Several methods exist for computing cross correlation from data which are related but not identical. However, their implementations for practical application have been limited by the high hardware costs. Thus, the choice of suitable methods compromising accuracy and economy requirements is of particular importance.

The purpose of this correspondence is to provide qualitative estimates of the magnitude of the error of the measured delay time. Computer simulation seems to be a particularly attractive method because it allows empirical comparison of the variance of the time delay estimator for all the methods under the same circumstances.

II. Definitions and Method of Comparison

Different time delay estimators have been proposed, discussed, and used in specific applications [7]–[10]. In this correspondence, we shall report the simulation studies concerning variance of the time delay estimators as a function of signal-to-noise ratio (SNR), number of samples (N), and quantization accuracy.

The computer experiment starts with the generation of the random sequence $s(i\Delta t)$ with autocorrelation function and spectral density shown in Fig. 1. The signal-plus-noise sequences $x_1(i\Delta t)$ and $x_2(i\Delta t)$ are formed by adding two independent Gaussian sequences $n_1(i\Delta t)$ and $n_2(i\Delta t)$ to $s(i\Delta t)$ and its delayed version

$$x_1(i\Delta t) = s(i\Delta t) + n_1(i\Delta t)$$
$$x_2(i\Delta t) = s(i\Delta t - D) + n_2(i\Delta t) \tag{1}$$

where D is the time delay and Δt denotes the sampling interval. For each realization of $x_1(i\Delta t)$ and $x_2(i\Delta t)$ the location of the correlation peak was determined by the following methods.

Manuscript received August 12, 1985; revised February 13, 1986.

The authors are with the Institute of Microwave Technology, P.O. Box 70033, S-100 44 Stockholm, Sweden.

IEEE Log Number 8609042.

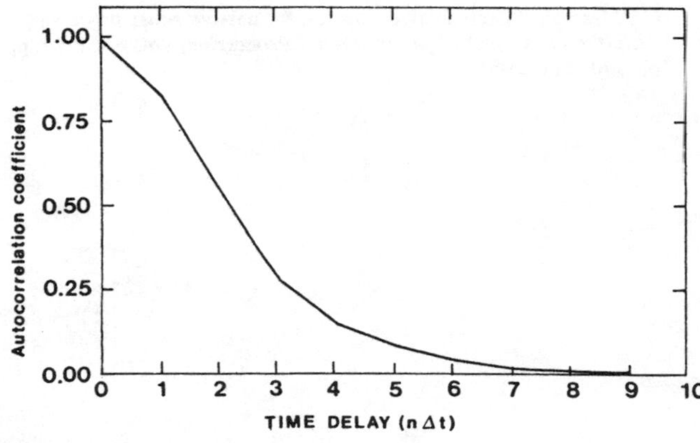

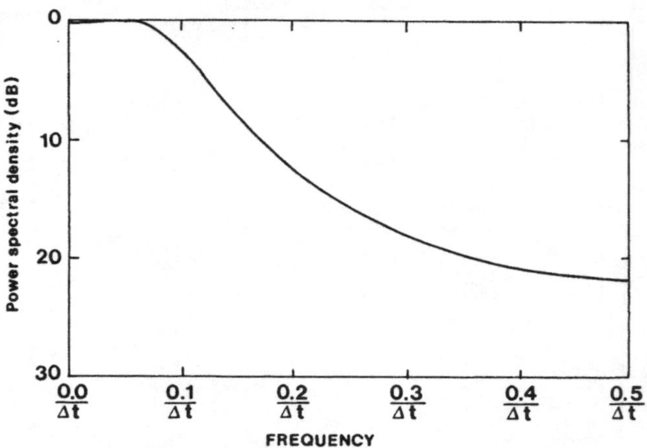

Fig. 1. Sampled autocorrelation function (ACF) and power spectral density (PSD) of the random time series $s(i\Delta t)$.

1) Direct correlation [10]:

$$R_{DC}(\tau) = \frac{1}{N} \sum x_1(i\Delta t) \cdot x_2(i\Delta t + \tau). \tag{2}$$

2) Hybrid-sign correlation [10]:

$$R_{HS}(\tau) = \frac{1}{N} \sum x_1(i\Delta t) \cdot \text{sign}\,(x_2(i\Delta t + \tau)). \tag{3}$$

3) Polarity-coincidence correlation [8], [10]:

$$R_{PC}(\tau) = \frac{1}{N} \sum \text{sign}\,(x_1(i\Delta t)) \cdot \text{sign}\,(x_2(i\Delta t + \tau)). \tag{4}$$

4) Average magnitude difference function [9]:

$$R_{AMDF}(\tau) = \frac{1}{N} \sum |x_1(i\Delta t) - x_2(i\Delta t + \tau)|. \tag{5}$$

Reprinted from *IEEE Trans. Acoust., Speech, Signal Processing*, vol. 34, no. 5, pp. 1329–1330, October 1986.

5) Meyr–Spies method [7]:

$$R_{MS}(\tau) = \frac{1}{N} \sum [-x_1(i\Delta t) + x_1((i-2)\Delta t) \cdot x_2((i-1)\Delta t + \tau). \quad (6)$$

Time delay variance was estimated from 300 independent simulations of $x_1(i\Delta t)$, $x_2(i\Delta t)$. The number of samples varied from 100 to 1000, the SNR varied from 0 to 40 dB, and the analog-to-digital conversion accuracy was 2, 4, and 8 bits. The computer wordlength was 24 bits, i.e., large enough to neglect the influence of roundoff error on the correlation coefficient.

III. Results

The variances of the time delay estimates normalized to the sampling interval ($V(\tau_{peak})\Delta t^2$) for different methods are plotted in Fig. 2 as a function of SNR. The input signals, each consisting of 500 samples, are quantized to 128 levels (8 bits). Simulations with various combinations of sample size (N) and analog-to-digital conversion accuracy show similar results. The variances of the DC, HS, PC, and MS estimators approach a constant value as the SNR approaches infinity. This value depends on the actual number of samples and the quantization error and is different for each method.

This is not the case for the AMDF estimator since the difference signal is always zero at a delay equal to D. Therefore, one may observe the linear decrease of the variance for high SNR.

Another interesting result is that the variance of the MS estimator increases considerably when the SNR decreases. Indeed, when the SNR is poor and the observation time is not sufficiently large, ambiguous zero-crossings may occur, which strongly influences the system performance.

Fig. 3 shows the variances of all five time delay estimators versus number of samples with the SNR fixed at 10 dB. The input signals are quantized to 8 bit accuracy. The results can be used to determine the value of N for various estimation methods required to obtain the same time delay variance. Referring to Fig. 3, it may be seen that the number of samples required to achieve a time delay variance equal to about $3 \cdot 10^{-3}$ is: 100 for the DC estimator, 130 for the AMDF estimator, 230 for the MS estimator, 320 for the HS estimator, and over 1000 for the PC estimator. This example shows that polarity correlation requires more samples compared to other methods than was assumed [4].

Examination of Fig. 2 suggests that the increase of observation time (measured in number of samples) depends on SNR. Hence, PC may be too slow for some of the practical applications.

In this respect, AMDF is the most advantageous method since it can give as accurate an estimate as direct correlation with the only disadvantage being that the observation time is slightly increased.

IV. Conclusions and Discussion

The computer simulation results presented here are useful to demonstrate the relative accuracy of different time delay estimates. The performance of five methods which are usually used in practical applications has been studied.

The interpretation of the results reveals that AMDF is almost as accurate as direct correlation, i.e., that the variance of the time delay estimator is comparable to that of the DC estimator for the same number of samples. The influence of the finite time of observation (N) on variance is of the same order for AMDF as for direct correlation. Moreover, AMDF requires no multiplication, which is the most significant practical advantage over the other methods (except polarity coincidence).

It should be noted that the variance of the polarity coincidence estimator is several times greater than that of AMDF, depending on the SNR. To compensate for this, the number of samples must be increased by an appropriate factor.

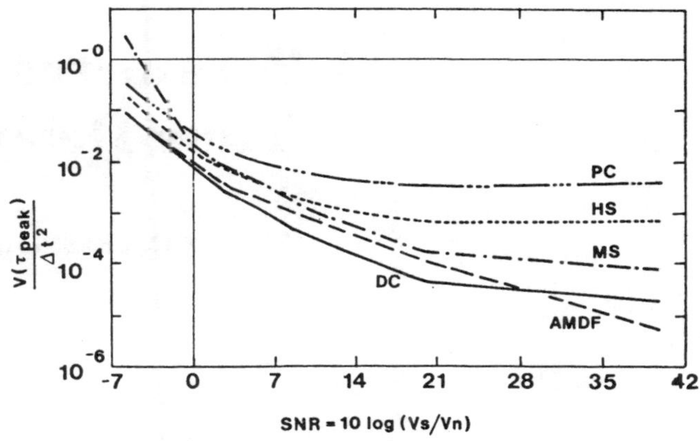

Fig. 2. Time delay estimate variances for various methods as a function of SNR for 500 samples. The input signals are quantized to 8 bit accuracy.

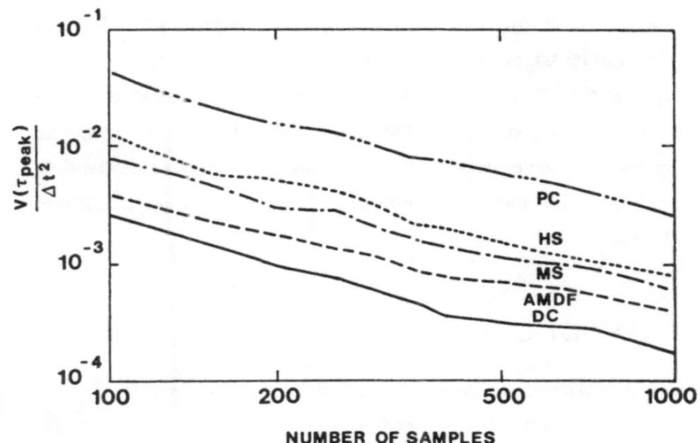

Fig. 3. Time delay estimate variances for various methods as a function of number of samples for SNR = 10 dB. The input signals are quantized to 8 bit accuracy.

References

[1] C. C. Carter, "Time delay estimation for passive sonar signal processing," *IEEE Trans. Acoust., Speech, Signal Processing*, vol. ASSP-29, pp. 463–470, June 1981.

[2] K. R. Godfrey, "The theory of the correlation method of dynamic analysis and its application to industrial processes and nuclear power plant," *Meas. Contr.*, vol. 2, pp. T65–T72, 1969.

[3] K. R. Godfrey and M. Devenish, "An experimental investigation of continuous gas chromatography using pseudo-random-binary sequence," *Meas. Contr.*, vol. 2, pp. 228–234, 1969.

[4] M. S. Beck, "Correlation in instruments: Cross correlation flowmeters," *J. Phys. E: Sci. Instrum.*, vol. 14, pp. 7–19, 1981.

[5] A. M. Featherstone, R. G. Green, and M. E. Shackleton, "Yarn velocity measurement," *J. Phys. E: Sci. Instrum.*, vol. 16, pp. 462–464, 1983.

[6] I. Andermo and A. Sjölund, "Laser velocity meter with correlation technique," *Acta IMEKO*, pp. 23–29, 1979.

[7] J. Bohmann, H. Meyr, R. Peters, and G. Spies, "A signal processor for a noncontact speed measurement system," *IEEE Veh. Technol.*, vol. VT-33, pp. 14–22, Feb. 1984.

[8] J. R. Jordan and K. G. Kelly, "Integrated circuit correlator for flow measurement," *Meas. Contr.*, vol. 9, pp. 267–270, 1976.

[9] M. Ross, H. Shaffer, A. Cohen, R. Freudberg, and H. Manley, "Average magnitude difference function pitch extractor," *IEEE Trans. Acoust., Speech, Signal Processing*, vol. ASSP-22, pp. 353–362, Oct. 1974.

[10] K. J. Gabriel, "Comparison of three correlation coefficient estimates for Gaussian stationary process," *IEEE Trans. Acoust., Speech, Signal Processing*, vol. ASSP-31, pp. 1023–1025, Aug. 1983.

A Parametric Technique for Time Delay Estimation

BENJAMIN FRIEDLANDER AND BOAZ PORAT

The problem of estimating the time difference of arrival of a signal with unknown spectrum to two receivers is treated. A signal model containing both spectral and delay parameters is derived. The model parameters are computed by a two-step procedure: (1) the modified Yule–Walker equations are used to estimate the autoregressive spectrum of the source signal, and (2) a frequency domain squared error criterion is minimized to provide the delay estimate. The performance of the proposed technique is illustrated by simulation results.

I. INTRODUCTION

The estimation of time delays between signals received at two or more sensors is an important issue in active and passive sonar signal processing, communications and radar systems, and in other fields. In its simplest form the problem is formulated as follows: Let $r_1(t)$ and $r_2(t)$ be signals received at two sensors, where

$$r_1(t) = s(t) + n_1(t) \tag{1a}$$

$$r_2(t) = cs(t-D) + n_2(t). \tag{1b}$$

The source signal $s(t)$ and the measurement noises $n_1(t)$ and $n_2(t)$ are Gaussian, stationary, and mutually uncorrelated processes. The objective is to estimate the time delay D given the observations $\{r_1(t), r_2(t) : 0 \leq t \leq T\}$. Numerous techniques have been proposed to solve this problem and some of its variations. For an overview of the extensive literature on time delay estimation see [1–3]. In the case of more than 2 sensors, say m, (1b) will be replaced by $r_i(t) = c_i s(t-D_i) + n_i(t)$, $i = 2, \ldots, m$, where D_i are the delays relative to the first sensor.

Most of the techniques proposed for time-delay estimation are based on a generalized correlation method [4, 5]. The two signals are filtered and then cross-correlated. The delay estimate is obtained by searching for a peak of the correlation function. The filters are derived so as to optimize a performance criterion. One performance criterion maximizes the expected peak relative to the total background noise; another criterion minimizes the difference between the incoming signal at the time delay and its estimated value. Approximate maximum likelihood estimates of the time delay can also be computed by the generalized correlator. The processing is almost always implemented in the frequency domain. Another approach is to use phase measurement data and to estimate the time delay from the slope of the phase versus frequency curve [6]. This approach was reported to have improved performance relative to standard correlation methods in the presence of strong scattering at the receiver locations and when the noise at the receivers is correlated. For the ideal case where the receiver noise is uncorrelated and no scattering is present, the phase measurement procedure provides delay estimates with an accuracy similar to other optimal estimation procedures.

Knowledge of the signal $s(t)$ and noise $n_1(t)$, $n_2(t)$ statistics is usually needed to achieve optimal processing. For example, the optimal filters used in the generalized correlation method depend on the signal and noise spectra. When these statistics are unknown, they have to be estimated from the available data. In other words, it is necessary to solve a spectral estimation problem prior to the delay estimation. Typically, the unknown spectra are estimated using the periodogram or the correlation methods [7]. Improved spectral estimates can often be obtained by using parametric models (e.g., autoregressive (AR) or autoregressive moving-average (ARMA) [8, 9]).

Manuscript received November 24, 1983; revised January 12, 1984.

This work was supported by the Office of Naval Research, Washington, D.C., under Contracts N00014-79-C-0743 and N00014-81-C-0300.

Authors' addresses: B. Friedlander, Systems Control Technology, Inc., 1801 Page Mill Road, Palo Alto, CA 94304; B. Porat, Department of Electrical Engineering, Technion, Haifa, Israel.

Reprinted from *IEEE Trans. Aerosp. Electron. Syst.*, vol. 20, no. 6, pp. 729–735, November 1984.

In this paper we present a parametric technique for time-delay estimation for the case of unknown signal spectrum. The proposed technique employs a model that contains both spectral parameters (an AR representation of the signal $s(t)$) and time delay D, as is described in Section II. Maximum likelihood estimation of these parameters involves a considerable amount of computations. In Section III we present a relatively simple suboptimal two-step procedure: spectral parameters are estimated by the modified Yule–Walker technique [10], followed by nonlinear least squares delay estimation. In Section IV we present some simulation results illustrating the performance of the proposed technique. Comparisons are made to the Cramer–Rao lower bound and to the performance of parametric techniques developed by Chan, Riley, and Plant [11, 12] (CRP) and by Hannan and Thomson [13].

II. THE PARAMETRIC MODEL

Many source signals of interest can be represented as an AR process,

$$s(t) = -\sum_{i=1}^{n} a_i s(t-i) + v(t) \tag{2}$$

where $v(t)$ is a white noise process with variance σ^2. We assume that the received signals are given by (1) with $n_1(t)$, $n_2(t)$ being white noise processes with variances σ_1^2 and σ_2^2. (This assumption can be relaxed to include finitely correlated noise.) The elements of the spectral density matrix of the observations $[G_{ij}(\omega)]$ are

$$G_{11}(\omega) = K_{11}/A(e^{j\omega})A(e^{-j\omega}) + \sigma_1^2,$$

$$K_{11} = \sigma^2 \tag{3a}$$

$$G_{22}(\omega) = K_{22}/A(e^{j\omega})A(e^{-j\omega}) + \sigma_2^2,$$

$$K_{22} = c^2\sigma^2 \tag{3b}$$

$$G_{12}(\omega) = G_{21}^*(\omega) = K_{12}e^{-j\omega D}/A(e^{j\omega})A(e^{-j\omega}),$$

$$K_{12} = c\sigma^2 \tag{3c}$$

where

$$A(e^{j\omega}) = 1 + a_1 e^{-j\omega} + \cdots + a_n e^{-jn\omega}. \tag{4}$$

Notice that the time-delay information D is contained in the off-diagonal elements G_{12}, G_{21} while the diagonal elements contain the information about the source spectral parameters $\{a_i\}$. Next we present an algorithm for estimating the $(n+1)$ model parameters.

III. A PARAMETER ESTIMATION ALGORITHM

Simultaneous estimation of all the model parameters leads to a complicated nonlinear optimization problem. A much simpler procedure is obtained by decomposing the problem into two parts: estimation of the spectral parameters, followed by delay estimation.

Estimating $\{a_i\}$

The two received signals $r_1(t)$, $r_2(t)$ provide two independent measurements for estimating the source spectrum. We will use a technique based on the modified Yule–Walker equations to estimate the AR part of an ARMA process [10, 14]. The algorithm presented in [10] was changed somewhat to take into account the two-channel structure of the problem treated here. The first step in the algorithm is the computation of the sample covariance sequence

$$\hat{R}_{ij}(k) = \frac{1}{T-k}\sum_{t=k}^{T} r_i(t)\, r_j(t-k), \qquad i,j = 1,2. \tag{5}$$

The true correlation coefficients can be shown to satisfy the following relationships, for $l \geq 1$:

$$R_{ii}(l+n) + a_1 R_{ii}(l+n-1) + \cdots + a_n R_{ii}(1) = 0, \qquad i = 1,2. \tag{6}$$

From this relationship we form a set of linear equations whose solution gives an estimate of $\{a_i\}$. Let

$$\rho_l = \begin{bmatrix} R_{11}(l) \\ R_{22}(l) \end{bmatrix} \tag{7}$$

Then

$$\begin{bmatrix} \rho_n & \cdots & \rho_1 \\ \rho_{n+1} & \cdots & \rho_2 \\ \vdots & & \vdots \\ \rho_{N-1} & \cdots & \rho_{N-n} \end{bmatrix} \begin{bmatrix} a_1 \\ \vdots \\ a_n \end{bmatrix} = -\begin{bmatrix} \rho_{n+1} \\ \rho_{n+2} \\ \vdots \\ \rho_N \end{bmatrix}. \tag{8}$$

These equations are usually called the (overdetermined) modified Yule–Walker (MYW) equations. Since in practice the true correlation coefficients $R_{ij}(k)$ are replaced by their estimates $\hat{R}_{ij}(k)$, this set of linear equations will hold only approximately. It is therefore necessary to solve for $\{a_i\}$ using a least squares procedure. In our simulation we used the singular value decomposition technique [15] which is known to be robust. Alternatively, it is possible to use the square-root normalized covariance lattice form [16]. Notice that when more than two sensors are available, (8) can be still used with $\rho_l = [R_{11}(l), r_{22}(l), \ldots, R_{mm}(l)]^T$, where $R_{ii}(l)$ are the autocorrelation coefficients of the signal received at the ith sensor.

The accuracy of the spectral estimates obtained by the MYW method depends on the choice of the number of equations $(N-n)$ and the number of parameters (n). A detailed discussion of these issues is presented in [10, 23]. Here we note only that the number of parameters n needs to be large compared with the true order of the source signal, in order to get accurate estimates at low signal-to-noise ratios. The choice of a high model order introduces spurious noise related modes into the estimated spectrum. These superfluous modes usually have low energies and do not seriously affect the quality of the spectral estimate. Furthermore, it is possible to systematically remove these modes by the procedure described in [10]. The choice of the number of equations will depend on the nature of the signal (see [23]).

Estimating the Time Delay D

To estimate the time delay we fit a parametric model of the cross spectrum to a nonparametric spectral estimate. The nonparametric estimate of the cross-spectral density function $\hat{G}_{12}(\omega)$ is obtained by Fourier transformation of the sample cross-covariance function $R_{12}(k)$,

$$\hat{G}_{12}(\omega_l) = R_{12}(0) + 2\sum_{k=1}^{M} w_k R_{12}(k)\cos k\omega_l, \qquad 0 \le l \le M \tag{9}$$

where w_k is some standard windowing function (see e.g., [7]). The number (M) of correlation coefficients used to compute $\hat{G}_{12}(\omega_l)$ determines the amount of spectral smoothing introduced in the estimation process. The number M should be chosen to be large enough not to introduce too much smoothing (the choice $M = T$ gives the nonsmoothed spectral estimate), but not too large, to avoid excessive computations.

To this cross-spectral estimate we want to fit the parametric model defined in (3c). Since the spectral estimate $\hat{G}_{12}(\omega)$ is a windowed version of the underlying cross spectrum, it is necessary to introduce the windowing effect in the parametric cross spectrum as well. This is done as follows: Let

$$P(\omega) \triangleq \sum_{k=-\infty}^{\infty} p_k e^{-jk\omega} = 1/A(e^{j\omega})A(e^{-j\omega}) \tag{10}$$

denote the unwindowed (parametric) source spectrum. The coefficients p_k can be computed using the algorithm in [17]. The windowed version of the (parametric) source spectrum will be computed by windowed Fourier transformation of the sequence p_k, i.e.,

$$\hat{P}(\omega_l) = p_0 + 2\sum_{k=1}^{M} w_k p_k \cos k\omega_l, \qquad 0 \le l \le M. \tag{11}$$

The windowed version of the (parametric) cross spectrum will, therefore, be given by (cf. (3c))

$$\hat{G}_{12}(\omega_l) \cong K_{12}\exp(-j\omega_l D)\,\hat{P}(\omega_l), \qquad 0 \le l \le M. \tag{12}$$

The delay D is chosen so as to minimize the squared error

$$e^2(D) = \sum_{l=0}^{M} W(\omega_l)\,|(\hat{G}_{12}(\omega_l) - K_{12}\exp(-j\omega_l D)\,\hat{P}(\omega_l))|^2 \tag{13}$$

where $W(\omega)$ is a properly chosen frequency weighting function. In our tests we used no frequency weighting, i.e. $W(\omega) = 1$. Alternatively, the delay can be estimated by maximizing the cost function

$$V(D) = \sum_{l=0}^{M} W(\omega_l)\hat{G}^*(\omega_l)\hat{P}(\omega_l)\exp(-j\omega_l D) \tag{14}$$

which can be computed efficiently for different values of D, using a fast Fourier transform (FFT).

Next we discuss how to estimate the gain K_{12} which appears in (13). From (3a) and (3b) it follows that

$$\begin{aligned}\hat{G}_{ii}(\omega_l) &\triangleq R_{ii}(0) + 2\sum_{k=1}^{M} w_k R_{ii}(k)\cos k\omega_l \\ &\cong K_{ii}\hat{P}(\omega_l) + \sigma_i^2, \qquad i = 1,2.\end{aligned} \tag{15}$$

The left-hand side of (15) is the nonparametric autospectral estimate, and the right-hand side is the parametric autospectral density function. Inserting (11) in (15) and comparing coefficients of $\cos k\omega_l$ leads to the conclusion that

$$R_{ii}(k) = K_{ii}p_k, \qquad 1 \le k \le M;\ i = 1,2. \tag{16}$$

The least squares estimate of K_{ii} associated with (16) is given by

$$\hat{K}_{ii} = \sum_{k=1}^{M} p_k R_{ii}(k) \Big/ \sum_{k=1}^{M} p_k^2, \qquad i = 1,2. \tag{17}$$

From (3) it follows that $K_{12} = \sqrt{K_{11}K_{22}}$. Therefore the estimate $\hat{K}_{12}$ can be computed from the estimates $\hat{K}_{11}$, $\hat{K}_{22}$ by

$$\hat{K}_{12} = \sqrt{\hat{K}_{11}\hat{K}_{22}}. \tag{18}$$

The minimization of the squared-error function $e^2(D)$ can be performed using various optimization routines. We found it more useful to work with the equation

$$\begin{aligned}\frac{\partial e^2(D)}{\partial D} &= 0 \\ &= \sum_{l=0}^{M} 2W(\omega_l)\,\omega_l K_{12}\hat{P}(\omega_l)[(\mathrm{Re}\{\hat{G}_{12}(\omega_l)\} \\ &\quad - K_{12}\hat{P}(\omega_l)\cos\omega_l D)\sin\omega_l D \\ &\quad + (\mathrm{Im}\{\hat{G}_{12}(\omega_l)\} \\ &\quad + K_{12}\hat{P}(\omega_l)\sin\omega_l D)\cos\omega_l D]\end{aligned} \tag{19}$$

or

$$\sum_{l=0}^{M} W(\omega_l)\,\omega_l\hat{P}(\omega_l)\,(\mathrm{Re}\{\hat{G}_{12}(\omega_l)\}\sin\omega_l D + \mathrm{Im}\{\hat{G}_{12}(\omega_l)\}\cos\omega_l D) = 0. \tag{20}$$

The value of D for which this equation is satisfied is chosen as the delay estimate $\hat{D}$. The equation is solved numerically using a secant method [18]. Notice that K_{12} does not appear in (20), and thus its computation via

(17), (18) is not needed in this case. The same is true when $\hat{D}$ is computed via (14).

The amount of computation required to estimate the delay is proportional to the number (M) of correlation coefficients and to the number of steps required by the secant method to solve (20) (which depends on various factors such as the distance of the initial solution from the true solution).

In the case of more than two sensors the cost function can be modified to include all of the cross-spectral terms $\{\hat{G}_{ij}(\omega_l)\}$ and the relative delays $\{D_i - D_j\}$. However, a simpler approach appears to be to estimate delays for each sensor pair and then to combine these estimates to obtain $\{\hat{D}_i,\ i = 1, \ldots, m\}$. See [24] for a discussion of nonparametric delay estimation in the multisensor case.

IV. SIMULATION RESULTS

To evaluate the performance of the proposed technique a simulation study was carried out. Two types of experiments were performed, as described below.

Experiment 1

In this experiment we compared the performance of the estimator to the case where the spectrum of the signal is known a priori. This was achieved by running the time delay estimation part of the algorithm (20) twice: once using the spectral parameters $\hat{A}(z)$ estimated as described in Section III, and once using the true parameters $A(z)$. The theoretical Cramer–Rao lower bound for the variance of the time delay $\sigma^2(\hat{D})$ was also computed, using the following formula [19]:

$$\sigma^2(\hat{D}) = \left[\frac{T}{4\pi}\int_0^{\pi} \frac{(2S(\omega)/N(\omega))^2}{1 + 2S(\omega)/N(\omega)}\,\omega^2 d\omega\right]^{-1} \tag{21}$$

where $S(\omega)$, $N(\omega)$ are the autospectra of the signal and noise processes (assuming that $n_1(t)$, $n_2(t)$ have the same spectra and $c = 1$ in (1)). In our case

$$S(\omega) = \frac{\sigma^2/2\pi}{A(e^{j\omega})A(e^{-j\omega})} \tag{22a}$$

$$N(\omega) = \sigma_1^2/2\pi = \sigma_2^2/2\pi. \tag{22b}$$

We define the signal-to-noise ratio S/N as the ratio of the signal energy to the noise variance. The source signal was generated according to (2) with

$$A(z) = 1 - 0.458z^{-1} + 0.55z^{-2}. \tag{23}$$

The received signal was generated according to (1) with $c = 1$, $\sigma_1 = \sigma_2$. The time delay was $D = 1.0$ and the number of data points $T = 1024$. The AR estimation procedure was used with $N = 48$, $n = 12$. A mode selection procedure [10] reduced the order of $\hat{A}(z)$ to the correct model order $n = 2$. The delay estimation procedure was used with $M = 60$. Table I summarizes the results obtained by 20 independent simulation runs for each case. Notice the close agreement of the error variance between the cases of known and unknown spectrum and the Cramer–Rao lower bound, except at the lowest S/N.

TABLE I
Experiment 1

	Estimation of Spectrum and Delay		Known Spectrum Estimated Delay		Theoretical Bound
S/N	Sample Mean	Sample Standard Deviation	Sample Mean	Sample Standard Deviation	Standard Deviation
1.00	0.997	0.043	0.997	0.037	0.032
0.10	1.015	0.168	1.025	0.186	0.151
0.01	1.047	0.783	0.930	0.547	1.134

Experiment 2

In this experiment we compared the performance of the estimator to that of other parametric delay estimation techniques that have appeared in the literature. One such method proposed by Chan et al. [11, 12] involves the fitting of a moving-average type model to the received signals. Another technique was proposed by Hannan and Thomson [13], involving the fitting of a two-channel AR model to the observatoin vector. The estimated AR parameters are used to evaluate a frequency weighting function which is then inserted into the Hamon–Hannan (HH) estimation procedure [20]. In [13] a comparison is made between the original HH procedure [20], the modified HH procedure described above, and the CRP procedure [11].

The source signal was generated in this experiment according to (2) with $\sigma^2 = 1$ and

$$A(z) = 1 - 1.77z^{-1} + 1.593z^{-2} - 0.7047z^{-3}. \tag{24}$$

The received signal was generated according to (1) with $c = 1$. The two values of D were $D = 0.5$ and $D = 10.5$. The signal to noise ratios were 1 and 0.25 (i.e., $\sigma_1^2 = \sigma_2^2 = 8.97995, 35.9198$, respectively). Simulations were performed for $T = 1024$ and $T = 256$. These test conditions are identical to those used by Hannan and Thomson [13].

The technique described in Section III was used to estimate the time delay in 20 independent simulation runs for each case. The AR estimation procedure was used with $N = 48$, $n = 16$. A mode selection procedure [10] reduced the order of $\hat{A}(z)$ to the correct model order $n = 3$. The delay estimation procedure was used with $M = 80$. The simulation results are summarized in Tables II and III under the heading "Parametric technique." The other entries in the table are a replication of Tables I and II in [13], to facilitate a comparison between the proposed technique and the HH, modified HH, and CRP methods.

An examination of Tables II and III indicates that the proposed technique tends to have a lower sample standard

TABLE II
Experiment 2, Sample Size 1024

		Hamon–Hannan			Modified Hamon–Hannan			Chan–Riley–Plant			Parametric Technique	
S/N	D	Sample Mean	Sample Std. Dev.	Theoretical Std. Dev.	Sample Mean	Sample Std. Dev.	Theoretical Std. Dev.	Sample Mean	Sample Std. Dev.	Theoretical Std. Dev.	Sample Mean	Sample Std. Dev.
1.0	0.5	0.4609	0.1404	0.0708	0.4971	0.1220	0.0708	0.4576	0.2651	0.2846	0.568	0.068
1.0	10.5	10.4719	0.1472	0.0708	10.4773	0.1221	0.0708	10.5023	0.2391	0.2846	10.558	0.069
0.25	0.5	0.4131	0.3802	0.1735	0.5541	0.2861	0.1735	0.3362	0.3802	0.6581	0.671	0.211
0.25	10.5	10.4488	0.3608	0.1735	10.4696	0.3309	0.1735	10.4616	0.3696	0.6581	10.414	0.432

TABLE III
Experiment 2, Sample Size 256

		Hamon–Hannan			Modified Hamon–Hannan			Chan–Riley–Plant			Parametric Technique	
S/N	D	Sample Mean	Sample Standard Deviation	Theoretical Standard Deviation	Sample Mean	Sample Standard Deviation	Theoretical Standard Deviation	Sample Mean	Sample Standard Deviation	Theoretical Standard Deviation	Sample Mean	Sample Standard Deviation
1.0	0.5	0.5588	0.3363	0.1416	0.5032	0.2146	0.1416	0.5875	0.4719	0.5693	0.499	0.138
1.0	10.5	10.4312	0.3812	0.1416	10.4945	0.3460	0.1416	10.3961	0.3894	0.5693	10.476	0.187
0.25	0.5	0.6749	0.5893	0.3470	0.6904	0.5916	0.3470	0.6287	0.6599	1.3163	0.615	0.446
0.25	10.5	10.6070	0.6989	0.3470	10.3732	0.5800	0.3470	10.4788	0.4884	1.3163	10.361	0.514

deviation than the other parametric techniques. The mean error of the delay estimate seems comparable to that of the other techniques.

The criterion that is being optimized (cf. (13)) has multiple minima. To see the behavior of this criterion we computed the function $e^2(0)/e^2(\hat{D} - D)$ for the case $S/N = 1$, $D = 0.5$, $T = 1024$. The result is depicted in Fig. 1. Similar plots for the HH and CRP procedures are given in [13] (Figs. 6 and 7). We note that the "sidelobes" of this function can be reduced by using a smoother (nonrectangular) windowing function $W(\omega)$.

V. CONCLUSIONS

A parametric technique was presented for estimating the time delay between two signals with unknown spectra. Limited simulation tests of the proposed technique seem to indicate that its performance compares favorably with that of other delay estimation algorithms described in the literature. The use of models with a small number of parameters may have some advantages over the widely used nonparametric generalized correlator.

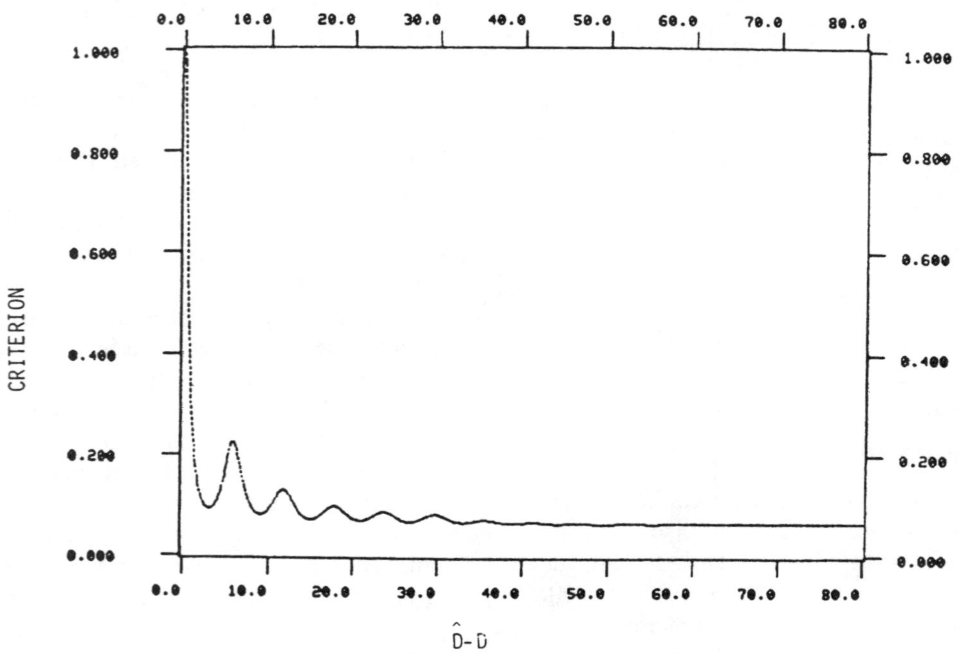

Fig. 1. Form of the criterion $e^2(0)/e^2(\hat{D} - D)$ for relatively high signal-to-noise ratio.

Finally we note that this parametric technique can be extended in a straightforward manner to more general spectral models, such as the ARMA model. See for example, [21, 22].

REFERENCES

[1] *Special issue on Time Delay Estimation.*
IEEE Transactions on Acoustics, Speech, and Signal Processing, ASSP-29, 3 (June 1981).

[2] *Proceedings of the Time Delay Estimation and Applications Conference.*
Naval Postgraduate School, Monterey, CA, May 1–3, 1979.

[3] Carter, G.C., et al. (1980)
Coherence estimation.
Naval Underwater Systems Center, 1980.

[4] Knapp, C.H., and Carter, G.C. (1976)
The generalized correlation method for estimation of time delay.
IEEE Transactions on Acoustics, Speech, and Signal Processing, ASSP-24, 4 (Aug. 1976), 320–327.

[5] Hassab, J.C., and Boucher, R.E. (1979)
Optimum estimation of time delay by a generalized correlator.
IEEE Transactions on Acoustics, Speech, and Signal Processing, ASSP-27, 6 (Aug. 1979), 373–380.

[6] Piersol, A.G. (1981)
Time delay estimation using phase data.
IEEE Transactions on Acoustics, Speech, and Signal Processing, ASSP-29, 3 (June 1981), 471–477.

[7] ——— (1979)
Programs for Digital Signal Processing.
New York: IEEE Press, 1979.

[8] Childers, D.G. (Ed.) (1978)
Modern Spectral Analysis.
New York: IEEE Press, 1978.

[9] Kay, S.M., and Marple, S.L. (1981)
Spectrum analysis—A modern perspective.
Proceedings of the IEEE, 69, 11 (Nov. 1981), 1380–1419.

[10] Friedlander, B., and Porat, B. (1984)
The modified Yule-Walker method of ARMA spectral estimation.
IEEE Transactions on Aerospace and Electronic Systems, AES-20, 2 (Mar. 1984), 158–173.

[11] Chan, Y.T., Riley, J.M., and Plant, J.B. (1980)
A parameter estimation approach to time-delay estimation and signal detection.
IEEE Transactions on Acoustics, Speech, and Signal Processing, ASSP-28 (Feb. 1980), 8–16.

[12] Chan, Y.T., Riley, J.M., and Plant, J.B. (1981)
Modeling of time delay and its application to estimation of nonstationary delays.
IEEE Transactions on Acoustics, Speech, and Signal Processing, ASSP-29, 3 (June 1981), 577–581.

[13] Hannan, E.J., and Thomson, P.J. (1981)
Delay estimation and the estimation of coherence and phase.
IEEE Transactions on Acoustics, Speech, and Signal Processing, ASSP-29, 3 (June 1981), 485–490.

[14] Cadzow, J.A. (1980)
High performance spectral estimation—a new ARMA method.
IEEE Transactions on Acoustics, Speech, and Signal Processing, ASSP-28, 5 (Oct. 1980), 524–529.

[15] Lawson, C.L., and Hanson, R.J. (1974)
Solving Least-Squares Problems.
Englewood Cliffs, N.J.: Prentice-Hall, 1974.

[16] Friedlander, B. (1983)
An efficient algorithm for ARMA spectral estimation.
Proceedings of the IEE, 130, pt. F, 3 (Apr. 1983), 195–201.

[17] Mullis, C.T., and Roberts, R.A. (1976)
Synthesis of minimum roundoff noise fixed point digital filters.
IEEE Transactions on Circuits and Systems, CAS-23, 9 (Sept. 1976), 551–562.

[18] Wolfe, P. (1959)
The secant method for simultaneous non-linear equations.
Communications of the ACM, 2, 1959.

[19] Chow, S.K., and Schultheiss, P.M. (1981)
Delay estimation using narrowband data.
IEEE Transactions on Acoustics, Speech, and Signal Processing, ASSP-29, 3 (June 1981), 478–484.

[20] Hamon, B.V., and Hannan, E.J. (1974)
Spectral estimation of time delay for dispersive and non-dispersive systems.
J. Royal Statist. Soc. Ser. C (Appl. Statist.), 23 (1974), 134–142.

[21] Porat, B., and Friedlander, B. (1983)
Estimation of spectral and spatial parameters of multiple sources.
IEEE Transactions on Information Theory, IT-29, 3 (May 1983), 412–425.

[22] Friedlander, B., and Porat, B. (1984)
A spectral matching technique for ARMA parameter estimation.
IEEE Transactions on Acoustics, Speech, and Signal Processing, ASSP-32, 2 (April 1984), 338–343.

[23] Friedlander, B., and Sharman, K.C.
Performance analysis of the modified Yule-Walker estimator.
IEEE Transactions on Acoustics, Speech, and Signal Processing, to appear.

[24] Hahn, W.R., and Tretter, S.A. (1973)
Optimum processing for delay-vector estimation in passive signal arrays.
IEEE Transactions on Information Theory, IT-19, 5 (1973), 608–614.

Delay Estimation and the Estimation of Coherence and Phase

EDWARD J. HANNAN AND PETER J. THOMSON

Abstract—The method of *delay* estimation given in Hamon and Hannan [1] is reviewed and a modification involving *autoregressive model* fitting is proposed. It is shown how numerical problems associated with the optimization of the criterion used in both the original and modified procedures can be circumvented. The modified procedure and a procedure due to Chan, Riley, and Plant [2] are compared to the Hamon-Hannan procedure by *simulations*. For moderate to large signal-to-noise ratios, the modified Hamon-Hannan procedure appears to provide better estimates. In the case of low signal-to-noise ratios, with the smaller size, the Chan, Riley, and Plant procedure performed best.

I. Introduction

CONSIDER two recorders receiving a delayed form of a common signal $s(t)$ together with noise. The signal is assumed to be a zero mean *stationary process* and thus has a spectral representation [5, p. 41]

$$s(t) = \int_{-\infty}^{\infty} e^{-jt\omega}\, dz(\omega) \qquad (-\infty < t < \infty).$$

We assume that $s(t)$ has an absolutely continuous spectrum with spectral density $G_s(\omega)$. The processes received at the two recorders are

$$x(t) = s(t) + n_1(t), y(t) = \int_{-\infty}^{\infty} e^{-j(t - D(\tau,\omega))\omega} a(\omega)\, dz(\omega) + n_2(t) \tag{1}$$

where $D(\tau, \omega)$ is the relative *delay*, $a(\omega)$ is the attenuation relative to the first recorder, and the $n_j(t)$ $(j = 1, 2)$ are zero mean *stationary* noise *processes* incoherent with each other and $s(t)$. $D(\tau, \omega)$ is assumed to be a known function of parameters $\tau_1, \cdots, \tau_r$ and ω. In the *nondispersive case* $D(\tau, \omega) = \tau$ where τ is the pure *delay*.

There is a considerable body of literature concerning the estimation of *delay* (see [4] for some references). Here we review the procedure in [1] and introduce a procedure based on *autoregressive model* fitting. These are together compared with a procedure in [2]. A numerical problem associated with the procedure in [1], and the modified procedure, is discussed.

We assume that $x(t)$, $y(t)$ have been sampled at equal time intervals, that this interval is the time unit which is sufficiently small for aliasing effects to be ignored. If the $n_i(t)$ $(i = 1, 2)$ have spectral densities $G_i(\omega)$ $(i = 1, 2)$, the spectral densities of $x(t), y(t)$ are $G_x(\omega) = G_s(\omega) + G_1(\omega), G_y(\omega) = |a(\omega)|^2 G_s(\omega) + G_2(\omega)$ and the cross-spectral density is $G_{yx}(\omega) = a(\omega)G_s(\omega) \exp jD(\tau, \omega)\omega$, $0 \leq \omega \leq \pi$. Then the coherence and phase at frequency ω are $\sigma(\omega) = |G_{yx}(\omega)|/\{G_x(\omega)G_y(\omega)\}^{1/2}$, $\phi(\omega) = \omega D(\tau, \omega)$.

Manuscript received March 4, 1980; revised September 26, 1980.

E. J. Hannan is with the Department of Statistics, Institute of Advanced Studies, Australian National University, Canberra, ACT 2600, Australia.

P. J. Thomson is with the Department of Mathematics and Statistics, Massey University, Palmerston North, New Zealand.

II. The Procedures

A detailed account of the Hamon-Hannan procedure (hereafter referred to as HH) is given in [1]. Estimates $\hat{\sigma}$, $\hat{\phi}$ are formed at the M frequencies $\lambda_u = (2u - 1)\pi/(2M)$ $(u = 1, \cdots, M)$. Then the vector of delay parameters is estimated by $\hat{\tau}$, which maximizes

$$Q_H(\tau) = \frac{1}{M} \sum_{\lambda_u \in B} \hat{W}(\lambda_u) \cos(\hat{\phi}(\lambda_u) - D(\tau, \lambda_u)\lambda_u). \tag{2}$$

Here $\hat{W}(\lambda_u) = \hat{\sigma}^2(\lambda_u)/(1 - \hat{\sigma}^2(\lambda_u))$ estimates $W(\lambda_u) = \sigma^2(\lambda_u)/(1 - \sigma^2(\lambda_u))$. M and the band B are chosen by the experimenter. The estimates of $\hat{\sigma}$, $\hat{\phi}$ can be formed from the discrete Fourier transforms

$$w_x(\omega_k) = \frac{1}{\sqrt{2\pi T}} \sum_{t=1}^{T} x(t) \exp jt\omega_k,\ w_y(\omega_k) = \frac{1}{\sqrt{2\pi T}} \sum_{t=1}^{T} \cdot y(t) \exp jt\omega_k$$

where $\omega_k = 2\pi k/T$ $(k = 1, \cdots, [\frac{1}{2}T])$ and T is the number of data points. A fader could be used to compute $w_x(\omega_k), w_y(\omega_k)$. (See [1], [5] for details.) Unver general conditions $\hat{\tau}$ converges (almost surely) to τ and $\sqrt{T(\hat{\tau} - \tau)}$ has an asymptotic limiting normal distribution with zero means and covariance matrix V^{-1} where V is estimated by

$$\hat{V}_{jk} = M^{-1} \sum_{\lambda_u \in B} \lambda_u^2 \hat{W}(\lambda_u) \frac{\partial D(\tau, \lambda_u)}{\partial \tau_j} \frac{\partial D(\tau, \lambda_u)}{\partial \tau_k}.$$

The weight function, $W(\lambda_u)$, is optimal in that it makes the covariance matrix V a minimum (see [1] and [6]). As pointed out in [1], $\hat{\sigma}$ may tend to be much too high when σ is low so that frequencies containing little information are given too much weight. The following procedure is suggested as a means of improving the estimate of W. If

$$\begin{bmatrix} G_x(\omega) & G_{xy}(\omega) \\ G_{yx}(\omega) & G_y(\omega) \end{bmatrix} \simeq \frac{1}{2\pi} \left[\sum_{i=0}^{p} A_i e^{ji\omega}\right]^{-1} \cdot K \left[\sum_{i=0}^{p} A_i e^{-ji\omega}\right]^{-1} \tag{3}$$

Reprinted from *IEEE Trans. Acoust., Speech, Signal Processing*, vol. 29, no. 3, pt. 2, pp. 485–490, June 1981.

where the A_i are 2×2 matrices and K is the covariance matrix of the "innovations," then estimates of p, K and the A_i will yield estimates of G_x, G_y and G_{xy} which yield estimates of σ and W. Since such estimates (say $\tilde{\sigma}$) are available for any frequency ω, an estimate of the vector of delay parameters is $\hat{\tau}$, which maximizes

$$\tilde{Q}_H(\tau) = \frac{1}{N} \sum_{0 < \omega_k < \pi} \tilde{W}(\omega_k) \cos (\hat{\phi}(\omega_k) - D(\tau, \omega_k)\omega_k) \tag{4}$$

where $\tilde{W}(\omega_k) = \tilde{\sigma}^2(\omega)/\{(1 - \tilde{\sigma}^2(\omega)\}$. *Now* $\hat{\phi}(\omega_k)$ *is the argument of* $w_y(\omega_k)\overline{w_x(\omega_k)}$. There is a large literature on the fitting of finite parameter models such as *autoregressive* (AR) *models*, i.e., of the form (3), (see [7]) and on the determination of the order p. One technique (see [3]) consists of fitting (3) to the data for a range of orders p and then selecting the p to minimize

$$\text{AIC}(p) = \log \det \hat{K}_p + 8p/T. \tag{5}$$

Here $\hat{K}_p$ is the estimated residual covariance matrix for the fitted model of order p (see also [8] and [9]). The method of [10] involving the recursive solution of the Yule-Walker equations provides a computationally efficient method for determining estimates of the A_j and K in (3).

For $D(\tau, \omega) = \tau$ the method of Chan, Riley, and Plant [2] (here called the CRP procedure) involves the fitting of a distributed lag relationship between $y(t)$ and $x(t)$ of the form

$$y(t) = \sum_{-p}^{p} \zeta_i x(t-i) + \epsilon(t)$$

where $\epsilon(t)$ is a *stationary* noise *process*. Estimates of the ζ_i (say $\hat{\zeta}_i$) are obtained by ordinary least squares and the estimate of the delay is then determined as the value $\hat{\tau}$ maximizing

$$Q_{\text{CRP}}(\tau) = \frac{1}{2p+1} \sum_{-p}^{p} \hat{\zeta}_j \frac{\sin \pi(\tau - j)}{\pi(\tau - j)}. \tag{6}$$

The criterion (6) is not, in general, symmetric with regard to the choice of regressor and regressand. If p is allowed to increase with T it is said in [2] that this criterion is asymptotically equivalent to the criterion given by (4) with $\tilde{W}(\omega_k)$ replaced by an estimate of $\sigma(\omega_k)\sqrt{G_y(\omega_k)/G_x(\omega_k)}$. Thus, this procedure then yields an estimator of τ whose asymptotic variance exceeds that of the $\hat{\tau}$ obtained using either (2) or (4).

If $W(\omega)$ were known, the HH procedure would be close to a maximum likelihood procedure. It is for this reason that we have sought to improve the estimation $W(\omega)$ by the use of a finite parameter model. The CRP procedure uses a finite parameter model also (effectively) to estimate phase, but the weighting of frequencies in this procedure is not optimal. A third alternative not investigated here would be to use a finite parameter model also for the noise in the distributed lag relationship so as to recover some of this loss due to incorrect weighting of frequencies.

III. Simulations: Pure Delay

To compare the three procedures in the pure *delay* case a *simulation* study was carried out with $T = 1024$ and 256. The data were generated according to (1) with $a(\omega) = 1$, $G_1(\omega) = G_2(\omega) = \alpha^2/(2\pi)$ and $s(t)$ was generated via an autoregression with spectral density

$$G_s(\omega) = \frac{1}{2\pi} |1 - 1.77 \exp j\omega + 1.593 \exp j2\omega - 0.7047 \cdot \exp j3\omega|^{-2}.$$

The two values of τ were $\tau = 0.5$, $\tau = 10.5$. The S/N ratios (the ratio of the signal variance to the noise variance) were 1 and 0.25, i.e., $\alpha^2 = 8.97995, 35.9198$, respectively. The coherence function is shown in Figs. 1 and 2.

For the HH procedure m, the number of ω_k in the band at λ_u, was 9 for $T = 1024$ and 5 for $T = 256$. Thus, M was effec-

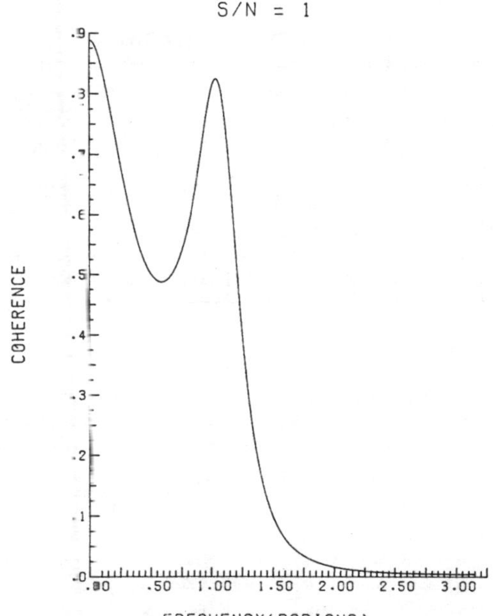

Fig. 1. Coherence. Relatively high signal-to-noise ratio.

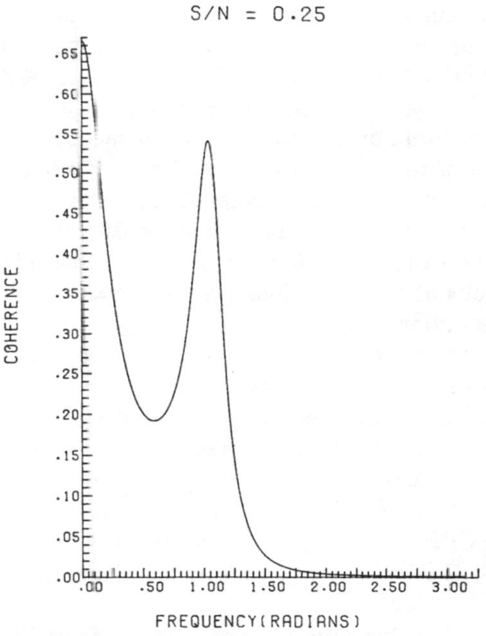

S/N = 0.25

COHERENCE

FREQUENCY(RADIANS)

Fig. 2. Coherence. Relatively low signal-to-noise ratio.

TABLE I
NONDISPERSIVE CASE; SAMPLE SIZE = 1024

$\frac{S}{N}$	τ	Hamon-Hannan			Modified Hamon-Hannan			Chan-Riley-Plant		
		Sample mean	Sample std. dev.	Theoretical std. dev.	Sample mean	Sample std. dev.	Theoretical std. dev.	Sample mean	Sample std. dev.	Theoretical std. dev.
1	0.5	0.4609	0.1404	0.0708	0.4971	0.1220	0.0708	0.4576	0.2651	0.2846
1	10.5	10.4719	0.1472	0.0708	10.4773	0.1221	0.0708	10.5023	0.2391	0.2846
0.25	0.5	0.4131	0.3802	0.1735	0.5541	0.2861	0.1735	0.3362	0.3802	0.6581
0.25	10.5	10.4488	0.3608	0.1735	10.4696	0.3309	0.1735	10.4616	0.3696	0.6581

TABLE II
NONDISPERSIVE CASE; SAMPLE SIZE = 256

$\frac{S}{N}$	τ	Hamon-Hannan			Modified Hamon-Hannan			Chan-Riley-Plant		
		Sample mean	Sample std. dev.	Theoretical std. dev.	Sample mean	Sample std. dev.	Theoretical std. dev.	Sample mean	Sample std. dev.	Theoretical std. dev.
1	0.5	0.5588	0.3363	0.1416	0.5032	0.2146	0.1416	0.5875	0.4719	0.5693
1	10.5	10.4312	0.3812	0.1416	10.4945	0.3460	0.1416	10.3961	0.3894	0.5693
0.25	0.5	0.6749	0.5893	0.3470	0.6904	0.5916	0.3470	0.6287	0.6599	1.3163
0.25	10.5	10.6070	0.6989	0.3470	10.3732	0.5800	0.3470	10.4788	0.4884	1.3163

tively 56 and 25. A taper was not used and may have helped the HH procedure. For the modified HH procedure, p was selected using (5) for $5 \leqslant p \leqslant 20$ for $\tau = 0.5$; $10 \leqslant p \leqslant 20$ for $\tau = 10.5$. In general, p should be large enough to ensure adequate resolution of the two component spectra and the cross-spectrum of the observed process. For large τ (e.g., $\tau = 10.5$), it is not hard to see that an order of at least the delay τ should be chosen. For the CRP procedure, similar considerations apply. For the CRP procedure, p was 5 ($\tau = 0.5$) and 15 ($\tau = 10.5$). The results of the simulations given were based on 20 independent replications.

For $T = 1024$, the results are given in Table I. The modified HH procedure performed best, followed by the HH procedure. However, the HH procedures gave sample standard deviations (sd's) that were significantly greater than the theoretical (asymptotic) values. Indeed, the smallest ratio of the former to the latter was 1.6, which is significant at any reasonable level. For the CRP procedure, the sample sd's were not significantly different ($S/N = 1$) or significantly less than ($S/N = 0.25$) the theoretical values.

For $T = 256$ the results are as given in Table II. Now, for $S/N = 1$ in terms of sample sd's (and mean-square errors), the modified HH procedure performed best, followed by the HH procedure, but for $S/N = 0.25$ the CRP procedure performed best, followed by the modified HH procedure.

The following may explain these results. The band B was taken as $(0, \pi)$. The theoretical sd for HH does not reflect inaccuracy in $\hat{W}$, since for large enough T this effect will be of lower order. However, the effect may not be negligible, particularly for low σ. Figs. 3 and 4 show $\hat{\phi}$ for two *simulations* and indicate how inaccurate is $\hat{\phi}$ for low σ. (The two $\hat{\tau}$ values were 0.475 for Fig. 3 and 0.543 for Fig. 4.) *Undoubtedly, better results would have been achieved had frequencies above about 1.25 rads been eliminated* (see [1, p. 141]). The theoretical weight function for the HH procedures for $S/N = 1$ is given in Fig. 5. That for the CRP procedure is $\sigma(\omega)$, exhibited for $S/N =$ 1 in Fig. 1. It may be noted that when $\sigma = 0$, $(m - 1)\hat{W}$ is (approximately) distributed as F with 2 and $2(m - 1)$ degrees of freedom (see [5, p. 261] for details). For $m = 9$ this must be above 0.22 (i.e., $\hat{\sigma} > 0.16$) for significance even at the 20 percent level.

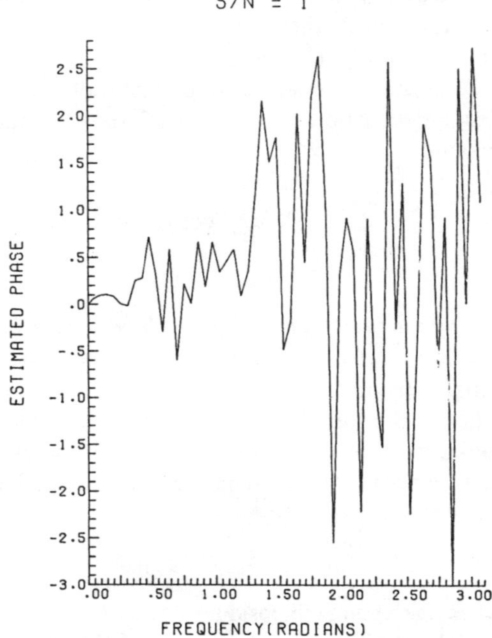

Fig. 3. Estimated phase for a simulation with relatively high signal-to-noise ratio.

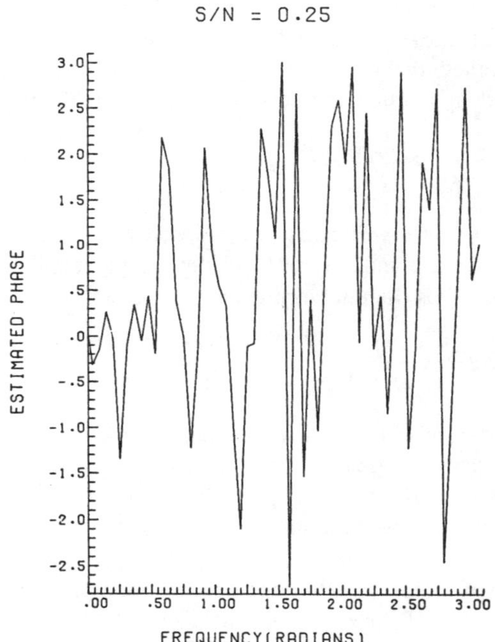

Fig. 4. Estimated phase for a simulation with relatively low signal-to-noise ratio.

The criterion being optimized will approach a limiting form as $T \to \infty$. These forms are shown in Figs. 6 and 7, for $S/N = 1$. Each is shown as a function of $\tau - \tau_0$ where τ_0 is the true value. In each case there is a main lobe and smaller sidelobes. The theoretical variances take no account of the possibility that a sample maximum may be near a sidelobe since, in the limit,

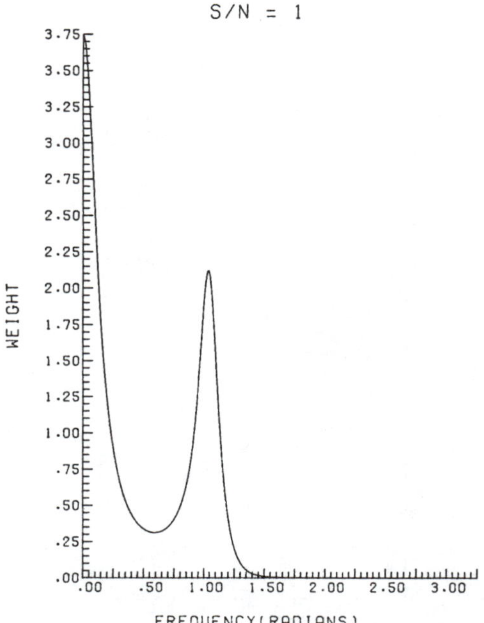

Fig. 5. Weight function $W(\omega)$ for relatively high signal-to-noise ratio.

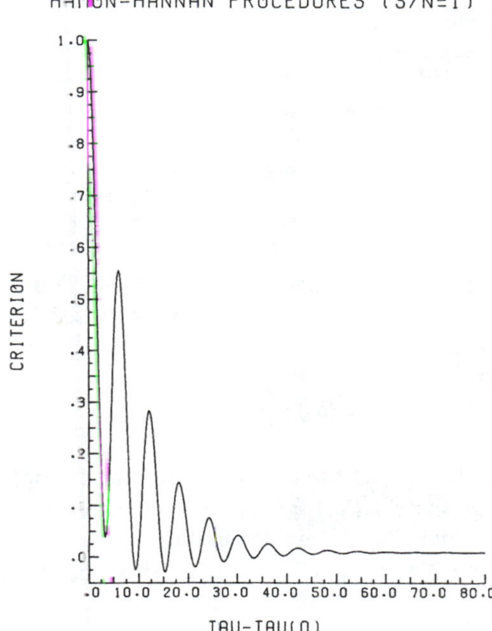

Fig. 6. Limiting form of the HH criterion for relatively high signal-to-noise ratio.

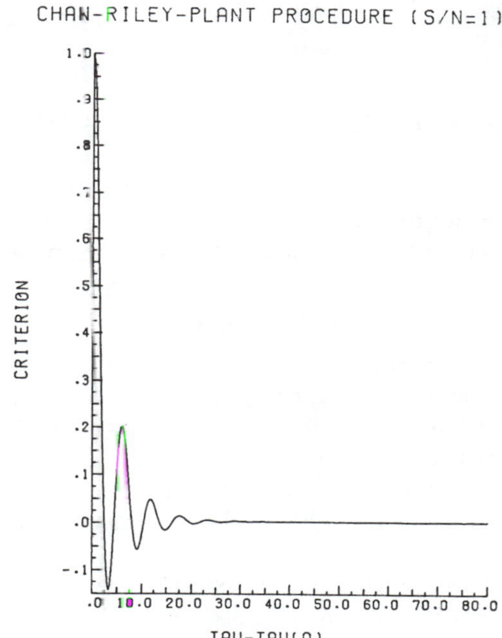

Fig. 7. Limiting form of the CRP criterion for relatively high signal-to-noise ratio.

that will not happen. However, again, it may happen in practice. (See [1, p. 141] for a discussion.) For $T = 1024$, this happened only once for the modified HH procedure, and then it happened for all three procedures. It happened twice for HH. It did not happen at $S/N = 1$ for $T = 256$, but for $T = 256$ and $S/N = 0.25$ it happened a number of times, especially for the modified HH procedure. This was due to the difficulty of fitting G_x, G_y, G_{xy} by a model of the form of (3) for $T = 256$ and α^2 relatively large. (It should be kept in mind that a distributed lag model may not work well except for large p, i.e., large T, when $G_x \neq G_y$ and the noise spectra are not white and equal.) For the HH procedures the false global maxima, when they occurred, were for large $\tau - \tau_0$. The technique of optimization used commenced iterations from τ_0 and thus located the nearest relative maximum to τ_0. Thus there was no inflation of the observed sd's from a false location of a global maximum for the HH procedures. It could be argued that this optimization procedure was reasonable, since values of $\hat{\tau}$ far away from τ_0 would be eliminated as spurious on prior grounds. For CRP there was a tendency for both a global and a relative maximum to be very near to τ_0, and the procedure on three or four occasions chose the relative maximum when $S/N = 0.25$, $T = 256$. This may have contributed to the reduction of the sd for CRP.

There is a problem with the location of the absolute maximum for HH because of the large number of relative maxima. However, this is easily overcome. Consider $Q_H(\tau)$ for example. This has period M, so consider τ of the form $t/2, t = 0, \pm 1, \cdots, \pm M$. Then, for such τ values, $Q_H(\tau)$ is the real part of

$$\frac{1}{T'} \sum_{k=1}^{T'} C_k \exp(j2\pi kt/T'), \qquad T' = 2T \tag{7}$$

where $C_k = T'M^{-1}\hat{W}(\lambda_u) \exp(-j\hat{\theta}(\lambda_u))$ for $\lambda_u = 2\pi k/T$ and zero otherwise. Now (7) may be evaluated *by fast Fourier transformation*, a global maximum located and a standard function optimization routine used to determine $\hat{\tau}$.

IV. Simulations: Dispersive Case

The simulations for the *dispersive case* were for the same S/N ratios, etc., the delay $D(\tau; \omega)$ being $\tau\omega$, so that $\phi(\omega) = \tau\omega^2$, with $\tau = 1.5$ and $T = 1024$. Again, 20 independent simulations were generated. The results are given in Table III. Since the CRP procedure does not apply to the *dispersive case*, the HH procedures only were simulated. The modified HH procedure performed better, but the significant discrepancies between the sample and theoretical sd persisted.

The limiting form of the criterion for $S/N = 1$ is shown in

TABLE III
DISPERSIVE CASE; SAMPLE SIZE = 1024

		Hamon-Hannan			Modified Hamon-Hannan		
$\frac{S}{N}$	τ	Sample mean	Sample std. dev.	Theoretical std. dev.	Sample mean	Sample std. dev.	Theoretical std. dev.
1	1.5	1.5615	0.1633	0.0697	1.5098	0.0914	0.0697
0.25	1.5	1.3542	0.3937	0.1718	1.4124	0.3304	0.1718

Fig. 8. A procedure similar to that based on (7) is again possible. Note that both criteria can be approximated as

$$\frac{1}{\pi}\int_0^{\pi} \overline{W}(\omega)\cos(\overline{\phi}(\omega) - \tau\omega^2)\,d\omega$$

where, in the case of the HH procedure, for example, $\overline{W}(\omega)$, $\overline{\phi}(\omega)$ are given by $\hat{W}(\lambda_u)$, $\hat{\phi}(\lambda_u)$ for ω a member of the narrow band containing λ_u. Consider evaluating the criterion over values of $\tau = t/4$ $(t = 0, \pm 1, \cdots, 2T)$. Then the above can be written as

$$\frac{1}{\pi}\int_0^{\pi} \omega^{-1/2}\, \overline{W}(2\sqrt{\omega})\cos t\omega\, d\omega$$

where we have defined $\overline{W}(\omega)$ to be zero for $\omega > \pi$. But this is approximately the real part of

$$\left\{\frac{1}{T}\sum_{k=1}^{T} (\pi k/T)^{-1/2}\, \overline{W}(2\sqrt{\pi k/T}\,)\exp j2\pi kt/(2T)\right\}$$

which can be evaluated by the fast Fourier transform. This technique can be generalized to deal with any dispersion situation where $D(\tau;\omega) = \tau D(\omega)$ and $D(\omega)$ is a monotone function of ω.

V. CONCLUSIONS

These simulations favor the HH procedures, and the modified HH procedure especially, except in the low S/N case for $T = 256$ and $\tau = 10.5$. A number of factors might make these results less relevant to a real estimation situation.

1) In a narrow-band situation, the variation of the coherence across the band might be small and optimal weighting less important.

2) The high experimental, as compared to theoretical, sd for the HH procedures is probably mainly due to overweighting of high frequencies where σ is low. As said in [1, Section 6.2], this could be reduced in effect by aggregating bands where $\hat{\sigma}$ is low with a smaller number of wider bands. It might be better to omit such bands from the calculations. In practice, this problem may be less important, since the frequencies where the signal is sensibly present may be known beforehand.

3) It is possible that, for the CRP procedure, the sample sd is too low because, by starting iterations at the true value, a subsidiary local maximum of the criterion has been formed. In practice, this subsidiary maximum could be at a wrong value, and this emphasizes the usefulness of scanning the criteria. This effect would increase as S/N falls and p rises. However, this may not be the only or major reason for the lower sd's, and an alternative explanation may be the advantage gained from a simple finite parameter model in the circumstances of the simulations.

4) The HH procedures could easily be misused to locate the estimate near a maximum of a sidelobe. The scanning procedures based on fast Fourier transformation avoid this.

In retrospect, many of these phenomena could have been investigated by a more extensive set of simulations involving a) a weighting of the frequencies with the true (asymptotic) weight function for the HH procedures and b) omission of frequencies above $\pi/2$ for the HH procedures. It also would have been interesting to examine the estimate of $\phi(\omega)$ got from the autoregression procedure and from $\Sigma\hat{\xi}_i \exp ij\omega$, and to examine more realistic models for the noise processes.

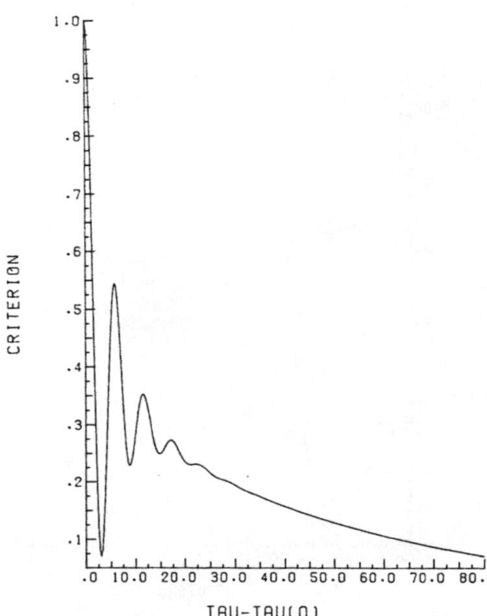

Fig. 8. Limiting form of the HH criterion in the dispersive case for relatively high signal-to-noise ratio.

REFERENCES

[1] B. V. Hamon and E. J. Hannan, "Spectral estimation of time delay for dispersive and non-dispersive systems," *J. Royal Statist. Soc. Ser. C (Appl. Statist.)*, vol. 23, pp. 134–142, 1974.

[2] Y. T. Chan, J. M. Riley, and J. B. Plant, "A parameter estimation approach to time delay estimation and signal detection," *IEEE Trans. Acoust., Speech, Signal Processing*, vol. ASSP-28, pp. 8–16, 1980.

[3] H. Akaike, "Fitting autoregressive models for prediction," *Ann. Inst. Statist. Math.*, vol. 21, pp. 243–247, 1969.

[4] J. C. Hassab and R. E. Boucher, "Optimum estimation of time delay by a generalized correlator," *IEEE Trans. Acoust., Speech, Signal Processing*, vol. ASSP-27, pp. 373–380, 1979.

[5] E. J. Hannan, *Multiple Time Series*. New York: Wiley, 1970.

[6] C. H. Knapp and G. C. Carter, "The generalized correlation method for estimation of time delay," *IEEE Trans. Acoust., Speech, Signal Processing*, vol. ASSP-24, pp. 320–327, 1976.

[7] G. E. P. Box and G. M. Jenkins, *Time Series Analysis Forecasting and Control*. San Francisco, CA: Holden-Day, 1970.

[8] E. J. Hannan and B. G. Quinn, "The determination of the order of an autoregression," *J. Royal. Statist. Soc. Ser. B*, vol. 41, pp. 190–195, 1979.

[9] B. G. Quinn, "Order determination for a multivariate autoregression," *J. Royal Statist. Soc. Ser. B*, vol. 42, pp. 182–185, 1980.

[10] P. Whittle, "On the fitting of multivariate autoregressions, and the approximate canonical factorization of a spectral density matrix," *Biometrika*, vol. 50, pp. 129–134, 1963.

A New Generalized Cross Correlator

ALFRED O. HERO, MEMBER, IEEE, AND STUART C. SCHWARTZ, MEMBER, IEEE

Abstract—A new generalized cross correlator (GCC) for the passive time delay estimation problem is presented. The interpretation of this GCC is that of estimating the cross-correlation function by cross correlating the least mean-square estimates of the signal component in each of the observed waveforms. The implementation is simply a GCC with the weighting filter equal to the magnitude coherency squared. Numerical evaluation of the performance of this processor and a robust version indicate that they compare favorably to some of the well-known GCC procedures.

I. Introduction

THE estimation of propagation delay in a common signal arriving at two spatially separated sensors is a problem which has received much attention in the literature. Cross-correlation methods, among which the standard cross correlator (CC) is the most basic, are particularly popular because of the richness and variety of processors within this class and the general ease of implementation. Of these, the Hannan-Thomson (HT) [1], [5], the Eckart (EK) [1], and the Hassab-Boucher (HB) [2] are examples of "optimum" processors which maximize some performance criteria. On the other hand, the SCOT [5] is an example of an "ad hoc" or intuitive correlation-type processor. In general, the optimum processors are very sensitive to deviations from the assumed signal and noise characteristics. By way of contrast, the CC and SCOT appear to be more robust to these deviations from the nominal model. However, these latter processors can have very poor performance at the nominal point.

In the following development, a different approach to the problem yields another type of (generalized) cross correlator. This is the Wiener Processor (WP), which has a simple form. Yet, preliminary results indicate that it outperforms most of the other above-named processors when compared under various performance criteria for the few important cases considered in this paper. A "robust" version of the WP also indicates good performance relative to the others under spectral uncertainty.

II. Problem Statement and Background

We first consider a system model generating the observations in Fig. 1. We observe Gaussian, ergodic, wide-sense stationary processes $x_1(t)$ and $x_2(t)$ over a time interval $[0, T]$ which contain uncorrelated broad-band noises $n_1(t)$ and $n_2(t)$ and signals $s(t)$ and $s_0(t)$, respectively. We assume that $c(t)$ is a linear time-invariant channel having a transfer function $C(\omega)$ with unknown linear phase so that $s_0(t)$ is a delayed but possibly distorted version of $s(t)$. Furthermore, we assume that the noises are uncorrelated with the signal and that T is much greater than the correlation time T_c of $x_1(t)$ and $x_2(t)$. The object is then to estimate the time delay D associated with the channel.

We define the sample cross correlation

$$\hat{R}_{12}(\tau) = \frac{1}{T}\int_0^T x_1(\sigma)\,x_2(\sigma+\tau)\,d\sigma \tag{1}$$

or for $T \gg T_c$ in the frequency domain,

$$\hat{R}_{12}(\tau) = \frac{1}{2\pi}\int_{-\infty}^{\infty} \hat{G}_{12}(\omega)\,e^{j\omega\tau}\,d\omega. \tag{2}$$

Here $\hat{G}_{12} = (1/T)\,X_1^* X_2$ where X_1 and X_2 are the finite time Fourier transforms of x_1 and x_2. For large observation time, $\hat{R}_{12}(\tau)$ is a good approximation to the true cross-correlation function which has a global peak at D. In fact, if $c(t)$ is pure delay and $s(t)$ is white, the cross-correlation function is a delta function at the true delay. For finite observation time, we can decompose $\hat{R}_{12}(\tau)$ into the sum of four terms:

$$\hat{R}_{12}(\tau) = c(\tau) * \hat{R}_{ss}(\tau) + c(\tau) * \hat{R}_{n_1 s}(\tau) + \hat{R}_{sn_2}(\tau) + \hat{R}_{n_1 n_2}(\tau) \tag{3}$$

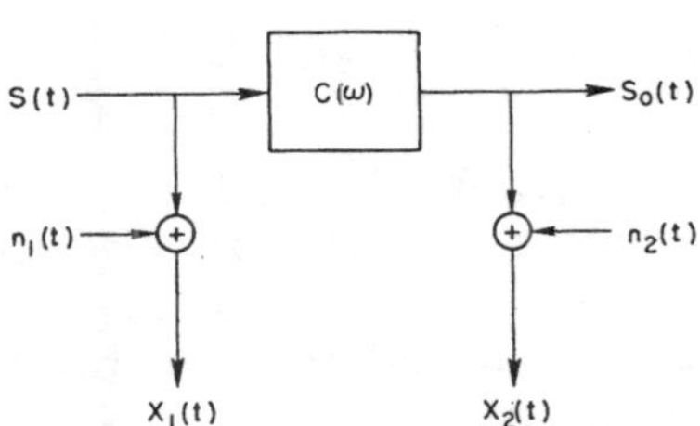

Fig. 1. Model governing the observations $x_1(t)$ and $x_2(t)$. $C(w)$ is the transfer function of a linear phase intersensor channel. The random signal $s(t)$ is independent of the noises $n_1(t)$ and $n_2(t)$.

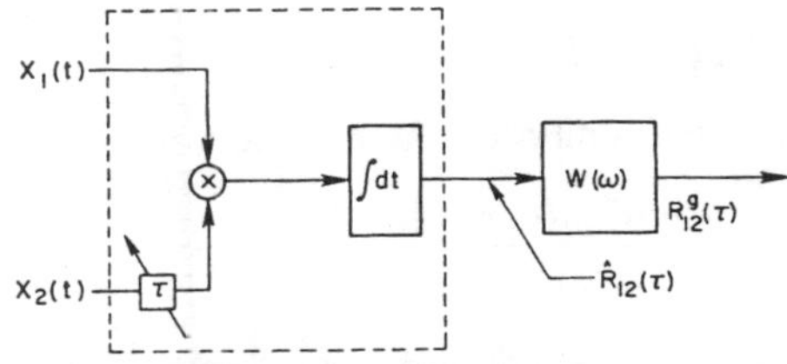

Fig. 2. Generalized cross correlator implemented as a filter on the sample cross-correlation function $\hat{R}_{12}$.

Manuscript received April 28, 1983; revised May 30, 1984. This work was supported in part by the Office of Naval Research under Contract N00014-81-K-0146.

A. O. Hero is with the Department of Electrical Engineering and Computer Science, University of Michigan, Ann Arbor, MI 48109.

S. C. Schwartz is with the Department of Electrical Engineering and Computer Science, Princeton University, Princeton, NJ 08544.

Reprinted from *IEEE Trans. Acoust., Speech, Signal Processing*, vol. 33, no. 1, pp. 38–45, February 1985.

where "$*$" denotes convolution. Here $\hat{R}_{ss}(\tau)$ is an estimate of the signal autocorrelation function $R_{ss}(\tau)$ and $\hat{R}_{n_1 s}(\tau)$, $\hat{R}_{sn_2}(\tau)$, and $\hat{R}_{n_1 n_2}(\tau)$ are estimates of the cross correlation between the respective signal and noise terms in the observations. In the limit, the sample cross correlation converges to $c(\tau) * R_{ss}(\tau)$ which displays an absolute maximum at D. Thus, it is the last three terms in (3) which constitute zero-mean disturbances affecting peak resolution of the first term. This suggests prefiltering the sample cross correlation with a filter $W(\omega)$ to obtain better resolution of the peak at D, where $W(\omega)$ has zero phase. This scheme is referred to as the generalized cross-correlation method or the generalized cross correlator (GCC) and is illustrated in Fig. 2. We denote the GCC output waveform $R_{12}^g(\tau)$. Therefore, we have

$$R_{12}^g(\tau) = \frac{1}{2\pi}\int_{-\infty}^{\infty} \hat{G}_{12}(\omega)\, W(\omega)\, e^{j\omega\tau}\, d\omega. \tag{4}$$

When $W(\omega)$ is unity, the resulting GCC is called the simple cross correlator (CC). Considering the first term in (3) as a "signal" in additive noise, classical optimal filtering theory can be applied to derive filters $W(\omega)$ which maximize signal-to-noise ratio.

Letting the last three terms of (3) be characterized as "noise," we can define a signal-to-noise ratio at the output of the GCC as the magnitude squared of the global peak of the "signal" term divided by the power of the "noise" which generates false peaks in $R_{12}^g(\tau)$. We will denote this SNR_1. For a sufficiently broad-band signal $s(t)$, the variance of the cross-correlation estimate (1) outside of the immediate vicinity of the true delay is given by

$$\text{var}\,(\hat{R}_{12}(\tau)) = \frac{1}{T2\pi}\int_{-\infty}^{\infty} G_{11}(\omega)\, G_{22}(\omega)\, d\omega \tag{5}$$

and the variance of $\hat{R}_{ss}$ is given by

$$\text{var}\,(\hat{R}_{ss}(\tau)) = \frac{1}{T2\pi}\int_{-\infty}^{\infty} |G_{12}(\omega)|^2\, d\omega \tag{6}$$

for $\tau \neq D$ [12]. $G_{11}(\omega)$ and $G_{22}(\omega)$ are the power spectral densities of the observations $x_1(t)$ and $x_2(t)$, respectively, and $G_{12}(\omega)$ is the cross spectrum. Using the above results, it is straightforward to derive the cross-correlation noise power which is given by

$$\sigma_n^2(\tau) = \frac{1}{2\pi T}\int_{-\infty}^{\infty} G_{11}(\omega)\, G_{22}(\omega)(1 - |\gamma_{12}(\omega)|^2) \cdot |W(\omega)|^2\, d\omega, \quad \tau \neq D. \tag{7}$$

$|\gamma_{12}(\omega)|^2$ is the magnitude coherency squared

$$|\gamma_{12}(\omega)|^2 = \frac{|G_{12}(\omega)|^2}{G_{11}(\omega)\, G_{22}(\omega)}. \tag{8}$$

Then from the defining relation

$$\text{SNR}_1 = \frac{[E\{R_{12}^g(\tau)|_{\tau=D}\}]^2}{\sigma_n^2}, \tag{9}$$

we obtain

$$\text{SNR}_1 = \frac{\left[\frac{1}{2\pi}\int_{-\infty}^{\infty} |G_{12}(\omega)|\, W(\omega)\, d\omega\right]^2}{\frac{1}{2\pi T}\int_{-\infty}^{\infty} G_{11}(\omega)\, G_{22}(\omega)(1 - |\gamma_{12}(\omega)|^2)|W(\omega)|^2\, d\omega}. \tag{10}$$

The maximum is obtained through the Schwarz inequality and yields the HT processor for the pure delay channel. The same result is derived in [4] as the result of minimizing the local variance of the delay estimate over the entire GCC class, and in [1] as the result of maximum likelihood estimation. The filter is

$$W_{\text{HT}}(\omega) = \frac{1}{|G_{12}(\omega)|}\, \frac{|\gamma_{12}(\omega)|^2}{1 - |\gamma_{12}(\omega)|^2}. \tag{11}$$

Neglecting the effect of the signal and noise cross terms, $c(\tau) * \hat{R}_{sn_1}(\tau)$ and $\hat{R}_{sn_2}(\tau)$ in (3) gives another characterization of the noise in the cross-correlation domain. With this definition of noise, another signal-to-noise ratio is defined in [6], SNR_2, which is shown for pure delay to be maximized by the Eckart Processor $W_{\text{EK}}(\omega)$

$$\text{SNR}_2 = \frac{\left[\frac{1}{2\pi}\int_{-\infty}^{\infty} |G_{12}(\omega)|\, W(\omega)\, d\omega\right]^2}{\frac{1}{2\pi T}\int_{-\infty}^{\infty} G_{n_1}(\omega)\, G_{n_2}(\omega)\, |W(\omega)|^2\, d\omega} \tag{12}$$

$$W_{\text{EK}}(\omega) = \frac{|G_{12}(\omega)|}{G_{n_1}(\omega)\, G_{n_2}(\omega)} \tag{13}$$

where $G_{n_1}(\omega)$ and $G_{n_2}(\omega)$ are the autospectra of the noises $n_1(t)$ and $n_2(t)$, respectively. Note that in terms of the spectra of the observables $x_1(t)$ and $x_2(t)$, the filter takes the form

$$W_{\text{EK}}(\omega) = \frac{|G_{12}(\omega)|}{(G_{11}(\omega) - |G_{12}(\omega)|)\,(G_{22}(\omega) - |G_{12}(\omega)|)}.$$

Hassab and Boucher [2] take the approach of maximizing a singal-to-noise ratio SNR_3 defined as the ratio of the expected peak energy at the true delay to the total statistical variation of the output of the GCC. This, in a sense, lumps the "signal" $c(\tau) * \hat{R}_{ss}(\tau)$ variation into the noise terms and yields the HB filter $W_{\text{HB}}(\omega)$

$$\text{SNR}_3 = \frac{\left[\frac{1}{2\pi}\int_{-\infty}^{\infty} |G_{12}(\omega)\,|\, W(\omega)\, d\omega\right]^2}{\frac{1}{2\pi T}\int_{-\infty}^{\infty} G_{11}(\omega)\, G_{22}(\omega)\, |W(\omega)|^2\, d\omega} \tag{14}$$

$$W_{\text{HB}}(\omega) = \frac{|G_{12}(\omega)|}{G_{11}(\omega)\, G_{22}(\omega)}. \tag{15}$$

The HB is similar to the SCOT introduced by Carter *et al.* [5] in that, for highly dynamic spectra, in addition to suppressing the cross-spectral estimate in ω-regions of low signal-to-noise ratio, high signal-to-noise ratio regions are also suppressed in an attempt to reject strong tonals in the observations.

Note that the above performance criteria impose equal penalty on small and large errors. That is, the location of the false peak in the GCC output exerts no influence on the signal-to-noise ratios defined in (9), (12), and (14). Therefore, one can only rely on these criteria if the signal-to-noise ratio is sufficiently high to guarantee a low probability of large error. The behavior of this probability as a function of signal-to-noise ratio, observation time, and signal bandwidth is investigated elsewhere [7], [13].

III. The Wiener Processor

Here a different approach is taken to derive an optimal filter. We deal directly with the quantities in the observation time domain (i.e., Fig. 1). The procedure is motivated by the following argument. If we knew the signal $s(t)$ and the filtered version $s_0(t)$ exactly, then from the linearity of the phase of the channel, the time delay could be estimated exactly by detecting the peak of the sample cross correlation of $s(t)$ and $s_0(t)$. Therefore, we simply try to estimate the signal $s(t)$ as best we can from the observations $x_1(t)$ and the channel output signal $s_0(t)$ from $x_2(t)$ by minimizing the mean-square errors:

$$E\{(s(t)-\hat{s}(t))^2\} = \min \tag{16}$$

$$E\{(s_0(t)-\hat{s}_0(t))^2\} = \min \tag{17}$$

where

$$\hat{s}(t) = \int_{-T}^{T} x_1(\sigma)\, h_1(t-\sigma)\, d\sigma \tag{18}$$

$$\hat{s}_0(t) = \int_{-T}^{T} x_2(\sigma)\, h_2(t-\sigma)\, d\sigma. \tag{19}$$

The above procedure is illustrated in Fig. 3. Given the channel characteristic $C(\omega)$, the solutions to (16) and (17) are the Wiener filters $H_1(\omega)$ and $H_2(\omega)$:

$$H_1(\omega) = \frac{G_{ss}(\omega)}{G_{ss}(\omega) + G_{n_1}(\omega)} \tag{20}$$

$$H_2(\omega) = \frac{G_{ss}(\omega)\, |C(\omega)|^2}{G_{ss}(\omega)\, |C(\omega)|^2 + G_{n_2}(\omega)} \tag{21}$$

Noting that $G_{12}(\omega) = C(\omega)\, G_{ss}(\omega)$, we can express the above filters in terms of the quantities derived from the observables

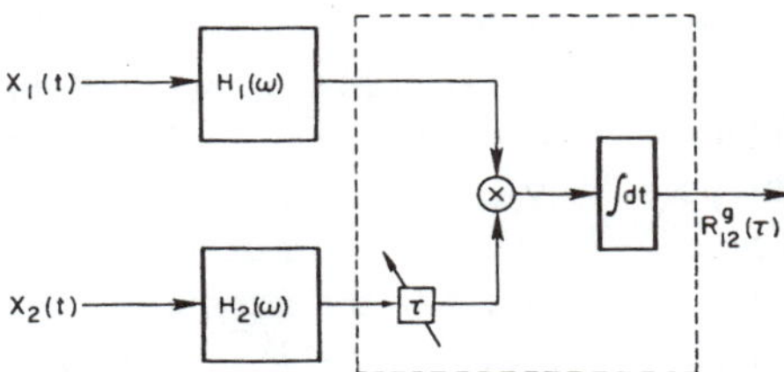

Fig. 3. Implementation of the generalized cross correlator as a preprocessor on the sensor waveforms. H_1 is the complex conjugate of H_2.

$$H_1(\omega) = \frac{1}{C(\omega)} \frac{G_{12}(\omega)}{G_{11}(\omega)} \tag{22}$$

$$H_2(\omega) = C^*(\omega) \frac{G_{12}(\omega)}{G_{22}(\omega)} \tag{23}$$

where $C^*(\omega)$ is the complex conjugate of $C(\omega)$.

With these filters, the sample cross correlation of the least mean-square error estimates of $s(t)$ and $s_0(t)$ yields the estimate of the cross-correlation function:

$$R_{12}^{WP}(\tau) = \frac{1}{2\pi} \int_{-\infty}^{\infty} \frac{1}{T} \hat{S}^*(\omega)\, \hat{S}_0(\omega)\, e^{j\omega\tau}\, d\omega \tag{24}$$

where

$$\hat{S}(\omega) = H_1(\omega)\, X_1(\omega) \tag{25}$$

$$\hat{S}_0(\omega) = H_2(\omega)\, X_2(\omega). \tag{26}$$

Regrouping terms in (24), we obtain

$$R_{12}^{WP}(\tau) = \frac{1}{2\pi} \int_{-\infty}^{\infty} \hat{G}_{12}(\omega) \frac{|G_{12}(\omega)|^2}{G_{11}(\omega)\, G_{22}(\omega)} e^{j\omega\tau}\, d\omega. \tag{27}$$

Comparing (27) to (4), we have the result that the WP is equivalent to using a generalized cross correlator with the filter $W(\omega)$ equal to the magnitude coherency squared.

It should be emphasized that even though the Wiener filters H_1 and H_2 involve the knowledge of the channel $C(\omega)$ itself, the GCC equivalent processor does not impose this requirement. In fact, as far as the cross-correlation estimate of time delay is concerned, the actual channel is immaterial to the peak detection procedure in the cross-correlation domain. Hence, the Wiener filter implementation (Fig. 3) with $C(\omega)$ arbitrarily set to unity in (22) and (23) is equivalent to any other choice of $C(\omega)$ for the time delay estimation problem.

From (5), the variance of the cross-spectrum estimate $\hat{G}_{12}(\omega)$ is proportional to the product of the autospectra of the observations $G_{11}(\omega)\, G_{22}(\omega)$. Fix the sample autocorrelation $\hat{R}_{ss}(\tau)$ in (3). Then the definition of "additive noise" leading to the signal-to-noise ratio SNR_1, (7), yields the interpretation of $1 - |\gamma_{12}(\omega)|^2$ as a measure of the cross-spectral estimator variance about the "desired signal" $c(\tau) * \hat{R}_{ss}(\tau)$. Thus, the WP deemphasizes those ω regions where the sample cross spectrum is likely to be a highly inaccurate estimate of the true cross spectrum. This is not surprising given the raison d'être of the WP which is to accurately estimate the smoothed sample autocorrelation $c(\tau) * \hat{R}_{ss}(\tau)$.

The WP does not, of course, maximize the signal-to-noise ratio in general. If we examine the optimal processor for SNR_1, the HT (11), we see that it has the additional ability to overemphasize as well as to deemphasize the cross-spectral estimate according to the function $|\gamma_{12}(\omega)|^2/(1-|\gamma_{12}(\omega)|^2)$. (Actually, in [4], the above function is shown to be inversely proportional to the variance of the phase estimate $\hat{G}_{12}(\omega)/|\hat{G}_{12}(\omega)|$ with respect to the true phase of the cross spectrum.) However, in situations where the coherence is low and the signal spectrum is nearly flat, the HT and the WP are virtually identical and exhibit identical performance [(11) becomes proportional to $|\gamma_{12}|^2$].

It is also observed that the WP is equivalent to the HB for nearly flat signal spectra, and also to the Eckart if we add a low signal-to-noise ratio condition

$$W_{\rm HB}(\omega) = \frac{1}{|G_{12}(\omega)|} |\gamma_{12}(\omega)|^2 = \frac{1}{|G_{12}(\omega)|} W_{\rm WP}(\omega) \quad (28)$$

$$W_{\rm EK}(\omega) = \frac{G_{ss}(\omega)}{G_{n_1}(\omega) G_{n_2}(\omega)} \approx \frac{1}{|G_{12}(\omega)|} |\gamma_{12}(\omega)|^2. \quad (29)$$

The above signal-to-noise ratio condition is that $G_{ss}(\omega)$ be uniformly small as compared to $G_{n_1}(\omega)$ and $G_{n_2}(\omega)$.

IV. The Robust Wiener Processor for Unknown Spectra

The optimal GCC's all require knowledge of the signal and noise spectra underlying the observations. When the spectral quantities used in the filter function for the GCC do not match the true spectra, there is a consequent deterioration in performance. Two approaches to the problem of unknown spectra are of interest. We either estimate the spectra and substitute the estimates into the aforementioned filters (totally unknown spectra) or we search for a robust solution over a range of spectra perturbed from some nominal point (partially unknown spectra).

For the estimation approach, the sensitivity of the GCC filter to small deviations in the estimated spectra may be an important consideration. As applied to the HT, the substitution method yields a procedure which weights the phase of the sample cross spectrum $\hat{G}_{12}(\omega)$ with the function $\hat{W}_{\rm HT}(\omega) = |\hat{\gamma}_{12}(\omega)|^2/(1 - |\hat{\gamma}_{12}(\omega)|^2)$, $|\hat{\gamma}_{12}(\omega)|^2$, a magnitude coherency squared estimate. A simple local analysis of the estimation error associated with $\hat{W}_{\rm HT}$ yields the variance

$$\text{var}(\hat{W}_{\rm HT}(\omega)) \approx \frac{1}{(1 - |\gamma_{12}(\omega)|^2)^4} \text{var}(|\hat{\gamma}_{12}(\omega)|^2). \quad (30)$$

$\hat{W}_{\rm HT}$ may critically underweight the phase estimate over frequencies where $|\gamma_{12}|^2$ is high, that is, where the phase estimate is apt to be the most accurate. On the other hand, substitution of $|\hat{\gamma}_{12}|^2$ for the WP gives only as much error as the estimation error of $|\hat{\gamma}_{12}|^2$ itself. An improved filter estimate could translate into improved performance of the time delay estimate. Naturally, these comments must be verified through a future simulation study.

In practice, the spectra may be only partially unknown, and a different strategy can be used to design the GCC. This is the "robust" approach which has been applied to classical matched and Wiener filtering with some success [8]-[10]. The resultant filters are robust in the maximim sense, e.g., the filter maximizes the minimum output signal-to-noise ratio as the spectra are allowed to vary over their regions of uncertainty. A more precise formulation is as follows.

It is assumed that the uncertainty corresponding to our partial knowledge of signal and noise spectra G_{ss}, G_{n_1}, and G_{n_2} is such that these quantities can be described as belonging to the uncertainty classes of spectra σ, η_1, and η_2, respectively. Frequently, the above classes represent neighborhoods centered around some nominal spectra G_{ss}^0, $G_{n_1}^0$, and $G_{n_2}^0$ where the size of the classes is chosen to reflect our overall confidence in the nominals as underlying the observations. Two common classes for the signal spectrum are the $\epsilon d - d$ mixture for $\epsilon > 0$ [9],

$$\sigma = \left\{G_{ss}: G_{ss} = (1 - \epsilon) G_{ss}^0 + \epsilon G_{ss}' \text{ and } \int_{-\infty}^{\infty} G_{ss}(\omega)\, d\omega = 2\pi v_s^2\right\} \quad (31)$$

and the total-variation class for $\delta > 0$ [9],

$$\sigma = \left\{G_{ss}: \int_{-\infty}^{\infty} |G_{ss}(\omega) - G_{ss}^0(\omega)|\, d\omega \leqslant \delta \text{ and } \int_{-\infty}^{\infty} G_{ss}(\omega)\, d\omega = 2\pi v_s^2\right\} \quad (32)$$

where, in (31), G_{ss}' is an arbitrary spectrum and v_s^2 is a known signal power. Analogous expressions can hold for the noise spectra.

Let W be a function in the class of filter functions $\boldsymbol{W}$. Now define ρ, a GCC performance criterion, e.g., signal-to-noise ratio, as a function of the three spectral quantities and the filter

$$\rho \equiv \rho(G_{ss}, G_{n_1}, G_{n_2}, W). \quad (33)$$

It is desired to find a $W(\omega)$ that maintains good performance as the spectra vary over the uncertainty classes. The idea is to design an optimum filter under adversity, i.e., one that assures us of a minimum level of "good" performance regardless of the underlying spectra. The *robust processor* for ρ is defined as a W, say W^R, that arises from the solution of

$$\max_{\boldsymbol{W}} \min_{\sigma, \eta_1, \eta_2} \rho(G_{ss}, G_{n_1}, G_{n_2}, W). \quad (34)$$

Related to robust processing is the concept of least favorable spectra discussed by Lehmann [14]. These are points in the uncertainty classes where the worst possible performance occurs subject to optimal filtering strategy. Specifically, let W^l be a filter which maximizes ρ for the set of spectra $(G_{ss}^l, G_{n_1}^l, G_{n_2}^l) \in \sigma \times \eta_1 \times \eta_2$. Then $(G_{ss}^l, G_{n_1}^l, G_{n_2}^l)$ is a *least favorable set of spectra* for ρ if, for any other set $(G_{ss}, G_{n_1}, G_{n_2}) \in \sigma \times \eta_1 \times \eta_2$,

$$\rho(G_{ss}, G_{n_1}, G_{n_2}, W^l) \geqslant \rho(G_{ss}^l, G_{n_1}^l, G_{n_2}^l, W^l). \quad (35)$$

Note that if (35) holds, we have a saddlepoint solution of (34) and we can take $W^R = W^l$.

Unfortunately, no results are known to us concerning the solution of the above robust time delay estimation problem, i.e., (34). Short of this, the only known published result in maximin filters for time delay is that of Kassam and Hussaini [11] for the pure delay case. In [11], they used the fact that the Eckart processor maximizes a classically defined signal-to-noise ratio [see (12)] to relate the filtering problem to robust hypothesis testing. This is achievable only by associating uncertainty classes with the spectral product $G_{n_1}(\omega) G_{n_2}(\omega)$ rather than with the individual noise spectra themselves.

An alternate approach to performance degradation under

spectral uncertainty is suggested by recent work in robust Wiener filtering [9], [10] when applied to the WP. Assume that the uncertainty classes σ, η_1, and η_2 govern the individual spectra as above, and that $|C(w)| = 1$. Here, one simple constructs two robust Wiener filters for the signal s, each acting on x_1 and x_2, respectively. This procedure can then be related to a GCC in the same manner as the Wiener filters were related to the WP in Section III.

Consider first the waveform $x_1(t)$ in Fig. 1. Define the mean-square error of an estimate of $s(t)$ obtained by linearly filtering $x_1(t)$ with $H_1 \in \boldsymbol{H}$, a class of filter functions, by

$$e \equiv e(G_{ss}, G_{n_1}, H_1). \tag{36}$$

Following Poor [9], $H_1 = H_1^R$ is called a *robust Wiener filter* if it is the solution of

$$\min_{\boldsymbol{H}} \max_{\sigma, \eta_1} e(G_{ss}, G_{n_1}, H_1). \tag{37}$$

Let H_1^l be the Wiener filter for $(G_{ss}^l, G_{n_1}^l)$, i.e., H_1^l minimizes the mean-square error for these spectra. The pair $(G_{ss}^l, G_{n_1}^l) \in \sigma \times \eta_1$ is called *least favorable for Wiener filtering* if, for any other spectra $(G_{ss}, G_{n_1}) \in \sigma \times \eta_1$,

$$e(G_{ss}, G_{n_1}, H_1^l) \leqslant e(G_{ss}^l, G_{n_1}^l, H_1^l). \tag{38}$$

If the uncertainty classes given by (31) and (32) govern the signal and noise spectra, it is shown in [9] that least favorable spectra exist and that H_1^l is the robust Wiener filter

$$H_1^R = \frac{G_{ss}^l}{G_{ss}^l + G_{n_1}^l}. \tag{39}$$

The same discussion carries over to the waveform $x_2(t)$ with the analogous robust Wiener filter

$$H_2^R = \frac{G_{ss}^l}{G_{ss}^l + G_{n_2}^l}. \tag{40}$$

With regard to the original formulation of the WP, we can use the robust filters H_1^R and H_2^R in place of H_1 and H_2 [see (22) and (23)]. Finally, we implement these filters in the cross-correlation domain as a GCC, a scheme which we will call the *Robust Wiener Processor*, or the RWP. In the next section, we give a detailed example of least favorable spectra and the resulting robust filters. For the example considered, there is a 3 dB processing gain at low signal-to-noise ratio using the RWP.

V. Numerical Comparisons

At the present time, no simulation results concerning the experimental performance of the WP and RWP as opposed to the other GCC's are available. In their absence, a preliminary investigation of the relative merits of the above processors was performed based on the various signal-to-noise ratio criteria defined in Section II for some specific observation spectra and for the pure delay channel.

Example 1

Figs. 6-10 show the relative performance of the HT, HB, Eckart, SCOT, and CC under the criteria SNR_1, SNR_2, SNR_3, and local variance var_L of the time delay estimate [3] for a third-order Markov signal in first-order Markov noises with the noise 3 dB bandwidth a factor of ten greater than that of the signal (see Figs. 4 and 5). Specifically, the signal spectrum has the form

$$G_{ss}(\omega) = S \frac{3}{2\pi B_s} \frac{1}{1 + \left(\dfrac{\omega}{2\pi B_s}\right)^6} \tag{41}$$

with $B_s = 10$ Hz, and the noise spectra $G_{n_1}(\omega)$ and $G_{n_2}(\omega)$ are of identical form:

$$G_n(\omega) = N \frac{1}{2\pi B_n} \frac{1}{1 + \left(\dfrac{\omega}{2\pi B_n}\right)^2} \tag{42}$$

with $B_n = 100$ Hz. Here S and N are the input signal and noise powers, respectively. These spectra were chosen for their tail behavior to avoid certain degeneracies in the local variance criterion. The interesting thing to note is that under SNR_1 and SNR_2, the WP exhibits better performance than all of the other suboptimum GCC's for that particular definition of SNR. Under SNR_3, it is a close second next to the MLE. In fact, under the criterion SNR_1, performance of the WP is virtually identical to the optimal HT processor. Although the local variance ranks the WP behind the HT, HB, and Eckart (see Figs. 9 and 10), it only marginally disfavors the WP at low signal-to-noise ratios. (It is to be noted from Fig. 9 that the SCOT and the CC have local variance orders of magnitude worse than the WP and are off scale. They are shown in Fig. 10.)

Example 2

Here the performance of the RWP, WP, and other GCC's are compared using SNR_1 for the ϵ-contaminated uncertainty class on the specific spectra in the example outlined in Kassam and Lim's paper on robust Wiener filtering [9]. Specifically, under the nominal assumption, at each sensor we have a signal with the flat band-limited spectrum $G_{ss}^0(\omega)$ in first-order Markov noise with the spectrum $G_n^0(\omega)$ where the signal and noises are of comparable bandwidths (see Figs. 11 and 12). The uncertainty on the signal and noise spectra are modeled as the ϵ-mixtures

$$(1 - \epsilon_1) G_{ss}^0(\omega) + \epsilon_1 G_{ss}'(\omega) \tag{43}$$

and

$$(1 - \epsilon_2) G_n^0(\omega) + \epsilon_2 G_n'(\omega), \tag{44}$$

respectively, with $G_{ss}'(\omega)$ and $G_n'(\omega)$ arbitrary spectra having the same mass as the nominal and ϵ_1 and ϵ_2 lying in the interval $[0, 1]$. When $\epsilon_1 = 0.2$ and $\epsilon_2 = 0.1$, the least favorable signal and noise spectra are plotted in Figs. 13 and 14, respectively. This corresponds to the case where one may have more confidence in the nominal noise than in the nominal signal. The least favorable spectra for this example illustrate a typical attribute of least favorables in that the worst performance of a Wiener filter occurs when the signal masquerades as the noise and vice versa, i.e., when we get a minimum separation of hypotheses concerning the presence or absence of the signal

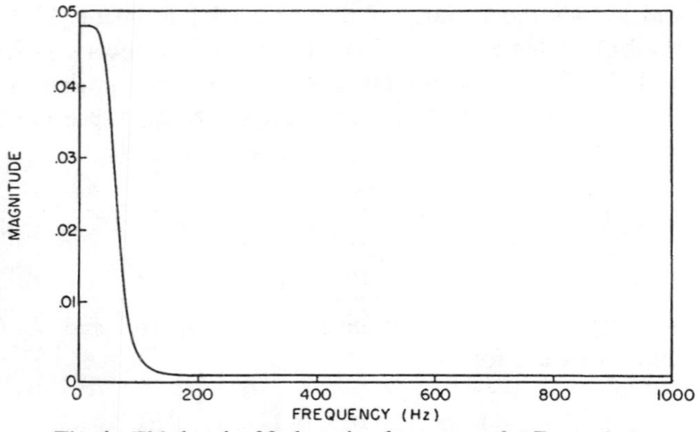

Fig. 4. Third-order Markov signal spectrum for Example 1.

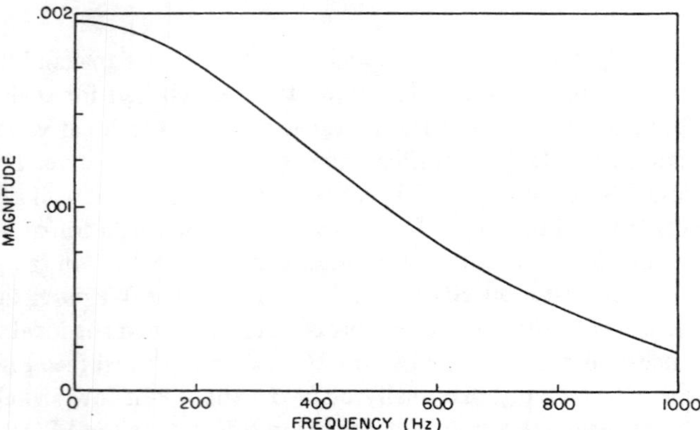

Fig. 5. First-order Markov noise spectra for Example 1.

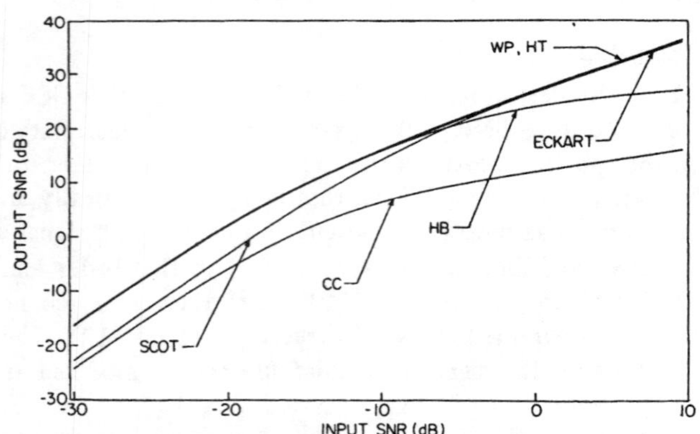

Fig. 6. Relative performance of various processors under criterion SNR_1.

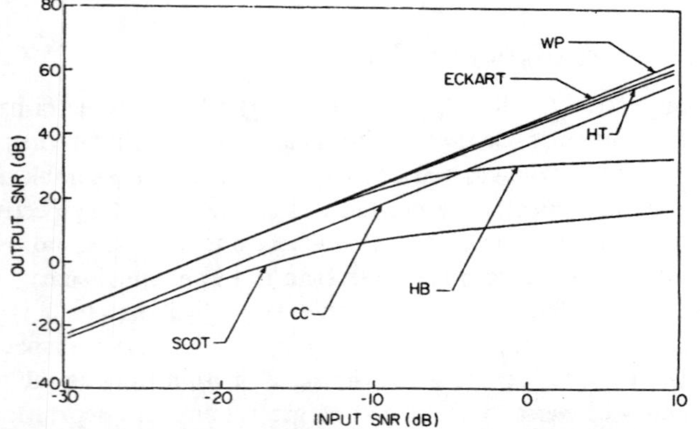

Fig. 7. Relative performance of various processors under criterion SNR_2.

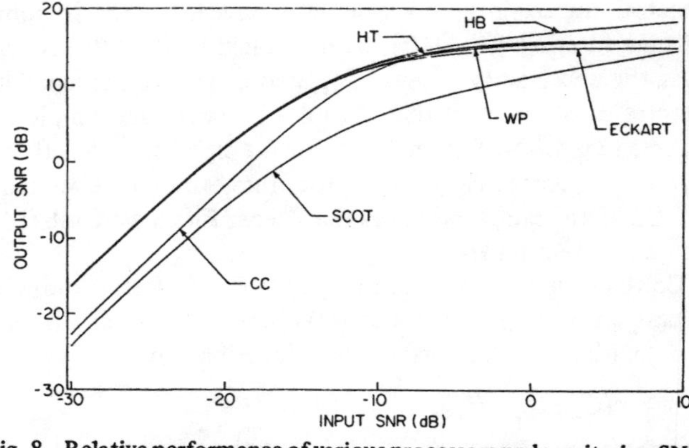

Fig. 8. Relative performance of various processors under criterion SNR_3.

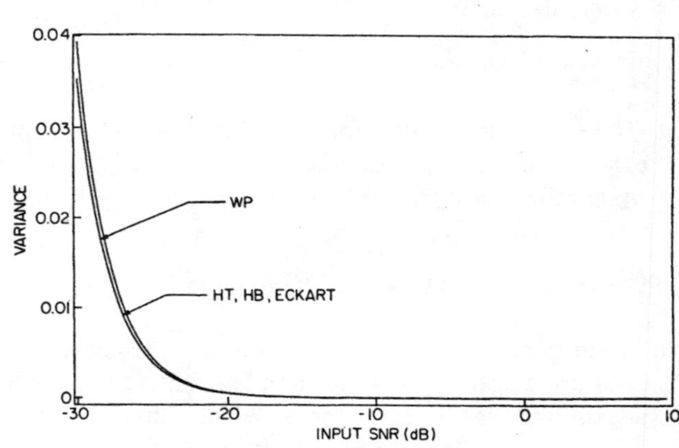

Fig. 9. Local variance for various processors.

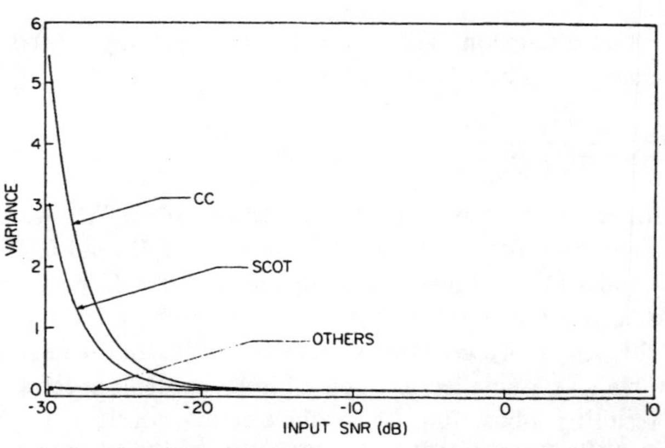

Fig. 10. Local variance for various processors (magnification of Fig. 9).

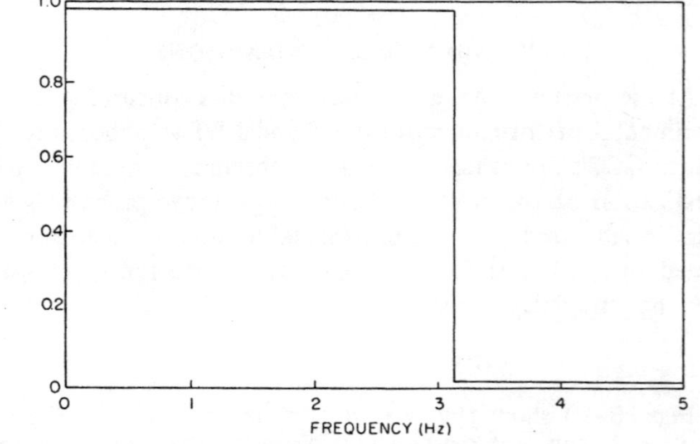

Fig. 11. Signal spectrum for Example 2.

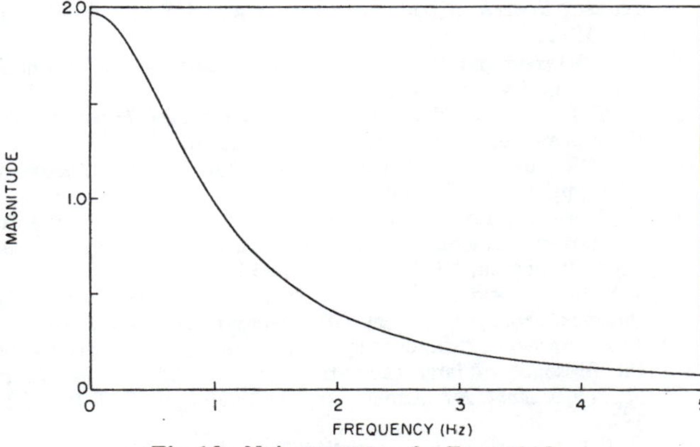

Fig. 12. Noise spectrum for Example 2.

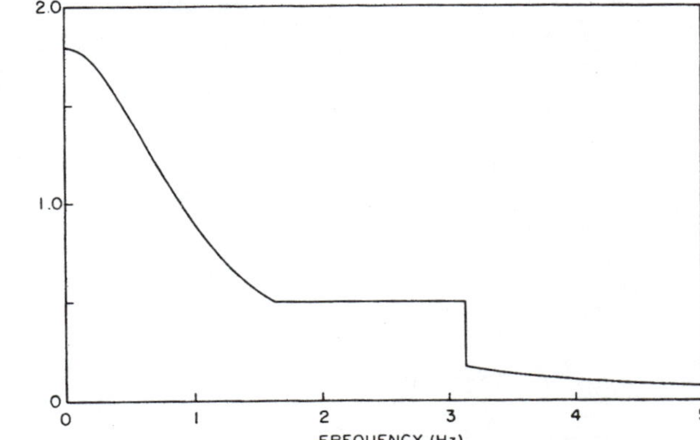

Fig. 13. Least favorable signal spectrum for Example 2.

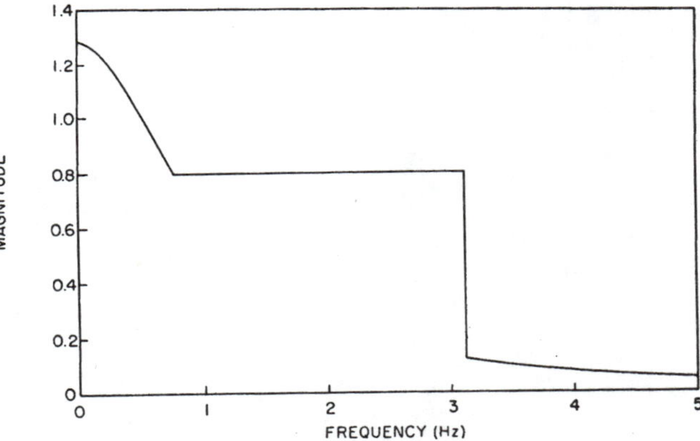

Fig. 14. Least favorable noise spectrum for Example 2.

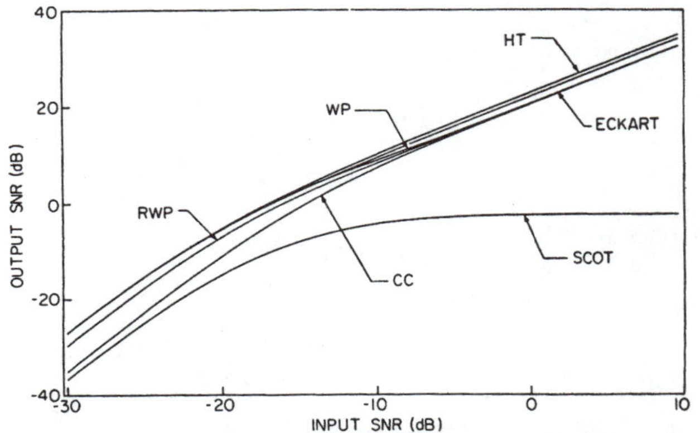

Fig. 15. Output SNR_1 for various processors at the nominal spectrum.

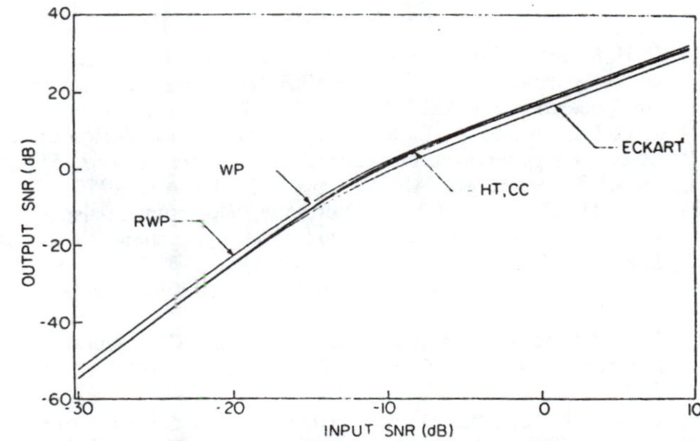

Fig. 16. Output SNR_1 for various processors at the least favorable spectrum.

within the uncertainty classes (43) and (44). Fig. 15 shows the relative performance for the nominal spectra and Fig. 16 the performance for the least favorable signal and noise spectra for Wiener filtering. Looking at the nominal case, we note that the use of the RWP entails a loss of about 3 dB at low SNR (below about 0 dB) over the optimal for the least favorable pair. However, when the true signal and noise spectra are least favorable for Wiener filtering, the RWP displays uniformly better relative performance, gaining about 3 dB over the other processors at low signal-to-noise ratios. Note that the pair in Figs. 13 and 14 is not necessarily the least favorable pair for HT filtering, so that no conclusive result is indicated here. However, Fig. 16 does suggest that, at least for some spectra in the above uncertainty class, we can expect better performance with the RWP than with the optimal scheme for the nominal spectra.

VI. Conclusion

We have outlined the development of several of the most popular GCC implementations and have introduced a simple GCC, the WP, which is believed to be new. Although the new GCC was motivated by intuition rather than any overall optimality considerations, it has been shown that the WP can perform well with respect to the other optimal schemes, as measured by their respective optimality criteria. For cases where the spectral uncertainty can be modeled as restricted to a specific class, classical robustness theory leads to a simple modification of the WP, which we called the RWP. The evaluation of its theoretical performance for a typical uncertainty class of signal and noise spectra was performed, which indicated a potential gain in output signal-to-noise ratio relative to the other GCC implementations. The above results suggest that the WP and the RWP may be viable alternatives to existing time delay estimation schemes. Further experimental- and simulation-based performance evaluation is required before any general conclusions can be drawn.

Acknowledgment

We thank the anonymous reviewers for their thorough review and constructive comments.

REFERENCES

[1] C. H. Knapp and G. C. Carter, "The generalized correlation method for estimation of time delay," *IEEE Trans. Acoust., Speech, Signal Processing*, vol. ASSP-24, pp. 320-327, Aug. 1976.

[2] J. C. Hassab and R. E. Boucher, "Optimum estimation of time delay by a generalized correlator," *IEEE Trans. Acoust., Speech, Signal Processing*, vol. ASSP-27, pp. 373-380, Aug. 1979.

[3] V. H. MacDonald and P. M. Schultheiss, "Optimum passive bearing estimation in a spatially incoherent noise environment," *J. Acoust. Soc. Amer.*, vol. 46, no. 1, pp. 37-43, 1969.

[4] E. J. Hannan and P. J. Thomson, "Estimating group delay," *Biometrika*, vol. 60, pp. 241-253, 1973.

[5] G. C. Carter, A. H. Nuttall, and P. G. Cable, "The smoothed coherence transform," *Proc. IEEE*, vol. 61, pp. 1497-1498, Oct. 1973.

[6] C. Eckart, "Optimal rectifier systems for the detection of steady signals," Scripps Inst. Oceanography, Marine Physics Lab., Univ. California, Rep. S10 12692, S10 Ref. 52-11, 1952.

[7] A. J. Weiss and E. Weinstein, "Fundamental limitations in passive time delay estimation–Part I: Narrow-band systems," *IEEE Trans. Acoust., Speech, Signal Processing*, vol. ASSP-31, pp. 472-485, Apr. 1983.

[8] S. A. Kassam and T. L. Lim, "Robust Wiener filters," *' Franklin Inst.*, vol. 304, pp. 171-185, 1977.

[9] H. V. Poor, "On robust Wiener filtering," *IEEE Trans. Automat. Contr.*, vol. AC-25, pp. 531-536, June 1980.

[10] —, "Robust matched filters," *IEEE Trans. Inform. Theory*, vol. IT-29, pp. 677-687, Sept. 1983.

[11] E. K. Al-Hussaini and S. Kassam, "Robust filters for time delay estimation problems," in *Proc. 16th Annu. Conf. ISS* Princeton Univ., Princeton, NJ, Mar. 1982, pp. 540-545.

[12] J. S. Bendat and A. G. Piersol, *Random Data: Analysis and Measurement Procedures*. New York: Wiley-Interscience Series, 1971.

[13] J. P. Ianniello, "Time delay estimation via cross-correlation in the presence of large estimation errors," *IEEE Trans. Acoust., Speech, Signal Processing*, vol. ASSP-30, pp. 998-1003, Dec. 1982.

[14] E. L. Lehmann, *Testing Statistical Hypotheses*. New York: Wiley, 1959.

Time Delay Estimation by Combining Efficient Algorithms and Generalized Cross-Correlation Methods

DAVID HERTZ

Abstract—**This paper analyzes the performance of the two efficient algorithms (presented by Stein and Cabot) for time delay of arrival estimation (TDE) between two signals. It is shown that these estimators are unbiased, and explicit expressions for the TDE mean-square error (MSE) are presented. It is also shown how to improve the performance of these algorithms by combining them with generalized cross-correlation (GCC) methods.**

In the analysis, we only assume stationary signals which are not necessarily Gaussian.

The first algorithm (Stein) is indirect and uses the symmetry of the cross-correlation function between the two signals. It is shown here that the TDE–MSE depends on the unknown delay. The performance of this algorithm can be improved by combining it with the GCC method, and the pertinent TDE–MSE expressions are presented.

The second algorithm (Cabot) is based on finding the zero of the cross-correlation function between one signal and the Hilbert transform of the other signal. Here, too, the pertinent TDE–MSE expressions are presented.

This algorithm is also combined with the GCC method, and the optimal weight function for which the TDE–MSE expression coincides with the Cramer–Rao lower bound is found.

I. Introduction

TIME delay of arrival estimation (TDE) has been the theme of a recent special issue [1], and numerous papers were published thereafter.

The popular methods for TDE are based on cross-correlation (CC) and generalized cross-correlation (GCC) methods which are implemented in the time domain or frequency domain [2]–[10].

In [8], TDE by CC and GCC methods is reexamined for the whole class of stationary signals. In this paper, the performance of two efficient algorithms for TDE described in [6] and [7] is presented. The analysis is carried out by extending the results in [8] and assuming stationary signals and very large observation time as in [8]. These two algorithms [6], [7] are based on computing the CC function in the time domain. By the algorithm of [6], the peak of the CC function is searched for indirectly.

This algorithm of [6] is indeed a practical fast approximation to the computation, starting from a base of sampled data, with an objective of reducing the number of interpolations needed to estimate location of the peak. Underlying this is the assumption that the sample interval is smaller than that required by Nyquist considerations for the input signals (i.e., sampling rate above the Nyquist rate).

By the algorithm of [7], the zero of the CC function between one signal and the Hilbert transform of the other signal is searched for. It will be shown in Section III that this algorithm is also a fast approximation to the computation, starting from a base of sampled data, with a similar objective as above of reducing the number of interpolations needed to estimate location of the zero and a similar assumption as above on the sampling rate. The algorithm of [7] has also been dealt with by other authors in more detail (see [4], [9], and the references cited in [7]).

These two algorithms also have importance for time variable delays, and they are computationally advantageous over directly searching the peak of the CC function.

In this paper, only the local variations in the neighborhood of the unknown delay D are studied. It is pointed out in [9] that the performance of an estimate of D is not sufficiently characterized by the local variations, and the occurrence of false peaks must be taken into account. The probability of such ambiguous peaks in the correlation function is derived in [9].

It will be shown that these two algorithms should be preferred over the CC method when the noise bandwidth is much larger than the signal's bandwidth and at low SNR's.

It will also be shown that the TDE–MSE for the algorithm in [6] depends on the unknown delay D and the sampling rate T_s^{-1}, and for signal and noise possessing flat power spectral densities we have

$$\max_D \text{MSE}/\min_D \text{MSE} = (4/\pi)^2 \simeq 1.6.$$

It will be shown that the errors associated with the zero crossing of the Hilbert transform cross correlation are not the same as the errors associated with the cross-correlation peak and that they are independent of the unknown delay.

Note that the TDE–MSE for searching the peak of the CC function directly does not depend on the unknown delay (see [2], [3], and [8]).

These two algorithms will be combined with the GCC method, and a performance analysis will be carried out.

Manuscript received June 26, 1984; revised July 5, 1985.

The author is with the Department of Electrical Engineering, Technion—Israel Institute of Technology, Haifa 32000, Israel.

IEEE Log Number 8406030.

Reprinted from *IEEE Trans. Acoust., Speech, Signal Processing*, vol. 34, no. 1, pp. 1–7, February 1986.

The GCC method consists of linear prefiltering of the two received signals and finding the maximum of the cross-correlation function of the filtered signals. It is well known [2], [3], [5], [8], [9] that there exist appropriate prefilters for which the MSE is minimized (MMSE).

Simulation results presented in [10] reveal the benefits of optimum prefiltering.

Assume the model

$$\begin{cases} x(t) = s(t) + n_1(t) \\ y(t) = s(t - D) + n_2(t) \end{cases} \quad -T/2 \leq t \leq T/2 \tag{1}$$

where the signal $s(t)$ and noise $n_1(t)$, $n_2(t)$ are real baseband signals. D is the unknown delay, T is the observation time, and the following assumptions hold.

Assumption A: The signal $s(t)$ and noise $n_1(t)$, $n_2(t)$ are stationary zero mean band-limited signals uncorrelated with each other.

Assumption B: The unknown delay and the correlation durations of the signal $s(t)$ and noise $n_1(t)$, $n_2(t)$: τ_s, τ_{n_1}, τ_{n_2}, respectively, are very small compared to the observation time T. Weak ergodicity of $s(t)$ is also assumed.

In this paper, we will show that for both algorithms the estimator $\hat{D}$ of D is unbiased, and we will derive general expressions for their TDE–MSE. We shall now present TDE–MSE expressions which were derived by using the following simplifying assumption.

Assumption C: The noises $n_1(t)$ and $n_2(t)$ possess flat power spectral density (PSD) of frequency bandwidth $B_{n_1} = B_{n_2} = B_n$ (single sided) and their PSD is denoted by $\overline{G}_{n_1}$ and $\overline{G}_{n_2}$, respectively. Let the bandwidth of the signal $s(t)$ be B_s (single sided) and its PSD not necessarily flat. Also assume that the "leakage" of the PSD's of $s(t)$, $n_1(t)$ and $n_2(t)$ outside of their bandwidths is negligible, and that $B_n > B_s$.

These simplified TDE–MSE expressions are convenient working formulas [6], [8] and therefore are presented in this section. The general expressions for TDE–MSE and their detailed derivations are presented in Sections II and III for the algorithms of [6] and [7], respectively. The performance of both algorithms combined with the GCC method is presented in Section IV.

For the algorithm of [6] we will show

$$\text{MSE} = E(\hat{D} - D)^2$$
$$\doteq \frac{1}{2B_n T \gamma_S \left\{ \left(\dfrac{R_s'(D - \tau_k)}{R_s(0)} \right)^2 \Big/ \left[1 - \dfrac{R_s(2D - 2\tau_k)}{R_s(0)} \right] \right\}} \tag{2}$$

where

$R_s(\tau)$ autocorrelation function of $s(t)$
B_n frequency bandwidth of $n_1(t)$ and $n_2(t)$ (single sided)
γ_s equivalent signal-to-noise ratio defined by

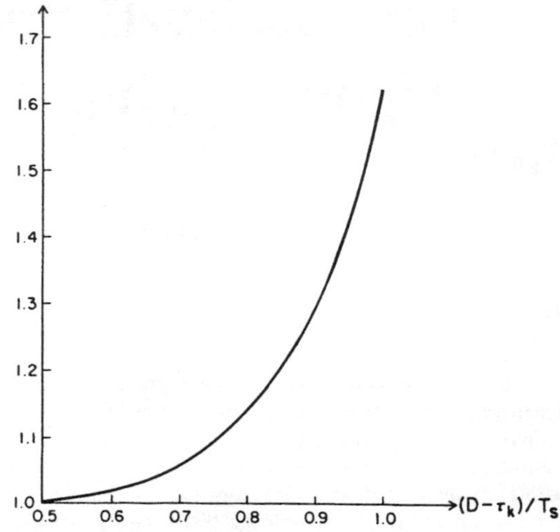

Fig. 1. Illustration of the dependence of the MSE in the algorithm of [6] on the unknown delay.

$$\frac{1}{\gamma_S} = \frac{1}{2} \left(\frac{1}{\gamma_1} + \frac{1}{\gamma_2} + \frac{1}{\gamma_1 \gamma_2} \frac{1 - \text{sinc}\,[4B_n(D - \tau_k)]}{[1 - R_s(2D - 2\tau_k)/R_s(0)]} \right) \tag{3}$$

where γ_1 and γ_2 [6] are the signal-to-noise ratios associated with $x(t)$ and $y(t)$ [see (1)], respectively, and sinc $(x) \triangleq \sin(\pi x)/(\pi x)$.

$\hat{D}$ the estimator of D
τ_k is given by kT_s (see Fig. 2); T_s is the sampling period.

If the signal $s(t)$ also possesses flat power spectral density and $T_s = 1/(2B_s)$, then (2) renders

$$\max_D E(\hat{D} - D)^2 / \min_D E(\hat{D} - D)^2 = (4/\pi)^2 \simeq 1.6. \tag{4}$$

Fig. 1 illustrates the terms max and min in (4) as well as the intermediate values.

For the algorithms of [7], we will show

$$\text{MSE} = E(\hat{D} - D)^2 = \frac{1}{B_n T \gamma_C b_s^2} \tag{5}$$

where

$$b_s \triangleq \frac{\int_{-\infty}^{\infty} df |2\pi f| \, G_s(f)}{\int_{-\infty}^{\infty} df \, G_s(f)} \tag{6}$$

and

$$\frac{1}{\gamma_C} \triangleq \frac{1}{2} \left[\frac{1}{\gamma_1} + \frac{1}{\gamma_2} + \frac{1}{\gamma_1 \gamma_2} \right]. \tag{7}$$

γ_C is an equivalent signal-to-noise ratio for this algorithm, and $G_s(f)$ is the Fourier transform of $R_s(\tau)$.

A condition for the validity of (5) is

$$B_s T \gamma_C \gg \tfrac{1}{3} \tag{8}$$

where B_s is the signal bandwidth (single sided).

Note that a similar result to (5) has been derived for the cross-correlation method in [8], however, instead of $1/\gamma_C$, we have in [8]

$$1/\gamma = 1/2(1/\gamma_1 + 1/\gamma_2 + (\beta_n/\beta_s)^2/(\gamma_1 \cdot \gamma_2))$$

where

$$\beta_\alpha^2 = \int_{-\infty}^{\infty} df\,(2\pi f)^2\, G_\alpha(f) \Big/ \int_{-\infty}^{\infty} G_\alpha(f)\, df$$

$\alpha = s$ or n, and instead of b_s^2 we have in [8] β_s^2. Considering (5), and the corresponding result in [8], it can be deduced that the algorithm of [7] should be preferred over the cross-correlation method for $\beta_n/\beta_s >> 1$ and at low SNR's.

In this paper, the above two algorithms are extended by combining them with the GCC method, and the pertinent TDE–MSE expressions are presented in (53) and (55).

The algorithm of [6] is proposed to be used with the weight function corresponding to the maximum likelihood estimator (MLE), thus yielding a suboptimal solution.

For the algorithm of [7], it will be shown that the weight function which yields MMSE coincides with the MLE's weight function multiplied by $2\pi f$. ($2\pi f$ is actually not part of the weight function. This point will be further cleared out in the sequel.)

II. Performance of the Algorithm Described in Stein [6]

We assume the model (1) and assumptions A and B.

In this algorithm, we estimate D by $\hat{D}$ which maximizes

$$\phi(\tau) = \int_{-T/2}^{T/2} dt\, x(t)\, y(t + \tau) = \phi_s(\tau) + \phi_N(\tau) \quad (9)$$

where the signal term $\phi_s(\tau)$ is defined by

$$\phi_s(\tau) \triangleq \int_{-T/2}^{T/2} dt\, s(t)\, s(t + \tau - D) \quad (10)$$

and the noise term $\phi_N(\tau)$ is defined by

$$\phi_N(\tau) \triangleq \int_{-T/2}^{T/2} dt\, s(t)\, n_2(t + \tau) + \int_{-T/2}^{T/2} dt\, n_1(t)\, s(t + \tau - D) + \int_{-T/2}^{T/2} dt\, n_1(t)\, n_2(t + \tau). \quad (11)$$

The pertinent algorithm is as follows.

Find the three largest values $\phi(\tau_k)$, $\phi(\tau_{k+1})$ and $\phi(\tau_{k+2})$ (see Fig. 2) where $\tau_k \triangleq kT_s$ (T_s is the sampling period).

Because of assumption B, it can be assumed that $\phi_s(\tau)$ is symmetric with respect to $\tau = D$.

The symmetry of the function $\phi_s(\tau)$ with respect to $\tau = D$ ensures that the intermediate value of the above three values of $\phi(\tau)$ is on one side of the peak, and the other two values are on the other side of the peak.

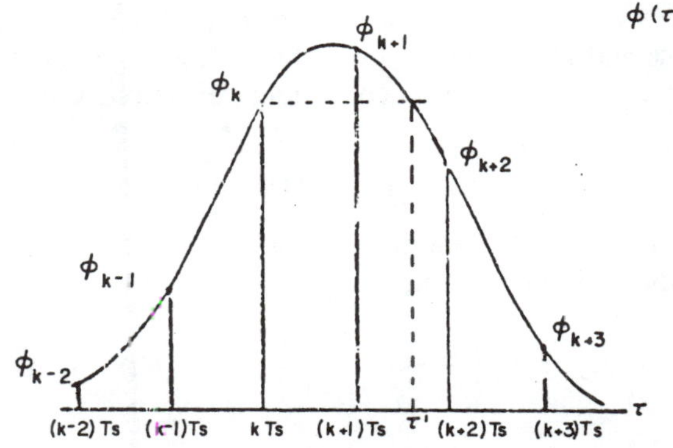

Fig. 2. Time delay of arrival estimation by Stein's method.

The estimator $\hat{D}$ of D maximizes $\phi(\tau)$ (see Fig. 2) and is given by

$$\hat{D} = \frac{\tau_k + \tau_k'}{2} \quad (12)$$

where τ_k' is the solution of the following equation:

$$\phi(\tau_k) = \phi(\tau_k'). \quad (13)$$

In [6], (13) is solved by a binary search combined with interpolation of the computed samples of $\phi(\tau_l)$.

Expanding the signal term $\phi_s(\tau)$ into a Taylor series in the neighborhood of τ' which equals τ_k' in the absence of noise, i.e., $n_1(t) = n_2(t) = 0$, $t\epsilon[-T/2, T/2]$, and retaining the linear part, we get

$$\tau' = 2D - \tau_k \quad (14)$$

and

$$\phi_s(\tau_k') \cong \phi_s(\tau') + \phi_s'(\tau')(\tau_k' - \tau). \quad (15)$$

Since $\phi_s(\tau)$ is symmetric with respect to $\tau = D$ (assumption B) we have

$$\phi_s(\tau') = \phi_s(\tau_k) \quad (16)$$

and

$$\phi_s'(\tau') = -\phi_s'(\tau_k). \quad (17)$$

Using (9) and (13)–(17), we get

$$\tau_k' - \tau' = -\frac{\phi_N(\tau_k) - \phi_N(\tau_k')}{\phi_s'(\tau_k)}. \quad (18)$$

Using (12) and (14), we get

$$\hat{D} - D = \tfrac{1}{2}(\tau_k' - \tau'). \quad (19)$$

Substituting (11) and (19) into (18) renders

$$\hat{D} - D = -\frac{1}{2\phi_s'(\tau_k)} \Bigg\{ \int_{-T/2}^{T/2} dt\, [s(t)\, n_2(t + \tau_k) + n_1(t)\, s(t + \tau_k - D) + n_1(t)\, n_2(t + \tau_k)] - \int_{-T/2}^{T/2} \cdot dt\, [s(t)\, n_2(t + \tau_k') + n_1(t)\, s(t + \tau_k' - D) + n_1(t)\, n_2(t + \tau_k')] \Bigg\}. \quad (20)$$

Now we show that $\hat{D}$ is unbiased (note that τ_k is a known variable and τ_k' is a random variable). Combining (20) with the following theorem on conditional expected values [11, p. 208]:

$$E[U] = E[E(U|V)] \tag{21}$$

where U and V are random variables and E or $\overline{\quad}$ denote the expectation operator, we get

$$\begin{aligned} E(\hat{D} - D) = E_s \Bigg\{ - \frac{1}{2\phi_s'(\tau_k)} \Bigg[\int_{-T/2}^{T/2} dt \, [s(t) \overline{n_2(t + \tau_k)} \\ + \overline{n_1(t)} \, s(t + \tau_k - D) \\ + \overline{n_1(t) \, n_2(t + \tau_k)}] - \int_{-T/2}^{T/2} \\ \cdot dt \, [s(t) \, E_{\tau_k'} \overline{[n_2(t + \tau_k') | \tau_k']} \\ + E_{\tau_k'} \overline{[n_1(t) \, s(t + \tau_k' - D | \tau_k']} \\ + E_{\tau_k'} \overline{[n_1(t) \, n_2(t + \tau_k') | \tau_k']]} | s(t), \\ t \epsilon [-T/2, T/2] \Bigg] \Bigg\} = 0. \end{aligned} \tag{22}$$

The fact that the right-hand side of (22) is zero follows from assumption A (zero mean, stationary signals).

Note that (22) has been derived without using the assumption that the signal $s(t)$ is weak ergodic (this assumption will be needed later on in deriving the TDE-MSE expression).

We now proceed to compute the TDE-MSE expression, i.e., $E(\hat{D} - D)^2$.

For $T >> \tau_s$, $|D|$ and using the assumption that $s(t)$ is weak ergodic, $\phi_s(\tau)$ can be approximated by

$$\phi_s(\tau) \simeq TR_s(D - \tau) \tag{23}$$

where $R_s(\tau)$ is the autocorrelation function of the signal $s(t)$. Therefore, $\phi_s'(\tau_k)$ is now no longer random. Following [8], and noting that τ_k' is a random variable, (20) and (23) render

$$\begin{aligned} E(\hat{D} - D)^2 = \frac{1}{2T[R_s'(D - \tau_k)]^2} \Bigg[\int_{-T}^{T} d\zeta \left(1 - \frac{|\zeta|}{T}\right) \\ \cdot [R_s(\zeta) \, R_{n_2}(\zeta) + R_{n_1}(\zeta) \, R_s(\zeta) \\ \cdot R_{n_1}(\zeta) \, R_{n_2}(\zeta) - R_s(\zeta) \, \overline{R_{n_2}(\zeta + \tau_k - \tau_k')} \\ - R_{n_1}(\zeta) \, \overline{R_s(\zeta + \tau_k - \tau_k')} \\ - R_{n_1}(\zeta) \, \overline{R_{n_2}(\zeta + \tau_k - \tau_k')}] \Bigg]. \end{aligned} \tag{24}$$

Expanding $R_\alpha(\xi + \tau_k - \tau_k')$, $\alpha = n_1, n_2$ or s into a Taylor series in the neighborhood of $\xi + \tau_k - \tau'$ (τ' equals τ_k' in the noiseless case), and retaining the linear part, we get

$$\begin{aligned} \overline{R_\alpha(\xi + \tau_k - \tau_k')} \cong R_\alpha(\xi + \tau_k - \tau') \\ - R_\alpha'(\xi + \tau_k - \tau') \, \overline{(\tau_k' - \tau')}. \end{aligned} \tag{25}$$

Using (19) and (22), (25) renders

$$\overline{R_\alpha(\xi + \tau_k - \tau_k')} \cong R_\alpha(\xi + \tau_k - \tau'). \tag{26}$$

To derive the MSE, we substitute (26) in (24), we use (14), because of assumption B we neglect in (24) $|\xi|/T$ as compared to 1, we extend the integration limits to $\pm\infty$ and we get

$$\begin{aligned} E(\hat{D} - D)^2 = \frac{1}{2T[R_s'(D - \tau_k)]^2} \Bigg[\int_{-\infty}^{\infty} d\xi \, [R_s(\xi) \, R_{n2}(\xi) \\ + R_{n1}(\xi) \, R_s(\xi) + R_{n1}(\xi) \, R_{n2}(\xi) \\ - R_s(\xi) \, R_{n2}(\xi + 2\tau_k - 2D) - R_{n1}(\xi) \\ \cdot R_{n2}(\xi + 2\tau_k - 2D) \\ - R_{n1}(\xi) \, R_{n2}(\xi + 2\tau_k - 2D)] \Bigg]. \end{aligned} \tag{27}$$

Note that a somewhat tighter result can be derived by also retaining in (25) the quadratic term.

Equation (27) in the frequency domain is

$$\begin{aligned} E(\hat{D} - D)^2 = \frac{1}{2T\left[\int_{-\infty}^{\infty} df (j2\pi f) \, G_s(f) \, e^{j2\pi f(D - \tau_k)}\right]^2} \\ \times \Bigg[\int_{-\infty}^{\infty} df [[G_s(f) \, G_{n2}(f) + G_{n1}(f) \\ \cdot G_s(f) + G_{n1}(f) \, G_{n2}(f)] \\ \times (1 - e^{j4\pi f(D - \tau_k)})] \Bigg]. \end{aligned} \tag{28}$$

Equations (27) and (28) are general expressions for the TDE-MSE as a function of observation time and the autocorrelation (auto-PSD's) of the signal $s(t)$ and noise $n_1(t)$, $n_2(t)$.

Using assumption C (see Section I), (27) and (28) render (2) (the simplified TDE-MSE expression).

III. Performance of the Algorithm Described in Cabot [7]

Assume the model (1) and assumptions A and B.

In this algorithm, we estimate D by $\hat{D}$, which is the solution of the following equation:

$$\phi_{x\check{y}}(\hat{D}) = 0 \tag{29}$$

where

$$\phi_{x\check{y}}(\tau) = \int_{-T/2}^{T/2} dt \, x(t) \, \check{y}(t + \tau) \tag{30}$$

and $\check{y}(t)$ is the Hilbert transform of $y(t)$

$$\check{y}(t) = \frac{1}{\pi t} \circledast y(t) \tag{31}$$

($\circledast$ denotes linear convolution). This algorithm is based on the fact that $R_{x\check{x}}(\tau)$ is an odd function and therefore satisfies

$$R_{x\check{x}}(0) = 0. \tag{32}$$

A fast approximation to the computation can be obtained as follows.

1) Compute $\phi_{x\check{y}}(\tau)$ at a discrete set of equispaced points τ_l, $l = 1, \cdots, L$ such that $\tau_1 << D << \tau_L$ (it is assumed that $\tau_{l+1} - \tau_l$ is above the Nyquist rate).

2) Find two successive values of $\phi_{x\check{y}}(\tau)$, say, $\phi_{x\check{y}}(\tau_k)$ and $\phi_{x\check{y}}(\tau_{k+1})$, such that

$$\phi_{x\check{y}}(\tau_k)\, \phi_{x\check{y}}(\tau_{k+1}) < 0. \tag{33}$$

The desired zero is therefore in the interval $[\tau_k, \tau_{k+1}]$.

3) Since the Hilbert transform does not affect the bandwidth of $y(t)$ (contrary to differentiation), one can compute intermediate values of $\phi_{x\check{y}}(\tau)$, $\tau \in [\tau_k, \tau_{k+1}]$ by the well-known interpolation formula for band-limited signals

$$\phi_{x\check{y}}(\tau) \simeq \sum_{l=1}^{L} \phi_{x\check{y}}(\tau_l) \operatorname{sinc}(\tau - \tau_l). \tag{34}$$

Now, starting from the interval $[\tau_k, \tau_{k+1}]$ and using (34), one can rapidly find by a binary search $\hat{D}$ the desired zero of $\phi_{x\check{y}}(\tau)$.

Let us now turn to the analysis of this algorithm. Denote by

$$\phi_{x\check{y}}(\tau) = \phi_s(\tau) + \phi_N(\tau) \tag{35}$$

where the signal term is defined by

$$\phi_s(\tau) = \int_{-T/2}^{T/2} dt\, s(t)\, \check{s}(t + \tau - D) \tag{36}$$

and the noise term is defined by

$$\phi_N(\tau) = \int_{-T2}^{T/2} dt\, s(t)\, \check{n}_2(t + \tau) + \int_{-T/2}^{T/2} dt\, n_1(t)\, \check{n}_2(t + \tau) + \int_{-T/2}^{T/2} dt\, \check{s}(t + \tau - D)\, n_1(t). \tag{37}$$

Expanding the signal term $\phi_s(\tau)$ into a Taylor series in the neighborhood of $\tau = D$, and retaining the linear part, we get

$$\phi_s(\tau) \simeq \phi_s(D) + \phi_s'(D)(\tau - D). \tag{38}$$

Using the assumption that $s(t)$ is weak ergodic, (32) and (38) render

$$\phi_s(\tau) \cong -TR'_{s\check{s}}(0)\,(\tau - \mathrm{D}). \tag{39}$$

The estimator $\hat{D}$ of D satisfies

$$\phi_{x\check{y}}(\hat{D}) = \phi_s(\hat{D}) + \phi_N(\hat{D}) = 0. \tag{40}$$

Substituting (39) in (40) we get

$$\hat{D} - D \cong -\frac{1}{TR'_{s\check{s}}(0)}\, \phi_N(\hat{D}). \tag{41}$$

Similarly to Section II it can be shown that

$$E(\hat{D} - D) = 0. \tag{42}$$

The expression for the TDE–MSE can be derived similarly to Section II by using the following relation from [11, p. 357]:

$$R_{\check{x}}(\tau) = \check{R}_x(\tau), \tag{43}$$

and we get

$$E(\hat{D} - D)^2 = \frac{1}{T[R'_{s\check{s}}(0)]^2} \left\{ \int_{-\infty}^{\infty} d\zeta [R_s(\zeta)\, R_{n2}(\zeta) + R_{n_1}(\zeta)\, R_s(\zeta) + R_{n_1}(\zeta)\, R_{n_2}(\zeta)] \right\}. \tag{44}$$

Equation (44) in the frequency domain reads

$$E(\hat{D} - D)^2 = \frac{1}{T\left[\int_{-\infty}^{\infty} df |2\pi f|\, G_s(f)\right]^2} \cdot \left\{ \int_{-\infty}^{\infty} df [G_s(f)\, G_{n_2}(f) + G_{n_1}(f)\, G_s(f) + G_{n_1}(f)\, G_{n_2}(f)] \right\}. \tag{45}$$

Note that the MSE in (43) does not depend on the unknown delay since here we have computed $\phi_{x\check{y}}(\hat{D})$, whereas in Section II we have computed $\phi(\tau_k')$ which is aside from the peak, and the location of τ_k' relative to the peak is a function of D and the sampling period T_s.

Here, too, using assumption C (see Section I), (45) renders (5) (the simplified TDE–MSE expression).

Now, using "wide-band assumptions," we shall justify the Taylor expansion in (38), and therefore the validity of the expressions derived for the bias and MSE.

Assume for simplicity that assumption C holds and, in addition, $s(t)$ possesses flat PSD and $B_{n_1} = B_{n_2} = B_s$ (a wide-band assumption).

The quadratic term in Taylor's series expension is zero and the cubic term can be neglected when compared to the linear term if

$$\left| \frac{\frac{1}{6}\phi_s'''(D)\,(\hat{D} - D)^3}{\phi_s'(D)\,(\hat{D} - D)} \right| << 1. \tag{46}$$

It can be argued that, if (46) is satisfied, the whole tail of the Taylor series expansion is negligible.

Similarly to the derivation in [8] applying (5) to (46), we get

$$B_s T \gamma_C >> \tfrac{1}{3} \tag{47}$$

which is the condition for the validity of the derivations, i.e., (38) cannot be used unless (47) is satisfied.

Remark: A result similar to (47) can be derived for the algorithm of [6].

IV. The Two Algorithms Combined with the GCC Method

Assume the model (1) and assumptions A and B.

By applying linear prefilters possessing impulse responses $h_1(t)$ and $h_2(t)$ to $x(t)$ and $y(t)$, respectively, we get

$$\begin{cases} \tilde{x}(t) = h_1(t) \circledast x(t) = \tilde{s}_1(t) + \tilde{n}_1(t) \\ \tilde{y}(t) = h_2(t) \circledast y(t) = \tilde{s}_2(t - D) + \tilde{n}_2(t) \end{cases} \quad (48)$$

where

$$\begin{cases} \tilde{s}_1(t) = h_1(t) \circledast s(t) & \tilde{n}_1(t) = h_1(t) \circledast n_1(t) \\ \tilde{s}_2(t - D) = h_2(t) \circledast s(t - D) & \tilde{n}_2(t) = h_2(t) \circledast n_2(t). \end{cases} \quad (49)$$

We denote the GCC function by

$$\phi_{\tilde{x}\tilde{y}}(\tau) = \int_{-T/2}^{T/2} dt\, \tilde{x}(t)\, \tilde{y}(t + \tau)\, dt \quad (50)$$

and define a weight function $W(f)$ by

$$W(f) = H_1(f)\, H_2^*(f) \quad (51)$$

where $H_1(f)$ and $H_2(f)$ are the Fourier transforms of $h_1(t)$ and $h_2(t)$, respectively.

A. Stein's Algorithm Combined with the GCC Method

To improve the TDE–MSE of the algorithm of [6], we can use the ML estimator that has the following weight function [2], [3], [8]:

$$W_{\text{ML}}(f) = \frac{G_s(f)}{G_s(f)\, G_{n_1}(f) + G_s(f)\, G_{n_2}(f) + G_{n_1}(f)\, G_{n_2}(f)}. \quad (52)$$

Using (48)–(51) and carrying out a similar derivation as in Section II, we get

$$E(\hat{D} - D)^2 = \frac{\int_{-\infty}^{\infty} df [G_s(f)\, G_{n_1}(f) + G_s(f)\, G_{n_2}(f) + G_{n_1}(f)\, G_{n_2}(f)]\, W^2(f)\, (1 - e^{j4\pi f(D - \tau_k)})}{2T\left[\int_{-\infty}^{\infty} dfj\, 2\pi f W(f)\, G_s(f)\, e^{j2\pi f(D - \tau_k)}\right]^2}. \quad (53)$$

B. Cabot's Algorithm Combined with the GCC Method

Here the estimator $\hat{D}$ of D is obtained by solving the following equation [see (50)]:

$$\phi_{\tilde{x}\tilde{y}}(\hat{D}) = 0. \quad (54)$$

A similar derivation as in Section III renders that $\hat{D}$ is unbiased and the MSE is given by

$$E(\hat{D} - D)^2 = \frac{1}{T\left[\int_{-\infty}^{\infty} df\, |2\pi f|\, W(f)\, G_s(f)\right]^2} \cdot \left\{\int_{-\infty}^{\infty} df\, W^2(f)\, [G_s(f)\, G_{n_2}(f) + G_{n_1}(f)\, G_s(f) + G_{n_1}(f)\, G_{n_1}(f)]\right\}. \quad (55)$$

Similarly to the derivation in [8], we get $W(f) = W_H(f)$ which leads to an MMSE of (55) given by

$$W_H(f) = 2\pi f W_{\text{ML}}(f) \quad (56)$$

where $W_{\text{ML}}(f)$ is given by (52).

The factor $2\pi f$ corresponds to differentiation, and, therefore, here one has to search for the zero of the GCC function rather than the peak of the GCC function.

Substituting (56) in (55) renders the MMSE obtained by this method and by the ML method [8].

The result arrived at here can be obtained by differentiating the GCC function corresponding to the ML estimator and thus replacing maximum detection by search for a zero. Therefore, the factor $2\pi f$ in (58) should not be interpreted as part of the weight function (see [9]). However, for the algorithm of [7], this observation does not hold (see Section III).

V. Discussion

The performance of two efficient algorithms for time delay estimation (TDE) [6], [7] has been analyzed. It has been shown that both algorithms are unbiased. Expressions for the time delay estimation mean-square error have been presented for both algorithms (their simplified versions are presented in the Introduction section). It has been shown that the MSE for the algorithm of [6] depends on the unknown delay D.

Next, both algorithms were combined with the GCC method. It has been shown that they are unbiased, and general expressions for the mean-square error have been presented.

It has been shown that a suboptimal estimator for the algorithm of [6] is to use the weight function corresponding to the maximum likelihood estimator.

An optimal estimator (the MSE coincides with the Cramer–Rao lower bound) for the algorithm of [7] is to use the weight function corresponding to the MLE multiplied by $2\pi f$.

In this paper, only the local variations in the neighborhood of the unknown delay were studied. The performance of any estimate of D is not fully characterized by the local variations, and the occurrence of false peaks must be taken into account (see [9]).

Acknowledgment

The author is grateful to the reviewers for their profound review and their useful remarks and suggestions.

References

[1] G. C. Carter, Ed., Special issue on time delay estimation, *IEEE Trans. Acoust., Speech, Signal Processing*, vol. ASSP-29, Part II, June 1981.

[2] C. Y. Knapp and G. C. Carter, "The generalized correlation method for estimation of time delay," *IEEE Trans. Acoust., Speech, Signal Processing*, vol. ASSP-24, pp. 320–327, Aug. 1976.

[3] G. C. Carter, "Time delay estimation," Ph.D. dissertation, Univ. Connecticut, pp. 67–70, 1976.

[4] H. Meyr, "Delay lock tracking of stochastic signals," *IEEE Trans. Commun.*, vol. COM-24, pp. 331–339, Mar. 1976.

[5] J. C. Hassab and R. E. Boucher, "Optimum estimation of time delay by a generalized correlator," *IEEE Trans. Acoust., Speech, Signal Processing*, vol. ASSP-27, pp. 373–380, Aug. 1979.

[6] S. Stein, "Algorithms for ambiguity function processing," *IEEE Trans. Acoust., Speech, Signal Processing*, vol. ASSP-29, Part II, pp. 588–599, June 1981.

[7] R. C. Cabot, "A note on the application of the Hilbert transform to time delay estimation," *IEEE Trans. Acoust., Speech, Signal Processing*, vol. ASSP-29, Part 2, pp. 607–609, June 1981.

[8] M. Azaria and D. Hertz, "Time delay estimation by generalized cross-correlation methods," *IEEE Trans. Acoust., Speech, Signal Processing*, vol. ASSP-32, pp. 280–285, Apr. 1984.

[9] H. Meyr and G. Spies, "The structure and performance of estimators for real-time estimation of randomly varying time delay," *IEEE Trans. Acoust., Speech, Signal Processing*, vol. ASSP-32, pp. 81–94, Feb. 1984.

[10] K. Scarbrough, N. Ahmed, and G. C. Carter, "On the simulation of a class of time delay algorithms," *IEEE Trans. Acoust., Speech, Signal Processing*, vol. ASSP-29, Part II, pp. 534–540, June 1981.

[11] A. Papoulis, *Probability, Random Variables and Stochastic Processes*. New York: McGraw-Hill, 1965.

Time Delay Estimation Via Cross-Correlation in the Presence of Large Estimation Errors

JOHN P. IANNIELLO

Abstract—The estimate of the difference in time of arrival of a common random signal received at two sensors, each of which also receives uncorrelated noise, is examined for both small and large estimation errors. It is shown that as the post-integration signal-to-noise ratio decreases, the correlator exhibits a thresholding effect; that is, the probability of a large error (an anomalous estimate) increases rapidly. Approximate theoretical results for the probability of an anomaly are presented and are verified experimentally. The variance of the time delay estimate is examined for both a gated mode, in which the time delay corresponding to the correlation peak closest to the true time delay is used as the estimate of time delay, and an ungated mode, in which the time delay corresponding to the largest peak over the full range of the correlator delay times is used as the estimate. The observed variance for both modes is compared with the theoretical variance based on a small error analysis. For the gated modes, the signal-to-noise ratio below which the observed variance begins to differ significantly from the small error theory can be reliably predicted from a linearity criterion. It is shown, however, that the expected variance for the ungated mode can depart from the small error theory at a higher signal-to-noise ratio than for the gated modes; thus the variance due to anomalies can be the most important factor in determining the region of applicability of the small error analysis.

I. Introduction

THE estimate of the difference in the time of arrival of a common random signal received at two or more sensors, each of which also receives uncorrelated noise, is a problem of considerable practical interest in underwater acoustics. One common method for estimating the time of arrival difference is to cross-correlate the two signals and to select the peak of the correlogram as the estimate of the time difference. The variance of the resulting time delay estimate has been derived by several authors [1], [2]. As discussed in [1] these results require that the estimation errors be sufficiently small for the estimate to remain within the linear region of the derivative of the signal autocorrelation function. It is also well established that the cross-correlator, with proper filtering, is the optimum time delay estimator in the sense that the variance of its estimate reaches the absolute minimum variance—the Cramer-Rao lower bound (CRLB) [1]-[3]. Again the correlator is optimum only for small estimation errors. This may be seen by noting that the cross-correlator, with proper filtering, is a maximum likelihood estimator of time delay [2]; maximum likelihood estimators, however, only reach the CRLB for nonlinear estimation problems such as the one we consider here, when estimation errors are small [4, p. 71]. Recent experimental studies have verified some of these theoretical conclusions [5], [6].

Manuscript received July 14, 1981; revised June 11, 1982.

The author is with the Naval Underwater Systems Center, New London, CT 06320.

Time delay estimation in the presence of large estimation errors has not been thoroughly examined. In particular it is not known under what conditions the variance of the time delay estimate begins to depart significantly from that given by the small error analysis. A rule of thumb is that when the standard deviation of the estimate is on the order of the inverse signal bandwidth, for low-pass signals, or the inverse center frequency for narrow-band signals, then typical estimates surely exceed the linear region of the derivative of the signal autocorrelation function, thus the assumption required to derive the variance expression is violated, implying that the errors will be larger than predicted. A related discussion concerning the applicability of the CRLB is given in [4, p. 70]. This argument applies only to local errors (when estimates are near the true value), it does not address the possibility of large errors or anomalous estimates. It is appropriate to consider only local errors when a tracking gate or an *a priori* knowledge of the time delay dynamics can be used to limit the magnitude of the error excursion. On the other hand, when *a priori* information about the time delay is lacking, cross-correlators are often used ungated in the sense that any time delay in the observed range of the correlogram delay times is allowed. It is important then to understand the behavior of this ungated mode.

As we show below, a fundamentally more interesting phenomenon than a simple increase in variance occurs as the post-integration signal-to-noise ratio (SNR) is decreased; namely, below a certain post-integration SNR the probability of an anomalous estimate increases suddenly and precipitously. This is a threshold phenomenon similar to that observed in frequency modulation systems [7, p. 661], pulse position modulation systems (PPM) [7, p. 627], or more generally for any nonlinear estimation or communications system [4, p. 273; 7, p. 617]. As discussed by Woodward [8, p. 90] a threshold effect is liable to occur whenever a message of low dimensionality (in our case time delay, which is one dimensional) is "encoded" in a signal having higher dimensionality or more degrees of freedom. Thus, the threshold effect is fundamental and unavoidable.

In this paper we first present an approximate analysis for the probability of an anomalous estimate, and hence for the onset of threshold. We then describe experimental results verifying the theory, and then discuss theoretical and experimental results for the variance of the time delay estimate in both the gated and ungated modes.

Reprinted from *IEEE Trans. Acoust., Speech, Signal Processing,* vol. 30, no. 6, pp. 998–1003, December 1982.

II. Analysis for Probability of Anomaly

We assume that we have two sensors which receive the two random signals $x_1(t)$ and $x_2(t)$ given by

$$x_1(t) = s(t) + n_1(t)$$
$$x_2(t) = s(t - D) + n_2(t) \tag{1}$$

where s, n_1, and n_2 are uncorrelated Gaussian random processes and D is the difference in arrival time. D is estimated by forming the correlator output $z(\lambda)$ via

$$z(\lambda) = \frac{1}{T}\int_0^T x_1(t)x_2(t+\lambda)\,dt, \qquad 0 \leq \lambda < T_0$$
$$= \frac{1}{T}\int_0^T x_1(t+\lambda)x_2(t)\,dt, \qquad -T_0 < \lambda < 0 \tag{2}$$

and selecting the value of λ which maximizes $z(\lambda)$ as the estimate $\hat{D}$ of D. We will assume that s, n_1, and n_2 have identical spectra. We also assume that a $T + T_0$ second record of $x_1(t)$ and $x_2(t)$ is available, although we only average for T seconds. By processing in this manner we avoid end effects and can form unbiased estimates which have the same variance (for noise only inputs) at each lag value. If we processed only a T second record, taper scaling [9, p. 182] would be required to avoid bias. Then, however, each lag estimate would have a different variance. The difference between the two methods will be small for large bandwidth time products.

To determine the probability of an anomalous estimate we follow the approximate analysis used for PPM [7, p. 627]; as for PPM the approximate analysis will be shown to agree remarkably well with experiments. We first recognize that there is a signal correlation time T_c (to be defined exactly later) over which the signal is reasonably well correlated and beyond which the signal correlation falls to zero. Thus the observed range of lag values, $\pm T_0$, may be expected to contain roughly $M = 2T_0/T_c$ independent values of $z(\lambda)$, which we denote by z_m. We assume that the true delay is located at one of the time delays λ_m corresponding to the z_m, say λ_0. We define an anomalous event as one for which the estimate of time delay is farther than $\pm T_c/2$ from the true value. If we define the event E as

$$E = [z_m > z_0 \text{ for at least one } \lambda_m] \tag{3}$$

then we can argue as in [7, p. 629] that E is a reasonable approximation to what we mean by an anomaly (the validity of the approximation is to be tested by experiment). Since the z_m are assumed independent, the probability of the event E is close to the probability of error in the communication of M equally likely messages or [4, p. 262]

$$P_r[\text{anomaly}] \simeq P_r[E] = 1 - \int_{-\infty}^{\infty} p(z_0) \cdot \left[\int_{-\infty}^{z_0} p(z_m)\,dz_m\right]^{M-1} dz_0 \tag{4}$$

where $p(z_0)$ is the probability density function of z_0 and $p(z_m)$ is the probability density function for any of the z_m, all of which are assumed to have identical probability density functions.

To proceed we must find appropriate forms for $p(z_0)$ and $p(z_m)$. Since z_0 and z_m are simply cross-correlation function estimates, their mean values and variances are known [9, p. 183]. Thus, if the autocorrelation functions of the signal $R_s(\tau)$ and the noises $R_N(\tau)$ are given by

$$R_S(\tau) = S\rho(\tau)$$
$$R_N(\tau) = N\rho(\tau) \tag{5}$$

where S and N are the signal and noise variance and $\rho(\tau)$ is the normalized correlation function of both signal and noise, and if the time bandwidth product is large, i.e., $T/T_c \gg 1$, then

$$\bar{z}_0 = S, \bar{z}_m = 0 \tag{6a}$$
$$\text{Var}[z_0] \simeq [(S+N)^2 + S^2]/(B_sT) \tag{6b}$$
$$\text{Var}[z_m] \simeq (S+N)^2/(B_sT) \tag{6c}$$
$$B_sT \equiv \left[\frac{1}{T}\int_{-\infty}^{\infty} \rho^2(\tau)\,d\tau\right]^{-1} \tag{6d}$$

where B_s is the statistical bandwidth [9, p. 278].

Now assuming that z_0 and z_m are Gaussianly distributed with means and variances given by (6), we can substitute these approximations into (4) and, after normalizing variables, we find

$$P_r[\text{anomaly}] \simeq 1 - \int_{-\infty}^{\infty} \frac{dx}{\sqrt{2\pi}} \exp\left[-\frac{1}{2}(x-A)^2\right] \cdot \left\{\int_{-\infty}^{Bx} \frac{dy}{\sqrt{2\pi}} \exp\left[-\frac{1}{2}y^2\right] dy\right\}^{M-1} \tag{7a}$$

where

$$A \equiv \frac{S}{\text{Var}[z_0]^{1/2}} = (B_sT)^{1/2} \frac{(S/N)}{[(S/N)^2 + (1+S/N)^2]^{1/2}} \tag{7b}$$

$$B \equiv \left[\frac{\text{Var}[z_0]}{\text{Var}[z_m]}\right]^{1/2} = \left[1 + \frac{(S/N)^2}{(1+S/N)^2}\right]^{1/2}. \tag{7c}$$

Note that A is the post-integration SNR and B is a scaling factor which takes on values between 1 and $2^{1/2}$. Equation (7a) will have to be evaluated numerically.

As in the communication analogy, an approximate analytic expression establishing an upper bound on P_r [anomaly] can be found. Thus as in [4, p. 264]

$$P_r[\text{anomaly}] \leq (M-1)\int_{-\infty}^{\infty} p(z_0)\left[\int_{z_0}^{\infty} p(z_m)\,dz_m\right] dz_0. \tag{8}$$

Then, again assuming that the underlying probability distributions are Gaussian with means and variances given by (6), it can be shown that

$$P_r[\text{anomaly}] \leq (M-1)\frac{1}{\sqrt{2\pi}\,C}\exp\left[-\frac{1}{2}C^2\right] \tag{9a}$$

where

$$C \equiv (B_s T)^{1/2} \frac{S/N}{[(S/N)^2 + 2(1 + S/N)^2]^{1/2}}. \tag{9b}$$

III. Description of Experiment

A computer experiment simulating the problem of interest was run by generating two unit variance, independent Gaussian random number sequences, U_i and V_i, and combining these to generate the sequences X_{1i} and X_{2i} according to the rule [10, p. 953]

$$X_{1i} = (1 + S/N)^{1/2} U_i \tag{10a}$$

$$X_{2i} = (1 + S/N)^{1/2} [rU_i + (1 - r^2)^{1/2} V_i] \tag{10b}$$

where $r = (S/N)/[1 + S/N]$. Thus both X_1 and X_2 have total variance $(1 + S/N)$, and the correlation of the two channels, $\overline{X_1 X_2}$, is S/N. This procedure thus simulates the problem described by (1) and (5) with N in (5) set equal to unity, with S in (5) set equal to (S/N), and with $D = 0$.

The sequences X_{1i} and X_{2i} were then low-pass filtered by a finite length Gaussian filter with sampled transfer function $h(n\Delta T)$ given by

$$h(n\Delta T) = [2(\pi a)^{1/2}]^{-1} \exp[-(n\Delta T)^2/(4a)], n \leqslant 16 \tag{11a}$$

resulting in sequences with sampled autocorrelation function

$$R(n\Delta T) \cong \frac{(1 + S/N)}{2(2\pi a)^{1/2}} \exp[-(n\Delta T)^2/(8a)]. \tag{11b}$$

where ΔT is the time between samples (ΔT is effectively set equal to unity below). These simulate time series with the spectral density

$$G(2\pi f) \cong (1 + S/N) \exp[-2a(2\pi f)^2]. \tag{11c}$$

This form of filter was chosen since it is a low-pass filter whose autocorrelation function has no sidelobes; bandpass filters or filters with high sidelobes could introduce added complications which we do not wish to deal with now. We discuss this again briefly in the final section.

A simulated $T + T_0$ second record of filtered data was formed by the processing described above. With $T = L\Delta T$ and $T_0 = P\Delta T$ this gives two, $L + P$ point records. The cross-correlation of these two sequences was estimated by dividing these $L + P$ point sequences into $(L + P)/P$ contiguous segments and proceeding via the FFT technique outlined in [11, p. 560]. This generates a correlogram estimate at lag values between $\pm P(\pm T_0$ s).

To determine which events constitute an anomaly we first must decide on a reasonable definition for the correlation time T_C of the process. We use the definition

$$T_C \equiv R^{-1}(0) \int_{-\infty}^{\infty} R(\tau)\, d\tau. \tag{12}$$

Note that if the correlation function were a boxcar function with constant value in $-T_1/2 \leqslant \tau \leqslant T_1/2$ and zero elsewhere, then (12) would give $T_C = T_1$. If, as another example, the process had a spectrum that was uniform in $-B \leqslant f \leqslant B$ then (12) would give a value of $T_C = 1/2B$. For the process that we are considering we find, after substituting the continuous version of (11b) into (12) and evaluating the integral,

$$T_C = 2(2\pi a)^{1/2}. \tag{13}$$

Other definitions for T_C could have been chosen. Equation (17) is reasonable and will be shown to yield satisfactory results. M, the number of independent values of $z(\lambda)$, is then given by

$$M = \frac{2T_0}{T_C} = \frac{2P\Delta T}{2(2\pi a)^{1/2}} = \frac{P}{(2\pi)^{1/2} b} \tag{14a}$$

where

$$a = b^2 \Delta T^2. \tag{14b}$$

The region R_e corresponding to a correct (nonanomalous) value of $\hat{D}$ is

$$R_e: -\frac{T_C}{2} \leqslant \hat{D} \leqslant \frac{T_C}{2} \text{ or } -(2\pi)^{1/2} b \leqslant \frac{\hat{D}}{\Delta T} \leqslant (2\pi)^{1/2} b. \tag{15}$$

Finally, to completely specify the problem we must choose a value for the bandwidth-time product given by (6d). For the spectrum of (11) $B_s = [2(\pi a)^{1/2}]^{-1}$ so with (14b)

$$B_s T = \frac{L}{2(\pi)^{1/2} b}. \tag{16}$$

To specify our problem we can choose M and the FFT size P; this determines the spectrum parameter b from (14b). Then we must choose L, the total record length, to insure that $B_s T$ is large.

Having specified the parameters of the problem a large number (several thousand) of separate realizations of the $L + P$ point filtered sequences were generated and correlated. For each realization the lag region $\pm P(\pm T_0)$ was searched to find the location of the correlation peak (largest positive value). If the peak fell within R_e then a refined estimate of the location of the peak was generated and stored using the parabolic interpolation formula

$$\frac{\hat{D}}{\Delta T} = i_p + \frac{1}{2} \frac{z(i_{p+1}) - z(i_{p-1})}{2z(i_p) - z(i_{p+1}) - z(i_{p-1})}. \tag{17}$$

Here i_p is the value of delay corresponding to the peak of the correlation function estimate and $i_{p\pm 1}$ are the adjacent delay values, separated by one sample interval (ΔT) from i_p. If i_p did not fall within R_e for a given realization, the estimate was counted as an anomaly. A refined estimate was still generated and stored using (17). In addition, when an anomalous event occurred, an estimate of the location of the peak value of the correlogram closest to the true value was also generated. Thus for every realization two delay estimates were generated. For nonanomalous events these estimates were identical. At the end of all the realizations the total number of anomalous events and the variances of the two delay estimates were generated. For each set of parameters (M, b, L) data were generated as a function of SNR.

IV. Comparison of Theoretical and Experimental Results for Probability of Anomaly

A test example with parameters $M = 16$, $b = (2/\pi)^{1/2}$, $P = 32$, $L = 320$ (hence $B_s T = 113$) was chosen. The theoretical results for the probability of anomaly plotted against the post-integra-

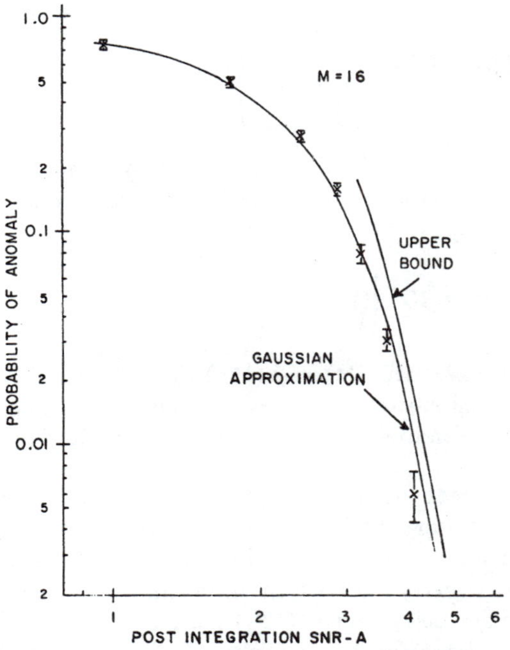

Fig. 1. Probability of anomaly versus post-integration SNR, $M = 16$.

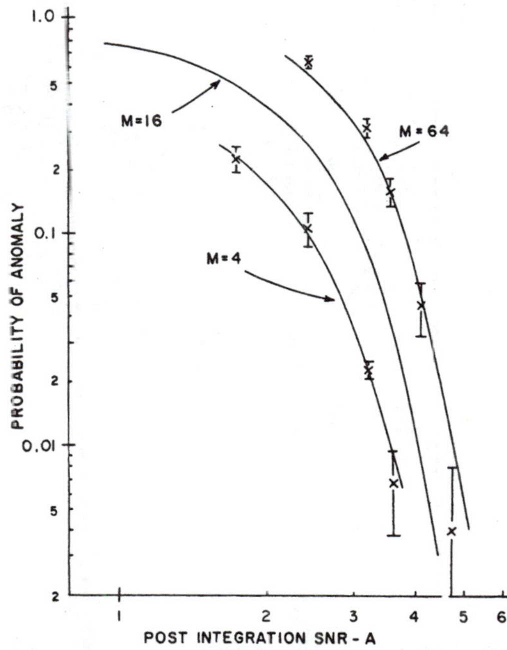

Fig. 2. Probability of anomaly versus post-integration SNR, $M = 4, 16, 64$.

tion SNR, A, are shown in Fig. 1. The curve labeled "Gaussian approximation" was obtained from (7). The curve labeled "Upper bound" was obtained from (9). The upper bound is close to, but does not converge to, the Gaussian approximation for the range of values shown.

The experimental results for the probability of anomaly are plotted as x's in Fig. 1. A variable number of total realizations (generally from 2000-5000) were used to obtain these data. The confidence regions indicated are 95 percent confidence intervals and are based on standard techniques for the estimation of proportions [12, p. 185]. The experimental results agree remarkably well with the theoretical results over the entire range of A shown. The data for very small P_r [anomaly] do appear to fall slightly below the theoretical predictions, however.

In Fig. 2 we show theoretical and experimental results for $M = 4$ [$(b = (2/\pi)^{1/2}, P = 8, L = 320)$], $M = 64$ [$(b = 0.4(2/\pi)^{1/2}, P = 128, L = 128)$] and the previous theoretical result for $M = 16$. Both new cases have $B_sT = 113$ as for the $M = 16$ case; thus Fig. 2 reflects only the effects of varying M. The experimental results again confirm the applicability of the theory. From Fig. 2 we see that as M is increased while holding the post-integration SNR, A fixed, the probability of an anomaly increases. This is similar to the effect seen in PPM and occurs because as M increases, with fixed SNR, there are simply more chances for a noise peak to exceed the peak at the true value of delay, since there are more noise bins.

P_r [anomaly] decreases, at first, with increasing SNR. As we see from (7b), however, A and hence P_r[anomaly] eventually become independent of SNR. Thus, for B_sT fixed, there is still a chance of making large errors even at infinite SNR. This effect occurs because the peak selecting correlator, as implemented, includes new data in the estimate of the cross-correlation at each lag value; thus it cannot be guaranteed that, even at infinite SNR, the peak value of the correlation output will be at the true value of time delay. It should be noted, though, that except for small B_sT, the error rate at very high S/N will be quite small and hence the resulting ungated variance will be less than the variance predicted by the linear theory. This suggests that for a small B_sT and a very large S/N the peak selecting correlator may not be the best instrumentation.

V. Comparison of Theoretical and Experimental Results for the Time Delay Estimation Variance

Theoretical values for the variance of the time delay estimate, for small estimation errors, can be obtained from the results of [2] using the Gaussian spectral shape given by (11c). Evaluating the integrals we find

$$\text{Var}\,[\hat{D}] = \frac{1}{(2\pi)(B_sT)B_s^2}\left[\frac{(S/N)^2}{1 + 2(S/N)}\right]^{-1}. \tag{18}$$

Normalizing Var $[\hat{D}]$ by ΔT^2 and recalling the definitions $a = b^2 \Delta T^2$, $T = L\Delta T$, $B_s^{-1} = 2(\pi a)$, (18) becomes

$$\frac{\text{Var}\,[\hat{D}]}{\Delta T^2} = \frac{4\pi^{1/2}b^3}{L}\left[\frac{(S/N)^2}{1 + 2S/N}\right]^{-1}. \tag{19}$$

Equation (19) is plotted, as a function of SNR, and labeled "Linear theory" in Fig. 3 using the same parameters [$(b = (2/\pi)^{1/2}, L = 320)$] as were used for the $M = 16$ example shown in Fig. 1. Experimental results for three different initial gate widths are also shown. The data labeled "Gate $\pm T_C/2$" were determined by selecting the peak value of the correlogram within the region $\pm T_C/2$ (the true value is at zero delay), then interpolating using (17), and computing the variance of the resulting time delay estimates. Note that a given estimate could be outside $\pm T_C/2$ after interpolation if the maximum value is not interior to $\pm T_C/2$. The data labeled "Gate $\pm T_C$" were computed similarly except that the initial range searched for a peak was $\pm T_C$. These two estimation procedures will yield identical

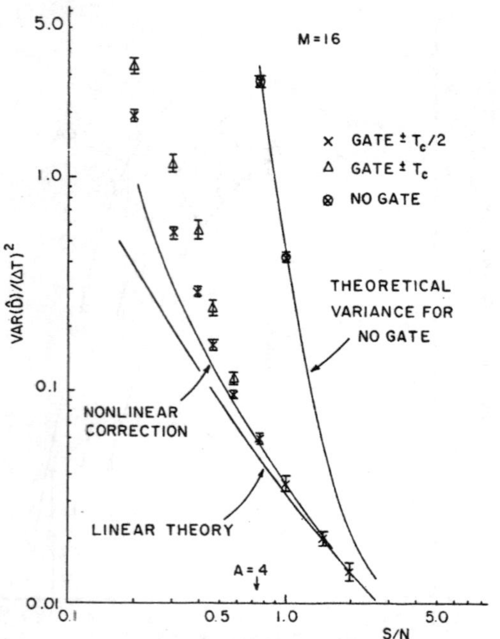

Fig. 3. Normalized variance of time delay estimate versus preintegration SNR, gated modes.

estimates except when there is a relative maximum interior to $\pm T_C/2$ and a larger peak value between $T_C/2$ and T_C (or $-T_C/2$ and $-T_C$). Finally, the data labeled "No gate" correspond to time delay estimates for which the entire correlogram ($\pm T_0$) was searched for an initial peak.

We first discuss the results for the two gated modes. From Fig. 3 we see that for high SNR, theory and experiment agree quite well as has previously been shown [5], [6]. For $S/N = 0.6$ or greater both gate widths gave identical results. As SNR is decreased the experimental results begin to exceed the theoretical results and at low SNR experiment and theory differ significantly. The low SNR behavior depends on the initial search gate width; as is reasonable, the larger the inital search gate the larger the variance. For SNR = 0 the "Gate $\pm T_C/2$" data have an average normalized variance of 11.6 or a standard deviation of 0.85 times the initial gate width ($T_C = 4$); the "Gate $\pm T_C$" data for SNR = 0 have an average normalized variance of 14.5 or 0.48 times the initial gate width.

We now examine at what point the measured variances for the gated modes begin to depart significantly from the theoretical prediction. The error expansion used in the derivation of (18) retains only terms to the order of $d^2R_s(\tau)/d\tau^2$ [1]. It can be shown that (18) is then valid as long as

$$\text{Var}[\hat{D}] \ll [(d^2R_s/d\tau^2)/(d^4R_s/d\tau^4)]_{\tau=0}. \tag{20}$$

Equation (20) thus gives a criterion for determining when the estimates remain on the linear region of $dR_s(\tau)/d\tau$; For the parameters of our example the right hand side of (20) equals 0.85; thus we might expect (18) to be accurate as long as $\text{Var}[\hat{D}] \leqslant 0.08$. From Fig. 3 we see that this is a reasonable estimate.

A simple correction to the theoretical result of (18) can be derived as follows. Equation (18) is computed from an expression of the form [1], [4, p. 70]

$$\text{Var}[\hat{D}] \equiv V_0^2 = \frac{\text{Var}\left(\dfrac{dz}{d\tau}\right)}{\left(\left.\dfrac{d^2R_s}{d\tau^2}\right|_{\tau=0}\right)^2}. \tag{21}$$

If instead of evaluating the second derivative at $\tau = 0$ we evaluate it at the value V_0 given by (21) we can find a corrected formula of the form

$$\text{Var}[D]_C = V_0^2 \left\{\left(1 - \frac{\pi}{8} V_0^2\right) \exp[-\pi V_0^2/16]\right\}^{-2} \tag{22}$$

where we have used $R(\tau)$ from (16b). $\text{Var}[\hat{D}]_C/\Delta T^2$ is plotted in Fig. 3 and labeled "Nonlinear correction." This simple correaction is seen to work reasonably well for SNR's down to say 0.5.

The preceeding discussion applies to the performance of the gated processing modes; to be practically usable the gated mode requires some *a priori* information about the true value of the time delay. If no such information is available the full correlogram delay region ($\pm T_0$) must be searched. The expected variance for the ungated mode is approximately equal to the variance, given no anomaly, times the probability of no anomaly, plus the variance, given an anomaly, times the probability of an anomaly. This function is plotted and labeled "Theoretical variance for no gate" on Figs. 3 and 4. The variance, given an anomaly, is approximately $T_0^2/3$ since this is the variance of a variable uniformly distributed in $\pm T_0$. The observed values are indicated by the x's in Fig. 4, and are seen to agree quite well with the theory.

It is clear from Figs. 3 and 4 that the variance for the ungated mode departs from the linearized theory at a higher value of SNR than does the variance for the gated modes. It is of interest to have a general criterion determining which mode of processing, gated or ungated, departs from the linearized theory at the higher SNR. First, it can be shown using the criterion of (20) that, for large bandwidth time products, the gated modes begin to depart significantly from the linear theory for SNR's below the value of $(S/N)_G \simeq (30/B_sT)^{1/2}$. Next we compare the value of the variance from the linear theory, evaluated at $(S/N)_G$, with the variance for the gated mode, also evaluated at $(S/N)_G$. This yields a criterion

$$M^3 \underset{UG}{\overset{G}{\lesseqgtr}} \frac{\sqrt{2}}{10\sqrt{\pi}} C_G \exp\left(\frac{1}{2} C^2\right) \tag{23}$$

where C_G is C from (9b) evaluated at $(S/N)_G$. Equation (23) shows that if M^3 is greater than the critical value on the right hand side then the variance for the ungated mode departs from the linear analysis at a higher SNR than does the variance for the gated mode; the opposite is true if M^3 is less than the right hand side. Hence, as stated earlier the larger the value of M the more likely is anomalous behavior to be important. If B_sT is very large then $C_G \simeq (15)^{1/2}$, independent of SNR, and the criterion becomes simply

$$M \underset{UG}{\overset{G}{\lesseqgtr}} 8.$$

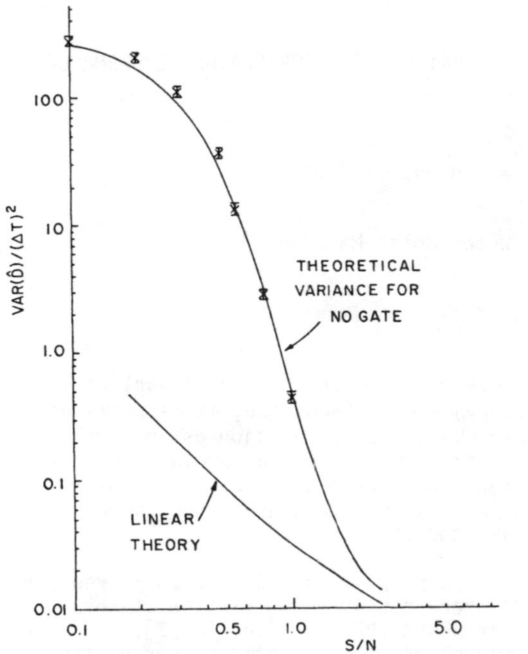

Fig. 4. Normalized variance of time delay estimate versus preintegration SNR, ungated mode.

VI. Final Comments

The experimental results shown here were obtained using a low-pass spectrum shose autocorrelation function has no sidelobes. Signals with narrow-band spectra or with large sidelobes will have a greater probability of an anomalous estimate due to the relatively large values of the autocorrelation function at time delays removed from zero. Such spectra will have to be considered individually. One would expect the mean square error for the ungated mode to be greater than that shown here at the higher SNR's due to the effect of the sidelobes or narrowband peaks (assuming the same bandwidth). At the lower SNR's these effects will probably not be so important and hence the results will be similar to those given here.

For small estimation errors, the standard of comparison for a given instrumentation is the CRLB. Further, it is known that cross-correlators with appropriate filters achieve the CRLB. [For the Gaussian signal and noise spectra used here, however, the CRLB, found after inserting the optimum filters, is zero, since an integral over all frequency is specified. This, of course, is physically unrealistic since some other noise process will eventually become important and limit performance.] For large estimation errors other bounds such as the Barankin bound [13] or the Ziv-Zakai bound [14] are more appropriate. These bounds must be evaluated and compared with the ungated mean-square estimation error as in Fig. 4, to establish how nearly optimum the cross-correlator is in the presence of large estimation errors.

Acknowledgment

I wish to thank P. Abraham, P. Cable, C. Carter, A. Nuttall, J. Pearson, and E. Weinstein for their comments and assistance at various stages of this work, and A. J. Meloney and A. J. Brault for typing the manuscript.

References

[1] V. H. MacDonald and P. M. Schultheiss, "Optimum passive bearing estimation in a spatially incoherent noise environment," *JASA*, vol. 46, pp. 37-43, July 1969.

[2] W. R. Hahn, "Optimum signal processing for passive sonar range and bearing estimation," *JASA*, vol. 58, pp. 201-207, July 1975.

[3] C. H. Knapp and G. C. Carter, "The generalized correlation method for estimation of time delay," *IEEE Trans. Acoust., Speech, Signal Processing*, vol. ASSP-24, pp. 320-327, Aug. 1976.

[4] H. L. Van Trees, *Detection, Estimation and Modulation Theory*. New York: Wiley, 1978.

[5] J. C. Hassab and R. E. Boucher, "A quantitative study of optimum and sub-optimum filters in the generalized correlator," in *Proc. 1979 IEEE Int. Conf. Acoust., Speech, Signal Processing*, 1979, pp. 124-127.

[6] K. Scarbrough, N. Ahmed, and G. C. Carter, "On the simulation of a class of time-delay estimation algorithms," *IEEE Trans. Acoust., Speech, Signal Processing*, vol. ASSP-29, pp. 534-539, June 1981.

[7] J. M. Wozencraft and I. M. Jacobs, *Principles of Communication Engineering*. New York: Wiley, 1965.

[8] P. M. Woodward, *Probability and Information Theory, with Applications to Radar*. Oxford: Pergamon, 1953.

[9] J. S. Bendat and A. G. Piersol, *Random Data: Analysis and Measurement Procedures*. New York: Wiley-Interscience, 1971.

[10] M. Abramowitz and I. A. Stegun, *Handbook of Mathematical Functions*. New York: Dover, 1965.

[11] A. V. Oppenheim and R. W. Schafer, *Digital Signal Processing*. Englewood Cliffs, NJ: Prentice-Hall, 1975.

[12] I. Miller and J. E. Freund, *Probability and Statistics for Engineers*. Englewood Cliffs, NJ: Prentice-Hall, 1965.

[13] S. K. Chow and P. M. Schultheiss, "Delay estimation using narrowband processes," *IEEE Trans. Acoust., Speech, Signal Processing*, vol. ASSP-29, pp. 478-484, June 1981.

[14] A. Weiss and E. Weinstein, "Passive time delay estimation from ambiguity prone signals-Part 1," submitted to *IEEE Trans. Acoust., Speech, Signal Processing*, 1982.

COMPARISON OF THE ZIV-ZAKAI LOWER BOUND ON TIME DELAY ESTIMATION WITH CORRELATOR PERFORMANCE

John P. Ianniello
Naval Underwater Systems Center, New London, CT 06320

Ehud Weinstein
Woods Hole Oceanographic Institution, Woods Hole, MA 02543

Anthony Weiss
School of Engineering, Tel-Aviv University, Tel-Aviv, Israel 69978

ABSTRACT

The Ziv-Zakai Lower Bound (ZZLB) on the mean square error (m.s.e.) of time delay estimators is compared with theoretical and computer simulation results for time delay estimation via cross-correlation. Comparisons are made for both lowpass and narrowband signal spectra. For both signal spectra it is shown that for sufficiently large time-bandwidth product the correlator performance is very close to the ZZLB in both the small and large error regions. This establishes the cross-correlator as an optimal instrumentation (in the m.s.e. sense) while demonstrating that the ZZLB is an extremely tight lower bound. The ZZLB is further used to accurately predict the threshold signal-to-noise ratio (SNR) above which the cross-correlator performance is closely characterized by the Cramer-Rao Lower Bound (CRLB).

INTRODUCTION

Time delay estimation between signals radiated from a common point source and observed at two spatially separated sensors, each of which also receives uncorrelated noise, is a problem of considerable practical interest in underwater acoustics. A common method for estimating the differential time delay is to cross-correlate the two sensor outputs, average for T seconds, and select as the estimate the delay associated with the peak of the correlogram. It is also well known that the cross-correlator, with proper filtering, is the Maximum Likelihood (ML) estimator of the sensor-to-sensor delay. Approximate theory and extensive computer simulation of the cross-correlator m.s.e. performance are reported in [1], where it is shown that a threshold SNR exists such that for SNR's greater than threshold the resulting m.s.e. approaches the absolute minimum characterized by the CRLB. For SNR's below threshold the cross-correlator performance rapidly deteriorates due to large anomalous or ambiguous estimates.

While the anomaly and ambiguity effects are fundamental and unavoidable features of the delay estimation problem, independent of the signal processing technique, it is still unclear whether the cross-correlation estimator is optimal for SNR's below threshold. Thus, it is important to compare the cross-correlator performance with some useful lower bound for the low SNR mode of operation.

A first step in this direction is reported in [2] where a simplified version of the Barankin Bound (BB), derived in [3], is compared with the correlator performance for narrowband signals. It is shown there that the threshold SNR achieved by the correlator is significantly larger than that predicted by the BB.

Another bounding technique based on a modified (improved) version of the ZZLB is developed in [4]. A direct comparison between the simplified version of the BB and the improved version of the ZZLB is carried out in [4] where it is indicated that the ZZLB is a greater (thus tighter) lower bound than the simplified BB, at least for narrowband signals. The important contribution of this paper is the direct comparison of the ZZLB and the correlator for both lowpass and narrowband signal spectra, thus allowing a joint assessment of how tight the ZZLB is and of how nearly optimal the cross-correlator is.

PROBLEM DESCRIPTION: ANOMALIES AND AMBIGUITIES

We consider two sensors one of which receives a random signal S(t) plus uncorrelated noise $N_1(t)$, and the other of which receives a time delayed version of the signal S(t - D) plus uncorrelated noise, $N_2(t)$. Both sensor outputs are observed during a common time period of T seconds. An estimate $\hat{D}$ of the time delay is to be formed. An a priori search region $\pm D_0/2$ for D is known; this comes about, perhaps, from the known sensor separation and the known sound velocity. To obtain some feeling for the source of the anomalous estimates assume that the estimate, $\hat{D}$, is formed by selecting the peak value of the cross-correlation estimate of the two sensor outputs. Figure 1 shows the correlator output $Z(\hat{D})$ for a lowpass signal spectrum. Assume the

Reprinted from *Proc. ICASSP '83*, vol. 2, pp. 875–878, April 1983.

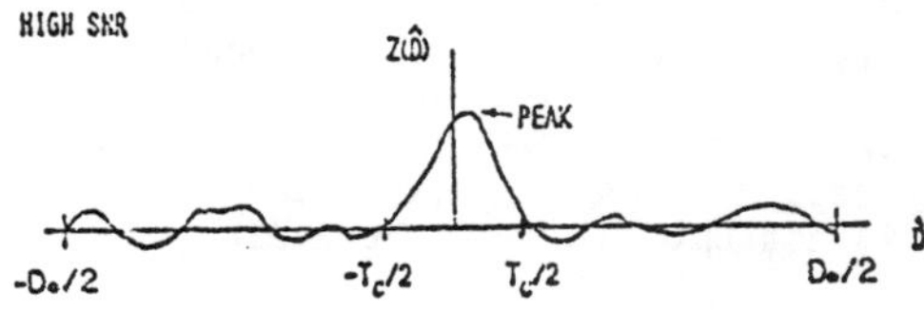

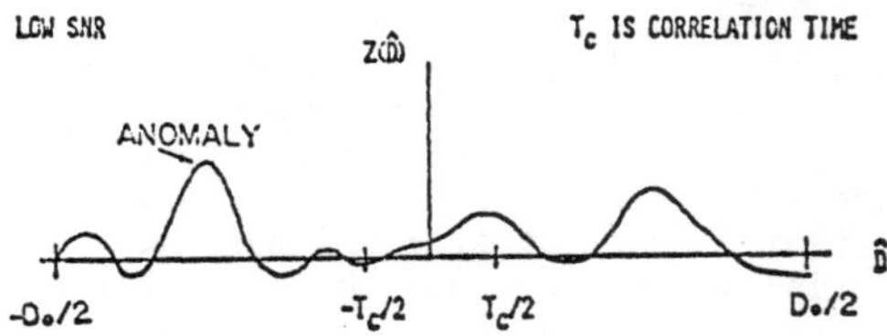

Figure 1. Correlator Output for Lowpass Signals

true value of D is zero. $Z(\hat{D})$ for high SNR is shown in the top panel. Here the errors are small and the correlation peak occurs close to the true value, say within a correlation time $\pm T_c/2$. This is the situation that must apply for the CRLB to be reachable. As SNR decreases the peak near D = 0 exhibits larger and larger variance until suddenly, at a certain SNR, large peaks occur far from the true value. These peaks will be uniformly spread throughout the a priori interval and at this point the estimates are useless. These may be referred to as anomalous estimates.

Figure 2 shows the correlator output for a narrowband signal spectrum. Now the envelope of $Z(\hat{D})$ is filled with a cosine carrier. The CRLB applies only when the peak value occurs within plus or minus one half cycle of the true value. At low SNR two types of large errors can occur. As in the lowpass case, anomalous estimates well removed from the true value can occur. Further, a second type of large error can now occur, as indicated, even if only the smaller region $\pm T_c/2$ is searched, since one of the highly oscillatory correlation peaks within $\pm T_c/2$ can exceed the peak value at D = 0. These errors are referred to as ambiguous estimates; they may still yield some useful information about the time delay, however.

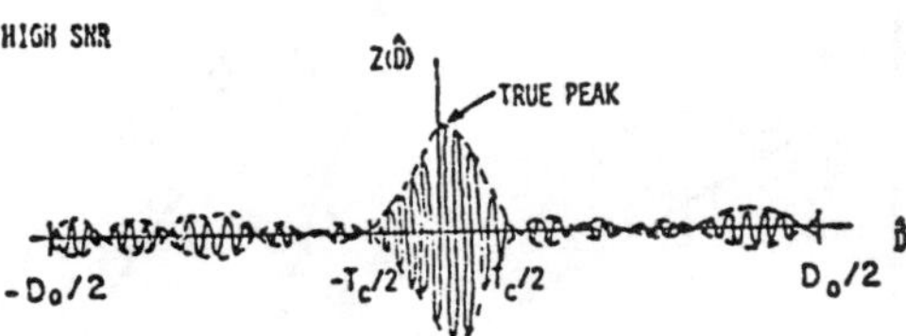

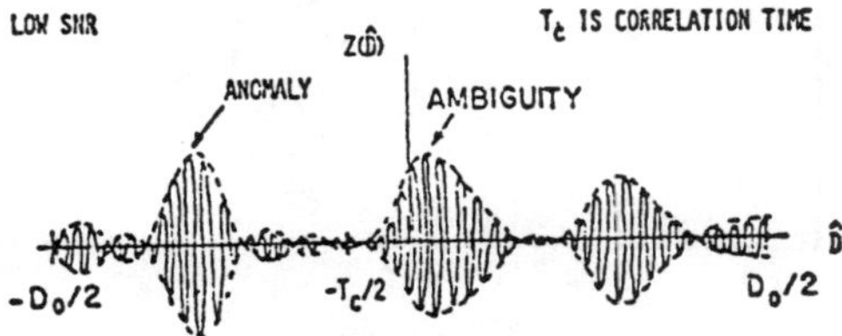

Figure 2. Correlator Output for Narrowband signals

In the next section we shall compare the cross-correlator performance above and below the threshold with the ZZLB. The section is divided into two sub-sections: first, we consider lowpass signals (and the associated anomaly effect); second, we consider narrowband signals (and the associated ambiguity effect).

COMPARISON OF THE ZZLB WITH CORRELATOR PERFORMANCE

Lowpass Signals

In this sub-section we consider signals whose power spectra is distributed about zero frequency. For computational convenience we shall assume that both signal and noise are spectrally flat within $|f| \leq B/2$ Hz and zero outside.

In Figure 3 we show the mean square estimation error, normalized by $D_0^2/12$ (the mean square value when the error is uniformly distributed in $\pm D_0/2$), as a function of the system SNR, defined by

$$\text{SNR} = \frac{(S/N)^2}{1 + 2S/N} \tag{1}$$

where S/N is the SNR at the individual sensors. The plot is for BT = 128 and BD_0 = 16. The

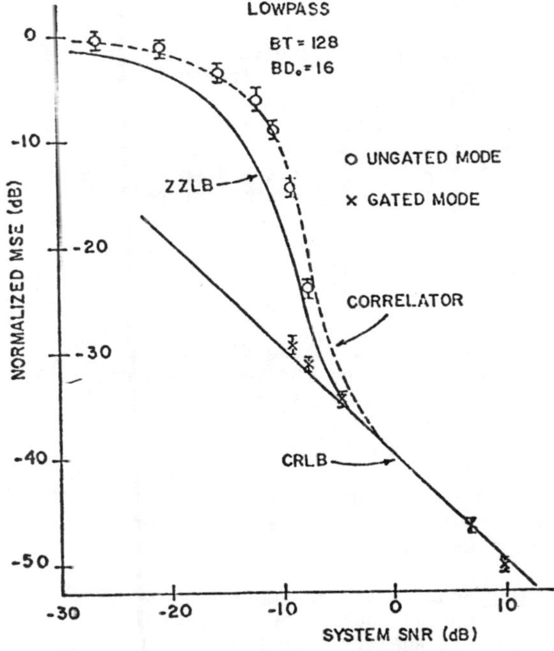

Figure 3. Normalized MSE vs. System SNR - Lowpass Signal

lowest solid curve is the CRLB while the upper solid curve is the ZZLB calculated from the exact results of [4]. The dashed line is an approximate theoretical result for the variance of the peak selecting correlator for large errors, calculated as described in [1].

As described in [1] and [2] a computer simulation was run to verify the theoretical results. Cross-correlation was performed via FFT processing. Data were acquired in both a gated and an ungated mode. In the gated mode the correlation peak closest to the known true delay that also fell within a correlation distance of the true value was used to calculate the variance. This restricts the variance calculation to local errors only and allows a comparison with the CRLB. These data are shown as the x's in Figure 3; the bars represent 95% confidence intervals. The agreement with the CRLB at high SNR is good, as expected; these data serve mainly as a check on the quality of the simulation. In the ungated mode the full a priori time delay region was searched for a peak. These data are shown as the circles in Figure 3. Note the close agreement between the ungated mode data and the approximate theory shown by the dashed line.

The most important observation to be made from Figure 3 is the close agreement between the correlator and the ZZLB. Indeed in the threshold region the two curves are displaced by roughly 1 dB in input S/N at the sensors. This establishes the ZZLB as a very tight bound and at the same time shows that the cross-correlator is very nearly an optimal instrumentation in the minimum m.s.e. sense.

The threshold SNR, denoted by SNR_{Th}, above which an essentially anomaly-free estimate can be achieved is of theoretical and practical importance. An extensive analysis of the ZZLB for lowpass spectra is carried out in [5] where it is shown that SNR_{Th} for the lowpass case is given to an excellent approximation by

$$SNR_{Th} \cong \frac{2}{BT}\left(F_{+}^{-1}\left[\left(\frac{3}{\pi BD_0}\right)^2\right]\right)^2 \qquad (2)$$

Here $F(x) = x^2\Phi(x)$ where $\Phi(x)$ is the integral from x to ∞ of the Gaussian probability density function. $F_{+}^{-1}(x)$ denotes the larger of the two solutions of $F^{-1}(x)$. SNR_{Th} is an approximation to the SNR at which the ZZLB is 3 dB above the CRLB. Thus for BT = 128 and BD_0 = 16, $SNR_{Th} \cong -7.4$ dB which is within 1 dB of the threshold point of the cross-correlator as observed from Figure 3.

Comparisons have also been made for other BT products. For BT equal 1280 the correlator performance is again within 1 dB of the ZZLB near threshold. For BT equal 64 the correlator is within 2 dB of the ZZLB. For BT equal 32 and 16 the correlator performance is 5 to 10 dB to the right of the ZZLB near threshold. Thus, at present we assert that, for lowpass signals, the ZZLB is tight and the correlator is nearly optimal for BT greater than, say, 50. For BT less than 50 we can only say, at present, that either the correlator is not optimal and/or the ZZLB is not tight.

Narrowband Signals

We now consider signals whose power spectra are narrowly distributed about some center frequency f_0. For computational convenience we shall assume that both signal and noise are spectrally flat within $|f \pm f_0| \leq B/2$ and zero elsewhere. In Figure 4 we show the mean square estimation error, normalized by $1/f_0^2$, as a function of system SNR. The plot is for $f_0/B = 10$, BT = 16 and $BD_0 = 2$. The lowest solid curve is the CRLB; the next solid curve is the simplified BB calculated from [3] and included here for reference; the upper solid curve is the ZZLB calculated from the exact results of [4]. The dashed line is an approximate theoretical result for the variance of the peak selecting correlator. This was found by an improved version of the method given in [2] and is based on the calculation of the probability that either of the two correlation peaks immediately adjacent to the peak at the true value of time delay are greater than the true peak. Unlike the method employed in [2] this technique does not assume that the two adjacent peaks are uncorrelated. The calculation is only valid for large SNR.

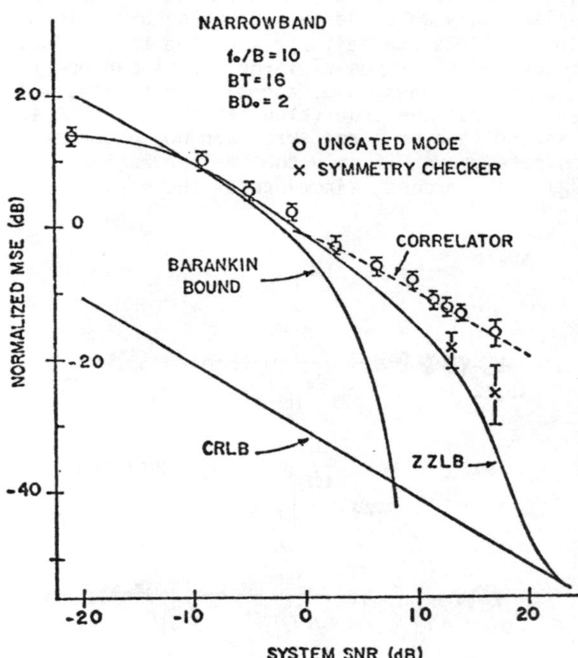

Figure 4. Normalized MSE vs. System SNR Narrowband Signals

A computer simulation was run as described above. The gated mode data were again within the 95% confidence limits of the CRLB for high SNR; these have not been included in Figure 4 for simplicity. The ungated data are shown by the circles. They agree quite well with the theoretical result for the correlator but the correlator variance is much larger than the ZZLB at high SNR. In fact, for BT = 16, the correlator, as implemented, will continue to make occasional large errors even at infinitely high SNR. This occurs because the correlator includes new data in the estimate at each lag value; as a result it cannot be guaranteed that, even at infinite SNR, the peak value will be at the true value of time delay.

Despite this, one feels that at least at infinite SNR some processing scheme which makes no errors ought to be realizable. One way to ensure that no errors are made at infinite SNR is as follows. For infinite SNR, and for a true time delay of zero, the correlator output will always be symmetrical about the true value of time delay. Thus, if one checks the symmetry of the three largest peaks one can always choose the correct peak. We employed this symmetry checking technique at non-infinite SNR to see if some improvement still resulted. This scheme is not practical but it does give some insight into the best one may be able to achieve. The data for the symmetry checker are shown by the x's in Figure 4. These data are quite a bit closer to the ZZLB than those for the simple peak selecting correlator. Since these data come from a realizable processor they also establish an upper bound; since they are within 2-3 dB in system SNR of the ZZLB they show that the ZZLB is at least this tight for small BT.

The situation is considerably simpler for larger BT products. In Figure 5 we show theoretical results for the normalized variance for BT = 1000, $f_o/B = 10$ and $BD_o = 2$ as before. Now we see that the threshold SNR for the correlator is within 1 dB of the ZZLB. Thus, for narrowband signals and large BT the ZZLB is shown to be very tight. Again we are interested in the threshold SNR above which an essentially ambiguity-free estimate of the sensor-to-sensor delay can be obtained. An extensive analysis of the ZZLB for narrowband signals is carried out in [4] where it is shown that for narrowband signals the threshold point is given to an excellent approximation by

$$SNR_{Th} \cong \frac{6}{\pi^2}\frac{1}{BT}\left[\Phi^{-1}\left(\frac{1}{24}\left[\frac{B}{f_o}\right]^2\right)\right]^2\left(\frac{f_o}{B}\right)^2 \quad (3)$$

where as in Eq. (2), SNR_{Th} is defined as the point at which the ZZLB is 3 dB above the CRLB. Thus for BT = 1000 and $f_o/B = 10$, $SNR_{Th} \cong -1.7$. dB which is within 1 dB of the threshold point of the cross-correlator, as observed from Figure 5.

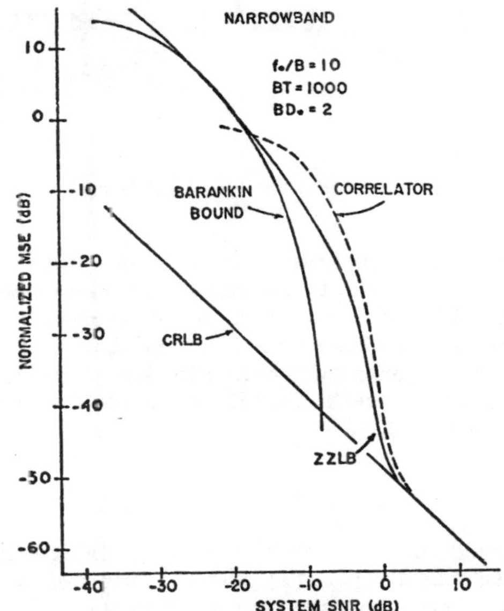

Figure 5. Normalized MSE vs. System SNR - Narrowband

SUMMARY

By comparing actual correlator performance with the ZZLB we have shown that the ZZLB is indeed a tight bound. For lowpass signals and BT greater than 50 the correlator threshold SNR is within 1 dB of the ZZLB. For narrowband signals and large BT the correlator threshold again is within 1 dB of the ZZLB; for small BT (BT = 16) the correlator (as implemented here) does not reach the ZZLB but an alternate scheme, based on a symmetry check, comes within 2-3 dB of the ZZLB near threshold.

REFERENCES

[1] J. P. Ianniello, "Time Delay Estimation via Cross-Correlation in the Presence of Large Estimation Errors", IEEE Trans. Acoust., Speech, Signal Proc., vol. ASSP-30, pp. 998-1003, 1982.

[2] J. P. Ianniello, "Threshold Effects in Time Delay Estimation Using Narrowband Signals", ICASSP 82 Conference Proc., vol. 1, pp. 375-378.

[3] S-K. Chow and P. M. Schultheiss, "Delay Estimation Using Narrowband Processors", IEEE Trans. Acoust., Speech, Signal Proc., vol. ASSP-29, pp. 478-484, 1981.

[4] A. Weiss and E. Weinstein, "Fundamental Limitations in Passive Time Delay Estimation - Part I: Narrowband Systems", to appear in IEEE Trans. Acoust., Speech, Signal Proc., 1982.

[5] E. Weinstein, "Performance Analysis of Time Delay Estimators", WHOI Technical Report, August 1982.

Delay Estimation Simulations and a Normalized Comparison of Published Results

R. LYNN KIRLIN AND JONATHAN N. BRADLEY

Abstract—Three recent publications have reported delay-estimate variances from simulations using the generalized cross-correlator method on both stationary random signals with noise and deterministic signals with noise. Further simulations with finite energy signals are reported here. A method of comparing the three reported data sets is developed, and a remarkable consistency is observed even though the simulation parameters have varied widely.

Introduction

The generalized correlator method of determining delay has been promoted recently [1], [2], and the results of extensive simulations have been published in three papers [3], [4], [8]. Although many algorithms for delay estimation have been proposed, the generalized correlator, shown in Fig. 1, has perhaps received the most widespread attention. It utilizes a frequency-domain windowing applied to the raw cross spectrum in order to produce a resulting cross-correlation function whose peak occurs at the time of delay of one signal with respect to the other. Various windows have been proposed under some optimality or heuristic criterion, and many of these were tested in simulations by Hassab and Boucher [4], [8]. Of these, the maximum-likelihood window W_I derived by Knapp and Carter [1] produces the most accurate results unless the signal spectrum contains a significant level of tonal power, in which case other windows such as the maximum-peak-to-variance window W_{II} perform better [5].

The three published simulation studies have considered both stationary random and deterministic signals and noise. However, there are other variations which make it difficult to compare results. For example, the sampling interval Δt, signal and noise spectral frequency bands, and record lengths all cause variations in the reported delay-estimate variances. The variances have been recorded in dimensionless sample-interval increments squared. Perhaps foreseeing this difficulty in comparison, Quazi [6] formulated the Cramer–Rao lower bound on this variance (when the W_I window is used) in terms of signal-to-noise ratio (SNR), record length T, and upper and lower frequencies f_2 and f_1 demarcing the ideally flat signal and noise spectra. The formula was most easily expressed at either high or low SNR:

$$\sigma_0^2 = \frac{3}{4\pi^2 T(f_2^3 - f_1^3)} \cdot \begin{cases} \dfrac{1}{2\,\mathrm{SNR}^2}, & \mathrm{SNR} \ll 1 \quad (1a) \\[2ex] \dfrac{1}{\mathrm{SNR}}, & \mathrm{SNR} \gg 1. \quad (1b) \end{cases}$$

Manuscript received June 17, 1981; revised December 1, 1981.
R. L. Kirlin was with the Department of Electrical Engineering, University of Wyoming, Laramie, WY 82071. He is now with the Auditory and Electroacoustics Research Group, University of Wyoming, Laramie, WY 82071.
J. N. Bradley is with the Department of Electrical Engineering, University of Wyoming, Laramie, WY 82071.

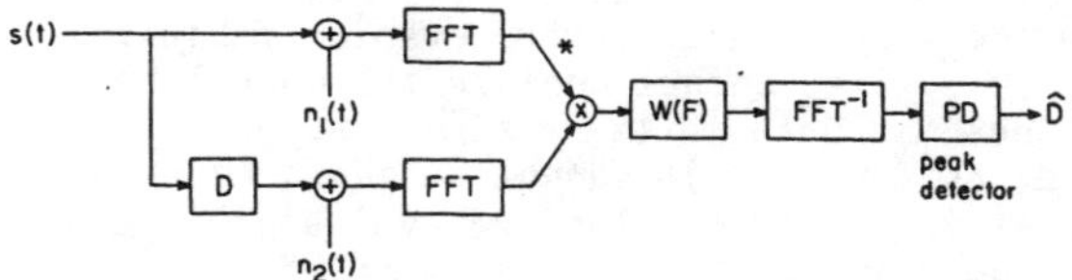

Fig. 1. Generalized correlator.

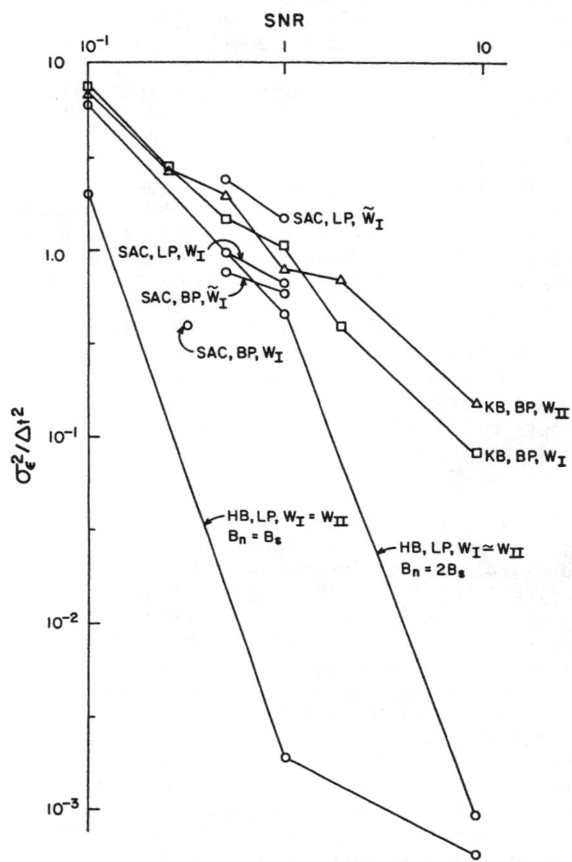

Fig. 2. Comparison of error variances with published data from similar situations. *HB*–N = 256, M = 300, D = 10, ΔD = 5, first-order signal noise. *SAC*–N = 512, M = 32, D = 4, ΔD = 10, white signal plus noise. *KB*–N = 128, M = 50, D = 20 and 1, ΔD = 5, Ricker and white noise. Various forms of low-pass (*LP*) and bandpass (*BP*) were used. N = block length, M = number of averages.

Discussion and Method of Comparison

Referring to Table I, it may be seen that only Scarbrough, Ahmed, and Carter used the ideally flat signal and noise spectra in their simulations. Other dissimilarities are noted as follows. The signal spectrum of the authors (K and B) is actually an energy spectrum which corresponds to a Ricker wavelet. (The Ricker wavelet is the second derivative of a Gaussian pulse, of duration approximately 40 ms and frequency content essentially in $20 \leqslant f \leqslant 50$ Hz.) The noise

Reprinted from *IEEE Trans. Acoust., Speech, Signal Processing*, vol. 30, no. 3, pp. 508–511, June 1982.

TABLE I
REPORTED DATA USING $W_I(f)$

Case #	f_2 Hz	f_1 Hz	B_s Hz	B_n Hz	experiment SNR (dB)	equivalent SNR in B_s(dB)
1	0.102	0	0.102	0.102	-10	-10
2	0.102	0	0.102	0.102	0	0
3	0.102	0	0.102	0.102	10	10
4	0.102	0	0.102	0.204	-10	-7
5	0.102	0	0.102	0.204	0	3
6	0.102	0	0.102	0.204	10	13
7	0.102	0	0.102	0.5	-10	-3
8	0.102	0	0.102	0.5	0	7
9	0.102	0	0.102	0.5	10	17
10	0.102	0	0.102	0.102	-10	-10
11	0.102	0	0.102	0.102	0	0
12	0.102	0	0.102	0.102	10	10
13	0.102	0	0.102	0.204	-10	-7
14	0.102	0	0.102	0.204	0	3
15	0.102	0	0.102	0.204	10	13
16	0.102	0	0.102	0.5	-10	-3
17	0.102	0	0.102	0.5	0	7
18	0.102	0	0.102	0.5	10	17
19	100	0	100	100	-3	-3
20	100	0	100	100	0	0
21	100	0	100	100	10	10
22	200	100	100	100	-3	-3
23	200	100	100	100	0	0
24	200	100	100	100	10	10
25	50	20	100	30	-10	-4.77
26	50	20	100	30	-3	2.23
27	50	20	100	30	0	5.23
28	50	20	100	30	3	8.23
29	50	20	100	30	10	15.23

spectrum of K and B is flat, low-pass out to 100 Hz. The K and B SNR was calculated according to total signal energy and expected total noise energy in the $T = 128$ ms records. Delay estimates include a parabolic fitting [see 7].

H and B spectra are fourth-order rolloff, low-pass, and the noise spectrum corner frequency is adjusted to give "noise bandwidth (B_n) equal to, twice, and 'much greater' than signal bandwidth (B_s)." We have interpreted "much greater than" to mean white out to the folding frequency (1/2 Hz) of the discrete-signal process.

Also tabulated in Table I are the reported results. Typically, a $\pm\delta D$ "acceptance region" of delay estimates is chosen, and this is given as so many sampling intervals. Estimates of delay falling within this acceptance region around the true delay D are experimentally found to do so with probability P. The delay error variance is also typically reported in number squared of sampling intervals, and is tabulated here as $\sigma_\epsilon^2/\Delta t^2$. Table I data as well as data on W_{II} and approximations to W_I from [3] and [4] are shown in Fig. 2.

In order to make use of the formulas of (1), the SNR must be measured in an equivalent, ideally flat signal bandwidth. That SNR will differ from the one reported except for SA and C. An equivalent bandwidth might be calculated. Alternately, the half-power or approximate half-power points might be used as has been done here; thus, we have the f_2 and f_1 values entered into Table II. Using these f_2 and f_1, the SNR in signal band (B_s) may be computed as tabulated in Table II. Finally, incorporating Δt^2 into the reported σ_ϵ^2, all simulation results can be compared on the same basis.

Specifically, this is done by normalizing the variance formula, moving all quantities except SNR to the left-hand side of (1). What is plotted in Fig. 3 then is

$$\sigma_\epsilon^2 \cdot \frac{4\pi^2 T(f_2^3 - f_1^3)}{3} \cdot \begin{cases} 2, & \text{SNR} \ll 1 \\ 1, & \text{SNR} \gg 1 \end{cases} = \begin{cases} \dfrac{1}{\text{SNR}^2}, & \text{SNR} \ll 1 \\ \dfrac{1}{\text{SNR}}, & \text{SNR} \gg 1. \end{cases} \tag{2}$$

In effect, the plot can be thought of as measured variances being used to estimate SNR in the signal band.

RESULTS

The asymptotes of (2) are plotted as a continuous line broken at SNR = 0 dB in Fig. 3. The line represents the Cramer–Rao lower bound on the delay-estimate variance, multiplied by $4\pi^2 T(f_2^3 - f_1^3)/3$. All estimated variances, when properly scaled, should fall above this line; allowances are, of course, to be made for the fact that all experiments do not use ideally flat signal and noise spectra. In fact, further simulations with $\sin x/x$-shaped spectra show that results are quite sensitive to "equivalent bandwidth" specifications.

Not plotted in Fig. 3 are three points of SA and C which show $\sigma_\epsilon^2 = 0$. In order to take into account the fact that only the asymptotic formulas are used, calculations have been made with the low-SNR expression, (2a). A transition region factor, linear from 0 dB to 3 dB, has been used to correct values in $10^{-1} < \text{SNR} < 10$. Although the various experimental results are scattered around the theoretical line, many of the points lie very close to it, and suggest that this method of comparison of results among simulations with widely varying parameters is useful.

TABLE II
NORMALIZING PARAMETERS

	Signal and Noise Spectra	SNR dB	T sec	Δt sec	$\frac{D}{\Delta t}$	$\frac{\delta D}{\Delta t}$	P	$\sigma_\varepsilon^2/\Delta t^2$	Case No.
Hassab and Boucher (H&B) (Deterministic)	$\Phi_s(\omega)=\frac{1}{(1+\omega^2)^2}$	-10					0.10	2.06	1
		0	256	1	20	±5	1.00	0.002	2
	$\Phi_n(\omega)=\frac{1}{(1+\omega^2)^2}$	10					1.00	0.0006	3
	$\Phi_s(\omega)=\frac{1}{(1+\omega^2)^2}$	-10					0.143	6.21	4
		0	256	1	20	±5	0.98	0.459	5
	$\Phi_n(\omega)=\frac{1}{(1+(\omega/2)^2)^2}$	10					1.00	0.001	6
	$\Phi_s(\omega)=\frac{1}{(1+\omega^2)^2}$	-10					0.69	3.37	7
		0	256	1	20	±5	1.00	0.175	8
	$\Phi_n(\omega)\simeq\sqcap(\frac{\omega}{2\pi})$	10					1.00	0.001	9
Hassab and Boucher (H&B) (Random)	$\Phi_n(\omega)=\frac{0.111}{(0.111+\omega^2)^2}$	-10					0.116	3.91	10
		0	256	1	20	±5	0.996	0.00294	11
	$\Phi_n(\omega)=\frac{1}{(0.111+\omega^2)^2}$	10					1.000	0.00047	12
	$\Phi_s(\omega)=\frac{0.111}{(0.111+\omega^2)^2}$	-10					0.103	4.79	13
		0	256	1	20	±5	0.900	0.565	14
	$\Phi_n(\omega)=\frac{1}{(0.111+(\omega/2)^2)}$	10					1.000	0.00151	15
	$\Phi_s(\omega)=\frac{0.111}{(0.111+\omega^2)^2}$	-10					0.613	3.41	16
		0	256	1	20	±5	1.000	0.230	17
	$\Phi_n(\omega)=\sqcap(\omega/2\pi)$	10					1.000	0.0219	18
Scarbrough Ahmed & Carter (SA&C) (Random)	$\Phi_s(f)=\sqcap(\frac{f}{200})$	-3					1.0	0.98	19
	$=\Phi_n(f)$	0	0.25	$\frac{1}{2048}$	4	±10	1.0	0.67	20
	(low pass, white)	10					1.0	0.0	21
	$\Phi_s(f)=\sqcap(\frac{\lvert f-150\rvert}{100})$	-3					1.0	0.40	22
	$=\Phi_n(f)$	0	0.25	$\frac{1}{2048}$	4	±10	1.0	0.0	23
	(band pass, white)	10					1.0	0.0	24
Kirlin & Bradley (K&B) (Deterministic)	$\Phi_s(\omega)=(0.005)^6(2\pi)\cdot$	-10					0.26	7.50	25
	$\omega^4\exp[-(0.005)^2\omega^2]$	-3					0.88	1.49	26
	(Ricker wavelet)	0	0.128	0.001	1	±5	0.96	1.07	27
	$\Phi_n(\omega)\simeq\sqcap(\frac{\omega}{2\pi\cdot 200})$	3					1.00	0.40	28
		10					1.00	0.08	29

Note: $\sqcap(\frac{x}{2B}) = \begin{cases}1, & B\leq x\leq B\\ 0, & \lvert x\rvert>B\end{cases}$

CONCLUSIONS

A formula [(1)] by Quazi has been shown to be useful for comparing and measuring accuracy of delay-estimation simulation results. Two simulations with stationary random signals and two with transient signals have been shown to fall close to an expected measure, even though simulation parameters vary widely. As previously only *SA* and *C*[3] compared their results to the Cramer–Rao lower bound, it is felt that the comparison method demonstrated here is useful in unifying all such experimental data inasmuch as the signal may be thought to have an equivalent ideally flat bandwidth. Viewing Figs. 2 and 3 verifies the comparability feature; in Fig. 2 it appears that the results are quite different, but Fig. 3 shows them very compatible.

Lastly, we report that many other simulations of our transient signal case with variations on the window (W_1, W_2, no window, estimated W_1) and methods of computing it involving signal and noise spectral estimation have yielded similar results.

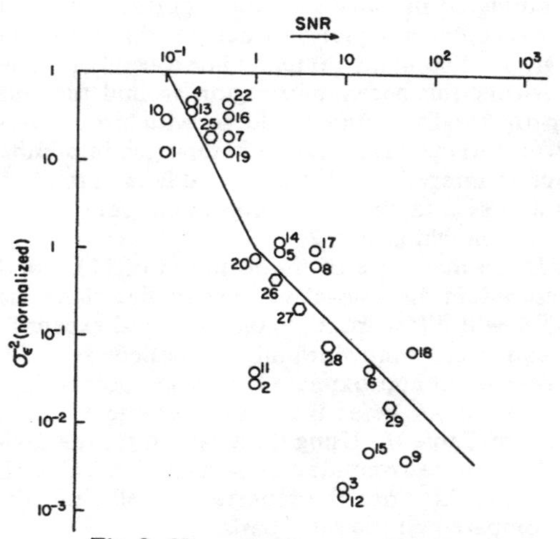

Fig. 3. Normalized error variances.

REFERENCES

[1] C. H. Knapp and G. C. Carter, "The generalized method for estimation of time delay," *IEEE Trans. Acoust., Speech, Signal Processing*, vol. ASSP-24, pp. 320–327, Aug. 1976.

[2] J. C. Hassab and R. E. Boucher, "Optimum estimation of time delay by a generalized correlator," *IEEE Trans. Acoust., Speech, Signal Processing*, vol. ASSP-27, pp. 373-380, Aug. 1979.

[3] K. Scarbrough, N. Ahmed, and G. C. Carter, "On the simulation of a class of time delay estimation algorithms," *IEEE Trans. Acoust., Speech, Signal Processing*, vol. ASSP-29, pp. 534-540, June 1981.

[4] J. C. Hassab and R. E. Boucher, "A quantitative study of optimum and sub-optimum filters in the generalized correlator," in *Proc. Int. Conf. Acoust., Speech, Signal Processing*, Washington, DC, 1979, IEEE Pub. 79Ch1379-7ASSP, pp. 124-127.

[5] —, "Performance of the generalized correlator in the presence of a strong spectral peak in the signal," *IEEE Trans. Acoust., Speech, Signal Processing*, vol. ASSP-29, pp. 549-555, June 1981.

[6] A. H. Quazi, "An overview on time delay estimation in active and passive systems for target localization," *IEEE Trans. Acoust., Speech, Signal Processing*, vol. ASSP-29, pp. 527-533, June 1981.

[7] R. E. Boucher and J. C. Hassab, "Analysis of discrete implementation of generalized cross-correlator," *IEEE Trans. Acoust., Speech, Signal Processing*, vol. ASSP-29, pp. 609-611, June 1981.

[8] J. C. Hassab and R. E. Boucher, "An experimental comparison of optimum and sub-optimum filters' effectiveness in the generalized correlator," *J. Sound Vibration*, vol. 76, pp. 117-128, May 1981.

THE ROLE OF THIRD ORDER SPECTRUM IN MAXIMUM LIKELIHOOD TIME DELAY ESTIMATION OF A RANDOM MULTI-TONE SIGNAL IN NOISE

by

Doron Kletter and Hagit Messer

Faculty of Engineering,Department of Electronic Systems,

Tel-Aviv University, Tel-Aviv, Israel 69978

ABSTRACT

The maximum likelihood (ML) processor for estimating the differential time delay between two noisy versions of a random signal is known to be the generalized cross correlator (GCC), when the signal and noise processes are Gaussian. For non-Gaussian signals, time delay estimation (TDE) in higher order spectra (HOS) domain was suggested, and indeed better performance was already reported with both empirical and simulated data. However, to the best of our knowledge, no claim for optimality of HOS based processors was ever made. Therefore, even-though promising, it is still considered to be an ad-hoc procedure. In this paper we present the ML time delay estimator for a special class of non-Gaussian signals, where the radiated signal is an harmonically related random multi-tone signal. For this case we show that when the signal has a non-zero bispectrum (i.e. in case of phase coupling) - the ML time delay estimator consists of the non-coherent estimator plus an extra processor which is, at least for low SNR, direcly related to the signal bispectrum. The performance of the ML TDE is analyzed by using the Cramer-Rao lower bound, and it is shown that the improved processor is superior.

1. Introduction and basic model

The most common method for estimating the pairwise delay is to cross-correlate one sensor output with a delayed version of the other sensor output. The delay value at which the cross-corrlation peaks - is the estimator of the sensor-to-sensor delay, [1,2]. A veraity of possible pre-filters is often used to better shape the generalized cross correlation (GCC) function. The derivation of the GCC estimator is strongly based on the assumption of Gaussian signal and noise processes, under which it was possible to show that the GCC estimator, with the appropriate pre-filtering, is the maximum likelihood (ML) estimator of the delay, [1]. The GCC is therefore optimal in the sense that its estimation variance approaches asymptotically to the absolute Cramer-Rao lower bound (CRLB).

When the signal is not Gaussian, however, the GCC estimator is no longer optimal, resulting in poor performance. There are many practical problems where the radiated signal is considered to be non-Gaussian. These include broad areas such as radar, sonar, oceanography, geophysics, etc.

It was mentioned recently that a considerable improvement in estimation performance is possible for random non-Gaussian signals, using a third order spectrum (bispectrum) technique, [3]. Such improvement was reported using computer simulations [5-8], and even demonstrated with real data [4], but no formal optimality proof nor any analytic performance evaluation is available. The conditions necessary to maintain a certain performance level are not well understood either.

A general random multi-tone signal consists of tones at different frequencies, with independent and/or harmonically related phases. Phase coupled signals are generated as a result of non linear interaction on multi-tones. We derive the ML estimator for a triple-tone signal, with and without phase coupling between two of its components. The resulting ML estimator in presence of phase coupling is similar to the uncoupled estimation scheme, but added to it there is a compensating block to take advantage of the existing phase coupling. This extra block, for low SNR, is closely related to the bispectral analysis. This should not be unreasonable, at least intuitively, since the power spectrum suppresses all phase information and therefore one is forced to use higher order analysis to utilize the coupling. Interestingly, for low SNR, only the second- (i.e. power spectral) and third- (i.e. bispectral) order analyses are needed, and it is unnecessary to use any higher order analysis than that.

The resulting estimation scheme is modular, because the bispectral analyzer provides zero output for uncoupled signals. It can also be shown that the above estimation scheme is robust in the sense that it is insensitive to a detailed knowledge of the sinusoidal amplitudes.

The main conclusion is that the non-Gaussian characteristics of the signal (or noises) can be used in order to better separate it from the noise, and estimate its parameters. This will generally be achieved at the expense of receiver complexity.

The basic system of interest consists of a stationary source emitting a sinusoidal signal, which is received at two spatially separated sensors. The observed signals are:

$$\begin{cases} x_1(t) = s_1(t) + n_1(t) \\ x_2(t) = s_2(t) + n_2(t) \end{cases} \qquad -T/2 \leq t \leq T/2 \qquad (1)$$

The noise at each sensor $n_m(t)$, m=1,2, is assumed to be a sample function from independent, zero mean, stationary white Gaussian random process, with (double sided) spectral density level N_m. T is the available observation time, and $s_m(t)$ is the received signal at the m-th sensor, which is assumed to be a triple-tone:

$$s_1(t) = \sum_{i=1}^{3} A_i \cos(\omega_i t - \phi_i)$$

$$s_2(t) = s_1(t+D) \qquad (2)$$

where D is the relative delay of interest between the sensors. The three frequencies are assumed to be harmonically related, that is $\omega_3 = \omega_1 + \omega_2$. To establish phase coupling, it is required that $\phi_3 = \phi_1 + \phi_2$ as well. The rest of the phases are assumed to be independent, uniformly distributed random variables.

Our main goal is to estimate D from the (finite length) observed data, and to set a lower bound on the estimation performance.

2. ML estimation of the time delay

Suppose that each sensor output is frequency bandlimited, with bandwidth W, and then sampled at Nyquist rate. The conditional joint distribution function (assuming WT >> 1) is given by:

$$P(\underline{X}/D,\underline{\phi}) = \qquad (3)$$

$$= (4 N_1 N_2)^{-L/2} \exp\left\{ - \sum_{m=1}^{2} \frac{1}{2N_m} \int_{-T/2}^{T/2} \left[x_m(t) - s_m(t)\right]^2 dt \right\}$$

Reprinted from *Proc. ICASSP '89*, vol. 4, pp. 2310–2313, May 1989.

where L is the total number of samples, i.e. $L = TW/\pi$.
Define:

$$\rho_m = \int_{-T/2}^{T/2} x_m^2(t)\, dt \qquad m=1,2 \tag{4a}$$

$$E = \int_{-T/2}^{T/2} s_m^2(t)\, dt = \sum_{i=1}^{3} A_i^2 T/2 \tag{4b}$$

$$C = (4\, N_1\, N_2)^{-L/2} \tag{4c}$$

and

$$C_{mi} = \int_{-T/2}^{T/2} x_m(t) \cos[\, \omega_i t + (m-1)\omega_i D\,]\, dt \qquad m=1,2$$

$$S_{mi} = \int_{-T/2}^{T/2} x_m(t) \sin[\, \omega_i t + (m-1)\omega_i D\,]\, dt \qquad i=1,2,3 \tag{5}$$

then eq.(3) can be re-written as:

$$P(\underline{X}/D,\underline{\phi}) = C \exp(J_1) \exp(J_2) \tag{6}$$

where

$$J_1 = -\sum_{m=1}^{2} \frac{\rho_m}{2N_m} - \sum_{m=1}^{2}\sum_{i=1}^{3} \frac{A_i^2 T}{4N_m}$$

$$J_2 = \sum_{m=1}^{2}\sum_{i=1}^{3} \frac{A_i}{N_m} [\, C_{mi}\cos(\phi_i) + S_{mi}\sin(\phi_i)\,] \tag{7}$$

It is clear that all the information about D is contained in the second exponent of eq.(6). Obviously, the marginal distribution of the data given D is:

$$P(\underline{X}/D) = \frac{1}{(2\pi)^k} \int_{-\pi}^{\pi} \cdots \int_{-\pi}^{\pi} P(\underline{X}/D,\underline{\phi})\, d\phi_1 \ldots d\phi_k \tag{8}$$

where k represents the number of independent phases to be averaged over. k=3 for the uncoupled signal (all phases are independent), while k=2 for a coupled one (since $\phi_3 = \phi_2 + \phi_1$).

The value of D which maximizes eq.(8) is the ML estimator of the delay, that is:

$$\hat{D}_{ML} = \max_{D} \{ \ln P(\underline{X}/D) \} \tag{9}$$

For the well known case of a single sinusoid (monochromatic) signal with unknown random phase the optimal ML estimator is achieved by constructing the non-coherent receiver, [9,chap.4]. The generalization of this result to a triple tone uncoupled signal under low SNR conditions is straightforward, and forms the basic processor shown in Fig.1 (thin lines only). It is given by:

$$\hat{D}_{ML} \Big|_{uc} = \max_{D} \left\{ \sum_{i=1}^{3} A_i^2 R_i^2 \right\} \tag{10}$$

where uc stands for uncoupled, and

$$R_i = \sqrt{X_i^2 + Y_i^2} \qquad i=1,2,3 \tag{11}$$

where

$$\left\{ \begin{array}{l} X_i = \dfrac{C_{1i}}{N_1} + \dfrac{C_{2i}}{N_2} \\[2ex] Y_i = \dfrac{S_{1i}}{N_1} + \dfrac{S_{2i}}{N_2} \end{array} \right. \qquad i=1,2,3 \tag{12}$$

To satisfy the low SNR conditions, it is required that $A_i\, R_i << 1$ for every i.

The derivation of the ML estimator for the coupled case is somewhat long, but follows in a similar manner. The result is:

$$\hat{D}_{ML} \Big|_{c} = \max_{D} \{ \sum_{i=1}^{3} A_i^2 R_i^2 + $$

$$+ A_1 A_2 A_3 [\, X_1 (X_2 X_3 + Y_2 Y_3) + Y_1 (X_2 Y_3 - X_3 Y_2)] \} \tag{13}$$

Comparing eq.(13) with (10), it is immediately clear that the presence of such quadratic phase coupling introduces an extra term (to the right of (13)) in the optimal (low SNR) ML estimator. The bispectrum of a random signal is defined as the Fourier transform of the third order cumulant function, i.e.-

$$B_s(\omega_1,\omega_2) = F\{ C_s(\tau,\rho) \} = F\{ E[\, s(t)s(t-\tau)s(t-\rho)\,] \} \tag{14}$$

where E( ) stands for expectation. Since $s_1(t)$ (the original signal) has zero mean, the third-order cumulant equals the third-order moment, which is known to be [3]:

$$C_{s_1}(\tau,\rho) = 1/4\, A_1 A_2 A_3 \{ \cos(\omega_1\tau+\omega_2\rho) + \cos(\omega_2\tau+\omega_1\rho) + $$
$$+ \cos(\omega_3\tau-\omega_1\rho) + \cos(\omega_1\tau-\omega_3\rho) + $$
$$+ \cos(\omega_3\tau-\omega_2\rho) + \cos(\omega_2\tau-\omega_3\rho) \} \tag{15}$$

Inserting $Cs_1(\tau,\rho)$ into eq.(14), and under the assumptions of large WT product and low SNR, it can be shwon that the bispectrum approximately equals four times the right side of eq.(13). Since the bispectrum of an uncoupled multi-tone is zero, adding this block is always suitable in any case. The new block is shown (with thick lines) in Fig. 1., thus the whole picture (the uncoupled estimation scheme together with the bispectral block) forms the complete estimaton scheme for a coupled signal.

3. Performance evaluation

It is clear from the previous discussion that the basic uncoupled estimation scheme is no longer optimal when the coupling exists. Hence, there must be some performance improvement when the phase coupling block is added. Intuitively, it is obvious that if there is prior knowledge that such coupling exists, then, rather than being an extra nuisance parameter, the third phase provides additional information now about the other two.

The Cramer-Rao lower bound on estimation variance is given by:

$$\mathrm{Var}\left(\hat{D}_{ML} \right) \geq J_D^{-1} = \left[-E\left\{ \frac{d^2 \ln P(\underline{X}/D)}{dD^2} \right\} \right]^{-1} \tag{14}$$

Again, the derivation of the CRLB for an uncoupled multi-tone is an extension to the single tone case considered in [9]. The result is:

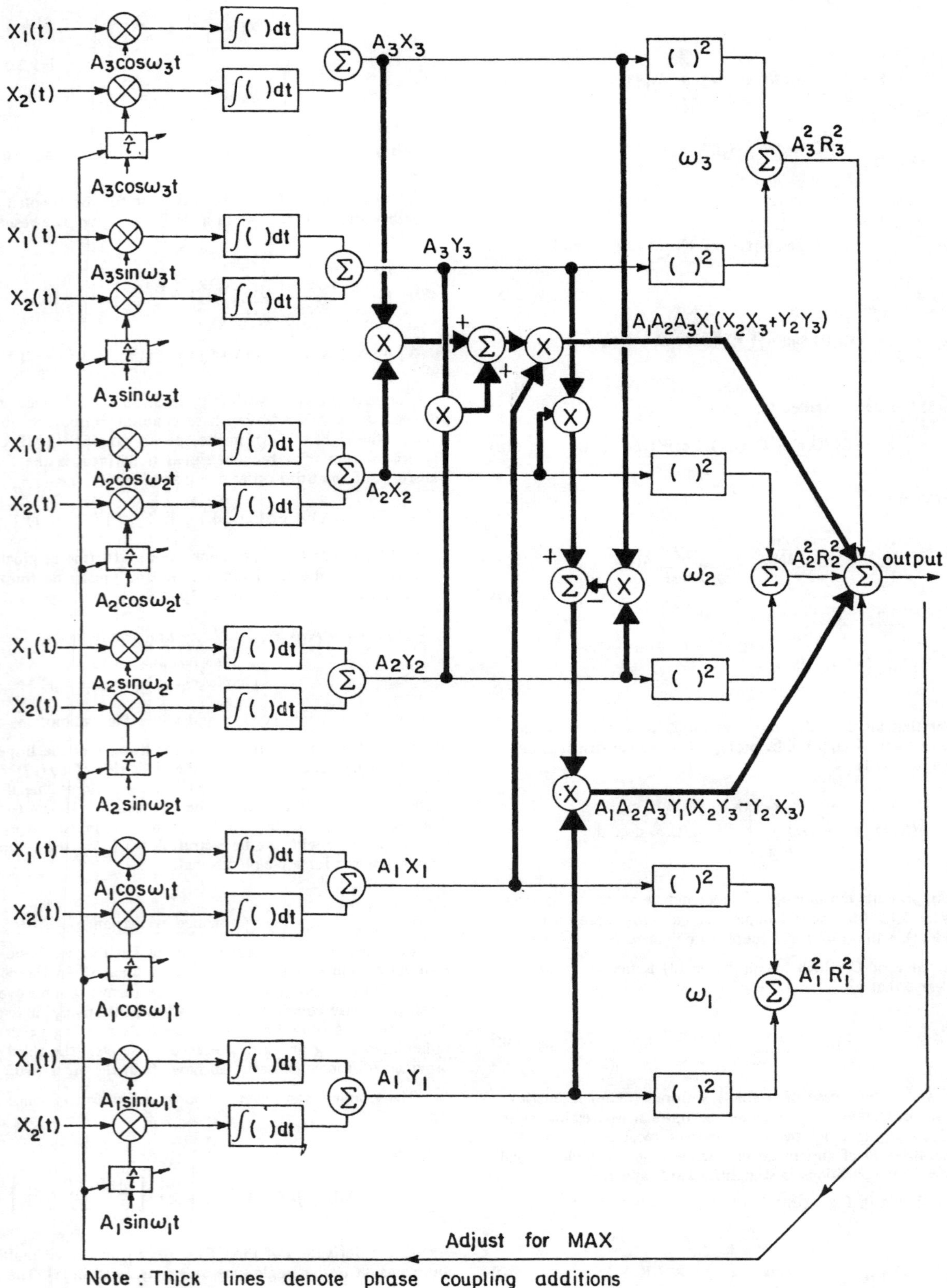

Fig. 1: The optimal (low SNR) ML estimator for random triple-tone signal

$$\mathrm{Var}\left\{ \hat{D}_{ML} \Big|_{uc} \right\} \geq \frac{2}{\sum_{i=1}^{3} \omega_i^2 \frac{A_i^2 T}{2N_1} \frac{A_i^2 T}{2N_2}} \quad (15)$$

where $A_i^2T/2N_m$ is actually the SNR of the i-th tone, in the m-th sensor. The performance can therefore be improved by either increasing the SNR, or the observation time. The low SNR CRLB in the coupled case (obtained following some tedious computations) is given by:

$$\mathrm{Var}\left\{ \hat{D}_{ML} \Big|_{c} \right\} \geq \left[J_D \Big|_{uc} + \Delta J_D \right]^{-1} \quad (16)$$

where $J_D \Big|_{uc}$ is the uncoupled Fisher information given by the inverse of the right side of eq.(15), and ΔJ_D is the coupling improvement, given by:

$$\Delta J_D = \frac{1}{2} \prod_{i=1}^{3} \frac{A_i^2 T}{2N} \left(\omega_1^2 + \omega_2^2 + \omega_1 \omega_2 \right) \quad (17)$$

We have assumed $N_1 = N_2 = N$ for simplicity. Note that ΔJ_D is always non negative, thus the total estimation error is always reduced when the phase coupling is present. This improvement is achieved by a non linear operation, namely - the bispectral analyzer.

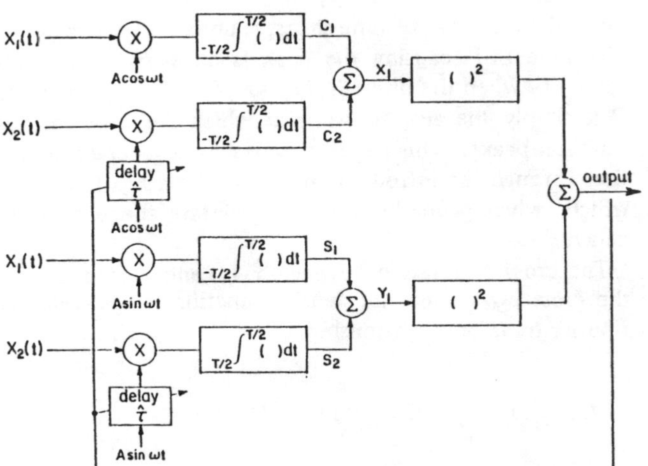

Fig. 2: The non-coherent ML estimator for monochromatic signal

4. Conclusions

In this paper an optimal ML time delay estimation procedure for a low SNR, phase coupled, multi-tone random signal was derived, and its asymptotic performance was analyzed. The resulting estimation scheme consists of a non-coherent estimator matched to the uncoupled signal, plus an additional block (introduced by the coupling) to be maximized together. This extra block is found to be directly related to the bispectrum of the coupled signal, thus the optimal estimation scheme is based only on the second (power spectral) and third (bispectral) order analyses. It is demonstrated, by evaluating the Cramer-Rao lower bound, that the presence of such quadratic phase coupling can be used to reduce the total estimation error. The theoretical results are in agreement with the empirical and simulated results previously reported. It can also be shown that this estimation method is insensitive to inexact knowledge of the tone amplitudes and frequencies. It must be emphasized that the non-Gaussian nature of the multi-tone signal (as expressed by its bispectrum) can effectively be used in order to better separate or estimate its parameters.

REFERENCES

[1] C.H. Knapp and G.C. Carter, "The Generalized Correlation Method for Estimation of Time Delay." IEEE Trans. Acoust., Speech, Signal Processing, vol. ASSP-24, No. 4, pp. 320-327, Aug. 1976.

[2] C.H. Knapp and G.C. Carter, "Estimation of Time Delay in Presence of Source or Receiver Motion". J. Acoust. Soc. Amer., vol. 61, No. 6, pp. 1545-1549, 1979.

[3] C.L. Nikias and M.R. Raghuveer, "Bispectrum Estimation: A Digital Signal Processing Framework," Proc. of the IEEE, Vol. 75, No. 7, pp. 869-892, July 1987.

[4] T. Sato, K. Sasaki, and Y. Nakamura, "Real Time Bispectral Analysis of Gear Noise and its Application to Contactless Diagnostics," J. Acoust. Soc. Amer., vol. 62, pp. 382-387, 1977.

[5] K. Sasaki, T. Sato, and Y. Nakamura, "Holographic Passive Sonar," IEEE Trans. Sonics., Ultrason., vol. SU-24, No. 3, pp. 193-200, May 1977.

[6] T. Sato and K. Sasaki, "Bispectral Holography," J. Acoust. Soc. Amer., vol. 62, pp. 404-408, 1977.

[7] C.L. Nikias and R. Pan, "Time Delay Estimation in Unknown Gaussian Spatially Correlated Noise," in Proc. ICASSP'88, pp. 2638-2641, 1988.

[8] M.R. Raghuveer and C.L. Nikias, "Bispectrum Estimation for Short Length Data," in Proc. ICASSP'85, pp. 1352-1355, 1985.

[9] Van Trees, H.L. "Detection, Modulation and Estimation Theory", Vol. I., Wiley 1969.

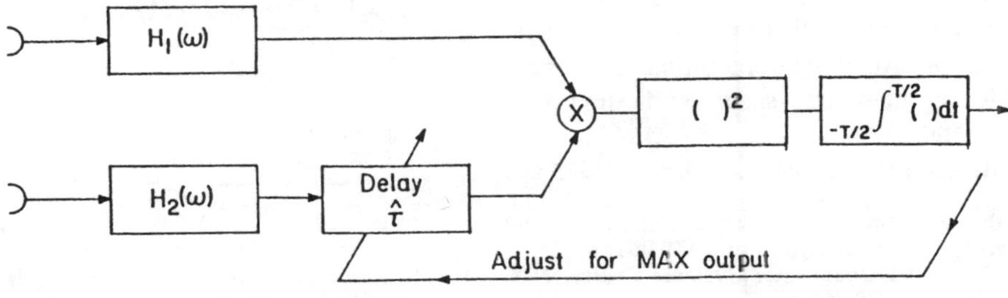

Fig. 3: The generalized-cross-correlator for time delay estimation

The Generalized Correlation Method for Estimation of Time Delay

CHARLES H. KNAPP, MEMBER, IEEE, AND G. CLIFFORD CARTER, MEMBER, IEEE

***Abstract*—A maximum likelihood (ML) estimator is developed for determining time delay between signals received at two spatially separated sensors in the presence of uncorrelated noise. This ML estimator can be realized as a pair of receiver prefilters followed by a cross correlator. The time argument at which the correlator achieves a maximum is the delay estimate. The ML estimator is compared with several other proposed processors of similar form. Under certain conditions the ML estimator is shown to be identical to one proposed by Hannan and Thomson [10] and MacDonald and Schultheiss [21].**

Qualitatively, the role of the prefilters is to accentuate the signal passed to the correlator at frequencies for which the signal-to-noise (S/N) ratio is highest and, simultaneously, to suppress the noise power. The same type of prefiltering is provided by the generalized Eckart filter, which maximizes the S/N ratio of the correlator output. For low S/N ratio, the ML estimator is shown to be equivalent to Eckart prefiltering.

INTRODUCTION

A SIGNAL emanating from a remote source and monitored in the presence of noise at two spatially separated sensors can be mathematically modeled as

$$x_1(t) = s_1(t) + n_1(t) \tag{1a}$$

$$x_2(t) = \alpha s_1(t + D) + n_2(t), \tag{1b}$$

where $s_1(t)$, $n_1(t)$, and $n_2(t)$ are real, jointly stationary random processes. Signal $s_1(t)$ is assumed to be uncorrelated with noise $n_1(t)$ and $n_2(t)$.

There are many applications in which it is of interest to estimate the delay D. This paper proposes a maximum likelihood (ML) estimator and compares it with other similar techniques. While the model of the physical phenomena presumes stationarity, the techniques to be developed herein are usually employed in slowly varying environments where the characteristics of the signal and noise remain stationary only for finite observation time T. Further, the delay D and attenuation α may also change slowly. The estimator is, therefore, constrained to operate on observations of a finite duration.

Another important consideration in estimator design is the available amount of *a priori* knowledge of the signal and noise statistics. In many problems, this information is negligible. For example, in passive detection, unlike the usual communications problems, the source spectrum is unknown or only known approximately.

Manuscript received July 24, 1975; revised November 21, 1975 and February 23, 1976.

C. H. Knapp is with the Department of Electrical Engineering and Computer Science, University of Connecticut, Storrs, CT 06268.

G. C. Carter is with the Naval Underwater Systems Center, New London Laboratory, New London, CT 06320.

One common method of determining the time delay D and, hence, the arrival angle relative to the sensor axis [1] is to compute the cross correlation function

$$R_{x_1x_2}(\tau) = E[x_1(t)\,x_2(t-\tau)], \tag{2}$$

where E denotes expectation. The argument τ that maximizes (2) provides an estimate of delay. Because of the finite observation time, however, $R_{x_1x_2}(\tau)$ can only be estimated. For example, for ergodic processes [2, p. 327], an estimate of the cross correlation is given by

$$\hat{R}_{x_1x_2}(\tau) = \frac{1}{T-\tau}\int_{\tau}^{T} x_1(t)\,x_2(t-\tau)\,dt, \tag{3}$$

where T represents the observation interval. In order to improve the accuracy of the delay estimate $\hat{D}$, it is desirable to prefilter $x_1(t)$ and $x_2(t)$ prior to the integration in (3). As shown in Fig. 1, x_i may be filtered through H_i to yield y_i for $i = 1, 2$. The resultant y_i are multiplied, integrated, and squared for a range of time shifts, τ, until the peak is obtained. The time shift causing the peak is an estimate of the true delay D. When the filters $H_1(f) = H_2(f) = 1, \forall f$, the estimate $\hat{D}$ is simply the abscissa value at which the cross-correlation function peaks. This paper provides for a generalized correlation through the introduction of the filters $H_1(f)$ and $H_2(f)$ which, when properly selected, facilitate the estimation of delay.

The cross correlation between $x_1(t)$ and $x_2(t)$ is related to the cross power spectral density function by the well-known Fourier transform relationship

$$R_{x_1x_2}(\tau) = \int_{-\infty}^{\infty} G_{x_1x_2}(f)\,e^{j2\pi f\tau}\,df. \tag{4}$$

When $x_1(t)$ and $x_2(t)$ have been filtered as depicted in Fig. 1, then the cross power spectrum between the filter outputs is given by [3, p. 399]

$$G_{y_1y_2}(f) = H_1(f)\,H_2^*(f)\,G_{x_1x_2}(f), \tag{5}$$

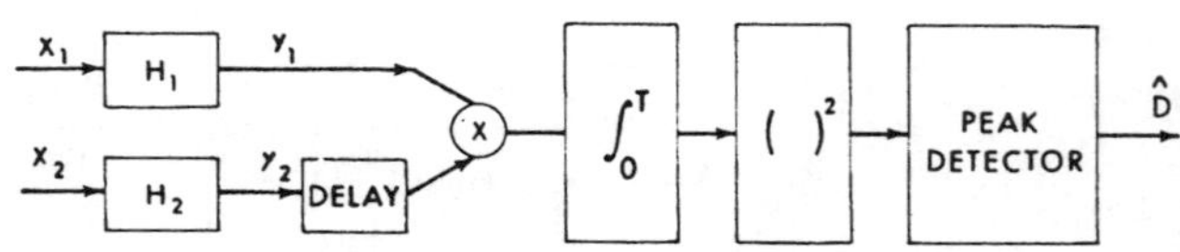

Fig. 1. Received waveforms filtered, delayed, multiplied, and integrated for a variety of delays until peak output is obtained.

Reprinted from *IEEE Trans. Acoust., Speech, Signal Processing,* vol. 24, no. 4, pp. 320–327, August 1976.

where $*$ denotes the complex conjugate. Therefore, the generalized correlation between $x_1(t)$ and $x_2(t)$ is

$$R^{(g)}_{y_1 y_2}(\tau) = \int_{-\infty}^{\infty} \psi_g(f)\, G_{x_1 x_2}(f)\, e^{j2\pi f\tau}\, df, \tag{6a}$$

where

$$\psi_g(f) = H_1(f)\, H_2^*(f) \tag{6b}$$

and denotes the general frequency weighting.

In practice, only an estimate $\hat{G}_{x_1 x_2}(f)$ of $G_{x_1 x_2}(f)$ can be obtained from finite observations of $x_1(t)$ and $x_2(t)$. Consequently, the integral

$$\hat{R}^{(g)}_{y_1 y_2}(\tau) = \int_{-\infty}^{\infty} \psi_g(f)\, \hat{G}_{x_1 x_2}(f)\, e^{j2\pi f\tau}\, df \tag{6c}$$

is evaluated and used for estimating delay. Indeed, depending on the particular form of $\psi_g(f)$ and the *a priori* information, it may also be necessary to estimate $\psi_g(f)$ in (6a)-(6b). For example, when the role of the prefilters is to accentuate the signal passed to the correlator at those frequencies at which the signal-to-noise (S/N) ratio is highest, then $\psi_g(f)$ can be expected to be a function of signal and noise spectra which must either be known *a priori* or estimated.

The selection of $\psi_g(f)$ to optimize certain performance criteria has been studied by several investigators. (See, for example, [4]-[12].) This paper will derive the ML estimator for delay D in the mathematical model (1a) and (1b), given signal and noise spectra. The results will be shown to be equivalent to (6a)-(6c) with an appropriate $\psi(f)$. This weighting turns out to be equivalent to that proposed in [12] and under simplifying assumptions to that proposed in [21]. The development presented here does not presume initially that the estimator has the form (6c). Rather, it is shown that the ML estimator may be realized by choosing τ that maximizes (6c) with proper weighting, $\psi_g(f)$, and proper estimate, $\hat{G}_{x_1 x_2}(f)$. The weighting $\psi_g(f)$ yielding the ML estimate will be compared to other weightings that have been proposed. Under certain conditions the ML estimator is shown to be equivalent to other processors.

Processor Interpretation

It is informative to examine the effect of processor weightings on the shape of $R_{y_1 y_2}(\tau)$ under ideal conditions. For models of the form of (1), the cross correlation of $x_1(t)$ and $x_2(t)$ is

$$R_{x_1 x_2}(\tau) = \alpha R_{s_1 s_1}(\tau - D) + R_{n_1 n_2}(\tau). \tag{7}$$

The Fourier transform of (7) gives the cross power spectrum

$$G_{x_1 x_2}(f) = \alpha G_{s_1 s_1}(f)\, e^{-j2\pi f D} + G_{n_1 n_2}(f). \tag{8}$$

If $n_1(t)$ and $n_2(t)$ are uncorrelated ($G_{n_1 n_2}(f) = 0$), the cross power spectrum between $x_1(t)$ and $x_2(t)$ is a scaled signal power spectrum times a complex exponential. Since multiplication in one domain is a convolution in the transformed domain (see, for example, [13]), it follows for $G_{n_1 n_2}(f) = 0$ that

$$R_{x_1 x_2}(\tau) = \alpha R_{s_1 s_1}(\tau) \circledast \delta(t - D), \tag{9}$$

where $\circledast$ denotes convolution.

One interpretation of (9) is that the delta function has been spread or "smeared" by the Fourier transform of the signal spectrum. If $s_1(t)$ is a white noise source, then its Fourier transform is a delta function and no spreading takes place. An important property of autocorrelation functions is that $R_{ss}(\tau) \leq R_{ss}(0)$. Equality will hold for certain τ for periodic functions (see, for example, [3, pp. 323-326]). However, for most practical applications, equality does not hold for $\tau \neq 0$, and the true cross correlation (9) will peak at D regardless of whether or not it is spread out. The spreading simply acts to broaden the peak. For a single delay this may not be a serious problem. However, when the signal has multiple delays, the true cross correlation is given by

$$R_{x_1 x_2}(\tau) = R_{s_1 s_1}(\tau) \circledast \sum_i \alpha_i \delta(\tau - D_i). \tag{10}$$

In this case, the convolution with $R_{s_1 s_1}(\tau)$ can spread one delta function into another, thereby making it impossible to distinguish peaks or delay times. Under ideal conditions where $\forall f$, $\hat{G}_{x_1 x_2}(f) \cong G_{x_1 x_2}(f)$, $\psi_g(f)$ should be chosen to ensure a large sharp peak in $R_{y_1 y_2}(\tau)$ rather than a broad one in order to ensure good time-delay resolution. However, sharp peaks are more sensitive to errors introduced by finite observation time, particularly in cases of low S/N ratio. Thus, as with other spectral estimation problems, the choice of $\psi_g(f)$ is a compromise between good resolution and stability.

The preceding discussion sets the background for the role that $\psi_g(f)$ is to play. Now the five generalizations of the cross-correlation function listed in Table I will be examined individually.

The Roth Processor

The weighting proposed by Roth [9],

$$\psi_R(f) = \frac{1}{G_{x_1 x_1}(f)} \tag{11}$$

[where the subscript R is to distinguish the choice of $\psi_g(f)$], yields[1]

$$\hat{R}^{(R)}_{y_1 y_2}(\tau) = \int_{-\infty}^{\infty} \frac{\hat{G}_{x_1 x_2}(f)}{G_{x_1 x_1}(f)}\, e^{j2\pi f\tau}\, df. \tag{12}$$

Equation (12) estimates the impulse response of the optimum linear (Wiener-Hopf) filter

$$H_m(f) = \frac{G_{x_1 x_2}(f)}{G_{x_1 x_1}(f)}, \tag{13}$$

which "best" approximates the mapping of $x_1(t)$ to $x_2(t)$ (see, for example, [14], [15]). If $n_1(t) \neq 0$, as is generally the case for (1), then

$$G_{x_1 x_1}(f) = G_{s_1 s_1}(f) + G_{n_1 n_1}(f), \tag{14}$$

[1] As discussed earlier, $\psi(f)$ may have to be estimated for this processor and those which follow, because of a lack of *a priori* information. In this case, (11) must be modified by replacing $G_{x_1 x_1}(f)$ with $\hat{G}_{x_1 x_1}(f)$.

and

$$R^{(R)}_{y_1 y_2}(\tau) = \delta(\tau - D) \circledast \int_{-\infty}^{\infty} \frac{\alpha G_{s_1 s_1}(f)}{\{G_{s_1 s_1}(f) + G_{n_1 n_1}(f)\}} \cdot e^{j2\pi f \tau}\, df. \tag{15}$$

Therefore, except when $G_{n_1 n_1}(f)$ equals any constant (including zero) times $G_{s_1 s_1}(f)$, the delta function will again be spread out. The Roth processor has the desirable effect of suppressing those frequency regions where $G_{n_1 n_1}(f)$ is large and $\hat{G}_{x_1 x_2}(f)$ is more likely to be in error.

The Smoothed Coherence Transform (SCOT)

Errors in $\hat{G}_{x_1 x_2}(f)$ may be due to frequency bands where $G_{n_2 n_2}(f)$ is large, as well as bands where $G_{n_1 n_1}(f)$ is large. One is, therefore, uncertain whether to form $\psi_R(f) = 1/G_{x_1 x_1}(f)$ or $\psi_R(f) = 1/G_{x_2 x_2}(f)$; hence, the SCOT [11] selects

$$\psi_s(f) = 1/\sqrt{G_{x_1 x_1}(f) G_{x_2 x_2}(f)}. \tag{16}$$

This weighting gives the SCOT

$$\hat{R}^{(s)}_{y_1 y_2}(\tau) = \int_{-\infty}^{\infty} \hat{\gamma}_{x_1 x_2}(f)\, e^{j2\pi f \tau}\, df, \tag{17}$$

where the coherence estimate[2]

$$\hat{\gamma}_{x_1 x_2}(f) \triangleq \frac{\hat{G}_{x_1 x_2}(f)}{\sqrt{G_{x_1 x_1}(f) G_{x_2 x_2}(f)}}. \tag{18}$$

For $H_1(f) = 1/\sqrt{G_{x_1 x_1}(f)}$ and $H_2(f) = 1/\sqrt{G_{x_2 x_2}(f)}$, the SCOT can be interpreted through Fig. 1 as prewhitening filters followed by a cross correlation. When $G_{x_1 x_1}(f) = G_{x_2 x_2}(f)$, the SCOT is equivalent to the Roth processor. If $n_1(t) \neq 0$ and $n_2(t) \neq 0$, the SCOT exhibits the same spreading as the Roth processor. This broadening persists because of an apparent inability to adequately prewhiten the cross power spectrum.

The Phase Transform (PHAT)

To avoid the spreading evident above, the PHAT uses the weighting [16]

$$\psi_P(f) = \frac{1}{|G_{x_1 x_2}(f)|}, \tag{19}$$

which yields

$$\hat{R}^{(P)}_{y_1 y_2}(\tau) = \int_{-\infty}^{\infty} \frac{\hat{G}_{x_1 x_2}(f)}{|G_{x_1 x_2}(f)|}\, e^{j2\pi f \tau}\, df. \tag{20}$$

For the model (1) with uncorrelated noise (i.e., $G_{n_1 n_2}(f) = 0$),

$$|G_{x_1 x_2}(f)| = \alpha G_{s_1 s_1}(f). \tag{21}$$

Ideally, when $\hat{G}_{x_1 x_2}(f) = G_{x_1 x_2}(f)$,

$$\frac{\hat{G}_{x_1 x_2}(f)}{|G_{x_1 x_2}(f)|} = e^{j\theta(f)} = e^{j2\pi f D} \tag{22}$$

has unit magnitude and

$$R^{(P)}_{y_1 y_2}(\tau) = \delta(t - D). \tag{23}$$

The PHAT was developed purely as an ad hoc technique. Notice that, for models of the form of (1) with uncorrelated noises, the PHAT (20), ideally, does not suffer the spreading that other processors do. In practice, however, when $\hat{G}_{x_1 x_2}(f) \neq G_{x_1 x_2}(f)$, $\theta(f) \neq 2\pi f D$ and the estimate of $R^{(P)}_{y_1 y_2}(\tau)$ will not be a delta function. Another apparent defect of the PHAT is that it weights $\hat{G}_{x_1 x_2}(f)$ as the inverse of $G_{s_1 s_1}(f)$. Thus, errors are accentuated where signal power is smallest. In particular, if $G_{x_1 x_2}(f) = 0$ in some frequency band, then the phase $\theta(f)$ is undefined in that band and the estimate of the phase is erratic, being uniformly distributed in the interval $[-\pi, \pi]$ rad. For models of the form of (1), this behavior suggests that $\psi_P(f)$ be additionally weighted to compensate for the presence or absence of signal power. The SCOT is one method of assigning weight according to signal and noise characteristics. Two remaining processors also assign weights or filtering according to S/N ratio: the Eckart filter [5] and the ML estimator or Hannan-Thomson (HT) processor [12].

The Eckart Filter

The Eckart filter derives its name from work in this area done in [5]. Derivations in [7], [8], [17], and [18] are outlined here briefly for completeness. The Eckart filter maximizes the deflection criterion, i.e., the ratio of the change in mean correlator output due to signal present to the standard deviation of correlator output due to noise alone. For long averaging time T, the deflection has been shown [8] to be

$$d^2 = \frac{L \left[\int_{-\infty}^{\infty} H_1(f) H_2^*(f)\, G_{s_1 s_2}(f)\, df \right]^2}{\int_{-\infty}^{\infty} |H_1(f)|^2 |H_2(f)|^2\, G_{n_1 n_1}(f)\, G_{n_2 n_2}(f)\, df}, \tag{24}$$

where L is a constant proportional to T, and $G_{s_1 s_2}(f)$ is the cross power spectrum between $s_1(t)$ and $s_2(t)$. For the model (1), $G_{s_1 s_2}(f) = \alpha G_{s_1 s_1}(f)$ exp $(j2\pi f D)$. Application of Schwartz's inequality to (24) indicates that

$$H_1(f) H_2^*(f) = \psi_E(f)\, e^{+j2\pi f D} \tag{25}$$

maximizes d^2 where

$$\psi_E(f) = \frac{\alpha G_{s_1 s_1}(f)}{G_{n_1 n_1}(f)\, G_{n_2 n_2}(f)}. \tag{26}$$

Notice that the weighting (26), referred to as the Eckart filter, possesses some of the qualities of the SCOT. In particular, it acts to suppress frequency bands of high noise,

[2] A more standard coherence estimate is formed when the auto spectra must also be estimated, as is usually the case.

as does the SCOT. Also note that the Eckart filter unlike the PHAT attaches zero weight to bands where $G_{s_1 s_1}(f) = 0$. In practice, the Eckart filter requires knowledge or estimation of the signal and noise spectra. For (1), when $\alpha = 1$ this can be accomplished by letting

$$\psi_E(f) = |\hat{G}_{x_1 x_2}(f)| \{\hat{G}_{x_1 x_1}(f) - |\hat{G}_{x_1 x_2}(f)|] \cdot [\hat{G}_{x_2 x_2}(f) - |\hat{G}_{x_1 x_2}(f)|]\}. \tag{27}$$

The first five processors in Table I can be justified on the basis of reasonable performance criteria, whether heuristic or mathematical.

In the next section, the ML estimator of the parameter D is derived. It is shown to be identical to that proposed by Hannan and Thomson [12].

The HT Processor

To make the model (1) mathematically tractable, it is necessary to assume that $s_1(t)$, $n_1(t)$, and $n_2(t)$ are Gaussian. Denote the Fourier coefficients of $x_i(t)$ as in [3, eq. (3.8)] by

$$X_i(k) = \frac{1}{T} \int_{-T/2}^{T/2} x_i(t)\, e^{-jkt\omega_\Delta}\, dt, \tag{28a}$$

where

$$\omega_\Delta = \frac{2\pi}{T}.$$

Note that the linear transformation $X_i(k)$ is Gaussian since $x_i(t)$ is Gaussian. Further, from [3, eq. (3.13)], as $T \to \infty$ and $K \to \infty$ such that $K\omega_\Delta = \omega$ is constant

$$\tilde{X}_i(\omega) = \lim_{T \to \infty} TX_i(k) \tag{28b}$$

$$= \int_{-\infty}^{\infty} x_i(t) e^{-j\omega t}\, dt, \tag{28c}$$

where $\tilde{X}_i$ is the Fourier transform of $x_i(t)$. A more complete discussion on Fourier transforms and their convergence is given in [3, p. 381], [4, pp. 23-25], [19, ch. 1], and [20, p. 11]. From [21], it follows for T large compared to $|D|$ plus the correlation time of $R_{s_1 s_1}(\tau)$, that

TABLE I
CANDIDATE PROCESSORS

Processor Name	Weight $\psi(f) = H_1(f) H_2^*(f)$	Text Reference
Cross Correlation	1	[2]-[4], [19]
Roth Impulse Response	$1/G_{x_1 x_1}(f)$	[9]
SCOT	$1/\sqrt{G_{x_1 x_1}(f)\, G_{x_2 x_2}(f)}$	[11], [16]
PHAT	$1/\lvert G_{x_1 x_2}(f)\rvert$	[16]
Eckart	$G_{s_1 s_1}(f)/[G_{n_1 n_1}(f)\, G_{n_2 n_2}(f)]$	[5], [7], [8] [17], [18]
ML or HT	$\dfrac{\lvert\gamma_{12}(f)\rvert^2}{\lvert G_{x_1 x_2}(f)\rvert[1 - \lvert\gamma_{12}(f)\rvert^2]}$	[12]

$$E[X_1(k) X_2^*(l)] \cong \begin{cases} \frac{1}{T} G_{x_1 x_2}(k\omega_\Delta), & k = l \\ 0, & k \neq l. \end{cases} \tag{29}$$

Now let the vector

$$X(k) = [X_1(k), X_2(k)]', \tag{30}$$

where $'$ denotes transpose. Define the power spectral density matrix Q such that

$$E[X(k) X^{*\prime}(k)] = E \begin{bmatrix} X_1(k) X_1^*(k) & X_1(k) X_2^*(k) \\ X_2(k) X_1^*(k) & X_2(k) X_2^*(k) \end{bmatrix} \tag{31a}$$

$$= \frac{1}{T} \begin{bmatrix} G_{x_1 x_1}(k\omega_\Delta) & G_{x_1 x_2}(k\omega_\Delta) \\ G_{x_1 x_2}^*(k\omega_\Delta) & G_{x_2 x_2}(k\omega_\Delta) \end{bmatrix} \tag{31b}$$

$$\triangleq \frac{1}{T} Q_x(k\omega_\Delta). \tag{31c}$$

Properties of $Q_x(k\omega_\Delta)$ can be used to prove

$$0 \leqslant |\gamma_{x_1 x_2}(k\omega_\Delta)|^2 \leqslant 1, \forall\, k\omega_\Delta$$

[4, p. 467]. The vectors $X(k)$, $k = -N, -N+1, \cdots, N$ are, as a consequence of (29), uncorrelated Gaussian (hence, independent) random variables. More explicitly, the probability for $X \equiv X(-N), X(-N+1), \cdots, X(N)$, given the power spectral density matrix Q (or the delay, attenuation, and spectral characteristics of the signal and noises necessary to determine Q) is

$$p(X|Q) = p(X|\alpha, G_{s_1 s_1}, G_{n_1 n_1}, G_{n_2 n_2}, G_{n_1 n_2}, D) = c \exp(-\tfrac{1}{2} J_1) \tag{32}$$

where

$$J_1 = \sum_{k=-N}^{N} X^{*\prime}(k)\, Q_x^{-1}(k\omega_\Delta)\, X(k)\, T \tag{33}$$

and c is a function of $|Q_x(k\omega_\Delta)|$ [15, p. 185]. Replacing $X(k)$ by $\frac{1}{T}\tilde{X}(k\omega_\Delta)$ from (28),

$$J_1 = \sum_{k=-N}^{N} \tilde{X}^{*\prime}(k\omega_\Delta)\, Q_x^{-1}(k\omega_\Delta)\, \tilde{X}(k\omega_\Delta) \frac{1}{T}. \tag{34}$$

For ML estimation (see, for example, [4] or [15]), it is desired to choose D to maximize $p(X|Q, D)$.

In general, the parameter D affects both c and J_1 in (32). However, under certain simplifying assumptions, c is constant or is only weakly related to the delay. Specifically, from (1) and (31), suppressing the frequency argument $k\omega_\Delta$,

$$|Q_x| = (G_{s_1 s_1} + G_{n_1 n_1})(\alpha^2 G_{s_1 s_1} + G_{n_2 n_2}) - (G_{n_1 n_2} + \alpha G_{s_1 s_1} e^{-j2\pi f D}) \cdot (G_{n_1 n_2}^* + \alpha G_{s_1 s_1} e^{+j2\pi f D}), \tag{35}$$

which is independent of D if $G_{n_1 n_2} = 0$ (i.e., the noises are uncorrelated).

For large T, (34) becomes

$$J_1 \cong \int_{-\infty}^{\infty} \tilde{X}^{*\prime}(f)\, Q_x^{-1}(f)\, \tilde{X}(f)\, df. \tag{36}$$

From (31),

$$Q_x^{-1}(f) = \frac{\begin{bmatrix} G_{x_2x_2}(f) & -G_{x_1x_2}(f) \\ -G^*_{x_1x_2}(f) & G_{x_1x_1}(f) \end{bmatrix}}{G_{x_1x_1}(f)\, G_{x_2x_2}(f) - |G_{x_1x_2}(f)|^2} \tag{37a}$$

$$= \frac{1}{[1 - |\gamma_{12}(f)|^2]} \cdot \begin{bmatrix} 1/G_{x_1x_1}(f), -G_{x_1x_2}(f)/\{G_{x_1x_1}(f) \cdot G_{x_2x_2}(f)\} \\ -G^*_{x_1x_2}(f)/\{G_{x_1x_1}(f)\, G_{x_2x_2}(f)\},\ 1/G_{x_2x_2}(f) \end{bmatrix}, \tag{37b}$$

which will exist provided $|\gamma_{12}(f)|^2 \neq 1$; i.e., $x_1(t)$ and $x_2(t)$ cannot be obtained perfectly from one another by linear filtering [14], or equivalently for the model (1) that observation noise *is* present.

When $G_{n_1n_2}(f) = 0$,

$$G_{x_1x_1}(f) = G_{s_1s_1}(f) + G_{n_1n_1}(f), \tag{38}$$

$$G_{x_2x_2}(f) = \alpha^2 G_{s_1s_1}(f) + G_{n_2n_2}(f), \tag{39}$$

$$G_{x_1x_2}(f) = \alpha G_{s_1s_1}(f)\, e^{-j2\pi fD}, \tag{40}$$

and it follows that

$$J_1 = \int_{-\infty}^{\infty} \tilde{X}^{*\prime}(f)\, Q_x^{-1}(f)\, \tilde{X}(f)\, df = J_2 + J_3, \tag{41}$$

where

$$J_2 = \int_{-\infty}^{\infty} \left[\frac{|\tilde{X}_1(f)|^2}{G_{x_1x_1}(f)} + \frac{|\tilde{X}_2(f)|^2}{G_{x_2x_2}(f)}\right] \cdot \frac{1}{[1 - |\gamma_{12}(f)|^2]}\, df, \tag{42a}$$

$$-J_3 = \int_{-\infty}^{\infty} A(f) + A^*(f)\, df, \tag{42b}$$

$$A(f) = \tilde{X}_1(f)\, \tilde{X}_2^*(f) \cdot \frac{\alpha G_{s_1s_1}(f)\, e^{j2\pi fD}}{G_{x_1x_1}(f)\, G_{x_2x_2}(f)[1 - |\gamma_{12}(f)|^2]}. \tag{42c}$$

In order to relate these results to [12] and interpret how to implement the ML estimation technique, note that for $x_1(t)$ and $x_2(t)$ real, $A^*(f) = A(-f)$. Then (42b) can be rewritten as

$$-J_3 = \int_{-\infty}^{\infty} A(f)\, df + \int_{-\infty}^{\infty} A(-f)\, df = 2\int_{-\infty}^{\infty} A(f)\, df. \tag{43}$$

Letting $T\hat{G}_{x_1x_2}(f) \triangleq \tilde{X}_1(f)\ \tilde{X}_2^*(f)$, (43) and (42c) can be rewritten as

$$-J_3 = 2T\int_{-\infty}^{\infty} \hat{G}_{x_1x_2}(f) \frac{1}{|G_{x_1x_2}(f)|} \frac{|\gamma_{12}(f)|^2}{[1 - |\gamma_{12}(f)|^2]} \cdot e^{j2\pi fD}\, df. \tag{44}$$

Notice that the ML estimator for D will minimize $J_1 = J_2 + J_3$, but the selection of D has no effect on J_2. Thus, D should maximize $-J_3$. Equivalently, when $\tilde{X}_1(f)\ \tilde{X}_2^*(f)$ is viewed as T times the estimated cross power spectrum, $T\hat{G}_{x_1x_2}(f)$, the ML estimator selects as the estimate of delay the value of τ at which

$$R_{y_1y_2}^{(\mathrm{HT})}(\tau) = \int_{-\infty}^{\infty} \hat{G}_{x_1x_2}(f) \frac{1}{|G_{x_1x_2}(f)|} \cdot \frac{|\gamma_{12}(f)|^2}{[1 - |\gamma_{12}(f)|^2]}\, e^{j2\pi f\tau}\, df \tag{45a}$$

achieves a peak.

The weighting in (6),

$$\psi_{\mathrm{HT}}(f) = \frac{1}{|G_{x_1x_2}(f)|} \cdot \frac{|\gamma_{12}(f)|^2}{[1 - |\gamma_{12}(f)|^2]}, \tag{45b}$$

where (as required for Q_x^{-1} to exist) $|\gamma_{12}(f)|^2 \neq 1$, achieves the ML estimator. When $|G_{x_1x_2}(f)|$ and $|\gamma_{12}(f)|^2$ are known, this is exactly the proper weighting. When the terms in (45b) are unknown, they can be estimated via techniques of [22]. Substituting estimated weighting for true weighting is entirely a heuristic procedure whereby the ML estimator can approximately be achieved in practice.

Note that, like the HT processor, the PHAT computes a type of transformation on

$$\frac{\hat{G}_{x_1x_2}(f)}{|G_{x_1x_2}(f)|} = \exp\,[j\hat{\theta}\,(f)].$$

However, the HT processor, like the SCOT, weights the phase according to the strength of the coherence.

From [4, p. 379],

$$\mathrm{var}\,[\hat{\theta}\,(f)] \cong \frac{1 - |\gamma|^2}{|\gamma|^2} \cdot \frac{1}{L_1}, \tag{46a}$$

where L_1 is a proportionality constant dependent on how the data are processed. Thus,

$$R_{y_1y_2}^{(\mathrm{HT})}(\tau) \cong \frac{1}{L_1} \int_{-\infty}^{\infty} e^{j\hat{\theta}(f)} \cdot \frac{1}{\mathrm{var}\,[\hat{\theta}(f)]}\, e^{j2\pi f\tau}\, df. \tag{46b}$$

Comparison of (46b) and (20) with (21) reveals that the ML estimator is the PHAT inversely weighted according to the variability of the phase estimates.

In interpreting the similarity of the HT processor to the other processors listed in Table I, it can be shown that if $G_{n_1n_1}(f) = G_{n_2n_2}(f) = G_{nn}(f)$ is equal to a constant times $G_{s_1s_1}(f)$, then the last five processors in Table I are the same except for a constant, but the cross-correlation processor ($\psi(f) = 1$, $\forall f$) is a delta function smeared out by the Fourier transform of the signal (noise) power spectrum.

Interpretation of Low S/N Ratio of ML Estimator

Good delay estimation is most difficult to achieve in the case of low S/N ratios. In order to compare estimates under low S/N ratio conditions, let $\alpha = 1$. Then,

$$\psi_{\mathrm{HT}}(f) = \frac{1}{G_{s_1 s_1}(f)} \cdot \frac{G^2_{s_1 s_1}(f)}{\{[G_{s_1 s_1}(f) + G_{n_1 n_1}(f)] \cdot [G_{s_1 s_1}(f) + G_{n_2 n_2}(f)] - G^2_{s_1 s_1}(f)\}} \tag{47a}$$

$$= \frac{\dfrac{G_{s_1 s_1}(f)}{G_{n_1 n_1}(f)\, G_{n_2 n_2}(f)}}{\left[1 + \dfrac{G_{s_1 s_1}(f)}{G_{n_2 n_2}(f)} + \dfrac{G_{s_1 s_1}(f)}{G_{n_1 n_1}(f)}\right]}, \tag{47b}$$

which agrees with [21, eq. (28)] if in (47b) $G_{n_1 n_1}(f) = G_{n_2 n_2}(f)$.

For low S/N ratio,

$$\frac{G_{s_1 s_1}(f)}{G_{n_1 n_1}(f)} << 1 \quad \text{and} \quad \frac{G_{s_1 s_1}(f)}{G_{n_2 n_2}(f)} << 1,$$

it follows that

$$\psi_{\mathrm{HT}}(f) \cong \frac{G_{s_1 s_1}(f)}{G_{n_1 n_1}(f)\, G_{n_2 n_2}(f)} = \psi_E(f). \tag{48}$$

That is, for $\alpha = 1$ and low S/N ratio, the HT processor is identical to the Eckart filter. Similarly, for low S/N ratio,

$$\psi_s(f) \cong 1/\sqrt{G_{n_1 n_1}(f)\, G_{n_2 n_2}(f)}. \tag{49}$$

Therefore, if $\alpha = 1$,

$$\psi_{\mathrm{HT}}(f) \cong \frac{G_{s_1 s_1}(f)}{\sqrt{G_{n_1 n_1}(f)\, G_{n_2 n_2}(f)}} [\psi_s(f)]. \tag{50a}$$

Furthermore, for $G_{n_1 n_1}(f) = G_{n_2 n_2}(f) = G_{nn}(f)$,

$$\psi_{\mathrm{HT}}(f) \cong \frac{G_{s_1 s_1}(f)}{G_{nn}(f)} [\psi_s(f)] = \left[\frac{G_{s_1 s_1}(f)}{G_{nn}(f)}\right]^2 \psi_p(f). \tag{50b}$$

Thus, under low S/N ratio approximations with $\alpha = 1$, both the Eckart and HT prefilters can be interpreted either as SCOT prewhitening filters with additional S/N ratio weighting or PHAT prewhitening filters with additional S/N ratio squared weighting.

Variance of Delay Estimators

It can be shown, by extending a result from [21], that the variance of the time-delay estimate in the neighborhood of the true delay for general weighting function $\psi(f)$ is given by

$$\mathrm{var}\,[\hat{D}] = \frac{\displaystyle\int_{-\infty}^{\infty} |\psi(f)|^2 (2\pi f)^2 G_{x_1 x_1}(f)\, G_{x_2 x_2}(f) [1 - |\gamma(f)|^2]\, df}{T \left[\displaystyle\int_{-\infty}^{\infty} (2\pi f)^2 |G_{x_1 x_2}(f)|\, \psi(f)\, df\right]^2}. \tag{51}$$

The variance of the HT processor (substituting from (45b) and using the definition of coherence) is

$$\mathrm{var}^{\mathrm{HT}}[\hat{D}] = \left\{2T \int_0^{\infty} (2\pi f)^2 |\gamma(f)|^2 / [1 - |\gamma(f)|^2]\, df\right\}^{-1}. \tag{52}$$

In particular, the HT processor achieves the Cramér–Rao lower bound (see Appendix). It should be pointed out that (51) and (52) evaluate the local variation of the time-delay estimate and thus do not account for ambiguous peaks which may arise when the averaging time is not large enough for the given signal and noise characteristics. Indeed, when T is not sufficiently large, local variation may be a poor indicator of system performance and the envelope of the ambiguous peaks must be considered [21, p. 40], [23], [24, p. 41]. Further, (51) and (52) predict system performance when signal and noise spectral characteristics are known; for T sufficiently large, these spectra can be estimated accurately. However, in general, (51) and (52) must be modified to account for estimation errors; alternatively, system performance can be evaluated by computer simulation. Empirical verification of expressions for variance has not been undertaken by simulation, because to do so without special purpose correlator hardware would be computationally prohibitive. For example, for a given $G_{s_1 s_1}(f)$,

$G_{n_1 n_1}(f)$, $G_{n_2 n_2}(f)$, α, and averaging time T, an estimated generalized cross-correlation function can be computed, from which only one number (the delay estimate) can be extracted. To empirically evaluate the statistics of the delay estimate (which would be valid *only* for these particular signal and noise spectra) many such trials would need to be conducted. We have conducted one such trial (with T large) and verified that useful delay estimates can be obtained by inserting estimates $|\hat{G}_{x_1 x_2}(f)|$ and $|\hat{\gamma}_{12}(f)|^2$ in place of the true values. (This might have been expected since the estimated optimum weighting will converge to the true weighting as $T \to \infty$. In practice, T may be limited by the stationary properties of the data and (52) may be an overly optimistic prediction of system performance when signal and noise spectra are unknown.)

Conclusions and Discussion

The HT processor has been shown to be an ML estimator for time delay under usual conditions. Under a low S/N ratio restriction, the HT processor is equivalent to Eckart prefiltering and cross correlation. These processors have been compared with four other candidate processors to demonstrate the interrelation of all six estimation techniques. The derivation of the ML delay estimator, together with its relation to various ad hoc techniques of intuitive appeal, suggests the practical significance of HT processing for determination of delay and, thence, bearing. Finally, interpretation of the results leads one to believe that, if the coherence is slowly changing as a function of time, the ML estimation of the source bearing will still be a cross correlator preceded by prefilters that must also vary according to time-varying estimates of coherence.

Appendix

Cramér-Rao Lower Bound on Variance of Delay Estimators

The Cramér-Rao lower bound is given [15, p. 72] by

$$\sigma_{\hat{D}}^2 \geqslant \left. \frac{-1}{E\left\{\dfrac{\partial^2 \ln p(X|Q,\tau)}{\partial \tau^2}\right\}} \right|_{\tau=D}. \tag{A1}$$

The only part of the log density which depends on τ, the hypothesized delay, is J_3 of (44). More explicitly,

$$E\left\{\frac{\partial^2}{\partial \tau^2} \ln p(X|Q,\tau)\right\} = \frac{\partial^2}{\partial \tau^2} E\left(\frac{-1}{2} J_3\right). \tag{A2}$$

If $G_{x_1 x_2}(f) = |G_{x_1 x_2}(f)| e^{-j2\pi f D}$, then since $E[\hat{G}_{x_1 x_2}(f)] = G_{x_1 x_2}(f)$, we have

$$E\left(\frac{-1}{2} J_3\right) = T \int_{-\infty}^{\infty} e^{j2\pi f(\tau - D)} \frac{|\gamma_{12}(f)|^2}{[1 - |\gamma_{12}(f)|^2]} \, df. \tag{A3}$$

Hence the minimum variance is

$$\text{minimum var}(\hat{D}) = \left[T \int_{-\infty}^{\infty} (2\pi f)^2 \frac{|\gamma_{12}(f)|^2}{[1 - |\gamma_{12}(f)|^2]} \, df\right]^{-1}. \tag{A4}$$

But this is the variance which the HT processor achieves [see (52)].

References

[1] A. H. Nuttall, G. C. Carter, and E. M. Montavon, "Estimation of the two-dimensional spectrum of the space-time noise field for a sparse line array," *J. Acoust. Soc. Amer.*, vol. 55, pp. 1034-1041, 1974.

[2] A. Papoulis, *Probability, Random Variables and Stochastic Processes*. New York: McGraw-Hill, 1965.

[3] W. B. Davenport, Jr., *Probability and Random Processes*. New York: McGraw-Hill, 1970.

[4] G. M. Jenkins and D. G. Watts, *Spectral Analysis and Its Applications*. San Francisco, CA: Holden-Day, 1968.

[5] C. Eckart, "Optimal rectifier systems for the detection of steady signals," Univ. California, Scripps Inst. Oceanography, Marine Physical Lab. Rep SIO 12692, SIO Ref 52-11, 1952.

[6] H. Akaike and Y. Yamanouchi, "On the statistical estimation of frequency response function," *Ann. of Inst. Statist. Math.*, vol. 14, pp. 23-56, 1963.

[7] C. H. Knapp, "Optimum linear filtering for multi-element arrays," Electric Boat Division, Groton, CT, Rep. U417-66-031, Nov. 1966.

[8] A. H. Nuttall and D. W. Hyde, "A unified approach to optimum and suboptimum processing for arrays," Naval Underwater Systems Center, New London Lab., New London, CT, Rep. 992, Apr. 1969.

[9] P. R. Roth, "Effective measurements using digital signal analysis," *IEEE Spectrum*, vol. 8, pp. 62-70, Apr. 1971.

[10] E. J. Hannan and P. J. Thomson, "The estimation of coherence and group delay," *Biometrika*, vol. 58, pp. 469-481, 1971.

[11] G. C. Carter, A. H. Nuttall, and P. G. Cable, "The smoothed coherence transform," *Proc. IEEE* (Lett.), vol. 61, pp. 1497-1498, Oct. 1973.

[12] E. J. Hannan and P. J. Thomson, "Estimating group delay," *Biometrika*, vol. 60, pp. 241-253, 1973.

[13] A. V. Oppenheim and R. W. Schafer, *Digital Signal Processing*. Englewood Cliffs, NJ: Prentice-Hall, 1975.

[14] G. C. Carter and C. H. Knapp, "Coherence and its estimation via the partitioned modified chirp-Z-transform," *IEEE Trans. Acoust., Speech, Signal Processing (Special Issue on 1974 Arden House Workshop on Digital Signal Processing)*, vol. ASSP-23, pp. 257-264, June 1975.

[15] H. L. Van Trees, *Detection, Estimation and Modulation Theory, Part I*. New York: Wiley, 1968.

[16] G. C. Carter, A. H. Nuttall, and P. G. Cable, "The smoothed coherence transform (SCOT)," Naval Underwater Systems Center, New London Lab., New London, CT, Tech. Memo TC-159-72, Aug. 8, 1972.

[17] D. W. Hyde and A. H. Nuttall, "Linear pre-filtering to enhance correlator performance," Naval Underwater Systems Center, New London Lab., New London, CT, Tech. Memo 2020-34-69, Feb. 27, 1969.

[18] A. H. Nuttall, "Pre-filtering to enhance clipper correlator performance," Naval Underwater Systems Center, New London Lab., New London, CT, Tech. Memo TC-11-73, July 10, 1973.

[19] L. H. Koopmans, *The Spectral Analysis of Time Series*. New York: Academic, 1974.

[20] R. K. Otnes and L. Enochson, *Digital Time Series Analysis*, New York: Wiley, 1972.

[21] V. H. MacDonald and P. M. Schultheiss, "Optimum passive bearing estimation," *J. Acoust. Soc. Amer.*, vol. 46, pp. 37-43, 1969.

[22] G. C. Carter, C. H. Knapp, and A. H. Nuttall, "Estimation of the magnitude-squared coherence function via overlapped fast Fourier transfrom processing," *IEEE Trans. Audio Electroacoust.*, vol. AU-21, pp. 337-344, Aug. 1973.

[23] D. Signori, J. L. Freeh, and C. Stradling, personal communication.

[24] B. V. Hamon and E. J. Hannan, "Spectral estimation of time delay for dispersive and non-dispersive systems," *J. Royal Stat. Soc. Ser. C (Appl. Statist.)*, vol. 23, pp. 134-142, 1974.

ARRIVAL TIME ESTIMATION USING ITERATIVE SIGNAL RECONSTRUCTION FROM THE PHASE OF THE CROSS-SPECTRUM *

by Y.T. Li ** and A.L. Kurkjian

Research Laboratory for Electronics
Massachusetts Institute of Technology
Cambridge MA 02139

Abstract

A situation encountered in such applications as seismics, ocean acoustics, radar, sonar and others, is that of an unknown wavelet propagating nondispersively in a reverberatory or multipath environment. Two receivers placed in such an environment will each record the arrival of this wavelet numerous times and each time with a different attenuation factor. In this paper, we present method for estimating the arrival times and attenuations at each receiver. The unknown wavelet can be estimated by deconvolving the receiver data by the estimated arrival sequences.

The method presented here involves the construction of a finite length sequence whose phase equals that of the cross-spectrum of the two received signals. The reconstructed signal, under certain conditions, can be uniquely "inverted" to recover the desired arrival time and attenuation information at each receiver.

A comparison of this method with generalized correlation time delay estimation techniques is presented and an example using synthetically generated data is provided. Comments on the performance of this method in the presence of noise are included.

Introduction

A situation encountered in such applications as seismics, ocean acoustics, radar, sonar and others, is that of an unknown wavelet propagating nondispersively in a reverberatory or multipath environment. Two receivers placed in such an environment will record the arrival of this wavelet numerous times and each time with a different attenuation factor. In this paper, we present a method for estimating the arrival times and attenuations at each receiver. This information is useful for eliminating the multiple arrivals, or for deducing the structure of the propagation channel. The unknown wavelet can be estimated by deconvolving the receiver data by the estimated arrival sequences.

The method involves the construction of a finite length sequence whose phase equals that of the cross-spectrum of the two received signals[1-4]. An iterative algorithm is used to perform the construction[1]. The convergence of this algorithm[2] is guaranteed provided that the length we choose for the constructed sequence is sufficiently long. However, if we choose this length to be longer than required, the reconstruction algorithm may converge to an undesired sequence. This sequence is related to the desired sequence through a convolution with a zero-phase (even) sequence. By factoring the zero-phase terms out of the sequence polynomial, a unique solution can always be obtained. The reconstructed signal, under certain conditions, can be uniquely "inverted" to recover the desired arrival time and attenuation information at each receiver.

Conventional generalized correlation methods[5] for estimating the time delay from one receiver to another do not perform well in the severely overlapped multiple arrival problem. These methods produce the desired arrival time and attenuation information convolved (smeared) with an unknown signal. This causes closely spaced arrivals to go unresolved and undesired sidelobes to be mistakenly interpreted as arrivals. The iterative signal reconstruction method presented here is shown to fit into the generalized correlator framework and corresponds to a choice of the cross-spectrum weighting function which results in no smearing of the desired answer.

An example of this method applied to noise-free synthetic data is provided. The result of this processing is compared with the PHAT generalized correlation method and is seen to be superior. Comments on the effect which noise, dispersion or other arrivals will have on the processing are included.

Signal Reconstruction from Phase

We model our two received signals as

$$x_1(t) = s(t) * \sum_{i=1}^{P} \gamma_i \, \delta(t-\zeta_i)$$

$$x_2(t) = s(t) * \sum_{i=1}^{Q} \beta_i \, \delta(t-\tau_i)$$

With $x_1(t)$ and $x_2(t)$ given, we wish to determine $s(t)$, P, Q, and the γ's, ζ's, β's and τ's. The Fourier transform of these signals are given by

$$X_1(f) = S(f) \sum_{i=1}^{P} \gamma_i \, e^{-j2\pi f \zeta_i}$$

$$X_2(f) = S(f) \sum_{i=1}^{Q} \beta_i \, e^{-j2\pi f \tau_i}$$

and the cross-spectrum is given by

$$G_{x_1x_2}(f) = X_2(f)\, X_1^*(f) = S(f)\, S^*(f) \sum_{i=1}^{P} \sum_{k=1}^{Q} \gamma_i \beta_k \, e^{-j2\pi f(\tau_k - \zeta_i)}$$

* This work has been supported in part by the National Science Foundation under grant No. ECS-8007102 and by Schlumberger-Doll Research, Ridgefield CT.
** Currently at the Department of Automation, Tsinghua University, Beijing, The Peoples's Republic of China

Reprinted from *Proc. of the 2nd International Symposium on Computer-Aided Seismic Analysis and Discrimination*, pp. 87–91, August 1981.

$$= G_{ss}(f) \sum_{n=1}^{N} \alpha_n \, e^{-j2\pi f D_n}$$

In this last equation, we have rewritten the double summation as a single summation with appropriate re-indexing. We now define $A(f)$ and $\varphi(f)$ and another sequence, y_m, as follows

$$A(f)\, e^{j\varphi(f)} = \sum_{n=1}^{N} \alpha_n e^{-j2\pi f D_n}$$

$$= \sum_{m=0}^{M-1} y_m \, e^{-j\frac{2\pi}{M}km}$$

Here, $k = Mf / \Delta t$ where Δt is small enough that it divides each D_n an integral number of times. We assume that no two D_n values are exactly equal so that there will be N distinct values of α_n. The sequence y_m will equal zero except at values of m such that $m\Delta t$ equals one of the D_n values. At this point, y_m equals the value of α_n associated with that particular D_n value.

The essence of our method is as follows. Because $G_{ss}(f)$ is a positive real-valued function, the phase of the cross-spectrum, $G_{x_1x_2}(f)$ is $\varphi(f)$, as is the phase of the finite length sequence y_m. Therefore, given the phase of the cross-spectrum, we can consider constructing the sequence y_m. Following that is the problem of the interpretation of the y_m sequence in terms of the desired attenuations and arrival times at each receiver. First let us treat the reconstruction issue.

It has recently been shown that if a sequence is of finite length and no zeros of the sequence occur in conjugate reciprocal pairs, then the signal can be uniquely reconstructed from the phase of its DFT[3]. A sequence with zeros in conjugate reciprocal locations can be decomposed into the convolution of an even sequence and a second sequence which has no zeros in conjugate reciprocal locations. Both one-step and iterative algorithms have been developed for the reconstruction of a finite length sequence from its phase[1-4]. The one-step algorithms typically require the inversion of a matrix whose dimension is equal to the data length. The iterative algorithms do not require a matrix inversion but may converge rather slowly. In this paper, we have elected to use an iterative algorithm.

The y_m sequence is certainly finite in length, but it may contain an even convolutional component. We use whatever apriori knowledge we have concerning the range over which the sequence may take on non-zero values to obtain a candidate for the y_m sequence. If this sequence is not the intended solution, we will have to factor a zero-phase component out of the sequence polynomial to obtain the desired result. An incorrect y_m solution will ultimately lead to an inconsistency in the interpretation step.

The reconstruction algorithm begins by associating an arbitrary (but non-zero) magnitude function with the phase of the cross-spectrum and transforming the result into the time domain. We then impose our finite length constraint by setting to zero those points in the sequence which we know to be zero. Again, this physically corresponds to apriori knowledge that the delay of the wavelet from one receiver to another is not larger than some number. We then transform back to the frequency domain and set the phase of the result equal to the phase of the cross-spectrum; the magnitude of the transform is left unchanged. The iterative process continues by repeatedly transforming from one domain to the other, demanding a finite length restriction in the time domain and a phase specification in the frequency domain. The convergence of this procedure is guaranteed provided the finite length constraint is not so restrictive that no sequence of that length or smaller exists with the desired phase. A unique sequence can always be obtained by factoring any zero-phase components out of the result.

The Interpretation of the y_m Sequence

The iterative signal reconstruction method just described provides us with the sequence y_m for $m = 0,1,...M-1$. In this section we discuss the interpretation of the y_m sequence in terms of the arrival times, τ's and ζ's, and attenuations, γ's and β's.

The y_m sequence is equal to the convolution of the arrival sequence at the second receiver with the time reversed arrival sequence at the first receiver. Thus, we wish to find two unknown sequences which convolve to produce the given sequence. There is no unique solution to this problem in general, however, in this application, we can uniquely determine the arrival sequences from the y_m sequence. The reason we are able to do so lies in our assumption that no two values of the D_n's are exactly equal. This implies that when convolving the two arrival sequences, at most one term in the convolution sum is non-zero. If two or more D_n values are equal, then the "inversion" is not unique.

The y_m sequence is zero except at N distinct points so that the determination of N is made by inspection. Because all the D_n are assumed to be distinct, we know that $N = PQ$. We begin the inversion process by guessing P and Q such that $N = PQ$. There may or may not be many reasonable choices for P and Q, depending on the application. As we will see, an incorrect guess will ultimately manifest itself. In this event, we keep guessing until the correct choice is found.

For a particular choice for P and Q, we now consider the extraction of the τ's and ζ's from the D_n's. We have PQ values of D_n and $P+Q$ unknown values to solve for. Therefore, we must solve a generally overdetermined set of linear equations in order to extract the arrival times at each receiver from the interarrival delays. As an example, suppose the signal reconstruction method results in $N=4$. Assuming that each receiver contains multiple arrivals, this implies that both P and Q equal 2. We then have to solve the following set of linear equations to obtain our arrival times.

$$\begin{bmatrix} 1 & -1 & 0 & 0 \\ 1 & 0 & 0 & -1 \\ 0 & -1 & 1 & 0 \\ 0 & 0 & 1 & -1 \end{bmatrix} \begin{bmatrix} \tau_1 \\ \zeta_1 \\ \tau_2 \\ \zeta_2 \end{bmatrix} = \begin{bmatrix} D_1 \\ D_2 \\ D_3 \\ D_4 \end{bmatrix}$$

The resulting solution for the arrival times may not be in ascending order (e.g. we may have $\tau_1 > \tau_2$), but they will be the unique values consistent with our choice for P and Q.

The determination of the attenuation factors is straightforward. We assume that $\gamma_1=1$ as a reference. Then, the β_k's are the values of α_n at $n=\tau_k-\zeta_1$. Having found the β's, the γ's are found similarly.

In summary, given N from the reconstruction of the y_m sequence, we guess P and Q such that $N=PQ$. We then solve a set of linear equations to obtain the values for τ_i, $i=1,2,\ldots,Q$ and ζ_i, $i=1,2,\ldots,P$. The values for β_i, $i=1,2,\ldots,Q$ and γ_i, $i=1,2,\ldots,P$ are then found assuming $\gamma_1=1$. If the assumption that $N=PQ$ is correct and the correct values for P and Q were chosen, the derived arrival sequences should each deconvolve their respective receiver signals to yield the same wavelet estimate. If this does not occur, then either P and Q are wrong, or the iterative signal reconstruction algorithm converged to an undesired solution, or the inter-arrival delays are not unique. If the latter is the case, then the interpretation problem is not unique. We do not treat this situation here.

The Relationship Generalized Correlation Methods

The generalized correlator structure is the following [5]

$$R_{x_1x_2}(\tau) = \int_{-\infty}^{\infty} W(f)\, G_{x_1x_2}(f)\, e^{j2\pi f\tau}\, df$$

Here, $W(f)$ is a weighting function we may choose. The intent is that through a judicious choice for $W(f)$, $R_{x_1x_2}(\tau)$ will have peaks at $\tau=D_n$ and no appreciable peaks for other values of τ. Clearly, if $W(f)=1$, then $R_{x_1x_2}(\tau)$ is simply the cross-correlation of the second received signal with a time reversed version of the first. Such a choice for $W(f)$ will produce intollerable results in the overlapped arrival case.

Another choice for $W(f)$ is $1/|G_{x_1x_2}|$. This is referred to as the Phase Transform (PHAT) method [5]. Here we obtain

$$\begin{aligned} R_{x_1x_2}(\tau) &= \int_{-\infty}^{\infty} e^{-j\varphi(f)}\, e^{j2\pi f\tau}\, df \\ &= \int_{-\infty}^{\infty} \left[\frac{1}{A(f)} \sum_{n=1}^{N} \alpha_n\, e^{-j2\pi f D_n}\right] e^{j2\pi f\tau}\, df \\ &= \sum_{n=1}^{N} \alpha_n \int_{-\infty}^{\infty} \frac{1}{A(f)} e^{j2\pi f(\tau-D_n)}\, df \\ &= U(\tau) * \sum_{n=0}^{N-1} \alpha_n\, \delta(\tau-D_n) \end{aligned}$$

where

$$U(\tau) = \int_{-\infty}^{\infty} \frac{1}{A(f)} e^{j2\pi f\tau}\, df$$

Thus, $R_{x_1x_2}(\tau)$ can be viewed as the desired answer smeared by $U(\tau)$. Other choices for $W(f)$ will also result in some sort of smearing interpretation. The following choice results in no smearing.

$$W(f) = \frac{A(f)}{|G_{x_1x_2}(f)|}$$

The problem with this choice is that $A(f)$ is unknown. Thus, we may not directly realize this function. The iterative phase-only method presented here is equivalent to such a choice for $W(f)$.

It is interesting to note that if we initially associate a magnitude function of unity at all frequencies to the phase of the cross-spectrum in our iterative signal reconstruction algorithm, then result after the first iteration is the PHAT estimate. Subsequent iterations then refine the estimate by removing the smearing.

An Example

A comparision of the signal reconstruction method with the PHAT method ($W(f)=1/|G_{x_1x_2}(f)|$) is shown in Figures 1-10. In this example, a synthetically generated wavelet, 30 samples in duration "arrived" twice at each of two "receivers". The arrival times and attenuations at each receiver are shown in Figures 1 and 2 and the signals themselves, $x_1(t)$ and $x_2(t)$, are shown in Figures 3 and 4. The received signals are each 50 samples in duration. The y_m sequence is shown in Figure 5. This is equal to the convolution of Figure 2 with a time reversed version of Figure 1

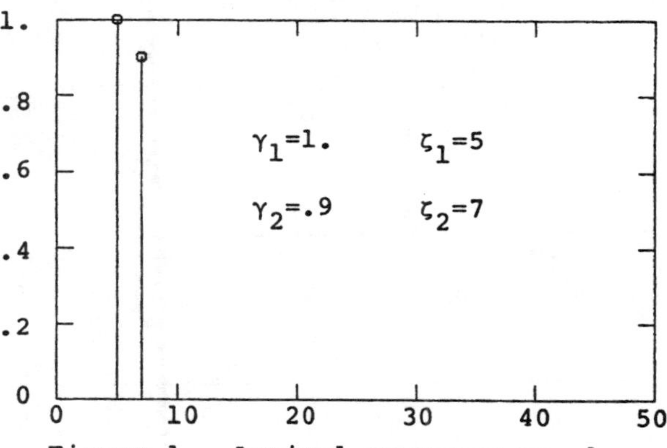

Figure 1. Arrival sequence no. 1

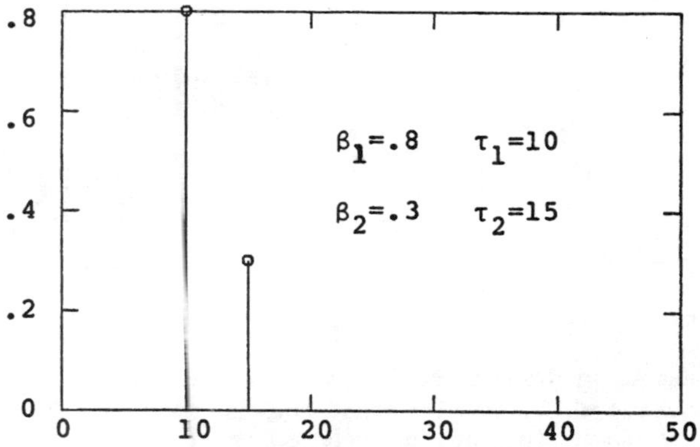

β1=.8 τ1=10

β2=.3 τ2=15

Figure 2. Arrival sequence no. 2

The results produced by the iterative signal reconstruction method are shown in Figures 6-10. We initially selected a unity magnitude function to associate with the phase of the cross-spectrum. The result of transforming this back to the time domain, shown in Figure 6, is the PHAT estimate as well as the zero-th iteration result of our method. We see that the correct peaks are unresolved and ambiguous peaks are

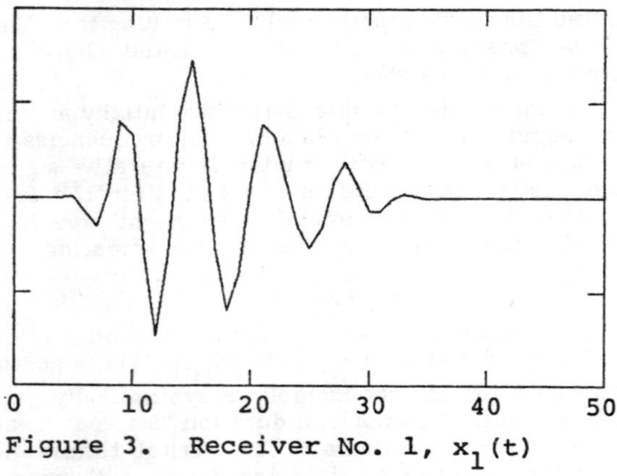

Figure 3. Receiver No. 1, $x_1(t)$

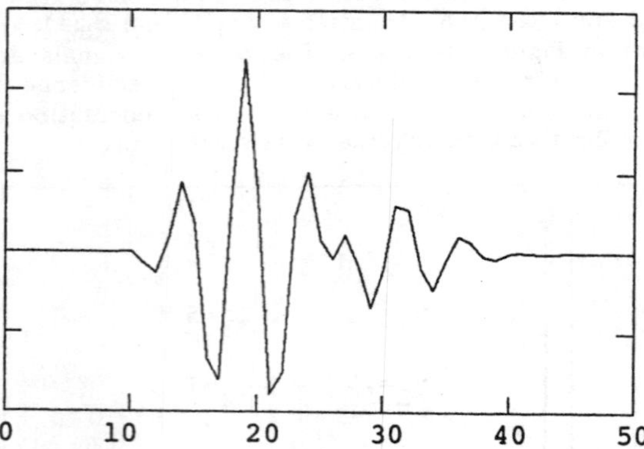

Figure 4. Receiver No. 2, $x_2(t)$

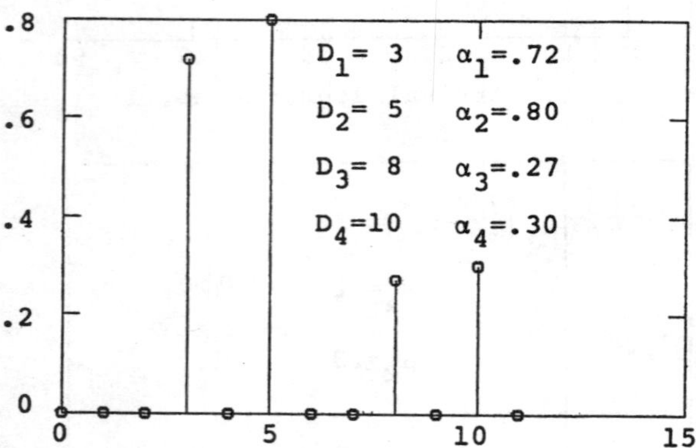

Figure 5. The y_m sequence.

present. In the time domain, we required that the y_m sequence will be zero for $m>11$ and also for $m<0$. This corresponds to our finite length restriction. The results of the iterative procedure after 25, 50 and 100 iterations are shown in Figures 7-9. After 50 iterations, the correct result becomes apparent. Note that the method only determines the y_m sequence to within a scale factor. We carried the method out for 100 iterations to illustrate nature of the convergence process. The computation involved in each pass of the iteration is dominated by the two FFT's. In this example the entire 100 iterations using 128 point FFT's, required approximately 9 seconds of cpu time on a PDP 11/45 computer with a UNIX operating system.

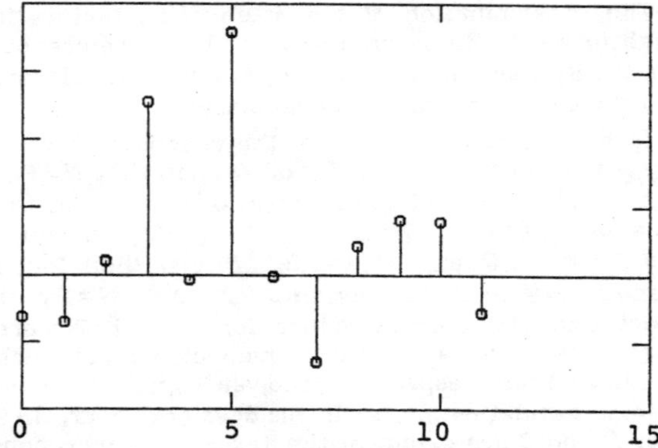

Figure 6. The PHAT estimate.

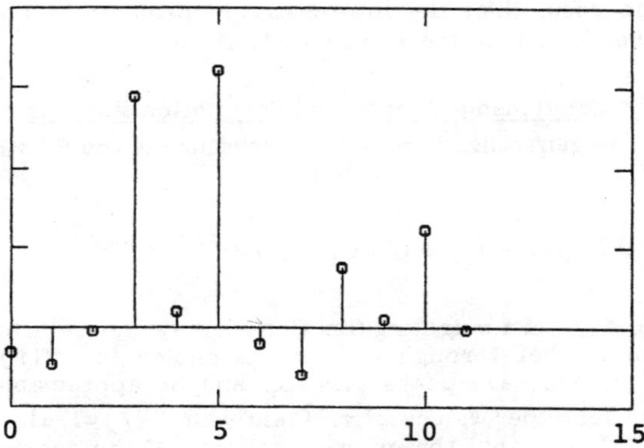

Figure 7. Iteration No. 25

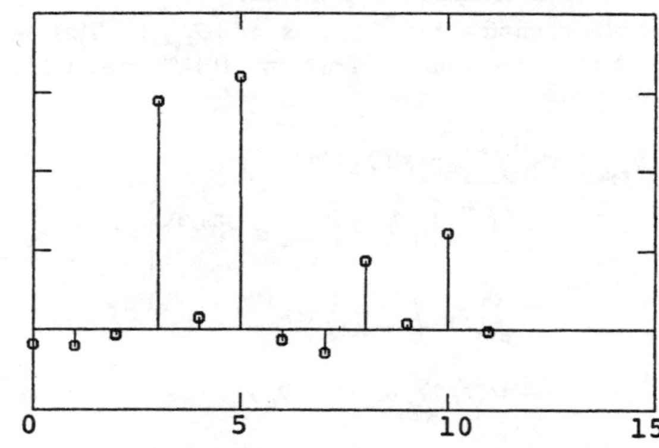

Figure 8. Iteration No. 50

The interpretation of the result of the iterative signal reconstruction method is straightforward and is not included here. The estimation of the unknown wavelet from the arrival sequences is, in principle, straightforward and is also not included.

In Figure 10 we show the result of 100 iterations of this iterative procedure for the same example. The difference here is that the finite length restriction was relaxed by only one point so that y_m is zero for $m>12$ instead of $m>11$. Clearly, the algorithm converged to an incorrect solution. It is easy to see that if we convolve the correct answer (Figure 9) with an even sequence of the form {...,0,g,0,1,0,g,0,...}, where g is a

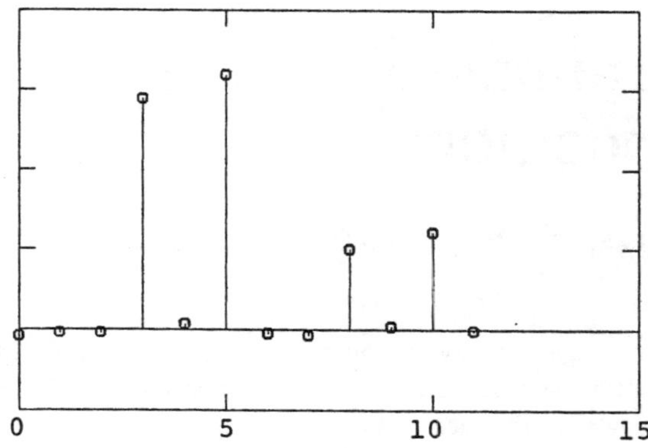

Figure 9. Iteration No. 100

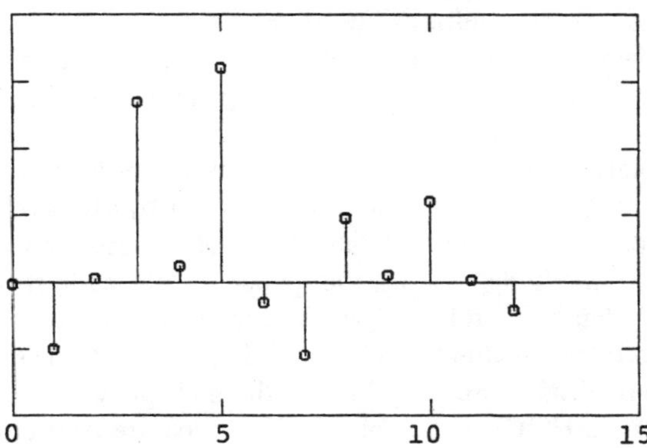

Figure 10. An example of convergence to an undesired sequence

small negative number, we will obtain the sequence shown in Figure 10. Thus, the sequence in Figure 10 contains zero's in conjugate reciprocal pairs. The algorithm was permitted to choose such a solution because we allowed non-zero values over too large a range.

Performance in the Presence of Noise

The iterative signal reconstruction method relies entirely on the phase of the cross-spectrum of the data. If the nondispersive multiple wavelet arrival model is accurate and the signal-to-noise ratio is very high, then the phase-only method can be expected to yield excellent results. However, if the wavelet does change shape from arrival to arrival, or if there are other undesired wavelets present in the data, or if there is significant measurement noise present, etc., then the phase of the cross-spectrum will differ from the desired phase by some significant amount and the method may breakdown.

In the presence of noise, we must estimate the phase of the cross-spectrum as well as the variance of the phase estimate, as a function of frequency. This information can be incorporated into the signal reconstruction algorithm by specifying the phase function only in the region where variance of the estimate is known to be low. Such an approach has only been partially successful in practice and requires further study.

Conclusions

We have considered the situation of an unknown wavelet propagating nondispersively in a multi-path or reverberatory environment. Such a situation is encountered in such applications as seismics, ocean acoustics, radar, sonar and others. Two receivers placed in such and environment each record the arrival of this wavelet an unknown number of times and each time with a different attenuation factor. The problem is to estimate the arrival times and attenuations at each receiver as well as the unknown wavelet.

The method presented here uses only the phase of the cross-spectrum of the two received signals to iteratively construct an intermediate sequence, y_m. Issues of convergence of the algorithm and uniqueness of the result have been explored. In the event that the algorithm does not converge, the length of the y_m sequence must be increased. If the algorithm converges to an undesired sequence, a contradiction will ultimately result. If this occurs, zero phase factors must be removed from the sequence polynomial to obtain the desired solution.

The intermediate y_m sequence can be "inverted" to obtain the desired arrival times and attenuations, provided that the interarrival delays, the D_n's, are distinct. If these delays are not distinct, the inversion is not unique.

This processing scheme is shown to fit into the generalized correlation structure for time delay estimation and corresponds to a choice for the cross-spectral weighting function which results in no smearing of the y_m sequence. Previous weighting functions all smear the y_m sequence to some degree.

The performance of this method in the presence of significant noise needs to be carefully explored.

Acknowledgement

The authors would like to acknowledge both Professor Alan V. Oppenheim and Douglas R. Mook for their support and helpful discussions.

References

[1] A.V. Oppenheim, M.H. Hayes and J.S. Lim, "Iterative Procedures for Signal Reconstruction from Phase", presented at the SPIE Conference, Washington DC, April 1980.

[2] T.F. Quatieri, V.T. Tom, M.H. Hayes and J. McClellan, "Convergence of Iterative Signal Reconstruction Algorithms", presented at the International Conference on Acoustics, Speech and Signal Processing, Atlanta GA, March 30, 1981.

[3] M.H. Hayes, J.S. Lim and A.V. Oppenheim, "Signal Reconstruction from Phase or Magnitude", IEEE Trans. on Acoustics Speech and Signal Processing, ASSP-28, 672-680, 1980.

[4] A.V. Oppenheim and J.S. Lim, "The Importance of Phase in Signals", Proceedings of the IEEE, 529-541, May 1981.

[5] C.H. Knapp and G.C. Carter, "The Generalized Correlation Method for Estimation of Time Delay", IEEE Trans. on Acoustics Speech and Signal Processing, ASSP-24, 320-327, 1976.

Estimation of Time Differences of Arrival by Pole Decomposition

ARYE NEHORAI, MEMBER, IEEE, GUANING SU, MEMBER, IEEE, AND MARTIN MORF, MEMBER IEEE

Abstract—Time differences of arrival (TDOA's) of emitter wave fronts to a spatially distributed array of sensors can be used to determine the source location. In this paper, we suggest a new method of TDOA estimation for multiple unknown autoregressive moving average (ARMA) sources and additive noise that may be correlated between the sensors. We derive a theoretical formula that only uses the receiver cross spectra and the source poles for the TDOA determination. The poles are estimated by a least squares technique, and two methods are suggested for the estimation of the cross spectra which allow tradeoffs between computational complexity and accuracy. A new time delay model is derived and used to show the applicability of the methods for noninteger TDOA's. Results from simulations illustrate the performance of the algorithm by Monte Carlo trials and compare it to the Cramer-Rao bound.

I. Introduction

THE location of electromagnetic, acoustic, or vibration wave sources by passive array systems is currently a topic of considerable interest in many areas, such as astronomy, defense, and geophysics [1]. The array sensors may comprise a group of antennas, microphones, or geophones, depending on the application. The time series output is usually modeled as a sum of delayed and amplitude scaled versions of the original source signals and additive noise components that are typically correlated among the array elements. The set of the delays, or the time differences of arrival (TDOA's), associated with each of the source wave fronts between the sensors can be used to determine the source location by solving a nonlinear or linear system of equations [2]. In this paper, we consider a new method for TDOA estimation of multiple sources whose spectra are unknown (see also [3]).

Existing TDOA estimation algorithms are usually devised for a single source. The generalized correlation method of Knapp and Carter [4] gives the maximum likelihood (ML) solution and unifies a large part of the available procedures as prefilters followed by cross correlators. These methods often require *a priori* knowledge of the source and noise spectra. For the case of unknown signal parameters, Chan *et al.* [5] and Hannan and Thomson [6] have suggested two different model fitting procedures, followed by an appropriate minimization function for the delay estimation. All of these methods are designed for the single source case.

The presence of many sources complicates the estimation problem, particularly when their spectral properties are unknown. Most traditional techniques employ beamformers to estimate the bearing angle of each of the distant emitters (see, for example, [7] and [8]). The spatial resolution of the beamformers is frequently limited by sidelobe effects, which unavoidably increase for narrower beams. Schweppe [9] derived an optimal decoupled beam processor by a least squares criterion. There is a considerable body of literature on array processing methods. Since this paper mainly addresses the TDOA estimation problem, we shall mention only the maximum likelihood technique of Capon [10], the multiple source location method of Reddi [11], and the multiple signal characterization (MUSIC) of Schmidt [12], which are restricted to narrow-band source signals. The signal subspace method used in [12] has been recently extended to band-limited sources by Coker and Ferrara [13], and for wide-band signals by Su and Morf [14] and Wax *et al.* [15].

In 1979, a new system identification approach for the TDOA estimation problem was suggested by Morf [16], using a multichannel rational model for the array signals. This model, which is useful for random sources with various spectra, was later applied by Porat and Friedlander [17] to design a new TDOA estimation technique for unknown multiple sources which can be modeled as autoregressive moving average (ARMA) processes. In their method, it is first necessary to estimate all the unknown signal parameters, such as the zeros and poles of the signal spectra as well as the amplitude attenuations from the sources to the receivers. A cost function is then minimized by a nonlinear optimization technique to estimate the desired delays. The receiver noises are assumed to be white and spatially uncorrelated.

In the next section of this paper, we first extend the ARMA model of [16] and [17] to include receiver noise that may be correlated among the sensors and in time. This model is still restricted to the integer delays case, but we generalize it to fractional delays later. In order to determine the unknown TDOA's, we derive a new formula that requires only the knowledge of the receiver cross spectra and the poles of the sources.

Based upon the new formula, we set up two new TDOA estimation procedures in Section III. In the first one, which we call the direct method, the receiver cross spectra are estimated by a windowed z-transform of the sensor sample cross correlations. The source poles are determined by a least squares technique which extends the ones of [17] and [18].

Manuscript received June 14, 1982; revised June 2, 1983. This work was supported by the Advanced Research Projects Agency under Contracts MDA903-80-C-0331 and MDA903-82-K-0382. The work of G. Su was also supported by a DSO scholarship from the Government of Singapore. This paper was presented in part at the 21st IEEE Conference on Decision and Control, Orlando, FL, December 1982.

A. Nehorai is with the Information Systems Laboratory, Department of Electrical Engineering, Stanford University, Stanford, CA 94305.

G. Su was with the Information Systems Laboratory, Stanford University, Stanford, CA 94305. He is now with the Ministry of Defence, Singapore.

M. Morf is with the Departments of Electrical Engineering and Computer Science, Yale University, New Haven, CT 06520.

Reprinted from *IEEE Trans. Acoust., Speech, Signal Processing*, vol. 31, no. 6, pp. 1478–1492, December 1983.

Then the TDOA formula is applied to estimate the desired TDOA's. This method based upon the TDOA formula was first presented by Nehorai and Morf in [3].

The second procedure, which we call the residue method, is similar to the first one, except that instead of the receiver cross spectra, we use their residues at the source poles. These residues are estimated by a special least squares method, first suggested by Su in [19]. The residue method requires more computations than the direct method, but it is asymptotically exact and yields more accurate results for overlapping spectra sources. The two methods are described, first for receiver noise of moving average (MA) type, and then we show how they can be applied for the more general ARMA noise.

In Section IV, we introduce a new ARMA array model for the case of noninteger delays. Our analysis is different from what is usually used in other TDOA estimation methods. Considering the appropriate modification for the model, it is shown that the TDOA formula is applicable for this case also.

Section V illustrates the performance of the algorithm by simulation results from Monte-Carlo experiments and comparisons to Cramer-Rao lower bound. Section VI summarizes the results and concludes the paper.

Appendix A relates our problem formulation to the multichannel rational one of Morf *et al.* [16]. This also facilitates the comparison to other estimation methods of multivariable type. Appendix B proves the TDOA estimation formula of Section II for the case of sources with common poles, assuming different multiplicity.

The major advantage of the proposed technique lies in its computational simplicity. It does not require the prior estimation of the amplitude attenuation factors, nor does it need estimation of the zeros of the source and noise spectra. In fact, the user does not have to know if the sources are of autoregressive (AR) or ARMA type.

Compared to the generalized cross-correlation technique of [4], the proposed method does not require prior knowledge of the signal and noise spectral densities. In comparison to [17], our technique has the advantage that it is completely independent of the attenuation factors, emitter and noise zeros, and requires significantly fewer computations. Our simulation results include a comparison with [17], exhibiting similar performance.

II. Problem Formulation and the TDOA Formula

A. Problem Formulation

We consider an array consisting of p omnidirectional sensors that receive signals originating from m uncorrelated sources which are modeled as finite order stationary ARMA processes. Thus, each of the sampled source signals is assumed to satisfy a stochastic recursion

$$x_k(nT_s) = \sum_{i=0}^{p_k} c_{ki} w_k[(n-i)T_s] - \sum_{i=1}^{q_k} d_{ki} x_k[(n-i)T_s] \qquad 1 \leqslant k \leqslant m \tag{1}$$

where n and k denote the time and source index, respectively, T_s is the sampling interval, and $\{w_k(nT_s)\}$ are uncorrelated unit variance white noise processes. (Note that any stationary random process can be modeled in this way by choosing sufficiently high order.) The z-transform of each source signal can be expressed from (1) as

$$x_k(z) = \frac{c_k(z)}{d_k(z)} w_k(z). \tag{2}$$

The transfer functions $\{c_k(z)/d_k(z)\}$ are assumed to be minimum phase and stable, i.e., with zeros and poles inside the unit circle.

Supposing nondispersive media as well as the absence of Doppler and multipath effects, the z-transform of the receiver signals can be written

$$y_i(z) = \sum_{k=1}^{m} G_{ik} z^{-\tau_{ik}} \frac{c_k(z)}{d_k(z)} w_k(z) + v_i(z) \qquad 1 \leqslant i \leqslant p \tag{3}$$

where the real scalars G_{ik} and τ_{ik} describe the unknown amplitude attenuation and normalized time delay, respectively, that a signal undergoes in traveling between the source k and receiver i. The receiver noise components $\{v_i(nT_s)\}$ may be correlated among the sensors.

The relation (3) is, strictly speaking, true only for integer delays. The appropriate modification to the more realistic noninteger delay situations will be given in Section IV.

All of the relations of this section can be put in a global matrix (or multichannel) form, as shown in Appendix A. In the matrix formulation of the Appendix, the z-transforms of this section become entries of rational matrix transfer functions, and therefore, the lower case letters are chosen for their notation.

B. The TDOA Formula

Denote the cross-correlation sequence between the receivers i and j by

$$R_{ij}(q) = E\,[y_i(nT_s) y_j^*(n-q)T_s] \qquad -\infty \leqslant q \leqslant \infty \tag{4}$$

where E is the expectation operator and * denotes the complex conjugate.[1] Then, under the above assumptions, it is easy to check that the cross spectrum (or the z-transform of the cross correlation) between the receivers i and j is given by

$$S_{ij}(z) = \sum_{k=1}^{m} G_{ik} G_{jk} z^{-\Delta_{ijk}} \frac{c_k(z)\, c_k^*(1/z^*)}{d_k(z)\, d_k^*(1/z^*)} + S_{ij}^v(z) \tag{5}$$

where $\Delta_{ijk} = \tau_{ik} - \tau_{jk}$ is the desired TDOA of the signal emitted from the kth source to the receivers i and j, and $S_{ij}^v(z)$ represents the possible correlation between the noise components $v_i(nT_s)$ and $v_j(nT_s)$.

Using the above relations, we establish now the following key formula [3]:

$$\lim_{z \to \lambda_{kl}} \frac{S_{ij}(z)}{S_{ij}^*(1/z^*)} = \lambda_{kl}^{-2\Delta_{ijk}} \tag{6}$$

[1] The complex signal notation is used for the sake of generality. It also can be used with some modifications for prefiltered signals.

where λ_{kl} denotes the lth pole of the source k.

Proof of (6): For simplicity, we shall give here the proof for only the special case where the source transfer functions are strictly proper and their poles are different and simple. The appropriate extension to the more general situations is given in Appendix B.

To prove (6), it is useful to expand each of the source transfer functions into the partial fraction decomposition

$$\frac{c_k(z)}{d_k(z)} = \sum_{r=1}^{p_k} \frac{C_{kr}}{1 - \lambda_{kr} z^{-1}} \tag{7}$$

where $\{C_{kr}\}$ are the residues of $\{c_k(z)/d_k(z)\}$ at the corresponding poles $\{\lambda_{kr}\}$. Now, substituting the expansion (7) into (5), we have

$$S_{ij}(z) = \sum_{k=1}^{m} G_{ik} G_{jk} z^{-\Delta_{ijk}} \sum_{r=1}^{p_k} \sum_{s=1}^{p_k} \frac{C_{kr}}{1 - \lambda_{kr} z^{-1}} \frac{C_{ks}^*}{1 - \lambda_{ks}^* z} + S_{ij}^v(z). \tag{8}$$

Consider the desired limit of (6). Multiplying both the numerator and denominator by $1 - \lambda_{kl} z^{-1}$ before taking the limit, we obtain

$$\lim_{z \to \lambda_{kl}} \frac{S_{ij}(z)}{S_{ij}^*(1/z^*)} = \lim_{z \to \lambda_{kl}} \frac{G_{ik} G_{jk} z^{-\Delta_{ijk}} C_{kl} \dfrac{c_k^*(1/z^*)}{d_k^*(1/z^*)}}{G_{ik} G_{jk} z^{+\Delta_{ijk}} C_{kl} \dfrac{c_k^*(1/z^*)}{d_k^*(1/z^*)}} = \lim_{z \to \lambda_{kl}} z^{-2\Delta_{ijk}} = \lambda_{kl}^{-2\Delta_{ijk}}. \tag{9}$$

The limit in (6) can also be interpreted as the residue ratios of $S_{ij}(z)$ and $S_{ij}^*(1/z^*)$. (Notice that $S_{ij}^*(1/z^*) = S_{ji}(z)$.) To prove this, note that $S_{ij}(z)$ can be written

$$S_{ij}(z) = \sum_{k=1}^{m} \sum_{l=1}^{p_k} \left(\frac{F_{ijkl}}{1 - \lambda_{kl} z^{-1}} + \frac{F_{ijkl}^*}{1 - \lambda_{kl}^* z} \right) + S_{ij}^v(z) \tag{10}$$

where $\{F_{ijkl}\}$ are the residues of the receiver cross spectra $S_{ij}(z)$ at the source poles. Computing the limit as in (9), one finds

$$\lim_{z \to \lambda_{kl}} \frac{S_{ij}(z)}{S_{ij}^*(1/z^*)} = \frac{F_{ijkl}}{F_{jikl}} = \lambda_{kl}^{-2\Delta_{ijk}}. \tag{11}$$

In Appendix B, we show that (6) still holds for more general conditions than in the above proof. More specifically, the source transfer functions may be nonproper and/or with repeated poles. If a pole is common to two or more sources and the multiplicities are different, (6) is applicable to the source with the highest multiplicity. These facts are proven using the Laurent expansion (see, e.g., [20]) of $S_{ij}(z)$ near each pole, as is done in Appendix B, or by L'Hôpital's rule.

All the proofs of (6) are based on the following four special features of $S_{ij}(z)$.

1) Its poles coincide with the source poles.

2) Its value near each pole is a function of only a single source.

3) All its signal components are symmetrical with respect to the unit circle except for $z^{-\Delta_{ijk}}$ [cf. (5)].

4) The effect of the delay is only on its residues (or the MA part).

It is also helpful to note that (6) is satisfied for any sources and values of z for which the source k dominates. To see this, note that here

$$S_{ij}(z) \approx S_k(z) S_k^*(1/z^*) z^{-\Delta_{ijk}} \tag{12}$$

where $S_k(z)$ denotes the spectrum of the source k, and therefore

$$\frac{S_{ij}(z)}{S_{ij}^*(1/z^*)} \approx z^{-2\Delta_{ijk}}. \tag{13}$$

In this sense, the above relationships are nonparametric, and the procedures below can be applied to more general situations than the ARMA modeling.

III. The Estimation Procedure

Using (6), we now establish our procedure for the TDOA estimation and consider some practical issues and extensions.

A. The Main Steps of the Procedure

The results of the previous section reveal that the receiver cross spectra or their residues and the source poles provide the necessary information for the TDOA determination. Since, in practice, these may not be available, we have to estimate them. The estimation procedure below first evaluates the source poles by solving a set of least squares equations. Then the TDOA formula is applied to estimate the desired TDOA's in two alternative ways: the first one uses direct estimates of the receiver cross spectra at the source poles by a windowed sum, while the second uses estimates of the corresponding residues. This allows tradeoffs between computational complexity and accuracy.

Source Poles Estimation: Before we describe the source pole estimation procedure, we introduce the assumption that the noise components are of the MA type. Thus, we write

$$y_i(z) = \sum_{k=1}^{m} G_{ik} z^{-\tau_{ik}} \frac{c_k(z)}{d_k(z)} w_k(z) + \sum_{k=1}^{m_v} b_{ik}(z) e_k(z) \tag{14}$$

where $\{b_{ik}(z)\}$ are finite order polynomials of z, and the signals $\{e_k(nT_s)\}$ associated with $\{e_k(z)\}$ are unit variance white processes, assumed to be uncorrelated both among themselves and the source signals. As we shall see later, our method is also applicable for ARMA noise. The present MA noise model is useful to characterize undesired nonwhite wideband signals and enables us to describe a possible correlation among the receiver noise components.

Let

$$a(z) = \prod_{k=1}^{m} d_k(z); \tag{15}$$

the roots of $a(z)$ coincide with the poles of the sources, and we shall denote its order by M.

For the estimation of the source poles, observe that the following relation can be inferred from (14):

$$a(z)y_i(z) = \sum_{k=1}^{m} z^{-\tau_{ik}} p_{ik}(z) w_k(z) + a(z) \sum_{k=1}^{m_v} b_{ik}(z) e_k(z) \tag{16}$$

where the polynomials $\{p_{ik}(z)\}$ are of order not larger than M. Now we can express the time domain version of (16) as

$$a(n) * y_i(nT_s) = \sum_{k=1}^{m} p_{ik}(n) * w_k[(n - \tau_{ik})T_s] + a(n) * \sum_{k=1}^{m_v} b_{ik}(n) * e_k(nT_s). \tag{17}$$

Define

$$\nu_{ij} = \max_k [\{\tau_{ik} - \tau_{jk}\}, \quad \{\text{order of } b_{ik}(z)\}]. \tag{18}$$

Multiplying both sides of (17) by $y_j^*[(n - M - \nu_{ij} - \mu)T_s]$ for $\mu \geqslant 0$, and upon taking expectation, one obtains

$$R_{ij}(M + \nu_{ij} + \mu) + a_1 R_{ij}(M + \nu_{ij} + \mu - 1) + \cdots + a_M R_{ij}(\nu_{ij} + \mu) = 0. \tag{19}$$

For notational convenience, we shall use the variable

$$\nu = \max_{ij} \{\nu_{ij}\} \tag{20}$$

instead of all $\{\nu_{ij}\}$. Now combining all the relations in the form of (19), we get the set of equations

$$\begin{bmatrix} \rho_{M+\nu} & \cdots & \rho_{\nu+1} \\ \rho_{M+\nu+1} & & \rho_{\nu+2} \\ \vdots & & \vdots \\ \rho_{N-1} & \cdots & \rho_{N-M} \end{bmatrix} \begin{bmatrix} a_1 \\ \vdots \\ a_M \end{bmatrix} = - \begin{bmatrix} \rho_{M+\nu+1} \\ \rho_{M+\nu+2} \\ \vdots \\ \rho_N \end{bmatrix} \tag{21}$$

where

$$\rho_q = [R_{11}(q), R_{12}(q), \cdots, R_{pp}(q)], \tag{22}$$

i.e., ρ_q is a p^2 dimension vector whose entries are $R_{ij}(q)$ for all i, j. The relation (21) has to be satisfied for the true receiver correlations. Since, in practice, they are not known, we replace them by their estimates

$$\hat{R}_{ij}(q) = \frac{1}{T - q} \sum_{n=q+1}^{T} y_i(nT_s) y_j^*[(n - q)T_s] \qquad 0 \leqslant q \leqslant N \tag{23}$$

where T denotes the available data length, and N has to be chosen large enough compared to the significant part of the true source correlation sequences. For negative lag estimates, use $\hat{R}_{ij}(-q) = \hat{R}_{ij}^*(q)$. Then the set of equations (21) is solved in the least squares sense for the coefficients of $a(z)$. Note that (21) extends the method of [17] as it uses the cross correlation where $i \neq j$, and it is applicable for correlated receiver noise of MA type. Equations (21) can also be viewed as an extension of the (single-input single-output) ones used by [18].

Next, we estimate the desired source poles by factoring $\hat{a}(z)$ into

$$\hat{a}(z) = \prod_{k,l} (1 - \hat{\lambda}_{kl} z^{-1}). \tag{24}$$

In other words, we extract the roots of $\hat{a}(z)$ to find the poles of the sources.

As may seem from (18), one must have some *a priori* knowledge about the TDOA's as well as the noise orders in order to apply the above least squares method. One *a priori* knowledge that is always available is the maximum possible TDOA between sensors, given by the propagation time between two sensors. Moreover, this restriction can be relaxed as follows. First, note that we can apply (21) with only part of the correlation terms that appear in (22). For instance, we may use only the correlation terms for which $i = j$. Here

$$\rho_q = [R_{11}(q), R_{22}(q), \cdots, R_{pp}(q)] \tag{25}$$

and the condition (18) is accordingly narrowed down to

$$\nu_{ii} = \max_k \{\text{order of } b_{ik}(z)\}. \tag{26}$$

This means that in this case, one does not need to have *a priori* information on the TDOA's to solve the set of equations (21). A sufficiently large value for ν can then be used to satisfy (19). Thus, one can apply the procedure only for the diagonal correlation terms, estimate the TDOA's, and then use (21) in its full version to obtain better estimates. The incorporation of the cross correlation for which $i \neq j$ can also be useful to separate the source poles from undesired noise poles, as we explain later.

In the next part of this section, we shall discuss how to choose the correct model order M. Here we note that, in practice, we usually use a higher order $\hat{M}$ in solving (21) than the true M. As a result, the lower bounds needed for the algorithm are less than the ones in the RHS of (18) and (26) by $\hat{M} - M$. This also severely diminishes the restriction of the above requirements.

TDOA Estimation–Direct Method: The direct method estimates the receiver cross spectra by the Blackman-Tukey-type windowed sum

$$\hat{S}_{ij}(z) = \sum_{q=-N}^{N} \hat{R}_{ij}(q) W(q) z^{-q} \tag{27}$$

where $\{W(q)\}$ denotes the window weights applied to reduce the sidelobe effects, cf. [21].

Given the estimates of the receiver cross spectra as well as the source poles, we are able to apply (6) to estimate the TDOA's. From the phase part of (6), we write the following estimation result at λ_{kl}:

$$\hat{\Delta}_{ijk} = \frac{\phi\{\hat{S}_{ij}^*(1/\hat{\lambda}_{kl}^*)\} - \phi\{\hat{S}_{ij}(\hat{\lambda}_{kl})\} + 2n\pi}{2\phi\{\hat{\lambda}_{kl}\}} \tag{28}$$

where

$$\phi(x) = tg^{-1}\left(\frac{\text{Im}\{x\}}{\text{Re}\{x\}}\right). \tag{29}$$

The amplitude part of (6) yields

$$\hat{\Delta}_{ijk} = \frac{\log|\hat{S}_{ij}^*(1/\hat{\lambda}_{kl}^*)| - \log|\hat{S}_{ij}(\hat{\lambda}_{kl})|}{2\log|\hat{\lambda}_{kl}|}. \tag{30}$$

The ambiguity term in (28) is resolvable by (30), provided that the sources are not too narrow-band (i.e., the poles are not too close to the unit circle). If the sources are of sinusoidal type, the ambiguity is inherent in the problem and can be prevented only by placing a limit on sensor spacing to be less than half a wavelength.

It should be clear that although the theoretical value of $\lim_{z\to\lambda_{kl}} S_{ij}(z)$ is infinity, the above windowed sum estimate of $\hat{S}_{ij}(\lambda_{kl})$ is always bounded since it approximates the infinite series by a finite sum.

Note that the cross-spectrum estimate (27) is best fitted to signals of the MA type. For the AR or ARMA considered here, this results in a lower resolution, and the above method may not be able to separate the sources at the poles when their spectra are strongly overlapping. The residue method below is suggested for solving this problem.

TDOA Estimation–Residue Method: From the results of the previous section, the residues of the receiver cross spectra can also be used to evaluate the TDOA's by the TDOA formula. The residues can be estimated from the sample receiver cross correlations by least squares fitting of the partial fraction decomposition (10). Comparing coefficients of powers of z^{-r}, where $r > \nu_{ij}$ in the power series expansion of (10) on the unit circle using the impulse response expansion

$$\frac{1}{1-\lambda_{kl}z^{-1}} = \sum_{r=0}^{\infty} \lambda_{kl}^r z^{-r} \tag{31}$$

we get the overdetermined set of equations for residues $F_{ij11}, \cdots, F_{ijmp_m}$

$$\begin{bmatrix} \lambda_{11}^{\nu_{ij}+1} & \cdot\ \lambda_{mp_m}^{\nu_{ij}+1} \\ \vdots & \vdots \\ \lambda_{11}^{N} & \cdot\ \lambda_{mp_m}^{N} \end{bmatrix} \begin{bmatrix} F_{ij11} \\ \vdots \\ F_{ijmp_m} \end{bmatrix} = \begin{bmatrix} R_{ij}(\nu_{ij}+1) \\ \vdots \\ R_{ij}(N) \end{bmatrix}. \tag{32}$$

The set of equations (32) can be solved in a least squares sense to give estimates for the residues. For more details of this residue estimation method and its application to signal subspace algorithms, see [19]. Here we note that the structure of (32) is similar to the structure in the Prony method (see, e.g., [22]).

Similarly to (28), we can write the estimation result

$$\hat{\Delta}_{ijk} = \frac{\phi\{\hat{F}_{jikl}\} - \phi\{\hat{F}_{ijkl}\} + 2n\pi}{2\phi\{\hat{\lambda}_{kl}\}}. \tag{33}$$

An additional equation analogous to (30) to solve the phase ambiguity by the residue amplitude gives

$$\hat{\Delta}_{ijk} = \frac{\log|\hat{F}_{jikl}| - \log|\hat{F}_{ijkl}|}{2\log|\hat{\lambda}_{kl}|}. \tag{33a}$$

Note that as the amount of data increases, the sample correlations used in (21) and (32) become exact asymptotically. This leads to exact pole and residue estimates and, hence, yields asymptotically exact TDOA estimation results independently of the SNR.

B. Some Practical Issues and Extensions

For the practical application of the above estimation procedure, it is first necessary to estimate the model order M when the sources are unknown. In addition, it is recommended to use an order larger than the true value of M (see, e.g., [17] and [18]). This yields better estimates for the source poles, but also introduces spurious poles. Therefore, it is also desired to know how to distinguish between the correct poles and the superfluous ones. The same problem arises when we apply our technique to receiver noise of the ARMA type. In what follows, we consider several alternative solutions to these problems (see also [17] for some related topics), as well as some possible extensions to our estimation procedures.

Choice of the Model Order M: One possibility for choosing the best model order M is based upon the fact that the least squares matrix in (21) does not have full rank (theoretically) for dimensions larger than the true model order M, cf. [22]. This suggests that M can be estimated by monitoring the determinant of this matrix for increasing dimensions until it becomes sufficiently small.

Another possibility is to use the Akaike information criterion (AIC) [23] where the best order is postulated by minimizing an information theoretical criterion.

Using the fact that the number of stable poles of $S_{ij}(z)$ is equal to M, we can estimate the model order by evaluating the corresponding number of relative maxima for the cross-spectra estimates $\hat{S}_{ij}(z)$ inside the unit circle. This can be done, for example, with the aid of the chirp z-transform [21]. Note that the resulting maxima are also useful to estimate the source poles, instead of the least squares technique described above.

ARMA Receiver Noise: From the discussion in Section II-B, the theoretical equation (6) is also applicable for noise that includes ARMA-type components, provided their poles are different from the source poles or have lower multiplicity if they coincide. When the noise has ARMA components, we apply the pole estimation equations (21) for the source and the noise signals together. Then, in order to estimate the TDOA's, we have to distinguish between the resulted source and noise poles. This can be done by utilizing available *a priori* spatial and/or temporal knowledge of the received signals. We give guidelines below.

Selection of the Correct Poles: In order to be able to distinguish between the desired and the spurious poles that result from (21), one should first note that the extra poles usually tend to approximate the spectra of the receiver noise components, particularly in low signal-to-noise ratio. Therefore, in the simplest situation of white receiver noise, the additional poles are expected to appear uniformly spaced within the unit circle. In low-pass noise, they will most likely appear in low frequency regions, and vice versa for high-pass noise.

If the receiver noise components are known to have low correlation between the sensors, it is recommended to apply the least squares (21) first for each receiver separately. In other words, apply (21) with $\rho_q = R_{ii}(q)$, $1 \leq i \leq m$. The correct poles can then be chosen as the consistent ones in the different receivers. Another possibility, in this case, is to use (21) with cross-correlation terms for which $i \neq j$ since they essentially have only signal components.

When the noise exhibits a relatively short temporal correlation, first estimating the poles for different time intervals is suggested. The desired poles can then be selected as the common ones for the different intervals. Another useful method is to first estimate the poles for different model orders, and then decide on the correct ones by a comparison.

It is important to note that the estimation results (28), (30), (33), and (34) can be used to find the unknown TDOA's before the selection of the true source poles has been carried out, since they are applicable to each pole independently of the others. This implies that we can apply them first to all the estimated poles, including the fictitious ones, and then use the obtained TDOA's as additional spatial information for the decision on the correct poles.

The source poles can also be selected by using a different estimation method than the least squares technique (21). For instance, as we mentioned above, the chirp z-transform can be employed for this purpose. In this approach, we first estimate the spectral peaks of $\hat{S}_{ij}(z)$ in desired regions inside the unit circle, and then apply the TDOA estimation results at the resulting points.

Yet another method for selecting the source poles is based on their corresponding gains (see, e.g., [17]). Here, the desired poles are chosen as the ones with the higher gains. In our method, the gains can be found by the residue estimation technique above.

Source Zeros Estimation: Although the zeros of the source transfer functions were not required for the TDOA evaluation, they can also be estimated by our pole decomposition procedure, as is illustrated below for the simple poles case. These zeros are useful for complete spectral estimation of the source signals.

To find the source zeros, it is sufficient to apply the above method to only one of the receiver signals. The results give the poles $\{\hat{\lambda}_{kl}\}$ and the residues $\hat{F}_{iik}$. Next, write

$$\sum_{l=1}^{p_k} \left(\frac{\hat{F}_{iikl}}{1 - \hat{\lambda}_{kl} z^{-1}} + \frac{\hat{F}^*_{iikl}}{1 - \hat{\lambda}^*_{kl} z} \right) = \hat{G}^2_{ik} \frac{\hat{c}_k(z)}{\hat{d}_k(z)} \frac{\hat{c}^*_k(1/z^*)}{\hat{d}^*_k(1/z^*)}. \tag{34}$$

In this equation, we have available all the terms of its LHS, and the RHS has to be formed so that the polynomial factors $\hat{c}_k(z)$ and $\hat{d}_k(z)$ have their roots inside the unit circle. These roots of $\hat{c}_k(z)$ provide the desired estimates for the zeros of the source transfer functions.

It should be noted that in applying the above zero estimation procedure, it is required to know how to associate each of the estimated poles with its corresponding source. This can be done by utilizing the spatial estimates obtained in the previous steps. In other words, estimated poles that yield similar TDOA's are associated with the same source.

Observe that in this way, we have also suggested a new method for spectral estimation of multivariable ARMA processes, based on our method of pole decomposition.

IV. Analysis of the Noninteger Delays Case

In this section, we consider the appropriate extension of the ARMA array model of Section II, as well as the one of [16] and [17] for the case of noninteger TDOA's. The approach we use was first suggested by Nehorai and Morf [3]. Since the analysis here is different from what is usually used in TDOA modeling methods, we give it in a rather detailed manner. We shall then use the results to show that the fundamental equation (6) is applicable also for this case.

A. The Modified ARMA Model

The problem here can be phrased as the evaluation of the sampled ARMA source signal model after a delay of a noninteger type. For the solution of this problem, it is useful to remodel the discrete source signal (1) back into its original continuous analog version. Then we are able to evaluate the effect of a noninteger delay on the z-transform of the received sampled signals. For simplicity, we consider only the nonrepeated poles case, although the description can be generalized to the more complicated situations.

Consider first the relation between a general discrete sampled signal model and its original continuous time version. Using Laplace and z-transforms, the result is illustrated in Fig. 1. In this figure, $H(z)$ denotes the z-transform of a linear causal and time invariant discrete model system whose impulse response sequence is denoted by $h(n)$ $(n \geq 0)$. $H^c(s)$ is the Laplace transform of the corresponding continuous time system whose impulse response is $h^c(t)$ $(t \geq 0)$. The input of the continuous system is the signal $w(t)$, multiplied by an impulse train $\sum_{n=-\infty}^{\infty} \delta(t - nT_s)$. The input of the discrete system is the sampled version of $w(t)$, namely, $w(nT_s)$.

Now note that if the above system models satisfy the relationship

$$h^c(nT_s) = h(n) \qquad n \geq 0, \tag{35}$$

then their corresponding outputs will be similarly related by

$$x^c(nT_s) = x(nT_s) \qquad -\infty \leq n \leq \infty. \tag{36}$$

This means that a system $H^e(s)$ satisfying (35) can be used to model the original continuous time version of $H(z)$. Note that in filter design, the relation (35) is sometimes called impulse invariance [21].

For our special purpose, the input $w(t)$ has to be a random signal with correlation length less than the sampling interval T_s and of unit variance. In this case, the inputs appearing in Fig. 1 become unit variance white signals, which is in accordance with our basic assumptions of Section II.

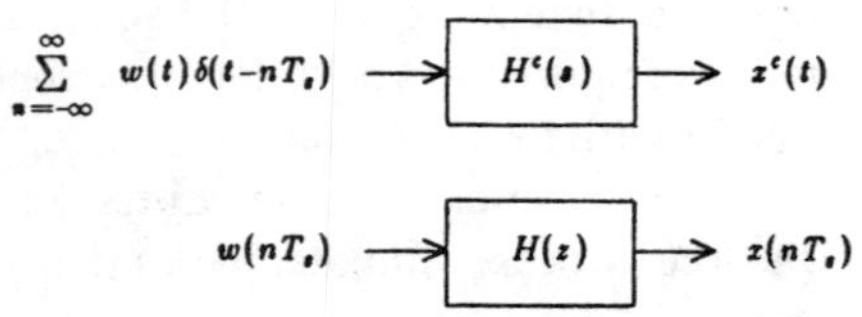

$\sum_{n=-\infty}^{\infty} w(t)\delta(t-nT_s) \longrightarrow$ [$H^c(s)$] $\longrightarrow x^c(t)$

$w(nT_s) \longrightarrow$ [$H(z)$] $\longrightarrow x(nT_s)$

Fig. 1. Relation between continuous and discrete time signal models.

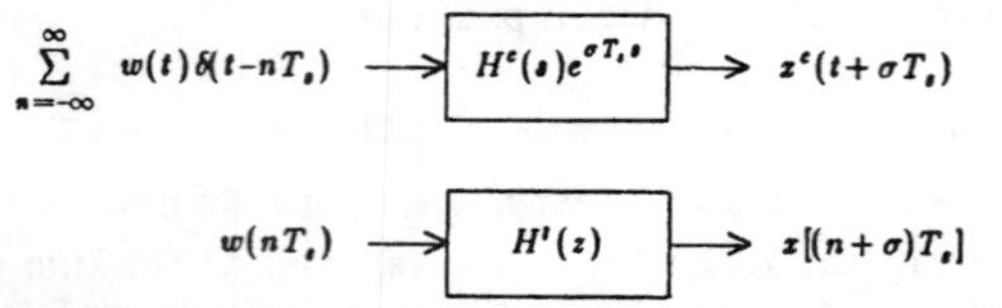

$\sum_{n=-\infty}^{\infty} w(t)\delta(t-nT_s) \longrightarrow$ [$H^c(s)e^{\sigma T_s s}$] $\longrightarrow x^c(t+\sigma T_s)$

$w(nT_s) \longrightarrow$ [$H'(z)$] $\longrightarrow x[(n+\sigma)T_s]$

Fig. 2. Relation between delayed continuous signal and its corresponding discrete time model.

Consider now the time invariance relationship for a system consisting of a single pole. With an obvious notation, it is well known that in this case, the following relation exists (cf. [21] or [24]):

$$H^c(s) = \frac{1}{s+a} \longleftrightarrow \frac{1}{1-\lambda z^{-1}} = H(z) \tag{37}$$

where

$$\lambda = e^{-aT_s}. \tag{38}$$

This relationship can be used to relate rational spectra of discrete type to their original continuous analog and vice versa. Note that while the usual impulse invariance relation is used for deterministic systems or filters, here we use it for stochastic processes.

Let us now examine the effect that a normalized fractional delay σ ($0 \leqslant \sigma < 1$) has on the above single pole system. The problem is displayed in Fig. 2. In this figure, the inputs and notations are the same as in Fig. 1. We are interested in the z-transform of the output sequence $x[(n+\sigma)T_s]$. The solution for this problem can be obtained by the modified z-transform first introduced by Barker in [25], in the context of control systems, cf. also [24]. For the single pole system, it is straightforward to show that

$$H^c(s)e^{\sigma T_s s} = \frac{e^{\sigma T_s s}}{s+a} \longleftrightarrow \frac{\lambda^{\sigma}}{1-\lambda z^{-1}} = H'(z) \tag{39}$$

where λ is the same as in (38). Notice that the effect of the delay is now expressed in the factor λ^{σ} of the RHS above. For the more general circumstances where the normalized delay is a noninteger larger than unity, write

$$\tau = [\tau] - \sigma \tag{40}$$

where $[\tau]$ denotes the next higher integer after τ. Then, for a single pole system, one can obtain

$$H^c(s)e^{-\tau T_s s} = \frac{e^{-\tau T_s s}}{s+a} \longleftrightarrow z^{-[\tau]} \frac{\lambda^{\sigma}}{1-\lambda z^{-1}} = H'(z). \tag{41}$$

B. The Noninteger Delay ARMA Array Model

From the above discussion, we are now able to express the required modification for the received array signal model, when the delays are not integers. Each source-receiver delay is written

$$\tau_{ik} = [\tau_{ik}] - \sigma_{ik} \tag{42}$$

where the notation is the same as for (40). Then it is evident from (41) and (42) that each receiver z-transform is now given by

$$y_i(z) = \sum_{k=1}^{m} G_{ik} z^{-[\tau_{ik}]} \sum_{l=1}^{p_k} \frac{\lambda_{kl}^{\sigma_{ik}} C_{kl}}{1-\lambda_{kl} z^{-1}} w_k(z) + v_i(z). \tag{43}$$

This result models the effect of the delay for each pole individually. Observe that since usually $\sigma_{ik} \neq \sigma_{jk}$ for $i \neq j$, (43) indicates that the received zero spectra associated with each source are slightly different from sensor to sensor, as a result of noninteger delays. Also, the denominators of (43) reveal that the received signal poles are unaltered by the fractional delays. This implies that the corresponding pole estimation procedure of Section II is still applicable for this case.

Notice from (43) that our result for the source-receiver delay operator differs from the usual model $e^{-j\omega\tau_{ik}}$ traditionally used for TDOA estimation, even when evaluated on the unit circle. Rather, it is a function that depends on each pole separately. This difference is due to the fact that the continuous time ARMA signal sources considered here are, in general, of wide-band type, while the regular delay model $e^{-j\omega\tau_{ik}}$ is valid only for band-limited signals. (Note that the two functions coincide for poles which are on the unit circle.)

The above model is also useful for prefiltered receiver signals, assuming that the filter transfer function $F(s)$ is rational. In this case, it is easy to check that the above discussion is still applicable by replacing $H^c(s)$ with $H^c(s)F(s)$.

C. Examination of (6) for Noninteger Delays

Using the result (43), we can now examine our TDOA (6) and verify that it is approximately useful also for the noninteger delay estimation case. From (43), we have

$$S_{ij}(z) = \sum_{k=1}^{m} G_{ik} G_{jk} z^{-([\tau_{ik}]-[\tau_{jk}])} \sum_{r=1}^{p_k} \sum_{s=1}^{p_k} \cdot \frac{C_{kr}\lambda_{kr}^{\sigma_{ik}}}{1-\lambda_{kr} z^{-1}} \frac{C_{ks}^{*}(\lambda_{ks}^{*})^{\sigma_{jk}}}{1-\lambda_{ks}^{*} z} + S_{ij}^{v}(z). \tag{44}$$

Now, following the same lines as in the proof of (6), we can get

$$\lim_{z\to\lambda_{kl}} \frac{S_{ij}(z)}{S_{ij}^{*}(1/z^{*})} = \lambda_{kl}^{-2([\tau_{ik}]-[\tau_{jk}])+(\sigma_{ik}-\sigma_{jk})} \frac{\sum_{r=1}^{p_k} \frac{C_{kr}^{*}(\lambda_{kr}^{*})^{\sigma_{jk}}}{1-\lambda_{kr}^{*}\lambda_{kl}}}{\sum_{r=1}^{p_k} \frac{C_{kr}^{*}(\lambda_{kr}^{*})^{\sigma_{ik}}}{1-\lambda_{kr}^{*}\lambda_{kl}}}. \tag{45}$$

This result is usually different from (6), unless $\sigma_{ik} = \sigma_{jk}$. But noting that usually

$$1 - \lambda_{kl}\lambda_{kl}^* << |1 - \lambda_{kr}\lambda_{kl}^*| \quad \text{for all} \quad r \neq l, \tag{46}$$

one can approximate each of the sums in (45) by the corresponding single term for which $r = l$. (Note also that the approximation (46) becomes better for poles which are closer to the unit circle and/or poles having sufficiently large central frequency distance.) As a result, we obtain

$$\lim_{z \to \lambda_{kl}} \frac{S_{ij}(z)}{S_{ij}^*(1/z^*)}$$

$$\approx \lambda_{kl}^{-2}{}^{(|\tau_{ik}| - |\tau_{jk}|) + (\sigma_{ik} - \sigma_{jk})}(\lambda_{kl}^*)^{(\sigma_{jk} - \sigma_{ik})}. \tag{47}$$

The importance of the last expression is that it yields the same TDOA estimation equations as (28) and (33). In other words, we have found that equations (28) and (33) are useful for the estimation of noninteger delays under the above conditions, which are usually satisfied.

V. Simulation Results

In this section, we demonstrate the performance of the two TDOA estimation procedures given in Section III. The experiments consist of Monte Carlo trials from which the sample statistics are computed. 20 trials are run for all experiments except the Cramer-Rao bound comparisons, for which 40 trials are performed. The resulting sample statistics are plotted against TDOA separation and signal-to-noise ratio (SNR) in a graph of expected range (mean ± standard deviation) with the mean shown as crosses. The results are also given in table form. For the Cramer-Rao bound comparisons, the quantity $10 \log (1/\sigma^2)$ is plotted against SNR where σ^2 is the variance of the estimates.

Source Spectra: Two sets of sources are used in the simulations. The first set consists of two second-order AR sources which are strongly overlapping. The source parameters are

transfer function of source 1

$$\frac{c_1(z)}{d_1(z)} = \frac{1}{1 - 1.096z^{-1} + 0.87z^{-2}}$$

transfer function of source 2

$$\frac{c_2(z)}{d_2(z)} = \frac{1}{1 - 0.899z^{-1} + 0.87z^{-2}}$$

This set of sources is used to test the capability of the residue method to separate strongly overlapping sources and is chosen to facilitate performance comparison with [17]. The power spectra of the above sources are plotted in Fig. 3.

The second set consists of two second-order ARMA sources. The source parameters are

transfer function of source 1

$$\frac{c_1(z)}{d_1(z)} = \frac{1 - 0.59z^{-1} + 0.93z^{-2}}{1 - 1.51z^{-1} + 0.87z^{-2}}$$

transfer function of source 2

$$\frac{c_2(z)}{d_2(z)} = \frac{1 - 1.84z^{-1} + 0.93z^{-2}}{1 - 1.10z^{-1} + 0.87z^{-2}}.$$

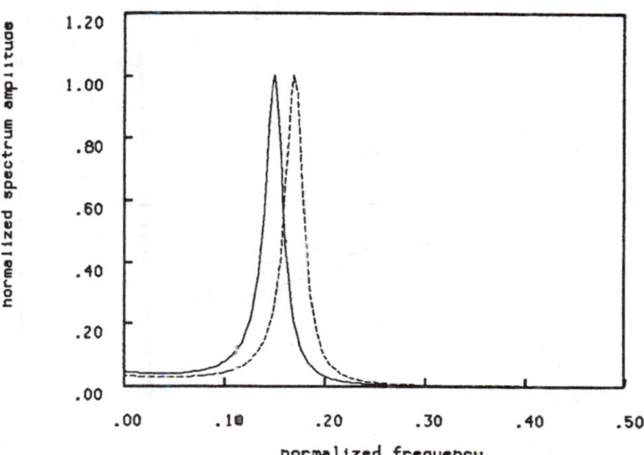

Fig. 3. Overlapping AR source spectra.

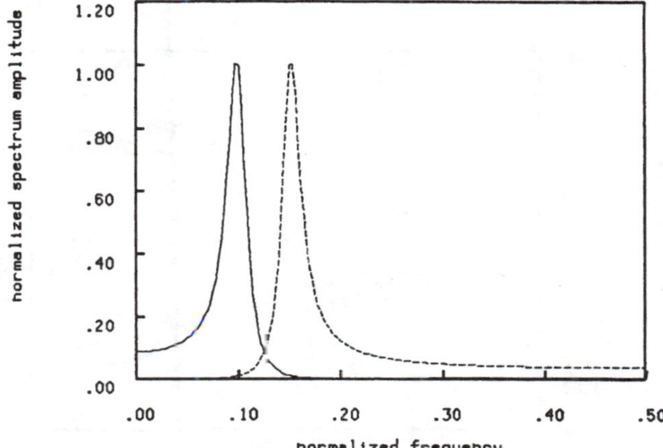

Fig. 4. ARMA source spectra.

The power spectra of the above sources are plotted in Fig. 4.

Noise Spectra: Two types of additive noise are considered in the simulations. The first is uncorrelated white noise at each sensor with equal power. The second is correlated colored noise with spectral factor

$$b_{ij}(z) = \begin{cases} 1 + z^{-1} & i = j \\ z^{-1} & i \neq j. \end{cases}$$

The algorithms are shown to be robust with respect to spatially correlated colored noise of moving average type.

Simulation Parameters: For each of the simulations, 1000 data samples were used. 40 correlation lags were computed for the direct method and 80 correlation lags for the residue method. The value of ν_{ij} used in (32) was 1 for uncorrelated white noise and 2 for moving average correlated noise, except for the Cramer-Rao bound computations, when it is 0, since the actual TDOA is 0. The overdetermined model order is 17 in all cases where poles are estimated, except for the overlapping AR sources, where an order of 19 was found to give better performance, and the Cramer-Rao bound comparisons, where an order of 13 was used, since there is only one source.

The phase ambiguity noted in (28) caused occasional step errors in the estimates. We have eliminated these ambiguous cases from the sample statistics. The number of ambiguities is

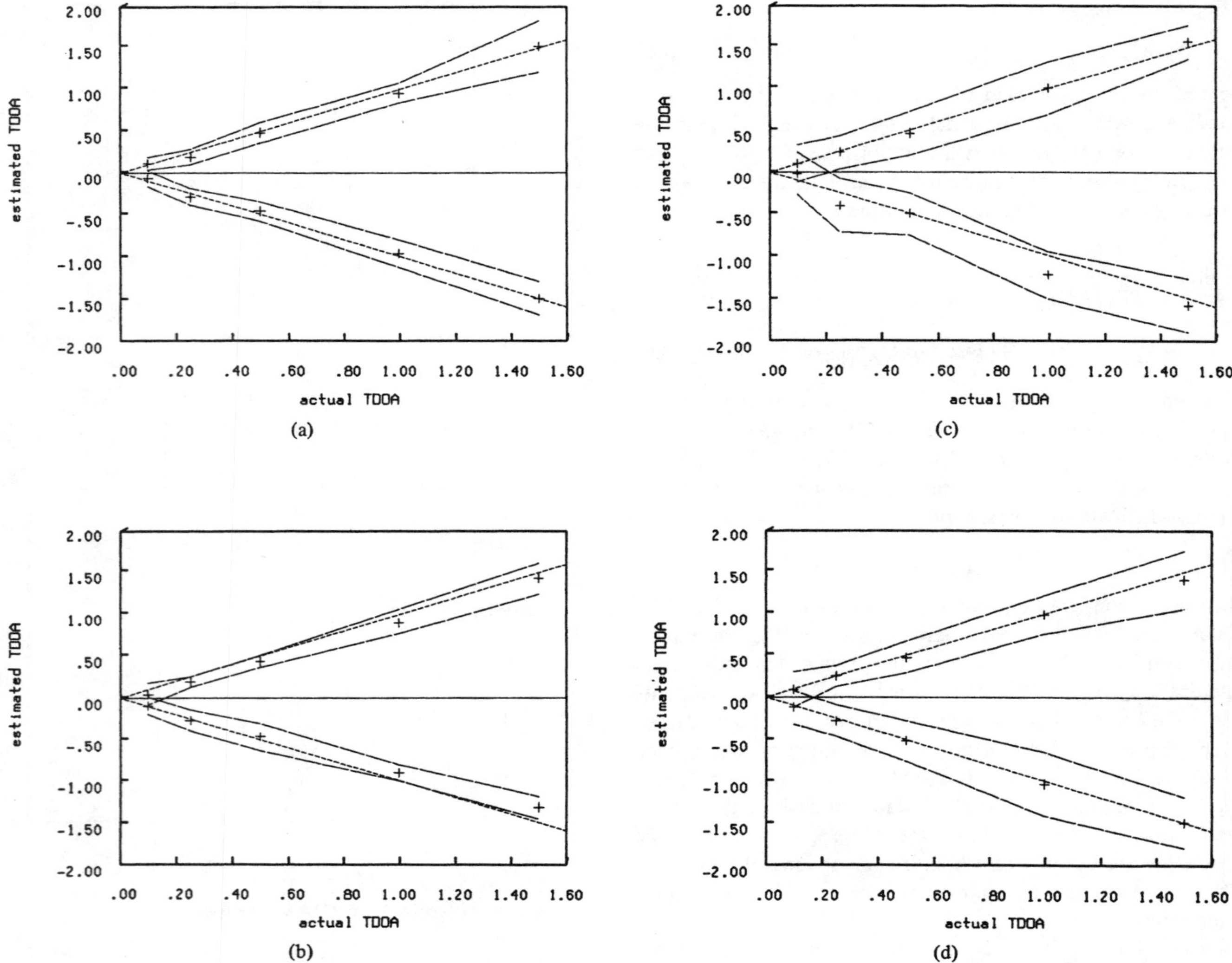

Fig. 5. (a) Direct method, uncorrelated white noise, SNR = 0 dB. (b) Direct method, correlated colored noise, SNR = 0 dB. (c) Residue method, uncorrelated white noise, SNR = –6 dB. (d) Residue method, correlated colored noise, SNR = –6 dB.

indicated by superscripts on the mean estimates in the tables and is relatively low (not more than three cases per 20 trials). The Cramer-Rao bound investigations, for example, exhibit no ambiguities at all.

Robustness with Respect to Spatially Correlated Colored Noise: Fig. 5(a) and (b) compares the results of the direct method applied to the ARMA sources for uncorrelated [5(a)] and correlated and colored [5(b)] additive noise at an SNR of 0 dB. The TDOA estimates are plotted as a function of TDOA separation in the figures. The corresponding statistics are given in Table I(a). Figs. 5(c) and 1(d) repeat the experiments, using the residue method at an SNR of -6 dB. The statistics are given in Table I(b). The results demonstrate the usefulness of the two methods at low SNR and their robustness with respect to correlated and colored noise spectra.

Effect of Signal-to-Noise Ratio: Figs. 6 and 7 give the results for both methods applied to the ARMA sources, with correlated colored noise at SNR's of 10 and 20 dB, respectively. The statistics are given in Tables II and III. The results clearly demonstrate the improved accuracy available at high SNR's. Fig. 8 plots the TDOA estimates as a function of signal-to-noise ratio between -10 and 20 dB, while the actual TDOA's are fixed at ± 1.0. The corresponding statistics are given in Table IV. The two methods behave differently at low SNR's. The direct method becomes more biased with moderate variance, while the residue method remains unbiased but has a larger variance.

Overlapping AR Sources: Fig. 9 demonstrates the applicability of the residue method to strongly overlapping sources at low SNR's and provides a comparison with [17]. The experiments are conducted at an SNR of -6 dB with white and uncorrelated additive noise. The results are also given in Table V. Our results show comparable performance with [17], although our method requires significantly fewer computations and is applicable with no additional complexity for more complicated situations such as ARMA sources and correlated colored noise.

Cramer-Rao Bound Comparison: Figs. 10 and 11 compare the TDOA estimate accuracy to the Cramer-Rao bound for one source. The Cramer-Rao bound is computed using results in the literature (see, e.g., [26]). The source spectrum is the same as the first source in Fig. 3, and the additive noise is un-

TABLE I
TDOA ESTIMATES: ARMA SOURCES, EFFECT OF NOISE CORRELATION AND SPECTRA

(a) Direct Method, SNR = 0 dB

	Uncorrelated White Noise				Correlated Colored Noise			
	Source 1		Source 2		Source 1		Source 2	
TDOA	Mean	St. Dev.	Mean	St. Dev.	Mean	St. Dev.	Mean	St. Dev.
±1.50	-1.499(1)	0.199	1.522(1)	0.305	-1.324(1)	0.134	1.434(1)	0.190
±1.00	-0.964(1)	0.163	0.956(1)	0.118	-0.909(2)	0.099	0.903(2)	0.131
±0.50	-0.465(1)	0.116	0.483(1)	0.123	-0.464	0.164	0.438	0.078
±0.25	-0.290(2)	0.101	0.193(2)	0.091	-0.262	0.125	0.190	0.060
±0.10	-0.070	0.099	0.117	0.074	-0.089	0.107	0.047	0.130

(b) Residue Method, SNR = -6 dB

	Uncorrelated White Noise				Correlated Colored Noise			
	Source 1		Source 2		Source 1		Source 2	
TDOA	Mean	St. Dev.	Mean	St. Dev.	Mean	St. Dev.	Mean	St. Dev.
±1.50	-1.585	0.312	1.566	0.203	-1.498	0.306	1.410	0.342
±1.00	-1.228	0.280	1.000	0.321	-1.038(1)	0.389	0.992(1)	0.231
±0.50	-0.504	0.247	0.461	0.262	-0.514	0.239	0.480	0.185
±0.25	-0.398	0.317	0.238	0.197	-0.273	0.184	0.249	0.122
±0.10	-0.016	0.253	0.107	0.219	-0.113	0.199	0.091	0.213

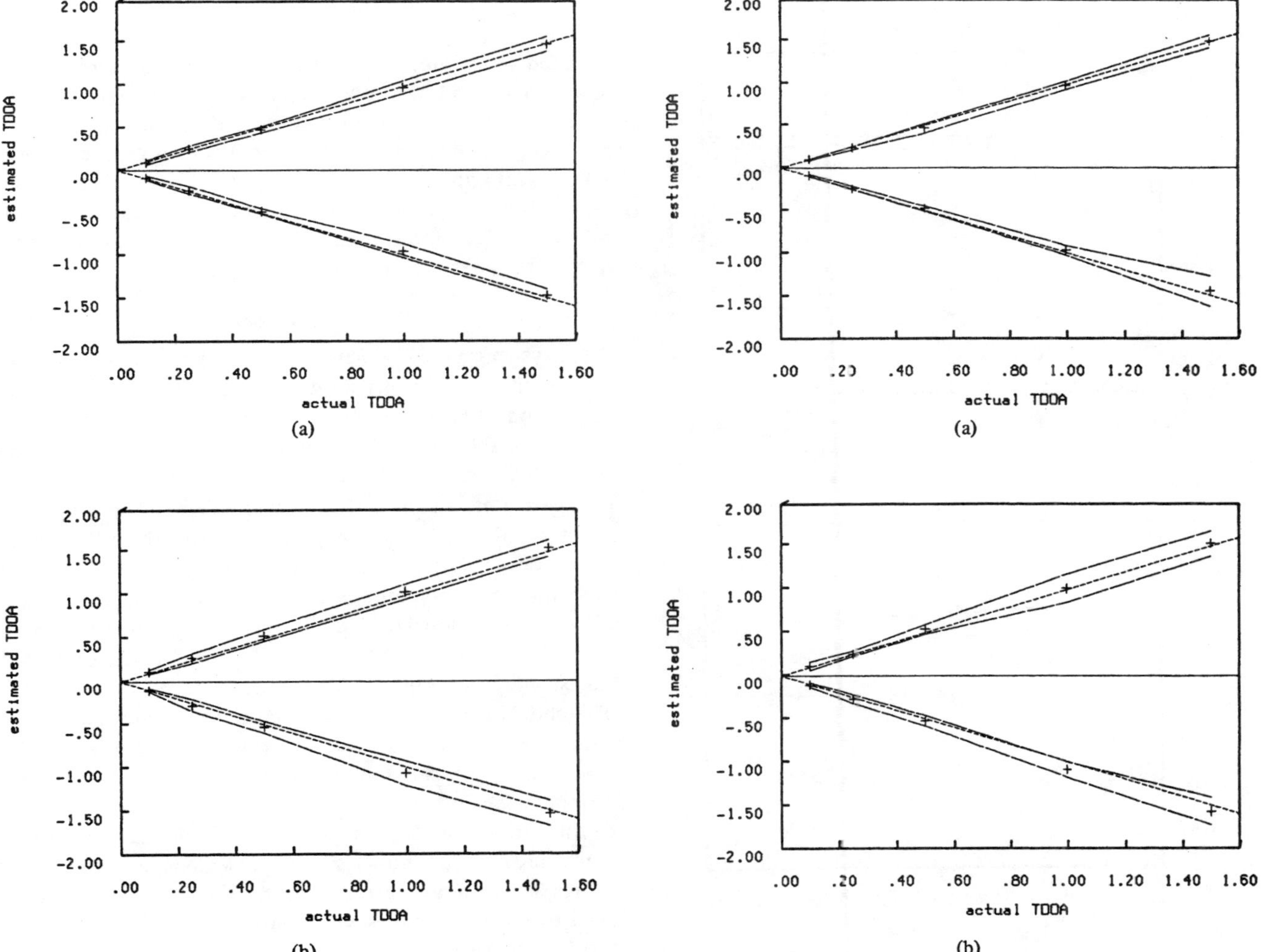

Fig. 6. TDOA estimates: ARMA sources, correlated colored noise, SNR = 10 dB. (a) Direct method. (b) Residue method.

Fig. 7. TDOA estimates: ARMA sources, correlated colored noise, SNR = 20 dB. (a) Direct method. (b) Residue method.

TABLE II
TDOA ESTIMATES: ARMA SOURCES AND CORRELATED COLORED NOISE, SNR = 10 dB

	Direct Method				Residue Method			
	Source 1		Source 2		Source 1		Source 2	
TDOA	Mean	St. Dev.	Mean	St. Dev.	Mean	St. Dev.	Mean	St. Dev.
±1.50	−1.470	0.069	1.495	0.083	−1.532	0.146	1.543	0.100
±1.00	−0.948	0.082	0.995	0.075	−1.070	0.137	1.040	0.090
±0.50	−0.481(1)	0.031	0.489(1)	0.038	−0.528	0.067	0.539	0.066
±0.25	−0.235	0.039	0.252	0.041	−0.272	0.065	0.272	0.053
±0.10	−0.095	0.025	0.092	0.022	−0.099	0.022	0.115	0.024

TABLE III
TDOA ESTIMATES: ARMA SOURCES AND CORRELATED COLORED NOISE, SNR = 20 dB

	Direct Method				Residue Method			
	Source 1		Source 2		Source 1		Source 2	
TDOA	Mean	St. Dev.	Mean	St. Dev.	Mean	St. Dev.	Mean	St. Dev.
±1.50	−1.451	0.183	1.507	0.069	−1.564	0.161	1.531	0.149
±1.00	−0.969	0.057	0.990	0.049	−1.087	0.088	1.016	0.154
±0.50	−0.478	0.028	0.468	0.059	−0.521	0.064	0.543	0.059
±0.25	−0.237	0.018	0.239	0.018	−0.267	0.051	0.254	0.031
±0.10	−0.094	0.009	0.099	0.008	−0.108	0.024	0.115	0.046

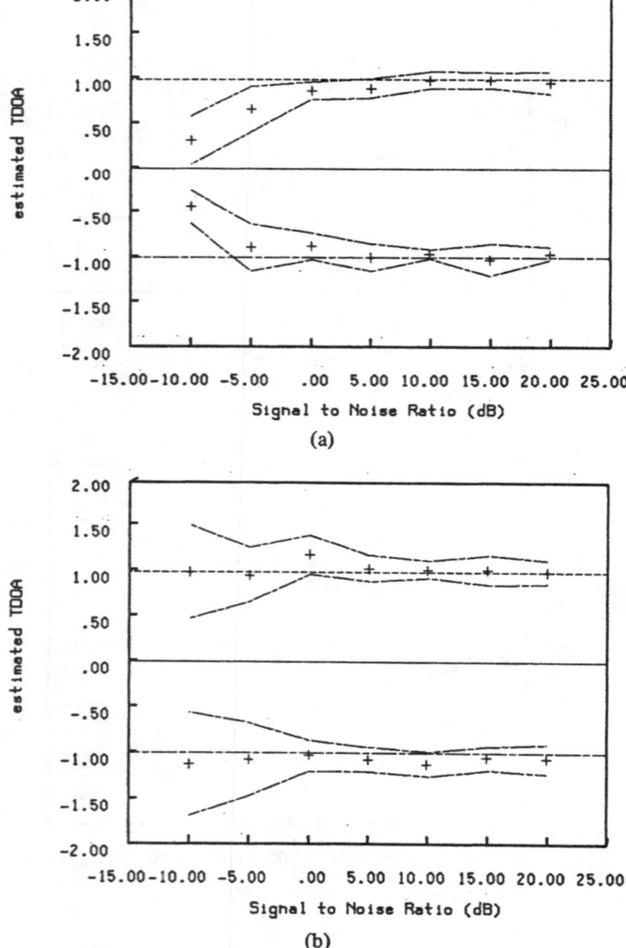

Fig. 8. TDOA estimates as a function of SNR, ARMA sources, correlated colored noise, TDOA = ±1.0. (a) Direct method. (b) Residue method.

correlated and white. In Fig. 10 the source spectra are assumed known; in Fig. 11 the source spectra are estimated. A noteworthy feature of the comparisons is that there is very little difference in performance whether the source spectrum is known or unknown, except at very low SNR's. The residue method has somewhat lower variance for both cases. At moderate SNR's, the standard deviations of the estimates are two to four times the Cramer–Rao bound.

VI. CONCLUSION

We have presented a new TDOA estimation method for unknown multiple ARMA sources and correlated receiver noise. The method utilizes the receiver cross spectra near the source poles to extract the TDOA information. Two algorithms were presented to estimate the TDOA's, one using direct receiver cross spectra estimates, and the other using residue estimates. The main advantage of the proposed algorithms lies in the computational simplicity for complicated situations. Both the direct and residue methods are independent of the source-receiver attenuation factors, as well as the source and noise zeros, so that they are equally applicable for AR or ARMA sources, white or correlated MA noise.

The choice between the direct and residue methods depends on the practical case considered. While the residue method is more accurate for overlapping source spectra, the direct method is computationally more efficient.

Results from simulations demonstrated the applicability of the methods for unknown sources, low SNR, and robustness with respect to noise characteristics. The residue method was shown to separate sources which have very close poles and low SNR. Comparisons to the Cramer–Rao bound were also given to facilitate comparisons to other methods, for example, the generalized correlation method of Knapp and Carter [4]. Note, however, that our method is applicable for multiple

TABLE IV
TDOA ESTIMATES AS A FUNCTION OF SNR, ARMA SOURCES, AND CORRELATED COLORED NOISE, TDOA = ±1.0

SNR (dB)	Direct Method Source 1 Mean	Direct Method Source 1 St. Dev.	Direct Method Source 2 Mean	Direct Method Source 2 St. Dev.	Residue Method Source 1 Mean	Residue Method Source 1 St. Dev.	Residue Method Source 2 Mean	Residue Method Source 2 St. Dev.
20	$-0.964^{(1)}$	0.070	$0.966^{(1)}$	0.120	−1.071	0.171	1.001	0.122
15	$-1.030^{(2)}$	0.171	$0.994^{(2)}$	0.086	−1.057	0.130	1.027	0.160
10	−0.967	0.050	0.994	0.093	−1.127	0.139	1.032	0.092
5	$-1.008^{(1)}$	0.148	$0.907^{(1)}$	0.106	−1.076	0.138	1.044	0.142
0	−0.875	0.148	0.878	0.096	−1.033	0.163	1.198	0.217
−5	$-0.890^{(3)}$	0.256	$0.676^{(3)}$	0.256	−1.074	0.402	0.972	0.298
−10	$-0.439^{(3)}$	0.185	$0.328^{(3)}$	0.272	$-1.128^{(2)}$	0.565	$1.003^{(2)}$	0.519

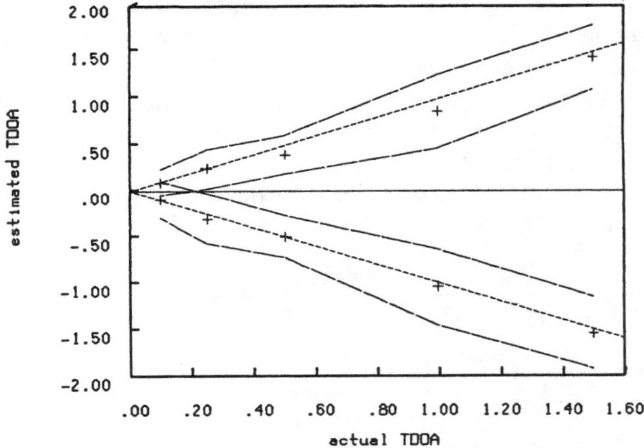

Fig. 9. Residue method performance with overlapping AR sources, SNR = -6 dB.

TABLE V
TDOA ESTIMATES: OVERLAPPING AR SOURCES, RESIDUE METHOD, SNR = −6 dB

TDOA	Source 1 Mean	Source 1 St. Dev.	Source 2 Mean	Source 2 St. Dev.
±1.50	$-1.540^{(1)}$	0.389	$1.444^{(1)}$	0.347
±1.00	$-1.046^{(2)}$	0.415	$0.870^{(2)}$	0.392
±0.50	−0.495	0.225	0.399	0.205
±0.25	−0.300	0.267	0.249	0.214
±0.10	−0.093	0.198	0.096	0.143

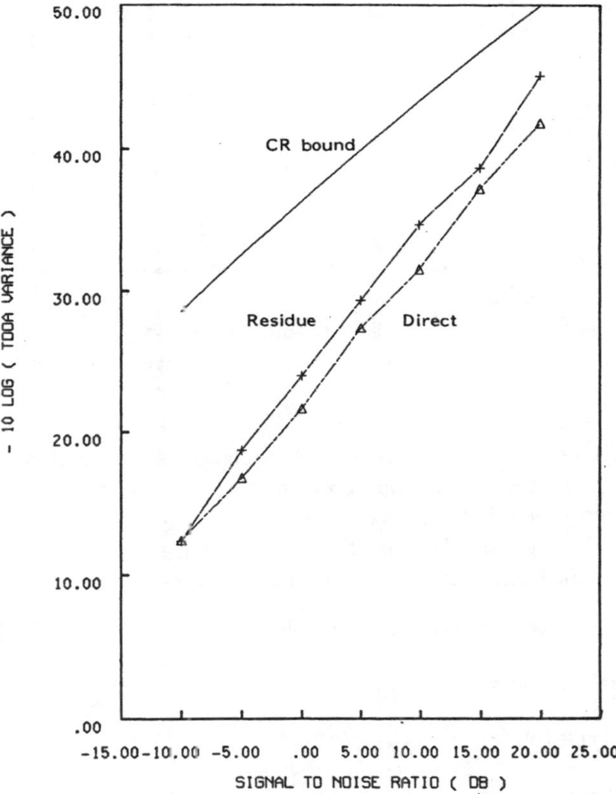

Fig. 10. Cramer–Rao bound comparison, source spectra known.

unknown sources, unlike most other time delay estimation methods. The simulations show that the performance of the proposed technique is in the same order of magnitude as the Cramer-Rao bound. Furthermore, the residue method is asymptotically exact.

We have also introduced a new ARMA array model for the practical situations of noninteger delays. The model relates the continuous time rational model of the source signals to the discrete sensor outputs. It is generally applicable for source location problems and useful for wide-band ARMA signals passed through rationally modeled filters. This model was used here to verify the TDOA formula for noninteger delays.

Our TDOA estimation approach is also useful for sources which are not necessarily of ARMA type, provided it is applied at peaks of cross spectra at points in the unit circle where one source dominates.

The pole decomposition method can be considered a new multivariable spectral estimation technique. More analysis is needed for comparison of this technique with other spectral estimation methods.

Finally, it is worth emphasizing the two main features presented in this paper which are different from other TDOA estimation procedures 1) the use of cross spectra inside the unit circle, and 2) the use of a new noninteger delay model based on rational modeling. The results are promising and useful for future research on the source location problem.

APPENDIX A
THE MULTICHANNEL ARMA ARRAY MODEL

The multichannel ARMA array model, first suggested in [16] for the source location problem, can be written as the

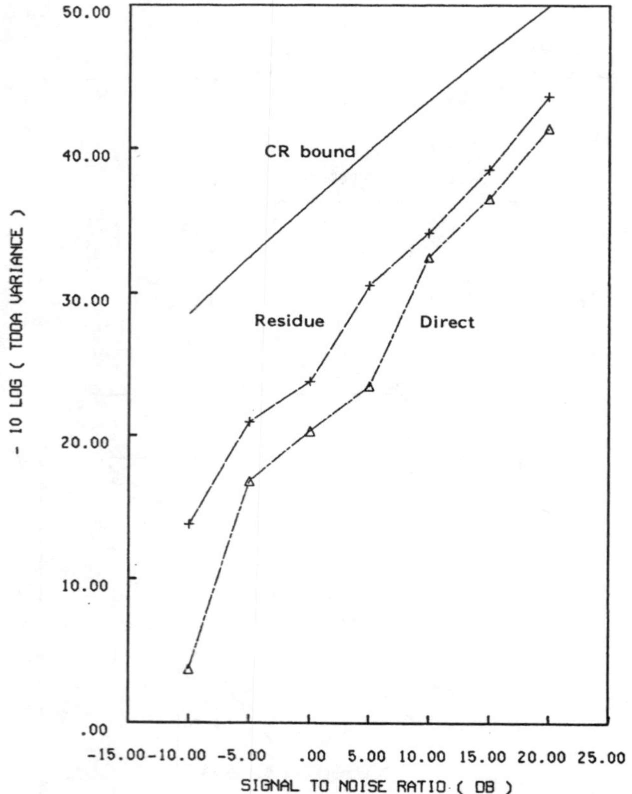

Fig. 11. Cramer–Rao bound comparison, source spectra unknown.

matrix version of the source-receiver relations from Section II. In this way, we relate our problem formulation to [16], as well as other general estimation methods.

Combining the relations of Section II to their matrix form, the z-transform of the vector of sensor signals can be written

$$Y(z) = N(z)C(z)D^{-1}(z)W(z) + V(z) \tag{A1}$$

where

$$Y(z) = [y_1(z), \cdots, y_p(z)]^T$$

$$W(z) = [w_1(z), \cdots, w_m(z)]^T$$

$$V(z) = [v_1(z), \cdots, v_p(z)]^T$$

$$C(z) = \text{diag}\{c_k(z)\} \quad m \times m$$

$$D(z) = \text{diag}\{d_k(z)\} \quad m \times m$$

$$N(z) = \begin{bmatrix} G_{11}z^{-\tau_{11}} & \cdots & G_{1m}z^{-\tau_{1m}} \\ \vdots & & \vdots \\ G_{p1}z^{-\tau_{p1}} & \cdots & G_{pm}z^{-\tau_{pm}} \end{bmatrix} \quad p \times m.$$

The spectral matrix of the received signals is

$$S(z) = N(z)C(z)D^{-1}(z)D^{-H}(1/z^*)C^H(1/z^*)N^H(1/z^*) + S^v(z) \tag{A2}$$

where H denotes the Hermitian transpose, $D^{-H}(z) \triangleq [D(z)^H]^{-1}$, and $S^v(z)$ is the receiver noise spectral matrix.

The relation (A1) suggests that an appropriate multivariable system identification algorithm can be used to estimate the source-receiver transfer function. This transfer function from the source inputs to the receivers, in terms of the physical parameters, is of right matrix fraction description type. While available algorithms are of left matrix fraction description, we recently designed an algorithm of this special structure for the given input case in [27] and [28] which can be applied to the present problem after some modifications.

Appendix B
Proof of (6) for Sources with Common Poles

We consider the case where some of the sources have a common pole and their transfer functions are not necessarily proper. Denote the common pole by λ_0. We show below that (6) is valid for λ_0 and the source whose corresponding order is the highest among the others, providing no other source has the same multiplicity.

The Laurent expansion of each of the source transfer functions about the pole λ_0 can be written

$$\frac{c_k(z)}{d_k(z)} = \frac{\psi_k(\lambda_0)}{(z-\lambda_0)^{m_k}} + \frac{\psi_k'(\lambda_0)}{(z-\lambda_0)^{m_k-1}} + \cdots + \frac{\psi_k^{(m_k-1)}(\lambda_0)}{(m_k-1)!}\,\frac{1}{z-\lambda_0} + \sum_{l=m_k}^{\infty} \frac{\psi_k^l(\lambda_0)}{l!}(z-\lambda_0)^{l-m_k}$$

where

$$\psi_k(z) = (z-\lambda_0)^{m_k}\,\frac{c_k(z)}{d_k(z)}$$

and m_k is the order of the pole λ_0 for the source k.

Assume now that among all the sources, the $\bar{k}$th source has highest order for the pole λ_0. Similarly to the proof of (6), upon multiplying the corresponding denominator and numerator by $(z-\lambda_0)^{m_{\bar{k}}}$ in the limit, one obtains

$$\lim_{z\to\lambda_0} \frac{S_{ij}(z)}{S_{ij}^*(1/z^*)} = \lim_{z\to\lambda_0} \frac{z^{-\Delta_{ij\bar{k}}}\psi_{\bar{k}}(\lambda_0)\dfrac{c_{\bar{k}}^*(1/z^*)}{d_{\bar{k}}^*(1/z^*)}}{z^{+\Delta_{ij\bar{k}}}\psi_{\bar{k}}(\lambda_0)\dfrac{c_{\bar{k}}^*(1/z^*)}{d_{\bar{k}}^*(1/z^*)}} = \lambda_0^{-2\Delta_{ij\bar{k}}},$$

which proves the above assertion.

References

[1] G. C. Carter, Ed., *IEEE Trans. Acoust., Speech, Signal Processing*, Special Issue on Time Delay Estimation, vol. ASSP-29, June 1981.

[2] R. O. Schmidt, "A new approach to the geometry of range difference location," *IEEE Trans. Aerosp. Electron. Syst.*, vol. AES-8, Nov. 1972.

[3] A. Nehorai and M. Morf, "Estimation of time differences of arrival for multiple ARMA sources by a pole decomposition method," in *Proc. 21st IEEE Conf. Decision Contr.*, Orlando, FL, Dec. 1982, pp. 1000-1002.

[4] C. H. Knapp and G. C. Carter, "The generalized correlation method for estimation of time delay," *IEEE Trans. Acoust., Speech, Signal Processing*, vol. ASSP-24, pp. 320-327, Aug. 1976.

[5] Y. T. Chan, J. M. Riley, and J. B. Plant, "A parameter estimation approach to time-delay estimation and signal detection," *IEEE Trans. Acoust., Speech, Signal Processing*, vol. ASSP-28, pp. 8-16, Feb. 1980.

[6] E. J. Hannan and P. J. Thomson, "Delay estimation and the estimation of coherence and phase," *IEEE Trans. Acoust., Speech, Signal Processing*, vol. ASSP-29, pp. 485-490, June 1981.

[7] J. B. Lewis and P. M. Schultheiss, "Optimum and conventional detection using a linear array," *J. Acoust. Soc. Amer.*, vol. 49, pp. 1083-1091, 1971.

[8] R. A. Monzingo and T. W. Miller, *Introduction to Adaptive Arrays.* New York: Wiley, 1980.

[9] F. C. Schweppe, "Sensor-array data processing for multiple-signal sources," *IEEE Trans. Inform. Theory*, vol. IT-14, pp. 294-305, Mar. 1968.

[10] J. Capon, "High resolution frequency-wavenumber spectrum analysis," *Proc. IEEE*, vol. 57, pp. 1408-1418, Aug. 1969.

[11] S. S. Reddi, "Multiple source location–A digital approach," *IEEE Trans. Aerosp. Electron. Syst.*, vol. AES-15, pp. 95-105, Jan. 1979.

[12] R. O. Schmidt, "Multiple signal classification (MUSIC)," ESL, Inc. Tech. Memo. TM-1098, Mar. 1979. See also "A signal subspace approach to multiple emitter location and spectral estimation," Ph.D. dissertation, Dep. Elec. Eng., Stanford Univ., Stanford, CA, Nov. 1981.

[13] M. Coker and E. Ferrara, "A new method for multiple source location," in *Proc. IEEE Conf. Acoust., Speech, Signal Processing*, Paris, France, May 3-5, 1982.

[14] G. Su and M. Morf, "The signal subspace approach for multiple wide band emitter location," *IEEE Trans. Acoust., Speech, Signal Processing,* this issue, pp. 1502-1522. See also *Proc. 16th Asilomar Conf. Circuits, Syst., Comput.*, Pacific Grove, CA, Nov. 1982.

[15] M. Wax, T. J. Shan, and T. Kailath, "Location and special density estimation of multiple sources," in *Proc. 16th Asilomar Conf. Circuits, Syst., Comput.*, Pacific Grove, CA, Nov. 1982, pp. 332-336.

[16] M. Morf *et al.*, "Investigation of new algorithms for locating and identifying spatially distributed sources and receivers," Tech. Summary Rep. to DARPA, SEL Rep. M355-1, Mar. 1979.

[17] B. Porat and B. Friedlander, "Estimation of spatial and spectral parameters of multiple sources," *IEEE Trans. Inform. Theory*, vol. IT-29, pp. 412-425, May 1983.

[18] J. A. Cadzow, "High performance spectral estimation–A new ARMA method," *IEEE Trans. Acoust., Speech, Signal Processing*, vol. ASSP-28, pp. 524-529, Oct. 1980.

[19] G. Su, "Signal subspace algorithms for emitter location and multidimensional spectral estimation," Ph.D. dissertation, Dep. Elec. Eng., Stanford Univ., Stanford, CA, Oct. 1983.

[20] R. V. Churchill, *Complex Variables and Applications.* New York: McGraw-Hill, 1960.

[21] A. V. Oppenheim and R. W. Schafer, *Digital Signal Processing.* Englewood Cliffs, NJ: Prentice-Hall, 1975.

[22] S. M. Kay and S. L. Marple, "Spectrum analysis–A modern perspective," *Proc. IEEE*, vol. 69, pp. 1380-1419, Nov. 1981.

[23] H. Akaike, "A new look at the statistical model identification," *IEEE Trans. Automat. Contr.*, vol. AC-19, pp. 716-723, Dec. 1974.

[24] J. R. Ragazzini and G. F. Franklin, *Sampled-Data Control Systems.* New York: McGraw-Hill, 1958.

[25] R. H. Barker, "The pulse transfer function and its application to sampling servo systems," *Proc. IEE*, Part IV, Monog. 43, pp. 302-317, July 1952.

[26] E. Weinstein, "Optimal source localization and tracking from passive array measurements," *IEEE Trans. Acoust., Speech, Signal Processing*, vol. ASSP-30, pp. 69-76, Feb. 1982.

[27] A. Nehorai, "Algorithms for system identification and source location," Ph.D. dissertation, Dep. of Elec. Eng., Stanford Univ., Stanford, CA, June 1983.

[28] A. Nehorai and M. Morf, "Recursive identification algorithms for right matrix fraction description models," *IEEE Trans. Automat. Contr.*, to be published.

TIME DELAY ESTIMATION IN UNKNOWN GAUSSIAN SPATIALLY CORRELATED NOISE

Chrysostomos L. Nikias and Renlong Pan

Communications and Digital Signal Processing (CDSP)
Center for Research and Graduate Studies
Department of Electrical and Computer Engineering
Northeastern University
Boston, MA 02115

ABSTRACT

A new class of methods that estimate the difference in arrival time between two signals corrupted by spatially correlated Gaussian noise sources of unknown cross-correlation is presented. The methods are based on the idea of "comparing the similarities" bewteen the two sensor measurements in higher-order spectrum domains (bispectrum) rather than in the cross-correlation domain. It is demonstrated that the time delay estimation techniques based on higher-order spectra suppress the effect of correlated Gaussian noise sources and therefore exhibit improved performance over generalized cross-correlation methods. Studies are shown for different lengths of data records and signal-to-noise ratios (SNRs).

I. INTRODUCTION

Let us assume that $\{X(n)\}$ and $\{Y(n)\}$ are two sensor measurements satisfying

$$X(n) = S(n) + W_1(n) \tag{1.1}$$

$$Y(n) = S(n - D) + W_2(n) \tag{1.2}$$

where $\{S(n)\}$ is an unknown signal and $\{S(n-D)\}$ is the same signal, delayed (or advanced) in time, and $\{W_1(n)\}$, $\{W_2(n)\}$ are unknown noise sources. The problem is to find from finite length records $\{X(n)\}$ and $\{Y(n)\}$ an estimate of the time delay D. This problem arises in many application fields such as sonar (active or passive), radar, biomedicine, geophysics, etc. [1]-[8].

The basic approach to the solution of the time delay estimation problem is to shift X(n) with respect to Y(n) and compare similarities between the two records at each shift. Best match will occur at a shift equal to D. Assuming that the noise sources are zero-mean independent stationary random processes, the fundamental operation adopted to "compare similarities" between X(n) and Y(n) is the cross-correlation which ideally is

$$r_{xy}(\tau) = E\{X(n)\ Y(n+\tau)\} = r_{ss}(\tau-D), \quad -\infty<\tau<\infty \tag{2.1}$$

$$\text{where } r_{ss}(\tau) = E\{S(n)\ S(n+\tau)\} \tag{2.2}$$

the $r_{xy}(\tau)$ peaks at $\tau = D$. However, in practical application problems due to finite length data records and not exactly independent noise sources the $r_{xy}(\tau)$ does not necessarily show a peak at the time delay position. Various window functions (prefilters) have been suggested to better shape the cross-correlation function via convolution operation, namely ROTH, SCOT, PHAT, Eckart, Hannan-Thompson to name a few [1]-[5]. Parameter estimation approaches have also been suggested for time delay estimation based on autocorrelations and cross-correlations such as the least-squares and Wiener filtering methods [2],[3]. Each one of the aforementioned techniques has certain advantages and limitations depending upon the nature of the signal and noise sources.

In practical application problems such as active sonar where the signal $\{S(n)\}$ can be regarded as non-Gaussian stationary random process, and the noise sources $\{W_1(n)\}$, $\{W_2(n)\}$ zero-mean stationary Gaussian, then the similarities between $\{X(n)\}$ and $\{Y(n)\}$ could also be "compared" in higher-order spectrum domains such as the bispectrum or trispectrum [7]-[8]. Higher-order spectra, which are defined in terms of higher-order cumulants, have been given a lot of attention lately due to their ability to preserve information of non-Gaussian stationary random processes. The definitions of cumulants and higher-order spectra as well as their general motivations in digital signal processing problems can be found in the tutorial review paper [9].

One of the fundamental properties of higher-order spectra which will be explored in this paper is the fact that for Gaussian processes only, all polyspectra of order greater than two are identically zero (in theory). Hence, in those signal processing settings where the signal is non-Gaussian stationary process and the additive noise to the process is stationary Gaussian, there might be certain advantages estimating signal parameters in higher-order spectrum domains. In the time delay estimation problem described by (1.1) and (1.2) if the noise sources are <u>correlated</u> Gaussian, then the cross-correlation techniques fail to work well because they estimate the cross-correlation of both the signals and noises. On the other hand, "comparing similarities" of $\{X(n)\}$ and $\{Y(n)\}$ in higher-order spectrum domains the effect of correlated Gaussian noises is completely suppressed. The bispectrum domain also suppresses non-Gaussian noises with symmetric probability density function.

II. TIME DELAY IN THE BISPECTRUM DOMAIN

Let $\{S(n)\}$ be zero-mean, non-Gaussian stationary random process with a non-zero measure of skewness, i.e., $E\{S^3(n)\} \neq 0$. Assuming that the two available

This work was supported by the Office of Naval Research (ONR) under contract N00014-86-K-0219.

Reprinted from *Proc. ICASSP '88*, vol. 5, pp. 2638–2641, April 1988.

data sequences from sensor measurements are $\{X(n)\}$, $\{Y(n)\}$ described by (1.1) and (1.2) and $\{W_1(n)\}$ and $\{W_2(n)\}$ are zero-mean, correlated stationary random processes, statistically independent of $S(n)$. The cross-correlation of the noise sources

$$r_{12}(\tau) = E\{W_1(n)\, W_2(n+\tau)\} \tag{3}$$

is assumed to be unknown. Let us note that these signal conditions usually arise in active sonar problems. Based on these assumptions, the cross-correlation of $\{X(n)\}$, $\{Y(n)\}$ is given by

$$r_{xy}(\tau) = r_{ss}(\tau - D) + r_{12}(\tau) \quad -\infty < \tau < \infty \tag{4}$$

and therefore all the cross-correlation based methods will generally fail to work for this problem. On the other hand, the following relationships hold in the third moment domain for the data given by (1) and the assumptions described above.

$$R_{xxx}(\tau,\rho) = E\{X(n)\, X(n+\tau)\, X(n+\rho)\} = R_{sss}(\tau,\rho) \tag{5.1}$$

$$R_{xyx}(\tau,\rho) = E\{X(n)\, Y(n+\tau)\, X(n+\rho)\} = R_{sss}(\tau - D,\rho) \tag{5.2}$$

where $R_{sss}(\tau,\rho) = E\{S(n)\, S(n+\tau)\, S(n+\rho)\}$.

This is because the third moment sequence of a zero-mean Gaussian process is identical to zero [9].

The bispectrum is by definition the Fourier transform of the third moment sequence, viz: [8]

$$B_{xxx}(\omega_1,\omega_2) = FT\{R_{xxx}(\tau,\rho)\} = B_{sss}(\omega_1,\omega_2) \tag{6.1}$$

$$B_{xyx}(\omega_1,\omega_2) = FT[R_{xyx}(\tau,\rho)] = B_{sss}(\omega_1,\omega_2)\, e^{j\omega_1 D} \tag{6.2}$$

where $FT[\cdot]$ denotes the 2-D Fourier transform operation.

A. Conventional Methods

If we write

$$B_{xxx}(\omega_1,\omega_2) = |B_{xxx}(\omega_1,\omega_2)| \exp j\, \psi_{xxx}(\omega_1,\omega_2) \tag{7.1}$$

$$B_{xyx}(\omega_1,\omega_2) = |B_{xyx}(\omega_1,\omega_2)| \exp j\, \psi_{xyx}(\omega_1,\omega_2) \tag{7.2}$$

then we have the following ad-hoc equivalent expressions which can be used as generally different time-delay estimation methods in practice:

Method I: If $\phi(\omega_1,\omega_2) \triangleq \psi_{xyx}(\omega_1,\omega_2) - \psi_{xxx}(\omega_1,\omega_2)$ (8.1)

and define $I(\omega_1,\omega_2) \triangleq \exp j\, \phi(\omega_1,\omega_2)$, (8.2)

from (6), (7) and (8), it follows that

$$I(\omega_1,\omega_2) = \exp j\, (\omega_1 D) \tag{9}$$

and therefore

$$T(\tau) = \int \int_{-\pi}^{+\pi} I(\omega_1,\omega_2)\, e^{-j\omega_1\tau}\, d\omega_1\, d\omega_2 \tag{10}$$

peaks at $\tau = D$.

Method II: [6],[7] $I(\omega_1,\omega_2) = \dfrac{B_{xyx}(\omega_1,\omega_2)}{B_{xxx}(\omega_1,\omega_2)}$ (11)

Method III:

$$I(\omega_1,\omega_2) = \frac{|B_{xyx}(\omega_1,\omega_2)| \exp j\, \phi(\omega_1,\omega_2)}{\sqrt{|B_{xxx}(\omega_1,\omega_2)|}\ \sqrt{|B_{yyy}(\omega_1,\omega_2)|}} \tag{12}$$

In practice we are always given finite length records of data and therefore the bispectrum estimation procedures described in [9] have to be employed first in order to obtain estimates of $I(\omega_1,\omega_2)$.

B. Parametric Methods

From (1.1) and (1.2) we can always write

$$Y(n) = \sum_{i=-\infty}^{+\infty} a_i\, X(n-i) + W_2(n) - W_1(n-D) \tag{13}$$

where in theory $a_i = 0$ for all $\{i\}$ except $i = D$ and $a_D = 1$. In practice D is always finite and therefore

$$Y(n) = \sum_{i=-P}^{+P} a_i\, X(n-i) + W_2(n) - W_1(n-D) \tag{14}$$

where P is the largest possible delay we can expect. By multiplying both sides of (14) by $X(n)$ and $X(n+\rho)$ and taking expectations we get

$$E(\tau,\rho) = R_{xyx}(\tau,\rho) - \sum_{i=-P}^{+P} a_i\, R_{xxx}(\tau - i,\rho) = 0 \tag{15}$$

because $R_{xw_2x}(\tau,\rho)$ and $R_{xw_1x}(\tau-D,\rho)$ are identical to zero due to the fact that the noise sources are zero-mean Gaussian and independent from the signal. Estimates of $\{a_i\}$ are obtained by minimizing the index $\Sigma\Sigma[E(\tau,\rho)]^2$. The time delay D is chosen to be the index of the parameter $\{a_i\}$ which has maximum value. However, in the case where the time delay is non-integer then interpolation is performed between the parameters to find the time instant at which the function is maximum.

The computational algorithms that correspond to the conventional and parametric time delay estimation techniques are described in [10].

III. SIMULATION RESULTS

Here we present some computer simulation results we have conducted to demonstrate the effectiveness of the time delay bispectrum estimation techniques. Specifically, the conventional bispectrum method I as well as the parametric method described in the previous section are employed in the simulations. For comparison purposes, three cross-correlation based methods, i.e., ROTH, SCOT, and ML [3],[4] have also been employed in exactly the same simulations. Throughout these simulation experiments, we made the data 50% overlapped when it is segmented into records, as proposed in [1], and have chosen the time delay D = 16.

Three different lengths of data have been used for the simulation examples: 64x128, 128x128 and 256x128 (128 samples per record). The signal is zero-mean, non-Gaussian color process with non-zero skewness and the two additive noise sources are also zero-mean Gaussian, uncorrelated with the signal but spatially correlated with each other [10].

Figures 1-6 illustrate the results obtained by the conventional and parametric bispectrum techniques, and by the ROTH, SCOT and ML cross-correlation methods, respectively. From these figures, it is apparent that the time-delay bispectrum techniques do have the capability to suppress the Gaussian noises. As the length of the data increases, the noise suppression becomes more effective. On the other hand, as expected, the cross-correlation methods (ROTH, SCOT, ML) are incapable of suppressing the spatially correlated Gaussian noises. It is also apparent from these figures that the parametric bispectrum approach exhibits best performance. Comprehensive simulation examples can be found in [10].

REFERENCES

[1] C.H. Knapp and G.C. Carter, "The generalized correlation method for estimation of time delay," IEEE Trans. Acoust., Speech, and Signal Processing, Vol. ASSP-24, pp. 320-327, August 1976.

[2] Y.T. Chan, J.M. Riley, and J. B. Plant, "A parameter estimation approach to time-delay estimation and signal detection," IEEE Trans. Acoust., Speech, and Signal Processing , Vol. ASSP-28(1), pp. 8-16, February 1980.

[3] G.C. Carter, "Coherence and time delay estimation," Proc. IEEE, Vol. 75(2), pp. 236-255, February 1987.

[4] IEEE Trans. Acoust., Speech, and Signal Processing, Special Issue on Time Delay Estimation, Vol. 29(3), June 1981.

[5] J.P. Ianniello, "Time delay estimation via cross-correlation in the presence of large estimation errors," IEEE Trans. Acoust., Speech, and Signal Processing, Vol. ASSP-30, pp. 998-1003, December 1982.

[6] V.H. MacDonald and P.M. Schultheiss, "Optimum passive learning estimation in a spatially incoherent noise environment," J. Acoust. Soc. Amer., Vol. 46, pp. 37-43, 1969.

[7] K. Sasaki, T. Sato, and Y. Nakamura, "Holographic passive sonar," IEEE Trans. Sonics and Ultrasonics, Vol. SU-24(3), pp. 193-200, May 1977.

[8] T. Sato and K. Sasaki, "Bispectral holography," J. Acoust. Soc. Amer., Vol. 62, pp. 404-408, 1977.

[9] C.L. Nikias and M.R. Raghuveer, "Bispectrum estimation: a digital signal processing framework," Proc. IEEE, Vol. 75(7), pp. 869-891, July 1987.

[10] C.L. Nikias and R. Pan, "Time delay estimation in unknown Gaussian spatially correlated noise," IEEE Trans. Acoust., Speech, and Signal Processing, to appear, 1988.

SNR=0 DB, 50 PERCENT OVERLAPPING

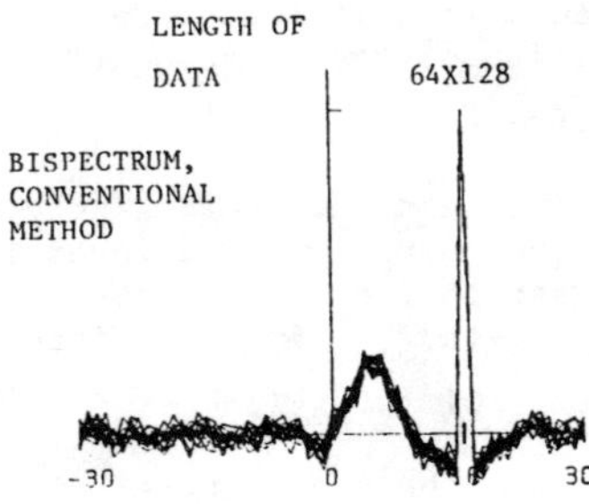

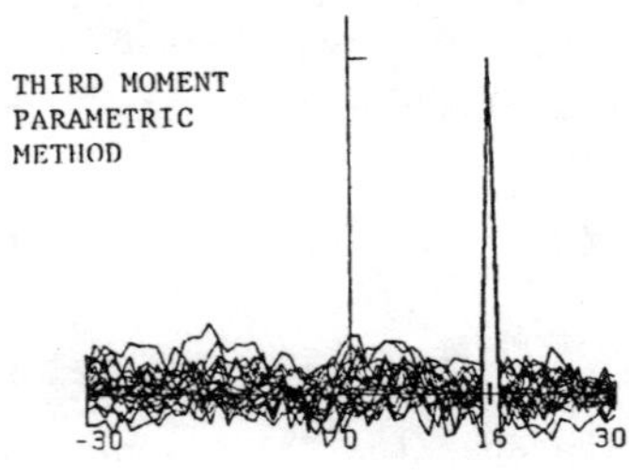

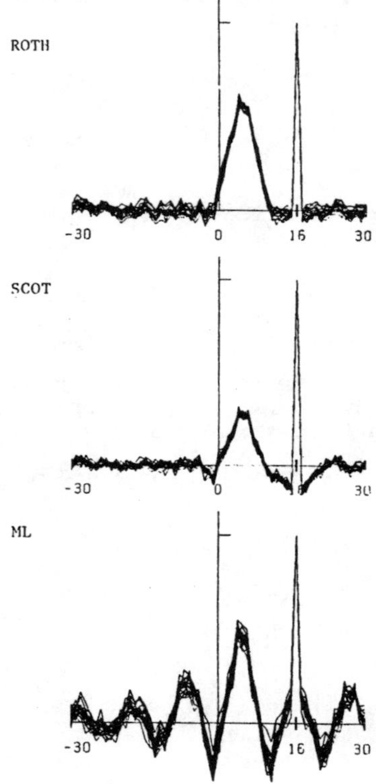

Fig.1

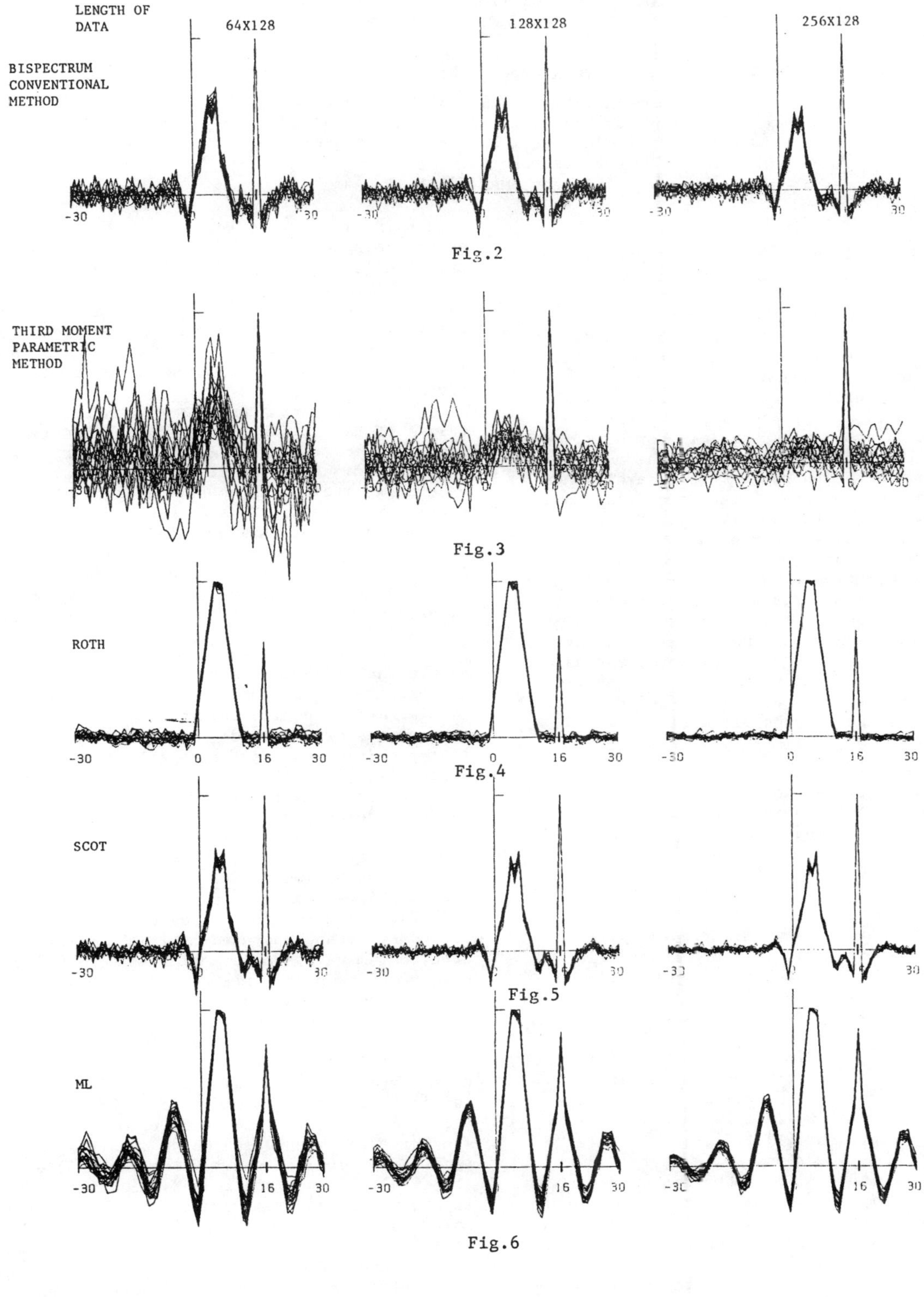

Fig.2

Fig.3

Fig.4

Fig.5

Fig.6

CUMULANT CEPSTRUM OF FM SIGNALS AND HIGH-RESOLUTION TIME DELAY ESTIMATION

Athina Petropulu, Chrysostomos L. Nikias and John G. Proakis

Communications and Digital Signal Processing (CDSP)
Center for Research and Graduate Studies
Department of Electrical and Computer Engineering
Northeastern University
Boston, MA 02115

ABSTRACT

One of the difficult sonar signal processing problems has been the case where the acoustic FM signal is embedded in reverberation noise, due to the presence of multipath and observation noise. The effects of multipath propagation have been shown to introduce biases and high variances in the bearing and range estimation of acoustic sources. It is therefore very critical to take into account multipath propagation in the time delay estimation problem and minimize or eliminate its effect. The impulse response of the multipath underwater communication channel is generally non-minimum phase and consists of the time delays and their attenuation constants. The specific problem we address in this paper is the one where the time delays are closely-spaced, i.e., the distance between two consecutive time delays is significantly less than the duration of the autocorrelation of the FM signal. We introduce in this paper a new high-resolution estimation method using the cumulant cepstra (polycepstra) of the received sensor data as the basic tool for reconstruction. The effectiveness of the method is demonstrated for different noise conditions and lengths of data.

I. INTRODUCTION

Suppose a superposition of attenuated and delayed versions of the same signal are received at a single sensor in the form

$$y(n) = \sum_{i=1}^{M} A_i\, s(n - D_i) + w(n) \qquad (1)$$

where $\{s(n)\}$ is the signal, $\{D_i\}$ are the signal time of arrivals (delays), $\{A_i\}$ are the scaling amplitudes and $\{w(n)\}$ is the additive noise component assumed to be independent of the signal. The problem is to estimate the scaling amplitudes $\{A_i\}$ and the time delays $\{D_i\}$ from a finite length record of $\{y(n)\}$. This is a multipath propagation problem that arises in active underwater propagation channels, radar, geophysics, telecommunications and in any other field where a signal and its replicas are present [1]-[15].

One of the key questions to be addressed in this paper is how to estimate the time delays $\{D_i\}$ and the scaling amplitudes $\{A_i\}$ in the case where the time delays are closely-spaced (non-resolvable), i.e., the distance between two consecutive time delays $\{D_{i+1} - D_i\}$ is less than the duration of the autocorrelation sequence of the signal $\{s(n)\}$. If the signals are separated in time by an interval greater than the length of the signal autocorrelation, then a traditional way of estimating the time delays is by using cross-correlation operations (matched filter) [1]. However, large errors are introduced in these estimates if the time delays are non-resolvable [2].

The estimation problem of non-resolvable time delays of deterministic signals in additive white noise was addressed in [3] as a nonlinear least-squares estimation problem which was partly based on cross-correlations among the signals and some of their derivatives. This method is a direct implementation of maximum likelihood (ML) and becomes computationally unattractive for practical implementations. The ML method has also been used in the case of additive color noise and known signals with resolvable and non-resolvable time delays [4],[5].

The non-resolvable time delay estimation problem has recently been addressed by transforming the maximum likelihood procedure into the frequency domain where the time delays become closely-spaced frequencies of sinusoidal signals. Specifically, in the noise-free case the spectrum of the multipath signal $\{y(n)\}$ is the spectrum of the signal $\{s(n)\}$ times a sum of cosine waves which have frequencies the time delays and their differences. Therefore, with known signal spectrum, the sequence of spectrum estimates of $\{y(n)\}$ in frequency can be regarded as a time series and the problem becomes that of high-resolution power spectrum estimation [6],[7]. Assuming noise-free deterministic signal, the Prony method was used in [8]. On the other hand, assuming random signal $\{s(n)\}$ with known power spectrum (low-pass flat) and additive white Gaussian noise, the Kumaresan-Tufts [9],[10] and autoregressive methods [6],[11] have been used.

Cepstrum analysis techniques have also been employed to solve this problem assuming that the noise $\{w(n)\}$ is a Gaussian process and the signal $\{s(n)\}$ is either deterministic or Gaussian process [12]-[14]. However, cepstrum techniques are very sensitive to observation noise when they are applied directly in the time-domain of the data [12]-[15]. On the other hand, cepstrum techniques in the power spectrum domain are less sensitive to noise but can only work for minimum phase channels [13].

Higher-order spectra, or polyspectra which are defined in terms of higher-order cumulants, have already become very useful tools in signal

This work was supported in part by the Office of Naval Research (ONR) under contract N00014-86-K-0219.

Reprinted from *Proc. ICASSP '88*, vol. 5, pp. 2642–2645, April 1988.

processing problems because of their ability to preserve the true phase character of signals and to suppress the effect of additive Gaussian noise sources. Several methodologies have been developed based on polyspectra that solve the aforementioned problem when the signal $\{s(n)\}$ is non-Gaussian white (see tutorial paper [16] and references therein). A new class of time delay estimation methods based on higher-order cumulants that suppress the effect of spatially correlated Gaussian noise sources was introduced in [17]. Finally, a new method for nonminimum phase system identification driven by non-Gaussian white noise has been developed using cumulant cepstra (bicepstrum and tricepstrum) [18].

II. PROBLEM ASSUMPTIONS AND PROPERTIES

This paper introduces a new method for estimating time delays $\{D_i\}$ and scaling amplitudes $\{A_i\}$ from a finite length record of $\{y(n)\}$ by extending the bicepstrum (cepstrum of third-order cumulants) approach originally introduced in [18]. The model and signal assumptions are the following: 1) the signal $\{s(n)\}$ is <u>known</u> FM, i.e., $s(n) = A \sin \omega_o(n)$; 2) the time delays $\{D_i\}$ are <u>non-resolvable</u>; 3) the scaling amplitudes, $\{A_i\}$, form a <u>non-minimum phase</u> FIR multipath channel; the minimum phase channel (attenuating amplitudes) is a special case; 4) the additive noise $\{w(n)\}$ is zero-mean, <u>color</u>, <u>Gaussian</u>, ergodic with <u>unknown</u> autocorrelation function and independent of the FM signal; 5) the sensor signal $\{y(n)\}$ is zero-mean (no loss of generality).

From (1) and the above assumptions the following identities hold:

$$y(n) = s(n) * h(n) + w(n) \quad \text{(time domain)} \tag{2.1}$$

where "*" denotes linear convolution and

$$h(n) = \sum_{i=1}^{M} A_i\, \delta(n - D_i) \tag{2.2}$$

is the channel impulse response,

$$P_y(\omega) = P_s(\omega) \cdot P_h(\omega) + P_w(\omega)\,, \tag{3}$$

(power spectrum domain)

$$R_y(\tau,\rho) = R_s(\tau,\rho) * R_h(\tau,\rho) \tag{4}$$

(third-order cumulant domain)

where $R_y(\tau,\rho)$, $R_s(\tau,\rho)$ and $R_h(\tau,\rho)$ are the third-order cumulants (or moments) of $\{y(n)\}$, $\{s(n)\}$ and $\{h(n)\}$, respectively. Let us note that $R_w(\tau,\rho) = 0$ (in theory) because the noise is zero-mean Gaussian. From (4) it follows that the bispectrum domain equation is

$$B_y(\omega_1,\omega_2) = B_s(\omega_1,\omega_2) \cdot B_h(\omega_1,\omega_2) \quad . \tag{5}$$

Since the FM signal $\{s(n)\}$ has pronounced resonances, the third moment sequence of $\{y(n)\}$ can be prefiltered by the operation

$$R_u(\tau,\rho) = R_e(\tau,\rho) * R_y(\tau,\rho) \tag{6.1}$$

or

$$B_u(\omega_1,\omega_2) = B_e(\omega_1,\omega_2) \cdot B_y(\omega_1,\omega_2) \tag{6.2}$$

where

$$B_e(\omega_1,\omega_2) \triangleq [B_s(\omega_1,\omega_2) * B_z(\omega_1,\omega_2)]/B_s(\omega_1,\omega_2) \tag{7}$$

to obtain a new third moment sequence $R_u(\tau,\rho)$. $B_z(\omega_1,\omega_2)$ is the bispectrum of the "window" signal $z(n) = a^n$, $0 < a < 1$. In other words, $R_u(\tau,\rho)$ is the third moment sequence of the time signal

$$u(n) = f(n) * h(n) \tag{8}$$

where

$$f(n) = z(n) \cdot s(n) \quad .$$

From (4)-(6) and [18] it follows that the bicepstrum of $\{u(n)\}$ is given by

$$c_u(m,n) = c_f(m,n) + c_h(m,n) \quad . \tag{9}$$

Let us note that $c_f(m,n)$ is assumed to be known. The bicepstrum $c_u(m,n)$ is related to $R_u(\tau,\rho)$ by [18]

$$m\, c_u(m,n) = F^{-1}\{F[mR_u(m,n)]/F[R_u(m,n)]\} \tag{10}$$

where $F[\cdot]$ denotes 2-d Fourier transform operation. Therefore, $c_h(m,n)$ can be computed by subtracting $c_f(m,n)$ from $c_u(m,n)$ and then $\{h(n)\}$ may be generated from $c_h(m,n)$ following the procedure described in [18].

III. THE HIGH-RESOLUTION METHOD

In summary, the high-resolution time delay estimation method proceeds as follows: 1) Since $s(n)$, $n = 0,1,\ldots,N-1$ is known, we generate $f(n) = a^n s(n)$ and estimate its third moment sequence $\hat{R}_f(\tau,\rho)$; using (10) with $\hat{R}_f(\tau,\rho)$ we estimate $\hat{c}_f(m,n)$ via 2-d FFT operations. 2) Given the sensor signal $y(n)$, $n = 0,1,\ldots,N-1$, we generate its third moment sequence $\hat{R}_y(\tau,\rho)$ which is prefiltered via (6.1) to generate $\hat{R}_u(\tau,\rho)$. 3) Using (10) we estimate $\hat{c}_u(m,n)$. 4) $\hat{c}_h(m,n) = \hat{c}_u(m,n) - \hat{c}_f(m,n)$. 5) We reconstruct $\{h(n)\}$ from $\hat{c}_h(m,n)$ via differential cepstrum operations as described in [18]. The scaling amplitudes and time delays follow from (2.2). Methods that estimate third-order moments are described in [16]; however, no data segmentation is performed here because $s(n)$ is a deterministic signal.

IV. SIMULATION RESULTS

We report Monte-Carlo simulation results in this section in order to demonstrate the ability of the bicepstrum technique in resolving closely-spaced time delays that arise in an underwater multipath channel. The FM signal used in the simulations is of the form [10] $s(n) = \cos[2\pi(6.28 \cdot 10^{-3})n^2]$, $n = 0, 1, \ldots, N-1$ and the multipath channel impulse response $h(n) = \delta(n) + 0.5\,\delta(n-3)$. Thus the distance between the two time delays $|D_3 - D_0| = 3$ which is significantly less than the length of the signal autocorrelation. The additive noise $\{w(n)\}$ (see eq. (2.1)) is zero-mean color Gaussian and is generated by passing a white noise process $N(0,Q)$ through a linear filter such that the autocorrelation of the noise output is either $r_w(m) = (0.8)^m$ (case I) or $r_w(m) = (0.8)^m \cdot \cos(0.56m)$ (case II). The signal-to-noise ratio (SNR) in our experiments is defined as SNR $\triangleq 10 \log_{10}(E_s/Q)$ where E_s is the total power of the signal $f(n)$ (see eq. (8)). The "window" signal is chosen $z(n) = \exp -0.15n$. For comparison purposes, we use the maximum likelihood (ML) method for the estimation of {h(n)} assuming that the noise is white (i.e., mismatch situation).

Table I illustrates the Monte-Carlo (50 runs) simulation results obtained by the bicepstrum method for two different lengths of data (N) and three SNRs. The bias and variance of the impulse response estimates $\hat{h}(0)$, $\hat{h}(1)$, $\hat{h}(2)$, $\hat{h}(3)$ have been computed for color noise case I (exponential autocorrelation). From this table, it is apparent that the bias and variance of the estimates increases as SNR decreases or the length of the data (N) decreases. However, we see that good reconstruction is achieved by the bicepstrum method without being necessary to know the autocorrelation of the additive white noise. For example, for SNR = 10 dB and N = 4096 the estimated impulse response values (sample mean) were $\hat{h}(0) = 1.027$, $\hat{h}(1) = 0.0226$, $\hat{h}(2) = 0.0383$, $\hat{h}(3) = 0.5182$.

Table II shows the results with color noise case II (damped sinusoidal autocorrelation) and SNR = 10 dB. The results obtained by the ML method in a mismatch situation are also shown. In other words, the ML method was implemented assuming that the additive noise was white. As expected, the bias and variance of the bicepstrum technique are less than the ML mismatch situation. For this specific example, the bicepstrum sample mean estimates were {1.002, -0.00614, -0.00557, 0.5009} whereas the ML-mismatch were {0.95, -0.0224, -0.0116, 0.523}. More simulation results will be presented at the conference.

REFERENCES

[1] J.E. Ehrenberg, T.E. Ewatt, and R.D. Morris, "Signal processing techniques for resolving individual pulses in multipath signal," J. Acoust. Soc. Am., 63(G), pp. 1861-1865, June 1978.

[2] J.P. Ianniello, "Large and small error performance limits for multipath time delay estimation," IEEE Trans. Acoust., Speech, and Signal Processing, Vol. ASSP-34(2), pp. 245-251, April 1986.

[3] R.J.P. deFigueiredo and A. Gerber, "Separation of superimposed signals by a cross-correlation method," IEEE Trans. Acoust., Speech, and Signal Processing, Vol. ASSP-31(5), pp. 1084-1089, October 1983.

[4] R.J. Tremblay, G.C. Carter, and D.W. Lytle, "A practical approach to the estimation of amplitude and time delay parameters of a composite signal in non-white Gaussian noise," Proc. ICASSP'87, pp. 467-470, Dallas, TX, April 1987.

[5] B.M. Bell and T.E. Ewart, "Separating multipaths by global optimization of a multidimensional matched filter," IEEE Trans. Acoust., Speech, and Signal Processing, Vol. ASSP-34(5), pp. 1029-1037, October 1986.

[6] Z. Hou and Z. Wu, "A new method for high-resolution estimation of time delay," Proc. ICASSP'82, pp. 420-423, Paris, France, May 1982.

[7] Y.T. Chan, J.G. Bryan and G.A. Lampropoulos, "Determining time delay via frequency estimation," Proc. ICASSP'86, pp. 2803-2806, Tokyo, Japan, April 1986.

[8] R.J.P. deFigueiredo and C.L. Hu, "Waveform feature extraction based on Tauberian approximation," IEEE Trans. Pattern Anal. Machine Intell., Vol. PAMI-4, pp. 105-116, March 1982.

[9] J.P. Ianniello, "High-resolution multipath time delay estimation for broadband random signals," Proc. ICASSP'87, pp. 447-450, Dallas, TX, April 1987.

[10] I.P. Kirsteins, "High resolution time delay estimation," Proc. ICASSP'87, pp. 451-454, Dallas, TX, April 1987.

[11] M.A. Pallas, N. Martin and J. Martin, "Time delay estimation by autoregressive modelization," Proc. ICASSP'87, pp. 455-458, Dallas, TX, April 1987.

[12] R.C. Kemerait and D.G. Childers, "Signal detection and extraction by cepstrum techniques," IEEE Trans. Inform. Theory, Vol. IT-18, pp. 745-759, November 1972.

[13] J.C. Hassab and R. Boucher, "A probabilistic analysis of time delay extraction by the cepstrum in stationary Gaussian noise," IEEE Trans. Inform. Theory, Vol. IT-22, pp. 444-454, July 1976.

[14] B.P. Bogert and J.F. Ossanna, "The heuristic of cepstrum analysis of a stationary complex echoed Gaussian signal in stationary Gaussian noise," IEEE Trans. Inform. Theory, Vol. IT-72, pp. 373-380, July 1966.

[15] D.G. Stone, "Wavelet estimation," Proc. IEEE, Vol. 10, pp. 1394-1402, October 1984.

[16] C.L. Nikias and M.R. Raghuveer, "Bispectrum estimation: a digital signal processing framework," Proc. IEEE, Vol. 75(7), pp. 869-891, July 1987.

[17] C.L. Nikias and R. Pan, "Time delay estimation in unknown Gaussian spatially correlated noise," IEEE Trans. Acoust., Speech, and Signal Processing, to appear, 1988.

[18] R. Pan and C.L. Nikias, "The complex cepstrum of higher-order cumulants and nonminimum phase system identification," IEEE Trans. Acoust., Speech, and Signal Processing, Vol. ASSP-36(2), February 1988.

Table I ($x10^{-3}$)

BICEPSTRUM METHOD: COLOR NOISE (CASE I) - RESULTS FROM 50 RUNS

SNR (dB)	N = 2048		N = 4096	
	bias	variance	bias	variance
20	1.13 0.06 12.00 6.20	0.193 9.31 5.34 0.59	0.19 0.92 8.41 4.0	0.07 4.45 2.59 0.25
15	7.00 9.60 31.60 14.00	1.31 40.44 21.20 3.44	2.06 1.13 16.20 8.00	0.38 14.30 9.08 1.17
10	110.00 130.00 38.00 20.00	83.4 214.0 179.0 42.6	27.00 22.60 38.30 18.20	7.46 92.40 33.10 8.43

Table II ($x10^{-3}$)

BICEPSTRUM METHOD AND ML WITH MISMATCH:
COLOR NOISE (CASE II) - RESULTS FROM 50 RUNS, LENGTH N = 4096

SNR (dB)	Bicepstrum		ML/Mismatch	
	bias	variance	bias	variance
10	2.03 6.14 5.57 0.90	0.215 1.60 1.41 0.04	50 22.4 11.6 23.0	1.69 6.45 0.318 0.929

Time Delay Estimation Using Phase Data

ALLAN G. PIERSOL

Abstract—The estimation of time delays between two received signals using phase measurements is discussed and the accuracy of such estimates is detailed. For the ideal case of statistically independent noise and no scattering at the receiver locations, it is shown that phase data regression lines yield time delay estimates with the same accuracy as other optimal time delay estimation procedures. For less ideal situations, the potential advantages of time delay estimation using phase data are discussed and illustrated. It is shown that regression analysis of phase estimates at properly selected frequencies can sometimes be employed to reduce bias errors in time delay estimates due to correlated receiver noise. It is also shown that the estimation errors due to scattering at the receiver location can often be assessed in nonparametric terms to provide time delay estimates with a realistic error bound.

I. Introduction

The accurate estimation of time delays between received signals at two or more transducer locations plays a dominant role in numerous engineering applications of signal processing [1]. Various *time delay estimation* procedures have been proposed and implemented over the years, including enveloped *cross-correlation functions*, unit impulse response calculations, smoothed coherence transforms, maximum likelihood estimates, and others as summarized in [2] and elsewhere. The general goal has been to minimize the variance (random error) of the resulting time delay estimates in the presence of statistically independent noise at the receiver locations.

Time delay estimation procedures such as those mentioned above generally can be viewed as inverse Fourier transforms of appropriately normalized and/or weighted *cross-spectral density function* measurements which yield delay estimates directly in time domain terms. There are instances, however, where it can be advantageous to base time delay estimates on properly interpreted phase data directly in the frequency domain. Common examples include situations where the noise at the receivers is correlated or there is strong *scattering* at the receiver location. This paper discusses such advantages for the simple case of a single path time delay estimate between two received signals with *nondispersive propagation*.

II. Background

Consider two received time history signals, $r_k(t)$ and $r_l(t)$, which originate from a common source with a wave form $s(t)$ at the closest receiver, as illustrated in Fig. 1(a). Assuming nondispersive propagation and a constant gain factor between the two received signals, it follows that

Manuscript received May 29, 1980; revised October 20, 1980.
The author is with Bolt Beranek and Newman, Inc., Canoga Park, CA 91305.

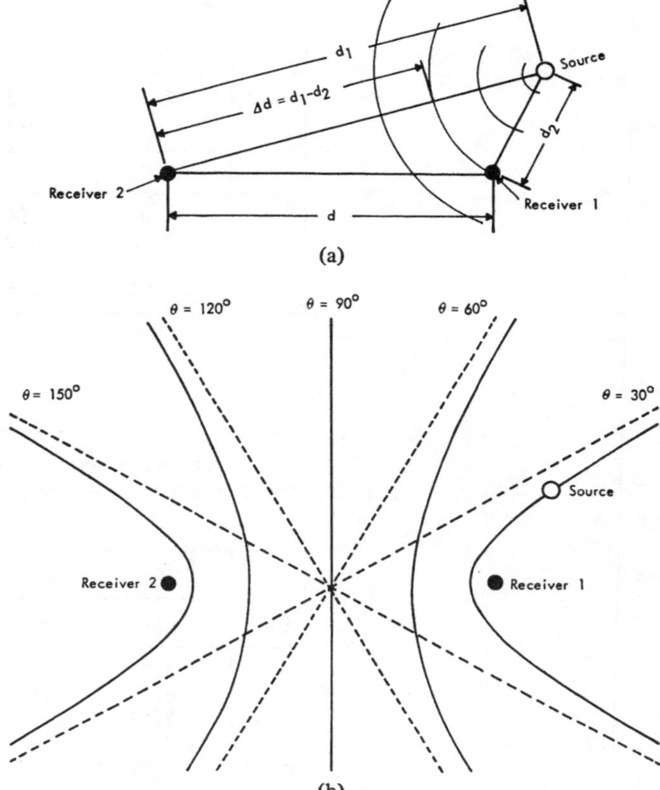

Fig. 1. Source location and angle of incidence based on measurements between two receivers. (a) Path length difference. (b) Source location.

$$r_k(t) = s(t) + n_k(t)$$
$$r_l(t) = as(t - D_{kl}) + n_l(t) \tag{1}$$

where $n_k(t)$ and $n_l(t)$ are the contaminating noise at the kth and lth receiver locations, respectively, and a is an attenuation factor. For the simple case where the contaminating noise terms are statistically independent, the cross-correlation function between the received signals is given by

$$R_{kl}(\tau) = E[r_k(t)\, r_l(t+\tau)] = aR_{ss}(\tau - D_{kl}) \tag{2}$$

where $R_{ss}(\tau - D_{kl})$ is the autocorrelation function of $s(t)$ transposed to have its peak value at $\tau = D_{kl}$. It is clear that the peak value of the cross correlation occurs at the time delay

$$D_{kl} = \Delta d / P \tag{3}$$

where $\Delta d = d_1 - d_2$ is the path length difference and P is the propagation velocity of the source signal $s(t)$. Knowing P, the delay value D_{kl} defines a hyperbolic surface on which the source must lie, as illustrated in Fig. 1(b).

Reprinted from *IEEE Trans. Acoust., Speech, Signal Processing,* vol. 29, no. 3, pt. 2, pp. 471–477, June 1981.

In frequency domain parameters, the Fourier transform of (2) yields

$$G_{kl}(f) = 2\int_{-\infty}^{\infty} R_{kl}(\tau)\, e^{-j2\pi f\tau}\, d\tau = aG_{ss}(f)\, e^{-2\pi fD_{kl}} \tag{4}$$

where $G_{ss}(f)$ is the one-sided autospectrum of $s(t)$. Hence, the time delay D_{kl} appears in the cross-spectrum as a *phase function*

$$\phi_{kl}(f) = 2\pi fD_{kl}. \tag{5}$$

For nondispersive propagation where P = constant, the phase shift $\phi_{kl}(f)$ in (5) is linear, as illustrated in Fig. 2. However, for dispersive propagation where $P(f)$ varies with frequency, $\phi_{kl}(f)$ will be nonlinear. For the special case where the source is sufficiently distant to be approximated by a plane wave at the receiver locations, the time delay D_{kl} can be interpreted as an angle of incidence for the received signal given by

$$\theta_{kl}(f) = \cos^{-1}[PD_{kl}/d] = \cos^{-1}\left[\frac{P\phi_{kl}(f)}{2\pi fd}\right] \tag{6}$$

where d is the separation distance between the receivers and $\phi_{kl}(f)$ is in radians. For nondispersive propagation, (6) reduces to

$$\theta_{kl}(f) = \cos^{-1}\left[\frac{Pb_{kl}}{2\pi d}\right] \tag{7}$$

where $b_{kl} = \phi_{kl}(f)/f$, the slope of the phase curve in radians per Hz. This angle of incidence is shown in Fig. 1(b). Note that the source will not actually fall on the line defined by $\theta_{kl}(f)$ except for the case where $d \ll d_2$.

III. Accuracy of Time Delay Estimates

There has been considerable study of the errors in time delay estimates, particularly as they relate to the *sonar* problem, as summarized, for example, in [2], [3]. The variance of the delay estimates provided by the various procedures often can be well defined in parametric terms, assuming the receiver noise is statistically independent and there are no other biasing effects. However, if these assumptions are violated, a more accurate estimate of the time delay error may be provided by nonparametric calculations on properly interpreted phase data using conventional regression analysis procedures. To demonstrate these points, it is first desirable to relate the error deduced by a regression analysis of the phase data between two receivers to the errors associated with other time delay estimation procedures assuming statistically independent receiver noise.

Consider a phase estimate $\hat{\phi}_{kl}(f)$ computed from two received signals $r_k(t)$ and $r_l(t)$. Specifically, assuming a total record length T is divided into n_d disjoint (independent) segments, each of length $T_d = T/n_d$, the one-sided cross-spectrum $\hat{G}_{kl}(f)$ is estimated by

$$\hat{G}_{kl}(f) = \frac{2}{T_d n_d}\sum_{i=1}^{n_d} R_{ki}^*(f) R_{li}(f) \tag{8}$$

where

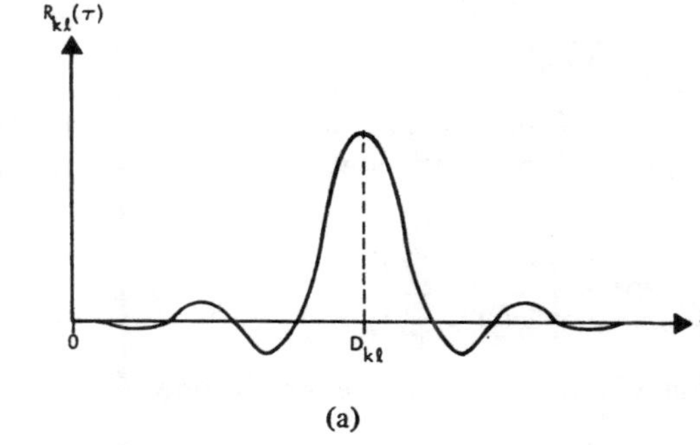

(a)

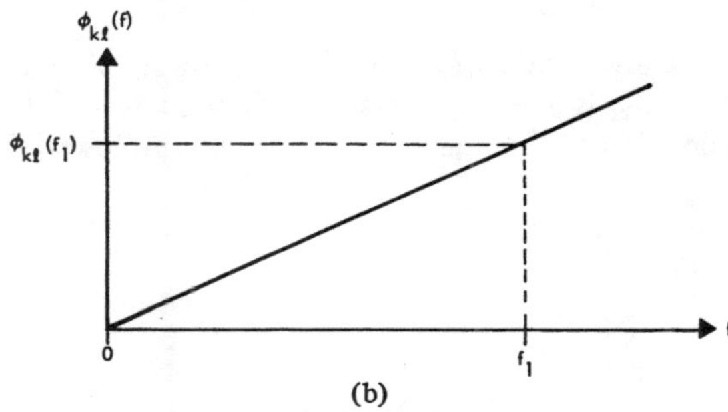

(b)

Fig. 2. Time delay estimation procedures. (a) Cross-correlation function estimate. (b) Cross-spectrum phase function estimate.

$$R_{ki}^*(f) = \int_0^{T_d} r_{ki}(t)\, e^{j2\pi ft}\, dt;$$

$$R_{li}(f) = \int_0^{T_d} r_{li}(t)\, e^{-j2\pi ft}\, dt. \tag{9}$$

The estimate $\hat{G}_{kl}(f)$ will generally be a complex number such that

$$\hat{G}_{kl}(f) = \hat{G}_{\mathrm{Re}}(f) - j\hat{G}_{\mathrm{Im}}(f) = |\hat{G}_{kl}(f)|\, e^{-j\hat{\phi}_{kl}(f)} \tag{10}$$

where

$$|\hat{G}_{kl}(f)| = [\hat{G}^2_{\mathrm{Re}}(f) + \hat{G}^2_{\mathrm{Im}}(f)]^{1/2};$$
$$\hat{\phi}_{kl}(f) = \tan^{-1}[\hat{G}_{\mathrm{Im}}(f)/\hat{G}_{\mathrm{Re}}(f)]. \tag{11}$$

From (9), it is clear that the spectral estimate will be at discrete frequencies separated by $\Delta f = 1/T_d$. For digital data processing with a sampling interval of Δt and N data points per segment, $\Delta f = 1/(\Delta tN)$ and the upper frequency limit of the analysis is $f_c = 1/(2\Delta t)$. Hence, the number of frequencies for nonredundant spectral estimates is $f_c/\Delta f = N/2$. Assuming the signals are stationary, the spectral estimates will be statistically independent of one another and have a standard deviation approximated by [4]

$$\sigma[|\hat{G}_{kl}(f_i)|] \approx \frac{|G_{kl}(f_i)|}{\sqrt{n_d C_{kl}(f_i)}}$$

$$\sigma[\hat{\phi}_{kl}(f_i)] \approx \sin^{-1}\left[\frac{1 - C_{kl}(f_i)}{2n_d C_{kl}(f_i)}\right]^{1/2} \qquad i = 1, 2, \cdots, N/2 \tag{12}$$

where $C_{kl}(f_i)$ is the *coherence function* (coherency squared) between $r_k(t)$ and $r_l(t)$ at frequency f_i defined by

$$C_{kl}(f_i) = \frac{|G_{kl}(f_i)|^2}{G_{kk}(f_i)\, G_{ll}(f_i)}. \tag{13}$$

For the case of small errors where $\sin \sigma \approx \sigma$, the standard deviation of the phase estimates can be further approximated by

$$\sigma[\hat{\phi}_{kl}(f_i)] \approx \left[\frac{1 - C_{kl}(f_i)}{2n_d C_{kl}(f_i)}\right]^{1/2}. \tag{14}$$

Now from (5), for nondispersive propagation,

$$\hat{D}_{kl} = \hat{\phi}_{kl}(f)/2\pi f = b_{kl}/2\pi. \tag{15}$$

Since we know that $\phi_{kl}(f) = 0$ at $f = 0$, the minimum error estimate of b_{kl} is given by the slope of the regression line for $\hat{\phi}_{kl}(f)$ on f forced through a zero intercept as follows [5].

$$b_{kl} = \frac{\sum_{i=1}^{N/2} f_i \hat{\phi}_{kl}(f_i)}{\sum_{i=1}^{N/2} f_i^2}. \tag{16}$$

The standard deviation of the slope estimate b_{kl} is approximated by [5]

$$\hat{\sigma}[b_{kl}] = \left\{\sum_{i=1}^{N/2} \frac{f_i^2}{\sigma^2[\hat{\phi}_{kl}(f_i)]}\right\}^{-1/2}. \tag{17}$$

Substituting from (14) and (15), it follows that

$$\hat{D}_{kl} = \frac{1}{2\pi}\left\{\sum_{i=1}^{N/2} f_i \hat{\phi}_{kl}(f_i)\right\} \Big/ \left\{\sum_{i=1}^{N/2} f_i^2\right\} \tag{18}$$

$$\hat{\sigma}[\hat{D}_{kl}] = \left\{8\pi^2 n_d \sum_{i=1}^{N/2} \frac{f_i^2\, C_{kl}(f_i)}{1 - C_{kl}(f_i)}\right\}^{-1/2}. \tag{19}$$

In practice, the estimated coherence function $\hat{C}_{kl}(f_i)$ would be used in place of the unknown true coherence $C_{kl}(f_i)$.

The error expression in (19) is written in a form appropriate for conventional *regression analysis* by digital computations. The result can be converted to analog form by noting from the relationships following (11) that the frequency increment $\Delta f = 1/T_d$ and the total record length $T = n_d T_d$. The sum expression in (19) then converts to the integral expression

$$\sigma[\hat{D}_{kl}] = \left[T \int_{-\infty}^{\infty} \frac{(2\pi f)^2\, C_{kl}(f)}{1 - C_{kl}(f)}\, df\right]^{-1/2} \tag{20}$$

where $C_{kl}(f)$ is the continuous, two-sided coherence function defined over both positive and negative frequencies. Note that the result in (20) is identical (after adjustments for notation) to the least squares error expression for time delay estimates deduced independently in [6] and noted there to be equivalent to the maximum likelihood estimate error. The result in (20) is also equivalent to the "Cramér-Rao lower bound" on the variance of time delay estimation errors presented in [7] and expanded upon in [2]. Hence, the results provided by a regression analysis of phase data are fully consistent with other optimum estimation procedures.

No matter how the error expression in (19) or (20) is arrived at, the result assumes proper analysis procedures are employed; specifically, 1) the instrumentation is calibrated to eliminate relative phase errors in the measured signals; 2) the received signals are aligned as required to eliminate time delay bias errors in the coherence calculations [1]; and 3) the number of averages n_d is sufficiently large to suppress small sample bias errors in the coherence calculations [1]. Perhaps most important, the error expression assumes that dispersion of the phase estimates about their regression line is due only to statistical sampling errors. There are various situations in practice where phase errors occur due to factors other than sampling considerations, for example, due to scattering at the receiver location. In such cases, (19) or (20) will underestimate the standard deviation associated with the time delay estimate. However, if it can be assumed that there is no bias in the estimate of (18), i.e., $E[\hat{D}_{kl}] = D_{kl}$, then the accuracy of $\hat{D}_{kl}$ can be approximated as follows.

Assume the standard deviation of the phase estimates given by (14) is a constant independent of frequency, i.e., $\sigma[\hat{\phi}_{kl}(f)] = \sigma_e$. The standard deviation of the calculated slope of the regression line per (16) is now estimated by

$$\hat{\sigma}[b_{kl}] = \sigma_e \left\{\sum_{i=1}^{m} f_i^2\right\}^{-1/2} \tag{21}$$

where m is the number of spectral components used in the regression calculation (not necessarily all $N/2$ values) and σ_e is estimated by [5]

$$\hat{\sigma}_e = \left\{\frac{1}{m-1}\left[\sum_{i=1}^{m} \hat{\phi}_{kl}^2(f_i) - \frac{\left[\sum_{i=1}^{m} f_i \hat{\phi}_{kl}(f_i)\right]^2}{\sum_{i=1}^{m} f_i^2}\right]\right\}^{1/2}. \tag{22}$$

Hence, from (15), the standard deviation of the time delay estimate $\hat{D}_{kl}$ is approximated by

$$\hat{\sigma}[\hat{D}_{kl}] = \frac{\hat{\sigma}_e}{2\pi}\left\{\sum_{i=1}^{m} f_i^2\right\}^{-1/2} \tag{23}$$

where $\hat{\sigma}_e$ is given by (22).

The virtues of estimating time delays and their accuracy using phase data via (18) and (23) are as follows: 1) the error in the resulting estimate is minimized by using the knowledge that the phase must be zero for infinite wavelengths; 2) the error is established by parameters of the analysis; and 3) phase estimates can be omitted from the time delay calculation at those frequencies where bias errors are anticipated. This last factor is particularly helpful for time delay estimates when there is *correlated noise* at the receiver locations, as discussed in the next section.

IV. Applications with Correlated Receiver Noise

The results in (2)-(7) assume that the contaminating noise terms in the receiver outputs are statistically independent, that is, $E[n_k(t)\, n_l(t)] = 0$. This is a reasonable assumption for noise which originates in the transducers and instrumentation. However, situations arise in practice where the received signals include contaminating noise terms of an outside origin which are correlated. A common example is the noise due to numerous secondary sources which surround the receiver locations. Such noise can often be modeled as an infinite collection of statistically independent monopole sources on a spherical surface surrounding the receiver, called diffuse or volume noise, as illustrated in Fig. 3.

It can be shown [8] that the cross-spectrum between any two points in a diffuse noise field is given by

$$G_{n_k n_l}(f) = G_{nn}(f)\frac{\sin Kd}{Kd} \tag{24}$$

where $K = 2\pi f/P$ (wave number), $G_{nn}(f)$ is the autospectrum of the received noise, and d is the separation distance between the receivers. From (4), the cross-spectrum between the two receivers due to a propagating signal from a single dominate source may be written as

$$G_{s_k s_l}(f) = aG_{ss}(f)\,[\cos K_t d - j\sin K_t d] \tag{25}$$

where $K_t = K\cos\theta$ (trace wave number) and θ is the angle of incidence defined previously in (6). It follows that the cross-spectrum between the total received signals is given by the sum of (24) and (25) as

$$G_{kl}(f) = G_{nn}(f)\frac{\sin Kd}{Kd} + aG_{ss}(f)\cos K_t d - jaG_{ss}(f)\sin K_t d. \tag{26}$$

Use of (26) yields the coherence and phase functions

$$C_{kl}(f) = (1+R)^{-2}\left[\left(\frac{R(f)\sin Kd}{Kd} + \cos K_t d\right)^2 + \sin^2 K_t d\right] \tag{27}$$

$$\phi_{kl}(f) = \tan^{-1}\left(\frac{\sin K_t d}{R(f)\dfrac{\sin Kd}{Kd} + \cos K_t d}\right) \tag{28}$$

where $R(f) = G_{nn}(f)/aG_{ss}(f)$. Plots of (27) and (28) for various values of R = constant and an arbitrarily selected angle of incidence $\theta = \pi/4$ are shown in Fig. 4.

Referring to (28), it is seen that the phase angle $\phi_{kl}(f) = i\pi$; $i = 1, 2, 3, \cdots$, at frequencies where $\sin K_t d = 0$, which occur where $K_t d = i\pi$; $i = 1, 2, 3, \cdots$, independent of the diffuse noise at the receiver locations. Hence, from (5), the correct (unbiased) time delay for the propagating component will be given by

$$D_{kl} = \frac{i}{2f_i}; \quad i = 1, 2, 3, \cdots \tag{29}$$

where f_i are those frequencies which satisfy

$$\phi_{kl}(f_i) = i\pi; \quad i = 1, 2, 3, \cdots. \tag{30}$$

It is also known that $\phi_{kl}(f) = 0$ at $f = 0$. Of course, in practice, only an estimate $\hat{\phi}_{kl}(f)$ will be available for the phase, so it is not possible to identify the exact frequencies $\hat{f}_i$ which satisfy (3). However, an estimate for the time delay $\hat{D}_{kl}$ and its variance $\hat{\sigma}^2[\hat{D}_{kl}]$ can be obtained by calculating the regression line for the phase estimates at those frequencies where $\hat{\phi}_{kl}(f_i) = i\pi$; $i = 1, 2, 3, \cdots$, using (18) and (23).

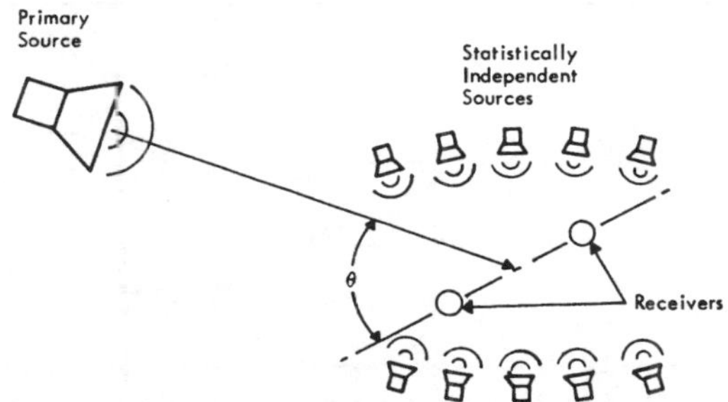

Fig. 3. Combined diffuse and propagating noise at receiver locations.

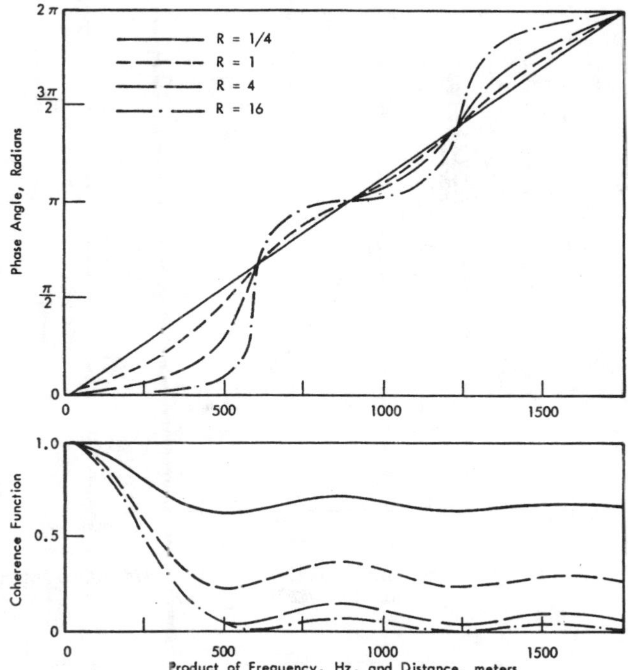

Fig. 4. Phase and coherence functions between two receivers for combined diffuse and propagating noise with an angle of incidence of $\theta = \pi/4$.

To illustrate this procedure for estimating time delays in the presence of diffuse noise at the receiver, consider the experiment outlined in Fig. 5 which involved coherence and phase measurements between two microphones in the test section of a wind tunnel [9]. The noise levels in the test section during tunnel operation include the effects of boundary layer turbulence and reverberation along the test section walls, which appear much like diffuse noise at the receiver locations, plus the contributions of the fan and perhaps other sources which propagate through the test section along its longitudinal axis. At frequencies below 1000 Hz, propagating components enter the test section primarily from the diffuser side of the tunnel. A typical autospectrum of the test section acoustic levels during tunnel operation is presented in Fig. 6. The measured phase and coherence between the two receivers in the test section during tunnel operation with a test section flow velocity of P_1 = 80 m/s are shown in Fig. 7. Note the strong similarity in the general form of the measured data in Fig. 7 and theoretical predictions for diffuse plus propagating noise in Fig. 4.

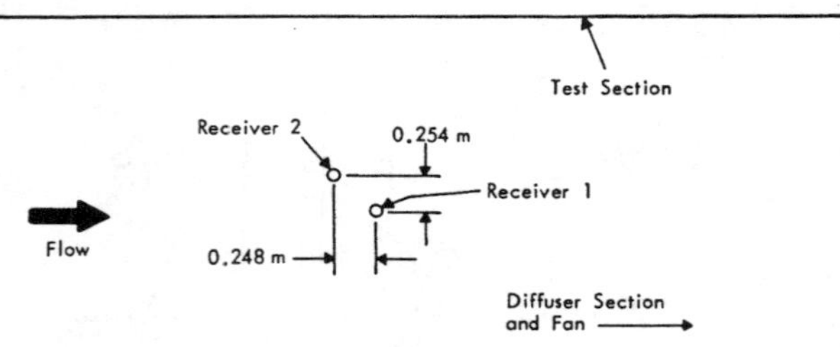

Fig. 5. Receiver locations for wind tunnel acoustic experiment.

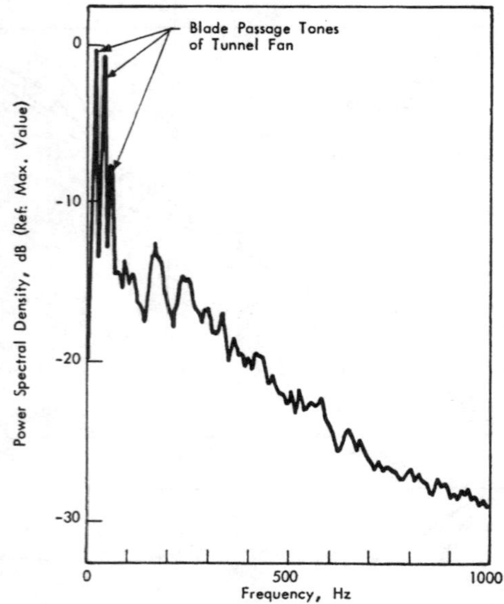

Fig. 6. Typical autospectrum of received signal in wind tunnel acoustic experiment ($\Delta f = 2$ Hz, $n_d = 128$).

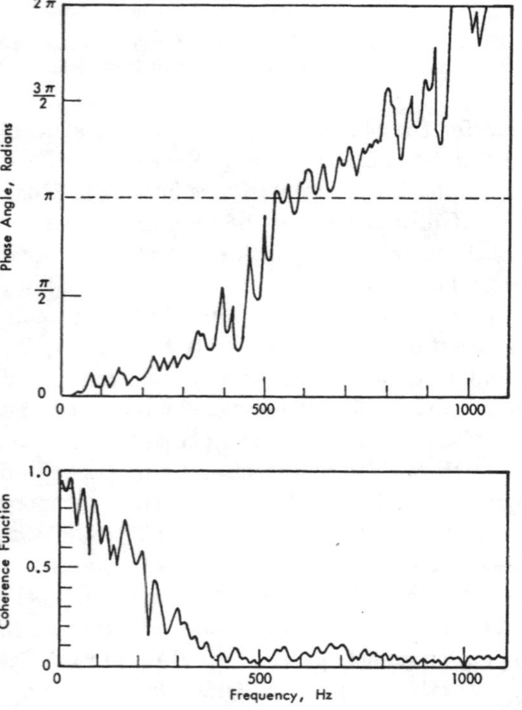

Fig. 7. Phase and coherence functions measured between two receivers in wind tunnel acoustic experiment ($\Delta f = 5$ Hz, $n_d = 128$).

The phase data in Fig. 7 cover a range of about 2π rads (360°), meaning there are two frequencies which satisfy (30). However, because of the limited record length and the low coherence between the two received signals, the phase data in Fig. 7 display substantial uncertainty fluctuations which make these frequencies ambiguous. Specifically, the frequencies where $\hat{\phi}_{kl}(f) = i\pi; i = i, 2$, are as follows:

$$\hat{\phi}_{kl}(f) = \begin{cases} \pi & \text{when } f = 515, 530, 545, 560, \text{and } 580 \text{ Hz} \\ 2\pi & \text{when } f = 945, 990, \text{and } 1040. \end{cases}$$

From (18) and (23), the best least squares estimate for the time delay and its standard deviation based upon the above eight frequencies are then given by

$$\hat{D}_{kl} = 0.976 \text{ ms} \qquad \hat{\sigma}[\hat{D}_{kl}] = 0.022 \text{ ms}.$$

From Fig. 5, noting that the path length difference along the longitudinal axis is $\Delta d = 0.248$ m and the propagation speed in air is $P = 340$ m/s, the true time delay between the two receivers for a propagating wave traveling upstream is given by

$$D_{kl} = \Delta d/(P - P_1) = 0.248/(340 - 80) = 0.95 \text{ ms}$$

which agrees well with the experimentally determined result (within about one standard deviation).

The time delay was also calculated from the wind tunnel data using the cross correlation function and the unit impulse response function between the two receivers. Since the autospectra of the received signals were essentially identical, the unit impulse response measurement is equivalent to a smoothed coherence transform (SCOT) calculation [2]. The cross correlation calculation produced $\hat{D}_{kl} = 0.27$ ms, which is in dramatic error from the correct delay of $D_{kl} = 0.95$ ms. This error obviously results from the fact that a simple cross correlation calculation gives greatest weight to the frequencies where the data are most intense, in this case at the low frequencies where the distortion of the phase data by the diffuse noise contribution is greatest. The unit impulse response function corrects this defect by essentially prewhitening the data. Nevertheless, the unit impulse response calculation produced a result of $\hat{D}_{kl} = 0.80$ ms, which is still in substantial error. The bias error in this calculation evolves from the fact that the unit impulse response gives greatest weight to the frequencies where the coherence is highest, i.e., it interprets high coherence as good signal-to-noise ratio. Of course, this would be correct for statistically independent noise, but not for correlated noise, in particular, diffuse noise where the coherence is highest at the low frequencies where the bias in the phase data is greatest. This problem will arise in other common time delay estimation procedures which are designed to optimize the signal relative to assumed independent noise.

It is important to note in the foregoing example that good results were obtained from the regression analysis on selected phase data only because one could anticipate that the background noise in the wind tunnel would be approximately diffuse in character. There are situations where correlated receiver noise of a different form may occur. One common example is underwater noise during a high sea state where a surface noise model assuming a plane of independent sources

[8] might be more appropriate. The frequencies at which the phase data may be considered unbiased are determined by the form of the assumed noise model. It should also be noted that the omission of phase data at selected frequencies in the time delay calculation per (18) will always yield a time delay estimate with a higher variance than would be obtained by the more conventional time delay estimation techniques. The procedure is designed to reduce bias errors only for those cases where the receiver noise is correlated via a well defined noise model.

V. Applications with Scattering at Receiver

Equations (2)-(7) assume a homogeneous path between the source and receivers and anechoic conditions at the receiver location. Conditions other than these will generally distort the correlation and phase data between two receivers and lead to erroneous time delay estimates. The most extreme distortions occur when the source and receivers are inside an enclosure which is sufficiently rigid to reflect incident waves and cause reverberation. In the limiting case where the reverberation time becomes very long and the density of acoustic modes in the enclosure becomes very large, the cross-spectral density function between the two receivers will approach the form [10]

$$G_{kl}(f) = G_{rr}(f) \frac{\sin Kd}{Kd} \tag{31}$$

where $G_{rr}(f)$ is the autospectrum of the reverberant noise at the receivers. In other words, reverberant noise in the limit will approach the same form as diffuse noise in (24) and, hence, the phase data between two receivers will not identify the source location. Even in partial enclosures, the phase data may be more influenced by anechoic openings in the enclosure than by the location of the source [11].

On the other hand, situations often arise where the receivers are surrounded by rigid objects with dimensions generally less than the wave length of the propagating signal. In such cases, scattering occurs which distorts the correlation and phase data between the receivers, but not to the extent that occurs from reverberation in a rigid enclosure. In fact, time delay estimates using phase data in accordance with (18) and (23) will often yield acceptable results with a well defined degree of error. This occurs because the phase errors due to a large number of scattering objects tend to be somewhat random as a function of frequency.

To illustrate this point, consider the experiment outlined in Fig. 8 involving an underwater noise source and an array of receivers located near various rigid objects. A square wave signal with a fundamental frequency of 236 Hz was generated by the source, and the phase between various receiver pairs along the two legs of the array were measured at the first four harmonics of the signal with the results shown in Figs. 9 and 10. Also shown in these figures are the regression lines for the phase data, b_{kl}, determined from (16), and 95 percent confidence intervals on the regression lines calculated from

$$b_{95} = b_{kl} \pm t_{\alpha/2:m-1}\, \hat{\sigma}\,[b_{kl}] \tag{32}$$

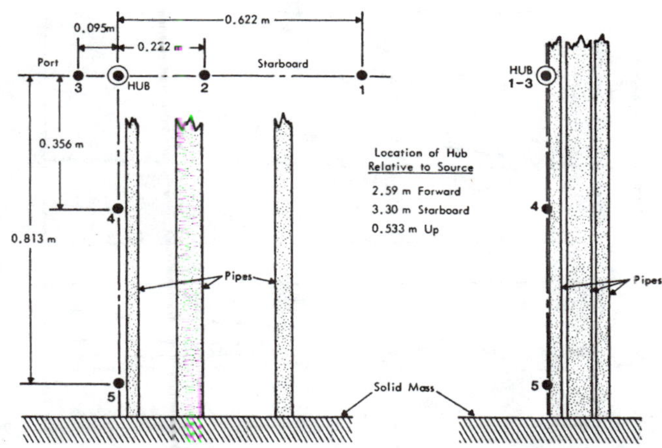

Fig. 8. Receiver locations for underwater acoustic experiment.

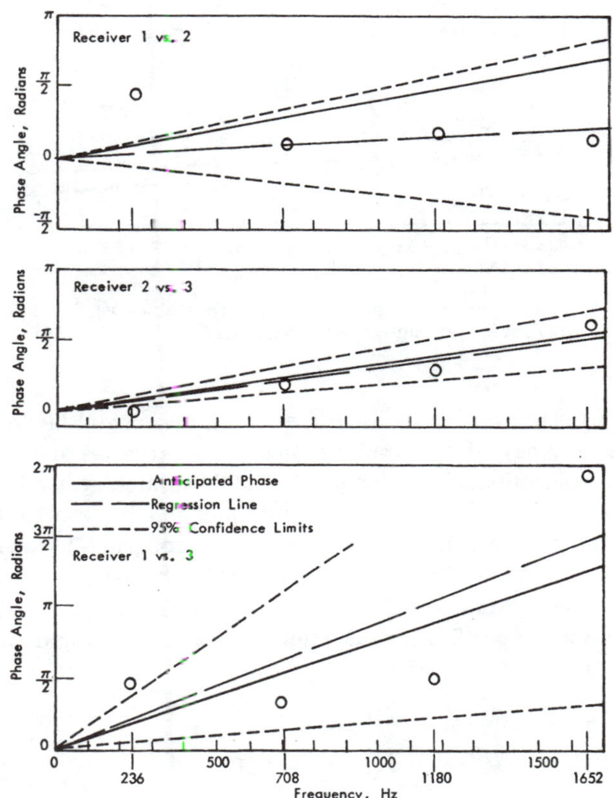

Fig. 9. Phase measurements between receivers along lateral axis of array for 236 Hz square wave source ($\Delta f = 4$ Hz, $n_d = 256$).

where $t_{\alpha/2;m-1}$ = the $\alpha/2 = 0.025$ percentage point of the student "t" variable with $m - 1 = 3$ degrees of freedom, and $\hat{\sigma}\,[b_{kl}]$ is given by (21). Finally, superimposed on the data in Figs. 9 and 10 are the true phase values which would be anticipated for anechoic propagation from the source.

The phase data for most of the receiver pairs in Figs. 9 and 10 display substantial discrepancies from their regression lines. The smallest discrepancies are in the data for transducers 2 versus 3 along the lateral axis and 5 versus 4 along the vertical axis. In all cases, however, the anticipated phase angles fall well within the 95 percent confidence intervals for the measured results. This is important, since it means that the phase

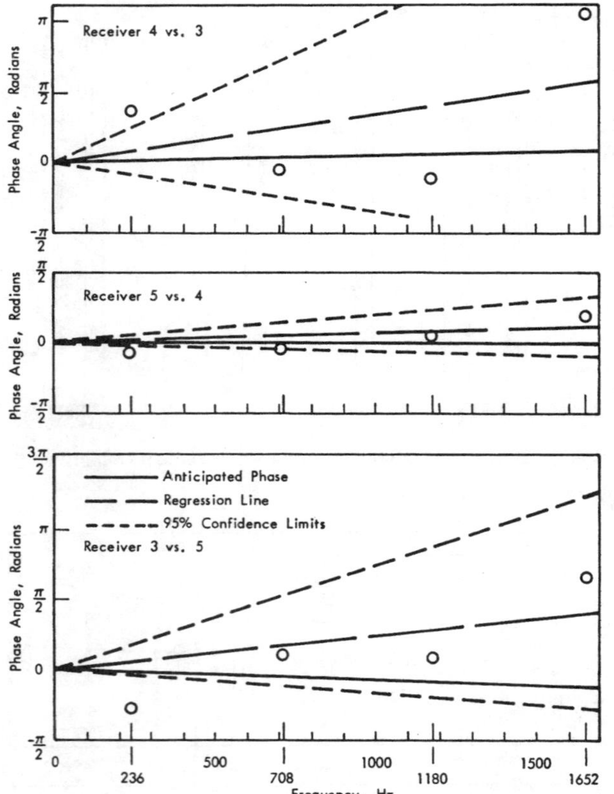

Fig. 10. Phase measurements between receivers along vertical axis of array for 236 Hz square wave source ($\Delta f = 4$ Hz, $n_d = 256$).

data themselves can be used to establish meaningful limits on the accuracy of the resulting time delay estimates and final source location. For example, in the illustration at hand, the phase data in Figs. 9 and 10 would probably be ignored due to the wide confidence intervals, except for pairs 2 versus 3 and 5 versus 4. These two pairs provide reasonably accurate time delay estimates using (15) in spite of the scattering at the receiver. Specifically, the estimated time delays and their standard deviations are

$$\hat{D}_{23} = 0.160 \text{ ms} \qquad \hat{\sigma}[\hat{D}_{23}] = 0.020 \text{ ms}$$

$$\hat{D}_{54} = 0.036 \text{ ms} \qquad \hat{\sigma}[\hat{D}_{54}] = 0.020 \text{ ms}.$$

The path length differences for propagation from the source past the two transducer pairs are $\Delta d_{23} = 0.249$ m and $\Delta d_{54} = 0.005$ m. Hence, assuming a propagation velocity of 1500 m/s in water, the anticipated time delays at the two transducer pairs are

$$D_{23} = 0.249/1500 = 0.17 \text{ ms} \qquad D_{54} = 0.005/1500 = 0.003 \text{ ms}$$

which agree with the experimental results to within 1.7 standard deviations. In terms of an angle of incidence θ as defined in (7), the discrepancy between the estimated and anticipated results is 0.048 rads (2.8°) for the 2-3 transducer pair and 0.11 rads (6.3°) for the 5-4 transducer pair. In summary, although the errors in the time delay estimates due to scattering effects are significant, the nonparametric estimate of the variance per (23) provides a realistic assessment of the resulting errors.

VI. Conclusions

Given two received signals where the receiver noise is correlated, the use of straightforward regression analysis procedures on phase estimates at properly selected frequencies can substantially reduce bias errors that would otherwise be present in the resulting time delay estimates. Furthermore, in situations where there is scattering from numerous objects near the receiver location, regression analysis procedures on phase estimates can yield time delay estimates with realistic error assessments based on nonparametric variance calculations. For the ideal case where the receiver noise is uncorrelated and no scattering is present, the phase data regression analysis procedure provides time delay estimates with an accuracy similar to other optimal estimation procedures.

References

[1] J. S. Bendat and A. G. Piersol, *Engineering Applications of Correlation and Spectral Analysis*. New York: Wiley-Interscience, 1980.

[2] C. H. Knapp and G. C. Carter, "The generalized correlation method for estimation of time delay," *IEEE Trans. Acoust., Speech, Signal Processing*, vol. ASSP-24, no. 4, pp. 320-327, 1976.

[3] G. C. Carter, "Sonar signal processing for source state estimation," IEEE Pub. 79CH1476-1-AES, pp. 386-395, 1979.

[4] J. S. Bendat, "Statistical errors in measurement of coherence functions and input/output quantities," *J. Sound and Vibration*, vol. 59, no. 3, pp. 405-421, 1978.

[5] K. A. Brownlee, *Statistical Theory and Methodology in Science and Engineering*, 2nd ed. New York: Wiley, 1965.

[6] Y. T. Chan, R. V. Hattin, and J. B. Plant, "The least squares estimation of time delay and its use in signal processing," *IEEE Trans. Acoust., Speech, Signal Processing*, vol. ASSP-26, no. 3, pp. 217-222, 1978.

[7] H. L. Van Trees, *Detection, Estimation and Modulation Theory*, Part 1. New York: Wiley, 1968.

[8] B. F. Cron and C. H. Sherman, "Spatial-correlation functions of various noise models," *J. Acoust. Soc. Amer.*, vol. 34, no. 11, pp. 1732-1736, 1962.

[9] A. G. Piersol, "Use of coherence and phase data between two receivers in evaluation of noise environments," *J. Sound and Vibration*, vol. 56, no. 2, pp. 215-228, 1978.

[10] C. T. Morrow, "Point-to-point correlation of sound pressures in reverberant chambers," *Shock and Vibration Bull.*, no. 39, part 2, pp. 87-97, 1969.

[11] W. K. Blake and R. V. Waterhouse, "The use of cross-spectral density measurements in partially reverberant sound fields," *J. Sound and Vibration*, vol. 54, no. 4, pp. 589-599, 1977.

Time-Delay Estimation Performance in a Scattering Medium

SURENDRA PRASAD, MEMBER, IEEE, M. S. NARAYANAN, AND SAMPATH R. DESAI

Abstract—The effects of a scattering medium on the performance of time-delay estimation are considered. The medium is assumed to exhibit angular scattering, causing angular (as well as delay) dispersion and hence loss of signal coherence across the array aperture. Both the variance (via the Cramer-Rao bound) and the bias introduced in the time-delay estimates are studied. The results have been converted to bearing and range error standard deviation and bias. It is shown that there is an optimum range of values for the separation distance between the sensors in the design of an array for time-delay estimation, for range and bearing measurements.

I. Introduction

The time-delay estimation (TDE) problem in Gaussian noise has been extensively studied in the literature. The measurement of time delays of a signal received at several locations is particularly important for source localization [1]. A number of workers have therefore analyzed the performance of the range and bearing estimators based on the delay estimates and the geometry of the problem, in terms of variance and bias [2]-[4]. Some studies have also reported on the effects of moving targets and/or platforms on the performance of time-delay estimators.

The purpose of this paper is to present some results on the effects of a scattering medium on the estimation performance. For simplicity, the medium is assumed to exhibit angular scattering, causing angular dispersion and hence loss of signal coherence across the array aperture. The performance in terms of variance is studied via the evaluation of the Cramer-Rao lower bound (CRLB), which is known to be achieved asymptotically (i.e., if the processing time is large enough) by the maximum likelihood estimator presented by Carter [5]. However, there are some cases when the CRLB will not be achieved, e.g., the case studied by Scarbrough *et al.* [8] and Ianniello *et al.* [9] when the processing time is not long enough. The results of this paper are therefore applicable only when the bandwidth-observation time product is large enough. The bias introduced by the TDE is also studied here. These results have been converted to the standard deviation and bias of the resulting bearing and range estimates. Numerical results are presented for different signal-to-noise ratios (SNR's), and sensor separation distances as a function of a "spatial coherence loss coefficient" introduced in the sequel, for typical signal spectra and observation intervals.

It is found from these studies that there is an "optimum" value for the separation distance in the design of an array used for time-delay estimation. The optimum distance depends not only on the scattering loss coefficient, but also whether we intend to minimize the bearing error variance or the range error variance, or the corresponding biases. Fortunately, however, the performance curves are quite flat near the minima, so that a reasonable compromise is easy to obtain. It should be noted here that the results obtained pertain to a particular scattering model. Hence the validity of these results as applied to different problems may vary in detail, although the general behavior is expected to be similar.

Manuscript received November 13, 1983; revised June 18, 1984. This work was supported by the Department of Electronics, National Radar Council, Government of India.

S. Prasad is with the Department of Electrical Engineering, Indian Institute of Technology, Delhi, New Delhi 110016, India.

M. S. Narayanan is with the Defence Research and Development Organization, Sena Bhavan, New Delhi 110001, India.

S. R. Desai is with the Indian Navy.

II. Model for the Scattering Medium [10]

A. Generalized Scattering Function

In a general scattering medium, the transmission characteristics of the medium depend on space and time parameters. We introduce the five-dimensional vector

$$\vec{p}_i = [f_i, t_i, x_i, y_i, z_i]^T \tag{1}$$

where f_i and t_i are frequency and time parameters, respectively, and (x_i, y_i, z_i) are space parameters. Thus, in the most general case, the transfer function $H(\cdot)$ between two points in the medium depends on the associated vectors $\vec{p}_i$.

We assume here that the transfer function process is stationary in time, frequency, and space. Defining the difference vector

$$\vec{p} = \vec{p}_1 - \vec{p}_2 = [\Delta f, \Delta t, \Delta x, \Delta y, \Delta z]^T \tag{2}$$

we can then define the space-time correlation function of the medium as

$$R_H(\vec{p}) = E\{H(\vec{p}_1) H(\vec{p}_2)\}. \tag{3}$$

We next define a dual, "transform domain" vector $\vec{q}$:

$$\vec{q} \triangleq [\tau, \varphi, u, v, w]^T \tag{4}$$

where

τ = delay

φ = Doppler

and where u, v, and w are variables in the angular domain related to the direction cosines $\cos\alpha$, $\cos\beta$, and $\cos\gamma$ as shown below:

Reprinted from *IEEE Trans. Acoust., Speech, Signal Processing*, vol. 33, no. 1, pp. 50–60, February 1985.

$$u \triangleq \frac{1}{\lambda}\cos\alpha = \frac{1}{\lambda}\sin\theta\cos\phi$$

$$v \triangleq \frac{1}{\lambda}\cos\beta = \frac{1}{\lambda}\sin\theta\sin\phi$$

$$w = \frac{1}{\lambda}\cos\gamma = \frac{1}{\lambda}\cos\theta. \tag{5}$$

Here λ is the wavelength and the angles θ, ϕ are with reference to a spherical coordinate system. The generalized scattering function $L(\vec{q})$ is then obtained as the five-dimensional Fourier transform of the space-time correlation function:

$$L(\vec{q}) = \int R_H(\vec{p})\, e^{j2\pi\vec{p}\cdot\vec{q}}\, d\vec{p} \tag{6}$$

where $\vec{p}\cdot\vec{q}$ represents the scalar product between the vectors $\vec{p}$ and $\vec{q}$ and $d\vec{p}$ is the five-dimensional volume element

$$d\vec{p} = [d(\Delta f), d(\Delta t), d(\Delta x), d(\Delta y), d(\Delta z)]^T. \tag{7}$$

The inverse transform of $L(\vec{q})$ yields $R_H(\vec{p})$:

$$R_H(\vec{p}) = \int L(\vec{q})\, e^{-j2\pi\vec{p}\cdot\vec{q}}\, d\vec{q}. \tag{8}$$

The scattering function $L(\vec{q})$ describes the distribution of the signal power with respect to channel delay, Doppler φ, and the three angular coordinates u, v, and w.

B. Special Case of Angular Scattering

Here we consider a simple example of a hypothesized medium, which has only angular scattering in one plane. In other words, the space-time correlation function is assumed to depend only on the separation distance Δx, so that the vector $\vec{p}$ becomes a scalar variable Δx, and the vector $\vec{q}$ contains the scalar variable $u = 1/\lambda \cos\alpha$. Let $\theta = 90 - \alpha$, so that $u = 1/\lambda \sin\theta$. We then have

$$L(\theta) = \int R_H(\Delta x)\, e^{(j2\pi/\lambda)\Delta x \sin\theta}\, d(\Delta x). \tag{9}$$

In order to get a clear picture of the above relation, it is instructive to obtain (9) from elementary principles for the case of two sensors separated from each other by a distance Δx, as shown in Fig. 1(a). The scattering model is summarized in Fig. 1(b).

Assume that the source is transmitting an impulse $\delta(t)$. Due to angular scattering, the two sensors receive energy from an angular sector, say between $\{\theta_o - \theta_m \text{ to } \theta_o + \theta_m\}$, where θ_o is the mean direction and θ_m is usually small. Assume first, that the energy comes via, say N, different directions within this sector, i.e., from N different plane waves. The kth plane wave arriving from direction θ_k has the associated delays $\tau_k \pm \Delta x \sin\theta_k/2c$ at the two sensors, where τ_k is the delay of the plane wave to the midpoint along the line of the two sensors, relative to an unscattered plane wave in the mean direction. Thus the impulse response of the medium to the two sensors is given by

$$h\left(t, \pm\frac{\Delta x}{2}\right) = \sum_{k=1}^{N} a_k\, \delta\left(t - \tau_k \pm \frac{\Delta x}{2c}\sin\theta_k\right) \tag{10}$$

where a_k is the strength associated with the kth plane wave. The corresponding transfer functions are given by

$$H\left(f, \pm\frac{\Delta x}{2}\right) = \sum_{k=1}^{N} a_k e^{j2\pi f[-\tau_k \pm (\Delta x/2c)\sin\theta_k]}. \tag{11}$$

In the limiting case, when we assume a continuum of plane waves arriving from all directions $[-\theta_m \text{ to } \theta_m]$ where θ_m is small, we have

$$H\left(f, \pm\frac{\Delta x}{2}\right) = \int_{-\theta_m}^{\theta_m} a(\theta)\, e^{j2\pi f[-\tau(\theta) \pm (\Delta x/2c)\sin\theta]} \tag{12}$$

where the mean direction θ_o is taken to be the broadside direction, without loss of generality, and where $a(\theta)$ and $\tau(\theta)$ now represent the path strength and relative delay, respectively, associated with the plane wave arriving at an angle θ.

The space-time correlation function for this case, then becomes

$$R_H(\Delta x) = E\left[H\left(f, \frac{\Delta x}{2}\right) H^*\left(f, -\frac{\Delta x}{2}\right)\right]$$

$$= \iint_{-\theta_m}^{\theta_m} E[a(\theta)\, a(\theta')\, e^{j2\pi[\tau(\theta') - \tau(\theta)]}]$$

$$\cdot\, e^{-j2\pi f(\Delta x/2c)(\sin\theta' + \sin\theta)}\, d\theta'\, d\theta. \tag{13}$$

Assuming uncorrelated scattering, i.e., assuming that the wavefronts coming from different directions arise from scatterers having uncorrelated cross sections, we have

$$E[a(\theta)\, a(\theta')\, e^{j2\pi f(\{\tau(\theta') - \tau(\theta)\}}] \triangleq L(\theta)\, \delta(\theta' - \theta) \tag{14}$$

where $L(\theta)$ is defined to be the angular scattering function giving the distribution of energy in the angular domain. This definition is motivated by the analogous uncorrelated scattering assumption of Bello for a time-varying channel [11]. From (13) and (14), it follows that

$$R_H(\Delta x) = \iint_{-\theta_m}^{\theta_m} L(\theta)\, \delta(\theta' - \theta)\, e^{-(j2\pi f \Delta x/2c)(\sin\theta' + \sin\theta)}\, d\theta$$

$$= \int_{-\theta_m}^{\theta_m} L(\theta)\, e^{-j2\pi f(\Delta x/c)\sin\theta}\, d\theta \tag{15}$$

which is the inverse transform of (9), as required.

C. Remarks

1) The one-dimensional scattering function $L(\theta)$ introduced above presents a highly simplified picture of the real medium

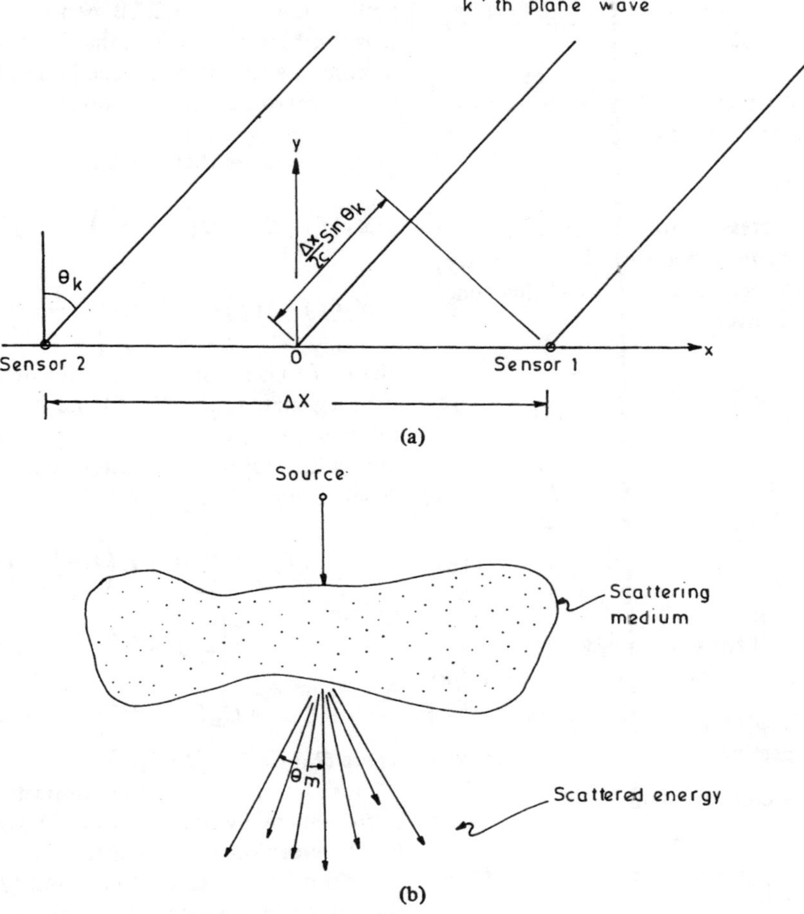

Fig. 1. (a) The array geometry for time-delay estimation. (b) The scattering model. Arrows in the scattered energy indicate directions of plane wavefronts.

characteristics in that it does not reflect the loss in coherence between the signals at the two sensors (or across the array aperture, as the case may be) due to other effects of scattering, viz. spreads in the delay and Doppler domains, etc. However, the simplicity of the resulting model not only permits easy evaluation of its effect on the time-delay estimation performance, but also makes the interpretation of the results simpler. It is, of course, of interest to generalize the results to be presented in the next section for more general scattering models.

2) It has been shown above that the space-time autocorrelation function and the scattering function are Fourier transform pairs. Thus if the scattering surface is assumed to be associated with a Gaussian, space-time autocorrelation function, then the resulting scattering function also has a Gaussian shape [7]. Thus, if we assume

$$R_H(\Delta x) = e^{-[(2\pi f/c)\Delta x \sigma]^2/2} \tag{16}$$

we then have

$$L(\theta) = \frac{1}{\sqrt{2\pi}\,\sigma} e^{-\theta^2/2\sigma^2} \tag{17}$$

where σ is defined to be "*spatial coherence loss coefficient*." The reciprocal of σ is a measure of the "coherence distance."

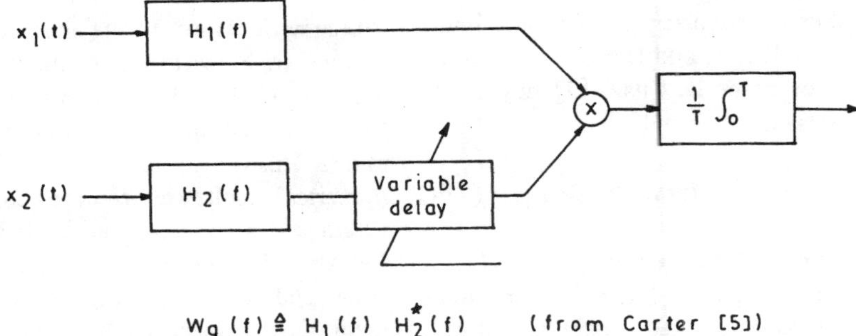

Fig. 2. Generalized cross-correlator for time-delay estimation (from [5]).

It is this correlation function which is used in the next section for typical performance calculations.

III. Performance of Time-Delay Estimators in Incoherent Media

A. Cramer-Rao Bound for TDE

Carter [5] has derived an expression for the variance of the time-delay estimate in the neighborhood of the true delay, for a general cross correlation receiver with a weighting function $W_g(f)$ (see Fig. 2). This is given by

$$\operatorname{var}^g D = \frac{\int_0^{\infty} |W_g(f)|^2 G_{x_1x_1}(f) G_{x_2x_2}(f)[1 - C_{12}(f)]\, f^2\, df}{8\pi^2 T\left[\int_0^{\infty} |G_{x_1x_2}(f)|\, W_g(f) f^2\, df\right]^2} \tag{18}$$

where

$G_{x_1x_2}(f)$ = cross spectrum of the two signals $x_1(t)$ and $x_2(t)$ received at the two sensors (19a)

$G_{x_1x_1}(f), G_{x_2x_2}(f)$ = autospectra of $x_1(t)$ and $x_2(t)$, respectively (19b)

$C_{12}(f)$ = magnitude square coherency function

$$\triangleq \left|\frac{G_{x_1x_2}(f)}{[G_{x_1x_1}(f) G_{x_2x_2}(f)]^{1/2}}\right|^2 \tag{19c}$$

T = observation interval. (19d)

In particular it has been shown that for the maximum likelihood (ML) processor, with

$$W_{\mathrm{ML}}(f) = \frac{C_{12}(f)}{[G_{x_1x_1}(f)|\,[1 - C_{12}(f)]} \tag{20}$$

the variance is given (under high SNR conditions) by

$$\operatorname{var}^{\mathrm{ML}} D = \left\{2T \int_{-\infty}^{\infty} (2\pi f)^2\, C_{12}(f)/(1 - C_{12}(f))^2\, df\right\}^{-1} \tag{21}$$

which is identical with the Cramer-Rao lower bound giving the minimum obtainable variance for delay estimation. This expression has been extensively evaluated in the literature for various special cases. It takes a particularly simple form, being a function only of the bandwidth, T, and the SNR, when the signals have flat spectra. For example, Quazi [3] has shown that for the ideal medium variance is given by

$$\operatorname{var}(D) = \left(\frac{3}{4\pi^2 T}\right) \frac{1}{\mathrm{SNR}} \frac{1}{f_2^3 - f_1^3} \qquad \text{for SNR} \gg 1 \tag{21a}$$

when the signal autospectrum is flat between f_1 to f_2 Hz. It should be noted, however, that even the ML estimator attains the CRLB performance of (21) only under certain conditions. It has been shown by Scarbrough *et al.* [8] that for a given bandwidth-observation time product, there is a threshold SNR below which CRLB performance will not be attained. It is this bound, however, that has been evaluated in the sequel to compare the performance in an ideal medium with that for a nonideal or scattering medium.

It is easy to see that, for our scattering model,

$$X_1(f) = S(f)\, H\left(f, -\frac{\Delta x}{2}\right) + N_1(f) \tag{22}$$

$$X_2(f) = S(f)\, e^{-j2\pi f D} H\left(f, \frac{\Delta x}{2}\right) + N_2(f) \tag{23}$$

where $S(f)$ is the spectrum of the desired signal, D is the delay to be estimated, and $N_1(f)$ and $N_2(f)$ are the spectra of the noises in the two sensors, assumed to be statistically independent with respect to (w.r.t.) each other as well as with the signal of interest. It follows that

$$\begin{aligned} G_{x_1x_2}(f) &= E\left\{\left[S(f) H\left(f, -\frac{\Delta x}{2}\right) + N_1(f)\right] \cdot \left[S(f)\, e^{j2\pi f D} H\left(f, \frac{\Delta x}{2}\right) + N_2(f)\right]^*\right\} \\ &= G_{ss}(f)\, e^{-j2\pi f D} R_H(\Delta x) \end{aligned} \tag{24}$$

where $G_{ss}(f) \triangleq E\{|S(f)|^2\}$ is the autospectrum of the signal.

Thus the value of the cross-spectrum $G_{x_1x_2}(f)$ and hence of the coherency function $C_{12}(f)$ would depend on the separation between the two sensors.

A close examination of (16) and (24) clearly shows that for the scattering medium considered in the previous section, as Δx increases, $G_{x_1x_2}(f)$ decreases and it follows from (21) that the variance of the delay error increases.

B. Bearing and Range Error Variances

The bearing error variance is related to the time-delay error variance by the relation [6]

$$\sigma_\theta^2 = \left(\frac{c}{\Delta x \cos\theta}\right)^2 \sigma_D^2 \tag{25}$$

where c is the velocity of sound in the medium, σ_D^2 is the variance of delay estimation error, and θ is the true angle of the source, with respect to the perpendicular to the line joining the two sensors.

For an ideal medium, therefore, the bearing error variance decreases as the separation distance between the sensors is increased, even though it is very high for small angles of the source with respect to the array line.

Since in a scattering medium σ_D is expected to increase with separation distance due to coherence loss, it follows that there will be an optimum distance for which σ_θ is minimum. This is demonstrated in the next section, where detailed numerical results are presented for the performance.

Range can be estimated with a minimum of three sensors. Let the distance between both adjacent pairs of sensors be equal and let the bearing of the source from the central element with respect to line perpendicular to the array be θ. Assuming that σ_D as obtained from both pairs is also same, it can be shown that [6]

σ_R^2 = range error variance

$$= \left(\frac{\sqrt{2} R^2 c}{(\Delta x)^2 \cos^2 \theta} \right)^2 \sigma_D^2 \tag{26}$$

where R is the true range.

Once again, unlike in an ideal medium where the performance is expected to improve as the fourth power of the separation distance, a range-dependent optimum separation distance would yield the minimum value of the range error variance for a scattering medium.

C. Bias Effects

Quazi [6] has derived expressions for the bias introduced in the bearing and range calculations from time-delay measurements. The results are reproduced here for convenience:

$$\theta_B = \text{bearing bias} = (-\sigma_\theta^2 \tan \theta)/2 \tag{27}$$

where θ is the true angle, and

$$R_B = \text{range bias} = \frac{\sigma_R^2}{R} \tag{28}$$

where R is the true range.

It is seen that bias in angle measurement is very high for grazing angles close to end fire and is negative or positive depending on θ. The range bias also becomes significant when the range variance is of the order of true range. It may be recalled from (26) that $\sigma_R^2 \propto R^4$ so that the range variance can quickly become greater than or equal to R as R increases.

IV. Typical Performance Calculations and Numerical Results

In this section we present typical performance curves for the estimation of bearing and range from time-delay measurements in scattering media. The parameters selected for study are the effects of SNR, separation distance between the sensors, and the "spatial coherence loss coefficient" introduced in Section II.

The signal and noise spectra are assumed to be flat and band limited between 3500 and 4500 Hz. It was felt that the performance of the ML estimator is mainly dependent on the SNR, rather than the signal and noise spectra, hence spectrum shape was not taken as a parameter in these studies. The observation time is taken to be 50 s, and has not been varied since σ_D^2 is known to be proportional to $1/T$. The speed of sound has been assumed to be 1500 m/s.

In order to provide a "benchmark" of comparison, the asymptotic performance of the maximum likelihood estimator in an ideal medium and additive Gaussian noise is first briefly summarized.

A. Performance of the TDE in Additive Noise in an Ideal Medium

Bearing Errors: Figs. 3 and 4 illustrate the important performance features of the bearing standard deviation (BSD) σ_θ in an ideal medium, with respect to SNR and sensor separation distance, respectively. Although bearing bias θ is not plotted here, its behavior is very similar to that of σ_θ^2 in view of its direct dependance on the latter via (27). Following are some of the important observations from Figs. 3 and 4 and (27).

1) BSD (σ_θ) varies inversely with the separation distance Δx, even though σ_D is independent of Δx in an ideal medium (Fig. 4).

2) The value of σ_θ also depends on the true bearing, being as low as 9.6×10^{-3} degrees for $\theta = 0^\circ$ (broadside direction) at an SNR of -20 dB and a separation distance of 150 m (Fig. 3).

3) The bearing bias θ_B is zero for $\theta = 0^\circ$. For small grazing angles with respect to the line of the array, however, the bias becomes very high in view of the $\tan \theta$ dependence (27).

Range Errors: The behavior of the range standard deviation (RSD) with respect to the SNR and Δx is also summarized in Figs. 3 and 4, respectively. The range bias R_B depends directly on σ_R^2 (28). Some of the important features of σ_R and R_B are as follows.

1) RSD, like BSD, displays an inverse relationship with respect to the SNR (Fig. 3).

2) RSD decreases more rapidly than BSD as the separation distance between the sensors is increased. This is because of the inverse square-law relationship between σ_R and Δx (Fig. 4).

3) The range bias R_B depends both on range and bearing. The bias, like the RSD σ_R, is least for a broadside direction ($\theta = 0^\circ$).

4) The range bias, however, has an inverse square-law dependence on the SNR, an inverse fourth-law dependence on the separation distance Δx, and a direct R^3 dependence on range. In order to obtain a reasonable value for R_B, therefore, it is required to choose an appropriately large separation distance. For example, with a separation distance of 150 m and an SNR of -20 dB, the value of the bias is only 0.316 m for a true range of 5000 m.

B. Performance of TDE in Scattering Media

For the following results, the spatial correlation function of (16) is used for calculating the cross-spectrum between the two received signals $x_1(t)$ and $x_2(t)$ via (24). The time-delay estimation error is computed via numerical evaluation of the integrals involved in (21), which in turn is used for the computation of bearing and range error standard deviations and biases. In order to study the effect of the medium, two representative values of σ are chosen [10], viz. $\sigma = 0.001$ and $\sigma = 0.01$. Figs. 5-8 demonstrate the important performance features of range and bearing measurement in a scattering medium. The following important observations can be made.

Bearing Standard Deviation (BSD):

1) Although the bearing standard deviation decreases with the SNR as in the ideal medium case, it can be seen that for a given SNR, BSD is considerably larger in the scattering medium. For example, for an SNR of -20 dB, the BSD is 9.6×10^{-3} degrees in an ideal medium (Fig. 3) as against 55.3×10^{-3} degrees in the scattering or incoherent medium (Fig. 5).

2) The behavior of the BSD in the scattering medium with respect to the sensor separation distance is illustrated in Fig. 6. As expected, we have an optimum separation distance for the

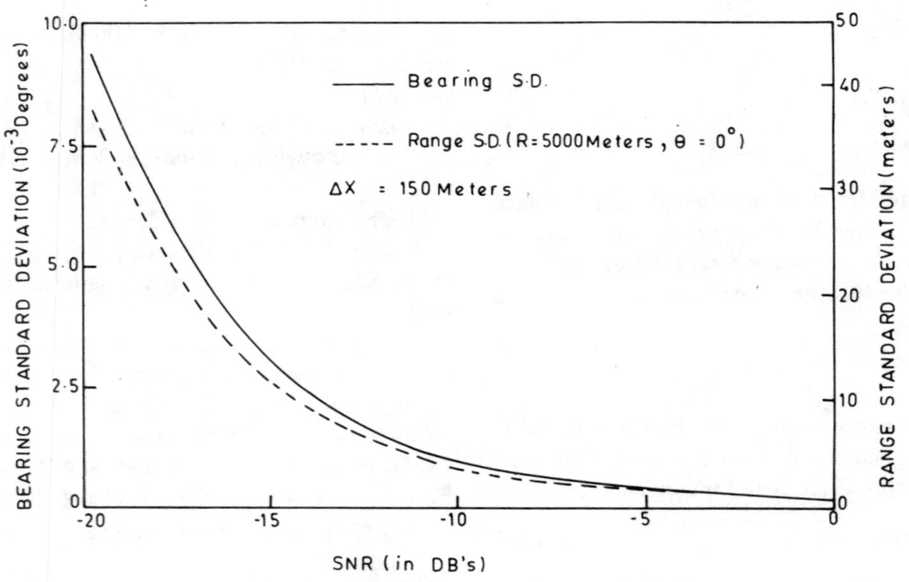

Fig. 3. SNR dependence of range and bearing standard deviations: ideal medium.

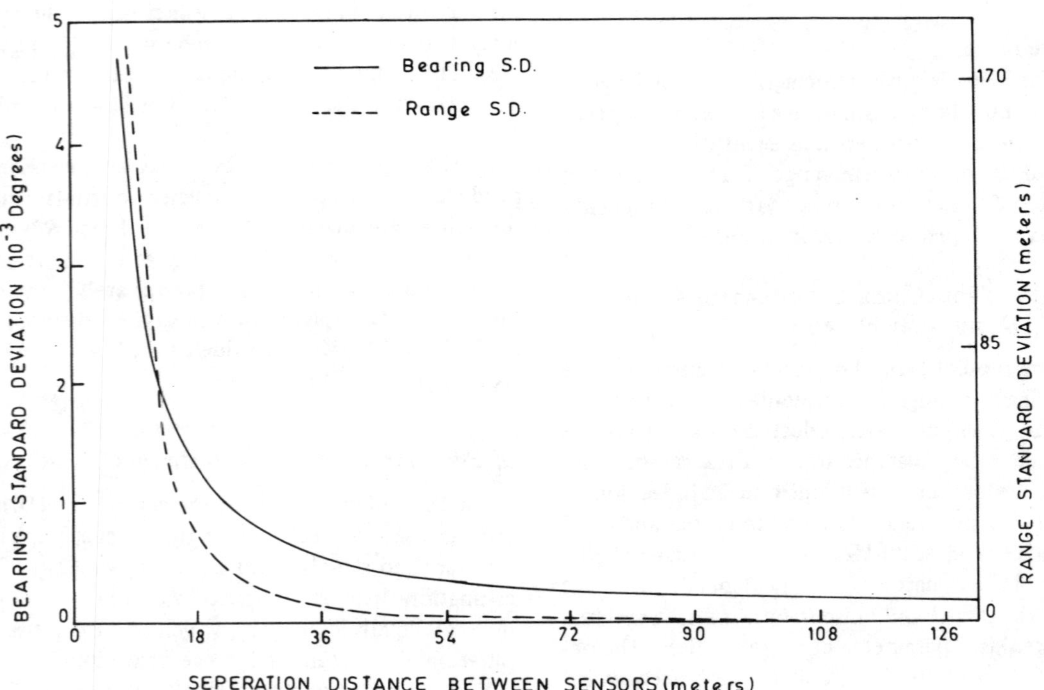

Fig. 4. Dependence of range and bearing standard deviations on sensor separation distance: ideal medium.

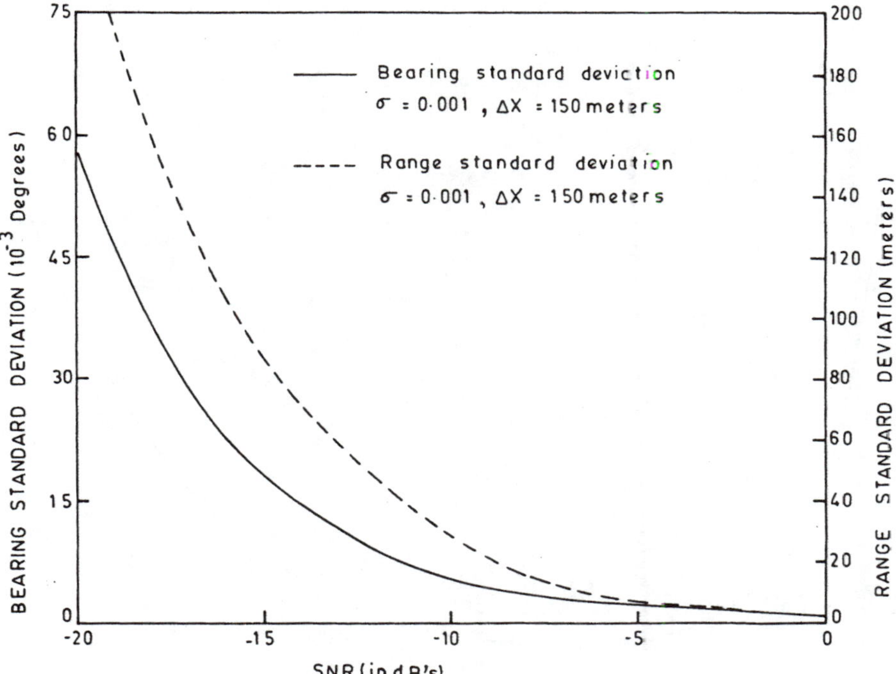

Fig. 5. SNR dependence of range and bearing standard deviations: nonideal medium ($\sigma = 0.001$).

minimum value of BSD. For the two cases shown here at an SNR of 0 dB, i.e., $\sigma = 0.01$ and $\sigma = 0.001$, the optimum distances are 8 and 75 m, respectively. The corresponding minimum values of BSD are 6.7×10^{-3} and 0.673×10^{-3} degrees, respectively.

It is interesting to observe, however, that for these cases the performance remains nearly constant (equal to the optimum value) when the separation distance is varied around the optimum value.

Range Standard Deviation (RSD):

1) A comparison of Figs. 3 and 5 shows a considerable loss in SNR performance in the nonideal medium. Thus, with a true range $R = 5000$ m and a sensor separation distance of 150 m, the range standard deviation is 11.4 m in the ideal medium ($\sigma = 0.001$) at an SNR of -15 dB.

2) The behavior of RSD with respect to the sensor separation distance is also similar to the corresponding behavior of BSD. Fig. 7 shows the existence of optimum values for the separation distance, the optimum values being 11 m for $\sigma = 0.01$ and 110 m for $\sigma = 0.001$. The corresponding values of RSD are 376 and 3.7 m, respectively. Once again, however, the curves are reasonably flat near the origin, although not as flat as for BSD.

Biases: Similar conclusions can also be drawn for bias in the range and bearing measurements. The results for range bias are summarized in Fig. 8. In fact, the behavior of the biases is very similar to the corresponding variances, except for scaling factors which depend on the true range and bearing [e.g., see (27) and (28)].

Choice of Sensor Separation Distance: A comparison of Figs. 6 through 8 clearly shows that the exact values of the separation distance are different for the minimum values of BSD, RSD, and the corresponding biases. However, for a given value of σ, these values are reasonably close. For example for $\sigma = 0.01$, the minimum value of BSD is obtained for $\Delta x = 8$ m, whereas the minimum value of RSD is obtained for $\Delta x = 11$ m. This fact, coupled with the flatness of the associated curves, enables us to obtain a compromise value of Δx. Thus, for the examples presented here, it may be seen that a reasonable choice of $\Delta x = 10$ m for $\sigma = 0.01$ and $\Delta x = 100$ m for $\sigma = 0.001$ can be made, without undue degradations in the BSD, RSD, and the corresponding biases.

V. Conclusions

It can be concluded that the concept of an optimum separation distance is a very important parameter in the design of an array for range and bearing estimation based on time-delay

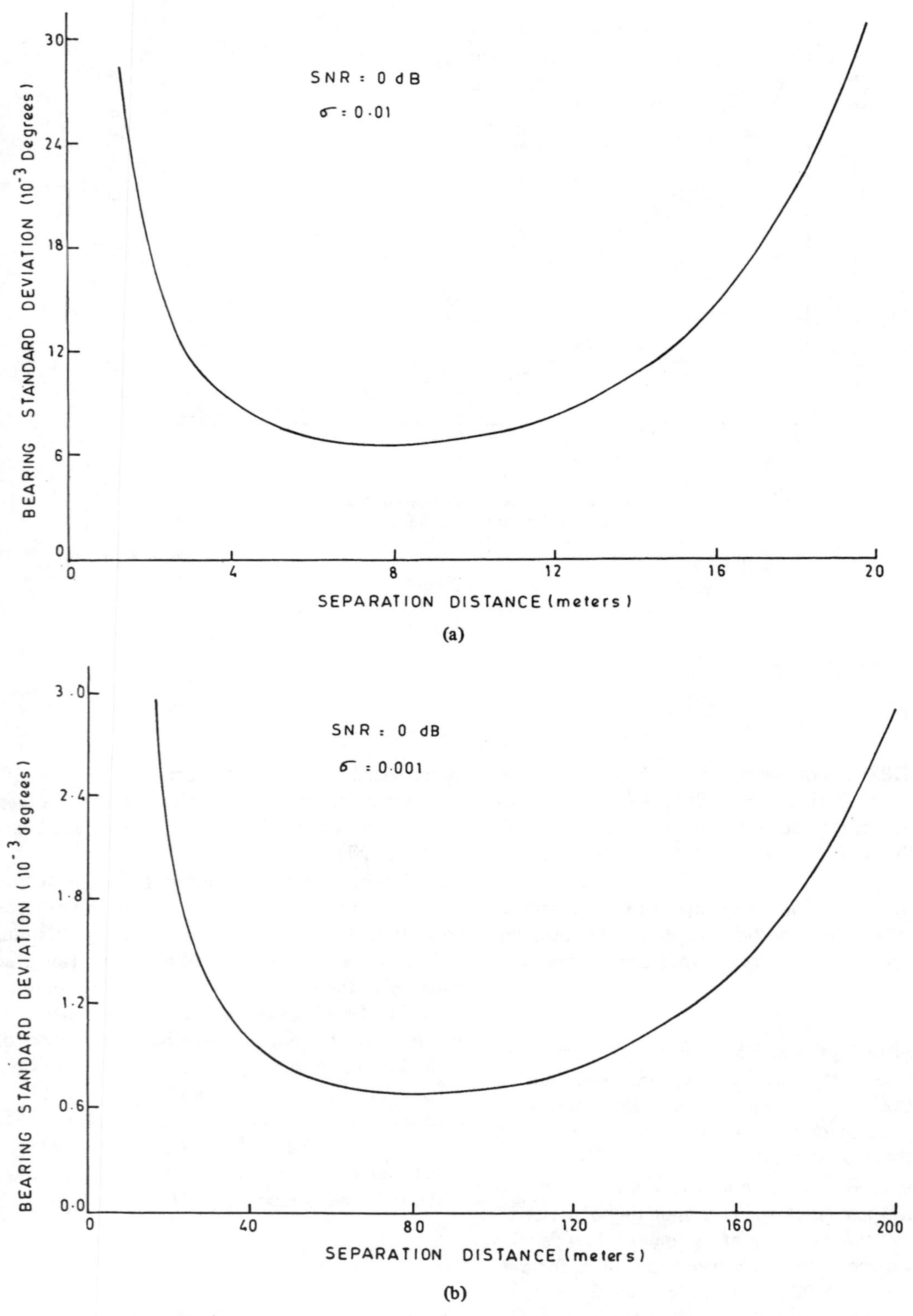

Fig. 6. Dependence of bearing standard deviation on sensor separation distance: nonideal medium. (a) $\sigma = 0.01$. (b) $\sigma = 0.001$.

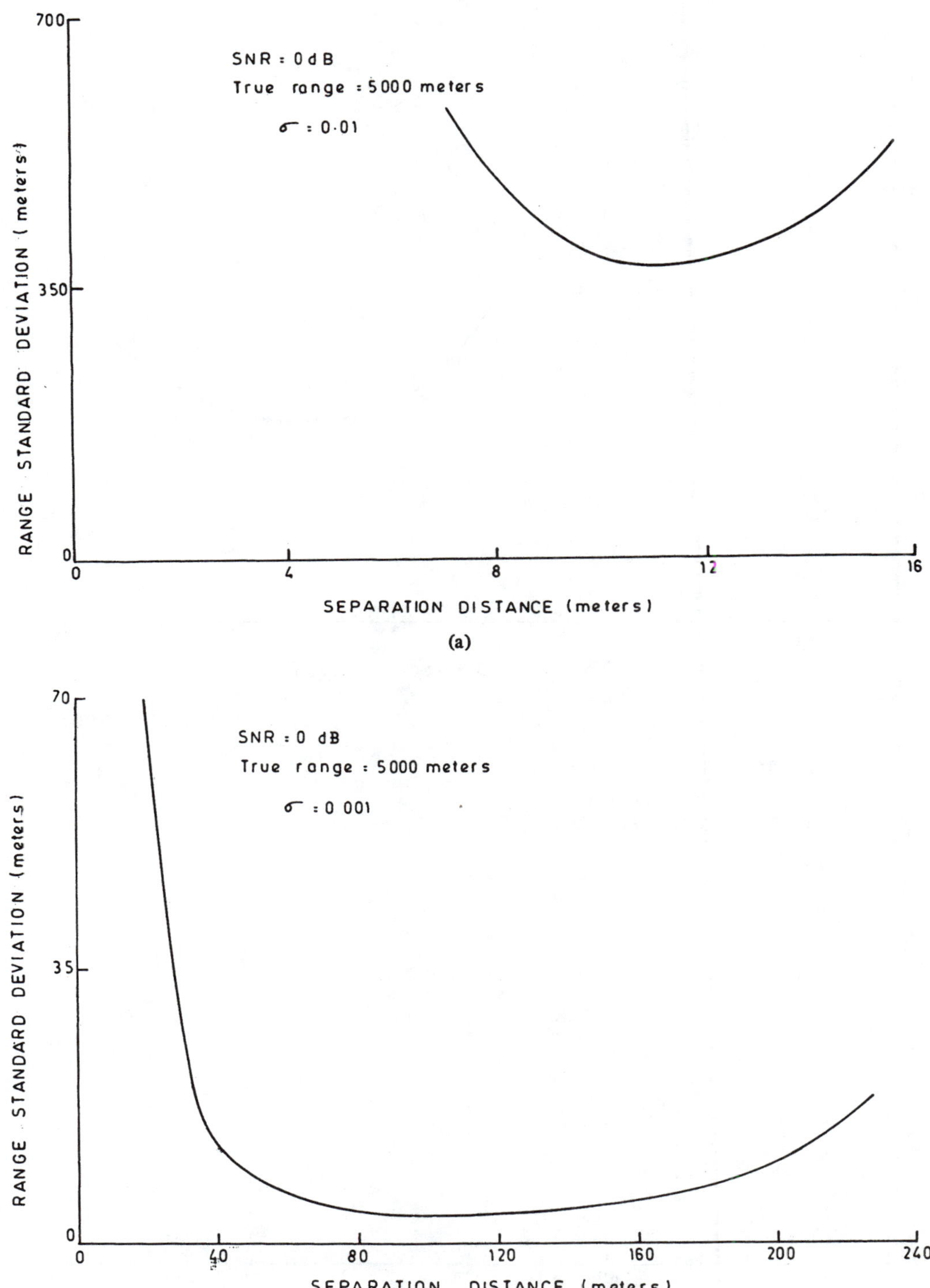

Fig. 7. Dependence of range standard deviation on sensor separation distance: nonideal medium. (a) $\sigma = 0.01$. (b) $\sigma = 0.01$.

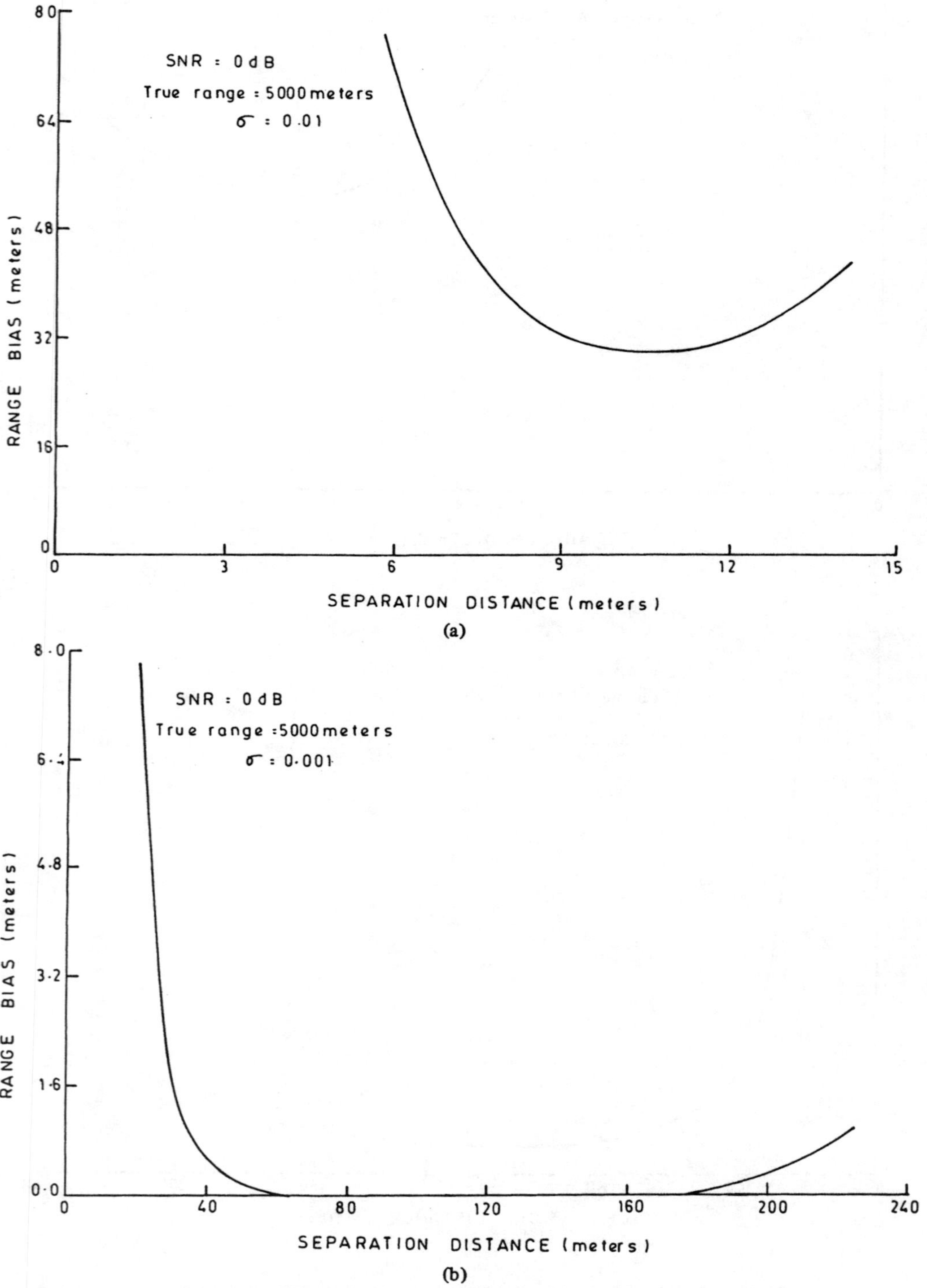

Fig. 8. Dependence of range bias on separation distance. (a) $\sigma = 0.01$. (b) $\sigma = 0.001$.

measurements. The optimum separation, however, depends not only on the scattering loss coefficient but also whether we intend to minimize the bearing error or range error standard deviation, or the corresponding biases. Hence the design of the array calls for an appropriate compromise. Fortunately, however, the performance curves are quite flat near the minima, so that a reasonable compromise is easy to obtain.

References

[1] Special Issue on Time Delay Estimation, *IEEE Trans. Acoust., Speech, Signal Process*, vol. ASSP-29, June 1981.

[2] C. H. Knapp and G. C. Carter, "The generalized correlation method for estimation of time delay," *IEEE Trans. Acoust., Speech, Signal Processing*, vol. ASSP-24, pp. 320–327, Aug. 1976.

[3] A. H. Quazi, "An overview of the time delay estimate in active and passive systems for target localization," *IEEE Trans. Acoust., Speech, Signal Processing*, vol. ASSP-29, pp. 527–533, June 1981.

[4] P. M. Schultheiss and E. Weinstein, "Lower bounds on the localization errors of a moving source observed by a passive array," *IEEE Trans. Acoust., Speech, Signal Processing*, vol. ASSP-29, pp. 600–607, June 1981.

[5] G. C. Carter, "Time delay estimation," Ph.D. dissertation, Univ. Connecticut, Storrs, 1976.

[6] A. H. Quazi, "Lower bounds on target localization errors for various signal and noise characteristics," presented at *Neuvieme Colloque sur le Trait du Signal et Ses Applications*, Nice, France, May 16–20, 1983.

[7] A. N. Venetsanopoulos, "Modelling of the sea surface scattering channel and underwater communications," in *Communication Systems and Random Process Theory*, J. K. Skwirzynski, Ed. The Netherlands: Sijthoff and Noordhoff, 1978.

[8] K. Scarbrough, R. J. Tremblay, and G. C. Carter, "Performance predictions for coherent and incoherent processing techniques of time delay estimation," *IEEE Trans. Acoust., Speech, Signal Processing*, vol. ASSP-31, pp. 1191–1196, Oct. 1983.

[9] J. P. Ianniello, E. Weinstein, and A. Weiss, "Comparison of the Ziv-Zakai lower bound on time delay estimation with correlator performance," in *Proc. 1983 Int. Conf. Acoust., Speech, Signal Processing*, 1983, pp. 875–878.

[10] A. Wasiljeff, "Spatial horizontal coherence of acoustic signals in shallow water," SACLANT ASW Res. Cen., La Spezia, Italy, SACLANTCEN Memo. SM-68, May 15, 1975.

[11] P. A. Bello, "Characterization of randomly time-variant linear channels," *IEEE Trans. Commun. Syst.*, vol. COM-16, pp. 360–393, 1963.

Optimistic and Pessimistic Approximations to Variance of Time Delay Estimators

J. ERIC SALT, MEMBER, IEEE, AND ARTHUR G. WACKER

Abstract—Estimators used for time delay estimation are sufficiently complex so that exact expressions for the variance of the estimators are generally not available. Existing approximate expressions, based on small error analysis, unfortunately give no indication of the accuracy of the approximation or even if the approximation yields variances that are larger or smaller than the actual variance of the estimator.

This paper presents a general theorem that can be used to find an optimistic or, if preferred, a pessimistic expression for the variance of a time delay estimator. The theorem concerns the relative location of the maxima of two functions which are contaminated with identical additive noise. The utility of the method developed is demonstrated by applying the approach to the generalized cross correlator to obtain both an optimistic expression and pessimistic expression for variance of the time delay.

I. INTRODUCTION

THERE are applications, for example, radar and sonar, where it is of interest to estimate the time delay between two noisy versions of the same signal received at spatially separated sensors. A common approach to estimating this delay is to define a similarity function between the two noisy signals and compute the similarity for different relative shifts between the two signals. The relative shift which maximizes or minimizes, whichever is appropriate, the similarity function is taken as an estimate of signal delay. A number of estimators of this type have been described in the literature [1]–[17].

Since a variety of similarity measures exist, it is of interest to be able to compare the performance of these estimators. The performance index generally chosen is the variance in the estimated delay. Unfortunately, most similarity measures of interest are sufficiently complex that it is very difficult, if not impossible, to determine a closed form expression for estimator variance in terms of signal and system parameters.

One approach that has been utilized to circumvent the complexity problem is a so-called small error analysis approach. In this approach, the similarity function is approximated by a simpler function to simplify the analysis. The variance of the time delay estimate computed for the simpler function is used as an approximation for the variance in the estimated time delay. Examples of this approach for the one-dimensional case are in the work of Helstrom [18] and Hertz [19], while McGillem and Svedlow [20] and Steding and Smith [21] consider the two-dimensional (image registration) case.

The principle difficulty with the small error analysis is that no indication of the accuracy of the approximate analysis is available. All that is known is that the estimation error of the approximate similarity function approaches the estimation error of the original similarity function under conditions where the estimation errors are infinitesimal. In situations where the received signals are contaminated with reasonable amounts of noise, i.e., the signal-to-noise ratio is above threshold, the estimation error may be considered small but not infinitesimal. In these situations, the accuracy of the approximation is unknown. In fact, it is not even known whether the approximation yields an optimistic or pessimistic value for estimation error.

The purpose of this paper is to provide a method of assessing the accuracy of variance expressions that are derived using approximations. The variance of interest could well be for a suboptimum estimator. The main contribution lies in the development of a general theorem concerning the relative location of the maximum of a loosely constrained class of functions. While this theorem may have broader applicability, it can be applied to determine error limits for a variance expression that is derived from a parabolic approximation to the similarity function. This includes variance expressions arrived at through small error techniques where the analysis is based on a parabolic approximation.

The error of any variance expression can be established by Monte Carlo simulation, but this requires a significant amount of computer time. At present there is no known way, outside of simulation, to establish error limits, at least a high side error limit, for the variance expression. A low side error limit could be established using the Cramer–Rao lower bound, since it is known that the variance must be greater than this bound. An error limit established in this way may be reasonably tight for near optimum estimators but may be somewhat loose for suboptimum estimators. The method of calculating error limits presented in this paper provides both a high side and a low side error limit. The low side limit is tighter than the low side limit derived from the Cramer–Rao lower bound when the es-

Manuscript received October 26, 1987; revised September 5, 1988. This work was supported by the Natural Science and Engineering Research Council of Canada.

The authors are with the Communications Systems Research Group, Electrical Engineering Department, University of Saskatchewan, Saskatoon, Sask., Canada S7N 0W0.

IEEE Log Number 8926666.

Reprinted from *IEEE Trans. Acoust., Speech, Signal Processing*, vol. 37, no. 5, pp. 634–641, May 1989.

timator of interest is suboptimum and the method is much more computer efficient than running Monte Carlo simulations.

The organization of this paper is as follows. First the theorem is stated and proven. The theorem is then applied to the similarity function of the commonly used time delay estimator known as the generalized cross correlator. Parabolic approximations which yield both optimistic and pessimistic variance estimates for the generalized cross correlator are discussed. Specific numeric results are presented for a particular signal and noise spectrum.

II. Theorem on the Relative Location of Maxima of Two Functions

There are situations where the location of the global maximum of some function is of interest but only a noisy version of the function is available for analysis. It is common to use the location of the global maximum of the noisy function as an estimate of the location of interest. If two noisy functions are available, one of which may well be contrived and only available in theory, and these functions satisfy certain conditions, then the noisy function that yields the better estimate can be identified.

Suppose the location of the global maximum of interest occurs at $\tau = \tau_0$. Consider noisy functions $m_1(\tau)$ and $m_2(\tau)$ with unique global maxima at $\tau = \tau_1$ and $\tau = \tau_2$, respectively. The distance $|\tau_1 - \tau_0|$ is less than or equal to distance $|\tau_2 - \tau_0|$ under the following conditions.

1) τ_1 is in the interval $\tau_0 - e_m \leq \tau_1 < \tau_0 + e_m$, where e_m is a constant.

2) The two functions can be represented as

$$m_1(\tau) = m_{S1}(\tau) + m_n(\tau) \tag{1}$$

and

$$m_2(\tau) = m_{S2}(\tau) + m_n(\tau), \tag{2}$$

where $m_n(\tau)$, $m_{S1}(\tau)$, and $m_{S2}(\tau)$ are differentiable for τ in the interval $\tau_0 - e_m \leq \tau \leq \tau_0 + e_m$.

3) Functions $m_{S1}(\tau)$ and $m_{S2}(\tau)$ are even about $\tau = \tau_0$.

4) The derivatives of the even functions $m_{S1}(\tau)$ and $m_{S2}(\tau)$ satisfy

$$m'_{S1}(\tau) \leq m'_{S2}(\tau), \qquad \text{for all } \tau, \qquad \tau_0 \leq \tau \leq \tau_0 + e_m. \tag{3}$$

Proof: This theorem is proved using the method of contradiction. The proof is accomplished by considering separately four cases: a) $\tau_0 \leq \tau_1 \leq \tau_0 + e_m$, $\tau_2 \geq \tau_0$; b) $\tau_0 - e_m \leq \tau_1 < \tau_0$, $\tau_2 \geq \tau_0$; c) $\tau_0 - e_m \leq \tau_1 < \tau_0$, $\tau_2 < \tau_0$; and d) $\tau_0 \leq \tau_1 \leq \tau_0 + e_m$, $\tau_2 < \tau_0$. These four cases cover all possible values for the pair (τ_1, τ_2).

Proof for Case a): The proof starts by assuming that $|\tau_1 - \tau_0|$ is greater than $|\tau_2 - \tau_0|$. Since case a) has both τ_1 and τ_2 greater than or equal to τ_0, the assumption implies $\tau_1 > \tau_2$. From the definition of interval integration, it follows that

$$m_1(\tau_1) - m_1(\tau_2) = \int_{\tau_2}^{\tau_1} m'_1(\tau)\, d\tau \tag{4}$$

and

$$m_2(\tau_1) - m_2(\tau_2) = \int_{\tau_2}^{\tau_1} m'_2(\tau)\, d\tau. \tag{5}$$

For this case, the limits of integration are in the interval $[\tau_0, \tau_0 + e_m]$. The fourth condition, together with (1) and (2), therefore implies the integrand in (4) is less than or equal to the integrand in (5) throughout the region of integration. Thus,

$$\int_{\tau_2}^{\tau_1} m'_1(\tau)\, d\tau \leq \int_{\tau_2}^{\tau_1} m'_2(\tau)\, d\tau, \tag{6}$$

which from (4) and (5) implies

$$m_1(\tau_1) - m_1(\tau_2) \leq m_2(\tau_1) - m_2(\tau_2). \tag{7}$$

However, τ_1 maximizes $m_1(\tau)$ so $m_1(\tau_1) - m_1(\tau_2) > 0$ and τ_2 maximizes $m_2(\tau)$ so $m_2(\tau_1) - m_2(\tau_2) < 0$, which is in disagreement with (7). Thus, the assumption that the displacement of the peak, $|\tau_1 - \tau_0|$, incurred by $m_1(\tau)$ is greater than the displacement of the peak, $|\tau_2 - \tau_0|$, incurred by $m_2(\tau)$ leads to a mathematical contradiction and therefore must be false.

Proof for Case b): The proof starts by assuming $|\tau_1 - \tau_0|$ is greater than $|\tau_2 - \tau_0|$. Case b) has τ_1 less than τ_0 and τ_2 greater than or equal to τ_0 so the assumption implies $-(\tau_1 - \tau_0) > (\tau_2 - \tau_0)$. From the definition of interval integration, it follows that

$$\begin{aligned} m_1(\tau_2) - m_1(\tau_1) &= \int_{\tau_1}^{\tau_2} m'_1(\tau)\, d\tau = \int_{\tau_1}^{\tau_0 - (\tau_2 - \tau_0)} m'_1(\tau)\, d\tau \\ &\quad + \int_{\tau_0 - (\tau_2 - \tau_0)}^{\tau_0 + (\tau_2 - \tau_0)} m'_1(\tau)\, d\tau \end{aligned} \tag{8}$$

and

$$\begin{aligned} m_2(\tau_2) - m_2(\tau_1) &= \int_{\tau_1}^{\tau_2} m'_2(\tau)\, d\tau = \int_{\tau_1}^{\tau_0 - (\tau_2 - \tau_0)} m'_2(\tau)\, d\tau \\ &\quad + \int_{\tau_0 - (\tau_2 - \tau_0)}^{\tau_0 + (\tau_2 - \tau_0)} m'_2(\tau)\, d\tau. \end{aligned} \tag{9}$$

Since $m_{S1}(\tau)$ is even about $\tau = \tau_0$, its derivative, $m'_{S1}(\tau)$, is odd about $\tau = \tau_0$ and does not contribute to the second term of (8) because the region of integration is centered on $\tau = \tau_0$. Thus, utilizing (1), the second term of (8) can be expressed as

$$\int_{\tau_0 - (\tau_2 - \tau_0)}^{\tau_0 + (\tau_2 - \tau_0)} m'_1(\tau)\, d\tau = \int_{\tau_0 - (\tau_2 - \tau_0)}^{\tau_0 + (\tau_2 - \tau_0)} m'_n(\tau)\, d\tau. \tag{10}$$

Similarly, the second term of (9) is given by

$$\int_{\tau_0 - (\tau_2 - \tau_0)}^{\tau_0 + (\tau_2 - \tau_0)} m'_2(\tau)\, d\tau = \int_{\tau_0 - (\tau_2 - \tau_0)}^{\tau_0 + (\tau_2 - \tau_0)} m'_n(\tau)\, d\tau, \tag{11}$$

which is the same as the second term of (8). Consider the first terms in (8) and (9). The limits of integration, τ_1 and $\tau_0 - (\tau_2 - \tau_0)$, are in the interval $[\tau_0 - e_m, \tau_0]$. Since $m_{S1}(\tau)$ and $m_{S2}(\tau)$ are even functions about $\tau = \tau_0$, their derivatives are odd functions about $\tau = \tau_0$, and so (3) implies

$$m'_{S1}(\tau) \geq m'_{S2}(\tau), \qquad \tau_0 - e_m \leq \tau \leq \tau_0. \quad (12)$$

From the derivatives of (1) and (2), it follows that $m'_1(\tau) \geq m'_2(\tau)$ and

$$\int_{\tau_1}^{\tau_0 - (\tau_2 - \tau_0)} m'_1(\tau)\, d\tau \geq \int_{\tau_1}^{\tau_0 - (\tau_2 - \tau_0)} m'_2(\tau)\, d\tau. \quad (13)$$

This, together with the fact established above that the second term in (8) and (9) are equal, implies

$$m_1(\tau_2) - m_1(\tau_1) \geq m_2(\tau_2) - m_2(\tau_1). \quad (14)$$

However, τ_1 maximizes $m_1(\tau)$ so $m_1(\tau_2) - m_1(\tau_1) < 0$ and τ_2 maximizes $m_2(\tau)$ so $m_2(\tau_2) - m_2(\tau_1) > 0$, which is in disagreement with (14). Thus, the assumption that the displacement of the peak, $|\tau_1 - \tau_0|$, incurred by $m_1(\tau)$ is greater than the displacement of the peak, $|\tau_2 - \tau_1|$, incurred by $m_2(\tau)$ leads to a contradiction and therefore must be false.

The proofs of cases c) and d) are similar to those of a) and b), respectively, and are, therefore, omitted. This completes the proof of the theorem.

III. Application of Theorem

A. The Generalized Cross Correlator

A similarity function, known as the generalized cross correlator, is frequently used in time delay estimation [7]. The theorem of Section I can be applied to this similarity function to get both optimistic and pessimistic estimates of the variance of estimation error. A block diagram of the models for the two noisy signals and the generalized cross correlator is given in Fig. 1. The generalized cross correlator estimates the delay between two noisy signals designated in Fig. 1 as

$$y_1(t) = x(t) + n_1(t) \quad (15)$$

and

$$y_2(t) = x(t - \tau_0) + n_2(t), \quad (16)$$

where τ_0 is the delay, $x(t)$ is an ergodic signal, and $n_1(t)$ and $n_2(t)$ are stationary signals that are independent of $x(t)$ and each other. The similarity function of the generalized cross correlator has been derived by Knapp and Carter [7] and is given by

$$\hat{R}_{12}(\tau) = \frac{1}{2\pi} \int_{-\infty}^{\infty} H_1(\omega)\, H_2^*(\omega)\, \hat{S}_{12}^T(\omega)\, e^{j\omega\tau}\, d\omega, \quad (17)$$

where the similarity function $\hat{R}_{12}(\tau)$ is an estimate of the cross-correlation function of the two received signals, ω is angular frequency with units radians per second, $H_1(\omega)$ and $H_2(\omega)$ are the frequency responses of the two filters

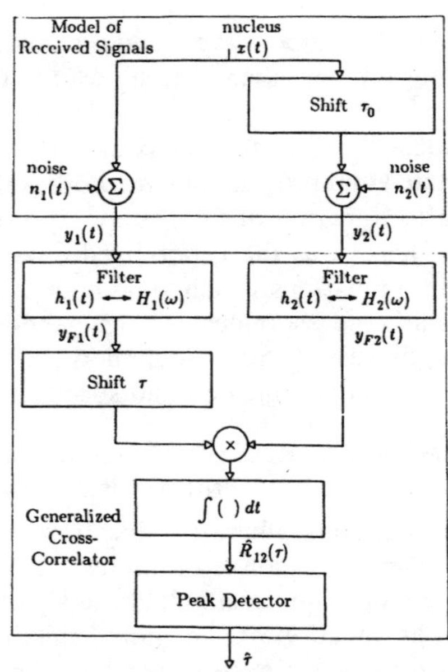

Fig. 1. Block diagram of the models for the received signals and the generalized cross correlator.

in the generalized cross correlator, $H_2^*(\omega)$ is the complex conjugate of $H_2(\omega)$, and $\hat{S}_{12}^T(\omega)$ is an estimte of the cross power spectrum of the two received signals. The superscript T in $\hat{S}_{12}^T(\omega)$ signifies that the spectral estimate is based on observations of received signals $y_1(t)$ and $y_2(t)$ from $t = 0$ to $t = T$ seconds.

There are a variety of ways that the cross power spectrum can be estimated. For example, it can be estimated by the Blackman and Tukey [22] method or by Welch's [23] overlapping method. The Blackman and Tukey method is often implemented by computing the discrete cross-correlation function using the entire lengths of the unweighted received signals, windowing the discrete cross-correlation function with a tapered window, and computing the discrete Fourier transform (DFT) of the discrete cross-correlation function. Welch's method involves partitioning the two received signals into segments and computing the cross power spectrum by averaging the cross power spectrums computed on the segments using the DFT.

Both of the methods outlined above estimate the cross power spectrum with a uniformly spaced line spectrum (discrete frequency spectrum). Since uniformly spaced line spectra are associated with periodic signals, both these methods must implicitly transform the received signals into periodic signals. The periodicity of the signals is the full record length of T seconds in the case of the Blackman and Tukey method, while it is somewhat shorter (the segment length) in the case of the Welch method.

The similarity function resulting from (17) depends on the method used to estimate the cross power spectrum. Thus, many different functions can be obtained from (17), although these functions will differ only slightly in the region of τ near $\tau_0 \ll T$. In the analysis that follows, only one similarity function is considered. This function is obtained by explicitly expanding the nucleus $x(t)$ and noise functions into periodic signals and computing a time average estimate of the cross-correlation function. The signal expansion is accomplished by repeating the T seconds signals *ad infinitum*. Thus, the expanded signals are periodic with period T and are given by

$$\tilde{x}(t) = x(t - mT), \quad mT \le t < (m + 1)T, \quad m = 0, \pm 1, \pm 2, \cdots, \tag{18}$$

$$\tilde{n}_1(t) = n_1(t - mT), \quad mT \le t < (m + 1)T, \quad m = 0, \pm 1, \pm 2, \cdots, \tag{19}$$

and

$$\tilde{n}_2(t) = n_2(t - mT), \quad mT \le t < (m + 1)T, \quad m = 0, \pm 1, \pm 2, \cdots, \tag{20}$$

where $\tilde{}$ is used to distinguish the expanded periodic signals from the T second duration signals. The similarity function is given by

$$\tilde{R}_{12}(\tau) = \frac{1}{T} \int_0^T \tilde{y}_{F1}(t + \tau)\, \tilde{y}_{F2}(t)\, dt, \tag{21}$$

where $\tilde{y}_{F1}(t)$ and $\tilde{y}_{F2}(t)$ are the periodic signals that result from filtering the extended received signals, $\tilde{y}_1(t)$ and $\tilde{y}_2(t)$. Specifically, $\tilde{y}_{F1}(t)$ is given by

$$\tilde{y}_{F1}(t) = \int_{-\infty}^{\infty} h_1(t - \lambda)\, [\tilde{x}(\lambda) + \tilde{n}_1(\lambda)]\, d\lambda, \tag{22}$$

which can be expressed as

$$\tilde{y}_{F1}(t) = \int_{-\infty}^{\infty} h_1(t - \lambda)\, \tilde{x}(\lambda)\, d\lambda + \int_{-\infty}^{\infty} h_1(t - \lambda)\, \tilde{n}_1(\lambda)\, d\lambda. \tag{23}$$

Letting $\tilde{x}_{F1}(t)$ and $\tilde{n}_{1F}(t)$ denote the filtered form of $\tilde{x}(t)$ and $\tilde{n}_1(t)$, this becomes

$$\tilde{y}_{F1}(t) = \tilde{x}_{F1}(t) + \tilde{n}_{1F}(t). \tag{24}$$

Similarly, $\tilde{y}_{F1}(t)$ is given by

$$\tilde{y}_{F2}(t) = \tilde{x}_{F2}(t - \tau_0) + \tilde{n}_{2F}(t), \tag{25}$$

where $\tilde{x}_{F2}(t - \tau_0)$ and $\tilde{n}_{2F}(t)$ denote $\tilde{x}(t - \tau_0)$ and $\tilde{n}_2(t)$ after they pass through the filter with impulse response $h_2(t)$.

Under conditions where the differential shift between the received signals is small compared to the observation interval, i.e., $\tau_0 \ll T$, the similarity function given by (21) will differ only slightly from each of the many functions that may be obtained from (17). The difference between the function of (21) and any of the functions given by (17) is a function of τ. There is essentially no difference for $\tau \ll T$. For $\tau_0 \ll T$, $\hat{R}_{12}(\tau)$ is only of interest in the region of $\tau \ll T$ and the similarity function given by (21) is representative of the many possible similarity functions of the generalized cross correlator.

B. Optimistic Approximation to $\hat{R}_{12}(\tau)$

The objective of this subsection is to approximate the similarity function $\hat{R}_{12}(\tau)$ given by (21) such that the variance of the estimation error incurred by the approximate similarity function is both calculable and less than or equal to the variance of the estimation error incurred by $\hat{R}_{12}(\tau)$. Such an approximate similarity function is obtained in two steps. First $\hat{R}_{12}(\tau)$ is separated into the two components—a component due to the received signals in the absence of noise and a component due to the noise corrupting the received signals. The component of $\hat{R}_{12}(\tau)$ that is due to the signals in the absence of noise is approximated by a parabola. The curvature of this parabola is carefully chosen in accordance with the theorem on the relative location of the global maxima of two functions to ensure the approximate similarity function incurs an equal or smaller variance of estimation error.

The similarity function given by (21) is separated into two components by using (24) and (25) to represent functions $\tilde{y}_{F1}(t + \tau)$ and $\tilde{y}_{F2}(t)$, cross multiplying, and expressing the integration of a sum as the sum of two integrations. This results in the expression

$$\hat{R}_{12}(\tau) = \hat{R}_{s12}(\tau) + \hat{R}_{n12}(\tau), \tag{26}$$

where $\hat{R}_{s12}(\tau)$ is the component due to the signal in the absence of noise and is given by

$$\hat{R}_{s12}(\tau) = \frac{1}{T} \int_0^T \tilde{x}_{F1}(t + \tau)\, \tilde{x}_{F2}(t - \tau_0)\, dt \tag{27}$$

and $\hat{R}_{n12}(\tau)$ is the component due to noise and is given by

$$\hat{R}_{n12}(\tau) = \frac{1}{T} \int_0^T [\tilde{x}_{F1}(t + \tau)\, \tilde{n}_{2F}(t) + \tilde{n}_{1F}(t + \tau)\, \tilde{x}_{F2}(t - \tau_0) + \tilde{n}_{1F}(t + \tau)\, \tilde{n}_{2F}(t)]\, dt. \tag{28}$$

The next step is to approximate the component due to the signal in the absence of noise with a parabola to get the approximate similarity function

$$m_o(\tau) = R_{s12}(\tau_0) + C_o(\tau - \tau_0)^2 + \hat{R}_{n12}(\tau), \tag{29}$$

where the curvature of the parabola, C_o, is determined in accordance with the theorem to ensure the approximate similarity function incurs the smaller error. It is necessary for C_o to be a constant in order to arrive at an expression for the variance of estimation error. Having to choose a fixed C_o means the parabola $R_{s12}(\tau_0) + C_o(\tau - \tau_0)^2$ can only approximate a single estimate of the cross-correlation function unless the nucleus $\tilde{x}(t)$ is an ergodic signal, in which case each estimate $\hat{R}_{s12}(\tau)$ yields the true function $R_{s12}(\tau)$. It is therefore necessary to assume the nucleus is an ergodic signal to assure $m_o(\tau)$ represents $\hat{R}_{s12}(\tau)$ for the ensemble of nucleus functions. (This is in fact the underlying reason why the nucleus is assumed to be ergodic in analyses that are premised on the estimation error being infinitesimal.) Under the assumption of ergodicity, it can be shown that $\hat{R}_{s12}(\tau)$ is the deterministic function of τ given by

$$\hat{R}_{s12}(\tau) = \sum_{k=0}^{K} \left|H_1\left(\frac{2\pi k}{T}\right)\right| \left|H_2\left(\frac{2\pi k}{T}\right)\right| \tilde{X}\left(\frac{2\pi k}{T}\right) \cdot \cos\left(\frac{2\pi k(\tau - \tau_0)}{T}\right), \tag{30}$$

where $\tilde{X}(2\pi k/T)$ denotes the power in $\tilde{x}(t)$ at $\omega = 2\pi k/T$, K denotes the highest nonzero frequency component of $\tilde{x}(t)$, and vertical bars $|\cdot|$ indicate absolute value.

It is also necessary to assume that the noise corrupting the two extended received signals, i.e., $\tilde{n}_1(t)$ and $\tilde{n}_2(t)$, originate from stationary processes in order to arrive at an expression for the variance of estimation error of the approximate similarity function. The expression for the variance can be obtained after straightforward but lengthy calculations using either the method of McGillen and Svedlow [19] or the method of Salt [24] with the result

$$\sigma_o^2 = \frac{1}{8C_o^2} \sum_{k=0}^{K} \left(\frac{2\pi k}{T}\right)^2 \left|H_1\left(\frac{2\pi k}{T}\right)\right|^2 \left|H_2\left(\frac{2\pi k}{T}\right)\right|^2 \cdot \left\{\tilde{N}_1\left(\frac{2\pi k}{T}\right)\tilde{X}\left(\frac{2\pi k}{T}\right) + \tilde{X}\left(\frac{2\pi k}{T}\right)\tilde{N}_2\left(\frac{2\pi k}{T}\right) + \tilde{N}_1\left(\frac{2\pi k}{T}\right)\tilde{N}_2\left(\frac{2\pi k}{T}\right)\right\}, \tag{31}$$

where σ_o^2 is the variance of estimation error for the approximate similarity function and $\tilde{N}_1(2\pi k/T)$ and $\tilde{N}_2(2\pi k/T)$ denote the power at angular frequency $\omega = 2\pi k/T$ in $\tilde{n}_1(t)$ and $\tilde{n}_2(t)$, respectively. In the derivation of this expression, it is assumed that the two filters in the generalized cross correlator, $H_1(\omega)$ and $H_2(\omega)$, do not introduce differential group delay.

An optimistic variance of estimation error for the similarity function can be obtained from the expression above, providing C_o is such that the approximation to the similarity function, $m_o(\tau)$, and the similarity function $\hat{R}_{12}(\tau)$ qualify for the two similarity functions of the theorem $m_1(\tau)$ and $m_2(\tau)$, respectively. Both functions can be separated into two components—one that is common to both functions and another that is an even function about $\tau = \tau_0$, as required by the theorem. This is evident from (26) and (29). Thus, $R_{s12}(\tau_0) + C_o(\tau - \tau_0)^2$, $\hat{R}_{s12}(\tau)$, and $\hat{R}_{n12}(\tau)$ assume the roles of the theorem functions $m_{S1}(\tau)$, $m_{S2}(\tau)$, and $m_n(\tau)$, respectively. For $m_o(\tau)$ to yield an optimistic variance, the theorem requires that

$$\frac{d}{d\tau}\left[R_{s12}(\tau_0) + C_o(\tau - \tau_0)^2\right] \le \frac{d}{d\tau}\tilde{R}_{s12}(\tau) \quad \text{for } \tau_0 \le \tau \le \tau_0 + e_m, \tag{32}$$

where e_m is a positive constant sufficiently large for the range $\tau_0 - e_m \le \tau \le \tau_0 + e_m$ to include the point of the global maximum of $m_o(\tau)$. Substituting (30) into the left side and taking the derivative of both sides yields

$$2C_o(\tau - \tau_0) \le -\sum_{k=0}^{K}\left[\frac{2\pi k}{T}\right]\left|H_1\left(\frac{2\pi k}{T}\right)\right|\left|H_2\left(\frac{2\pi k}{T}\right)\right| \cdot \tilde{X}\left(\frac{2\pi k}{T}\right)\sin\left(\frac{2\pi k(\tau - \tau_0)}{T}\right), \quad \text{for } \tau_0 \le \tau \le \tau_0 + e_m. \tag{33}$$

Realizing that this inequality is satisfied for any C_o when $\tau = \tau_0$, the range of τ can be reduced to exclude τ_0 and both sides of (33) can be divided by the nonzero quantity $2(\tau - \tau_0)$ with the result

$$C_o \le \frac{-1}{2}\sum_{k=0}^{K}\left[\frac{2\pi k}{T}\right]^2\left|H_1\left(\frac{2\pi k}{T}\right)\right|\left|H_2\left(\frac{2\pi k}{T}\right)\right| \cdot \tilde{X}\left(\frac{2\pi k}{T}\right)\frac{\sin\left(\frac{2\pi k(\tau - \tau_0)}{T}\right)}{\frac{2\pi k(\tau - \tau_0)}{T}}, \quad \text{for } \tau_0 < \tau \le \tau_0 + e_m. \tag{34}$$

The right side of this inequality will be referred to as the curvature function.

The largest of the possible optimistic variances is of most value as it is least optimistic. Since the variance is inversely proportional to the square of C_o, this variance is obtained when C_o is the smallest in magnitude. Since C_o must be less than or equal to the curvature function for τ in the range $\tau_0 < \tau \le \tau_0 + e_m$, C_o is given by the minimum of the curvature function over this range. From (34) it is seen that C_o is given by the value of the curvature function for τ approaching zero and this value is given by

$$C_o = \frac{-1}{2}\sum_{k=0}^{K}\left[\frac{2\pi k}{T}\right]^2\left|H_1\left(\frac{2\pi k}{T}\right)\right|\left|H_2\left(\frac{2\pi k}{T}\right)\right|\tilde{X}\left(\frac{2\pi k}{T}\right) \tag{35}$$

and does not depend on e_m.

The optimistic variance of estimation error for the similarity function of the generalized cross correlator is obtained by substituting (35) into the variance expression of (31) and is given by

$$\sigma_o^2 = \frac{\sum_{k=0}^{K}\left(\frac{2\pi k}{T}\right)^2\left|H_1\left(\frac{2\pi k}{T}\right)\right|^2\left|H_2\left(\frac{2\pi k}{T}\right)\right|^2\left\{\tilde{N}_1\left(\frac{2\pi k}{T}\right)\tilde{X}\left(\frac{2\pi k}{T}\right) + \tilde{X}\left(\frac{2\pi k}{T}\right)\tilde{N}_2\left(\frac{2\pi k}{T}\right) + \tilde{N}_1\left(\frac{2\pi k}{T}\right)\tilde{N}_2\left(\frac{2\pi k}{T}\right)\right\}}{2\left[\sum_{k=0}^{K}\left[\frac{2\pi k}{T}\right]^2\left|H_1\left(\frac{2\pi k}{T}\right)\right|\left|H_2\left(\frac{2\pi k}{T}\right)\right|\tilde{X}\left(\frac{2\pi k}{T}\right)\right]^2}. \tag{36}$$

This expression is the discrete form of the expression for an approximate variance obtained by both Hahn [25] and Knapp and Carter [7], suggesting their results yield optimistic estimates for the variance of the time delay. The discrete nature of the expression is a consequence of extending the received signals in a way that makes them periodic.

It is interesting to note that this discrete expression is greater than or equal to the Cramer–Rao lower bound for periodic signals [26]. The two become equal when the filters, $H_1(2\pi k/T)$ and $H_2(2\pi k/T)$, are the optimum maximum likelihood filters derived by Knapp and Carter [7]. This suggests the optimistic variance provides a tighter low side error limit than does the Cramer–Rao lower bound when the estimator of interest is a suboptimum estimator.

C. Pessimistic Approximation to $\hat{R}_{12}(\tau)$

A pessimistic value for the variance of the generalized cross correlator can be obtained in the same way that the optimistic variance was obtained. To find a pessimistic variance, the approximate similarity function and the actual similarity functions must again satisfy the conditions on theorem functions but the roles of these functions are reversed. The generalized cross-correlator similarity function must assume the role of $m_1(\tau)$, while the approximate similarity function assumes the role of $m_2(\tau)$. The value of the pessimistic curvature, C_p, yielding the smallest pessimistic variance is given by the maximum of the curvature function expressed by

$$C_p = \max_{\tau_0 < \tau \le \tau_0 + e_m} \frac{-1}{2} \sum_{k=0}^{K} \left[\frac{2\pi k}{T}\right]^2 \left|H_1\left(\frac{2\pi k}{T}\right)\right| \cdot \left|H_2\left(\frac{2\pi k}{T}\right)\right| \bar{X}\left(\frac{2\pi k}{T}\right) \frac{\sin\left(\dfrac{2\pi k(\tau - \tau_0)}{T}\right)}{\dfrac{2\pi k(\tau - \tau_0)}{T}}, \tag{37}$$

where the positive constant e_m is sufficiently large so that the location of the global maximum of the similarity function of the generalized cross correlator, $\hat{R}_{12}(\tau)$, is in the interval $\tau_0 - e_m \le \tau \le \tau_0 + e_m$. The pessimistic variance is given by the variance expression (31) with C_p substituted for C_o. Since the analysis of the approximate similarity function that led to the expression for its variance was premised on the parabola being concave down, should (37) yield a positive value for C_p, the variance expression is invalid and no pessimistic variance can be obtained.

Unlike C_o, C_p is a function, in fact a nondecreasing function, of the interval over which the maximum of (37) is sought. It is desirable to have this interval small so that the magnitude of C_p is as large as possible and the pessimistic variance is as small as possible. Thus, e_m should be chosen as small as possible within the constraint that the interval $\tau_0 - e_m \le \tau \le \tau_0 + e_m$ encompass the global maximum of the similarity function $\hat{R}_{12}(\tau)$. If the system is operating above the threshold signal-to-noise ratio, then e_m can be established from knowledge of the autocorrelation function of the nucleus $\bar{x}(t)$.

Example: Consider an example where the nucleus $\bar{x}(t)$ has a flat low-pass line spectrum with lines of amplitude 1 unit of power at angular frequencies $0, 2\pi/T, \cdots, 2\pi K/T$, where K is an integer constant. The filters in the generalized cross correlator are taken to be flat all pass filters. $\hat{R}_{s12}(\tau)$, which is the cross correlation of the nucleus, $\bar{x}(t)$, with a delayed version of itself, $\bar{x}(t - \tau_0)$, given by (30), simplifies to

$$\hat{R}_{s12}(\tau) = \frac{1}{2} + \frac{\sin\left((2K+1)\pi\dfrac{\tau - \tau_0}{T}\right)}{2\sin\left(\pi\dfrac{\tau - \tau_0}{T}\right)}. \tag{38}$$

It is again pointed out that (30) yields a deterministic function. This is a consequence of the nucleus $\bar{x}(t)$ being an ergodic signal. $\hat{R}_{s12}(\tau)$, with $K = 100$, is normalized and plotted against normalized shift $(\tau - \tau_0/T)$ in Fig. 2, where it is referred to, perhaps somewhat loosely, as normalized autocorrelation.

The curvature function, whose minimum yields the curvature of the optimistic parabola and whose maximum over the interval $\tau_0 \le \tau \le \tau_0 + e_m$ yields the curvature of the pessimistic parabola, given by the right side of (34) reduces to

$$\frac{\pi(2K+1)\cos\left((2K+1)\pi\dfrac{\tau-\tau_0}{T}\right)}{4T^2\left(\dfrac{\tau-\tau_0}{T}\right)\sin\left(\pi\dfrac{\tau-\tau_0}{T}\right)} - \frac{\pi\sin\left((2K+1)\pi\dfrac{\tau-\tau_0}{T}\right)\cos\left(\pi\dfrac{\tau-\tau_0}{T}\right)}{4T^2\left(\dfrac{\tau-\tau_0}{T}\right)\left(\sin\left(\pi\dfrac{\tau-\tau_0}{T}\right)\right)^2}. \tag{39}$$

This curvature function, with $K = 100$, is also normalized and plotted against normalized shift $(\tau - \tau_0/T)$ in Fig. 2. Fig. 2 demonstrates that the optimistic curvature is indeed the value of the curvature function at $\tau = \tau_0$, since it is at $\tau = \tau_0$ that the curvature function is minimum. Fig. 2 also demonstrates that the pessimistic curvature is a function of the range over which the maximum of the curvature function is sought. For the range extending across the mainlobe of the autocorrelation function, which has τ constrained by $\tau_0 \le \tau \le \tau_0 + 0.005T$, the curvature function is increasing. Thus, for $e_m < 0.005T$, the pessimistic curvature is obtained by evaluating the curvature function at range boundary $\tau = \tau_0 + e_m$, since it is at this value of τ that the curvature function is maximum.

It is desirable to have very little spread between the optimistic and pessimistic variance as spread leaves un-

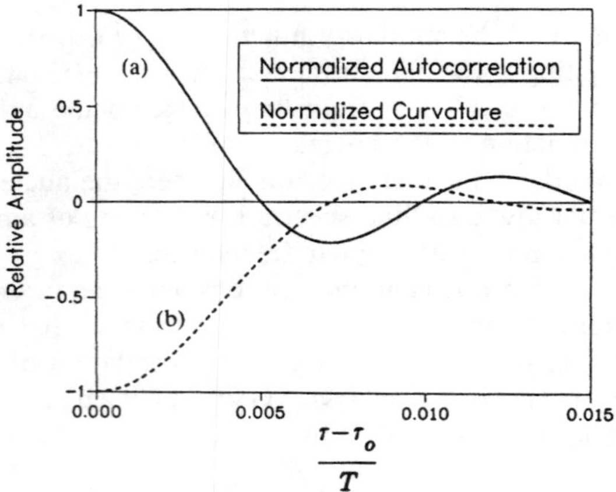

Fig. 2. Graph of: (a) normalized correlation of the nucleus with a delayed version of itself, and (b) normalized curvature versus normalized shift $(\tau - \tau_0/T)$.

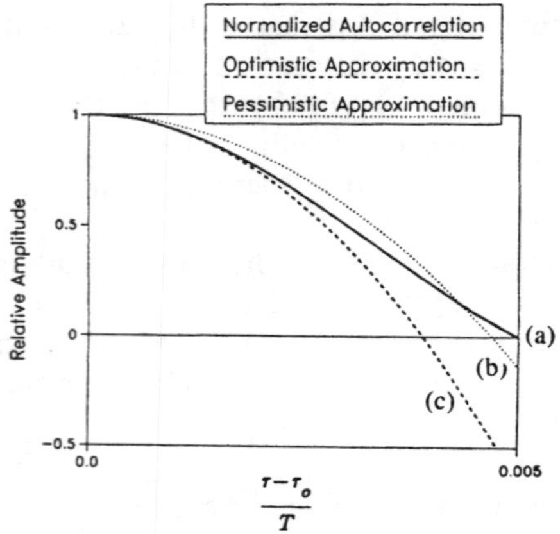

Fig. 3. Graph of: (a) normalized autocorrelation, (b) normalized optimistic approximation; and (c) normalized pessimistic approximation.

certainty as to the actual variance. The spread between the optimistic and pessimistic variances can be measured by their ratio, where a ratio close to 1 indicates little spread. Since the pessimistic variance is given by the variance expression (31) with C_p substituted for C_o, the ratio of pessimistic to optimistic variance is C_o^2/C_p^2. This is just the square of the reciprocal of the curvature function evaluated at $\tau = \tau_0 + e_m$. It is clear from Fig. 2 that the ratio approaches 1 as e_m approaches 0. If $e_m = 0.003\,T$, which places the end of the range marked by $\tau = \tau_0 + e_m$ at the point where the normalized autocorrelation function equals 0.5, then $C_o^2/C_p^2 = 1.45^2 = 2.1$. Thus, for $e_m = 0.003\,T$, there is considerable uncertainty as to the actual variance, however, the availability of optimistic and pessimistic values facilitates worse case and best case analyses of a system.

Coincidental graphs of the normalized autocorrelation function, the normalized optimistic parabola, and the normalized pessimistic parabola for $e_m = 0.003\,T$ are shown in Fig. 3. These curves show that the optimistic parabola is the better approximation over the range $\tau_0 \leq \tau \leq \tau_0 + 0.003\,T$. It can be concluded from Fig. 3 that, at least for this example, the optimistic variance is closer to the actual variance than is the pessimistic variance.

IV. Summary and Conclusions

Time delay is often estimated by the shift that maximizes some similarity function. The similarity functions that are commonly used are so complex that the variance of the estimators cannot be calculated. Approximate variances have been calculated using small error analysis techniques. These techniques, in essence, approximate the information bearing component of the similarity function with a parabola. The curvature of the parabola is chosen so that the approximate variance approaches the actual variance as the estimation errors became vanishingly small. Unfortunately, when the estimation errors are of practical size, the approximate variance differs from the actual variance. Currently, Monte Carlo simulations are used to establish the difference between the approximate and actual variances, however, these simulations require significant amounts of computer time.

With the aid of the theorem presented in this paper, it is shown that the curvature of the parabola used in the approximation can be chosen to make the approximate variance optimistic (less than the actual variance). It is also shown that the curvature can be chosen to make the approximate variance pessimistic. The optimistic and pessimistic variances establish limits for the true variance and facilitate a worst case analysis of a system. These variances can also be used to establish limits for the error in existing variance expressions. Both optimistic and pessimistic variances are calculated in an example where the estimator is based on the generalized cross correlator.

References

[1] C. Eckart, "Optimum rectifier systems for the detection of steady signals," Univ. California, Scripps Inst. Oceanography, Marine Physical Lab. Rep. SIO 12692, Ref. SIO 52-11, 1952.

[2] P. R. Roth, "Effective measurements using digital signal analysis," *IEEE Spectrum*, vol. 8, pp. 62–70, Apr. 1971.

[3] G. C. Carter, A. H. Nuttal, and P. G. Cable, "The smoothed coherence transform (SCOT)," Naval Underwater Syst. Center, New London Lab., New London, CT, Tech. Memo TC-159-72, Aug. 8, 1972.

[4] E. J. Hannan and P. J. Thomson, "Estimating group delay," *Biometrika*, vol. 60, pp. 241–253, 1973.

[5] G. C. Carter, A. H. Nuttal, and P. G. Cable, "The smoothed coherence transform," *Proc. IEEE* vol. 61, pp. 1497–1498, Oct. 1973.

[6] B. V. Hamon and E. J. Hannan, "Spectral estimation of time delay for dispersive and non-dispersive systems," *Appl. Stat.*, vol. 23, pp. 134–142, 1974.

[7] C. H. Knapp and G. C. Carter, "The generalized correlation method for estimation of time delay," *IEEE Trans. Acoust., Speech, Signal Processing*, vol. ASSP-24, pp. 320–327, Aug. 1976.

[8] Y. T. Chan, R. V. Hattin, and J. B. Plant, "The least squared estimation of time delay and its use in signal detection," *IEEE Trans. Acoust., Speech, Signal Processing*, vol. ASSP-26, pp. 217–222, June 1978.

[9] J. C. Hassab and R. E. Boucher, "Optimum estimation of time delay by a generalized correlator," *IEEE Trans. Acoust., Speech, Signal Processing*, vol. ASSP-27, pp. 373–380, Aug. 1979.

[10] Y. T. Chan, J. M. Riley, and J. B. Plant, "A parameter estimation approach to time-delay estimation and signal detection," *IEEE Trans. Acoust., Speech, Signal Processing*, vol. ASSP-28, pp. 8–16, Feb. 1980.

[11] A. F. Hassan, E. K. Al-Hassainy, and M. Bakry, "Nonparametric detectors for signal detection time delay estimation," in *Proc. 9th*

IEEE Conf. Acoust., Speech, Signal Processing (ICASSP 82), Paris, France, May 1982, pp. 387-390.

[12] B. Friedlander and B. Porat, "A parametric technique for time delay estimation," in *Proc. 7th IEEE Conf. Acoust., Speech, Signal Processing (ICASSP 82)*, Paris, France, May 1982, pp. 416-419.

[13] D. H. Youn, N. Ahamed, and G. C. Carter, "On using the LMS algorithm for time delay estimation," *IEEE Trans. Acoust., Speech, Signal Processing*, vol. ASSP-30, pp. 798-801, Oct. 1982.

[14] V. T. Chan and R. K. Miskowicz, "An ARMA modeling method for estimation of coherence and time delay," in *Proc. 8th IEEE Conf. Acoust., Speech, Signal Processing (ICASSP 83)*, Boston, MA, 1982, pp. 567-570.

[15] Z. Zhen and H. Zi-qiang, "The generalized phase spectrum method for time delay estimation," in *Proc. 9th IEEE Conf. Acoust., Speech Signal Processing (ICASSP 84)*, San Diego, CA, Mar. 19-21, 1984, pp. 46.2.1-46.2.4.

[16] A. Hero and S. Schwartz, "Alternative to the generalized cross correlator for time delay estimation,"in *Proc. 9th IEEE Conf. Acoust., Speech, Signal Processing (ICASSP 84)*, San Diego, CA, Mar. 19-21, 1984, pp. 15.4.1-15.4.4.

[17] M. Simaan, "A frequency-domain method for time-shift estimation and alignment of seismic signal," *IEEE Trans. Geosci. Remote Sensing*, vol. GE-23, pp. 132-138, Mar. 1985.

[18] C. W. Helstrom, *Statistical Theory of Signal Detection*. New York: Pergamon, 1960.

[19] D. Hertz, "Time delay estimation by combining efficient algorithms and generalized cross-correlation methods," *IEEE Trans. Acoust., Speech, Signal Processing*, vol. ASSP-34, pp. 1-7, Feb. 1986.

[20] C. D. McGillem and M. Svedlow, "Image registration error as a measure of overlay quality," *IEEE Trans. Geosci. Electron.*, vol. GE-14, pp. 44-49, Jan. 1976.

[21] T. L. Steding and F. W. Smith, "Optimum filters for image registration," *IEEE Trans. Aerosp. Electron. Syst.*, vol. AES-15, Nov. 1979.

[22] R. B. Blackman and J. W. Tukey, *The Measurement of Power Spectra*. New York: Dover, 1958.

[23] P. D. Welch, "The use of the fast Fourier transform for estimation of power spectra: A method based on time averaging over short modified periodograms," *IEEE Trans. Audio Electro-Acoust.*, vol. AU-15, pp. 70-73, 1967.

[24] J. E. Salt, Ph.D. dissertation, Univ. Saskatchewan, Saskatoon, Sask., Canada, 1987.

[25] W. R. Hahn, "Optimum signal processing for passive sonar range and bearing estimation," *J. Acoust. Soc. Amer.*, vol. 58, pp. 201-207, July 1975.

[26] A. J. Weiss and E. Weinstein, "Fundamental limitations in passive time delay estimation—Part 1: Narrow-band systems," *IEEE Trans. Acoust., Speech, Signal Processing*, vol. ASSP-31, pp. 472-486, Apr. 1983.

On the Simulation of a Class of Time Delay Estimation Algorithms

KENT SCARBROUGH, STUDENT MEMBER, IEEE, NASIR AHMED, SENIOR MEMBER, IEEE, AND G. CLIFFORD CARTER, SENIOR MEMBER, IEEE

Abstract—The time delay between signals received at two (or more) sensors has proven to be a useful parameter in passive sonar for estimating the location of an acoustic source. This paper presents the results of a simulation comparing the smoothed coherence transform and maximum likelihood estimation methods to the basic cross correlation technique for time delay estimation. Band-limited random signals which are corrupted by white noise and received at two sensors are considered at various signal-to-noise ratios. The variance of the time delay estimates are compared to the minimum variance obtainable in theory as given by the Cramér-Rao lower bound.

I. Introduction

THE problem of estimating the time delay between noisy signals which are received at two or more remote sensors has attracted considerable attention in recent years due to a variety of applications. One important application is in the localization of an underwater acoustic source. Many of the methods proposed for *time delay estimation* are related through a *generalized cross correlation* (GCC) approach [1]. Some of the related schemes are: 1) the Roth processor [2], 2) the *smoothed coherence transform* (SCOT) [3], 3) the phase transform (PHAT) [4], and 4) the Eckart processor [5]. In addition, a *maximum likelihood* (ML) *estimator* has been derived [1], and it has been shown that the ML estimator is identical to that proposed by Hannan and Thomson [6].

The chief objective of this paper is to evaluate the performance of the SCOT and ML estimators as compared to the basic *cross correlation* method via a digital simulation. The experimental results are compared to the *Cramér-Rao lower bound*, which gives a theoretical bound on the performance of these estimators, and to the results of a similar simulation which have been previously published [7].

II. Some Theoretical Considerations

A simple but useful model of the two sensor problem is

$$\begin{aligned} r_1(t) &= s(t) + n_1(t) \\ r_2(t) &= s(t-D) + n_2(t) \end{aligned} \tag{1}$$

Manuscript received June 1, 1980; revised October 31, 1980. This work was supported in part by the Naval Underwater Systems Center, New London, CT.

K. Scarbrough and N. Ahmed are with the Department of Electrical Engineering, Kansas State University, Manhattan, KS 66506.

G. C. Carter is with the Naval Underwater Systems Center, New London, CT 06320.

where $r_1(t)$ and $r_2(t)$ denote the sensor outputs, $s(t)$ is the source signal, and $n_1(t)$ and $n_2(t)$ are uncorrelated, zero-mean, Gaussian random processes. The problem is to estimate the time delay parameter D, which in turn enables the source bearing to be estimated. With three sensors the range of the source as well as the bearing can be estimated.

In the GCC approach, the sensor outputs $r_1(t)$ and $r_2(t)$ are linearly prefiltered as depicted in Fig. 1. The cross-power spectrum between the filter outputs $p_1(t)$ and $p_2(t)$ is then computed as [8]

$$G_{p_1p_2}(f) = H_1(f)H_2^*(f)G_{r_1r_2}(f) \tag{2}$$

where $*$ denotes the complex conjugate. The GCC function is then obtained by taking the inverse Fourier transform (IFT) of (2), i.e.,

$$R_{r_1r_2}^G(\tau) = \int_{-\infty}^{\infty} W(f)G_{r_1r_2}(f)\, e^{j2\pi f\tau}\, df \tag{3}$$

where $W(f) = H_1(f)H_2^*(f)$. In practice, the weighting function $W(f)$ is real valued. Thus, the filters H_1 and H_2 have the same phase. From (3) it is seen that the GCC method can be alternately viewed as applying a weighting or window function to the cross-power spectrum before computing the IFT. The argument τ that maximizes $R_{r_1r_2}^G(\tau)$ is then the desired estimate of the time delay D. Note that multiplying $W(f)$ by a frequency independent scalar does not affect the location at which the GCC function peaks occur.

Due to the limitation of finite observation time, only an estimate, say $\hat{G}_{r_1r_2}(f)$, of the cross-power spectrum can be obtained. Also, many of the weighting functions that are commonly used in (3) depend on the auto- or cross-power spectral estimates. Therefore, only an estimate $\hat{R}_{r_1r_2}^G(\tau)$ can be computed for GCC function $R_{r_1r_2}^G(\tau)$.

The basic cross correlation method is obtained by setting $W(f) = 1$ in (3). In an effort to reduce the error in the time delay estimate, various other weighting functions have been proposed. This discussion will be confined to the weighting functions used in the SCOT and maximum likelihood methods.

The SCOT algorithm uses the weighting function

$$W(f) = [G_{r_1r_1}(f)G_{r_2r_2}(f)]^{-1/2} \tag{4}$$

where $G_{r_1r_1}(f)$ and $G_{r_2r_2}(f)$ denote the autopower spectra of $r_1(t)$ and $r_2(t)$, respectively. This method was developed to

Reprinted from *IEEE Trans. Acoust., Speech, Signal Processing*, vol. 29, no. 3, pt. 2, pp. 534–540, June 1981.

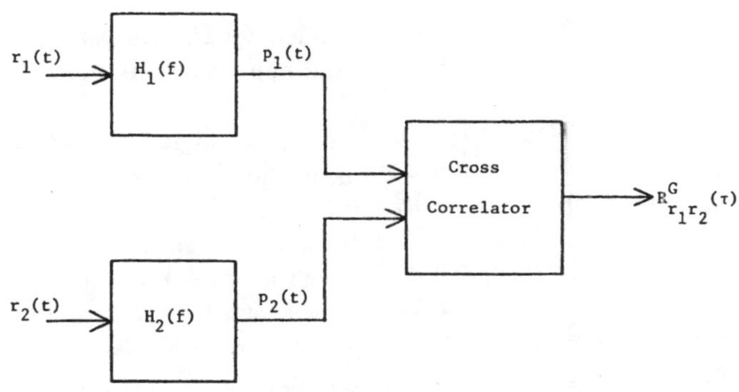

Fig. 1. Generalized cross correlation (GCC) method.

counteract the undesirable effects of strong tonals in broadband signals. Returning to the realization in Fig. 1, if $|H_1(f)| = [G_{r_1r_1}(f)]^{-1/2}$ and $|H_2(f)| = [G_{r_2r_2}(f)]^{-1/2}$, the SCOT method can be interpreted as two prewhitening filters followed by a cross correlator. Substituting (4) into (3) one obtains

$$R^{\mathrm{SCOT}}_{r_1r_2}(\tau) = \int_{-\infty}^{\infty} \gamma_{r_1r_2}(f)\, e^{j2\pi f\tau}\, df \tag{5}$$

where

$$\gamma_{r_1r_2}(f) = \frac{G_{r_1r_2}(f)}{[G_{r_1r_1}(f)G_{r_2r_2}(f)]^{1/2}} \tag{6}$$

is the complex coherence between $r_1(t)$ and $r_2(t)$.

The ML estimator applies the weighting function

$$W(f) = \frac{C_{r_1r_2}(f)}{|G_{r_1r_2}(f)|\,[1 - C_{r_1r_2}(f)]} \tag{7}$$

where

$$C_{r_1r_2}(f) = |\gamma_{r_1r_2}(f)|^2 \tag{8}$$

is the *magnitude squared coherence* (MSC). Substitution of this weighting function (7) into (3) yields

$$R^{\mathrm{ML}}_{r_1r_2}(\tau) = \int_{-\infty}^{\infty} \frac{C_{r_1r_2}(f)}{[1 - C_{r_1r_2}(f)]}\, e^{j\phi(f)}\, e^{j2\pi f\tau}\, df \tag{9}$$

where

$$e^{j\phi(f)} = \frac{G_{r_1r_2}(f)}{|G_{r_1r_2}(f)|}. \tag{10}$$

Since the additive noises in (1) are assumed to be uncorrelated, we have

$$e^{j\phi(f)} = e^{-j2\pi fD}. \tag{11}$$

Thus, the phase $\phi(f)$ is a measure of the time delay. The ML estimator assigns greater weight to the phase in regions of the frequency domain where the MSC is relatively large. It is easily seen that the ML correlation function in (9) attains its maximum value when $\tau = D$, that is, when $e^{j\phi(f)} \cdot e^{j2\pi f\tau} = 1$. Note that due to the necessity of estimating the pertinent power spectra, only an estimate of (9) is obtainable in practice. As such only an *approximate maximum likelihood* (AML) *estimator* can be implemented.

An important question to ask is how closely does the performance of the AML estimator compare with that of the ML estimator. It is known that the theoretical variance of the time delay estimate for a general weighting function $W(f)$ is given by [1]

$$\sigma^2_{\Delta D} = \frac{\int_{-\infty}^{\infty} (2\pi f)^2 |W(f)|^2 G_{r_1r_1}(f) G_{r_2r_2}(f)[1 - C_{r_1r_2}(f)]\, df}{T\left[\int_{-\infty}^{\infty} (2\pi f)^2 |G_{r_1r_2}(f)| W(f)\, df\right]^2} \tag{12}$$

where T denotes the observation time in seconds. Substitution of the ML weighting function from (7) into (12) and simplifying yields

$$\sigma^2_{\Delta D} = \left[T \int_{-\infty}^{\infty} (2\pi f)^2 \frac{C_{12}(f)}{1 - C_{12}(f)}\, df\right]^{-1} \tag{13}$$

which is the Cramér–Rao lower bound (CRLB) for the variance of the time delay estimate [1]. Thus, the theoretical performance of a GCC processor can be computed from (12), and (13) allows the experimental performance of the AML processor to be compared with the theoretical performance of the ML processor.

For the band-limited white Gaussian noise signals considered in this paper, the MSC is a constant over the signal bandwidth and zero elsewhere. The value of the MSC depends on the signal-to-noise ratio (SNR). For this case the CRLB in (13) can be simplified to

$$\sigma^2_{\Delta D} = \frac{K}{T(f_2^3 - f_1^3)} \tag{14}$$

where K is a constant dependent on the SNR and f_1 and f_2 are the lower and upper frequency limits of the signal power spectrum, respectively. Equivalently, (14) can be written as

$$\sigma^2_{\Delta D} = \frac{K'}{TBf_0^2(1 + B^2/12f_0^2)} \tag{15}$$

where K' is again a constant dependent on the SNR, $B = f_2 - f_1$ is the signal bandwidth, and $f_0 = (f_2 + f_1)/2$ is the center or "carrier" frequency. As one would intuitively expect, the CRLB decreases with increasing observation time T and with increasing signal bandwidth B. However, it is interesting to note that, for a given bandwidth, the CRLB also decreases as the carrier frequency f_0 increases. The explanation for this has to do with the properties of the Fourier transform. For white band-limited signals the GCC function corresponding to the ML estimator has a $\sin x/x$ envelope where the envelope width depends on the signal bandwidth. The width of the local peak, however, depends on the carrier frequency and narrows as the carrier frequency increases. Thus, the CRLB for the variance of the time delay estimate decreases as the carrier frequency increases. It can be shown that the other GCC processors considered in this paper behave similarly.

III. Simulation Details

The simulation considered the two sensor model given in (1). The time delay parameter D was 4 samples and the sampling interval was 1/2048 s. The simulation was done for two different source spectra. To obtain the representations for the sensor outputs, three real white Gaussian noise sequences were generated and checked to ensure they were mutually incoherent. Two of these sequences were used to represent the additive noises. The third sequence was used to generate two different (low-passed and bandpassed) source signal spectra by passing it through a pair of tenth-order Butterworth digital filters which had the following characteristics:

$$|H_1(f)| \simeq \begin{cases} 1, & 0 \leq f \leq 100 \text{ Hz} \\ 0, & \text{elsewhere} \end{cases} \tag{16}$$

$$|H_2(f)| \simeq \begin{cases} 1, & 100 \leq f \leq 200 \text{ Hz} \\ 0, & \text{elsewhere.} \end{cases} \tag{17}$$

The resulting outputs were thus low-passed and bandpassed sequences with the initial transients discarded. The signal sequences were then scaled to obtain the desired SNR and added to the noise sequences as in (1). SNR's of approximately 10, 0, -3, and -10 dB were considered, where

$$\text{SNR} \equiv \frac{\int_{f_1}^{f_2} G_{ss}(f)\, df}{\int_{f_1}^{f_2} G_{nn}(f)\, df}. \tag{18}$$

In (18), $G_{ss}(f)$ and $G_{nn}(f)$ denote the autopower spectra of the signal and noise sequences, respectively, and f_1 and f_2 define the frequency band where the signal power is nonzero, i.e., the integrals in (18) were from 0 to 100 Hz for the low-passed signal and from 100 to 200 Hz for the bandpassed signal. Note that the SNR definition in (18) is the SNR in the signal band and that the signal is present in only about ten percent of the frequency band (in particular 100 out of 1024 Hz). Thus, the total noise power is about ten times greater than indicated in (18).

Next, the digital data sequences $r_1(t)$ and $r_2(t)$ were processed using the SCOT, AML and basic cross correlation techniques, using the computer algorithm developed by Carter *et al.* [9]. In addition, an estimator that applies a weighting function, which is optimum in the sense that it is matched to the known power spectra of the signal sequence, was implemented. This was done to determine the effect of the nonzero weighting of the other processors in frequency bands where the signal power is zero. Thus, this optimum estimator uses the weighting functions

$$W_1(f) = \begin{cases} 1, & 0 \leq f \leq 100 \text{ Hz} \\ 0, & \text{elsewhere} \end{cases}$$

$$W_2(f) = \begin{cases} 1, & 100 \leq f \leq 200 \text{ Hz} \\ 0, & \text{elsewhere.} \end{cases} \tag{19}$$

In (19), $W_1(f)$ and $W_2(f)$ correspond to the low-passed and bandpassed cases, respectively. It should also be noted that, for the signals used in this simulation, these weighting functions are the same as those for the ML processor except for a scalar factor. Thus, this optimum processor is essentially the ML estimator, with the exception that the cross-power spectrum $G_{r_1 r_2}(f)$ is still estimated from the data.

The data sequences for each trial were processed using 16 disjoint segments of 512 points each. Each trial represented 4 s (i.e., $16 \times 512/2048$) of input data, and a total of 32 independent trials were conducted. The time argument which maximized the GCC function was taken to be the time delay estimate. In this manner, 32 estimates of the time delay parameter D were obtained. This was done for the low-passed as well as the bandpassed signal at each of the four SNR's: 10, 0, -3, and -10 dB.

Values for the time delay estimates which fell outside the range $(D \pm 10)$ samples, where D is the true time delay, were omitted when calculating statistics related to the time delay estimates. This was done to avoid biasing the statistics. In practice, estimates that are widely variant with the average values are normally discarded as "false" estimates by a tracking algorithm. The choice of ±10 samples to define the region of acceptable estimates is arbitrary, but seems reasonable for the problem of interest. Also note that the expression in (12) gives the variance of the time delay estimates in the neighborhood of the true delay. As such, experimental and theoretical variances can still be compared.

IV. Results and Discussion

The results of the simulation for the cases of SNR approximately equal to 10, 0, and -3 dB are summarized in Table I. When the SNR $\simeq$ -10 dB, approximately half of the trials failed to meet the time delay tracker requirement that the time delay estimates be within the range $(D \pm 10)$ samples for the optimum processor. An even greater number of "false" estimates was obtained when the other processors were used.

TABLE I
SIMULATION RESULTS

		LOW-PASSED CASE				BANDPASSED CASE			
		SCOT	Cross Corr.	AML	Optimum	SCOT	Cross Corr.	AML	Optimum
SNR ≈ 10 dB	$\sigma_{\Delta D}$ (exp)	1.53	0.56	0.49	0.0	0.61	0.0	0.0	0.0
	$\sigma_{\Delta D}$ (theo)	7.2	0.65	>0.09	0.09	1.00	0.10	>0.03	0.03
	"False" Estimates (%)	0.0	0.0	0.0	0.0	0.0	0.0	0.0	0.0
SNR ≈ 0 dB	$\sigma_{\Delta D}$ (exp)	2.17	1.59	1.53	0.67	0.76	0.59	0.61	0.0
	$\sigma_{\Delta D}$ (theo)	13.1	3.28	>0.35	0.35	1.88	0.95	>0.13	0.13
	"False" Estimates (%)	0.0	0.0	0.0	0.0	C.0	0.0	0.0	0.0
SNR ≈ -3 dB	$\sigma_{\Delta D}$ (exp)	2.95	2.32	2.48	0.98	1.28	0.83	0.80	0.40
	$\sigma_{\Delta D}$ (theo)	19.8	6.56	>0.57	0.57	2.83	1.90	>0.21	0.21
	"False" Estimates (%)	6.25	0.0	0.0	0.0	21.8	15.6	12.5	0.0

* $\sigma_{\Delta D}$ values represent number of samples.

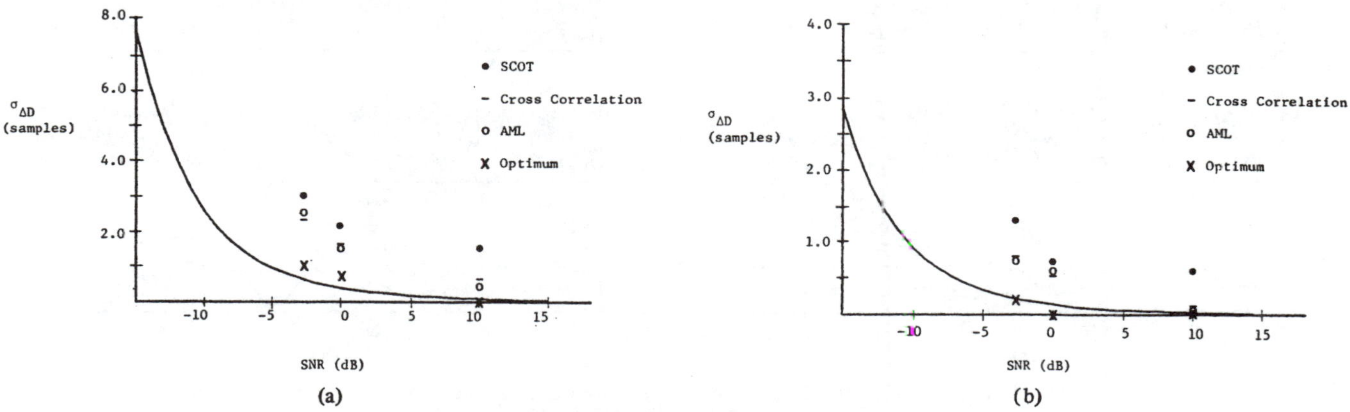

Fig. 2. Plots of the CRLB and experimental values of $\sigma_{\Delta D}$ versus SNR. (a) Low-passed case. (b) Bandpassed case.

Due to the limited number of trials remaining, little can be said statistically about the time delay estimates, and these results have been omitted from the tables. Plots of the CRLB versus SNR and the values obtained experimentally are shown in Fig. 2(a) and (b) for the low-passed and bandpassed cases, respectively. Typical plots of the correlation functions for three of the 32 trials using the low-passed signal for the four SNR's considered are shown in Fig. 3(a)-(d). Similar plots for the bandpassed case are shown in Fig. 4(a)-(d).

For the integration time used in this simulation, the AML estimator exhibited little or no improvement in performance over the basic cross correlation method. In fact, the performance of these two methods is nearly identical. However, for the signal and noise waveforms considered in this work, the AML and basic cross correlation methods perform significantly better than the SCOT technique in all cases. As the SNR decreases, the performance of the SCOT approaches that of the AML and cross correlation methods. As analytically predicted, all three processors gave better estimates for the bandpassed case than the low-passed case. The degradation due to nonzero weighting in frequency bands where the signal power is zero is clearly shown by the much improved performance of the optimum processor compared to the other methods. Also, comparison of the optimum processor, which essentially uses ML weighting, with the AML processor reveals the deterioration in performance caused by estimating the weighting function from the data. This result indicates that it may be beneficial to make use of *a priori* information about the signal power spectrum in determining the weighting function, even if that information is not entirely correct.

Comparison of the CRLB for the variance of the time delay estimates with that obtained experimentally for the optimum

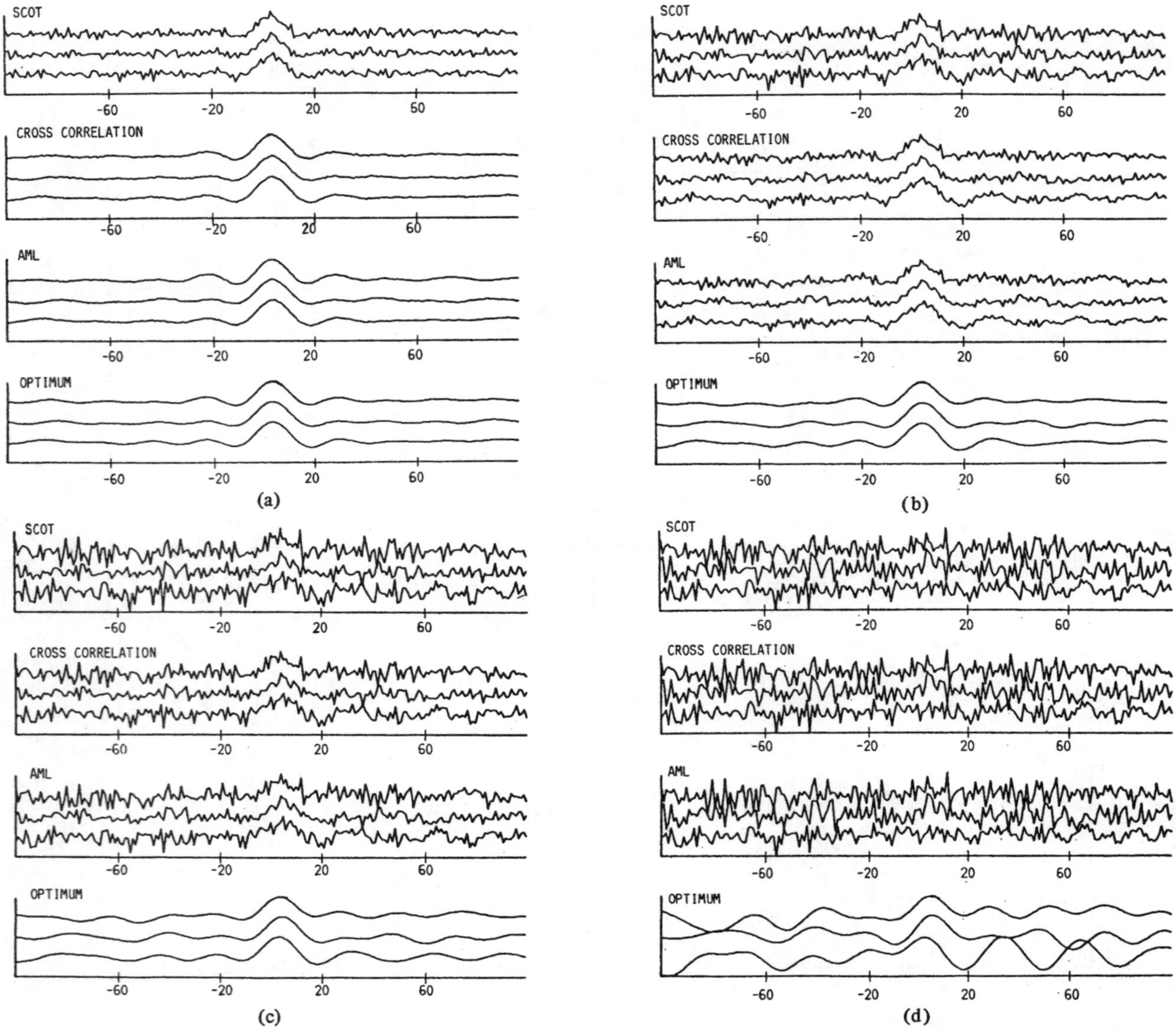

Fig. 3. Example GCC functions for 3 of 32 trials for the low-passed case. (a) SNR $\simeq$ 10 dB. (b) SNR $\simeq$ 0 dB. (c) SNR $\simeq$ -3 dB. (d) SNR $\simeq$ -10 dB.

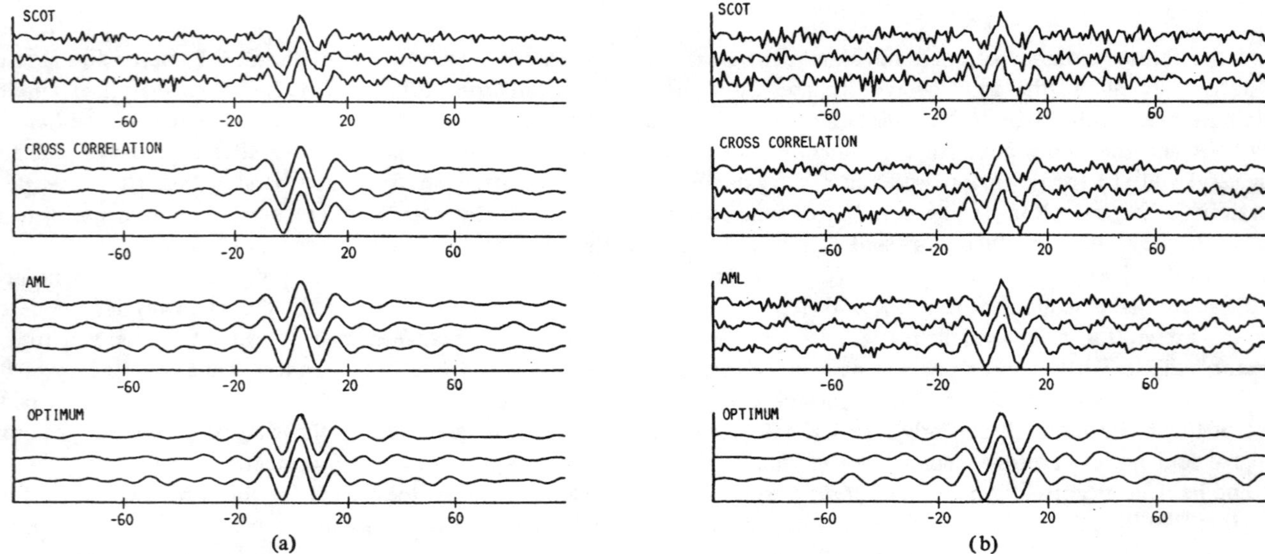

Fig. 4. Example GCC functions for 3 of 32 trials for the bandpassed case. (a) SNR $\simeq$ 10 dB. (b) SNR $\simeq$ 0 dB.

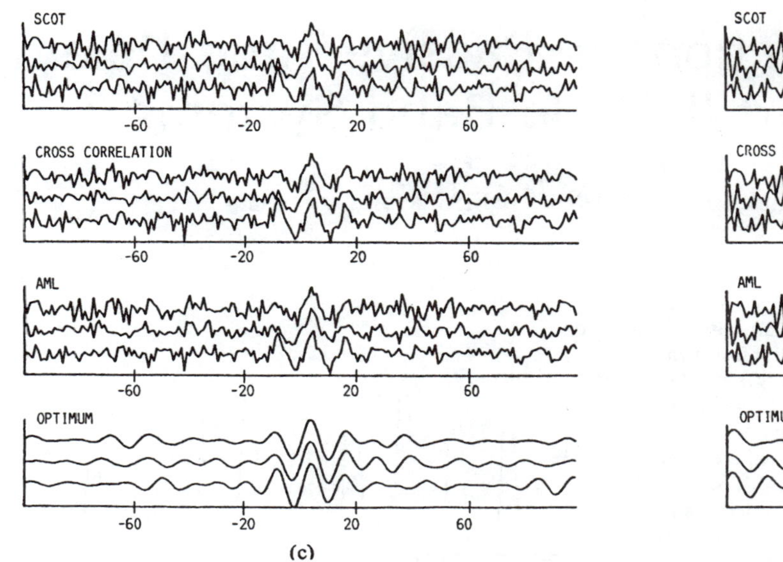

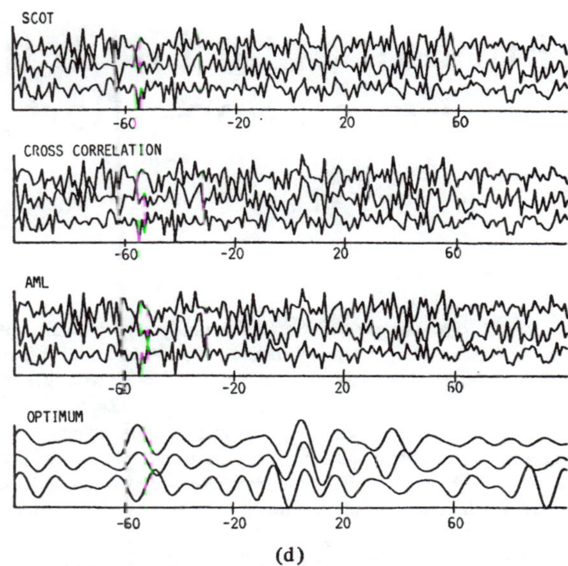

Fig. 4. (Continued). Example GCC functions for 3 of 32 trials for the bandpassed case. (c) SNR $\simeq -3$ dB. (d) SNR $\simeq -10$ dB.

processor shows good agreement. However, (12) is only marginally useful in predicting the experimental results for the other processors. While fairly good agreement between the experimental and theoretical values was obtained for the cross correlation method, this agreement tended to worsen as the SNR decreased. Moreover, the theoretical variances for the SCOT method were consistently higher than those found experimentally. As expected, the experimental variances of the AML estimator were higher than the CRLB.

For the case of a band-limited Gaussian noise signal which contains no strong sinusoids and which is embedded in white Gaussian noise, the basic cross correlation algorithm has shown significantly better performance than the SCOT algorithm. The AML estimator also performs better than the SCOT, but shows no improvement compared to the cross correlation method. The optimum processor using the ML weighting gave the best performance, as expected.

Hassab and Boucher have conducted a similar simulation using somewhat different signal and noise sequences [7]. Their results agree substantially with those of this work in that they found the ML method performed better than the cross correlation method, which performed better than the SCOT. Hassab and Boucher also noted that the differences in the variances of the time delay estimates of the various processors become statistically significant at moderate to high SNR's. This simulation has found that the performance of the SCOT tends to converge to that of the AML and cross correlation methods as the SNR decreases.

Finally, it is noted that while the SCOT algorithm exhibited the poorest performance for this simulation, it has been shown that for certain types of data (e.g., when tonals are present), the SCOT algorithm achieves better performance than the basic cross correlation method [3].

References

[1] C. H. Knapp and G. C. Carter, "The generalized correlation method for estimation of time delay," *IEEE Trans. Acoust., Speech, Signal Processing*, vol. ASSP-24, pp. 320-327, Aug. 1976.

[2] P. R. Roth, "Effective measurements using digital signal analysis," *IEEE Spectrum*, vol. 8, pp. 60-62, Apr. 1971.

[3] G. C. Carter, A. H. Nuttall, and P. G. Cable, "The smoothed coherence transform," *Proc. IEEE*, vol. 61, pp. 1497-1498, Oct. 1973.

[4] —, "The smoothed coherence transform (SCOT)," Naval Underwater Syst. Center, New London Lab., New London, CT, Tech. Memo. TL-159-72, Aug. 8, 1972.

[5] C. Eckhart, "Optimal rectifier systems for detection of steady signals," Scripps Inst. Oceanography, Marine Physical Lab., Univ. California, Rep. SIO 12692, SIO Ref. 52-11, 1952.

[6] E. J. Hannan and P. J. Thomson, "The estimation of coherence and group delay," *Biometrika*, vol. 58, pp. 469-481, 1971.

[7] J. C. Hassab and R. E. Boucher, "A quantitative study of optimum and suboptimum filters in the generalized correlator," in *Proc. 1979 IEEE Int. Conf. Acoust., Speech, Signal Processing*, pp. 124-127, 1979.

[8] W. B. Davenport, Jr., *Probability and Random Processes.* New York: McGraw-Hill, 1970.

[9] IEEE ASSP Digital Signal Processing Committee, *Programs for Digital Signal Processing.* New York: IEEE Press, 1979.

Fundamental Limitations in Passive Time-Delay Estimation—Part II: Wide-Band Systems

EHUD WEINSTEIN, MEMBER, IEEE, AND ANTHONY J. WEISS, STUDENT MEMBER, IEEE

Abstract—This is the second part of a study which deals with the problem of passive time delay estimation. The focus here is on systems employing wide-band signals and/or arrays of very widely separated receivers. A modified (improved) version of the Ziv-Zakai lower bound (ZZLB) is used to analyze the effect of additive noise and signal ambiguities on the attainable mean-square estimation errors. When the lower bound is plotted as a function of signal-to-noise ratio (SNR), one observes two distinct threshold phenomena dividing the SNR domain into three disjointed segments. At high SNR, the lower bound coincides with the Cramér-Rao lower bound (CRLB). This is the ambiguity-free mode of operation where differential delay estimation is subject only to local errors. At moderate SNR (between the two thresholds), the lower bound exceeds the CRLB by a factor of $12(\omega_0/w)^2$ where ω_0 and w are, respectively, the center frequency and signal bandwidth. In this region, the ambiguities in the received signal phases cannot be resolved; however, a useful estimate of the differential delay can still be obtained using the received signal envelopes. At low SNR, the lower bound approaehes a constant level depending only on the *a priori* search domain of the unknown delay parameter. In this region, signal observations are subject to envelope ambiguities as well, and are thus essentially useless for the delay estimation.

I. INTRODUCTION

A. Ambiguity Phenomena in Time Delay Estimation

ESTIMATION of the time difference of arrival of a noise-like random signal observed at two or more spatially separated receivers is a problem of considerable practical interest in many disciplines such as underwater acoustics, geophysics, and radio astronomy, to mention a few. Consequently, numerous procedures have been proposed for passive time-delay estimation (e.g., [1]-[7]). In most of the systems which have been analyzed, the Cramér-Rao inequality has been used to set a lower bound on the attainable mean-square estimation errors. Its use was justified by invoking a well-known theorem in statistics asserting that the maximum likelihood (ML) estimator is asymptotically unbiased, and that its error variance approaches the Cramér-Rao lower bound (CRLB) arbitrarily close for sufficiently long observation times. There remains the question: How long is "long enough"? Clearly, the observation time T must be large compared to the correlation time (inverse bandwidth) of signal and noise ($WT/2\pi >> 1$), a condition which presents very little difficulty in practice. However,

Manuscript received August 31, 1983. This work was supported by the Naval Underwater Systems Center under Contract 00140-83-C-KA35.

E. Weinstein is with the Department of Ocean Engineering, Woods Hole Oceanographic Institution, Woods Hole, MA 02543, and with the Department of Electronic Systems, Faculty of Engineering, Tel-Aviv University, Ramat-Aviv 69978, Tel-Aviv, Israel.

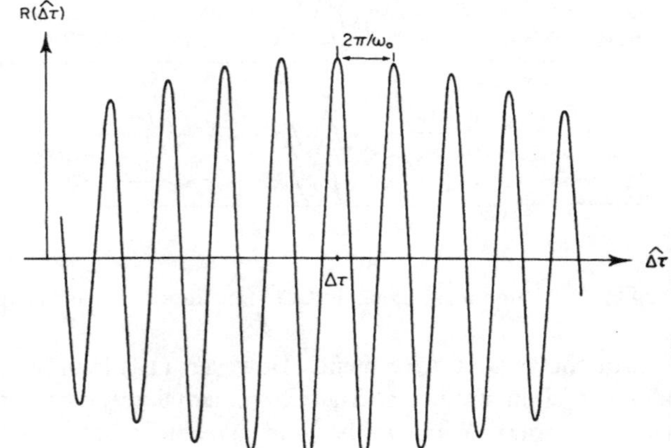

Fig. 1. Typical narrow-band signal cross correlation.

if one examines the problem more closely, one finds a second condition: the ML estimator (which approaches the CRLB asymptotically) must not be subject to ambiguities.

Perhaps the most common setting in which this difficulty occurs is in delay estimation using very narrow-band signals. Consider the extreme case of observations at only two receivers so that only one differential delay can be estimated. The ML estimate of that delay cross correlates the received signals, averages for time T, and obtains the desired delay from the peak of the cross-correlation function. In the narrow-band case, the differential delay causes essentially a phase shift between the received signals, and thus generates a formidable ambiguity problem. This is illustrated in Fig. 1. The cross-correlation output peaks at $\Delta\tau$, the true differential delay, but it is quasiperiodic with a period of $2\pi/\omega_0$ where ω_0 is the center frequency of the signal. To come close to the CRLB, one must be able to distinguish unambiguously between adjacent peaks of the correlation function. If the signal bandwidth is only a small fraction of its center frequency (i.e., $w/\omega_0 <<$ 1), adjacent peaks have very nearly equal height, and identification of the largest one will require either very large SNR or exceedingly long observation times. In many important practical situations, therefore, the attainable mean-square error (MSE) may be very drastically inferior to that predicted by the CRLB.

In [8], a new lower bound, based on a modified (improved) version of the Ziv-Zakai lower bound (ZZLB), is developed to analyze the attainable MSE in delay estimation schemes. The resulting lower bound is then applied to investigate the effect of ambiguity on delay estimation using narrow-band signals

Reprinted from *IEEE Trans. Acoust., Speech, Signal Processing*, vol. 32, no. 5, pp. 1064–1077, October 1984.

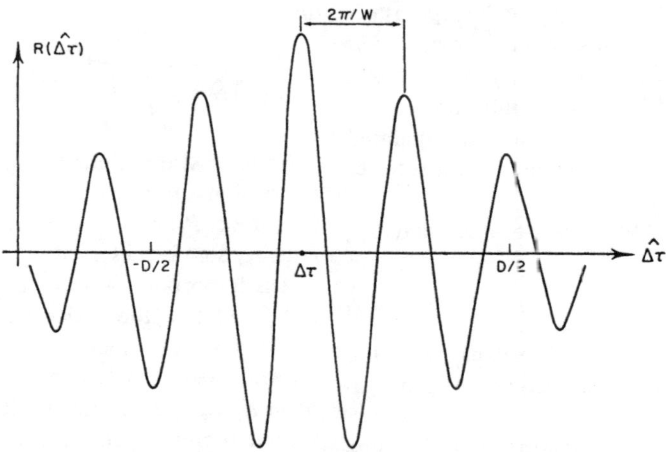

Fig. 2. Typical baseband signal cross correlation.

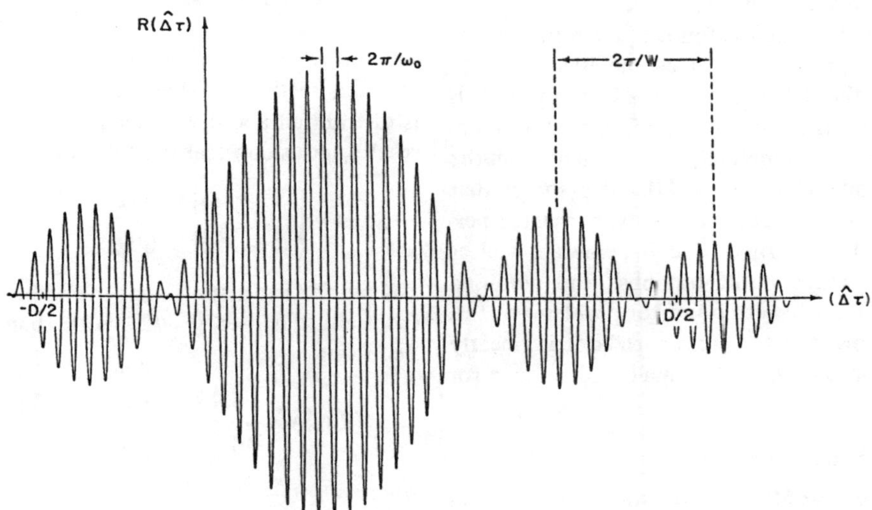

Fig. 3. Typical bandpass signal cross correlation.

[9]. When the lower bound is plotted as a function of SNR, one observes a strong threshold phenomenon. Above a critical SNR, the lower bound coincides with the CRLB. This is the ambiguity-free mode of operation. Below the threshold, the lower bound exceeds the CRLB by a large factor, indicating that in this region, differential delay estimation is subject to unresolved phase ambiguities. The point at which the threshold occurs as well as its magnitude are calculated in [9] as a function of SNR and the available WT product. Such information is of obvious practical interest since system performance predictions generally are based on the smaller error assumption (i.e., the CRLB) and may, therefore, be extrapolated incorrectly below the threshold, yielding extremely optimistic performance predictions.

In the second part of the study, we broaden the range of investigation to wide-band signals which, for the delay estimation problem, will be divided into two categories: baseband signals and bandpass signals. Both of these signal classes are subject to ambiguity and threshold phenomena. In the baseband case, a typical signal cross correlation is illustrated in Fig. 2. The cross-correlation function peaks at $\Delta\tau$, the true differential delay, but it is an oscillatory function of the estimated delay, thus generating an ambiguity problem. In the bandpass case, the situation is even more complicated. A typical signal cross-correlation response is illustrated in Fig. 3. One observes two types of ambiguities. The more critical ambiguity problem results from the highly oscillatory nature of the phase of the observed cross correlation function. A secondary ambiguity phenomenon results from the oscillatory nature of the envelope of the cross-correlation function.

The purpose of this paper is to analyze the effect of ambiguities on the attainable MSE for these two classes of signals.

B. Problem Formulation and Assumptions

The basic system of interest here consists of a stationary source radiating a noise-like signal towards two spatially separated receivers. Each receiver also receives an additive noise component so that the actual waveforms observed at the receiver outputs are given by

$$\begin{aligned} r_1(t) &= s(t) + n_1(t) \\ r_2(t) &= s(t - \Delta]) + n_2(t) \end{aligned} \qquad -T/2 \leqslant t \leqslant T/2. \tag{1}$$

We shall assume that $s(t)$, $n_1(t)$, and $n_2(t)$ are sample functions from uncorrelated zero-mean Gaussian random processes with spectral densities $S(\omega)$, $N_1(\omega)$, and $N_2(\omega)$, respectively. Since we are primarily interested in the ambiguity problem in time delay estimation, implying the use of widely separated receivers, the assumption of noise incoherence from receiver to receiver is likely to be satisfied.

We shall further assume that $\Delta\tau$, the receiver-to-receiver delay, is confined to the interval

$$-D/2 \leqslant \Delta\tau \leqslant D/2. \tag{2}$$

This *a priori* domain may come about, perhaps, from the known receiver separation and the known velocity of propagation in the medium.

Finally, we shall assume that the observation time T is large compared to the correlation time (inverse bandwidth) of signal and noise, i.e., $WT/2\pi >> 1$. This condition is very generally satisfied in problems of practical interest here.

The problem may now be stated as follows: given the data at the receiver outputs (i.e., $v_i(t)$ $i = 1, 2$), characterize the minimum MSE estimate of the delay parameter. Our approach is to set a lower bound on the attainable MSE using the modified ZZLB. This approach is completely independent from the actual estimation method. However, in [10] it is shown that for a sufficiently large WT product, the cross-correlator performance comes close to the lower bound below as well as above the threshold. These results establish the modified ZZLB as an extremely tight lower bound for problems of this type, while demonstrating that the cross correlator is a nearly optimal instrumentation in both the small and large error regimes.

C. The Modified Ziv-Zakai Lower Bound

The modified ZZLB on the MSE of any estimate $\widehat{\Delta\tau}$ of $\Delta\tau$ is given by [8], [9]

$$\overline{\epsilon^2} \geqslant \frac{1}{D} \int_0^D x G[(D-x) P_e(x)]\, dx \tag{3}$$

where $\overline{\epsilon^2}$ is the MSE averaged over the *a priori* parameter domain. $G[\ ]$ is a nonincreasing function of x obtained by filling the valleys in the function $(D - x)\, P_e(x)$ (see Fig. 12). $P_e(x)$ is the minimum probability of error (achievable by the likelihood ratio test) for deciding whether the true value of the parameter is $\Delta\tau_0$ or $\Delta\tau_1$ where $\Delta\tau_1 - \Delta\tau_0 = x$. In general, a closed analytical form for $P_e(x)$ cannot be found. However, in [9, Appendix A], it has been shown that for $WT/2\pi >> 1$, $P_e(x)$ is very closely approximated by

$$P_e(x) \approx e^{a(x) + b(x)} \phi(\sqrt{2b(x)}) \tag{4}$$

where

$$a(x) = -\frac{T}{2\pi} \int_0^\infty \ln\,[1 + \mathrm{SNR}(\omega) \cdot \sin^2 \omega x/2]\, d\omega \tag{5}$$

$$b(x) = \frac{T}{2\pi} \int_0^\infty \frac{\mathrm{SNR}(\omega) \sin^2 \omega x/2}{1 + \mathrm{SNR}(\omega) \cdot \sin^2 \omega x/2}\, d\omega \tag{6}$$

$$\mathrm{SNR}(\omega) = \frac{[S(\omega)/N_1(\omega)]\,[S(\omega)/N_2(\omega)]}{1 + S(\omega)/N_1(\omega) + S(\omega)/N_2(\omega)} \tag{7}$$

and

$$\phi(y) \triangleq \frac{1}{\sqrt{2\pi}} \int_y^\infty e^{-\mu^2/2}\, d\mu. \tag{8}$$

We have also included Appendix A to demonstrate that the expression on the right-hand side of (4) is, in fact, a *lower bound* on $P_e(x)$. Hence, by substituting (4) into (3), the inequality sign is preserved.

It is important to observe that since (7) uses arbitrary spectral functions, the lower bound can be applied to investigate a wide class of signals. As pointed out before, the study will be separated into two parts. In Section II we consider baseband signals. Analytically, it appears to be a simpler case and should therefore be understood first. In Section III we consider bandpass signals. In that context, all the results derived in [9] concerning narrow-band signals will be included and referred to as a special case.

II. Baseband Systems

In this section we shall concentrate on signals whose power is distributed about zero frequency. To simplify the form of results, let us consider the following special case:

$$S(\omega) = \begin{cases} S & |\omega| \leqslant W/2 \\ 0 & |\omega| > W/2. \end{cases} \tag{9}$$

If we further assume that the additive noise components are spectrally flat over $[-W/2, W/2]$, than (7) assumes the simplified form

$$\mathrm{SNR}(\omega) = \begin{cases} \mathrm{SNR} & |\omega| \leqslant W/2 \\ 0 & |\omega| > W/2 \end{cases} \tag{10}$$

where

$$\mathrm{SNR} = \frac{(S/N_1)\,(S/N_2)}{1 + S/N_1 + S/N_2}. \tag{11}$$

S/N_i is the in-band signal-to-noise ratio at the ith receiver output. Substitution of (10) into (5) and (6) immediately yields

$$a(x) = -\frac{T}{2\pi} \int_0^{W/2} \ln\,(1 + \mathrm{SNR} \sin^2 \omega x/2)\, d\omega \tag{12}$$

$$b(x) = \frac{T}{2\pi} \int_0^{W/2} \frac{\mathrm{SNR} \cdot \sin^2 \omega x/2}{1 + \mathrm{SNR} \cdot \sin^2 \omega x/2} \cdot d\omega. \tag{13}$$

To obtain the lower bound, one must substitute (12) and (13) into (4) and (4) into (3) successively, and carry out the indicated algebraic operations. Since we are primarily interested in the ambiguity phenomenon, we shall further assume that the *a priori* search domain of the unknown delay parameter contains at least several peaks of the signal cross correlation, i.e., $WD/2\pi >> 1$ (see Fig. 2). In that case, following some rather extensive algebra manipulations outlined in Appendix B, it is shown that the lower bound exhibits a distinct threshold effect, dividing the entire SNR domain into essentially two disjointed segments as suggested by the following

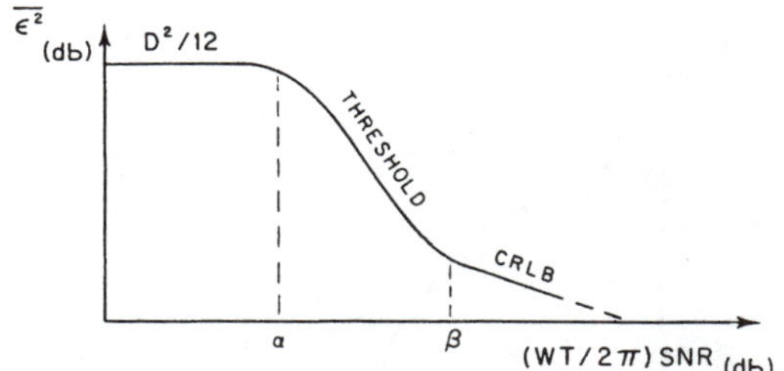

Fig. 4. Composite bound on $\overline{\epsilon^2}$—baseband systems.

equation:

$$\overline{\epsilon^2} \geq \begin{cases} D^2/12 & (WT/2\pi)\ \text{SNR} < \alpha \\ \text{threshold} & \alpha < (WT/2\pi)\ \text{SNR} < \beta \\ \dfrac{12}{W^2} \cdot \dfrac{1}{(WT/2\pi)\ \text{SNR}} & \beta < (WT/2\pi)\ \text{SNR} \end{cases} \tag{14}$$

where, in the threshold region, the lower bound varies essentially exponentially with $(WT/2\pi)$ SNR. This result is illustrated in Fig. 4. It is interesting to observe that (14) depends on $WT/2\pi$ and SNR only through their product. This is true only if $WT/2\pi >> 1$. The quantity $(WT/2\pi)$ SNR is usually referred to as the postintegration SNR.

α and β are the lower and upper limits of the threshold region. In the analysis carried out in Appendix B, α is defined as the point at which the lower bound is 3 dB (a factor of 2) *below* the $D^2/12$ performance level. It is given approximately by

$$\alpha \approx 0.92 = -0.36 \text{ dB}. \tag{15}$$

Similarly, β is defined as the point at which the lower bound is 3 dB above the performance level indicated by the third line of (14). That point is perhaps the most important factor in the composite result illustrated above, since it determines the boundary between small and large estimation errors. Analytical considerations outlined in Appendix B indicate that β is approximately the solution to the following transcendental equation:

$$(\beta/2)\phi(\sqrt{\beta/2}) = (6/WD)^2. \tag{16}$$

Note that (16) has two solutions. We only consider the larger one. In Fig. 5, β is plotted and tabulated as a function of $WD/2\pi$ for the convenience of the reader. Thus, in a logarithmic (dB) scale, β varies from about 13 dB for moderate WD products to about 16 dB for exceedingly large WD products.

To put these results into perspective, we first observe that a MSE of $D^2/12$ corresponds to a random variable uniformly distributed in $[-D/2, D/2]$. Such a performance level can always be achieved, regardless of signal observations. The first line in (14), therefore, reads that for a combination of $WT/2\pi$ and SNR such that their product does not exceed ~1 (i.e., if the postintegration SNR does not exceed ~0 dB), signal observations are completely dominated by noise, and thus are essentially useless for the delay estimation.

We next observe that MSE predictions based on the Cramér-Rao inequality yield the following lower bound [9, Appendix B]:

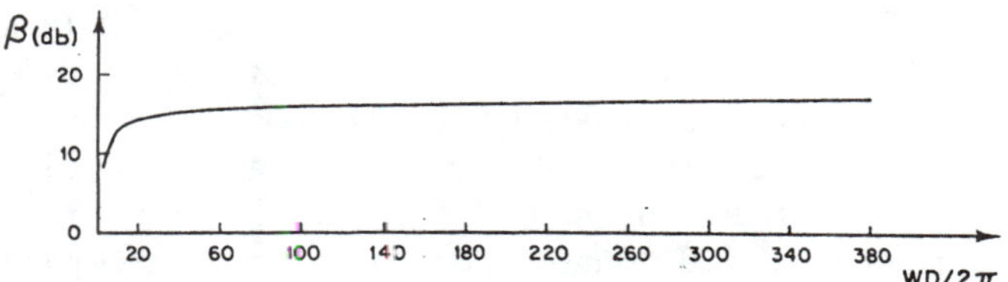

$WD/2\pi$	β	β(db)
3	7.38	8.68
10	19.33	12.86
20	25.44	14.06
60	34.98	15.44
100	39.34	15.95
140	42.17	16.25
180	44.30	16.46
220	45.91	16.62
260	47.31	16.75
300	48.60	16.87
340	49.56	16.95
380	50.53	17.04

Fig. 5. β versus $WD/2\pi$.

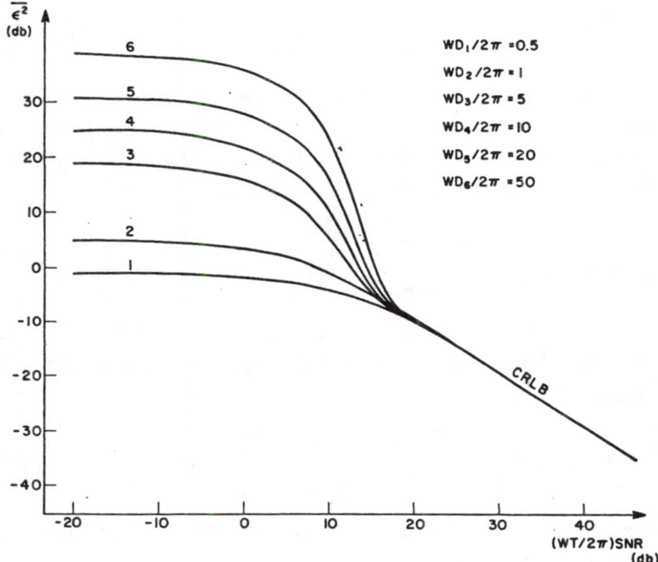

Fig. 6. $\overline{\epsilon^2}$ versus postintegration SNR for baseband signals.

$$\overline{\epsilon^2} \geq \left[\frac{T}{2\pi}\int_0^\infty \omega^2\ \text{SNR}(\omega)\, d\omega\right]^{-1}. \tag{17}$$

Substituting (10) into (17) and carrying out the indicated integration, one immediately obtains the third line in (14). Hence, if the postintegration SNR exceeds β, the ambiguities in the differential delay observations can essentially be resolved. By cross correlating the received signals, one obtains a differential delay estimate whose MSE is closely characterized by the CRLB.

In the derivation of (14) we have assumed that $WD/2\pi >> 1$. This condition means that the correlation time (inverse bandwidth) of the signal is small compared to the maximum expected receiver-to-receiver delay, or that the spacing between receivers is large compared to the half-wavelength of the highest frequency component. Only in that case we are dealing with the possibility of a significant ambiguity problem. There remains the question: How are the results stated by (14) affected when this condition is not satisfied? In Fig. 6 the lower bound is generated numerically, by exact integration of

(3), and plotted as a function of the postintegration SNR (in a logarithmic (dB) scale) for different values of $WD/2\pi$. Only the upper set of curves (curves 3-6) exhibits a significant threshold phenomenon. In case $WD/2\pi < 1$, one observes a rather smooth transition from the CRLB to the $D^2/12$ asymptote. The point at which the transition occurs is given to a very good approximation by simply intersecting the first and third lines of (14).

We finally note that in the derivation of (14), it is assumed that the source signal and the additive noises are spectrally flat over the receiver frequency band. A spectrally flat signal, however, has a highly oscillatory correlation function, indicating a serious ambiguity problem. For signal spectra whose correlation function (inverse Fourier transform) is smoothly varying, the ambiguity phenomenon may not be as critical. This effect can be studied by analyzing the lower bound for various signal and noise spectra.

III. Bandpass Systems

In this section we concentrate on signals whose power is distributed about some center frequency ω_0. To simplify the form of results, let us consider, in complete analogy with the baseband case, the following special case:

$$S(\omega) = \begin{cases} S & |\omega \pm \omega_0| \leqslant W/2 \\ 0 & |\omega \pm \omega_0| > W/2. \end{cases} \tag{18}$$

We shall further assume that the additive noise components are spectrally flat over the signal frequency band so that (7) assumes the form

$$\mathrm{SNR}(\omega) = \begin{cases} \mathrm{SNR} & |\omega \pm \omega_0| \leqslant W/2 \\ 0 & |\omega \pm \omega_0| > W/2 \end{cases} \tag{19}$$

where SNR is defined in (11). Substitution of (19) into (5) and (6) immediately yields

$$a(x) = -\frac{T}{2\pi} \int_{\omega_0 - W/2}^{\omega_0 + W/2} \ln(1 + \mathrm{SNR} \sin^2 \omega x/2)\, d\omega \tag{20}$$

$$b(x) = \frac{T}{2\pi} \int_{\omega_0 - W/2}^{\omega_0 + W/2} \frac{\mathrm{SNR} \sin^2 \omega x/2}{1 + \mathrm{SNR} \sin^2 \omega x/2}\, d\omega. \tag{21}$$

The lower bound is now obtained by successive substitutions of (20) and (21) into (4) and (4) into (3), and carrying out the indicated algebra operations.

Let us first assume that $W/\omega_0 << 1$. Since we are concerned with the joint effect of envelope and phase ambiguities on the attainable MSE (see Fig. 3), we shall further assume that $WD/2\pi >> 1$. In this setting, therefore, we are dealing with narrowband signals and very widely separated receivers. Following some rather extensive algebraic manipulations outlined in Appendix C, it is shown that the lower bound consists of essentially two distinct threshold effects dividing the entire SNR domain into three disjointed segments as suggested by the following equation:

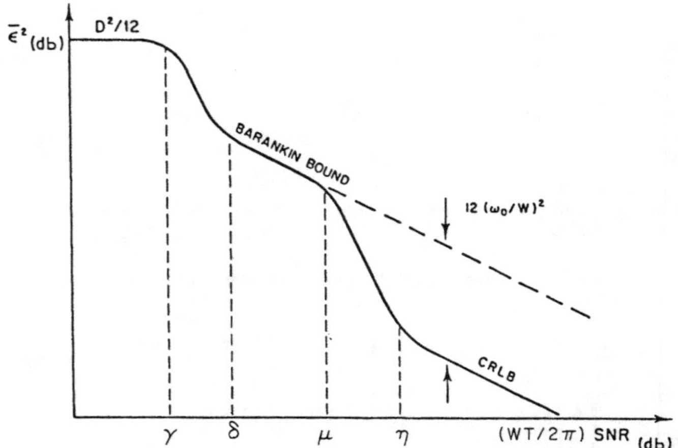

Fig. 7. Composite bound on $\overline{\epsilon^2}$—bandpass systems.

$$\overline{\epsilon^2} \geqslant \begin{cases} D^2/12 & (WT/2\pi)\,\mathrm{SNR} < \gamma \\ \text{threshold} & \gamma < (WT/2\pi)\,\mathrm{SNR} < \delta \\ \dfrac{6}{W^2} \dfrac{1}{(WT/2\pi)\,\mathrm{SNR}} & \delta < (WT/2\pi)\,\mathrm{SNR} < \mu \\ \text{threshold} & \mu < (WT/2\pi)\,\mathrm{SNR} < \eta \\ \dfrac{1}{2\omega_0^2} \cdot \dfrac{1}{(WT/2\pi)\,\mathrm{SNR}} & \eta < (WT/2\pi)\,\mathrm{SNR} \end{cases} \tag{22}$$

where, in the threshold regions, the lower bound varies essentially exponentially with $(WT/2\pi)$ SNR. Equation (22) depends on $WT/2\pi$ and SNR only through their product. This result is illustrated in Fig. 7. Analytical information concerning the various threshold points γ, δ, μ, and η will be given shortly.

To put this composite result into perspective, we observe that MSE predictions based on the Cramér-Rao inequality yield the following lower bound:

$$\overline{\epsilon^2} \geqslant \frac{1}{2\omega_0^2} \frac{1}{(WT/2\pi)\,\mathrm{SNR}}. \tag{23}$$

Equation (23) is obtained by substituting (19) into (17) and carrying out the indicated integration. Thus, if $(WT/2\pi)\,\mathrm{SNR} > \eta$, the CRLB coincides with the modified ZZLB. This is the ambiguity-free mode of operation.

If $\delta < (WT/2\pi)\,\mathrm{SNR} < \mu$, the lower bound exceeds the CRLB by a factor of $12\,(\omega_0/W)^2$. In this region, signal observations are subject to unresolved phase ambiguities. However, a useful estimate of the differential delay can still be obtained from the envelope of the cross-correlation function. Note that this segment of the lower bound coincides with MSE predictions based on a simplified version of the Barankin lower bound [11].

Finally, if $(WT/2\pi)\,\mathrm{SNR} < \gamma$, the lower bound is essentially characterized by the constant level of $D^2/12$. This is the noise-dominated region where signal observations are subject to envelope ambiguities as well, and are thus essentially useless for the delay estimation.

The 3 dB points of the more critical threshold are given, respectively, by

$$\mu = (2.76/\pi^2)(\omega_0/W)^2 \tag{24}$$

$$\eta = (6/\pi^2)(\omega_0/W)^2 [\phi^{-1}(W^2/24\,\omega_0^2)]^2 \tag{25}$$

where $\phi^{-1}(\ )$ denotes the inverse of $\phi(\ )$. Equations (24) and (25) depend only on ω_0/W, the ratio of center frequency to signal bandwidth. Equation (25) is of particular interest since it represents the minimum amount of postintegration SNR required to achieve the CRLB. Thus, for example, if $\omega_0/W = 10.0$ (i.e., 10 percent signal bandwidth), $\mu = 14.5$ dB and $\eta = 28.3$ dB. If $\omega_0/W = 100$ (i.e., 1 percent signal bandwidth), $\mu = 34.5$ dB and $\eta = 50.8$ dB. One further observes that the threshold region is not infinitely small, as may be interpreted from the analysis based on the Barankin lower bound [11]. For 10 percent signal bandwidth, it is a segment of 28.3 - 14.5 = 13.8 dB. For 1 percent signal bandwidth, it is a segment of 50.8 - 34.5 = 16.3 dB. Further discussion concerning this unavoidable threshold phenomenon and its relation to the threshold effect predicted by the Barankin lower bound is given in [9].

The 3 dB points of the secondary threshold are given, respectively, by

$$\gamma = \alpha/2 \tag{26}$$

$$\delta = \beta/2 \tag{27}$$

where α and β are defined by (15) and (16), respectively. Thus, $\gamma \approx -3.36$ dB where δ depends, to some extent, on the WD product and it varies from about 10 dB for moderate WD products to about 13 dB for exceedingly large WD products.

The derivation of (22) is based on the assumption that $W/\omega_0 << 1$. In that case, the lower bound exhibits two distinct threshold effects. The phase ambiguities and the envelope ambiguities occur at essentially disjointed segments of the SNR domain and are therefore, in a sense, strictly additive. As the signal bandwidth increases, the two threshold regions come close (i.e., δ converges to μ) and the two ambiguity phenomena cannot be considered separately in the SNR domain. This effect is illustrated in Figs. 8-10 where the various performance characteristics were generated for $WD/2\pi = 20$ and different values of ω_0/W by numerical integration of the exact lower bound (3). Note that $\overline{\epsilon^2}$ is normalized by $D^2/12$ so that we actually measure the relative efficiency of the attainable MSE to the *a priori* variance. The dashed lines denote the various 3 dB points calculated numerically. The sign † denotes the various 3 dB points calculated using the analytical results (24)-(27). Note the close agreement between the two sets as can be observed in Figs. 8 and 9.

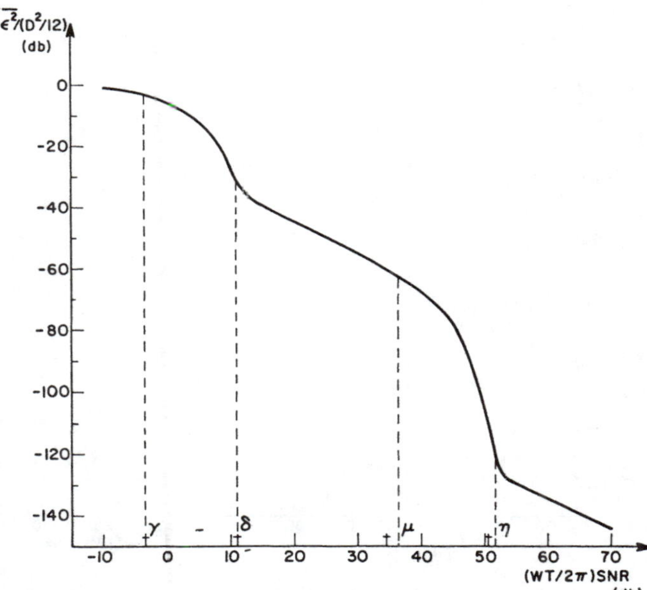

Fig. 8. Normalized $\overline{\epsilon^2}$ versus postintegration SNR for $\omega_0/W = 100$.

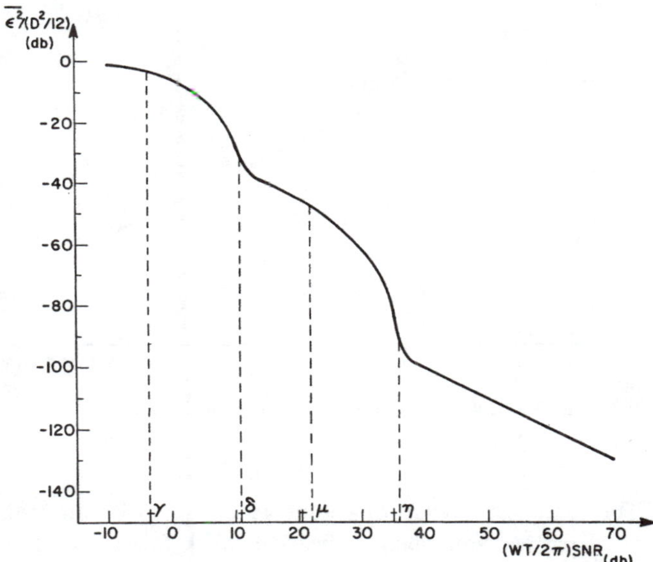

Fig. 9. Normalized $\overline{\epsilon^2}$ versus postintegration SNR for $\omega_0/W = 20$.

Narrow-Band Systems

It is not difficult to relate the composite result given by (22) to the analysis of the narrow-band systems carried out in [9]. In [9], the lower bound is derived under the assumption that the signal bandwidth is so narrow that its correlation time (inverse bandwidth) exceeds the maximum expected delay (i.e., $WD/2\pi < 1$). The envelope-type ambiguities, therefore, are completely eliminated from consideration. Instead of having a threshold effect, there is a smooth transition from the first line of (22) to its third line as illustrated in Fig. 11. The point at which the transition occurs (denoted by ρ in the figure) is obtained by simply intersecting the two indicated lines. One finds

$$\rho = (18/\pi^2)/(WD/2\pi)^2. \tag{28}$$

With this modification, (22) reduces to the result derived in [9].

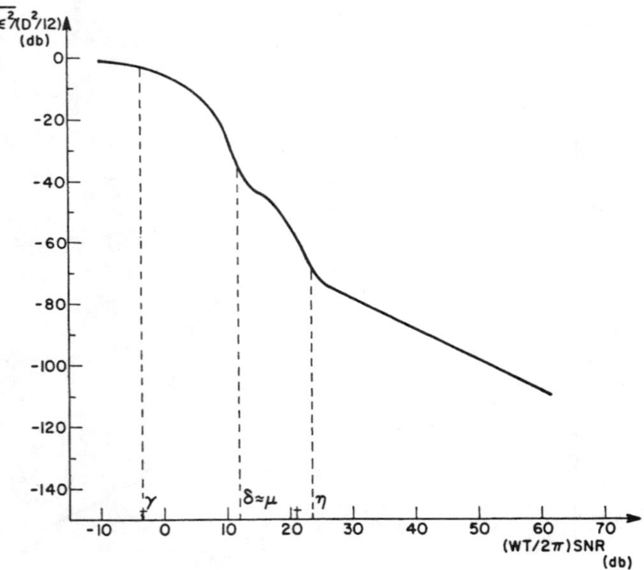

Fig. 10. Normalized $\overline{\epsilon^2}$ versus postintegration SNR for $\omega_0/W = 5$.

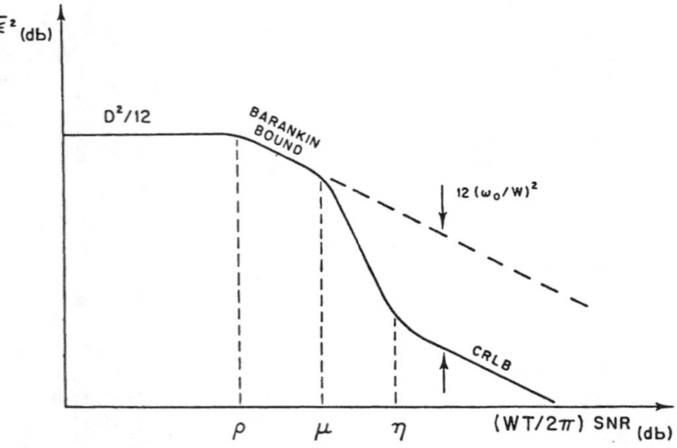

Fig. 11. Composite bound on $\overline{\epsilon^2}$—narrow-band systems.

IV. Conclusions

This is the second part of a study which deals with the problem of passive time delay estimation. The focus here is on systems employing wide-band signals and/or arrays of very widely separated receivers. A modified (improved) version of the ZZLB is used to analyze the effect of additive noise and signal ambiguities on the attainable mean-square estimation errors.

We first concentrated on baseband signals. The analysis is carried out in Appendix B. The lower bound is presented by (14). We observe a distinct threshold phenomenon. Above a certain postintegration SNR, the lower bound converges to the CRLB. This is the small error region where the attainable MSE varies as the first inverse power of the postintegration SNR. Below the threshold, the lower bound quickly approaches a constant level, depending only on the *a priori* search domain. In that region, signal observations are completely dominated by noise and are thus essentially useless for the delay estimation. The threshold level, which determines the boundary between small and large estimation errors, depends on WD, the product of signal bandwidth by the *a priori* search domain, and it varies from about 13 dB for moderate WD products to about 16 dB for exceedingly large WD products.

Delay estimation using bandpass signals is analytically a more complicated problem. The analysis is carried out in Appendix C. The lower bound is presented by (22). Here we observe two distinct threshold effects. The location of the more critical threshold depends only on ω_0/W, the ratio of center frequency to signal bandwidth. The location of the secondary threshold depends only on the WD product. These threshold effects divide the entire SNR domain into essentially three disjointed segments indicating three distinct modes of operation: at high SNR the lower bound coincides with the CRLB. This is the ambiguity-free mode of operation where differential delay estimation is subject only to local errors. At moderate SNR (between the two thresholds), the lower bound exceeds the CRLB by a factor of $12\,(\omega_0/W)^2$. In that region, the ambiguities in the received signal phases cannot be resolved. However, a useful estimate of the differential delay can still be obtained from the received signal envelopes. At low SNR the lower bound approaches a constant level depending only on the *a priori* search domain. In that region, signal observations are subject to envelope ambiguities as well, and are thus essentially useless for the delay estimation.

Appendix A

Lower Bounds on $P_e(x)$

The binary decision problem under consideration here is given by

$$H_0: \Delta\tau = \Delta\tau_0 = a$$

$$H_1: \Delta\tau = \Delta\tau_1 = a + x. \tag{A-1}$$

The log likelihood ratio test (log LRT) between H_0 and H_1 is defined by

$$l = \log \frac{p(v|H_1)}{p(v|H_0)} \tag{A-2}$$

where $p(v|H_i)$ is the conditional probability density of v under H_i hypothesis, and v represents the data vector (e.g., time samples or Fourier coefficients of the received signals). Assuming that H_0 and H_1 are equally likely to occur (i.e., $P(H_0) = P(H_1) = \frac{1}{2}$), a decision rule which minimizes the probability of error compares the log LRT to a zero threshold. If $l \geqslant 0$, we decide on H_1; if $l < 0$, we decide on H_0. Hence, the minimum attainable probability of error is given by

$$P_e(a, a+x) = \tfrac{1}{2} \int_0^\infty p(l|H_0)\, dl + \tfrac{1}{2} \int_{-\infty}^0 p(l|H_1)\, dl. \tag{A-3}$$

Let $\psi_i(s)$ denote the characteristic function associated with $p(l|H_i)$. $p(l|H_i)$ and $\psi_i(s)$ are, by definition, a Fourier transform pair satisfying

$$R(l|H_0) = \frac{1}{2\pi j} \int_{c_0 - j\infty}^{c_0 + j\infty} \psi_0(s)\, e^{-sl}\, ds \tag{A-4}$$

and

$$p(l|H_1) = \frac{1}{2\pi j}\int_{c_1-j\infty}^{c_1+j\infty} \psi_1(s)\, e^{-sl}\, ds \tag{A-5}$$

where it will be shown that the integral in (A-4) can be evaluated for any c_0 in the open interval (0, 1), and the integral in (A-5) can be evaluated for any c_1 in (-1, 0). Substituting (A-4) and (A-5) into (A-3), one obtains

$$\begin{aligned} P_e(a, a+x) &= \frac{1}{2}\cdot\frac{1}{2\pi j}\int_{c_0-j\infty}^{c_0+j\infty} \psi_0(s)\, ds \int_0^{\infty} e^{-sl}\, dl \\ &\quad + \frac{1}{2}\cdot\frac{1}{2\pi j}\int_{c_1-j\infty}^{c_1+j\infty} \psi_1(s)\, ds \int_{-\infty}^{0} e^{-sl}\, dl \\ &= \frac{1}{2}\cdot\frac{1}{2\pi j}\int_{c_0-j\infty}^{c_0+j\infty} [\psi_0(s)/s]\, ds \\ &\quad - \frac{1}{2}\cdot\frac{1}{2\pi j}\int_{c_1-j\infty}^{c_1+j\infty} [\psi_1(s)/s]\, ds \end{aligned} \tag{A-6}$$

where the transition from the first version of (A-6) to its second version can be carried out only if $c_0 > 0$ and $c_1 < 0$. In this setting, $P_e(a, a+x)$ is expressed in terms of $\psi_i(s)$, $i = 0, 1$. Now, by definition,

$$\psi_i(s) \triangleq E\{e^{sl}|H_i\} \qquad i = 0, 1 \tag{A-7}$$

where $E\{\ \}$ denotes the statistical expectation of the bracketed quantity. Substituting (A-2) into (A-7), and observing that $l = l(\boldsymbol{v})$ (so that the statistical expectation operation can be performed with respect to the probability density of $\boldsymbol{v}$), one immediately obtains

$$\psi_0(s) = \int_{-\infty}^{\infty} [R(\boldsymbol{v}|H_1)]^s\, [p(\boldsymbol{v}|H_0)]^{1-s}\, d\boldsymbol{v} \tag{A-8}$$

$$\psi_1(s) = \int_{-\infty}^{\infty} [p(\boldsymbol{v}|H_1)]^{1+s} [p(\boldsymbol{v}|H_0)]^{-s}\, d\boldsymbol{v}. \tag{A-9}$$

We shall now generate the data vector $\boldsymbol{v}$ by Fourier analyzing $v_k(t)$ in (1). Since signal and noise are assumed to be zero-mean Gaussian processes, and the components of $\boldsymbol{v}$ are generated by linear operations on these time functions, $\boldsymbol{v}$ has a multivariate Gaussian distribution

$$p(\boldsymbol{v}|H_i) = \frac{1}{\det(\pi K_i)}\cdot \exp(-\boldsymbol{v}^* K_i^{-1} \boldsymbol{v}) \qquad i = 0, 1 \tag{A-10}$$

where

$$K_i = E\{\boldsymbol{v}\boldsymbol{v}^*|H_i\} \qquad i = 0, 1 \tag{A-11}$$

and $\boldsymbol{v}^*$ denotes the conjugate transpose of $\boldsymbol{v}$. To obtain $\psi_0(s)$, one must substitute (A-10) into (A-8) and carry out the indicated algebraic operations. For observation time T large compared to the correlation time (inverse bandwidth) of signal and noise (i.e., $WT/2\pi >> 1$), the Fourier coefficients (at each receiver output) associated with different frequencies are statistically uncorrelated. In that case, the required computations become relatively easy. Details are contained in [9, Appendix A]. The result is

$$\psi_0(s) = \prod_k\, [1 + 4s(1-s)\,\Gamma(\omega_K, x)]^{-1} \tag{A-12}$$

where

$$\omega_k = 2\pi k/T \tag{A-13}$$

and

$$\Gamma(\omega_k, x) = \mathrm{SNR}\,(\omega_k)\, \sin^2 \omega_k x/2. \tag{A-14}$$

The index k varies over all components in the signal frequency band. Following very similar considerations for $\psi_1(s)$, one obtains

$$\psi_1(s) = \prod_k\, [1 - 4s(1+s)\,\Gamma(\omega_k, x)]^{-1}. \tag{A-15}$$

Note that $\psi_0(s)$ and $\psi_1(s)$ depend only on x so that $P_e(a, a+x) = P_e(x)$. One must now substitute (A-12) and (A-15) into (A-6) and carry out the indicated integration. It can be easily shown that all the poles (singular points) of $\psi_0(s)$ are located on the real axis outside the interval [0, 1], and all the poles of $\psi_1(s)$ are located on the real axis outside the interval [-1, 0]. Hence, the first integral in (A-6) can be evaluated for any value of c_0 in the open interval (0, 1) without affecting the desired result. Similarly, the second integral in (A-6) can be evaluated for any value of c_1 in (-1, 0). We shall find it most convenient to choose $c_0 = \frac{1}{2}$, $c_1 = -\frac{1}{2}$. We shall further make the change of variables $s = \frac{1}{2} + jy$ in the first integral and $s = -\frac{1}{2} + jy$ in the second integral. In this setting, one obtains

$$\begin{aligned} P_e(x) &= \frac{1}{2}\frac{1}{2\pi}\int_{-\infty}^{\infty} \frac{1}{\frac{1}{2}+jy}\left\{\prod_k \left[1 + 4\left(\frac{1}{2}+jy\right)\left(\frac{1}{2}-jy\right)\cdot \Gamma(\omega_k, x)\right]^{-1}\right\} dy \\ &\quad - \frac{1}{2}\frac{1}{2\pi}\int_{-\infty}^{\infty} \frac{1}{-\frac{1}{2}+jy}\left\{\prod_k \left[1 + 4\left(\frac{1}{2}+jy\right)\left(\frac{1}{2}-jy\right)\cdot \Gamma(\omega_k, x)\right]^{-1}\right\} dy \\ &= \frac{1}{\pi}\int_{-\infty}^{\infty} \left\{\prod_k\, [1 + 1 + 4y^2)\,\Gamma(\omega_k, x)]^{-1}\right\} \frac{dy}{1+4y^2} \\ &= \frac{2}{\pi}\left\{\prod_k\, [1 + \Gamma(\omega_k, x)]^{-1}\right\} \\ &\quad \cdot \int_0^{\infty} \left\{\prod_k \left[1 + \frac{\Gamma(\omega_k, x)}{1+\Gamma(\omega_k, x)}\, 4y^2\right]^{-1}\right\} \frac{dy}{1+4y^2}. \end{aligned} \tag{A-16}$$

This is an exact expression for $P_e(x)$. To generate a lower bound on $P_e(x)$ we shall use the inequality $(1+x)^{-1} \geqslant e^{-x}$ whenever $x \geqslant 0$. Thus,

$$\left[1 + \frac{\Gamma(\omega_k, x)}{1+\Gamma(\omega_k, x)} 4y^2\right]^{-1} \geqslant \exp\left\{-\frac{\Gamma(\omega_k, x)}{1+\Gamma(\omega_k, x)} 4y^2\right\}. \tag{A-17}$$

Substituting (A-17) into (A-16), one obtains

$$P_e(x) \geqslant \left\{\prod_k [1+\Gamma(\omega_k, x)]^{-1}\right\} \frac{2}{\pi} \int_0^\infty \cdot \exp\left\{-4y^2 \sum_k \frac{\Gamma(\omega_k, x)}{1+\Gamma(\omega_k, x)}\right\} \frac{dy}{1+4y^2}$$

$$= e^{a(x)} \frac{2}{\pi} \int_0^\infty e^{-4y^2 b(x)} \frac{dy}{1+4y^2} \tag{A-18}$$

where we defined

$$a(x) = -\sum_k \ln [1+\Gamma(\omega_k, x)] \tag{A-19}$$

$$b(x) = \sum_k \frac{\Gamma(\omega_k, x)}{1+\Gamma(\omega_k, x)}. \tag{A-20}$$

The integral in (A-18) can be evaluated analytically using [12, p. 338, formula 3,466-1]. The result is given by

$$P_e(x) \geqslant e^{a(x)+b(x)} \phi(\sqrt{2b(x)}), \tag{A-21}$$

which is the desired result. Substituting (A-14) into (A-19) and (A-20), one obtains

$$a(x) = -\sum_k \ln [1 + \mathrm{SNR}(\omega_k) \cdot \sin^2 \omega_k x/2]$$

$$= -\frac{T}{2\pi} \sum_k \ln [1 + \mathrm{SNR}(\omega_k) \cdot \sin^2 \omega_k x/2] \cdot \Delta\omega \tag{A-22}$$

$$b(x) = \sum_k \frac{\mathrm{SNR}(\omega_k) \cdot \sin^2 \omega_k x/2}{1 + \mathrm{SNR}(\omega_k) \cdot \sin^2 \omega_k x/2}$$

$$= \frac{T}{2\pi} \sum_k \frac{\mathrm{SNR}(\omega_k) \cdot \sin^2 \omega_k x/2}{1 + \mathrm{SNR}(\omega_k) \cdot \sin^2 \omega_k x/2} \cdot \Delta\omega \tag{A-23}$$

where we define $\Delta\omega = 2\pi/T$. For large WT product and smoothly varying signal and noise spectra, the function SNR (ω) $\sin^2 \omega x/2$ changes only insignificantly over the frequency increment of $\Delta\omega$, so that the sums in (A-22) and (A-23) can be converted into the integrals given by (5) and (6), respectively.

Equation (A-21) can further be used to generate weaker lower bounds on $P_e(x)$, which will become useful in the proceeding analysis. We shall start from (A-19) and (A-20). Since

$$\frac{\Gamma(\omega_k, x)}{1+\Gamma(\omega_k, x)} \leqslant \Gamma(\omega_k, x) \tag{A-24}$$

and since

$$\ln [1+\Gamma(\omega_k, x)] - \frac{\Gamma(\omega_k, x)}{1+\Gamma(\omega_k, x)} \leqslant \frac{1}{2} \Gamma^2(\omega_k, x), \tag{A-25}$$

it immediately follows that

$$2b(x) \leqslant 2 \sum_k \Gamma(\omega_k, x) \tag{A-26}$$

and

$$a(x) + b(x) \geqslant -\tfrac{1}{2} \sum_k \Gamma^2(\omega_k, x) \tag{A-27}$$

where $2b(x)$ and $[a(x)+b(x)]$ are the terms appearing in (A-21). Replacing the first term by its upper bound, and the second term by its lower bound, $P_e(x)$ can further be bounded by

$$P_e(x) \geqslant e^{-d(x)} \phi(\sqrt{c(x)}) \tag{A-28}$$

where $c(x)$ and $d(x)$ are given by

$$c(x) = 2 \sum_k \Gamma(\omega_k, x) = \frac{T}{\pi} \sum_k \mathrm{SNR}(\omega_k) \sin^2 (\omega_k x/2)\, \Delta\omega \xrightarrow[WT/2\pi >> 1]{} \frac{T}{\pi} \int_0^\infty \mathrm{SNR}(\omega) \cdot \sin^2 (\omega x/2)\, d\omega \tag{A-29}$$

$$d(x) = \frac{1}{2} \sum_k \Gamma^2(\omega_k, x) = \frac{T}{4\pi} \sum_k [\mathrm{SNR}(\omega_k) \cdot \sin^2 (\omega_k x/2)]^2 \, \Delta\omega \cdot \xrightarrow[WT/2\pi >> 1]{} \frac{T}{4\pi} \int_0^\infty [\mathrm{SNR}(\omega) \cdot \sin^2 \omega x/2]^2 \, d\omega. \tag{A-30}$$

Using the inequality $\sin^2 (\omega x/2) \leqslant (\omega x/2)^2$, $c(x)$ and $d(x)$ can further be bounded by

$$c(x) \leqslant c^2 x^2 \tag{A-31}$$

$$d(x) \leqslant d^4 x^4 \tag{A-32}$$

where

$$c^2 = \frac{T}{4\pi} \int_0^\infty \omega^2 \, \mathrm{SNR}(\omega)\, d\omega \tag{A-33}$$

and

$$d^4 = \frac{T}{64\pi} \int_0^\infty [\omega^2 \, \mathrm{SNR}(\omega)]^2 \, d\omega. \tag{A-34}$$

Substituting $c(x)$ and $d(x)$ by their upper bounds, one immediately obtains

$$P_e(x) \geqslant e^{-d^4 x^4} \phi(cx). \tag{A-35}$$

We note, in passing, that the lower bound given by (A-35) is useful only for small x where the inequalities in (A-31) and (A-32) are tight.

Appendix B
Analysis of the Modified ZZLB for Baseband Signals

The modified ZZLB is given by (3), and is rewritten here for reference:

$$\overline{\epsilon^2} \geq \frac{1}{D} \int_0^D xG[(D-x) P_e(x)]\, dx. \tag{B-1}$$

The function $G[(D-x)\, P_e(x)]$ assumes a simple form if $P_e(x)$ contains several well-defined peaks. This is illustrated in Fig. 12 for a typical baseband case. Thus, if $x_0, x_1, x_2, \cdots$ are the local maxima of $P_e(x)$, then at each segment $[x_{n-1}, x_n]$ the function $G[(D-x) P_e(x)]$ is closely bounded by

$$G[(D-x) P_e(x)] \geq (D-x_n) P_e(x_n) \qquad x_{n-1} \leq x \leq x_n. \tag{B-2}$$

Note that since $G[(D-x) P_e(x)]$ is a nonincreasing function of x, (B-2) holds for arbitrary set of x_n's. We further note that the local maxima of $P_e(x)$ are, in fact, the local maxima (ambiguity points) of the cross-correlation function. These occur, approximately, at (see Fig. 2)

$$x_n = (2\pi/W)\, n \qquad n = 0, 1, 2, \cdots. \tag{B-3}$$

$P_e(x_n)$, required for the computation of (B-2), is closely approximated using the expression in (4). Following the derivation in Appendix A, that expression is shown to be a lower bound. Thus,

$$P_e(x_n) \geq e^{a(x_n) + b(x_n)}\, \phi(\sqrt{2b(x_n)}) \tag{B-4}$$

where $a(x_n)$ and $b(x_n)$ are obtained by substituting $x = x_n$ into (12) and (13), respectively. One obtains

$$a(x_n) = -\frac{T}{2\pi} \int_0^{W/2} \ln(1 + \text{SNR} \cdot \sin^2 \omega x_n/2)\, d\omega$$

$$= -\frac{WT}{2\pi} \frac{1}{Wx_n/2} \int_0^{Wx_n/4} \ln(1 + \text{SNR} \sin^2 \Omega)\, d\Omega \tag{B-5}$$

and

$$b(x_n) = \frac{T}{2\pi} \int_0^{W/2} \frac{\text{SNR} \cdot \sin^2 \omega x_n/2}{1 + \text{SNR} \cdot \sin^2 \omega x_n/2}\, d\omega$$

$$= \frac{WT}{2\pi} \frac{1}{Wx_n/2} \int_0^{Wx_n/4} \frac{\text{SNR} \cdot \sin^2 \Omega}{1 + \text{SNR} \cdot \sin^2 \Omega}\, d\Omega \tag{B-6}$$

where $Wx_n/4 = n\pi/2$. Since the integrands in (B-5) and (B-6) are even and periodic functions of Ω with a period of π, and since we are integrating over $[0, n\pi/2]$, these equations become independent of n, i.e., $a(x_n) = a$ and $b(x_n) = b$, where

$$a = -\frac{WT}{2\pi} \cdot \frac{1}{\pi} \int_0^{\pi/2} \ln(1 + \text{SNR} \sin^2 \Omega)\, d\Omega$$

$$= -\frac{WT}{2\pi} \ln \frac{1 + \sqrt{1 + \text{SNR}}}{2} \tag{B-7}$$

$$b = \frac{WT}{2\pi} \frac{1}{\pi} \int_0^{\pi/2} \frac{\text{SNR} \cdot \sin^2 \Omega}{1 + \text{SNR} \cdot \sin^2 \Omega}\, d\Omega$$

$$= \frac{WT}{4\pi} \frac{\sqrt{1 + \text{SNR}} - 1}{\sqrt{1 + \text{SNR}}}. \tag{B-8}$$

The integral appearing in (B-7) can be found in [12, p. 594, formula 4.399]. The integral appearing in (B-8) can be modified to a form found in [12, p. 152, formula 2.562]. Thus,

$$P_e(x_n) \geq e^{a+b}\, \phi(\sqrt{2b}). \tag{B-9}$$

Substituting (B-3) and (B-9) into (B-2), one obtains

$$G[(D-x) P_e(x)] \geq (D-x_n)\, e^{a+b} \phi(\sqrt{2b})$$
$$x_n - 2\pi/W \leq x \leq x_n. \tag{B-10}$$

Since $D - x_n \geq D - 2\pi/W - x$ for all $x \in [x_n - 2\pi/W, x_n]$, it immediately follows that

$$G[(D-x) P_e(x)] \geq \begin{cases} (D - 2\pi/W - x)\, e^{a+b} \phi(\sqrt{2b}) \\ \qquad 0 \leq x \leq D - 2\pi/W \\ 0 \qquad D - 2\pi/W \leq x \leq D. \end{cases} \tag{B-11}$$

The first line in (B-11) is illustrated by the dashed line in Fig. 12. The second line in (B-11) indicates that zero is a better bound. For $WD/2\pi \gg 1$, the lower bound presented by (B-11) is very tight except for values of x in the vicinity of $x = 0$ where $P_e(x)$ changes rapidly.

To take this effect into account, we observe that for small x, $P_e(x)$ is closely bounded using (A-35). In that region, therefore, $G[(D-x) P_e(x)]$ is closely bounded by

$$G[(D-x) P_e(x)] \geq G[(D-x)\, e^{-d^4 x^4} \phi(cx)]$$
$$= (D-x)\, e^{-d^4 x^4} \phi(cx) \tag{B-12}$$

where, in the transition from the first version of (B-12) to its second version, we observe that $G[f(x)] = f(x)$ whenever $f(x)$ is a nonincreasing function of x. The parameters c and d are obtained by substituting (10) into (A-33) and (A-34), respectively. One obtains

$$c^2 = \frac{T}{4\pi} \int_0^{W/2} \omega^2\, \text{SNR}\, d\omega = \frac{W^3 T\, \text{SNR}}{96\pi} \tag{B-13}$$

and

$$d^4 = \frac{T}{64\pi} \int_0^{W/2} [\omega^2\, \text{SNR}]^2\, d\omega = \frac{W^5 T\, \text{SNR}^2}{5 \cdot 2^{11} \cdot \pi}. \tag{B-14}$$

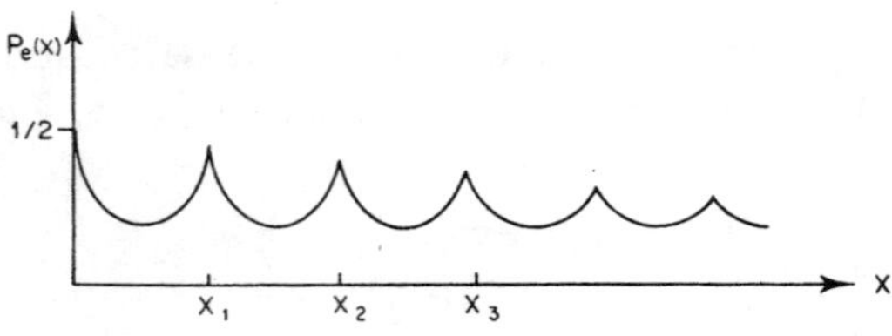

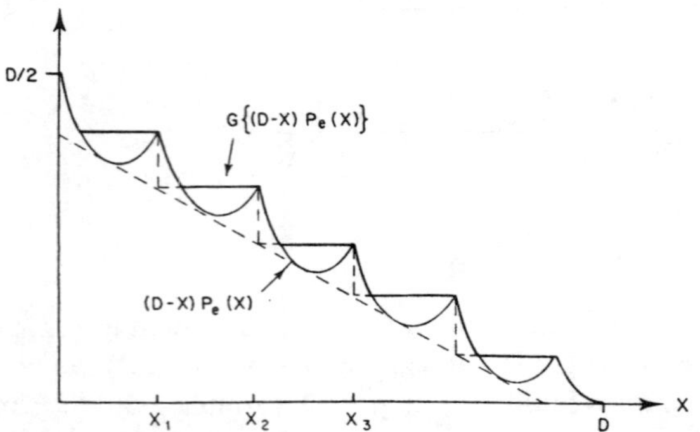

Fig. 12. $P_e(x)$ and $G[(D-x)P_e(x)]$ for typical baseband case.

For future manipulations we note that

$$(c/d)^4 = \frac{20}{9}(WT/2\pi). \tag{B-15}$$

Combining (B-11) and (B-12), $G[(D-x)P_e(x)]$ is tightly bounded by

$$G[(D-x)P_e(x)] \geqslant \begin{cases} (D-x)\,e^{-d^4x^4}\phi(cx) & 0 \leqslant x \leqslant z \\ (D-2\pi/W-x)\,e^{a+b}\phi(\sqrt{2b}) & z < x \leqslant D-2\pi/W \\ 0 & D-2\pi/W < x \leqslant D \end{cases} \tag{B-16}$$

where z is defined by

$$cz = \sqrt{2b}. \tag{B-17}$$

Note that z is approximately the point at which the first and second lines in (B-16) intersect. Substituting (B-16) into (B-1) one obtains the following lower bound:

$$\overline{\epsilon^2} \geqslant \frac{1}{D}\int_0^z x(D-x)\,e^{-d^4x^4}\phi(cx)\,dx + \left[\frac{1}{D}\int_z^{D-2\pi/W} x(D-2\pi/W-x)\,dx\right] e^{a+b}\phi(\sqrt{2b}). \tag{B-18}$$

Carrying out some straightforward algebra manipulations, it can easily be shown that $z < 2\pi/W$. Since we are assuming that $2\pi/W \ll D$, the first term on the right-hand side of (B-18) is closely approximated by

$$\frac{1}{D}\int_0^z x(D-x)\,e^{-d^4x^4}\phi(cx)\,dx \approx \int_0^z x e^{-d^4x^4}\phi(cx)\,dx$$

$$= \frac{1}{c^2}\int_0^{cz} x\exp\left\{-\frac{x^4}{(c/d)^4}\right\}\phi(x)\,dx$$

$$= \frac{1}{c^2}\int_0^{\sqrt{2b}} x\cdot\exp\left\{-\frac{9x^4}{20(WT/2\pi)}\right\}\phi(x)\,dx \tag{B-19}$$

where, in the transition from the second version of (B-19) to its third version, we have substituted (B-15) and (B-17). Similarly, the integral appearing in the second term of (B-18) is closely approximated by

$$\frac{1}{D}\int_z^{D-2\pi/W} x(D-2\pi/W-x)\,dx \approx D^2/6. \tag{B-20}$$

Substituting (B-19) and (B-20) into (B-18), one obtains

$$\overline{\epsilon^2} \geqslant \sim \frac{1}{c^2}\int_0^{\sqrt{2b}} x\cdot\exp\left\{-\frac{9x^4}{20(WT/2\pi)}\right\}\phi(x)\,dx + \frac{D^2}{6}e^{a+b}\phi(\sqrt{2b}). \tag{B-21}$$

Numerical integration indicates that (B-21) is an excellent approximation to the exact result (B-1). The lower bound presented by (B-21) exhibits two distinct asymptotes: as SNR $\to$ 0, the variables a, b, and c approach zero. In that limit, the second term in (B-21) becomes the dominant term and the lower bound approaches

$$\overline{\epsilon^2} \approx D^2/12. \tag{B-22}$$

As SNR $\to \infty$, $a + b \to -\infty$ and $b \to WT/4\pi$. In that limit, the first term in (B-21) becomes the dominant term in the sum and the lower bound approaches

$$\overline{\epsilon^2} \geqslant \frac{1}{c^2}\int_0^{\sqrt{WT/2\pi}} x\cdot\exp\left\{-\frac{9x^4}{20(WT/2\pi)}\right\}\phi(x)\,dx \xrightarrow[WT/2\pi \gg 1]{} \frac{1}{c^2}\int_0^\infty x\phi(x)\,dx = \frac{1}{4c^2}$$

$$= \frac{12}{W^2}\cdot\frac{1}{(WT/2\pi)\,\mathrm{SNR}}. \tag{B-23}$$

One immediately identifies (B-22) with the first line of (14), and (B-23) with its third line.

The transition from the $D^2/12$ asymptote to the $1/4c^2$ asymptote essentially starts when

$$e^{a+b}\phi(\sqrt{2b}) = \tfrac{1}{4} \tag{B-24}$$

and is essentially completed when

$$\frac{D^2}{6}e^{a+b}\phi(\sqrt{2b}) = 1/4c^2 \tag{B-25}$$

where, in the transition region (the so-called threshold region), the lower bound varies essentially as $\phi(\sqrt{2b})$. One must now substitute (B-7), (B-8), and (B-13) into (B-24) and (B-25), and solve these equations with respect to SNR in order to obtain the corresponding 3 dB points. Suppose, for the moment, that these solutions are obtained at SNR $\ll 1$. In that case, (B-7) and (B-8) can be approximated, without incurring any significant errors, by

$$a \approx -\frac{WT}{2\pi} \ln (1 + \text{SNR}/4) \approx -\frac{WT}{2\pi} \cdot \frac{\text{SNR}}{4} \tag{B-26}$$

$$b \approx \frac{WT}{4\pi} \cdot \frac{\text{SNR}}{2}. \tag{B-27}$$

With these approximations, (B-24) and (B-25) assume the simplified forms

$$\phi(\sqrt{R/2}) = \tfrac{1}{4} \tag{B-28}$$

and

$$(R/2)\phi(\sqrt{R/2}) = (6/WD)^2 \tag{B-29}$$

where R is the postintegration SNR defined by

$$R = (WT/2\pi)\ \text{SNR}. \tag{B-30}$$

Denoting by $R = \alpha$ the solution to (B-28), and substituting $2[\phi^{-1}(\frac{1}{4})]^2 = 0.92$, one immediately obtains (15). Denoting by $R = \beta$ the solution to (B-29), one immediately obtains (16). We further observe that in terms of SNR, the solution to (B-28) reads $\text{SNR} = \alpha/(WT/2\pi)$, and the solution to (B-29) reads $\text{SNR} = \beta/(WT/2\pi)$. Thus, for $WT/2\pi \gg 1$, the simplifying approximations made in (B-26) and (B-27) may affect these results only insignificantly.

Appendix C
Analysis of the Modified ZZLB for Bandpass Signals

The modified ZZLB is given by (3) and is rewritten here for reference:

$$\overline{\epsilon^2} \geq \frac{1}{D} \int_0^D xG[(D - x) P_e(x)]\, dx. \tag{C-1}$$

For values of x in the vicinity of $x = 0$, the function $G[(D - x) P_e(x)]$ is closely bounded using (B-12) where the parameters c and d are obtained by substituting (19) into (A-33) and (A-34), respectively. Assuming that $\omega_0/W \gg 1$ (so that one can ignore terms on the order of $(W/\omega_0)^2$ relative to 1), c^2 and d^4 are given, respectively, by

$$c^2 = \frac{WT}{4\pi} \omega_0^2\ \text{SNR} \tag{C-2}$$

$$d^4 = \frac{WT}{64\pi} \omega_0^4\ \text{SNR}^2. \tag{C-3}$$

For future manipulations, we note that

$$(c/d)^4 = 8(WT/2\pi). \tag{C-4}$$

For values of x away from $x = 0$, a tighter lower bound on $G[(D - x) P_e(x)]$ can be generated from the local maxima of $P_e(x)$. This is illustrated in Fig. 12 for a typical baseband case. In the bandpass case, however, $P_e(x)$ has two sets of local maxima. One set, $x_n = (2\pi/W)\, n$, $n = 1, 2, \cdots$, is associated with the ambiguities in the envelope of the cross-correlation function, and the other set, $\tilde{x}_n = (2\pi/\omega_0)\, n$, is associated with the ambiguities in the phase of the cross-correlation function.

We shall assume, without any significant loss in generality, that $\omega_0/W = k$ is an integer. In that case, using the set $x_n = (2\pi/W)\, n$ and following the same considerations outlined in Appendix B, $G[(D - x) P_e(x)]$ is bounded using (B-11), where a and b are given by

$$a = -\frac{WT}{\pi} \ln \frac{1 + \sqrt{1 + \text{SNR}}}{2} \tag{C-5}$$

$$b = \frac{WT}{2\pi} \frac{\sqrt{1 + \text{SNR}} - 1}{\sqrt{1 + \text{SNR}}}. \tag{C-6}$$

Note that there is a factor of 2 difference between (C-5) and (B-7), and between (C-6) and (B-8).

We shall now use the set $\tilde{x}_n = (2\pi/\omega_0)\, n$ to generate another lower bound on $G[(D - x) P_e(x)]$ using similar considerations. Making use of (A-28), $P_e(\tilde{x}_n)$ is bounded by

$$P_e(\tilde{x}_n) \geq e^{-d(\tilde{x}_n)} \phi(\sqrt{c(\tilde{x}_n)}) \tag{C-7}$$

where $c(\tilde{x}_n)$ and $d(\tilde{x}_n)$ are obtained by substituting (19) into (A-29) and (A-30), and calculating these functions for $x = \tilde{x}_n$. Following some straightforward algebra manipulations, one obtains

$$c(\tilde{x}_n) = \frac{WT}{2\pi} \frac{1}{W\tilde{x}_n/2} \int_{-W\tilde{x}_n/4}^{W\tilde{x}_n/4} \text{SNR} \cdot \sin^2 \Omega\, d\Omega \tag{C-8}$$

$$d(\tilde{x}_n) = \frac{WT}{4\pi} \frac{1}{W\tilde{x}_n/2} \int_{-W\tilde{x}_n/4}^{W\tilde{x}_n/4} [\text{SNR} \cdot \sin^2 \Omega]^2\, d\Omega. \tag{C-9}$$

Using the inequality $\sin^2 \Omega \leq \Omega$, $c(\tilde{x}_n)$ and $d(\tilde{x}_n)$ can be bounded by

$$c(\tilde{x}_n) \leq \tilde{c}^2 \tilde{x}_n^2 \tag{C-10}$$

$$d(\tilde{x}_n) \leq \tilde{d}^4 \tilde{x}_n^4 \tag{C-11}$$

where

$$\tilde{c}^2 = \frac{W^3 T\, \text{SNR}}{48\pi} \tag{C-12}$$

$$\tilde{d}^4 = \frac{W^5 T\, \text{SNR}^2}{5 \cdot 2^{10} \cdot \pi}. \tag{C-13}$$

For future manipulations, we note that

$$(\tilde{c}/\tilde{d})^4 = \frac{40}{9}(WT/2\pi). \tag{C-14}$$

Substituting $c(\tilde{x}_n)$ and $d(\tilde{x}_n)$ by their upper bounds, one obtains

$$P_e(\tilde{x}_n) \geq e^{-\tilde{d}^4 \tilde{x}_n^4} \phi(\tilde{c} \tilde{x}_n). \tag{C-15}$$

Now, since

$$G[(D - x) P_e(x)] \geq (D - \tilde{x}_n) P_e(\tilde{x}_n) \qquad \tilde{x}_n - 2\pi/\omega_0 \leq x \leq \tilde{x}_n, \tag{C-16}$$

and since the right-hand side of (C-15) is a monotonically decreasing function, it immediately follows that

$$G[(D-x)P_e(x)] \geqslant (D - 2\pi/\omega_0 - x)\, e^{-\tilde{d}^4(x+2\pi/\omega_0)^4} \cdot \phi[\tilde{c}(x+2\pi/\omega_0)]. \tag{C-17}$$

Combining (B-12), (C-17), and (B-11) (in the given order), $G[(D-x)P_e(x)]$ is tightly bounded by

$$G[(D-x)P_e(x)] \geqslant \begin{cases} (D-x)\,e^{-d^4x^4}\,\phi(cx) & 0 \leqslant x < z_1 \\ (D - 2\pi/\omega_0 - x)\,e^{-\tilde{d}^4(x+2\pi/\omega_0)^4}\,\phi[\tilde{c}(x+2\pi/\omega_0)] & z_1 \leqslant x < z_2 \\ (D - 2\pi/W - x)\,e^{a+b}\,\phi(\sqrt{2b}) & z_2 \leqslant x < D - 2\pi/n \\ 0 & D - 2\pi/W \leqslant x \leqslant D \end{cases} \tag{C-18}$$

where z_1 and z_1 are defined by

$$cz_1 = \tilde{c}(z_1 + 2\pi/\omega_0) \tag{C-19}$$

$$\tilde{c}(z_2 + 2\pi/\omega_0) = \sqrt{2b}. \tag{C-20}$$

Note that z_1 is approximately the point at which the first and second lines in (C-18) intersect, and z_2 is approximately the point at which the second and third lines intersect. Substituting (C-18) into (C-1), one obtains the following lower bound:

$$\begin{aligned}\overline{\epsilon^2} \geqslant &\frac{1}{D}\int_0^{z_1} x(D-x)\,e^{-d^4x^4}\,\phi(cx)\,dx \\ &+ \frac{1}{D}\int_{z_1}^{z_2} x(D - 2\pi/\omega_0 - x)\,e^{-\tilde{d}^4(x+2\pi/\omega_0)^4} \\ &\cdot \phi[\tilde{c}(x+2\pi/\omega_0)]\,dx \\ &+ \left[\frac{1}{D}\int_{z_2}^{D-2\pi/W} x(D - 2\pi/W - x)\,dx\right] e^{a+b}\,\phi(\sqrt{2b}).\end{aligned} \tag{C-21}$$

Following some straightforward algebra manipulations, it can easily be shown that $z_1 \ll 2\pi/\omega_0$, and that $z_2 < 2\pi/W$. Since we are assuming that $2\pi/W \ll D$, the lower bound can be approximated, without incurring any significant errors, by

$$\begin{aligned}\overline{\epsilon^2} \geqslant\sim &\int_0^{z_1} xe^{-d^4x^4}\,\phi(cx)\,dx \\ &+ \int_{z_1}^{z_2} xe^{-\tilde{d}^4(x+2\pi/\omega_0)^4}\,\phi[\tilde{c}(x+2\pi/\omega_0)]\,dx \\ &+ \frac{D^2}{6}e^{a+b}\,\phi(\sqrt{2b}) \\ = &\frac{1}{c^2}\int_0^{cz_1} x \cdot \exp\left\{-\frac{x^4}{(c/d)^4}\right\}\phi(x)\,dx \\ &+ \frac{1}{\tilde{c}^2}\int_{\tilde{c}(z_1+2\pi/\omega_0)}^{\tilde{c}(z_2+2\pi/\omega_0)} (x - \tilde{c}\cdot 2\pi/\omega_0) \\ &\cdot \exp\left\{-\frac{x^4}{(\tilde{c}/\tilde{d})^4}\right\}\phi(x)\,dx + \frac{D^2}{6}e^{a+b}\,\phi(\sqrt{2b}) \\ = &\frac{1}{c^2}\int_0^{\tilde{c}(z_1+2\pi/\omega_0)} x \cdot \exp\left\{-\frac{x^4}{8(WT/2\pi)}\right\}\phi(x)\,dx \\ &+ \frac{1}{\tilde{c}^2}\int_{\tilde{c}(z_1+2\pi/\omega_0)}^{\sqrt{2b}} (x - \tilde{c}2\pi/\omega_0) \\ &\cdot \exp\left\{-\frac{9x^4}{40(WT/2\pi)}\right\}\phi(x)\,dx + \frac{D^2}{6}e^{a+b}\,\phi(\sqrt{2b}) \\ \approx &\frac{1}{c^2}\int_0^{\tilde{c}2\pi/\omega_0} x \cdot \exp\left\{-\frac{x^4}{8(WT/2\pi)}\right\}\phi(x)\,dx \\ &+ \frac{1}{\tilde{c}^2}\int_{\tilde{c}2\pi/\omega_0}^{\sqrt{2b}} (x - \tilde{c}2\pi/\omega_0) \\ &\cdot \exp\left\{-\frac{9x^4}{40(WT/2\pi)}\right\}\phi(x)\,dx + \frac{D^2}{6}e^{a+b}\,\phi(\sqrt{2b})\end{aligned} \tag{C-22}$$

where, in the transition from the second version of (C-22) to its third version, we have substituted (C-4), (C-14), (C-19), and (C-20).

Substituting (C-2), (C-12), (C-5), and (C-6) into (C-22), one can now generate MSE predictions for any prespecified SNR. We further note that the second and third terms in (C-22) contribute significantly to the sum only when $\text{SNR} \ll 1$. In that region, a and b can be approximated, without incurring any significant errors, by

$$a \approx -\frac{WT}{\pi}\cdot\frac{\text{SNR}}{4} \tag{C-23}$$

$$b \approx \frac{WT}{2\pi}\cdot\frac{\text{SNR}}{2}. \tag{C-24}$$

With these considerations, the lower bound assumes the form

$$\begin{aligned}\overline{\epsilon^2} \geqslant &\frac{1}{(\omega_0^2/2)R}\int_0^{\sqrt{(\pi^2/6)(W/\omega_0)\sqrt{R}}} x \cdot \exp\left\{-\frac{x^4}{8(WT/2\pi)}\right\} \\ &\cdot \phi(x)\,dx \\ &+ \frac{1}{(W^2/24)R}\int_{\sqrt{(\pi^2/6)(W/\omega_0)\sqrt{R}}}^{\sqrt{R}} x \\ &\cdot \exp\left\{-\frac{9x^4}{40(WT/2\pi)}\right\}\phi(x)\,dx + \frac{D^2}{6}\phi(\sqrt{R})\end{aligned} \tag{C-25}$$

where R is defined as the product of $WT/2\pi$ by SNR (the so-called postintegration SNR).

We shall now examine a few limiting cases. If $R \ll 1$, the third term on the right-hand side of (C-25) becomes the dominant term in the sum, and the lower bound approaches

$$\overline{\epsilon^2} \geq \sim \frac{D^2}{6} \phi(\sqrt{R}) \rightarrow \frac{D^2}{12}. \tag{C-26}$$

If $1 << R << (\omega_0/W)^2$, the second term becomes the dominant term in the sum, and the lower bound approaches

$$\overline{\epsilon^2} \geq \sim \frac{1}{(W^2/24)R} \int_{\sqrt{(\pi^2/6)}(W/\omega_0)\sqrt{R}}^{\sqrt{R}} x \cdot \exp\left\{-\frac{9x^4}{40(WT/2\pi)}\right\} \phi(x)\, dx \xrightarrow[(WT/2\pi)>>1]{} \frac{1}{(W^2/12)R} \phi\left(\sqrt{\frac{\pi^2}{6}} \frac{W}{\omega_0} \sqrt{R}\right) \approx \frac{1}{(W^2/6)R}. \tag{C-27}$$

Finally, if $R >> (\omega_0/W)^2$, the first term in the sum becomes the dominant one, and the lower bound approaches

$$\overline{\epsilon^2} \geq \sim \frac{1}{(\omega_0^2/2)R} \int_0^{\sqrt{(\pi^2/6)} \cdot (W/\omega_0)\sqrt{R}} x \cdot \exp\left\{-\frac{x^4}{8(WT/2\pi)}\right\} \phi(x)\, dx \xrightarrow[WT/2\pi>>1]{} \frac{1}{2\omega_0^2 R}\left[1 - 2\phi\left(\sqrt{\frac{\pi^2}{6}} \frac{W}{\omega_0} \sqrt{R}\right)\right] \approx \frac{1}{2\omega_0^2 R}. \tag{C-28}$$

One immediately identifies (C-26), (C-27), and (C-28) with the first, third, and fifth lines of (22), respectively.

The transition from the first line of (22) to its third line essentially starts when

$$\phi(\sqrt{R}) = \tfrac{1}{4} \tag{C-29}$$

and is essentially completed when

$$\frac{D^2}{6} \phi(\sqrt{R}) = \frac{1}{(W^2/6)R}, \tag{C-30}$$

where, in the transition region, the lower bound varies essentially as $\phi(\sqrt{R})$. Denoting by $R = \gamma$ the solution to (C-29), and using the similarity between (C-29) and (B-28), one immediately obtains (26). Denoting by $R = \delta$ the solution to (C-30), and using the similarity between (C-30) and (B-29), one immediately obtains (27).

The transition from the third line of (22) to its fifth line essentially starts when

$$\phi\left(\sqrt{\frac{\pi^2}{6}} \frac{W}{\omega_0} \sqrt{R}\right) = \frac{1}{4} \tag{C-31}$$

and is essentially completed when

$$\frac{1}{(W^2/12)R} \phi\left(\sqrt{\frac{\pi^2}{6}} \frac{W}{\omega_0} \sqrt{R}\right) = \frac{1}{2\omega_0^2 R} \tag{C-32}$$

where, in the transition region, the lower bound varies essentially as $\phi(\sqrt{(\pi^2/6)}W/\omega_0 \sqrt{R})$. Denoting by $R = \mu$ and $R = \eta$ the solutions tc (C-31) and (C-32), and following some straightforward algebra manipulations, one immediately obtains (24) and (25), respectively.

ACKNOWLEDGMENT

The authors would like to thank Prof. P. M. Schultheiss and Dr. J. P. Ianniello for their helpful comments and discussions throughout the course of study, and to the Word Processing Department at Tel-Aviv University for their patient and excellent secretarial assistance.

REFERENCES

[1] V. H. MacDonald and P. M. Schultheiss, "Optimum passive bearing estimation in a spatially incoherent noise environment," *J. Acoust. Soc. Amer.*, vol. 46, pp. 37–43, 1969.

[2] W. R. Hahn, "Optimum signal processing for passive sonar range and bearing estimation," *J. Acoust. Soc. Amer.*, vol. 58, no. 1, pp. 201-207, 1981.

[3] G. C. Carter, "Time delay estimation for passive sonar signal processing," *IEEE Trans. Acoust., Speech, Signal Processing*, vol. ASSP-29, pp. 463–470, June 1981.

[4] Y. T. Chan, R. V. Hattin, and J. B. Plant, "The least squares estimation of time delay and its use in signal detection," *IEEE Trans. Acoust., Speech, Signal Processing*, vol. ASSP-26, no. 3, pp. 217–222, 1978.

[5] W. R. Hahn and S. A. Tretter, "Optimum processing for delay-vector estimation in passive signal arrays," *IEEE Trans. Inform. Theory*, vol. IT-19, no. 5, pp. 608–614, 1973.

[6] B. V. Hamon and E. J. Hannan, "Spectral estimation of time delay for dispersive and nondispersive systems," *J. Appl. Statist.*, vol. 23, no. 2, pp. 134–142, 1974.

[7] C. H. Knapp and G. C. Carter, "The generalized correlation method for estimation of time delay," *IEEE Trans. Acoust., Speech, Signal Processing*, vol. ASSP-24, no. 4, pp. 320–327, 1976.

[8] A. J. Weiss and E. Weinstein, "Composite bound on the attainable mean-square error in passive time-delay estimation from ambiguity prone signal," *IEEE Trans. Inform. Theory*, vol. IT-28, pp. 977–979, Nov. 1982.

[9] —, "Fundamental limitations in passive time delay estimation–Part I: Narrow-band systems," *IEEE Trans. Acoust., Speech, Signal Processing*, vol. ASSP-31, pp. 472–486, Apr. 1983.

[10] J. P. Ianniello, E. Weinstein, and A. Weiss, "Comparison of the Ziv-Zakai lower bound on time delay estimation with correlator performance," in *Proc. IEEE Int. Conf. Acoust., Speech, Signal Processing (ICASSP'83)*, Apr. 1983.

[11] S. K. Chow and P. M. Schultheiss, "Delay estimation using narrow-band processes," *IEEE Trans. Acoust., Speech, Signal Processing*, vol. ASSP-29, pp. 478–484, June 1981.

[12] I. S. Gradshteyn and I. M. Ryzhik, *Tables of Integrals, Series and Products.* New York: Academic, 1980.

Composite Bound on the Attainable Mean-Square Error in Passive Time-Delay Estimation from Ambiguity Prone Signals

ANTHONY WEISS and EHUD WEINSTEIN

Abstract—The location of a radiating source can be determined by measuring the relative time-delay of its signal wavefront to several spatially separated receivers. A new technique is presented to investigate the performance of time-delay measurement schemes based on a modified version of the Ziv–Zakai lower bound. This technique is shown to yield a tight bound on the attainable mean-square measurement error for any prespecified signal-to-noise ratio conditions.

The location of a radiating source can be determined by observation of its signal at three or more spatially separated receivers. In the absence of detailed knowledge concerning the signal waveshape, all information about source location (i.e., its bearing and range) is contained in the relative (differential) delay of the signal wavefront to the various receiver pairs. Differential-delay estimation has therefore attracted a great deal of interest in recent literature (e.g., [1]–[7]).

In most of the systems which have been analyzed, the Cramér–Rao inequality has been used to set bounds on the attainable mean-square estimation error. Its use was justified by invoking an asymptotic theorem which asserts that the maximum likelihood estimator comes arbitrarily close to reaching the lower bound for sufficiently long observation times. The practically important question is therefore: How long is "long enough?" Clearly, the observation time T must be large compared with the inverse bandwidth of the signal and noise (i.e., $WT \gg 1$), a condition which presents very little difficulty in practice. However if one examines the problem more closely, one finds a second condition: the maximum likelihood estimate (which approaches the Cramér–Rao bound asymptotically) must not be subject to ambiguities. For differential-delay estimation with narrow-band signals this means that one must be able to distinguish unambiguously between adjacent maxima of the cross correlation function for a given receiver pair. This, in turn, implies either a very high signal-to-noise ratio (SNR) or a very long observation time. If the signal bandwidth is only a small fraction of the center frequency (i.e., $W/\omega_0 \ll 1$), the requirement of SNR or observation time quickly becomes unreasonable, indicating that the actual performance may be drastically inferior to that predicted by the Cramér–Rao lower bound (CRLB).

A technique which has been used with some success to address the ambiguity problem is based on the Barankin bound [8]. The result derived in [8] indicates a distinct threshold phenomenon: above a critical SNR the lower bound on the mean-square error is given by Cramér–Rao inequality, which is approached by the Barankin inequality under these conditions. Below the threshold the Barankin bound exceeds the CRLB by a factor proportional to $(\omega_0/W)^2$.

Manuscript received October 23, 1981; revised December 28, 1981. This work was supported in part by the Office of Naval Research under Contract N 00014-82-C-0152. This research is documented by the Woods Hole Institute under contribution number 5229.

A. Weiss is with the Department of Electronic Systems, Tel-Aviv University, Tel-Aviv, 69978, Israel.

E. Weinstein is with the Department of Ocean Engineering, Woods Hole Oceanographic Institution, Woods Hole, MA 02543.

This approach, however, suffers from two fundamental limitations. 1) For a fixed choice of test points the Barankin bound diverges for SNR's in excess of the indicated threshold (i.e., one obtains the trivial equation $\overline{\epsilon^2} \geq 0$, where $\epsilon = \Delta\hat{\tau} - \Delta\tau$ is the estimation error). The exact transition from the Barankin bound to the CRLB cannot be clearly specified. 2) In the formulation of the Barankin bound one completely ignores *a priori* information about the parameter under investigation. In particular, the receiver-to-receiver delay cannot exceed certain limits depending on the spacing between the receivers. Such prior information concerning $\Delta\tau$ is rather critical since the spacing between receivers determines the number of local solutions of the maximum likelihood equation or the order of ambiguity. As a result of (1) and (2), the approach based on the Barankin theory appears to give useful numerical information only for a limited range of SNR conditions.

In this correspondence, a new technique to analyze estimator performance is presented based on the Ziv–Zakai lower bound (ZZLB) [9]. This technique does not suffer from limitations (1) and (2) thus can be applied to a wide range of SNR's. Furthermore, by using an improved version of the ZZLB [10], the suggested technique is suitable for dealing with ambiguity-prone signals.

For simplicity let us consider a two-receiver array so that only one differential delay can be estimated. If $s(t)$ is the signal observed at one of the receivers in the absence of noise, then the actual waveforms at two spatially separated receivers are

$$
\begin{aligned}
r_1(t) &= s(t) + n_1(t), \qquad 0 \leq t \leq T \\
r_2(t) &= s(t - \Delta\tau) + n_2(t).
\end{aligned} \tag{1}
$$

$s(t)$, $n_1(t)$, and $n_2(t)$ are assumed to be sample functions from uncorrelated zero-mean stationary Gaussian processes with spectral densities $S(\omega)$, $N_1(\omega)$, and $N_2(\omega)$, respectively. Suppose $\Delta\tau$, the receiver-to-receiver delay, can take values only in the interval

$$-D/2 \leq \Delta\tau \leq D/2, \tag{2}$$

If L is the spacing between receivers and c is the propagation velocity then $D = 2L/c$.

The basic bound on the attainable mean-square error, averaged over all possible values of $\Delta\tau$ in the *a priori* domain defined in (2), is given by

$$\overline{\epsilon^2} \geq \frac{1}{D}\int_0^D xG[(D-x)P_e(x)]\,dx, \tag{3}$$

where $G[\cdot]$ is a nonincreasing function of x obtained by filling the valleys in the function $(D-x)P_e(x)$ (see [10], Fig. 1). $P_e(x)$ is the minimum attainable probability of error in deciding between two hypothesized values of the differential-delay parameter, say $\Delta\tau = a$ and $\Delta\tau = a + x$. For $WT \gg 1$ (i.e., a large sample size) $P_e(x)$ can be closely approximated by [11]

$$P_e(x) = \frac{1}{2}\exp\left\{\mu(s_m) + \frac{s_m^2}{2}\ddot{\mu}(s_m)\right\}\phi\left(s_m\sqrt{\ddot{\mu}(s_m)}\right)$$

Reprinted from *IEEE Trans. Inform. Theory*, vol. 28, no. 6, pp. 977–979, November 1982.

$$+\frac{1}{2}\exp\left\{\mu(s_m)+\frac{(1-s_m)^2}{2}\ddot{\mu}(s_m)\right\}\phi\left((1-s_m)\sqrt{\ddot{\mu}(s_m)}\right), \tag{4}$$

where $\phi(x)=\int_x^\infty dt\, e^{-t^2/2}/\sqrt{2\pi}$ is the error function. For the binary detection problem under consideration here one finds (after some rather extensive computations)

$$s_m = 1/2, \tag{5a}$$

$$\mu(1/2) = -\frac{T}{2\pi}\int_0^\infty \ln[1+\gamma(\omega)]\,d\omega, \tag{5b}$$

and

$$\ddot{\mu}(1/2) = \frac{4T}{\pi}\int_0^\infty \frac{\gamma(\omega)}{1+\gamma(\omega)}\,d\omega, \tag{5c}$$

where

$$\gamma(\omega) = \frac{[S(\omega)/N_1(\omega)][S(\omega)/N_2(\omega)]}{1+S(\omega)/N_1(\omega)+S(\omega)/N_2(\omega)}\sin^2\frac{\omega x}{2}. \tag{5d}$$

Substituting (4) into (5) one can now generate mean-square error predictions for any prespecified SNR conditions. Furthermore, since (5d) uses arbitrary spectral functions the resulting bound is equally applicable for wide-band as well as narrow-band systems.

Discussion

Our ultimate interest is to analyze the performance of time-delay estimators for narrow-band signals. To further investigate the bound consider the following illustration

$$\frac{S(\omega)}{N_i(\omega)} = \begin{cases} \text{SNR}_i \text{ (a constant)}, & |\omega \pm \omega_0| \leq W/2, \\ 0, & |\omega \pm \omega_0| > W/2. \end{cases} \tag{6}$$

In the narrow-band case $W/\omega_0 \ll 1$. The resulting bound is plotted in Fig. 1 as a function of R, where R is given by

$$R = \frac{\text{SNR}_1 \cdot \text{SNR}_2}{1+\text{SNR}_1+\text{SNR}_2}. \tag{7}$$

One immediately observes the entire domain of R is divided into several disjoint segments (which follow) indicating several distinct modes of operation.

If $R < R_1$ the signal observations from the receiver outputs are essentially useless in differential-delay estimation. This is the noise dominated region. The attainable mean-square error is bounded only by the *a priori* parameter domain (2).

If $R_1 < R < R_2$ the modified ZZLB coincide with the Barankin bound. In this regime time-delay information from the signal observations is subject to ambiguities. As already indicated in [8], all the essential information is contained in the received signal envelope, not in the signal phase. A near optimal $\Delta\tau$ estimate is therefore obtained by forming the cross correlation between the received signal envelopes.

If $R > R_3$ the ambiguity in the differential-delay estimation can be completely resolved. Cross correlating the received signal waveforms yields a $\Delta\tau$ estimate whose performance is characterized by the CRLB. The transition from the ambiguity dominated mode of operation to the ambiguity free mode of operation starts at R_2 and ends at R_3. This is the threshold phenomenon in the differential-delay estimation.

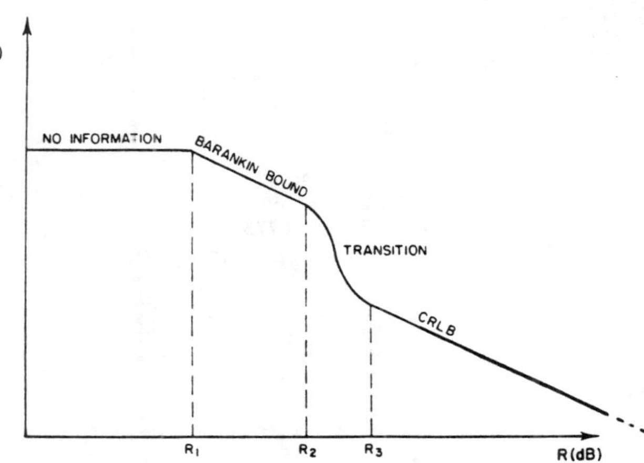

Fig. 1. Bound on the attainable mean-square error.

Information concerning the various segments of the composite bound and the various deflection points R_1, R_2, and R_3 as functions of such important parameters as WT, W/ω_0, and D is contained in [12]. With this information the composite bound illustrated in Fig. 1 provides the most complete analytical characterization of the attainable performance level under any prespecified SNR conditions.

References

[1] V. H. MacDonald and P. M. Schultheiss, "Optimum passive bearing estimation in a spatially incoherent noise environment," *J. Acous. Soc. Amer.*, vol. 46, pp. 37–43, 1969.

[2] W. R. Hahn, "Optimum signal processing for passive sonar range and bearing estimation," *J. Acous. Soc. Amer.*, vol. 58(1), pp. 201–207, 1975.

[3] G. C. Carter, "Time delay estimation for passive sonar signal processing," *IEEE Trans. Acous., Speech, Signal Processing*, vol. ASSP-29, no. 3, June 1981.

[4] Y. T. Chan, R. V. Hattin, and J. B. Plant, "The least squares estimation of time delay and its use in signal detection," *IEEE Trans. Acous., Speech, Signal Processing*, vol. ASSP-26, no. 3, pp. 217–222, 1978.

[5] W. R. Hahn and S. A. Tretter, "Optimum processing for delay-vector estimation in passive signal arrays," *IEEE Trans. Inform. Theory*, vol. IT-19, no. 5, pp. 608–614, 1973.

[6] B. V. Hamon and E. J. Hannan, "Spectral estimation of time delay for dispersive and non-dispersive systems," *J. Appl. Statist.*, vol. 23, no. 2, pp. 134–142, 1974.

[7] C. H. Knapp and G. C. Carter, "The generalized correlation method for estimation of time delay," *IEEE Trans. Acous., Speech, Signal Processing*, vol. ASSP-24, no. 4, pp. 320–327, 1976.

[8] S. K. Chow and P. M. Schultheiss, "Delay estimation using narrow-band processes," *IEEE Trans. Acous., Speech, Signal Processing*, vol. ASSP-29, no. 3, pp. 478–484, June 1981 (special issue).

[9] D. Chazan, M. Zakai, and J. Ziv, "Improved lower bounds on signal parameter estimation," *IEEE Trans. Inform. Theory*, vol. IT-21, pp. 90–93, Jan. 1975.

[10] S. Bellini and G. Tartara, "Bounds on error in signal parameter estimation," *IEEE Trans. Commun.*, pp. 340–342, Mar. 1974.

[11] H. L. Van Trees, *Detection, Estimation and Modulation Theory, Part 1*. New York: Wiley 1968, ch. 2, eq. 484.

[12] A. Weiss and E. Weinstein, "Fundamental limitations in passive time-delay estimations—Part I: Narrow-band systems," accepted for publication in *IEEE Trans. Acous., Speech, Signal Processing*.

On Time Delay Estimation Involving Received Signals*

by

C. Y. Wuu
Bell Laboratories
Holmdel, NJ 07733

A. E. Pearson
Division of Engineering
Brown University
Providence, RI 02912

Abstract

The estimation of time delay between two received signals is re-examined for the three basic approaches to this problem: (1) Phase Data, (2) Generalized Cross Correlation, and (3) Parameter Estimation. Special consideration is given to the problem of estimating the time delay when signals are received through transmitting media with unknown transfer functions.

I. Introduction

The time delay estimation between two received signals has several important applications such as locating and tracking the position of a signal source [1], indicating a transit time, or estimating some other process time. Given two received signals, $r_1(t)$ and $r_2(t)$, a widely used mathematical model can be expressed as follows:

$$r_1(t) = s(t)+n_1(t)$$
$$r_2(t) = ws(t-\alpha)+n_2(t), \qquad (1)$$

where "w" is a constant and (n_1,n_2) are the contaminating noises. In spite of its simple structure, several approaches based on quite different points of view have been proposed and studied starting with model (1). These existing approaches can be classified into three categories:

(1) Phase Data [2, 3]
(2)(Generalized) Cross Correlation [4, 5, 6]
(3)Parameter Estimation [7-12].

Section II will analyze the three approaches in a way that reveals some comparative features between them. Section III will discuss the effect on estimating the delay when the transmitting media are modeled by unknown channel transfer functions, i.e., the transmitting media introduce different phase shifts and attenuations for different frequency components. The problem is then reformulated such that the delay can be estimated via a system identification algorithm. A sequel report will include simulation data.

* This work was supported in part by the National Science Foundation under grant ECS-8111219, and in part by the Air Force Office of Scientific Research under grant AFOSR-82-0230.

II. Some Review and Analysis of Existing Approaches

Assuming the signal s(t) and noises $(n_1(t), n_2(t))$ are mutually independent processes, the cross-correlation function between the received signals is given by

$$R_{12}(\tau) = E\left[r_1(t)r_2(t+\tau)\right]$$
$$= wR_{ss}(\tau-\alpha) \qquad (2)$$

where $R_{ss}(\cdot)$ is the autocorrelation function of s(t). Given a finite observation interval and assuming the observed signals are ergodic processes, the cross correlation is estimated by

$$\hat{R}_{12}(\tau) = \frac{1}{T}\int_0^T r_1(t)r_2(t+\tau)dt \qquad (3)$$

where T is the length of the observation interval and has to be sufficiently long for the estimated cross correlation to be accurate.

The one-sided cross spectrum $G_{12}(f)$ is obtained by performing the Fourier Transform of (2):

$$G_{12}(f) = 2\int_{-\infty}^{\infty} R_{12}(\tau)e^{-j2\pi f\tau}d\tau$$
$$= wG_{ss}(f)e^{-j2\pi f\alpha} \qquad (4)$$

where $G_{ss}(f)$ is the one-sided (real) spectral density function of s(t). Hence, the phase function is given by

$$P_{12}(f) = 2\pi f\alpha \qquad (5)$$

and is linear in the frequency "f". The estimation of delay can thus be based on the phase data. Usually, a regression or least squares method is used to estimate the delay α [2 , 3]. The Cross Correlation method can also be used to estimate the delay between the received signals. In view of the important property of autocorrelation that $R_{ss}(\tau) \leq R_{ss}(0)$, the delay can be estimated by locating the peak of $\hat{R}_{12}(\tau)$ in (3). In order to improve the estimation accuracy, it is usually suggested that $r_1(t)$ and $r_2(t)$ be prefiltered prior to the integral in (3). The cross correlation function of the filtered $\tilde{r}_1(t)$ and $\tilde{r}_2(t)$ is then the "generalized" cross correlation function $\tilde{R}_{12}(\tau)$. The equivalent representation in the frequency domain of this situation is

$$\tilde{R}_{12}(\tau) = \int_{-\infty}^{\infty} W(f)\hat{G}_{12}(f)e^{j2\pi f\tau}df \qquad (6)$$

Reprinted from *Proc. ICASSP '83*, vol. 2, pp. 871–874, April 1983.

where $\hat{G}_{12}(f)$ is the estimated cross spectral density function and W(f) denotes the frequency domain weighting function. The delay is then estimated by locating the peak of the generalized cross correlation function $\hat{R}_{12}(\tau)$. There exist many publications investigating the design and effect of this weighting function W(f) [4,5,6]. A sufficient (though not necessary) condition for this estimation to be unbiased is that W(f) be a real function and, therefore, introduce no phase distorsion into the transformation (6). For example, the Roth weighting function, $W(f) = G_{11}^{-1}(f)$, will yield an unbiased estimation.

With the considerable progress in parameter estimation algorithms made during the last decade, some researchers are motivated to reformulate the delay estimation problem in terms of a parameter estimation problem [7-12]. Although differences exist in their formulations, the fundamental idea is to approximate the time shift operator as a finite impulse response filter with unknown parameters. Compared with Generalized Cross Correlation, this method is very closely related to the Roth weighting function [4 , 1] in which W(f) in (6) is chosen according to

$$W(f) = \frac{1}{G_{11}(f)}. \quad (7)$$

This will be further delineated below.

Consider the two received signals observed in discrete time:

$$r_1(k) = s(k)+n_1(k)$$
$$r_2(k) = ws(k-\alpha)+n_2(k) \quad (8)$$

where α is the delay, $n_1(k)$ and $n_2(k)$ are noises and the sampling period is normalized as unity. If the actual delay α is an integral multiple of the sampling period, the received signals can be further related by

$$r_2(k) = \sum_{i=0}^{M} w_i[r_1(k-i)-n_1(k-i)]+n_2(k) \quad (9)$$

where $w_i=w$ if $i = \alpha$, and $w_i=0$ otherwise. If the actual delay is not an integral multiple of the sampling period, the expression in (9) can be modified as given in [7,11].

Various parameter estimation approaches can be proposed to estimate the unknown w_i. Since the disturbance term e(k) is correlated with $r_1(k)$ and $r_2(k)$, the estimated parameter w_i would be biased when the usual least mean square error $E\{e^2(k)\}$ is used. The final estimation of delay is also biased as long as the order M is finite.

A different derivation is given in the following which shows how the optimal estimate of w_i relates to the unknown noises. First of all, (9) can be expressed in a vector form as

$$Z_2(k) = [Z_1(k)-D_1(k)]\,V+D_2(k) \quad (10)$$

where

$$Z_2(k) = \begin{bmatrix} r_2(k+N-1) \\ r_2(k+N-2) \\ \vdots \\ r_2(k) \end{bmatrix}, \quad V = \begin{bmatrix} w_0 \\ w_1 \\ \vdots \\ w_M \end{bmatrix}$$

$$Z_1(k) = \begin{bmatrix} r_1(k+N-1) & \cdots & r_1(k+N-M-1) \\ \vdots & & \vdots \\ r_1(k) & \cdots & r_1(k-M) \end{bmatrix}$$

while the matrix D_1 and the vector D_2 are respectively defined in the same way as Z_1 and Z_2 but in terms of the noises $n_1(k)$ and $n_2(k)$. Then the unknown vector parameter V can be solved exactly if the inverse matrix exists in the following expression:

$$V^* = [(Z_1'Z_1-Z_1'D_1-D_1'Z_1+D_1'D_1)/N]^{-1}[(Z_1'Z_2 - Z_1'D_2-D_1'Z_2+D_1'D_2)/N] \quad (11)$$

Since the noises $n_1(t)$ and $n_2(t)$ are unobserved, the above exact solution cannot be obtained. However, assuming that the signal and noises are ergodic processes, the exact solution can be further simplified as follows:
Assume the signal and noises are mutually uncorrelated and the noises have zero mean values. Then the following approximations are valid when N is large enough

$$(D_2'D_1/N \cong 0$$

$$(Z_1'D_2)/N \cong 0$$

$$(Z_1'D_1)/N \cong (D_1'D_1)/N.$$

In this case the unbiased solution of (11) can be approximated as

$$V^* \cong [(Z_1'Z_1-D_1'D_1)/N]^{-1}[Z_1'Z_2/N]. \quad (12)$$

Since the information in D_1 is unavailable, the following least mean square estimation $E[e^2(k)]$ is often used [7-12]:

$$\tilde{V} = [(Z_1'Z_1)/N]^{-1}\;[Z_1'Z_2/N]. \quad (13)$$

Compared with V* in (12), the estimate $\tilde{V}$ in (13) would be biased even when the order M approaches to infinity. However, since this $\tilde{V}$ vector will approach the Roth Generalized Cross Correlation, $[\tilde{R}_{12}(\tau_1), \tilde{R}_{12}(\tau_2)\ldots\tilde{R}_{12}(\tau_M)]$, cf. (6) and (7):

$$\tilde{R}_{12}(\tau) = F^{-1}\left\{\frac{G_{12}(f)}{G_{11}(f)}\right\}$$

Furthermore, since the above Roth Generalized Cross Correlation has an unbiased peak corresponding to the actual delay, the estimated delay via (13) will tend to become unbiased as M approaches infinity. Since only the estimation of the delay is desired, it is suggested that the above ambiguity in biasedness can be resolved by emphasizing the parameter estimation scheme (13) as a time domain implementation of the Roth processor rather than emphasizing the filter response model for the time delay operator as in [7-12]. Moreover, one also should note that the vector $(Z_1'Z_2')/N$ is an approximated cross correlation function between the received signals, i.e.,

$$(\tilde{Z}_1'Z_2)/N = \begin{bmatrix} R_{12}(0) \\ \vdots \\ R_{12}(M) \end{bmatrix}$$

Under the ideal condition that both the signal and the noises ($n_1(t)$, $n_2(t)$) are independent white processes, (13) is replaced by

$$\tilde{\tilde{V}} = \frac{1}{s^2+d^2}[(Z_1'Z_2)/N] \quad (14)$$

where s^2 and d^2 are the mean square powers of $s(t)$ and $n_1(t)$ respectively. Estimating the peak of $\hat{V}$ in (14), i.e., the largest component of the vector $\tilde{V}$, is thus equivalent to that of the approximated cross correlation function $Z_1'Z_2/N$. Thus the superiority of this parameter estimation method cannot be justified with the simulations under the above ideal (or an almost ideal) condition [9-12], since the usual cross correlation method can yield at least equivalently good estimation under the same ideal conditions.

III. Delay Estimation with Unknown Channel Transfer Function

In contrast with the model (1), consider the more general measurement model:

$$r_1(t) = G_1(D)\, s(t)+n_1(t)$$
$$r_2(t) = G_2(D)s(t-\alpha)+n_2(t) \quad (15)$$

where $G_1(D)$ and $G_2(D)$ are ratios of polynomials in the differential operator $D=\frac{d}{dt}$ representing the dynamic transfer functions of the transmitting media and sensor response. The terms $n_1(t)$ and $n_2(t)$ are the contaminating noises and the delay is to be estimated. Clearly, the model (15) reduces to (1) under the condition $G_1(D) = G_2(D) = G(D)$ and by denoting $G(D)s(t)$ as the source signal.

With the assumption that the signal and noises are mutually uncorrelated processes, the cross-spectral density function $G_{12}(f)$ between the received signals can be expressed as

$$G_{12}(f) = G_1(f)G_2{}^*(f)G_{ss}(f)e^{-j2\pi f\alpha} \quad (16)$$

and the cross correlation function $R_{12}(\tau)$ is then

$$R_{12}(\tau) = \int_{-\infty}^{\infty} G_1(f)G_2{}^*(f)G_{ss}(f)e^{j2\pi f(\tau-\alpha)}df. \quad (17)$$

Since the product $G_1(f)G_2{}^*(f)$ is usually a complex valued function with nonzero phase shift, the phase function of $G_{12}(f)$ in (16) is not a linear function of frequency as in (5). The estimated delay which corresponds to the peak of $R_{12}(\tau)$ is usually a biased estimation because of the phase distorsion in the transformation of (17). For example, when the signal is a combination of sinusoidal signals,

$$s(t) = \sum_i S_i(t)$$
$$= R(\sum_i c_i e^{j2\pi f_i t}) \quad (18)$$

where R denotes the "real part" operator, and

$$G_1(f)G_2{}^*(f) = A(f)e^{j2\pi f\theta(f)},$$

then the cross correlation function $R_{12}(\ )$ can be expressed as

$$R_{12}(\tau) = \sum_i A(f_i)R_i(\tau) \quad \delta(\tau-\alpha+\theta(f_i)) \quad (19)$$

where $R_i(\tau)$ is the autocorrelation of the ith component $S_i(t)$ and has its peak at $\tau = 0$. After the convolution in (19), the peak of $R_{12}(\tau)$ would (in general) be biased from the actual delay α.

However, under the extreme case that the signal $s(t)$ is a white process, a sufficient condition for the estimated delay to be unbiased is that the impulse response functions of $G_1(D)$ and $G_2(D)$ are both positive and decreasing. In order to verify this, denote by $h_1(\cdot)$ and $h_2(\cdot)$ the impulse reponses of $G_1(D)$ and $G_2(D)$ respectively. Then the cross correlation function can be expressed as

$$R_{12}(\tau) = \int_{-\infty}^{\infty}\int_{-\infty}^{\infty} h_1(v)(h_2(u)E\ s(t-v)s(t+\tau-\alpha-u)\ dvdu$$
$$= \int_{-\infty}^{\infty}\int_{-\infty}^{\infty} h_1(v)h_2(u)\delta(\tau-\alpha+v-u)dvdu \quad (20)$$
$$= \int_{-\infty}^{\infty} h_1(v)h_2(\tau-\alpha+v)dv.$$

When the above sufficient condition is satisfied, the following relationship is seen to hold:

$$R_{12}(\alpha) > R_{12}(\tau) \text{ for all } \tau \neq \alpha.$$

It is noted that the function $R_{12}(\tau)$ is no longer symmetric about its peak value. Compared with the discussion given in 14. These analyses have been experimentally justified via the simulations to be reported in a sequel report.

Now suppose the transfer functions $G_1(D)$ and $G_2(D)$ are unknown, or only very poor information in available. Then the received signals can be related as follows:

$$r_2(t) = G_1^{-1}(D)G_2(D)r_1(t-\alpha)-G_1^{-1}(D)n_1(t-\alpha)+n_2(t) \quad (21)$$
$$= G_1^{-1}(D)G_2(D)r_1(t-\alpha)+d(t).$$

With the form given in (21), the delay can be estimated with some identification algorithm designed for systems with time delay by taking $r_1(t)$ as input and $r_2(t)$ as output. Since, statistically, the disturbance term $d(t)$ would be a colored noise and correlated with $r_1(t)$ and $r_2(t)$, the identification algorithm proposed by Gabay and Merhav [18], which requies that the input $r_1(t-\alpha)$ and noise $d(t)$ be independent processes, cannot be applied to this case. Then, the delay can be estimated with the algorithm proposed by Kurz and Gredecke [15] or another algorithm by Pearson and Wuu [16, 17]. These details will be discussed in the sequel.

IV. Summary

It has been shown that several existing approaches for estimating the delay between received signals can be closely related. The rederivation of the Parameter Estimation Approach showed some insights about its relationship with the (Generalized) Cross Correlation method. With the confusing simulation results published in the literature, a fair experimental evaluation is considered important. When the more complicated model involving different channel dynamics is considered, the Phase Data approach cannot be applied and the optimal weighting function in the Generalized Cross Correlation method cannot be designed. The Cross Correlation method still yields unbiased estimation under a sufficient condition as explained. Along with the discussion, it is illustrated how the channel dynamics would affect the final estimation. When the difference in channel dynamics is significant, the use of a system identification approach to estimate the delay may be justified.

References

[1] G. Clifford Carter, "Time Delay Estimation for Passive Sonar Signal Processing", IEEE Trans. on Acoustics, Speech and Signal Processing, Vol. ASSP-29, no. 3, pp. 463-470, 1981.

[2] Y.T. Chan, Richard V. Hattin and J.B. Plant, "The Least Squares Estimation of Time-Delay and Its Use in Signal Detection", IEEE Trans. on Acoustics, Speech and Signal Processing, Vol. ASSP-26, no. 3, pp. 217-222, 1978.

[3] Allan G. Piersol, "Time Delay Estimation Using Phase Data", IEEE Trans. on Acoustics, Speech and Signal Processing, Vol. ASSP-29, no. 3, pp. 471-477, 1981.

[4] Charles H. Knapp and G. Clifford Carter, "The Generalized Correlation Method for Estimation of Time Delay", IEEE Trans. on Acoustics, Speech and Signal Processing, Vol. ASSP-24, no. 4, pp. 320-327, 1976.

[5] Joseph C. Hassab and Ronald E. Boucher, "Optimum Estimation of Time Delay by a Generalized Correlator", IEEE Trans. on Acoustics, Speech and Signal Processing, Vol. ASSP-27, no. 4, pp. 373-380, 1979.

[6] William H. Haas and Claude S. Lindquist, "A Synthesis of Frequency Domain Filters for Time Delay Estimation", IEEE Trans. on Acoustics, Speech and Signal Processing, Vol. ASSP--29, no. 3, pp. 540-548, 1981.

[7] Y.T. Chan, J.M.F. Riley and J.B. Plant, "A Parameter Estimation Approach to Time-Delay Estimation and Signal Detection", IEEE Trans. on Acoustics, Speech and Signal Processing, Vol. ASSP-28, no. 1, pp. 8-15, 1980.

[8] Y.T. Chan, J.M.F. Riley and J.B. Plant, "Modeling of Time-Delay and Its Application to Estimation of Nonstationary Delays", IEEE Trans. on Acoustics, Speech and Signal Processing, Vol. ASSP-29, no. 3, pp. 577-581, 1981.

[9] Francis A.Reed, Paul L. Feintuch and Neil J. Bershad, "Time Delay Estimation Using the LMS Adaptive Filter -Static Behavior", IEEE Trans. on Acoustics, Speech and Signal Processing, Vol. ASSP-29, no. 3, pp. 561-571, 1981.

[10] Francis A. Reed, Paul L. Feintuch and Neil J. Bershad, "Time Delay Estimation Using the LMS Adaptive Filter -Dynamic Behavior", IEEE Trans. on Acoustics, Speech and Signal Processing, Vol. ASSP-29, no. 3, pp. 571-576, 1981.

[11] B. Friedlander, "Multi-target Tracking Studies: System Description and Preliminary Evaluation", TM 5334-04, Systems Control Inc., Palo Alto, CA.

[12] D.H. Youn, N. Ahmed and G.C. Carter, "On Using the LMS Algorithm for Time Delay Estimation", IEEE Trans. on Acoustics, Speech and Signal Processing, Vol. ASSP-30, no. 5, pp. 798-801, 1982.

[13] Peter R. Roth, "Effective Measurements Using Digital Signal Analysis", IEEE Spectrum, pp. 62-70, April 1971.

[14] G.D. Lassahn and A.G. Baker, "Errors in Cross-Correlation Peak Location", Trans. ASME, Journal of Dynamic Systems, Measurement and Control, Vol. 104, pp. 194-199, June 1982.

[15] H. Kurz and W. Gredecke, "Digital Parameter-Adaptive Control of Processes with Unknown Dead Time", Automatica, Vol. 17, no. 1, pp. 245-252, 1981.

[16] A.E. Pearson and C.Y. Wuu, "System Identification of Pure Time Delay with Finite Time Data", Proc. of 19th IEEE Conference on Decision and Control, Albuquerque, New Mexico, pp. 739-740, 1980.

[17] C.Y. Wuu, "A Finite Time Parameter Estimation Scheme for System Identification and Signal Processing with Unknown Delays", Ph.D. Dissertation, Brown University, June 1982.

[18] E. Gabay and S.J.Merhav, "Identification of Linear Systems with Time Delay Operating in a Closed Loop in the Presence of Noise", IEEE Trans. on Auto. Control, AC-21, pp 711-716, 1976.

THE GENERALIZED PHASE SPECTRUM METHOD FOR TIME DELAY ESTIMATION

Zhao Zhen and Hou Zi-qiang

Institute of Acoustics Academia Sinica
Beijing China

ABSTRACT

The concept of the Generalized Phase Spectrum (GPS) TDE is put forward. The relation between the GPS TDE and the GCC TDE is derived. A multipath signal model is considered. The method of Amplitude Square (AS) weighting for TDE is proposed and the comparison between the AS TDE and the Phase Data (PD) TDE is made in the multipath environment. The results of theoretical calculation and computer simulation experiment show that the performance of the AS weighting TDE is superior to that of the PD TDE.

I. INTRODUCTION

A rapid development of Time Delay Estimation (TDE) theory has been made in recent years, many different methods for TDE based on the Crosscorrelation have been put forward for different situations, these methods can be included in an integral theoretical frame by using the Generalized Crosscorrelation (GCC) TDE theory. Because we must interpolate the crosscorrelation curve to acquire accurate TDE, the computation complexity will increase. A.G. Piersol estimates the time delay by using Phase Data(PD) method, this method is based on FFT Processing, so it is simple in computation, however it has not taken the problem of frequency spectrum weighting into account, when signal power spectrum curve has considerable fluctuation, the accuracy of TDE will decrease significantly.

This paper developes the method proposed by Piersol. The theoretical frame of the Generalized Phase Spectrum(GPS) TDE corresponding to the GCC TDE theory is set up. The computation of variance of the GPS TDE shows that the Cramer-Rao lower bound(CRLB) of variance of TDE can be attained. Further, the relation between the GPS and the GCC TDE has been found, and the different versions of the GPS TDE using different weighting function have been given.

This paper discusses the case in which the GPS TDE method is used in the multipath environment. In the underwater acoustical system, the interference produced by multipath channel will make the signal power spectrum fluctuating violently within the frequency band, the SNR also fluctuates with the frequency, therefore the TDE performance of the PD method fails in the multipath situation. According to the theory of the GPS TDE, we should adopt the weight of coherence function, so as to improve the estimation accuracy and achieve the CRLB. In order to realize easily, the Amplitude Square(AS) of cross power spectrum is adopted as weighting function. Essentially, this new approach takes SNR square as weight to modify the PD method, therefore the variance of AS weighting method, $\sigma[\hat{\tau}_A]$, would be smaller than $\sigma[\hat{\tau}_p]$ of PD method, especially in the situation of low SNR. For the ideal case, theoretical analysis and computer simulation experiment show that the variance of TDE errors using AS weighting method is very close to the CRLB.

II. DERIVATION FOR THE GPS TDE

The method for TDE using Phase Data(PD) [2] is

$$\hat{\tau}_p = \frac{\sum_i f_i \hat{\phi}_{x_1x_2}(f_i)}{2\pi \sum_i f_i^2} = \frac{b_{12}}{2\pi} \tag{1}$$

where $\hat{\phi}_{x_1x_2}(f_i)$ is the smooth phase spectrum estimate, b_{12} is the slop of the regression line for $\hat{\phi}_{x_1x_2}(f_i)$, we obtain the variance of b_{12}

$$D[b_{12}] = \frac{\sum_i f_i^2 D[\hat{\phi}_{x_1x_2}(f_i)]}{\left(\sum_i f_i^2\right)^2} \tag{2}$$

However, the reference [2] gives

$$D[b_{12}] = \left\{\sum_i \frac{f_i^2}{D[\hat{\phi}_{x_1x_2}(f_i)]}\right\}^{-1} \tag{3}$$

Note that only in the case that $D[\hat{\phi}_{x_1x_2}(f_i)]$ has nothing to do with i, can eq.(3) hold true. In physics meaning, only in the condition that the power spectrum is flat, can the variance of PD

Reprinted from *Proc. ICASSP '84*, vol. 3, pp. 46.2/1–4, March 1984.

TDE achieve the CRLB. However, the fluctuation of the amplitude of cross spectrum versus frequency is considerably large in the multipath environment. The SNRs are very low at some frequencies where the cross spectrum amplitudes are small, therefore the deviation of $\hat{\phi}_{x_1x_2}(f_i)$ from the ideal $\phi_{x_1x_2}(f_i)$ are very large at those frequencies. Because the regression line in PD method is equally weighting essentially, the deviation of $\hat{\phi}_{x_1x_2}(f_i)$ at those frequencies would cause the increase of the error of TDE. From eq.(2) we have

$$D[\hat{\tau}_p]=\left\{8\pi^2\cdot\frac{T\cdot\left(\int_0^\infty f^2\,df\right)^2}{\int_0^\infty f^2\frac{[1-C_{x_1x_2}(f)]}{C_{x_1x_2}(f)}\,df}\right\}^{-1} \tag{4}$$

where $C_{x_1x_2}(f)$ is magnitude-square coherence, T is the total record length. From eq.(4), we can see that if there are some frequencies at which $C_{x_1x_2}(f)$ is equal zero approximately, the $D[\hat{\tau}_p]$ will increase notably. The PD TDE corresponds to the PHAT TDE in the GCC TDE method [3] , in fact

$$\begin{aligned}R^p_{x_1x_2}(\tau)&=\int\frac{\hat{G}_{x_1x_2}(f)}{|\hat{G}_{x_1x_2}(f)|}\cdot e^{j2\pi f\tau}\cdot df\\&=\int 1\cdot e^{j\hat{\phi}_{x_1x_2}(f)}\cdot e^{j2\pi f\tau}\cdot df\end{aligned} \tag{5}$$

Because the PHAT method in GCC may be regarded as using weight $W(f_i)=1$ to $\hat{\phi}_{x_1x_2}(f_i)$, we revise the PD TDE method, the Generalized weighting function $W(|G_{x_1x_2}(f)|, G_{x_1x_1}, G_{x_2x_2})$ is introduced into eq.(1), and define this new method as the Generalized Phase Spectrum TDE, shown in Fig. 1

$$\hat{\tau}^G=\frac{\sum_i f_i\cdot W(f_i)\cdot\hat{\phi}_{x_1x_2}(f_i)}{2\pi\sum_i W(f_i)\cdot f_i^2} \tag{6}$$

when the observation time T is long enough, i.e. $TB>8$, where B is the bandwidth of signal, in the condition of Gaussian assumption, we obtain the expression of variance of GPS TDE in discrete form

$$D[\hat{\tau}^G]=\frac{1}{4\pi^2}\cdot\frac{\sum_i W^2(f_i)\cdot f_i^2\cdot D[\hat{\phi}_{x_1x_2}(f_i)]}{\left(\sum_i W(f_i)\cdot f_i^2\right)^2} \tag{7}$$

III. RELATION BETWEEN THE GPS AND THE GCC TDE

The GPS TDE is equivalent to the GCC TDE. The reasons are as follows. First, in the GCC TDE procedure the time delay is estimated with weighted cross correlation, which is the inverse Fourier transform of weighted cross spectral density function. The information of time delay is involved entirely in the cross spectral density function. On the basis of duality relation between frequency domain and time domain, it is entirely equivalent to estimate time delay using the weighted cross spectral function directly in frequency domain. Second, because the slope of a phase spectrum regression line is the measure of time delay, we can use the weighted phase spectrum to estimate time delay.

GCC TDE weightes cross spectrum with $\psi(f)$, GPS TDE weightes phase spectrum with $W(f)$. Since the cross spectrum differs from the phase spectrum only with the magnitude of cross spectrum, we must multiply $\psi(f)$ with the magnitude of cross spectrum, so as to convert $\psi(f)$ into weighting for phase spectrum, i.e.

$$W(f)=\psi(f)\cdot|G_{x_1x_2}(f)| \tag{8}$$

this is the relation between the two weighting functions for TDE. In reality we can deduce eq.(8) from the following expression

$$\begin{aligned}R^G(\tau)&=\int\psi(f)\cdot G_{x_1x_2}(f)\cdot e^{j2\pi f\tau}\cdot df\\&=\int\{\psi(f)\cdot|G_{x_1x_2}(f)|\}\cdot e^{j\phi_{x_1x_2}(f)}\cdot e^{j2\pi f\tau}\cdot df\end{aligned} \tag{9}$$

using eq.(8), we list the corresponding relation between the GPS and GCC time delay estimators, as illustrated in Table 1.

IV. GPS TDE IN MULTIPATH ENVIRONMENT

We consider the following multipath signal model, as illustrated in Fig.2

$$\begin{aligned}x_1(t)&=s(t)/\sqrt{2}+\eta\cdot s(t+D)/\sqrt{2}+n_1(t)\\x_2(t)&=s(t+\tau)/\sqrt{2}+\eta\cdot s(t+D+\tau)/\sqrt{2}+n_2(t)\end{aligned} \tag{10}$$

where η is the multipath coefficient, τ is the time delay between $x_1(t)$ and $x_2(t)$, D is the multipath time delay. From eq.(7) and (10), we can obtain the discret expression of $\sigma[\hat{\tau}_A]$, $\sigma[\hat{\tau}_p]$ and CRLB in the case of multipath environment.

$$\sigma[\hat{\tau}_A]=\frac{N}{2\sqrt{2}\pi G}\cdot\frac{1+\eta^2}{2}\cdot\frac{T_s}{\sqrt{n_d}}\times\frac{\sqrt{\sum_{i=1}^{N/2-1} i^2\left(\frac{1+\eta^2}{2}+\eta\cos\frac{2\pi i d}{N}\right)^2\left[1+2G\left(1+\frac{2\eta}{1+\eta^2}\cdot\cos\frac{2\pi i d}{N}\right)\right]}}{\sum_{i=1}^{N/2-1}\left(\frac{1+\eta^2}{2}+\eta\cdot\cos\frac{2\pi i d}{N}\right)^2\cdot i^2} \tag{11}$$

$$\sigma[\hat{\tau}_p]=\frac{N}{2\sqrt{2}\,\pi G}\cdot\frac{T_s}{\sqrt{n_d}}\times$$

$$\times\frac{\sqrt{\sum_{i=1}^{N/2-1} i^2\frac{\left[1+2G\left(1+\frac{2\eta}{1+\eta^2}\cdot\cos\frac{2\pi i d}{N}\right)\right]}{\left(1+\frac{2\eta}{1+\eta^2}\cdot\cos\frac{2\pi i d}{N}\right)^2}}}{\sum_{i=1}^{N/2-1} i^2} \quad (12)$$

$$CRLB=\frac{1}{2\sqrt{2}\,\pi G}\cdot\frac{T_s}{\sqrt{N}\cdot\sqrt{n_d}}\times \quad (13)$$

$$\times\left\{\int_0^b\frac{x^2\left(1+\frac{2\eta}{1+\eta^2}\cdot\cos 2\pi d x\right)^2}{\left[1+2G\left(1+\frac{2\eta}{1+\eta^2}\cdot\cos 2\pi d x\right)\right]}\cdot dx\right\}^{-\frac{1}{2}}$$

Where N is the point number of FFT, n_d is the average number of times, B=b/Ts, D=d·Ts, Ts is the sampling period, G is the average SNR within the bandwidth of signal. If d=10, b=0.25, we calculate the $\sigma[\hat{\tau}_A]$, $\sigma[\hat{\tau}_p]$, and CRLB for different G and η, the two results are illustrated in Fig.3 and 4. From these two figures we know that $\sigma[\hat{\tau}_A]$ is lower than $\sigma[\hat{\tau}_p]$, and fairly good agreement between $\sigma[\hat{\tau}_A]$ and CRLB was obtained, esp. in low SNR case, in fact if SNR≪1, then $\sigma[\hat{\tau}_A]$=CRLB. Along with the increase of η, $\sigma[\hat{\tau}_p]$ increases rapidly. The variances are inversely proportional to the square root of the average number of times n_d. The curve in Fig.5 illustrates the relation between the standard deviation and η, when SNR=0 dB, N=64. It is clear that $\sigma[\hat{\tau}_p]$ increases rapidly with η increased. The variations of $\sigma[\hat{\tau}_A]$ and CRLB with η are small, and $\sigma[\hat{\tau}_A]$ approaches the CRLB gradually with η increased.

V. COMPUTER SIMULATION EXPERIMENT

In our computer simulation experiment we adopt the following multipath signal model

$$\begin{aligned} x_1(t)&=s(t)+s(t+D)+n_1(t)\\ x_2(t)&=s(t+\tau)+s(t+D+\tau)+n_2(t)\end{aligned} \quad (14)$$

Let the point number of FFT N=64, D=10 Ts, τ=0.5 Ts, the bandwidth of low pass filter B=0.25/Ts. For SNR = 3, 6 and 12 dB, n_d=64, for SNR = 0 dB, n_d= 128. In this manner, 32 estimates of the time delay parameter $\hat{\tau}_A$ and $\hat{\tau}_p$ are obtained at each of the four SNR's : 12, 6, 3 and 0 dB, then $\sigma[\hat{\tau}_A]$, $\sigma[\hat{\tau}_p]$, $\bar{\hat{\tau}}_A$ and $\bar{\hat{\tau}}_p$ are calculated as well. The result of the simulation shown in Fig.6, from the diagram we know that the performance of AS TDE is better than that of PD TDE. Typical plots of the phase spectrum $\phi(f_i)$ and amplitude square spectrum logarithm $\ln[A^2(f_i)]$ for one trial in the computer simulation experiment for the two SNR's : 12 and 0 dB, are shown in Fig. 7 and 8, the line forced through the origin in phase spectrum is the theoretical phase line for τ = 0.5 Ts. From Fig. 7 and 8, we can find that the amplitude of cross spectrum fluctuates versus frequency. The more the average SNR in bandwidth is low, the more the fluctuation of the phase spectrum is large. At some spectral bins where the amplitudes of the cross spectrum are small and local SNRs are low, the deviations of phase spectrum from the theoretical phase line are large, but at other spectral bins where the amplitudes of the cross spectrum are large and SNRs are high, the deviations are small.

VI. CONCLUSIONS

The methods for TDE using the cross power spectrum can be summarized in an integral theory frame of GPS TDE. In the model discussed above, the GPS TDE can achieve the CRLB. The GPS TDE is equivalent to the GCC TDE, and a definite relation exists between both weighting functions. The AS weighting method is a practical, effective and suboptimal means for TDE. The main advantage of GPS TDE is that single chip FFT processor such as AMIs2814, NEC μpd7720 etc. can be utilized to simplify the signal processing equipment and raise the estimate speed for time delay.

ACKNOWLEDGEMENT

This paper was completed under the direction of Prof. Wang Te-chao, the authors wish to express their deep gratitude to him for his instruction.

REFERENCES

1. Wang Te-chao, Shang Er-chang, Underwater Acoustics, Science Press, China, 1981.
2. A.G. Piersol, "Time Delay Estimation Using Phase Data" IEEE Trans. Acoust., Speech, Signal Processing, Vol. ASSP-29, No.3, June 1981.
3. C.H. Knapp and G.C. Carter, "The Generalized Correlation Method for Estimation of Time Delay" IEEE Trans. Vol. ASSP-24, No.4, 1976.
4. J.S. Bendat, "Statistical Errors in Measurement of Coherence Functions and Input/Output Quantities" Journal of Sound and Vibration (1978) 59 (3), 405-421.

Table 1.

Processor Name	GPS	Processor Name	GCC
Phase Data	$W_p = 1$	PHAT	$\psi_P=\frac{1}{\lvert G_{x_1x_2}(f)\rvert}$
Amplitude Square	$W_A=\lvert G_{x_1x_2}(f)\rvert^2$	Magnitude	$\psi_M=\lvert G_{x_1x_2}(f)\rvert$
Amplitude Noise Ratio	$W_{AN}=\frac{\lvert G_{x_1x_2}(f)\rvert^2}{G_{n_1n_1}(f)\cdot G_{n_2n_2}(f)}$	Eckart	$\psi_E=\frac{G_{ss}(f)}{G_{n_1n_1}(f)\cdot G_{n_2n_2}(f)}$
Optimum Coherence Function	$W_{OC}=\frac{C_{x_1x_2}(f)}{1-C_{x_1x_2}(f)}$	ML	$\psi_{ML}=\frac{1}{\lvert G_{x_1x_2}(f)\rvert}\cdot\frac{C_{x_1x_2}(f)}{[1-C_{x_1x_2}(f)]}$

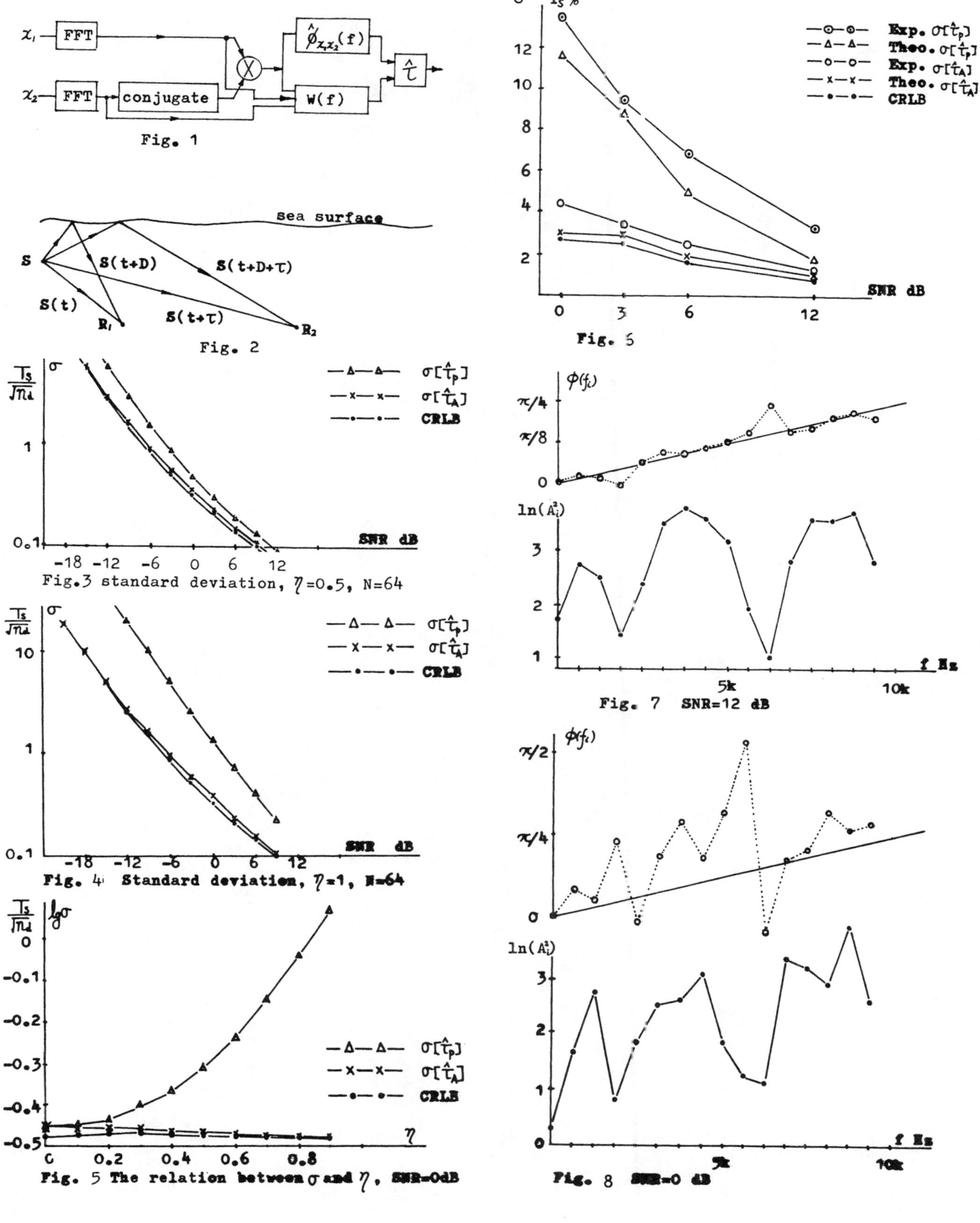

x_1
FFT
$\hat{\phi}_{x_1x_2}(f)$
$\hat{\tau}$
x_2
FFT
conjugate
W(f)
Fig. 1
sea surface
S
S(t+D)
S(t+D+τ)
S(t)
R1
S(t+τ)
R2
Fig. 2
Ts/√nd
σ
σ[t̂p]
σ[t̂A]
CRLB
1
0.1
SNR dB
-18 -12 -6 0 6 12
Fig.3 standard deviation, η=0.5, N=64
10
1
0.1
Fig. 4 Standard deviation, η=1, N=64
lgσ
0
-0.1
-0.2
-0.3
-0.4
-0.5
η
0 0.2 0.4 0.6 0.8
Fig. 5 The relation between σ and η, SNR=0dB
σ Ts %
Exp. σ[t̂p]
Theo. σ[t̂p]
Exp. σ[t̂A]
Theo. σ[t̂A]
CRLB
12
10
8
6
4
2
0 3 6 12
SNR dB
Fig. 6
φ(fi)
π/4
π/8
0
ln(Ai²)
3
2
1
f Hz
5k
10k
Fig. 7 SNR=12 dB
π/2
π/4
Fig. 8 SNR=0 dB

Part 4
Selected Papers on Time-Varying Time Delay Estimation

Correlator Compensation Requirements for Passive Time-Delay Estimation with Moving Source or Receivers

WILLIAM B. ADAMS, JOHN P. KUHN, AND WILLIAM P. WHYLAND, MEMBER, IEEE

Abstract—An analysis is given of the effects of source or receiver motions on the output of a cross correlator used to estimate the source time-delay difference across a baseline. For both narrow-band and wide-band correlators, the need for time-varying correlator delay compensation is quantified for the case of a quadratic signal delay-difference variation. Two useful concepts which emerge are 1) the three-dimensional delay/delay-rate/delay-acceleration mean ambiguity function of the source signal, and 2) the equivalent τ-domain filter, which transforms the source autocorrelation function into the mean output of the mismatched correlator.

The required correlator compensation is approximately a quadratic delay modulation matched to the input delay-difference function. For 3 dB peak output loss with a narrow-band signal, the maximum allowable delay-rate mismatch will produce 158° of phase rotation at the centroid frequency f_0 during the correlation integration time T, while the maximum allowable delay-acceleration mismatch will produce 156° of quadratic phase rotation at f_0 between the center ($t = 0$) and each edge ($t = \pm T/2$) of the correlator integration window. For broad-band signals, the mismatch tolerances become about 11 percent tighter.

I. Introduction

GIVEN two sensors separated by a baseline, passive estimation of the time-delay difference between signals received from a common source may be accomplished by cross correlating the appropriately filtered [1] sensor outputs. When source or receiver motions cause the delay difference to vary, during the correlator integration time T, by more than the correlation time width of the source signal, then the output correlogram is degraded (smeared) unless the correlator implements a compensating delay modulation during T. In many cases the simple expedient of reducing T causes an unacceptable reduction in processing gain.

For a linear delay-difference variation during T, the optimum correlator delay modulation is also linear in time during T, so that one receiver output is time scaled (compressed or expanded) by a constant factor before cross correlating [2]. For a narrow-band signal, this compensation can be approximated by a constant frequency shift [3], [6], [7].

For a quadratic delay-difference variation during T, the optimum correlator delay modulation is approximately quadratic in time during T, so that one receiver output is time scaled by a linearly varying factor before cross correlating. For a narrow-band signal, this compensation can be approximated by a linearly varying frequency shift [3]. If the correlator can only implement a linear delay modulation (or its narrow-band approximation by a constant frequency shift), then the maximum allowable value of T for a specified correlogram degradation becomes a function of the geometry and motion scenario, together with the centroid and width of the source spectrum [3], [4].

All these effects for either broad-band or narrow-band correlators are treated in a unified manner by employing analytic-signal representation, assuming a stochastic source emission, and examining the mean (ensemble average) correlogram output as a function of errors (mismatches) in the correlator delay-modulation schedule. Correlator compensation requirements, for a specified degradation of 3 dB in the mean peak correlator output, are then obtained as error budgets on the allowable mismatches in compensating for delay-difference rates and accelerations (which at any reference frequency, and in particular at the centroid frequency f_0 of the source spectrum, may be translated into allowable values of frequency difference and frequency-difference rate).

II. Mean Correlator Output with Quadratic Delay Variations

A. General Approach

Assume the source emits a waveform $s(t)$, which is a sample function of a zero-mean stationary random process with autocorrelation function (acf)

$$R_s(\tau) = E_s\{s(t)\, s(t+\tau)\}. \tag{1}$$

Neglecting medium distortion and attenuation, the signal received at sensor "k" ($k = 1, 2$) is delay-modulated because of source or receiver motion

$$s_k(t) = s[t - \tau_k(t)]. \tag{2}$$

If the propagation delays $\tau_k(t)$ are assumed to vary quadratically with time during the observation interval T_{obs},

$$\tau_k(t) = \tau_{ok} + \tau'_{ok} t + \tau''_{ok} t^2/2, \qquad |t| \leq T_{\text{obs}}/2, \tag{3}$$

then the delay difference is also quadratic in time

$$\begin{aligned}\Delta(t) &= \tau_1(t) - \tau_2(t) \\ &= \Delta_0 + \Delta'_0 t + \Delta''_0 t^2/2\end{aligned} \tag{4}$$

Manuscript received June 11, 1979; revised October 15, 1979. This work was supported by Naval Systems Command Code 06H1.

The authors are with the Heavy Military Equipment Department, Advanced Development Engineering, Advanced Sonar Concepts, General Electric Company, Syracuse, NY 13221.

Reprinted from *IEEE Trans. Acoust., Speech, Signal Processing*, vol. 28, no. 2. pp. 158–168, April 1980.

where

$$\Delta_0 = \tau_{01} - \tau_{02}$$
$$\Delta_0' = \tau_{01}' - \tau_{02}'$$
$$\Delta_0'' = \tau_{01}'' - \tau_{02}''. \quad (5)$$

The cross correlator must then implement a time-varying delay-modulation schedule during its integration time T to compensate for the moving correlation point. We assume a quadratic[1] correlator delay-modulation schedule given by

$$\Delta_c(t) = \tau_c + \lambda_c t + \alpha_c t^2/2, \qquad |t| \leq T/2 \quad (6)$$

so that when the correlator inputs, including additive noises, are

$$x(t) = s[t - \tau_1(t)] + n_1(t)$$
$$y(t) = s[t - \tau_2(t)] + n_2(t), \quad (7)$$

then the correlator output is

$$F = F(\tau_c, \lambda_c, \alpha_c) = \frac{1}{T} \int_{-T/2}^{T/2} x(t)\, y[t - \Delta_c(t)]\, dt. \quad (8)$$

The correlator attempts to compensate for Δ_0' and Δ_0'' by a proper choice of λ_c and α_c, after which it performs (e.g., by parallel processing) a scan in τ_c to find the delay τ_{cm} at which F is a maximum. The value τ_{cm} is its estimate of the delay difference Δ_0 between the received signals at the instant $t = 0$ (taken as the center of its integration window).

Since F is a random function of τ_c because s, n_1 and n_2 are sample functions of random processes (assumed independent), we require a large product $(T \cdot B_{\text{signal}})$ so that the time average performed by the correlator gives a useful approximation to the ensemble average over the noise and signal distributions, i.e., so that F versus τ_c is a good sampling image of the signal acf with a delay shift. The correlator compensation requirements may then be determined by studying the mean (ensemble average) correlator output [5], [6] $R_d = E_{s, n_1, n_2}\{F\}$ as a function of the mismatches in choosing τ_c, λ_c, and α_c. Combining (7) and (8) and averaging over n_1 and n_2 yields $R_d = E_s\{F | n_1 = n_2 = 0\}$, which expresses the well-known [5], [11] result that additive zero-mean independent noises do not affect the *mean* cross-correlator output. Thus for our purposes the additive noise will be ignored.

B. Analytic-Signal (Preenvelope) Notation

In discussing both wide-band and narrow-band correlators it is convenient to use analytic-signal or preenvelope notation [6], [8], [9], where $(\tilde{\ })$ denotes $[(\,) + j(\hat{\ })]$ with $(\hat{\ })$ the Hilbert transform of ( ).

Thus the real source waveform is

$$s(t) = \text{Re}[\tilde{s}(t)] \quad (9)$$

where

$$\tilde{s}(t) = s(t) + j\hat{s}(t) \quad (10)$$

[1] This delay schedule can give perfect compensation for a constant or linearly-varying delay difference during T, but can only give a good approximation to perfect compensation when $\tau_{02}'' \neq 0$ (Appendix B).

and the complex acf of $\tilde{s}(t)$ may be shown to be [6], [8], [9]

$$R_{\tilde{s}}(\tau) = E_{\tilde{s}}\{\tilde{s}^*(t)\, \tilde{s}(t + \tau)\}$$
$$= 2\tilde{R}_s(\tau) = 2[R_s(\tau) + j\hat{R}_s(\tau)] \quad (11)$$

which is twice the preenvelope of $R_s(\tau)$.

With a source mean power density spectrum $W_s(f)$, then

$$R_s(\tau) = \int_{-\infty}^{\infty} W_s(f)\, e^{j2\pi f\tau}\, df \quad (12)$$

and [8]

$$\tilde{R}_s(\tau) = 2 \int_0^{\infty} W_s(f)\, e^{j2\pi f\tau}\, df. \quad (13)$$

C. Mean Output of Delay-Modulated Correlator

Using analytic-signal notation, the basic expression for the mean correlator output $\tilde{R}_d$ is now derived. Referring to the functional diagram shown in Fig. 1, the analytic source signal $\tilde{s}(t)$ experiences the time-varying propagation delays $\tau_k(t)$, $k = 1, 2$, and appears at correlator input port "k" as $\tilde{s}_k(t) = \tilde{s}(t - \tau_k(t))$. The correlator imposes the compensating delay modulation $\Delta_c(t)$ on $\tilde{s}_2(t)$ to yield $\tilde{s}_3(t) = \tilde{s}_2[t - \Delta_c(t)] = \tilde{s}\{t - \Delta_c(t) - \tau_2[t - \Delta_c(t)]\}$, and then evaluates the time average over $(-T/2, T/2)$ of the product $\tilde{s}_1(t)\tilde{s}_3^*(t)$. The mean correlator output $\tilde{R}_d$ is then half the ensemble average (over $\tilde{s}$) of this finite time average, which from (11) is seen to be

$$\tilde{R}_d = \frac{1}{2T} \int_{-T/2}^{T/2} E_{\tilde{s}}\{\tilde{s}^*(t - \Delta_c(t) - \tau_2[t - \Delta_c(t)])$$
$$\cdot\, \tilde{s}[t - \tau_1(t)]\}\, dt$$
$$= \frac{1}{2T} \int_{-T/2}^{T/2} R_{\tilde{s}}[\epsilon(t)]\, dt$$
$$= \frac{1}{T} \int_{-T/2}^{T/2} \tilde{R}_s[\epsilon(t)]\, dt \quad (14)$$

where

$$\epsilon(t) = \Delta_c(t) - \{\tau_1(t) - \tau_2[t - \Delta_c(t)]\} \quad (15)$$

is the negative of the delay difference at the input to the multiplier.

Evaluating (15) with (3) and (6), we obtain after some minor approximations (Appendix B)

$$\epsilon(t) \approx (\beta_2\tau_c - \Delta_0) + (\beta_2\lambda_c - \Delta_0')\, t + (\beta_2\alpha_c - \Delta_0'')\, t^2/2$$
$$= \beta_2(\tau + \lambda t + \alpha t^2/2) \quad (16)$$

where we define

$$\beta_2 = 1 - \tau_{02}'$$
$$\tau = \tau_c - \Delta_0/\beta_2$$
$$\lambda = \lambda_c - \Delta_0'/\beta_2$$
$$\alpha = \alpha_c - \Delta_0''/\beta_2 \quad (17)$$

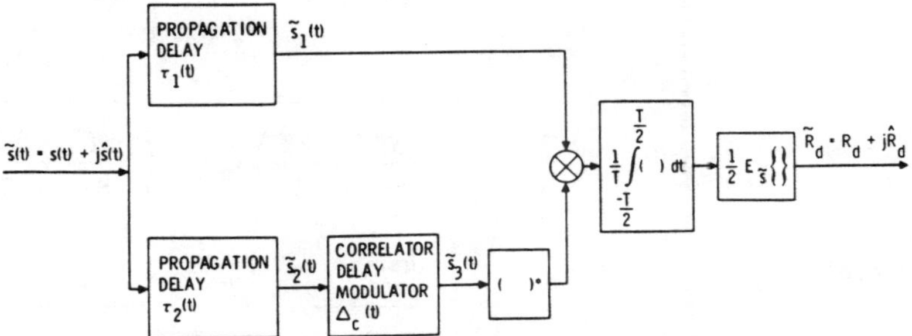

Fig. 1. Schematic for mean output of delay-modulated correlator.

so that (14) is

$$\tilde{R}_d = \tilde{R}_d(\tau, \lambda, \alpha) = \frac{1}{T}\int_{-T/2}^{T/2} \tilde{R}_s[\beta_2(\tau + \lambda t + \alpha t^2/2)]\, dt, \tag{18}$$

which may be interpreted [7] as a three-dimensional mean ambiguity function of the source waveform. For our purposes it is simply the mean preenvelope of the correlator output, as a function of correlator mismatches in delay (τ), delay rate (λ), and delay acceleration (α). With perfect compensation in delay rate and delay acceleration, $\lambda = \alpha = 0$ so that

$$\tilde{R}_d(\tau, 0, 0) = \tilde{R}_s(\beta_2\tau). \tag{19}$$

For a wide-band correlator, the mean real output is given by the real part of $\tilde{R}_d(\tau, \lambda, \alpha)$ from (18). For a narrow-band correlator this output versus τ has the form of a modulated carrier, and since the delay as measured by the "carrier" phase is too ambiguous to be useful, we rely instead on the modulated source bandwidth for delay determination [7] by finding that τ_{cm} which maximizes the envelope $|\tilde{R}_d(\tau, \lambda, \alpha)|$.

D. Equivalent Filter for Mismatched Correlator

A useful concept in understanding the effects of correlator mismatches on the mean output $\tilde{R}_d$ is that of the equivalent filter [5]. If we put (13) in (18), interchange the order of integration, and change variable

$$\beta_2 f \to f, \tag{20}$$

then we obtain

$$\tilde{R}_d(\tau, \lambda, \alpha) = \int_0^\infty \left[\frac{2}{\beta_2} W_s\left(\frac{f}{\beta_2}\right)\right] H_d(f, \lambda, \alpha)\, e^{j2\pi f\tau}\, df \tag{21}$$

where

$$H_d(f, \lambda, \alpha) = \frac{1}{T}\int_{-T/2}^{T/2} e^{j2\pi f(\lambda t + \alpha t^2/2)}\, dt. \tag{22}$$

The bracketed term in (21) is the Fourier transform of $\tilde{R}_s(\beta_2\tau)$, and $H_d(f, \lambda, \alpha)$ is an equivalent filter in the following sense: with τ analogous to a time variable, then $\tilde{R}_d(\tau, \lambda, \alpha)$ considered as a function of τ is the result of passing the "time" waveform $\tilde{R}_s(\beta_2\tau)$ through a filter with frequency response $H_d(f, \lambda, \alpha)$ and τ-domain impulse response

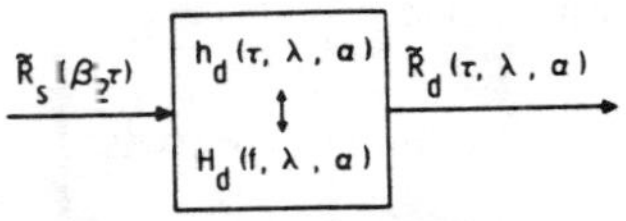

Fig. 2. Equivalent filter for delay-modulated correlator.

$$h_d(\tau, \lambda, \alpha) = \int_{-\infty}^{\infty} H_d(f, \lambda, \alpha)\, e^{j2\pi f\tau}\, df. \tag{23}$$

Note that H_d is *not* a physical filter, but rather a mathematical analogy, as indicated by Fig. 2.

By completing the square in the exponent in (22), we obtain

$$H_d(f, \lambda, \alpha) = \frac{e^{-j\pi(\lambda^2 f/\alpha)}}{2x}\left\{Z\left[\left(1 + \frac{2\lambda}{\alpha T}\right)x\right] + Z\left[\left(1 - \frac{2\lambda}{\alpha T}\right)x\right]\right\} \tag{24}$$

where

$$x = T\sqrt{f\alpha/2}$$

and

$$Z(u) = C(u) + jS(u) = \int_0^u e^{j(\pi/2)t^2}\, dt \tag{25}$$

is the complex Fresnel integral, obeying

$$Z(-u) = -Z(u)$$
$$Z(ju) = jZ^*(u). \tag{26}$$

For special cases we obtain directly from (22),

$$H_d(f, 0, 0) = 1 \tag{27}$$

$$H_d(f, \lambda, 0) = \text{sinc}\,(\lambda f T),$$
$$\text{sinc}\,(u) = \frac{\sin \pi u}{\pi u}, \tag{28}$$

while from (24) we obtain

$$H_d(f, 0, \alpha) = \frac{Z(x)}{x}. \tag{29}$$

The corresponding τ-domain impulse response $h_d(\tau, \lambda, \alpha)$ is obtained by inverting (22), and differs according to whether $\lambda \gtrless \alpha T/2$:

$$h_d(\tau, \lambda, \alpha)]_{\lambda > \alpha T/2} = \begin{cases} \dfrac{1}{T\sqrt{\lambda^2 - 2\alpha\tau}}, \\ 0, \end{cases}$$

and

$$h_d(\tau, \lambda, \alpha)]_{\lambda < \alpha T/2} = h_d(\tau, \lambda, \alpha)]_{\lambda > \alpha T/2} + h_2(\tau, \lambda, \alpha)$$

where

$$h_2(\tau, \lambda, \alpha) = \begin{cases} \dfrac{2}{T\sqrt{\lambda^2 - 2\alpha\tau}}, & \left(\dfrac{-\alpha T^2}{8} + \dfrac{\lambda T}{2}\right) \leqslant \tau < \dfrac{\lambda^2}{2\alpha} \\ 0, & \text{otherwise.} \end{cases} \tag{31}$$

For special cases this reduces to

$$h_d(\tau, 0, 0) = \delta(\tau) \tag{32}$$

$$h_d(\tau, \lambda, 0) = \begin{cases} \dfrac{1}{\lambda T}, & |\tau| \leqslant \dfrac{\lambda T}{2} \\ 0, & \text{otherwise} \end{cases} \tag{33}$$

and

$$h_d(\tau, 0, \alpha) = \begin{cases} \dfrac{2}{T\sqrt{-2\alpha\tau}}, & \dfrac{-\alpha T^2}{8} \leqslant \tau < 0 \\ 0, & \text{otherwise.} \end{cases} \tag{34}$$

III. Correlator Compensation Requirements for Delay-Difference Rates and Accelerations

A. General Approach

For both narrow-band and broad-band source spectra, we consider peak loss, resolution, and possible bias in the mean correlator output as a function of mismatches in the correlator delay-compensation parameters. Most of the results are well known when expressed in terms of range variation, rather than (as here) delay-variation coefficients. For physical interpretation and comparisons, we assume that the range from the source to sensor "k" ($k = 1, 2$) varies quadratically during the observation interval

$$r_k(t) = r_{ok} + r'_{ok} t + r''_{ok} t^2/2, \qquad |t| \leqslant T_{\text{obs}}/2. \tag{35}$$

Thus, to obtain consistency with our model (3) of quadratic time variation of the propagation delays $\tau_k(t)$, we use the results given in Appendix A to obtain the correspondences shown in Table I.

B. Delay-Rate Mismatch with Perfect Delay-Acceleration Match

1) Derivation of Results: From (18) with $\alpha = 0$,

$$\tilde{R}_d(\tau, \lambda, 0) = \frac{1}{T} \int_{-T/2}^{T/2} \tilde{R}_s[\beta_2(\tau + \lambda t)]\, dt$$

$$= \frac{1}{\beta_2 \lambda T} \int_{\beta_2(\tau - (\lambda T/2))}^{\beta_2(\tau + (\lambda T/2))} \tilde{R}_s(x)\, dx \tag{36}$$

$$\left(\frac{-\alpha T^2}{8} - \frac{\lambda T}{2}\right) \leqslant \tau \leqslant \left(\frac{-\alpha T^2}{8} + \frac{\lambda T}{2}\right) \tag{30}$$

otherwise

with an equivalent expression given by (21) and (28) as

$$\tilde{R}_d(\tau, \lambda, 0) = \int_0^{\infty} \frac{2}{\beta_2} W_s\left(\frac{f}{\beta_2}\right) \text{sinc}\,(\lambda T f)\, e^{j2\pi f \tau}\, df. \tag{37}$$

The frequency functions pertinent to (37) are shown in Fig. 3 for a block source spectrum of width B centered at f_0, normalized so that

$$\tilde{R}_s(\tau) = \text{sinc}\,(B\tau)\, e^{j2\pi f_0 \tau}. \tag{38}$$

Since $H_d(f, \lambda, 0)$ is purely real, then (refer to Fig. 2) by the principle of stationary phase [12] the τ-centroid of $\tilde{R}_d(\tau, \lambda, 0)$ is the same as that of $\tilde{R}_s(\beta_2\tau)$, which is at $\tau = 0$. Thus $H_d(f, \lambda, 0)$ introduces no τ-bias which varies with λ, and we need consider only the main-axis responses $\tilde{R}_d(\tau, 0, 0)$ and $\tilde{R}_d(0, \lambda, 0)$.

From (36) and (38),

$$\tilde{R}_d(\tau, 0, 0) = \tilde{R}_s(\beta_2\tau) = \text{sinc}\,(\beta_2 B\tau)\, e^{j2\pi f_0 \beta_2 \tau}, \tag{39}$$

while from (37) and Fig. 3,

TABLE I
Relations between Delay and Range Coefficients

ITEM	KINEMATIC EQUIVALENT: MOVING SOURCE, FIXED RECEIVERS (*)	KINEMATIC EQUIVALENT: FIXED SOURCE, MOVING RECEIVERS
$\beta_k = 1 - \tau'_{0k}$, $k = 1, 2$	$1/(1 + r'_{0k}/c)$	$1 - r'_{0k}/c$
τ_{0k}	$\beta_k r_{0k}/c$	r_{0k}/c
$\Delta_0 = \tau_{01} - \tau_{02}$	$(\beta_1 r_{01} - \beta_2 r_{02})/c$	$(r_{01} - r_{02})/c$
$\Delta'_0 = \tau'_{01} - \tau'_{02}$	$\beta_2 - \beta_1$	$\beta_2 - \beta_1$
$\Delta''_0 = \tau''_{01} - \tau''_{02}$	$(\beta_1^3 r''_{01} - \beta_2^3 r''_{02})/c$	$(r''_{01} - r''_{02})/c$

(*) EXACT FOR $r''_{0k} \to 0$.

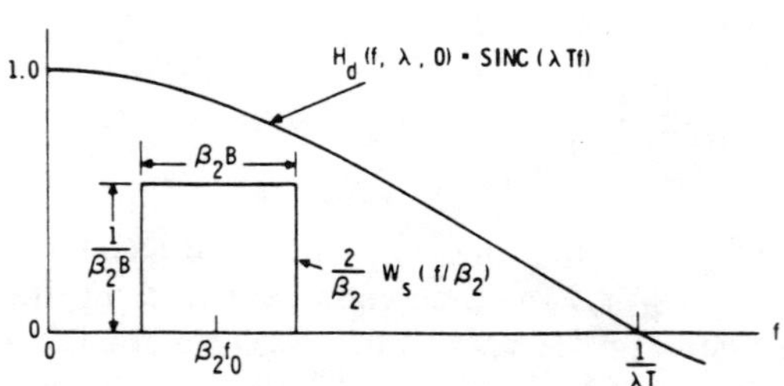

Fig. 3. Scaled source spectrum and equivalent filter response for delay-rate mismatch λ and correlator integration time T.

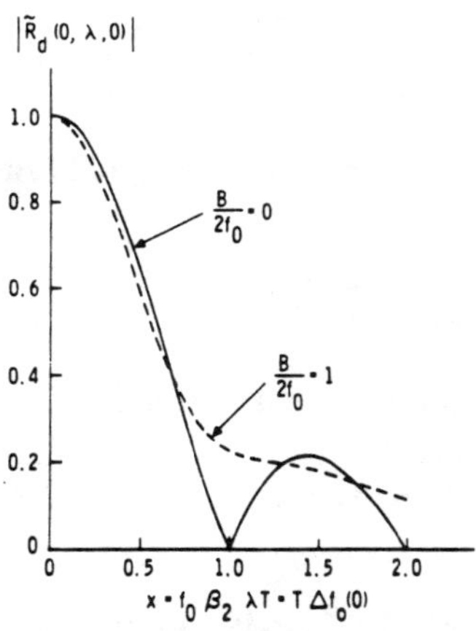

Fig. 4. Delay-rate mismatch response at $\tau = 0$ (block source spectrum, bandwidth B, centroid f_0).

$$\tilde{R}_d(0, \lambda, 0) = \frac{1}{\beta_2 B}\int_{\beta_2(f_0-B/2)}^{\beta_2(f_0+B/2)} \text{sinc}\,(\lambda T f)\,df$$

$$= \frac{Si[\pi x(1+\mu)] - Si[\pi x(1-\mu)]}{2\pi x\mu} \tag{40}$$

where

$$\mu = \frac{B}{2f_0}$$

$$x = \beta_2 f_0 \lambda T$$

$$Si(x) = \int_0^x \frac{\sin t}{t}\,dt.$$

For a source bandwidth $B \to 0$, the spectrum in Fig. 3 goes over to $\delta(f - \beta_2 f_0)$ and (40) becomes

$$\tilde{R}_d(0, \lambda, 0]_{B\to 0} = \text{sinc}\,(\beta_2 f_0 \lambda T), \tag{41}$$

while at the other extreme of a low-pass block spectrum of video bandwidth $B = 2f_0$,

$$\tilde{R}_d(0, \lambda, 0)]_{B/2=f_0} = \frac{Si(2\pi x)}{2\pi x}, \quad x = \beta_2 B\lambda T/2. \tag{42}$$

Equations (41) and (42) are plotted in Fig. 4.

We note from (40) that $\tilde{R}_d(0, \lambda, 0)$ is purely real. Thus from Fig. 4 the 3 dB λ-width (λ_3) of either $R_d(0, \lambda, 0)$ for a wide-band correlator, or of $|\tilde{R}_d(0, \lambda, 0)|$ for a narrow-band correlator, satisfies

$$f_0\beta_2\lambda_3 = \frac{0.88}{T} \quad \text{for } B \to 0 \tag{43}$$

and is only about 11 percent smaller for $B \to 2f_0$, where f_0 is the centroid of the source spectrum. Thus the allowable mismatch (λ) in delay-rate compensation for a peak[2] correlator loss of at most 3 dB is essentially

$$|\lambda| \leq \frac{0.44}{f_0\beta_2 T}. \tag{44}$$

To translate this tolerance into that of an allowable frequency difference, we observe that, for the component of the source signal at the centroid frequency $f = f_0$, the instantaneous frequency difference at $t = 0$ between the signals at the input to the multiplier in Fig. 1 is obtained from (15)–(17) as

$$\Delta f_0(0) = f_0 \epsilon'(0)$$

$$\approx f_0(\beta_2\lambda_c - \Delta_0') = f_0\beta_2\lambda \tag{45}$$

so that the abscissa of Fig. 4 may be interpreted as

$$x = T \cdot \Delta f_0(0) \tag{46}$$

where $\Delta f_0(0)$ is the uncompensated frequency difference at $t = 0$ at frequency ($\beta_2 f_0$). The mismatch tolerance specification given by (44) is then

$$|\Delta f_0(0)| < \frac{0.44}{T} \tag{47a}$$

which, for 3 dB peak loss, corresponds to a linear phase rotation at frequency ($\beta_2 f_0$) during time T of

$$2\pi T \cdot \Delta f_0(0) = 2\pi(0.44) \simeq 158°. \tag{47b}$$

If *no* delay-rate compensation is attempted, then $\lambda_c = 0$ in (17) so that

$$\lambda = -\Delta_0'/\beta_2 \tag{48}$$

and the abscissa in Fig. 4 then has a magnitude of

$$x = |Tf_0\Delta_0'|, \tag{49}$$

which from Table I is seen to be

$$x = |Tf_0(\beta_2 - \beta_1)|$$

$$= |Tf_0\beta_1\beta_2(r_{01}' - r_{02}')/c| \tag{50a}$$

for moving source and fixed receivers

$$= |Tf_0(r_{01}' - r_{c2}')/c| \tag{50b}$$

for fixed source and moving receivers.

Since β_1 and β_2 are within 1 percent of unity, (50a) and (50b) are essentially equal.

2) Comparisons With Published Results [2], [3], [5], [13]: From (50b), Fig. 4 is seen to be equivalent to Fig. 4 of [3], but with a more general abscissa.

The approach used and the general results obtained[3] are identical to those given in [13] for the case where the delay-acceleration mismatch is zero and where real rather than complex signal notation is used. Thus the real part of (18), with $\alpha = 0$ and $\beta_2\lambda \to \delta$, is the same as equation (3.10.2-33), p. 111 of [13]; and for the block source spectrum, the further notational changes $\beta_2 f_0 \to f_c$, $\beta_2 B \to B$ reveal that our equations (40), (41), and (42) are identical to equations (3.10.2-35, -36, -38), p. 113, *op. cit.*

[2]This assumes λT is small enough so that the mainlobe responses versus τ of $R_d(\tau, \lambda, 0)$ for the broad-band, and $|\tilde{R}_d(\tau, \lambda, 0)|$ for the narrow-band correlator are unimodal with global maxima at $\tau = 0$. For the block source spectrum, calculations of $\tilde{R}_d(\tau, \lambda, 0)$ show that this requires that $x = |\beta_2 f_0 \lambda T|$ be less than 0.6 to 1.0, with the exact bound depending on $B/2f_0$ and the correlator type. At larger values of x, the mainlobe responses remain even in τ but become bimodal, with a local minimum at $\tau = 0$.

[3]We thank a reviewer for pointing this out.

The equivalent filter, when no delay-rate compensation is used, is given by (28) and (48) and Table I

$$H_d(f, -\Delta_0'/\beta_2, 0) = \text{sinc}\,(fT\Delta_0'/\beta_2) = \text{sinc}\,[(1 - \beta_1/\beta_2) fT] \tag{51}$$

which is slightly different, both from the result given in (21) of [5], and also from the result implied by (26b) of [2]. It appears that the difference from [5] stems from the definition of correlator clock time used there, while the difference from [2] rises from an approximation neglecting spectral compression made there [p. 1548, paragraph preceding (26a)].

As a final comparison we note that the "bias" in the time-delay estimate suggested in [2] as a consequence of ignoring delay rate is an artifact which vanishes if we agree that the correlator-delay estimate applies at the *center* of its data window. The peak of $\tilde{R}_d(\tau, \lambda, 0)$ occurs at $\tau = 0$ for *any* λ[4], so that from (17) and Table I the mean correlator-delay estimate is

$$\tau_{cm} = \frac{\Delta_0}{\beta_2}, \tag{52}$$

which for the moving source/fixed receiver scenario is

$$\begin{aligned}\tau_{cm} &= \left(\frac{\beta_1}{\beta_2} r_{01} - r_{02}\right) \Big/ c \\ &= \left(\frac{r_{01} - r_{02}}{c}\right) - (1 - \beta_1/\beta_2)\frac{r_{01}}{c} \\ &= \left(\frac{r_{01} - r_{02}}{c}\right) - \left(\frac{r_{01}' - r_{02}'}{c}\right)\beta_1 \frac{r_{01}}{c} \\ &= \left(\frac{r_{01} - r_{02}}{c}\right) - \left(\frac{r_{01}' - r_{02}'}{c}\right)\tau_{01} \\ &\cong [r_1(-\tau_{01}) - r_2(-\tau_{01})]/c \end{aligned} \tag{53}$$

if we neglect the acceleration terms $r_{0k}''\tau_{01}^2/2c$.

Thus in this case the mean correlator *time*-delay difference estimate for *any* λ is a measure of the *range*-difference delay at a time earlier than $t = 0$ by the propagation time (at $t = 0$) from the source to sensor "1." This geometric, or kinematic, bias is a time-lag effect arising from finite propagation velocity, and is almost always negligible.

The corresponding result for the fixed-source/moving-receiver scenario is

$$\tau_{cm} = (r_{01} - r_{02})/c\beta_2 \tag{54}$$

which has a scale-factor geometric error of up to 1 percent, again almost always negligible.

C. Delay-Acceleration Mismatch with Perfect Delay-Rate Match

1) Derivation of Results: From (18) with $\lambda = 0$, we have

[4]Assuming as before that λT is not so large that the mainlobe of $\tilde{R}_d(\tau, \lambda, 0)$ becomes bimodal.

$$\tilde{R}_d(\tau, 0, \alpha) = \frac{1}{T}\int_{-T/2}^{T/2} R_s[\beta_2(\tau + \alpha t^2/2)]\, dt \tag{55}$$

with an equivalent expression given by (21) and (29) as

$$\tilde{R}_d(\tau, 0, \alpha) = \int_0^\infty \frac{2}{\beta_2} W_s\left(\frac{f}{\beta_2}\right) \frac{Z(T\sqrt{f\alpha/2})}{T\sqrt{f\alpha/2}} e^{j2\pi f\tau}\, df \tag{56}$$

where Z is the complex Fresnel integral (25).

In this case both $|\tilde{R}_d(\tau, 0, \alpha)|$ and $\text{Re}[\tilde{R}_d(\tau, 0, \alpha)]$ have their peaks at (different) nonzero values of τ which vary with α

$$\max_{\{\tau\}} |\tilde{R}_d(\tau, 0, \alpha)| = |\tilde{R}_d(\tau_e(\alpha), 0, \alpha)| \tag{57}$$

$$\max_{\{\tau\}} \text{Re}[\tilde{R}_d(\tau, 0, \alpha)] = \text{Re}[\tilde{R}_d(\tau_r(\alpha), 0, \alpha)] \tag{58}$$

so that the peak loss must be evaluated at either τ_e or τ_r, rather than at $\tau = 0$. These α-varying τ-biases arise because the time shift function in the argument of the integrand in (55) is not odd about zero. As a consequence the equivalent filter (in the sense of Fig. 2) has a nonlinear phase response over frequency, the study of which allows a good understanding of these effects.

From (25) and (29), the equivalent filter is

$$H_d(f, 0, \alpha) = \frac{C(x) + jS(x)}{x}, \quad x \triangleq T\sqrt{\frac{f\alpha}{2}} \tag{59}$$

which we write in amplitude/phase response form as

$$H_d(f, 0, \alpha) = A(f, 0, \alpha)\, e^{j\phi(f, 0, \alpha)} \tag{60}$$

where

$$A(f, 0, \alpha) = \frac{1}{x}\sqrt{C^2(x) + S^2(x)} \tag{61}$$

$$\phi(f, 0, \alpha) = \tan^{-1}[S(x)/C(x)]. \tag{62}$$

The phase-delay (T_ϕ) and group-delay (T_g) functions for this filter are [10]

$$T_\phi(f) = \frac{-\phi(f, 0, \alpha)}{2\pi f} = \frac{-\alpha T^2}{4\pi x^2}\tan^{-1}[S(x)/C(x)] \tag{63}$$

$$\begin{aligned}T_g(f) &= \frac{-1}{2\pi}\frac{\partial\phi(f, 0, \alpha)}{\partial f} \\ &= \frac{-\alpha T^2}{8\pi x}\left[\frac{C(x)\sin(\pi/2)x^2 - S(x)\cos(\pi/2)x^2}{C^2(x) + S^2(x)}\right],\end{aligned} \tag{64}$$

with limiting small-argument values given by

$$\lim_{x\to 0}[T_\phi(f)] = \lim_{x\to 0}[T_g(f)] = \frac{-\alpha T^2}{24} \tag{65}$$

which is the average of the function $(-\alpha t^2/2)$ over the interval $(-T/2, T/2)$.

The amplitude, phase, and delay responses are plotted in Fig. 5 versus the scaled frequency variable $x^2 = f\alpha T^2/2$.

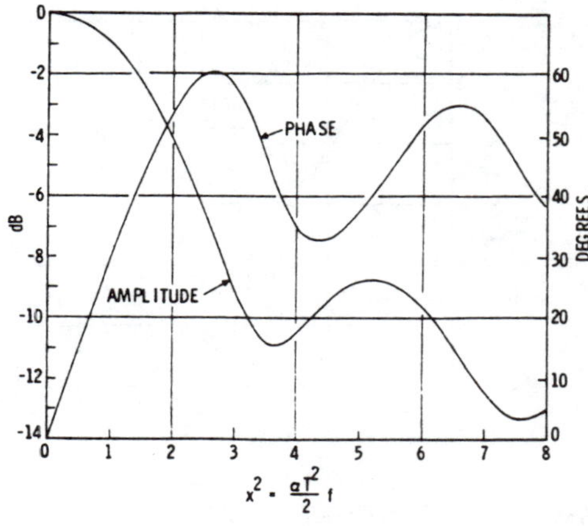

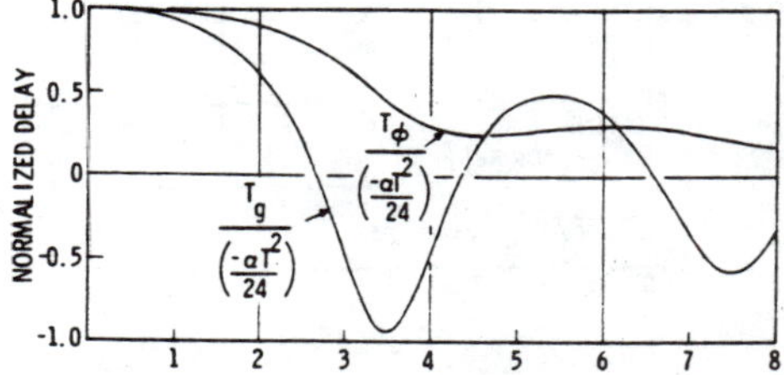

Fig. 5. Amplitude, phase, phase-delay, and group-delay functions for

Now consider the block source spectrum of width B centered at f_0. The frequency domain relations pertinent to (56) are similar to Fig. 3, but with sinc (λTf) replaced by $H_d(f, 0, \alpha)$.

For a very narrow-band signal $(B \to 0)$, the compressed source spectrum goes over to $\delta(f - \beta_2 f_0)$, and (56) gives

$$\tilde{R}_d(\tau, 0, \alpha)]_{B=0} = e^{j2\pi f_0 \beta_2 \tau} \left[\frac{C(y_0) + jS(y_0)}{y_0}\right]$$

$$\text{where } y_0 = T\sqrt{\frac{f_0\beta_2\alpha}{2}}. \tag{66}$$

Here the envelope of the mean correlator output is

$$|\tilde{R}_d(\tau, 0, \alpha)| = \frac{1}{y_0}\sqrt{C^2(y_0) + S^2(y_0)} \tag{67}$$

which is independent of τ since $B = 0$, while the real part of the mean correlator output is

$$\begin{aligned}\text{Re}\,[\tilde{R}_d(\tau, 0, \alpha)] &= \frac{C(y_0)\cos 2\pi f_0\beta_2\tau - S(y_0)\sin 2\pi f_0\beta_2\tau}{y_0} \\ &= \frac{1}{y_0}\sqrt{C^2(y_0) + S^2(y_0)} \\ &\quad \cdot \cos 2\pi f_0\beta_2[\tau - T_\phi(f_0\beta_2)]\end{aligned} \tag{68}$$

where $T_\phi(f_0\beta_2)$ is the phase delay of the equivalent filter at frequency $(f_0\beta_2)$, as given by (63) with $x = y_0$. From (65), (66), and (68) and Fig. 5, we see that for

$$y_0^2 = f_0\beta_2\alpha T^2/2 \leqslant 1.73 \tag{69}$$

then both the envelope and the peak real response of the correlator output are attenuated by less than 3 dB, while

$$T_\phi(f_0\beta_2) \approx \frac{-\alpha T^2}{24}. \tag{70}$$

Thus, for a very narrow-band signal with centroid frequency f_0, the 3 dB tolerance value for delay-acceleration mismatch is, from (69),

$$|\alpha| \leqslant \frac{3.46}{\beta_2 f_0 T^2}. \tag{71}$$

For the component of the source signal at its centroid frequency f_o, the difference-frequency rate between the signals

$$H_d(f, 0, \alpha) = \frac{C\left(T\sqrt{\frac{f\alpha}{2}}\right) + jS\left(T\sqrt{\frac{f\alpha}{2}}\right)}{T\sqrt{\frac{f\alpha}{2}}}.$$

at the input to the multiplier in Fig. 1 is obtained from (15)-(17) as

$$\Delta f_0'(0) = f_0\epsilon''(0) \simeq f_0\beta_2\alpha \tag{72}$$

so that the equivalent 3 dB tolerance for a mismatch in difference-frequency rate is

$$|\Delta f_0'(0)| \leqslant \frac{3.46}{T^2} \tag{73}$$

which, at the allowable maximum for 3 dB peak loss, corresponds to a quadratic phase rotation at the edge of the correlator window of

$$\begin{aligned}\Delta\phi_{\text{edge}} &= 2\pi \int_0^{T/2} \Delta f_0'(0) \cdot t\, dt \\ &= \frac{\pi}{2}\,\Delta f_0'(0) \cdot T^2/2 \\ &= \frac{\pi}{2}\,(1.73) \approx 156^\circ\end{aligned} \tag{74}$$

where $\Delta f_0'$ is the difference-frequency rate at frequency $\beta_2 f_0$.

If now we allow the signal bandwidth B to increase from zero, then the input to the equivalent filter of Fig. 2 is given by (39), which has the form of a "carrier" at frequency $(f_0\beta_2)$, modulated by sinc $(\beta_2 B\tau)$. If $\mu = (B/2f_0)$ is small enough that $H_d(f, 0, \alpha)$ has essentially constant amplitude and linear phase over the band $(\beta_2 B)$ centered at $(\beta_2 f_0)$, then [10] the output envelope is delayed by $T_g(\beta_2 f_0)$, while the output "carrier" is delayed by $T_\phi(\beta_2 f_0)$, which is the delay for the peak of Re $\{\tilde{R}_d(\tau, 0, \alpha)\}$. In either case, the output peak is approximately reduced by the "narrow-band" factor given by (67).

As $\mu = (B/2f_0)$ increases further, the concepts of phase delay and group delay lose meaning [10], and we must evaluate both $|\tilde{R}_d|$ and Re $\{\tilde{R}_d\}$ versus τ to find first the τ-biases, and then the peak attenuations at these values of τ. For the block spectrum this requires numerical integrations which are simpler if we use (38) and (55) to obtain

$$\tilde{R}_d(\tau, 0, \alpha) = \frac{1}{T}\int_{-T/2}^{T/2} \frac{\sin \pi B\beta_2(\tau + \alpha t^2/2)}{\pi B\beta_2(\tau + \alpha t^2/2)} \cdot e^{j2\pi f_0\beta_2(\tau + \alpha t^2/2)}\, dt$$

$$= 2\int_0^{1/2} \frac{\sin 2\pi\mu(x + a\xi^2)}{2\pi\mu(x + a\xi^2)} e^{j2\pi(x+a\xi^2)}\, d\xi \quad (75)$$

with

$$\xi = t/T$$
$$x = f_0\beta_2\tau$$
$$a = f_0\beta_2\alpha T^2/2$$
$$\mu = B/2f_0.$$

Equation (75) was evaluated numerically for various μ and a in the region around $x = -a/12$ (i.e., $\tau = -(\alpha T^2/24)$), and the peak delays and corresponding peak responses obtained for both $\mathrm{Re}\{\tilde{R}_d(\tau_r, 0, \alpha)\}$ and $|\tilde{R}_d(\tau_e, 0, \alpha)|$. The results are shown in Fig. 6, in which the results for $B = 0$ are simply the values given in Fig. 5 for $f = f_0\beta_2$. It is seen that the delay-acceleration mismatch tolerance for a 3 dB peak loss is very nearly independent of bandwidth for both wide-band and narrow-band correlators, and is essentially given by (71) or (73).

We note from (72) and (73) that the abscissa in Fig. 6 may be expressed as

$$a = \Delta f_0'(0) \cdot T^2/2 \quad (76)$$

and that if *no* delay-acceleration compensation is attempted, then $\alpha_c = 0$ in (17) and

$$\alpha = -\Delta_0''/\beta_2 \quad (77)$$

so that

$$|a| = |f_0\Delta_0'' T^2/2| \quad (78)$$

and the algebraic signs of τ_r and τ_e are reversed in Fig. 6. From Table I, (78) is seen to be

$$|a| = \left| f_0 \frac{T^2}{2} (\beta_1^3 r_{01}'' - \beta_2^3 r_{02}'')/c \right|$$

for moving source/fixed receivers (79a)

$$= \left| f_0 \frac{T^2}{2} (r_{01}'' - r_{02}'')/c \right|$$

for fixed source/moving receivers (79b)

and for $|1 - \beta_k| < 0.01$, these are roughly equal.

2) Comparisons with Published Results [3], [4]: For the narrow-band ($B/2f_0 \to 0$) case, the peak loss for no delay-acceleration compensation given in (4-32) of [4] is essentially (68) with

$$y_0 = T\sqrt{-f_0\Delta_0''/2} = jT\sqrt{f_0\Delta_0''/2} \quad (80)$$

so that by (26), the sign of the phase delay $T_\phi(f_0\beta_2)$ in (68) is reversed. In [4] the small-argument approximation (70) is used.

In (44) of [3] the degradation factor for $B/2f_0 \to 0$ is given as $C(y_0)/y_0$, which is erroneous since from (68) this equals $\mathrm{Re}[\tilde{R}_d(0, 0, \alpha)]$, which is neither the real output peak nor the peak of the envelope of the output. This error appears to have arisen because the term involving the Fresnel sine integral was ignored in passing from (43) to (44), *op. cit.*

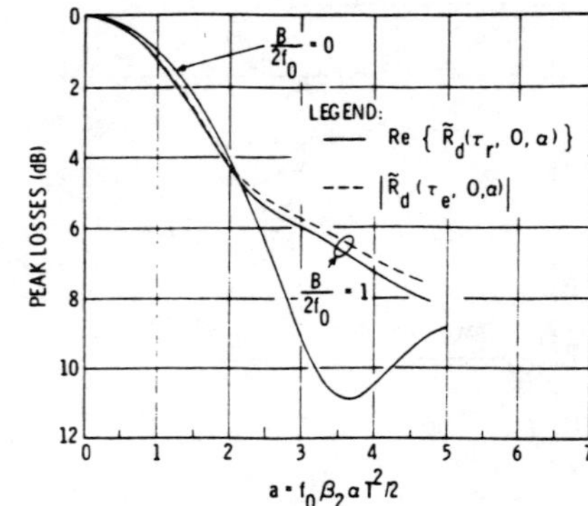

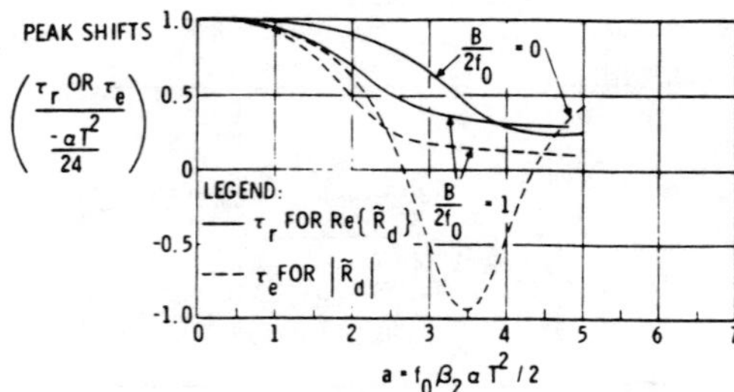

Fig. 6. Peak shifts and losses for delay-acceleration mismatch with wide-band [$\mathrm{Re}(\tilde{R}_d)$] and narrow-band [$|\tilde{R}_d|$] correlators (block source spectrum, bandwidth B, centroid f_0).

These results are plotted in Fig. 7 for comparisons.

IV. Summary

We have considered the problem of how, and how well, to compensate for the effects of source or receiver motions on the output of a cross correlator used to estimate delay difference across a baseline. For a quadratic delay difference $\Delta(t)$ during the correlator integration period T given by

$$\Delta(t) = \Delta_0 + \Delta_0' t + \Delta_0'' t^2/2, \quad |t| \leq T/2, \quad (4)$$

the required correlator compensation was found to be essentially a quadratic delay modulation $\Delta_c(t)$ inserted in one input channel before multiplication

$$\Delta_c(t) = \tau_c + \lambda_c t + \alpha_c t^2/2, \quad |t| \leq T/2. \quad (6)$$

The approach used was to assume a stochastic source signal and to determine the mean (ensemble average) correlator out-

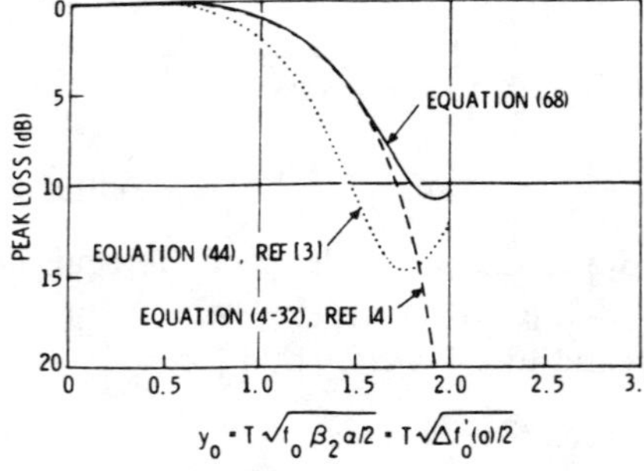

Fig. 7. Comparison of results for peak output loss of correlator with quadratic-delay mismatch and negligible signal bandwidth. $\left(\frac{B}{2f_0} = 0\right)$.

put as a function of the mismatches

$$\lambda = \lambda_c - \Delta_0'/\beta_2$$
$$\alpha = \alpha_c - \Delta_0''/\beta_2, \qquad \beta_2 = 1 - \tau_{02}' \tag{17}$$

in compensating for delay-difference rate and acceleration. The use of analytic signal representation allowed broad-band and narrow-band correlators to be treated simultaneously.

For no more than 3 dB reduction in the peak of the mean correlator output with a *narrow-band* signal and a perfect delay-acceleration match ($\alpha = 0$), the delay-rate mismatch (λ) must satisfy

$$|\lambda T| \leqslant \frac{0.44}{\beta_2 f_0} \quad \text{or} \quad |T\Delta f_0(0)| \leqslant 0.44, \tag{44, 47}$$

while with a perfect delay-rate match ($\lambda = 0$), the delay-acceleration mismatch (α) must satisfy

$$|\alpha T^2/2| \leqslant \frac{1.73}{\beta_2 f_0} \quad \text{or} \quad \left| \frac{T^2}{2} \Delta f_0'(0) \right| \leqslant 1.73 \tag{71, 73}$$

where f_0 is the centroid frequency of the source spectrum, T is the correlator integration time, and $\Delta f_0(0)$, $\Delta f_0'(0)$ are the frequency difference and frequency-difference rate at frequency $\beta_2 f_0$ and $t = 0$, at the input to the correlator multiplier. For *broad-band* signals, these mismatch tolerances become about 11 percent tighter.

For given values of λ and α, these mismatch tolerances prescribe the maximum allowable correlator integration time $T_{\max}$ as the smaller of

$$T_{m_\lambda} = \frac{0.44}{\beta_2 f_0 |\lambda|}$$

and

$$T_{m_\alpha} = \sqrt{\frac{3.46}{\beta_2 f_0 |\alpha|}}.$$

If the correlator does not attempt to compensate for delay-difference accelerations, then $\alpha_c = 0$ and α is determined by the kinematics Δ_0'' and may be large, leading to a small value of $T_{\max}$. A similar conclusion holds for a correlator which does not compensate for delay-difference rates.

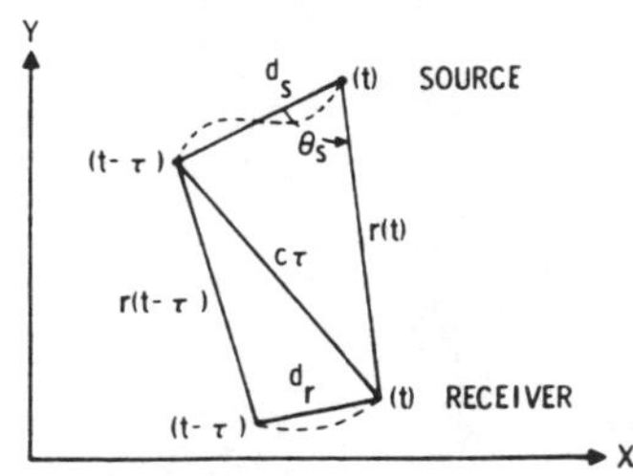

Fig. 8. Assumed propagation geometry.

Appendix A
Signal Range-Delay Variation with Moving Source or Receiver

The aim here is to describe the time-varying signal delay $\tau = \tau(t)$ in terms of the time-varying range $r(t)$ when the source or receiver, or both, move. Assuming direct-path propagation in a horizontal plane in an idealized medium with constant speed of propagation c, the geometry is indicated in Fig. 8.

The source transmits a waveform $s(t - \tau)$ at time $(t - \tau)$ when the range is $r(t - \tau)$. Neglecting attenuation and distortion in the medium, this point value of the source waveform travels along the path $c\tau(t)$ and impinges on the receiver at time t when the range is $r(t)$, so that the received waveform at time t is

$$s_r(t) = s(t - \tau(t)). \tag{A1}$$

For a fixed source and moving receiver (FS, MR) then $d_s = 0$ in Fig. 8, and

$$c\tau(t) = r(t). \tag{A2}$$

For a moving source and fixed receiver (MS, FR) then $d_r = 0$ so that

$$c\tau(t) = r(t - \tau(t)). \tag{A3}$$

When both source and receiver move (MS, MR), then by the law of cosines

$$c\tau(t) = r(t)\sqrt{1 - 2d_s \cos\theta_s/r(t) + (d_s/r(t))^2} \tag{A4}$$

where $d_s = d_s(t, \tau)$ is the straight line connecting the source positions at times $(t - \tau)$ and (t), and $\theta_s = \theta_s(t, \tau)$ is the angle as indicated between d_s and $r(t)$. With an instantaneous source velocity schedule given by $v_{sx}(t)$, $v_{sy}(t)$ during $(t - \tau, t)$, then the average source speed over this interval is

$$V_s = \sqrt{V_x^2 + V_y^2} = V_s(t, \tau) \tag{A5}$$

with

$$V_x = \frac{1}{\tau} \int_{t-\tau}^{t} v_{sx}(\xi)\, d\xi = V_x(t, \tau) \tag{A6}$$

and similarly for V_y. The length d_s is then

$$d_s = V_s \cdot \tau(t) \tag{A7}$$

where

$$\tau(t) \simeq \frac{r(t)}{c} \tag{A8}$$

so that

$$\frac{d_s}{r(t)} \simeq \frac{V_s}{c} < 0.01 \tag{A9}$$

and a first-order expansion of (A4) yields

$$c\tau(t) \cong r(t) - (V_s \cos\theta_s)\, \tau(t). \tag{A10}$$

Equations (A2), (A3), and (A10) relate $\tau(t)$ to $r(t)$ for the (FS, MR), (MS, FR), and (MS, MR) cases. We now concentrate on the first two cases on the assumption that the range $r(t)$ varies quadratically with time over the observation interval

$$r(t) = r_0 + r_0' t + \frac{r_0''}{2} t^2, \quad |t| \leq T_{obs}/2. \tag{A11}$$

1) FS, MR–From (A2) and (A11),

$$\tau(t) = \frac{r(t)}{c} = \tau_0 + \tau_0' t + \tau_0'' \frac{t^2}{2} \tag{A12}$$

where

$$\tau_0 = \frac{r_0}{c}$$

$$\tau_0' = \frac{r_0'}{c}$$

$$\tau_0'' = \frac{r_0''}{c}. \tag{A13}$$

2) MS, FR–From (A3) and (A11),

$$c\tau = r(t - \tau) = r - \tau r' + \frac{\tau^2}{2} r'' \tag{A14}$$

where we temporarily suppress the explicit time-function notation. The solution of (A14) for τ yields

$$\tau = \frac{(c + r')}{r''}\left[1 - \sqrt{1 - \frac{2r''r}{(c + r')^2}}\right] \simeq \frac{r}{c + r'} \tag{A15}$$

since the second term under the radical in (A15) is of the order of c^{-1} times the change in r' during τ, and is thus much less than 1.

Successive differentiation of (A14) yields

$$\tau' = \frac{\dfrac{r' - \tau r''}{c}}{1 + \left(\dfrac{r' - \tau r''}{c}\right)} \tag{A16}$$

$$\tau'' = \frac{\dfrac{r''}{c}(1 - \tau')}{\left[1 + \left(\dfrac{r' - \tau r''}{c}\right)\right]^2} = \frac{\dfrac{r''}{c}}{\left[1 + \left(\dfrac{r' - \tau r''}{c}\right)\right]^3}. \tag{A17}$$

The signal delay function for the (MS, FR) case is then approximately quadratic for a quadratic range variation

$$\tau(t) \simeq \tau_0 + \tau_0' t + \frac{\tau_0''}{2} t^2 \tag{A18}$$

where from (A11), (A15), $\cdots$, (A17),

$$\tau_0 \simeq \frac{r_0/c}{1 + r_0'/c}$$

$$\tau_0' = \left(\frac{r_0' - r_0''\tau_0}{c}\right) \Big/ \left[1 + \left(\frac{r_0' - r_0''\tau_0}{c}\right)\right] \cong \frac{r_0'/c}{1 + r_0'/c}$$

$$\tau_0'' = \frac{r_0''/c}{\left[1 + \left(\dfrac{r_0' - r_0''\tau_0}{c}\right)\right]^3} \cong \frac{r_0''/c}{[1 + r_0'/c]^3} \tag{A19}$$

and where we have assumed that

$$r_0' - r_0''\tau_0 = r'(-\tau_0) \cong r'(0) = r_0'.$$

Equations (A13) and (A19) relate the coefficients of the quadratic signal-delay function to the coefficients of the quadratic range function for the two scenarios (FS, MR) and (MS, FR). Note that if $r'' = 0$ all approximations become exact for the (MS, FR) case.

Appendix B

Delay-Difference Compensation Approximations

With $\tau_k(t)$ and $\Delta(t) = \tau_1(t) - \tau_2(t)$ assumed quadratic in time (3), (4), then for an *arbitrary* delay-compensation modulation $\Delta_c(t)$, the delay-difference function at the input to the multiplier in Fig. 1 is the negative of $\epsilon(t)$, as given by (15). Combining (3), (4), and (15) gives

$$\epsilon(t) = \frac{\tau_{02}''}{2} \Delta_c^2(t) + [1 - \tau_{02}' - \tau_{02}'' t]\, \Delta_c(t) - \Delta(t). \tag{B1}$$

For perfect compensation we require $\epsilon(t) = 0$ over $(-T/2, T/2)$, so that $\Delta_c(t)$ is the root of a quadratic

$$\Delta_c(t) = \frac{-(\beta_2 - \tau_{02}'' t)}{\tau_{02}''}\left[1 - \sqrt{1 + \frac{2\tau_{02}''\Delta(t)}{(\beta_2 - \tau_{02}'' t)^2}}\right],$$

$$\beta_2 \triangleq 1 - \tau_{02}'. \tag{B2}$$

In nearly all practical cases the second term under the radical in (B2) is $\ll 1$, so that

$$\Delta_c(t) \approx \frac{\Delta(t)}{\beta_2 - \tau_{02}'' t} \approx \frac{\Delta(t)}{\beta_2}; \tag{B3}$$

therefore, if $\Delta(t)$ is quadratic in time, then the optimum $\Delta_c(t)$ is very nearly quadratic in time.[5] Thus we assume that $\Delta_c(t)$ has the quadratic form given by (6), and when (3), (4), (6), (15) are combined, we obtain

$$\epsilon(t) = A + Bt + \frac{D}{2} t^2 + Et^3 + Ft^4 \tag{B4}$$

where

$$A = \beta_2 \tau_c - \Delta_0 + [\tau_{02}'' \tau_c^2/2]$$

$$B = \beta_2 \lambda_c - \Delta_0' - [\tau_{02}''(1 - \lambda_c)\tau_c]$$

$$D = \beta_2 \alpha_c - \Delta_0'' - [\tau_{02}''\{\lambda_c(2 - \lambda_c) - \tau_c \alpha_c\}]$$

$$E = [-\tau_{02}''(1 - \lambda_c)\alpha_c/2]$$

$$F = [\tau_{02}'' \alpha_c^2/8]. \tag{B5}$$

[5] This is exact if $\tau_{02}'' = 0$.

By neglecting the terms in brackets in (B5), we obtain (16). In any particular scenario the effect of these approximations can be estimated by expanding $\tilde{R}_s[\epsilon(t)]$ to low order in $\epsilon(t)$ and integrating (14). In nearly all cases the approximations leading from (B5) to (16) are very good.

Acknowledgment

We thank S. M. Garber, D. W. Winfield, and S. E. Robison for support, encouragement, and many useful discussions.

References

[1] C. H. Knapp and G. C. Carter, "The generalized correlation method for estimation of time delay," *IEEE Trans. Acoust., Speech, Signal Processing*, vol. ASSP-24, pp. 320–327, 1976.

[2] —, "Estimation of time delay in the presence of source or receiver motion," *J. Acoust. Soc. Amer.*, vol. 61, pp. 1545–1549, 1977.

[3] J. T. Patzewitsch, M. D. Srinath, and C. I. Black, "Nearfield performance of passive correlation processing sonars," *J. Acoust. Soc. Amer.*, vol. 64, pp. 1412–1423, 1978.

[4] A. A. Gerlach, "Motion induced coherence degradation in passive systems," *IEEE Trans. Acoust., Speech, Signal Processing*, vol. ASSP-26, pp. 1–15, 1978.

[5] W. R. Remley, "Correlation of signals having a linear delay," *J. Acoust. Soc. Amer.*, vol. 35, no. 1, pp. 65–69, 1963.

[6] —, "Doppler dispersion effects in matched filter detection and resolution," *Proc. IEEE*, vol. 54, no. 1, pp. 33–39, 1966.

[7] E. J. Kelly and R. P. Wishner, "Matched-filter theory for high-velocity accelerating targets," *IEEE Trans. Mil. Electron.*, vol. MIL-9, pp. 56–69, 1965.

[8] M. Schwartz, W. R. Bennett, and S. Stein, *Communication Systems and Techniques.* New York: McGraw-Hill, 1966, pp. 29–45.

[9] J. Dugundji, "Envelopes and pre-envelopes of real waveforms," *IRE Trans. Inform. Theory*, vol. IT-4, pp. 53–57, Mar. 1958.

[10] A. Papoulis, *The Fourier Integral and Its Applications.* New York: McGraw-Hill, 1962, pp. 134–136.

[11] G. A. Korn, *Random-Process Simulation and Measurements.* New York: McGraw-Hill, 1966, pp. 5-11–5-12.

[12] S. Goldman, *Frequency Analysis, Modulation and Noise.* New York: McGraw-Hill, 1948, pp. 112–113.

[13] A. A. Gerlach, *Theory and Applications of Statistical Wave-Period Processing.* New York: Gordon & Breach, Science Publishers, Inc., 1970, vol. I, pp. 110–113.

An Optimum First-Order Time Delay Tracker

ROY E. BETHEL AND ROBERT G. RAHIKKA

The key to the general estimation problem is the computation of the posterior probability density function (PDF) of the desired parameter, or PDF conditioned on knowing all available observations. Any optimum estimate can be computed from this PDF. The theory and implementation of a numerical solution is presented for the computation of the posterior PDF in real time for time delay tracking with a first-order Markov assumption. Benefits include a lack of threshold effects usually associated with nonlinear trackers as well as theoretically optimum performance.

I. INTRODUCTION

Real-time estimation, or tracking, of a time varying parameter in a noisy environment is one of the most challenging problems in signal processing. The observations consist of a wanted signal component and an unwanted noise component. The signal component contains information about a desired parameter. If this signal component could be perfectly isolated, the desired parameter could be measured with deterministic accuracy. The noise component represents the corrupting influence on the observations and inhibits accurate measurement. Errors are inherent in any estimation technique. The task is further complicated by the sometimes unpredictable time varying nature of the desired parameter. The objective is to devise an estimation algorithm which computes periodic estimates in real time of the desired parameter with minimum errors.

A probabilistic characterization is the accepted mathematical model for this problem. The unwanted noise component is usually assumed to be additive Gaussian noise (observation model). The desired parameter is usually assumed to be a Markov random process (dynamic model). With a probabilistic characterization, the theory of applied statistical inference can be used. Optimum estimates are mathematically defined. Two such optimum estimates are MMSE (Minimum Mean Square Error) and MAP (Maximum A Posteriori) [1].

The posterior probability density function (PDF) is defined as the PDF of the desired parameter conditioned on knowing all available observed data and any other useful information. In a real-time system, the posterior PDF envolves over time as new observed data become available. The posterior PDF is required for the computation of the above or any other optimum estimates. The MMSE and MAP estimates are the mean and mode (peak) of the posterior PDF. Furthermore, the posterior PDF contains all the information about the desired parameter. In light of these facts, the preferred design rationale is to propagate the posterior PDF in time if at all feasible. This is the philosophy adopted here. McGarty [2], Bryson and Ho [3], Snyder [4], Lee [5], DeGroot [6], and Box and Tiao [7] are strongly recommended for a fuller discussion and development. They provide the necessary background material for this paper.

Consider the special case of linear Gaussian observation and dynamic models. For this case, a linear Kalman filter is both an analytical solution and a practical implementation for the propagation of the posterior PDF. The posterior PDF is known to be Gaussian so that the mean and covariance matrix are modified over time. Unfortunately, this case is a limited subset of real problems. An analytical solution, much less a practical implementation of the analytical solution, for the general nonlinear problem is impossible. However, this does not lessen the attractiveness of this approach.

In summary, the current state of the art for the general nonlinear tracking problem is that the mathematical models, objective (including optimization criteria), and technical approach have been identified. Specifically, the technical approach is to compute, or propagate, the posterior PDF of the desired parameter over time as a function of the observed data and mathematical models. Any desired output is computed from the posterior PDF. The remaining difficulty, a practical implementation of the technical approach, is addressed here and a solution is offered.

The proposed solution is to let both time and parameter space be discrete in the mathematical models. In [2], the development of a solution for general

Manuscript received September 19, 1986; revised February 26 and May 28, 1987.

Authors' addresses: Bethel, IBM FSD, 9500 Godwin Dr., 400/041, Manassas, VA 22110; Rahikka, Naval Surface Weapons Center, Silver Springs, MD 20910.

Reprinted from IEEE Trans. Aerosp. Electron. Syst., vol. 23, no. 6, pp. 718–725, November 1987.

continuous mathematical models is pursued and results in an unsolvable nonlinear partial-differential integral equation. For discrete mathematical models, a closed-form solution for the propagation of the posterior PDF is not found in a mathematical sense. Instead, a recursive numerical solution is found, which can be implemented on high speed digital computers. This solution is presented here.

Hatsell and Nolte [8, 9] have arrived at a similar approach from the investigation of the detection problem. That is, the propagation of the posterior PDF is required for the computation of the optimum detection statistic, the true (not generalized) likelihood ratio. Thus, the same algorithm is simultaneously the optimum estimator and detector. The use of the presented algorithm as a detector is under investigation.

This approach was first applied to frequency tracking [10]. In this paper, the selected problem to demonstrate the proposed solution is time delay tracking, which, like frequency tracking, is of considerable interest. Although this paper concentrates on the time delay tracking problem, much of the discussion generalizes to other problems. An expanded version of this paper is contained in [11]. Extensions of the results in this paper to other cases such as multisignal problems, higher dimensional problems, and second-order Markov processes are found in [12].

Meyr and Spies [13] have given an overview of the time delay tracking problem and a discussion of two methods used to date: 1) peak picker on the output of a cross-correlation, and 2) delay-lock tracker which is similar in concept to a phase-lock loop for frequency tracking.

The primary difficulty with the peak picking method is the tendency to select "false peaks". Ianniello [14] has also analyzed this problem. The primary difficulty with the delay-lock tracker is its threshold or "out-of-lock" behavior. Meyr and Spies offer an alternative approach because of these difficulties. Their approach uses a peak picking front end to generate inputs to a linear Kalman filter with a data rejection mechanism to detect and eliminate "false peaks." Kirlin, Moore, and Kubichek [15] also investigate the use of a linear Kalman filter as a post processor. Compared with traditional nonlinear trackers, the presented tracker is robust. It does not suffer from these two problems, which are shared by other nonlinear trackers. The following sections describe in detail the theory and implementation.

II. OBSERVATION MODEL

The assumed observation model for two channels is

$$r_1(t) = s(t) + n_1(t)$$

$$r_2(t) = s(t - d(t)) + n_2(t). \quad (1)$$

The observed analog time series data in (1) is sampled faster than the Nyquist rate. The observed sampled time series data is processed in contiguous blocks of length T seconds each as depicted in Fig. 1. There are K time series data points per block for each channel. Within a particular block j

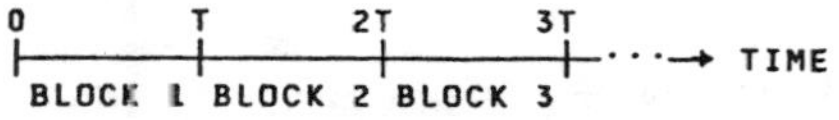

0 T 2T 3T ... → TIME
BLOCK 1 BLOCK 2 BLOCK 3

Fig. 1. Time series data.

$$r_{1j}(t_j(k)) = s_j(t_j(k)) + n_{1j}(t_j(k))$$

$$r_{2j}(t_j(k)) = s_j(t_j(k) - d_j) + n_{2j}(t_j(k))$$

$$k = 1, \ldots, K \quad (2)$$

where

$r_{1j}(t_j(k)), r_{2j}(t_j(k))$ = observed sampled time series data for the two channels during block j.

$n_{1j}(t_j(k)), n_{2j}(t_j(k))$ = additive zero mean Gaussian noise during block j.

$s_j(t_j(k))$ = true signal in each channel during block j. Assumed zero mean Gaussian.

d_j = time delay (desired parameter) of the signal between the two channels during block j. Assumed constant during a block, but permitted to change over blocks subject to a stochastic dynamic model. T is selected according to expected time variations in $d(t)$ [13].

III. TECHNICAL APPROACH

The design goal is to find a computational procedure to determine the posterior PDF of the desired parameter d_j in real time. Rather than force any particular parametric density (i.e., Gaussian), d_j is assumed to be discrete. That is, d_j can take on one of M values. These values, or bins, are uniformly spaced with sufficient resolution over a finite selected range. The Bayesian identity is invoked.

$$p(d_j = d(m) \mid \mathbf{R}_j) = \frac{p(\mathbf{r}_j \mid d_j = d(m), \mathbf{R}_{j-1})\, p(d_j = d(m) \mid \mathbf{R}_{j-1})}{p(\mathbf{r}_j \mid \mathbf{R}_{j-1})},$$

$$m = 1, \ldots, M. \quad (3)$$

The parameters and terms of (3) are

1) The scalar $d(m)$ is one of M discrete time delays that d_j can assume.
2) The vector $\mathbf{r}_j$ is the observed time series data of both channels for block j.
3) The vector $\mathbf{R}_j$ is the set of all observed time series vectors for all blocks up to and including block j. $\mathbf{R}_j = \{\mathbf{r}_j, \mathbf{r}_{j-1}, \ldots\}$.
4) The term $p(d_j = d(m) \mid \mathbf{R}_j)$ is the posterior probability that d_j equals $d(m)$ given all observed data up to and including block j. These quantities constitute the required discrete posterior PDF.

5) The term $p(\mathbf{r}_j | d_j = d(m), \mathbf{R}_{j-1})$ is the conditional PDF of the observed data during block j given that $d(m)$ is the true time delay.
6) The term $p(d_j = d(m) | \mathbf{R}_{j-1})$ is the prior probability that d_j equals $d(m)$ given all observed data prior to block j.

The vector $\mathbf{R}_{j-1}$ is dropped from the conditional PDF under the assumption that the signal and noise are independent over blocks. Since $p(\mathbf{r}_j | \mathbf{R}_{j-1})$ is not a function of time delay, it can be represented as a normalization constant. Equation (3) can be written as

$$p(d_j = d(m) | \mathbf{R}_j) = C\, p(\mathbf{r}_j \,| d = d(m))\, p(d_j = d(m) | \mathbf{R}_{j-1}), \quad m = 1, \ldots, M. \quad (4)$$

In implementation, these discrete posterior probabilities are maintained in an M element array. The tracking algorithm consists of two procedures executed at the end of each block of data.

1) *Correction (Bayesian):* This procedure is a direct implementation of (4). The prior PDF is available in an M element array. It is known from the projection procedure at the previous block. For the first block, the prior PDF must be initialized, usually uniform. The conditional PDF is computed from the current block of data for each of the M values that the time delay can assume. The unnormalized desired posterior PDF is computed and stored in an M element array by a multiplication of the corresponding values of the prior and conditional PDFs. The posterior PDF is normalized by dividing each value by the sum of all values. Any desired outputs, such as the mean (true MMSE estimate) and standard deviation (true MMSE error), are computed from the posterior PDF.

2) *Projection:* The posterior PDF of the current block is modified to become the prior PDF of the next block. This procedure takes into account the change in time delay over blocks. The time delay dynamic model is assumed to be a first-order Markov random process.

Both of these procedures may be carried out in place so that only a single M element array is needed for storage. The M element array of probabilities is the state of the time delay tracker. On the basis of the first-order Markov assumption of dynamics and the assumption of independent signal and noise over blocks, this PDF is a sufficient statistic of history. The implementation of the tracker is shown in Fig. 2. These two procedures are identical in concept to a linear Kalman filter but with the Gaussian restriction on the prior and posterior PDFs removed.

IV. COMPUTATION OF THE CONDITIONAL PDF

The conditional PDF $p(\mathbf{r}_j | d_j = d(m))$ is required for the correction procedure in (4). The conditional PDF is the likelihood function required in the computation of the maximum likelihood estimate under the assumption of a

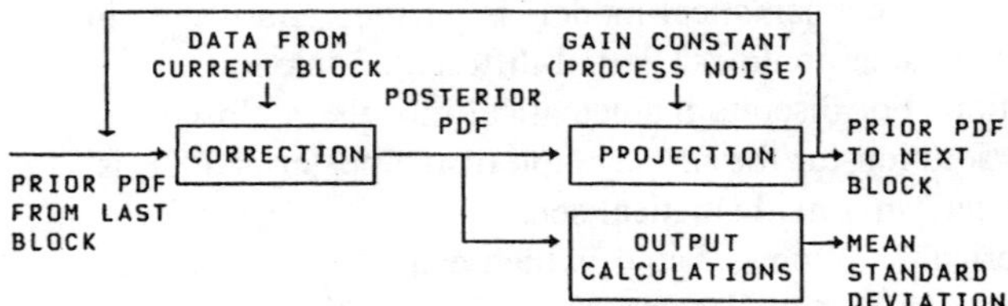

Fig. 2. Tracker block diagram.

constant time delay over a finite time interval. This is the same assumption as in this paper. As such, it has been the subject of many previous works. Knapp and Carter [16], Hassab and Boucher [17], and Haas and Lindquist [18] are cited as examples. This paper follows the work of Bradley and Kirlin [19].

$$p(\mathbf{r}_j | d_j = d(m)) = C \exp(\lambda), \qquad m = 1, \ldots, M \qquad (5)$$

where

$$\lambda = 2T \int_{-\infty}^{+\infty} X_{1j}^*(f)\, X_{2j}(f)\, \psi(f) \exp(j2\pi f d(m))\, df$$

$X_{1j}(f), X_{2j}(f)$ = Fourier transform of received time series data in each channel for block j.

$\psi(f)$ = Filter based on signal and noise statistics. Signal and noise terms assumed mutually independent.

$$\psi(f) = \frac{S_{ss}(f)}{2S_{ss}(f)S_{nn}(f) + S_{nn}^2(f)}$$

$S_{ss}(f), S_{nn}(f)$ = Signal, noise power spectrum.

Under the assumption that the signal and noise power spectrums are flat over a known bandwidth and the time series data are independently sampled, (5) reduces to

$$p(\mathbf{r}_j | d_j = d(m)) = C \exp(\lambda), \qquad m = 1, \ldots, M \qquad (6)$$

where

$$\lambda = \frac{S}{2SN + N^2} \sum_{k=1}^{K} r_{1j}(t_j(k))\, r_{2j}(t_j(k) - d(m))$$

S = Signal power, or variance.

$= E[s_j^2(t_j(k))]$

N = Noise power, or variance.

$= E[n_{1j}^2(t_j(k))] = E[n_{2j}^2(t_j(k))]$.

The commonly used additive Gaussian noise observation model ((2)) is assumed in the derivation of (5) and (6). The presented tracker does not depend on this model. The correction procedure requires only a computable function of the observed data in a block for the conditional PDF. A useful extension would be the investigation of other observation models.

V. PROJECTION PROCEDURE

Time delay dynamics are modeled as a discrete-discrete (discrete in both time and parameter space) first-order Markov random process [20]. The general model of a first-order Markov projection is

$$\mathbf{p}_{j+1} = P\,\mathbf{p}_j \tag{7}$$

where $\mathbf{p}_j$ is the input $M \times 1$ vector of the discrete PDF of the state parameter at time j, P is the $M \times M$ state transition matrix which is assumed constant over time, and $\mathbf{p}_{j+1}$ is the output $M \times 1$ vector of the discrete PDF of the state parameter at time $j + 1$.

The element $p(m)$ of the vectors $\mathbf{p}_j$ and $\mathbf{p}_{j+1}$ is the probability of the state parameter assuming the mth state. The element $p(m,k)$ of the state transition matrix P is the probability of the state parameter assuming state m at time $j + 1$ given that it is in state k at time j. To ensure that $\mathbf{p}_{j+1}$ is a valid PDF, the state transition matrix must satisfy the necessary and sufficient conditions

$$p(m,k) \geq 0, \qquad m,k = 1, \ldots, M \tag{8}$$

$$\sum_{m=1}^{M} p(m,k) = 1, \qquad k = 1, \ldots, M. \tag{9}$$

A desirable property of the state transition matrix is that its stationary PDF be uniform. The necessary and sufficient condition for this is

$$\sum_{k=1}^{M} p(m,k) = 1, \qquad m = 1, \ldots, M. \tag{10}$$

For this application, the state parameter is time delay. The times are the previously defined blocks. The states are the constant values, or bins, that the time delay is allowed to assume during a block. The input PDF $\mathbf{p}_j$ is the known posterior PDF from the correction procedure at block j. The output PDF $\mathbf{p}_{j+1}$ is the desired prior PDF required for the correction procedure at block $j + 1$. The state transition matrix P must be assumed.

With the proper choice of the state transition matrix, the projection procedure could be directly implemented as (7). However, (7) is inefficient in terms of computation and storage requirements. Also, a rationale must be developed for selecting the state transition matrix. Equations (11)–(17) are an efficient implementation of the projection procedure that can be easily tuned via a gain constant. The operations are executed sequentially and in-place. There is an option in (15) depending on whether a linear or circular projection is desired. A circular projection is required, as an example, for bearing or phase shift tracking applications. A linear projection is required for time delay and frequency tracking applications.

$$p(1) \leftarrow \alpha p(1), \qquad \text{Boundary Condition} \tag{11}$$

$$\{p(m) \leftarrow p(m-1) + \alpha(p(m) - p(m-1))\},$$
$$m = 2, \ldots, M. \qquad \text{Forwards Filter} \tag{12}$$

$$p(M) \leftarrow \frac{1}{2-\alpha}\, p(M), \qquad \text{Boundary Condition} \tag{13}$$

$$\{p(m) \leftarrow p(m+1) + \alpha(p(m) - p(m+1))\},$$
$$m = M-1, \ldots, 1. \qquad \text{Backwards Filter} \tag{14}$$

$$r \leftarrow p(1); \quad s \leftarrow p(M) \qquad \text{Linear}$$
$$r \leftarrow p(M); \quad s \leftarrow p(1) \qquad \text{Circular,}$$
$$\text{Boundary Condition} \tag{15}$$

$$r \leftarrow \frac{r}{1-(1-\alpha)^M}; \quad s \leftarrow \frac{s}{1-(1-\alpha)^M},$$
$$\text{Finite Range Compensation} \tag{16}$$

$$\{p(m) \leftarrow p(m) + r(1-\alpha)^m\}$$
$$\{p(M+1-m) \leftarrow p(M+1-m) + s(1-\alpha)^m\},$$
$$m = 1, \ldots, M. \qquad \text{Fold} \tag{17}$$

where α is the selected gain constant ($0 < \alpha < 1$). Equations (11)–(17) essentially follow the random walk dynamic model in (18), which is a special case of (7).

$$d_{j+1} = d_j + w_j \tag{18}$$

where w_j is the independent additive process noise.

The process noise variance σ^2 is

$$\sigma^2 = E[w_j^2] = E[(d_{j+1} - d_j)^2] \qquad \text{bins}^2$$

$$= \sum_{m=-\infty}^{+\infty} m^2 q(m) \qquad \text{bins}^2$$

$q(m)$ is the zero mean discrete PDF of w_j.

$$= \frac{2(1-\alpha)}{\alpha^2} \qquad \text{bins}^2 \tag{19}$$

The rationale and derivation of (11)–(17) are contained in [11]. The projection procedure is a mathematical identity of (7). In particular, there exists a unique effective state transition matrix. The kth column of the effective state transition matrix is obtained by setting the input PDF to all zeros except for a one in the kth state and executing (11)–(17) for a selected value of α ($0 < \alpha < 1$). The effective state transition matrix satisfies (8), (9), and (10). Equation (19) is used to relate process noise variance σ^2 to gain constant α. The gain constant α is selected on the basis of expected time delay dynamics, not signal-to-noise ratio (SNR) or tracker time constant. The choice of α determines the effective state transition matrix. The projection procedure broadens the PDF (with a uniform PDF in the limit), indicating increased uncertainty. Neglecting finite range effects, the variance of the prior PDF is the sum of the variances of the posterior PDF and noise PDF ($q(m)$). The projection procedure also induces a fading memory into the tracker. For sufficient separation in time, the current PDF is independent of a PDF in the past, including initialization.

VI. TEST RESULTS

The following simulated scenario is selected to demonstrate performance. The scenario consists of seven segments.

1) Noise only (no signal).
2) The signal is turned on. A constant time delay is applied.
3) Discontinuous jump (step function) to a different but also constant time delay.
4) A time delay rate (ramp function) is applied.
5) A change in time delay rate is applied (same magnitude but opposite sign).
6) The time delay rate is removed (constant time delay).
7) The signal is turned off (noise only).

This scenario is run under the following conditions.

1) The signal and noise time series samples in the two channels are Monte Carlo zero mean Gaussian random variables. The sampling rate is 4096 Hz. The signal and noise time series samples are mutually independent of each other and independent over time.

2) The SNR per channel (see below) is -14 dB in all signal-plus-noise segments. The magnitude of the time delay rate in segments 4 and 5 is 0.5 ms/s.

3) The time delay range is -0.12085 s to $+0.12085$ s. There are 991 time delay bins with a resolution of 0.24414 ms per bin, which is the time series sampling interval (1/4096 Hz.). The block length T is 1.0 s.

4) The process noise standard deviation is 2.048 bins. This corresponds to a gain constant α of 0.49212 ((19)) in the projection procedure ((11)–(17)). The process noise standard deviation matches the known rate in segments 4 and 5.

5) The conditional PDF in the correction procedure is computed as (6). The constants S and N are known. The signal power in the signal-plus-noise segments is S. The noise power in all segments is N. SNR $= 10 \log_{10} S/N$.

Fig. 3 is the raw cross-correlation required in (6). The magnitude of the cross-correlation is intensity modulated. Fig. 4 is the ground truth. The seven segments can be visually identified on these two figures. The tracker MMSE estimates (mean of the posterior PDF) are shown in Fig. 5. Figs. 6–10 are the tracker errors (solid line) with respect to ground truth. The plus and minus tracker predicted errors (standard deviation of the posterior PDF) are shown on the same figures. Fig. 11 is the maximum likelihood estimates for the same test data. In particular, the time delay bin with the maximum cross-energy is selected and plotted for each block. The number of correct selections is shown in Table I. As expected, the percentage of correct selections decreases for the segments with a time delay rate because of smearing [13]. The following observations are offered.

1) These test results demonstrate the operational characteristics in Section VII. In particular, attention is drawn to the following:

Fig. 3. Raw cross correlation.

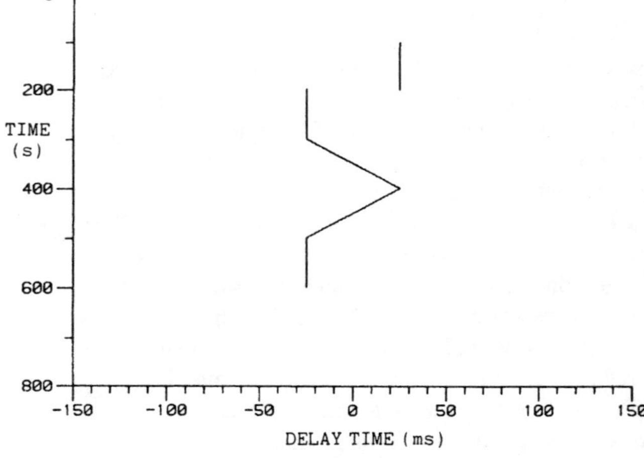

Fig. 4. Ground truth.

a) The convergence in segment 2 from a noise-only condition with no initialization or acquisition. This demonstrates the claims that neither initialization nor an acquisition stage are required.

b) The response to the step and ramp functions in time delay with no "loss-of-lock" usually associated with nonlinear trackers. The step function would be well outside the "tracking region" of a delay-lock tracker. A delay-lock tracker would have to maintain the ramp functions in this "tracking region" not to fall "out-of-lock". This is by no means assured. This ability is a

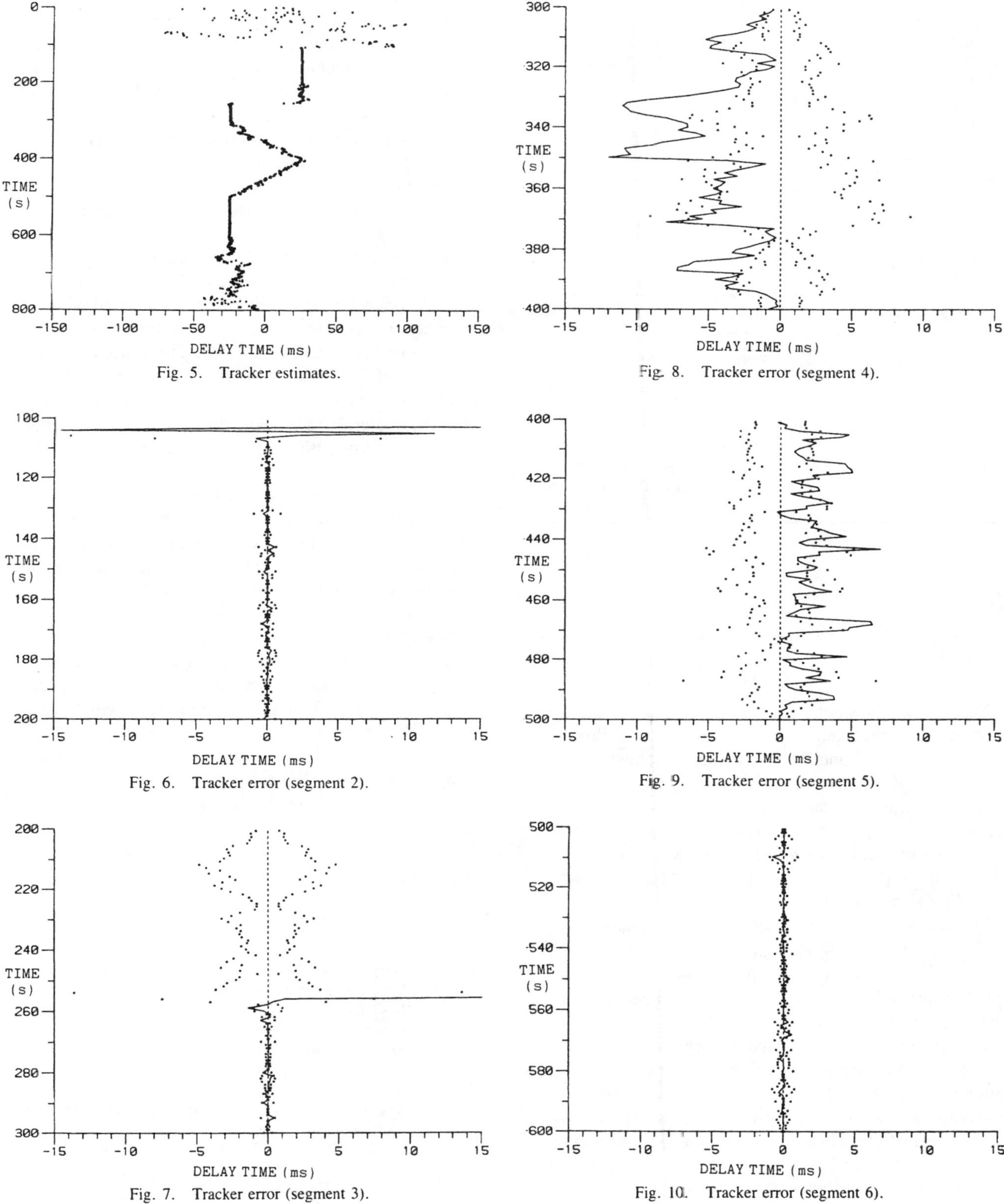

Fig. 5. Tracker estimates.

Fig. 6. Tracker error (segment 2).

Fig. 7. Tracker error (segment 3).

Fig. 8. Tracker error (segment 4).

Fig. 9. Tracker error (segment 5).

Fig. 10. Tracker error (segment 6).

function of SNR and the severity of the dynamics [13]. There are no such restrictions on the presented tracker. This response also demonstrates the claims that no reinitialization or reacquisition logic is required.

c) The successful performance of the tracker with no threshold effects despite a high probability of "false peaks" [13, 14]. The percentages in Table I are compared with the 80 percent correct selection rate required for successful performance of the tracker in [13]. There is no peak picking procedure in the presented tracker. Therefore, its performance does not depend on the ability of such a procedure to select the correct peak.

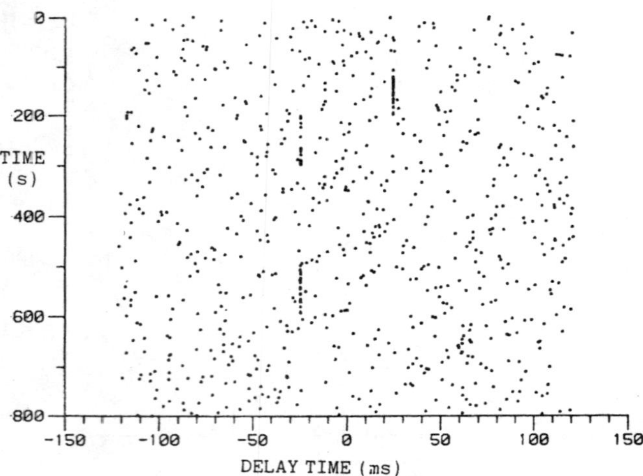

Fig. 11. Maximum likelihood estimates.

TABLE I
Maximum Likelihood Estimator Performance

Segment Number	Number of Correct Selections (Out of 100)
2	26
3	15
4	6
5	8
6	19

These results are obtained with no external monitoring or interference, thus demonstrating the appeal of this approach in an environment where tracker robustness, or reliability, is of utmost concern.

2) The lag in segments 4 and 5 and lack of an overshoot between segments are a result of the first-order (as opposed to second- or higher order) Markov assumption. In [12], results are presented for a second-order Markov assumption.

3) The amount of time to converge in segments 2 and 3 and the amount of lag in segments 4 and 5 are a function of SNR for a fixed gain constant α in the projection procedure. Thus, the tracker is inherently adaptive in response time.

4) The predicted error (standard deviation of the posterior PDF) is consistent with the actual error quantitatively and qualitatively. That is, the magnitude of the actual error is correctly predicted, and there is a high positive correlation between the standard deviation and actual error. The standard deviation of the posterior PDF is the identical MMSE error. Many potential applications require an accurate indication of estimate quality.

VII. SUMMARY AND CONCLUSIONS

The presented time delay tracker is a direct consequence of the assumed mathematical models for time series observations and time delay dynamics and the design goal of propagating the time delay posterior PDF. Both time and parameter space must be discrete in the mathematical models to achieve this. In the realization of the time delay tracker, no compromises are made, such as linear approximations in an extended Kalman filter. The propagation procedures (correction and projection) are exact on the basis of the discrete mathematical models. The strict adherence to this philosophy [1–7] makes the benefits below possible. In the development of the projection procedure, attention has been given to an efficient implementation.

There are two benefits from this design approach.

1) *Error Performance:* The true MMSE estimate and the true (not a bound or an approximation) minimum root mean square error are computed over time based on the modeled mathematical conditions. These are the mean and standard deviation of the posterior PDF. The true MAP and other optimum estimates may also be computed from the posterior PDF. This error performance is not subject to any qualification, such as linear approximations.

2) *Operational Characteristics:* This approach is robust in that 1) no initialization is required, 2) no acquisition stage is required, 3) there are no threshold effects or ''out-of-lock'' condition, 4) no reinitialization or reacquisition logic is required, and 5) the tracker can be used to implement a detector [8, 9].

These benefits are without precedent among realizable nonlinear trackers. The operational characteristics are more important than the error performance. No operator attention is required. Algorithm reliability will become increasingly critical as the degree of automation increases in future systems. The pursuit of this approach for more complex and higher dimensional problems [12] would be an ideal application for the new generations of high performance computers.

REFERENCES

[1] Van Trees, H.L. (1968)
Detection, Estimation, and Modulation Theory, Part I.
New York: John Wiley and Sons, Inc., 1968.

[2] McGarty, T.P. (1974)
Stochastic Systems and State Estimation.
New York: Wiley, 1974.

[3] Bryson, A.E., Jr., and Ho, Yu-Chi (1975)
Applied Optimal Control.
New York: Halsted Press, 1975, pp. 296–389.

[4] Snyder, D.L. (1969)
The state-variable approach to continuous estimation with applications to analog communication theory.
Resarch Monograph No. 51, Cambridge, Mass.: MIT Press, 1969.

[5] Lee, R.C.K. (1964)
Optimum Estimation, Identification, and Control.
Research monograph 28, M.I.T. Press, Cambridge, Mass., 1964.

[6] DeGroot, M.H. (1986)
Probability and Statistics (2nd ed.)
Reading, Mass.: Addison-Wesley, 1986.

[7] Box, G.E.P., and Tiao, G.C. (1973)
Bayesian Inference in Statistical Analysis.
Reading, Mass.: Addison-Wesley, 1973.

[8] Hatsell, C.P., and Nolte, L.W. (1974)
Optimal detection of a signal with time-varying carrier phase.
IEEE Transactions on Aerospace and Electronic Systems, AES-10, 6 (Nov. 1974), 788–794.

[9] Hatsell, C.P. (1970)
Optimum tracking detectors.
Technical Report 5, Adaptive Signal Detection Laboratory, Dept. of Electrical Engineering, Duke University, Durham, N.C., Mar. 1970.

[10] Bethel, R.E., and Gauss, J.A. (1981)
A Bayesian approach to frequency line tracking.
IEEE Eascon 1981 Proceedings, pp. 286–290.

[11] Bethel, R.E. (1986)
A probability density function time delay tracker.
IBM FSD technical report TR63.0143, Manassas, Va., Jan. 1986.

[12] Bethel, R.E. (1986)
PDF (probability density function) line and target tracking.
IBM FSD technical report TR63.0144, Manassas, Va., Jan. 1986.

[13] Meyr, H., and Spies, G. (1984)
The structure and performance of estimators for real-time estimation of randomly varying time delay.
IEEE Transactions on Acoustics, Speech, and Signal Processing, ASSP-32, 1 (Feb. 1984), 81–94.

[14] Ianniello, J.P. (1982)
Time delay estimation via cross-correlation in the presence of large estimation errors.
IEEE Transactions on Acoustics, Speech, and Signal Processing, ASSP-30, 6 (Dec. 1982), 988–1003.

[15] Kirlin, R.L., Moore, D.F., and Kubichek, R.F. (1981)
Improvement of delay measurements from sonar arrays via sequential state estimation.
IEEE Transactions on Acoustics, Speech, and Signal Processing, ASSP-29, 3 (June 1981), 514–519.

[16] Knapp, C.H., and Carter, G.C. (1976)
The generalized correlation method for estimation of time delay.
IEEE Transactions on Acoustics, Speech, and Signal Processing, ASSP-24, 4 (Aug. 1976), 320–327.

[17] Hassab, J.C., and Boucher, R.E. (1979)
Optimum estimation of time delay by a generalized correlator.
IEEE Transactions on Acoustics, Speech, and Signal Processing, ASSP-27, 4 (Aug. 1979), 373–380.

[18] Haas, W.H., and Lindquist, C.S. (1981)
A synthesis of frequency domain filters for time delay estimation.
IEEE Transactions on Acoustics, Speech, and Signal Processing, ASSP-29, 3 (June 1981), 540–548.

[19] Bradley, J.N., and Kirlin, R.L. (1984)
Delay estimation by expected value.
IEEE Transactions on Acoustics, Speech, and Signal Processing, ASSP-32, 1 (Feb. 1984), 19–27.

[20] Larson, H.J., and Shubert, B.O. (1979)
Probabilistic Models In Engineering Sciences, Vol. II. New York: Wiley, 1979.

PERFORMANCE OF THE DESKEWED SHORT-TIME CORRELATOR

John W. Betz

RCA Automated Systems
Burlington, Massachusetts

and

Department of Electrical Engineering
Northeastern University
Boston, Massachusetts

ABSTRACT

A technique is described for time-delay estimation of linearly varying time delay using coherent processing of short-time correlograms. Results of a computer simulation of this technique are compared to analytical predictions and theoretical bounds, in terms of output signal/noise ratio and estimation accuracy. The deskewed short-time correlator is seen to provide near-optimal performance, implementational simplicity, and little sensitivity to the number of short time segments into which the data is segmented.

INTRODUCTION

Signal detection and estimation of differential time delay (DTD) between two broadband signals in broadband noise is degraded when the signal exhibits relative time companding (RTC). A simple model of the two received waveforms, assuming linear RTC, is given below [1, 2]:

$$r_1(t) = s(t) + n_1(t)$$
$$r_2(t) = s((1+q)t+D) + n_2(t) \qquad (1)$$

where $s(t)$, $n_1(t)$ and $n_2(t)$ are assumed to be independent stationary Gaussian random processes, with corresponding power spectral densities $G_s(f)$, $G_1(f)$, $G_2(f)$, and standard deviations σ_s^2, σ_1^2, σ_2^2. In (1), D is the DTD and q the RTC introduced by nature, and it is assumed that $q \ll 1$.

When signal and noise are not white, appropriate filtering of the received waveforms improves the detection of s(t) and the estimation of D and q [3, 4, 5].

The present paper is concerned with the mechanism for estimating and compensating for RTC and DTD, and thus assumes that the $r_i(t)$ in (1) are the outputs of selected filters. Crosscorrelation of $r_1(t)$ and $r_2(t)$ is then generalized crosscorrelation.

Given these conventions, the time-companding correlator (TCC) [6] is a maximum-likelihood (ML) estimator of D and q. The TCC operation is expressed by

$$C_T(\tau, \alpha) = \frac{1}{T} \int_{-\frac{T}{2}}^{\frac{T}{2}} r_1(t) r_2(t/(1+\alpha)+\tau)dt \qquad (2)$$

where τ and α are DTD and RTC inserted by the TCC to compensate for D and q inserted by nature. With no prior knowledge of D or q, a broadband passive ambiguity surface (PAS) can be formed by evaluating (2) over a range of τ and α values. The coordinates of the PAS maximum correspond to ML estimates of D and q under assumptions of large correlator time-bandwidth product.

Direct implementation of (2) to compute a PAS is undesirable because the entire crosscorrelation operation must be repeated for each value of α, a computationally intensive procedure.

The deskewed short-time correlator (DSTC) provides an approximation to (2) while requiring fewer arithmetic operations to compute a PAS [7,8,9].

The DSTC process is defined by

$$C_D(\tau, \alpha) = \frac{1}{N_T} \sum_{i=1}^{N_T} c_i(\tau + \gamma_i (\tfrac{\alpha}{1+\alpha})) \qquad (3a)$$

where the short-time correlograms are defined by

$$c_i(\tau) = \frac{N_T}{T} \int_{\ell_i}^{u_i} r_i(t) \ r_2(t-\tau)dt \qquad (3b)$$

and are computed without RTC compensation. The upper and lower limits in (3b) are given by

$$u_i = \frac{-T}{2} + iT/N_T$$

$$\ell_i = u_i - T/N_T,$$

the deskewing term is $\gamma_i(\frac{\alpha}{1+\alpha})$ with α the RTC estimate and

$$\gamma_i = (\ell_i + u_i)/2,$$

Reprinted from *Proc. ICASSP '84*, vol. 1, pp. 15.5/1–4, March 1984.

and N_T is the number of short-time correlograms into which the time aperture T is segmented.

Implementing (3) involves computing the short-time correlograms only once, then deskewing and summing the $c_i(\tau)$. In order to compute a PAS, only the deskewing and summing need be repeated for each value of α. This simplification in computational complexity facilitates implementation of the DSTC.

The DSTC provides a piecewise constant approximation to the linear delay versus time trend in equation (1), with improvement in the approximation as N_T increases. The DSTC has been shown analytically to provide more than an order of magnitude reduction in the number of arithmetic operations, compared to the TCC, when many values of RTC must be searched [7].

The following sections of this paper compare analytical expressions for DSTC performance with results obtained using a computer simulation of the DSTC.

OUTPUT SNR

The output signal/noise ratio (SNR) of the DSTC is defined as the mean squared of the PAS divided by the variance of the PAS, at $\tau = D$ [7]. A useful standard with which to compare DSTC output SNR is the output SNR of a correlator without RTC compensation, while an upper bound on output SNR is that for the TCC with $\alpha = q$ (equivalently, the DSTC output SNR with $\alpha = q = 0$).

The output SNR without RTC compensation is given by

$$SNR_1 = \frac{\left[\int_{-\infty}^{\infty} G_s(f)\,\mathrm{sinc}[\pi f q T]\,df\right]^2}{\iint_{-\infty}^{\infty} B(f_1, f_2)\,\mathrm{sinc}^2[\pi(f_1+f_2)T]\,df_1 df_2} \tag{4}$$

where

$$B(f_1, f_2) \triangleq G_s(f_1)G_2(f_2) + G_1(f_1)G_s(f_2) + G_1(f_1)G_2(f_2)$$

and $\mathrm{sinc}\,[x] \triangleq \sin[x]/x$. The upper bound on output SNR is

$$SNR_2 = \frac{\left[\int_{-\infty}^{\infty} G_s(f)\,df\right]^2}{\iint_{-\infty}^{\infty} B(f_1,f_2)\,\mathrm{sinc}^2[\pi(f_1+f_2)T]\,df_1 df_2} \tag{5}$$

The DSTC output SNR when $\alpha = q$ is given by [7]

$$SNR_D = \frac{\left[\int_{-\infty}^{\infty} G_s(f)\,\mathrm{sinc}\,[\pi f q T/N_T]\,df\right]^2}{\left(\frac{1}{N_T}\right)\iint_{-\infty}^{\infty} B(f_1,f_2)\,\mathrm{sinc}^2[\pi(f_1+f_2)T/N_T]\,df_1 df_2} \tag{6}$$

For lowpass white signal and noise with one-sided bandwidth W, equations (4), (5) and (6) reduce to

$$SNR_1 = 2WT\ \psi^2\ [Si(\pi qTW)/(\pi qTW)]^2 \tag{7}$$

$$SNR_2 = 2WT\ \psi^2 \tag{8}$$

$$SNR_D = 2WT\ \psi^2\ [Si(\pi qTW/N_T)/(\pi qTW/N_T)]^2 \tag{9}$$

where $\psi^2 \triangleq \sigma_s^4[2\sigma_s^2 + \sigma_s^2(\sigma_1^2 + \sigma_2^2) + \sigma_1^2\sigma_2^2]^{-1}$,

$Si[x] \triangleq \int_0^x \mathrm{sinc}\,[y]\,dy$, and it is assumed that

$TW >> 1$ and $N_T \leq \frac{TW}{3.5}$. (This latter assumption ensures that adjacent correlograms are uncorrelated [7].) It is seen in (7) that the loss in output SNR with no compensation is a function of the dimensionless term $|qTW|$.

A computer simulation of the DSTC was developed to confirm analytical expressions for the case of lowpass white signals and noise [10]. Time series were generated and signals were time companded using numerical interpolation. Short-time correlograms were computed, deskewed using numerical interpolation and summed.

Define compensation gain $G = SNR_D/SNR_1$, to provide a measure of the benefit in using the DSTC over using no compensation. Figure 1 shows compensation gain computed analytically, and measured using the simulation. The simulation used TW = 100, input SNR of $\psi^2 = 0.2$, i.e., $\sigma_s^2 = \sigma_1^2 = \sigma_2^2$, and 25 waveform realizations per data point. The simulation and theoretical results agree closely. For large N_T, the DSTC compensates nearly optimally. Even for small values of N_T, substantial compensation gain is achieved.

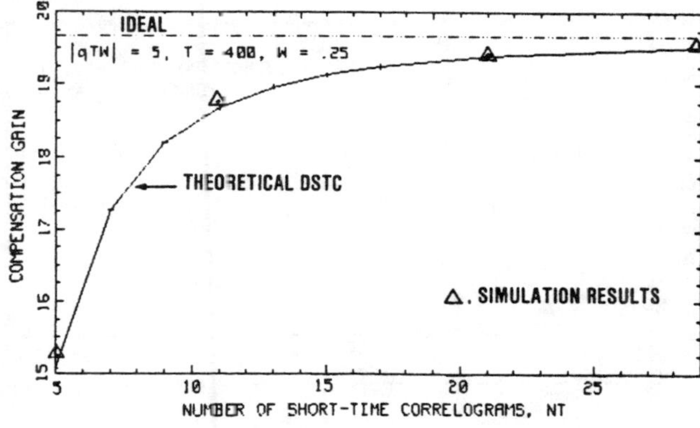

Figure 1. Compensation Gain with DSTC

LOCAL ESTIMATION ERRORS

The Cramer-Rao lower bounds (CRLB) on the error estimating DTD and RTC have been derived and shown to be statistically uncorrelated [11, 12]. For the case of interest, with lowpass white signals and noise, the expressions are [13, 7]

$$\text{CRLB}(D) = 3\,[8\,(\pi\rho W)^2\,(TW)]^{-1} \qquad (10)$$

$$\text{CRLB}(\hat{q}) = 9\,[2\,(\rho\pi)^2\,(TW)^3]^{-1}$$

$$\text{for } \rho^2 \triangleq \sigma_s^4\,[\sigma_s^2(\sigma_1^2 + \sigma_2^2) + \sigma_1^2\,\sigma_2^2]^{-1}$$

Derivation of the corresponding variances for the DSTC under these assumptions yields [7]

$$\sigma_D^2 = \Omega^2[\pi qTW/N_T]\,[24(\rho\pi W)^2(TW)]^{-1} \qquad (12)$$

$$\sigma_q^2 = \Omega^2[\,qTW/N_T]\left(\frac{N_T^2}{N_T^2-1}\right) 2(\rho\pi)^2(TW)^3{}^{-1} \qquad (13)$$

$$\text{where } \Omega^2(x) \triangleq x^6/[\sin(x) - x\cos(x)]^2$$

The simulation evaluation of estimation accuracy used a time-bandwidth product of TW = 500, ρ^2 = .0083, i.e., 10 $\sigma_s^2 = \sigma_1^2 = \sigma_2^2$ and 100 waveform realizations. Numerical interpolation of the PAS was used to estimate the coordinates of the peak to a higher accuracy than the discrete (τ, a) grid provided.

The parameters were selected to ensure that the probability of global errors was very low, so that the CRLB applies. (This topic is discussed in the next section.)

Comparison of simulation results with equations (10), (11), (12), and (13) are presented in Figures 2 and 3.

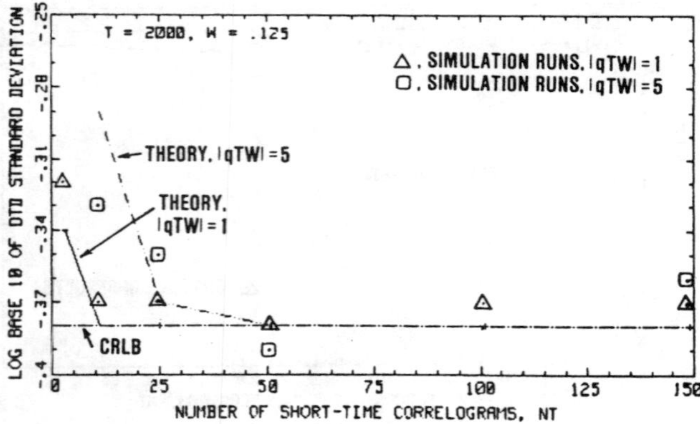

Figure 2. DTD Estimation Accuracy with DSTC

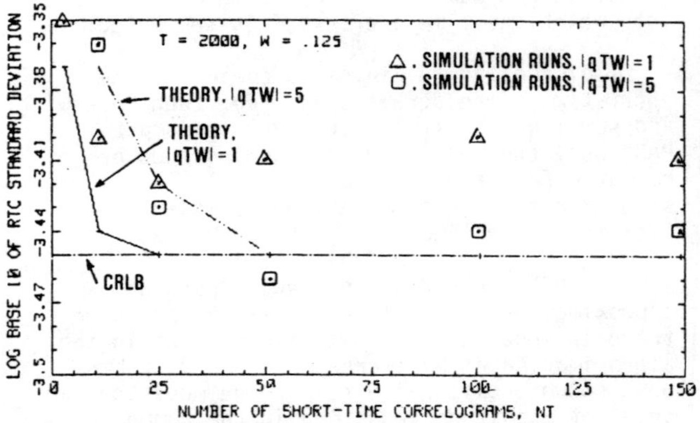

Figure 3. RTC Estimation Accuracy with DSTC

Close agreement is observed between simulation and theory and near-optimal DSTC performance is shown.

GLOBAL ERRORS IN DTD ESTIMATION

A theory has been developed to account for global errors in DTD estimation which occur when a maximum caused by noise-only exceeds the maximum produced by the signal [14, 15, 16]. To this author's knowledge, this theory has not yet been extended to RTC estimation, so the discussion is limited to DTD estimation errors, with no RTC present.

The effect of global errors is expressed by the correlator performance estimate (CPE) [17]. The DSTC, being a coherent process, is expected to achieve the CPE.

To experimentally evaluate DSTC performance, simulation runs were made for parameter sets where global errors in DTD estimation occur with non-zero probability. Results are virtually identical for N_T = 1 and N_T = TW/5, as expected [10].

CONCLUSION

A computer simulation of the DSTC has been used to quantitively observe performance compared both to analytical predictions and to measures of ideal performance. For lowpass white waveforms, DSTC performance is seen to be near optimal. The simplicity of implementing the DSTC is furthered by the observation that most of the DSTC performance is retained when the time aperature is segmented into relatively few short-time segments. In fact, N_T = TW/10 is satisfactory for the cases examined.

It is concluded that the deskewed short-time correlator is a readily implementable technique for differential time delay estimation in the presence of relative time companding.

ACKNOWLEDGEMENT

Work was supported by the RCA Graduate Studies Program, and simulations were performed under NUSC Contract No. N00140-83-M-NN24. The author expresses his gratitude to Dr. G. C. Carter for his encouragement and support.

REFERENCES

1. Remley, W.R., "Correlation of Signals Having a Linear Delay, " J. Acoust. Soc. America, Vol. 35, No. 1 , pp 65-69, Jan. 1963.

2. Adams, W.B., et al, "Correlator Compensation Requirements for Passive Time Delay Estimation with Moving Source or Receivers," IEEE Trans. Acoust., Speech, Signal Processing, Vol. ASSP-28, No. 2, pp 158-168, 1980.

3. Knapp, C.H. and Carter, G.C., "The Generalized Correlation Method for Estimation of Time Delay', IEEE Trans. Acoust., Speech, Signal Processing, Vol. ASSP-24, No. 4, pp 320-327, August 1976.

4. Scarbrough, K., et al, "On the Simulation of a Class of Time Delay Estimation Algorithms", IEEE Trans. Acoust., Speech, Signal Processing, Vol. ASSP-29, No 3, June 1981.

5. Schwartz, S.C. and Hero, A., "A New Generalized Cross-Correlator", Report Number 11 by Information Sciences and Systems Laboratory, Princeton University, April 1983.

6. Knapp, C.H. and Carter, G.C., "Estimation of Time Delay in the Presence of Source or Receiver Motion", J. Acoust. Soc. America, Vol. 61, No. 6, pp 1545-1549, 1977.

7. Betz, J.W., "Comparison of the Deskewed Short-Time Correlator and the Maximum Likelihood Correlator", to appear in IEEE Trans. Acoust., Speech, Signal Processing.

8. Kuhn, J.P., et al, "Time Delay Detection and Estimation Techniques for Moving Sources", Proc. 24th Midwestern Symp. Circuits and Syst., June 29-30, 1981.

9. Johnson, G.W., et. al, "Time Delay Detection and Estimation Techniques for Moving Sources", Proc. 1983 Int. Conf. Acoust. Speech Signal Processing, pp. 583-586, 1983.

10. "Computer Simulation of the Deskewed Short-Time Correlator", Report Prepared by RCA for Naval Underwater Systems Center under Contract No. N00140-83-M-NN24, January 1984.

11. Carter, G.C., "Time Delay Estimation", NUSC Technical Report 5335, 9 April 1976.

12. Schultheis, P.M. and Weinstein, E., "Estimation of Differential Doppler Shifts", J. Acoust. Soc. America, Vol. 65, No. 5, pp 1412-1419, November 1979.

13. Quazi, A.H., "An Overview on the Time Delay Estimate in Active and Passive Systems for Target Localization", IEEE Trans. Acoust., Speech, Signal Processing, Vol. ASSP-29, No. 3, June 1981.

14. Ianniello, J.P., "Time Delay Estimation Via Cross-Correlation in the Presence of Larger Estimation Errors, IEEE Trans. Acoust., Speech, Signal Processing, Vol. 30, No. 6, pp 988-1003, December 1982.

15. Ianniello, J.P., Weinstein, E., and Weiss, A., "Comparison of Ziv-Zakai Lower Bound on Time Delay Estimation with Correlator Performance", Proc. 1983 Int. Conf. Acoust. Speech Signal Proc., pp. 875-878, 1983.

16. Weiss, A. and Weinstein, E., "Composite Bound on the Attainable Mean-Square Error in Passive Time-Delay from Ambiguity Prone Signals", IEEE Trans. Info. Theory (Corr.), Vol. 29, No. 3, pp. 478-484, June 1981.

17. Scarbrough, K., et al, "Performance Predictions for Coherent and Incoherent Processing Techniques of Time Delay Estimation", IEEE Trans. Acoust., Speech, Signal Processing, November 1983.

A New Method for Adaptive Time Delay Estimation for Non-Gaussian Signals

HSING-HSING CHIANG, MEMBER, IEEE, AND CHRYSOSTOMOS L. NIKIAS, SENIOR MEMBER, IEEE

Abstract—**A new adaptive scheme for time delay estimation is introduced for signal environments where the signal is non-Gaussian and the additive noise sources are spatially correlated Gaussian with unknown power spectrum characteristics. The new scheme is based on parametric modeling between two sensor measurements and employs higher-order statistics (third- or fourth-order) of the data. It is demonstrated by means of extensive simulations that the new adaptive scheme works well for both stationary and nonstationary cases. As expected, it outperforms the cross-correlation-based gradient method for time delay adaptation in spatially correlated Gaussian noises. The new scheme is also compared to the overdetermined recursive instrumental variable method, and is shown to exhibit substantially less computational complexity.**

I. Introduction

THE problem of estimating and tracking the time delay in a signal between two sensor measurements arises in many application fields such as sonar (active and passive), radar, biomedicine, geophysics, etc. [1]–[11].

One common method of estimating the time delay is to find the peak location of the cross-correlation function of the two sensor measurements. Various window functions (prefilters) have been suggested to better shape the cross-correlation function via convolution operation; namely, ROTH, SCOT, PHAT, Eckart, and Hannan–Tompson just to name a few [2], [6]. Parameter estimation approaches have also been proposed for time delay estimation based on autocorrelations and cross correlations such as the least-squares and Wiener filtering methods [3], [4].

The least-mean-square (LMS) adaptive filter was applied to determine the time delay in a signal between two split-array outputs [8], [9]. A recursive least-squares algorithm and a peak detection algorithm have also been suggested for adaptive time delay estimation [4], [10]. Another adaptive technique that has been proposed was based on the assumption that the cross-correlation function is unimodal or periodically unimodal [11]. Each one of the aforementioned techniques has certain advantages and limitations depending upon the nature of the signal and noise sources.

In those practical application problems where the signal can be regarded as non-Gaussian and the additive noises can be regarded as zero-mean, stationary, Gaussian processes, then the time delay can be estimated in higher-order spectrum domains such as the bispectrum or trispectrum [12]–[15]. Higher-order spectra (or polyspectra), which are defined in terms of higher-order cumulants, have been given a lot of attention lately due to their ability to preserve information of non-Gaussian stationary random processes. The definition of cumulants and higher-order spectra as well as their general motivations for digital signal processing problems can be found in the tutorial review paper [13]. The estimation of stationary delays from two sensor measurements using polyspectra (or cumulants) has been addressed in [12]–[15]. One of the fundamental properties of cumulants which will be explored in this paper is the fact that for Gaussian processes only, all cumulants of order greater than two are identically zero. Hence, if the additive noise sources are correlated Gaussian, their effect may be suppressed in higher-order cumulant domains. On the other hand, the cross-correlation techniques estimate the cross correlation of both the signal and noises. The third-order cumulant domain also suppresses non-Gaussian noises with symmetric probability density function [13]–[18].

The objective of this paper is to introduce a new adaptive time delay estimation method for signal environments where the signal is non-Gaussian and the additive noises are spatially correlated Gaussian with unknown correlation functions. The new method is developed by using a parametric bispectrum model and a gradient-type algorithm to adapt the time delay at each time instant for both stationary and nonstationary cases.

The organization of the paper is as follows. The problem definition and preliminaries are given in Section II. Section III describes the new adaptive time delay estimation method in the third-order moment domain (it is equivalent to third-order cumulant domain for zero-mean signals). The convergence properties are analyzed in Section IV. Section V discusses the computational complexity of the new algorithm. Extensive simulation results are presented in Section VI. Conclusions are drawn in Section VII.

Manuscript received June 2, 1988; revised March 13, 1989. This work was supported by the Office of Naval Research (ONR) under Contract N00014-88-K-0062.

H.-H. Chiang is with Biometrak Corporation, Cambridge, MA 02139.

C. L. Nikias is with the Communications and Digital Signal Processing (CDSP) Center for Research and Graduate Studies, Department of Electrical and Computer Engineering, Northeastern University, Boston, MA 02115.

IEEE Log Number 8932765.

Reprinted from *IEEE Trans. Acoust., Speech, Signal Processing*, vol. 38, no. 2, pp. 209–219, February 1990.

II. Problem Definition and Preliminaries

Let us assume that $\{X(n)\}$ and $\{Y(n)\}$ are two sensor measurements satisfying

$$X(n) = S(n) + W_1(n) \tag{1.1}$$

$$Y(n) = S(n - D(n)) + W_2(n) \tag{1.2}$$

where $\{S(n)\}$ is an unknown signal and $\{S(n - D(n))\}$ is the delayed signal, $D(n)$ changes slowly with time, and $\{W_1(n)\}$, $\{W_2(n)\}$ are unknown noises. Moreover, we also assume that $\{S(n)\}$ is zero-mean non-Gaussian stationary random process with nonzero skewness, i.e., $E\{S^3(n)\} \neq 0$, and that $\{W_1(n)\}$ and $\{W_2(n)\}$ are zero-mean, Gaussian, possibly correlated stationary random processes, statistically independent of $S(n)$. The problem is to track the delay $D(n)$ at each time instant from the two sensor measurements $\{X(n)\}$ and $\{Y(n)\}$.

For the stationary case, i.e., $D(n) = \hat{D}$, and from (1.1), we can always write $S(n - \hat{D}) = X(n - \hat{D}) - W_1(n - \hat{D})$. By substituting into (1.2), we have

$$Y(n) = X(n - \hat{D}) - W_1(n - \hat{D}) + W_2(n) \tag{2}$$

or, in a more general form,

$$Y(n) = \sum_{i=-\infty}^{\infty} a_i X(n - i) - W_1(n - \hat{D}) + W_2(n) \tag{3}$$

where, in theory, $a_i = 0$ for all $\{i\}$ except $i = \hat{D}$, and $a_{\hat{D}} = 1$. In practice, $\hat{D}$ is always finite and therefore

$$Y(n) = \sum_{i=-P}^{P} a_i X(n - i) - W_1(n - \hat{D}) + W_2(n) \tag{4}$$

where P is much greater than the largest possible delay we can expect. By multiplying both sides of (4) by $X(n + \tau)\, X(n + \rho)$ and taking expectations, we get [14]

$$\begin{aligned} E\{Y(n)\, X(n + \tau)\, X(n + \rho)\} &= \sum_{i=-P}^{P} a_i E\{X(n - i)\, X(n + \tau)\, X(n + \rho)\} \\ &\quad - E\{W_1(n - \hat{D})\, X(n + \tau)\, X(n + \rho)\} \\ &\quad + E\{W_2(n)\, X(n + \tau)\, X(n + \rho)\} \end{aligned}$$

or

$$\begin{aligned} R_{YXX}(\tau, \rho) &= \sum_{i=-P}^{P} a_i R_{XXX}(\tau + i, \rho + i) \\ &\quad - R_{W_1XX}(\tau + \hat{D}, \rho + \hat{D}) + R_{W_2XX}(\tau, \rho). \end{aligned} \tag{5}$$

However, $R_{W_1XX}(\tau, \rho)$ and $R_{W_2XX}(\tau, \rho)$ are identically zero for all (τ, ρ) due to the fact that i) the signal and noises are zero mean, ii) the signal is independent from the noises, and iii) the noises are Gaussian. So, even in the case where the noises are spatially correlated with unknown cross-correlation function, (5) becomes

$$R_{YXX}(\tau, \rho) = \sum_{i=-P}^{P} a_i R_{XXX}(\tau + i, \rho + i). \tag{6}$$

The basic idea in this paper is to introduce a method for adaptive time delay estimation, $D(n)$, based on the parametric model described by (6), i.e., the adaptive version of

$$R_{YXX}(n, \tau, \rho) = \sum_{i=-P}^{P} a_i(n)\, R_{XXX}(n, \tau + i, \rho + i) \tag{7}$$

where n is the iteration (or time) index.

III. The Adaptive Time Delay Estimation Scheme

In order to develop the adaptive time delay estimation scheme, we need to define first the "criterion of goodness." Assuming that $R_{XXX}(\tau, \rho)$ are the third-order cumulants of $\{X(n)\}$, then from Fig. 1, we have that the output of the FIR filter with coefficients $\{a_i\}$ is given by

$$R_{ZXX}(\tau, \rho) = \sum_{i=-P}^{P} a_i R_{XXX}(\tau + i, \rho + i).$$

The "criterion of goodness" is defined as the sum of the squared errors between the desired output $R_{YXX}(\tau, \rho)$ and the actual output $R_{ZXX}(\tau, \rho)$, viz.

$$\xi_{\text{general}} \equiv \sum_{\tau} \sum_{\rho} [R_{ZXX}(\tau, \rho) - R_{YXX}(\tau, \rho)]^2, \tag{8.1}$$

where (τ, ρ) may be defined to include the whole 2-D plane. However, if we choose $\tau - k = \rho = -P, -P + 1, \cdots, P$, $k = 0, 1, \cdots, N_s - 1$, the criterion of goodness, ξ, becomes one special case of (8.1)

$$\begin{aligned} \xi &= \sum_{k=0}^{N_s - 1} \sum_{\rho=-P}^{P} \left[\sum_{i=-P}^{P} a_i R_{XXX}(\rho + k + i, \rho + i) - R_{YXX}(\rho + k, \rho) \right]^2 \\ &= (\boldsymbol{R}_{XXX}\boldsymbol{A} - \boldsymbol{R}_{YXX})^T (\boldsymbol{R}_{XXX}\boldsymbol{A} - \boldsymbol{R}_{YXX}) \end{aligned} \tag{8.2}$$

where

$$\boldsymbol{R}_{XXX} = [\boldsymbol{R}_1^T(0), \boldsymbol{R}_1^T(1), \cdots, \boldsymbol{R}_1^T(N_s - 1)]^T$$

is $N_s(2P + 1) \times (2P + 1)$

$$\boldsymbol{R}_1(k) = \begin{pmatrix} R_{XXX}(k, 0) & R_{XXX}(k + 1, 1) & \cdots & R_{XXX}(k + 2P, 2P) \\ R_{XXX}(k - 1, -1) & R_{XXX}(k, 0) & \cdots & R_{XXX}(k + 2P - 1, 2P - 1) \\ \vdots & \vdots & \ddots & \\ R_{XXX}(k - 2P, -2P) & R_{XXX}(k - 2P + 1, -2P + 1) & \cdots & R_{XXX}(k, 0) \end{pmatrix}$$

is $(2P + 1) \times (2P + 1)$

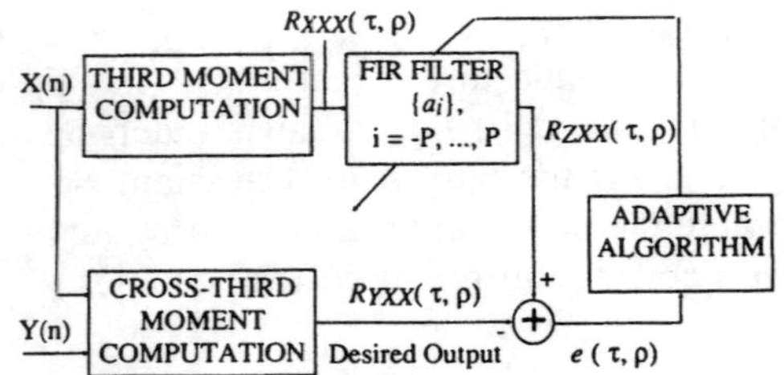

Fig. 1. The configuration of the adaptive time delay estimation method based on cumulants (ATDC).

$$A = [a_{-P}, a_{-P+1}, \cdots, a_0, \cdots, a_{P-1}, a_P]^T$$

$$\text{is } (2P + 1) \times 1$$

$$R_{YXX} = [R_2^T(0), R_2^T(1), \cdots, R_2^T(N_s - 1)]^T$$

$$\text{is } N_s(2P + 1) \times 1,$$

$$\begin{aligned} R_2(k) = [&R_{YXX}(k + P, P), R_{YXX}(k + P - 1, P - 1), \\ &\cdots, R_{YXX}(k, 0), \cdots, \\ &R_{YXX}(k - P + 1, -P + 1), \\ &R_{YXX}(k - P, -P)]^T \quad \text{is } (2P + 1) \times 1. \end{aligned}$$

The gradient of ξ is

$$\nabla \equiv \frac{\partial \xi}{\partial A} = 2(R_{XXX}^T R_{XXX} A - R_{XXX}^T R_{YXX}). \tag{9}$$

Let us note that, in practice, the third-order moments $R_{XXX}(\tau, \rho)$ and cross moments $R_{YXX}(\tau, \rho)$ in (8) and (9) need to be substituted by their estimates $\hat{R}_{XXX}(\tau, \rho)$ and $\hat{R}_{YXX}(\tau, \rho)$. The gradient-type algorithm is then employed to develop the adaptive time delay estimation scheme as follows. First, we take

$$\begin{aligned} \hat{\xi}(n) &\equiv \sum_{k=0}^{N_s-1} \sum_{\rho=-P}^{P} \left[\sum_{i=-P}^{P} \hat{a}_i(n) \hat{R}_{XXX}(n, \rho + k + i, \right. \\ &\qquad \left. \rho + i) - \hat{R}_{YXX}(n, \rho + k, \rho) \right]^2 \\ &\equiv (\hat{R}_{XXX}(n) \hat{A}(n) - \hat{R}_{YXX}(n))^T \\ &\qquad \cdot (\hat{R}_{XXX}(n) \hat{A}(n) - \hat{R}_{YXX}(n)) \end{aligned} \tag{10}$$

as an estimate of ξ at the nth iteration where $\hat{A}(n) = [\hat{a}_{-P}(n), \hat{a}_{-P+1}(n), \cdots, \hat{a}_0(n), \cdots, \hat{a}_{P-1}(n), \hat{a}_P(n)]^T$, and $\hat{R}_{XXX}(n)$ and $\hat{R}_{YXX}(n)$ are estimates of R_{XXX} and R_{YXX} at the nth iteration. Therefore, the gradient estimate becomes

$$\begin{aligned} \hat{\nabla}(n) \equiv \frac{\partial \hat{\xi}(n)}{\partial \hat{A}(n)} &= 2[\hat{R}_{XXX}^T(n) \hat{R}_{XXX}(n) \hat{A}(n) \\ &\quad - \hat{R}_{XXX}^T(n) \hat{R}_{YXX}(n)]. \end{aligned} \tag{11}$$

Then, the parameter update equation takes the form

$$\hat{A}(n + 1) = \hat{A}(n) - \mu(n) \hat{\nabla}(n), \tag{12.1}$$

where $\mu(n)$ is chosen as

$$0 < \mu(n) < \frac{1}{\text{tr}\left\{\hat{R}_{XXX}^T(n) \hat{R}_{XXX}(n)\right\}} \tag{12.2}$$

in order to satisfy the stability requirement without knowing the spread of the eigenvalues of $\hat{R}_{XXX}^T(n) \hat{R}_{XXX}(n)$.

In order to iterate the algorithm, assuming that we have data

$$\{X(0), \cdots, X(n - 1), X(n), X(n + 1), \cdots, X(n + 4P)\}$$

and

$$\{Y(0), \cdots, Y(n - 1), Y(n), Y(n + 1), \cdots, Y(n + 4P)\}$$

at the nth iteration, the third moments $\hat{R}_{XXX}(n, \tau, \rho)$ and cross moments $\hat{R}_{YXX}(n, \tau, \rho)$ are estimated as

$$\begin{aligned} \hat{R}_{XXX}(n, \tau, \rho) &= \frac{1}{n + 1} \sum_{i=0}^{n} f^{n-i} X(i + 2P) \\ &\quad \cdot X(i + 2P + \tau) X(i + 2P + \rho) \end{aligned} \tag{13.1}$$

$$\begin{aligned} \hat{R}_{YXX}(n, \tau, \rho) &= \frac{1}{n + 1} \sum_{i=0}^{n} f^{n-i} Y(i + 2P) \\ &\quad \cdot X(i + 2P + \tau) X(i + 2P + \rho) \end{aligned} \tag{13.2}$$

where $0 < f \leq 1$, and f is a forgetting factor which controls the shape of the window of data being used at each iteration. We choose $f = 1$ for the stationary case.

For the nonstationary case, if the time delay $D(n)$ slowly changes with time, we may follow the variation of $D(n)$ by properly choosing the forgetting factor, $f(<1)$, and applying the same algorithm (10)–(13) to track the time delay. In this case, values of the parameters, $\hat{a}_i(n)$, are changing with time as $D(n)$ slowly changes. However, the condition for good tracking is when $D(n)$ slowly changes with time so enough data can be included in a window for the estimation of third-order moments.

Table I summarizes our Adaptive Time Delay estimation method based on Cumulants (ATDC) for both stationary and nonstationary cases. It also provides simple recursive expressions for the computation of cumulants at each iteration.

IV. A Note on Convergence Analysis

The convergence property of the parameters obtained by the adaptive time delay estimation method (ATDC) is studied in this section for the case of stationary and ergodic signals. It is shown that the model parameters asymptotically converge to their true values provided that P is chosen sufficiently large.

Let us first define

$$\tilde{A}(n) \equiv [\hat{R}_{XXX}^T(n) \hat{R}_{XXX}(n)]^{-1} \hat{R}_{XXX}^T(n) \hat{R}_{YXX}(n). \tag{14}$$

TABLE I

SUMMARY OF THE ATDC METHOD

Initialization

$C_{XXX}(\rho + k, \rho) = 0 \qquad k = 0, 1, \ldots, N_s - 1, \quad \rho = -2P, -2P + 1, \ldots, 2P$

$C_{YXX}(\rho + k, \rho) = 0 \qquad k = 0, 1, \ldots, N_s - 1, \quad \rho = -P, -P + 1, \ldots, P$

$\underline{\hat{A}}(0) = \underline{A}_I, \quad \underline{A}_I$ is a initial guess, $\qquad (2 + P) \times 1$

For $n = 0, 1, 2, \ldots, \quad n_1 = n + 2P$

for $k = 0, 1, \ldots, N_s - 1, \quad \rho = -2P, -2P + 1, \ldots, 2P$

$$C_{XXX}(\rho + k, \rho) = f C_{XXX}(\rho + k, \rho) + X(n_1) X(n_1 + \rho + k) X(n_1 + \rho)$$

$$\hat{R}_{XXX}(n, \rho + k, \rho) = \tfrac{1}{n+1} C_{XXX}(\rho + k, \rho)$$

for $k = 0, 1, \ldots, N_s - 1, \quad \rho = -P, -P + 1, \ldots, P - 1, P$

$$C_{YXX}(\rho + k, \rho) = f C_{YXX}(\rho + k, \rho) + Y(n_1) X(n_1 + \rho + k) X(n_1 + \rho)$$

$$\hat{R}_{YXX}(n, \rho + k, \rho) = \tfrac{1}{n+1} C_{XXX}(\rho + k, \rho)$$

$$\underline{\hat{R}}_{XXX}(n) = [\underline{\hat{R}}_1^T(0), \underline{\hat{R}}_1^T(1), \ldots, \underline{\hat{R}}_1^T(N_s - 1)]^T$$

where $\underline{\hat{R}}_1(k) =$

$$\begin{pmatrix} \hat{R}_{XXX}(n, k, 0) & \hat{R}_{XXX}(n, k + 1, 1) & \cdots & \hat{R}_{XXX}(n, k + 2P, 2P) \\ \hat{R}_{XXX}(n, k - 1, -1) & \hat{R}_{XXX}(n, k, 0) & \cdots & \hat{R}_{XXX}(n, k + 2P - 1, 2P - 1) \\ \vdots & \vdots & \ddots & \vdots \\ \hat{R}_{XXX}(n, k - 2P, -2P) & \hat{R}_{XXX}(n, k - 2P + 1, -2P + 1) & \cdots & \hat{R}_{XXX}(n, k, 0) \end{pmatrix}$$

$$\underline{\hat{R}}_{YXX}(n) = [\underline{\hat{R}}_2^T(0), \underline{\hat{R}}_2^T(1), \ldots, \underline{\hat{R}}_2^T(N_s - 1)]^T$$

where $\underline{\hat{R}}_2(k) = [\hat{R}_{YXX}(n, k + P, P), \hat{R}_{YXX}(n, k + P - 1, P - 1), \ldots, \hat{R}_{YXX}(n, k - P, -P)]^T$

$$\hat{\nabla}(n) = 2[\underline{\hat{R}}_{XXX}^T(n) \underline{\hat{R}}_{XXX}(n) \underline{\hat{A}}(n) - \underline{\hat{R}}_{XXX}^T(n) \underline{\hat{R}}_{YXX}(n)]$$

$$\underline{\hat{A}}(n + 1) = \underline{\hat{A}}(n) - \mu(n) \hat{\nabla}(n), \qquad 0 < \mu(n) < 1/tr[\underline{\hat{R}}_{XXX}^T(n) \underline{\hat{R}}_{XXX}(n)]$$

We made ergodic assumption for cumulants as in (13) for the stationary case ($f = 1.0$), i.e.,

$$\lim_{n \to \infty} \hat{R}_{XXX}(n, \tau, \rho) = R_{XXX}(\tau, \rho)$$

$$\lim_{n \to \infty} \hat{R}_{YXX}(n, \tau, \rho) = R_{YXX}(\tau, \rho)$$

and

$$\lim_{n \to \infty} \hat{R}_{W_i XX}(n, \tau, \rho) = R_{W_i XX}(\tau, \rho) = 0 \qquad i = 1, 2$$

where $\hat{R}_{W_i XX}(n, \tau, \rho)$ is estimated using $\{W_i(n)\}$ instead of $\{Y(n)\}$ in (13). In this method, we also assume that

$$\lim_{n \to \infty} \hat{R}_{XXX}^T(n) \hat{R}_{XXX}(n) \qquad \text{is nonsingular.}$$

By using the assumptions we made and the results in [21], we get

$$\lim_{n \to \infty} \tilde{A}(n) = A^* = (R_{XXX}^T R_{XXX})^{-1} R_{XXX}^T R_{YXX} \qquad (15)$$

provided that P is sufficiently large.

The ATDC method is like using the LMS algorithm [8], [9], with the third-order cumulant estimates as its input instead of the data themselves. Therefore, the convergence rate of the ATDC method depends on the step size, $\mu(n)$, the order of the parametric model, P, and the eigenvalues of $\hat{R}_{XXX}^T(n)\ \hat{R}_{XXX}(n)$. When the eigenvalue spread of $\hat{R}_{XXX}^T(n)\ \hat{R}_{XXX}(n)$ is large, the ATDC method slows down and it requires a large number of iterations to converge.

V. COMPUTATIONAL COMPLEXITY

The computational complexity of an algorithm is an important aspect in the adaptation process and therefore should be taken into account. We use the number of multiplications per iteration as a figure of merit. The required complexity of the new adaptive time delay estimation method (ATDC) is illustrated in Table IV. For comparison purposes, the required complexity of the Cross-Correlation-Based Gradient method (CCBG) [22] and the Overdetermined Recursive Instrumental Variable method (ORIV), which is also based on cumulants [19], [23], is also included in Table IV for both stationary and nonstationary cases. The CCBG and ORIV methods are summarized in Tables II and III.

From Table IV, it is apparent that the number of multiplications required by the ORIV method is the largest one. The number of multiplications required by the adaptive time delay estimation method (ATDC) requires approximately $7(2P + 1)^2$ less than that required by the ORIV method. However, it is $3N_s$ times larger than that

TABLE II
SUMMARY OF THE CCBG METHOD

Initialization

$C_{XX}(\tau) = 0 \qquad \tau = 0, 1, \ldots, 2P$

$C_{XY}(\rho) = 0 \qquad \rho = -P, -P+1, \ldots, P$

$\underline{\hat{A}}(0) = \underline{A}_I, \quad A_I$ is a initial guess. $\qquad (2+P) \times 1$

For $n = 0, 1, 2, \ldots$ *and* $n_1 = n - 2P$

$C_{XX}(\tau) = fC_{XX}(\tau) - X(n_1)X(n_1 - \tau) \qquad \tau = 0, 1, \ldots, 2P$

$C_{XY}(\rho) = fC_{XY}(\rho) - X(n_1)Y(n_1 - \rho) \qquad \rho = -P, -P+1, \ldots, P$

$$r_{XX}(n,\tau) = \frac{1}{n+1}C_{XX}(\tau), \qquad \tau = 0, 1, \ldots, 2P$$

$$r_{XY}(n,\rho) = \frac{1}{n+1}C_{XY}(\rho), \qquad \rho = -P, -P+1, \ldots, P$$

$$\tilde{R}_{XX}(n) = \begin{pmatrix} r_{XX}(n,0) & r_{XX}(n,1) & \ldots & r_{XX}(n,2P) \\ r_{XX}(n,1) & r_{XX}(n,0) & \ldots & r_{XX}(n,2P-1) \\ \vdots & \vdots & \ddots & \vdots \\ r_{XX}(n,2P) & r_{XX}(n,2P-1) & \ldots & r_{XX}(n,0) \end{pmatrix}$$

$$\tilde{R}_{XY}(n) = [r_{XY}(n,-P), r_{XY}(n,-P+1), \ldots, r_{XY}(n,P)]^T$$

$$\nabla(n) = 2[\tilde{R}_{XX}(n)\underline{\hat{A}}(n) - \tilde{R}_{XY}(n)]$$

$$\underline{\hat{A}}(n-1) = \underline{\hat{A}}(n) - \mu(n)\nabla(n), \qquad 0 < \mu(n) < \frac{1}{tr[\tilde{R}_{XX}(n)]}$$

TABLE III
SUMMARY OF THE ORIV METHOD

Initialization

$S_0 = \mu[I_{2P+1}\|0]$.	$(2P+1) \times N_s(2P+1)$
$\underline{L}_0 = \underline{0}$,	$N_s(2P+1) \times 1$
$P_0 = \mu^{-2}I_{2P+1}$,	$(2P+1) \times (2P+1)$
$\underline{\hat{A}}(0) = \underline{A}_I$, $\underline{A}_I$ is a initial guess,	$(2P+1) \times 1$
For $n = 0, 1, 2, \ldots$	
$\underline{Z}_{n+1} = [\underline{Z}_1^T(0), \underline{Z}_1^T(1), \ldots, \underline{Z}_1^T(N_s - 1)]^T$,	$N_s(2P+1) \times 1$
where $\underline{Z}_1(k) = [X(n+k+2P-1)X(n+2P+1), X(n+k+2P)X(n+2P),$ $\ldots, X(n-k+1)X(n+1)]^T$,	$(2P+1) \times 1$
$\underline{X}_{n+1} = [X(n-2P-1), X(n+2P), \ldots, X(n+1)]^T$,	$(2P+1) \times 1$
$Y_{n+1} = Y(n+P-1)$,	1×1
$\underline{W}_{n+1} = S_n\underline{Z}_{n+1}$,	$(2P+1) \times 1$
$S_{n+1} = \lambda S_n + \underline{X}_{n+1}\underline{Z}_{n+1}^T$,	$(2P+1) \times N_s(2P+1)$
$\phi_{n+1} = [\underline{W}_{n+1}\|\underline{X}_{n+1}]$,	$(2P+1) \times 2$
$\lambda^2\Lambda_{n+1} = \begin{pmatrix} -\underline{Z}_{n+1}^T\underline{Z}_{n+1} & \lambda \\ \lambda & 0 \end{pmatrix}$	2×2
$K_{n+1} = P_n\phi_{n+1}(\lambda^2\Lambda_{n+1} + \phi_{n+1}^T P_n \phi_{n+1})^{-1}$,	$(2P-1) \times 2$
$P_{n+1} = \lambda^{-2}(P_n - K_{n+1}\phi_{n+1}^T P_n)$,	$(2P+1) \times (2P+1)$
$\underline{V}_{n+1} = \begin{pmatrix} \underline{Z}_{n+1}^T\underline{L}_n \\ Y_{n+1} \end{pmatrix}$	2×1
$\underline{L}_{n+1} = \lambda\underline{L}_n + \underline{Z}_{n+1}Y_{n+1}$,	$N_s(2P+1) \times 1$
$\underline{\hat{A}}(n+1) = \underline{\hat{A}}(n) + K_{n+1}(\underline{V}_{n+1} - \phi_{n+1}^T\underline{\hat{A}}(n))$,	$(2P+1) \times 1$

TABLE IV
THE NUMBER OF MULTIPLICATIONS PER ITERATION OF THE CCBG, ATDC, AND ORIV METHODS

type of method	No. of multiplications per iteration	P		
		15	30	60
CCGB	$(2P+1)^2 + 3(2P+1) + 3$	1057	3907	15007
ORIV	$3N_s(2P+1)^2 + (5N_s+12)(2P+1) + 7(2P+1)^2 + 7$	16220	61190	237530
ATDC	$3N_s(2P+1)^2 + (12N_s+1)(2P+1) - 3$	9793	35743	136243

Note that $N_s = 3$ is used in the calculation.

required by the CCBG method. We therefore conclude that the improved performance of the adaptive time delay estimation scheme over the cross-correlation-based gradient method demonstrated in the next section is achieved at the expense of more computations.

VI. Simulation Results

We consider six examples to illustrate the performance of the adaptive time delay estimation scheme (ATDC) introduced in this paper. For comparison purposes, the cross-correlation-based gradient method (CCBG) and the overdetermined recursive instrumental variable method (ORIV) are included in the simulations.

A. Stationary Case ($f = 1.0$)

In the following four examples, the stationary time delay $D(n) = 16$ for all n and an initial guess $\hat{a}_i(0) = 0$ for all i are used. The signal-to-noise ratio (SNR) is defined as $20 \log_{10} (\sigma_s / \sigma_n)$ where σ_s and σ_n are the standard deviations of signal and noise, respectively. The SNR in each example is 0 dB. The range of search for the time delay estimate is from -30 to 30 ($P = 30$) with a step length equal to one.

Example A.1—White Signal and Spatially Uncorrelated White Noises: The signal is zero-mean non-Gaussian (one-sided exponentially distributed) white process and is generated by the GGEXN subroutine of the IMSL library [24] with $E\{S^2(n)\} = 1$ and $E\{S^3(n)\} = 2$. There is no special significance for choosing one-sided exponential probability density function (pdf). The two additive noises are also zero mean, but are Gaussian and uncorrelated with each other and with the signal. They are generated by the GGNML subroutine of the IMSL library [24] using different seeds to ensure their independence.

Figs. 2(a), 3(a), and 4(a) show the values of the coefficients, $\hat{a}_i(n)$, $i = -30, \cdots, 0, \cdots, 30$, for each iteration n using the ATDC, CCBG, and ORIV methods, respectively. Figs. 2(b), 3(b), and 4(b) illustrate the magnitudes of $\hat{a}_i(n)$, $i = -30, \cdots, 0, \cdots, 30$, at iteration $n = 3000$. For this specific example, all three methods

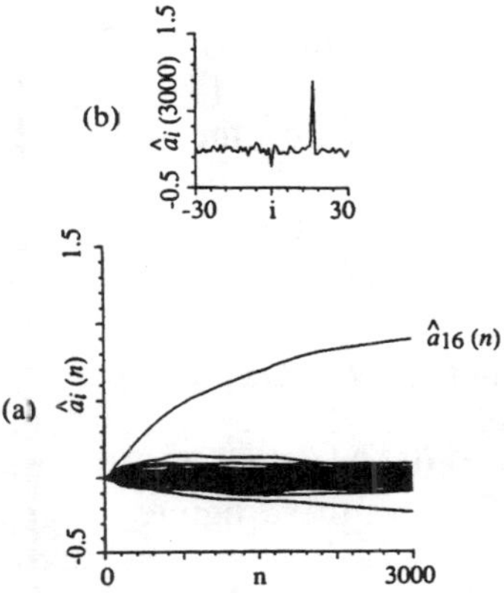

Fig. 2. Stationary case: white signal and spatially uncorrelated white noises. Results obtained by the ATDC method: (a) the values of the coefficients, $\{\hat{a}_i(n)\}$, $i = -30, \cdots, 0, \cdots, 30$, for each iteration n, (b) the values of $\{\hat{a}_i(3000)\}$.

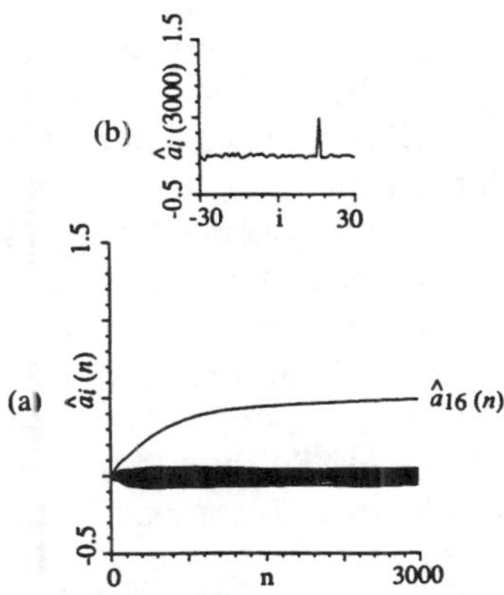

Fig. 3. Stationary case: white signal and spatially uncorrelated white noises. Results obtained by the CCBG method: (a) the values of the coefficients, $\{\hat{a}_i(n)\}$, $i = -30, \cdots, 0, \cdots, 30$, for each iteration n, (b) the values of $\{\hat{a}_i(3000)\}$.

converge to time delay $\hat{D} = 16$ because all coefficients converge to zero except $\hat{a}_{16}(n)$ that converges to one.

Example A.2—White Signal and Spatially Correlated Noises: The signal is the same as before, but the noises are no longer uncorrelated. The Gaussian noise process

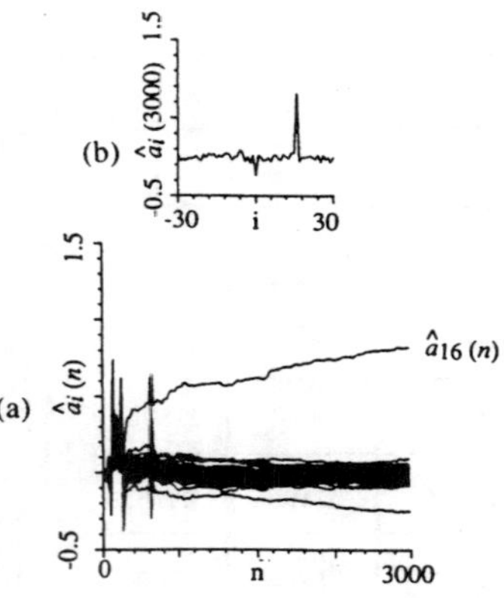

Fig. 4. Stationary case: white signal and spatially uncorrelated white noises. Results obtained by the ORIV method: (a) the values of the coefficients, $\{\hat{a}_i(n)\}$, $i = -30, \cdots, 0, \cdots, 30$, for each iteration n, (b) the values of $\{\hat{a}_i(3000)\}$.

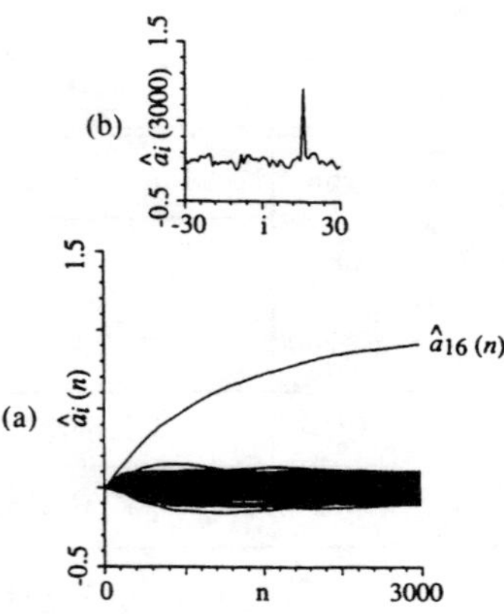

Fig. 5. Stationary case: white signal and spatially correlated noises. Results obtained by the ATDC method: (a) the values of the coefficients, $\{\hat{a}_i(n)\}$, $i = -30, \cdots, 0, \cdots, 30$, for each iteration n, (b) the values of $\{\hat{a}_i(3000)\}$.

$\{W_1(n)\}$ in (1.1) is generated by the GGNML subroutine of the IMSL library, whereas $\{W_2(n)\}$ in (1.2) is generated from $\{W_1(n)\}$ using the following FIR system equation:

$$W_2(n) = \sum_{i=0}^{10} b_i W_1(n-i) \tag{16}$$

where the parameters $\{b_i\}$ take values $\{0.0337, 0.0675, 0.1012, 0.1349, 0.1514, 0.9466, 0.1514, 0.1181, 0.0843, 0.0506, 0.0169\}$. Figs. 5–7 illustrate the results of this example. From these figures, it appears that the ATDC and ORIV methods (cumulant-based methods) exhibit better performance than the CCBG method. As expected, the CCBG method is incapable of suppressing the correlated Gaussian noises and therefore fails to work well for this signal environment.

Example A.3—Colored Signal and Spatially Correlated Colored Noises (Case I): The signal $\{S(n)\}$ is generated by passing the signal used in Examples A.1 and A.2 through a low-pass filter with digital cutoff frequency 0.1π. The noise $\{W_1(n)\}$ is generated the same way by passing through the same low-pass filter a Gaussian white process, and $\{W_2(n)\}$ is generated using (16). Again, the noises are spatially correlated. Thus, signal and noises have flat spectra in the band $|\omega| \leq 0.1\pi$. Figs. 8–10 illustrate the results of this example. As expected, the CCBG method does not suppress correlated Gaussian colored noises while the ATDC and ORIV methods do so.

Example A.4—Colored Signal and Spatially Correlated Colored Noises (Case II): The signal $\{S(n)\}$ and noise $\{W_1(n)\}$ are the same as in Example A.3. The spatial correlation function between $\{W_1(n)\}$ and $\{W_2(n)\}$ is modeled as a time delay (directional noise), i.e.,

$$W_2(n) = W_1(n-8). \tag{17}$$

Thus, for this specific example, the signal time delay is $\hat{D} = 16$, and the noise spatial correlation is another time delay at 8. Figs. 11–13 illustrate the results of this example. These results are consistent with those of Example A.3, i.e., the CCBG method shows two peaks at 8 and 16, whereas the ATDC and ORIV methods show one peak at 16 $[\hat{a}_{16}(n)]$.

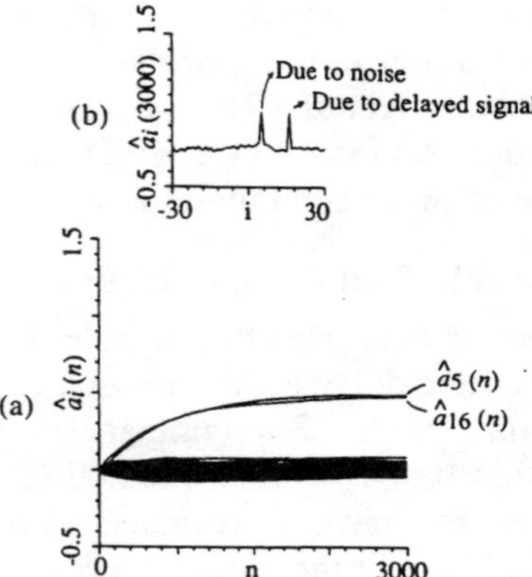

Fig. 6. Stationary case: white signal and spatially correlated noises. Results obtained by the CCBG method: (a) the values of the coefficients, $\{\hat{a}_i(n)\}$, $i = -30, \cdots, 0, \cdots, 30$, for each iteration n, (b) the values of $\{\hat{a}_i(3000)\}$.

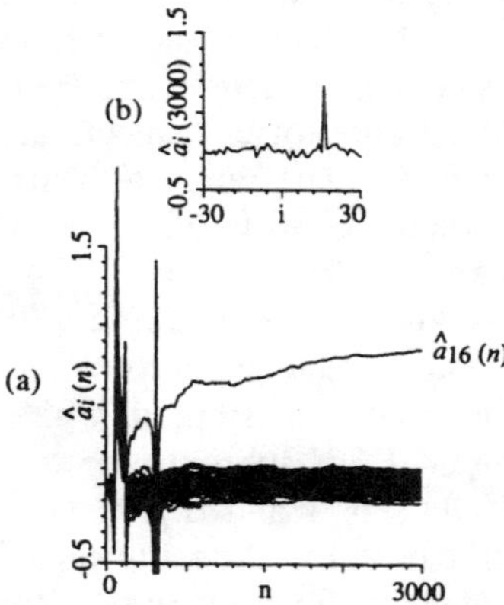

Fig. 7. Stationary case: white signal and spatially correlated noises. Results obtained by the ORIV method: (a) the values of the coefficients, $\{\hat{a}_i(n)\}$, $i = -30, \cdots, 0, \cdots, 30$, for each iteration n, (b) the values of $\{\hat{a}_i(3000)\}$.

B. Nonstationary Case

In the following two examples, $D(n)$ changes slowly with time.

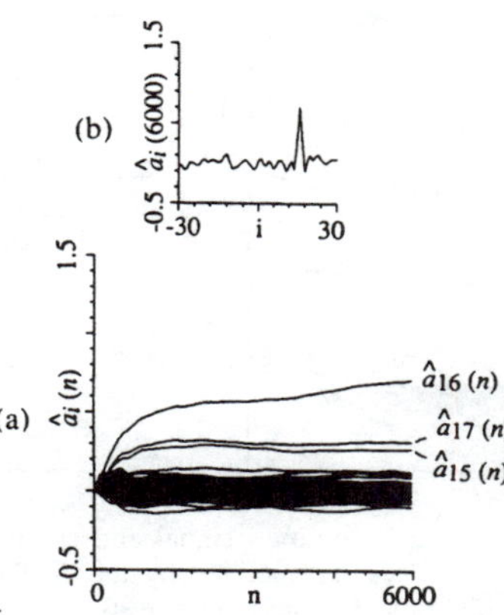

Fig. 8. Stationary case: colored signal and spatially correlated colored noises (case I). Results obtained by the ATDC method: (a) the values of the coefficients, $\{\hat{a}_i(n)\}$, $i = -30, \cdots, 0, \cdots, 30$, for each iteration n, (b) the values of $\{\hat{a}_i(6000)\}$.

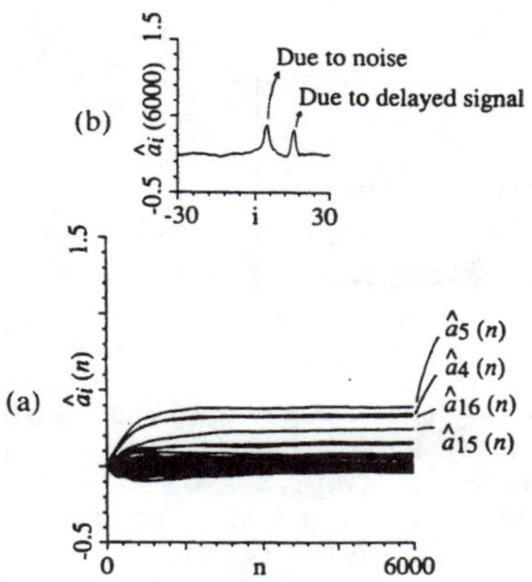

Fig. 9. Stationary case: colored signal and spatially correlated colored noises (case I). Results obtained by the CCBG method: (a) the values of the coefficients, $\{\hat{a}_i(n)\}$, $i = -30, \cdots, 0, \cdots, 30$, for each iteration n, (b) the values of $\{\hat{a}_i(6000)\}$.

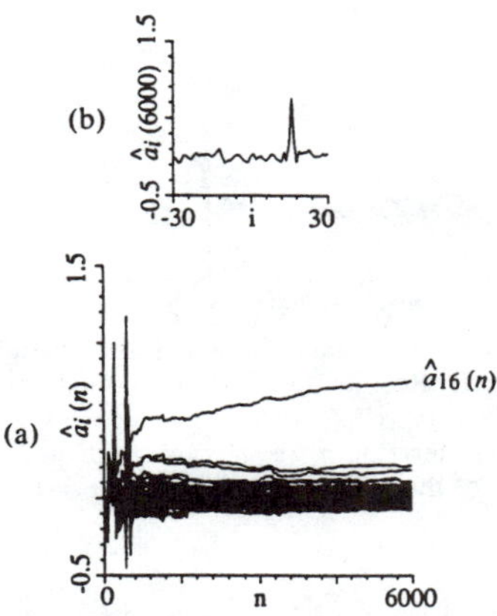

Fig. 10. Stationary case: colored signal and spatially correlated colored noises (case I). Results obtained by the ORIV method: (a) the values of the coefficients, $\{\hat{a}_i(n)\}$, $i = -30, \cdots, 0, \cdots, 30$, for each iteration n, (b) the values of $\{\hat{a}_i(6000)\}$.

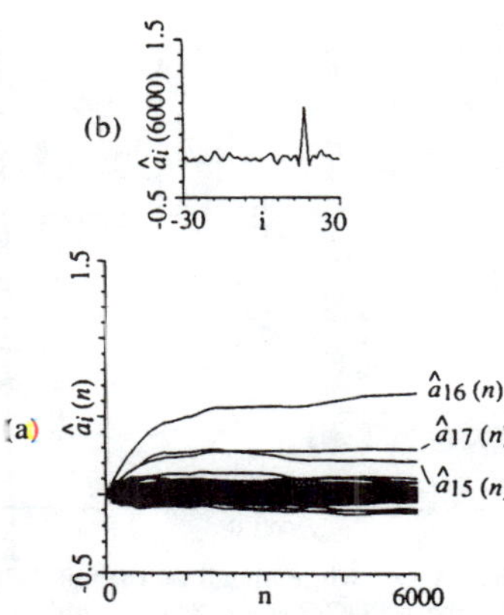

Fig. 11. Stationary case: colored signal and spatially correlated colored noises (case II). Results obtained by the ATDC method: (a) the values of the coefficients, $\{\hat{a}_i(n)\}$, $i = -30, \cdots, 0, \cdots, 30$, for each iteration n, (b) the values of $\{\hat{a}_i(6000)\}$.

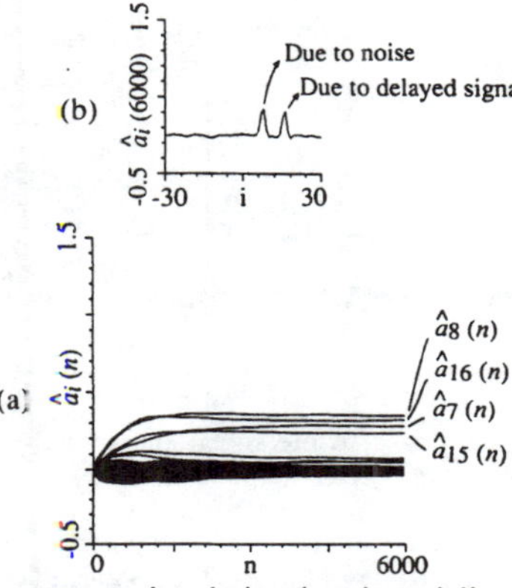

Fig. 12. Stationary case: colored signal and spatially correlated colored noises (case II). Results obtained by the CCBG method: (a) the values of the coefficients, $\{\hat{a}_i(n)\}$, $i = -30, \cdots, 0, \cdots, 30$, for each iteration n, (b) the values of $\{\hat{a}_i(6000)\}$.

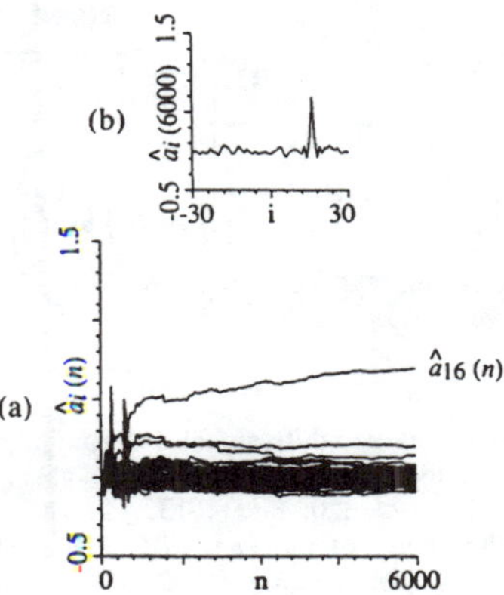

Fig. 13. Stationary case: colored signal and spatially correlated colored noises (case II). Results obtained by the ORIV method: (a) the values of the coefficients, $\{\hat{a}_i(n)\}$, $i = -30, \cdots, 0, \cdots, 30$, for each iteration n, (b) the values of $\{\hat{a}_i(6000)\}$.

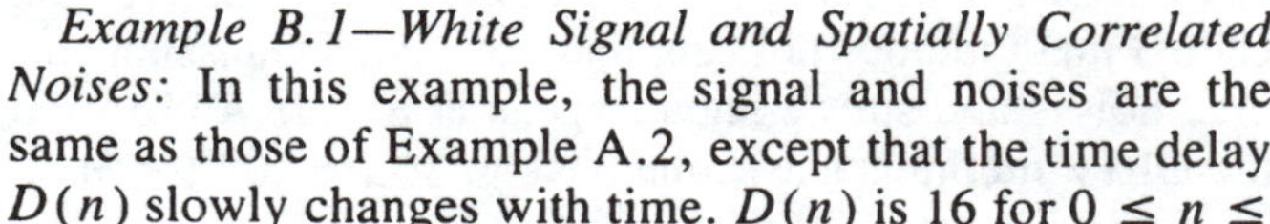

Example B.1—White Signal and Spatially Correlated Noises: In this example, the signal and noises are the same as those of Example A.2, except that the time delay $D(n)$ slowly changes with time. $D(n)$ is 16 for $0 \leq n \leq 4000$, 17 for $4000 < n \leq 8000$, and 14 for $8000 < n \leq 12000$. From Figs. 14–16, it is apparent that the ATDC method with $f = 0.995$, the CCBG method with $f = 0.995$, and the ORIV method with $\lambda = 0.999$ do follow

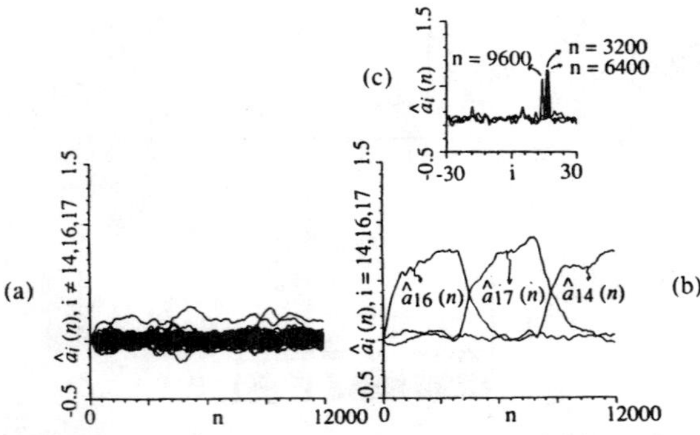

Fig. 14. Nonstationary case: white signal and spatially correlated noises. Results obtained by the ATDC method: (a) the values of the coefficients, $\{\hat{a}_i(n)\}$, $i = -30, \cdots, 0, \cdots, 13, 15, 18, \cdots, 30$, for each iteration n, (b) the values of $\{\hat{a}_{16}(n), \hat{a}_{17}(n), \hat{a}_{14}(n)$ for each iteration n, (c) the values of $\{\hat{a}_i(n)\}$ at n = 3200, 6400, 9600.

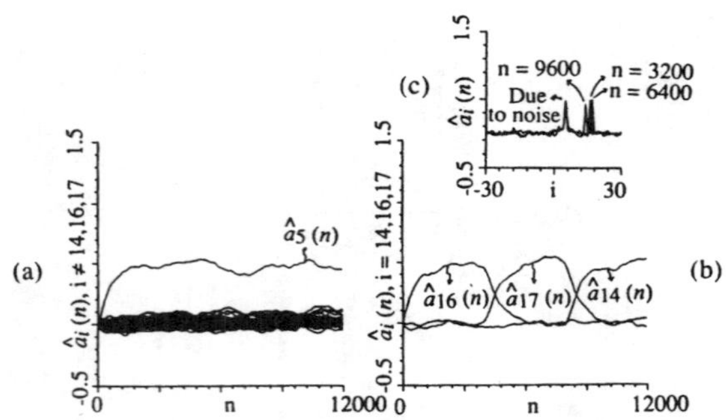

Fig. 15. Nonstationary case: white signal and spatially correlated noises. Results obtained by the CCBG method: (a) the values of the coefficients, $\{\hat{a}_i(n)\}$, $i = -30, \cdots, 0, \cdots, 13, 15, 18, \cdots, 30$, for each iteration n, (b) the values of $\{\hat{a}_{16}(n), \hat{a}_{17}(n), \hat{a}_{14}(n)$ for each iteration n, (c) the values of $\{\hat{a}_i(n)\}$ at n = 3200, 6400, 9600.

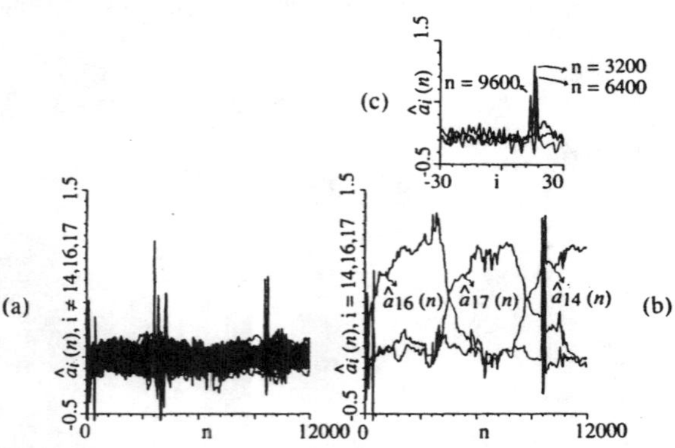

Fig. 16. Nonstationary case: white signal and spatially correlated noises. Results obtained by the ORIV method: (a) the values of the coefficients, $\{\hat{a}_i(n)\}$, $i = -30, \cdots, 0, \cdots, 13, 15, 18, \cdots, 30$, for each iteration n, (b) the values of $\{\hat{a}_{16}(n), \hat{a}_{17}(n), \hat{a}_{14}(n)$ for each iteration n, (c) the values of $\{\hat{a}_i(n)\}$ at n = 3200, 6400, 9600.

the variation of the time delay. However, the CCBG method does not suppress the correlated Gaussian noises and always shows a second large peak (stationary) at 5 [Fig. 15(c)]. From Fig. 17, it appears that the ORIV method exhibits higher variability than the other two adaptive schemes.

Example B.2—Colored Signal and Spatially Correlated Colored Noises (Case II): The signal and noises in this

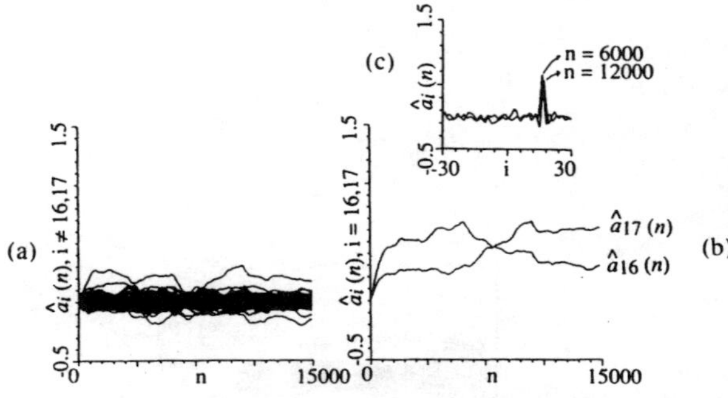

Fig. 17. Nonstationary case: colored signal and spatially correlated colored noises (case II). Results obtained by the ATDC method: (a) the values of the coefficients, $\{\hat{a}_i(n)\}$, $i = -30, \cdots, 0, \cdots, 15, 18, \cdots, 30$, for each iteration n, (b) the values of $\{\hat{a}_{16}(n)$ and $\hat{a}_{17}(n)$ for each iteration n, (c) the values of $\{\hat{a}_i(n)\}$ at n = 6000, 12 000.

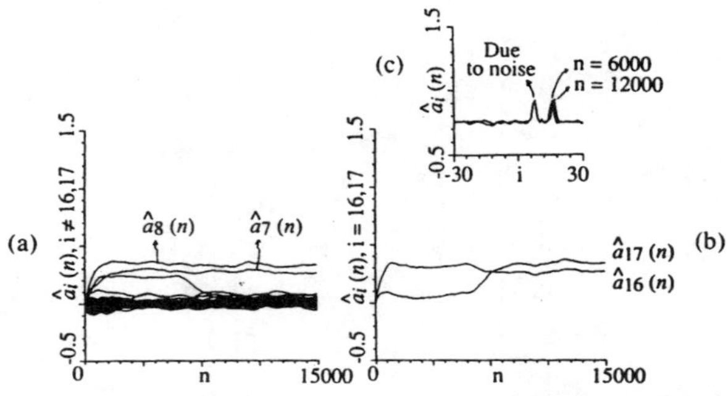

Fig. 18. Nonstationary case: colored signal and spatially correlated colored noises (case II). Results obtained by the CCBG method: (a) the values of the coefficients, $\{\hat{a}_i(n)\}$, $i = -30, \cdots, 0, \cdots, 15, 18, \cdots, 30$, for each iteration n, (b) the values of $\{\hat{a}_{16}(n)$ and $\hat{a}_{17}(n)$ for each iteration n, (c) the values of $\{\hat{a}_i(n)\}$ at n = 6000, 12 000.

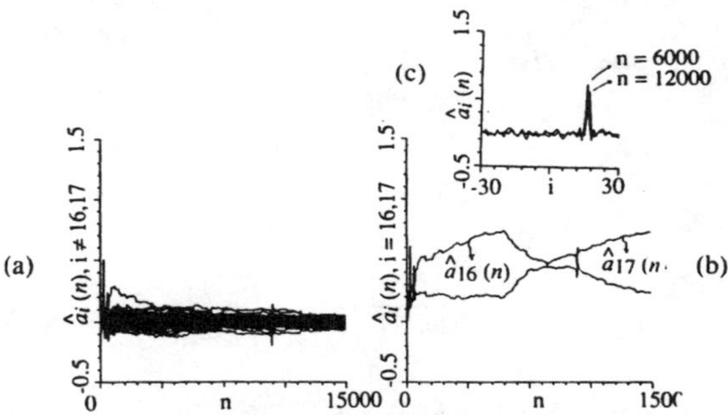

Fig. 19. Nonstationary case: colored signal and spatially correlated colored noises (case II). Results obtained by the ORIV method: (a) the values of the coefficients, $\{\hat{a}_i(n)\}$, $i = -30, \cdots, 0, \cdots, 15, 18, \cdots, 30$, for each iteration n, (b) the values of $\{\hat{a}_{16}(n)$ and $\hat{a}_{17}(n)$ for each iteration n, (c) the values of $\{\hat{a}_i(n)\}$ at n = 6000, 12 000.

example are the same as in Example A.4 except for the fact that time delay $D(n)$ slowly changes with time. $D(n)$ is 16 for $0 \leq n \leq 6000$ and 17 for $6000 < n \leq 15\,000$. Figs. 17–19 illustrate the results of this example using the ATDC with $f = 0.999$, the CCBG method with $f = 0.999$, and the ORIV method with $\lambda = 0.9999$. As expected, the CCBG method does not suppress correlated Gaussian colored noises and shows another peak at 8. The ATDC and the ORIV methods show one peak at 16 for $1000 < n < 7000$ and 17 for $9000 < n \leq 15\,000$.

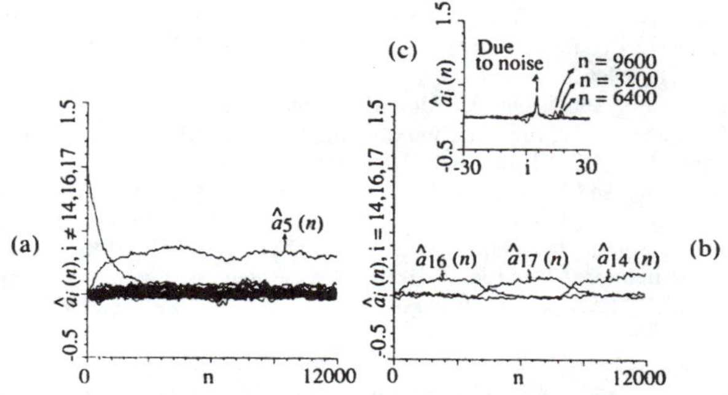

Fig. 20. Nonstationary case: white signal and spatially correlated noises. Results obtained by the ATDC method: (a) the values of the coefficients, $\{\hat{a}_i(n)\}$, $i = -30, \cdots, 0, \cdots, 13, 15, 18, \cdots, 30$, for each iteration n, (b) the values of $\{\hat{a}_{16}(n), \hat{a}_{17}(n), \hat{a}_{14}(n)$ for each iteration n, (c) the values of $\{\hat{a}_i(n)\}$ at $n = 3200, 6400, 9600$.

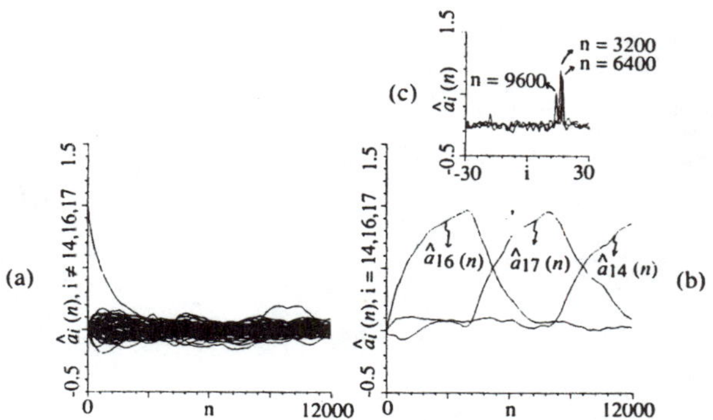

Fig. 21. Nonstationary case: white signal and spatially correlated noises. Results obtained by the ATDC method: (a) the values of the coefficients, $\{\hat{a}_i(n)\}$, $i = -30, \cdots, 0, \cdots, 13, 15, 18, \cdots, 30$, for each iteration n, (b) the values of $\{\hat{a}_{16}(n), \hat{a}_{17}(n), \hat{a}_{14}(n)$ for each iteration n, (c) the values of $\{\hat{a}_i(n)\}$ at $n = 3200, 6400, 9600$.

We also use the ATDC method with $f = 0.99$ and $f = 0.999$ in Example B.1 to demonstrate the effects of the forgetting factor f. Comparing Figs. 20, 21, and 14, we can see that when f decreases, the ATDC method does not suppress the correlated Gaussian noises and shows a second large peak at 5 [Fig. 20(c)]; this is because there is not enough data in the window decided by f to get good third-order moment estimates. But when f increases, the transition time needed to track the change of one time delay to another time delay increases [Fig. 21(b)].

In all the examples, $\mu(n) = \mu_0/\mathrm{tr}\ \{\hat{R}_{XXX}^T(n)\ \hat{R}_{XXX}(n)\}$. The reason is that we need to normalize step size to ensure the stability of the algorithm because of the large spread of the eigenvalues of $[\hat{R}_{XXX}^T(n)\ \hat{R}_{XXX}(n)]$.

From the simulation results, it is apparent that the cumulant-based methods, i.e., ATDC and ORIV, can track the time delay changes even when the additive noises are spatially correlated Gaussian with unknown correlations. The behavior of the ORIV method, however, appears to be irregular during the initial phase. The ATDC method results in much smoother trajectories for the parameters with less computational complexity.

VII. Concluding Remarks

A new adaptive method is introduced for time delay estimation for non-Gaussian signals in the third-order cumulant domain based on parametric modeling. The adaptive scheme has been tested and compared to the cross-correlation-based method and the overdetermined recursive instrumental variable method (cumulant-based too) using different sets of signal and noise conditions. It has been demonstrated that in the case where the signal is non-Gaussian and the additive noises are correlated Gaussian with unknown correlation function, the adaptive time delay estimation methods based on cumulants (both ATDC and ORIV) track the time delay changes correctly and suppress the effect of additive noises. The new method presented here is ad hoc, and therefore no claims of optimality are made.

Although this paper has dealt with third-order statistics which require that the signal has nonzero skewness, the same methods and techniques described in this paper can be extended in a straightforward manner to the case of fourth-order statistics to deal with signals that have nonzero kurtosis and symmetric probability density function.

References

[1] B. Y. Hamon and E. J. Hannan, "Spectral estimation of time delay for dispersive and non-dispersive systems," *Appl. Statist.*, vol. 23, no. 2, pp. 134–142, 1974.

[2] C. H. Knapp and G. C. Carter, "The generalized correlation method for estimation of time delay," *IEEE Trans. Acoust., Speech, Signal Processing*, vol. ASSP-24, pp. 320–327, Aug. 1976.

[3] Y. T. Chan, R. V. Hattin, and J. B. Plant, "The least squares estimation of time delay and its use in signal detection," *IEEE Trans. Acoust., Speech, Signal Processing*, vol. ASSP-26, pp. 217–222, June 1978.

[4] Y. T. Chan, J. M. Riley, and J. B. Plant, "A parameter estimation approach to time-delay estimation and signal detection," *IEEE Trans. Acoust., Speech, Signal Processing*, vol. ASSP-28, pp. 8–16, Feb. 1980.

[5] G. C. Carter, "Time delay estimation for passive sonar signal processing," *IEEE Trans. Acoust., Speech, Signal Processing*, vol. ASSP-29, pp. 463–470, June 1981.

[6] —, "Coherence and time delay estimation," *Proc. IEEE*, vol. 75, pp. 236–255, Feb. 1987.

[7] V. H. MacDonald and P. M. Schultheiss, "Optimum passive bearing estimation in a spatially incoherent noise environment," *J. Acoust. Soc. Amer.*, vol. 45, pp. 37–43, 1969.

[8] F. A. Reed, P. L. Feintuch, and N. J. Bershad, "Time delay estimation using the LMS adaptive filter—Static behavior," *IEEE Trans. Acoust., Speech, Signal Processing*, vol. ASSP-29, pp. 561–571, June 1981.

[9] P. L. Feintuch, N. J. Bershad, and F. A. Reed, "Time delay estimation using the LMS adaptive filter—Dynamic behavior," *IEEE Trans. Acoust., Speech, Signal Processing*, vol. ASSP-29, pp. 571–576, June 1981.

[10] Y. T. Chan, J. M. F. Riley, and J. B. Plant, "Modeling of time delay and its application to estimation of nonstationary delays," *IEEE Trans. Acoust., Speech, Signal Processing*, vol. ASSP-29, pp. 577–581, June 1981.

[11] D. M. Etter and S. D. Stearns, "Adaptive estimation of time delays in sampled data systems," *IEEE Trans. Acoust., Speech, Signal Processing*, vol. ASSP-29, pp. 582–587, June 1981.

[12] K. Sasaki, T. Sato, and Y. Nakamura, "Holographic passive sonar," *IEEE Trans. Sonics Ultrason.*, vol. SU-24, pp. 193–200, May 1977.

[13] C. L. Nikias and M. R. Raghuveer, "Bispectrum estimation: A digital signal processing framework," *Proc. IEEE*, vol. 75, pp. 869–891, July 1987.

[14] C. L. Nikias and R. Pan, "Time delay estimation in unknown Gaussian spatially correlated noise," *IEEE Trans. Acoust., Speech, Signal Processing*, vol. 36, Nov. 1988.

[15] A. Petropulu, C. L. Nikias, and J. G. Proakis, "Cumulant cepstrum of FM signals and high resolution time delay estimation," in *Proc. ICASSP'88*, New York, NY, Apr. 1988, pp. 2642–2645.

[16] C. L. Nikias, "ARMA bispectrum estimation to the identification of nonminimum phase system," *IEEE Trans. Acoust., Speech, Signal Processing*, vol. 36, pp. 513–524, Apr. 1988.

[17] C. L. Nikias and H. H. Chiang, "Noncausal autoregressive bispectrum estimation and deconvolution," in *Proc. ICASSP'87*, Dallas, TX, Apr. 1987, pp. 1557–1560.

[18] H. H. Chiang and C. L. Nikias, "Adaptive filtering via cumulants and LMS algorithm," in *Proc. ICASSP'88*, New York, NY, Apr. 1988, pp. 1479–1482.

[19] B. Friedlander and B. Porat, "Adaptive IIR algorithms based on higher-order statistics," in *Proc. Amer. Contr. Conf.*, Atlanta, GA, June 1988.

[20] H. Akaike, "On the use of non-Gaussian process in the identification of a linear dynamic system," *Ann. Inst. Statist. Math.*, vol. 18, pp. 269–276, 1966.

[21] K. Y. Wong and E. Polak, "Identification of linear discrete time systems using the instrumental variable method," *IEEE Trans. Automat. Contr.*, vol. AC-12, pp. 707–718, Dec. 1967.

[22] J. F. Yang and M. Kaveh, "Adaptive eigensubspace algorithms for direction or frequency estimation and tracking," *IEEE Trans. Acoust., Speech, Signal Processing*, vol. 36, pp. 241–251, Feb. 1988.

[23] B. Friedlander, "The overdetermined recursive instrumental variable method," *IEEE Trans. Automat. Contr.*, vol. AC-29, pp. 353–356, Apr. 1984.

[24] The IMSL Library, IMSL Inc., Houston, TX, 1982.

includes cardiac and cerebral bispectral analysis. Her research interests lie in the area of digital signal processing with applications.

Adaptive Estimation of Time Delays in Sampled Data Systems

DELORES M. ETTER, MEMBER, IEEE, AND SAMUEL D. STEARNS, SENIOR MEMBER, IEEE

Abstract—An adaptive technique is developed which iteratively determines the time delay between two sampled signals that are highly correlated. Although the procedure does not require *a priori* information on the input signals, it does require that the signals have a unimodal or periodically unimodal cross-correlation function. The adaptive delay algorithm uses a gradient technique to find the value of the adaptive delay that minimizes the mean-squared (MS) error function. This iterative algorithm is similar to the adaptive filter coefficient algorithm developed by Widrow. However, the MS error function for the adaptive delay is not quadratic, as it is in the adaptive filter. A statistical analysis determines the value of the convergence parameter which effects rapid convergence of the adaptive delay. This convergence parameter is a function of the power of the input signal. Computer simulations are presented which verify that the adaptive delay correctly estimates the time delay difference between two sinusoids, including those in noisy environments. The adaptive delay is also shown to perform correctly in a time delay tracking application.

I. Introduction and Summary

The problem of estimating and tracking the delay between two highly correlated signals is encountered in many applications such as localization and tracking of a signal source, determination of bearing and elevation angles, and in sonar or radar detection. Most techniques for *time delay estimation* are based on some form of cross correlation technique [6], [7]. These techniques generally require *a priori* statistics about the signals and require peak detecting algorithms [3]. The technique introduced in this paper develops an adaptive process which synchronizes two highly correlated signals with a time-domain iterative algorithm. The algorithm requires no previous statistics on the input signals; instead it uses an error value and a gradient approximation to continually update the time delay estimate until the process has minimized the error function. The technique used is a steepest-descent *gradient technique* and thus requires that the error function be *unimodal*, or periodically unimodal over the range of values being considered. Gradient estimation techniques employ a *convergence parameter* that is very sensitive to the performance of the system [4], [5]. Improper selection of the convergence parameter can cause either very slow adaptations or situations where the *adaptive delay* does not converge to the desired value at all. Therefore, in addition to developing an upper bound for this parameter in terms of the input signal power, we also illustrate the practical selection of this parameter to optimize the convergence rate. Examples of six different configurations that include white noise sources are given to illustrate the performance of the adaptive delay in estimating the time delay between sinusoids. Learning curves of the adaptations are also shown in examples where the time delay is a time-varying function. In these examples the adaptive delay successfully tracks the changes in time delay between sinusoids.

Manuscript submitted June 1, 1980; revised October 24, 1980. This work was supported in part by the Sandia University Research Program, Sandia National Laboratories, under Contract 42-7947-I.

D. M. Etter is with the Department of Electrical Engineering and Computer Engineering, University of New Mexico, Albuquerque, NM 87131.

S. D. Stearns is with Sandia National Laboratories, Albuquerque, NM 87185.

II. Development of an Adaptive Delay Element

A. MS Error Function Properties

The system model used in the development of the adaptive delay element for time delay estimation is given in Fig. 1. One channel contains the signal s_m, m being the time index or sample number, and the other channel contains the signal s_{m-D}, a delayed version of s_m. We assume for the present that s_m is a stationary time series.

In order to drive the error to a minimum, we modify the variable delay until it "locks" onto a value that minimizes the error e_m. This value of the variable delay is then the estimate of the time delay between the two signals. From Fig. 1, we have

$$e_m = s_{m-D} - s_{m-d}. \tag{1}$$

Then,

$$\begin{aligned} e_m^2 &= (s_{m-D} - s_{m-d})^2 \\ &= s_{m-D}^2 - 2s_{m-D}s_{m-d} + s_{m-d}^2. \end{aligned}$$

Applying the expected value operator to both sides of this result gives

$$E[e_m^2] = 2R_{ss}(0) - 2R_{ss}(DT - dT) \tag{2}$$

where R_{ss} is the continuous autocorrelation function of the signal s, and T is the sampling time step. Thus, in Fig. 1, the *MS error* [9], [10] $E[e_m^2]$ is determined completely by the autocorrelation function of the input signal.

If we assume that the autocorrelation function is unimodal with a single maximum over the range of values allowed for the variable delay d, then the MS error is also unimodal with a single minimum. We now develop an iterative algorithm that modifies the variable delay d until it reaches this minimum value. A continuous variable $\hat{d}$ will be used in the iteration,

Reprinted from *IEEE Trans. Acoust., Speech, Signal Processing*, vol. 29, no. 3, pt. 2, pp. 582–587, June 1981.

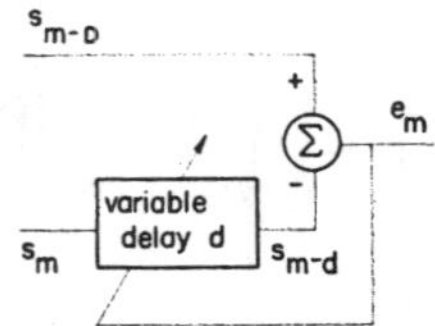

Fig. 1. Model of the adaptive delay system.

but the integer nearest $\hat{d}$ will be the estimated time delay between the two input signals and will be designated by d.

B. Gradient Techniques for Optimization

The general form for an optimization algorithm which uses gradient information is

$$\hat{d}_{m+1} = \hat{d}_m - \mu \nabla_m \tag{3}$$

where μ is a convergence parameter such that $\mu > 0$ and ∇_m is the gradient (derivative) of the MS error with respect to $\hat{d}$. The gradient can sometimes be determined by direct differentiation of the MS error expression, as in Widrow's adaptive filter coefficient algorithm [11]. If the derivative cannot be determined in this way, other approximation methods must be used.

From (2), the gradient of the error function with respect to the delay $\hat{d}$ is

$$\nabla_m = \frac{\partial E[e_m^2]}{\partial \hat{d}} \tag{4}$$

$$= -2 \frac{\partial R_{ss}(DT - \hat{d}t)}{\partial \hat{d}}. \tag{5}$$

We can approximate ∇_m by assuming that $E[e_m^2]$ can be approximated by individual values of e_m^2 [1]. Using (1) and (4), and using continuous forms of the sampled signals e and s, the gradient is thus approximated by

$$\tilde{\nabla}_m = 2 e_m \frac{\partial e(mT)}{\partial \hat{d}}$$

$$= -2 e_m \frac{\partial s(mT - \hat{d}T)}{\partial \hat{d}}.$$

A closed form for the derivative of s with respect to the delay cannot be determined without specific information about the signal s. Therefore, using the symmetric difference as an approximation to the derivative, our gradient approximation becomes

$$\tilde{\nabla}_m = -2 e_m \frac{(s_{m-d-1} - s_{m-d+1})}{(d+1) - (d-1)}$$

$$= -e_m (s_{m-d-1} - s_{m-d+1}). \tag{6}$$

C. Bias and Variance of the Gradient Estimate

Since we are using an approximation to the real gradient, it is necessary to determine the quality of the estimator $\tilde{\nabla}_m$. Hence, we will examine the bias and variance of this estimator.

To examine first the bias of our estimator $\tilde{\nabla}_m$, we use (1) and (6) to obtain the following:

$$\tilde{\nabla}_m = -e_m(s_{m-d-1} - s_{m-d+1})$$

$$= -(s_{m-D}s_{m-d-1} - s_{m-d}s_{m-d-1} - s_{m-D}s_{m-d+1} + s_{m-d}s_{m-d+1}).$$

The expected value of the gradient approximation then follows:

$$E[\tilde{\nabla}_m] = -[R_{ss}(DT - dT - T) - R_{ss}(DT - dT + T)]. \tag{7}$$

From (4), we also have

$$\nabla_m = -2 \lim_{T \to 0} \frac{R_{ss}(DT - (d+1)T) - R_{ss}(DT - (d-1)T)}{(d+1) - (d-1)}$$

$$= -\lim_{T \to 0} [R_{ss}(DT - dT - T) - R_{ss}(DT - dT + T)]. \tag{8}$$

Thus, from (7) and (8), the estimate is unbiased as T approaches zero. For nonzero T, we are primarily concerned with the bias as the gradient approaches zero, because this is the point at which we determine the estimate of the time delay between the two signals. The maximum of the autocorrelation function R_{ss} in (5) occurs when $DT = \hat{d}T$. As dT approaches DT in (7), $E[\tilde{\nabla}_m]$ also approaches zero because of the symmetry of the autocorrelation about its maximum. Thus, the bias of our estimate of the gradient approaches zero as the gradient approaches zero.

For a specific example, let

$$s_m = \sin(4\pi m T)$$

where T is 0.025 s and let D, the fixed delay, be three time steps. The autocorrelation can be shown to be

$$R_{ss}(nT) = \tfrac{1}{2} \cos(n\pi/10).$$

This autocorrelation is periodic, but it is still unimodal within each period, and thus will still be suitable for the "steepest descent" gradient technique. Also, from (1), (2), (5), and (6), we have

$$E[e_m^2] = 1 - \cos((3 - d)\pi/10),$$

$$\nabla_m = -\pi/10 \sin((3 - d)\pi/10),$$

and

$$\tilde{\nabla}_m = -\tfrac{1}{2}[\cos((2 - d)\pi/10) - \cos((4 - d)\pi/10)].$$

Fig. 2 is a plot of ∇_m and $\tilde{\nabla}_m$, and verifies that the bias of the gradient estimate for this example approaches zero as the gradient approaches zero.

Having discussed the bias of the gradient estimator $\tilde{\nabla}_m$, we now turn to its variance. Instead of calculating the variance directly, we will determine its upper bound. This upper bound v will be used in the next section to determine a range for the parameter μ which assures convergence to the correct time delay value. To obtain an exact result, we will assume that the input signal s_m is Gaussian with zero mean. The examples will suggest, however, that the result is also valid for non-Gaussian inputs.

The variance is the expected squared deviation from the mean,

$$\sigma_{\tilde{\nabla}}^2 = E[(\tilde{\nabla}_m - E[\tilde{\nabla}_m])^2] = E[\tilde{\nabla}_m^2] - E^2[\tilde{\nabla}_m].$$

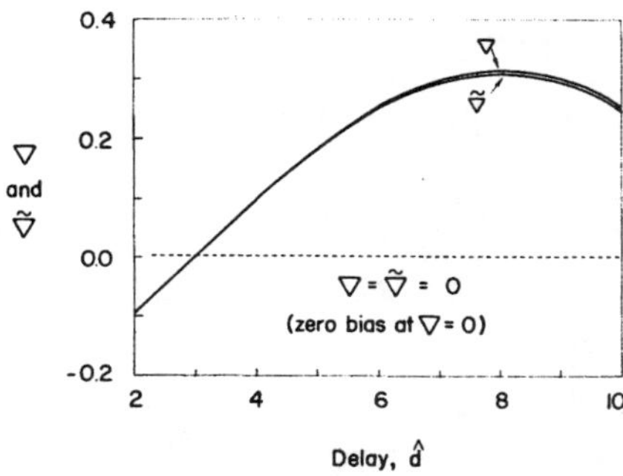

Fig. 2. Gradient estimate that approaches zero as the gradient approaches zero.

Since $E^2[\tilde{\nabla}_m]$ is positive, an upper bound v of the variance can be calculated, using (6) and the theorem of Isserlis [8], to yield the following:

$$\begin{aligned} v &= E[\tilde{\nabla}_m^2] \\ &= E[(-e_m(s_{m-d-1} - s_{m-d+1}))^2] \\ &= 24\,R_{ss}(0). \end{aligned}$$

(All details for obtaining this final result may be obtained from [2].) Since v is an upper bound on the variance, an upper bound SD of the standard deviation is

$$\text{SD} \leqslant 5\,R_{ss}(0).$$

Thus, the SD of the gradient estimate is at most five times the input signal power.

D. Adaptive Delay Algorithm and Associated Convergence Parameter

Combining the general gradient optimization formula (3) with the gradient estimate (6) gives the following iterative formula for determining the optimal delay:

$$\hat{d}_{m+1} = \hat{d}_m + \mu e_m(s_{m-d-1} - s_{m-d+1}). \tag{9}$$

In this recursive algorithm for modifying the continuous delay $\hat{d}_m$, the sign of the gradient will keep modifying $\hat{d}_m$ in the proper direction, but must be small enough to allow convergence to the integer nearest the optimum delay value. (Recall that d_m, the actual delay value, is the integer nearest $\hat{d}_m$.) A large μ could cause the delay to continually jump back and forth across the optimum value without converging. Thus, as the delay converges to the optimum value, we would like the increment in the delay to be less than one. That is, we specify

$$|\mu\tilde{\nabla}_m| < 1, \quad \text{or} \quad 0 < \mu < |1/\tilde{\nabla}_m|. \tag{10}$$

If we assume that the gradient estimate has a Gaussian distribution about the true value of the gradient, then approximately 95 percent of the values for the gradient estimate fall within two standard deviations of the correct value. Thus, as the gradient approaches zero, 95 percent of our estimates will be within two standard deviations of zero. Therefore, from the previous result, i.e., SD $\leqslant 5\,R_{ss}(0)$, the condition $\mu < |1/\tilde{\nabla}_m|$ will hold 95 percent of the time as ∇_m approaches zero if

$$0 < \mu < 1/2\text{SD}, \quad \text{or} \quad 0 < \mu < 1/(10\,R_{ss}(0)). \tag{11}$$

The upper bound for the convergence parameter is thus expressed in terms of the power of the input signal. A large number of simulations have shown that convergence occurs when μ is calculated in this manner. In the examples that follow, μ was set at one-fourth of the upper bound in (11).

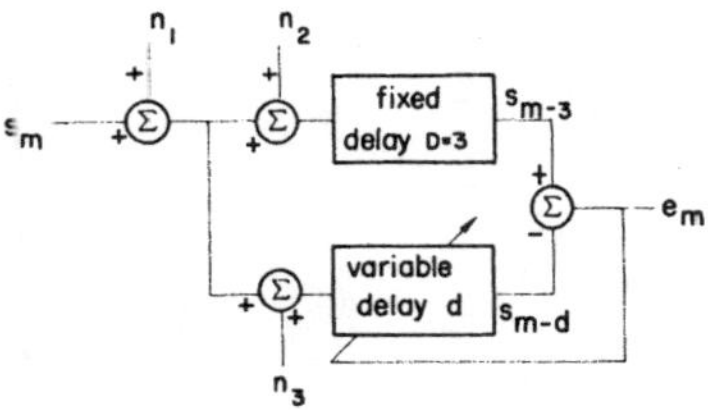

Fig. 3. Delay-lock loop configuration.

III. Time Delay Estimation With the Adaptive Delay

Using the recursive algorithm for the adaptive delay $\hat{d}$ given in (9), we examine examples of the iterative process that minimizes the error signal e_m and thereby determines the estimate of the time delay between two signals. We will consider six variations of the *delay-lock loop* configuration in Fig. 3. The input s_m will be the example used earlier,

$$s_m = \sin 4\pi(m-1)T; \qquad m = 1, 2, \cdots, N$$

with $T = 0.025$ s, a fixed delay of $D = 3$ time steps, and an amplitude that gives a signal power of 40.

Case 1: Assume there is no noise in the system; that is, $n_1 = n_2 = n_3 = 0$. The MS error for this configuration is plotted in Fig. 4(a). As previously mentioned, this error surface is not unimodal, but all the minimum points have the same function value. Therefore, if the adaptive delay is initialized to zero, the adaptive delay should increment to a value of three which is the correct time delay and also drives the error to zero. Plots of the learning curves of the continuous variable delay $\hat{d}$ and the integer variable delay d for a particular computer run are also shown in Fig. 4(a).

Case 2: Assume there is white noise only at n_1; that is, $n_2 = n_3 = 0$. The noise power at n_1 is equal to the signal power. The error function of this system is given in Fig. 4(b), along with typical learning curves of the adaptive delay. As expected, the time delay estimate converges to the correct value of three. Since the same noise exists on both channels, the minimum error is zero when $d = D$.

Case 3: Assume there is white noise only at n_2; that is, $n_1 = n_3 = 0$. The noise power at n_2 is equal to the signal power. Fig. 4(c) shows the MS error of this system and an adaption of the time delay estimate which converges to the correct value of three. The minimum value of the MS error is nonzero, but the correct value of the time delay is still obtained when the error is minimized.

Case 4: Assume there is white noise only at n_3; that is, $n_1 = n_2 = 0$. The noise power at n_3 is equal to the signal power. Fig. 4(d) contains the error function and the adaption of the time delay estimate. Note that Cases 3 and 4 have the same error function.

Case 5: Assume there is white noise at both n_2 and n_3; that is, $n_1 = 0$. The noise samples are independent of each other. The noise power at each of the two noise sources is equal to

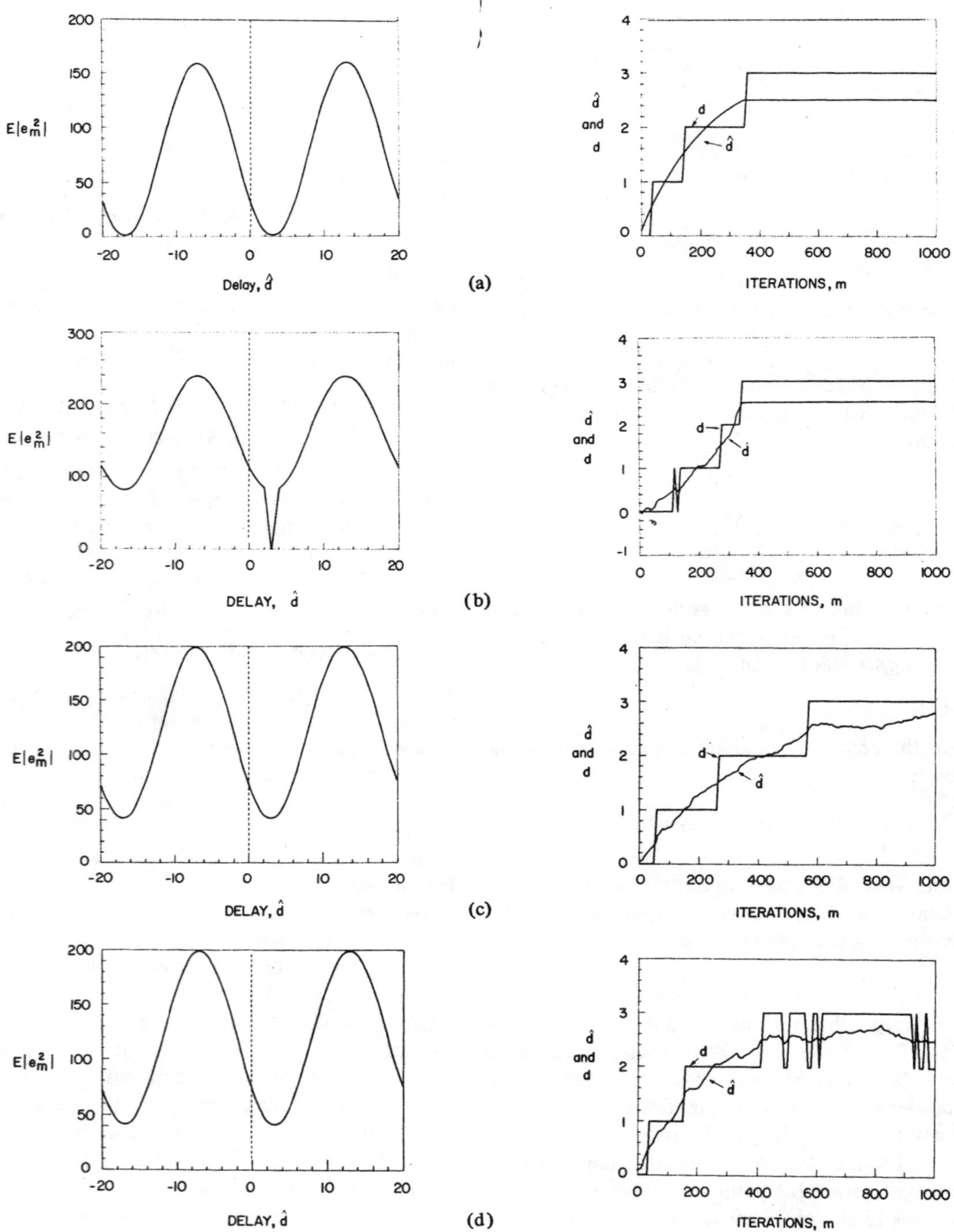

Fig. 4. Simulations of time delay estimation. (a) Case 1. No noise in delay-lock loop. (b) Case 2. Noise in channel n_1 of delay-lock loop. (c) Case 3. Noise in channel n_2 of delay-lock loop. (d) Case 4. Noise in channel n_3 of delay-lock loop.

the signal power. The minimum value of the error function is again nonzero. Fig. 4(e) shows the MS error and the learning curves of the adaptive process, which again converges to the correct time delay estimate.

Case 6: Assume there is white noise at n_1, n_2, and n_3. All three noise samples are independent of each other. The noise power at each of the three noise sources is equal to the signal power. The minimum value of the error function is nonzero. Fig. 4(f) shows the MS error and the learning curves of the adaptation process.

IV. Time Delay Tracking With the Adaptive Delay

The previous section illustrates correct convergence of the adaptive delay element to the time delay difference between two signals which have unimodal or periodically unimodal correlation functions. This section examines the performance of the adaptive delay in *time delay tracking* configurations. The simulations were performed in a noiseless environment, and then in a noisy environment with a signal-to-noise ratio of 1 : 1. Fig. 5 contains the learning curve of a simulation using a sinu-

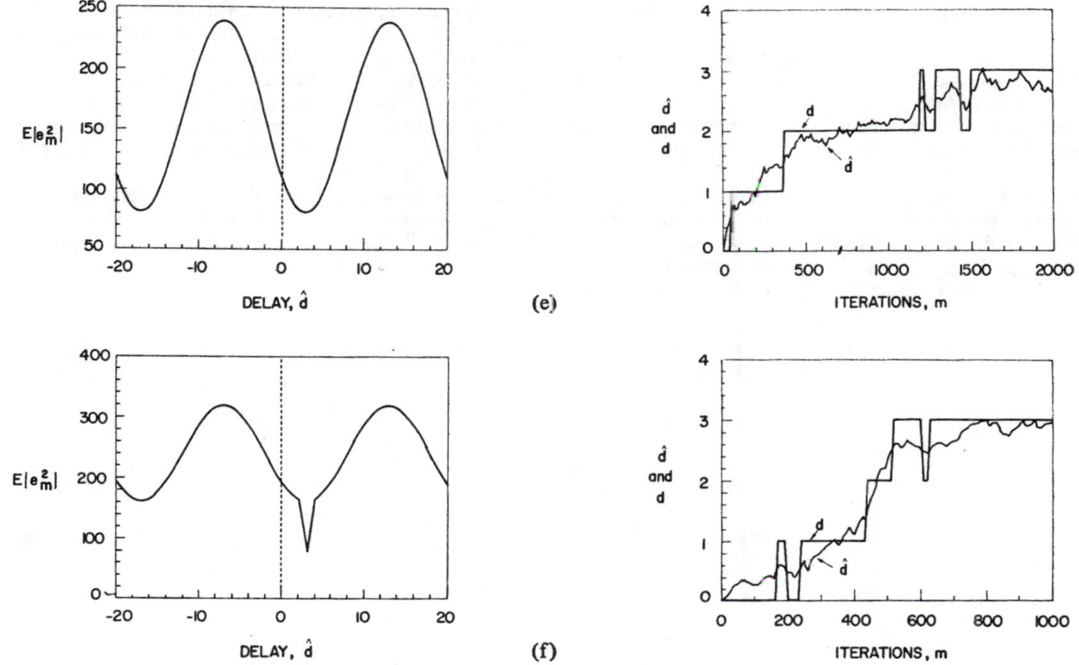

Fig. 4 (Continued). Simulations of time delay estimation. (e) Case 5. Noise in channels n_2 and n_3 of delay-lock loop. (f) Case 6. Noise in channels n_1, n_2, and n_3 of delay-lock loop.

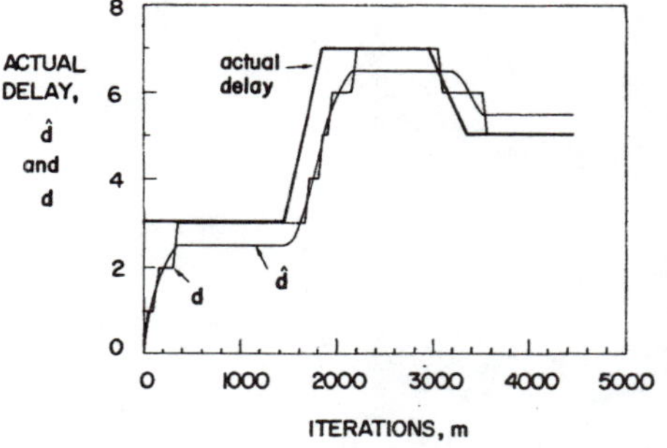

Fig. 5. Time delay tracking with no noise in system.

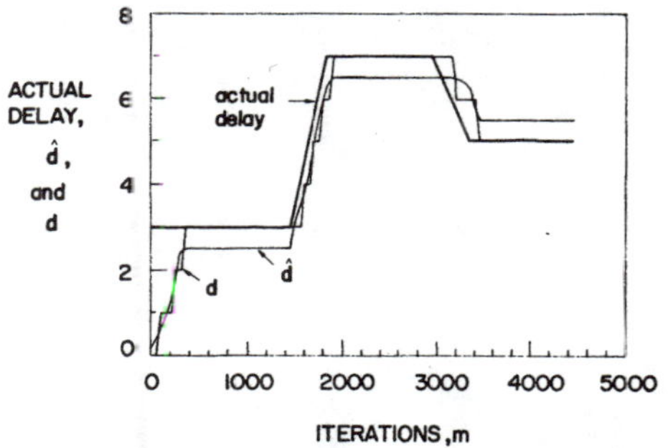

Fig. 6. Time delay tracking with noise in system.

soid that was initially three sampling intervals behind the original signal. The time delay between the signals was increased gradually from three to seven time steps, and then similarly decreased to five steps. Thus, the delay changes represent continuous changes rather than abrupt changes. The adaptive delay correctly tracked the changing time delay. This example involved the adaptive delay moving both forward and backward on the error surface to obtain the optimum value of d that minimized the error e_m. Depending on the setting of μ, if the time delay changes occur too rapidly, the adaptive delay may not adapt to the previous final value before beginning to track the new time delay value. Fig. 6 contains the same configuration, but noise has been added on the n_1 channel. Again the time delay tracking performs correctly.

V. Conclusions

The adaptive delay element has proven to be an effective means for *synchronizing* two *sampled signals* that are highly correlated by estimating the time delay between the signals. However, the requirement for the proper convergence of the adaptive delay is that the MS error of the configuration be a unimodal function with one minimum, or periodic with one minimum per period. With these requirements, the adaptive delay will estimate the delay between two signals with no *a priori* information. The adaptive delay performs correctly in noise environments, as shown in the examples with additive noise. The convergence factor which controls the rate of convergence has been shown to depend on the input signal power.

The simplicity of the adaptive algorithm is a significant characteristic of the adaptive delay. It does not require squaring, differentiating, averaging, or the use of transforms. With so many applications now based in microprocessors and minicomputers, the reduced computation time and storage requirements of the adaptive delay are advantages worthy of note.

References

[1] D. M. Etter and S. D. Stearns, "Convergence properties of the adaptive delay element in delay-lock loops and frequency tracking," presented at the 13th Ann. Asilomar Conf. Circuits, Syst., Comput., Pacific Grove, CA, Nov. 1979.

[2] D. M. Etter, "Digital signal processing with an adaptive delay element," Ph.D. dissertation, Univ. New Mexico, Albuquerque, 1979.

[3] R. Fletcher and M.J.D. Powell, "A rapidly convergent descent

method for minimization," *Comput. J.*, vol. 6, pp. 163–168, 1963.

[4] L. J. Griffiths, "Rapid measurement of digital instantaneous frequency," *IEEE Trans. Acoust., Speech, Signal Processing*, vol. ASSP-23, pp. 207–222, Apr. 1975.

[5] H. L. Groginsky *et al.*, "Adaptive detection of statistical signals in noise," *IEEE Trans. Inform. Theory*, vol. IT-12, pp. 337–348, July 1966.

[6] J. C. Hassab and R. E. Boucher, "Optimum estimation of time delay by a generalized correlator," *IEEE Trans. Acoust., Speech, Signal Processing*, vol. ASSP-27, pp. 373–380, Aug. 1979.

[7] C. H. Knapp and G. C. Carter, "The generalized correlation method for estimation of time delay," *IEEE Trans. Acoust., Speech, Signal Processing*, vol. ASSP-24, pp. 320–327, Aug. 1976.

[8] L. H. Koopmans, *The Spectral Analysis of Time Series*. New York: Academic, 1974.

[9] H. W. Sorenson, "Least-squares estimation: From Gauss to Kalman," *IEEE Spectrum*, pp. 63–67, July 1970.

[10] B. Widrow *et al.*, "Stationary and nonstationary learning characteristics of the LMS adaptive filter," *Proc. IEEE*, vol. 64, pp. 1151–1162, Aug. 1976.

[11] B. Widrow and J. M. McCool, "A comparision of adaptive algorithms based on the methods of steepest descent and random search," *IEEE Trans. Antennas Propagat.*, vol. AP-24, pp. 615–637, Sept. 1976.

Time Delay Estimation Using the LMS Adaptive Filter—Dynamic Behavior

PAUL L. FEINTUCH, MEMBER, IEEE, NEIL J. BERSHAD, MEMBER, IEEE, AND FRANCIS A. REED, MEMBER, IEEE

Abstract—**The LMS adaptive filter is used to estimate a linearly moving time delay between two broad-band waveforms. The tracking behavior of the mean weights is analyzed and is compared with simulations of the actual device.**

I. Introduction

THE concept of using an *LMS adaptive filter*, configured as a canceller to estimate the time delay difference between two waveforms, is studied in [1] for a static time delay. The estimate of delay is obtained by interpolating on the weights in the filter to select the point in the tapped delay line that corresponds to the peak weight. The resulting estimate is unbiased in the nonmoving input case, and the variance is within 0.5 dB of the Cramér–Rao lower bound for the variance of any delay estimate using the same data. In addition, it was shown that if assumed *a priori* statistics concerning the input power spectra differ from what is actually the case, then the adaptive filter tracker can dramatically outperform a fixed parameter conventional tracker being operated in an environment other than that for which it was designed.

Since the *delay estimation* is performed on the peak value of the weights in the adaptive filter, the time constant for the selection of the peak is the significant transient response. The peak weight can be correctly selected much more quickly than the time required for either the weights or the mean square error to converge [1]. This suggests that this processor has the potential to track rapidly changing time delays by just observing the peak weight move through the adaptive filter tapped delay line.

This paper extends the results of [1] to consider a *linearly time-varying delay.* The *mean weights* are derived for both spectrally white signals and band-limited broad-band signals with exponential correlation function and with small signal-to-noise ratio. The mean weights in steady state are derived and shown to be time varying and lagging the ramp input. Simulations are then presented, demonstrating good agreement with the analytical results.

II. Analysis

A. Mean Weight Behavior of Adaptive Filter Tracker

In this section, the time-varying mean weights of the adaptive filter are derived for a linearly time-varying change of delay. Two signal models are considered—a broad-band spectrally white signal and a broad-band nonwhite signal.

For a spectrally white process, the mean filter weights are shown to be a traveling wave with a decaying exponential envelope. Tracking the time delay involves estimating the delay location of the leading edge of the weights. For the broadband nonwhite signal, the mean weights behave similarly except, because of correlation between the taps, the peak of the traveling wave is much larger and the decaying exponential envelope is dependent on the signal dynamics and correlation.

1) Mean Weights for Broad-Band White Signals in Uncorrelated Noise: Let the time delay between the two input channels be given by $D(t) = bt$. The input signal and noise are assumed to be independent zero-mean Gaussian random processes that are spectrally white over the band corresponding to the sampling frequency, with power σ_s^2 and σ_n^2, respectively. One channel, $d(n)$, is the desired signal for the adaptive process, and the other is the input $x(n)$ to the adaptive filter.

The algorithm for changing the weights in the adaptive filter is given by [2]

$$\begin{aligned} W(n+1) &= W(n) + \mu\,[d(n) - X^T(n)\,W(n)]\,X(n) \\ &= W(n) + \mu\,[d(n)\,X(n) - X(n)\,X^T(n)\,W(n)] \end{aligned} \tag{1}$$

where

$W(n)$ = filter weight vector at time sample n
$d(n)$ = desired signal at time sample n
$X(n)$ = observed data vector of samples
μ = LMS algorithm step size.

T denotes vector transpose. Fig. 1 shows the data vector $X(n)$ in the tapped line. The scalar $d(n)$ is the delayed and sampled signal plus noise,

$$d(n) = s(nT_s - D(nT_s)) + n_1(nT_s) \tag{2}$$

and the vector $X^T(n)$ is

$$X^T(n) = (s(nT_s), s((n-1)\,T_s) \cdots s((n-M)\,T_s)) + N_2^T(nT_s) \tag{3}$$

where

T_s = time delay between taps = algorithm sampling time
M = number of taps.

It is first assumed that the signal is white, so that

Manuscript received May 9, 1980; revised September 30, 1980. This work was supported by the Naval Sea Systems Command Code 63R under Contract N00024-77-C-6251.

P. L. Feintuch and F. A. Reed are with Hughes Aircraft Company, Fullerton, CA 92634.

N. J. Bershad is with the Department of Electrical Engineering, University of California, Irvine, CA 92664, and is a consultant to Hughes Aircraft Company, Fullerton, CA 92634.

Reprinted from *IEEE Trans. Acoust., Speech, Signal Processing*, vol. 29, no. 3, pt. 2, pp. 571–576, June 1981.

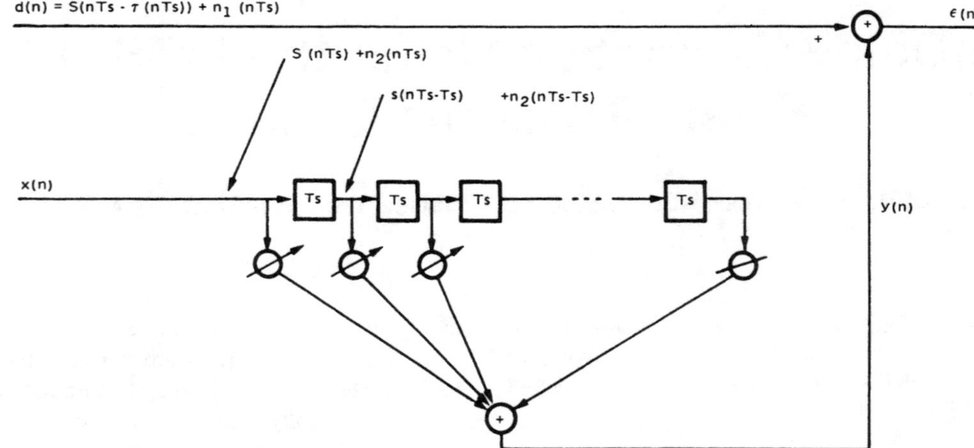

Fig. 1. Data in the delay line.

$$E[s(n)\,s(m)] = \sigma_s^2 \delta(n-m) \tag{4}$$

where $\delta(n-m) = 1$ if $n = m$ and $\delta(n-m) = 0$ otherwise. Taking expectations of (1) and making the assumption [3] that the data and weights at time n are uncorrelated yields a difference equation for the mean of the adaptive filter weight,[1]

$$E[W(n+1)] = [I - \mu R_{xx}(n)]\, E[W(n)] + \mu R_{dx}(n) \tag{5}$$

where

$I = M \times M$ identity matrix

$R_{dx}(n) = E[d(n)\,X(n)]$

$R_{xx}(n) = E[X(n)\,X^T(n)]$.

Since $R_{xx}(n)$ is independent of n, and the initial mean weight can be $E[W(0)] = 0$, then (5) can be rewritten as

$$E[W(n)] = \mu \sum_{k=0}^{n-1} [I - \mu R_{xx}(k)]^{n-k-1} R_{dx}(k). \tag{6}$$

The second-order statistics can be calculated from the input waveforms as follows:

$$R_{xx}(n) = (\sigma_s^2 + \sigma_n^2)\, I \tag{7}$$

$$R_{dx}^T(n) = (\delta(bnT_s), \delta(bnT_s - T_s) \cdots \delta(bnT_s - MT_s)). \tag{8}$$

Using (6), the mean weight vector at the nth iteration is given by

$$E[W(n)] = \mu\sigma_s^2 \sum_{k=0}^{n-1} [1 - \mu(\sigma_s^2 + \sigma_n^2)]^{n-k-1} R_{dx}(k). \tag{9}$$

In the static case, $b = 0$ and only the first weight is nonzero, with mean value

$$E[W_0(n)] = \frac{\sigma_s^2}{\sigma_s^2 + \sigma_n^2}\,[1 - (1 - \mu(\sigma_s^2 + \sigma_n^2))^n] \tag{10}$$

which converges to the Wiener filter [4] for the broad-band stationary case as $n \to \infty$.

[1]The time compression parameters, as discussed in [6] for example, are imbedded in the cross correlation properties of the taps and hence are contained directly in $R_{dx}(n)$ and $R_{xx}(n)$.

In the dynamic case, the weights are a moving set of spikes that are changing with amplitude as the signal moves and as the weights converge. The total weight vector is the sum of the vectors in (9). The weights can be viewed as a sliding window of exponentially growing responses, or as a moving weight at the leading edge that leaves behind it an exponentially decaying wake.

This can be seen by examining the weight at the leading edge of the response. In (9), the leading edge will occur at the latest time. If the filter is sufficiently long so that the response still falls within the tapped delay line, i.e., $M > n$, then the amplitude and location of the leading weight are found by examining the term in the summation for which $k = n - 1$. The amplitude of the leading edge is $\mu\sigma_s^2$ and its location is at tap number $b(n-1)$. If $b = 1$ then the signal moves one tap per iteration; if the signal changes more slowly, then $b < 1$ and the leading edge moves more slowly than the iteration rate.

For the special case where $b = 1$ the weight vector in (9) can be readily expanded. Letting $r = 1 - \mu(\sigma_s^2 + \sigma_n^2)$,

$$E[W^T(n)] = \mu\sigma_s^2(r^{n-1}, r^{n-2}, \cdots, r^1, 1, 0, \cdots) \tag{11}$$

which shows the decaying wake behind the leading weight which shifts along the delay line as n increases, as pictured in Fig. 2.

For $b < 1$ the weights move more slowly and the basic model herein tends to become less realistic. The signal sequence used is uncorrelated in time. In general this is not the case. The impact of this assumption is to have weights respond only at the exact correct alignment of input delays and tap delay values. In the band-limited, nonwhite signal case, correlation will exist even at noninteger delay shifts, and larger weight responses should be expected at the leading edge for slower moving signals. It is shown in the next section that the amplitude of the leading edge decreases monotonically with b from $\sigma_s^2/(\sigma_s^2 + \sigma_n^2)$ (the value for $b = 0$) to $\mu\sigma_s^2$ (the value for which movement is so fast that the signal samples essentially decorrelate totally at each time sample). The extent of the wake and the height of the leading edge will depend on signal dynamics.

2) Mean Weights for Band-Limited Broad-Band (Correlated) Signal in Uncorrelated Noise for Small Signal-to-Noise Ratios:

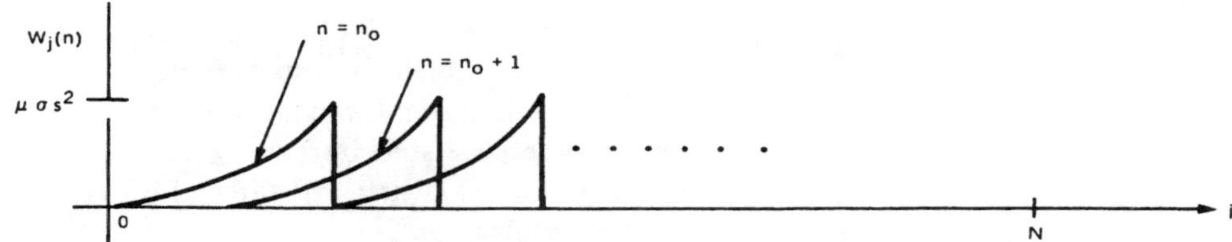

Fig. 2. Envelope of the tracker weights as they move in time.

The conditions of the previous section hold with one extension. The signal spectrum is no longer white. Equations (2) and (3) are still valid. However, $R_{dx}(n)$ and $R_{xx}(n)$ [(7) and (8)] are now replaced by

$$R_{dx}^T(n) = (R_s(bnT_s), R_s(bnT_s - T_s) \cdots R_s(bnT_s - MT_s)) = \sigma_s^2(\rho^{|bnT_s|}, \rho^{|bnT_s - T_s|} \cdots \rho^{|bnT_s - MT_s|}) = \sigma_s^2 V^T(n) \quad (12)$$

$$R_{xx}(n) = \sigma_n^2 I + \begin{bmatrix} \sigma_s^2 R_s(T_s) \cdots R_s(MT_s) \\ \text{SYM} \quad \ddots \\ \sigma_s^2 \end{bmatrix} = \sigma_n^2 I + \sigma_s^2 \Lambda \quad (13)$$

where

$$\Lambda = \begin{bmatrix} 1 & \rho^{T_s} \rho^{2T_s} \rho^{MT_s} \\ & \ddots \\ \text{SYM} & \end{bmatrix}. \quad (14)$$

Since $R_{xx}(n)$ is independent of n, for $E[W(0)] = 0$, (6) simplifies to

$$E[W(n)] = \mu\sigma_s^2 \sum_{k=0}^{n-1} R^{n-k-1} V(k) \quad (15)$$

where

$$R = (1 - \mu\sigma_n^2) I - \mu\sigma_s^2 \Lambda. \quad (16)$$

In order to proceed further, it is necessary to assume small signal-to-noise ratio, i.e., $\sigma_s^2 << \sigma_n^2$. Then $R_{xx}(n) \approx \sigma_n^2 I, R \approx (1 - \mu\sigma_n^2) I$. The mean weight vector, for small signal-to-noise ratio, is approximately

$$E[W(n)] \approx \mu\sigma_s^2 \sum_{k=0}^{n-1} (1 - \mu\sigma_n^2)^{n-k-1} V(k). \quad (17)$$

The jth weight becomes

$$E[W_j(n)] = \mu\sigma_s^2 \sum_{k=0}^{n-1} (1 - \mu\sigma_n^2)^{n-k-1} \rho^{|bk-j| T_s}. \quad (18)$$

Assume that the filter is long enough so that $M = n - 1$, and let $j = Mb$ so that we are at the end of the filter. For $r = 1 - \mu\sigma_n^2$,

$$E[W_{Mb}(n)] = \mu\sigma_s^2 r^M \sum_{k=0}^{M} r^{-k} \rho^{-(k-M)bT_s} = \mu\sigma_s^2 \frac{r^M \rho^{(M+1)bT_s} - 1}{r\rho^{bT_s} - 1}. \quad (19)$$

Equation (19) describes the transient behavior of the mean weight for small signal-to-noise ratios. In steady state, i.e., M arbitrarily large, one can readily see the relationships between signal correlation ρ, signal dynamics b, and algorithm dynamics μ:

$$\lim_{M \to \infty} E[W_{Mb}(\infty)] = \frac{\mu\sigma_s^2}{1 - r\rho^{bT_s}} = \frac{\mu\sigma_s^2}{1 - [1 - \mu\sigma_n^2]\rho^{bT_s}}. \quad (20)$$

There are several cases of interest. If $\rho = 1$, then the signal is totally correlated from tap to tap, regardless of signal dynamics. For this case $E[W_{Mb}(\infty)] = \sigma_s^2/\sigma_n^2$. This is a small signal-to-noise ratio approximation to $\sigma_s^2/(\sigma_n^2 + \sigma_s^2)$ which is what one would expect in the static case. If $\rho = 0$, the signal is uncorrelated and again signal dynamics should not affect the result. For this case, $E[W_{Mb}(\infty)] = \mu\sigma_s^2$ which agrees with (11) where dynamics tended to decorrelate the signal. If there are no signal dynamics, i.e., $b = 0$, then $E[W_{Mb}(\infty)] = \sigma_s^2/\sigma_n^2$ independent of ρ, which is again the expected small signal-to-noise ratio static result. The tradeoff between dynamics and signal correlation can be seen by examining the term ρ^{bT_s} in the expression for $E[W_{Mb}(\infty)]$. As b decreases, ρ^{bT_s} looks more like unity and the signal looks more correlated. As b increases, ρ (which is less than one) is raised to a higher power and the signal has become less correlated.

B. Steady-State Mean Weights

Equation (18) is an expression for the mean weights of the adaptive tracker, for small signal-to-noise ratios, when the input signal is a broad-band correlated process with linearly time-varying delay, buried in uncorrelated noise. Equation (18) can be expressed, in steady state, in closed form as [5, Appendix G]

$$\lim_{n \to \infty} E[W_j(n)] = \mu\sigma_s^2 \left[\frac{(1 - \mu\sigma_n^2)^{n-l-1}}{1 - (1 - \mu\sigma_n^2)\rho^{bT_s}} + \frac{\rho^{bT_s(n-l-1)} - (1 - \mu\sigma_n^2)^{n-l-1}}{1 - (1 - \mu\sigma_n^2)\rho^{-bT_s}} \right],$$

$$j \leqslant b(n-1)$$

$$= \frac{\mu\sigma_s^2}{1 - (1 - \mu\sigma_n^2)\rho^{bT_s}} \rho^{-bT_s[n-l-1]},$$

$$j \geqslant b(n-1) \tag{21}$$

where

$$l = \text{integer such that } l \leqslant \frac{j}{b} < l + 1, \text{ and } |j - bl| T_s \ll 1.$$

The weights have an exponentially decaying leading and trailing edge. For $\mu\sigma_n^2 < 1 - \rho^{bT_s}$, the trailing edge of the weights has a decay factor $1 - \mu\sigma_n^2$. For $\mu\sigma_n^2 > 1 - \rho^{bT_s}$, the trailing edge of the weights has a decay factor ρ^{bT_s}. The leading edge of the weights has decay factor ρ^{bT_s} independent of the ratio. Hence, for $\mu\sigma_n^2 > 1 - \rho^{bT_s}$, the shape of the trailing edge of the weights is determined by the input signal dynamics, whereas for $\mu\sigma_n^2 < 1 - \rho^{bT_s}$, the shape of the trailing edge of the weights is determined by the system dynamics.

The *lag* in the peak of the weights in (21) in comparison to the true delay is given by

$$\text{tap lag} = \left[\frac{b}{b\alpha T_s + \ln(1 - \mu\sigma_n^2)}\right] \cdot \ln\left\{\frac{\alpha b T_s[1 - (1 - \mu\sigma_n^2)\rho^{bT_s}]}{-\ln(1 - \mu\sigma_n^2)\,[(1 - \mu\sigma_n^2)(\rho^{-bT_s} - \rho^{bT_s})]}\right\} \tag{22}$$

$$\approx \frac{b}{b\alpha T_s - \mu\sigma_n^2} \ln\left[\frac{1}{2} + \frac{b\alpha T_s}{2\mu\sigma_n^2}\right] \tag{23}$$

where $\rho = e^{-\alpha}$. Thus the lag increases with the rate of the signal b, and decreases with larger feedback coefficient, μ.

III. Simulations

Simulations were run of a 16-tap adaptive filter operating at a sample rate 2400 Hz with broad-band (i.e., spectrally flat) inputs band-limited to 800 Hz, and a delay between signals that is linearly varying with time. The band-limited inputs were generated by passing computer generated white noise through a digital FIR filter. In Section II, it was shown that the mean weight vector, as a function of time, is a peak that follows the instantaneous delay between array halves, moving through the filter at the delay rate of change with lag and with an exponentially decaying trailing edge. This analysis is verified by Fig. 3, which shows the weight vector at 5000, 6000, and 7000 iterations for a 10 dB signal-to-noise ratio (SNR) and a delay changing at 1.0 ms/s.

A more quantitative assessment of the tracking behavior of the *adaptive tracker* can be had from Fig. 4, showing the delay estimate as a function of time for a linearly increasing delay of 261.8 μs/s, with $\mu = 2^{-10}$ for an SNR of 0 dB and $\mu = 2^{-14}$ for an SNR of -10 dB. Note that the delay estimate lags the time but has the correct time-averaged rate of change. The losses of track in Fig. 4(b) occur when weight fluctuations obscure the peak of the mean weights. The instantaneous estimates of delay while tracking will contain two sources of error-dynamics and random noise. The noise errors will cause the instantaneous delay estimates to deviate from a straight line. The dynamic errors will cause the estimate to lag. If one time averages with a moving window, the random noise errors will be reduced (although not dramatically because the fluctuations are probably highly correlated—the LMS algorithm has already performed significant averaging). The remaining error will be due to a fixed lag which is generally supported by Fig. 4. Evaluating the lag using (23) for this example produces a value of 0.0008 s, which is greater than the average lag of 0.003 s in Fig. 4(b). This difference may be due to the different spectral shape in the analysis and the simulations, causing the bandwidth parameters to disagree.

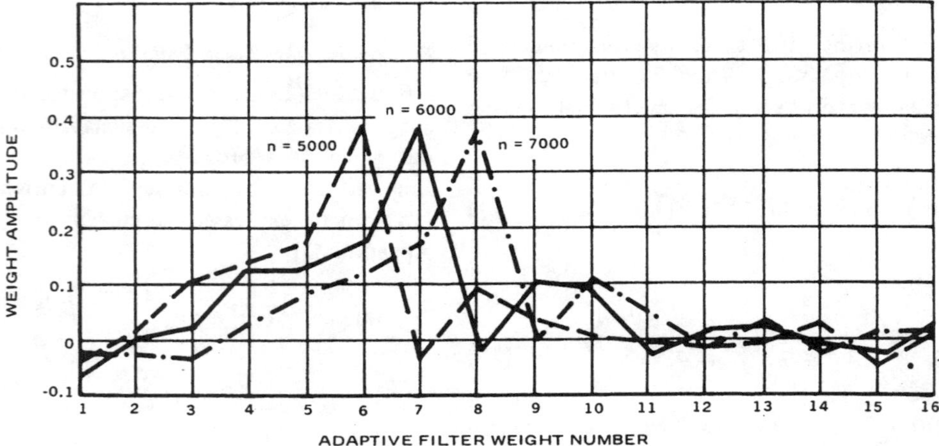

Fig. 3. Weight vector as a function of time (n) for linearly varying delay and broad-band input, N = number of taps = 16, SNR = 10 dB, $\mu = 2^{-10}$, b = 1.0 ms/s.

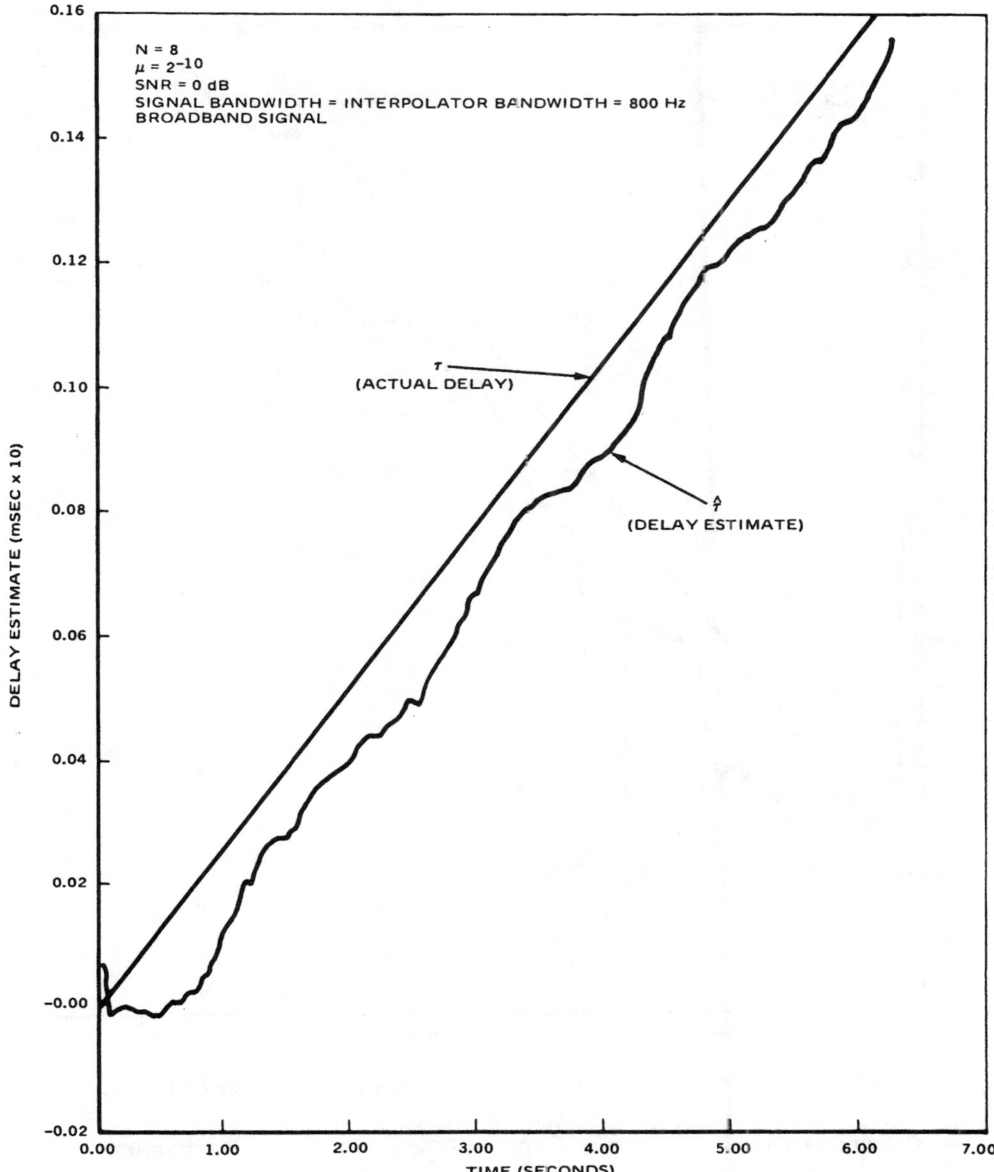

Fig. 4. (a) Delay estimate versus time for linearly varying delay (261.8 μs/s).

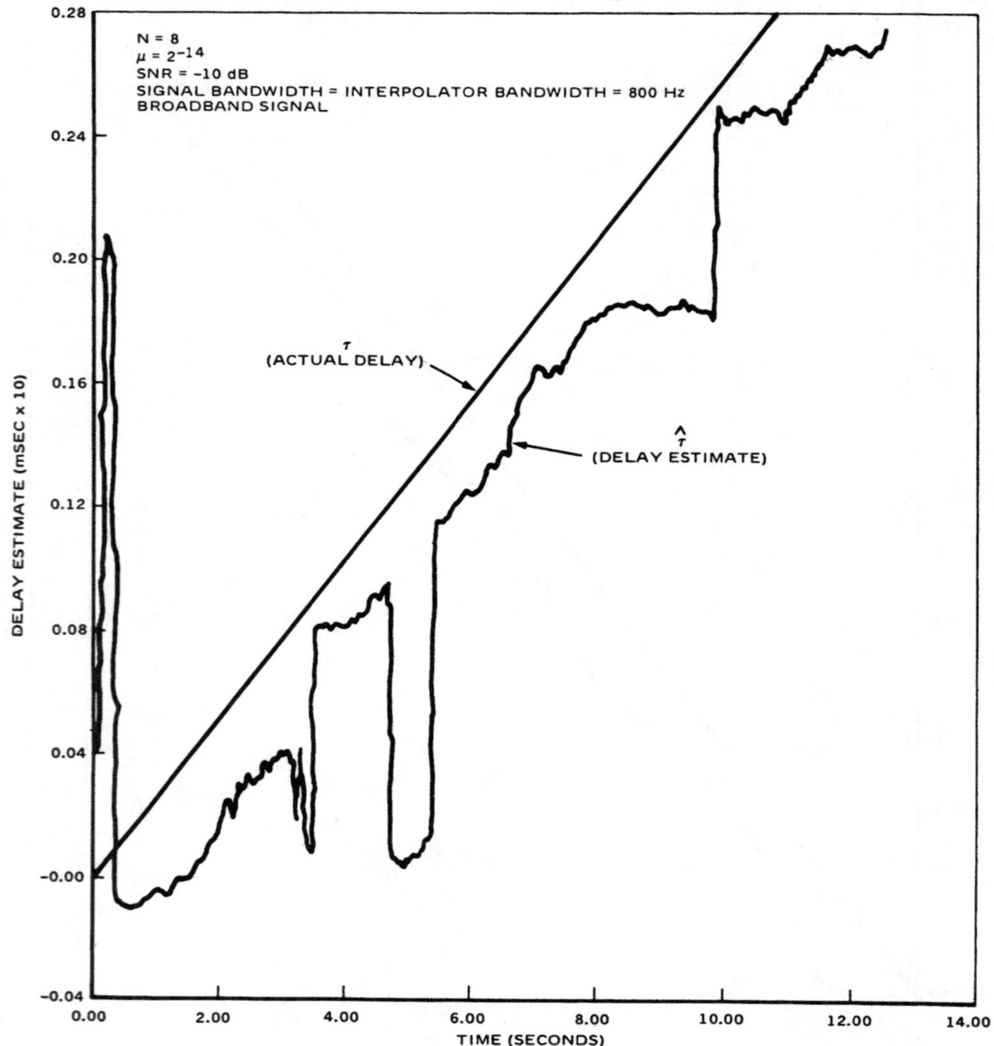

Fig. 4 (Continued). (b) Delay estimate versus time for linearly varying delay (261.8 μs/s).

IV. Conclusions

The transient and steady tracking behavior of the mean value of the LMS adaptive tracker weights have been derived for both spectrally white and nonwhite signal processes with linearly time-varying delays. The mean filter weights were shown to be a traveling wave which leaves behind an exponentially decreasing wake. It was shown that the trailing edge of the adaptive filter mean weight is determined by the algorithm dynamics for small weight update adjustments and by the input signal dynamics for large weight update adjustments.

It was also shown that the location of the peak of the weights lags the true time delay by an amount that depends on the delay rate, the signal correlation function, and the adaptive filter time response. For a large adaptive filter time constant, the lag is linear with respect to delay rate and inversely proportional to the adaptive filter time constant.

References

[1] F. A. Reed, P. L. Feintuch and N. J. Bershad, "Time delay estimation using the LMS adaptive filter–Static behavior," this issue, pp. 561–571.

[2] B. Widrow *et al.*, "Adaptive noise cancelling: Principles and applications," *Proc. IEEE*, vol. 63, pp. 1692–1716, Dec. 1975.

[3] J. R. Treichler, "Transient and convergent behavior of the adaptive line enhancer," *IEEE Trans. Acoust., Speech, Signal Processing*, vol. ASSP-27, p. 53, Feb. 1979.

[4] A. Papoulis, *Probability, Random Variables, and Stochastic Processing.* New York: McGraw-Hill, 1965.

[5] P. L. Feintuch, F. A. Reed, N. J. Bershad, and C. M. Flynn, "Adaptive tracking system study–Phase 2," Hughes Aircraft Co., Fullerton, CA, Final Rep., Oct. 1979, prepared for NAVSEA Code 63R, Contract N00024-77-C-6251.

[6] C. H. Knapp and G. C. Carter, "Estimation of time delay in the presence of source and receiver motion," *J. Acoust. Soc. Amer.*, vol. 61, June 1977.

A New Constrained Least Mean Square Time-Delay Estimation System

K. C. HO AND P. C. CHING

Abstract —**A novel adaptive filter for determining the time difference in a signal between two split-array outputs is described. A least mean square (LMS) algorithm is used to adapt the filter coefficients which are constrained to samples of a sinc function. The newly configured LMS time-delay estimation model has a faster convergence speed and a superior ability to track time-varying delays.**

I. Introduction

The estimation of the time delay between signals arriving at two spatially separated sensors is a problem of considerable practical interest in many applications such as radar, seismology and sonar [1]. Let $x(k)$ and $y(k)$ be the output sequences of the two sensors, and

$$x(k) = s(k) + \theta(k) \tag{1}$$

$$y(k) = s(k-D) + \phi(k) \tag{2}$$

where the source $s(k)$ is uncorrelated with the corrupting noises $\theta(k)$ and $\phi(k)$ and D is the time difference between the receiver outputs. Signals $s(k)$, $\theta(k)$, and $\phi(k)$ are all assumed to be real, jointly stationary, random processes with uniform distribution. The problem of time-delay estimation (TDE) is to determine D from $x(k)$ and $y(k)$. The conventional approach for TDE involves finding the location of the peak of the generalized correlation function of the outputs of the two sensors [2]. An alternative approach for TDE with limited *a priori* knowledge of the source spectrum employs adaptive filtering technique. The basic method, known as least mean square time-delay estimation (LMSTDE) [3], models the time difference between the two signals as a FIR filter in one of the receiver channels. Widrow's LMS algorithm is used to update the filter coefficients and a rough delay estimate is given by the filter tap number corresponding to the maximum weight. More recently, Ching and Chan [4] have shown that by imposing the condition that filter coefficients of a time delay model are samples of a sinc function, the adaptation can be made shorter and less computations are involved.

Manuscript received July 3, 1989; revised November 8, 1989 and November 30, 1989. This paper was recommended by Associate Editor D. Graupe.

The authors are with the Department of Electronic Engineering, The Chinese University of Hong Kong, Shatin, N.T., Hong Kong.

IEEE Log Number 9036292.

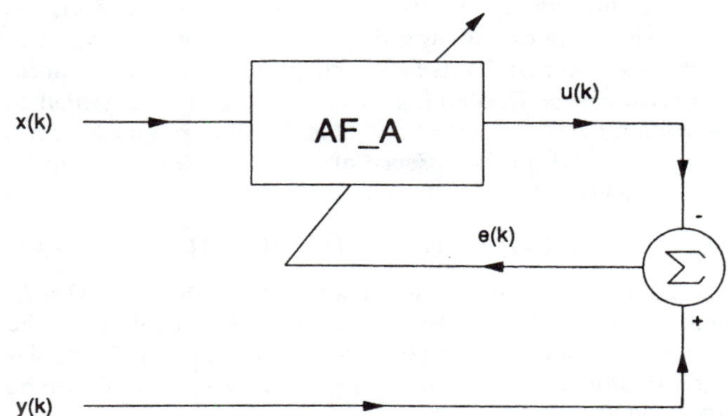

Fig. 1. Basic adaptive TDE configuration.

In this paper, a novel configuration for constrained adaptive TDE is proposed. The formulation, analysis and practical implementation of this model will first be discussed in Sections II and III. The adaptive algorithm together with simulation results will then be presented in Section IV and finally, conclusions are drawn in Section V.

II. The Adaptive Filter for Constrained TDE

The basic configuration for time delay estimation using LMS adaptive filter is shown in Fig. 1, where $x(k)$ and $y(k)$ are output samples of two split-array sensors. Determination of the inter-array delay requires estimation of the peak coefficient from the weight vector a_i of the transversal adaptive filter AF_A. For constrained adaptation [4], only the filter tap with the largest amplitude of AF_A is updated using the LMS algorithm in each iteration. All other coefficients are, however, assigned to samples of a sinc function accordingly through a table look-up operation. The expectation of the mean-square error (MSE), $E[e^2(k)]$, is given by

$$E[e^2(k)] = E[\{y(k)-u(k)\}^2] = E\left[\left\{y(k) - \sum_{i=-N}^{N} a_i x(k-i)\right\}^2\right] \tag{3}$$

which will be minimized by the stochastic gradient method. Now, $2N$ is the filter order and a_i are constrained to values of a

Reprinted from *IEEE Trans. Circuits Syst.*, vol. 37, no. 8, pp. 1060–1064, August 1990.

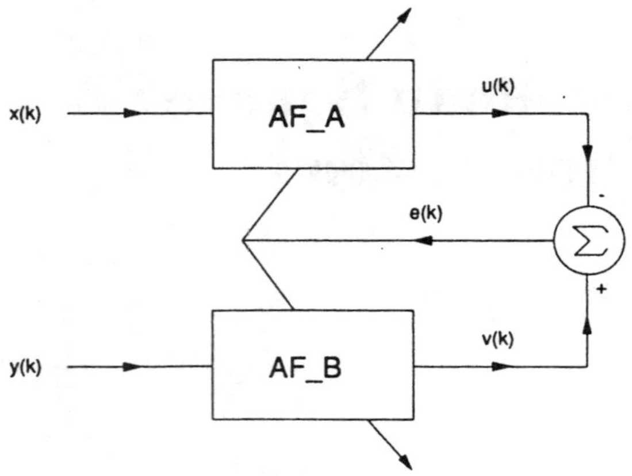

Fig. 2. The new adaptive TDE configuration.

sinc function, say, $a_i = \mathrm{sinc}(i - \Omega)$. When Ω is an integer, the filter weights are essentially a δ function with non-zero value at tap Ω. Assume that $2N$ is large enough such that filter truncation error can be ignored [5], then, AF_A can be regarded as inserting a delay Ω to $x(k)$ and thus the filter output becomes $u(k) = x(k - \Omega)$. In the absence of noise, substitute (1) and (2) into (3) yields

$$E\left[e^2(k)\right] = E\left[\{s(k-D) - s(k-\Omega)\}^2\right]. \tag{4}$$

At steady state, $E[e^2(k)]$ will tend to zero and thus $\Omega = D$. Hence, the interarray delay D can be derived easily from the filter weights, and no interpolation is necessary even for nonintegral (multiples of sampling period, T) delays. Since D can be expressed as

$$D = D_I \cdot T + D_F \tag{5}$$

where D_I is an integer and $|D_F| \leqslant 0.5T$, therefore, only cases in which $|D| \leqslant 0.5T$ needed to be considered because D_I can simply be deduced from the tap position of the largest filter coefficient.

It is trivial that the two input signals of the basic LMSTDE model can be interchanged. In this case, the filter will provide a time shift of $-D$, or a time advance of D, to $y(k)$. Consider two adaptive filters are being used in the model instead. Fig. 2 shows the block diagram of the system in which the two filters, AF_A and AF_B, are introduced on the two separate paths. $E[e^2(k)]$ is now given by

$$\begin{aligned} E\left[e^2(k)\right] &= E\left[\{v(k) - u(k)\}^2\right] \\ &= E\left[\{s(k - D - \Omega_B) - s(k - \Omega_A)\}^2\right] \end{aligned} \tag{6}$$

where Ω_A and Ω_B are time delays inserted by AF_A and AF_B onto $x(k)$ and $y(k)$ respectively and assuming D is a fractional value. Of course, the coefficients of the two filters must also satisfy the same constraint as before. In order to make $E[e^2(k)]$ equal to null at equilibrium, one possible solution is

$$\Omega_A = -\Omega_B = D/2. \tag{7}$$

The inter-array delay can, again, be determined directly from the peak coefficients of the two filters. There are two advantages of this new LMSTDE model: 1) the same filter length can be used to measure a larger delay, and 2) it has a faster convergence and a better tracking ability for nonstationary delays.

III. Analysis and Filter Implementation

It is apparent that the steady state solution for $E[e^2(k)]$ to become zero in (6) is not unique. In fact, as long as $\Omega_A = D + \Omega_B$ is satisfied, $E[e^2(k)]$ will be diminished. However, if $\Omega_A = -\Omega_B$, the delay estimation process can be simplified and the implementation of the adaptive filter can also be made more efficiently and economically. Thus, necessary procedures are included in the algorithm to ensure that the system will converge to the desired solution as given by (7).

Suppose the coefficients of AF_A and AF_B are being updated at alternate iteration and constrained adaptation is used after normal LMS adaptation has converged to a coarse time delay estimate. Let k and h denote the sampling and iteration count, respectively, where $h = (k \,\mathrm{div}\, 2)$ and div is an integer division operation. Then, for $|D| \leqslant 0.5T$, the coefficient with the largest amplitude of the two filters will appear in the center tap position. In the absence of noise, the two peak weights of AF_ and AF_B are updated using the LMS stochastic gradient method as follows:

$$\begin{aligned} a_{0,h+1} &= a_{0,h} + 2\mu\{v(k) - u(k)\}x(k) \\ &= a_{0,h} + 2\mu\{s(k - D - \Omega_{B,h}) - s(k - \Omega_{A,h})\}s(k) \end{aligned} \tag{8}$$

$$\begin{aligned} b_{0,h+1} &= b_{0,h} + 2\mu\{u(k+1) - v(k+1)\}y(k+1) \\ &= b_{0,h} + 2\mu\{s(k+1-\Omega_{A,h}) \\ &\quad - s(k+1-D-\Omega_{B,h})\}s(k+1-D) \end{aligned} \tag{9}$$

where $a_{0,h}$ and $b_{0,h}$ denote the center tap of AF_A and AF_B at iteration h. They are then matched to the nearest values in the look-up table and therefore, are samples of a sinc function $\Omega_{A,h}$ and $\Omega_{B,h}$ are, respectively, the time delays, and μ is a preselected parameter that controls the rate of convergence and stability of the adaptation. Now, taking expectation of (8) and (9) yields

$$E[a_{0,h+1}] = E[a_{0,h}] + 2\mu\{R_{ss}(D + \Omega_{B,h}) - R_{ss}(\Omega_{A,h})\} \tag{10}$$

$$E[b_{0,h+1}] = E[b_{0,h}] + 2\mu\{R_{ss}(D - \Omega_{A,h}) - R_{ss}(\Omega_{B,h})\} \tag{11}$$

where $R_{ss}(t)$ is the low-pass autocorrelation of sequence $s(k)$, which is given by

$$R_{ss}(t) = E[s(k)s(k-t)] = \sum_{j=-\infty}^{\infty} \mathrm{sinc}(j-t)E[s(k)s(k-j)].$$

It can be seen from (10) and (11) that if the time shifts, $\Omega_{A,h}$ and $\Omega_{B,h}$, at iteration h are equal but opposite in sign, then $a_{0,h} = \mathrm{sinc}(\Omega_{A,h})$ and $b_{0,h} = \mathrm{sinc}(\Omega_{B,h})$ are equal and the expected amount of weight adjustments for them in the next update will be identical, thereby still making $\Omega_{A,\,h+1} = -\Omega_{B,\,h+1}$. Hence, if initially $\Omega_{A,0} = -\Omega_{B,0}$, then upon further adaptation, whenever an expected value of delay Ω_A is inserted on $x(k)$ by AF_A, a similar time shift of $\Omega_B = -\Omega_A$ will also be introduced on $y(k)$ by AF_B. Subsequently, when steady state is reached $\Omega_A = -\Omega_B = D/2$, which is the ideal solution.

With alternate adaptation of the two filters, it can be shown that the number of iterations for the two filters to reach equilibrium are almost the same and the convergence rate between the new and old LMSTDE model are approximately equal. However, it is intriguing to note that adapting AF_B with LMS is not necessary so long as the shifts provided by AF_A and AF_B are always kept equal but opposite. In this case, constrained adaptation is first applied to AF_A, and the respective delay at each iteration is estimated. The sign of the shift is then reversed and the related filter weight values are assigned to AF_B correspondingly. This procedure can further be simpli-

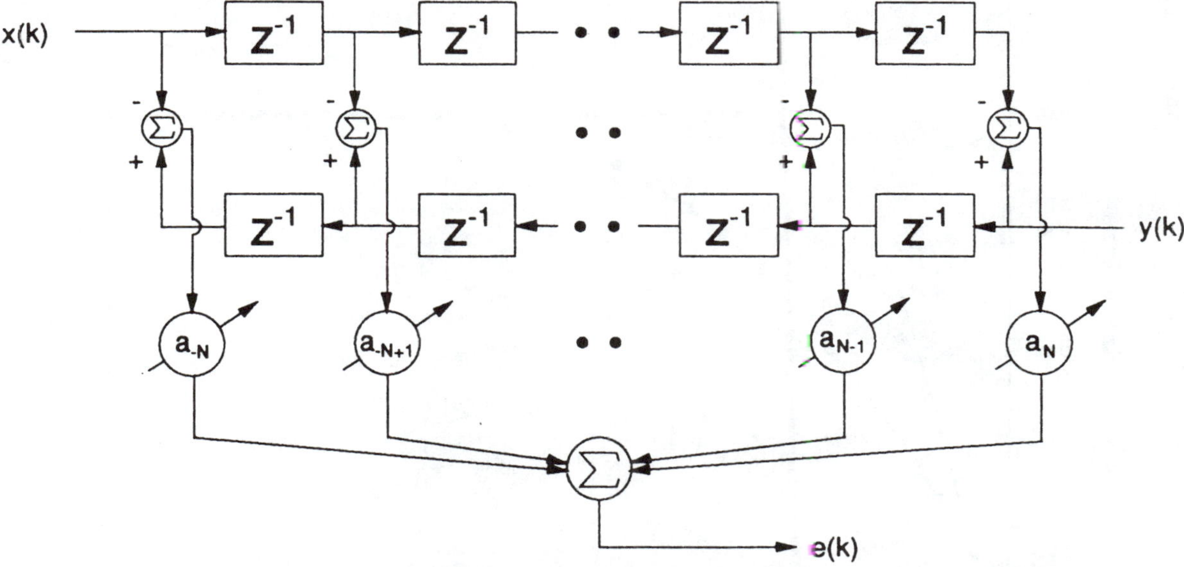

Fig. 3. A practical implementation of the new LMSTDE model.

fied by making use of the fact that $\text{sinc}(i-\Omega)=\text{sinc}(-i+\Omega)$. The coefficients of AF_A and AF_B are actually related to each other by

$$b_{i,h}=a_{-i,h}, \qquad i=-N,\cdots,N. \tag{12}$$

Hence, updating of AF_B can simply be facilitated by a weight copying operation which involves no extra computation. Since adaptation of AF_B is not required, updating of AF_A can be carried out in every sampling instant and thus speed up the convergence rate of the new LMSTDE model. Indeed, eigenvalue analysis of the input correlation matrix shows that the new system has a smaller time constant in the learning curve while keeping the same steady state excess MSE. The detail convergence analysis will be reported elsewhere. Now, since each of the two filters only provides half of the expected delay and thus if the same filter length is maintained, the truncation error induced is comparatively smaller. It can also be shown that if the delay to be estimated is large, then the filter length required can be reduced and the savings between the old and new LMSTDE approaches in terms of number of filter taps would be significant.

From (6), it seems that one additional convolution is needed in each iteration in order to obtain $e(k)$ which obviously increases the computational complexity. However, by using (12), $e(k)$ can be simplified to

$$\begin{aligned} e(k) &= \sum_{i=-N}^{N} b_{i,k}y(k-i)-\sum_{i=-N}^{N} a_{i,k}x(k-i) \\ &= \sum_{i=-N}^{N} a_{i,k}\{y(k+i)-x(k-i)\}. \end{aligned} \tag{13}$$

Equation (13) shows that by subtracting the two receiver signals prior to performing convolution, additional multiplication is not needed although there are $2N$ extra subtractions in each iteration. As AF_B is neither directly involved in coefficient updating nor error computing, therefore, a simplified configuration of the new LMSTDE adaptive filter can be devised which is shown in Fig. 3. When comparing with the old LMSTDE system, the new model has a penalty of requiring N additional shift registers.

IV. The Adaptive Algorithm and Simulation Results

In this study, it is assumed that the input correlation matrix R_{ss} is unknown *a priori*. The adaptive algorithm for the proposed LMSTDE system is summarized as below:

a) Relax all filter coefficients at the beginning.

b) Adapt AF_A for the first 200 iterations using normal LMS algorithm.

c) After that, determine the peak tap weight, a_p, of AF_A. Pre-shift $x(k)$ by p samples correspondingly. The initial delay inserted by AF_A is set to $0.125T$ if $a_{p-1}<0$ and $a_{p+1}>0$. Otherwise, the delay is set to $-0.125T$. Then assign filter weights, which are values of a sinc function, to AF_A according to the initial set delay with maximum at a_0. The value of D_I is now given by p.

d) Constrained adaptation by using (8) and (13) is next carried out until steady state is reached. At each iteration, updating of AF_A is subject to the satisfaction of some predetermined sign conditions [6] which confine the movement of delay estimate in a proper direction.

e) From a_0, derive the delay, Ω_A, introduced by AF_A. Now, if D is not a fractional value, then from (5), $p=D_I$ and $\Omega_A T=D_F/2$. Apparently, the overall delay estimate, $\hat{D}$, can be computed from

$$\hat{D}=(p+2\Omega_A)T\approx D. \tag{14}$$

Computer simulations were run to evaluate the performance of the new LMSTDE model. The source $s(k)$ and, noises $\theta(k)$ and $\phi(k)$ were all uniformly distributed white sequences generated by a random noise generator with source power equal to unity. The delayed signal was obtained by passing $s(k)$ through a FIR filter of order 40 with its impulse response given by samples of a sinc function. The order of the adaptive transversal filter AF_A was set to 20. The look-up table [4] for time difference and filter weights conversion had a size of 512×21. Note that this look-up table only needs to contain delay ranging from 0 to $0.25T$, that is the resolution of delay estimate is approximately $0.001T$. Simulation results were the average of 10 independent ensembles to reduce point-to-point fluctuation.

Fig. 4 compares the adaptive behaviour of the basic and the new LMSTDE model in the absence of noise. Convergence

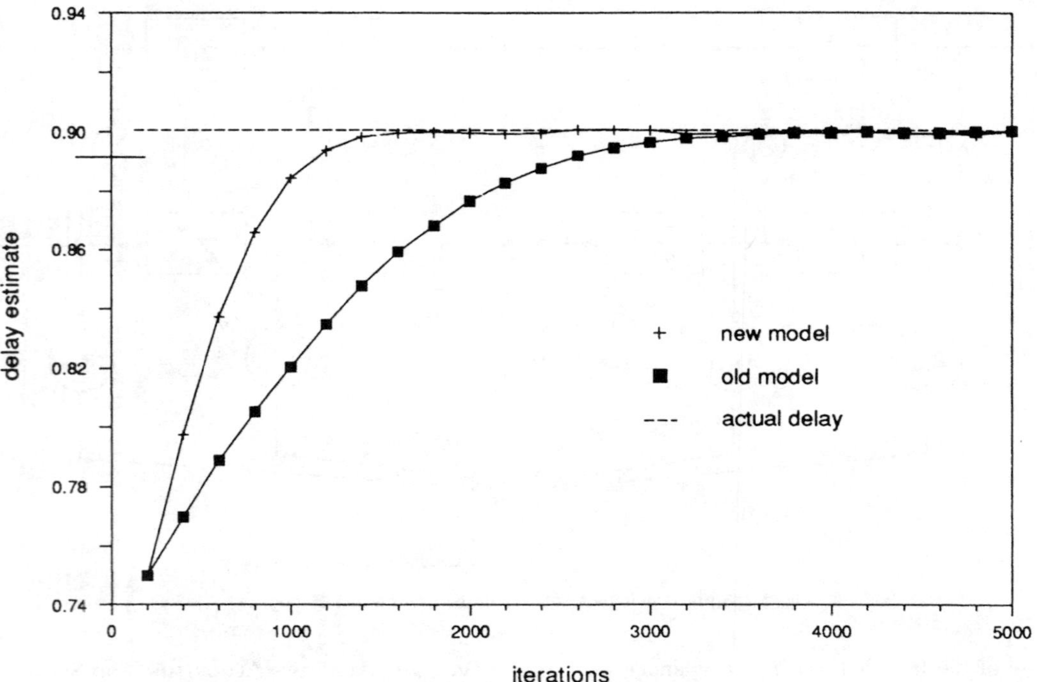

Fig. 4. Delay estimation in the absence of noise.

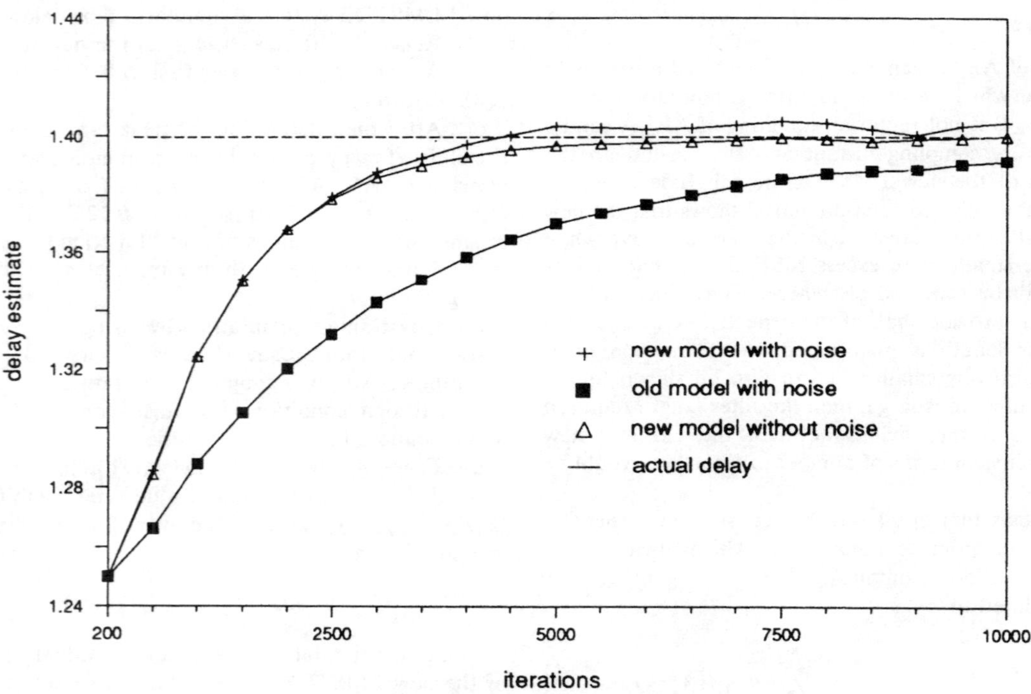

Fig. 5. Delay estimation with noise at 20-dB SNR.

factor μ was set to 0.0004 and the time delay D was chosen as $0.9T$. Both systems started with the same initial conditions. Let the time at which the estimate enters and stays within $\pm 1.0\%$ of the actual delay be the capture time. As shown in Fig. 4, the capture time for the new and old system are roughly $1200T$ and $2500T$, respectively. Thus the improvement of convergence speed of the proposed model in terms of capture time is about $1300T$. A similar test has also been run to study the performance of the two models in the presence of white noise. Fig. 5 shows a typical result in this regard. Uniformly distributed random noise at SNR of 20 dB were added to both $x(k)$ and $y(k)$ while μ and D were set to 0.0002 and 1.4T, respectively. As noise was added, so a smaller μ was chosen in order to maintain the same MSE which in turn give rise to a longer lapse for the system to reach equilibrium as compared to Fig. 4. It is apparent from the graph that only marginal degradation in performance was recorded and, again, the new configuration outperforms the basic system. For both new and old LMSTDE system, deviation of the delay estimate from the desired value occured if SNR decreased. However, even for an SNR as low as 10 dB, both systems will

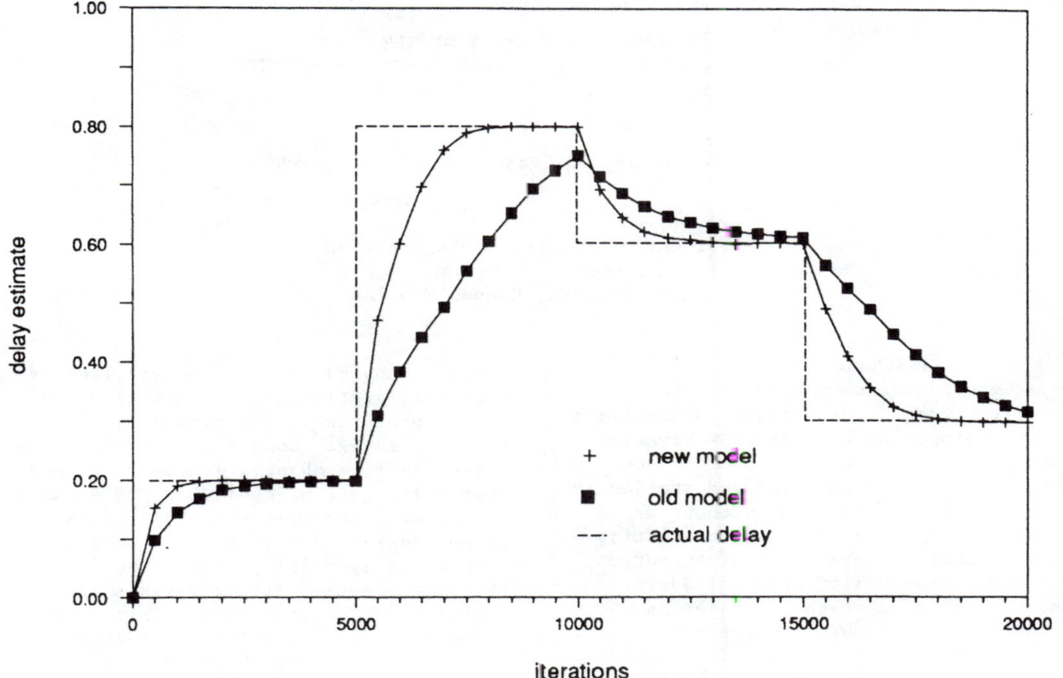

Fig. 6. Estimation of nonstationary delay without noise.

still converge though with a discrepancy of around 10%.

The faster convergence speed obtained by the proposed LMSTDE model is quite significant especially when signal statistics is time varying. Fig. 6 depicts the trajectories of the two adaptive systems with nonstationary delays. In this example, the unknown time delay was given a series of step offsets and adaptive algorithms were applied to estimate and track the dynamic changes of the time difference parameter. It is seen that both systems responded to the step changes and were able to give accurate estimates after the transients. The new LMSTDE system, however, is able to trace time-varying delays more effectively.

V. Conclusions

A novel configuration with constrained adaptive algorithm is proposed for TDE. This new LMSTDE model has the advantages of having a faster convergence speed and a better ability to track nonstationary delays. In addition, the same filter length can be used to measure a longer delay without giving rise to a larger truncation error. The only disadvantages of the system are the requirement of some extra inexpensive shift registers and a moderate increase in computation.

References

[1] Special issue on time delay estimation, *IEEE Trans. Acoust., Speech, Signal Processing*, vol. ASSP-29, June 1981.

[2] C. H. Knapp and G. C. Carter, "The generalized correlation method for estimation of time delay," *IEEE Trans. Acoust., Speech, Signal Processing*, vol. ASSP-24, pp. 320–327, Aug. 1976.

[3] D. H. Youn, N. Ahmed, and G. C. Carter, "On using the LMS algorithm for time delay estimation," *IEEE Trans. Acoust., Speech, Signal Processing*, vol. ASSP-30, pp. 798–801, Oct. 1982.

[4] P. C. Ching and Y. T. Chan, "Adaptive time delay estimation with constraints," *IEEE Trans. Acoust., Speech, Signal Processing*, vol. ASSP-36, pp. 599–602, Apr. 1988.

[5] Y. T. Chan, J. Riley, and J. B. Plant, "Modeling of time delay and its application to estimation of nonstationary delays," *IEEE Trans. Acoust., Speech, Signal Processing*, vol. ASSP-29, pp. 577–581, June 1981.

[6] P. C. Ching, Y. T. Chan, and K. M. Tse, "Constrained adaptive estimation of time delay with multipath," in *Proc. 11th GRETSI Symp. on Signal and Image Processing*, Nice, France, vol. 1, pp. 305–308, June 1–5, 1987.

A COMPARATIVE STUDY OF THE LMS ADAPTIVE FILTER VERSUS GENERALIZED CORRELATION METHODS FOR TIME DELAY ESTIMATION

J. Krolik, M. Joy, S. Pasupathy, M. Eizenman

Department of Electrical Engineering
University of Toronto
Toronto, Toronto, Canada M5S 1A4

ABSTRACT

This paper provides a comparison of the LMS adaptive filter versus conventional Generalized Correlation (GC) methods of time delay estimation in terms of their mean-square-error performance. The treatment is restricted to broadband stationary inputs with finite observation time without a priori knowledge of the processor input statistics. The time delay estimator probability distribution, variance, and error probability are derived for the LMS adaptive filter approach. Further, a performance index is given which is optimized with respect to choice of step size. The simulation results presented indicate that without a priori knowledge of the input statistics, both approaches yield similar sub-optimal results. On the other hand, optimal processing of the adaptive filter weights can yield an estimator with variance similar to that of the minimum variance GC method which approaches the Cramer-Rao Lower Bound.

1. INTRODUCTION

Generalized correlation (GC) methods [1,6,7] and LMS adaptive filtering [2-5] offer two different approaches to the problem of time delay estimation. A comparison of adaptive and conventional schemes can be made by considering their mean-square-error performance given a fixed observation time in the static delay situation. In this paper, the probability distribution of the time delay estimate obtained via the LMS adaptive filter approach (LMSTDE) is derived, leading to closed form expressions for the estimator variance and the probability of estimation error. A performance index based on the probability of estimation error is maximized with respect to choice of step size for a fixed observation time. Finally, the effect of lack of a priori knowledge of the input statistics is demonstrated by the results of a comparative simulation study employing a lowpass broadband source in white noise.

2. THEORETICAL PERFORMANCE OF THE LMSTDE

The simple model considered in this paper for discrete-time signals received in the presence of noise at two spatially separated sensors is given by:

$$r_1(k) = s(k) + n_1(k) \qquad (1)$$
$$r_2(k) = s(k-D) + n_2(k)$$

where the source signal, $s(k)$, and the noises $n_1(k)$ and $n_2(k)$ are sampled versions of real, Gaussian, stationary, and mutually uncorrelated random processes. The time delay, D (assumed here to be an integer constant), is the parameter to be estimated from observation of $r_1(k)$ and $r_2(k)$ over a finite time equivalent to N samples.

The mean-square-error (MSE) of the LMSTDE is now derived for broadband stationary sources with an integer delay, under the assumption that the adaptive filter weights are uncorrelated Gaussian random variables. This assumption is known to be approximately correct under a wide range of conditions [8]. For instructive purposes, consider the following ideal low pass band-limited source signal and white additive noises:

$$S_{ss}(f) = \begin{cases} P_S & , 0 < f < B \\ 0 & , \text{otherwise} \end{cases} \qquad (2a)$$

and

$$S_{nn}(f) = P_N \quad , |f| < 0.5 \text{ for } i=1,2 \qquad (2b)$$

where $S_{ss}(f)$ and $S_{nn}(f)$ are the power spectral densities of $s(k)$ and $n_i(k)$, i=1,2, respectively and $2\pi f$ is the normalized frequency in radians.

To obtain the variance of the LMSTDE, expressions for the mean and variance of the adaptive filter weights are required. Denote the adaptive filter weights at iteration n by $h(m,n)$, $(m=-L, -L+1, \ldots L-1, L)$. Assume that the length of the filter is much longer than the decorrelation time of the source signal and the delay D. Consideration of the eigenvalues and eigenvectors of the converging adaptive filter weights [9] yields an approximate expression for the expected value of $h(m,n)$ with the inputs of (2):

$$E(h(m,n)) \cong (1-(1-2\mu P_N(1+SNR))^n) \cdot \frac{2B \cdot SNR}{1+SNR} \cdot \frac{\sin 2\pi B(m-D)}{2\pi B(m-D)} \qquad (3)$$

where $SNR=P_S/P_N$ is the signal-to-noise ratio in the signal band and μ is the step size. Observe that provided $0<\mu<1/2(P_N+P_S)$, the adaptive filter weights will converge to the discrete Roth GC function [10]. The adaptive filter weights may now be written as:

$$h(m,n) = E(h(m,n)) + w(m,n) \qquad (4)$$

where $w(m,n)$ is the zero-mean noise on the m-th weight at iteration n and E( ) denotes expectation. Under the conditions that $h(m,n)$ is near the converged solution, the covariance of the weight noise is approximately [8]:

Reprinted from *Proc. ICASSP '84*, vol. 1, pp. 15.11/1–4, March 1984.

$$E(w(j,n)w(k,n)) = \begin{cases} \sigma_w^2 & \text{, for } j=k \\ 0 & \text{, otherwise} \end{cases} \quad (5a)$$

where

$$\sigma_w^2 \cong \mu P_N \quad (5b)$$

under conditions of low signal-to-noise ratio.

The LMSTDE is defined as the argument m for which h(m,n) is maximized. Hence the probability distribution of the LMSTDE, $P_{LMS}(m)$, is given by:

$$P_{LMS}(m)=\text{Prob}[\varepsilon_{m,-L},\ldots,\varepsilon_{m,m-1},\varepsilon_{m,m+1},\ldots,\varepsilon_{m,L}] \quad (6)$$

where $\varepsilon_{m,i}$= the event$\{h(m,n) \geq h(i,n)\}$. Equivalently, using (4),

$$\varepsilon_{m,i}= \text{the event}\{w(i) \leq d(m,i)+w(m)\} \quad (7)$$

where d(m,i)=E(h(m,n))-E(h(i,n)) and w(i)=w(i,n), with the dependence on the iteration number n is suppressed to simplify the notation. Then,

$$P_{LMS}(m)= \int_{-\infty}^{\infty} \text{Prob}[\bigcap_{i\neq m} \varepsilon_{m,i}/w(m)=x]f_{w(m)}(x)dx \quad (8)$$

where $f_{w(m)}(x)$ is the probability density function (PDF) of w(m). Since w(i) are uncorrelated, zero-mean Gaussian random variables with variance σ_w^2,

$$P_{LMS}(m)= \frac{1}{\sigma_w\sqrt{2\pi}} \int_{-\infty}^{\infty} \prod_{i\neq m} Q\left(\frac{d(m,i)+x}{\sigma_w}\right)e^{-x^2/2\sigma_w^2}\, dx \quad (9)$$

where $Q(a)=(1/\sqrt{2\pi}) \int_{-\infty}^{a} \exp(-x^2/2)dx$. Observe that $P_{LMS}(m)$ is a function of the ratio of mean filter weight differences, d(m,i), to weight noise standard deviation, σ_w. Using (3) and (5), $P_{LMS}(m)$ may be calculated for the inputs of (2) by numerical integration of (9). Figure 1 shows $P_{LMS}(m-D)$ for three SNR values with B=0.05, $\mu=1.2 \times 10^{-4}$, N=8192 and P_N=1.0 (corresponding to a 100 Hz. lowpass source in unit variance white noise with a 4 second observation time and sampling rate of 2048 Hz.). Clearly, $P_{LMS}(m)$ becomes more sharply peaked around the true delay D as SNR increases.

The probability distribution of the LMSTDE can be used to derive various performance measures. The probability of error, P(e), in estimating the time delay D equals 1 - $P_{LMS}(m=D)$. Note that equation (17) in [5] represents the union bound [11] on P(e) and yields approximately correct results in this application only in cases of high SNR and large source bandwidth. The variance of the LMSTDE is also easily expressed as

$$\text{Var(LMSTDE)}= \sum_{m=-L}^{L} (m-D)^2 P_{LMS}(m).$$

A useful performance index for the LMSTDE is suggested by considering the case of a white source signal in white additive noises (i.e. let B=0.5 in (2)). The mean filter weights are now indentically zero except for m=D hence d(D,i)= E(h(D,n)) for all $i\neq D$. The probability of error is given by:

$$P(e)=1-P_{LMS}(m)\big|_{d(D,i)=E(h(D,n)),\quad i\neq D} \quad (10)$$

Defining the performance index $R=E(h(D,n))/\sigma_w$, inspection of (10) shows that P(e) decreases with increasing R. For a fixed observation time, P(e) may be minimized by choosing the step size, μ, to maximize R. Using (3) and (5), the performance index R at iteration number N for white inputs and low SNR is given by:

$$R = \frac{(1-(1-2\mu P_N(1+SNR))^N}{\sqrt{\mu P_N}} \cdot \frac{SNR}{1+SNR} \quad (11)$$

In figure 2, R is plotted versus step size for different SNR values (where N = 8192, P_N=1.0). The choise of μ which maximizes R in figure 2 is relatively insensitive to SNR. Reed, Feintuch, and Bershad have implicity derived the optimal choice of μ by minimizing the ratio of the variance of the LMSTDE to the CRLB ((18) in [2]). Their results suggest choosing μ equal to $1/NP_N$. This is consistent with the present derivation since $\mu=1/(1)(8192)=1.2\text{x}10^{-4}$ is near the choice of μ which maximizes R for all plots in fig. 2. Empirically, values of $R \lesssim 1$ indicate that anomalous time delay estimates may result. The index is thus useful in predicting the threshold SNR,below which supurious behavior may be evident using the LMSTDE with a fixed observation time.

3. A COMPARATIVE SIMULATION STUDY

Simulations using computer-generated data give experimental verification of the theoretical LMSTDE variance and allow a direct comparison of LMSTDE and GC techniques. The study reported here has been modelled after that of Scarbrough Ahmed, and Carter [7]. The two sensor model given in (1) was realized with a sampling interval, T_S, of 1/2048 s. and delay parameter of 4 samples. The representation for the sensor outputs was obtained by using three mutually uncorrelated unit variance Gaussian pseudo-random noise sequences. Two of these sequences were used to represent additive noises, while the third was passed through a tenth-order Chebyshev lowpass digital filter with a cut-off frequency of 100 Hz. and became the source signal representation. The simulated sensor outputs were then processed by a 64-tap LMS adaptive filter with $r_2(k)$ used as the reference input and $r_1(k)$ used as the filter input. A constant delay of 32 sampling intervals was placed in the reference channel so that the maximum of the converged filter weights would correspond to taps near the middle of the filter for small D. The adaptive filter was allowed to run for a 4 second period at the end of which time delay estimates were made. A step size, $\mu=1.2 \times 10^{-4}$, was used, as discussed in the previous section. Thirty-two independent trials were conducted to determine the mean square error of the estimator for each of several signal-to-noise ratios.

For each trial, three different time delay estimates were obtained from the LMS adaptive filter weights. The first of these, the "unsmoothed LMS" TDE, is simply the lag corresponding to the largest filter weight (i.e. precisely the LMSTDE as defined previously). To obtain the other two, the sequence of filter weights were used to form the function, $h_s(t,n)$, defined by:

$$h_s(t,n)=\sum_{m=-L}^{L} h(m,n)\,\frac{\sin 2\pi\beta T_s(t-m)}{2\pi\beta T_s(t-m)} \qquad (12)$$

This function was employed in [2, eq.19] for the purpose of interpolating h(m,n) to obtain non-integer delays where β was chosen to match the source signalbandwidth. Strictly speaking, for interpolation alone, β should be chosen so as to match the sampling frequency since matching to the source signal bandwidth clearly implies a priori knowledge of the signal bandwidth. In this paper, delay estimates obtained by taking the nearest integer lag corresponding to the maximum value of $h_s(m,n)$, with β=100 Hz., are called "smoothed LMS" TDE estimates. The precise non-integer lag associated with the peak of $h_s(t,n)$, also with β=100 Hz, result in "interpolated LMS" estimates. As in [7], time delay estimates which fell outside the range (D±10) samples were omitted in calculating mean-square-error to avoid biasing the statistics.

The results of applying the LMS adaptive filter to time delay estimation are given in Table 1 for three low SNR values. Theoretical values of LMSTDE mean-square-error (MSE) were obtained as discussed in sec. 2. The "unsmoothed LMS" TDE variances are in agreement with the theoretical LMSTDE variance. The values of the Cramer-Rao Lower Bound were calculated using expression (A4) in [1]. The unsmoothed LMSTDE yields an MSE significantly greater than the CRLB, while the smoothed and interpolated LMSTDE gives performance approaching the CRLB. A more complete description is given by fig. 3, where the results of the LMS adaptive filter approach are compared with the Cramer-Rao Lower Bound. Mean-square error performance is clearly improved by application of the interpolator matched to the source signal bandwidth. In effect, a priori knowledge of the interpolator bandwidth corresponds to optimal filtering in the GC processor. The mean square error performance of four different GC estimators may be compared with the above results in Fig. 4. (The GC functions were estimated as in [7] with 16 disjoint segments and an FFT length of 512 points.) The results indicate that the LMS adaptive filter with proper smoothing of the filter weights can yield a performance equivalent to that of the optimum/ML GC processor. Both these estimators make use of a priori knowledge of the source bandwidth to obtain significantly improved time delay estimates. On the other hand, in this simulation the unsmoothed LMS time delay estimates offer little advantage over approximate maximum likelihood (AML), Smoothed Coherence Transform (SCOT), and basic cross-correlation methods, each of which estimate receiver input statistics directly from the available data.

CONCLUSION

In this paper, theoretical and experimental results for the MSE performance of the LMSTDE have been presented. A simulation study comparing the LMSTDE and conventional GC methods indicates that without a priori knowledge both methods offer similar performance, with variances significantly larger than the Cramer-Rao Lower Bound, particularly at low signal-to-noise ratios. These results suggest that noise on the adaptive filter weights can degrade the performance of the LMSTDE as much as errors in estimating the GC function can in conventional methods. For the simulation considered here, a priori knowledge may be used by both approaches to produce time delay estimates with variance approaching the Cramer-Rao Lower Bound. This work is directed towards the implementation of robust post-processing of weight vectors derived from the LMS adaptive filter for time delay estimation. Design methods which account for uncertainty in a priori knowledge of the processor input statistics are currently under development [12].

ACKNOWLEDGEMENT

This work was supported in part by the Defence Research Establishment Atlantic, Dartmouth, N.S., Canada, under Supply and Services (DND/DREA) contract no. 8SC83-00274.

REFERENCES

1. C.H. Knapp and G.C. Carter, "The generalized correlation method for estimation of time delay", IEEE Trans. ASSP, pp. 320-327, 1976.
2. F.A. Reed, P.L. Feintuch, and N.J. Bershad, "Time delay estimation using the LMS adaptive filter - static behavior", IEEE Trans. ASSP, pp. 561-571, June 1981.
3. F.A. Reed, P.L. Feintuch, and N.J. Bershad, "Time delay estimation using the LMS adaptive filter - dynamic behavior", IEEE Trans. ASSP, pp. 571-576, June 1981.
4. D.H. Youn, N. Ahmed, and G.C. Carter,, "On Using the LMS Algorithm for time delay estimation", IEEE Trans. ASSP, pp. 798-801, Oct. 1982.
5. D.H.Youn, N. Ahmed and G.C. Carter, "An adaptive approach for time delay estimation of band-limited signals", IEEE Trans. ASSP, June 1983.
6. J.C. Hassab and R.E.Boucher, "Optimum estimation of time delay by a generalized correlator", IEEE Trans. ASSP, pp. 373-380, 1979.
7. K. Scarborough, N. Ahmed, and G.C. Carter, "On the simulation of a class of time delay estimation algorithms", IEEE Trans. ASSP, pp. 534-540, June 1981.
8. B. Widrow et.al., "Adaptive noise cancelling: Principles and applications", Proc. of IEEE, pp. 1692-1716, December 1975.
9. J.R. Treichler, "Transient and convergent behavior of the adaptive line enchancer", IEEE Trans. ASSP, February 1979.
10. K. Scarborough, N. Ahmed, and G.C. Carter,"On the SCOT and Roth algorithms for time delay estimation", in Proc. ICASSP-82,Paris,May 1982.
11. J.M. Wozencraft and I.M. Jacobs, "Principles of Communication Engineering", Wiley, 1965.
12. J.L. Krolik, "Constrained time delay estimattion via zer-crossing methods", M.A.Sc. thesis, Dept. of Elec. Engg., University of Toronto, 1983.

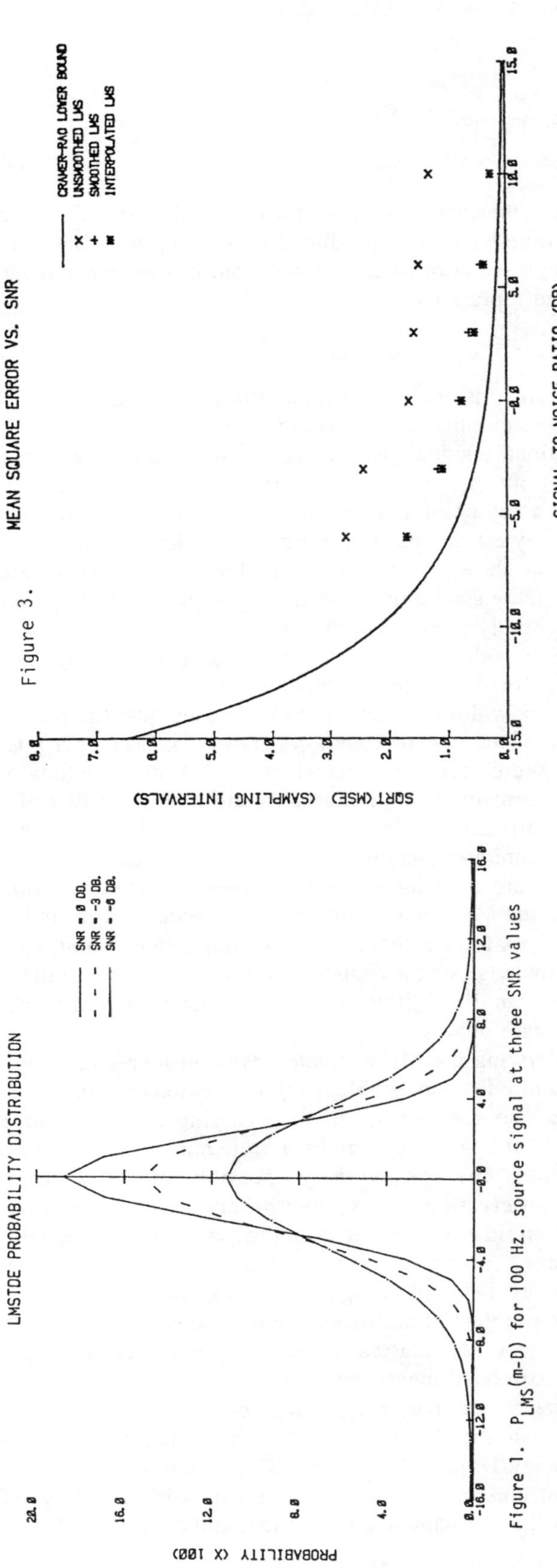

Figure 3.

Figure 1. $P_{LMS}(m-D)$ for 100 Hz. source signal at three SNR values

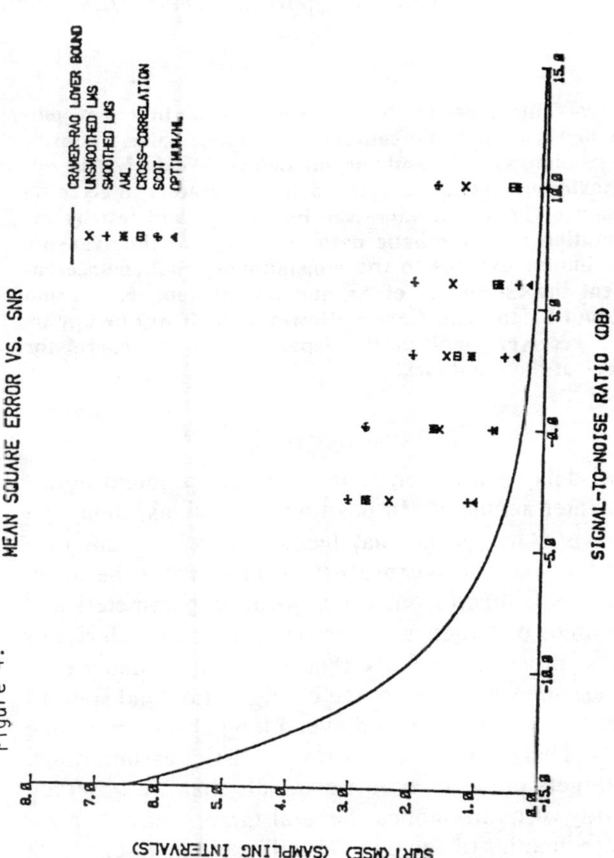

Figure 4.

Figure 2.

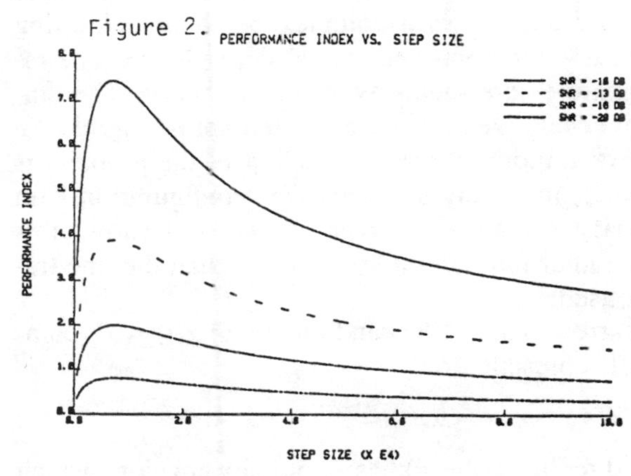

TABLE 1. LMSTDE Low SNR Simulation Results

SNR (dB)	unsmoothed LMSTDE expt.	unsmoothed LMSTDE theo.	smoothed LMSTDE expt.	interpolated LMSTDE expt.	CRLB
	SQRT(MSE) (SAMPLING INTERVALS)				
0	1.85	2.07	.637	.527	.35
-3	2.43	2.62	.829	.783	.57
-6	3.71	3.89	1.41	1.43	1.2

Multisource Delay Estimation: Nonstationary Signals

Isabel M. G. Lourtie, *Member, IEEE*, and José M. F. Moura, *Senior Member, IEEE*

***Abstract*—This paper studies delay estimation in a very general framework: multiple sources, correlated noises, nonstationary random signals, and varying delays. We derive the optimal maximum likelihood (ML) delay estimator, analyze its performance via the Cramer–Rao inequality, and test by experimentation with synthetic data. The present time-varying delay estimator extends to the nonstationary/multisource environment the estimators of Ng and Bar-Shalom, Kirlin and Dewey, and Knapp and Carter. However, as it will be apparent, our receiver significantly departs from the correlator structures of these authors.**

I. Introduction

Time delay estimation is an important problem in underwater acoustics. In positioning systems, delay estimation is a first stage that feeds into subsequent processing blocks. The postprocessing blocks use the delay estimates as features from which location parameters and the dynamics of targets are then recovered. In underwater acoustics, most of the work reported on time delay estimation assumes a single source configuration and stationary signal processes observed over a long observation time interval. These are not always realistic assumptions. Noisy targets emanate acoustic signals that are observed at an array of hydrophones. Several targets may be present in the vicinity of the one being located or tracked. At the array of sensors we receive then the superposition of all the radiated signals. Over time, the received signal cannot be assumed to be stationary. The transmission medium, the ocean, shows a complex behavior exhibiting simultaneously temporal and spatial dependence. The radiated signatures are themselves nonstationary. The emitter/receiver relative motions are often not negligible. In this context, a more accurate modeling of the problem is often needed, justifying a multisource configuration with nonstationary signals. However, as we will show, this leads to a rather more complex receiver than the one traditionally used.

For a narrow-band (NB) random signal $x(t)$ with complex signal representation

$$\tilde{x}_b(t)e^{j2\pi f_c t} \tag{1}$$

the delayed replica $x(t - \tau)$ has a complex envelope which is approximately given by

$$\tilde{x}_b(t)e^{-j2\pi f_c \tau}. \tag{2}$$

Manuscript received May 6, 1988; revised June 25, 1990.

I. M. G. Lourtie is with CAPS, Complexo, Instituto Superior Técnico, P-1096 Lisboa Cedex, Portugal.

J. M. F. Moura is with LASIP, Department of Electrical and Computer Engineering, Carnegie-Mellon University, Pittsburgh, PA 15213.

IEEE Log Number 9042718.

In (2), we used the NB approximation

$$\tilde{x}_b(t - \tau) \simeq \tilde{x}_b(t). \tag{3}$$

On the other hand, for a stationary random signal with a large time bandwidth product BT, working with the Fourier representation, we have the Fourier transform of the delayed replica $x(t - \tau)$ as

$$X(f)e^{-j2\pi f\tau} \tag{4}$$

where $X(f)$ is the Fourier transform of $x(t)$. Equations (2) and (4) show that under assumptions of either NB signals or stationary signals with large BT, the delay dependence factors out of the structure of the signal. This factorization is a key point determining the structure of the present day delay estimators, see, for example, Knapp and Carter [3] who showed that the optimal maximum likelihood (ML) delay estimator for stationary signals with large BT is essentially a cross correlator.

We consider in this paper the problem of delay estimation for nonstationary random signals with arbitrary time bandwidth product BT. For these signals, the factorization of the type of equation (2) or (4) is no longer possible. We derive the structure of the ML delay estimator which turns out to be remarkably different from that of a cross correlator. Lacking stationarity, the Fourier transform techniques that underly the design of these cross correlators are no longer useful. Rather, the maximization step of the ML delay estimator is preceded by a recursive filter. This filter estimates the random signal conditioned on knowledge of the values of the delays. It generalizes the Kalman–Bucy filter to the case of estimation of waveforms with delays.

In deriving the ML estimator, we consider a very general setup for the problem: i) nonstationary correlated multiple source signals, ii) time-varying delays (relative or absolute) parameterized by a finite number of unknown quantities, iii) nonstationary, possibly mutually correlated, observation noises, iv) arbitrary (short or long) observation time interval. In doing so, we extend to the nonstationary, arbitrary BT case, the works of Knapp and Carter [3], [4], and of Ng and Bar-Shalom [1], who studied the problem of multisource environment, and of Kirlin and Dewey [2], that assumed a single source with spatially correlated measurement noise.

When the sources are moving, besides delay, we need to estimate the Doppler shift. We derive the joint optimal ML delay/Doppler estimator. This extends to the more general class of nonstationary signals with arbitrary BT the works of Knapp and Carter [5], and of Kirlin, Moore,

Reprinted from *IEEE Trans. Sig. Proc.*, vol. 39, no. 5, pp. 1033–1048, May 1991.

and Kubichek [6]. The latter work develops a suboptimal strategy to deal with signals where the nonstationarity is due to time-varying delays. This is achieved by sequentially processing the data into continuous blocks of length T seconds. Within each block the delay is assumed to be time invariant, but allowed to change over successive blocks. In [6], a linear Kalman filter is used as a postprocessor that filters the sequence of delay measurements. These delay estimates result from processing the most recent received data block by means of a delay estimator such as the one described by Hassab and Boucher [7]. For narrow-band signals, and when the delays are themselves sample functions of a random process Bucy, Moura, Mallinckrodt [8], Leitão and Moura [9], Bethel and Rahikka [10] have developed Bayes law type receivers to estimate the delay waveform. Moura and Baggeroer [11] have used the estimators of [9] with real data underwater signals that propagated under the Arctic ice crust.

In Section II, we formulate and model the delay estimation problem. A multisource environment is considered. The signals are sample functions of Gaussian nonstationary stochastic processes, described via a state space model, and the delays are continuous real functions. The maximum likelihood (ML) delay estimator for the above context is derived in Section III. The ML delay estimator maximizes, over the unknown delay parameters, the log-likelihood function (LLF). When the source signals $y_l(t)$ are not deterministic known functions, but rather sample functions of nonstationary random processes, the construction of the likelihood function involves the minimum mean-square error estimate (MMSE) $\hat{y}_l(t - D_l^s)$ of the delayed source signals $y_l(t - D_l^s)$ given the observation set, and conditioned on the values of the delays D_l^s. This is the first step of the ML receiver. We develop the causal MMSE recursive filter for $\hat{y}_l(t - D_l^s)$ which exhibits the structure of a Kalman–Bucy filter for signals with delays. These MMSE estimates $\hat{y}_l(t - D_l^s)$, which are functions of the unknown delays D_l^s's, are used in the LLF whose maxima yield the delay estimates. In this sense, the ML-delay receiver here studied generalizes to unknown nonstationary random signals, the FSK-type receiver of radar contexts. To evaluate the estimator performance, in Section IV we develop analytically the time evolution of the Cramer–Rao bound. Under the general framework considered herein, the Fisher information matrix requires the gradient with respect to the delays of the estimate of the signal process. Our parameterization in terms of the GKBF, promptly provides algorithmically the means to compute these gradients. To gain insight into the general ML delay estimator developed herein, in Section V we analyze its asymptotic structure under stationary signals, long observation time interval (SLOT approximation), and time-invariant delay assumptions. For the SLOT case, we provide a frequency interpretation showing that the new delay estimator recovers the structure of a generalized cross correlator, [1]–[4]. Assuming a multisource environment with spatially uncorrelated observation noise and SLOT conditions we obtain the processor of Ng and Bar-Shalom [1], while considering a single source configuration we recover either the estimator of Kirlin and Dewey [2], for spatially correlated measurement noise, or the generalized cross correlator of Knapp and Carter [3], [4], when the observation noise is uncorrelated among sensors. In Section VI, we carry out a simulation study that illustrates the capabilities of the general ML processor and compare it with the classical generalized cross-correlator structure. Finally, Section VII presents the paper's conclusions.

Notation: The general setup of the problem involves a detailed notation that is presented here. The following symbols are used throughout:

i) $\otimes$: Kronecker product [12] which, for example, for $X, Y \in \mathcal{R}^2 \times \mathcal{R}^2$, is defined by

$$X \otimes Y = \begin{bmatrix} x_{11}Y & x_{12}Y \\ x_{21}Y & x_{22}Y \end{bmatrix} \in \mathcal{R}^4 \times \mathcal{R}^4. \tag{5}$$

ii) diag $\{\bullet\}$: diagonal matrix of block-matrix elements, e.g.,

$$\text{diag}\,\{A_l\} = \begin{bmatrix} A_1 & & & 0 \\ & A_2 & & \\ & & \ddots & \\ 0 & & & A_L \end{bmatrix}, \qquad l = 1, 2, \cdots, L \tag{6}$$

where A_l are matrices of appropriate dimensions.

iii) I_S: identity matrix of order $S \times S$.

iv) $\mathbf{1}_S = [1 \cdots 1]^T$: one vector of order S.

v) $\nabla_{tr_1r_2}(*)$, $\nabla_{tr}(*)$, and $\nabla_t(*)$: differential operators defined by

$$\nabla_{tr_1r_2}(*) = \gamma(t)\left[\frac{\partial}{\partial t}(*)\right]\gamma(t) + \beta(t, r_1)\left[\frac{\partial}{\partial r_1}(*)\right]\gamma(t) + \gamma(t)\left[\frac{\partial}{\partial r_2}(*)\right]\beta(t, r_2) \tag{7}$$

$$\nabla_{tr}(*) = \gamma(t)\left[\frac{\partial}{\partial t}(*)\right] + \beta(t, r)\left[\frac{\partial}{\partial r}(*)\right] \tag{8}$$

$$\nabla_t(*) = \frac{d}{dt}(*) \tag{9}$$

where t, r_1, r_2 are scalars representing time and delay variables, γ and β are diagonal matrices.

vi) $\langle *, \bullet \rangle_R$: weighted inner product, i.e.,

$$\langle *, \bullet \rangle_R = *^T R^{-1} \bullet \tag{10}$$

where R is a covariance matrix.

vii) $\text{Tr}_N(\bullet)$: matricial trace operator, defined as follows: given a block square matrix

$$X = [X_{ij}] \in \mathcal{R}^{MN} \times \mathcal{R}^{MN} \tag{11}$$

with the blocks $X_{ij} \in \mathcal{R}^N \times \mathcal{R}^N$, $\mathrm{Tr}_N(X)$ is the $M \times M$ matrix

$$\mathrm{Tr}_N(X) = [\mathrm{tr}\, X_{ij}] \in \mathcal{R}^M \times \mathcal{R}^M. \tag{12}$$

viii) On occasion, a short notation is used for functions, e.g.,

$$P_{tr_1r_2} = P(t, r_1, r_2). \tag{13}$$

When no ambiguity arises, arguments may also be omitted.

II. Problem Formulation

A. Statement of the Problem

We now formulate the *multiple source* delay estimation problem, when the processor array has $S \geq 1$ sensors, and the receiving signal at each sensor is a superposition of $L \geq 1$ sources. Specifically, let $\{z(t, s) \in \mathcal{R}, t \geq 0\}$ be a sample function of the observed process at sensor s, modeled by

$$z(t, s) = \sum_{l=1}^{L} \alpha_l^s(t) y_l(t - D_l^s(t)) + v(t, s), \qquad t \in \boldsymbol{T}, \quad s \in \boldsymbol{S}, \quad l \in \boldsymbol{L} \tag{14}$$

where $t \in \boldsymbol{T} = [0, T] \subseteq \mathcal{R}^+$ is the time variable, $s \in \boldsymbol{S} = \{1, 2, \cdots, S\}$ is the sensor index, and $l \in \boldsymbol{L} = \{1, 2, \cdots, L\}$ is the source index. Functions $D_l^s(t)$ and $\alpha_l^s(t)$ represent the *time-varying* delay and the attenuation associated to each pair (s, l). The observation noise $\{v(t, s) \in \mathcal{R}, t \in \boldsymbol{T}\}$ is a nonstationary zero mean Gaussian temporally white *spatially correlated* noise process with cross-covariance $E\{v(t, s)v(\sigma, m)\} = R_{sm}(t)\delta(t - \sigma)$, $(s, m \in \boldsymbol{S})$. The signal $\{y_l(t) \in \mathcal{R}, t \geq t_0\}$ emanating from source $l \in \boldsymbol{L}$ is a stochastic *nonstationary* process assumed to be uncorrelated with the noise $\{v(t, s)\}$. Throughout the paper, the signals are assumed scalar. Generalization to vector signals is trivial.

At each sensor of the receiving array, the observed signal is the superposition of delayed (travel time) and attenuated replicas of the emitted signals. Inherent to the problem description is a propagation effect introduced through the travel time delays $(D_l^s(t))$ and the attenuation factors $(\alpha_l^s(t))$. These are functions of both the transmission channel impulse response, and of the sources and sensors geometry and relative motion.

Equation (14) describes the problem. It encompasses a *multisource* configuration with a single direct acoustic path from each source $l \in \boldsymbol{L}$ to each sensor $s \in \boldsymbol{S}$. It also includes the single source *multipath* geometry if we take a single signal $y(t) = y_l(t)$ propagating through multiple paths $l \in \boldsymbol{L}$ from the source to each sensor $s \in \boldsymbol{S}$ of the array. Fig. 1 shows the multisource configuration for two sources and an array of three hydrophones.

In our formulation, the time-varying delay function $D_l^s(t)$, $\forall t \in \boldsymbol{T}$, is assumed to satisfy the following conditions:

i) it is a continuous function of t;

ii) it has first derivative with respect to t;

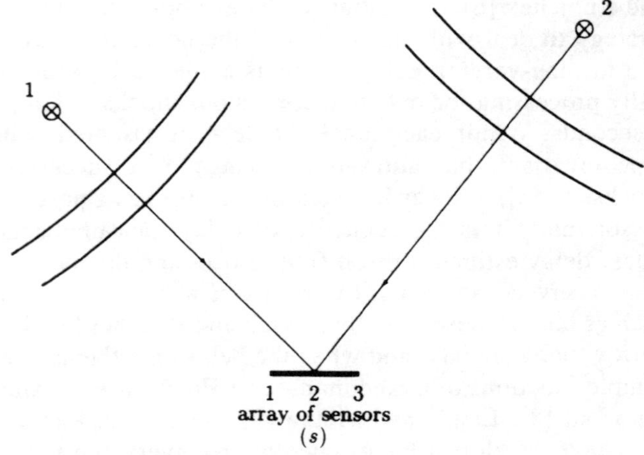

Fig. 1. Multisource, single direct acoustic path geometry.

iii) it is described by a deterministic time-varying function which is parameterized by a finite number of unknown parameters, i.e.,

$$D_l^s(t) = \tau_l(t, \theta_l) \tag{15}$$

where θ_l is the vector of unknown parameters.

Under i) above, for bounded $\boldsymbol{T}$, we can define

$$D_{\min} \leq \min_{t \in \boldsymbol{T}, s \in \boldsymbol{S}, l \in \boldsymbol{L}} D_l^s(t) \tag{16}$$

and

$$D_{\max} \geq \max_{t \in \boldsymbol{T}, s \in \boldsymbol{S}, l \in \boldsymbol{L}} D_l^s(t). \tag{17}$$

The values $D_{\min}$ and $D_{\max}$ are, respectively, lower and upper bounds to all the time-varying delay functions $D_l^s(t)$. In the sequel, we need the technical assumption

$$t_0 \leq -D_{\max} \tag{18}$$

where t_0 represents a time reference with respect to which all sources start radiating the signals $y_l(t)$, $l \in \boldsymbol{L}$. Equation (18) guarantees that the process $y_l(t - D_l^s(t))$ in (14) exists for every $t \in \boldsymbol{T}$, that is, all signal sources are present during the entire observation time interval $\boldsymbol{T}$. For a linear array, and depending upon the case under study, absolute or relative delay, the parameter $D_{\min}$ is defined accordingly. For an absolute delay, since

$$\forall t \in \boldsymbol{T}, \quad s \in \boldsymbol{S}, \quad l \in \boldsymbol{L}, \quad D_l^s(t) \geq 0$$

$D_{\min}$ has the trivial value

$$D_{\min} = 0.$$

For the relative delay case, i.e., when the differential delays $D_l^s(t)$ can be negative, $D_{\min}$ is a function of the intersensors separation and/or of the maximum range of the time-varying delays. If there are no constraints on the source location, the relative delay D_l^s is minimum or maximum for the endfire configuration.

The work presented in this paper does not assume any particular parameterization $\tau_l(t, \theta_l)$ of the delay functions. It remains valid for a general parameterization. A discussion of parameterizations for a planar source/receiver geometry, with a linear array shape and a target following a linear uniform motion can be found in Moura and Baggeroer [13], and Moura [14], [15]. The ML estimation of the time-varying delays $D_l^s(t)$ is accomplished after the ML estimates of the parameters are performed.

To model general nonstationary source signatures, the signals $y_l(t)$ are described as sample functions of Gauss Markov processes, which are the output of a linear system driven by noise. In formal terms, the source signature is described by

$$\frac{d}{dt}x_l(t) = A_l(t)x_l(t) + B_l(t)u_l(t), \qquad t \geq t_0, \; l \in L, \tag{19}$$

$$y_l(t) = C_l(t)x_l(t). \tag{20}$$

The initial condition $x_l(t_0) \in \mathcal{R}^{n_l}$ is a Gaussian random vector with mean value $\bar{x}_l(t_0) \in \mathcal{R}^{n_l}$ and cross-covariance matrix $E\{[x_l(t_0) - \bar{x}_l(t_0)][x_k(t_0) - \bar{x}_k(t_0)]^T\} = \Sigma_{lk}(t_0, t_0)(l, k \in L)$. The dynamics disturbance $\{u_l(t) \in \mathcal{R}^{m_l}, t \geq t_0\}$ is a Gaussian white noise vector process, independent of the observation noise $\{v(t, s)\}$ and of the random initial condition $x_k(t_0)(k \in L)$, with cross-covariance matrix $E\{u_l(t)u_k^T(\sigma)\} = Q_{lk}(t)\delta(t - \sigma)$. For $l \neq k$, the noises $u_l(t)$ and $u_k(t)$ may be statistically correlated. In this case, the signatures $y_l(t)$ and $y_k(t)$, emitted by sources l and k, are correlated. If the sources are mutually uncorrelated, we simply take

$$Q_{lk}(t) = 0 \quad \text{and} \quad \Sigma_{lk}(t_0, t_0) = 0, \qquad \forall l \neq k. \tag{21}$$

Depending on the system matrices $A_l(t)$, $B_l(t)$, and $C_l(t)$, (19), (20) model *nonstationary* narrow- or broad-band processes. When the system matrices are time invariant, the matrix A is asymptotically stable, and the observation interval is large, model (19), (20) leads then to signals which are asymptotically stationary. Equations (19), (20) represent a paradigm which is alternative to Fourier representation of signals. The latter describes signals as linear superposition of complex exponentials, and in the context of delay estimation leads, under a large BT assumption, to batch cross-correlation techniques. Equations (19), (20) represent the signals as output of linear systems driven by noise. They are well-suited representations to handle nonstationary environments and lead to estimators that emphasize time recursiveness. Equations (14) and (19), (20) are to be interpreted in a formal sense. A rigorous writing requires the Ito calculus formulation, e.g., see [16].

B. Modeling

As referred to in Section I, the log-likelihood function involves the minimum mean-square error (MMSE) estimate of the received signals. Because at different sensors, these correspond to delayed replicas of sample functions of nonstationary random processes, we need to develop filtering structures for nonstationary stochastic processes with delays. The complexity of the problem is brought about not so much by the nonstationarity but more so by the necessity of filtering with delay of random signals. We achieve this by reformulating the problem into the context of linear filtering of signals described by partial differential equations (distributed parameter systems). As we will see, the present problem corresponds, however, to a very specific context within this very general topic.

To be able to obtain a compact description we work with vector and matrix notations. For that, define the extended processes that collect the snapshots over the array of sensors:

i) Received signal vector:

$$Z(t) = [z(t, 1) \vdots z(t, 2) \vdots \cdots \vdots z(t, S)]^T. \tag{22}$$

ii) Measurement noise vector:

$$V(t) = [v(t, 1) \vdots v(t, 2) \vdots \cdots \vdots v(t, S)]^T. \tag{23}$$

iii) Delay line vector $D_l(t)$ for source l:

$$D_l(t) = [D_l^1(t) \vdots D_l^2(t) \vdots \cdots \vdots D_l^S(t)]. \tag{24}$$

iv) The $L \times S$ delay matrix:

$$D(t) = [D_1^T(t) \vdots D_2^T(t) \vdots \cdots \vdots D_L^T(t)]^T. \tag{25}$$

v) State vector:

$$X(t, r) = [X^{1^T}(t, r) \vdots X^{2^T}(t, r) \vdots X^{S^T}(t, r)]^T. \tag{26}$$

vi) State vector for sensor s:

$$X^s(t, r) = [x_1^T(\xi_1^s(t, r)) \vdots x_2^T(\xi_2^s(t, r)) \vdots \cdots \vdots x_L^T(\xi_L^s(t, r))]^T. \tag{27}$$

vii) Equivalent auxiliary time variable:

$$\xi_l^s(t, r) = t - D_{\min} - r(D_l^s(t) - D_{\min}). \tag{28}$$

viii) Input noise vector:

$$U(t) = [u_1^T(t - D_{\min}) \vdots u_2^T(t - D_{\min}) \vdots \cdots \vdots u_L^T(t - D_{\min})]^T. \tag{29}$$

The parameter $r \in [0, 1]$ in (26)–(28) is useful in describing the evolution in the time interval $[t - D_l^s(t), t - D_{\min}]$ of the state vectors $x_l (l \in L)$ that model the emitted signals y_l (see (19), (20)). It introduces gradually and implicitly the nonstationarity imposed on the signals by the time-varying delays. In (28), $\xi_l^s(t, r)$ is an auxiliary time variable that represents, for fixed observation time t, the time evolution of the emitted signal when propagating along path (l, s) from source l to sensor s. For example, if $D_l^s(t)$ is an absolute delay, we have

$$\xi_l^s(t, r) = t - rD_l^s(t). \tag{30}$$

In the above context, at fixed time t, along path (l, s) from source l to sensor s the emitted signal is at source l

$$y_l(t) = y_l(\xi_l^s(t, 0)) \tag{31}$$

while at the receiver is

$$y_l(t - D_l^s(t)) = y_l(\xi_l^s(t, 1)). \tag{32}$$

In between, $y_l(\xi_l^s(t, r))$ is a nonobservable signal at a fictitious sensor located a distance $rd_l^s(t)$ apart from the emitter (Fig. 2). When $r = 1$, $X(t, 1)$ collects the state vectors $x_l(t - D_l^s(t))$ used in modeling the signals received at each sensor of the array.

Under definitions (22)–(29), the multisource delay problem is now compactly described. The vector process $\{Z(t) \in \mathcal{R}^S, t \in T\}$ observed at the receiving array is

$$Z(t) = Y(t) + V(t) \tag{33}$$

where the observation noise $\{V(t) \in \mathcal{R}^S, t \in T\}$ is a nonstationary Gaussian white noise vector process with covariance matrix $E\{V(t)V^T(\sigma)\} = R(t)\delta(t - \sigma)$. When the noise is spatially uncorrelated, $R(t)$ is diagonal. The received signal $\{Y(t) \in \mathcal{R}^S, t \in T\}$

$$Y(t) = C(t)X(t, 1) \tag{34}$$

is a stochastic nonstationary vector process, uncorrelated with the noise $\{V(t)\}$. It is interpreted as the output of a linear distributed parameter system. From (26), (27) and (19), the state vector $X(t, r)$ is described by the partial differential equation

$$\nabla_{tr} X(t, r) = 0, \qquad r \in [0, 1], \qquad t \geq 0 \tag{35}$$

where $\nabla_{tr}(\cdot)$ is the differential operator introduced in (8), with

$$\gamma(t) = \text{diag } \{\gamma^s(t)\} \tag{36}$$

where

$$\gamma^s(t) = \text{diag } \{[D_l^s(t) - D_{\min}]I_{nl}\} \tag{37}$$

and

$$\beta(t, r) = \text{diag } \{\beta^s(t, r)\} \tag{38}$$

where

$$\beta^s(t, r) = \text{diag } \{[1 - r\nabla_t D_l^s(t)]I_{nl}\}. \tag{39}$$

The boundary condition for (35) is at $r = 0$

$$\nabla_t X(t, 0) = A(t)X(t, 0) + B(t)U(t). \tag{40}$$

Matrices A, B, and C in model (34)–(40) are given by

$$A(t) = I_S \otimes \text{diag } \{A_l(t - D_{\min})\} \tag{41}$$

$$B(t) = 1_S \otimes \text{diag } \{B_l(t - D_{\min})\} \tag{42}$$

and

$$C(t) = \text{diag } \{C^s(t)\} \tag{43}$$

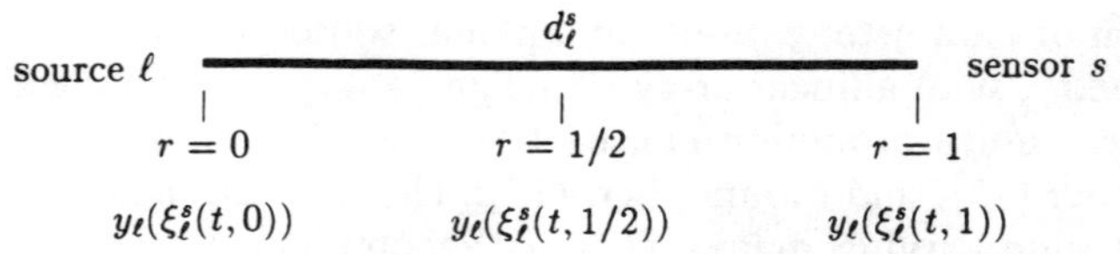

source ℓ —— d_ℓ^s —— sensor s

$r = 0$ $\quad$ $r = 1/2$ $\quad$ $r = 1$

$y_\ell(\xi_\ell^s(t,0))$ $\quad$ $y_\ell(\xi_\ell^s(t,1/2))$ $\quad$ $y_\ell(\xi_\ell^s(t,1))$

Fig. 2. Path (l, s) between source l and sensor s at a fixed time t.

with

$$C^s(t) = [\alpha_1^s(\xi_1^s)C_1(\xi_1^s) \vdots \alpha_2^s(\xi_2^s)C_2(\xi_2^s) \vdots \cdots \vdots$$

$$\vdots \alpha_L^s(\xi_L^s)C_L(\xi_L^s)] \tag{44}$$

where ξ_l^s, defined by (28), is taken at $r = 1$. Expressions (34)–(44) model the received signal $Y(t)$. Although not explicitly specified, the signal $Y(t)$ and the state vector $X(t, r)$ are functions of the delay matrix $D(t)$.

The partial differential equation (35) describes the propagation of the state vectors $x_l(\xi_l^s)$ of the radiated signals along each propagation path (l, s). The boundary equation (40) describes the signals at the source locations. They correspond to time $t - D_{\min}(r = 0)$, rather than $t = 0$. This is a useful gimmick that allows us to consider relative delays. In this case, $D_{\min} < 0$ and the boundary condition is advanced relative to the signal received at the reference sensor. This procedure represents a time-invariant translation which does not modify the emitted signal characterization.

III. Time-Varying Delay Estimation

A. Estimation Criterion: Maximum Likelihood

The observations along the time interval $[0, t]$ and array sensor set S are collected by

$$Z^t = \{Z(\sigma): 0 \leq \sigma \leq t\}. \tag{45}$$

We assume that the time delays, if time varying, are described by deterministic functions whose structure is specified up to a finite number of unknown parameters collected in the vector θ. Let the time-varying delay realization be

$$D^t(\theta) = \{D(\sigma) = \tau(\sigma, \theta): 0 \leq \sigma \leq t\}. \tag{46}$$

The vector θ is estimated by maximum likelihood (ML) techniques [17], [18], i.e., the ML-estimate $\hat{\theta}(t)$ of θ given the observation set Z^t is

$$\hat{\theta}(t) = \arg \max_{\theta \in \Theta} J(t; D^t(\theta)) \tag{47}$$

where $J(t; D^t)$ is the log-likelihood function (LLF). The maximization operation in (47) is carried out on the compact domain $\Theta \subseteq \mathcal{R}^\nu$ where the vector θ is assumed to lie.

B. Log-Likelihood Function (LLF)

In this subsection, we present the LLF when the time variable t is continuous. To get insight into the structure of the LLF, we derive it in Appendix A for the discrete time case. For continuous time t the LLF structure is established in Appendix B. The evaluation of the LLF involves the minimum mean-square error (MMSE) causal estimate of the received signal $Y(t)$, conditioned on both the observation set Z^t and the time-varying delay realization $D^t(\theta)$.

The log-likelihood function (LLF) is given by (B.13) of Appendix B

$$J(t; D^t) = \int_0^t [\langle \hat{Y}_\sigma, Z_\sigma \rangle_R - \tfrac{1}{2} \langle \hat{Y}_\sigma, \hat{Y}_\sigma \rangle_R - \tfrac{1}{2} \operatorname{tr}(R^{-1}P_Y)] \, d\sigma \tag{48}$$

where $\hat{Y}(\sigma)$ is the minimum mean-square error (MMSE) estimate of the signal process $Y(\sigma)$, and $P_Y(\sigma)$ is the signal error estimation covariance matrix. Both $\hat{Y}(\sigma)$ and $P_Y(\sigma)$ are functions of the time delay realization $D^\sigma(\theta)$. The first and second terms on the RHS of expression (48) depend on the observation process: the first term computes the cross correlation, normalized by the observation noise covariance, between the observation process $Z(t)$ and the MMSE signal estimate $\hat{Y}(t)$; the second term represents the energy of the MMSE signal estimate. The third term on the RHS of (48) is a correction term. It depends only on the signal estimate error covariance $P_Y(t)$. The next subsection shows that $P_Y(t)$ does not depend on the observation process, and thus can be precomputed.

To conclude, we emphasize that the ML delay estimate $\hat{D}(t)$, which is obtained by maximizing the LLF given by (48), involves the MMSE estimate of the received signal $\hat{Y}(t)$. If the emitted signals were deterministic, that is, the received signal $Y(t)$ was known except for the finite dimensional vector of parameters θ that describes the delay realization $D^t(\theta)$, then it would be trivial to show that

$$\hat{Y}(t) = Y(t) \tag{49}$$

and

$$P_Y(t) = 0 \tag{50}$$

leading to the well-known expression for the LLF for a deterministic signal (see [17, ch.4])

$$J(t, D^t) = \int_0^t [\langle Y_\sigma, Z_\sigma \rangle_R - \tfrac{1}{2} \langle Y_\sigma, Y_\sigma \rangle_R] \, d\sigma. \tag{51}$$

In the general case of random $Y(t)$, we describe in the next subsection how to construct the MMSE estimate $\hat{Y}(t)$ of $Y(t)$ conditioned on the value of the delay.

C. Generalized Kalman–Bucy Filter (GKBF)

In the previous subsection, the LLF was expressed in terms of the minimum mean-square error (MMSE) estimate $\hat{Y}(t)$ of $Y(t)$. Because the signals are delayed sample functions of linear random processes, the MMSE estimates are obtained by generalization of the Kalman–Bucy filter to the distributed parameter context of Section II.

The optimal filtering theory of Kalman and Bucy [19] was first extended by Kwakernaak [20] to include linear systems with multiple time-invariant delays in both the dynamics and the observations. Yu *et al.* [21] generalized Kwakernaak's results to nonlinear distributed parameter systems. The structure of [21] is too general. The generalized Kalman–Bucy filter (GKBF) that we need for the delay estimator is obtained by deriving the causal MMSE filter when delays (possibly time varying) are present only in the observations. The details of the derivation are lengthy. They are available in [18].

Under the general linear framework stated in Section II, the signal MMSE estimate $\hat{Y}(t)$ and the error covariance matrix $P_Y(t)$, appearing in (48), are given by

$$\hat{Y}(t) = C(t)\hat{X}(t, 1) \tag{52}$$

and

$$P_Y(t) = C(t)P(t, 1, 1)C^T(t) \tag{53}$$

where

$$\hat{X}(t, r) = E\{X_{tr} | Z^t, D^t\} \tag{54}$$

and $P(t, r_1, r_2)$ is the error covariance conditioned on D^t

$$P(t, r_1, r_2) = E\{[X_{tr_1} - \hat{X}_{tr_1}][X_{tr_2} - \hat{X}_{tr_2}]^T | D^t\} \tag{55}$$

with $t \geq 0$, $r_1, r_2 \in [0, 1]$. Denoting the filter gain by

$$K(t, r) = P_{tr_1} C_t^T R_t^{-1} \tag{56}$$

the state vector MMSE estimate $\hat{X}(t, r)$ is given by the partial differential equation

$$\nabla_{tr} \hat{X}_{tr} = \gamma_t K_{tr} \tilde{Z}_t, \qquad t \in \boldsymbol{T}, \qquad r \in [0, 1] \tag{57}$$

with the boundary condition

$$\nabla_t \hat{X}_{t0} = A_t \hat{X}_{t0} + K_{t0} \tilde{Z}_t \tag{58}$$

and the initial condition

$$\hat{X}_{0r} = \overline{X}_{0r} \tag{59}$$

where

$$\tilde{Z}(t) = Z(t) - \hat{Y}(t) \tag{60}$$

is the innovations process, and matrix $\gamma(t)$ is defined in (36), (37).

The state estimate error covariance matrix $P(t, r_1, r_2)$ is given by the partial differential equation

$$\nabla_{tr_1r_2} P_{tr_1r_2} = -\gamma_t K_{tr_1} R_t K_{tr_2}^T \gamma_t, \qquad t \in \boldsymbol{T}, \qquad r_1, r_2 \in [0, 1] \tag{61}$$

with the boundary conditions

$$\nabla_{tr} P_{tr0} = \gamma_t P_{tr0} A_t^T - \gamma_t K_{tr} R_t K_{t0}^T \tag{62}$$

$$P_{t0r} = P_{tr0}^T \tag{63}$$

$$\nabla_t P_{t00} = A_t P_{t00} + P_{t00} A_t^T + B_t Q_t B_t^T - K_{t0} R_t K_{t0}^T \tag{64}$$

and the initial condition

$$P_{0r_1r_2} = \Sigma_{0r_1r_2} \tag{65}$$

where the matrix $\Sigma(0, r_1, r_2)$ stands for the cross covariance of the state vectors $X(t, r_1)$ and $X(t, r_2)$.

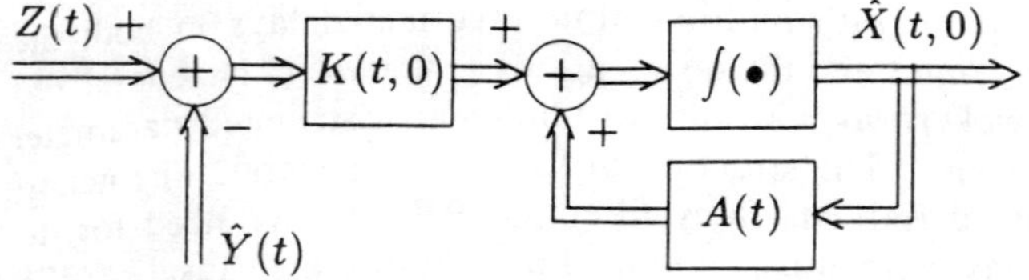

Fig. 3. Generalized Kalman-Bucy filter: boundary condition.

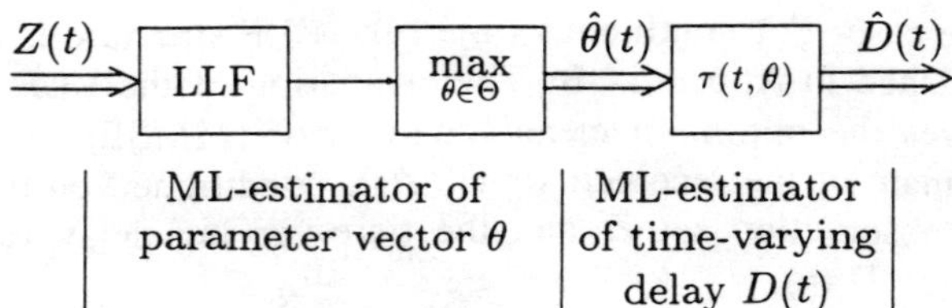

Fig. 4. Time-varying delay estimator.

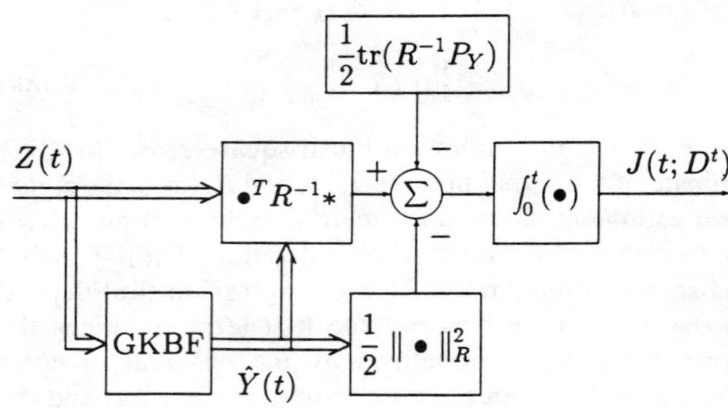

Fig. 5. Log-likelihood function (LLF) block.

Equations (52)–(65) describe the generalized Kalman-Bucy filter (GKBF), which extends the Kalman-Bucy filter [19] to systems with multiple time-varying delays in the observation process. The GKBF is a recursive system that reproduces the dynamics structure of the signal, (35) and (40), driven by the innovations process.

The received signal $Y(t)$ contains different delayed replicas of the emitted signals. The boundary condition (58) computes the estimates of the state vectors x_l, $\forall l \in L$, describing each radiating source, at time instant $t - D_{\min}$. In the above context, the boundary condition (58), Fig. 3, performs a prediction operation in the sense that it gives estimates of the state vectors x_l at a time instant

$$t - D_{\min} \geq t - D_l^s(t), \quad t \in T, \qquad l \in L, \qquad s \in S. \tag{66}$$

On the other hand, the propagation equation (57) evaluates the estimates of the state vectors x_l in the time interval $[t - D_l^s(t), t - D_{\min}]$, that is, it carries out a smoothing operation over the boundary state estimate.

The filter gain requires the solution of the coupled partial differential equations (61)–(64). These represent the extension to filtering with time-varying delays in the observation process of the Riccati equation of the Kalman-Bucy theory [19]. Because of the linearity assumption on the signal process model, the covariance matrix $P(t, r_1, r_2)$ does not depend on the observation process being precomputable. As mentioned, the interested reader is referred to [18] for the complete derivation.

D. Estimator Structure

As referred to before, $\hat{Y}(t)$ depends on the time delay realization $D^t(\theta)$. Based on the same set of data Z^t, distinct delay realizations $D^t(\theta)$ lead to different signal estimates $\hat{Y}(t)$. The ML processor chooses for the delay estimate the delay realization that is most likely to have caused the particular set of observations Z^t, that is, the one that accomplishes a better matching between signal modeling and data. The ML time-varying delay estimator developed herein is conceptually equivalent to two blocks (Fig. 4). It first computes the ML estimate of the parameter vector $\hat{\theta}(t)$. The LLF block generates the log-likelihood function (LLF—expression (48), Fig. 5). It constructs the MMSE estimate of the received signal, conditioned on both the observation set Z^t and the time-varying delay realization $D^t(\theta)$, via the generalized Kalman-Bucy filter. The maximization of the LLF is carried out on the compact domain Θ where the parameter vector θ that describes $D(t)$ is assumed to lie. Thereafter, the time-varying delay estimate $\hat{D}(t)$ is constructed based on both the parameter vector estimate $\hat{\theta}(t)$, and the assumed delay function $\tau(t, \theta)$.

When the second and third terms on the LLF (expression (48)) are weakly dependent on the delay matrix D, for example, when the problem geometry assumes a single stationary source configuration, and the observation noise is uncorrelated among sensors, we have

$$\hat{D}(t) = \arg \max_{\theta \in \Theta} \int_0^t \langle Z(\sigma), \hat{Y}(\sigma) \rangle_R \, d\sigma. \tag{67}$$

IV. Cramer–Rao Bound

A. Introduction

In this section we establish the Cramer–Rao bound (CRB) for the parameter vector estimate $\hat{\theta}(t)$. It is well known that the ML method provides asymptotically optimal Gaussian estimates which are consistent (asymptotically unbiased) and whose variance achieves the CRB (efficiency) [17]. For any unbiased estimator, the CRB is given by the inverse of the Fisher information matrix $\Upsilon(Z^t, \theta^a)$ [17], i.e.,

$$E\{[\hat{\theta}(t) - \theta^a][\hat{\theta}(t) - \theta^a]^T\} \geq \Upsilon^{-1}(Z^t, \theta^a) \tag{68}$$

where $\theta^a \in \mathbb{R}^\nu$ is the actual value of the parameter vector. The $\nu \times \nu$ Fisher information matrix is

$$\Upsilon(Z^t, \theta^a) = -E\left\{ \left. \frac{\partial^2 J(t; \theta)}{\partial\theta\partial\theta^T} \right|_{\theta^a} \right\} \tag{69}$$

and $J(t; \theta)$ is the log-likelihood function (LLF) given by (48).

B. Fisher Information Matrix

Denote by

$$\nabla_a \hat{Y}(t) = \frac{\partial \hat{Y}(t)}{\partial \theta}\bigg|_{\theta^a} \tag{70}$$

the $S\nu$ dimensional gradient of the MMSE signal estimate $\hat{Y}(t)$ with respect to θ, taken at $\theta = \theta^a$. By expanding in a Taylor series $\hat{Y}(t)$ about the actual parameter vector θ^a, noting that $\hat{Y}(t)$ satisfies the orthogonal projection lemma, expression (70) can be written after lengthy algebraic manipulations (which again are omitted here but can be found in [18])

$$\Upsilon(Z^t, \theta^a) = \int_0^t \mathrm{Tr}_S \{[I_\nu \otimes R_\sigma^{-1}][P^a_{\nabla Y}(\sigma) + \overline{\nabla_a \hat{Y}_\sigma}\, \overline{\nabla_a \hat{Y}_\sigma^T}]\}\, d\sigma \tag{71}$$

where $\otimes$ is the Kronecker product introduced in (5)

$$\overline{\nabla_a \hat{Y}(t)} = E\{\nabla_a \hat{Y}(t)\} \tag{72}$$

and

$$P^a_{\nabla Y}(t) = E\{[\nabla_a \hat{Y}(t) - \overline{\nabla_a \hat{Y}(t)}][\nabla_a \hat{Y}(t) - \overline{\nabla_a \hat{Y}(t)}]^T\} \tag{73}$$

represent, respectively, the mean value and the covariance matrix of the gradient stochastic process $\nabla_a \hat{Y}(t)$.

Fig. 6 shows the evaluation in block diagram of the Fisher information matrix. The block GM generates the first- and the second-order moments of the gradient vector process $\nabla_a \hat{Y}(t)$. To obtain the gradient process $\nabla_a \hat{Y}(t)$ two approaches can be considered: i) Numerical solution: $\nabla_a \hat{Y}(t)$ is computed numerically from $\hat{Y}(t)$; ii) Analytical solution: obtain, from the GKBF model, a dynamical equation for the gradient

$$\nabla_a \hat{X}(t, r) = \frac{\partial \hat{X}(t, r)}{\partial \theta}\bigg|_{\theta^a}. \tag{74}$$

Then describe the gradient $\nabla_a \hat{Y}(t)$ by means of the state vector $X^a(t, r)$, its estimate $\hat{X}^a(t, r)$, and the gradient $\nabla_a \hat{X}(t, r)$. The superscript a means that the process at which it is applied is taken at the actual parameter vector θ^a. For the analytical approach, it is shown in [18] that $\nabla_a \hat{Y}(t)$ is modeled as the output of a dynamic distributed parameter system with the Gaussian white noise inputs $\{U(t)\}$ and $\{V(t)\}$.

V. SLOT Approximation: Time-Invariant Delay, Stationary Processes, Long Observation Time Interval

A. Introduction

In Sections III and IV, we presented the structure of the ML delay estimator, of the GKBF used to construct the causal MMSE filters for random signals with delays, and derive expressions for the Cramer–Rao bounds for the delays. Here, we show that the ML estimator extends to the nonstationary multisource contexts, the more often encountered cross-correlator techniques used in delay estimation when the signals are assumed stationary. We do this by carrying out the asymptotic analysis of the ML estimator of Section III under stationary long observation time interval (SLOT approximation), and time-invariant delay assumptions. A further technical condition needed is that the signal processes are completely controllable and observable [19].

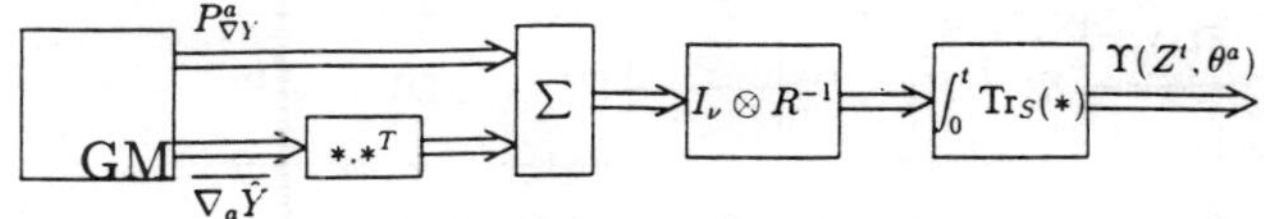

Fig. 6. Fisher information matrix.

In the above context, the covariance matrix $P(t, r_1, r_2)$ (61)–(64) is asymptotically time invariant, i.e.,

$$\lim_{t \to \infty} P(t, r_1, r_2) = P_\infty(r_1, r_2). \tag{75}$$

The steady-state error covariance matrix $P_\infty(r_1, r_2)$ gives rise to an asymptotically time-invariant generalized Kalman–Bucy filter (IGKBF).

We proceed now by factorizing this asymptotic Riccati equation, showing that the IGKBF corresponds to a generalized Wiener filter, and then interpreting this as the usual cross-correlator structures arising in the delay estimators of, for example, [1]–[3].

B. The Generalized Wiener Filter (GWF)

Denote by $\hat{Y}_I(\omega)$ and $Z_I(\omega)$ the integrated Fourier transforms of the vectorial processes $\hat{Y}(t)$ and $Z(t)$, [17], [22], [23]

$$\hat{Y}(t) = \frac{1}{2\pi} \int_{-\infty}^{+\infty} e^{j\omega t}\, d\hat{Y}_I(\omega) \tag{76}$$

and accordingly for $Z(t)$. The IGKBF equivalent frequency domain representation is (see Appendix C)

$$d\hat{Y}_I(\omega) = F_W(\omega)\, dZ_I(\omega) \tag{77}$$

with

$$F_W(\omega) = I - H(\omega, D) \tag{78}$$

where

$$H^{-1}(\omega, D) = I_S + Ce^{-j\omega\gamma}\left[(j\omega I - A)^{-1}K_0 + \int_0^1 e^{j\omega\gamma\sigma}\gamma K_\sigma\, d\sigma\right]. \tag{79}$$

The transfer matrix $F_W(\omega)$ of the IGKBF corresponds to a generalized Wiener filter extended to systems which have delays in the observation process. This is a linear, causal system that, for a stationary signal process $Y(t)$, minimizes the mean-square error between $Y(t)$ and its estimate $\hat{Y}(t)$. We can show (see [18] for details) that the generalized Wiener filter (GWF) (77)–(79) is conceptually equivalent to two cascade filters (Fig. 7). The first

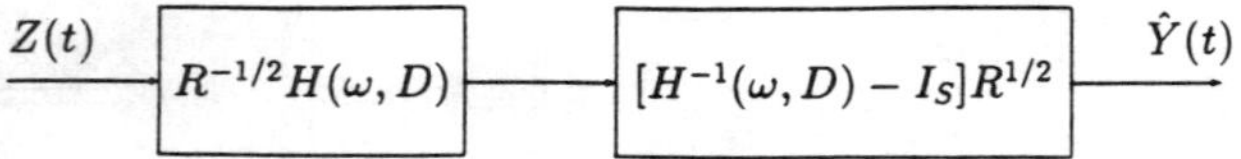

Fig. 7. Generalized Wiener filter.

one, with transfer matrix $R^{-1/2}H(\omega, D)$, is a realizable, minimum phase, whitening filter. The second one, with transfer matrix $[H^{-1}(\omega, D) - I_S]R^{1/2}$, is a realizable filter that gives at its output the signal steady-state MMSE estimate $\hat{Y}(t)$.

C. Log-Likelihood Function Asymptotic Analysis

To be able to compare the general ML processor developed in this paper with the solutions available in the literature, e.g., [1]–[3], we need to express the log-likelihood function (LLF) (48), and the corresponding time delay processor structure, in the frequency domain. These are obtained by working with truncated versions of the processes to a bounded interval T. Under the hypothesis of stationarity, assuming T is large compared to the correlation times of the processes plus the maximum delay magnitude $|D|_{\max}$ (SLOT approximation), the estimate of the (truncated version) of the received signal $\hat{Y}_T(t)$ is the output given by the GWF when the input is the (truncated) observed process $Z_T(t)$. Processes $\hat{Y}_T(t)$ and $Z_T(t)$ being limited to T seconds are finite energy processes, which then admit Fourier transforms, $\hat{Y}_T(\omega) = d\hat{Y}_{T_1}(\omega)/d\omega$ and $Z_T(\omega) = dZ_{T_1}(\omega)/d\omega$ [22]. From expression (77), it follows

$$\hat{Y}_T(\omega) = F_W(\omega)Z_T(\omega). \tag{80}$$

As referred to before, the LLF has two parts

$$J(T; D) = J_o(T; D) + J_c(T; D). \tag{81}$$

The first one, $J_o(T; D)$, first and second terms on the RHS of expression (48), depends on the observation process $Z_T(t)$; the second part, $J_c(T; D)$, is the third term on the RHS of (48). From (48), under the SLOT approximation

$$J_c(T; D) = -\tfrac{1}{2} T \operatorname{tr} [R^{-1}P_{Y_\infty}] \tag{82}$$

where P_{Y_∞} is the steady-state signal estimate error covariance matrix, while the first term of the log-likelihood function is

$$J_o(T; D) = \int_{-\infty}^{+\infty} [\langle Z_T(t), \hat{Y}_T(t)\rangle_R - \tfrac{1}{2} \langle \hat{Y}_T(t), \hat{Y}_T(t)\rangle_R]\, dt \tag{83}$$

where the integration over $]-\infty, +\infty[$ is valid given the time limitation imposed to the processes $\hat{Y}_T(t)$ and $Z_T(t)$.

Application of Parseval's theorem to the term $J_o(T; D)$ gives

$$J_o(T; D) = \int_{-\infty}^{+\infty} Z_T^T(-\omega)[R^{-1} - G_{ZZ}^{-1}(\omega)]Z_T(\omega)\, \frac{d\omega}{4\pi} \tag{84}$$

where, it is shown in Appendix D, the power spectral density of the observation process is

$$G_{ZZ}(\omega) = H^{-1}(\omega, D)RH^{-T}(-\omega, D). \tag{85}$$

The term $J_o(T; D)$, expression (84), is a generalized correlation function that computes the autocorrelation of the prefiltered observation process.

Until now the signals were assumed to be mutually correlated, as well as the observation noise processes. In the next subsection, we are going to consider the simpler case of mutually uncorrelated signals.

D. Mutually Uncorrelated Signals

Assume that the emitted signals $y_l(t)$, $l \in \mathbf{L}$, are mutually uncorrelated. In other words, the covariance matrix Q is diagonal. The power spectral density of the observation process is then

$$G_{ZZ}(\omega) = R + \sum_{l=1}^{L} \mathcal{V}_l(\omega)G_l(\omega)\mathcal{V}_l^T(-\omega) \tag{86}$$

where

$$\mathcal{V}_l(\omega) = [\alpha_l^1 e^{-j\omega D_l^1} \vdots\ \alpha_l^2 e^{-j\omega D_l^2} \vdots \cdots \vdots\ \alpha_l^S e^{-j\omega D_l^S}]^T \tag{87}$$

is the steering vector, and $G_l(\omega)$ is the power spectral density of the signal $y_l(t)$ radiated by the source l, $l = 1, \cdots, L$.

An ML estimate of the $L \times (S - 1)$ differential delay matrix

$$\Delta D = \begin{bmatrix} \Delta D_1^1 & \Delta D_1^2 & \cdots & \Delta D_1^{S-1} \\ \Delta D_2^1 & \Delta D_2^2 & \cdots & \Delta D_2^{S-1} \\ \vdots & \vdots & \ddots & \vdots \\ \Delta D_L^1 & \Delta D_L^2 & \cdots & \Delta D_L^{S-1} \end{bmatrix}, \tag{88}$$

where the relative delays are

$$\Delta D_l^{s-1} = D_l^s - D_l^{s-1}, \tag{89}$$

is obtained by solving the root equation

$$\nabla J(T; D) = \frac{\partial J(T; D)}{\partial(\Delta D)} = 0 \tag{90}$$

where ∇ is the gradient operator. By differentiation of $J(T; D)$ (expressions (81), (82) and (84)) with respect to ΔD_l^s, $l \in \mathbf{L}$, $s \in \mathbf{S}$, assuming the signal estimation error is small, we get approximately

$$\frac{\partial J(T; D)}{\partial \Delta D_l^s} \cong \int_0^{+\infty} j\omega\{\Gamma_l^{-1} Z_T^{*^T} \mathcal{G}_l^{-1} V_l \Lambda_l^s V_l^{*^T} \cdot \mathcal{G}_l^{-1} Z_T - Tb_l^s\}\, \frac{d\omega}{2\pi} \tag{91}$$

where the superscript * stands for complex conjugate

$$V_l(\omega) = \operatorname{diag} \{\alpha_l^s e^{-j\omega D_l^s}\} \tag{92}$$

$$\Lambda_l^s(\omega) = \Phi_s - b_l^s \mathbf{1}_S \mathbf{1}_S^T \tag{93}$$

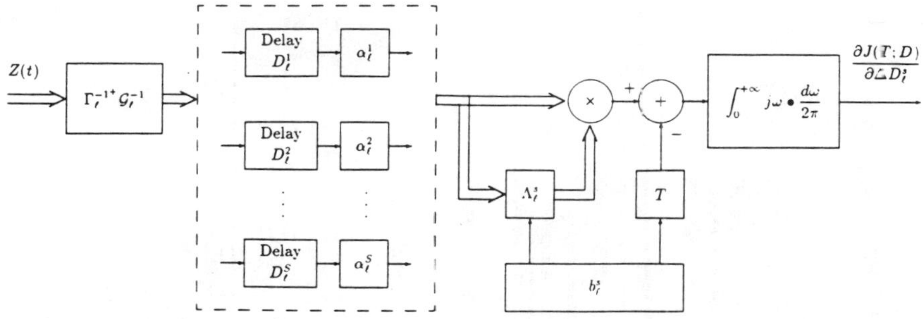

Fig. 8. Generation of $\partial J(T; D)/\partial \Delta D_l^s$: expression (91).

$$\Phi_s = \sum_{m=1}^{S} [-\mathbf{1}_S e_m^T + e_m \mathbf{1}_S^T] \tag{94}$$

$$b_l^s(\omega) = \Gamma_l^{-1} \operatorname{tr} [\mathcal{G}_l^{-1} V_l \Phi_s V_l^{*^T}] \tag{95}$$

$$\Gamma_l(\omega) = \mathcal{V}_l^{*^T} \mathcal{G}_l^{-1} \mathcal{V}_l + G_l^{-1} \tag{96}$$

$$\mathcal{G}_l(\omega) = R + \sum_{k=1, k \neq l}^{L} \mathcal{V}_k G_k \mathcal{V}_k^{*^T} \tag{97}$$

$$e_m = [0 \cdots 0 \;\; \underbrace{1}_{\text{column } m} \;\; 0 \cdots 0]^T \in \mathbb{R}^S \tag{98}$$

and $\mathbf{1}_S$ is the one vector introduced in Section I.

The delay processor assembles $L(S - 1)$ parallel processing channels that perform the derivative $\partial J(T; D)/\partial \Delta D_l^s$. Fig. 8 shows the path (l, s) processing channel block diagram. The causal filter $\Gamma_l^{-1}(\omega)^+ \mathcal{G}_l^{-1}(\omega)$, where $\Gamma_l^{-1}(\omega)^+$ contains all poles and zeros of $\Gamma_l^{-1}(\omega)$ that lie on the left-half s plane (stable, minimum phase), is a function of the delays regarding the first l sources. The term $b_l^s(\omega)$ is a correction term due to either the presence of a number of sources L greater than 1 and/or the spatially correlated observation noise.

For a multisource configuration, under the SLOT approximation with spatially uncorrelated observation noise, i.e., diagonal covariance matrix R, Ng and Bar-Shalom [1] developed a ML processor to estimate the delay matrix ΔD. These authors modeled the observation noises and the signals as zero mean band-limited Gaussian processes. For a white noise assumption it is straightforward to show that the structure of Ng and Bar-Shalom (see [1, eq. (9), (10)], and the processor stated above (see (91)–(98)) are equivalent. The general ML processor presented in this paper (48), (52)–(65) generalizes the estimator of Ng and Bar-Shalom to *nonstationary* mutually correlated signal processes with possibly mutually spatially correlated observation noises.

Assuming a single source configuration ($L = \{1\}$), i.e., taking $l = 1$ in (88)–(98), the ML differential delay estimate is given by the solution of the system of $S - 1$ equations (90), (91), with

$$\mathcal{G}(\omega) = R. \tag{99}$$

The delay receiver involves $S - 1$ parallel processing channels. Each channel block diagram is obtained from the one in Fig. 8, with $\mathcal{G}(\omega)$ defined in (99). For this case, the prefilter $\Gamma^{-1}(\omega)^+ R^{-1}$, although dependent on the delay vector D, becomes a causal system. The term $b^s(\omega)$ is the correction regarding the nonzero observation noise cross covariance. This structure is now equivalent to that of Kirlin and Dewey (see [2, eq. (6)]). A similar argument also recovers the generalized cross correlator of Knapp and Carter [3], when we specialize the structure of (90), (91) to a stationary single source geometry under a spatially uncorrelated observation noise assumption.

VI. Simulation Results

To illustrate the general ML-processor behavior we present an experimental study based on synthetic data. In [18], as well as [24]–[26], several other examples have been described, where we consider the performance of the novel delay estimator when the source signal is nonstationary, in presence of a directional interference, for a two sources configuration, Doppler estimation, and joint estimation of both delay and signal spectrum. Here we do not repeat these experiments that show that the ML estimator developed in this paper achieves asymptotically the Cramer–Rao bound. Rather, we present a comparison with the classical cross-correlator structure when the signals are not stationary. This is, of course, a situation where the classical estimator does not strictly apply.

To implement the delay processor, the parameter domain Θ is restricted to a grid of allowable values. The ML estimator reduces to a bank of parallel processing blocks, each one computing the log-likelihood function (LLF) tuned to each feasible solution $\theta \in \Theta$. This scheme converges uniformly to the optimal algorithm when the mesh of the discretization grid goes to zero.

A Monte Carlo experiment is analyzed next. We carry out the comparison study of our processor with that of Knapp and Carter [3]. For both structures, we evaluate their statistical mean-square error performance, and compare them to that predicted by the Cramer–Rao theory of Section IV.

The array of hydrophones has 2 sensors. The observations are modeled by (14), with

$$\alpha^s = 1, \quad D^s = (s-1)D^a, \quad s = 1, 2 \tag{100}$$

where the actual delay value is $D^a = 0.5$ ms. The delay processor does not know the actual delay D^a, but bounds

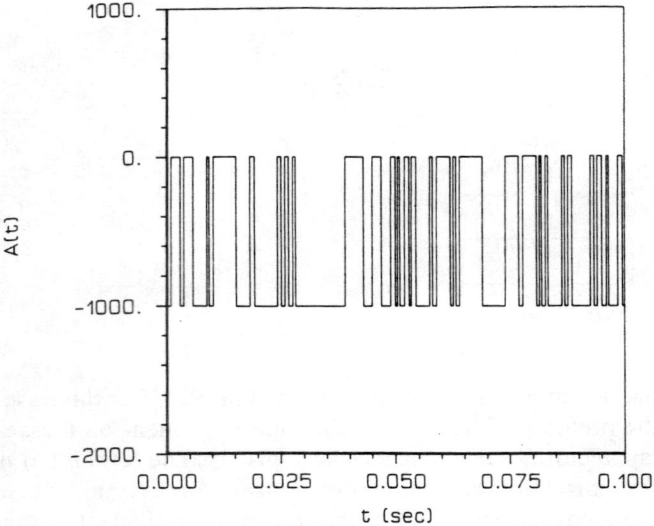

Fig. 9. Time evolution of $A(t)$.

the *a priori* region of uncertainty to

$$D \in [-2.5, +2.5] \quad \text{ms.} \tag{101}$$

This constraint is established from the prior knowledge concerning the problem geometry. The observation noise is spatially and temporally white with spectral height

$$R(s) = 0.001, \qquad s = 1, 2. \tag{102}$$

The single signal process $y(t)$ is a zero mean nonstationary process modeled by the linear dynamical system (19), (20) with $A(t)$ shown in Fig. 9, and

$$B = 10, \quad C = 1, \quad Q = 20, \quad \text{cov}\, x_0 = 1. \tag{103}$$

The time dependence on $A(t)$ may model, for example, a fast maneuvering target subject to sudden changes in its trajectory.

Taking a discretization step of 0.25 ms for the delay domain, both the ML processor developed herein and the generalized cross correlator (GCC) of Knapp and Carter [3] are implemented through a bank of 21 filters working in parallel. We compare both estimators for different observation time interval durations T, where we take

$$T = 5n \quad \text{ms}, \quad n = 1, 2, \cdots, 20. \tag{104}$$

For each value of T, we average the results of 150 Monte Carlo simulations. The time evolution of both the sample mean (SM) and the mean-square error (MSE) of the delay estimate are plotted in Figs. 10 and 11, respectively. From Fig. 10 we see that, for small t, the GCC suffers a larger bias on the delay estimate than the ML processor. Furthermore, as t increases, Fig. 10 shows that the ML estimate is asymptotically unbiased while the GCC retains a residual bias. Fig. 11 shows that the mean-square error (MSE) of the GCC is much larger than the MSE of the ML estimator developed in this paper. In particular, as t increases, Fig. 11 shows that the ML estimate is asymptotically efficient, i.e., its MSE converges to the Cramer–Rao bound (CRB).

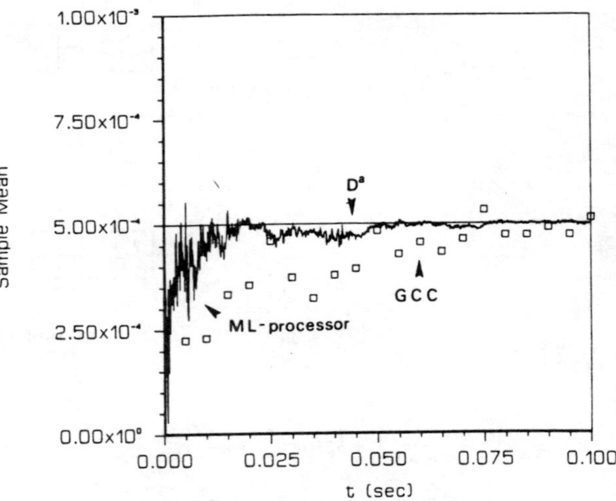

Fig. 10. Delay estimate sample mean.

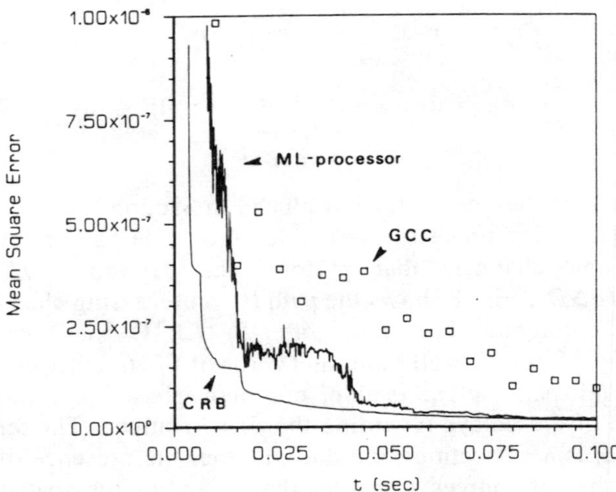

Fig. 11. Delay estimate mean-square error.

VII. Conclusion

The paper reports on maximum likelihood time-varying delay estimation with stochastic nonstationary signals. The framework includes multiple sensors with possibly spatially correlated noise, for either multisources in a single path, or a single source in a multipath environment.

The maximum likelihood (ML) delay estimator maximizes the log-likelihood function (LLF). This function involves estimates of each of the source signals received at each array sensor. To obtain these estimates, we reformulated the nonstationary multisource/multisensor delay problem in the framework of a distributed parameter system model. The signal estimation was now constructed as the causal minimum mean-square error (MMSE) filter for the distributed parameter signal model.

The paper presented the structure of the ML-delay estimator. Due to the nonstationarity assumed, the estimator emphasizes time domain techniques. We provided an interpretation of the delay estimator by studying it under SLOT conditions. We showed that under stationary and long observation intervals, working in the frequency domain, the Riccati equation associated with our generalized Kalman–Bucy filter (GKBF) can be factored, leading to the transfer function of a corresponding generalized Wiener filter (GWF). Using this GWF into the LLF recovers the more common cross-correlator delay estimators of [1]–[3]. The statistical analysis of the delay estimator is carried out in terms of the Cramer–Rao bound. The paper presents an experimental study based on synthetic data that illustrates the behavior of the ML-delay estimator and compares it with the classical generalized cross-correlator configuration.

Appendix A

In this Appendix we derive the LLF for discrete time t. Two steps are considered: in the first one, the model discretization is carried out, while in the second one, the LLF is established.

1. Model Discretization

Define in $[0, t]$ the partition

$$0, \tau, 2\tau, \cdots, N\tau = t \tag{A.1}$$

with τ small and positive. The discretization of the observation set (45) leads to the observation record

$$Z^t = \{Z(k\tau): k = 0, 1, \cdots, N\} \tag{A.2}$$

modeled by the algebraic equation (from (33))

$$Z(k\tau) = Y(k\tau) + V(k\tau), \qquad k = 0, 1, \cdots, N \tag{A.3}$$

where $\{V(k\tau), k = 0, 1, \cdots, N\}$ is a zero mean Gaussian white noise vector sequence, with normalized covariance matrix $R(k\tau)/\tau$.

To the observation record (A.2) corresponds the signal sequence

$$Y^t = \{Y(k\tau): k = 0, 1, \cdots, N\}. \tag{A.4}$$

Denote the corresponding delay sequence as

$$D^t = \{D(k\tau): k = 0, 1, \cdots, N\}. \tag{A.5}$$

2. Establishment of the LLF

The log-likelihood function (LLF) is

$$J(t, D^t) = \ln p(Z^t|D^t) \tag{A.6}$$

with

$$\begin{aligned} p(Z^t|D^t) &= E_{Y^t}[p(Z^t|Y^t; D^t)|D^t\} \\ &= E_{Y^t}[p(Z(t)|Y^t, Z^{t-\tau}, D^t)p(Z^{t-\tau}|Y^t, D^t)|D^t\} \end{aligned} \tag{A.7}$$

where E_{Y^t} denotes the expectation operation over all paths of the signal sequence Y^t.

From (A.3)

$$p(Z(t)|Y^t, Z^{t-\tau}, D^t) = p(Z(t)|Y(t), D(t)). \tag{A.8}$$

From (A.7), noting that

$$E_{Y^{t-\tau}}\{p(Z^{t-\tau}|Y^t, D^t)|D^t\} = p(Z^{t-\tau}|Y(t), D^t) \tag{A.9}$$

we have

$$p(Z^t|D^t) = E_{Y(t)}\{p(Z(t)|Y(t), D(t))p(Z^{t-\tau}|Y(t), D^t)|D^t\} \tag{A.10}$$

where $E_{Y(t)}$ denotes the expectation operation over all paths of the random variable $Y(t)$. Since

$$p(Z^{t-\tau}|D^t) = p(Z^{t-\tau}|D^{t-\tau}) \tag{A.11}$$

from Bayes law [22], we can write

$$p(Z^{t-\tau}|Y(t), D^t) = \frac{p(Y(t)|Z^{t-\tau}, D^t)p(Z^{t-\tau}|D^{t-\tau})}{p(Y(t)|D^t)}. \tag{A.12}$$

Substitution of (A.12) in (A.10) leads to

$$p(Z^t|D^t) = E_{Y(t)}\{p(Z(t)|Y(t), D(t))|Z^{t-\tau}, D^t\}p(Z^{t-\tau}|D^{t-\tau}). \tag{A.13}$$

From (A.3), we have

$$p(Z(t)|Y(t), D(t)) = \mathfrak{N}\left(Z(t) - Y(t), \frac{R(t)}{\tau}\right) \tag{A.14}$$

where $\mathfrak{N}(.,.)$ represents the normalized Gauss function. On the other hand, noting that $Y(t)$ is Gaussian

$$p(Y(t)|Z^{t-\tau}, D^t) = \mathfrak{N}(Y(t) - \hat{Y}(t), P_Y(t)) \tag{A.15}$$

where

$$\hat{Y}(t) = E\{Y(t)|Z^{t-\tau}, D^t\} \tag{A.16}$$

is the minimum mean-square error (MMSE) estimate of the received signal, and

$$P_Y(t) = E\{[Y(t) - \hat{Y}(t)][Y(t) - \hat{Y}(t)]^T|Z^{t-\tau}, D^t\} \tag{A.17}$$

is the signal error estimation covariance matrix.

Substitution of (A.14) in (A.13), considering (A.15), leads to

$$p(Z^t|D^t) = \mathfrak{N}\left(Z(t) - \hat{Y}(t), P_Y(t) + \frac{R(t)}{\tau}\right) p(Z^{t-\tau}|D^{t-\tau}). \tag{A.18}$$

From (A.18), recalling the partition (A.1) considered in the discretization of the time interval $[0, t]$, we obtain

$$\begin{aligned} \ln p(Z^t|D^t) = \sum_{k=0}^{N} \Bigg\{ & [Z(k\tau) - \hat{Y}(k\tau)]^T[P_Y(k\tau)\tau \\ & + R(k\tau)]^{-1}[Z(k\tau) - \hat{Y}(k\tau)]\tau \\ & - \frac{1}{2} \ln \Bigg\{(2\pi)^S \det\left[\frac{R(k\tau)}{\tau}\right] \\ & \cdot \det[I + R^{-1}(k\tau)P_Y(k\tau)\tau]\Bigg\}\Bigg\}. \end{aligned} \tag{A.19}$$

In (A.19), the term $(2\pi)^S \det [R(k\tau)/\tau]$ does not depend on the delay sequence D', its inclusion on the LLF being not necessary. Thus

$$J(t; D') = \sum_{k=0}^{N} \{[Z(k\tau) - \hat{Y}(k\tau)]^T[P_Y(k\tau)\tau + R(k\tau)]^{-1} \cdot [Z(k\tau) - \hat{Y}(k\tau)]\tau - \tfrac{1}{2} \ln \det [I + R^{-1}(k\tau)P_Y(k\tau)\tau]\}. \quad (A.20)$$

Expanding as a Taylor series, around $\tau = 0$, the second term on the RHS of expression (A.20), keeping $k\tau = t_k$ fixed, neglecting the second and higher order terms, leads to the LLF for discrete time

$$J(t; D') = \sum_{k=0}^{N} \{[Z(k\tau) - \hat{Y}(k\tau)]^T[P_Y(k\tau)\tau + R(k\tau)]^{-1} \cdot [Z(k\tau) - \hat{Y}(k\tau)] - \tfrac{1}{2} \operatorname{tr} [R^{-1}(k\tau)P_Y(k\tau)]\}\tau. \quad (A.21)$$

Appendix B

In this Appendix, the log-likelihood function (LLF) for continuous time t is presented.

As referred to in Section II-A, a rigorous description of the processes requires the Ito calculus formulation. Letting formally $Z(t) = d\zeta(t)/dt$ and $V(t) = dW(t)/dt$ in (33), the observed process is modeled by the stochastic differential equation

$$d\zeta(t) = Y(t)\, dt + dW(t) \quad (B.1)$$

where $\{W(t)\}$ is a Wiener vectorial process. Assume that i) the processes $\{Y(t)\}$ and $\{W(t)\}$ are independent; ii) $E\{\int_0^T \langle Y(t), Y(t)\rangle_R\, dt\} < \infty$, where $\langle *, *\rangle_R$ is the weighted inner product operator defined in (10), and $R(t)$ is the covariance matrix of the observation noise.

In the above context, the log-likelihood function (LLF) is nothing but the logarithm of the Radon–Nikodym derivative of the measure induced by the observation process with respect to the Wiener measure. This derivative is (see [27, ch. 7, theorem 7.13, note 3])

$$J(t; D') = \int_0^t \langle \hat{Y}_\sigma, d\zeta_\sigma\rangle_R - \tfrac{1}{2} \int_0^t \langle \hat{Y}_\sigma, \hat{Y}_\sigma\rangle_R\, d\sigma. \quad (B.2)$$

In (B.2), $\hat{Y}(\sigma)$ is the minimum mean-square error (MMSE) estimate of the signal process $Y(\sigma)$ given the observation process and the delay sequence

$$\hat{Y}(\sigma) = E[Y(\sigma) | \zeta_\vartheta, \vartheta \in [0, \sigma], D'(\theta)]. \quad (B.3)$$

The first term on the right-hand side (RHS) of expression (B.2) represents an Ito stochastic integral [16].

When implementing the LLF, one converts the Ito integral into a Stratonovich integral [16]. Denoting the innovations process as

$$d\tilde{\zeta}(t) = d\zeta(t) - \hat{Y}(t)\, dt \quad (B.4)$$

the LLF becomes

$$J(t; D') = \int_0^t \langle \hat{Y}_\sigma, d\tilde{\zeta}_\sigma\rangle_R + \tfrac{1}{2} \int_0^t \langle \hat{Y}_\sigma, \hat{Y}_\sigma\rangle_R\, d\sigma. \quad (B.5)$$

The first term in (B.5), the Ito integral, can now be written [16] as

$$\int_0^t \langle \hat{Y}_\sigma, d\tilde{\zeta}_\sigma\rangle_R = \int_0^t \langle \hat{Y}_\sigma, d\tilde{\zeta}_\sigma\rangle_R - \tfrac{1}{2} (\hat{Y}_t, \tilde{\zeta}_t)_R. \quad (B.6)$$

The integral in the RHS of (B.6) is a Stratonovich integral which is computed using the ordinary rules of calculus. The second term in the RHS of (B.6) is the quadratic variation [16] of the processes $\hat{Y}(t)$ and $\tilde{\zeta}(t)$. Define in $[0, t]$ the partition

$$\tau^n = \{t_0, t_1, \cdots, t_n\} \quad (B.7)$$

where

$$0 = t_0 < t_1 < \cdots < t_n = t. \quad (B.8)$$

The quadratic variation of the processes $\hat{Y}(t)$ and $\tilde{\zeta}(t)$ is defined by [16]

$$(\hat{Y}_t, \tilde{\zeta}_t)_R = \operatorname*{l.i.m.}_{|\tau^n| \to 0} \sum_{k=0}^{n-1} [\hat{Y}(t_{k+1}) - \hat{Y}(t_k)]^T R^{-1} [\tilde{\zeta}(t_{k+1}) - \tilde{\zeta}(t_k)] \quad (B.9)$$

where $\hat{Y}(t)$, the best linear estimate of $Y(t)$ in the mean-square error sense, is

$$\hat{Y}_t = E\{Y_t\} + \int_0^t P_{Y_\sigma} R_\sigma^{-1}\, d\tilde{\zeta}_\sigma. \quad (B.10)$$

Noting that

$$\tilde{\zeta}(t) = \int_0^t d\tilde{\zeta}_\sigma \quad (B.11)$$

and considering (B.10), the quadratic variation becomes [16]

$$(\hat{Y}_t, \tilde{\zeta}_t)_R = \int_0^t \operatorname{tr} [R_\sigma^{-1} P_{Y_\sigma}]\, d\sigma. \quad (B.12)$$

Using formally $Z(t) = d\zeta(t)/dt$ along with (B.6) and (B.12), we obtain the LLF

$$J(t; D') = \int_0^t [\langle \hat{Y}_\sigma, Z_\sigma\rangle_R - \tfrac{1}{2} \langle \hat{Y}_\sigma, \hat{Y}_\sigma\rangle_R - \tfrac{1}{2} \operatorname{tr} (R^{-1} P_Y)]\, d\sigma \quad (B.13)$$

where the integral is computed using the ordinary rules of calculus, and both $\hat{Y}(\sigma)$ and $P_Y(\sigma)$ are functions of the time delay realization $D^o(\theta)$. The third term on the RHS of (B.13) is the correction term regarding the corresponding Ito integral representation.

Appendix C

In this Appendix we derive the frequency domain representation of the time-invariant generalized Kalman-Bucy filter (IGKBF).

Let $\hat{X}_I(\omega, r)$ ($r \in [0, 1]$) and $Z_I(\omega)$ be the integrated Fourier transforms of, respectively, the state vector $\hat{X}(t, r)$ and the observation process $Z(t)$. From (57), (58) with

$$K(t, r) = K(r) = P_\infty(r, 1) C^T R^{-1} \tag{C.1}$$

the frequency domain representation of the IGKBF is

$$j\omega\gamma \, d\hat{X}_I(\omega, r) + \frac{\partial}{\partial r}[d\hat{X}_I(\omega, r)] = \gamma K(r)[dZ_I(\omega) - C\, d\hat{X}_I(\omega, 1)] \tag{C.2}$$

with the boundary condition

$$d\hat{X}_I(\omega, 0) = (j\omega I - A)^{-1} K(0)[dZ_I(\omega) - C\, d\hat{X}_I(\omega, 1)]. \tag{C.3}$$

For $r = 1$ the solution of (C.2) is

$$d\hat{X}_I(\omega, 1) = e^{-j\omega\gamma} d\hat{X}_I(\omega, 0) + e^{-j\omega\gamma} \int_0^1 e^{j\omega\gamma\sigma} \gamma K(\sigma) \cdot d\sigma [dZ_I(\omega) - C\, d\hat{X}_I(\omega, 1)]. \tag{C.4}$$

Substituting (C.3) into (C.4), premultiplying by C, and noting that

$$d\hat{Y}_I(\omega) = C\, d\hat{X}_I(\omega, 1) \tag{C.5}$$

where $\hat{Y}_I(\omega)$ is the integrated Fourier transform of the signal estimate $\hat{Y}(t)$, we get

$$d\hat{Y}_I(\omega) = [I_S - H(\omega, D)]\, dZ_I(\omega) \tag{C.6}$$

where

$$H^{-1}(\omega, D) = I_S + Ce^{-j\omega\gamma}\left[(j\omega I - A)^{-1} K(0) + \int_0^1 e^{j\omega\gamma\sigma} \gamma K(\sigma)\, d\sigma\right]. \tag{C.7}$$

Appendix D

In this Appendix we establish expression (85). The technique carries out a spectral factorization of the steady state (PDE) Riccati equation. To work with a compact notation, the subscript ∞ on the steady state covariance matrix $P_\infty(r_1, r_2)$ will be often omitted.

From (79), we have

$$\begin{aligned}
H^{-1}&(\omega, D) R H^{-T}(-\omega, D) \\
&= \left\{ I_S + Ce^{-j\omega\gamma}\left[(j\omega I - A)^{-1} K_0 + \int_0^1 e^{j\omega\gamma\sigma}\gamma K_\sigma\, d\sigma\right]\right\} R \\
&\quad \cdot \left\{ I_S + \left[K_0^T(-j\omega I - A^T)^{-1} + \int_0^1 K_\sigma^T \gamma e^{-j\omega\gamma\sigma}\, d\sigma\right] e^{j\omega\gamma} C^T\right\} \\
&= R + Ce^{-j\omega\gamma}(j\omega I - A)^{-1} K_0 R \\
&\quad + Ce^{-j\omega\gamma}\int_0^1 e^{j\omega\gamma\sigma}\gamma K_\sigma\, d\sigma R \\
&\quad + RK_0^T(-j\omega I - A^T)^{-1} e^{j\omega\gamma} C^T \\
&\quad + R\int_0^1 K_\sigma^T \gamma e^{-j\omega\gamma\sigma}\, d\sigma e^{j\omega\gamma} C^T \\
&\quad + Ce^{-j\omega\gamma}\left[(j\omega I - A)^{-1} K_0 R K_0^T(-j\omega I - A^T)^{-1} \right. \\
&\quad + (j\omega I - A)^{-1} K_0 R \int_0^1 K_\sigma^T \gamma e^{-j\omega\gamma\sigma}\, d\sigma \\
&\quad + \int_0^1 e^{j\omega\gamma\sigma}\gamma K_\sigma\, d\sigma R K_0^T(-j\omega I - A^T)^{-1} \\
&\quad \left. + \int_0^1 e^{j\omega\gamma\sigma}\gamma K_\sigma\, d\sigma R \int_0^1 K_\sigma^T \gamma e^{-j\omega\gamma\sigma}\, d\sigma\right] e^{j\omega\gamma} C^T.
\end{aligned} \tag{D.1}$$

Each one of the quadratic terms on the RHS of (D.1) can be obtained from the generalized Riccati equation that describes the steady-state covariance matrix $P_\infty(r_1, r_2)$, $r_1, r_2 \in [0, 1]$. From expressions (61)–(64) with $\partial P(t, r_1, r_2)/\partial t = 0$, $P_\infty(r_1, r_2)$ is given by the partial differential equation

$$\nabla_{r_1 r_2} P_\infty(r_1, r_2) = -\gamma K(r_1) R K^T(r_2)\gamma \tag{D.2}$$

with the boundary conditions

$$\nabla_r P_\infty(r, 0) = \gamma P_\infty(r, 0) A^T - \gamma K(r) R K^T(0) \tag{D.3}$$

$$\nabla_r P_\infty(0, r) = A P_\infty(0, r)\gamma - K(0) R K^T(r)\gamma \tag{D.4}$$

and

$$A P_\infty(0, 0) + P_\infty(0, 0) A^T + BQB^T - K(0) R K^T(0) = 0 \tag{D.5}$$

where the differential operators are

$$\nabla_{r_1 r_2}(*) = \left[\frac{\partial}{\partial r_1}(*)\right]\gamma + \gamma\left[\frac{\partial}{\partial r_2}(*)\right] \tag{D.6}$$

and

$$\nabla_r(*) = \frac{d}{dr}(*). \tag{D.7}$$

Multiplication of (D.2) on the left by $e^{j\omega\gamma r_1}$ and on the right by $e^{-j\omega\gamma r_2}$, and integration on r_1 and r_2 over the interval $[0, 1]$, leads to

$$\begin{aligned}
\int_0^1 & e^{j\omega\gamma\sigma}\gamma K_\sigma\, d\sigma R \int_0^1 K_\sigma^T \gamma e^{-j\omega\gamma\sigma}\, d\sigma \\
&= -e^{j\omega\gamma}\int_0^1 P_{1\sigma}\gamma e^{-j\omega\gamma\sigma}\, d\sigma + \int_0^1 P_{0\sigma}\gamma e^{-j\omega\gamma\sigma}\, d\sigma \\
&\quad - \int_0^1 e^{j\omega\gamma\sigma}\gamma P_{\sigma 1}\, d\sigma e^{-j\omega\gamma} + \int_0^1 e^{j\omega\gamma\sigma}\gamma P_{\sigma 0}\, d\sigma.
\end{aligned} \tag{D.8}$$

Noting that

$$e^{j\omega\gamma r}\nabla_r P_{r0} = \nabla_r(e^{j\omega\gamma r} P_{r0}) - j\omega\gamma e^{j\omega\gamma r} P_{r0} \tag{D.9}$$

multiplication of (D.3) on the left by $e^{j\omega\gamma r}$, and integration on r over the interval [0, 1], yields after manipulation

$$\int_0^1 e^{j\omega\gamma\sigma}\gamma K_\sigma \, d\sigma R K_0^T(-j\omega I - A^T)^{-1}$$

$$= -e^{j\omega\gamma}P_{10}(-j\omega I - A^T)^{-1}$$

$$+ P_{00}(-j\omega I - A^T)^{-1} - \int_0^1 e^{j\omega\gamma\sigma}\gamma P_{\sigma 0} \, d\sigma. \quad \text{(D.10)}$$

Noting that

$$\nabla_r P_{0r} e^{-j\omega\gamma r} = \nabla_r (P_{0r} e^{-j\omega\gamma r}) + P_{0r} e^{-j\omega\gamma r} j\omega\gamma \quad \text{(D.11)}$$

multiplication of (D.4) on the right by $e^{-j\omega\gamma r}$, and integration on r over the interval [0, 1], gives after manipulation

$$(j\omega I - A)^{-1}K_0 R \int_0^1 K_\sigma^T \gamma e^{-j\omega\gamma\sigma} \, d\sigma$$

$$= -(j\omega I - A)^{-1}P_{01}e^{-j\omega\gamma} + (j\omega I - A)^{-1}P_{00}$$

$$- \int_0^1 P_{0\sigma}\gamma e^{-j\omega\gamma\sigma} \, d\sigma. \quad \text{(D.12)}$$

From (D.5), it is straightforward to show that

$$(j\omega I - A)^{-1}K_0 R K_0^T(-j\omega I - A^T)^{-1}$$

$$= -(j\omega I - A)^{-1}P_{00} - P_{00}(-j\omega I - A^T)^{-1}$$

$$+ (j\omega I - A)^{-1}BQB^T(-j\omega I - A^T)^{-1}. \quad \text{(D.13)}$$

Substitution of (D.8), (D.10), (D.12), and (D.13) in (D.1), remarking that the IGKBF gain matrix is

$$K(r) = P_\infty(r, 1)C^T R^{-1} \quad \text{(D.14)}$$

yields after manipulation

$$H^{-1}(\omega, D)RH^{-T}(-\omega, D) = R + G_{YY}(\omega) \quad \text{(D.15)}$$

where

$$G_{YY}(\omega) = Ce^{-j\omega\gamma}(j\omega I - A)^{-1}BQB^T(-j\omega I - A^T)^{-1}e^{j\omega\gamma}C^T \quad \text{(D.16)}$$

is the power spectral density of the received signal $Y(t)$. The RHS of (D.15) represents the power spectral density $G_{ZZ}(\omega)$ of the observed process $Z(t)$, i.e.,

$$H^{-1}(\omega, D)RH^{-T}(-\omega, D) = G_{ZZ}(\omega). \quad \text{(D.17)}$$

References

[1] L. C. Ng and Y. Bar-Shalom, "Multisensor multitarget time delay vector estimation," *IEEE Trans. Acoust., Speech, Signal Processing*, vol. ASSP-34, pp. 669–678, Aug. 1986.

[2] R. L. Kirlin and L. A. Dewey, "Optimal delay estimation in a multiple sensor array having spatially correlated noise," *IEEE Trans. Acoust., Speech, Signal Processing*, vol. ASSP-33, pp. 1387–1396, Dec. 1985.

[3] C. H. Knapp and G. C. Carter, "The generalized correlation method for estimation of time delay," *IEEE Trans. Acoust., Speech, Signal Processing*, vol. ASSP-24, pp. 320–327, Aug. 1976.

[4] G. C. Carter, "Time delay estimation," Ph.D. dissertation, University of Connecticut, Storrs, CT, 1976.

[5] C. H. Knapp and G. C. Carter, "Estimation of time delay in the presence of source or receiver motion, *J. Acoust. Soc. Amer.*, vol. 61, no. 6, pp. 1545–1549, June 1977.

[6] R. L. Kirlin, D. F. Moore, and R. F. Kubichek, "Improvement of delay measurements from sonar arrays via sequential state estimation," *IEEE Trans. Acoust., Speech, Signal Processing*, vol. ASSP-29, pp. 514–519, June 1981.

[7] J. C. Hassab and R. E. Boucher, "Optimum estimation of time delay by a generalized correlator," *IEEE Trans. Acoust., Speech, Signal Processing*, vol. ASSP-27, pp. 373–380, Aug. 1979.

[8] R. S. Bucy, J. M. F. Moura, and A. J. Mallinckrodt, "A Monte Carlo study of absolute phase determination," *IEEE Trans. Inform. Theory*, vol. IT-29, pp. 509–520, July 1983.

[9] J. M. N. Leitão and J. M. F. Moura, "Algorithm structures for cyclic phase estimation," presented at GRETSI-81, Nice, France, 1981.

[10] R. E. Bethel and R. G. Rahikka, "An optimum first-order time delay tracker," *IEEE Trans. Aerosp. Electron. Syst.*, vol. AES-23, pp. 718–725, Nov. 1987.

[11] J. M. F. Moura and A. B. Baggeroer, "Phase unwrapping of signals propagated under the Arctic ice crust: A statistical approach," *IEEE Trans. Acoust., Speech, Signal Processing*, vol. ASSP-36, pp. 617–630, May 1988.

[12] A. Graham, *Kronecker Products and Matrix Calculus: with Applications.* Ellis Horwood Ltd., 1981.

[13] J. M. F. Moura and A. B. Baggeroer, "Passive systems theory with narrow-band and linear constraints: Part I—Spatial diversity," *IEEE J. Ocean. Eng.*, vol. OE-3, pp. 5–13, Jan. 1978.

[14] J. M. F. Moura, "Passive systems theory with narrow-band and linear constraints: Part II—Temporal diversity," *IEEE J. Ocean Eng.*, vol. OE-4, pp. 19–30, Jan. 1979.

[15] J. M. F. Moura, "Passive systems theory with narrow-band and linear constraints: Part III—Spatial/temporal diversity," *IEEE J. Ocean. Eng.*, vol. OE-4, pp. 113–119, July 1979.

[16] J. M. F. Moura, "Linear and nonlinear stochastic filtering," in *Signal Processing*, vol. 1, J. L. Lacoume, T. S. Durrani, and R. Storo, Eds. North-Holland, 1987, pp. 205–276.

[17] H. L. Van Trees, *Detection, Estimation, and Modulation Theory, Part I.* New York: Wiley, 1968.

[18] I. M. G. Lourtie, "Estimação Óptima de Tempos de Atraso com Sinais Não-Estacionários," Ph.D. dissertation, Instituto Superior Técnico, Portugal, Oct. 1987; also available as "Optimal estimation of time delays with nonstationary signals," Tech. Rep., LASIP, CMU, June 1988.

[19] R. E. Kalman and R. S. Bucy, "New results in linear filtering and prediction theory," *Trans. ASME J. Basic Eng.*, Mar. 1961.

[20] H. Kwakernaak, "Optimal filtering in linear systems with time delays," *IEEE Trans. Automat. Contr.*, vol. AC-12, pp. 169–173, Apr. 1967.

[21] T. K. Yu, J. H. Seinfeld, and W. H. Ray, "Filtering in nonlinear time delay systems," *IEEE Trans. Automat. Contr.*, vol. AC-19, pp. 324–333, Aug. 1974.

[22] A. Papoulis, *Probability, Random Variables, and Stochastic Processes.* New York: McGraw-Hill, 1965.

[23] A. M. Yaglom, *An Introduction to the Theory of Stationary Random Functions.* Englewood Cliffs, NJ: Prentice-Hall, 1962.

[24] I. M. G. Lourtie and J. M. F. Moura, "Time delay determination: Maximum likelihood and Kalman-Bucy type structures," in *Proc. 1985 IEEE Int. Conf. Acoust., Speech, Signal Processing*, Mar. 1985.

[25] I. M. G. Lourtie and J. M. F. Moura, "Optimal estimation of time-varying delay," in *Proc. 1988 IEEE Int. Conf. Acoust., Speech, Signal Processing*, Apr. 1988.

[26] I. M. G. Lourtie and J. M. F. Moura, "Joint delay and signal determination," in *Underwater Acoustic Data Processing*, Y. T. Chan, Ed. Kluwer Academic, 1989, pp. 531–535.

[27] R. S. Lipster and A. N. Shiryayev, *Statistics of Random Processes*, vol. I. New York: Springer, 1977.

REAL-TIME ESTIMATION OF MOVING TIME DELAY

H. Meyr, G. Spies, J. Bohmann

Aachen Technical University (RWTH), West-Germany
Templergraben 55, D-5100 Aachen

ABSTRACT

A two-step algorithm for the estimation of a rapidly varying time delay between two stochastic signals is described. In the first step the maximum-likelihood (ML) estimate is computed over an observation interval small enough to consider the delay to be approximately constant. Due to the short averaging interval the probability of ambiguous peaks is greatly increased. Therefore in the second step, a non-linear, adaptive postfiltering algorithm is presented that effectively suppresses 'outliers' of the ML-estimator.

INTRODUCTION

The maximum-likelihood (ML) technique applied in the estimation of a constant delay between two random signals yields a structure that performs the cross-correlation of the two prefiltered signals. The ML-estimate of the delay is the value $\hat{D}$ for which the cross-correlation assumes its maximum [1]. For the estimation of time variable delays prefilters have been proposed that performs a time compression-expansion on the signals prior to the correlation. This approach can be viewed as the simultaneous estimation of the two (constant) parameters $(D_o, \dot{D})$, where $\dot{D}$ is the derivative of the delay with respect to time and D_o is a constant delay. The ML-estimator consists of a bank of cross-correlators each preceeded by a signal compression-expansion device.

We consider in this paper the estimation of time variable delay where the delay D(t) is the output of a state-space model driven by a random signal. The practical application which leads to the problem is non-contact velocity measurement of railguided vehicles. As shown in [2], the velocity is obtained by determining the time delay between two stochastic signals. This delay, then, is inversely proportional to the velocity.

We describe here a 'structured' approach to the problem. The estimate $\hat{D}$ is obtained in a two-step algorithm. In the first step the ML-estimate is computed over an observation interval small enough to consider the delay to be approximately constant. Due to the small averaging interval the probability of detecting a false peak in the cross-correlation function can be in the order of magnitude of 0.1. In this case, it is characteristic for a sequence of estimates that they are either very close to the true delay or completely wrong, if a false peak is detected. Therefore, a post-filter to eliminate these 'outliers' is necessary. It has the structure of a state-estimator with the system matrix determined by the known state-space model of the delay D(t). The ML-estimates, then, can be viewed as noisy measurements for the state estimator. The state-estimator has the familiar structure of a Kalman-filter with the important difference that a measurement update is done only if the measurement is considered reliable, otherwise the filter acts as a predictor. This two-step approach is capable of handling signal-interruptions by simply operating the state-estimator as a predictor. This is of utmost importance in our application and might be of interest in other applications as well. The system was realized with standard digital hardware at low cost to operate in real-time.

SIGNAL MODEL FOR VELOCITY ESTIMATION

Two sensors S_1 and S_2 are moving relative to a surface a distance L apart along the direction of movement (fig.1).

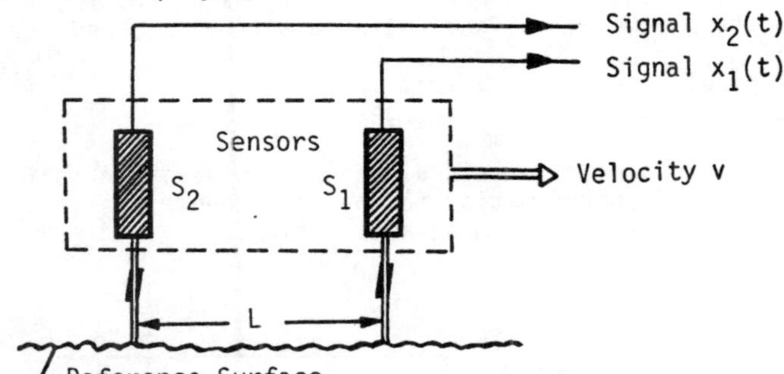

Fig. 1: Principle of Signal Generation

Due to surface irregularities, each sensor generates a stochastic signal whose amplitude in time is determined by the time dependent location of the sensor. If S_2 reaches at a time $t=t_2$ the same location reached by sensor S_1 at a time $t=t_1$ then the values $x_2(t=t_2)$ and $x_1(t=t_1)$ will be similar. The difference in time $D(t)=t_2-t_1$ (delay) depends on the translation velocity during the covering of the distance L.

This delay D(t) has two possible definitions,

Reprinted from *Proc. ICASSP '82*, vol. 1, pp. 383–386, May 1982.

depending on which signal is under consideration . For example the two definitions

$$\text{I:} \quad s_1(t)=s_2(t+D_I(t)) \tag{1}$$

$$\text{II:} \quad s_2(t)=s_1(t-D_{II}(t)) \tag{2}$$

yield different delays D_I, D_{II}. D_I gives the time that sensor S_2 will take to reach the present position of sensor S_1. On the other hand if sensor S_2 is at a fixed location, the sensor S_1 had already passed the same location at time $t-D_{II}(t)<t$. If we assume a translation velocity v linear in time with a constant term v_0 and with the acceleration a

$$v(t)=v_0+at \tag{3}$$

and using the relation between distance w and velocity v

$$w(t)=\int_0^t v(t)dt \tag{4}$$

we obtain the following expressions for the delays $D_I(t)$ and $D_{II}(t)$

$$D_I(t)=-\frac{v(t)}{a}+\frac{v(t)}{a}\sqrt{1+\frac{2aL}{v^2(t)}} \tag{5}$$

$$D_{II}(t)=\frac{v(t)}{a}-\frac{v(t)}{a}\sqrt{1-\frac{2aL}{v^2(t)}} \tag{6}$$

These eqs. are valid for positive and negative acceleration a, provided the square root has a real solution and the velocity is always positive.

For accelerations small with respect to the velocity ($aL<<v^2$) the eqs. (5) and (6) converge to the equation

$$D(t)\approx L/v(t) \tag{7}$$

which gives sufficient accuracy for a lot of applications. The maximum error for D(t) using eq. (7) instead of eqs. (5) and (6) occurs if $v=v_{min}$ and $a=a_{max}$.
Typical values for railway application:

$L=0.005m$; $a_{max}=2m/s^2$; $v_{min}=0.2m/s$

In fig.2 a model of signal generation is shown that is a good approximation to the real conditions in our velocity measurement application.

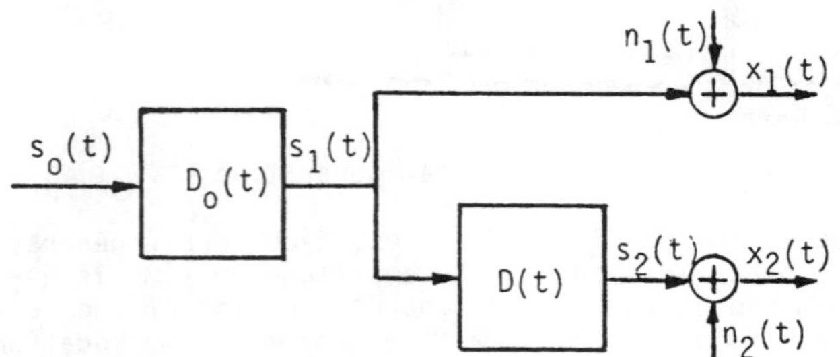

Fig. 2: Model of Signal Generation

The delay D(t) is given by the eqs. (5) to (7) respectively, $s_0(t)$ is a fictitious signal which would have been generated by sensor S1, moving with the constant reference velocity $v=v_{max}$. Therefore s_0 can be treated as a stationary lowpass process with the known noise bandwidth B_{00}. The signal s_0 affected by the delay $D_0(t)$ yields the compressed sensor signal $s_1(t)$. The delay $D_0(t)$ is related to the velocity $v=v_0+at$ by eq. (8) if we use $s_0'(t)=s_1(t+D_0(t))$

$$D_0(t)=-\frac{v}{a}+\frac{v_0}{a}\sqrt{1+\frac{2\cdot v_{max}\cdot a\cdot t}{v_0^2}} \tag{8}$$

The eq. (8) is valid, if the square root has a real solution and $D_0(t)>0$. The condition $v_{max}\cdot a\cdot t<<v_0{}^2$ gives the approximation

$$D_0(t)\approx t(\frac{v_{max}}{v_0}-1) \tag{9}$$

Eqs. (8) and (9) show that the time t is multiplied by a factor which corresponds to a compression/ expansion in time. This means the spectral properties are modified dependent on the velocity.

STRUCTURE OF THE ESTIMATOR

For constant acceleration a maximum-likelihood (ML) estimator for the simultaneous estimation of the two parameters (v_0,a) appearing in eqs. (5) and (6) is conceivable. As shown by Carter [3] and Adams et al. [4] such an estimator compensates a time-varying delay by signal compression/expansion. In our application, however, the acceleration is a stochastic process assuming values within the interval (a_{min},a_{max}). A typical realization of a(t) is characterized by piecewise constant segments of random duration. Rather than trying to find an 'optimum' estimator a 'structured' approach was taken. We assume that the delay does not change appreciably over the estimation interval T_e. Thus the ML-estimator has the familiar structure of a generalized cross-correlator [1]. The quasi-stationary assumption on the delay D(t) can be stated in the form of an inequality

$$\dot{D}\cdot T_e<\frac{1}{2B}\delta \quad <=> \quad 2T_e\cdot B\cdot\dot{D}<\delta \tag{10}$$

with T_e: estimation interval
B : bandwidth of the signal s(t)
δ : constant value, typical $\delta<0.5$

Eq. (10) implies that the change in D over the averaging interval T_e is smaller than δ times the correlation time (=1/(2·B)) of the signal s(t). In order to cope with large delay variations the estimation time T_e must be chosen very small. Worst case values are $|\dot{D}|<0.10$, $\delta=0.5$ which leads to a very small time-bandwidth product ($2\cdot T_e\cdot B$) of 5. The penalty to be paid for such extremly small products $T_e\cdot B$ is the greatly increased probability that the cross-correlation function occasionally peak for a value $\hat{D}$ which is entirely different from the true delay D. We denote the probability of occurence of such false peaks by $1-P_d$. More precisely, we define P_d as the probability that the maximum of the cross-correlation function is located within the interval $|D-\hat{D}|<1/(2\cdot B)$.

It can be shown [5] that P_d depends on the two parameters β and R.

$$R=2D_R\cdot B \tag{11}$$

$$\beta=\sqrt{2T_e\cdot B}\,\frac{\alpha\gamma}{1+\alpha\gamma} \tag{12}$$

$$\gamma=1-(2T_e\cdot B)\dot{D} \tag{13}$$

where D_R is the range of possible delays $\hat{D}$, α is the signal-to-noise ratio, $\dot{D}$ is the rate of change of the delay D and γ is a dynamic suppression factor. For simplicity it has been assumed that signal and noise band-width are equal. The parameter R is (roughly) the number of resolvable values of D in the interval D_R while β combines the signal-to-noise ratio α with the number of independent samples of the estimation function during the averaging interval T_e. The factor γ approximately takes into account the blurring of the correlation peak caused by the time-varying delay D(t). As can be seen from eq. (12) γ has the effect of reducing the SNR. Therefore γ has been called the dynamic suppression factor. The probability P_d of detecting the correct peak is plotted upon β with R as parameter.

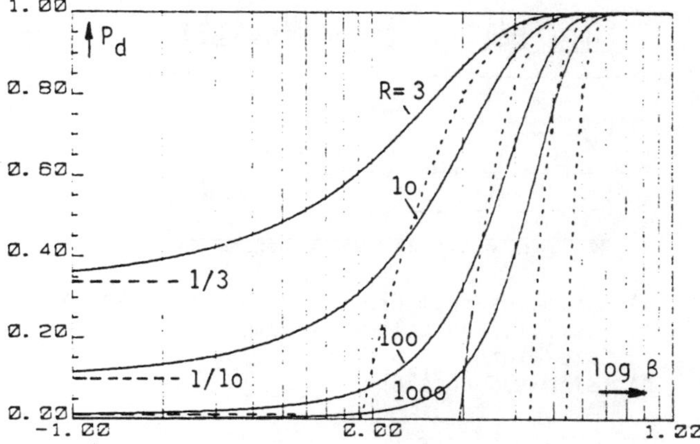

Fig. 3: Detection Probability

We observe a very pronounced threshold behaviour which occurs when

$$(R/\beta)\cdot\exp(-\frac{1}{2}\beta^2) \approx 1 \tag{14}$$

For very small values of β the function P_d asymptotically reaches a value of 1/R. This corresponds to the case where the location of the peak is uniformly distributed over the entire range D_R. For large values of P_d an asymptotically valid expression for P_d can be derived [5].

$$P_D = 1-\frac{1}{\sqrt{2\pi}}\cdot\frac{R}{\beta}\cdot\exp(-\frac{1}{2}\beta^2) \tag{15}$$

The detection probability P_d rapidly increases as a function of β. The resolution R only affects P_d in an essential way for small values. From the definition of β in eq. (12) we deduce the familiar feature of correlation analysis that signal-to-noise ratio α can be traded for averaging time (T_e B). It is worthwhile to note that for $\alpha\cdot\gamma\to\infty$ the parameter β assumes a finite value of $\sqrt{2\cdot T_e\cdot B}$ which means that the probability of detecting a false peak is not zero. This is perfectly plausible: even for vanishing noise processes $n_1(t)$ and $n_2(t)$ the short-time correlation function contains a randomly fluctuating part whose variance vanishes for $T_e\to\infty$ only.

Worst case values for the parameters in eq. (15) are

$$\alpha\cdot\gamma > 10 \quad ; \quad R=10 \quad ; \quad 2T_e B > 5 \tag{16}$$

which yields

$$\beta > 2 \quad \text{and} \quad P_D > 0.8 \tag{17}$$

The occurrence of false peaks cannot be neglected and a suitable postfilter must be used to eliminate these 'outliers'. At first glance such a postfilter seems to be heuristically justified only. This, however, is not the case. As will be seen shortly the postfilter is based on sound theoretical ground.

THE POSTFILTER

The basic task of a postfilter is to estimate the velocity v(t) of a vehicle from the noisy delay estimates $\hat{D}(k)$ obtained from the ML-estimator at discrete times $t=k\cdot T_e$.
The first step in tackling the problem is to derive a discrete time state-space model for the motion of the vehicle. If we choose a piecewise linear approximation of the velocity trajectory for segments of duration T_e the two-dimensional state equations read

$$x(k+1) = \begin{bmatrix} v(k+1) \\ a(k+1) \end{bmatrix} = \begin{bmatrix} 1 & T_e \\ 0 & 1 \end{bmatrix}\begin{bmatrix} v(k) \\ a(k) \end{bmatrix} + \begin{bmatrix} 0 \\ w_2(k) \end{bmatrix} \tag{18}$$

The random changes in acceleration are modelled by the white sequence $w_2(k)$. Since only the first component of the state vector can be measured, we obtain the measurement equation.

$$z(k) = \begin{bmatrix} 1 & 0 \end{bmatrix}\begin{bmatrix} v(k) \\ a(k) \end{bmatrix} + n(k) \tag{19}$$

The measurement z(k) is related to the ML-estimate $\hat{D}(k)$ by the equation

$$z(k) = L/\hat{D}(k) \tag{20}$$

Inserting the identity

$$\hat{D}(k) \equiv D(k) + (\hat{D}(k)-D(k)) \tag{21}$$

into (20) yields

$$z(k) = \frac{L}{D(k)+(\hat{D}(k)-D(k))} = \frac{L}{D(k)} + \frac{L[D(k)-\hat{D}(k)]}{D(k)\cdot\hat{D}(k)} \tag{22}$$

since v(k)=L/D(k) we can write eq. (20) in the form

$$z(k) = v(k) + n(k)$$
$$\text{with} \quad n(k) = \frac{L[D(k)-\hat{D}(k)]}{D(k)\cdot\hat{D}(k)} \tag{23}$$

where n(k) represents the measurement noise sequence. This sequence is white, but it is not Gaussian because of 'outliers' as has been demonstrated in the previous section. This fact will be of key importance in developing the postfilter algorithm. Suppose for the moment that w(k) and n(k) are white Gaussian noise sequences. Since the state equation of the vehicle dynamics (eqs. 18 and 19) are linear, the optimum postfilter is a conventional Kalman-filter with a gain matrix K determined by the error covariance matrix. Such a filter performs well if the probability of ambiguous peaks is small, i.e. the occurrence of 'outliers' is a rare event which is effectively filtered out. This is not true for non-zero probability of 'outliers' caused by the finite observation time T_e. In this case the performance of the Kalman-filter can be greatly improved by incorporating an adaptive non-linearity into the measurement path as shown in fig. 4.

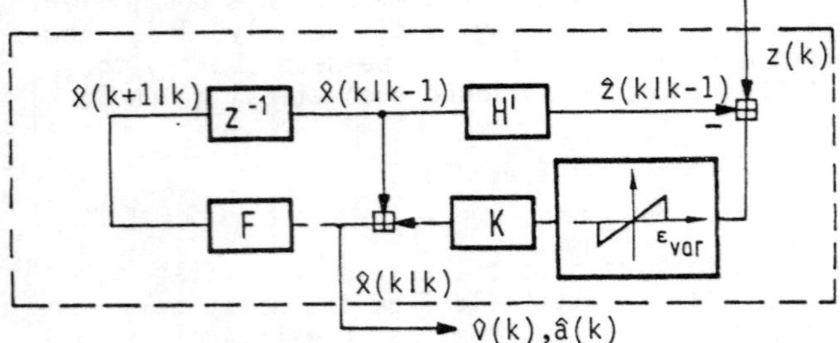

Fig. 4: Structure of the Postfilter

The function and design of this non-linearity can be briefly described as follows. Only if the difference between predicted value $\hat{z}(k|k-1)$ and the actual measurement z(k) (innovation) is small

$$|z(k)-\hat{z}(k|k-1)| < \varepsilon_{var} \tag{24}$$

is a measurement update done by the filter. In the other case no measurement update is done and the Kalman-filter operates as a predictor. To cope with multiple consecutive failures ('outliers') as well as with high dynamics of a(k) the width ε_{var} of the nonlinearity must be adaptively controlled. This procedure is illustrated in fig. 5.

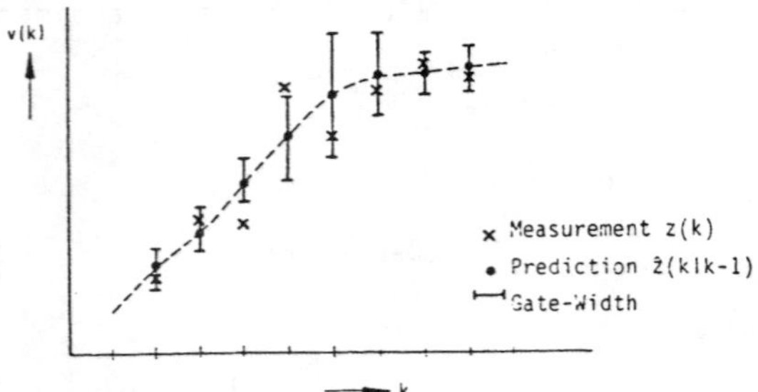

Fig. 5: Example of the Variable Gate-Width ε_{var}

After each 'outlier' the width ε_{var} of the nonlinearity is increased by a certain amount $\Delta\varepsilon$. The Kalman-filter then accepts larger differences between predicted and measured values that occasionally are caused by a sudden change in acceleration or a false estimate of the acceleration by the filter.

Conversely, the width $\varepsilon_{var}(k+1)$ is decreased if the difference at time k was smaller than $\varepsilon_{var}(k-1)$, provided ε_{var} is larger than the minimum width $\Delta\varepsilon$. The minimum width $\Delta\varepsilon$ is determined by the dynamics a(k) and the variance of the conditional estimate of D, conditioned by the fact that 'outliers' are absent. The constant Kalman-filter gain K is also computed under linear operating conditions, i.e. 'outliers' are absent.

Fig. 6 shows an example for the function of the postfilter.

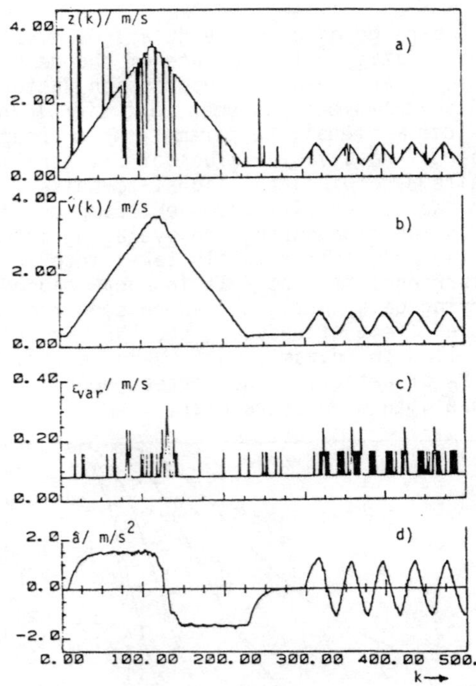

a) Estimates from Crosscorrelation (P_d=0.9)
b) Postfiltered Estimates $\hat{v}(k)$
c) Gate-Width ε_{var} ($\Delta\varepsilon$=0.08m/s)
d) Estimated Acceleration $\hat{a}$

Fig. 6: Function of the Postfilter (Simulation)

CONCLUSIONS

A two-step estimator for determining rapidly varying time delay was described. The system was successfully employed in a practical application under real-life conditions. The results leads one to believe that the approach taken is not far from the optimum achievable under the additional constraints of varying signal-to-noise ratio and possible signal failures.

REFERENCES

1 Knapp, Carter, "The Generalized Correlation Method for Estimation of Time-Delay", IEEE Trans. on ASSP, August 1976

2 J. Bohmann, H. Meyr, G. Spies, "A Digital Processor for High Precision Non-Contact Speed Measurement of Rail Guided Vehicles", 32nd IEEE Vehicular Technology Conference, San Diego, Calif., 1982

3 Carter, "Time Delay Estimation", NUSC Technical Report 5335, Naval Underwater Systems Center, New London, April 1976

4 Adams et al., "Correlator Compensation Requirements for Passive Delay Estimation . . . , IEEE Trans. on ASSP, Vol ASSP-28, No 2, April 1980

5 Bohmann, Meyr, Spies, "Performance of Estimators for Variable Time Delay Estimation", Internal Report, el. Regelungst. RWTH Aachen, Jan 1982

6 Meyr, "Delay-Lock Tracking of Stochastic Signals, IEEE Trans. on Com., Vol COM-24, No 3, March 1976

Time-Delay Estimation Using the Wigner Distribution

P. Rao C. Griffin F. J. Taylor

Department of Electrical Engineering
University of Florida
Gainesville, FL 32611

Abstract

A new approach is proposed and investigated for the high resolution estimation and tracking of non-stationary time delay between frequency modulated signals. The method, based on cross-correlation of the pseudo-Wigner Distributions of the noise-contaminated f.m. signals exploits the time-varying instantaneous frequency characteristic of such signals. The time-delay estimation algorithm is presented together with simulation results that show the performance of the technique in the estimation of time-varying delay.

1. Introduction.

The estimation of time-delay between signals is a problem of importance in areas such as sonar, radar, biomedicine, geophysics, etc. For example, in passive sonar signal processing the time difference of arrival of a signal observed at two or more spatially separated receivers is used to estimate the signal source range and bearing. Several techniques exist for the estimation of time-delay [1] under the various conditions that arise in practical applications such as low signal-to-noise ratios, motion of the signal source or receiver, and other commonly occurring propagation conditions. Of chief interest here is the estimation and tracking of time-varying delay due to source or receiver motion. Existing methods of dealing with the estimation of variable time-delay include modeling the time-delay esimation (t.d.e.) process as one of estimating the coefficients of an FIR filter [2], and adaptive techniques [3] to track slowly varying delay. These methods usually involve a trade-off between the noise-sensitivity and the tracking performance of the system. In this paper the t.d.e. of frequency modulated signals (or signals in which the variation of instantaneous frequency with time is a dominant characteristic) received at two spatially separated sensors is investigated using a new approach that exploits the time-varying frequency structure of the signals.

Section 2 provides a brief overview of the Wigner Distribution (WD) as a useful tool for the time-frequency analysis of signals. Section 3 describes the use of the WD in the estimation of time difference of arrival of f.m. signals in uncorrelated noise received at two spatially separated sensors. The time-delay estimation algorithm is presented in Section 4, together with simulation results of the t.d.e. of stationary and of time-varying delay .

2. Wigner Distribution.

The WD is a joint time-frequency representation that enables an effective description of the time-varying spectral content of non-stationary signals and systems.

The auto-WD (henceforth referred to as WD) of a signal x(t) is defined by

$$W_x(t, \omega) = \int_{-\infty}^{+\infty} x(t+\frac{\tau}{2})x^*(t-\frac{\tau}{2})e^{-j\omega\tau}\, d\tau \qquad (1)$$

The properties of the WD are well described in several papers [4,5] and are only summarized briefly here. The auto-WD is real-valued. The time and frequency moments of the WD convey useful information on the time-frequency behavior of the signal. The zeroth order local moments of the WD are equal to the instantaneous signal energy $|x(t)|^2$ and the power spectrum $|X(\omega)|^2$. The first-order local moments of the WD are proportional to the instantaneous frequency and the group delay of the signal. The signal energy, average frequency, and average time delay can be obtained from global zeroth- and first-order moments of the WD. The WD preserves the time and the frequency support of the signal and also reflects any time or frequency shifts of the signal as corresponding shifts in the time-frequency plane. Of particular relevance to the application studied in this paper is the property of the WD to yield in the case of monocomponent, frequency-modulated signals a high resolution spectral estimate or a time-frequency function that is concentrated along the curve which describes the instantaneous frequency as a function of time, of the signal. Due to its bilinear nature however, the WD of a real signal contains in addition to the spectral components of the signal itself, cross-products arising from the interaction of the positive and negative spectral components of the real signal. Since these cross-terms contain no useful information they are usually suppressed by using the analytic signal derived from the real signal in place of the real signal to compute the WD [4].

In practice, due to finite signal observation times a windowed version of the WD known as the pseudo-

Reprinted from *Proc. 22nd Asilomar Conference on Signals, Systems, and Computers*, pp. 726–729, November 1988.

WD (pWD) is used. If $h(t)$ denotes a window function such that $h(t) = 0$ for $|t| > T/2$ applied to the signal $x(t)$ to get $x(t) = x(t)h(t-t')$, then the pWD of $x(t)$ is given by

$$W_x(t,\omega) = \int_{-T/2}^{+T/2} x(t+\frac{\tau}{2})x^*(t-\frac{\tau}{2})h(\frac{\tau}{2})h^*(-\frac{\tau}{2})e^{-j\omega\tau}\, d\tau \quad (2)$$

The pWD is a frequency-smoothed version of the WD. However it does not suffer from the time- versus frequency-resolution trade-off inherent in the Short-time Fourier analysis of non-stationary signals [4]. Figure 1 displays the pseudo-WD of a linear f.m. signal.

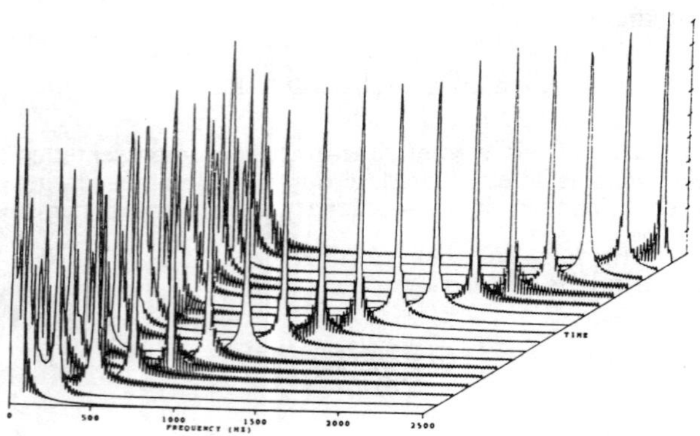

Figure 1
WD of a linear f.m. signal.

The WD has in recent years been applied to a large variety of signal processing problems [5] such as the analysis of non-stationary signals in speech and seismic applications, analysis of phase distortion in audio engineering problems, interpretation and realization of optimum detection in the time-frequency plane, activity detection in underwater passive acoustic surveillance, and pattern recognition in image analysis.

3. T.D.E. using WD Cross-correlation.

The WD satisfies the Moyal's formula [6] given by

$$\int_{-\infty}^{+\infty}\int_{-\infty}^{+\infty} W_x(t,\omega)W_y(t,\omega)\, dtd\omega = \left| \int_{-\infty}^{+\infty} x(t)y^*(t)\, dt \right|^2 \quad (3)$$

which relates cross-correlation of WDs in the time-frequency plane to cross-correlation of the signals in time. Based on the above property Flandrin [6] has suggested the use of the WD in the maximum-likelihood estimation of unknown signal parameters in an optimum detection-estimation problem in the following manner. If $r(t)$ is a reference signal and $f(t;\theta)$ is the received signal where θ is an unknown signal parameter, then the detection-estimation process is based on determining and maximizing with respect to the unknown parameter the cross-correlation of the signal WDs.

$$\hat{\theta} = \arg\{\max_\theta \int_{-\infty}^{+\infty}\int_{-\infty}^{+\infty} W_r(t,\omega)W_f(t,\omega;\theta)\, dtd\omega\} \quad (4)$$

This approach may be extended to the estimation of the time-difference of arrival of noise-contaminated signals received at two spatially separated sensors. If the received signals are modeled by

$$r_1(t) = s(t) + n_1(t) \qquad \frac{-T}{2} < t < \frac{T}{2} \quad (5)$$
$$r_2(t) = s(t-D) + n_2(t)$$

where $s(t)$ is the source signal and $n1(t)$ and $n2(t)$ are sample functions of independent noise processes, T is the observation time and D is the time delay between signals then for a fixed delay D the time delay estimate $\hat{D}$ can be obtained from i.e. a variable delay is

$$\hat{D} = \arg\{\max_d \int_{-\infty}^{+\infty}\int_{-\infty}^{+\infty} W_{r_1}(t,\omega)W_{r_2}(t,\omega;d)\, d\omega dt\} \quad (6)$$

where

$$W_{r_1}(t,\omega) = \int_{-T/2}^{T/2} r_1(t+\frac{\tau}{2})r_1(t-\frac{\tau}{2})e^{-j\omega\tau}\, d\tau \quad (7)$$

$$W_{r_2}(t,\omega;d) = \int_{-T/2}^{T/2} r_2(t+d+\frac{\tau}{2})r_2(t+d-\frac{\tau}{2})e^{-j\omega\tau}\, d\tau$$
$$= W_{r_2}(t+d,\omega)$$

inserted in one channel and the cross-correlation of the pWDs in the time-frequency plane computed. The estimate of time-delay D, is taken to be the delay for which the measured correlation is maximum. This, as can be seen from eqn. 3 is equivalent to using cross-correlation of signals in time to estimate the time delay.

The above formulation is extended here to the more general case of estimation of time-varying delay, i.e. D is a function of time. The estimate of time-delay at a specific time-instant t is obtained from the cross-correlation in frequency alone of the pWDs of the received signals computed at the time t over the observation interval T as in eqn. 7.

$$\hat{D}(t) = \arg\{\max_d \int_{-\infty}^{+\infty} W_{r_1}(t,\omega)W_{r_2}(t+d,\omega)\, d\omega\} \quad (8)$$

In practice, compensation for the relative Doppler frequency shift introduced between received signals due to source motion and which could be significant, needs to be incorporated into the t.d.e. procedure. This compensation is achieved by offsetting in frequency the pWD of one of the signals in eqn. 7 by the estimated Doppler shift at that time instant. (Note that this procedure is far simpler than the time-scaling necessary to achieve Doppler-shift compensation when the generalized cross-correlation technique of t.d.e. is employed in presence of source motion [7].)

4. T.D.E. Algorithm and Simulation Results.

The time-delay estimation algorithm based on a simplification of eqn. 8, is illustrated in the block diagram in Figure 2. The pWDs for the received signals

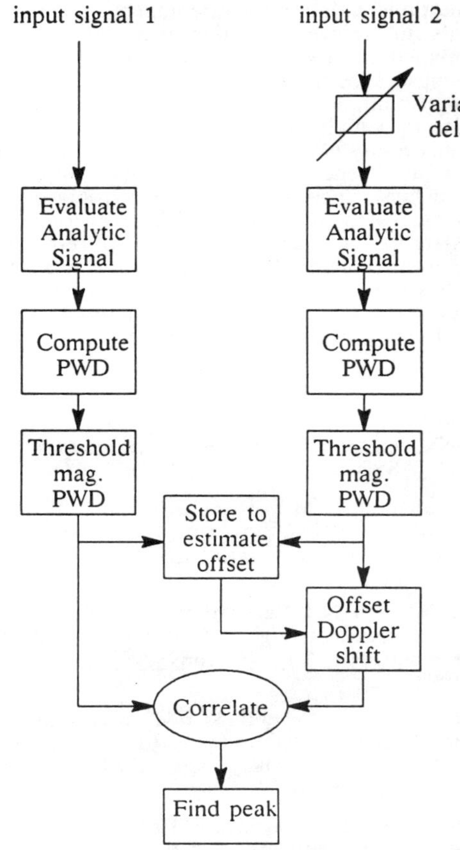

Figure 2
Time-Delay Estimation Algorithm

are computed after a variable delay is inserted in one of the channels. The delay is varied over the expected range (which is updated continually based on the current estimate of delay) to determine the delay value that maximizes the measured output cross-correlation of the pWDs after modifying to compensate for the estimated Doppler shift. The presence of noise introduces degradation in the pWD of the received signal due both, to noise components and cross-terms. A way to get around this and obtain accurate time delay estimates is to apply a threshold to the pWDs. The f.m. signal components in the pWD being concentrated in frequency tend to be of higher magnitude than the noise components and cross-terms. A further advantage of thresholding of the pWDs is that it simplifies the cross-correlation procedure so that simple polarity-coincidence technique may be used to estimate the cross-correlation.

The relative Doppler shift between the received signals is estimated by assuming that the time-varying delay can be closely approximated by a piece-wise linear function of time (that is, the relative source velocity is constant), so that $D(t) \approx ut$ for some u over an interval of time. The Doppler-shift for a frequency component f is then well approximated by $u \times f$ when $u << 1$ (which is usually true in practice). The quantity u can be estimated from the measured pWDs observed over a suitable time interval. For example, for the case of linear f.m. signals, it can be shown by simple geometrical derivation that $(1+u)/(1-u) \approx m1/m2$ where $m1$ and $m2$ are the slopes of the instantaneous frequency *vs* time curves of the received signals as estimated from the observed pWDs.

At low SNR's (around 0 dB) the noise-immunity of the t.d.e. may be improved greatly by using the analytic signal instead of the real signal to compute the pWD since this suppresses the cross-terms arising from the interaction of positive and negative spectral components of the signal plus noise. The time-smoothing of the pWD that results from using the analytic signal however leads to a slight loss in estimator resolution.

Implementation of the pWD of real sampled signals based on using the analytic signal can be carried out in real time by one of the several efficient algorithms available [8,9].

Figure 3 shows the estimation of a fixed delay between linear and sinusoidal f.m. signals respectively in white Gaussian noise at 0 dB SNR using 512 sample windows to compute the pWD. The analytic signal has been used to compute the pWD. The accuracy of the

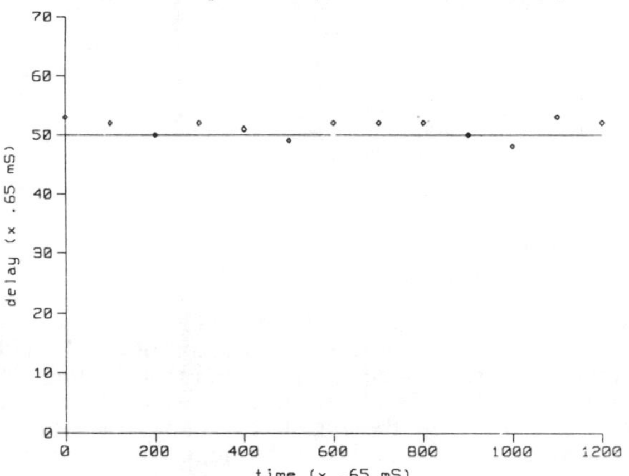

Figure 3 (a)
Estimation of fixed delay in
linear f.m. signals, SNR 0 dB.

delay estimate is similar to that of using cross-correlation of the signals in time over the same observation interval. The WD estimator performs better for the linear f.m. signal since the WD spectrum is inherently of higher resolution for this signal. Figure 4 shows the estimation of time-varying delay between linear f.m. signals at 0 dB SNR. The actual delay versus time

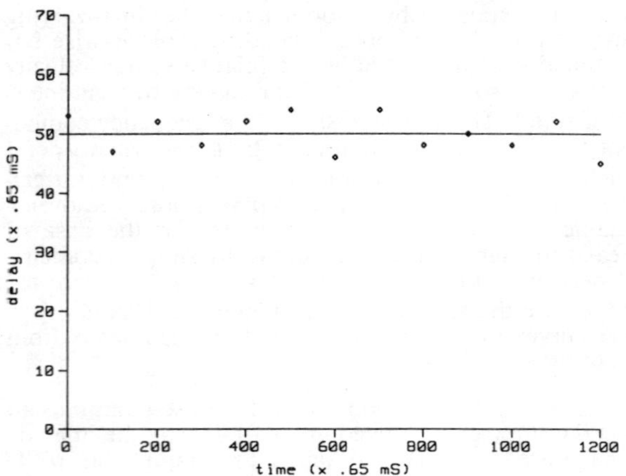

Figure 3 (b)
Estimation of fixed delay in sinusoidal f.m. signals, SNR 0 dB.

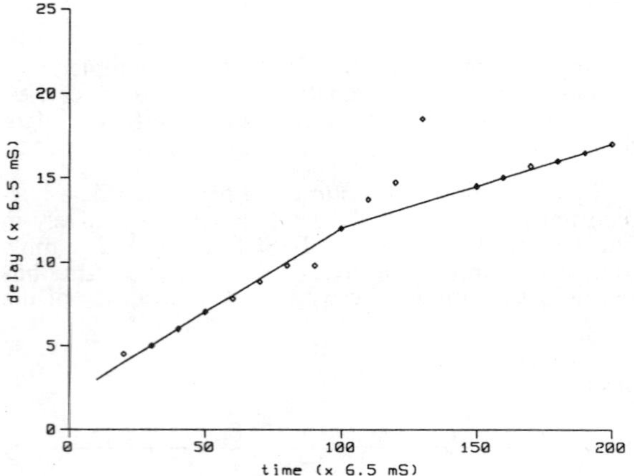

Figure 4
Estimation of varying delay in linear f.m. signals, SNR 0 dB.

curve is shown by the smooth line made up of the two segments with different rates of variation of time-delay or relative source velocities. It is seen that the linearly varying time-delay is closely tracked while there is a loss of accuracy in the region of change of relative source velocity. This is due to the finite time necessary for the system to update the Doppler-shift factor after the abrupt change.

5. Discussion.

A new technique has been proposed and investigated for the estimation of time-varying delay between f.m. signals contaminated with noise. The method is based on exploiting the variation of instantaneous frequency with time characteristic of such signals to derive a high resolution delay estimator. Simulation results demonstrate that the method is simple and robust in its applicability to the estimation and tracking of delay between f.m. signals at low SNRs. There are some shortcomings however in the technique as formulated here that merit further consideration and study. These are outlined below.

Since the approach presented here is based on the use of the time-variation of instantaneous frequency of the signals to achieve estimation of time-delay, it cannot be applied exactly as formulated here to stationary signals whose spectra do not change with time. The existence of multiple reflections in the propagation path giving rise to a multicomponent received signal would cause unnecessary cross-terms to appear in the WD of the signal and hence ambiguity in the estimation of multiple delays. This problem can possibly be overcome by time-smoothing of the pWD [10] to eliminate the cross-terms with some loss however of resolution of the resulting estimator. Linearity of the propagation channel has been assumed in the problem investigated here. Future work could consider including non-linear channel effects in the time-delay estimation problem.

References

[1]. Special issue on time-delay estimation, IEEE Trans. ASSP, vol. 29, no. 3, 1981.

[2]. Y.T. Chan and J.M.F. Riley, "Modeling of time delay and it's application to estimation of non-stationary delays, " IEEE Trans. ASSP, vol. 29, no. 3, 1981.

[3]. P.L. Feintuch, N.J. Bershad and F.A. Reed, "Time delay estimation using the LMS adaptive filter-Dynamic behavior, " IEEE Trans. ASSP, vol. 29, no. 3, 1981.

[4]. T.A.C.M. Claasen and W.F.G. Mecklenbrauker, "The Wigner distribution-A tool for time-frequency signal analysis. Parts I,II,III, " Philip's J. Res., vol. 35, 1980.

[5]. G.F. Boudreaux-Bartels, "Time-varying signal processing using the Wigner-distribution time-frequency signal representation, " Advances in Geophys. Data Processing, vol. 2, JAI Press Inc., 1985.

[6]. P.Flandrin, "A Time-frequency formulation of optimum detection, " IEEE Trans. ASSP, vol. 36, no. 9, 1988.

[7]. C.H. Knapp and G.C. Carter, "Estimation of time delay in the presence of source or receiver motion," J. Acoust. Soc. Amer., vol. 61., no. 6, 1977.

[8]. B. Boashash and P.J. Black, "An efficient real-time implementation of the Wigner-Ville distribution, " IEEE Trans. ASSP, vol. 35, no. 11, 1987.

[9]. J. Wilbur and F.J. Taylor, "High-speed Wigner processing based on a single modulus QRNS, " Proc. IEEE Intl. Conf. on ASSP, vol.2, 1985.

[10].P. Flandrin, "Some features of time-frequency representations of multicomponent signals, "Proc. IEEE Intl. Conf. on ASSP, vol.4, 1984.

This work was supported in part by an IBM grant and by ARO.

Adaptive Multipath Delay Estimation

Julius O. Smith and Benjamin Friedlander

Systems Control Technology
1801 Page Mill Road
Palo Alto, California 94303

Abstract

Two algorithms are proposed for the continuous tracking of multipath delay. The methods use an adaptive delay line interpolated by a first-order filter. The interpolation coefficient is explicitly estimated using the recursive Gauss-Newton algorithm.

1. Introduction

Multipath propagation is often encountered in a variety of communication and surveillance systems. Spurious propagation paths may degrade system performance, unless their effect is properly compensated. Given knowledge of the relative delays in the different propagation paths, it is possible to coherently recombine the received signals, alleviating fading effects, and in fact enhancing the performance of the system. Multipath delays are not known a priori, and are constantly changing due to motion of the signal source or the receiver, or due to the time-varying characteristics of the medium. It is therefore necessary to estimate these delays from the available data and track their changes over time.

In this paper, we propose two algorithms for adaptively tracking the multipath delay in a simple propagation model. We consider the case in which the signal received by a single sensor can be modeled as

$$y(n) = x(n) + g_0 x(n - \tau_0) + e(n) \tag{1}$$

where $x(n)$ is the source signal, $0 \le g_0 < 1$ is the attenuation of the secondary path relative to the primary path, τ_0 is the (real-valued) relative delay in samples, and $e(n)$ is noise at the receiver.

Given samples $y(n)$ for $n = 0, 1, \ldots, N-1$, the problem is to estimate the multipath delay τ_0. In some cases, it is also advantageous to estimate multipath attenuation g_0.

This work was supported by the Office of Naval Research under contract no. N00014-82-C-0476.

If the probability density function (PDF) of $y(n)$ given by (1) is known, then the *maximum likelihood (ML) method* can be used to estimate τ_0 and/or g_0 by maximizing the PDF with respect to τ, g. In the Gaussian case where $y(n)$, $x(n)$, and $e(n)$ are assumed to be stationary Gaussian processes, with $e(\cdot)$ and $x(\cdot)$ independent, the PDF is relatively simple to derive. As shown in [12], for large N, the ML estimate of τ_0 is found by minimizing

$$\hat{L}(\theta) = \frac{1}{N}\sum_{k=0}^{N-1}\left\{\ln S_{y_\theta}(\omega_k) + \frac{\hat{S}_y(\omega_k)}{S_{y_\theta}(\omega_k)}\right\}, \qquad \omega_k \triangleq \frac{2\pi}{N}k, \tag{2}$$

where $\theta^T \triangleq [g, \tau]$,

$$\hat{S}_y(\omega_k) \triangleq \frac{1}{N}\left|\sum_{n=0}^{N-1} y(n)e^{-j\omega_k n}\right|^2$$
$$S_{y_\theta}(\omega_k) \triangleq S_x(\omega_k)[1 + g^2 + 2g\cos(\omega_k\tau)] + S_e(\omega_k), \tag{3}$$

and $S_x(\omega_k), S_e(\omega_k)$ are the power spectral densities of the source $x(n)$ and noise $e(n)$, respectively. Thus $\hat{S}_y(\omega_k)$ is the sample power spectral density (PSD) of the observations and $S_{y_\theta}(\omega_k)$ is the PSD observed when the multipath parameters are given by θ. The objective function $\hat{L}(\theta)$ is the limiting form of the negative log likelihood function (NLLF) as N approaches infinity [11,12].

By dropping the term $\ln S_{y_\theta}(\omega_k)$ in (2), the NLLF takes the form of the output power of an *inverse filter*:

$$\hat{L}_e(\theta) = \frac{1}{N}\sum_{k=0}^{N-1}\frac{\hat{S}_y(\omega_k)}{S_{y_\theta}(\omega_k)}. \tag{4}$$

The dropped term can be defined as the sample entropy of $y_\theta(\cdot)$ [2]. If $x(n)$ and $e(n)$ are assumed to have constant power spectral densities S_x and S_e, then the inverse filter becomes simpler; $1/S_{y_\theta}(\omega_k)$ becomes the power response of a filter of the form

$$H_r(d) \triangleq \frac{a}{1 + bd^\tau}, \tag{5}$$

where d is the unit-sample delay operator:

$$d^m y(n) \triangleq y(n - m). \tag{6}$$

Reprinted from *Proc. ICASSP '84*, vol. 1, pp. 15.9/1–4, March 1984.

The coefficient b depends on signal-to-noise ratio, and approaches g as $S_e \to 0$. This simplified approximation to the NLLF is minimized asymptotically by the adaptive inverse filter estimator presented in the next section.

As a further approximation, we may truncate the Taylor series expansion (impulse response) of the inverse filter to obtain a finite impulse-response (FIR) filter:

$$\begin{aligned} H_r(d) &= \frac{1}{1+bd^\tau} \\ &= 1 - bd^\tau + b^2 d^{2\tau} - b^3 d^{3\tau} + \cdots \\ &\approx H_f(d) \triangleq 1 - bd^\tau . \end{aligned} \tag{7}$$

The objective function becomes (assuming b fixed)

$$\begin{aligned} \hat{L}_f(\tau) &\triangleq \frac{1}{N}\sum_{k=0}^{N-1} \hat{S}_y(\omega_k)\left| H_f(e^{-j\omega_k})\right|^2 \\ &= \frac{1}{N}\sum_{k=0}^{N-1} \hat{S}_y(\omega_k)[(1+b^2) - 2b\cos(\omega_k \tau)] \\ &= (1+b^2) r_y(0) - 2b r_y(\tau) , \end{aligned} \tag{8}$$

where $r_y(\tau)$ is the sample autocorrelation function. Since the first term is constant in τ, minimizing $\hat{L}_f(\tau)$ is the same (for $b > 0$) as maximizing the autocorrelation $r_y(\tau)$ with respect to τ. An adaptive correlator algorithm, minimizing (8), is also given in the next section.

The two adaptive algorithms in this paper are based on the recursive prediction error (RPE) algorithm for system identification [5]. An advantage of the RPE approach is that algorithms for the time-varying case are obtained having convergence properties which have been studied in detail [3,4,5].

A novel feature of the proposed algorithms is the manner in which interpolated delay estimation is handled. The interpolated delay is estimated implicitly by optimizing the coefficient of a first-order interpolation filter. Also, the integer part of the delay is maintained outside of the RPE algorithm so that the interpolation filter always works within a one-sample range of interpolated delay. More details on this approach are given in [9].

In section 3 we present some numerical examples illustrating the behavior of these algorithms. A more complete performance analysis is presented in [9].

2. The Adaptive Algorithms

The RPE method of system identification [5] employs a recursive Gauss-Newton (RGN) technique to minimize a least-squares error criterion. Since it converges locally like Newton's method, it is quadratically convergent near a local minimum of the error function, while gradient methods are only linearly convergent. For a discussion of the convergence of RGN, see [3,4,5]. The RGN parameter update is given by

$$\begin{aligned} \theta(n+1) &= \theta(n) - R^{-1}(n)\epsilon'(n)\epsilon(n) \\ R(n+1) &= \lambda R(n) + \epsilon'(n)\epsilon'(n)^T , \end{aligned} \tag{9}$$

where $\theta^T(n) \triangleq [\, g(n), \tau(n)\,]$, $\epsilon(n) \triangleq y(n) - \hat{y}(n)$ is the prediction error,

$$\epsilon'(n) \triangleq \frac{\partial \epsilon(n)}{\partial \theta(n)} \tag{10}$$

(a column vector), and $\epsilon'(n)^T$ denotes the transpose of $\epsilon'(n)$. The parameter $0 < \lambda \leq 1$ controls how much averaging over time is performed [5]. The time-constant of adaptation, in samples, can be approximated by $1/(1-\lambda)$. The two multipath delay estimation algorithms differ only in the way the prediction $\hat{y}(n)$ is formed.

In order to allow non-integer values of estimated delay τ, it is necessary to interpolate $\hat{y}(n)$ between samples which translates to the need for interpolating $y(n)$. Perhaps the simplest interpolation technique is linear interpolation. The linearly interpolated value $y(n-\tau)$ is obtained by passing $y(n-N)$ through a first-order FIR filter $H_\alpha(d)$, where $N \triangleq \lfloor \tau \rfloor$, and $\lfloor \tau \rfloor$ denotes the greatest integer less than or equal to τ. Letting $\alpha \triangleq \tau - N$, and $\overline{\alpha} \triangleq 1 - \alpha$, the filter equation can be written as

$$\begin{aligned} y(n-\tau) &\triangleq H_\alpha(d) y(n-N) \\ &\triangleq (\overline{\alpha} + \alpha d) y(n-N) \\ &= y(n-N) + \alpha[y(n-N-1) - y(n-N)] . \end{aligned} \tag{11}$$

An advantage to the linear interpolation approach is low complexity—only one multiply and two adds are needed per interpolation. In [9], a first-order allpass interpolation scheme having identical computational requirements is discussed.

When using linear interpolation, it is convenient to remove the integer part of τ from the RGN parameter vector. The integer $N(n)$ is the length at time n of a variable pure delay maintained so that

$$\tau(n) \triangleq N(n) + \alpha(n) ,$$

where $\alpha(n)$ is the fractional delay. While α is precisely the delay of the linear interpolation filter (11) only at $\omega = 0$, it is an excellent approximation elsewhere [9].

We confine $\alpha(n)$ to the interval $[\alpha_-, \alpha_+] = [0,1]$, which gives a range of one sample of continuous delay from the interpolation filter at 0Hz. When $\alpha(n) > \alpha_+$, we remove a sample of delay from the interpolation filter and add it to the delay-line by adding 1 to $N(n)$ and subtracting 1 from $\alpha(n)$. If $\alpha(n)$ is changing slowly, then an approximate reset is $\alpha(n) = \alpha_-$. Analogous resetting is done when $\alpha(n) \leq \alpha_-$.

The following is a summary of the proposed algorithm.

Initialization:

Assume that $y(n)$ is available for $n = 1, \ldots, T$, and let the algorithm begin at time n_0 where n_0 is sufficiently large so that all delayed quantities are available. If initial tracking accuracy is not important, then one can use $n_0 = 1$ and substitute zero for unavailable data.

$N(n_0) = \lfloor \tau(n_0) \rfloor$ should be as accurate as possible since the RGN algorithm attempts only to track the nearest local minimum of the corresponding loss function.

$\alpha(n_0) = \tau(n_0) - N(n_0)$, or $\alpha(n_0) = 0$.

$R(n_0) > 0$ controls the initial speed of adaptation. For a smooth startup, a reasonable setting is $\mathrm{diag}(R(n_0)) = \mathrm{E}\{\epsilon^2(n_0)\}/(1-\lambda)$.
Larger values inhibit initial tracking, while smaller values allow rapid initial variations in $\alpha(n)$.

Iteration for $n = n_0, n_0+1, \ldots, T$:

$$\begin{aligned}
\epsilon(n) &= y(n) - \hat{y}(n) \\
\theta(n+1) &= \theta(n) - R^{-1}(n)\epsilon'(n)\epsilon(n) \\
R(n+1) &= \lambda R(n) + \epsilon'(n)\epsilon'(n)^T \\
N(n+1) &= N(n) \\
\alpha(n+1) > \alpha_+ &\Rightarrow \begin{Bmatrix} \alpha(n+1) = \alpha_- \\ N(n+1) = N(n) + 1 \end{Bmatrix} \\
\alpha(n+1) < \alpha_- &\Rightarrow \begin{Bmatrix} \alpha(n+1) = \alpha_+ \\ N(n+1) = N(n) - 1 \end{Bmatrix}
\end{aligned} \tag{12}$$

where $\theta^T(n) = [g(n)\ \alpha(n)]$. For the inverse filter, $\epsilon(n)$ is given by

$$\begin{aligned}
\epsilon_r(n) &= y(n) - \hat{y}(n) \\
&= y(n) - g(n)\epsilon_r(n - \tau(n)) \\
&= y(n) - g(n)\{\epsilon_r(n - N(n)) \\
&\quad + \alpha(n)[\epsilon_r(n - N(n) - 1) - \epsilon_r(n - N(n))]\},
\end{aligned} \tag{13}$$

and

$$\begin{aligned}
\frac{\partial \epsilon_r}{\partial g(n)} &= [\overline{\alpha}(n) + \alpha(n)d]\tilde{\epsilon}_r(n) \\
\frac{\partial \epsilon_r}{\partial \alpha(n)} &= g(n)(d-1)\tilde{\epsilon}_r(n).
\end{aligned} \tag{14}$$

where

$$\tilde{\epsilon}_r(n) \triangleq - \frac{d^{N(n)}}{1 + g(n)d^{N(n)}[\overline{\alpha}(n) + \alpha(n)d]}\epsilon_r(n). \tag{15}$$

In the simulations presented in the next section, this form of the algorithm will be denoted RGN-I (inverse filter case), and the adaptive correlator (or FIR) version will be called RGN-F.

For the FIR model, $\epsilon(n)$ is given by

$$\begin{aligned}
\epsilon_f(n) &= y(n) - g(n)\hat{y}_f(n - \tau(n)) \\
&= y(n) - g(n)\{y(n - N(n)) \\
&\quad + \alpha(n)[y(n - N(n) - 1) - y(n - N(n))]\}.
\end{aligned} \tag{16}$$

and $\epsilon'(n)$ is given by

$$\begin{aligned}
\frac{\partial \epsilon_f}{\partial g(n)} &= \alpha(n)[y(n - N(n)) - y(n - N(n) - 1)] - y(n - N(n)) \\
\frac{\partial \epsilon_f}{\partial \alpha(n)} &= g(n)[y(n - N(n)) - y(n - N(n) - 1)].
\end{aligned} \tag{17}$$

Finally, the time-delay estimate is

$$\tau(n) \triangleq N(n+1) + \alpha(n+1).$$

The above algorithm estimates both multipath gain g and delay τ. It is straightforward to derive a version which estimates delay only with g fixed [12].

3. Simulation Results

In this example, the source signal $x(n)$ and receiver noise $e(n)$ are zero-mean, unit-variance, Gaussian white noise, and the signal to noise ratio is 0dB The received signal is

$$y(n) = x(n) + gx(n - \tau(n)) + e(n), \qquad n = 1, 2, \ldots, N,$$

where $g = 0.5$, $N = 4096$, and

$$\tau(n) = \begin{cases} 10, & 1 \le n < \frac{1}{3}N \\ 10 + \frac{n - \frac{1}{3}N}{\frac{1}{3}N}, & \frac{1}{3}N \le n < \frac{2}{3}N \\ 11, & \frac{2}{3}N \le n \le N \end{cases}$$

Thus the delay is a constant 10 samples for the first third of the record, linearly increases to 11 samples over the middle third, and remains fixed at 11 samples thereafter. The technique used to continuously delay the signal $x(n)$ is described in [13].

The algorithms are initialized with $n_0 = 13$, $R(n_0) = 10.0$, $\lambda = 0.995$, and correct initial delay τ. The initial gain, however, is set to $g = 0.234$, which is the expected value of b for zero-dB SNR. The recursively computed quantities in RGN-I are initially zero.

Figure 1 shows the tracking performance of the RGN-I algorithm. Figure 2 gives the results for RGN-F.

4. Conclusions

Two recursive algorithms suitable for tracking multipath delays were presented. These algorithms performed quite well in a limited number of simulations, giving performance close to that predicted by asymptotic theory [12]. There is need for analytical results on the "capture range" of the algorithms, i.e., how far can the initial guess be from the desired solution, and how fast can the solution vary and still be tracked. In addition, the convergence properties of identification algorithms with parameter "resets" such as employed here need further investigation.

References

[1] R. E. Williams and H. F. Battestin, "Coherent Recombination of Acoustic Multipath Signals Propagated in the Deep Ocean," *J. Acoust. Soc. Amer.*, vol. 50, no. 6, pt. 1, pp. 1433–1442, 1971.

[2] G. C. Goodwin and R. L. Payne, *Dynamic System Identification,* Academic Press, New York, 1977.

[3] L. Ljung, "Analysis of Recursive Stochastic Algorithms," *IEEE Trans. Automat. Contr.*, vol. AC–22, pp. 551–575, Aug. 1977.

[4] L. Ljung, "Convergence Analysis of Parameteric Identification Methods," *IEEE Trans. Automat. Contr.*, vol. AC–23, pp. 770–783, 1978.

[5] L. Ljung and T. L. Soderstrom, *Theory and Practice of Recursive Identification,* MIT Press, Cambridge MA, 1983.

[6] J. P. Ianniello, "The Variance of Multipath Time Delay Estimation using Autocorrelation," NUSC tech. memo. no. 831008, Jan. 24, 1983.

[7] J. P. Ianniello, "The Cramer-Rao Lower Bound on Multipath Time Delay Estimation," NUSC tech. memo. no. 831067, May 9, 1983.

[8] J. P. Ianniello, "The Ziv-Zakai Lower Bound on Multipath Time Delay Estimation," NUSC tech. memo. no. 831100, July 11, 1983.

[9] J. O. Smith and B. Friedlander, "Adaptive Interpolated Time-Delay Estimation," Asilomar, 1983. Full version to appear.

[10] B. Friedlander, "On the Cramer-Rao Bound for Delay and Doppler Estimation," *IEEE Tr. Info. Theory,* to appear.

[11] P. Whittle, "The Analysis of Multiple Stationary Time Series," *J. Royal Statistical Society,* vol. 15, pp. 125–130, 1953.

[12] B. Friedlander and J. O. Smith, "Multipath Delay Estimation," Tech. Rep. 5466-04, Systems Control Tech., Dec. 1983.

[13] J. O. Smith and P. Gossett, "A Flexible Sampling-Rate Conversion Method," IEEE Conf. Acoust. Sp. and Sig. Proc., San Diego, 1984.

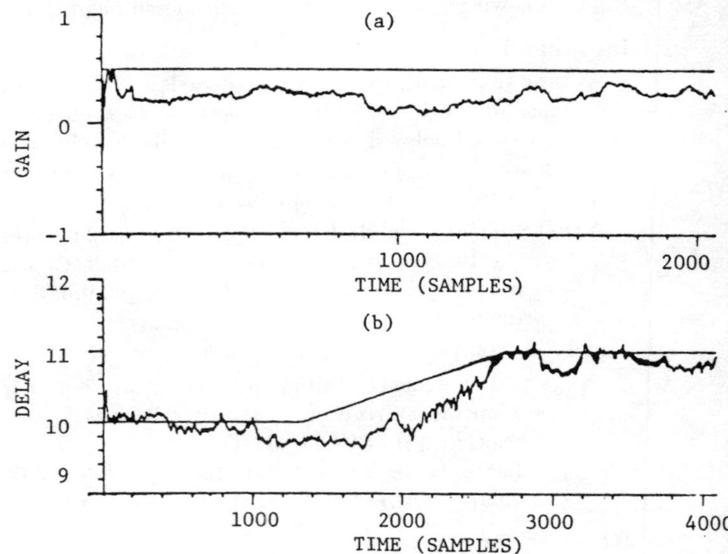

Figure 1. Performance of the RGN-I algorithm in tracking a) multipath gain and b) multipath delay at zero-dB SNR. While the true multipath gain is shown in part a), the correct asymptotic value of the gain estimate is ≈ 0.234 due to the presence of receiver noise.

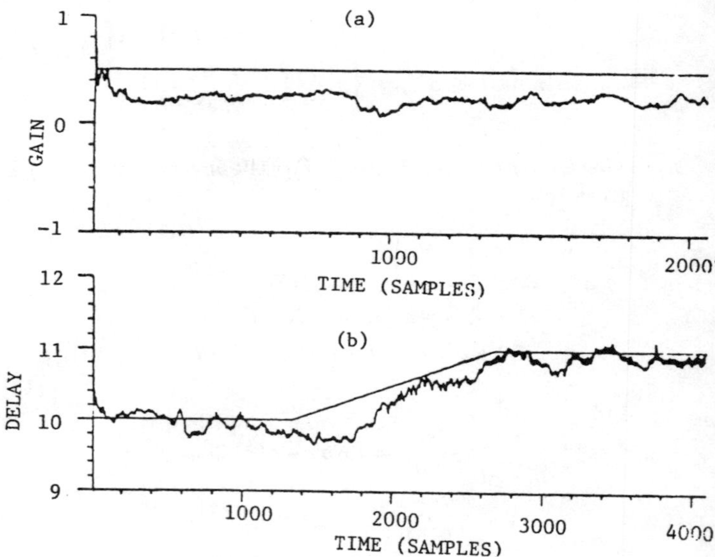

Figure 2. Performance of the RGN-F algorithm in tracking a) multipath gain and b) multipath delay at zero-dB SNR.

Maximum-Likelihood Estimation of Time-Varying Delay—Part I

JOHN A. STULLER, SENIOR MEMBER, IEEE

Abstract—**This paper presents, for the first time, the exact theoretical solution to the problem of maximum-likelihood (ML) estimation of time-varying delay $d(t)$ between a random signal $s(t)$ received at one point in the presence of uncorrelated noise, and the time-delayed, scaled version $\alpha s(t - d(t))$ of that signal received at another point in the presence of uncorrelated noise. The signal is modeled as a sample function of a nonstationary Gaussian random process and the observation interval is arbitrary. The analysis of this paper represents a generalization of that of Knapp and Carter [1], who derived the ML estimator for the case that the delay is constant, $d(t) = d_o$, the signal process is stationary, and the received processes are observed over the infinite interval $(-\infty, +\infty)$. We show that the ML estimator of $d(t)$ can be implemented in any of four canonical forms which, in general, are time-varying systems. We also show that our results reduce to a generalized cross correlator for the special case treated in [1].**

I. Introduction

In their classic paper, "The generalized correlation method for estimation of time delay" [1], Knapp and Carter presented the solution to the problem of maximum-likelihood (ML) estimation of constant delay d_o between signals received at two spatially separated sensors in the presence of uncorrelated noise. The received waveforms were modeled mathematically as

$$r_a(t) = s(t) + n_a(t), \quad -\infty < t < +\infty, \tag{1.1a}$$

$$r_b(t) = \alpha s(t - d_o) + n_b(t), \quad -\infty < t < +\infty \tag{1.1b}$$

where α was a relative attenuation constant and $s(t)$, $n_a(t)$, and $n_b(t)$ were uncorrelated, stationary Gaussian random processes. Knapp and Carter showed that the ML estimator of d_o can be realized by a pair of prefilters followed by a cross correlator. Their solution turned out to be identical to that proposed by Hannan and Thomson [2], [3] whose motivation for estimating delay was to improve estimates of the spectra, cross spectra, and coherence of stationary time series. The present paper presents a major generalization of these preceding analyses that includes: i) arbitrary time-varying delay $d(t)$; ii) nonstationary random signal process; and iii) arbitrary observation interval.

Previous attempts to extend the theoretical solution described in [1]–[3] have been relatively limited in scope. Assuming a stationary signal process and an infinite observation interval, Knapp and Carter [4] obtained an approximate ML estimator of both differential delay and Doppler for a source whose relative velocity is much less than the signal propagation velocity. Here the estimator structure included time companding before cross correlation to compensate for the Doppler time scaling of the waveform. Wax [5] generalized the analysis to include differential phase. Several other authors [6]–[9] have described the degradation in compensated and uncompensated cross-correlator outputs due to motion of the source. Beyond these theoretical studies, there has been both a need for and a continuing effort to develop practical algorithms that estimate time-varying delay more generally [10]–[16]. The general theory presented here may provide guidance to that effort, as well as new insights into previous theoretical results.

We model the problem of time-varying delay estimation as follows. A vector of real waveforms

$$r(t) = \begin{bmatrix} r_1(t) \\ r_2(t) \end{bmatrix} = \begin{bmatrix} s(t) \\ \alpha s(t - d(t)) \end{bmatrix} + \begin{bmatrix} w_1(t) \\ w_2(t) \end{bmatrix} \tag{1.2}$$

is observed on the interval $[T_i, T_f]$ where T_i and T_f denote initial and final observation time, respectively. For convenience, we define $r(t)$ as zero for t outside this interval. The signal $s(t)$ is a sample function of a zero-mean Gaussian random process having covariance function

$$R_s(t_1, t_2) = E\{s(t_1)\, s(t_2)\}. \tag{1.3}$$

The delayed and attenuated signal $\alpha s(t - d(t))$ is related to $s(t)$ through a nonrandom but unknown invertible linear operator

$$\mathcal{L}_{d(t),\alpha}\{s(t)\} = \alpha s(t - d(t)). \tag{1.4}$$

The noise waveforms $w_1(t)$ and $w_2(t)$ are sample functions of white Gaussian random processes having covariance functions

$$R_{w_1}(t_1, t_2) = R_{w_2}(t_1, t_2) = \frac{N_o}{2}\,\delta(t_1 - t_2). \tag{1.5}$$

The signal process and noise processes are mutually independent. The attenuation factor α and delay function $d(t)$ appearing in (1.2) and (1.4) are nonrandom but unknown. Since $d(t)$ represents delay, we will assume

Manuscript received July 20, 1985; revised September 17, 1986. This work was supported in part by the Graduate School, University of Missouri, Rolla, and in part by the Signal Processing Branch of the Naval Underwater Systems Center, New London, CT.

The author is with the Department of Electrical Engineering, University of Missouri, Rolla, MO 65401.

IEEE Log Number 8611977.

Reprinted from *IEEE Trans. Acoust., Speech, Signal Processing*, vol. 35, no. 3, pp. 300–313, March 1987.

throughout this paper that $d(t) \geq 0$. The attenuation constant α can be any nonzero real number. The problem is to estimate $d(t)$ and α.

The model (1.2)–(1.5) has been chosen to focus on the central issue of time-delay estimation. The techniques we will use to solve this problem, however, can be extended to more general problems. For example, our model assumes white noise processes and a nondispersive (frequency independent) propagation medium. One can include nonwhite noise by applying noise whitening techniques similar to those described in [17, p. 290]. One can also include dispersion by replacing the operation (1.4) by a more general invertible time-varying linear system. Hamon and Hannan [18] have previously described an approximate ML solution to the time-delay estimation problem for dispersive systems. Extensions involving time-varying attenuation and signal delay in both received waveforms are also possible. Because of the length of the development, this paper is concerned only with the *structure* of the ML estimator of time-varying delay. The important issues of practical realization and performance will be considered in subsequent papers.

Although the details of our derivation differ substantially from those in [1], the estimation criterion is identical. We seek a function, the log-likelihood function $\ln \Lambda(D(t), \mathrm{A})$, whose value is an indicator of the likelihood that a hypothetical delay function $D(t)$ and attenuation constant A caused a particular received vector waveform $\boldsymbol{R}(t)$, $T_i \leq t \leq T_f$. The function $\hat{D}(t)]_{\mathrm{ML}}$ and the constant $\hat{\mathrm{A}}]_{\mathrm{ML}}$ jointly maximizing $\ln \Lambda(D(t), \mathrm{A})$ are the maximum-likelihood estimates of $d(t)$ and α, respectively, when $\boldsymbol{R}(t)$ is the received vector waveform.

This paper is organized as follows. The log-likelihood function $\ln \Lambda(D(t), \mathrm{A})$ is derived in Section II. The log-likelihood function is shown to depend upon the minimum mean square error (MMSE) estimators of $s(t)$ and $\alpha s(t - d(t))$ from $\boldsymbol{r}(t)$ conditioned on given attenuation and delay. We show, in Section III, that these estimators can be implemented using the structure of Fig. 4. In Section IV we obtain the four alternative systems for computing $\ln \Lambda(D(t), \mathrm{A})$ shown in Figs. 5–8. In Section V we show that the general solution to the problem of ML estimation of $d(t)$ reduces to the generalized cross-correlator receiver of [1] for the special case that $d(t)$ is a constant, the signal process is stationary, and the observation interval is long.

The following symbols are used throughout this paper.

T_i: initial observation time.

T_f: final observation time.

MMSE and LMMSE: Unconstrained minimum mean square error and linear minimum mean square error, respectively. In this paper, these terms can be used interchangeably because the processes are Gaussian.

Subscript n: noncausal estimate or system.

Subscript c: causal estimate or system.

ML: maximum likelihood.

T (superscript): vector or matrix transposition.

$\mathrm{Tr}\{\cdot\}$: trace of matrix argument.

$|\boldsymbol{\Theta}|^2$: squared Euclidean norm, $|\boldsymbol{\Theta}|^2 = \boldsymbol{\Theta}^T\boldsymbol{\Theta}$.

$\delta(t)$: Dirac delta function.

δ_{ij}: Kronecker delta function.

$\mathrm{SGN}(\mathrm{A})$: algebraic sign function. $\mathrm{SGN}(\mathrm{A}) = 1$ for $\mathrm{A} > 0$, $\mathrm{SGN}(\mathrm{A}) = -1$ for $\mathrm{A} < 0$.

Random or unknown variables are denoted with lower case italic letters. Given or assumed values of random or unknown variables are denoted with upper case italic letters.

II. Derivation of the Log-Likelihood Function

The derivation of the log-likelihood function $\ln \Lambda(D(t), \mathrm{A})$ involves an extension of the material in [17, pp. 203–205, 221–223] and [19, pp. 22, 170–173] from scalar- to vector-valued random processes. This takes three steps. The first step, described in Section II-A, is to obtain a series form for $\ln \Lambda(D(t), \mathrm{A})$ by means of the generalized Karhunen–Loève expansion. The result is shown in (2.16)–(2.18). The second and third steps, described in Sections II-B and C, put the series (2.17) and (2.18) into the closed forms (2.19) and (2.36), respectively, which in turn depend upon the noncausal and the causal MMSE estimators of $s(t)$ and $\alpha s(t - d(t))$ from $\boldsymbol{r}(t)$ conditioned on given attenuation and delay.

A. Series Form

The problem of estimating attenuation and time-varying delay can be reframed as a parameter estimation problem. One way to do this is to represent the time-varying delay $d(t)$ by a generalized Fourier series

$$d(t) = \sum_{i=1}^{\infty} d_i \psi_i(t); \qquad T_i \leq t \leq T_f \tag{2.1}$$

where

$$d_i = \int_{T_i}^{T_f} d(t)\, \psi_i(t)\, dt; \tag{2.2}$$

and where $\{\psi_i(t)\}$ is any convenient set of basis functions that is complete and orthonormal (CON) over the interval $[T_i, T_f]$. Because $d(t)$ is nonrandom but unknown, the coefficients $d_1, d_2, d_3, \cdots$, are nonrandom but unknown. The substitution of (2.1) into $s(t - d(t))$ yields a function that depends upon the basis set $\{\psi_i(t)\}$, the vector of unknown coefficients $\boldsymbol{d} = (d_1, d_2, \cdots,)$ and t. To show the dependency on $\boldsymbol{d}$ explicitly, we will denote this function by $s(t; \boldsymbol{d})$:

$$s(t; \boldsymbol{d}) \triangleq s\left(t - \sum_{i=1}^{\infty} d_i \psi_i(t)\right). \tag{2.3}$$

It follows from notation (2.3) that

$$s(t; \boldsymbol{0}) = s(t). \tag{2.4}$$

We now write $\boldsymbol{r}(t)$ of (1.2) as

$$\boldsymbol{r}(t) = \boldsymbol{s}(t; \boldsymbol{d}, \alpha) + \boldsymbol{w}(t) \tag{2.5}$$

where

$$r(t) = \left(r_1(t) \quad r_2(t)\right)^T, \tag{2.6a}$$

$$s(t; d, \alpha) \triangleq \left(s(t; 0) \quad \alpha s(t; d)\right)^T, \tag{2.6b}$$

and

$$w(t) = \left(w_1(t) \quad w_2(t)\right)^T. \tag{2.6c}$$

The problem of estimating the unknown delay $d(t)$ and relative attenuation constant α from $r(t)$ in (1.2) is equivalent to that of estimating the unknown vector d and scalar α from $r(t)$ in (2.6). We have considered the generalized Fourier series representation of $d(t)$ (2.1) because it is both well known and general. Other techniques for representing $d(t)$ as a vector may be preferred in a particular application. For example, if it is known *a priori* that $d(t) = a_0 + a_1 t + a_2 t^2$ for $T_i \le t \le T_f$ (where a_0, a_1, and a_2 are unknown), one can simply set $d = (a_0, a_1, a_2)^T$ to obtain a four parameter estimation problem involving the unknown attenuation α and the physically meaningful constants a_0, a_1, and a_2.

Notice that if $d(t)$ is a known function $D(t)$ and if α is a known constant A, then $\alpha s(t - d(t))$ is related to $s(t)$ by a known linear transformation $\mathcal{L}_{D(t),A}\{s(t)\} = As(t - D(t))$. Therefore, for given α and $d(t)$, the signal $s(t)$ and its delayed, attenuated version $\alpha s(t - d(t))$ are jointly Gaussian random processes. Let D be the vector of coefficients corresponding $D(t)$. Then for $d = D$ and $\alpha = A$, $r(t)$ of (2.6) is a vector-valued Gaussian random process having mean zero and 2×2 matrix covariance function

$$\begin{aligned} K_{r;d,\alpha}(t, u; D, A) &\triangleq E\left\{r(t)\, r^T(u) \middle| d = D, \alpha = A\right\} \\ &= E\left\{s(t; D, A)\, s^T(u; D, A)\right\} + E\left\{w(t)\, w^T(u)\right\} \\ &= K_{s;d,\alpha}(t, u; D, A) \\ &\quad + \frac{N_o}{2} I\delta(t - u) \end{aligned} \tag{2.7}$$

where I is the 2×2 identity matrix.

We proceed by representing vector process $r(t)$ as an infinite-dimensional vector r using the generalized Karhunen–Loève expansion [17, pp. 221–223] and [20]. We define

$$r_N(t) \triangleq \sum_{i=1}^{N} r_i \varphi_i(t; D, A) \tag{2.8}$$

where

$$r_i = \int_{T_i}^{T_f} \varphi_i^T(t; D, A)\, r(t)\, dt; \qquad i = 1, 2, \cdots, N \tag{2.9}$$

and where the $\varphi_i(t; D, A)$ are the normalized vector eigenfunctions of the matrix covariance function $K_{s;d,\alpha}(t, u; D, A)$. We assume that $\{\varphi_i(t; D, A)\}$ is complete. The normalized vector eigenfunctions are 2 element column vectors that satisfy the equations

$$\lambda_i(D, A)\, \varphi_i(t; D, A) = \int_{T_i}^{T_f} K_{s;d,\alpha}(t, u; D, A)\, \varphi_i(u; D, A)\, du; \qquad T_i \le t \le T_f \tag{2.10}$$

and

$$\int_{T_i}^{T_f} \varphi_i^T(t; D, A)\, \varphi_j(t; D, A)\, dt = \delta_{ij} \tag{2.11}$$

where $\lambda_i(D, A)$ is the (scalar) eigenvalue associated with $\varphi_i(t; D, A)$. With the $\varphi_i(t; D, A)$ so specified, it follows that

$$r(t) = \underset{N\to\infty}{\text{l.i.m.}}\ r_N(t). \tag{2.12}$$

This is the generalized Karhunen–Loève expansion of $r(t)$.

Because $\{\varphi_i(t; D, A)\}$ is complete, one can represent $r(t)$ by (2.12) [using (2.8)–(2.9)] for hypothetical or assumed values of D and A. It is easy to show that if the assumed values of D and A are the true values of the unknown quantities d and α, respectively, then the r_i of (2.9) are statistically independent Gaussian random variables having zero means and variances

$$E\{r_i^2 | d = D, \alpha = A\} = \lambda_i(D, A) + \frac{N_o}{2}; \qquad i = 1, 2, \cdots, N. \tag{2.13}$$

The joint probability density function of the r_i conditioned on $d = D$ and $\alpha = A$ is, therefore,

$$p_{r_N;d,\alpha}(R_N; D, A) = \prod_{i=1}^{N} \frac{1}{\sqrt{2\pi\left[\lambda_i(D, A) + \frac{N_o}{2}\right]}} \cdot \exp\left\{-\frac{R_i^2}{2\left[\lambda_i(D, A) + \frac{N_o}{2}\right]}\right\} \tag{2.14}$$

where $r_N = (r_1 r_2 \cdots r_N)^T$ and $R_N = (R_1 R_2 \cdots R_N)^T$. To obtain the likelihood function associated with $r(t)$, we take the logarithm and the limit $N \to \infty$. This leads to a convergence difficulty which can be bypassed in the usual way [17, p. 274] of dividing (2.14) by the function

$$f(R_N) = \prod_{i=1}^{N} \frac{1}{\sqrt{\pi N_o}} \exp\left\{-\frac{R_i^2}{N_o}\right\}. \tag{2.15}$$

The result of this division is still a likelihood function because $f(R_N)$ does not depend upon D or A. After dividing (2.14) by (2.15), we take the logarithm and the limit $N \to \infty$. The result is the log-likelihood function

$$\ln \Lambda(D, A) = l_R(D, A) + l_B(D, A) \tag{2.16}$$

where

$$l_R(D, A) \triangleq \frac{1}{N_o} \sum_{i=1}^{\infty} \frac{\lambda_i(D, A)}{\lambda_i(D, A) + N_o/2} R_i^2 \tag{2.17}$$

and

$$l_B(\boldsymbol{D}, \mathrm{A}) \triangleq -\frac{1}{2}\sum_{i=1}^{\infty} \ln\left[1 + \frac{2}{N_o}\lambda_i(\boldsymbol{D}, \mathrm{A})\right]. \quad (2.18)$$

B. Closed Form for $l_R(\boldsymbol{D}, \mathrm{A})$

The first term in (2.16), $l_R(\boldsymbol{D}, \mathrm{A})$, can be written as

$$l_R(\boldsymbol{D}, \mathrm{A}) = \frac{1}{N_o}\int_{T_i}^{T_f}\int_{T_i}^{T_f} \boldsymbol{R}^T(t)\,\boldsymbol{H}_n(t, v; \boldsymbol{D}, \mathrm{A})\,\boldsymbol{R}(v)\,dt\,dv \quad (2.19)$$

where

$$\boldsymbol{H}_n(t, v; \boldsymbol{D}, \mathrm{A}) \triangleq \sum_{i=1}^{\infty} \frac{\lambda_i(\boldsymbol{D}, \mathrm{A})}{\lambda_i(\boldsymbol{D}, \mathrm{A}) + N_o/2}\,\boldsymbol{\varphi}_i(t; \boldsymbol{D}, \mathrm{A})\,\boldsymbol{\varphi}_i^T(v; \boldsymbol{D}, \mathrm{A}); \quad T_i \le t, v \le T_f \quad (2.20)$$

and $\boldsymbol{R}(t)$ is the sample vector function of $\boldsymbol{r}(t)$ corresponding to the vector $\boldsymbol{R}$:

$$\boldsymbol{R}(t) = \sum_{i=1}^{\infty} R_i\boldsymbol{\varphi}_i(t; \boldsymbol{D}, \mathrm{A}) \quad (2.21)$$

with

$$R_i = \int_{T_i}^{T_f} \boldsymbol{\varphi}_i^T(t; \boldsymbol{D}, \mathrm{A})\,\boldsymbol{R}(t)\,dt. \quad (2.22)$$

Equation (2.19) can be verified by substituting $\boldsymbol{H}_n(t, v; \boldsymbol{D}, \mathrm{A})$ of (2.20) into (2.19) and using (2.22). An interpretation of (2.19) is obtained by considering the following noncausal linear estimate of $\boldsymbol{s}(t; \boldsymbol{D}, \mathrm{A})$:

$$\hat{\boldsymbol{s}}_n(t; \boldsymbol{D}, \mathrm{A}) = \int_{T_i}^{T_f} \boldsymbol{H}_n(t, v; \boldsymbol{D}, \mathrm{A})\,\boldsymbol{r}(v)\,dv, \quad T_i \le t \le T_f, \quad (2.23)$$

where the subscript "n" is used to denote a noncausal estimate or system.

It follows from (2.23) and (2.7) that

$$E\left\{[\boldsymbol{s}(t; \boldsymbol{d}, \alpha) - \hat{\boldsymbol{s}}_n(t; \boldsymbol{D}, \mathrm{A}))]\,\boldsymbol{r}^T(u)\,\middle|\,\boldsymbol{d} = \boldsymbol{D}, \alpha = \mathrm{A}\right\}$$
$$= \boldsymbol{K}_{s;d,\alpha}(t, u; \boldsymbol{D}\,\mathrm{A}) - \frac{N_o}{2}\boldsymbol{H}_n(t, u; \boldsymbol{D}, \mathrm{A}) - \int_{T_i}^{T_f} \boldsymbol{H}_n(t, v; \boldsymbol{D}, \mathrm{A})\,\boldsymbol{K}_{s;d,\alpha}(v, u; \boldsymbol{D}, \mathrm{A})\,dv; \quad T_i \le t, u \le T_f. \quad (2.24)$$

According to the matrix version of Mercer's theorem [20]:

$$\boldsymbol{K}_{s;d,\alpha}(t, u; \boldsymbol{D}, \mathrm{A}) = \sum_{i=1}^{\infty} \lambda_i(\boldsymbol{D}, \mathrm{A})\,\boldsymbol{\varphi}_i(t; \boldsymbol{D}, \mathrm{A})\,\boldsymbol{\varphi}_i^T(u; \boldsymbol{D}, \mathrm{A}); \quad T_i \le t, u \le T_f. \quad (2.25)$$

By substituting (2.25) and (2.20) into the right-hand side of (2.24) and using (2.11), one obtains

$$\boldsymbol{K}_{s;d,\alpha}(t, u; \boldsymbol{D}, \mathrm{A}) - \frac{N_o}{2}\boldsymbol{H}_n(t, u; \boldsymbol{D}, \mathrm{A}) - \int_{T_i}^{T_f} \boldsymbol{H}_n(t, v; \boldsymbol{D}, \mathrm{A})\,\boldsymbol{K}_{s;d,\alpha}(v, u; \boldsymbol{D}, \mathrm{A})\,dv = \boldsymbol{0}; \quad T_i \le t_i, u \le T_f, \quad (2.26a)$$

which yields

$$E\left\{[\boldsymbol{s}(t; \boldsymbol{d}, \alpha) - \hat{\boldsymbol{s}}_n(t; \boldsymbol{D}, \mathrm{A})]\,\boldsymbol{r}^T(u)\,\middle|\,\boldsymbol{d} = \boldsymbol{D}, \alpha = \mathrm{A}\right\} = \boldsymbol{0}; \; T_i \le t, u \le T_f. \quad (2.26b)$$

Therefore, if $\boldsymbol{D}$ and A are the true values of $\boldsymbol{d}$ and α, then the estimation error

$$\boldsymbol{e}_n(t; \boldsymbol{d}, \alpha, \boldsymbol{D}, \mathrm{A}) \triangleq \boldsymbol{s}(t; \boldsymbol{d}, \alpha) - \hat{\boldsymbol{s}}_n(t; \boldsymbol{D}, \mathrm{A}) \quad (2.27)$$

is orthogonal to $\boldsymbol{r}(u)$, $T_i \le u \le T_f$. Consequently [21, p. 390], $\hat{\boldsymbol{s}}_n(t; \boldsymbol{D}, \mathrm{A})$ of (2.23) is the noncausal point LMMSE estimate of $\boldsymbol{s}(t; \boldsymbol{d}, \alpha)$ from $\boldsymbol{r}(u)$, $T_i \le u \le T_f$, given that $\boldsymbol{d} = \boldsymbol{D}$ and $\alpha = \mathrm{A}$, and $\boldsymbol{H}_n(t, v; \boldsymbol{D}, \mathrm{A})$ is the impulse response of the noncausal point LMMSE estimator. Note that the 2×2 matrix function $\boldsymbol{H}_n(t, v; \boldsymbol{D}, \mathrm{A})$ is the solution to (2.26a). As will be described in Section IV, the substitution of (2.23) into (2.19) results in a vector estimator-correlator realization for $l_R(\boldsymbol{D}, \mathrm{A})$.

The error covariance matrix of the noncausal point LMMSE estimate of $\boldsymbol{s}(t; \boldsymbol{d}, \alpha)$ conditioned on $\boldsymbol{d} = \boldsymbol{D}$ and $\alpha = \mathrm{A}$ is

$$\boldsymbol{E}_n(t; \boldsymbol{D}, \mathrm{A}) \triangleq E\left\{\boldsymbol{e}_n(t; \boldsymbol{d}, \alpha, \boldsymbol{D}, \mathrm{A})\,\boldsymbol{e}_n^T(t; \boldsymbol{d}, \alpha, \boldsymbol{D}, \mathrm{A})\,\middle|\,\boldsymbol{d} = \boldsymbol{D}, \alpha = \mathrm{A}\right\}. \quad (2.28)$$

By substituting (2.27) into (2.28) and using (2.23) and (2.26), one obtains

$$\boldsymbol{E}_n(t; \boldsymbol{D}, \mathrm{A}) = \frac{N_o}{2}\boldsymbol{H}_n(t, t; \boldsymbol{D}, \mathrm{A}). \quad (2.29)$$

C. Closed Form for $l_B(\boldsymbol{D}, \mathrm{A})$

The term $l_B(\boldsymbol{D}, \mathrm{A})$ of (2.18) can be written in closed form by noting that, according to (2.10)–(2.11), the eigenvalues $\lambda_i(\boldsymbol{D}, \mathrm{A})$ and the vector eigenfunctions $\boldsymbol{\varphi}_i(t; \boldsymbol{D}, \mathrm{A})$ depend upon the final observation time T_f. To indicate this dependency, we write

$$\lambda_i(\boldsymbol{D}, \mathrm{A}) = \lambda_i(\boldsymbol{D}, \mathrm{A}, T_f) \quad (2.30)$$

$$\boldsymbol{\varphi}_i(t; \boldsymbol{D}, \mathrm{A}) = \boldsymbol{\varphi}_i(t; \boldsymbol{D}, \mathrm{A}, T_f). \quad (2.31)$$

It follows that $l_B(\boldsymbol{D}, \mathrm{A})$ of (2.18) can be rewritten as

$$l_B(\boldsymbol{D}, \mathrm{A}) = -\frac{1}{2}\int_{T_i}^{T_f} dt\,\frac{d}{dt}\sum_{i=1}^{\infty} \ln\left[1 + \frac{2}{N_o}\lambda_i(\boldsymbol{D}, \mathrm{A}, t)\right] = -\frac{1}{N_o}\int_{T_i}^{T_f} dt \sum_{i=1}^{\infty} \frac{[d\lambda_i(\boldsymbol{D}, \mathrm{A}, t)]/dt}{1 + (2/N_o)\lambda_i(\boldsymbol{D}, \mathrm{A}, t)}, \quad (2.32)$$

where $\lambda_i(\boldsymbol{D}, \mathrm{A}, T_i) = 0$. It can be shown by a straightforward extension of the derivation in [17, pp. 204–205] that

$$\frac{d\lambda_i(\boldsymbol{D}, \mathrm{A}, t)}{dt} = \lambda_i(\boldsymbol{D}, \mathrm{A}, t)\, \mathrm{Tr}\left\{\boldsymbol{\varphi}_i(t; \boldsymbol{D}, \mathrm{A}, t)\, \boldsymbol{\varphi}_i^T(t; \boldsymbol{D}, \mathrm{A}, t)\right\}. \tag{2.33}$$

By using (2.33) and the fact that Tr $\{\cdot\}$ is a linear operator, (2.32) becomes

$$l_B(\boldsymbol{D}, \mathrm{A}) = -\frac{1}{2}\int_{T_i}^{T_f} \mathrm{Tr}\left\{\sum_{i=1}^{\infty} \frac{\lambda_i(\boldsymbol{D}, \mathrm{A}, t)}{\lambda_i(\boldsymbol{D}, \mathrm{A}, t) + N_o/2} \cdot \boldsymbol{\varphi}_i(t; \boldsymbol{D}, \mathrm{A}, t)\, \boldsymbol{\varphi}_i^T(t; \boldsymbol{D}, \mathrm{A}, t)\right\} dt. \tag{2.34}$$

A closed form for the quantity in the braces in (2.34) is recognized by rewriting (2.20) with the notation of (2.30) and (2.31)

$$\boldsymbol{H}_n(t, v; \boldsymbol{D}, \mathrm{A}, T_f) = \sum_{i=1}^{\infty} \frac{\lambda_i(\boldsymbol{D}, \mathrm{A}, T_f)}{\lambda_i(\boldsymbol{D}, \mathrm{A}, T_f) + N_o/2} \cdot \boldsymbol{\varphi}_i(t; \boldsymbol{D}, \mathrm{A}, T_f)\, \boldsymbol{\varphi}_i^T(v; \boldsymbol{D}, \mathrm{A}, T_f). \tag{2.35}$$

The quantity in the braces in (2.34) is $\boldsymbol{H}_n(t, t; \boldsymbol{D}, \mathrm{A}, t)$. Since $\boldsymbol{H}_n(t, v; \boldsymbol{D}, \mathrm{A}, t)$ is the matrix impulse response of the point LMMSE estimator of $\boldsymbol{s}(t; \boldsymbol{d}, \alpha)$ from $\boldsymbol{r}(v)$, $T_i \le v \le t$, given $\boldsymbol{d} = \boldsymbol{D}$ and $\alpha = \mathrm{A}$, then $\boldsymbol{H}_n(t, v; \boldsymbol{D}, \mathrm{A}, t)$ is, by definition, the matrix impulse response of the causal point LMMSE estimator of $\boldsymbol{s}(t; \boldsymbol{d}, \alpha)$ from $\boldsymbol{r}(v)$ given $\boldsymbol{d} = \boldsymbol{D}$ and $\alpha = \mathrm{A}$. Denoting the causal matrix impulse response by $\boldsymbol{H}_c(t, v; \boldsymbol{D}, \mathrm{A})$, (2.34) becomes

$$l_B(\boldsymbol{D}, \mathrm{A}) = -\tfrac{1}{2}\int_{T_i}^{T_f} \mathrm{Tr}\left[\boldsymbol{H}_c(t, t; \boldsymbol{D}, \mathrm{A})\right] dt. \tag{2.36}$$

All the previous equations describing noncausal estimation of $\boldsymbol{s}(t; \boldsymbol{d}, \alpha)$ from $\boldsymbol{r}(v)$, describe causal estimation of $\boldsymbol{s}(t; \boldsymbol{d}, \alpha)$ from $\boldsymbol{r}(v)$ when t is substituted for T_f. In particular, with $T_f = t$, (2.29) describes the error covariance matrix of the causal LMMSE estimate of $\boldsymbol{s}(t; \boldsymbol{d}, \alpha)$ given $\boldsymbol{d} = \boldsymbol{D}$ and $\alpha = \mathrm{A}$:

$$\boldsymbol{E}_c(t; \boldsymbol{D}, \mathrm{A}) = \frac{N_o}{2}\boldsymbol{H}_c(t, t; \boldsymbol{D}, \mathrm{A}). \tag{2.37}$$

The substitution of (2.37) into (2.36) yields an alternative expression for $l_B(\boldsymbol{D}, \mathrm{A})$:

$$l_B(\boldsymbol{D}, \mathrm{A}) = -\frac{1}{N_o}\int_{T_i}^{T_f} \mathrm{Tr}\left[\boldsymbol{E}_c(t; \boldsymbol{D}, \mathrm{A})\right] dt. \tag{2.38}$$

III. The Matrix Impulse Response $\boldsymbol{H}_n(t, v; \boldsymbol{D}, \mathrm{A})$

In this section we derive a simple explicit form for the matrix impulse response $\boldsymbol{H}_n(t, v; \boldsymbol{D}, \mathrm{A})$. It is relatively difficult to obtain this form by solving (2.26a). The constructive approach of Section III-B has the advantage of being both mathematically and conceptually simple. Before proceeding with the constructive solution, it will be helpful to derive the explicit form for the inverse of the operator (1.4).

A. Inverse of Operator (1.4)

By definition, the inverse operator satisfies

$$\mathcal{L}_{\boldsymbol{D},\mathrm{A}}^{-1}\left\{\boldsymbol{s}(t; \boldsymbol{D}, \mathrm{A})\right\} = \mathcal{L}_{D(t),\mathrm{A}}^{-1}\left\{\mathrm{A}s(t - D(t))\right\} = s(t). \tag{3.1}$$

Let $v(t)$ be an arbitrary waveform and try an inverse having the form

$$\mathcal{L}_{D(t),\mathrm{A}}^{-1}\left\{v(t)\right\} = \frac{1}{\mathrm{A}} v(\beta(t)) \tag{3.2}$$

where $\beta(t)$ is to be determined. Define

$$f(t) = t - D(t) \tag{3.3}$$

so that by the definition (3.1)

$$\mathcal{L}_{D(t),\mathrm{A}}^{-1}\left\{v(f(t))\right\} = \frac{1}{\mathrm{A}} v(t). \tag{3.4a}$$

Replacing t by $f(t)$ in (3.2) gives

$$\mathcal{L}_{D(t),\mathrm{A}}^{-1}\left\{v(f(t))\right\} = \frac{1}{\mathrm{A}} v(\beta(f(t))) \tag{3.4b}$$

which, when compared to (3.4a), yields

$$\beta(f(t)) = t. \tag{3.5}$$

Therefore, the inverse operator is given by (3.2) where $\beta(\cdot)$ is the inverse of the function $f(t)$. Since $\mathcal{L}_{D(t),\mathrm{A}}\{\cdot\}$ is invertible, the function $f(t)$ is one to one. We now proceed to the constructive derivation of $\boldsymbol{H}_n(t, v; \boldsymbol{D}, \mathrm{A})$.

B. Constructive Derivation of $\boldsymbol{H}_n(t, v; \boldsymbol{D}, \mathrm{A})$

The first step in the derivation of $\boldsymbol{H}_n(t, v; \boldsymbol{D}, \mathrm{A})$ is to (noncausally) transform $\boldsymbol{r}(t)$, $T_i \le t \le T_f$, into the vector process $\boldsymbol{r}'(u)$, $f(T_i) \le u \le T_f$, where

$$\boldsymbol{r}'(u) = \begin{bmatrix} \mathrm{A}^{-1}r_2(\beta(u)) \\ 0 \end{bmatrix}; \quad f(T_i) \le u \le T_i \tag{3.6a}$$

$$\boldsymbol{r}'(u) = \tfrac{1}{2}\begin{bmatrix} r_1(u) + \mathrm{A}^{-1}r_2(\beta(u)) \\ r_1(u) - \mathrm{A}^{-1}r_2(\beta(u)) \end{bmatrix}; \quad T_i < u \le f(T_f) \tag{3.6b}$$

$$\boldsymbol{r}'(u) = \begin{bmatrix} r_1(u) \\ 0 \end{bmatrix}; \quad f(T_f) < u \le T_f \tag{3.6c}$$

and $f(u)$ is defined by (3.3). In (3.6), A can be regarded as an assumed value for the unknown relative attenuation constant α, and $D(t)$ as an assumed function for the unknown delay function $d(t)$. (We naturally require $D(t) \ge 0$ which implies that $f(t) \le t$.) Notice that (3.6b) assumes that $f(T_f) \ge T_i$. This is equivalent to the assumption that the signal delay does not exceed the observation interval. Since this assumption is likely to be met in most applications, we will retain it in the following. It is possible to generalize our results to include the case $f(T_f) < T_i$.

The transformation $\mathbf{r}(t) \rightarrow \mathbf{r}'(u)$ is illustrated in Fig. 1 where, for simplicity in interpretation, the noise processes $w_1(t)$ and $w_2(t)$ have been drawn as small ripples. A system block diagram for the transformation is shown in Fig. 2. An examination of (3.6), Fig. 1, or Fig. 2 will reveal that the transformation from $\mathbf{r}(t)$ to $\mathbf{r}'(u)$ is linear and invertible. Thus, $\mathbf{r}(t)$, $T_i \leq t \leq T_f$ can be recovered from $\mathbf{r}'(u)$, $f(T_i) \leq u \leq T_f$, using a linear transformation. It follows from the reversibility theorem [17, p. 289] that the noncausal LMMSE estimate $\hat{s}_n(t; \mathbf{D}, \mathrm{A})$ of (2.23), given $\mathbf{d} = \mathbf{D}$ and $\alpha = \mathrm{A}$, can be obtained from $\mathbf{r}'(u)$. Before describing the structure of the LMMSE estimator, it will be helpful to observe that if $\mathbf{d} = \mathbf{D}$ and $\alpha = \mathrm{A}$, then, from (2.5), (2.6), and (3.6):

$$\mathbf{r}'(u) = \mathbf{0}; \qquad u < f(T_i), \tag{3.7a}$$

$$\mathbf{r}'(u) = \begin{bmatrix} s(u) \\ 0 \end{bmatrix} + \begin{bmatrix} n_1(u) \\ n_2(u) \end{bmatrix}; \qquad f(T_i) \leq u \leq T_f \tag{3.7b}$$

$$\mathbf{r}'(u) = \mathbf{0}; \qquad T_f < u, \tag{3.7c}$$

where

$$\begin{bmatrix} n_1(u) \\ n_2(u) \end{bmatrix} = \begin{bmatrix} \mathrm{A}^{-1} w_2(\beta(u)) \\ 0 \end{bmatrix}; \qquad f(T_i) \leq u \leq T_i, \tag{3.8a}$$

$$\begin{bmatrix} n_1(u) \\ n_2(u) \end{bmatrix} = \frac{1}{2} \begin{bmatrix} w_1(u) + \mathrm{A}^{-1} w_2(\beta(u)) \\ w_1(u) - \mathrm{A}^{-1} w_2(\beta(u)) \end{bmatrix}; \qquad T_i < u \leq f(T_f), \tag{3.8b}$$

$$\begin{bmatrix} n_1(u) \\ n_2(u) \end{bmatrix} = \begin{bmatrix} w_1(u) \\ 0 \end{bmatrix}; \qquad f(T_f) < u \leq T_f. \tag{3.8c}$$

We present the form of the LMMSE estimator of $s(t)$, $f(T_i) \leq t \leq T_f$, in the following theorem.

Theorem: The noncausal point LMMSE estimator of $s(t)$ from $\mathbf{r}'(u)$, $f(T_i) \leq t, u \leq T_f$, conditioned on $\mathbf{d} = \mathbf{D}$ and $\alpha = \mathrm{A}$, is given by the system in Fig. 3, where $f(t, u; \mathbf{D}, \mathrm{A})$ is the impulse response of the noncausal point LMMSE estimator $\hat{n}_1(t)$ of $n_1(t)$ from $n_2(u)$,

$$\hat{n}_1(t) = \int_{f(T_i)}^{T_f} f(t, u; \mathbf{D}, \mathrm{A})\, n_2(u)\, du; \qquad f(T_i) \leq t \leq T_f \tag{3.9}$$

and $g_n(t, u; \mathbf{D}, \mathrm{A})$ is the impulse response of the noncausal point LMMSE estimator $\hat{s}_n(t)$ of $s(t)$ from $s(u) + n_1(u) - \hat{n}_1(u)$,

$$\hat{s}_n(t) = \int_{f(T_i)}^{T_f} g_n(t, u; \mathbf{D}, \mathrm{A})\, [s(u) + n_1(u) - \hat{n}_1(u)]\, du; \qquad f(T_i) \leq t \leq T_f. \tag{3.10}$$

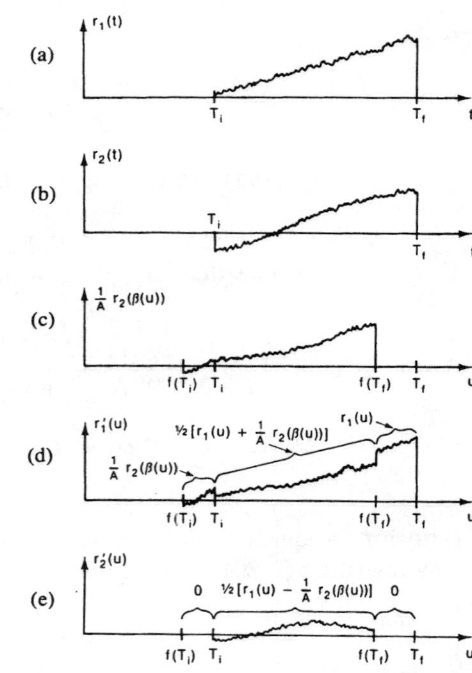

Fig. 1. Components of $\mathbf{r}(t) = (r_1(t)\, r_2(t))^T$ and $\mathbf{r}'(t) = (r_1'(t)\, r_2'(t))^T$. (a) $r_1(t)$, (b) $r_2(t)$, (c) output of inverse operator $\mathcal{L}_{\mathbf{D},\mathrm{A}}^{-1}$ of Fig. 2, (d) $r_1'(u)$, (e) $r_2'(u)$.

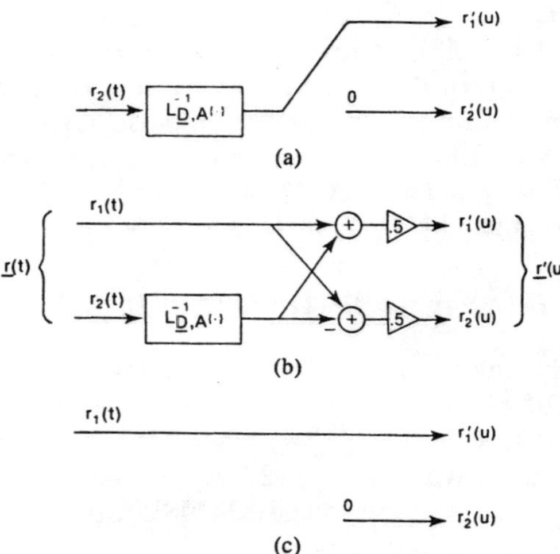

Fig. 2. System block diagram corresponding to transformation (3.6). (a) $f(T_i) \leq u \leq T_i$; (b) $T_i < u \leq f(T_f)$, (c) $f(T_f) < u \leq T_f$.

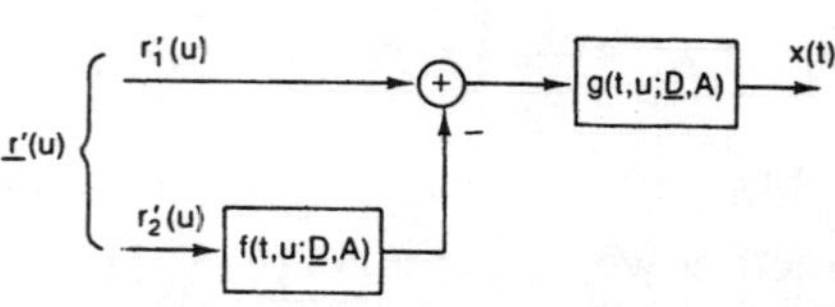

Fig. 3. Structure of the noncausal conditional LMMSE estimator of $s(t)$ from $\mathbf{r}'(u)$, $f(T_i) \leq u \leq T_f$. When $\mathbf{d} = \mathbf{D}$ and $\alpha = \mathrm{A}$, $x(t)$ equals the noncausal LMMSE estimate of $s(t)$. The impulse responses $f(t, u; \mathbf{D}, \mathrm{A})$ and $g_n(t, u; \mathbf{D}, \mathrm{A})$ are defined in the theorem of Section III. $f(t, u; \mathbf{D}, \mathrm{A})$ is specified by (3.30) and (3.31), and $g(t, u; \mathbf{D}, \mathrm{A})$ is specified by (3.44) and (3.41).

Proof: According to the orthogonality principle [21, p. 390] a linear functional $\hat{p} = L[\boldsymbol{q}]$ is the LMMSE estimate of a random variable p from data vector $\boldsymbol{q}(\xi)$, $\xi \in \mathfrak{D}$ (where $\mathfrak{D}$ is the domain of the data), if and only if the estimation error $p - \hat{p}$ is orthogonal to $\boldsymbol{q}$ for all $\xi \in \mathfrak{D}$

$$E\{(p - \hat{p})\boldsymbol{q}(\xi)\} = \boldsymbol{0};\ \xi \in \mathfrak{D}. \tag{3.11}$$

Therefore, a necessary and sufficient condition that $\hat{s}_n(t)$ be the LMMSE estimate of $s(t)$ from $\boldsymbol{r}'(u)$ given that $\boldsymbol{d} = \boldsymbol{D}$ and $\alpha = \mathrm{A}$ is that the vector

$$\boldsymbol{v}(t, u) \triangleq E\{[s(t) - \hat{s}_n(t)]\,\boldsymbol{r}'(u) \mid \boldsymbol{d} = \boldsymbol{D}, \alpha = \mathrm{A}\} \tag{3.12}$$

be identically zero for $f(T_i) \le t, u \le T_f$. We note from (3.7) that the components of $\boldsymbol{v}(t, u)$ are

$$v_1(t, u) = E\{[s(t) - \hat{s}_n(t)]\,[s(u) + n_1(u)]\} \tag{3.13}$$

and

$$v_2(t, u) = E\{[s(t) - \hat{s}_n(t)]\,n_2(u)\} \quad \text{for } f(T_i) \le t, u \le T_f. \tag{3.14}$$

By the definitions of $f(t, u; \boldsymbol{D}, \mathrm{A})$ and $g_n(t, u; \boldsymbol{D}, \mathrm{A})$,

$$E\{[n_1(t) - \hat{n}_1(t)]\,n_2(u)\} = 0; \quad f(T_i) \le t, u \le T_f \tag{3.15}$$

and

$$E\{[s(t) - \hat{s}_n(t)]\,[s(u) + n_1(u) - \hat{n}_1(u)]\} = 0; \quad f(T_i) \le t, u \le T_f. \tag{3.16}$$

The signal process $s(t)$ is orthogonal to the white noise processes $w_1(t)$ and $w_2(t)$. Therefore, $s(t)$ is also orthogonal to the noise processes $n_1(t)$ and $n_2(t)$ defined in (3.8). It follows that (3.14) simplifies to

$$v_2(t, u) = -E\{\hat{s}_n(t)\,n_2(u)\} \tag{3.17a}$$

which, with the aid of (3.10), becomes

$$\begin{aligned} v_2(t, u) &= -E\left\{\int_{f(T_i)}^{T_f} g_n(t, \sigma; \boldsymbol{D}, \mathrm{A}) \cdot [s(\sigma) + n_1(\sigma) - \hat{n}_1(\sigma)]\,d\sigma\, n_2(u)\right\} \\ &= -\int_{f(T_i)}^{T_f} g_n(t, \sigma; \boldsymbol{D}, \mathrm{A})\, E\{[n_1(\sigma) - \hat{n}_1(\sigma)]\,n_2(u)\}\,d\sigma \\ &= 0; \quad f(T_i) \le t, u \le T_f, \end{aligned} \tag{3.17b}$$

where the last step follows from (3.15). The result that $v_2(t, u)$ of (3.14) equals zero, and the fact that $\hat{n}_1(t)$ depends linearly upon $n_2(t)$, together imply

$$E\{[s(t) - \hat{s}_n(t)]\,\hat{n}_1(u)\} = 0 \quad f(T_i) \le t, u \le T_f. \tag{3.18}$$

Subtracting (3.18) from (3.13) leads to

$$v_1(t, u) = E\{[s(t) - \hat{s}_n(t)]\,[s(u) + n_1(u) - \hat{n}_1(u)]\}. \tag{3.19}$$

Comparing (3.19) to (3.16), we see that

$$v_1(t, u) = 0; \quad f(T_i) \le t, u \le T_f. \tag{3.20}$$

This completes the proof.

The LMMSE estimator of $\alpha s(t - d(t))$, $T_i \le t \le T_f$, from $\boldsymbol{r}(u)$, $T_i \le u \le T_f$, conditioned on $\boldsymbol{d} = \boldsymbol{D}$ and $\alpha = \mathrm{A}$, follows easily from the fact that $\alpha s(t - d(t))$ is a linear transformation of $s(t)$. Because all available data have been used to obtain $\hat{s}_n(t)$, $f(T_i) \le t \le T_f$, the noncausal LMMSE estimate of $\alpha s(t - d(t))$, given $\boldsymbol{d} = \boldsymbol{D}$ and $\alpha = \mathrm{A}$, is simply the scaled and delayed version of $\hat{s}_n(t)$ of (3.10), namely, $\mathrm{A}\hat{s}_n(t - D(t))$.

The explicit form for $f(t, u; \boldsymbol{D}, \mathrm{A})$ follows easily from (3.9) and (3.15), which together imply

$$E\{n_1(t)\,n_2(u)\} = \int_{f(T_i)}^{T_f} f(t, \sigma; \boldsymbol{D}, \mathrm{A})\, E\{n_2(\sigma)\,n_2(u)\}\,d\sigma, \quad f(T_i) \le t, u \le T_f. \tag{3.21}$$

This can be simplified by using (3.8) and (1.5), which imply

$$E\{n_2(t)\,n_2(u)\} = \begin{cases} \dfrac{N_o}{8}\left[\delta(t - u) + \mathrm{A}^{-2}\delta(\beta(t) - \beta(u))\right]; & T_i < u \le f(T_f) \\ 0; & \text{otherwise} \end{cases} \tag{3.22}$$

and

$$E\{n_1(t)\,n_2(u)\} = \begin{cases} \dfrac{N_o}{8}\left[\delta(t - u) - \mathrm{A}^{-2}\delta(\beta(t) - \beta(u))\right]; & T_i < u \le f(T_f) \\ 0; & \text{otherwise.} \end{cases} \tag{3.23}$$

Since $\beta(t)$ is a one-to-one function, then

$$\delta(\beta(t) - \beta(u)) = \frac{1}{|\dot{\beta}(u)|}\,\delta(t - u) \tag{3.24}$$

where the dot denotes the derivative of a function. It follows from (3.3) and (3.5) that

$$\dot{\beta}(u) = \frac{1}{1 - \dot{D}(\beta(u))}. \tag{3.25}$$

Note that $s(t - D(t))$ is locally reversed in time where $\dot{D}(t) > 1$ and frozen in time where $\dot{D}(t) = 1$. It is therefore reasonable to define $D(t)$ as a valid delay function if and only if

$$\dot{D}(t) \le 1, \tag{3.26}$$

where the equality holds only at isolated values of t. It is easy to see that the above condition guarantees that $f(t)$ is invertible. By combining (3.24), (3.25), and (3.26), we obtain

$$\delta(\beta(t) - \beta(u)) = [1 - \dot{D}(\beta(u))]\,\delta(t-u). \quad (3.27)$$

Therefore, (3.22) and (3.23) become, respectively,

$$E\{n_2(t)\,n_2(u)\} = \begin{cases} \dfrac{N_o}{8}\left[1 + \dfrac{1 - \dot{D}(\beta(u))}{A^2}\right]\delta(t-u); & T_i < t \le f(T_f) \\ 0; \quad \text{otherwise} \end{cases} \quad (3.28)$$

and

$$E\{n_1(t)\,n_2(u)\} = \begin{cases} \dfrac{N_o}{8}\left[1 - \dfrac{1 - \dot{D}(\beta(u))}{A^2}\right]\delta(t-u); & T_i < u \le f(T_f) \\ 0; \quad \text{otherwise.} \end{cases} \quad (3.29)$$

The substitution of (3.28) and (3.29) into (3.21) yields

$$f(t, u; \mathbf{D}, A) = k(u)\,\delta(t-u) \quad (3.30)$$

where

$$k(u) \triangleq \begin{cases} \dfrac{A^2 - [1 - \dot{D}(\beta(u))]}{A^2 + [1 - \dot{D}(\beta(u))]}; & T_i < u \le f(T_f) \\ 0; \quad \text{otherwise.} \end{cases} \quad (3.31)$$

Therefore,

$$\hat{n}_1(t) = k(t)\,n_2(t). \quad (3.32)$$

Note that the MMSE estimate of $n_1(t)$ from $n_2(t)$ requires only multiplication of $n_2(t)$ by a time-varying gain.

$$E\{n_1(t)\,n_1(u)\} = \begin{cases} \dfrac{N_o}{2A^2}[1 - \dot{D}(\beta(u))]\,\delta(t-u); & f(T_i) \le t \le T_i \\ \dfrac{N_o}{8}\left[1 + \dfrac{1 - \dot{D}(\beta(u))}{A^2}\right]\delta(t-u); & T_i < t \le f(T_f) \\ \dfrac{N_o}{2}\,\delta(t-u); & f(T_f) < t \le T_f. \end{cases} \quad (3.39)$$

It is interesting to observe from (3.29) that if

$$D(t) = D_0 + (1 - A^2)\,t, \quad (3.33)$$

then $n_1(t)$ and $n_2(t)$ become statistically independent and

$$\hat{n}_1(t) = 0. \quad (3.34)$$

Equation (3.33) is a necessary and sufficient condition for (3.34). A sufficient condition arises when the delay is constant

$$D(t) = D_0 \quad (3.35a)$$

and the magnitude of the attenuation constant is unity

$$A = \pm 1. \quad (3.35b)$$

These results are a consequence of the fact that the statistics of $w_2(t)$ are unchanged by the inverse operator (3.1) when $D(t)$ and A satisfy (3.33). The point is that if $d(t)$ and α are known *a priori* to satisfy (3.33), then the hypothetical quantities $D(t)$ and A can also be assumed to satisfy (3.33). This results in a simplified receiver because, under these conditions, $k(u)$ is identically zero.

An equation specifying $g_n(t, u; \mathbf{D}, A)$ can be obtained by using the fact that $g_n(t, u; \mathbf{D}, A)$ is the LMMSE estimator of $s(t), f(T_i) \le t \le T_f$, from

$$z(u) = s(u) + n(u); \quad f(T_i) \le u \le T_f, \quad (3.36)$$

where

$$n(u) \triangleq n_1(u) - \hat{n}_1(u); \quad f(T_i) \le u \le T_f. \quad (3.37)$$

The noise process $n(u)$ is zero mean and uncorrelated with the signal process $s(u)$. Its covariance function is

$$\begin{aligned} E\{n(t)\,n(u)\} &= E\{[n_1(t) - \hat{n}_1(t)]\,[n_1(u) - \hat{n}_1(u)]\} \\ &= E\{[n_1(t) - \hat{n}_1(t)]\,n_1(u)\} \\ &= E\{n_1(t)\,n_1(u)\} - k(t)\,E\{n_2(t)\,n_1(u)\}. \end{aligned} \quad (3.38)$$

By a derivation similar to that leading to (3.28), one finds (3.39).

The substitution of (3.29), (3.31), and (3.39) into (3.38) yields

$$E\{n(t)\,n(u)\} = Q(u)\,\delta(t-u); \quad f(T_i) \le u \le T_f \quad (3.40)$$

where

$$Q(u) = \begin{cases} \dfrac{N_o}{2}[1 - \dot{D}(\beta(u))]; & f(T_i) \le u \le T_i \\ \dfrac{N_o[1 - \dot{D}(\beta(u))]}{2[A^2 + [1 - \dot{D}(\beta(u))]]}; & T_i < u \le f(T_f) \\ \dfrac{N_o}{2}; & f(T_f) < u \le T_f; \end{cases} \quad (3.41)$$

and it follows that $n(t)$ is nonstationary white noise. The equation specifying $g_n(t, u; \mathbf{D}, A)$ is now obtained by substituting

$$\hat{s}_n(t) = \int_{f(T_i)}^{T_f} g_n(t, \sigma; \mathbf{D}, A)\, z(\sigma)\, d\sigma; \quad f(T_i) \le t \le T_f \tag{3.42}$$

into the orthogonality condition

$$E\{[s(t) - \hat{s}_n(t)]\, z(u)\} = 0; \quad f(T_i) \le t, u \le T_f. \tag{3.43}$$

This leads directly to

$$R_s(t, u) = \int_{f(T_i)}^{T_f} g_n(t, \sigma; \mathbf{D}, A)\, R_s(\sigma, u)\, d\sigma + Q(u)\, g_n(t, u; \mathbf{D}, A); \quad f(T_i) < t, u < T_f. \tag{3.44}$$

Note that the problem of finding $g_n(t, u; \mathbf{D}, A)$ is equivalent to the problem of deriving the noncausal LMMSE estimator of $s(t)$ from $s(t) + n(t)$ where the process $n(t)$ is nonstationary white noise uncorrelated with $s(t)$. If $s(t)$ is a state representable process, then $\hat{s}_n(t)$ can be obtained using optimal linear smoothing [22, ch. 5].

C. Explicit Form for Entries in $\mathbf{H}_n(t, u; \mathbf{D}, A)$

We have now specified the structure of $\mathbf{H}_n(t, u; \mathbf{D}, A)$. This structure is shown in Fig. 4, where the filter $g_n(t, u; \mathbf{D}, A)$ is given by the solution to (3.44) and where $k(t)$ is given by (3.31). Using this structure, we now derive the explicit form for the individual entries $h_{ij}(t, u; \mathbf{D}, A)$ in $\mathbf{H}_n(t, u; \mathbf{D}, A)$. This can be done by noting that, by definition, the output of $\mathbf{H}_n(t, u; \mathbf{D}, A)$ is, as in Fig. 4,

$$x(t) = \int_{T_i}^{T_f} h_{11}(t, u; \mathbf{D}, A)\, r_1(u)\, du + \int_{T_i}^{T_f} h_{12}(t, u; \mathbf{D}, A)\, r_2(u)\, du \tag{3.45}$$

$$y(t) = \int_{T_i}^{T_f} h_{21}(t, u; \mathbf{D}, A)\, r_1(u)\, du + \int_{T_i}^{T_f} h_{22}(t, u; \mathbf{D}, A)\, r_2(u)\, du. \tag{3.46}$$

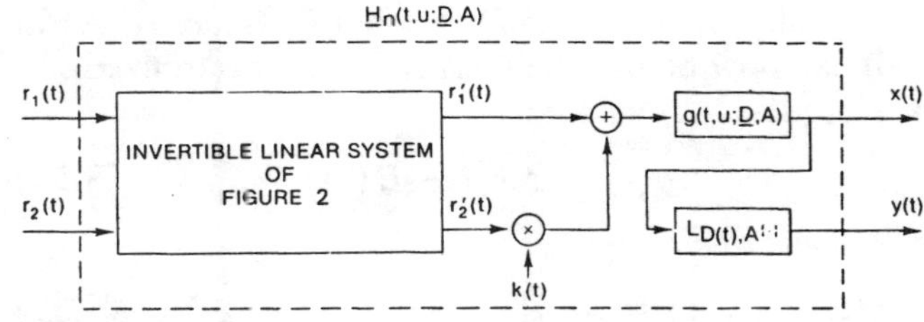

Fig. 4. The system $\mathbf{H}_n(t, u; \mathbf{D}, A)$. If $\mathbf{d} = \mathbf{D}$ and $\alpha = A$, then outputs $x(t)$ and $y(t)$ are the noncausal LMMSE estimates of $s(t)$ and $\alpha s(t - d(t))$, respectively.

On the other hand, by tracing the signals through the system in Fig. 4, one can express $x(t)$ and $y(t)$ in terms of $g_n(t, u; \mathbf{D}, A)$ and $k(u)$. To keep the notation simple, we will subsequently write $g_n(t, u; \mathbf{D}, A)$ as $g_n(t, u)$. An examination of Fig. 4 with the aid of Fig. 2 and (3.6) yields, after a little labor,

$$\begin{aligned} x(t) = &\int_{f(T_i)}^{T_i} g_n(t, u) \frac{1}{A} r_2(\beta(u))\, du \\ &+ \int_{T_i}^{f(T_f)} g_n(t, u) \frac{1}{2} [1 - k(u)]\, r_1(u)\, du \\ &+ \int_{T_i}^{f(T_f)} g_n(t, u) \frac{1}{2A} [1 + k(u)]\, r_2(\beta(u))\, du \\ &+ \int_{f(T_f)}^{T_f} g_n(t, u)\, r_1(u)\, du. \end{aligned} \tag{3.47}$$

By changing variables in the first and third integrals (set $\sigma = \beta(u)$), (3.47) becomes after a little more labor

$$\begin{aligned} x(t) = &\int_{T_i}^{\beta(T_i)} g_n(t, \sigma - D(\sigma)) \frac{1}{A} r_2(\sigma) [1 - \dot{D}(\sigma)]\, d\sigma \\ &+ \int_{T_i}^{f(T_f)} g_n(t, u) \frac{1}{2} [1 - k(u)]\, r_1(u)\, du \\ &+ \int_{\beta(T_i)}^{T_f} g_n(t, \sigma - D(\sigma)) \frac{1}{2A} [1 + k(\sigma - D(\sigma))] \\ &\cdot r_2(\sigma) [1 - \dot{D}(\sigma)]\, d\sigma \\ &+ \int_{f(T_f)}^{T_f} g_n(t, u)\, r_1(u)\, du. \end{aligned} \tag{3.48}$$

By comparing (3.48) to (3.45), one obtains, for $T_i \le t, u \le T_f$

$$h_{11}(t, u; \mathbf{D}, A) = \begin{cases} g_n(t, u) \frac{1}{2} [1 - k(u)]; & T_i \le u \le f(T_f) \\ g_n(t, u); & f(T_f) < u \le T_f \end{cases} \tag{3.49}$$

$$h_{12}(t, u; \mathbf{D}, A) = \begin{cases} g_n(t, u - D(u)) \frac{1}{A} [1 - \dot{D}(u)]; & T_i \le u \le \beta(T_i) \\ g_n(t, u - D(u)) \frac{1}{2A} [1 + k(u - D(u))] [1 - \dot{D}(u)]; & \beta(T_i) < u \le T_f. \end{cases} \tag{3.50}$$

The formulas for $h_{21}(t, u; \boldsymbol{D}, A)$ and $h_{22}(t, u; \boldsymbol{D}, A)$ in (3.46) can now be obtained easily by noting from Fig. 4 that

$$y(t) = Ax(t - D(t)) \tag{3.51}$$

which leads to

$$h_{21}(t, u; \boldsymbol{D}, A) = Ah_{11}(t - D(t), u; \boldsymbol{D}, A) \tag{3.52}$$

$$h_{22}(t, u; \boldsymbol{D}, A) = Ah_{12}(t - D(t), u; \boldsymbol{D}, A), \tag{3.53}$$

where $T_i \leq t, u \leq T_f$.

D. Explicit Form for Bias $l_B(\boldsymbol{D}, A)$

One can obtain the explicit form for the entries of the causal matrix impulse response $\boldsymbol{H}_c(t, u; \boldsymbol{D}, A)$ by noting that if $T_f = t$, then no data $\boldsymbol{r}(v)$, $T_i \leq v \leq T_f$, are future data. Thus, for $T_f = t$ and for $\boldsymbol{d} = \boldsymbol{D}$ and $\alpha = A$, $\boldsymbol{H}_n(t, v; \boldsymbol{D}, A)$ becomes the impulse response $\boldsymbol{H}_c(t, v; \boldsymbol{D}, A)$ of the causal LMMSE estimate $\hat{\boldsymbol{s}}_c(t; \boldsymbol{D}, A)$ of $\boldsymbol{s}(t; \boldsymbol{d}, \alpha)$ from $\boldsymbol{r}(v)$, $T_i \leq v \leq t$, and $g_n(t, \sigma)$ of (3.42)–(3.44) becomes the causal LMMSE estimator $g_c(t, \sigma)$ of $s(t)$ from $z(\sigma)$, $f(T_i) \leq \sigma \leq t$. One obtains the components of $\boldsymbol{H}_c(t, v; \boldsymbol{D}, A)$ by replacing $g_n(t, u)$, in (3.49), (3.50), (3.52), and (3.53), with $g_c(t, u)$ where $g_c(t, u)$ is the solution to (3.44) for $T_f = t$, with $g_c(t, u) = 0$ for $t < u$. With $\boldsymbol{H}_c(t, v; \boldsymbol{D}, A)$ so determined, $l_B(\boldsymbol{D}, A)$ can be obtained directly from (2.36). A little analytical simplification then yields

$$l_B(\boldsymbol{D}, A) = -\tfrac{1}{2} \int_{f(T_i)}^{T_f} g_c(\sigma, \sigma)\, d\sigma. \tag{3.54}$$

The minimum mean square error associated with $g_c(t, u)$ is

$$\begin{aligned} \xi_{oc}(t) &= E\left\{(s(t) - \hat{s}_c(t))^2\right\} \\ &= R_s(t, t) - \int_{f(T_i)}^{t} g_c(t, u)\, R_s(t, u)\, du \\ &= Q(t)\, g_c(t, t), \end{aligned} \tag{3.55}$$

where the last step follows from (3.44) with $T_f = t$. An alternative expression for $l_B(\boldsymbol{D}, A)$ can be obtained by substituting (3.55) into (3.54). This observation is important because if $s(t)$ is a state representable process, then $\xi_{oc}(t)$ can be obtained from the matrix Riccati equation [22, ch. 4.3].

IV. Canonical Realizations

Our formulation of the delay estimation problems leads naturally to four "canonical" realizations which are based upon well-known receiver structures for the detection of Gaussian signals in white Gaussian noise [19, sec. 2.1]. In this paper, we simply point out the potential application of these structures in delay estimation. A more detailed development and comparison of these structures, with the view to obtaining practical estimation algorithms, appears to be a fertile area for future research.

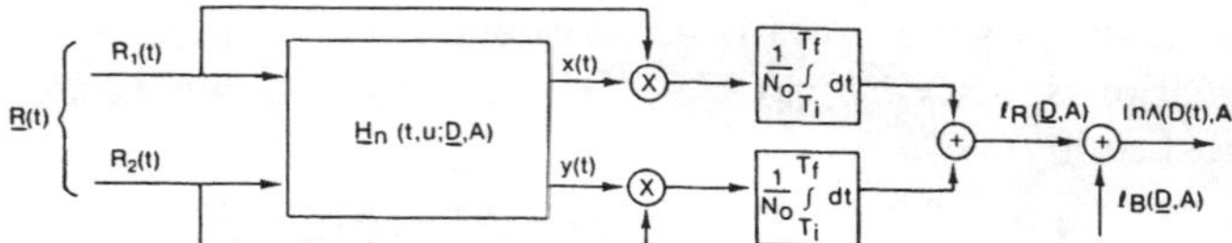

Fig. 5. Canonical realization number 1. Vector estimator-correlator receiver when $\boldsymbol{d} = \boldsymbol{D}$ and $\alpha = A$, $x(t)$ and $y(t)$ are the noncausal LMMSE estimates of $s(t)$ and $\alpha s(t - d(t))$, respectively.

The substitution of (2.23) into (2.19) [with $\boldsymbol{r}(\cdot)$ replaced by $\boldsymbol{R}(\cdot)$] yields

$$l_R(\boldsymbol{D}, A) = \frac{1}{N_o} \int_{T_i}^{T_f} \boldsymbol{R}^T(t)\, \hat{\boldsymbol{s}}_n(t; \boldsymbol{D}, A)\, dt. \tag{4.1}$$

The resulting ML estimator of $d(t)$ and α is shown in Fig. 5 where, following the terminology of Van Trees, it is referred to as canonical realization number 1. Observe that this realization is a vector estimator–correlator analogous to the scalar estimator–correlator in [19, Fig. 2.2]. Here is how it works. The system tentatively hypothesizes that the unknown delay $d(t)$ is $D(t)$ and that the unknown attenuation α is A, where $D(t)$ is a possible delay function and A is a possible relative attenuation constant. The received vector-valued waveform $\boldsymbol{R}(t)$ is input to the noncausal conditional LMMSE estimator of $\boldsymbol{s}(t; \boldsymbol{d}, \alpha)$, which is designed with the assumption that $\boldsymbol{D}$ and A represent the true values of delay vector $\boldsymbol{d}$ and attenuation scalar α. The output of the estimator is $\hat{\boldsymbol{s}}_n(t; \boldsymbol{D}, A)$ of (2.23). A possible realization of the estimator was shown in Fig. 4. The vector correlator then yields $l_R(\boldsymbol{D}, A)$ of (4.1) which, when added to $l_B(\boldsymbol{D}, A)$ of (2.36), (2.38), or (3.54), yields the value of the log-likelihood function $\ln \Lambda(D(t), \hat{A})$ for the assumed $D(t)$ and A. This process is repeated for all choices of $D(t)$ and A that are possible for the application in question. The particular $D(t)$ and A jointly maximizing $\ln \Lambda(D(t), A)$ are the ML estimates $\hat{D}(t)]_{ML}$ and $A]_{ML}$ of $d(t)$ and α.

The problem of finding the function $D(t)$ and the constant A that jointly maximize $\ln \Lambda(D(t), A)$ can be viewed as a generalization of the problem of finding the constant D_o that maximizes the output of the generalized cross correlator shown in Fig. 1 of [1]. This is an interesting problem in algorithm design in which prior constraints on $d(t)$ (such as finite bandwidth, parabolic form, etc.) must be exploited in order to achieve a practical result.

An alternative form for canonical realization number 1 can be obtained by noting that the lower integrator output of Fig. 5 can be written

$$\begin{aligned} \frac{1}{N_0} \int_{T_i}^{T_f} y(t)\, r_2(t)\, dt &= \frac{1}{N_0} \int_{T_i}^{T_f} Ax(t - D(t))\, r_2(t)\, dt \\ &= \frac{1}{N_0} \int_{f(T_i)}^{f(T_f)} Ax(\sigma)\, r_2(\beta(\sigma))\, [1 - \dot{D}(\beta(\sigma))]\, d\sigma. \end{aligned} \tag{4.2}$$

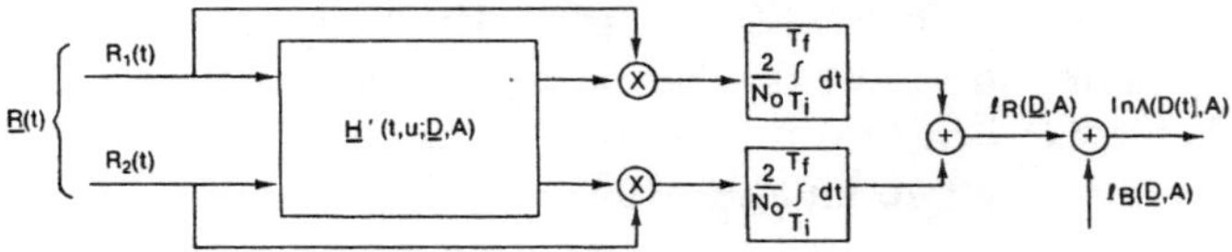

Fig. 6. Canonical realization number 2. Vector filter-correlator receiver.

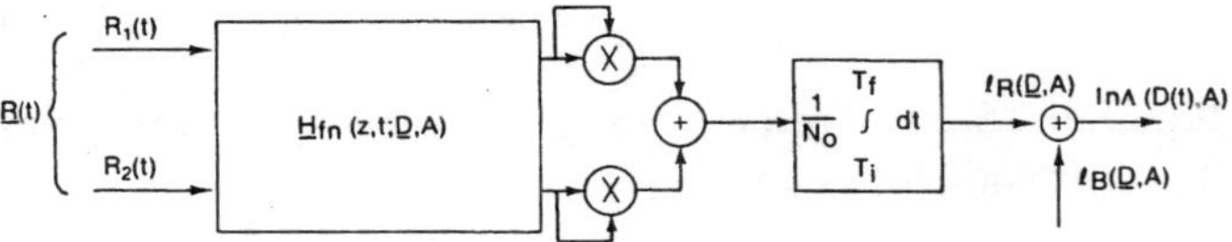

Fig. 7. Canonical realization number 3. Vector filter-squarer receiver.

The process $A^{-1} r_2(\beta(u))$ is available at the output of the inverse operator $\mathcal{L}_{D,A}^{-1}\{r_2(u)\}$ in $H_n(t, u; D, A)$ shown in Fig. 2, and $x(t)$ is the output of $g_n(t, u; D, A)$ shown in Fig. 4. Equation (4.2) therefore provides a means for eliminating the operator $\mathcal{L}_{D(t),A}\{\cdot\}$ from the system in Fig. 4.

Alternative estimator structures can be obtained by straightforward generalizations of the material in [19, pp. 15–23]. Canonical realization number 2, shown in Fig. 6, is a vector filter-correlator receiver. The matrix impulse response $H'(t, u; D, A)$ is defined by

$$H'(t, u; D; A) = \begin{cases} H_n(t, u; D, A); & t \geq u \\ 0; & t < u. \end{cases} \quad (4.3)$$

We point out that the output of the realizable filter $H'(t, u; D, A)$ is not the causal MMSE estimate of $s(t; d, \alpha)$ given $d = D$ and $\alpha = A$.

Canonical realization number 3, shown in Fig. 7, is a vector version of a noncausal filter-squarer receiver [19, Fig. 2.5]. The matrix impulse response $H_{fn}(t, u; D, A)$ is a noncausal solution to

$$H_n(t, u; D, A) = \int_{T_i}^{T_f} H_f^T(z, t; D, A)\, H_f(z, u; D, A)\, dz, \quad T_i \leq t, u \leq T_f. \quad (4.4)$$

As with the scalar case, there are an infinite number of noncausal solutions. Also as with the scalar case, it may be possible to obtain a causal solution $H_{fc}(t, u; D, A)$ to (4.4).

A straightforward but lengthy generalization of the material in [19, pp. 19–24] leads to the expression

$$l_R(D, A) = \frac{1}{N_o} \int_{T_i}^{T_f} \left\{ 2R^T(t)\, \hat{s}_c(t; D, A) - \left| \hat{s}_c(t; D, A) \right|^2 \right\} dt \quad (4.5)$$

where $\hat{s}_c(t; D, A)$ is the LMMSE causal estimate of $s(t; d, \alpha)$ from $R(t)$ given $d = D$ and $\alpha = A$. Equation (4.5) can be realized by the system shown in Fig. 8 which is referred to as canonical realization number 4. The system $H_c(t, u; D, A)$ in Fig. 8 is the matrix impulse of the casual LMMSE estimator encountered previously. Its structure can be obtained from that of $H_n(t, u; D, A)$ by setting $T_f = t$. If $s(t)$ is state representable, then $g_c(t, u)$ can be realized using the Kalman filter [22, ch. 4.].

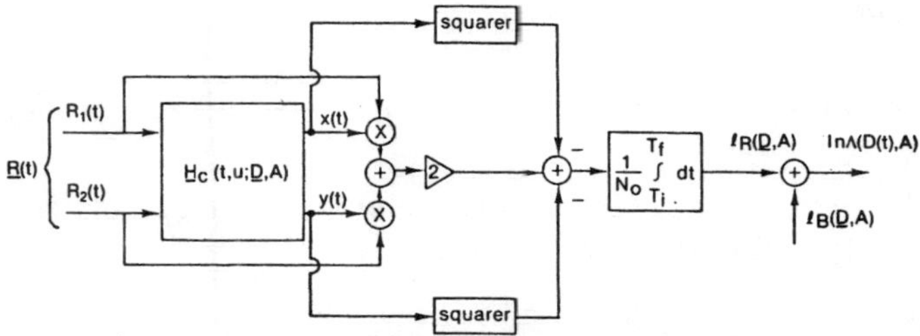

Fig. 8. Canonical realization number 4. Causal LMMSE estimator realization. When $d = D$ and $\alpha = A$, $x(t)$ and $y(t)$ are the causal LMMSE estimates of $s(t)$ and $\alpha s(t - d(t))$, respectively.

V. The Special Case of Constant Delay, Stationary Processes, and Long Observation Interval (CDSPLOT)

This section links the general solution to the problem of maximum-likelihood (ML) time-delay estimation presented above to the solution of Knapp and Carter [1] in which the delay is constant, $d(t) = d_o$, the signal process is stationary, and the observation interval is long. We call this the CDSPLOT case. This exercise has the benefit of providing additional insight into Knapp and Carters' solution as well as a more explicit description of the bias term $l_B(D_o, A)$.

As was shown in (2.16), the log-likelihood function ln $\Lambda(D, A)$ consists of the sum of a data dependent term $l_R(D, A)$ and a bias term $l_B(D, A)$. The forms of these terms under the CDSPLOT approximation are derived in the next two subsections.

A. Data Dependent Term $l_R(D, A)$ Under CDSPLOT Approximation

The data dependent term $l_R(D, A)$ is given in the general case by (2.19), where the entries of $H_n(t, v; D, A)$ are given by (3.49), (3.50), (3.52), and (3.53). It can be seen that the entries themselves are specified in terms of $g_n(t, u)$ and $k(u)$ of (3.44) and (3.31), respectively. Under the CDSPLOT approximation, we obtain $g_n(t, u)$ (approximately) by replacing T_i and T_f in (3.44) by $-\infty$ and $+\infty$, respectively, $R_s(t, u)$ by $R_s(t - u)$, and $Q(u)$ of (3.41) by

$$Q(u) = \frac{N_o'}{2} \qquad -\infty < u < \infty \quad (5.1a)$$

where

$$N_o' \triangleq \frac{N_o}{A^2 + 1}. \quad (5.1b)$$

With (3.44) so modified, we try a solution of the form $g_n(t, \sigma) = g_n(t - \sigma)$. This, and a change in variables, yields

$$R_s(\tau) = \int_{-\infty}^{+\infty} g_n(\tau - \lambda)\, R_s(\lambda)\, d\lambda + \frac{N_o'}{2} g_n(\tau); \quad (5.2)$$

for $-\infty < \tau < \infty$. The solution to (5.2) can be obtained easily by Fourier transforms and is

$$g_n(t) = \int_{-\infty}^{+\infty} G(f)\, e^{j2\pi ft}\, df \quad (5.3)$$

where

$$G(f) = \frac{S_s(f)}{S_s(f) + N_o'/2} \quad (5.4)$$

and where $S_s(f)$ is the power spectral density of $s(t)$.

It can be seen from (5.2) that $g_n(t)$ does not depend on D_o. Also, with $D(t) = D_o$, and with $T_i = -\infty$, $T_f = +\infty$, $k(u)$ of (3.31) reduces to

$$k(u) = \frac{A^2 - 1}{A^2 + 1} = k. \quad (5.5)$$

The substitution of the above with $[T_i, T_f] = [-\infty, \infty]$ into (3.49), (3.50), and (3.52), (3.53) yields the time-invariant impulse responses

$$h_{11}(t) = \frac{1}{A^2 + 1} g_n(t) \quad (5.6)$$

$$h_{12}(t) = \frac{A}{A^2 + 1} g_n(t + D_o) \quad (5.7)$$

$$h_{21}(t) = \frac{A}{A^2 + 1} g_n(t - D_o) \quad (5.8)$$

$$h_{22}(t) = \frac{A^2}{A^2 + 1} g_n(t), \quad (5.9)$$

which, when substituted into (2.19), yields

$$\begin{aligned} l_R(D_o, A) = {} & \frac{1}{N_o} \int_{-\infty}^{+\infty} \int_{-\infty}^{+\infty} \frac{1}{A^2 + 1} g_n(t - v)\, R_1(t)\, R_1(v)\, dt\, dv \\ & + \frac{1}{N_o} \int_{-\infty}^{+\infty} \int_{-\infty}^{+\infty} \frac{A}{A^2 + 1} g_n(t - v + D_o)\, R_1(t)\, R_2(v)\, dt\, dv \\ & + \frac{1}{N_o} \int_{-\infty}^{+\infty} \int_{-\infty}^{+\infty} \frac{A}{A^2 + 1} g_n(t - v - D_o)\, R_2(t)\, R_1(v)\, dt\, dv \\ & + \frac{1}{N_o} \int_{-\infty}^{+\infty} \int_{-\infty}^{+\infty} \frac{A^2}{A^2 + 1} g_n(t - v)\, R_2(t)\, R_2(v)\, dt\, dv. \end{aligned} \quad (5.10)$$

In (5.10) we have written the integration limits as $\pm\infty$ for convenience. Since $r(t)$ is *defined* as zero outside $[T_i, T_f]$, the integration is actually still over the long but finite interval $[T_i, T_f]$. Therefore, the integrals exist.

We are interested in finding the maximum-likelihood estimate of d_o, $\hat{D}_o]_{ML}$, which will be obtained by choosing D_o to maximize $l_R(D_o, A) + l_B(D_o, A)$. We will find in the next section that $l_B(D_o, A)$ does not depend on D_o under the CDSPLOT assumption. Thus, it will be equivalent to maximize $l_R(D_o, A)$ of (5.10). Only the two middle terms in (5.10) depend on D_o. It follows from (5.4) that $g_n(t)$ is an even function. Using this fact, one can combine the two middle terms of (5.10) to obtain

$$\begin{aligned} l_R'(D_o, A) = {} & \frac{2}{N_o} \frac{A}{A^2 + 1} \int_{-\infty}^{+\infty} \int_{-\infty}^{+\infty} g_n(t - \sigma)\, R_1(t) \\ & \cdot R_2(\sigma + D_o)\, dt\, d\sigma. \end{aligned} \quad (5.11)$$

Equation (5.11) can be written another way by introducing functions $h_1(t)$ and $h_2(t)$ satisfying the equation

$$\frac{2}{N_o} \frac{A}{A^2 + 1} g_n(t - \sigma) = \int_{-\infty}^{+\infty} h_1(z - t)\, h_2(z - \sigma)\, dz, \quad (5.12)$$

which, when substituted into (5.11), yields

$$\begin{aligned} l_R'(D_o, A) = {} & \int_{-\infty}^{+\infty} dz \int_{-\infty}^{+\infty} h_1(z - t)\, R_1(t)\, dt \\ & \cdot \int_{-\infty}^{+\infty} h_2(z - \sigma)\, R_2(\sigma + D_o)\, d\sigma. \end{aligned} \quad (5.13)$$

This confirms that $l_R(D_o, A)$ can be obtained from the generalized cross correlator shown in Fig. 1 of [1]. The Fourier transform of (5.12) is

$$\frac{2}{N_o} \frac{A}{A^2 + 1} G(f) = H_1(f)\, H_2^*(f) \equiv \psi(f) \quad (5.14)$$

where $\psi(f)$ is the "frequency weighting function" appearing in equation (6) of [1]. The substitution of (5.4) into (5.14) yields

$$\psi(f) = \frac{2}{N_o} \frac{A}{A^2 + 1} \frac{S_s(f)}{S_s(f) + N_o'/2}. \quad (5.15)$$

The function $\psi(f)$ can be written in terms of the coherence function of $r_1(t)$ and $r_2(t)$, defined as

$$\gamma_{12}(f) \triangleq \frac{S_{r_1 r_2}(f)}{\sqrt{S_{r_1}(f)\, S_{r_2}(f)}} \quad (5.16)$$

where $S_{r_1}(f)$ is the power density spectrum of $r_1(t)$

$$S_{r_1}(f) = S_s(f) + \frac{N_o}{2}, \tag{5.17}$$

$S_{r_2}(f)$ is the power density spectrum of $r_2(t)$

$$S_{r_2}(f) = A^2 S_s(f) + \frac{N_o}{2}, \tag{5.18}$$

and $S_{r_1 r_2}(f)$ is the cross-power density spectrum of $r_1(t)$ and $r_2(t)$

$$S_{r_1 r_2}(f) = A S_s(f)\, e^{+j2\pi f D_o}. \tag{5.19}$$

It can be verified that

$$\psi(f) = \frac{|\gamma_{12}(f)|^2 \text{ SGN } (A)}{|S_{r_1 r_2}(f)|\left[1 - |\gamma_{12}(f)|^2\right]} \tag{5.20}$$

which, for $A > 0$, is the frequency weighting function associated with the ML or HT (for Hannan/Thomson [2], [3]) processor in Table 1 of [1]. Knapp and Carter do not include the factor SGN (A) because their receiver has a square law device before the peak detector.

B. Bias Term $l_B(D, A)$ Under CDSPLOT Approximation

The bias term for the general case is given by (3.54), where $g_c(t, u)$ is the solution to (3.44) for $T_f = t$, with $g_c(t, u) = 0$ for $t < u$. Under the CDSPLOT approximation, we set $D(t) = D_o$, $R_s(t, u) = R_s(t - u)$, and $T_f = t$ in (3.44) to obtain

$$R_s(t - u) = \int_{T_i - D_o}^{t} g_c(t, \sigma)\, R_s(\sigma - u) \cdot d\sigma + Q(u)\, g_c(t, u); \tag{5.21}$$

for $T_i - D_o < u \le t$. Under the CDSPLOT approximation, the function $Q(u)$ of (3.41) becomes

$$Q(u) = \begin{cases} \dfrac{N_o}{2}; & T_i - D_o \le u \le T_i \\ \dfrac{N_o'}{2}; & T_i < u \le t - D_o \\ \dfrac{N_o}{2}; & t - D_o < u \le t \end{cases} \tag{5.22}$$

where N_o' was defined in (5.1b). Recall that $g_c(t, \sigma)$ is the impulse response of the casual LMMSE estimator of $s(t)$ from $z(t) = s(t) + n(t)$, where $n(t)$ has covariance function $Q(u)\,\delta(t - u)$ [see (3.36)–(3.40)]. Looking at (5.22), we see that under the CDSPLOT approximation $n(u)$ is "piecewise" stationary in the three intervals $[T_i - D_o, T_i]$, $(T_i, t - D_o]$, and $(t - D_o, t]$, but $Q(u)$ changes abruptly at the interval boundaries. If we let $T_i \to -\infty$, then $n(u)$ will be stationary for $-\infty < u \le t - D_o$ and $g_c(t, u)$ will be time invariant in this range. Thus,

$$g_c(t, u) = g_c^{(1)}(t - u) \quad \text{for} \quad -\infty < u \le t - D_o \tag{5.23}$$

where the function $g_c^{(1)}(v)$ is the solution to the Wiener–Hopf equation

$$R_s(\tau) = \int_0^{\infty} g_c^{(1)}(v)\, R_z(\tau - v)\, dv; \qquad 0 < \tau < \infty \tag{5.24}$$

with

$$R_z(\tau) = R_s(\tau) + \frac{N_o'}{2}\,\delta(\tau) \tag{5.25}$$

and

$$g_c^{(1)}(\tau) = 0; \qquad \tau < 0. \tag{5.26}$$

Since the statistics of $n(u)$ change abruptly at $u = t - D_o$, and since $g_c(t, u)$ operates in general over all past data, we cannot expect $g_c(t, u)$ to be time invariant for $t - D_o < u \le t$. Thus, under the CDSPLOT approximation, $g_c(t, u)$ is a time-varying casual impulse response that has the approximate time-invariant form $g_c^{(1)}(t - u)$ specified by (5.24) for $T_i < u \le t - D_o$, but not otherwise. Using these results in (3.54) with $f(t) = t - D_o$, one obtains

$$l_B(D_o, A) \simeq -\frac{1}{2}\int_{T_i - D_o}^{T_i} g_c(\sigma, \sigma)\, d\sigma - \frac{1}{2}\int_{T_i}^{t - D_o} g_c^{(1)}(0)\, d\sigma - \frac{1}{2}\int_{t - D_o}^{t} g_c(\sigma, \sigma)\, d\sigma. \tag{5.27}$$

Since D_o is finite, the values of the first and the third integrals in (5.27) become negligible compared to that of the second integral as $t - T_i$ increases. Consequently,

$$\begin{aligned} l_B(D_o, A) &\simeq -\tfrac{1}{2}[t - D_o - T_i]\, g_c^{(1)}(0) \\ &\simeq -\tfrac{1}{2}[t - T_i]\, g_c^{(1)}(0). \end{aligned} \tag{5.28}$$

We can obtain a more explicit form for $l_B(D_o, A)$ by referring to (3.55) which becomes, under the CDSPLOT approximation (for $T_i = -\infty < u < t - D_o$),

$$\begin{aligned} \xi_{oc}(u) &= E\left\{(s(u) - \hat{s}_c(u))^2\right\} \\ &= \frac{N_o'}{2}\, g_c^{(1)}(0) \\ &\equiv \xi_{oc}. \end{aligned} \tag{5.29}$$

If the signal spectrum $S_s(f)$ is rational with finite variance, then [17, p. 501]

$$\xi_{oc} = \frac{N_o'}{2}\int_{-\infty}^{+\infty} \ln\left[1 + \frac{2}{N_o'} S_s(f)\right] df. \tag{5.30}$$

Combining (5.28)–(5.30), we have

$$l_B(D_o, A) \simeq -\frac{1}{2}[t - T_i]\int_{-\infty}^{+\infty} \ln\left[1 + \frac{2}{N_o'} S_s(f)\right] df \tag{5.31}$$

which, as in [1], does not depend upon D_o.

Acknowledgment

It is a pleasure to acknowledge C. Knapp, C. Carter, and four anonymous reviewers for their encouragement and helpful suggestions.

References

[1] C. H. Knapp and G. C. Carter, "The generalized correlation method for estimation of time delay," *IEEE Trans. Acoust., Speech, Signal Processing*, vol. ASSP-24, pp. 320–327, Aug. 1976.

[2] E. J. Hannan and P. J. Thomson, "The estimation of coherence and group delay," *Biometrika*, vol. 58, no. 3, pp. 469–481, 1971.

[3] —, "Estimating group delay," *Biometrika*, vol. 60, no. 2, pp. 241–253, 1973.

[4] C. H. Knapp and G. C. Carter, "Estimation of time delay in the presence of source and receiver motion," *J. Acoust. Soc. Amer.*, vol. 61, pp. 1545–1549, June 1977.

[5] M. Wax, "The joint estimation of differential delay, Doppler, and phase," *IEEE Trans. Inform. Theory*, vol. IT-28, pp. 817–820, Sept. 1982.

[6] A. A. Gerlach, "Motion induced coherence degradation in passive systems," *IEEE Trans. Acoust., Speech, Signal Processing*, vol. ASSP-26, pp. 1–15, Feb. 1978.

[7] W. B. Adams, J. P. Kuhn, and W. P. Whyland, "Correlator compensation requirements for passive time-delay estimation with moving source or receivers," *IEEE Trans. Acoust., Speech, Signal Processing*, vol. ASSP-28, pp. 158–168, Apr. 1980.

[8] J. W. Betz, "Comparison of the deskewed short-time correlator and the maximum likelihood correlator," *IEEE Trans. Acoust., Speech, Signal Processing*, vol. ASSP-32, pp. 285–294, Apr. 1984.

[9] —, "Effects of uncompensated relative time companding on a broadband cross correlator," *IEEE Trans. Acoust., Speech, Signal Processing*, vol. ASSP-33, pp. 505–510, June 1985.

[10] J. O. Smith and B. Friedlander, "Adaptive interpolated time-delay estimation," *IEEE Trans. Aerosp. Electron. Syst.*, vol. AES-21, pp. 180–199, Mar. 1985.

[11] Y. T. Chan, J. M. F. Riley, and J. B. Plant, "Modeling of time delay and its application to estimation of nonstationary delays," *IEEE Trans. Acoust., Speech, Signal Processing*, vol. ASSP-29, pp. 577–581, June 1981.

[12] P. L. Feintuch, N. J. Bershad, and F. A. Reed, "Time delay estimation using the LMS adaptive filter-dynamic behavior," *IEEE Trans. Acoust., Speech, Signal Processing*, vol. ASSP-29, pp. 561–571, June 1981.

[13] —, "Time delay estimation using the LMS adaptive filter-dynamic behavior," *IEEE Trans. Acoust., Speech, Signal Processing*, vol. ASSP-29, pp. 571–576, June 1981.

[14] D. M. Etter and S. D. Stearns, "Adaptive estimation of time delay in sampled data systems," *IEEE Trans. Acoust., Speech, Signal Processing*, vol. ASSP-29, pp. 582–587, June 1981.

[15] D. H. Youn, N. Ahmed, and G. C. Carter, "On using the LMS algorithm for time delay estimation," *IEEE Trans. Acoust., Speech, Signal Processing*, vol. ASSP-30, pp. 798–801, Oct. 1982.

[16] A. N. Netravali and J. D. Robbins, "Motion—Compensated television coding: Part 1," *Bell Syst. Tech. J.*, vol. 58, pp. 631–670, Mar. 1979.

[17] H. L. Van Trees, *Detection, Estimation and Modulation Theory, Part I.* New York: Wiley, 1968.

[18] B. V. Hamon and E. J. Hannan, "Spectral estimation of time delay for dispersive and non-dispersive systems," *J. Roy. Stat. Soc.*, ser. C, vol. 23, no. 2, pp. 134–142, 1974.

[19] H. L. Van Trees, *Detection, Estimation and Modulation Theory, Part III.* New York: Wiley, 1971.

[20] E. J. Kelly and W. L. Root, "A representation of vector-valued random processes," M.I.T., Lincoln Lab., Cambridge, MA, revised Group Rep. 55-21, Apr. 22, 1960.

[21] A. Papoulis, *Probability, Random Variables and Stochastic Processes.* New York: McGraw-Hill, 1965.

[22] A. Gelb, *Applied Optimal Estimation.* Cambridge, MA: MIT Press, 1980.

Adaptive phase transform processors for time delay estimation[a)]

Dae Hee Youn
Department of Electronic Engineering, Yonsei University, Seoul, Korea

Shen-Neng Chiou
Department of Electrical Engineering, University of Southern California, Los Angles, California 90007

V. John Mathews
Department of Electrical Engineering, University of Utah, Salt Lake City, Utah 84112

(Received 10 January 1984; accepted for publication 10 February 1986)

This paper introduces two recursive realizations of the phase transform (PHAT) processor for time-delay estimation (TDE), using a simple one-pole low-pass filter and the least-mean-square (LMS) adaptive filter, respectively. It is shown that these adaptive methods are capable of tracking time-varying delay functions which correspond to moving sources or receivers, and are very effective in reducing the effect of interfering tonals which must be generated by the target as jamming signals to mask its movement. The performances of these methods are compared with those of other existing adaptive TDE algorithms via computer simulations.

PACS numbers: 43.60.Gk, 43.30.Vh

INTRODUCTION

The problem of estimating the time difference of arrival of the same signal at two spatially separated sensors arises in a variety of applications of sonar, radar, acoustics, geophysics, and biomedical engineering where we need to locate the signal source.[1–5]

Of interest in this paper are passive systems, in which, unlike the active systems, the source signal strength cannot be controlled. However, their covertness can be advantageous, since passive systems do not rely on self-generated energy that is reflected off the source or target. An important example of such systems is a passive sonar system which receives the signals generated by a source, possibly corrupted by noise, at an array of spatially separated sensors. It is well known[1] that the location of the source can be determined if the time delays between the arrival times of the signal at three sensors are available.

We consider the two-sensor time delay estimation (TDE) problem, where the signals received at the two sensors are given by

$$x_1(k) = s(k) + w_1(k) + p(k) \quad (1a)$$

and

$$x_2(k) = s(k-D) + w_2(k) + p(k-\tilde{D}), \quad (1b)$$

where k is the discrete time index, $s(k)$ is the source signal, $w_1(k)$ and $w_2(k)$ are the additive noises at sensors 1 and 2, $p(k)$ denotes interfering tonals which might be generated by a target as a jamming signal to mask its movement, and D and $\tilde{D}$ are delay parameters associated with the signal and interfering tonals, respectively. Also, it is assumed that the source signal $s(k)$ and additive noises $w_1(k)$ and $w_2(k)$ are mutually uncorrelated random processes with zero mean.

Most approaches for TDE have been shown to be related through generalized cross correlation (GCC) methods which involve prefiltering the received signals and estimating the time delay as the time lag where the cross correlation function of the prefiltered signals

$$R_{12}^{(g)}(m) = F^{-1}\{W^{(g)}(f)e^{j\theta_{12}(f)}\}, \quad |m|<M \quad (2)$$

is maximum.[6] In (2), $F^{-1}\{\cdot\}$ denotes the inverse Fourier transform of $\{\cdot\}$, $W^{(g)}(f)$ is a weighting function in the frequency domain that is determined by the prefilters, and $\theta_{12}(f)$ is the phase function of the cross-power density spectrum (cross-PDS) of $x_1(k)$ and $x_2(k)$. That is,

$$e^{j\theta_{12}(f)} = [G_{12}(f)]/|G_{12}(f)|, \quad (3)$$

where $G_{12}(f)$ is the cross-PDS of $x_1(k)$ and $x_2(k)$. If there are no interfering tonals in the received signals [i.e., $p(k)=0$ in (1)], the phase function in (3) is given by $\theta_{12}(f) = 2\pi f D$, which means that the phase function is directly proportional to the delay parameter D. The frequency domain weighting functions of the GCC methods of interest in this paper are summarized below:

$$W^{(B)}(f) = |G_{12}(f)|; \quad (4a)$$

BCC (basic cross correlation) method,[2]

$$W^{(R)}(f) = |G_{12}(f)|G_{22}(f); \quad (4b)$$

Roth processor,[7]

$$W^{(P)}(f) = |G_{12}(f)|/|G_{12}(f)| = 1; \quad (4c)$$

PHAT (phase transform).[2]

Recently, the BCC method and the Roth processor have been realized using a simple one-pole low-pass filter[8–10] and the LMS adaptive filter,[10–14] respectively. The main advantages of these recursive time-domain implementations are that they track time-varying delay functions and also avoid the difficulties encountered in spectral estimation with finite record lengths.

[a)] Part of this paper was presented at the International Conference on Acoustics, Speech and Signal Processing, San Diego, CA, March 1984.

Reprinted with permission from the *J. of the Acoustical Soc. of America*, vol. 80, no. 1, pp. 188–194, July 1986.

The phase transform processor was proposed as an *ad hoc* method to reduce the effect of strong tonals by uniformly weighting the phase function $e^{j\theta_{12}(f)}$ in the entire frequency band.[2] The purpose of this paper is to introduce two recursive methods which realize the PHAT processor. In these adaptive techniques the relevant GCC functions are updated using a simple one-pole low-pass filter[8–10] and the LMS adaptive filter,[10–16] respectively.

In Sec. I, adaptive realizations of the BCC and the Roth processors are briefly summarized, while Sec. II is devoted to the PHAT processor and its adaptive implementations. Experimental results and conclusions are presented in Secs. III and IV, respectively.

I. SOME THEORETICAL BACKGROUND

From (2) and (4a), the GCC function of the BCC method is given by the cross correlation function of the received signals without prefiltering. That is,

$$R_{12}^{(B)}(m) = F^{-1}\{G_{12}(f)\} = C_{12}(m), \quad |m| \leqslant M, \quad (5a)$$

where

$$C_{12}(m) = E\{x_1(k)x_2(k+m)\}, \quad (5b)$$

and $E\{\cdot\}$ denotes the statistical expectation of $\{\cdot\}$.

It has been shown[8–10] that the cross correlation function of $x_1(k)$ and $x_2(k)$ can be estimated using a bank of simple one-pole low-pass filters as

$$\hat{C}_{12}(m,k) = \beta \hat{C}_{12}(m,k-1) + (1-\beta)x_1(k)x_2(k+m), \quad |m| \leqslant M, \quad (6a)$$

where $\hat{C}_{12}(m,k)$ denotes an estimate of $C_{12}(m,k)$ in (5b) at time k and $0<\beta<1$ controls the time constant of the low-pass filter whose transfer function is given by

$$A(z) = (1-\beta)/(1-\beta z^{-1}) \quad (6b)$$

when $x_1(k)x_2(k+m)$ is applied as its input. The time constant of the above low-pass filter can be approximated as[9,10]

$$\tau_A \cong 1/(1-\beta) \text{ samples.} \quad (7)$$

From (5a)–(6a), we can see that taking the Fourier transform (FT) of $\hat{C}_{12}(m,k)$ with respect to m yields an estimate of the cross-PDS of $x_1(k)$ and $x_2(k)$ at time k. That is,

$$\hat{G}_{12}(f,k) = F\{\hat{C}_{12}(m,k)\}, \quad (8)$$

where $F\{\cdot\}$ represents the FT of $\{\cdot\}$ with respect to m.

The cross correlation function estimate $\hat{C}_{12}(m,k)$ in (6a) has been used to estimate the time delay parameters,[8–10] and the approach has been referred to as the ABCTDE (adaptive basic cross correlation for TDE) algorithm.[10] From (5a), (6a), and (8), we can see that the ABCTDE algorithm realizes the BCC method in a recursive way.

From (2) and (4b), the GCC function of the Roth processor is given by

$$R_{12}^{(R)}(m) = F^{-1}\{[G_{12}(f)]/[G_{22}(f)]\}, \quad |m| \leqslant M. \quad (9)$$

It is known that $R_{12}^{(R)}(m)$ represents the impulse response function $h_{12}(m)$ of the optimum (Weiner) filter which best approximates $x_1(k)$ as a weighted sum of $x_2(k-m)$ for $|m| \leqslant M$.

A class of adaptive filter algorithms has been developed to recursively update the optimum filter coefficients.[15–17] In this paper, we restrict our interest to the LMS adaptive filter[15,16] since it is computationally very simple but still very effective. The LMS adaptive filter algorithm updates the filter coefficients $h_{12}(m,k)$ to minimize the mean-squared error $E\{e^2(k)\}$ in Fig. 1, where $x_1(k)$ and $x_2(k)$ are applied as primary and reference inputs, respectively, and the M-sample delay is introduced to $x_1(k)$ to make the system causal. The LMS algorithm is summarized in the following:

$$\hat{h}_{12}(m,k+1) = \hat{h}_{12}(m,k) + 2\mu e(k)x_2(k-m), \quad |m| \leqslant M, \quad (10a)$$

where

$$e(k) = x_1(k) - \sum_{m=-M}^{M} \hat{h}(m,k)x(k-m). \quad (10b)$$

In (10a), μ controls the convergence rate and stability of the adaptive filter. The time constant of the LMS adaptive filter can be approximated as[15,16]

$$\tau \cong 1/2\mu\sigma_2^2, \quad (11)$$

where σ_2^2 is the variance of $x_2(k)$. From (9) and (10a), we can see that taking the Fourier transform of $\hat{h}_{12}(m,k)$ with respect to m yields

$$\hat{H}(f,k) \triangleq F\{\hat{h}(m,k)\}, \quad |m| \leqslant M \quad (12a)$$

$$= \widehat{\left\{\frac{G_{12}(f,k)}{G_{22}(f,k)}\right\}}, \quad (12b)$$

which is an estimate of $G_{12}(f)/G_{22}(f)$ in (9) at time k.

From (9), (12a), and (12b), we can see that the impulse response function of the LMS adaptive filter is an estimate of the GCC function of the Roth processor. This approach has been referred to as the LMSTDE (LMS for TDE) algorithm.[10,13,14]

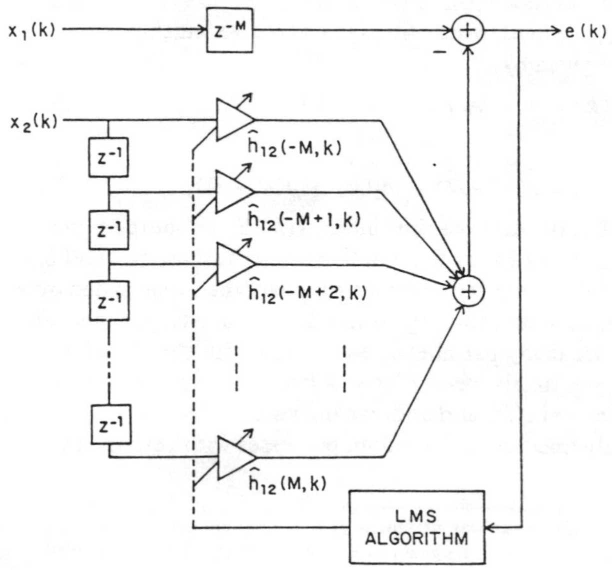

FIG. 1. Block diagram of the LMS adaptive filter algorithm.

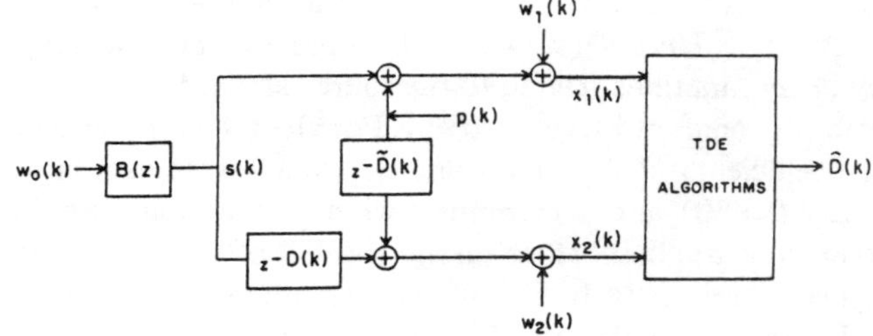

FIG. 2. Block diagram for generating signals for simulations.

II. THE PHASE TRANSFORM PROCESSOR AND ITS ADAPTIVE IMPLEMENTATIONS

The phase transform processor was proposed as an *ad hoc* method to obtain a clear indication of the peak and to remove the effect of interfering tonals of the pertinent GCC function by weighting the phase function in (3) uniformly over the entire frequency band. Thus, from (2) and (4c), the GCC function of the PHAT processor is given by

$$R_{12}^{(P)}(m) = F^{-1}\{[G_{12}(f)]/|G_{12}(f)|\}, \quad |m| \leqslant M$$
$$= F^{-1}\{e^{j\theta_{12}(f)}\}. \tag{13}$$

Introducing a time index k in (13) yields the time-varying GCC function of PHAT as

$$R_{12}^{(P)}(m,k) = F^{-1}\{[G_{12}(f,k)]/|G_{12}(f,k)|\}$$
$$= F^{-1}\{e^{j\theta_{12}(f,k)}\}. \tag{14}$$

Now, using (8) and (12b), the time-varying GCC function of the PHAT can be estimated using the one-pole low-pass filter in (6a) and the LMS adaptive filter algorithm in (10a) and (10b) as follows:

$$\hat{R}_{12}^{(P1)}(m,k) = F^{-1}\left\{\frac{\hat{G}_{12}(f,k)}{|\hat{G}_{12}(f,k)|}\right\}, \quad |m| \leqslant M \tag{15a}$$

and

$$\hat{R}^{(P2)}(m,k) = F^{-1}\left\{\frac{\hat{H}_{12}(f,k)}{|\hat{H}_{12}(f,k)|}\right\}$$
$$= F^{-1}\left\{\widehat{\left(\frac{G_{12}(f,k)}{G_{22}(f,k)}\right)}\left[\widehat{\left|\frac{G_{12}(f,k)}{G_{22}(f,k)}\right|}^{-1}\right]\right\},$$
$$|m| \leqslant M. \tag{15b}$$

The above approaches in (15a) and (15b) will be referred to as the APHAT-1 and APHAT-2, respectively, when the time-delay estimate is given by the argument $m = \hat{D}(k)$, where the relevant time-varying GCC functions $\hat{R}_{12}^{(P1)}(m,k)$ and $\hat{R}_{12}^{(P2)}(m,k)$ are maximum.

In many passive sonar signal processing problems, the received signals often include strong tonals $p(k)$ [see (1)]. One of the sources of the periodic components might be the engine or propeller of a target. Another important case of such signals can be encountered when the target transmits narrow-band jamming signals to hide its location and movement. In general, there may be more than one tonal involved. Computing the cross correlation function of $x_1(k)$ and $x_2(k)$ in (1), we have

$$C_{12}(m) = C_{ss}(m - D) + C_{pp}(m - \tilde{D}), \tag{16a}$$

where

$$C_{ss}(m) = E\{s(k)s(k+m)\} \tag{16b}$$

and

$$C_{pp}(m) = E\{p(k)p(k+m)\} \tag{16c}$$

represent the auto correlation functions of $s(k)$ and $p(k)$, respectively. If no periodic components are involved in the received signals, (16a) becomes

$$C_{12}(m) = C_{ss}(m - D) \tag{17}$$

and the time-delay parameter D can be estimated as the argument $m = D$, where $C_{12}(m)$ is maximum. However, in the presence of strong tonals, the cross correlation function $C_{12}(m)$ might yield peaks at several different places to estimate incorrect delay parameters, since the cross correlation functions of periodic signals are also periodic.

The PHAT processor in (13) is rather simple but performs very well in the presence of strong tonals when the source signal is white or broad bandlimited. If we consider the magnitude of the cross-PDS of $x_1(k)$ and $x_2(k)$ in the presence of strong tonals, the spectral components of the periodic signals are given by impulse functions at the relevant frequencies. Thus we see that normalizing the cross-PDS with its magnitude as in (13) or (15) produces an effect of de-emphasizing the strong tonals.

Now, consider the case of $G_{12}(f) = 0$ in some frequency band (i.e., bandlimited source signal). Then the phase function in (3) is undefined in that band and the estimate of the phase is erratic. Thus normalizing the cross-PDS with its magnitude or weighting the phase function uniformly in the entire frequency range introduces errors in estimating the time delay. Therefore, this behavior suggests that the phase

TABLE I. Summary of the parameters used for the simulations.

Case	$B(z)$	$P(k)$[a]	$D(k)$	$\tilde{D}(k)$	β	μ
1	$(1+z^{-1})/2$	0	4	0	0.9998	5×10^{-5}
2	b	0	4	0	0.9998	5×10^{-5}
3	1	0	4	0	0.9999	5×10^{-5}
4	$\frac{z^{-1}}{1-z^{-1}+0.8z^{-2}}$	$3P_1(k)+2P_2(k)$	4	9	0.9999	5.88×10^{-6}
5	1	$2P_1+P_2(k)$	$-8+0.002k$	0	0.99	1.11×10^{-3}
6	1	$3P_1(k)+2P_2(k)$	$-8+0.002k$	$4-0.001k$	0.998	1.18×10^{-4}

[a] $P_1(k) = \sin(0.46\cdot\pi\cdot k\cdot 0.5)$ and $P_2(k) = \sin(0.12\cdot\pi\cdot k\cdot 0.5)$.
[b] $B(z)$ for case 2 is the 6th-order Butterworth low-pass filter with cutoff frequency of 0.2 Hz, and sampling frequency of 2 Hz.

function $e^{j\theta_{12}(f)}$ be additionally weighted to compensate for the presence or absence of signal power as in the case of the Roth,[7] Scot,[8] and ML (maximum likelihood)[2] processors. Even though the APHAT algorithms, like the conventional PHAT, have the above problem, it will be shown that they are very effective when the source signal has broad bandwidth and when the received signals contain strong interfering tonals. This property will be demonstrated in the next section via computer simulations.

III. EXPERIMENTAL RESULTS

The properties of the APHAT algorithms will be discussed by comparing the performances of the APHAT-1 and -2 processors with those of the ABCTDE[9,10] and LMSTDE[10–14] algorithms through computer simulations.

The schematic diagram used to generate the received signals $x_1(k)$ and $x_2(k)$ is depicted in Fig. 2, where a white Gaussian random signal $w_0(k)$ is processed through $B(z)$ to generate the source signal $s(k)$. Also, the source signal $s(k)$ and the periodic signal $p(k)$ were passed through time-varying filters with the transfer functions of $e^{-j2\pi f D(k)}$ and $e^{-j2\pi f \tilde{D}(k)}$ to generate $s[k-D(k)]$ and $p[k-\tilde{D}(k)]$, respectively.[19] Here, $D(k)$ and $\tilde{D}(k)$ represent the time-varying delay functions related to the source signal $s(k)$ and interfering tonals $p(k)$, respectively. For all of the simulations, 61 coefficients of $\hat{C}_{12}(m,k)$ and $\hat{h}_{12}(m,k)$ were estimated (i.e., $M=30$) and a Hamming window function with 61 points was applied before taking the FT of $\hat{C}_{12}(m,k)$ and $\hat{h}_{12}(m,k)$, respectively. For all of the simulations except case 3, the source signals and additive noises were scaled to have unit variances, while the variances of $s(k)$ and $w_i(k)$ are given by 0.1 and 0.9 for case 3(a) (i.e., SNR = 1/9) and 0.0476 and 0.9524 for case 3(b) (i.e., SNR = 1/20). Other parameters for the simulations are summarized in Table I. The estimated GCC functions at $k=8000$ for cases 1–4 are displayed in Figs. 3–6, where the delay parameter of interest is constant [i.e., $D(k)=4$ samples]. Also, the estimated delay functions for cases 5 and 6 are presented in Figs. 7 and 8, respectively, where the delay function of the source signal linearly increases from -8 to 8 in 8000 samples as indicated by a dotted line, and the delay parameter was computed every 20 samples, starting from $k=80$ and ending at $k=8000$.

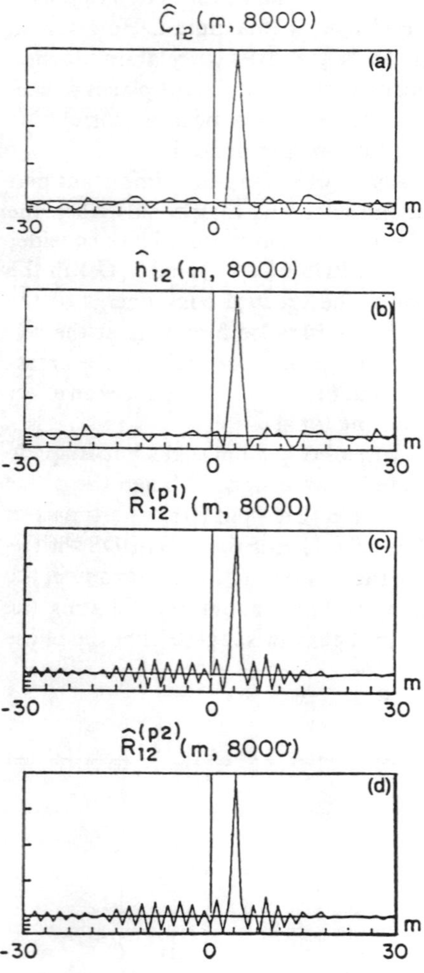

FIG. 3. Estimated GCC functions for broadband low-pass source signal with additive white noise: (a) ABCTDE; (b) LMSTDE; (c) APHAT-1; (d) APHAT-2.

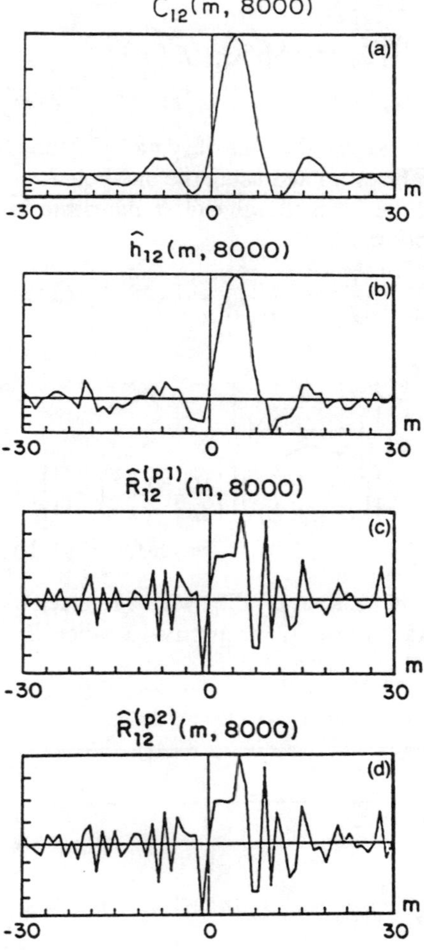

FIG. 4. Estimated GCC functions for narrow-band low-pass source signal with additive white noise: (a) ABCTDE; (b) LMSTDE; (c) APHAT-1; (d) APHAT-2.

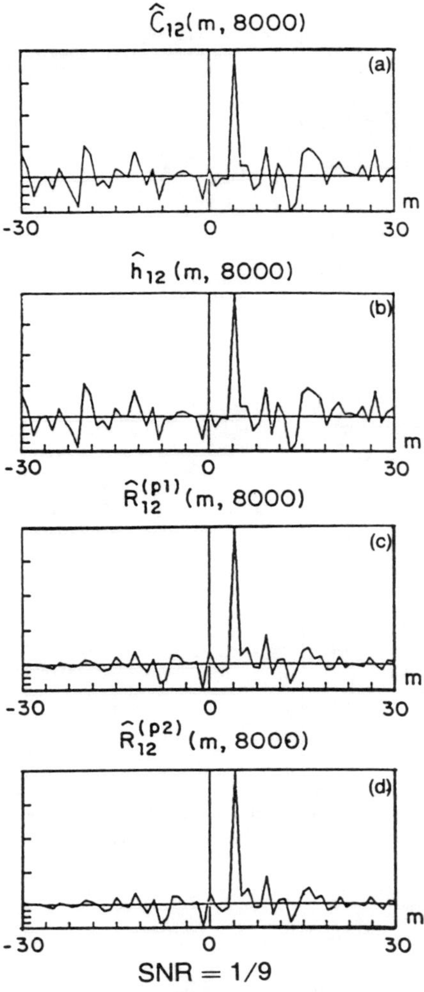

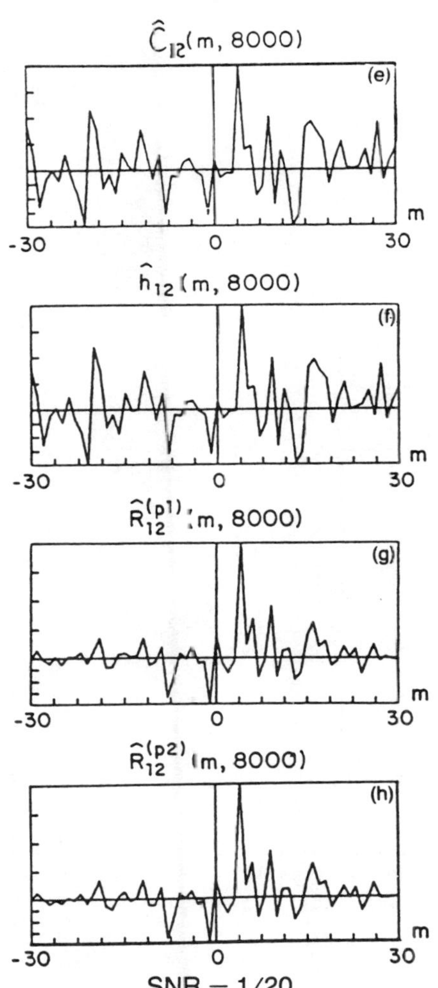

FIG. 5. Estimated GCC functions for white source signal with additive white noise: (a) ABCTDE; (b) LMSTDE; (c) APHAT-1; (d) APHAT-2; (e) ABCTDE; (f) LMSTDE; (g) APHAT-1; (h) APHAT-2.

IV. DISCUSSION

Cases 1 and 2: The estimated GCC functions in Fig. 3 demonstrate that the APHAT-1 and -2 algorithms perform as well as the ABCTDE and LMSTDE algorithms do, when the source signal has broad bandwidth. However, since the source signal for case 2 is narrow bandlimited, the phase information outside the frequency band of the source signal is not related to the time delay, but is given by a random-phase function. Therefore, uniformly weighting the phase function in the entire frequency range as in APHAT-1 and -2 results in emphasizing the frequency band where only spectral estimation errors exist, to yield noisy GCC function estimates as shown in Fig. 4(c) and (d).

The results for cases 1 and 2 suggest that the APHAT-1 and -2 are efficient methods to estimate time delay for the source signals with broad bandwidth, but fail to estimate correct delay parameter for narrow bandlimited source signals.

Case 3: The relevant GCC functions for the four adaptive time-delay estimation algorithms are displayed in Fig. 5 when the source signals are white and for two different SNR's (i.e., 1/9 and 1/20). These results show that the performances of the APHAT-1 and -2 are as good as those of the others. Here, the less noisy GCC function estimates for the APHAT-1 and -2 are due to the Hamming window functions applied before taking the Fourier transform of $\hat{C}_{12}(m,k)$ and $\hat{h}_{12}(m,k)$, respectively.

Case 4: The signals used in this set of simulations were obtained by passing white Gaussian signals through a second-order bandpass filter and then corrupting the output with interfering tonals as well as additive white noises, and the delay parameter of the source signal is given by $D(k)$ = four samples. The result in Fig. 6(a) shows that the GCC function for the ABCTDE algorithm is maximum at $m = 0$, which is the delay parameter relevant to interfering tonals. Similarly, the GCC function in Fig. 6(b) for the LMSTDE algorithm peaks at an incorrect position, even though the effect of the tonals is less than that of the ABCTDE algorithm. However, the GCC function estimates of the APHAT-1 and -2 are maximum at $m = 4$, and yield the correct time-delay estimate. We notice that the APHAT-2 performs better than the APHAT-1. This is because the periodic components have been already de-emphasized and the bandlimited source signal is whitened in the process of LMS adaptive filtering.

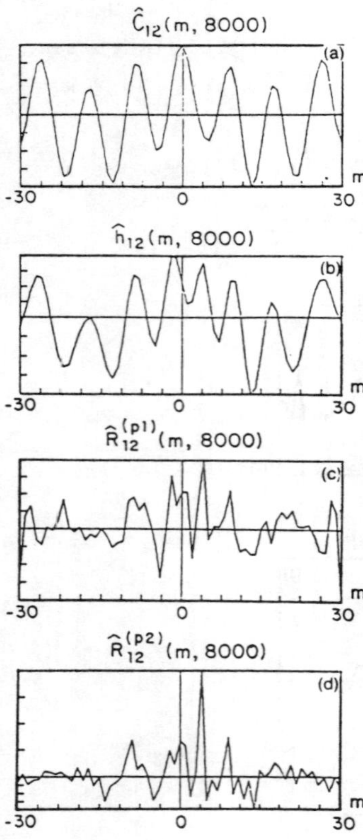

FIG. 6. Estimated GCC functions for bandpass source signal with additive white noise and interfering tonals: (a) ABCTDE; (b) LMSTDE; (c) APHAT-1; (d) APHAT-2.

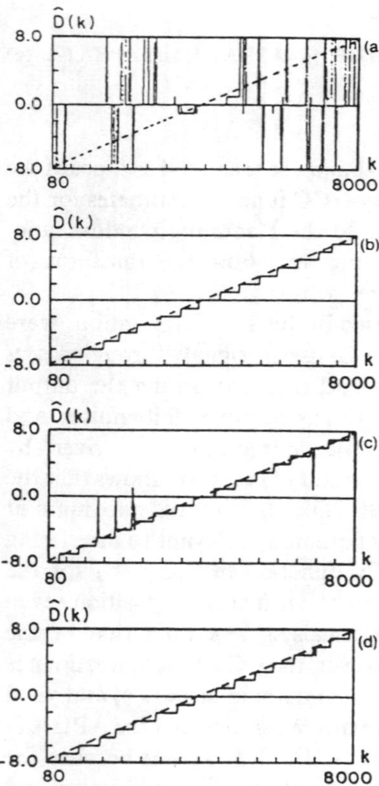

FIG. 7. Estimated time-varying delay functions for white source signal with additive white noise and interfering tonals (case 5): (a) ABCTDE; (b) LMSTDE; (c) APHAT-1; (d) APHAT-2.

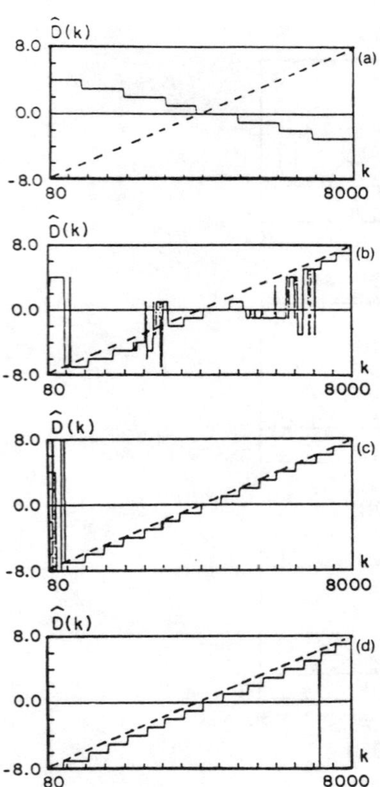

FIG. 8. Estimated time-varying delay functions for white source signal with additive white noise and interfering tonals (case 6): (a) ABCTDE; (b) LMSTDE; (c) APHAT-1; (d) APHAT-2.

From the above results, we see that the APHAT-1 and -2 are effective even when the source signal is narrow bandlimited but still has some power in a wide range of frequency bands.

Cases 5 and 6: The last two sets of simulations concern the problems of estimating time-varying delay functions, which correspond to moving source or receivers.[10–14,21–23] Here, the delay functions relevant to the source signals linearly increase from -8 to $+8$ in 8000 samples, while those of the interfering tonals are constant [i.e., $\tilde{D}(k) = 0$] and linearly decrease from $+4$ to -4 in 8000 samples; $\tilde{D}(k) = 4\text{–}0.001\,k$ for cases 5 and 6, respectively.

From the estimated delay functions in Figs. 7 and 8, we observe that the ABCTDE method estimates the delay function relevant to the interfering tonals [i.e., $\tilde{D}(k)$, see Figs. 7(a) and 8(a)], while the LMSTDE algorithm, APHAT-1, and APHAT-2 track the correct relay parameter relevant to the source signals [i.e., $D(k)$]. Also, the results show that the APHAT-1 and APHAT-2 perform superiorly to the LMSTDE algorithm.

V. CONCLUSIONS

Two adaptive implementations of the phase transform processor, using a bank of simple one-pole low-pass filters

(APHAT-1) and the LMS adaptive filter (APHAT-2), respectively, were introduced. It was demonstrated that these algorithms are more effective than the ABCTDE and LMSTDE algorithms in tracking constant and time-varying delay functions associated with broad bandlimited source signals in the presence of strong tonals. It was also shown that the APHAT algorithms should be used with caution when the source signals are narrow bandlimited.

The APHAT-1 algorithm is attractive because of its computational simplicity. However, as demonstrated in Fig. 6, the APHAT-2 processor may give more accurate time-delay estimates because of the inherent whitening of the source signals in the process of LMS adaptive filtering.

[1]G. C. Carter (Ed.), Special issue on time delay estimation, IEEE Trans. Acoust. Speech Signal Process. **ASSP-29**, 461–624 (1981).

[2]C. H. Knapp and G. C. Carter, "The generalized correlation method for estimation of time delay," IEEE Trans. Acoust. Speech Signal Process. **ASSP-24**, 320–327 (1976).

[3]W. R. Hahn, "Optimum signal processing for passive sonar range and bearing estimation," J. Acoust. Soc. Am. **58**, 201–207 (1975).

[4]V. H. MacDonald and P. M. Schultheiss, "Optimum passive bearing estimation in a spatially incoherent noise environment," J. Acoust. Soc. Am. **46**, 37–43 (1969).

[5]G. C. Carter, "Passive ranging errors due to receiving hydrophone position uncertainty," J. Acoust. Soc. Am. **65**, 528–530 (1979).

[6]K. Scarbrough, N. Ahmed, and G. C. Carter, "On the simulation of a class of time delay estimation algorithms," IEEE Trans. Acoust. Speech Signal Process. **ASSP-29**, 534–540 (1981).

[7]P. R. Roth, "Effective measurements using digital signal analysis," IEEE Spectrum **8**, 62–70 (1971).

[8]N. Ahmed and S. Vijayendra, "An algorithm for line enhancement," Proc. IEEE **70**, 1459–1460 (1982).

[9]D. L. Johnstone and O. L. Frost, "High resolution differential time-of-arrival and differential Doppler estimation," ARGO Systems, Inc., 1069 East Meadow Circle, Palo Alto, CA 94303.

[10]D. H. Youn, "A class of adaptive methods for estimating coherence and time delay functions," Ph.D. thesis, Department of Electrical Engineering, Kansas State University (1982).

[11]P. L. Feintuch, N. J. Bershad, and F. A. Reed, "Time delay estimation using the LMS adaptive filter—dynamic behavior," IEEE Trans. Acoust. Speech Signal Process. **ASSP-29**, 571–576 (1981).

[12]F. A. Reed, P. L. Feintuch, and N. J. Bershad, "Time delay estimation using the LMS adaptive filter—static behavior," IEEE Trans. Acoust. Speech Signal Process. **ASSP-29**, 561–571 (1981).

[13]D. H. Youn, N. Ahmed, and G. C. Carter, "On using the LMS algorithm for time delay estimation," IEEE Trans. Acoust. Speech Signal Process. **ASSP-30**, 798–801 (1982).

[14]D. H. Youn, N. Ahmed, and G. C. Carter, "An adaptive approach for time delay estimation of bandlimited signals," IEEE Trans. Acoust. Speech Signal Process. **ASSP-31**, 780–784 (1983).

[15]B. Widrow, J. R. Glover, Jr., J. M. McCool, J. Kaunitz, C. S. Williams, R. H. Hearn, J. R. Zeidler, E. Dong, Jr., and R. C. Goodlin, "Adaptive noise cancelling: Principles and applications," Proc. IEEE **63**, 1692–1716 (1975).

[16]B. Widrow, J. M. McCool, M. G. Larimore, and C. R. Johnson, Jr., "Stationary and nonstationary learning characteristics of the LMS adaptive filter," Proc. IEEE **64**, 1151–1162 (1976).

[17]S. R. Parker and L. J. Griffiths (Eds.), Joint Special Issue on Adaptive Signal Processing, IEEE Trans. Acoust. Speech Signal Process. **ASSP-29**, 625–775 (1981); IEEE Trans. Circuits Syst. **CAS-29**, 465–615 (1981).

[18]G. C. Carter, A. H. Nuttall, and P. G. Cable "The smoothed coherence transform (SCOT)," Proc. IEEE **61**, 1497–1498 (1973).

[19]D. H. Youn, N. Ahmed, and G. C. Carter, "A method for generating a class of time-delayed signals," in *Proceedings of ICASSP*, Atlanta, GA, 1981, pp. 1257–1260.

[20]P. C. Chestnut, "Emitter location accuracy using TDOA and differential Doppler," IEEE Trans. Aerosp. Electron. Syst. **AES-18**, 214–218 (1982).

[21]C. H. Knapp and G. C. Carter, "Estimation of time delay in the presence of source or receiver motion," J. Acoust. Soc. Am. **61**, 1545–1549 (1977).

[22]D. M. Etter and S. D. Stearns, "Adaptive estimation of time delays in sampled data systems," IEEE Trans. Acoust. Speech Signal Process. **ASSP-29**, 582–587 (1981).

[23]Y. T. Chan, J. M. F. Riley, and J. B. Plant, "Modeling of time delay and its application to estimation of nonstationary delays," IEEE Trans. Acoust. Speech Signal Process. **ASSP-29**, 577–582 (1981).

Part 5
Selected Papers on Detection, Localization, and Other Applications

The Least Squares Estimation of Time Delay and Its Use in Signal Detection

Y. T. CHAN, RICHARD V. HATTIN, AND J. B. PLANT, SENIOR MEMBER, IEEE

Abstract—This paper examines the use of two spatially separated receivers to determine the presence of a distant signal source and its relative bearing. Ideally, the phase shift between the receivers' output is proportional to the frequency with the time delay between outputs equal to the proportionality constant. Because of noise, the plot of phase against frequency is scattered along a straight line whose slope is the time delay. A least squares estimator of the slope turns out to be equivalent to the maximum likelihood estimator developed by Hamon and Hannan [1]. Since the goodness of fit of the least squares line is a function of the coherence between the receivers' output, the sum of the squared errors is used as a test statistic in detection. The proposed detector has a detection threshold that depends only on the probability of false alarm and not on the ambient noise level. It can also be simply extended to an array of receivers.

I. Introduction

THE estimation of time difference (or delay) between the outputs of two receivers at different locations has found applications in many areas. Specifically, let the two receiver outputs be

$$x_1(t) = s(t) + n_1(t)$$
$$x_2(t) = s(t+\tau) + n_2(t) \tag{1}$$

where the signal $s(t)$ is uncorrelated with the corrupting noises $n_1(t)$ and $n_2(t)$ and τ is the time difference between the signals received. When $x_1(t)$ and $x_2(t)$ are outputs of two receivers of known separation, an estimation of τ can give the bearing of the signal source. Other applications [1] deal with the determination of the lag time in mean sea level variations between two locations, drift velocity of ionospheric clouds, and speed of propagation of an impulse along a nerve fiber, to mention a few.

We present here a scheme that first computes an estimate of τ by minimizing a cost function. The resultant minimum cost is then used for detection purposes in a nonparametric detection technique. Hannan and Thomson [2] and Hamon and Hannan [1] have developed a maximum likelihood (ML) estimator for τ under very general conditions. Knapp and Carter [3], in a definitive analysis of several other estimators together with that in [1], showed that they can be generalized to a two prefilter, cross-correlator configuration. The time argument at which the correlator attains a maximum is the delay estimate. The properties of the estimators differ in the choice of the prefilters, which are in general nonlinear.

Manuscript received August 1, 1977; revised December 22, 1977.

The authors are with the Department of Electrical Engineering, Royal Military College of Canada, Kingston, Ont., Canada.

Implicit in this generalized cross-correlator approach is that the time argument obtained is always an integer multiple of the sampling interval. There are instances, for an example see Section III, where good bearing resolution requires time delay estimates whose resolution is less than the sampling interval. One must then interpolate between samples using standard interpolating techniques. However, this will increase the computation load.

It is easy to see, from the Fourier transform of (1), that the phase between $s(t)$ and $s(t+\tau)$ is proportional to frequency with τ as the proportionality constant. By plotting the phase of the cross-spectral density of $x_1(t)$ and $x_2(t)$ against frequency, an estimator of τ is given simply by the slope of the generalized, restricted least squares line [4]. The merits of using this estimator are: it is easily computed, it is the best linear unbiased estimator, and, most importantly, if the samples from $x_1(t)$ and $x_2(t)$ are Gaussian, it is also an ML estimator; further, since the least squares cost is itself a function of the signal-to-noise (S/N) ratio, the measure of how well the straight line fits the data can be used for detection purposes. Intuitively, if the S/N ratio is large, the phase distribution will crowd closely to the least squares line so that the squared error sum will be small. The derivation of the receiver operating characteristics (ROC) of this new detector is given in Section IV.

II. Preliminaries

Assuming that $s(t)$, $n_1(t)$, and $n_2(t)$ in (1) are jointly stationary, zero-mean Gaussian random processes and uncorrelated with each other, the problem is to determine 1), if $s(t)$ is present and 2), the value of τ.

We compute [5] the estimate of the magnitude squared coherence,

$$|\hat{\gamma}(k)|^2 = \frac{\left|\sum_{l=1}^{n} X_{1l}(k)\, X_{2l}^*(k)\right|^2}{\sum_{l=1}^{n} |X_{1l}(k)|^2 \sum_{l=1}^{n} |X_{2l}(k)|^2}, \quad k = 0, 1, \cdots, M+1 \tag{2}$$

and the phase of the cross-spectral density,

$$\theta_k = \tan^{-1} \frac{\operatorname{Im}\left\{\sum_{l=1}^{n} X_{1l}(k)\, X_{2l}^*(k)\right\}}{\operatorname{Re}\left\{\sum_{l=1}^{n} X_{1l}(k)\, X_{2l}^*(k)\right\}} \tag{3}$$

Reprinted from *IEEE Trans. Acoust., Speech, Signal Processing*, vol. 26, no. 3, pp. 217–222, June 1978.

where $X_{1l}(k)$ and $X_{2l}(k)$ denote the FFT of the lth weighted segments of samples from $x_1(t)$ and $x_2(t)$, respectively, at frequency ω_k, and Im$\{\cdot\}$ and Re$\{\cdot\}$ denote the imaginary and real components, respectively, of the bracketed complex quantity.

When there is no noise, it follows from (1) that

$$G_{x_1 x_2}(\omega) = G_{ss}(\omega)\, e^{j\omega\tau} \tag{4}$$

where the theoretical cross and auto power spectra are given by $G_{x_1 x_2}(\omega)$ and $G_{ss}(\omega)$, respectively. With noise, the actual phase angle becomes

$$\theta_k = \tau\omega_k + \epsilon_k \qquad k = 0, 1, \cdots, M+1$$

$$\omega_k = 0, \frac{\pi}{M+1}, \cdots, \pi. \tag{5}$$

The disturbance term ϵ_k, which is a random variable, accounts for the phase deviations caused by $n_1(t)$ and $n_2(t)$ and the effects of finite record length, computational inaccuracies, etc. Without loss of generality, we assume available from the FFT process $M+2$ discrete frequency components in the interval $[0, \pi]$ and the sampling frequency is 2π rad/s. The data at $k = 0$ and $M+1$, however, will not be used.

We now present several properties of the random variable ϵ_k which will be useful in the sequel.

1) If $s(t) = 0$, ϵ_k is uniformly distributed in the interval $-\pi$ to π [7] with variance

$$\sigma_0^2 = \frac{\pi^2}{3}. \tag{6}$$

This property follows from the assumption that $x_1(t)$ and $x_2(t)$ are zero-mean Gaussian processes so that $X_{1l}(k)$ and $X_{2l}(k)$ are also Gaussian random variables [7].

2) When K is large ($\geqslant 100$) or when $|\gamma(k)|^2$ is small ($\leqslant 0.1$), then ϵ_k has the probability density [8], [9]

$$f(\epsilon_k) = \frac{1}{2\pi} \frac{(1-s^2)^{1/2}}{(1 - s\cos 2\epsilon_k)} \cdot \exp\left[-\frac{Ks}{2(1+s)}\right]\left[1 + \sqrt{2\pi}\, r \exp\left(\frac{r^2}{2}\right)\Phi(r)\right]$$

where

$$r = \left[\frac{K s(1-s)}{(1+s)(1 - s\cos 2\epsilon_k)}\right]^{1/2} \cos\epsilon_k, \qquad s = |\gamma(k)|^2$$

$$\Phi(r) = \int_{-\infty}^{r} \frac{1}{\sqrt{2\pi}} \exp\left(\frac{-z^2}{2}\right) dz$$

and

$$K = \frac{2n}{\sum_{l=-n+1}^{n-1} \left(1 - \frac{|l|}{n}\right) \left|\frac{\phi_w(lS)}{\phi_w(0)}\right|^2} \tag{7}$$

with l and n as defined in (2) and (3), S the shift of each successive overlapped window, and $\phi_w(\cdot)$ is the autocorrelation of the window used. It is shown in [8], [9] that for $K \geqslant 300$, ϵ_k is approximately Gaussian and has variance

$$\sigma_k^2 = \frac{1}{\psi_k K} = \frac{\sigma_s^2}{\psi_k}. \tag{8}$$

The quantity ψ_k is a function of coherence and is given by

$$\psi_k = \frac{|\gamma(k)|^2}{1 - |\gamma(k)|^2}$$

and, following [3], we define

$$S/N \text{ ratio} = \frac{|\gamma(k)|}{1 - |\gamma(k)|}.$$

When (8) is not valid, $E\{\epsilon_k^2\}$ can be computed directly from (7).

3) If the Fourier coefficients from (1) are uncorrelated, that is, if $E\{X_{1l}(i)\, X_{2l}(j)\} = 0$, $E\{X_{2l}(i)\, X_{2l}(j)\} = 0$, then $E\{\epsilon_i\epsilon_j\} = 0$ for $i \neq j$. Hodgkiss and Nolte [10] have shown that the Fourier coefficients of a bandlimited Gaussian noise become uncorrelated if $WT_o > 8$, where W is the bandwidth in hertz and T_o is the observation time. This condition is easily met in practice.

III. The Least Squares Estimator

Consider the cost function

$$J = \sum \psi_k(\theta_k - \tau\omega_k)^2 \tag{9}$$

where the summation is assumed taken from $k = 1$ to M, inclusive, unless otherwise stated. The estimate $\hat{\tau}$ of τ that minimizes J is

$$\hat{\tau} = \frac{\sum \psi_k \theta_k \omega_k}{\sum \psi_k \omega_k^2}. \tag{10}$$

This estimator differs from the classical least squares (l.s.) estimator in two respects. First, it uses the well-known generalized l.s. [4] where ψ_k normalizes the disturbances, assuming (8) is valid, so that $E\{\psi_k\epsilon_k^2\} = \sigma_s^2$ for all k. Heuristically, by use of the weights ψ_k, we place more weight on the measurement that has less variance. Less well-known, however, is the so-called restricted l.s. that (10) incorporates. The restricted l.s. [4] makes use of the *a priori* information that the l.s. line should have zero intercept. For if there were no noise, $\theta_k = 0$ at $\omega_k = 0$. It is also shown in [4] that, in view of the properties 1)–3) of ϵ_k, (10) is an ML estimator and has a smaller variance than the classical l.s. estimator. Indeed, using (10) and (5) gives

$$E\{\hat{\tau}\} = \tau \tag{11}$$

and

$$E\{(\hat{\tau} - \tau)^2\} = \frac{\sum \psi_k^2\omega_k^2 E\{\epsilon_k^2\}}{(\sum \psi_k\omega_k^2)^2}. \tag{12a}$$

If (8) is valid, the variance becomes

$$E\{(\hat{\tau} - \tau)^2\} = \frac{\sigma_s^2}{\sum \psi_k\omega_k^2}. \tag{12b}$$

In [1], the asymptotic ($N \to \infty$, $M \to \infty$, $M/N \to 0$) variance of $\hat{\tau}$ is given by

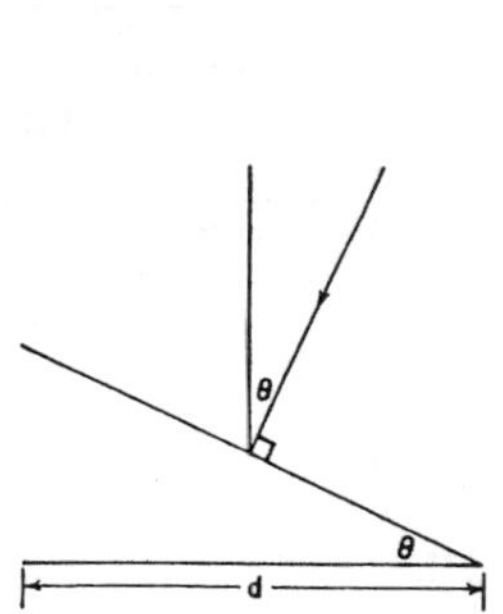

Fig. 1. Bearing estimation from time delay.

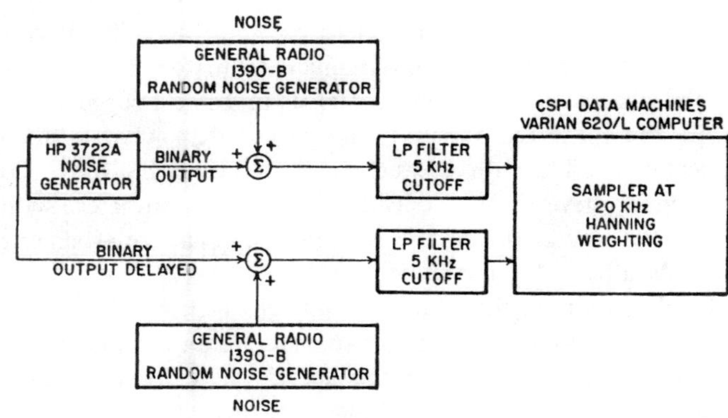

Fig. 2. Experimental setup.

TABLE I
TIME DELAY ESTIMATES

	S/N Ratio	1/4	1/1	4/1	16/1	64/1	∞
	Time Delay	0.2T	0.2T	0.2T	0.2T	0.2T	0.2T
	Var$\{\hat{\tau}\}$ from (12a) K = 13.37	4.10x10^{-3}T^2	1.24x10^{-3}T^2	3.36x10^{-4}T^2	1.50x10^{-4}T^2	1.08x10^{-4}T^2	—
	Var$\{\hat{\tau}\}$ from (12b) K = 300	1.92x10^{-4}T^2	2.40x10^{-5}T^2	4.50x10^{-6}T^2	1.03x10^{-6}T^2	2.52x10^{-7}T^2	—
L.S. Estimator	mean $\hat{\tau}$	0.12T	0.18T	0.21T	0.20T	0.19T	0.19T
	var$\{\hat{\tau}\}$	2.89x10^{-2}T^2	2.27x10^{-2}T^2	3.15x10^{-3}T^2	8.35x10^{-4}T^2	3.17x10^{-4}T^2	1.71x10^{-4}T^2
	Std. Dev.	0.17T	0.15T	0.06T	0.0289T	.0177T	0.0131T
H.H. Estimator	mean $\hat{\tau}$	0.12T	0.21T	0.20T	0.19T	0.19T	0.19T
	var $\{\hat{\tau}\}$	1.49x10^{-1}T^2	2.58x10^{-2}T^2	3.09x10^{-3}T^2	8.44x10^{-4}T^2	3.21x10^{-4}T^2	1.65x10^{-4}T^2
	Std. Dev.	0.39T	0.16T	0.0556T	0.0291T	0.0179T	0.0128T

$$\text{Var}\{\hat{\tau}\} = \frac{M}{N \sum \psi_k \omega_k^2} \tag{13}$$

where N is the number of time samples and $M + 1$ is the number of frequency points obtained by averaging. Thus, $N/M = 2n$, where n is the number of segments in (2) and (3). When there is no overlap, K also equals $2n$ [8]. Thus, (12b) and (13) are equivalent. This shows that the asymptotic variance given by (13) holds for $K \geqslant 300$. Suppose $n = 150$ segments, $M = 127$, and $\psi = 0.33$ (S/N ratio = 1); then Var$\{\hat{\tau}\} = 2.4 \times 10^{-5}\ T^2$ and the standard deviation is $4.9 \times 10^{-3}\ T$, where T is the sampling period. Thus, 95 percent of estimates fall within two standard deviations which is $\approx\pm\ 0.01T$. Hence, reliable estimates of delays less than T are obtainable. Accuracy in time delay estimates affects directly the bearing estimates. For example, if $d = 4.3$ m in Fig. 1 and using the formula $d \sin\theta/C = \tau$ to compute θ, where $C = 1494$ m/s is the speed of sound in water, the correlation technique [3], without interpolation, can only give a bearing estimate of $\pm 1^\circ$ accuracy for $T = 50\ \mu$s. In comparison, when using the l.s. estimator, the limiting resolution is $\pm 0.01^\circ$ under the preceding conditions.

Of course, the variance of $\hat{\tau}$ obtained in practice will be higher than that predicted by (12) because ψ_k is generally not known and has to be estimated from

$$\hat{\psi}_k = \frac{|\hat{\gamma}(k)|^2}{1 - |\hat{\gamma}(k)|^2}. \tag{14}$$

When $\hat{\psi}_k$ is used instead of ψ_k in (10), it only approximates the ML estimator. As an assessment of the estimator (10), when K is small and $\hat{\psi}_k$ is used, we have performed the following measurements. The experimental setup is illustrated in Fig. 2. The computer takes 1024 samples from each channel and divides them into seven segments of 256 points each (hence, a 50 percent overlap). It then applies a Hanning weighting to each segment before taking the FFT and producing 128 Fourier coefficients for frequencies from 0 to π/T. Using this information, K is found [8] to be 13.37. Next, (2) and (3) are evaluated, and finally $\hat{\tau}$ is obtained from (10). The variances, computed from 40 runs for each S/N ratio, together with the values from (12), are listed in Table I. As a comparison, we also calculated $\hat{\tau}$ using the Hamon-Hannan [1] estimator which maximizes $J_{HH} = \Sigma\, \psi_k \cos(\theta_k - \hat{\tau}\omega_k)$. A Newton-Raphson iterative procedure was used. Since J_{HH} is not quadratic, the procedure sometimes converged to a point which is not the global maximum. This happened when $|\theta_k - \hat{\tau}\omega_k|$ became larger than $\pi/2$ at a certain iteration and caused the cosine function to go negative. This is one disadvantage of the Hamon-Hannan estimator, in addition to the necessity for iteration. Values in Table I suggest that while the measured variances are higher than what is theoretically possible with a much larger K (requiring over 100 segments), nevertheless, reliable time delay estimates, with accuracies within 0.1 of the sampling interval, are attainable for S/N ratio of 4/1 and above.

Finally, we note that since (3) only gives angles in the range $-\pi$ to π, the θ_k can sometimes (depending on the magnitude of time delay) differ from the actual phase by an integral multiple of 2π. We must, therefore, unwrap the phase as given by (3), thereby producing a continuous phase, before using (10). Several so-called "phase-unwrapping" algorithms are available for this purpose, including a new one by Tribolet [11]. The same problem is nicely avoided in the Hamon-Hannan [1] estimator. For in J_{HH}, the addition or subtraction of an integral multiple of 2π in the argument does not affect the cosine function. However, there are applications in which a series of receiver pairs are arranged so that each pair covers only a specific small arc. In this situation, there is no need to unwrap the phase provided the largest bearing covered is equivalent to a time delay of less than T. For if θ_k from (3) is directly used in (10), any time delay larger than T will produce an incorrect $\hat{\tau}$, and hence a large cost J, in (9). In these applications our detector, discussed in the next section, because of the large resulting J, will not detect signals outside the bearing arc assigned to that receiver pair.

IV. The Detector

A measure of the goodness of fit of the least squares line is given by the sum of the residuals from

$$R = \sum (\theta_k - \hat{\tau}\omega_k)^2. \tag{15}$$

This sum is also a measure of the strength of signal present since R would be small for a large coherence value. Thus, R can be used as a sample for detection purposes. To develop a detection threshold as well as the receiver operating characteristics (ROC), it is necessary to compute the two probability density functions of R, $f(R/0)$ when $s(t) = 0$ and $f(R/s)$ when $s(t) \neq 0$. Since the ROC are meaningful only for a constant S/N ratio over the frequency band of interest, we have accordingly made the assumption that the coherence and ψ_k are constant so that (10) becomes

$$\hat{\tau} = \frac{\sum \theta_k \omega_k}{\sum \omega_k^2} \tag{16}$$

and $f(\epsilon_k)$ is independent of k. Substituting (16) into (15) and taking the expectation gives

$$E\{R\} = (M-1)\, E\{\epsilon_k^2\}. \tag{17}$$

The derivation of variance of R is more lengthy but straightforward. Through direct manipulation of (15), (16), and (17), we get

$$\mathrm{Var}\{R\} = E\{R^2\} - E^2\{R\}$$

TABLE II
ROC Table

Graph Number	P_{FA} \ S/N	0.32/1	0.29/1	0.25/1	0.21/1	0.16/1	0.11/1
(1)	2.866×10^{-7}	0.997	0.979	0.896	0.655	0.271	0.029
(2)	3.397×10^{-6}	≈1	0.996	0.971	0.841	0.488	0.089
(3)	3.167×10^{-5}	≈1	0.999	0.994	0.940	0.709	0.212
(4)	2.326×10^{-4}	≈1	≈1	0.999	0.987	0.871	0.401
(5)	1.350×10^{-3}	≈1	≈1	≈1	0.997	0.956	0.618

$$= E\{\epsilon_k^4\}\left[M - 2 + \frac{\sum \omega_k^4}{\left(\sum \omega_k^2\right)^2}\right] + E^2\{\epsilon_k^2\}\left[4 - M - \frac{3\sum \omega_k^4}{\left(\sum \omega_k^2\right)^2}\right]. \tag{18}$$

For large $M(>100)$, (18) simplifies, with negligible error, to

$$\mathrm{Var}\{R\} \approx E\{\epsilon_k^4\}\,[M-2] + E^2\{\epsilon_k^2\}\,[4-M]. \tag{19}$$

Now R is a sum of M random variables and since $M > 100$ normally, $f(R/0)$ and $f(R/s)$ are approximately Gaussian by the central-limit theorem [6] with mean and variance given by (17) and (19). When $s(t) = 0$, $E\{\epsilon_k^2\} = \pi^2/3$ and $E\{\epsilon_k^4\} = \pi^4/5$ from property 1) of Section II. When a signal is present, $f(\epsilon_k)$ is given by (7) so that $E\{\epsilon_k^2\}$ and $E\{\epsilon_k^4\}$ can be found by numerical integration. Or, if $K \geqslant 300$, $E\{\epsilon_k^2\}$ is given by (8) and $E\{\epsilon_k^4\} = 3\,\sigma_k^4$ using the Gaussian assumption of ϵ_k. To see the effectiveness of using R as a detection sample, let $M = 127$. Then $E\{R/0\} = 414.32$ and $\mathrm{Var}\{R/0\} = 1103.97$. For $K = 13.37$ and $|\gamma|^2 = 0.25$, which is equivalent to a S/N ratio of one at both receivers, $E\{R/S\} = 65.09$ and $\mathrm{Var}\{R/S\} = 191.30$. Thus, there is considerable separation between the means of the sample with signal and without.

We conceive our nonparametric detector as follows. For a given probability of false alarm P_{FA}, a detection threshold TH is computed from

$$P_{FA} = \int_0^{TH} f(R/0)\, dR. \tag{20}$$

Any $R < TH$ is considered a detection. The probability of detection P_D, which depends on the S/N ratio, is given by

$$P_D = \int_0^{TH} f(R/s)\, dR. \tag{21}$$

Strictly speaking, the lower limits of integration in (20) and (21) should be taken at $-\infty$. However, 0 is used because R is always a positive number. The error $\int_{-\infty}^{0} f(R/0)\, dR$ is in the order of 10^{-33}.

Using the same processing parameters as those in Section III, i.e., $K = 13.37$, $M = 127$, etc., we have computed the values of P_D under different S/N ratios (assuming same at both receivers) and they are listed in Table II. Fig. 3 is a plot of P_D versus S/N ratio with P_{FA} as a parameter. Of course, the P_D will improve with either an increase in K or M.

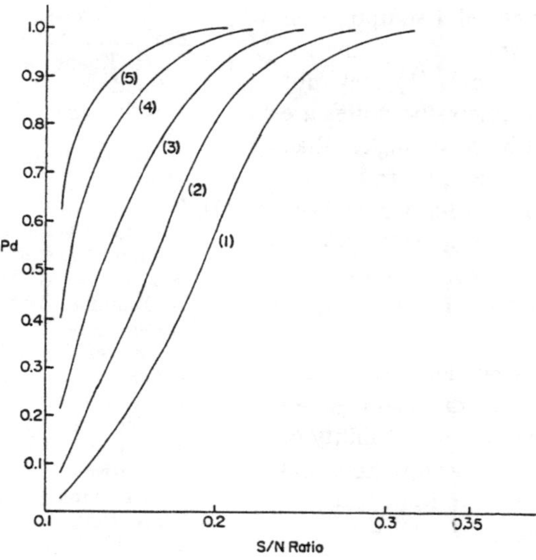

Fig. 3. P_D versus S/N ratio.

It is of interest to compare the performance of the new detector with the likelihood ratio detector for the general Gaussian problem [12] and the coherence detector by Gosselin [13]. At a S/N ratio of 0.21, the new detector, when using 1024 points, 7 segments, and $M = 127$, gives a P_D of about 0.9 for a P_{FA} of 10^{-4}. The coherence detector, which uses $|\hat{\gamma}|^2$ as a sample and assumes $|\hat{\gamma}|^2$ constant over a narrow band, would give the same result but requires 500 segments. The likelihood ratio detector, which uses the sum of squares of time samples as a sufficient statistic, would give a P_D of about 0.8 with 1024 samples.

To summarize the detector-estimation procedures, we first compute (2) and (3) from the time samples, unwrap θ_k if the maximum bearing the detector is covering may cause a delay greater than T, and then compute (16) and (15) and compare R with TH. If there is a detection, a better estimate of τ is available from (10) if $|\hat{\gamma}(k)|^2$ is not constant over the band.

In practice, $|\gamma(k)|^2$ is generally not constant over the band. A sensitivity analysis of the detector can be performed by considering the effects on P_D of small deviations in the values of $E\{\epsilon_k^2\}$ and $E\{\epsilon_k^4\}$ in (17) and (19). However, from the shapes of the ROC curves, we can see that the effects will be minimal provided the deviations in S/N ratio are above the knee. For example, in graph (1) of Fig. 3, the P_D will remain approximately 1 for any variations in S/N ratio across the band as long as each S/N ratio is $\geqslant 0.3$.

The detection scheme is readily extendable to a linear array of $2P$ receivers. The corrupting noise sources at the receivers are assumed independent, so that even though spacings between receivers can take on any values, their separation cannot be too small. We then form P pairs of receivers so as to obtain, as above, for each pair an

$$R_i = \sum (\theta_{k_i} - \hat{\tau}_i \omega_{k_i})^2 \tag{22}$$

with the subscript i denoting the measurements from the ith pair. Any two receivers can be used to form a pair. However, each receiver is only used once. This is to ensure that the disturbances ϵ_{k_i} in (5) are uncorrelated between pairs of receivers, i.e., $E\{\epsilon_{k_i}\epsilon_{k_j}\} = 0$ for $i \neq j$. Again, the assumption is made that the ψ_{k_i} are constant; thus, $E\{\epsilon_{k_i}^2\}$, $E\{\epsilon_{k_i}^4\}$ are constant for all k, i, so as to make the ROC meaningful. Now we take

$$R_s = \sum_{i=1}^{P} R_i \tag{23}$$

as a test statistic. As before, it can be shown that $f(R_s/0)$ and $f(R_s/s)$ are Gaussian with

$$E\{R_s\} = P(M - 1) E\{\epsilon_k^2\} \tag{24}$$

and

$$\text{Var}\{R_s\} \approx E\{\epsilon_k^4\} P(M - 2) + E^2 \{\epsilon_k^2\} P(4 - M). \tag{25}$$

These two equations are simply P times those of (17) and (19). Thus, the spread between the means of $f(R_s/0)$ and $f(R_s/s)$ are P times higher than the single pair case. It appears that the detection capability of P pairs is better than a single pair. However, a single pair can achieve the same capability if P measurements of R from the same pair are summed to form (23). The only benefit of using P pairs is that it allows simultaneous measurements and processing so that for a fixed interval, the performance is better than a single pair.

V. Conclusions

This paper has proposed a nonparametric detection scheme which uses the least squares errors from time delay estimation as a test statistic. The ambient noise level is not needed to establish a TH for a given P_{FA}, and this aspect is particularly useful in many practical situations. Also, the scheme can be easily extended to an array of $2P$ receivers. The detector characteristics compare favorably with two other detection schemes, although an optimum detector could also easily be derived since both $f(R/0)$ and $f(R/s)$ are Gaussian. Some comments on the capability of the scheme to detect at extremely small S/N ratio (thus low coherence) are in order. The

ROC are accurate as long as the uncorrelated assumption on $n_1(t)$ and $n_2(t)$ are valid. In practice, when $s(t) = 0$, the coherence is zero for uncorrelated noises, but (1) may not accurately model the physical situation unless the noises are slightly correlated. Thus, the P_{FA} will be much higher than predicted. A more reasonable approach might be to select a specified S/N ratio above which we wish to have detection. For example, suppose we choose a P_D of 0.98 at S/N ratio = 1 so that $P_D > 0.98$ for S/N ratio > 1. Then calculations show that $TH = 93.86$ (compare with $E\{R/0\} = 414.32$) and a P_{FA} in the order of 10^{-19}.

The l.s. time delay estimator has the same asymptotic properties as the Hamon–Hannan estimator. Its disadvantage lies in the requirement of an unwrapped phase. The reliability of the phase unwrapping algorithms under low signal-to-noise has yet to be investigated. But if delays are known to be less than the sampling period, it is simple to compute.

Acknowledgment

The authors wish to thank Dr. A. H. Nuttall who, in a private communication [9], provided (7).

References

[1] B. V. Hamon, and E. J. Hannan, "Spectral estimation of time delay for dispersive and non-dispersive systems," *Appl. Statist.*, vol. 23, no. 2, pp. 134-142, 1974.

[2] E. J. Hannan, and P. Thomson, "Estimating group delay," *Biometrika*, vol. 60, pp. 241–253, 1973.

[3] C. H. Knapp, and G. C. Carter, "The generalized correlation method for estimation of time delay," *IEEE Trans. Acoust., Speech, Signal Processing*, vol. ASSP-24, pp. 320–327, Aug. 1976.

[4] A. S. Goldberger, *Econometric Theory*. New York: Wiley, 1964.

[5] G. C. Carter, C. H. Knapp, and A. H. Nuttall, "Estimation of the magnitude-squared coherence function via overlapped fast Fourier transform processing," *IEEE Trans. Audio Electroacoust.*, vol. AU-21, pp. 337–344, Aug. 1973.

[6] A. Papoulis, *Probability, Random Variables and Stochastic Processes*. New York: McGraw-Hill, 1965.

[7] G. M. Jenkins and D. G. Watts, *Spectral Analysis and Its Applications*. San Francisco: Holden-Day, 1968.

[8] A. H. Nuttall, "Estimation of cross-spectra via overlapped fast fourier transform," Naval Underwater Systems Center, New London, CT, Rep. 4169-S, July 1975.

[9] —, private communication, Naval Underwater Systems Center, New London, CT, June 1977.

[10] W. S. Hodgkiss and L. W. Nolte, "Covariance between Fourier coefficients representing the time waveforms observed from an array of sensors," *J. Acoust. Soc. Amer.*, vol. 59, pp. 582–590, Mar. 1976.

[11] J. M. Tribolet, "A new phase unwrapping algorithm," *IEEE Trans. Acoust., Speech, Signal Processing*, vol. ASSP-25, pp. 170–177, Apr. 1977.

[12] H. L. Van Trees, *Detection, Estimation, and Modulation Theory, Part 1*. New York: Wiley, 1968.

[13] J. J. Gosselin, "Comparative study of two-sensor (magnitude-squared coherence) and single-sensor (square-law) receiver operating characteristics," in *Proc. IEEE Int. Conf. on Acoust., Speech, and Signal Processing*, Hartford, CT, May 1977.

A Parameter Estimation Approach to Time-Delay Estimation and Signal Detection

Y. T. CHAN, JACK M. RILEY, MEMBER, IEEE, AND J. B. PLANT, SENIOR MEMBER, IEEE

Abstract—Present techniques that estimate the difference in arrival time between two signals corrupted by noise, received at two separate sensors, are based on the determination of the peak of the generalized cross correlation between the signals. To achieve good resolution and stability in the estimates, the input sequences are first weighted. Invariably, the weights are dependent on input spectra which are generally unknown and hence have to be estimated. By approximating the time shift as a finite impulse response filter, estimation of time delay becomes one of determination of the filter coefficients. With this formulation, a host of techniques in the well-developed area of parameter estimation is available to the time-delay estimation problem—with the possibilities of reduced computation time as compared with present methods. In particular, it is shown that the least squares estimation of the filter coefficients is equivalent to estimating the Roth processor. However, the parameter estimation approach is expected to have a smaller variance since it avoids the need for spectra estimation. Indeed, experimental results from two examples show that the Roth processor, found by least squares parameter estimation, has a smaller variance than the approximate maximum likelihood estimator of Hannan-Thomson where spectral estimation is required. A detector that uses the sum of the estimated parameters as a test statistic is also given, together with its receiver operating characteristics.

I. Introduction

The estimation of time difference (or delay) between signals corrupted by noise, received at two sensors located at a known distance apart, have applications in many fields [1]. A familiar use is in passive sonar where the bearing of a signal source is related to the time delay [2], [3]. Other not so well-known examples make use of the time delay and known sensor separation to compute the speed of a ship, or the rolling speed of hot steel, or the flow rate of solids through a pneumatic conveyor [4].

The generalized correlation method [2], which unifies many of the existing time-delay estimators into a two-prefilter cross-correlator configuration, performs a weighting on the inputs through the prefilters. To compute the weights, the input spectra at the two sensors should ideally be known. The same is true of the method in [3]. Since the input spectra are not known in many problems, they must first be estimated in order to establish the weights prior to the estimation of time delay. Due to the inaccuracies associated with estimating spectrum and coherence [5], and hence the weights, these time-delay estimators fail to achieve in practice, where data length is finite, their theoretical performance (based on known spectra)—a difficulty recognized in [2].

Manuscript received December 18, 1978; revised April 30, 1979 and August 16, 1979.

The authors are with the Department of Electrical Engineering, Royal Military College of Canada, Kingston, Ont., Canada.

This paper presents an alternative approach to time-delay estimation by modeling the time delay as a finite impulse response filter (FIR). With this formulation, time-delay estimation becomes a parameter estimation problem, that of estimating the coefficients of the FIR filter. The literature on parameter estimation is extensive in the areas of control, economics, and speech [6], [7]. Many existing techniques are directly applicable to the present problem and it will be shown that the least squares estimation of the parameters is equivalent to the Roth processor in [2]. However, since spectra estimation is avoided in the least squares estimator, it should, in practice, have a lower variance than those realized via the method in [2]. While this claim is not proven in theory, it is at least substantiated by experimental results which show in two examples that for equal data length (1024 points) the Roth processor, realized by parameter estimation, has a smaller variance than the Hannan-Thomson processor [2]. The latter processor is a maximum-likelihood estimator, if the input spectra are known.

Section II contains the derivation that relates the time delay to the coefficients of the FIR filter. The delay is, in general, assumed to be a nonintegral multiple of the sampling period, otherwise the formulation is trivial. The least squares solution that gives the Roth processor is described in Section III, together with the experimental results. Section IV shows that the estimates themselves can be used in a detection scheme and goes on to present the development for a receiver operating characteristics (ROC) curve. The conclusions are in Section V.

II. Problem Formulation

Let $x(t)$ and $y(t) = x(t + \tau)$ represent, respectively, the signal and its delayed version, the delay being τ. Their corresponding sampled values are $x(iT)$ and $y(iT)$, and if the sampling interval T is adequately small for the bandwidth of $x(t)$, and assuming $x(t)$ is band limited, then [8]

$$x(t) = \sum_{i=-\infty}^{\infty} x(iT) \operatorname{sinc}(t - iT) \tag{1}$$

where

$$\operatorname{sinc}(\cdot) \triangleq \frac{\sin\left\{\dfrac{\pi(\cdot)}{T}\right\}}{\dfrac{\pi(\cdot)}{T}}. \tag{2}$$

Without loss of generality, let $T = 1$ and $\tau = (l + f)\,T$ where l is any integer and $0 < f < 1$, i.e., τ is a nonintegral multiple of T.

Reprinted from *IEEE Trans. Acoust., Speech, Signal Processing,* vol. 28, no. 1, pp. 8–15, February 1980.

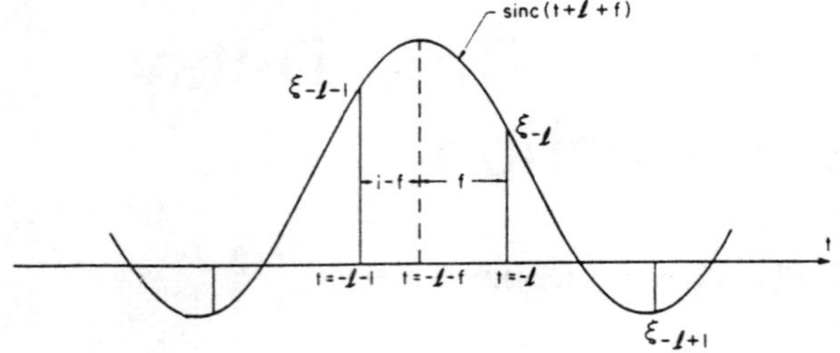

Fig. 1. The sinc $(t + e + f)$ function.

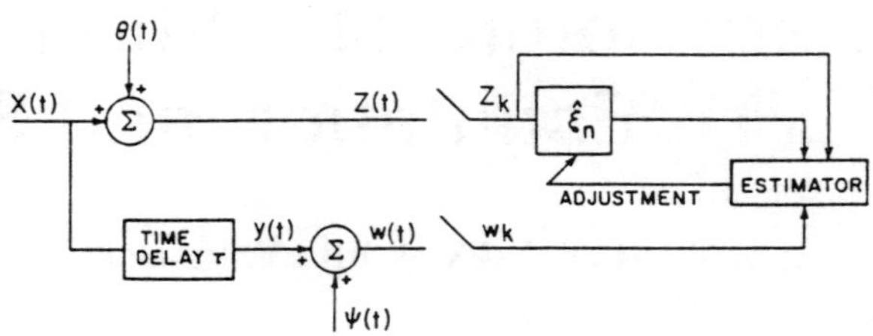

Fig. 2. Parameter estimation of time delay.

The delayed signal $y(t)$ can also be reconstructed from the samples $x(iT)$ by

$$y(t) = x[t + (l + f)] = \sum_{i=-\infty}^{\infty} x(i) \operatorname{sinc}(t + l + f - i), \tag{3}$$

so that for any integer k

$$y(k) = \sum_{i=-\infty}^{\infty} x(i) \operatorname{sinc}(k + l + f - i). \tag{4}$$

With the change of variable $k - n = i$, (4) becomes

$$y(k) = \sum_{k-n=-\infty}^{\infty} x(i) \operatorname{sinc}(n + l + f). \tag{5}$$

On defining

$$\zeta_n \triangleq \operatorname{sinc}(n + l + f) \tag{6}$$

and letting $y_k = y(k)$ and $x_k = x(k)$, (5) can be written as

$$y_k = \sum_{k-n=-\infty}^{\infty} \zeta_n x_{k-n}. \tag{7}$$

Since k is finite in (7) and the summation is from $-\infty$ to $+\infty$, (7) simplifies to

$$y_k = \sum_{n=-\infty}^{\infty} \zeta_n x_{k-n}. \tag{8}$$

Thus, the time series x_k is related to its delayed version y_k through a filter whose coefficients are ζ_n, the values of which are dependent on τ.

From (6), it is clear that ζ_n are the samples of the function sinc $(t + l + f)$, with the maximum at $t + l + f = 0$, as shown in Fig. 1. Hence, given the coefficients ζ_n, the delay τ is the value of t at which the maximum of $\zeta(t)$ given by

$$\zeta(t) = \sum_{n=-\infty}^{\infty} \zeta_n \operatorname{sinc}(t - n) \tag{9}$$

occurs. Alternatively, since

$$\zeta_n = \frac{(-1)^{n+l} \sin \pi f}{\pi(n + l + f)}, \tag{10}$$

the maximum value of ζ_n occurs at either ζ_{-l} (if $f < 0.5$) or ζ_{-l-1} (if $f > 0.5$) (see Fig. 1). Let ζ_j be the maximum value of ζ_n (i.e., $j = -l$ or $-l - 1$); then it is easy to verify, by direct substitution from (10), that

$$\tau = l + f = -j + \frac{\zeta_{j-1}}{\zeta_j + \zeta_{j-1}}. \tag{11}$$

Now with $x(t)$ as the signal, let $\theta(t)$ and $\psi(t)$ be the corrupting noise sources at the two sensors with the usual assumptions that the signal and noise sources are real, jointly stationary and independent random processes. The time-delay problem is as depicted in Fig. 2, where

$$z(t) = x(t) + \theta(t) \tag{12}$$

$$w(t) = y(t) + \psi(t) \tag{13}$$

and

$$z_k = x_k + \theta_k \tag{14}$$

$$w_k = y_k + \psi_k \tag{15}$$

are the samples (assuming an adequate sampling frequency) of $z(t)$ and $w(t)$, respectively. The estimator computes the estimates $\hat{\zeta}_n$, $n = -\infty$ to $+\infty$, using z_k and w_k, so that the delay τ can be obtained from either (9) or (11).

Clearly, it is not practical to estimate an infinite number of coefficients. However, since the function sinc $(n + l + f)$ approaches zero for large values of $n + l$, and since the maximum delay to be estimated (hence the largest possible value for l), is normally known, the series summation in (8) can be truncated at some predetermined, finite number p. Thus (8) changes to

$$y_k = \sum_{n=-p}^{p} \zeta_n x_{k-n}, \tag{16}$$

which models the time delay as an FIR filter. The modeling error introduced by (16) is examined in [9], which shows that, for an ideal low-pass [10] x_k and for $p = 4$, $l = 0$, the difference between the true delay and that produced by (16) varies, depending on the value of f, from a maximum of $-0.0231T$ at $f = 0.25T$ (about a 10 percent error) to a minimum of $-0.0004T$ at $f = 0.5T$. Similar results can be expected for sequences whose spectra are not ideal low pass. As an example in choosing p, suppose the desired accuracy is 10 percent and the maximum time difference that the estimator may encounter is $\pm 3T$. Then $p = 4 + 3 = 7$. It should be emphasized that the 10 percent error is the worst case, occurring at $\tau = 0.25T$. If $\tau = 1.25T$, the error is only $0.0231T/1.25T \times 100$ percent = 1.85 percent. Quite often, the errors due to the approximation in (16) are insignificant compared with the variances of the estimator.

To proceed with the estimation of ζ_n using z_k and w_k, their relationship is first developed via (15) and (8) to yield

$$w_k = \sum_{n=-\infty}^{\infty} \zeta_n x_{k-n} + \psi_k. \tag{17}$$

Substituting (14) into (17) gives

$$w_k = \sum_{n=-\infty}^{\infty} \zeta_n z_{k-n} - \sum_{n=-\infty}^{\infty} \theta_{k-n} + \psi_k \tag{18}$$

$$= \sum_{n=-p}^{p} \zeta_n z_{k-n} + d_k \tag{19}$$

where

$$d_k = \sum_{n=-\infty}^{-p-1} \zeta_n z_{k-n} + \sum_{n=p+1}^{\infty} \zeta_n z_{k-n} - \sum_{n=-\infty}^{\infty} \zeta_n \theta_{k-n} + \psi_k \tag{20}$$

is the disturbance that contains the noise terms and the modeling errors. For $k = p$ to N, (19) expands to the matrix equation

$$w = Z\zeta + d \tag{21}$$

where

$$w = \begin{bmatrix} w_p \\ w_{p+1} \\ \vdots \\ w_{p+N} \end{bmatrix}, \quad \zeta = \begin{bmatrix} \zeta_{-p} \\ \vdots \\ \zeta_0 \\ \vdots \\ \zeta_p \end{bmatrix}, \quad Z = \begin{bmatrix} z_{2p} & z_{2p-1} \cdots z_p \cdots z_1 z_0 \\ z_{2p+1} & z_{2p} \quad \cdots \quad z_1 \\ \vdots & \vdots \\ z_{2p+N} & \cdots \quad z_N \end{bmatrix}, \quad d = \begin{bmatrix} d_p \\ d_{p+1} \\ \vdots \\ d_{p+N} \end{bmatrix}. \tag{22}$$

The estimation problem is: given the measurements vector w and matrix Z and (21), find the estimates $\hat{\zeta}$ such that $\hat{\zeta}$ is close to ζ in some sense.

While (21) is a familiar equation in parameter estimation, it possesses some properties that are not normally present. The first one is that the measurements z_k are related with the disturbances d_k, so that $E\{Z^T d\} \neq 0$. Secondly, the disturbance sequence is not white, so that $E\{dd^T\}$ is not a diagonal matrix. Literature on parameter estimation refers to the first property as correlation between observation and disturbance and the second as nonspherical disturbances [11]. Together, they add complexities to the estimation of ζ. But, if the final objective is estimating τ and not ζ, analysis in the next section reveals that the standard parameter estimation procedures are still applicable.

III. The Time-Delay Parameter Estimator

As mentioned earlier, many techniques are available to solve the problem posed in Section II. Among them, the simplest one chooses $\hat{\zeta}_n$ to minimize $\{\Sigma_{k=p}^{N} (w_k - \Sigma_{n=-p}^{p} \hat{\zeta}_n z_{k-n})\}^2$, i.e., it minimizes the square of the differences between the output of the FIR filter and the observations w_k. This well-known solution [11] is given by

$$\hat{\zeta} = (Z^T Z)^{-1} Z^T w. \tag{23}$$

Now because $E\{Z^T d\} \neq 0$, the estimate $\hat{\zeta}$ is asymptotically biased [11], i.e., $\hat{\zeta}$ is not a consistent estimator of ζ. To confirm this, take the probability limit, as N approaches infinity, denoted by $p \lim_{N\to\infty}$, of (23). On using a corollary of Slutsky's theorem [11], the result is

$$\bar{\zeta} \triangleq \underset{N\to\infty}{p \lim}\, \hat{\zeta} = \left(\underset{N\to\infty}{p \lim} \frac{Z^T Z}{N}\right)^{-1} \underset{N\to\infty}{p \lim} \frac{Z^T w}{N}. \tag{24}$$

Let

$$R_{zz}(i) \triangleq E\{z_k z_{k+i}\}, \qquad R_{\theta\theta}(i) \triangleq E\{\theta_k \theta_{k+i}\} \tag{25}$$

and

$$R_{zw}(i) \triangleq E\{z_k w_{k+i}\}, \qquad R_{xx}(i) \triangleq E\{x_k x_{k+i}\}; \tag{26}$$

then,

$$\underset{N\to\infty}{p \lim} \frac{Z^T Z}{N} = \begin{bmatrix} R_{zz}(0) & R_{zz}(1) \cdots R_{zz}(-2p) \\ R_{zz}(1) & R_{zz}(0) \cdots \\ \vdots & \\ R_{zz}(2p) & \cdots \end{bmatrix} \triangleq \Sigma_{zz} \tag{27}$$

which is the covariance matrix of the sequence z_k. Similarly,

$$\underset{N\to\infty}{p \lim} \frac{Z^T w}{N} = [R_{zw}(-p) \cdots R_{zw}(p)]^T. \tag{28}$$

Thus, from (24),

$$\bar{\zeta} = \Sigma_{zz}^{-1} [R_{zw}(-p) \cdots R_{zw}(p)]^T. \tag{29}$$

Next, from (26), (17), and (14), and the fact that the signal x_k and the noise sources are uncorrelated with each other, one obtains

$$R_{zw}(i) = \sum_{n=-\infty}^{\infty} \zeta_n R_{xx}(i-n) \tag{30}$$

and, for sufficiently large p,

$$R_{zw}(i) \approx \sum_{n=-p}^{p} \zeta_n R_{xx}(i-n). \tag{31}$$

Therefore,

$$[R_{zw}(-p) \cdots R_{zw}(p)]^T = \Sigma_{xx}\zeta \tag{32}$$

where

$$\Sigma_{xx} \triangleq \begin{bmatrix} R_{xx}(0) & R_{xx}(-1) \cdots R_{xx}(-2p) \\ R_{xx}(1) & \vdots \\ \vdots & \\ R_{xx}(2p) & \cdots \quad R_{xx}(0) \end{bmatrix} \tag{33}$$

is the covariance matrix of x_k. From (14), (25), and (26), one easily deduces

$$\Sigma_{zz} = \Sigma_{xx} + \Sigma_{\theta\theta} \tag{34}$$

where $\Sigma_{\theta\theta}$ is the covariance matrix of θ_k. Finally, the combination of (29), (34), and (32) yields

$$\bar{\zeta} = (\Sigma_{xx} + \Sigma_{\theta\theta})^{-1} \Sigma_{xx} \zeta \tag{35}$$

showing that, in general, $\bar{\zeta} \neq \zeta$ unless $\Sigma_{\theta\theta} = 0$.

If x_k and θ_k are samples from ideal low-pass processes, Σ_{xx} and $\Sigma_{\theta\theta}$ become diagonal matrices with diagonal elements equal to σ_x^2, the signal power, and σ_θ^2, the noise power, respectively. Then, (35) simplifies to

$$\bar{\zeta} = \frac{\sigma_x^2}{\sigma_x^2 + \sigma_\theta^2} \zeta. \tag{36}$$

Thus, although $\bar{\zeta} \neq \zeta$, the use of $\hat{\zeta}_n$ in (11) to estimate τ will be unbiased in the asymptotic sense because the factors $\sigma_x^2/(\sigma_x^2 + \sigma_\theta^2)$ cancel each other in (11). Of course, the assumption of x_k being an ideal low-pass process is not always valid. To use (11), unbiased estimates of ζ must be obtained by some of the more complex estimation procedures (the instrumental variable method [11], for example). However, for purposes of time-delay estimation, it will next be shown that using the estimates from (23) in (9) will suffice.

First, recall from (29) that

$$\sum_{n=-p}^{p} \bar{\zeta}_n R_{zz}(i-n) = R_{zw}(i) \tag{37}$$

where $\bar{\zeta}_n$ is the nth element of $\bar{\zeta}$. But (37) is also an approximation to the discrete solution of the unrealizable Wiener-Hopf equation [12]

$$\sum_{n=-\infty}^{\infty} h_n R_{zz}(i-n) = R_{zw}(i) \qquad -\infty \leqslant i \leqslant \infty \tag{38}$$

where h_n are the coefficients of the unrealizable filter that minimizes $E\{(w_k - \sum_{n=-\infty}^{\infty} h_n z_{k-n})^2\}$. Further, since [12]

$$\sum_{n=-\infty}^{\infty} h_n e^{-j\omega n} = \frac{G_{zw}(\omega)}{G_{zz}(\omega)} \tag{39}$$

with $G_{zw}(\omega)$, $G_{zz}(\omega)$, and $G_{ww}(\omega)$ denoting the cross and auto spectra of z_k and w_k, the coefficients h_n are samples, at $t = nT$, of the function

$$h(t) = \int_{-\infty}^{\infty} \frac{G_{zw}(\omega)}{G_{zz}(\omega)} e^{j\omega t}\, d\omega. \tag{40}$$

The function $h(t)$ is recognized as the output of the Roth processor [2]. Noting the similarity between $\bar{\zeta}_n$ in (37) and h_n in (38), we conclude that the $\bar{\zeta}_n$ are samples of the Roth processor and $\hat{\zeta}_n$ from (23) are estimates of h_n. The time delay $\hat{\tau}$ is now found from (9) by substituting $\hat{\zeta}_n$ for ζ_n, $n = -p$ to p, and computing the value of t that maximizes (9). The Roth processor is unbiased [2], but this method of realization has modeling errors as discussed in the paragraph following (16). The theoretical variance is [2]

$$\operatorname{var}\{\hat{\tau}\} = \frac{2\pi \displaystyle\int_{-\infty}^{\infty} \frac{\omega^2\{G_{zz}(\omega)\, G_{ww}(\omega) - |G_{zw}(\omega)|^2\}}{G_{zz}^2(\omega)}\, d\omega}{T_l \left[\displaystyle\int_{-\infty}^{\infty} \frac{\omega^2 |G_{zw}(\omega)|}{G_{zz}(\omega)}\, d\omega\right]^2} \tag{41}$$

where T_l is the record length.

The computational requirement for the time-delay parameter estimator (TDPE) is rather modest. A recursive algorithm [13] is available for the implementation of (23). At the Nth sample, let $\hat{\zeta}^{(N)}$ be the estimate of ζ and $Z^{(N)}$ and $w^{(N)}$ be the measurement matrix and vector. Then at $(N+1)$, with

$$\underline{z}^{(N+1)} = [z_{2p+N+1} \cdots z_{N+1}], \tag{42}$$

the new estimate is

$$\hat{\zeta}^{(N+1)} = \hat{\zeta}^{(N)} + \frac{\Sigma^{(N)} \underline{z}^{(N)^T} (w_{N+1} - \underline{z}^{(N+1)} \hat{\zeta}^{(N)})}{D^{(N)}} \tag{43}$$

where

$$D^{(N)} = 1 + \underline{z}^{(N+1)} \Sigma^{(N)} \underline{z}^{(N+1)^T} \tag{44}$$

$$\Sigma^{(N+1)} = \Sigma^{(N)} - \frac{\Sigma^{(N)} \underline{z}^{(N+1)^T} \underline{z}^{(N+1)} \Sigma^{(N)}}{D^{(N)}}. \tag{45}$$

At the start of the algorithm, $\Sigma^{(0)}$ is a diagonal matrix with elements equal to some large numbers, 10^6 for example, and $\hat{\zeta}^{(0)} = 0$. This algorithm eliminates the need for matrix inversion and the requirement for storing a large quantity of data. In addition, because of its recursive nature, it is well suited for applications where data arrive sequentially, such as in sonar.

To determine the value of t that maximizes (9), a Newton-Raphson iterative scheme was used. In all the experiments below, the search converged to within a 0.0001 resolution in, at most, six iterations.

As an assessment of the TDPE, experiments were performed on a PDP 11/34 computer using Fortran IV and single precision arithmetic. Three random number generators produced the independent ideal low-pass sequences x_k, θ_k, and ψ_k. Then x_k was shifted via (16) to give y_k, with $p = 16$, so that the error due to truncation of series is negligible. The ζ_n in (16) were computed according to (6) to realize the desired delay. The estimator used an FIR filter of $p = 4$ in (21) to model the time delay which had a value of either $0.25T$ or $0.50T$. The mean and mean-squared errors (MSE), with respect to the time delay quoted below, were based on 50 independent runs.

Experiment 1: Since x_k is ideal low pass, after obtaining $\hat{\zeta}_n$, the time delay $\hat{\tau}$ can be estimated from either (9) or (11). Table I summarizes the results for $\tau = 0.25T$ and $\tau = 0.50T$. Comparisons between the mean values in the tables confirm that the error in using $p = 4$ is largest for $\tau = 0.25T$ at $-0.0231T$ and smallest for $\tau = 0.50T$ at $-0.0004T$. It was this error at $\tau = 0.25T$ that caused large MSE, compared with the theoretical values, at a signal-to-noise (S/N) ratio of 4. At lower S/N ratios, when the estimator variances dominate, the experimental MSE agrees well with the predicted values from (41).

Experiment 2: The signals x_k and y_k used in this experiment were nonwhite sequences obtained from their ideal low-

TABLE I
TDPE, x_k WHITE

	S/N Ratio	$\hat{\tau}$ from (11)						$\hat{\tau}$ from maximization of (9)					
		N = 500			N = 1000			N = 500			N = 1000		
		Mean	MSE Theo.	MSE Exp.	Mean	MSE Theo.	MSE Exp.	Mean	MSE Theo.	MSE Exp.	Mean	MSE Theo.	MSE Exp.
$\tau = 0.25$	4	.249	3.4×10^{-4}	4.1×10^{-4}	.249	1.7×10^{-4}	2.2×10^{-4}	.225	3.4×10^{-4}	1.1×10^{-3}	.226	1.7×10^{-4}	8.1×10^{-4}
	1	.262	1.8×10^{-3}	2.2×10^{-3}	.263	9.0×10^{-4}	1.6×10^{-3}	.236	1.8×10^{-3}	1.8×10^{-3}	.239	9.0×10^{-4}	1.2×10^{-3}
	0.25	.242	1.5×10^{-2}	1.5×10^{-2}	.245	7.3×10^{-2}	9.5×10^{-3}	.219	1.5×10^{-2}	1.4×10^{-2}	.215	7.3×10^{-3}	8.0×10^{-3}
$\tau = 0.50$	4	.500	3.4×10^{-4}	3.9×10^{-4}	.500	1.7×10^{-4}	1.7×10^{-4}	.498	3.4×10^{-4}	5.2×10^{-4}	.500	1.7×10^{-4}	2.2×10^{-4}
	1	.509	1.8×10^{-3}	1.7×10^{-3}	.509	9.0×10^{-4}	1.9×10^{-3}	.509	1.8×10^{-3}	2.3×10^{-3}	.510	9.0×10^{-4}	1.4×10^{-3}
	0.25	.505	1.5×10^{-2}	1.0×10^{-2}	.500	7.3×10^{-3}	7.3×10^{-3}	.504	1.5×10^{-2}	1.7×10^{-2}	.505	7.3×10^{-3}	9.6×10^{-3}

TABLE II
TDPE, x_k NONWHITE

	S/N Ratio	$\hat{\tau}$ from maximization of (9)					
		N = 500			N = 1000		
		Mean	MSE Theo.	MSE Exp.	Mean	MSE Theo.	MSE Exp.
$\tau = 0.25$	4	.262	6.4×10^{-4}	1.3×10^{-2}	.247	3.2×10^{-4}	9.1×10^{-4}
	1	.255	4.9×10^{-3}	9.6×10^{-3}	.266	2.5×10^{-3}	6.4×10^{-3}
	0.25	.258	3.1×10^{-2}	5.2×10^{-2}	.223	1.5×10^{-2}	5.0×10^{-2}
	0.125	.329	9.2×10^{-2}	2.0×10^{-1}	.347	4.6×10^{-2}	1.2×10^{-1}
$\tau = 0.50$	4	.504	6.4×10^{-4}	4.3×10^{-3}	.497	3.2×10^{-4}	1.0×10^{-3}
	1	.492	4.9×10^{-3}	1.1×10^{-2}	.507	2.5×10^{-3}	5.6×10^{-3}
	0.25	.495	3.1×10^{-2}	4.9×10^{-2}	.469	1.5×10^{-2}	4.0×10^{-2}
	0.125	.515	9.2×10^{-2}	1.2×10^{-1}	.560	4.6×10^{-2}	1.1×10^{-1}

TABLE III
TDPE VERSUS *HT* ESTIMATOR $\tau = 0.25$

		HT estimator N=1024			TDPE N=1000		
	S/N RATIO	Mean	MSE Theo.	MSE Exp.	Mean	MSE Theo.	MSE Exp.
White	1	.222	9.0×10^{-4}	2.3×10^{-3}	.239	9.0×10^{-4}	1.2×10^{-3}
	0.25	.063	7.3×10^{-3}	4.3×10^{-2}	.215	7.3×10^{-3}	8.0×10^{-3}
Nonwhite	1	.159	1.5×10^{-3}	1.2×10^{-2}	.266	2.5×10^{-3}	6.4×10^{-3}
	0.25	.036	1.2×10^{-2}	5.3×10^{-2}	.223	1.5×10^{-2}	5.0×10^{-2}

pass counter parts by passing each of them through the filter

$$a_k = b_k + b_{k-1} \tag{46}$$

where b_k is the input to the filter and a_k is its output. The resultant x_k and y_k sequences have the nonwhite spectrum [14] $[\sin^2 \omega]/[\sin^2 (\omega/2)]$. Since θ_k and ψ_k remained white, the S/N ratio was no longer constant across the spectrum and was taken as the ratio of total signal power-to-noise power. The results are in Table II. In comparison with Table I, the nonwhite case has higher MSE, as predicted.

Experiment 3: This experiment compares the performance between the TDPE and the Hannan-Thomson (HT) estimator, which was computed according to [1]. For spectral estimation, the program divided 1024 points into 7 segments of 256 points each (hence a 50 percent overlap). It then applied a Hanning weighting to each segment before taking the fast Fourier transform and producing 128 Fourier coefficients. The nonwhite signals were produced in the same manner as described in experiment 2. Results in Table III show that, although the HT estimator has theoretical variances equal to (when x_k white) or less than (when x_k nonwhite) the TDPE, the latter has smaller experimental MSE. This, as conjectured in Section I, is because the TDPE avoids the need for spectral estimation. Mean values in Table III also indicate that the HT estimator exhibits a larger bias.

Several comments are in order on the experimental results. First, it is well-known [11] that the sample, or experimental, MSE are unbiased and have a variance equal to $\sigma^4/25$ for Gaussian distribution and 50 independent samples, and a theoretical (or population) MSE of σ^2. Thus, in 68 percent of the time (for one standard deviation $= \sigma^2/5$) the experimental MSE is within ± 20 percent of the theoretical value. Second, in experiment 3, the choice of 128 Fourier coefficients in the HT estimator is ad hoc. This choice is taken to cover a reasonable variation in the signal spectrum. If it were known *a priori* that the signal spectrum was white, for example, a much smaller number of coefficients, say eight, could have been used. The resultant experimental MSE would be much closer to the theoretical value because more segments would then be available for averaging. Of course, in practice, the shape of signal spectrum is unknown and a reasonable number of frequency points have to be used to ensure a sufficient resolution in frequencies. Finally, it should be noted that the experimental MSE in the HT estimator is in part caused by a large bias. This bias could possibly be due to the bias associated with coherence estimation [16], which would then bias the location of the maximum of the cost function [1] used in the HT estimator.

IV. The Detector

The development of the signal detection scheme is rather straightforward. This scheme is nonparametric with the test statistic

$$\Lambda = \sum_{n=-p}^{p} \hat{\zeta}_n. \tag{47}$$

Assuming that x_k and the noise sources are ideal low pass (an assumption necessary for the derivation of the ROC curves) so that the S/N ratio is constant over the band, then it follows from (36), for large N,

$$E\{\Lambda\} = \frac{\sigma_x^2}{\sigma_x^2 + \sigma_\theta^2} \sum_{n=-p}^{p} \zeta_n. \tag{48}$$

In addition, for large N, the estimate $\hat{\zeta}$ is Gaussian [11] with the covariance (see Appendix I) given by

$$\frac{I}{N} \frac{1 + 2r}{(1 + r)^2},$$

where I is an identity matrix. Thus the $\hat{\zeta}_n$ are independent of each other and the variance of Λ is

$$\mathrm{var}\,\{\Lambda\} = \frac{(2p + 1)}{N} \frac{(1 + 2r)}{(1 + r)^2}. \tag{49}$$

It is shown in [15] that $\sum_{n=-\infty}^{\infty} \zeta_n = 1$ and for ζ_n as given by (6) with $p = 4$, $l = 0$, direct evaluation of the series $\sum_{n=-p}^{p} \zeta_n$ gives a sum of 0.992 for $f = 0.5$ and 0.997 for $f = 0.25$. Hence, with negligible error, (48) becomes

$$E\{\Lambda\} = \frac{\sigma_x^2}{\sigma_x^2 + \sigma_\theta^2} = \frac{r}{1 + r} \tag{50}$$

where $\sigma_x^2/\sigma_\theta^2 = r$ is the S/N ratio and is equal to zero when no signal is present.

Let $\Lambda/1$ denote the test statistic when a signal is present and $\Lambda/0$ when not; then (50) and (49) give

$$E\{\Lambda/0\} = 0, \qquad \mathrm{var}\,\{\Lambda/0\} = \frac{2p + 1}{N} \tag{51}$$

and

$$E\{\Lambda/1\} = \frac{r}{1 + r}, \qquad \mathrm{var}\,\{\Lambda/1\} = \frac{(2p + 1)}{N} \frac{(1 + 2r)}{(1 + r)^2}. \tag{52}$$

With the mean and variance of the Gaussian random variable Λ known, the probability of false alarm P_{FA} and detection P_D are given by

$$P_{FA} = \int_{TH}^{\infty} f(\Lambda/0)\, d\Lambda, \tag{53}$$

which is independent of the ambient noise levels and

$$P_D = \int_{TH}^{\infty} f(\Lambda/1)\, d\Lambda. \tag{54}$$

The detection threshold is TH, i.e., $\Lambda > TH$ is a detection and $f(\cdot)$ denotes the probability density function.

We have computed, for $p = 4$ and $N = 1000$, the values of P_D and P_{FA} as a function of r. The results are in Table IV and the ROC curves are plotted in Fig. 3.

This detection technique is easily extendable to a linear array of 2R receivers. The spacings between receivers can be arbitrary, but cannot be too close to invalidate the assumption of independent noise sources at the receivers. The test statistic for 2R receivers, γ, is simply the sum of the Λ_j statistic from each pair of receivers. Thus,

$$\gamma = \sum_{j=1}^{R} \Lambda_j. \tag{55}$$

Any two receivers can form a pair, but each is used only once to ensure the independence of the Λ_j. Since the signal is ideal low pass and the noise sources are independent, the Λ_j are independent Gaussian random variables so that, for constant S/N ratio in each receiver in the array, γ has mean

$$E\{\gamma\} = \sum_{j=1}^{R} E\{\Lambda_j\} = RE\{\Lambda\} \tag{56}$$

and variance

$$\mathrm{var}\,\{\gamma\} = \sum_{j=1}^{R} \mathrm{var}\,\{\Lambda_j\} = R\, \mathrm{var}\,\{\Lambda\}. \tag{57}$$

The ROC curves for the γ statistic can be established with the same procedures as before.

V. Conclusions

This paper has introduced a new formulation of the time-delay estimation problem by modeling the delay as an FIR filter. This formulation is most useful when the largest ex-

TABLE IV
P_D FOR $N = 1000$

r \ P_{FA}	.001	.005	.02	.035	.045
0.1	.02	.05	.15	.20	.23
0.25	.16	.32	.53	.61	.66
0.35	.36	.57	.76	.83	.86
0.5	.67	.84	.94	.96	.97

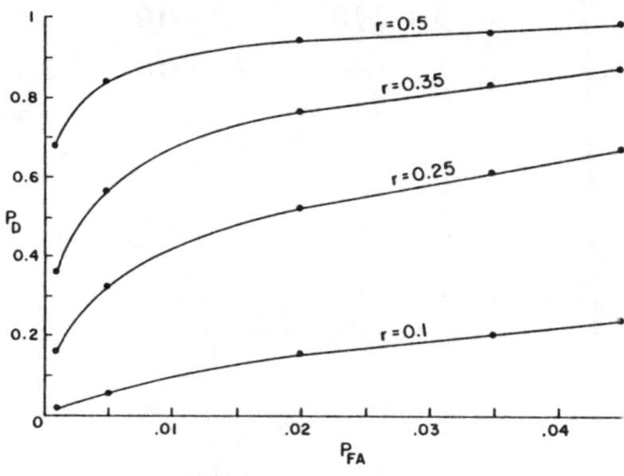

Fig. 3. ROC curves.

pected delay does not exceed several (3 or 4, say) times the sampling period, for then the order of the FIR filter can take on a reasonable number. While there are many methods available to estimate the coefficients of an FIR filter, only the least squares method is considered and is shown to be equivalent to the Roth processor. The effectiveness of this least squares algorithm has been demonstrated by experiments which also compare the TDPE with the HT estimator. The TDPE achieves better performance than the HT estimator in two examples in which one has a white input sequence and the other a nonwhite sequence. We attribute this outcome to the additional variances created by spectral estimations in the HT estimator. We have not given a comparison of computational times between the TDPE and those estimators in [2]. This is because the real time requirement can be very dependent on the special features of computing equipment. The choice of an estimator may well be decided by the particular application.

APPENDIX
THE ASYMPTOTIC COVARIANCE OF $\hat{\zeta}$

In [11] the asymptotic covariance (ACV) of $\hat{\zeta}$, as estimated from (23), is given for the case $E\{Z^T d\} = 0$. When elements of the measurement matrix Z are correlated with the disturbances d_k in (21), the derivation for the ACV is considerably more complex and is not available in the literature. The ACV for $\hat{\zeta}$ is needed for the construction of the ROC curves in Section IV, where it is assumed that x_k and the noise sources are all ideal low pass. Without this assumption, the development for the ACV of $\hat{\zeta}$ would be untractable.

From [11], the ACV of $\hat{\zeta}$ is defined as

$$\mathrm{ACV}(\hat{\zeta}) \triangleq \frac{1}{N} \lim_{N\to\infty} E\{\sqrt{N}(\hat{\zeta} - \bar{\zeta})\sqrt{N}(\hat{\zeta} - \bar{\zeta})^T\} \tag{A.1}$$

where $\bar{\zeta}$ is given by (36). Note that, since $\bar{\zeta} \neq \zeta$, this variance of $\hat{\zeta}$ is in respect to its own mean and not the true value. The evaluation of (A.1) is rather tedious and the reader is referred to [15] for complete details. The procedures involve considering the asymptotic distribution of the terms in $\sqrt{N}(\hat{\zeta} - \bar{\zeta})$ and frequent use of the assumptions that signal and noise sources are white and are uncorrelated with each other. The result is an $\mathrm{ACV}(\hat{\zeta})$ which has all diagonal elements equal to

$$\frac{1}{N}\frac{1+2r}{(1+r)^2} \tag{A.2}$$

and all off diagonal elements equal to

$$\frac{1}{N}\frac{r}{(1+r)^2}\sum_{n=-p}^{p-1} \zeta_n \zeta_{n+1} \tag{A.3}$$

where $r \triangleq \sigma_x^2/\sigma_\theta^2$ is the S/N ratio. It is further shown in [15] that $\Sigma_{n=-\infty}^{\infty} \zeta_n \zeta_{n+1} = 0$. For $p = 4$ and $f = 0.5$, direct evaluation of $\Sigma_{n=-4}^{4} \zeta_n \zeta_{n+1}$ gives 0.052. Hence, with negligible error, $\mathrm{ACV}(\hat{\zeta})$ is a diagonal matrix with diagonal elements equal to

$$\frac{1}{N}\frac{1+2r}{(1+r)^2}.$$

To verify this theoretical $\mathrm{ACV}(\hat{\zeta})$, we have computed in experiment 1 the covariance of $\hat{\zeta}$ for $N = 1000$, $f = 0.5$, and $r = 4$ and 1, and are reproduced below. Only the upper triangular elements are given because $\mathrm{ACV}(\hat{\zeta})$ is a symmetric matrix.

$$r = 4, \quad \frac{1}{N}\frac{1+2r}{(1+r)^2} = 3.6 \times 10^{-4}, \qquad \frac{1}{N}\frac{r}{(1+r)^2}\sum_{n=-p}^{p} \zeta_n \zeta_{n+1} = 8.3 \times 10^{-6}$$

Experimental ACV($\hat{\zeta}$):

4.8×10^{-4}	1.9×10^{-5}	2.9×10^{-5}	5.6×10^{-6}	-7.7×10^{-5}	-1.9×10^{-5}	-2.3×10^{-5}	6.5×10^{-5}	1.9×10^{-5}
	4.2×10^{-4}	-2.3×10^{-5}	-5.3×10^{-5}	-6.0×10^{-5}	-3.4×10^{-5}	-2.9×10^{-5}	-2.2×10^{-5}	-3.5×10^{-5}
		2.5×10^{-4}	6.4×10^{-5}	4.0×10^{-5}	-1.1×10^{-5}	-3.4×10^{-5}	4.6×10^{-5}	-3.4×10^{-5}
			4.6×10^{-4}	1.1×10^{-4}	-4.3×10^{-5}	-4.0×10^{-3}	6.5×10^{-5}	3.3×10^{-5}
				3.4×10^{-4}	1.3×10^{-5}	9.5×10^{-6}	-2.4×10^{-5}	1.5×10^{-5}
					3.9×10^{-4}	-3.5×10^{-5}	6.1×10^{-5}	9.4×10^{-5}
						2.2×10^{-4}	-9.4×10^{-5}	3.7×10^{-5}
							3.1×10^{-4}	3.4×10^{-6}
								3.7×10^{-4}

$$r = 1, \quad \frac{1}{N}\frac{1+2r}{(1+r)^2} = 7.5 \times 10^{-4}, \qquad \frac{1}{N}\frac{r}{(1+r)^2}\sum_{n=-p}^{p} \zeta_n \zeta_{n+1} = 1.3 \times 10^{-5}$$

Experimental ACV($\hat{\zeta}$):

8.7×10^{-4}	4.0×10^{-5}	-1.6×10^{-4}	-9.5×10^{-6}	-2.0×10^{-4}	1.7×10^{-4}	9.6×10^{-5}	-9.2×10^{-5}	-5.7×10^{-5}
	7.9×10^{-4}	-3.0×10^{-5}	7.9×10^{-5}	-3.4×10^{-5}	-1.2×10^{-4}	-8.8×10^{-5}	3.7×10^{-5}	7.5×10^{-5}
		9.6×10^{-4}	5.0×10^{-5}	1.1×10^{-4}	-7.5×10^{-5}	-3.7×10^{-5}	-6.2×10^{-5}	1.5×10^{-5}
			8.4×10^{-4}	-2.0×10^{-4}	-9.3×10^{-5}	1.3×10^{-4}	-1.1×10^{-4}	2.2×10^{-4}
				6.2×10^{-4}	-1.1×10^{-4}	-1.1×10^{-4}	1.3×10^{-4}	-9.3×10^{-5}
					7.9×10^{-4}	-3.5×10^{-5}	-7.1×10^{-5}	1.5×10^{-4}
						6.9×10^{-4}	9.9×10^{-5}	-1.8×10^{-4}
							7.3×10^{-4}	-3.9×10^{-5}
								8.7×10^{-4}

REFERENCES

[1] B. V. Hamon and E. J. Hannan, "Spectral estimation of time delay for dispersive and non-dispersive systems," *Appl. Statist.*, vol. 23, no. 2, pp. 134-142, 1974.

[2] C. H. Knapp and G. C. Carter, "The generalized correlation method for estimation of time delay," *IEEE Trans. Acoust., Speech, Signal Processing*, vol. ASSP-24, Aug. 1976.

[3] Y. T. Chan, R. V. Hattin, and J. B. Plant, "The least squares estimation of time delay and its use in signal detection," *IEEE Trans. Acoust., Speech, Signal Processing*, vol. ASSP-26, June 1978.

[4] H. K. Whitsel and L. W. Griswold, "Speed sensors for high speed surface craft," presented at 5th Ship Control Systems Symposium, Annapolis, MD, Nov. 1978.

[5] E. H. Scannell and G. C. Carter, "Confidence bounds for coherence estimates," presented at IEEE Intl. Conf. Acoust., Speech, Signal Processing, Tulsa, OK, April 1978.

[6] K. J. Astrom and P. Eykhoff, "System identification–A survey," *Automatica*, vol. 7, 1971.

[7] J. Makhoul, "Linear prediction–A tutorial review," *Proc. IEEE*, vol. 63, Apr. 1975.

[8] M. Schwartz and L. Shaw, *Signal Processing–Discrete Spectral Analysis, Detection, and Estimation.* New York: McGraw-Hill, 1975.

[9] Y. T. Chan, J. M. Riley, and J. B. Plant, "A method to shift a time series by a nonintegral multiple of sampling period," submitted to *IEEE Trans. Syst., Man, Cybern.*

[10] A. Papoulis, *Probability, Random Variables, and Stochastic Processes.* New York: McGraw-Hill, 1965.

[11] A. S. Goldberger, *Econometric Theory.* New York: Wiley, 1964.

[12] H. Freeman, *Discrete-Time Systems.* New York: Wiley, 1965.

[13] J. M. Mendel, *Discrete Techniques of Parameter Estimation.* New York: Marcel Dekker Inc., 1973.

[14] A. V. Oppenheim and R. W. Schafer, *Digital Signal Processing.* Englewood Cliffs, NJ: Prentice-Hall, 1975.

[15] J. M. Riley, "The least squares estimation of time delay," M.Eng. thesis, Dept. Elec. Eng., Royal Military College of Canada, Kingston, Ont., Canada, Oct. 1978.

[16] G. C. Carter and C. H. Knapp, "Coherence and its estimation via the partitioned modified chirp-*Z* transform," *IEEE Trans. Acoust., Speech, Signal Processing*, vol. ASSP-23, June 1975.

Analysis of Auditory Evoked Potentials by Magnitude-Squared Coherence*

Robert A. Dobie and Michael J. Wilson

Department of Otolaryngology-Head and Neck Surgery, University of Washington, Seattle, Washington

ABSTRACT

Evoked potentials are usually analyzed in the time domain (voltage versus time). The most familiar frequency-domain measure, the power spectral density function, displays power as a function of frequency but doesn't distinguish signal power from noise power. The coherence function estimates, for each frequency, the ratio of signal power to total (signal plus noise) power and, thus, indicates the degree to which system output (scalp potential) is determined by input (acoustic stimulus). Coherence ranges from 0 to 1; values above specified critical values can be accepted as demonstrating statistically significant system response. In this paper, we present coherence analysis of human scalp responses to clicks and amplitude-modulated tones. In both cases, this analytic method provides insight into the spectral character of the response (for example, assisting in specifying desirable filter characteristics). Threshold sensitivity is also improved: statistically significant responses can be detected at lower intensity by coherence analysis than by inspection of time-domain waveforms.

Auditory evoked potentials (AEPs) have usually been averaged and analyzed in the time domain, i.e., as waveforms representing scalp voltage as a function of time. Amplitudes, latencies, and thresholds—the latter most often estimated by inspection—have been the dependent variables of traditional interest. AEPs can also be analyzed in the frequency domain. Beyond its inherent descriptive interest, spectral analysis, by identifying the frequencies prominent in a particular AEP type, assists in the selection of response filter passbands (Fridman, John, Bergelson, Kaiser, & Baird, 1982). Comparison of the power spectrum measured during signal presentation (containing system response plus system noise) to that measured in a no-stimulus condition (noise only) permits a judgment of the presence or absence of a response. However, comparison of power spectra is difficult when a low-amplitude response is present, and is not easily amenable to statistical inference.

* This research was supported by a grant from NINCDS (No. NS23116).

The goal of this paper is to present and illustrate the use of the magnitude-squared coherence function (γ^2) as an alternative to "simple" spectral analysis of AEPs. Although the examples in this paper use AEPs, coherence analysis is equally applicable to evoked potentials from any sensory modality.

γ^2 is a measure of the degree to which system output (such as an AEP) is determined by a specified input, as a function of frequency (Bendat & Piersol, 1986). For a linear system, coherence measures, for each frequency of interest, the proportion of response power attributable to a given stimulus; it is thus a "signal to signal plus noise" power ratio, varying from 0 to 1. Multiple response spectral measures are required for a single "smoothed" and valid coherence estimate; these can be obtained by breaking up a single evoked potential average into multiple subaverages (e.g., 16 subaverages of 256 responses each instead of a single average of 4096), or by combining spectral estimates for adjacent frequencies (at the expense of frequency resolution). These two methods—smoothing across subaverages or frequencies—can also be used together to further reduce variance. Depending on the degree of smoothing, critical values can be determined for any desired significance level; coherence estimates above these values indicate statistically significant response.

The use of periodic signals, as is necessary in evoked potential work, poses some problems: the measured coherence values no longer are valid estimates of linear relationship between stimulus and response. However, the technique is still useful, since our main interest is not in actually measuring coherence per se, but in determining whether a response is present. An estimated coherence value above the appropriate critical value does this, whether the response is linear or not. An advantage of using periodic stimuli is that coherence can be very rapidly calculated using only the spectra of the subaverages and the grand average (mean) response: for a given frequency, coherence is then equal to the power of the mean response

Reprinted with permission from *Ear and Hearing*, vol. 10, no. 1, pp. 2–13, February 1989.

divided by the mean power (of the subaverages).

The basic principle is that during the averaging process, signal (phase-locked to the stimulus) remains while noise (uncorrelated to the stimulus) tends to cancel out. We periodically interrupt this process to measure subaverages in addition to the grand average response, then calculate power spectra for each. For a frequency containing only signal, the powers in the subaverages and the grand average will all be the same, and coherence will be 1. For any frequency containing only noise, the power in the grand average will be much less than the mean power in the subaverages, and estimated coherence will be much less than 1, tending toward 0 as the number of subaverages increases. When signal and noise are mixed, coherence will depend on the signal to noise ratio.

The following section, "Calculation of Coherence Estimates," is mathematically intuitive and nonrigorous, requiring no calculus, and should be understandable by both clinicians and investigators working with evoked potentials. However, readers willing to accept the preceding assertions on faith may skip directly to the "Methods" section.

CALCULATION OF COHERENCE ESTIMATES

Frequency Domain Analysis

Table 1 lists and briefly defines abbreviations and symbols used. The "system" of interest is a human or animal subject receiving an acoustic input, x(t), and producing a scalp voltage output y(t)—y(t) can be considered to be the sum of system response to the input, S(t), and random noise produced by the system, N(t).

Fourier analysis produces complex frequency domain representations of the input and output signals, X(f) and Y(f); for each of these, real and imaginary (orthogonal) components are calculated for each frequency. Amplitude A(f) and phase θ(f) can be derived by well-known trigonometric relationships (Fig. 1 compares the polar and complex representations of X(f)):

$$A(f) = \sqrt{Re[X(f)]^2 + Im[X(f)]^2} \quad (1)$$

$$\theta(f) = \tan^{-1}\left(\frac{Im[X(f)]}{Re[X(f)]}\right) \quad (2)$$

A(f) is sometimes indicated as | X(f) |, i.e., the "absolute value" of the complex number X(f) equals the length of its vector.

The power spectral density function, $G_{xx}(f)$, the frequency domain display most familiar to evoked potential workers, is a real-valued function equal to the square of A(f). For the input x(t):

$$G_{xx}(f) = A_x^2(f) = |X(f)|^2 \quad (3)$$

Note that $G_{xx}(f)$ can also be obtained by complex multiplication. X*(f), the complex conjugate of X(f), is a function whose real components are identical to those of X(f), but whose imaginary components have reversed signs (the phases are thus also reversed). Since the phase of the product of two complex numbers is the sum of the phases of the original numbers, X(f)X*(f) has zero phase [θ(f) − θ(f)], and is thus real-valued. Therefore, $G_{xx}(f)$ = X(f)X*(f).

Table 1. Abbreviations and symbols.

A (f)	Amplitude of Fourier component as a function of frequency f
cos	Cosine
f	Frequency (f_c = carrier frequency; f_m = modulating frequency)
G_{xx} (f)	Power spectral density function of x (t)
G_{xy} (f)	Cross spectral density function relating x (t) and y (t) as a function of frequency f
i, j	Subscripts indicating one of a series of measurements
Im []	Imaginary component of []
ℓ	Number of adjacent frequencies over which spectral estimates are averaged prior to calculation of coherence
m	Number of adjacent frequencies over which multiple coherence estimates are averaged
n	Number of measurements averaged in the time domain
N (t)	Portion of system response y (t) attributable to noise
p	Probability
PC	Phase coherence
q	Number of measurements averaged in the frequency domain
Re []	Real component of []
S (t)	Portion of system response y (t) attributable to input x (t)
sin	Sine
t	Time
tan	Tangent
x (t)	Input signal (time domain)
X (f)	Fourier transform (frequency domain) of input signal
y (t)	System output (time domain)
Y (f)	Fourier transform (frequency domain) of system output
z	Number of standard deviations encompassing a given proportion of a normal distribution
α	Desired level of significance for statistical inference
Δ	A small change in some quantity
γ^2 (f)	Magnitude-squared coherence relating x (t) and y (t) as a function of frequency f
θ (f)	Phase angle as a function of frequency f
Σ	Sum
($\bar{}$)	Superscript indicating an averaged measurement
($\hat{}$)	Superscript indicating an estimate
(*)	Indicates a complex conjugate (imaginary component multiplied by −1); thus, X = Re [X] + Im [X] X* = Re [X] − Im [X]
| |	Absolute value

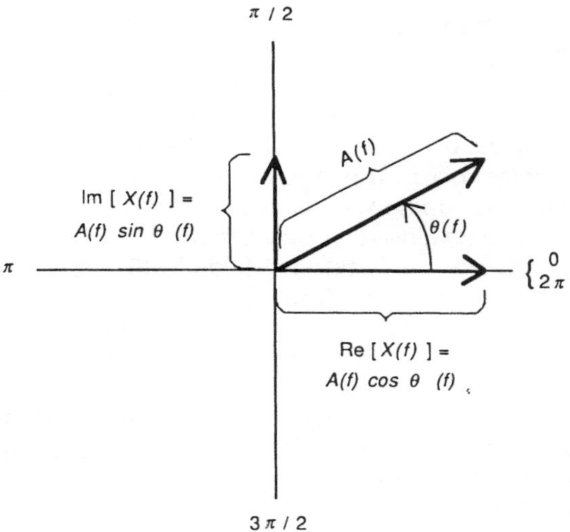

Figure 1. The amplitude and phase of a single frequency component (f) of a signal X(f) are shown in polar coordinates. Note that one can equivalently represent this vector by its Cartesian coordinates, the real and imaginary (orthogonal) parts of X(f).

We now consider the cross-spectral density function $G_{xy}(f)$, which relates x(t), the system input, to y(t), the system output, as a function of frequency f:

$$G_{xy}(f) = X^*(f)Y(f) \tag{4}$$

The amplitude of $G_{xy}(f)$ is the product of the amplitudes of X(f) and Y(f), while its phase will be the difference between output and input phase, i.e., the phase lead or lag introduced by the system. $G_{xy}(f)$ can be normalized by division by the square root of the product of the input and output power spectra, so that its real and imaginary components vary between −1 and +1. This normalized cross-spectrum is the complex coherence function, $\gamma_{xy}(f)$.

Coherence

γ^2_{xy} is the magnitude-squared coherence; assuming a linear system:

$$\gamma^2_{xy}(f) = \frac{|G_{xy}(f)|^2}{G_{xx}(f)G_{yy}(f)} \tag{5}$$

which varies from 0 to 1 and is real-valued. $\gamma^2(f)$ estimates the proportion of the power in y(t) which is attributable to input x(t), for frequency f. Thus,

$$\gamma^2(f) = \frac{|S(f)|^2}{|Y(f)|^2} = \frac{|S(f)|^2}{|S(f)|^2 + |N(f)|^2} \tag{6}$$

Algebraically manipulating this in terms of the signal to noise ratio, S/N (omitting the "f" for simplicity), yields

$$\gamma^2 = \frac{(S/N)^2}{(S/N)^2 + 1} \tag{7}$$

and

$$S/N = \sqrt{\frac{\gamma^2}{1 - \gamma^2}} \tag{8}$$

Expanding equation 5 yields an interesting result for the case in which all the necessary functions are estimated from a single pair of time segments $\hat{x}(t)$ and $\hat{y}(t)$. The superscript (ˆ) indicates an estimate of the "true" function based on a finite data sampling. Based on equations 3 and 4:

$$\hat{\gamma}^2(f) = \frac{|\hat{X}^*(f)\hat{Y}(f)|^2}{|\hat{X}(f)|^2\,|\hat{Y}(f)|^2} = \frac{|\hat{X}^*(f)|^2\,|\hat{Y}(f)|^2}{|\hat{X}(f)|^2\,|\hat{Y}(f)|^2} = 1 \tag{9}$$

Since $\hat{X}^*(f)$ and $\hat{X}(f)$ have identical amplitudes, their powers are also identical, and $\hat{\gamma}^2 = 1$. Coherence estimates from a single pair of samples will always be 1, regardless of the true coherence. This is analogous to estimating a correlation coefficient (e.g., for height and weight) from measurements of one individual; the result will be (spuriously) equal to 1.

For this reason, γ^2 estimates must always be smoothed, based on multiple independent samples. Most often, this is done by collecting multiple data segments for both x(t) and y(t), then averaging (for each frequency) the spectral estimates from the q segments, to yield the best overall estimates for $G_{xy}(f)$, $G_{xx}(f)$, and $G_{yy}(f)$:

$$\hat{G}_{xy}(f) = \frac{1}{q}\sum X_i^*(f)Y_i(f) \tag{10}$$

$$\hat{G}_{xx}(f) = \frac{1}{q}\sum |X_i(f)|^2 \tag{11}$$

$$\hat{G}_{yy}(f) = \frac{1}{q}\sum |Y_i(f)|^2 \tag{12}$$

$$\hat{\gamma}^2_{xy}(f) = \frac{|\hat{G}_{xy}|^2}{\hat{G}_{xx}(f)\,\hat{G}_{yy}(f)} \tag{13}$$

The averaging in equation 10 operates separately on the real and imaginary parts of $X_i^*(f)Y_i(f)$, not on amplitudes and phases per se. Thus, the real part of $\hat{G}_{xy}(f)$ is the average of the real parts of cross-spectra estimated from q separate segments of data. If there is no coherence between x(t) and y(t) at frequency f, the phases of the q separate estimates of $G_{xy}(f)$ will be randomly distributed, and both real and imaginary components will be as often negative as positive. Their averages and the magnitude of $\hat{G}_{xy}(f)$ will tend toward 0 as q increases. Conversely, the denominator of equation 13 is equal to the product of the mean powers of the q estimates of x(t) and y(t) and will not be reduced by averaging over data segments. Thus, when true coherence $\gamma^2 = 0$, estimated coherence $\hat{\gamma}^2$ will approach 0 and the variance of $\hat{\gamma}^2$ will decrease as q increases.

The spectral estimates can also be averaged across adjacent frequencies to reduce variance. For example, an FFT yielding 128 frequencies can be smoothed by calculating complex average values for groups of 4 frequencies, yielding a function with only 32 frequency values. If ℓ equals the number of frequencies in each group, and f now represents the center frequency for each group,

$$\hat{G}_{xy}(f) = \frac{1}{\ell} \sum X^*(f_j)Y(f_j) \qquad (14)$$

$\hat{G}_{xx}(f)$ and $\hat{G}_{yy}(f)$ are similarly smoothed. Frequency averaging and segment averaging can be used together, with the net reduction in variance of $\hat{\gamma}^2$ related to the product $q\ell$.

Our main interest in evoked potential work is in rejecting the null hypothesis $\gamma^2 = 0$, i.e., confirming the presence of a response. Critical values of $\hat{\gamma}^2$ for $\gamma^2 = 0$ (levels above which a response can be accepted) have been presented by various authors, based on theoretical calculations (Amos & Koopmans, 1963; Carter, Knapp, & Nuttall, 1973) and empirical simulations (Benignus, 1969). Carter et al. propose that for $q\ell > 32$, and γ^2 (true coherence) = 0, the bias of $\hat{\gamma}^2$ will be $1/q\ell$, and the variance $1/(q\ell)^2$. If the critical value for $\hat{\gamma}^2$ is equal to bias plus z standard deviations (determined by desired statistical significance level, α), then

$$\text{critical value}_\alpha\ (\hat{\gamma}^2) = \frac{1 + z}{q\ell} \qquad (15)$$

(This calculation assumes that $\hat{\gamma}^2$ is normally distributed; as will be seen, this assumption is unjustified and produces slightly too-liberal critical values.)

Recall from equation 7 the relation between γ^2 and the S/N ratio. For $S/N \ll 1$, as would be typical of most short-latency AEPs, γ^2 is approximately $(S/N)^2$. If $S/N = 0.1$ (i.e., -20 dB), $\gamma^2 = 0.01$. From equation 15, for a one-tailed test at a 99% confidence level ($\alpha = 0.01$), we would require $q \geq 333$ to reject the null hypothesis for a measured coherence value this low (assume for the moment that $\ell = 1$ and all smoothing is via segment averaging). At least for software-based systems, this would make inordinate demands on processing time and/or memory, and it is usually more practical to perform some time domain averaging prior to spectral analysis.

If each of q data segments is the result of averaging n separate samples, the resultant "subaverages" will have S/N ratios improved by a factor of $\sqrt{n}$, and γ^2 will increase approximately as n. Conversely, the required degree of smoothing will decrease by a similar factor, since (for large $q\ell$) the critical values of $\hat{\gamma}^2$ are inversely proportional to $q\ell$. For a fixed total number of samples (= qn), there is a trade-off between q and n. For high q and low n, $\hat{\gamma}^2$ and its critical value will be high. For low q and high n, both will be low.

Smoothing across frequencies ($\ell > 1$) decreases $\hat{\gamma}^2$ variance without requiring increased data collection time, but at the expense (often very acceptable) of decreased spectral resolution. Also, if there are rapid phase shifts in the system's transfer function for small frequency changes, cross-spectral values for adjacent frequencies will be out of phase and may tend to cancel one another, even if coherence is high for each.

We have so far ignored the nature of the stimulus x(t). We must assume that x(t) is a stationary time series, i.e., its average properties over successive segments of time do not change; the most appropriate stimulus for coherence analysis is Gaussian random noise. For such stimuli, $\gamma^2_{xy}(f)$ reflects the degree to which x(t) and y(t) are related by a linear process. It is often convenient to use periodic, deterministic inputs such as pseudorandom noise, particularly when extensive time domain averaging must be done to improve S/N ratios. Maki (1986) has shown that such stimuli permit nonlinearities present in the system to contribute to $\hat{\gamma}^2_{xy}(f)$. For example, harmonics generated in response to a given input frequency can result in apparently significant coherence for frequencies not present in the input. He suggests the use of pseudorandom stimuli composed only of prime-numbered harmonics, none of which are integer multiples of other components, to eliminate at least some kinds of nonlinearities from $\hat{\gamma}^2_{xy}(f)$. Even-order nonlinearities, such as are introduced by rectification, can be avoided by using inverse-repeat stimuli (Swerup, 1978); such stimuli have energy only at odd multiples of their fundamental frequency.

When a periodic stimulus is used (this would include not only pseudorandom noise, but any stimuli time-locked to the data collection period, such as clicks, tonebursts, etc.), and spectral estimates are averaged only across data segments, it can be shown that $\hat{\gamma}^2_{xy}(f)$ depends only upon output data. Since X(f) is invariant,

$$\gamma^2(f) = \frac{|X^*(f)|^2|\hat{Y}(f)|^2}{|X(f)|^2\hat{G}_{yy}(f)} = \frac{|\hat{Y}(f)|^2}{\hat{G}_{yy}(f)}$$

$$= \frac{\left|\frac{1}{q}\sum Y_i(f)\right|^2}{\frac{1}{q}\sum |Y_i(f)|^2} \qquad (16)$$

Restated, coherence in this case is equal to the power of the grand average response divided by the average power of the subaverages. When spectral estimates are averaged across adjacent frequencies ($\ell > 1$), however, this simplification is not applicable. Since the amplitudes and phases of adjacent frequencies in the stimulus may be quite different, the cross-spectra must be separately estimated for all frequencies prior to averaging.

Our software used algorithms based on equations 10 to 14. However, for any frequency not represented in the stimulus, this would lead to an undefined coherence since zeroes would appear in both numerator and denominator of equation 13. For these cases, equation 16 was used.

"Phase Coherence" and Other Measures

We must distinguish the magnitude-squared coherence $[\gamma^2(f)]$ from the "phase coherence" (PC) statistic which has recently been introduced to the AEP literature, for use in analysis of steady state evoked potentials (Jerger, Chmiel, Frost, & Coker, 1986; Picton, Vajsar, Rodriguez, & Campbell, 1987b; Stapells, Makeig, & Galambos, 1987). As for γ^2, PC analysis breaks up the data collection period into q subaverages, but instead of then carrying out complex averaging of the response or cross-spectrum vectors, the PC approach considers only the phases of each subaverage for a given frequency. These phase angles are projected onto a unit circle, and their sines and cosines are separately averaged. The length of the resulting vector (ranging from 0–1) is a measure of relative phase aggregation:

$$PC = \sqrt{\left(\frac{1}{q}\sum \cos\theta_i\right)^2 + \left(\frac{1}{q}\sum \sin\theta_i\right)^2} \quad (17)$$

Compare this to $\hat{\gamma}^2$, for the special case of an invariant input (equation 16):

$$\hat{\gamma}^2 = \frac{\left|\frac{1}{q}\sum Y_i(f)\right|^2}{\frac{1}{q}\sum |Y_i(f)|^2}$$

$$= \frac{\left(\frac{1}{q}\sum A_i\cos\theta_i\right)^2 + \left(\frac{1}{q}\sum A_i\sin\theta_i\right)^2}{\frac{1}{q}\sum A_i^2} \quad (18)$$

If we ignore amplitudes by setting all $A_i = 1$, $\hat{\gamma}^2 = PC^2$. Thus, the PC statistic is analogous to the square root of $\hat{\gamma}^2$, but disregards amplitude information. To the extent that AEPs result from phase reordering of EEG energy rather than the addition of new energy at specific frequencies, this is reasonable.

PC is identical to Goldberg and Brown's (1969) "vector strength," offered as a measure of phase-locking of single-unit discharges, and to the statistic $\bar{R}$ used in the Raleigh test of circular variance (see Mardia, 1972, Appendix 2.5, for a table of critical values for different numbers of subaverages, q). Fridman et al (1984) used a statistic they called "synchrony measure" (equal to PC^2) for objective detection of auditory brain stem responses.

Jervis, Nichols, Johson, Allen, and Hudson (1983) claim, on theoretical grounds, that PC (the Raleigh test) is less powerful than a statistic incorporating both amplitude and phase information, when the AEP actually adds energy to the EEG, rather than simply causing phase aggregation. Their "modified Raleigh" test statistic (nearly identical to $\hat{\gamma}^2$) proved superior to the Raleigh test in detecting long-latency AEPs.

Fridman et al (1982) computed power spectra for multiple subaverages and for a grand average response, then subtracted the spectrum of the grand average from the average spectrum of the subaverages. The difference spectrum value was large for frequencies not contributing to the ABR, and small for those contributing to the response. Their difference score (in dB), in fact, is equivalent to $10 \log_{10}\left(\frac{1}{\hat{\gamma}^2}\right)$, as can be seen from equation 16.

Sayers, Beagley, and Riha (1979) assessed phase aggregation using a simple "rotating Chi-square" test. The unit circle was divided into 24 bins of 15° width. For each of the 12 lines bisecting this circle, a Chi-square test determined whether there was a significant clustering of response phases in one or the other half-circles. Jervis et al (1983) describe an improved version of this test.

Picton, Skinner, Champagne, Kellett, and Maiste (1987) describe the use of the Hotelling T^2 test (Hotelling, 1931) to calculate a "confidence ellipse" for a set of vectors such as those shown in Figure 3: if the origin (0,0) is not included, the mean vector can be considered to represent a statistically significant response. Like γ^2 and Jervis' (1983) "modified Raleigh test," the T^2 test considers both magnitude and phase.

METHOD

Subjects

Subjects were young adults (21–33 years old) with normal hearing (15 dB HL or better for octave-interval pure tones from 250–8000 Hz, ANSI 1970). Otoscopic examinations and otologic histories were unremarkable. This paper presents results for individual subjects, but these were, for each experiment, representative of at least three subjects.

Stimuli

Two types of stimuli (clicks and amplitude-modulated tones) were digitally synthesized using a 256-point array (for details, see Wilson & Dobie, 1987), which was repetitively output at a D/A rate of 5.12 KHz. Thus, the interval between points was about 195.3 μsec, and the period of the entire stimulus waveform was 50 msec (256 × 0.1953); the fundamental frequency (f_0) of the stimulus was 20 Hz. A 500-Hz tone 100% amplitude-modulated at 40 Hz (AM tone) was synthesized by adding the 23rd, 25th, and 27th harmonics (= 460, 500, and 540 Hz, for f_0 = 20 Hz) in appropiate amplitude (1:2:1) and phase relationships. Spectral analysis of the transducer output showed no other spectral components within 70 dB of these three. A click (20/sec) was generated by simply outputting a positive pulse at the beginning of each cycle; alternating-polarity clicks (40/sec) were produced by adding a negative pulse at the midpoint of the cycle.

Both the AM tone and the alternating-polarity click are so-called inverse-repeat stimuli, containing energy only at odd harmonics of the fundamental frequency. (Consider, e.g., the 3rd harmonic. While phase at the end of the stimulus period will be the same as at the beginning, phase at the midpoint will be 180° opposite, 1½ cycles having been completed. Its waveform in the second half of the stimulus period is a mirror image of its first-half waveform, hence "inverse-repeat.") Such stimuli have the interesting property of permitting, in the frequency domain, separation of linear response components from even-order nonlinearities. For AEPs, stimulus artifact, cochlear microphonic, and frequency-following response (FFR) are at least primarily linear, and will contribute energy (and coherence) at the (odd) stimulus frequencies. Response coherence at even frequencies must represent nonlinearity introduced, for example, by rectification (this is presumably how the envelope of an AM tone becomes represented in auditory nerve discharge patterns). Auditory brain stem response (ABR) is itself primarily an even-order nonlinearity, since there is little difference between ABRs to condensation and rarefaction clicks. The common practice of adding responses to clicks of opposite polarity is equivalent to extracting only the even-order nonlinearities while suppressing linear (and odd-order nonlinear) response components (artifact, CM, FFR).

Stimuli were low-pass filtered (24 dB/octave) above 2 kHz, amplified, attenuated, and presented through a TDH-49 earphone in a MX-41/AR supra-aural cushion. Intensity is specified both in decibels sensation level (SL) and in either SPL (for the AM tone) or peSPL (for clicks). Naturally, SL-SPL differences varied slightly across subjects.

Recording and Analysis

Subjects were tested reclining, in a double-walled, sound-treated booth. They were instructed to remain alert but relaxed, and were spoken with by intercom at least every 5 minutes. Gold cup electrodes were attached to alcohol-cleansed skin at the vertex (positive), ipsilateral mastoid (negative), and contralateral mastoid (ground). Electrode impedances were less than 2 kΩ at 30 Hz. Scalp potentials were preamplified (10^5), sampled digitally at 5.12 kHz, and averaged. Preamplifier (Grass P511K) response passband (6 dB/octave) was 30 to 3000 Hz. While the low-pass setting (3 kHz) was greater than the Nyquist frequency (2.56 kHz), the next lower choice was 1 kHz, which would possibly have attenuated response frequencies of interest (ABR spectra roll off rapidly above about 1500 Hz). Time domain responses (except in Fig. 3) are usually presented as replicated averages (typically 2048 sweeps were averaged for each). Coherence functions were obtained as described in the "Calculation" section, with q subaverages, each of which was the average of n responses, and were sometimes further smoothed across groups of ℓ adjacent frequencies. Typically, q = 16 and n = 256, for a total (qn) of 4096 responses; this required about 3.5 minutes of recording time per coherence function (replicated time domain averages were simultaneously acquired). Critical values for $\hat{\gamma}^2$, from the literature, were tested by collecting data from human subjects under "no stimulus" conditions (where, by definition, $\gamma^2 = 0$) and obtaining empirical distributions of $\hat{\gamma}^2$ for various values of q, ℓ and α.

RESULTS

Distribution of Coherence Estimates when True Coherence = 0

For each of several values of q (4, 8, 16, 32, 64), eight coherence functions were calculated using scalp potentials from a subject receiving no stimulus. Each function contained coherence values for 127 frequencies, yielding 1016 (8 × 127) coherence estimates ($\hat{\gamma}^2$) for true coherence ($\hat{\gamma}^2$) = 0. Figure 2a shows the 99th percentile values for this distribution plotted with the theoretical distribution of Amos and Koopmans (1963) and the critical values estimated from Carter et al (1973), assuming a normal distribution (equation 15). Our data agree well with Amos and Koopmans, but the simple estimates based on the Carter et al mean and variance (for q ≤ 32) appear to be slightly too liberal. Figure 2b compares our 97.5th percentile data with the sources mentioned above, and also with the data of Benignus (1969), who carried out a Monte Carlo simulation similar to ours. Both sets of empirical data agree well with the Amos and Koopmans distribution.

We repeated this experiment, varying both q (as above) and ℓ (1, 2, 4, 8, 16). The results are not shown, but confirmed that the 99th and 97.5th percentile points in these distributions fit the Amos and Koopmans distributions very well, based on the ql product. For example, if qℓ = 8 and ℓ = 4, the 99th percentile was about 0.14, as for the case where q = 32 and ℓ = 1. For subsequent experiments, we used critical values from Amos and Koopmans (Table 2). Some of the qℓ values of interest to us are not tabulated in Amos and Koopmans and had to

(a)

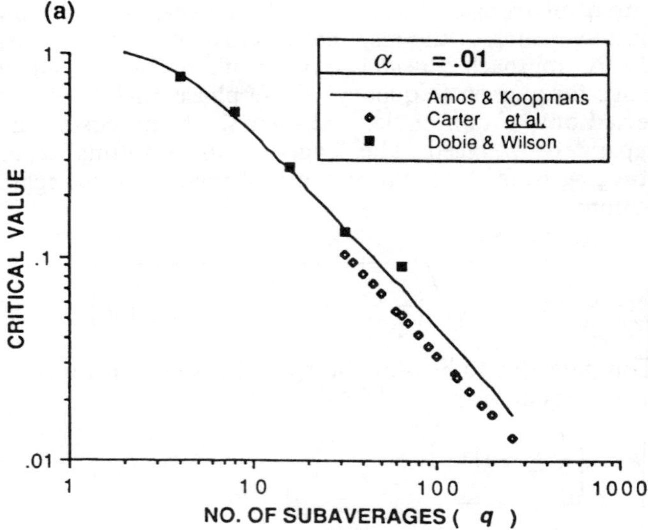

(b)

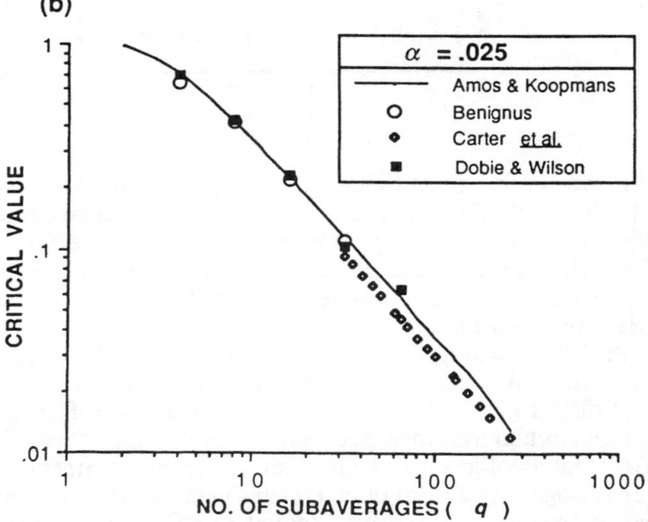

Figure 2. (a), Values (99th percentile) for coherence estimates obtained under no-stimulus conditions are plotted as a function of the number of subaverages (*filled squares*). Critical values for α = 0.01 using the bias and variance estimates of Carter et al (1973), and assuming normal distribution, are also shown (*open diamonds*). The continuous line represents the 99th percentile of coherence estimates for zero true coherence, according to Amos and Koopmans (1963). (b), The same data sources, as well as simulations by Benignus (1969) (*open circles*), for α = 0.025.

be obtained by log-linear interpolation (for example, they included q = 30 and 35, but not q = 32). Note also that their tables show distributions of $|\hat{\gamma}|$ rather than $\hat{\gamma}^2$; the values in Table 2 thus have been squared.

Responses to an AM Tone

Figure 3 shows in some detail how an actual coherence estimate (for one frequency) is obtained. The stimulus was the AM tone ($f_c = 500$ Hz; $f_m = 40$ Hz). At low to medium intensities, the primary scalp response follows the stimulus envelope: this is the amplitude-modulated following response (AMFR) described by Kuwada, Batra, and Maher (1986). Like the "40-Hz" response (Galambos, Makeig, & Talmachoff, 1981), it is maximal for modulation frequencies around 40 Hz, but it is less sensitive than response to 40 Hz transients (Palaskas, Wilson, & Dobie, 1988). The stimulus intensity in Figure 3 was 30 dB SL (58 dB SPL), just at the visual detection level of AMFR for this subject. The upper left portion of the figure (a) shows the q (=16) separate subaverages, with the grand average waveform shown below (b). The 40 Hz response appears as two cycles in this 50 msec interval. The upper right part of the figure (c) shows the 40 Hz Fourier components of the q subaverages, and the 40 Hz component of the grand average. The distribution of the powers of the q subaverages is shown in (d). Since a periodic stimulus was used, coherence can be calculated as the power of the mean response divided by the mean power of the subaverages (equation 16); these two power values are shown as scalars in Figure 3d. The result ($\hat{\gamma}^2 = 0.472$) far exceeds the 0.01

Table 2. Critical values for $\hat{\gamma}^2$ (when $\gamma^2 = 0$), from Amos and Koopmans (1963).

	Critical Value for Desired α Level (One-Tailed)		
qℓ	0.01	0.025	0.05
4	0.792	0.714	0.632
8	0.483	0.410	0.354
16	0.266	0.219	0.183
32*	0.139	0.114	0.094
64*	0.072	0.058	0.047
128*	0.036	0.029	0.023

* *By interpolation (see text).*

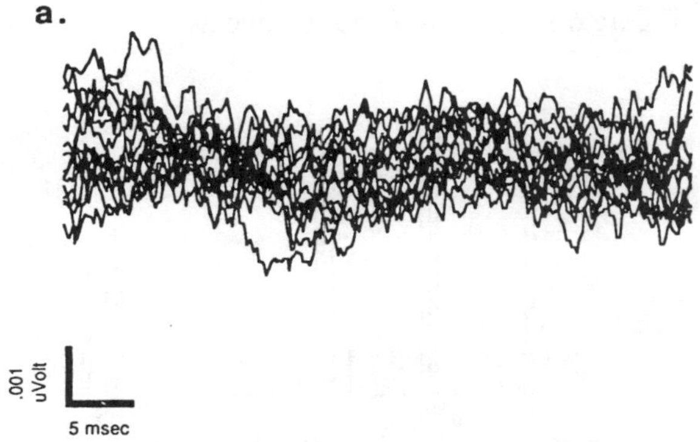

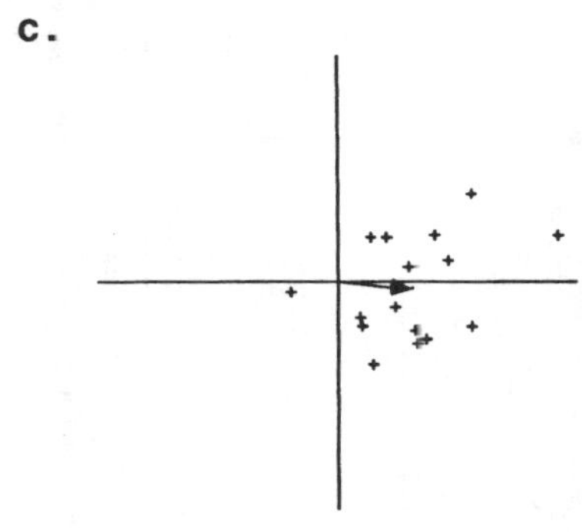

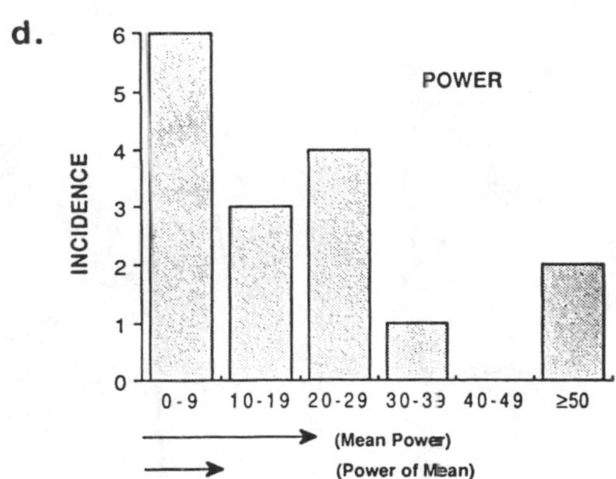

Figure 3. (a), Sixteen "subaverage" scalp responses ($n = 256$) to an AM tone (30 dB SL), superimposed. (b), The grand average of the waveforms in (a) shows an apparent 40-Hz component (2 cycles within the 50 msec response period). (c), The 40 Hz components of each of the 16 subaverage responses, obtained by FFT, are plotted in polar coordinates ("plus" signs), along with the 40 Hz component of the grand average (vector with arrowhead). (d), The histogram shows the distribution of 40-Hz power for the subaverages; the arrows (scalars) show the mean power of the subaverage responses and the power of the grand average. The ratio of the latter to the former is the estimated coherence.

critical value for q = 16 (Table 2), and indicates with more than 99% certainty that a response was present.

Figure 4 shows time domain response and the coherence functions for the same stimulus (a) at higher intensity (60 dB SL; 85 dB SPL). The 40 Hz envelope response is clearly evident, as is a 500 Hz FFR (Fig. 4b). Cross-correlation of the time domain response with the AM tone stimulus revealed a response latency of 5.96 msec, consistent with an FFR of brain stem origin. The coherence function (Fig. 4d) demonstrates statistically significant coherence for several frequencies: 40 Hz, its 2nd and 3rd harmonics, 500 Hz, the 460 and 540 Hz sidebands also present in the stimulus, and a few frequencies around 1000 Hz. The latter group of frequencies could be considered as harmonics of the stimulus components, but note that they are also present in the envelope of the stimulus (Fig. 4c; obtained by half-wave rectification, as at the hair cell afferent synapse). Rectification is an even-order nonlinearity; for an "inverse-repeat" stimulus containing only odd harmonics, like our AM tone, the new frequencies introduced by rectification are all even harmonics of the fundamental.

Responses to 40 Hz Alternating-Polarity Clicks

Figure 5 shows a series of time domain responses and coherence functions for 40 Hz alternating-polarity clicks (note that this too is an inverse-repeat stimulus). At 70 dB SL one can see stimulus artifact, ABR, and an apparent myogenic response. The coherence function displays statistically significant coherence across the frequency range, especially below 500 Hz and above 1000 Hz. In general, low coherence values alternate with high values. For example, 20 Hz coherence (the first harmonic) is about 0.17, while 40 Hz coherence (second harmonic) is 0.89. Below 460 Hz, all the significant coherences are for even harmonics; above 1560 Hz, only odd harmonics yield significant coherence. Since the stimulus is of the inverse-repeat (odd harmonic) type, most or all of the significant coherence above 1560 Hz is probably attributable to the large stimulus artifact present at this high intensity level.

At 50 dB SL, stimulus artifact is absent, and there is no significant odd-harmonic coherence. There is strong, even-harmonic coherence up to about 750 Hz, with a few values above the critical level for 1000 to 2500 Hz. Any of these isolated peaks could be spurious; since 127 coherence values are computed for each coherence function, there is a 0.72 probability that at least one will exceed the $p < 0.01$ critical value by chance alone ($1 - 0.99^{127}$). Note that 40 Hz coherence is statistically significant down to 10 dB SL, while visual detection of replicable time-domain response fails at about 30 dB SL in this case.

Figure 6 shows 40-Hz coherence as a function of inten-

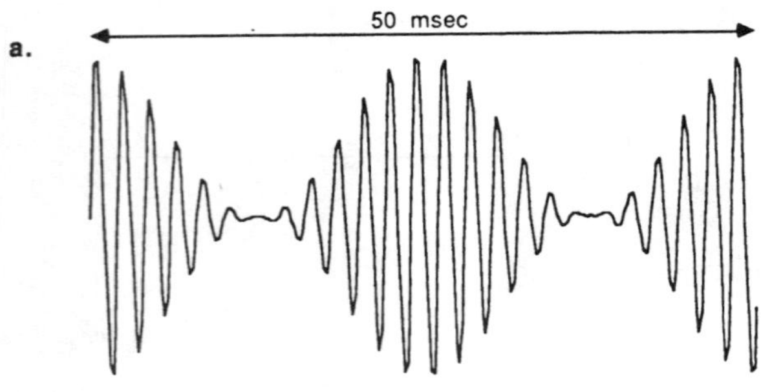

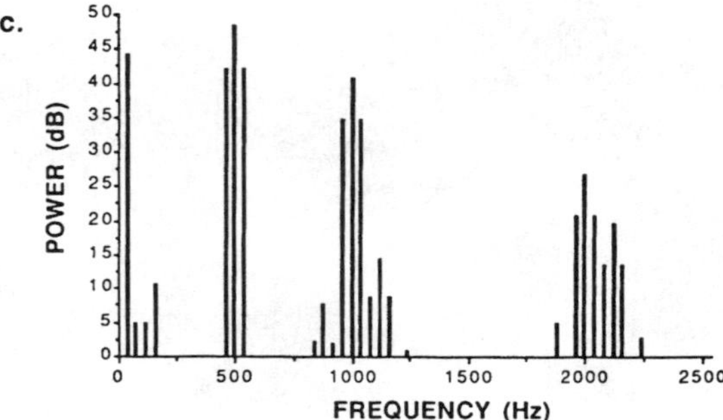

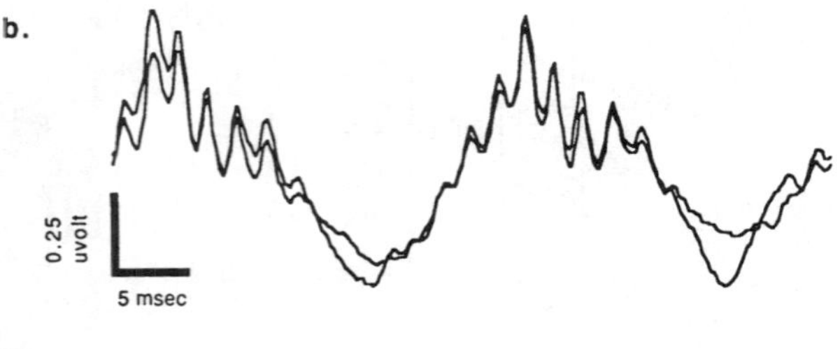

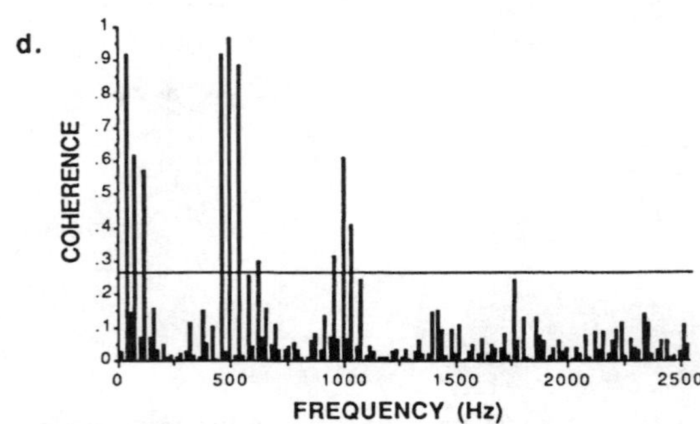

Figure 4. (a), AM tone; f_c = 500 Hz, f_m = 40 Hz. (b), Replicated scalp response showing responses to both f_c and f_m. (c), The envelope of the stimulus was obtained by half-wave rectification; the amplitude spectrum of that envelope is shown here. (d), The coherence function, calculated from the same scalp potential data in (b), shows peaks corresponding to the half-wave rectified stimulus, as well as harmonics of the principal envelope frequency (see text). The critical value at the 0.01 level of significance is indicated by the horizontal line at $\hat{\gamma}^2 = 0.266$ (see Table 2).

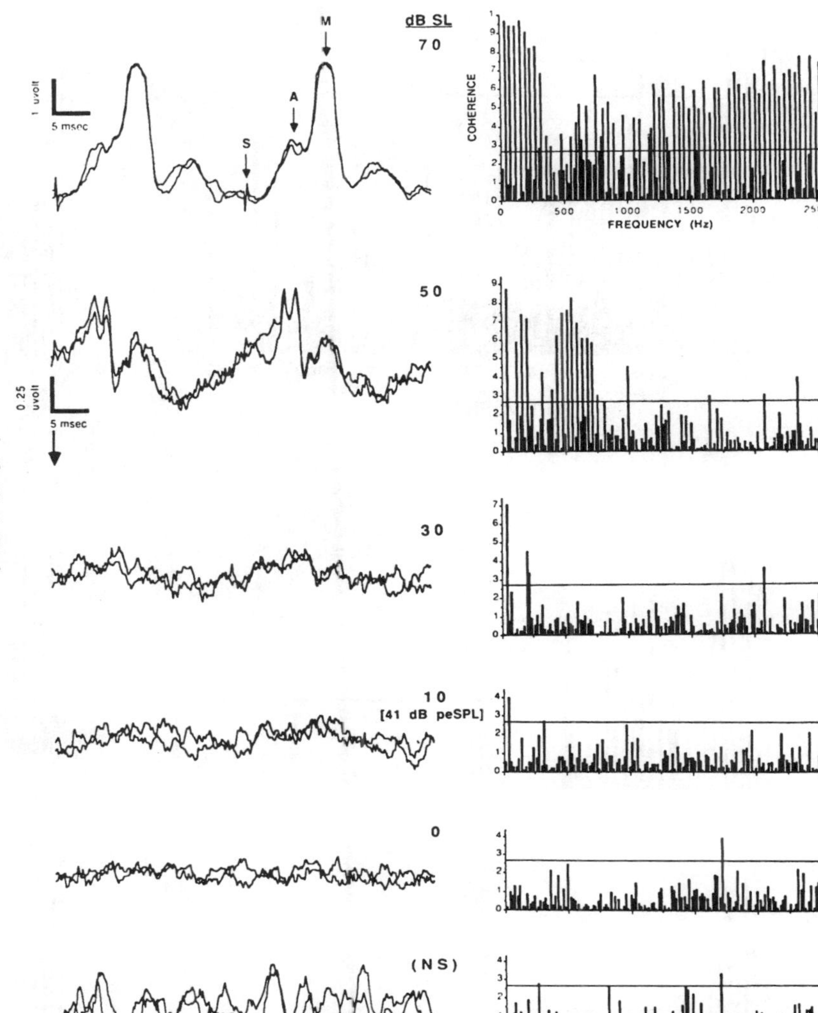

Figure 5. (*Left side*), Replicated scalp responses to alternating-polarity clicks. Intensity is indicated and decreases from top to bottom; NS, no stimulus; S, stimulus artifact; A, auditory brain stem response; M, myogenic response. (*Right side*) Corresponding coherence functions are shown (see text), with the critical value at the 0.01 level of significance.

sity, for two subjects. Significant coherence is present for all intensities down to 10 dB SL (equivalent to 41 and 44 dB peSPL for the two subjects), and is absent for 0 dB, −10 dB, and the no-stimulus condition.

Effects of Varying q, n, and ℓ

Recall (equation 9) that coherences estimated without-subaveraging will always (and spuriously) equal 1. Increasing the number of subaverages (q), while holding constant the number of responses per subaverage (n), should reduce all coherence estimates toward their true values. As Figure 7 illustrates (using responses to 40 Hz alternating-polarity clicks at a low intensity of 40 dB SL), this results in improved identification of those frequencies genuinely contributing to the evoked potential. Note that the critical values (from Table 2) decrease as q increases.

Alternatively, one can spend data acquisition time by

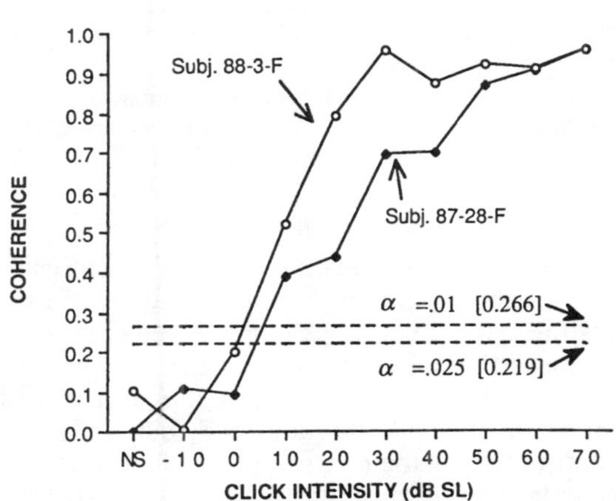

Figure 6. Forty Hz coherence (to alternating-polarity clicks) is shown for each of two subjects as a function of stimulus intensity. NS, no stimulus.

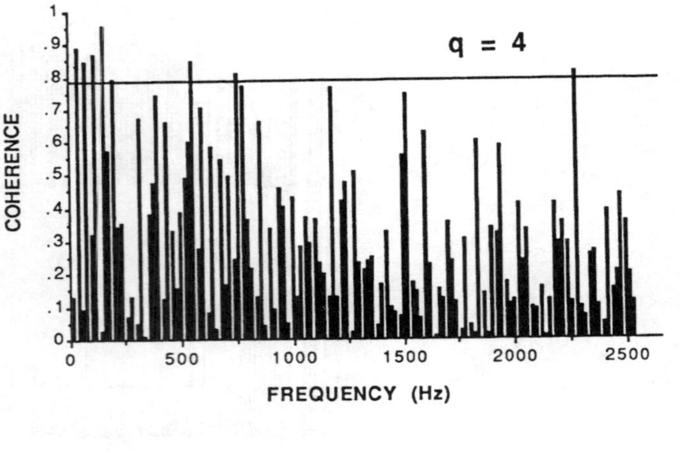

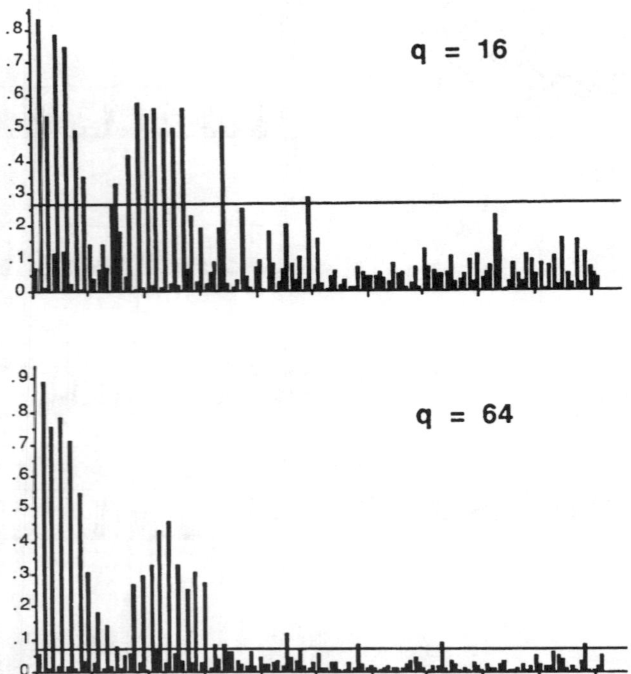

Figure 7. The effect of increasing the number of subaverages (q) on the coherence function is shown; the critical values at the 0.01 level of significance are included. Click intensity = 40 dB SL (74 dB peSPL).

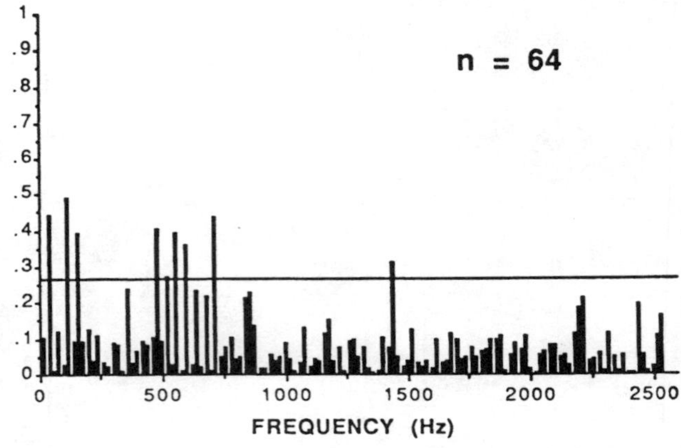

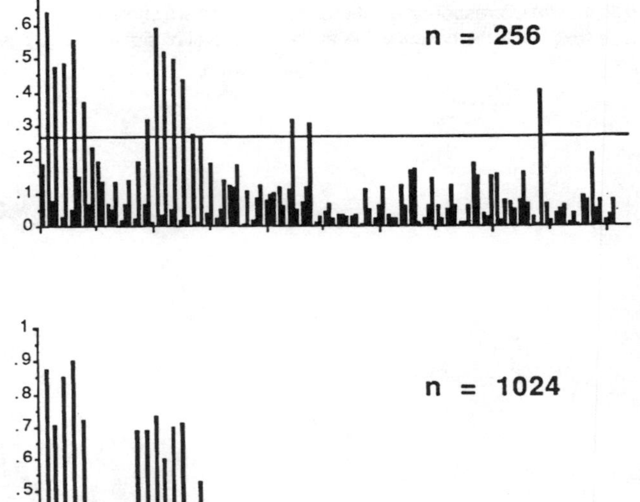

Figure 8. The effect of increasing the number of responses per subaverage (n) is shown; the critical value at the 0.01 level of significance is included. Click intensity = 40 dB SL (74 dB peSPL).

increasing n, leaving q constant. Since this increases S/N ratio, one expects (and finds) increasing coherence estimates, while critical values are unchanged (Fig. 8).

Increasing either q or n improves detection of significant coherence values and results in a greater number of contributing frequencies being identified, but both strategies require more time. One can improve coherence estimates without collecting more data by smoothing spectral estimates across frequency ($\ell > 1$). We initially tried this approach with alternating-polarity clicks. Since response to these stimuli, at moderate levels, shows significant coherence primarily for even harmonic frequencies, one would expect averaging across frequencies to be ineffective (unless the odd frequencies could be eliminated from calculations). This was indeed the case: for $\ell \geq 2$, all coherence estimates were below critical values. We then repeated the experiment with fixed-polarity clicks at 20/sec (Fig. 9). The coherence function ($\ell = 1$) is above the critical value to about 1100 Hz, and for increasing l, one finds both $\hat{\gamma}^2$ and critical values decreasing at about the same rate. At about 500 Hz, for example, $\hat{\gamma}^2$ is significant ($p < 0.01$) for $\ell = 1$ and $\ell = 4$, but not for $\ell = 16$.

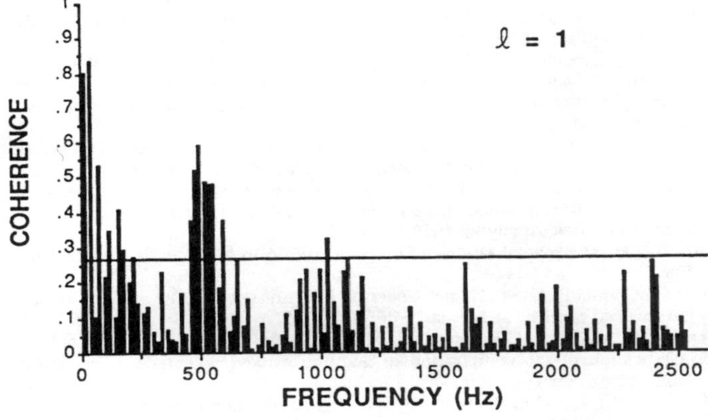

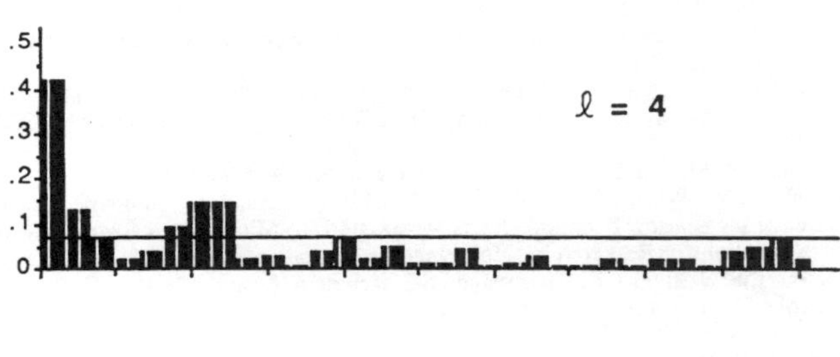

Figure 9. The effect of smoothing coherence estimates across frequency ($\ell > 1$) is shown; the critical values at the 0.01 level of significance are included. Click intensity = 40 dB SL (71 dB peSPL).

DISCUSSION

Spectral Content

The magnitude-squared coherence function identifies those frequencies contributing significantly to an evoked potential. This information can be useful in specifying analog and/or digital filter parameters and sampling frequencies for improving evoked potential detection in time domain waveforms. For example, while significant coherence is present (disregarding stimulus artifact) up to about 1500 Hz for responses to high intensity 40-Hz clicks, low-intensity responses have a much narrower coherence spectrum. This suggests that different low-pass cutoffs might be selected for optimizing either latency measures (high intensity) or threshold estimation (low intensity).

Several investigators (e.g., Boston, 1981; Suzuki, Sakabe, & Miyashita, 1982; Laukli & Mair, 1981) have studied power spectra of human ABR; there is general agreement that the upper limit of response power is between 1500 and 2000 Hz for high-intensity clicks, with lower frequencies predominating for lower intensities.

Threshold Estimation

Coherence analysis provides an objective estimate of response threshold with a known risk of a false positive response, depending on the α value chosen. As the data in Figures 5 and 6 suggest, this method may prove to be more sensitive than simple inspection of replicated responses; we have shown this to be true for AMFR (DL Tucci, MJ Wilson, RA Dobie, unpublished data). Coherence estimates should be tested for significance against the critical values tabulated by Amos and Koopmans (1963), a few of which are reproduced in Table 2.

Envelope Detection

Responses to AM tones showed clear FFR to the stimulus frequencies. In addition, significant coherence was present for the modulating frequency (40 Hz), its 2nd and 3rd harmonics, and the frequencies 960, 1000, and 1040 Hz. Response energy at overtones of the envelope and component frequencies has been noted before for both 40 Hz (Kavanagh & Domico, 1986) and FFR (Davis & Britt, 1984; Snyder & Schreiner, 1984, 1987). Interpretation of these nonlinearities requires some discussion of the concept of an envelope. Our AM tone, although synthesized by addition of sinusoids, was identical to the product of a carrier (500 Hz) and a modulator (40 Hz plus DC). In one obvious sense, the latter is the envelope, and can be derived from the AM tone via the Hilbert transform (Thrane, 1984). However, the mechanism by which the envelope is extracted in the cochlea is basically half-wave rectification and low-pass filtering. This produces a more complex envelope whose strongest spectral components are precisely those best seen in the coherence function. Thus, it seems that scalp response to an intense 500 Hz AM tone consists primarily of FFR plus even-order nonlinearities equivalent to rectification.

Some subjects showed significant coherence for other frequencies: 420, 580, and 620 Hz. These could represent combination tones of the form $2f_1$-f_2 or $2f_2$-f_1.

Coherence around 1000 Hz was seen only at 60 to 70 dB SL, and 500 Hz coherence disappeared at about 40 dB SL. As no acoustic distortion was measurable around 1000 Hz (phone output more than 70 dB down re:500 Hz), and because FFR sensitivity declines monotonically for frequencies above 500 Hz, it is quite unlikely that the 1000 Hz responses were FFRs to acoustic distortion.

Choice of Analysis Parameters

As expected, increasing either q or n improves the detection of significant coherence; both increase the total number of responses averaged. For a fixed data collection period, q and n will be inversely proportional, and there is no obvious way to optimize their values. For AEPs, q between 16 and 64 seems to produce acceptable coherence functions, and we used q = 16 for most of the experiments in this paper because computational time increases with q.

Smoothing coherence estimates across groups of adjacent frequencies is relatively ineffective for ABR, probably because response phase changes rapidly relative to the spectral resolution (20 Hz) in these studies. Finer resolution, achieved by use of longer stimulus/response periods, could improve the situation, but at the expense of longer recording times.

Alternatively, one could average groups of coherence estimates for adjacent frequencies. This differs from the frequency smoothing previously described, which requires the averaging of complex cross-spectral density functions prior to calculation of a coherence estimate and depends on a minimal rate of change of phase in the system transfer function. Averages of coherence estimates for groups of m frequencies will have narrower distributions than the single estimates (standard error of mean = standard deviation $\div \sqrt{m}$), and critical values will be correspondingly lower. This approach could be useful in analyzing responses such as ABR.

CONCLUSIONS

Our goal has been to introduce and illustrate the magnitude-squared coherence function as an AEP analysis tool. This technique appears promising for purposes of objectively identifying stimulus-response relationships in the frequency domain. It could be useful for threshold estimation for a variety of evoked potentials. Whether it offers any advantages over the other frequency domain analytic approaches described in the "Calculation" section remains to be seen. γ^2 can be calculated on-line, and is computationally about as complex as "phase coherence" and less complex than the T^2 test. It has the potential advantage over both of permitting smoothing across frequencies to reduce variance without increasing averaging time.

References

American National Standards Institute. American National Specifications for Audiometers, ANSI S3.6-1969. New York, 1970.

Amos DE and Koopmans LH. Tables of the Distribution of the Coefficient of Coherence for Stationary Bivariate Gaussian Processes [Monograph SCR-483]. Albuquerque: Sandia Corp, 1963.

Bendat JS and Piersol AG. Random Data: Analysis and Measurement Procedures, 2nd ed. New York: Wiley, 1986.

Benignus VA. Estimation of the coherence spectrum and its confidence interval using the fast Fourier transform. IEEE Trans Audio Electroacoust 1969;17:145–150.

Boston JR. Spectra of auditory brainstem responses and spontaneous EEG. IEEE Trans Biomed Engineer 1981;28:334–341.

Carter GC, Knapp CH, and Nuttall AH. Estimation of the magnitude-squared coherence function via overlapped fast Fourier transform processing. IEEE Trans Audio Electroacoust 1973;21:337–344.

Davis RL and Britt RH. Analysis of the frequency following response in the cat. Hear Res 1984;15:29–37.

Fridman J, John ER, Bergelson M, Kaiser JB, and Baird HW. Application of digital filtering and automatic peak detection to brain stem auditory evoked potential. Electroencephalogr Clin Neurophysiol 1982;53:405–416.

Fridman J, Zappulla R, Bergelson M, Greenblatt E, Malis L, Morrell F, and Hoeppner T. Application of phase spectral analysis for brain stem auditory evoked potential detection in normal subjects and patients with posterior fossa tumors. Audiology 1984;23:99–113.

Galambos R, Makeig S, and Talmachoff PJ. A 40 Hz auditory potential recorded from the human scalp. Proc Natl Acad Sci USA 1981;78:2643–2647.

Goldberg JM and Brown PB. Response of binaural neurons of dog superior olivary complex to dichotic tonal stimuli: some physiological mechanisms of sound localization. J Neurophysiol 1969;32:613–636.

Hotelling H. The generalization of Student's ratio. Ann Math Statist 1931;2:360–378.

Jerger J, Chmiel R, Frost JD, and Coker N. Effect of sleep on the auditory steady state evoked potential. Ear Hear 1986;7:240–245.

Jervis BW, Nichols MJ, Johnson TE, Allen E, and Hudson NR. A fundamental investigation of the composition of auditory evoked potentials. IEEE Trans Biomed Engineer 1983;30:43–49.

Kavanagh KT and Domico WD. High-pass digital filtration of the 40 Hz response and its relationship to the spectral content of the middle latency and 40 Hz responses. Ear Hear 1986;7:93–99.

Kuwada S, Batra R, and Maher VL. Scalp potentials of normal and hearing-impaired subjects in response to sinusoidally amplitude-modulated tones. Hear Res 1986;21:179–192.

Laukli E and Mair IWS. Early auditory-evoked responses: spectral content. Audiology 1981;20:453–464.

Maki BE. Interpretation of the coherence function when using pseudorandom inputs to identify non-linear systems. IEEE Trans Biomed Engineer 1986;33:775–779.

Mardia KV. Statistics of Directional Data. London: Academic Press, 1972.

Palaskas CW, Wilson MJ, and Dobie RA. Electrophysiologic assessment of low-frequency hearing: Sedation effects. Abstracts 11th Midwinter Research Meeting, Association for Research in Otolaryngology, Clearwater Beach, FL, 1988.

Picton TW, Skinner CR, Champagne SC, Kellett AJC, and Maiste AC. Potentials evoked by the sinusoidal modulation of the amplitude or frequency of a tone. J Acoust Soc Am1987;82:165–178.

Picton TW, Vajsar J, Rodriguez R, and Campbell KB. Reliability estimates for steady-state evoked potentials. Electroencephalogr Clin Neurophysiol 1987;68:119–131.

Sayers BM, Beagley HA, and Riha J. Pattern analysis of auditory-evoked EEG potentials. Audiology 1979;18:1–16.

Snyder RL and Schreiner CE. The auditory neurophonic: basic properties. Hear Res 1984;15:261–280.

Snyder RL and Schreiner CE. Auditory neurophonic responses to amplitude-modulated tones: transfer functions and forward masking. Hear Res 1987;31:79–92.

Stapells DR, Makeig S, and Galambos R. Auditory steady-state responses: threshold prediction using phase coherence. Electroencephalogr Clin Neurophysiol 1987;67:260–270.

Suzuki T, Sakabe N, and Miyashita Y. Power spectral analysis of auditory brain stem responses to pure tone stimuli. Scand Audiol 1982;11:25–30.

Swerup C. On the choice of noise for the analysis of the peripheral auditory system. Biol Cybern 1978;29:97–104.

Thrane N. The Hilbert transform. Brüel & Kjaer Tech Rev 1984;(3):3–15.

Wilson MJ and Dobie RA. Human short-latency auditory responses obtained by cross-correlation. Electroencephalogr Clin Neurophysiol 1987;66:529–538.

Acknowledgments: The authors wish to express their thanks and appreciation to Chris Prall, scientific programmer, who wrote most of the software used in our laboratory. Drs. Ed Rubel, William Lippe, Nina Kraus, Therese McGee, Richard Folsom, Ben Clopton, Douglas Keefe, Francis Spelman, and Jack Cullen read earlier versions of this manuscript and made helpful suggestions.

Address reprint requests to Robert A. Dobie, M.D., Department of Otolaryngology-Head and Neck Surgery (RL-30), University of Washington, Seattle, WA 98195.

Received August 5, 1988; accepted November 7, 1988.

TIME DELAY ESTIMATION : APPLICATION TO FLOW RATE MEASUREMENT OF COOLING FLUID IN NUCLEAR POWER PLANTS

J.-M. FAVENNEC*, B. GEORGEL* and J. MASSON**

*Electricité de France
DER, 6 Quai Watier
78400 CHATOU, France

**Institut National Polytechnique de Grenoble. CEPHAG. Domaine Universitaire. 38400 St Martin D'Hères

Abstract : *Precise measurement of the coolant flow rate is of utmost importance for the control and safety of PWR nuclear power plants. In this paper several methods based on generalized cross-correlation of γ-radioactivity signals are tested. We examine the influence of hydrodynamical parameters such as velocity profile and turbulence on the measurement. We show that recently proposed sophisticated estimates (PHAT-SCOT) are not efficient for our purpose and that the Hannan-Thomson (HT) estimate is just slightly better than classical cross correlation (CC).*

I. INTRODUCTION

The measurement of the coolant fluid flow-rate in a PWR-type nuclear reactor is very important in regard of the safety and control.
This can conveniently be achieved by estimating a time delay between two signals. The new method is easy-to-set-up, non-intrusive and at least as accurate as the indirect methods that have been used so far. Our paper presents the results of a study comparing various time-delay estimates applied to on-site recorded signals.

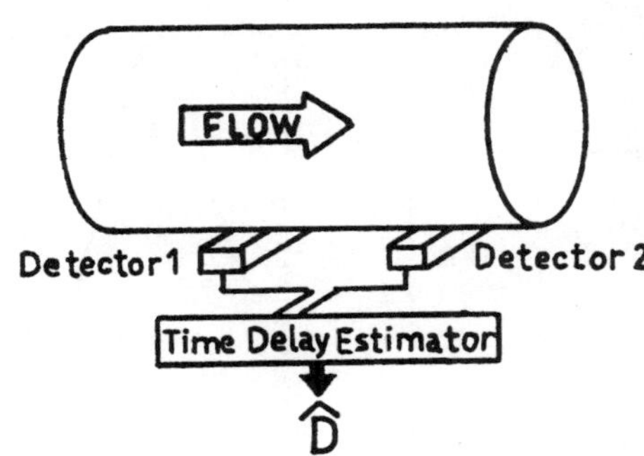

Fig. 1 : Principle of the method

II. PRINCIPLE OF THE METHOD (Fig. 1)

A. *The primary circuit*

In pressurized water reactors (PWR's) the thermal energy yielded in the core is supplied to the primary fluid (or coolant). This coolant is ordinary water at 155 bar pressure and 300°C (570°F) temperature which flows through three primary loops. We are interested in measuring the flow rate of the coolant in each loop.

B. *Origin of the signals*

When passing through the core the water molecules are activated by the fast neutrons and Oxygen-16 elements are turned into Nitrogen-16 elements (N16). N16 is a γ-ray emitting element which is carried by the fluid.

The method of time delay estimation mainly utilizes the time fluctuations of the γ-activity of the coolant.

C. *Signal Detection*

The highly energetic (6 MeV) γ-rays of N16 are detected outside of the pipe by two ionization chambers separated by a distance d (d = 1 m).
Each detector integrates the γ-activity of different elementary volumes of primary fluid located in the cross-section of the pipe.

D. *Flow characterization*

The primary fluid flows in 0.7m diameter steel pipes at an approximative velocity of 17 m/s. The value of the Reynolds number is 10^8 : the flow is therefore a fully developed turbulent one [1]. The parameter to be measured is the true mean velocity in the pipe defined as :

$$V_{mean} = \frac{1}{S} \iint_S V(r,\theta) r \, dr \, d\theta \qquad (1)$$

where $V(r,\theta)$ is the velocity profile (Fig.2).

Defining $D(r,\theta) = \frac{d}{V(r,\theta)}$ (time-delay profile) we obtain : $V_{mean} = \frac{1}{S} \iint_S \frac{d}{D(r,\theta)} r \, dr \, d\theta$.

The method of time-delay estimation yields : $V_{TDE} = \frac{d}{\hat{D}}$ (2), $\hat{D}$ being the estimate of the

Reprinted from *Proc. ICASSP '82*, vol. 1, pp. 379–382, May 1982.

transit time.

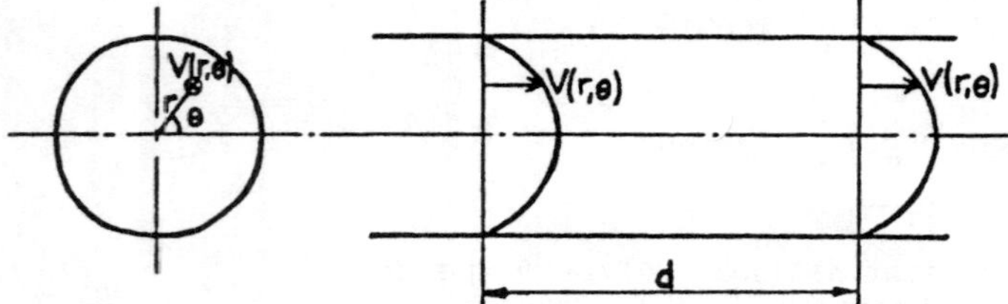

Fig. 2. Flow profile in the pipe.

E. *Influence of the time-delay profile on the measurement*

It has been shown [2] that the Cross-Correlation function can be written as follows :

$$R_{vu}(\tau) = \iint_S R_{uu}(\tau - D(r,\theta))r\,dr\,d\theta \tag{3}$$

where R_{uu} is the auto-correlation function of the signal originating from a cross-section of the pipe.

A non-constant time-delay profile gives a bias of $E\{\hat{D}\} - \frac{d}{V_{mean}}$. This is also apparent on the expression of the coherence function between u and v :

$$\Gamma_{vu}(f) = \frac{1}{S}\iint_S e^{-2\pi jfD(r,\theta)}\,r\,dr\,d\theta \tag{4}$$

Assuming various laws for $D(r,\theta)$, it can be shown that the velocity profile results in a degradation of coherence (Fig.3) and phase distortion (Fig.4). The values of V_{mean} (see (1)) and V_{TDE} (see (12)) are represented in Fig.5.

If one simulates a realistic velocity profile for the flow, the resulting bias does not exceed 0.5 % and the simulated phase and coherence are poorly affected. The comparison with the estimations (see II.F.) suggests that another phenomenon degrades the measurement : this can be regarded as the effect of turbulence.

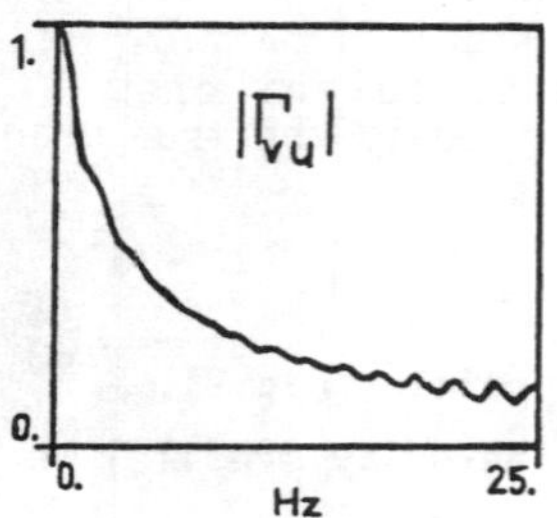

Fig.3 Simulated Coherence

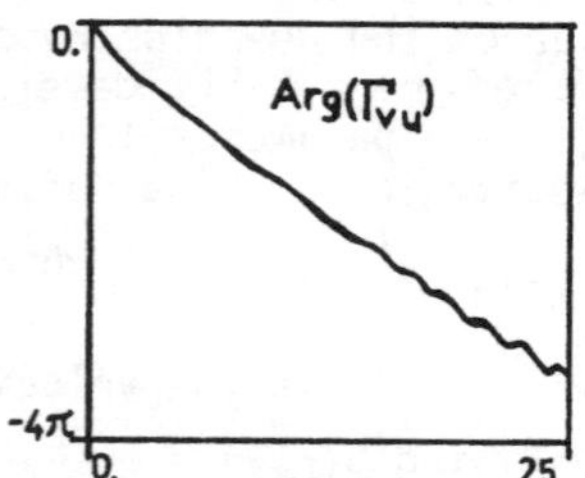

Fig.4 Simulated Phase

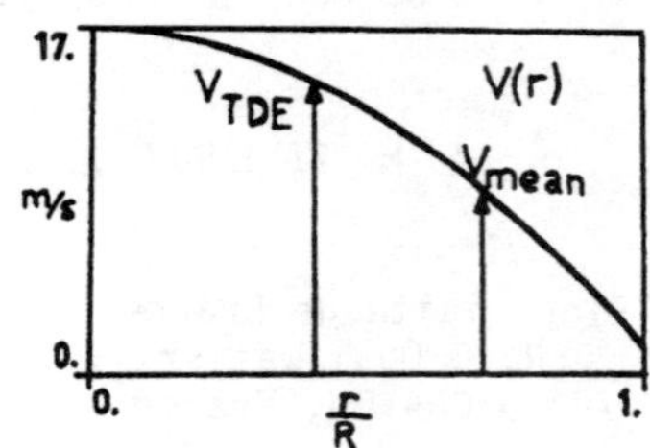

Fig.5 Velocity Profile

F. *Signal Characterization*

The signals x and y provided by the two ionization chambers were recorded during a long period on a 900 MW French reactor stabilized at its normal power level.

The estimates $\hat{G}_{xx}$, $\hat{G}_{yy}$, $\hat{\Gamma}_{yx} = \frac{\hat{G}_{yx}}{\sqrt{\hat{G}_{xx}\cdot\hat{G}_{yy}}}$ and $\hat{\rho}_{yx} = \frac{\hat{R}_{yx}(\tau)}{\hat{R}_{xx}(0.)}$ are shown in Fig.6 to Fig.9.

The useful bandwidth is about 20 Hz. It can be pointed out that the bias on the estimation of $C_{yx} = |\Gamma_{yx}|^2$ resulting from misalignment [3] is very small (-0.6 % C_{yx} in our case).

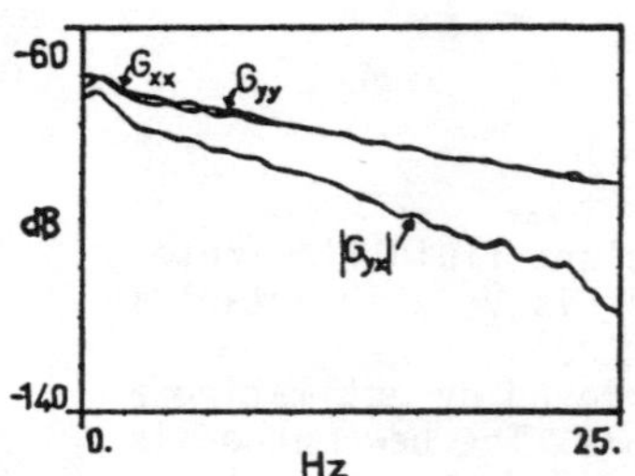

Fig.6 Estimated Spectra and cross spectrum

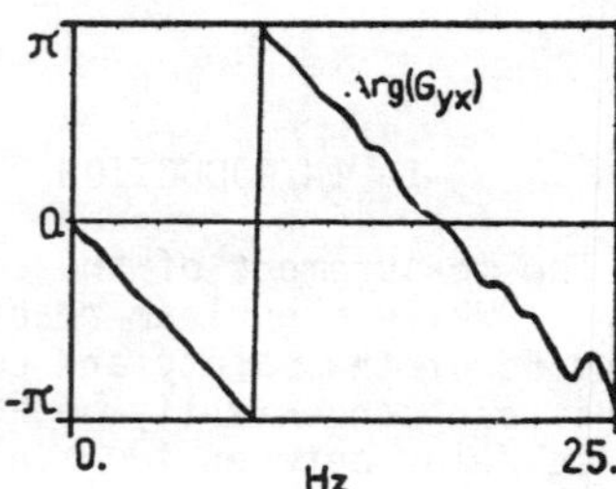

Fig.7 Estimated Phase of the cross spectrum

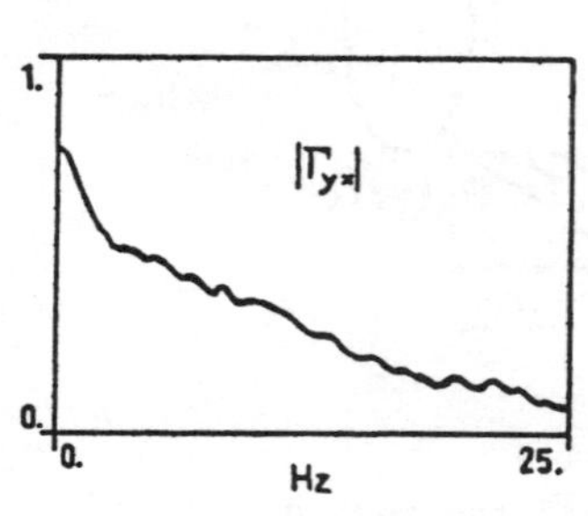

Fig.8 Estimated Coherence

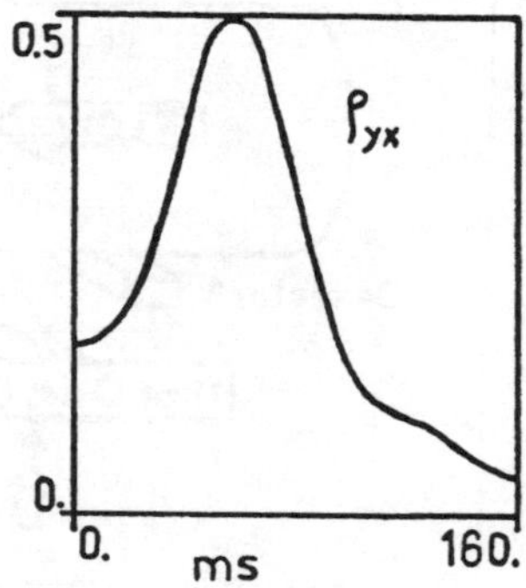

Fig.9 Estimated Normalized cross - Correlation

III. MODEL

Let the physical phenomena be modelled as (see Fig.10) :

$$(5)\quad \begin{cases} x(t) = S(t) + n_1(t) \\ y(t) = F(S(t)) + n_2(t) + n_3(t) \end{cases}$$

where :

- x and y are the output signals of the detection and processing channels.
- S represents the time fluctuation of the γ-activity.
- F is a linear transform whose gain is :

  $H(f) = \alpha(f)\ e^{-2\pi jfD}\ e^{j\varphi(f)}$ where $\varphi(f)$ represents the dispersive effect of the velocity profile.
- n_1 and n_2 represent the emission noise of the radioactive particles, the detection noise and the electronic noise of each channel.
- n_3 represents the non-coherent fluctuations of γ-activity between the two cross-sections due to the velocity profile.

Hypotheses :

H1. S, n_1, n_2 and n_3 are ergodic, stationnary and mutually independent random signals.

H2. The turbulence phenomena are not taken into account.

H3. The two channels are calibrated : their relative phase is zero.

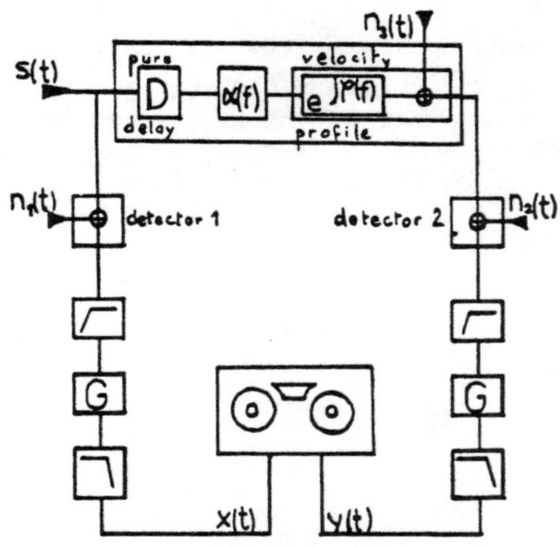

Fig.10 Block diagram of the model.

Equations (5) can be summed up in one simple equation : $y(t) = F(x(t)) + n(t)$ (6)

which leads to :

$$(7)\quad \begin{cases} G_{yy}(f) = \alpha^2(f)\ G_{xx}(f) + G_{nn}(f) \\ G_{yx}(f) = \alpha(f)\ G_{xx}(f)\ e^{-2\pi jfD}\ e^{j\varphi(f)} \end{cases}$$

The transit time is small enough to ensure that the spectral content remains unchanged.

Hence : $G_{xx} = G_{yy}$ (see Fig.6).

$$(8)\quad C_{yx} = \alpha^2 \quad ; \quad (9)\quad G_{nn} = (1-C_{yx})\ G_{xx}$$

IV. TIME DELAY ESTIMATION

It has been shown [4], [5], [6] that numerous methods can be used to estimate a time delay between two signals and that they could be written in the following short form :

$\hat{D}$ = abcissa of the maximum of the function $\hat{R}_{yx}$

$$\text{where } \hat{R}_{yx}(\tau) = \int_{-\infty}^{+\infty} W(f)\ \hat{G}_{yx}(f)\ e^{2\pi jf\tau}\, df \qquad (10)$$

W(f) is a weighting function defined as follows :

Estimate	W(f)
Cross-Correlation (CC)	1
Smoothed Coherence Transform (SCOT)	$\frac{1}{\sqrt{G_{xx}\cdot G_{yy}}} = \frac{1}{G_{xx}}$
Phase Transform (PHAT)	$\frac{1}{\lvert G_{yx}\rvert} = \frac{1}{\alpha\ G_{xx}}$
Hannan-Thomson (HT)	$\frac{1}{\lvert G_{yx}\rvert}\ \frac{C_{yx}}{1-C_{yx}} = \frac{1}{G_{xx}}\ \frac{\alpha}{1-\alpha^2}$

Table I. Time Delay Estimates.

A. *Bias*

All these estimates have a bias due to the presence of $\varphi(f)$ in the cross spectrum's phase. It can be shown using [7] that :

$$\text{Bias}(\hat{D}) = -\frac{\int_{-\infty}^{+\infty} W(f)\ \alpha(f)\ \sin(\varphi(f))\ G_{xx}(f)\, df}{\int_{-\infty}^{+\infty} f^2\alpha(f)\ W(f)\ \cos(\varphi(f))\ G_{xx}(f)\, df} \qquad (11)$$

It would thus be interesting to use the function :

(12) $\int_{-\infty}^{+\infty} W(f)\ e^{-j\varphi(f)}\ \hat{G}_{yx}(f)\ e^{2\pi jf\tau}\, df$. But the estimation of $\varphi(f)$ is rather difficult : one must know how to model phenomena such as non steady turbulent flow ...

B. *Variance*

The variance of the different estimates can be theoretically computed [5] by :

$$\mathrm{Var}(\hat{D}) = \frac{\int_0^\infty W^2(f)\ G_{xx}(f)\ (1-C_{yx}(f))\ f^2\,df}{8\pi^2 T\left[\int_0^\infty |G_{yx}(f)|\ W(f)\ f^2\,df\right]^2} \qquad (13)$$

In practice (13) is used with the estimations $\hat{G}_{xx}$, $\hat{G}_{yx}$, $\hat{C}_{yx}$, $\hat{W}$ computed on a long integration time.

The results for T = 50 s are presented in Table II and Fig.11 :

Estimate	PHAT	SCOT	CC	HT
Standard Deviation (ms)	1.65	1.38	1.24	1.22

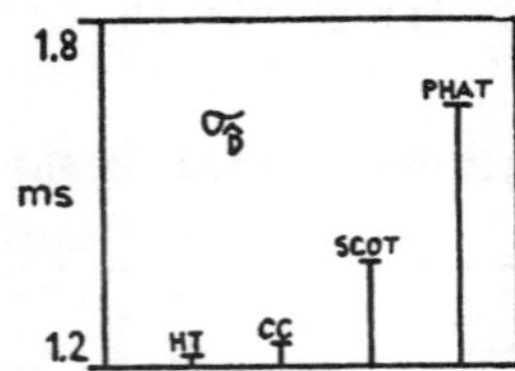

Table II and Fig.11

Theoretical Standard Deviations

It can be seen that SCOT and PHAT are quite unsuitable for our application. This is in accordance with Carter's results [8] which tend to prove that these estimates show a good performance only in the presence of "strong" periodic components.

On the other hand, the very slight difference between CC and HT is worth noticing (1.6 %).

To check this result we applied the two methods to the same signals and experimentally computed the standard deviations of series of 50 measurements (no overlapping, individual integration time of 50 s). The estimated time-delay was found equal to 58 ms.
The weighting function of the HT estimate was computed once for all with a 30 mn-long integration time.

The results are presented in Table III :

Estimate	Theoretical St. Deviation	Sample St.Deviation ($\hat{\sigma}_S$)	95 % Confidence Interval	$\hat{\sigma}_S / \hat{D}$
CC	1.24	1.5	1.24 - 1.84	2,1 %
HT	1.22	1.49	1.23 - 1.83	

Table III. Comparison of Theoretical and Sample Standard Deviations.

V. CONCLUSION

In this study we have been interested in various recently developed time delay estimates for an industrial application of flow-rate measurement.

Our paper shows that SCOT and PHAT are not suited to our problem because of a large variance. If HT does minimize the estimation error, CC is almost as efficient.

In regard of the complexity of implementing the HT estimate and the very slight resulting gain, the CC estimator appears to be the right compromise for our particular application.

REFERENCES

[1] S.I. Pai, Viscous Flow Theory Part II Turbulent Flow. D. Van Nostrand Company 1957.

[2] J. Masson, "Effet du profil des vitesses sur l'estimation du débit primaire par des méthodes de corrélation" Technical Report EDF-DER, HP/126/81-27. Oct. 81.

[3] G.C. Carter, "Bias in magnitude-Squared Coherence estimation due to misalignment", IEEE Trans. on A S S P, Vol 28, n°1, pp 97-99, 1980.

[4] G.C. Carter, A.H. Nuttall and P.G. Cable, "The smoothed Coherence Transform", Proceedings IEEE (Lett.), Vol 61, pp 1497-1498, 1973.

[5] C.H. Knapp and G.C. Carter, "The generalized Correlation method for estimation of time-delay" IEEE Trans. on A S S P, Vol 24, n° 4, pp 320-327, 1976.

[6] E.J. Hannan and P.J. Thomson, "The estimation of coherence and group delay", Biometrika, Vol 58, pp 469-481, 1971.

[7] Ph. Bolon, "Application de méthodes de corrélation de signaux optiques à des mesures de vitesse sans contact. Utilisation dans l'industrie papetière", Thèse de Docteur-Ingénieur, Institut National Polytechnique de Grenoble, Janv. 1981.

[8] K. Scarbrough, N. Ahmed and G.C. Carter, "an experimental comparison of the cross-correlation and SCOT techniques for time-delay estimation", Proceedings ICASSP 80, pp 807-810, 1980.

Delay Estimation of Disturbances on the Basilar Membrane

MARY MORTON GIBSON

Abstract–With current measurement techniques, it is impossible to directly observe the traveling wave on the basilar membrane over its entire length. Treating the mechanical propagation as a pure conduction delay, the travel time to a particular region can be inferred from the phase-frequency characteristics of the neural response.

INTRODUCTION

Changes in sound pressure around us are transmitted to the fluids of the inner ear where they are transduced into electrical impulses which are sent to the brain. The external ear directs the impinging sound pressures down the auditory canal and against the eardrum, causing it to move. This motion causes the bones of the middle ear to move, transmitting the motion to the oval window membrane. The middle ear behaves as an impedance matching device between the outside air and the inner ear fluid. When the oval window membrane moves, a differential pressure is established across the *basilar membrane* (BM), causing it to exhibit a *traveling wave* of displacement. The BM's mechanical properties vary greatly along its length. Near the base it is narrow and stiff; at the apex it is wider and more flaccid. Due to the elasticity changes with length, the region of maximum vibration of the BM changes with the frequency of the sound. High frequencies cause a maximal displacement at the basal end and lower frequencies cause a maximal displacement at the apical end. Motion of the BM is transduced into action potentials which propagate along the *auditory nerve* fibers. Each fiber makes contact with nerve cells (*neurons*) in the *cochlear nucleus*. The response of individual neurons in the anteroventral portion of the cochlear nucleus (AVCN) seems to be in several respects very similar to the response of single auditory nerve fibers [11].

Békésy was the first to measure the *travel time* of a mechanical disturbance to various loci on the BM [3]. His measurements were made in dead specimens of several species including man, using a light microscope with stroboscopic illumination. Observations were limited to the apical $\frac{1}{3}$ of the BM. To make the vibrations visible it was necessary to use very intense sounds (140 dB SPL, sound pressure level in dB re 20 μPa). The use of nonliving tissue and nonphysiological intensities almost certainly affected the results. Greenwood [7] used Békésy's data and a behaviorally determined frequency map [13] to develop an empirical function to describe the distance-frequency relation for several species:

$$X = A \log((f + B)/B)$$

Manuscript received May 26, 1980; revised October 24, 1980. This work was supported by NIH Grants NS06225, NS12732, NS0536, and NS30897.

The author is with the Department of Neurophysiology, University of Wisconsin, Madison, WI 53706.

where X is the distance in millimeters measured *from the apex* and f is *best frequency* in hertz. For the cat, $A = 10.48$ and $B = 419$.

The Mössbauer technique has been used to measure the squirrel monkey BM response to brief clicks. Robles [10] and his colleagues observed the response of a basal segment of the *cochlea* (6.0–7.8 kHz) where they found the travel time to be 300–390 μs.

Kiang [9] measured (102 cat auditory nerve fibers) response latency to brief clicks. He found latencies (presumably travel time plus any constant delays) of 1.3–1.8 ms for fibers with BF above 2–3 kHz and longer latencies for fibers with a lower BF. More recently, efforts have been made to infer the travel time of disturbances on the BM from the responses of either auditory nerve fibers or AVCN neurons to pure tones. Anderson *et al.* [2] were the first to infer travel time (58 squirrel monkey auditory nerve fibers) by the methods we used. In the 7 kHz region they found a travel time of about 500 μs, which is slightly longer than the estimate from Mössbauer data [10].

METHODS

Data were recorded from neurons in AVCN of anesthetized cats. Surgical preparations and recording techniques have been described previously [11]. In each animal we used a Kellogg probe-tube condenser microphone to calibrate the sound pressure level near the eardrum for frequencies from 20 to 25 600 Hz. Calibration data were stored by the LINC computer to allow delivery of a fixed SPL at any frequency under computer control.

When the electrical activity of a neuron was isolated, the frequency-intensity domain was explored systematically, from low to high, both for SPL's at each frequency and for successive frequencies. Stimulus duration and interstimulus interval were independently variable. Frequency steps of 200–300 Hz and intensity steps of 10 dB were used.

Two measures were used to evaluate the neuronal discharges. 1) Response areas: the number of discharges plotted as a function of stimulus frequency with intensity plotted as a parameter. 2) Period histograms: the number of discharges plotted as a function of stimulus phase. The abscissa, always equal to the stimulus period, is divided into a set of contiguous time periods (bins). The bin height indicates the number of discharges which occur within the time limits of that bin. The period histogram portrays a statistical estimate of the probability of firing as a function of the phase of the stimulus.

The amplitude R and angle ϕ of the radius vector [6] were computed for each histogram. This vector is the vector sum computed by considering each bin of the histogram to be a vector whose length is the number of spikes in the bin and whose angle is the midpoint of the bin. Each vector is normal-

Reprinted from *IEEE Trans. Acoust., Speech, Signal Processing*, vol. 29, no. 3, pt. 2, pp. 621–623, June 1981.

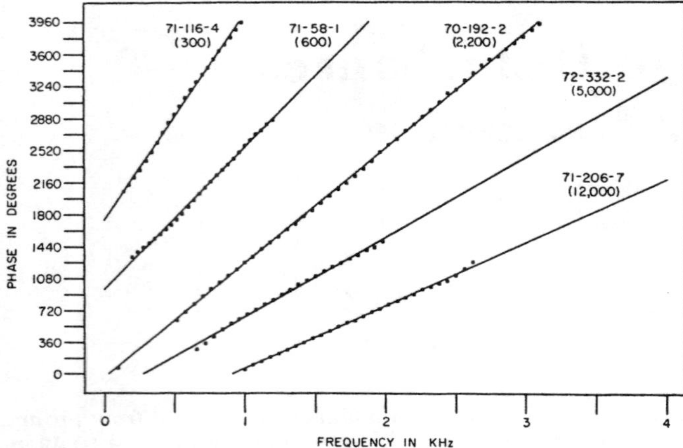

Fig. 1. Cumulative phase shift versus stimulus frequency plotted for five AVCN neurons. Note that the slope of the lines drawn through the points is dependent upon best frequency (shown in parentheses). Some of the lines have been arbitrarily shifted by multiples of 360° for clarity.

ized by dividing by the total number of spikes. The amplitude is a measure of how well the response is synchronized or phase-locked to the input sinusoid; the angle is the mean discharge phase. Generally, responses to stimuli below about 4 kHz are phase-locked ($R > 0.4$), while higher frequency responses are not. If the response of the neuron was found to be phase-locked, pure tones were presented at a fixed SPL (usually 80 dB) with the frequency incremented in very small steps (40 Hz) in order to determine the relation between phase and frequency.

Results

Complete response areas were obtained for 148 AVCN neurons which demonstrated phase locking. The best frequency (BF) of a neuron is defined as that frequency which activates it at the lowest intensity. Once the BF and frequency range were determined, the phase characteristics were determined. Cumulative phase shift versus input frequency is shown in Fig. 1 for five different neurons. Each phase-frequency relation can reasonably be approximated by a straight line. Such a linear relationship implies a time delay that is independent of the input frequency. The delay is the slope of the line. Note that with increasing BF, the slope of the line decreases, and thus delay decreases. Since high frequencies activate the basal portion of the BM, this result is expected.

Delays for each neuron were determined from a least squares fit to the phase-frequency line. In Fig. 2(a) these delays are illustrated as a function of distance along the BM (from the apex) as calculated from the BF using Greenwood's function. The delay estimates include not only travel time, but also any constant delays. These include the delay between the electrical signal and the resulting sound pressure wave at the eardrum, transmission delays through the middle ear, and the synaptic and conduction delays from the BM to the recording site. Greenwood [8] has suggested an exponential relation between travel time and distance along the BM of the form

$$D = A\,\{\exp(Bx) - \exp(22B)\} + C.$$

A Box algorithm [1] fit to the data yielded parameter values of $A = 7.11$ ms, $B = -0.154$/mm, and $C = 1.51$ ms. Parameter C presumably represents the sum of all constant delays, and may be subtracted to determine the travel time of the disturbance. (See Fig. 2(b).) Our data are consistent with Anderson's [2] in two respects: the travel times are close and the 0.5 ms difference in constant delay approximates the neural delay between the nerve and AVCN. Our values are slightly

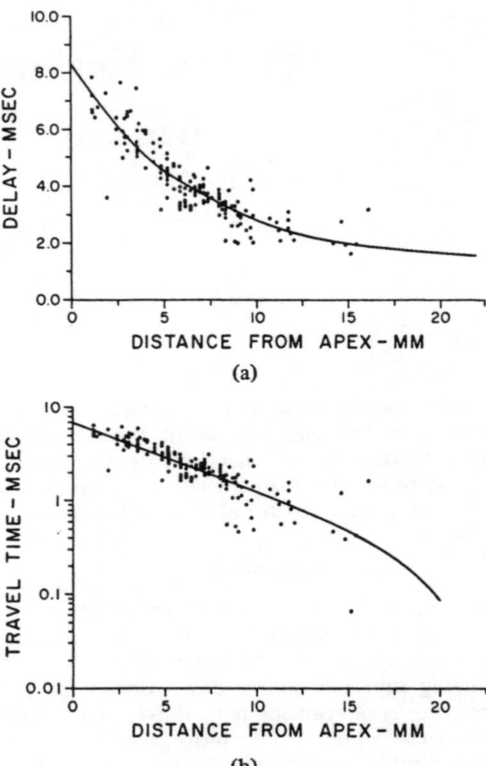

Fig. 2. Estimated delay (a) and travel time (b) for 148 AVCN neurons plotted versus cochlear distance. Fitted curve indicates an exponential relation between distance and travel time. See text for details. This relation is consistent with an exponentially varying velocity along the basilar membrane. (Modified from Gibson *et al.*, 1977, with permission from *Psychophysics and Physiology of Hearing*, E. F. Evans and J. P. Wilson, Eds. Copyright Academic Press Inc. (London) Ltd.)

larger than those of Robles *et al.* [10] or Kiang [9]. Recently, Ruggero [12] has argued that the values thus derived are larger than signal-front delay, but smaller than the average-weighted group delay. If so, the small discrepancy between the click data and the phase-frequency data is explained. An exponential relation between travel time and distance is consistent with a propagation velocity that varies exponentially along the BM. Some of these results have been reported previously [4], [5].

Acknowledgment

This investigation was done in collaboration with Dr. J. E. Hind, Dr. L. M. Kitzes, and Dr. J. E. Rose.

References

[1] K. P. Ambrose, "Programming note," *Decuscope*, vol. 9, pp. 11-12, 1970.

[2] D. J. Anderson, J. E. Rose, J. E. Hind, and J. F. Brugge, "Temporal position of discharges in single auditory nerve fibers within the cycle of a sine-wave stimulus: Frequency and intensity effects," *J. Acoust. Soc. Amer.*, vol. 49, pp. 1131-1139, 1971.

[3] G. von Békésy, *Experiments in Hearing*, E. G. Wever, Transl. and Ed. New York: McGraw-Hill, 1960.

[4] M. M. Gibson, L. M. Kitzes, J. E. Rose, and J. E. Hind, "Average phase angle of discharges of neurons in the anteroventral cochlear nucleus of the cat," *J. Acoust. Soc. Amer.*, vol. 55, p. 467, 1973.

[5] M. M. Gibson, J. E. Hind, L. M. Kitzes, and J. E. Rose, "Estimation of traveling wave parameters from the response properties of cat AVCN neurons," in *Psychophysics and Physiology of Hearing*, E. F. Evans and J. P. Wilson, Eds. London, England: Academic, 1977, pp. 57-68.

[6] J. M. Goldberg and P. B. Brown, "Response of binaural neurons of dog superior olivary complex to dichotic tonal stimuli; Some physiological mechanisms of sound localization," *J. Neurophysiol.*, vol. 32, pp. 613–636, 1969.

[7] D. D. Greenwood, "Critical bandwidth in man and some other species in relation to the traveling wave envelope," in *Sensation and Measurement*, H. R. Moskowitz, B. Scharf, and J. C. Stevens, Eds. Dordrecht, The Netherlands, and Boston, MA: Reidel, 1974, pp. 231–239.

[8] —, "Empirical travel time functions on the basilar membrane," in *Psychophysics and Physiology of Hearing*, E. F. Evans and J. P. Wilson, Eds. London, England: Academic, 1977, pp. 43–53.

[9] N. Y.-S. Kiang, *Discharge Patterns of Single Fibers in the Cat's Auditory Nerve.* Boston, MA: M.I.T. Press, 1965.

[10] L. Robles, W. S. Rhode, and C. D. Geisler, "Transient response of the basilar membrane measured in squirrel monkeys using the Mössbauer effect," *J. Acoust. Soc. Amer.*, vol. 59, pp. 926–939, 1976.

[11] J. E. Rose, L. M. Kitzes, M. M. Gibson, and J. E. Hind, "Observations on phase-sensitive neurons of anteroventral cochlear nucleus of the cat: Nonlinearity of cochlear output," *J. Neurophysiol.*, vol. 37, pp. 218–253, 1974.

[12] M. A. Ruggero, "Systematic errors in indirect estimates of basilar membrane travel times," *J. Acoust. Soc. Amer.*, vol. 67, pp. 707–710, 1980.

[13] H. F. Schuknecht, "Neuroanatomical correlates of auditory sensitivity and pitch discrimination in the cat," in *Neural Mechanisms of the Auditory and Vestibular Systems*, G. L. Rasmussen and W. F. Windle, Eds. Springfield, IL: Charles C Thomas, 1960, pp. 76–90.

Invariance of the Magnitude-Squared Coherence Estimate with Respect to Second-Channel Statistics

HERBERT GISH AND DOUGLAS COCHRAN

Abstract—**Properties of the magnitude-squared coherence (MSC) estimate in the presence of spherically symmetric noise are examined from a geometric point of view. Invariance to second-channel statistics is shown and the distribution function of the estimate is derived. A relationship between spherically symmetric and Gaussian sequences is also given.**

I. Introduction

A commonly used technique for detecting the presence of a common signal in two noisy channels is to compare the magnitude-squared coherence (MSC) estimate made with sample sequences from the two channels to a threshold [1]. Thresholds corresponding to particular false alarm probabilities are typically computed using the well-known probability distribution function of the MSC estimate derived under the assumption of Gaussian processes on both channels [2]. In [3], it was shown that this distribution function is valid, regardless of the statistics of the process on one of the two channels, provided the other sequence is Gaussian.

In this correspondence, a geometric argument is presented that shows the Gaussian assumption, even on one channel, is not necessary; spherical symmetry of the distribution function of one of the sequences is sufficient to ensure invariance of the distribution of the MSC estimate with respect to the statistics of the second sequence. The distribution function of the MSC estimate is derived as a corollary to the invariance argument, and it is also shown that spherically symmetric complex sequences are Gaussian if their real and imaginary parts are independent.

Finally, invariance is shown in one important nonspherically symmetric case in which each sample in one sequence has unit magnitude and uniformly distributed phase. In this case, the distribution of the MSC estimate is identical to the solution of a classical problem in random motion.

II. Geometry of the MSC Estimate

Let $a \triangleq \{x_n + iy_n\}_1^N$ and $b \triangleq \{u_n + iv_n\}_1^N$ denote the complex random sequences to be used in the MSC estimate. Then the estimate is given by

$$\gamma^2(a, b) \triangleq \frac{\left|\langle a, b\rangle\right|^2}{\|a\|^2 \cdot \|b\|^2} = \frac{\left|\sum_{n=1}^{N} (x_n + iy_n)(u_n + iv_n)^*\right|^2}{\sum_{n=1}^{N} |x_n + iy_n|^2 \cdot \sum_{n=1}^{N} |u_n + iv_n|^2} \tag{1}$$

$$= \frac{\left\{\sum_{n=1}^{N} (x_n u_n + y_n v_n)\right\}^2 + \left\{\sum_{n=1}^{N} (-x_n v_n + y_n u_n)\right\}^2}{\sum_{n=1}^{N} (x_n^2 + y_n^2) \cdot \sum_{n=1}^{N} (u_n^2 + v_n^2)} \tag{2}$$

where * denotes complex conjugation and $\langle\ ,\ \rangle$ denotes inner product. Define unit vectors in $\boldsymbol{R}^{2N}$ by

$$\boldsymbol{\alpha} \triangleq \frac{(x_1, \cdots, x_N, y_1, \cdots, y_N)}{\sum_{n=1}^{N} \sqrt{x_n^2 + y_n^2}},$$

$$\boldsymbol{\beta}_1 \triangleq \frac{(u_1, \cdots, u_N, v_1, \cdots, v_N)}{\sum_{n=1}^{N} \sqrt{u_n^2 + v_n^2}},$$

and

$$\boldsymbol{\beta}_2 \triangleq \frac{(-v_1, \cdots, v_N, u_1, \cdots, u_N)}{\sum_{n=1}^{N} \sqrt{u_n^2 + v_n^2}}$$

so that (2) reduces to

$$\gamma^2(\boldsymbol{a}, \boldsymbol{b}) = \langle \boldsymbol{\alpha}, \boldsymbol{\beta}_1\rangle^2 + \langle \boldsymbol{\alpha}, \boldsymbol{\beta}_2\rangle^2. \tag{3}$$

By construction, $\boldsymbol{\beta}_1$ and $\boldsymbol{\beta}_2$ are orthonormal. Thus, $\gamma^2(\boldsymbol{a}, \boldsymbol{b})$ is shown by (3) to be the square of the length of the projection of $\boldsymbol{\alpha}$ onto the plane defined by $\boldsymbol{\beta}_1$ and $\boldsymbol{\beta}_2$ (see Fig. 1). Hence, the distribution of γ^2 is the distribution of the square of the length of the projection of $\boldsymbol{\alpha}$ onto the plane formed by $\boldsymbol{\beta}_1$ and $\boldsymbol{\beta}_2$.

III. Invariance with Respect to Second-Channel Statistics

In the above construction, $\boldsymbol{\alpha}$ depends only on the statistics of $\boldsymbol{a}$, and the plane of $\boldsymbol{\beta}_1$ and $\boldsymbol{\beta}_2$ depends only on $\boldsymbol{b}$. If the probability density function $f_a(x_1, \cdots, x_N, y_1, \cdots, y_N)$ of $\boldsymbol{a}$ can be written as $f_a(x_1, \cdots, x_N, y_1, \cdots, y_N) = f_{ss}(\Sigma_{n=1}^N (x_n^2 + y_n^2))$ where f_{ss} is a one-dimensional function (i.e., if the value of the density depends only on the distance from the origin and not on the direction), then $\boldsymbol{a}$ is said to be *spherically symmetric.* With $\boldsymbol{a}$ spherically symmetric, then $\boldsymbol{\alpha}$ is also spherically symmetric. Furthermore, since $\boldsymbol{\alpha}$ is a vector of unit length, its spherical symmetry implies that it is uniformly distributed over the $2N$-dimensional unit sphere. If $\boldsymbol{a}$ is statistically independent of $\boldsymbol{b}$, the evaluation of the distribution of γ^2 becomes purely geometric in character, i.e., the probability that γ^2 is less than or equal to r is equal to the fraction of the $2N$-dimensional sphere that orthogonally projects into a disk of radius r centered at the origin in the plane formed by $\boldsymbol{\beta}_1$ and $\boldsymbol{\beta}_2$. Since this will not depend on the orientation of the plane (i.e., the specific occurrences of $\boldsymbol{\beta}_1$ and $\boldsymbol{\beta}_2$), we have the following.

Theorem 1: The distribution of the MSC estimate $\gamma^2(\boldsymbol{a}, \boldsymbol{b})$ does not depend on the statistics of $\boldsymbol{b}$ provided that $\boldsymbol{a}$ is spherically symmetric and statistically independent of $\boldsymbol{b}$.

Although statistical independence of $\boldsymbol{a}$ and $\boldsymbol{b}$ is assumed in the argument leading to Theorem 1, it is not a necessary condition for the result to hold. Suppose, for example, $\boldsymbol{b}$ is multiplied by the length of $\boldsymbol{a}$ to form a new sequence $\boldsymbol{b}'$. Then $\boldsymbol{b}'$ and $\boldsymbol{a}$ are not in-

Manuscript received September 16, 1986; revised July 21, 1987.
The authors are with BBN Laboratories, Inc., Cambridge, MA 02238.
IEEE Log Number 8717269.

Reprinted from *IEEE Trans. Acoust., Speech, Signal Processing,* vol. 35, no. 12, pp. 1774–1776, December 1987.

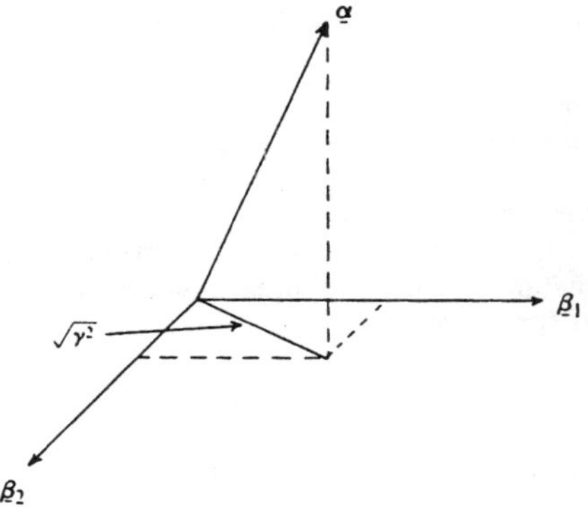

Fig. 1. The coherence estimate as a projection.

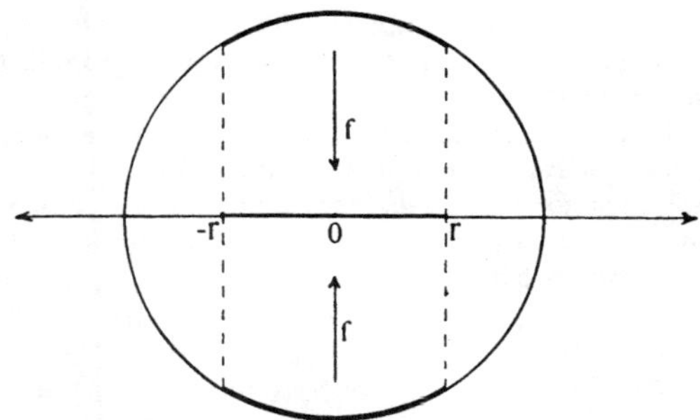

Fig. 2. For $0 \le r \le 1$, the probability $\Pr\{\gamma^2(a, b) \le r\}$ is the fraction of the unit spherical surface in $\mathbb{R}^{2N}$ that projects into a disk of radius r.

dependent, but $\gamma^2(a, b') = \gamma^2(a, b)$, so that the distribution of the MSC estimate still does not depend on the second-channel statistics. A stronger version of Theorem 1 could be formulated with the hypothesis $f_{a|b}$ is spherically symmetric for all b.

IV. Distribution of the MSC Estimate

In the previous section, the geometric character of the distribution of γ^2 under the assumptions of Theorem 1 was pointed out. In this section, the distribution of γ^2 will be derived from these geometric considerations as opposed to analytical methods that rely on the Gaussian distribution of the signals. This will be accomplished by explicitly computing the fraction of the surface of the unit sphere in $\boldsymbol{R}^{2N}$ that projects onto the annular region in $\boldsymbol{R}^2$ given in polar coordinates by $\{(r, \theta) : |R| < r \le 1 \text{ and } 0 \le \theta < 2\pi\}$, thereby implicitly determining the fraction of this sphere that projects into the disk of radius R centered at the origin in $\boldsymbol{R}^2$.

Consider an extension of $\{\beta_1, \beta_2\}$ to an orthonormal basis $\{\beta_1, \beta_2, \beta_3, \cdots, \beta_{2N}\}$ for $\boldsymbol{R}^{2N}$, and let $(x_1, \cdots, x_{2N})$ be coordinates in terms of this basis (see [4]). Then, assuming the hypotheses of Theorem 1, the probability that $\gamma^2(a, b) \le r$ for $0 \le r \le 1$ is the fraction of the area of the unit spherical surface $\{x \in \boldsymbol{R}^{2N} : x_1^2 + \cdots + x_{2N}^2 = 1\}$ that is situated over the disk $x_1^2 + x_2^2 \le r$ (Fig. 2). To determine this area, first define the projection mapping f: $\boldsymbol{R}^{2N} \rightarrow \boldsymbol{R}^2$ by $f(x_1, \cdots, x_{2N}) = (x_1, x_2)$ and consider the preimage $f^{-1}(\bar{x}_1, \bar{x}_2)$ of a point $(\bar{x}_1, \bar{x}_2) \in \boldsymbol{R}^2$. This consists of all points $(\bar{x}_1, \bar{x}_2, x_3, \cdots, x_{2N})$ and, because $x_3, \cdots, x_{2N}$ are arbitrary, is a $(2N-2)$-dimensional affine subspace of $\boldsymbol{R}^{2N}$ (defined in [4]).

Suppose now that $\bar{x}_1^2 + \bar{x}_2^2 = R \le 1$. Then the intersection of $f^{-1}(\bar{x}_1, \bar{x}_2)$ with the surface of the unit sphere in $\boldsymbol{R}^{2N}$ is not empty. In fact, a point $(x_3, \cdots, x_{2N})$ in the $(2N-2)$-dimensional space $f^{-1}(\bar{x}_1, \bar{x}_2)$ is in this intersection if and only if $x_3^2 + \cdots + x_{2N}^2 = 1 - R^2$. Thus, the points x on the surface of the unit sphere in $\boldsymbol{R}^{2N}$ for which $f(x) = (\bar{x}_1, \bar{x}_2)$ form a spherical surface of radius $\sqrt{1 - R^2}$ in the $(2N-2)$-dimensional space $f^{-1}(\bar{x}_1, \bar{x}_2)$.

Now, consider a line segment L_θ in $\boldsymbol{R}^2$ connecting the points (in polar coordinates) (R, θ) and $(1, \theta)$ where $R \le 1$ (see Fig. 3). Integrating the intersection of $f^{-1}(x_1, x_2)$ and the unit spherical surface in $\boldsymbol{R}^{2N}$ over $\{(x_1, x_2) \in L_\theta\}$ yields, by the above argument, the volume (by "spherical shells") of a solid sphere of radius $\sqrt{1 - R^2}$ in a $(2N-2)$-dimensional space [5]. This sphere has $(2N-2)$-dimensional volume $V = \pi^{N-1}(1 - R^2)^{N-1}/(N-1)!$. Define $A_R \triangleq \{(r, \theta) \in \boldsymbol{R}^2 : R < r \le 1\}$. Integrating V over $0 \le \theta < 2\pi$ yields the $(2N-1)$-dimensional volume of the intersection of $f^{-1}(A_R)$ and the surface of the unit sphere in $\boldsymbol{R}^{2N}$. This is $2\pi V = 2\pi^N(1 - R^2)^{N-1}/(N-1)!$. Since the $(2N-1)$-dimensional volume of the entire surface of the unit sphere in $\boldsymbol{R}^{2N}$ is $2\pi^N/(N-1)!$, the fraction of this spherical surface mapped by f into A_R is $(1 - R^2)^{N-1}$; the remainder of the sphere is mapped into the disk $\{(r, \theta) : 0 \le r \le R\}$. Hence, $\Pr\{\gamma^2(a, b) \le R\} = 1 - (1 - R^2)^{N-1}$. This, of course, is the same distribution function for the MSC estimate given in [2] assuming Gaussian sequences on both channels.

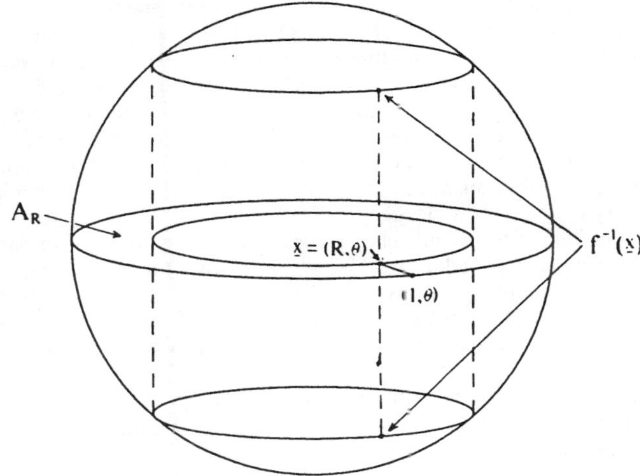

Fig. 3. The preimage of each point $(x_1, x_2) \in \boldsymbol{R}^2$ is a spherical surface in a $(2N-2)$-dimensional space.

V. A Comment on Spherically Symmetric Sequences

Let $a \triangleq \{x_n + iy_n\}_1^N$ be a spherically symmetric sequence. If the real and imaginary parts of a are statistically independent, then its probability density function $f(x_1, \cdots, x_N, y_1, \cdots, y_N)$ factors as $f(x_1, \cdots, x_N, y_1, \cdots, y_N) = g(x_1, \cdots, x_N) \cdot h(y_1, \cdots, y_N)$ where g and h are the density functions of the real and imaginary parts of a, respectively. Because a spherically symmetric random vector of dimension $2N$ is also spherically symmetric on any lower dimensional subspace, the real and imaginary parts of a are also spherically symmetric. Hence, $f_{ss}(\sum_{n=1}^N (x_n^2 + y_n^2)) = g_{ss}(\sum_{n=1}^N x_n^2) \cdot h_{ss}(\sum_{n=1}^N y_n^2)$. If f_{ss} is continuous, this relationship implies it must be Gaussian. The reason for this arises from the definition of the exponential function (see, for example, [6]). Hence, we have Theorem 2.

Theorem 2: If a sequence of complex random variables has a spherically symmetric distribution and statistically independent real and imaginary parts, then it has a Gaussian distribution.

VI. Sequences with Unit-Magnitude Samples

In some applications, the magnitudes of the elements of the two sequences may have undergone rather different distortions, and the only useful information in the sequences with respect to the presence of a common signal is contained in the phases of the sequence elements. In these circumstances, the amplitudes of the elements of both sequences are converted to a unit magnitude before the coherence estimate is evaluated. In this section, a condition is given for the distribution of the coherence estimate to be independent of second-channel statistics in this situation. Interestingly, the independence does not rely on the notion of spherical symmetry.

Assume that the samples in sequence $\boldsymbol{a}$ are statistically independent. Also, suppose $\boldsymbol{a}$ and $\boldsymbol{b}$ have the property that each sample has unit magnitude, i.e., $x_n^2 + y_n^2 = u_n^2 + v_n^2 = 1$ for $n = 1, \cdots, N$. Then they can be written as $\boldsymbol{a} = (e^{i\theta_1}, \cdots, e^{i\theta_N})$ and $\boldsymbol{b} = (e^{i\phi_1}, \cdots, e^{i\phi_N})$. The MSC estimate becomes $\gamma^2(\boldsymbol{a}, \boldsymbol{b}) = |\Sigma_{n=1}^N \exp[i(\theta_n - \phi_n)]|^2/N^2$.

If each sample $e^{i\phi_n}$ of $\boldsymbol{a}$ is uniformly distributed on the unit circle, then so is each term $\exp[i(\theta_n - \phi_n)]$ in this expansion of $\gamma^2(\boldsymbol{a}, \boldsymbol{b})$—regardless of the distribution of the samples of $\boldsymbol{b}$. This is easily proven by convolution of an arbitrary probability density function $p(\phi_n)$ with a uniform density function, modulo 2π. Furthermore, the terms $\exp[i(\theta_n - \phi_n)]$ will be statistically independent of one another. Thus, in this case, the distribution of the MSC estimate is $1/N^2$ times the distribution of the length squared of a vector formed by adding N unit vectors in a plane which have uniformly distributed direction. This problem has a long history as the problem of a random walk in the plane. The problem of determining this distribution was posed by Pearson in 1905 (see [7]) and its history is detailed in [8].

Note that if the elements of the $\boldsymbol{a}$ sequence are allowed to have nonconstant amplitudes (i.e., $A_n = |x_n + iy_n|$) that are independently distributed, then invariance with respect to second-channel statistics is still maintained. The distribution of the coherence estimate is, in this case, related to a random walk with a random step size.

It is also important to observe that although each term of the $\boldsymbol{a}$ sequence $A_n e^{i\theta_n}$ is spherically symmetric, the $\boldsymbol{a}$ sequence itself *is not spherically symmetric.*

VII. Conclusions

We feel the important aspect of this correspondence is the demonstration of the utility of a geometric approach to this probability problem, not just for the results obtained but also for the insight into why these results come about. Although we have focused on the problem of coherence estimation with no signal present, we anticipate that the geometric approach will prove useful in the case when signal is present.

References

[1] J. J. Gosselin, "Comparative study of two sensor (magnitude-squared coherence) and single sensor (square law) receiver operating chacteristics," in *Proc. IEEE Int. Conf. Acoust., Speech, Signal Processing*, Hartford, CT, May 1977, pp. 311-314.

[2] G. C. Carter, C. Knapp, and A. H. Nuttall, "Estimation of the magnitude-squared coherence function via overlapped fast Fourier transform processing," *IEEE Trans. Audio Electroacoust.*, vol. AU-21, pp. 337-344, Aug. 1973.

[3] A. H. Nuttall, "Invariance of distribution of coherence estimate to second-channel statistics," *IEEE Trans. Acoust., Speech, Signal Processing*, vol. ASSP-29, pp. 120-122, Feb. 1981.

[4] S. MacLane and G. Birkhoff, *Algebra.* New York: MacMillan, 1967, pp. 394-295 and pp. 420-433.

[5] W. Fleming, *Functions of Several Variables.* Reading, MA: Addison-Wesley, 1965, p. 183.

[6] W. Rudin, *Principles of Mathematical Analysis.* New York: McGraw-Hill, 1953, pp. 145-150 and p. 162.

[7] K. Pearson, "The problem of the random walk," *Nature*, vol. 72, pp. 294-203, 1905.

[8] J. A. Greenwood and D. Durand, "The distribution of length and components of the sum of *n* random unit vectors," *Ann. Math. Stat.*, vol. 26, pp. 233-246, 1955.

Large and Small Error Performance Limits for Multipath Time Delay Estimation

JOHN P. IANNIELLO, MEMBER, IEEE

Abstract—**The Cramer-Rao lower bound (CRLB) and a lower bound on the probability of large errors for multipath time delay estimation, for a single resolvable multipath, are developed. These theoretical results are compared to the performance of the maximum likelihood estimator (MLE) and an autocorrelator, via computer simulation. For small errors and large SNR, the MLE reaches the CRLB; the variance of the time delay estimate for the autocorrelator is much larger than the CRLB if the reflection coefficient for the delayed path is near unity. For small errors and small SNR, the MLE reduces to an autocorrelator, hence, their performance is identical. For large errors and bandwidth time products greater than 500, the error probability for the autocorrelator is within 1 dB in input SNR of the lower bound on large error probability. Thus, the autocorrelator is nearly an optimal instrumentation from the point of view of large errors, at least for large bandwidth time products and the flat, low-pass spectrum considered here.**

I. Introduction

Multipath propagation, in which one or more attenuated and delayed versions of the same radiated signal are received at a single sensor (or a beamformed array of sensors), is a frequent occurrence in the ocean [1, p. 95], and in other fields where a signal and its "echo" are present. In the ocean acoustics application, the time of arrival difference between the various multipath arrivals can provide useful source localization information, hence, this is a problem of practical concern.

We are interested here only in random signals, not signals of known waveshape. If such multipath arrivals are received on a single sensor, one straightforward way to estimate the time delays is to autocorrelate the output signal with itself. The resulting correlogram will have peaks corresponding to the various delay differences; if these peaks are resolvable, their locations can be used as estimates of the time delay. It is of interest therefore to analyze the performance of such an autocorrelation processor. More basically, however, it is of interest to establish the fundamental limits on multipath time delay estimation, and to see how closely the autocorrelator comes to this performance.

As has been pointed out elsewhere [2]–[6], nonlinear estimation problems of this type are unavoidably characterized by a threshold signal-to-noise ratio (SNR) such that above threshold the estimation errors are small (the estimates are close to the true value), while below threshold the errors are large (the estimates can be anywhere in the *a priori* known parameter region). In the small error region, the minimum achievable estimation variance is given by the Cramer–Rao lower bound (CRLB); if this performance can be realized, then it is reached by the maximum likelihood estimator (MLE). Below threshold SNR the CRLB cannot be reached, thus, some other lower bound is needed as a standard of comparison. In [19] the Ziv–Zakai lower bound (ZZLB) developed in [4] was applied to this problem. It was tentatively concluded there that, since the ZZLB and the autocorrelator variance were within 2 dB in input SNR in the threshold region, the autocorrelator was very nearly an optimal instrumentation. In [7], however, a different approach to characterizing the large error performance of estimators was developed. The ZZLB is a bound on mean-square error. One may, however, be more interested in the probability of a large error than in the actual value of the error, since the very occurrence of a large error may make the estimate useless. In this case, a bound on probability of large error, rather than a bound on mean-square error, will be more valuable. In [7] a lower bound on probability of large error, based on an *M*-ary hypothesis test, was developed. A valuable feature of the bound developed in [7] is that the large error probability increases linearly with increasing *a priori* parameter region, as one expects it should [3], [5]. This differs from the performance of the ZZLB developed in [4] and [19] which increases more slowly with increasing *a priori* region. Hence, this *M*-ary bound should be a tighter bound than the ZZLB developed previously for the multipath problem.

Several authors have considered various aspects of this or related problems. Levin [8] has derived the MLE and the CRLB for multipath time delay estimation for the noiseless case. Friedlander [9] has examined the CRLB for resolvable and nonresolvable delays. Much effort has been concentrated on cepstral processing for multipath problems [10]–[12]. Bogert and Ossanna [10] have considered cepstrum analysis for a single, resolvable multipath delay such as we will consider. They conclude via simulation that for irregular spectra the cepstrum is superior to autocorrelation for echo detectability. They point out that although the MLE is a natural approach, it requires detailed knowledge of signal and noise properties; the cepstrum requires no such information. Kemerait and Childers [11] point out that when two (or more) "echos" are present, erroneous peaks are observed at the sum and

Manuscript received April 1, 1985; revised October 9, 1985. This work was supported by the Office of Naval Research under Code 411 SP.

The author is with the Naval Underwater Systems Center, New London, CT 06320.

IEEE Log Number 8407031.

Reprinted from *IEEE Trans. Acoust., Speech, Signal Processing*, vol. 34, no. 2, pp. 245–251, April 1986.

difference of the echo arrival times and, further, that echo detectability suffers. We will not consider cepstral processing in this paper since we are interested in establishing fundamental limits; for this purpose, the MLE and autocorrelator are sufficient. Further, the cepstrum's advantage over autocorrelation depends on how irregular the spectrum is, and hence, general conclusions are difficult to reach. Finally, Kemerait and Childers' comment indicates that cepstral processing may be a poor choice for oceanic propagation since several multipaths are generally present (although we only consider two paths here).

In the next section, we describe the signal model and analyze the small and large error performance for the autocorrelator. Then we develop the MLE and the CLRB. Next we develop the probability of large error bound for multipath. Finally we compare our theoretical results to computer simulation of the autocorrelator and the MLE.

II. Model Description: Small and Large Error Autocorrelator Performance

We consider the single multipath situation where the received signal $r(t)$ is given by

$$r(t) = s(t) + as(t - D_0) + n(t), \qquad 0 \le t \le T \quad (1)$$

where $s(t)$ and $n(t)$ are uncorrelated Gaussian random processes, a is a frequency independent attenuation coefficient, and D_0 is the multipath time delay. The autocorrelation function of $r(t)$, $R_r(D)$, is

$$R_r(D) = (1 + a^2)\, R_s(D) + aR_s(D - D_0) + aR_s(D + D_0) + R_n(D) \quad (2)$$

where $R_s(D)$ and $R_n(D)$ are the autocorrelation functions of $s(t)$ and $n(t)$. Note that $R_r(D)$ has a peak at $D = 0$ and peaks at D close to $\pm D_0$. The latter peaks may not be exactly at $\pm D_0$ due to the slight biases caused by the superposition of the various correlograms. We denote the true location of these peaks by $\pm D_0'$. We will assume that the peaks at D_0' are well removed from 0, i.e., that the delays are highly resolvable.

We now determine the variance (small error performance) of the multipath time delay estimate via autocorrelation. This is a fairly straightforward, although tedious, analysis. The output of the autocorrelator $Z(D)$ is given by

$$Z(D) = \frac{1}{T}\int_0^T r(t)\, r(t - D)\, dt. \quad (3)$$

We estimate the location of the peak near D_0 by finding where $dZ/dD = 0$ in the vicinity of D_0; call this point $\hat{D}$. Expanding dZ/dD in a Taylor series about D_0' and retaining only linear terms, we find [13]

$$\text{var}\,[\hat{D}] \cong \overline{[dZ/dD]^2}|_{D_0'} / [d^2\overline{Z}/dD^2]|_{D_0'} \quad (4)$$

where the overbar denotes the expected value. The denominator of (4) is easily found by differentiation of (2). The numerator of (4) can be shown to be given by [13]

$$\overline{(dZ/dD)^2}|_{D_0'} = -\frac{1}{T}\int_{-T}^{T}\left(1 - \frac{|x|}{T}\right)[R_r(x)\, R_r''(x) + R_r'(x + D_0')\, R_r'(x - D_0')]\, dx \quad (5)$$

where the primes associated with the R_r's denote differentiation with respect to the argument. For large bandwidth time product, (4) can be written in terms of the spectrum of $r(t)$, $G_r(f)$ as

$$\text{var}\,[\hat{D}] \cong \frac{1}{T}\frac{\int_{-\infty}^{\infty}(2\pi f)^2\, G_r^2(f)\,[1 - \cos(4\pi f D_0')]\, df}{\left[\int_{-\infty}^{\infty}(2\pi f)^2\, G_r(f)\cos(2\pi f D_0')\, df\right]^2} \quad (6)$$

which can be evaluated numerically for arbitrary spectra.

A useful approximate result for var $[\hat{D}]$ can also be found if we assume that $s(t)$ and $n(t)$ have uniform, lowpass spectra in the band $|f| \le B/2$. For highly resolvable multipath ($BD_0 >> 1$), large observation time with respect to multipath delay ($T/D_0 >> 1$), and $BT >> 1$ we find

$$\text{var}\,[\hat{D}] \cong \frac{1}{BT}\frac{3}{\pi^2 B^2} \cdot \frac{[(1 + a^2)^2 + a^2]\, S^2 + 2(1 + a^2)\, SN + N^2}{a^2 S^2} \quad (7)$$

where S and N are the spectrum levels of $s(t)$ and $n(t)$.

Note that at infinite SNR the variance levels off at a nonzero value. This contrasts with the two channel time delay estimation case where the variance goes to zero as SNR increases. Referring to (5), we see that this occurs because, even at infinite SNR, there will be signal x signal terms contributing to the variance. We note, finally, that for low SNR and $a = 1$, the autocorrelator variance (for highly resolvable signals and large BT) is equal to the variance of a cross correlator for two channel time delay estimation (also for low SNR).

An approximate result for the probability of large error (or an anomalous estimate) can be obtained via a modification of the development in [3]. As in [3] we assume there are $M = B(D_U - D_L)$ independent autocorrelation values in an *a priori* known parameter range for D_0, $D_L \le D_0 \le D_U$. The probability of large error is given by

$$P_r[\text{large error}] \cong 1 - \int_{-\infty}^{\infty}\frac{dx}{\sqrt{2\pi}}\exp\left[-\frac{1}{2}(x - A_1)^2\right] \cdot \left\{\int_{-\infty}^{A_2}\frac{dy}{\sqrt{2\pi}}\exp\left[-\frac{1}{2}y^2\right]\right\}^{M-1} \quad (8)$$

where we have assumed that the autocorrelator output at the true value of delay Z_0, as well as the outputs at the other $M - 1$ delays Z_m, are Gaussianly distributed, and

$$A_1 = \frac{\overline{Z_0}}{\text{var}\,[Z_0]^{1/2}}, \qquad A_2 = \left\{\frac{\text{var}\,[Z_0]}{\text{var}\,[Z_m]}\right\}^{1/2}. \quad (9)$$

Assuming $BD_0 >> 1$, and the flat, low-pass spectrum described above, it is easily shown that

$$A_1 = (BT)^{1/2} \cdot \frac{aS}{\{[(1 + a^2)^2 + 3a^2]\, S^2 + 2(1 + a^2)\, SN + N^2\}^{1/2}}$$

$$A_2 = \left\{ \frac{[(1 + a^2)^2 + 3a^2]\, S^2 + 2(1 + a^2)\, SN + N^2}{[(1 + a^2)^2 + 2a^2]\, S^2 + 2(1 + a^2)\, SN + N^2} \right\}^{1/2}. \tag{10}$$

Thus, with (8)–(10) an approximate result for the probability of large error can be found numerically. The validity of the various approximations can only be checked via simulation; as discussed in [7] the approximations are reasonable for large BT. For large BT practically interesting error probabilities ($10^{-1} - 10^{-5}$, say) will correspond to small SNR. For small SNR we see from (10) that A_2 goes to unity while A_1 goes to $(BT)^{1/2}\, a(S/N)$. Thus, for small SNR, changing a is equivalent to changing input SNR.

III. Derivation of MLE and CRLB

The CRLB can be found in straightforward fashion by Fourier transforming the received data and performing the appropriate operations on the probability density function of the complex Fourier coefficients [14], [15]. For large BT and $T/D_0 >> 1$, it can be shown that [15]

$$\text{CRLB} = \left\{ \frac{T}{2\pi} \int_0^{\infty} \left[\frac{2a\omega \sin(\omega D_0)\, S(\omega)}{[1 + a^2 + 2a\cos(\omega D_0)]\, S(\omega) + N(\omega)} \right]^2 d\omega \right\}^{-1}. \tag{11}$$

Specializing to uniform, low-pass spectra in the band $|f| \leq B/2$, an approximate result, valid for $2\pi BD_0 >> 1$, can be found [15] by expanding (11) in a series in powers of $R = 2aS/((1 + a^2)\, S + N)$, integrating, and regrouping terms. The result is

$$\text{CRLB} = \frac{1}{BT} \frac{6}{\pi^2 B^2} \left[\frac{1}{(1 - R^2)^{1/2}} - 1 \right]^{-1}. \tag{12}$$

By comparing the approximate result in (12) to an exact result obtained by numerically integrating (11), the approximation was found to be accurate to better than 10 percent for $BD_0 > 10$.

For low SNR, $R \to 2aS/N$, and (12) becomes

$$\text{CRLB} \to \frac{1}{BT} \frac{3}{\pi^2 B^2} \frac{N^2}{a^2 S^2}. \tag{13}$$

Note that (13) agrees with (7), the variance for the autocorrelator, for low SNR. We can, in fact, show that an autocorrelator, with proper filtering [13], achieves the CRLB for low SNR and resolvable time delays. For the flat signal and noise spectra assumed here, the optimum filter is all-pass, hence, no filtering is required for the autocorrelator to achieve the CRLB.

For high SNR

$$R \to 2a/(1 + a^2)$$

and

$$\text{CRLB} \to \frac{1}{BT} \frac{3}{\pi^2 B^2} \frac{1 - a^2}{a^2}. \tag{14}$$

This agrees with Levin's [8] result with $a << 1$. From (14) we see that, except for $a \to 1$, the CRLB levels off at a nonzero value for infinite SNR (and finite BT). Thus, even an instrumentation that reaches the CRLB will not achieve zero variance for infinite SNR. We note that for large SNR and a near unity, the CRLB can be much less than the autocorrelator variance. For large SNR and small a, the autocorrelator variance also goes to the CRLB.

The MLE instrumentation for large BT and $T/D_0 >> 1$ also follows directly from the analysis in [15]. Thus, it can be shown that the MLE minimizes, through choice of $\hat{D}$, the log likelihood ratio (LLR)

$$\text{LLR} = \sum_{k=1}^{K} \frac{|R_k|^2}{(1 + a^2 + 2a\cos\omega_k \hat{D})\, S(\omega_k) + N(\omega_k)} + \sum_{k=1}^{K} \ln \{[(1 + a^2 + 2a\cos\omega_k \hat{D}) \cdot S(\omega_k) + N(\omega_k)]\} \tag{15}$$

where R_k is the complex Fourier coefficient of the data at frequency $\omega_k = (2\pi k)/T$ and $K = \frac{1}{2} BT$. The denominator of the first term in (15) is the spectrum of the received signal, given that $\hat{D}$ (and a) are the true parameter values. The numerator is the periodogram value at ω_k. The second term in (15) ensures that the MLE is (asymptotically) unbiased [15]. The MLE thus selects the $\hat{D}$ (and a, if unknown, we assume it is known here) which minimizes the power out of a whitening filter. We note, as in [8] and [9], that if $aS(\omega) << N(\omega)$ for all ω, then the denominator of the first term can be expanded in a series in aS and, retaining only the first term, we find that the MLE maximizes, by choice of $\hat{D}$,

$$\sum_{k=1}^{K} |R_k|^2 \cos \omega_k \hat{D}.$$

This is an estimate of the autocorrelation function of the data, and shows that the MLE reduces to selecting the peak of the autocorrelation function for $aS(\omega) << N(\omega)$. This agrees with our earlier conclusion that the autocorrelator reaches the CRLB for low SNR.

IV. Lower Bound on Large Error Probability

To characterize the large error performance, we develop a lower bound on the worst-case probability of large error. Following the development in [7], we assume that there is an *a priori* region $[D_L, D_U]$ in which the delay D_0 is known to occur. We assume that $M = (D_U - D_L)/2\Delta$ is an integer and that D_0 can only take on one of the M values D_i given by

$$D_i = D_L + \Delta + (i - 1)\, 2\Delta \qquad i = 1, M. \tag{16}$$

We consider the M hypothesis problem H_i: D_i $i = 1, M$. For equally likely hypotheses, the minimum probability of error decision scheme for choosing between the M hypotheses computes the likelihood function at each D_i and selects the hypothesis corresponding to the largest value [16, p. 50]. Denote this minimum probability of error by P_{LF}.

Now consider any other scheme for choosing between the M hypotheses. Denote the probability of error for this (possibly) suboptimum scheme by P_{SO}. Clearly, $P_{LF} \le P_{SO}$. Now by specializing to a particular decision scheme, namely, one which first estimates D_0 in any way (call the estimate $\hat{D}$) and then decides on H_i according to which D_i $\hat{D}$ is closest to, we can show [7]

$$P_{LF} \le \text{worst case } P_r[\text{large error}] \triangleq P_{wc}(\Delta)$$
$$D_L \le D_0 \le D_U \tag{17}$$

where a large error is defined as the event $|\hat{D} - D_0| > \Delta$ given that D_0 is the true value of the parameter. Equation (17) states that the worst-case (over all values of D_0 in D_L, D_U) probability of large error must be greater than P_{LF}. Thus, P_{LF} is a lower bound on the worst-case probability of large error.

Unfortunately, P_{LF} is difficult to calculate exactly. We have two options. First, we can always determine P_{LF} arbitrarily closely via simulation, if we can instrument the MLE. Second, we can calculate an approximate result for P_{LF} (actually an upper bound) as follows. Denote the likelihood processor output at each of the D_i by l_i. We can then show, using the union bound [7; 16; p. 263], that

$$P_{LF} \le \max_{H_j} \sum_{\substack{i=1 \\ i \ne j}}^{M} P_r[l_i > l_j \mid H_j]. \tag{18}$$

If $P_r[l_i > l_j \mid H_j]$ does not depend on i

$$P_{LF} \le (M - 1) \max_{H_j} P_r[l_i > l_j \mid H_j] \triangleq P_{LFU} \qquad i \ne j. \tag{19}$$

Since the union bound has been used to derive P_{LFU}, P_{LFU} will be a more accurate approximation to P_{LF} for small values of P_{LF}. This is so because the union bound neglects joint error probabilities such as $P_r[l_i \text{ and } l_k > l_j \mid H_j]$, but for small P_{LF} these joint probabilities become negligibly small, as is discussed more fully in [7].

To find P_{LFU}, we must evaluate $P_r[l_i > l_j \mid H_j]$. But, since l_i and l_j are MLE outputs, $P_r[l_i > l_j \mid H_j]$ is simply the error probability for the binary hypothesis test for deciding whether the delay is at D_i or D_j given that the true value is D_j. Thus, for equally likely hypotheses,

$$P_r[l_i > l_j \mid H_j] = \int_0^\infty P_{l|H_j}[L|H_j]\, dL. \tag{20}$$

where $P_{l|H_j}$ is the probability density function of the log likelihood ratio given that H_j is true [16, p. 117]. The integral in (20) can be evaluated numerically [18].

An approximate analytic result for $P_r[l_i > l_j \mid H_j]$, and hence for P_{LFU}, can be obtained by starting from the general formula for the Chernoff bound for $P_r[l_i > l_j \mid H_j]$ and developing a representation in terms of a power series in $R = 2as/[(1 + a^2)\, S + N]$ [19]. For small R and for the spacing $2\Delta = B^{-1}$, i.e., $|D_i - D_j| = kB^{-1}$ where k is an integer, the result for P_{LFU} reduces to

$$P_{LFU} \cong (M - 1)\, \Phi\left[\left(\frac{BT}{2}\right)^{1/2} a \frac{S}{N}\right] \tag{21}$$

where $\phi(y)$ is the integral from y to infinity of the standardized Gaussian density. This is interesting since it is identical with the result derived for two channel time delay estimation for small SNR in [7] (if we equate aS/N with S/N). Thus, for equal *a priori* regions, and for small SNR, the large error probabilities for both multipath and two channel time delay estimation are the same. Notice that for small SNR, $R = a(S/N)$ and thus changes in a are equivalent to changes in input SNR.

V. Comparison of Theory and Simulation

The received signal $r(t)$ [see (1)] was simulated by filtering white Gaussian noise with an 11th-order elliptic filter which had its 3 dB down points at one-eighth of the sampling frequency. $s(t)$ and $n(t)$ were generated from uncorrelated white sequences. The autocorrelator was implemented via straightforward digital techniques [17, p. 556]; the peak of the correlogram in the *a priori* region was selected as the estimate of the time delay. The MLE was implemented by Fourier transforming the data and performing the operations indicated in (15). For a given set of parameters [BT, a, SNR] a large number of independent realizations was run; each realization produced independent time delay estimates which were saved and then used to compute the relevant statistics [variance, bias, probability of large error]. The autocorrelator and the MLE (as implemented) produce output values at the sampling interval. The largest of those output values was used as a coarse estimate of the peak location. A refined estimate of the peak location was found by numerically determining the zero crossing of the derivative of the relevant function [autocorrelator or MLE] that was closest to the true peak.

A. Small Error Results

We first discuss the small error results. In Fig. 1 we show estimation variance (normalized by the sample interval squared) versus SNR for the autocorrelator and the MLE for different values of BT and a. The upper solid curve is the variance for the autocorrelator computed from (6) for $BT = 512$, $BD_0 = 8$, and $a = 0.9$. The x's are the simulation results for the autocorrelator. It is seen that theory and simulation results agree well for high SNR's

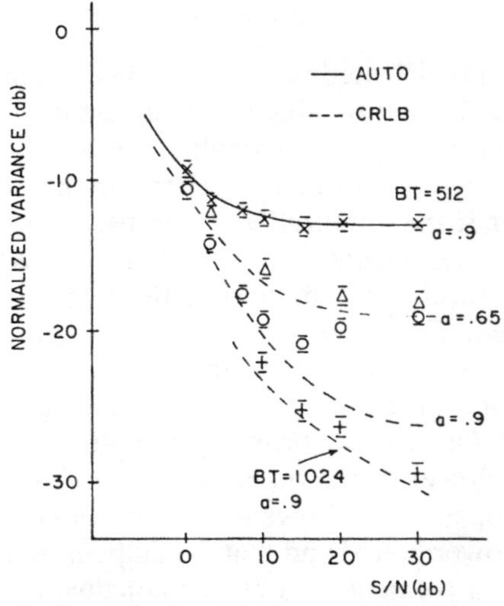

Fig. 1. Normalized variance versus SNR; $BD_0 = 8$.

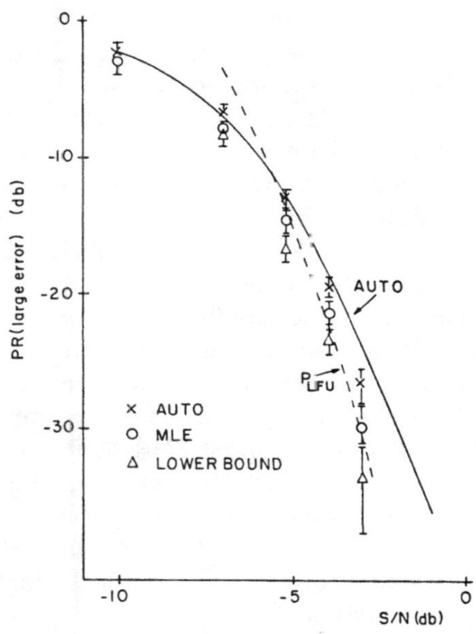

Fig. 2. Probability of large error versus SNR; $BT = 512$, $BD_0 = 16$, and $a = 0.9$.

where the linearizing approximations for the theoretical results are valid. The bars above and below the x's indicate 95 percent confidence intervals. Notice the leveling off of the variance with increasing SNR as discussed earlier. Similar close agreement was observed for other values of a.

The middle two dashed lines show the CLRB for $BT = 512$ and $a = 0.65$ and 0.9. Notice the large theoretical variance reduction at high SNR. The MLE simulation results for $a = 0.65$ are shown by the triangles, while the MLE results for $a = 0.9$ are shown by the circles. For $a = 0.65$ the agreement is fairly good for all SNR's, while for $a = 0.9$ the simulation results become much larger than the CRLB above SNR's of 10 dB. The bottom curve shows the CRLB for $BT = 1024$ and again $a = 0.9$. The crosses show the MLE simulation results. Here the simulation agrees fairly well with the theory for all SNR's shown. Thus, by increasing the BT product by a factor of two, the "breakdown" in the simulation results for $BT = 512$, $a = 0.9$ is overcome. This indicates the sensitivity of the MLE for high SNR, a near unity, and moderate BT. SNR's in the range of 20–30 dB are extremely large and probably have little practical significance, at least in underwater acoustics.

These small error simulation results serve mainly as a check on the quality of the simulation as well as giving some idea of where the various linearizing assumptions break down. One further point is worth noting. As mentioned above, the autocorrelator will generally be biased (how much bias depends on the superposition of the correlograms) while the MLE is asymptotically unbiased. The bias in the simulation results for the autocorrelator agreed well with theoretical predictions. The bias for the MLE was generally much smaller, on the order of 4–10 times smaller than the bias for the autocorrelator, at high SNR, for the cases run. For some applications, the bias reduction may be more important than the variance reduction.

B. Large Error Results

To illustrate the large error bound we show, in Fig. 2, theoretical and simulation results for the probability of large error versus SNR for $BT = 512$, $BD_0 = 16$, and $a = 0.9$. The solid line is the approximate probability of large error for the autocorrelator computed from (8), while the dashed line is P_{LFU} from (19). In calculating (8) we used $M = B(D_U - D_L) = 27$ since the *a priori* region searched was from 20 ΔT to 128 ΔT and $B = \frac{1}{4}\,\Delta T$ (ΔT is the sample interval, equal to unity in the simulation). As a large error criterion we used $2\Delta = B^{-1}$. This choice of Δ corresponds to the large error definition used in [3] and [7], $\pm T_c/2$, where T_c is a correlation length scale. When errors are greater than $\pm T_c/2$, the linearity assumption required for small error analysis to apply is surely violated. With this choice of Δ, (18) can be used since it was observed from the numerical calculation that $P_r[l_i > l_j \mid H_j]$ is independent of l_i for l_i's spaced at intervals of 2Δ from the true value ($D_0 = 68\ \Delta T$). As noted in [7], tighter bounds may be obtainable by other choices of Δ.

The x's in Fig. 2 represent the measured probability of large error for the autocorrelator. A large error was declared whenever the delay estimate was further than $\pm\Delta$ from the known true value. The simulated data agree quite well with the approximate theory except at very high SNR. This type of behavior was also observed in the two channel time delay estimation problem.

The circles in Fig. 2 represent the measured probability of error for the MLE. This is seen to be smaller than the probability of error for the autocorrelator for all SNR's. The improvement in performance for the MLE is especially apparent at the higher SNR's, as is reasonable, since at low SNR the MLE approaches an autocorrelator, hence, the major differences between the MLE and the autocorrelator should appear at the higher SNR's.

The triangles in Fig. 2 represent the probability of error for the lower bound. These data were obtained by comparing the MLE outputs at the delay values given by (10) (with $2\Delta = B^{-1}$ and $M = (D_U - D_L)/2\Delta = 27$). If the peak was not at the known, true value of delay a large error was declared. The lower bound is, of course, less than the probability of error for both the autocorrelator and the MLE for all SNR's, again with the largest difference at the higher SNR's. In comparing these results we must note that, due to the rapidity of the decrease in error probability in the threshold region, even large differences in error probability represent small differences in threshold SNR. Thus, for example, at an error probability of −30 dB, the difference in input SNR between the lower bound and the autocorrelator is seen from Fig. 2 to be roughly 1 dB. From this we can assert that for the case examined here, the lower bound that we have developed is nearly a greatest lower bound since it is close to a realizable instrumentation. On the other hand, the autocorrelator is nearly optimal from the point of view of large error probability since it is close to the lower bound. These conclusions are strictly applicable only to the flat, low-pass spectrum examined here.

Note that P_{LFU} (the dashed line in Fig. 2) is within 1 dB in SNR of the lower bound, represented by the triangles, for small error probability. Thus, P_{LFU} is a useful approximate analytic formula. The conclusions reached here apply strictly for a BT product of 512. The underlying assumption in the development of the theoretical results is that the BT product is large. We may, therefore, expect that the agreement between theory and simulation will be at least as good as that shown here for BT products greater than 512. Thus, we may infer that our conclusions about the tightness of the bound and the near optimality of the autocorrelator are valid for BT products greater than 512.

The straightforward implementation of the MLE that was used for the results described above assumes that a, S/N, and the spectral shape are all known. To obtain some idea of how sensitive the basic MLE structure is to these assumptions, simulations were run using estimates of the necessary quantities. From (15) it is seen that the two quantities $[(1 + a^2) S + N]$ and $2aS$ are needed to implement the MLE. From (2) we see that $[(1 + a^2) S + N]$ is the value of $R_r(0)$ while aS is the mean value of $R_r(D_0)$ (for highly resolvable time delays). Thus, as an estimate of $[(1 + a^2) S + N]$ the value of the sample autocorrelation function at $D = 0$ was used, while as an estimate for aS, the largest value of the autocorrelation function in the *a priori* region was used as an estimate. The performance of the approximate MLE obtained by using these estimated values in (15) was close to that of the exact MLE for both small and large errors and, in all cases, still outperformed the autocorrelator. Thus, we conclude that the MLE implementation is not overly sensitive to the exact values of the parameters used, at least for the flat, low-pass spectrum used here.

VI. Summary

We have investigated the limits to performance for single, resolvable, multipath time delay estimation for both small and large errors. For small errors and large SNR's, the MLE can have a much smaller variance than an autocorrelator if the amplitudes of the two paths are nearly equal. For small errors and small SNR, the MLE reduces to an autocorrelator and, hence, the two have identical performance. For large errors we have shown that for moderately large bandwidth time products ($BT > 500$), and for error probabilities greater than 10^{-4}, the error probability for an autocorrelator is within 1 dB in input SNR of the lower bound on large error probability that we have developed. This shows that the lower bound is nearly a greatest lower bound and that the autocorrelator is nearly an optimal instrumentation. These conclusions are strictly applicable only for the flat, low-pass spectrum we have examined. All of the theoretical results can, however, be evaluated for arbitrary signal and noise spectra.

References

[1] C. B. Officer, *Introduction to the Theory of Sound Transmission*. New York: McGraw-Hill, 1958.

[2] S.-K. Chow and P. M. Schultheiss, "Delay estimation using narrow-band signals," *IEEE Trans. Acoust., Speech, Signal Processing*, vol. ASSP-29, pp. 478–484, 1981.

[3] J. P. Ianniello, "Time delay estimation via cross-correlation in the presence of large estimation errors," *IEEE Trans. Acoust., Speech, Signal Processing*, vol. ASSP-30, pp. 998–1003, 1982.

[4] A. Weiss and E. Weinstein, "Fundamental limitations in passive time delay estimation—Part 1: Narrow-band systems," *IEEE Trans. Acoust., Speech, Signal Processing*, vol. ASSP-31, pp. 472–485, 1983.

[5] H. Meyr and G. Spies, "The structure and performance of estimators for real-time estimation of randomly varying time delay," *IEEE Trans. Acoust., Speech, Signal Processing*, vol. ASSP-32, pp. 81–94, 1984.

[6] J. P. Ianniello, E. Weinstein, and A. Weiss, "Comparison of the Ziv–Zakai lower bound on time delay estimation with correlator performance," in *ICASSP 83 Conf. Proc.*, Boston, MA, Apr. 1983, pp. 875–878.

[7] —, "Lower bounds on worst-case probability of large error for two channel time delay estimation," submitted to *Trans. Acoust., Speech, Signal Processing*, 1984.

[8] M. J. Levin, "Power spectrum parameter estimation," *IEEE Trans. Inform. Theory*, vol. IT-11, pp. 100–107, Jan. 1965.

[9] B. Friedlander, "On the Cramer-Rao bound for time delay and Doppler estimation," *IEEE Trans. Inform. Theory*, vol. IT-30, pp. 575–580, May 1984.

[10] B. P. Bogert and J. F. Ossanna, "The heuristics of cepstrum analysis of a stationary complex echoed Gaussian signal in stationary Gaussian noise," *IEEE Trans. Inform. Theory*, vol. IT-12, pp. 373–380, July 1966.

[11] R. C. Kemerait and D. G. Childers, "Signal detection and extraction by cepstrum techniques," *IEEE Trans. Inform. Theory*, vol. IT-18, pp. 745–759, Nov. 1972.

[12] J. C. Hassab and R. Boucher, "A probabilistic analysis of time delay extraction by the cepstrum in stationary Gaussian noise," *IEEE Trans. Inform. Theory*, vol. IT-22, pp. 444–454, July 1976.

[13] J. P. Ianniello, "The variance of multipath time delay estimation using autocorrelation," NUSC Tech. Memo 831008, Jan. 24, 1983.

[14] W. J. Bangs, "Array processing with generalized beamformers," Ph.D. dissertation, Yale Univ., New Haven, CT, 1971.

[15] J. P. Ianniello, "The Cramer-Rao lower bound on multipath time delay estimation," NUSC Tech. Memo 831067, May 9, 1983.

[16] H. L. Van Trees, *Detection, Estimation and Modulation Theory*. New York: Wiley, 1968.

[17] A. Oppenheim and R. Schafer, *Digital Signal Processing*. Englewood Cliffs, NJ: Prentice-Hall, 1975.

[18] J. P. Ianniello, "A numerical technique for evaluating the minimum error probability for the binary hypothesis test," NUSC Tech. Memo 841082, May 6, 1984.

[19] ——, "Comparison of the Ziv–Zakai lower bound on multipath time delay estimation with autocorrelator performance," in *ICASSP 84 Conf. Proc.*, San Diego, CA, Mar. 1984, pp. 15.3.1–15.3.4.

Matched Filters in Nerve Conduction Velocity Estimation

VIJAI S. JASROTIA AND PHILIP A. PARKER, MEMBER, IEEE

Abstract—The evoked response signal-to-noise ratio in peripheral sensory nerves is of the order of one or less. To reduce noise induced errors in the nerve conduction velocity measurement, signal averaging is employed. The number of responses required depends upon the signal-to-noise ratio and the acceptable error. It is proposed to use maximum likelihood estimators to improve the signal-to-noise ratio and hence, reduce the number of responses required. The matched filter and generalized cross correlator are studied and noise performance equations obtained. It is shown that the matched filter offers a 6.8 dB improvement over the conventional 5 kHz band-limiting filter. It is further shown that the matched filter as a prefilter in the generalized cross correlator gives the optimum correlator configuration for the low signal-to-noise ratio case. These results are verified with experiments.

I. Introduction

The conduction delay and velocity in nerves are well established diagnostic measures of neuromuscular disorders [1]. The clinical measurement techniques developed for these measures rely on evoked responses and on averaging to reduce the effects of additive noise. This paper considers the application of the matched filter and cross correlator to nerve conduction delay estimation. In particular, it obtains the delay estimation error and signal-to-noise ratio equations for these estimators.

The approach is to formulate the problem as the estimation of the position in time of a deterministic signal contaminated with noise. The deterministic signal is the surface recorded nerve evoked response and the stimulus application time gives a time origin. The additive noise is generated primarily by the skin–electrode interface and the electronics of the signal acquisition system. In practice, the signal-to-noise ratio for sensory nerves is of the order of one or less. The evoked response plus noise is observed by the estimator whose task it is to determine the conduction velocity of the nerve. As this is an estimation problem, the appropriate performance measure is the estimation mean-square error.

II. Nerve Signal Processing

Peripheral sensory nerve signal processing utilizes evoked response signals for signal-to-noise ratio considerations. The evoked response of a nerve trunk gives a larger signal than the asynchronous activity of voluntary signals. Also, the evoked response is repeatable and signal averaging is possible. The standard orthodromic technique of peripheral sensory nerve conduction delay measurement stimulates the nerve (Fig. 1) at a site distal to two surface signal sites [1]. The evoked responses $s_1(t)$ and $s_2(t)$ from the monopolar electrode configurations M_1 and M_2, respectively, give the conduction delay over the distance d and hence, the conduction velocity. This measurement is subject to error, due to the noise terms $n_1(t)$ and $n_2(t)$ which perturb $s_1(t)$ and $s_2(t)$, respectively. To ensure the maximum signal-to-noise ratio and hence, minimize this error, it is reasonable to choose the peaks of $s_1(t)$ and $s_2(t)$ as the observed points. It will be useful for later sections to define the signal-to-noise ratio SNR for the standard technique. The received signals $r_1(t)$ and $r_2(t)$ at M_1 and M_2 are given by

$$r_1(t) = s_1(t) + n_1(t) = s(t - \alpha_1) + n_1(t), \qquad 0 \leqslant t \leqslant T$$

and for small d

$$r_2(t) = s_2(t) + n_2(t) \simeq s(t - \alpha_2) + n_2(t), \qquad 0 \leqslant t \leqslant T$$

where $s(t)$ is the response waveform and α_1 and α_2 are time delays from stimulus to response.

For a conduction velocity v and with the above approximation

$$s_2(t) \simeq s_1(t - \Delta)$$

where

$$\Delta = d/v = \alpha_2 - \alpha_1.$$

The signal-to-noise ratio SNR is defined as

$$\text{SNR} = s_p^2/\delta_n^2 \tag{1}$$

where s_p is the peak value of $s_1(t)$ and δ_n is the root-mean-square (rms) value of $n_1(t)$. When signal averaging is used to reduce noise effects, we have an average result $\bar{r}_1(t)$ where

$$\bar{r}_1(t) = \frac{1}{N}\sum_{i=1}^{N} [s_{1i}(t) + n_{1i}(t)]$$

$$= s_1(t) + \frac{1}{N}\sum_{i=1}^{N} n_{1i}(t), \qquad 0 \leqslant t \leqslant T$$

and the signal-to-noise ratio SNR_0 is given by the well-known result

$$\text{SNR}_0 = N(\text{SNR}) = N s_p^2/\delta_n^2. \tag{2}$$

The delay estimation error is inversely proportional to SNR_0 [2].

In the subsequent sections, we shall be considering maximum likelihood estimators to improve SNR_0 for a given value of N.

Manuscript received October 21, 1981; revised March 4, 1982. This work was supported in part by the National Science and Engineering Research Council of Canada under Grant A-4445.

V. S. Jasrotia is with Northern Telecom, Ltd., Toronto, Ont., Canada.
P. A. Parker is with the Department of Electrical Engineering, University of New Brunswick, Fredericton, N.B. E3B 5A3, Canada.

Reprinted from *IEEE Trans. Biomed. Eng.*, vol. 30, no. 1, pp. 1–9, January 1983.

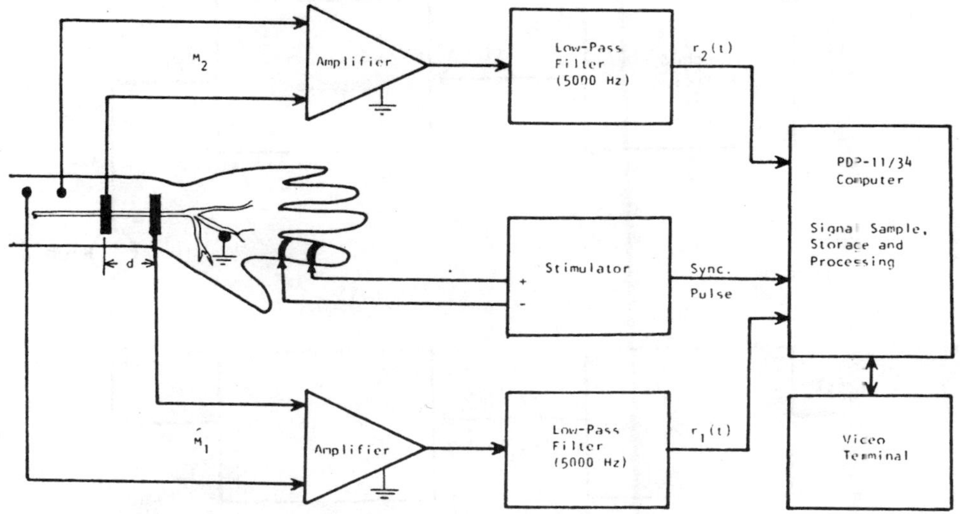

Fig. 1. Sensory nerve conduction velocity measurement.

The maximum likelihood estimator of the position in time of a deterministic waveform is the matched filter [2]. The generalized cross correlator [3] is a form of a maximum likelihood delay estimator. Parker and Kelly [4] demonstrated the possibility of using cross correlation and spectral analysis for nerve conduction delay estimation with surface electrodes. Van der Vliet *et al.* [5] applied these techniques to short segments of exposed nerve. The matched filter either as a maximum likelihood signal position estimator or as a prefilter in a generalized correlator needs to be investigated in the nerve conduction delay application. In order to proceed with this investigation, models for the signal and noise are required.

III. Noise and Signal Models

The following model is assumed to hold for the noise. $n_1(t)$ and $n_2(t)$ are both zero-mean band-limited white independent processes with

$$E[n_1(t)n_1(t+\tau)] = E[n_2(t)n_2(t+\tau)] = N_0 B \frac{\sin 2\pi B\tau}{2\pi B\tau}. \quad (3)$$

That these processes are zero mean follows from the zero dc gain of the acquisition system. That they are approximately white is easily demonstrated by measurement [6]. That they are independent follows by virtue of the fact that $n_1(t)$ and $n_2(t)$ come from two unrelated measurement channels.

In order to design the matched filter we must know *a priori* the form of $s(t)$. Further, it is necessary for optimum filter performance that the form of $s(t)$ be invariant with patient and interelectrode spacing up to 4 cm. The model chosen to represent $s(t)$ for the purpose of SNR calculations is given by

$$s(t) = \begin{cases} kt(2-ct)e^{-ct}, & 0 \leq t \leq T_0 \\ 0, & \text{otherwise} \end{cases} \quad (4)$$

where $T \geq T_0 + \alpha_2$, k is a scale factor, and c is a constant with units of s^{-1} which, from measurements, has a value of 2500. This model was arrived at by choosing a rational energy density spectrum which, on average, best fits the measured energy density spectra of many evoked nerve responses. This model form was also used by Shwedyk *et al.* [7] and Parker and Scott [8] for skeletal muscle fiber action potentials. The parameters k and c together determine the signal energy and the parameter c determines the signal bandwidth. Any signal model with the same energy as that of (4) would suffice as far as the matched filter SNR calculations are concerned. However, for the cross correlator SNR calculations, the signal waveform becomes important.

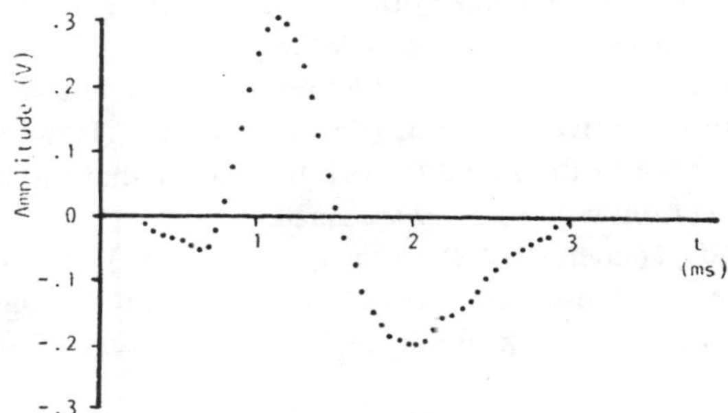

Fig. 2. Average evoked response.

In practice, $s(t)$ cannot be completely invariant over subject and interelectrode spacing. This is due to variations over subjects of the electrode positions with respect to the nerve and to velocity dispersion over fibers of a given nerve trunk. These two factors give rise to amplitude and phase distortions as discussed by Van der Vliet *et al.* [5].

For consistent electrode placement and for small d, the variations are expected to be small. As a check, the sensory nerve action potential of the median nerve was measured at two wrist locations separated by 1-4 cm for four normal subjects. Fig. 2 shows the average action potential averaged over all subjects and the two recording locations for a total of eight action potentials. The integral square value of the error between the average signal and the individual signals, expressed as a percentage of the average signal energy, is 10 percent.

It must be kept in mind that the model of (4) and the discussion of the previous paragraph are with respect to nerve

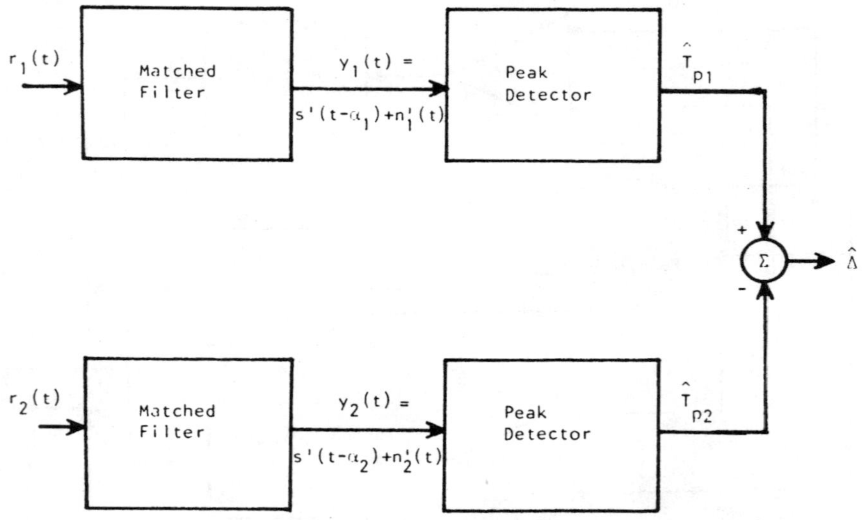

Fig. 3. Matched filter delay estimator.

action potentials from normals. The measurements to be reported in Sections IV and V were also made on normals. In the case of abnormal conduction velocities, changes in the action potential waveform can be expected. However, there is evidence in the literature [9] to indicate that the variations in waveform with velocity are small.

Two parameters of the signal model of (4) that will be of interest and used throughout the paper are the signal peak value s_p and the signal energy E. These are easily found to be given by

$$s_p = 0.46\, k/c \tag{5}$$

and

$$E = k^2/(4c^3). \tag{6}$$

Using (5) in (1) gives

$$\text{SNR} = 0.21\, k^2/(c^2 \sigma_n^2)$$

and for a measurement bandwidth B, σ_n^2 can be replaced by BN_o and

$$\text{SNR} = 0.21\, k^2/(c^2 BN_o). \tag{7}$$

Sections IV and V which follow consider the matched filter delay estimator and the cross correlator delay estimator, respectively. Since there are two distinct estimators, it is considered best from the reader's point of view to make the sections self-contained. As such, each section contains the theoretical development and experimental results for its respective estimator.

IV. Matched Filter

The maximum likelihood estimator for the time position of a known waveform in noise is the matched filter $h(t)$ followed by a peak detector, as shown in Fig. 3. For the noise and signal models of (3) and (4), the matched filter is given by

$$h(t) = \begin{cases} kg(T_0 - t)[2 - c(T_0 - t)]\, e^{-c(T_0 - t)}, & 0 \leq t \leq T_0 \\ 0, & \text{otherwise} \end{cases} \tag{8}$$

where g is an arbitrary gain factor.

The performance of the matched filter will be given in terms of the output signal-to-noise ratio, the mean-square error in the estimate of the output peak position T_p, and the mean-square error in the estimate of Δ. The signal-to-noise ratio SNR_M is defined as the ratio of the square of the peak output at time T_p due to evoked response to the variance of the noise output, i.e.,

$$\text{SNR}_M = \left. \frac{E[y(t)]^2}{E[y^2(t)] - E[y(t)]^2} \right|_{t=T_p}$$

where $E[x]$ denotes the expected value of the random variable x. Now, with $B \gg 1/T_0$, the signal-to-noise ratio for the matched filter output is given by

$$\text{SNR}_M = 2E/N_o. \tag{9}$$

Using (6) and (7) in (9) gives

$$\text{SNR}_M = 2.4B\, \text{SNR}/c.$$

For a typical bandwidth B of 5000 Hz and for c of 2500,

$$\text{SNR}_M = 4.8\ \text{SNR}. \tag{10}$$

Thus, the introduction of a matched filter in the peak position estimator improves the signal-to-noise ratio by 6.8 dB.

The mean-square error in the estimate $\hat{T}_p$ of T_p is a function of the signal-to-noise ratio. In the case of high SNR values, the mean-square error can be found from the Cramér–Rao bound. In the case of low SNR values, the estimate is subject to anomalies which give rise to large errors. By determining the probability of an anomaly, these two cases can be combined to give the mean-square estimation error σ_p^2. The details are given in Appendix A with the result

TABLE I
THEORETICAL AND MEASURED SNR_M VERSUS SNR FOR 12 SUBJECTS

Subject	$\sqrt{SNR}$	$\sqrt{SNR_M}$ Measured	$\sqrt{SNR_M}$ Theoretical
1	0.64	1.26	1.40
2	1.15	2.10	2.52
3	1.03	2.16	2.26
4	0.86	1.63	1.88
5	0.77	1.54	1.69
6	0.85	1.74	1.86
7	0.86	1.63	1.88
8	0.97	2.33	2.12
9	0.95	1.87	2.08
10	0.66	1.52	1.45
11	0.93	1.86	2.04
12	0.90	1.70	1.97

$$\sigma_p^2 \simeq \frac{0.14}{Bc\ \mathrm{SNR}} + \frac{T_1^2(2T_1W-1)}{6(7.4B\ \mathrm{SNR}/c)^{1/2}} \exp\left[-0.6B\ \mathrm{SNR}/c\right] \tag{11}$$

where W is the equivalent rectangular signal bandwidth and T_1 is the observation interval which is searched for the peak.

The time delay Δ and estimate $\hat{\Delta}$ are given by

$$\Delta = T_{p_1} - T_{p_2}$$

and

$$\hat{\Delta} = \hat{T}_{p_1} - \hat{T}_{p_2}.$$

The estimation variance $\sigma_{\hat{\Delta}}^2$ will, by virtue of the independence of $n_1(t)$ and $n_2(t)$, be given by

$$\sigma_{\hat{\Delta}}^2 = \sigma_{p_1}^2 + \sigma_{p_2}^2$$

and assuming the SNR at the two electrodes to be the same

$$\sigma_{\hat{\Delta}}^2 = 2\sigma_p^2. \tag{12}$$

The experimental setup used to verify the theoretical results of this section and the next is shown in Fig. 1. All measurements were made on the median nerve at the wrist with normal subjects. Evoked responses were obtained by stimulating with 0.2 ms monopolar constant voltage pulses of amplitudes of 15-30 V. The responses were amplified, bandlimited to 5000 Hz, and sampled and stored by the computer at a sample rate of 18 kHz. Various electrode spacings d of 1-4 cm were used.

A finite impulse response digital matched filter was designed from the average response of Fig. 2 and implemented on the computer. A series of 512 evoked responses was recorded for each electrode. The responses were averaged to determine T_{p1}, T_{p2}, Δ, and SNR. These values were taken as the true values as they were measured by the standard technique and with 512 responses averaged, the effect of noise is negligible. The responses were then processed by the matched filter and $\hat{T}_{p1}$, $\hat{T}_{p2}$, $\hat{\Delta}$, SNR_M, and σ_p^2 were obtained for various values of SNR which could be varied by averaging an appropriate number of responses at the matched filter input.

Table I gives the measured values of SNR_M for 12 subjects and the theoretical results according to (10). With 512 data points in each measurement, the statistical error is approximately ±5 percent. For comparison purposes, the theoretical signal-to-noise ratio improvement given by a first-order low-pass filter with a 3 dB bandwidth of 1500 Hz was calculated and found to be 4.1 dB, which is 2.7 dB poorer than the matched filter. Fig. 4(a) and (b) show typical measurement results and theoretical results for σ_p^2 versus SNR with observation intervals T_1 of 4.0 and 1.0 ms, respectively. The threshold effect due to anomalies is clearly evident in the 4.0 ms data. For comparison purposes, experimental results are also shown for a first-order low-pass filter with a 3 dB bandwidth of 1500 Hz.

The theoretical error variance $\sigma_{\hat{\Delta}}^2$ is given by (12) and the experimental results assuming equal SNR at the two electrodes will be simply the experimental results of Fig. 4(a) and (b) with a vertical shift of 3 dB.

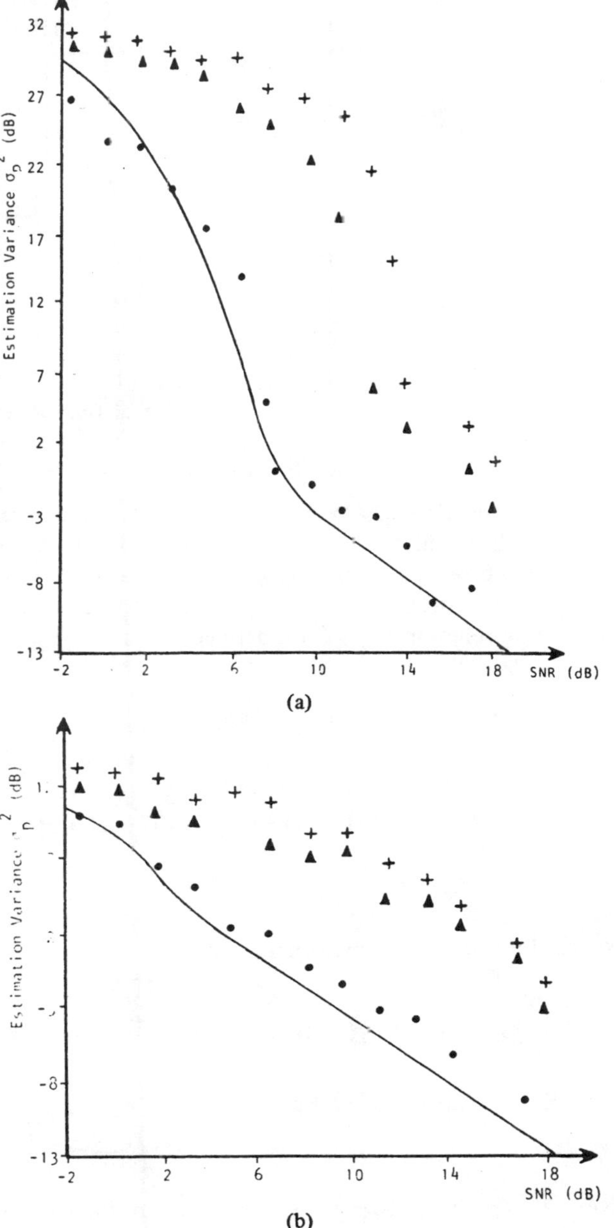

Fig. 4. Theoretical and measured peak estimation variance σ_p^2. (a) T_1 = 4.0 ms, (b) T_1 = 1.0 ms; measured: (+) 5 kHz band-limited filter, (▲) 1.5 kHz low-pass filter, (•) matched filter; theoretical: — matched filter (measured in units of sample intervals).

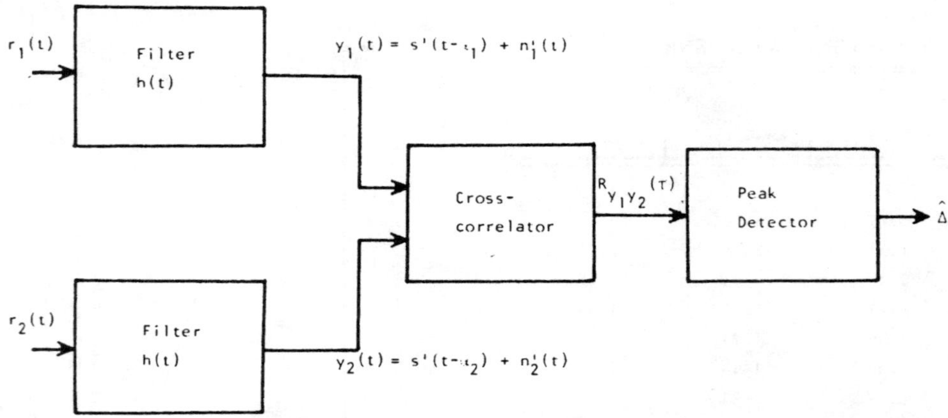

Fig. 5. Generalized cross correlator delay estimator.

V. Cross Correlation

Cross correlation is a technique by which the conduction delay can be estimated from a single peak position. A generalized cross correlator block diagram is given in Fig. 5 in which the filter $h(t)$ is left unspecified for the moment. The cross correlation function $R_{y_1 y_2}(\tau)$ is defined as

$$R_{y_1 y_2}(\tau) = \int_0^{T-\tau} y_1(t) y_2(t+\tau)\, dt$$

$$= \int_0^{T-\tau} [s'(t-\alpha_1) + n_1'(t)]\,[s'(t-\alpha_2+\tau) + n_2'(t+\tau)]\, dt$$

which, for $\tau \ll T$, is approximated by

$$\simeq \int_0^{T} [s'(t-\alpha_1) + n_1'(t)]\,[s'(t-\alpha_2+\tau) + n_2'(t+\tau)]\, dt$$

$$= R_{s's'}(\tau-\Delta) + \hat{R}_{s'n_2'}(\tau+\alpha_1) + \hat{R}_{n_1's'}(\tau-\alpha_2) + \hat{R}_{n_1'n_2'}(\tau). \quad (13)$$

Now $R_{s's'}(\tau-\Delta)$ is, except for a time shift Δ, the autocorrelation function for $s'(t)$, and its maximum value occurs at that value of τ for which the argument $\tau - \Delta$ is equal to zero. Thus, the delay Δ is estimated from the peak position of $R_{s's'}(\tau-\Delta)$. In the presence of noise which manifests itself in the last three terms of (13), the estimate is subject to error.

The signal-to-noise ratio SNR_C for the cross correlator output is defined as

$$SNR_C = \left.\frac{E[R_{y_1 y_2}(\tau)]^2}{E[R^2_{y_1 y_2}(\tau)] - E[R_{y_1 y_2}(\tau)]^2}\right|_{\tau=\Delta}$$

and since the noise processes are zero mean and uncorrelated, we have

$$SNR_C = \left.\frac{R^2_{s's'}(\tau-\Delta)}{E[\hat{R}^2_{s'n_2'}(\tau+\alpha_1) + \hat{R}^2_{n_1's'}(\tau-\alpha_2) + \hat{R}^2_{n_1'n_2'}(\tau)]}\right|_{\tau=\Delta}. \quad (14)$$

It is shown in Appendix B that for $T \gg 1/B$, (14) can be approximated as

$$SNR_C \simeq \frac{\left[\int_{-\infty}^{\infty} |S(f)H(f)|^2\, df\right]^2}{N_0^2 T/4 \int_{-\infty}^{\infty} |H(f)|^4\, df + N_0 \int_{-\infty}^{\infty} |S(f)|^2\, |H(f)|^4\, df} \quad (15)$$

where $S(f)$ and $H(f)$ are the Fourier transforms of $s(t)$ and $h(t)$, respectively. It is now useful to consider SNR_C for three cases.

Case 1:

$$H(f) = \begin{cases} 1, & |f| \leq 5000 \\ 0, & \text{otherwise.} \end{cases}$$

Recognizing that $\int_{-\infty}^{\infty} |S(f)|^2\, df = E$ and substituting (6) and (7) in (15) gives

$$SNR_C = \frac{2.35\, SNR^2}{SNR + 0.43\, cT}$$

and as before with $c = 2500$, we have

$$SNR_C = \frac{2.35\, SNR^2}{SNR + 1075T}. \quad (16)$$

Case 2:

$$H(f) = f_c/(f_c + if),$$

first-order low-pass filter with $f_c = 1500$ Hz.

Substituting for $H(f)$ in (15), evaluating the integrals, and substituting (6) and (7) with $c = 2500$ gives

$$SNR_C = \frac{2.10\, SNR^2}{SNR + 256T}. \quad (17)$$

Case 3:

$$H(f) = S^*(f)\, e^{-j2\pi f T_0}, \quad \text{matched filter.}$$

Substituting for $H(f)$ in (15), evaluating the integrals, and

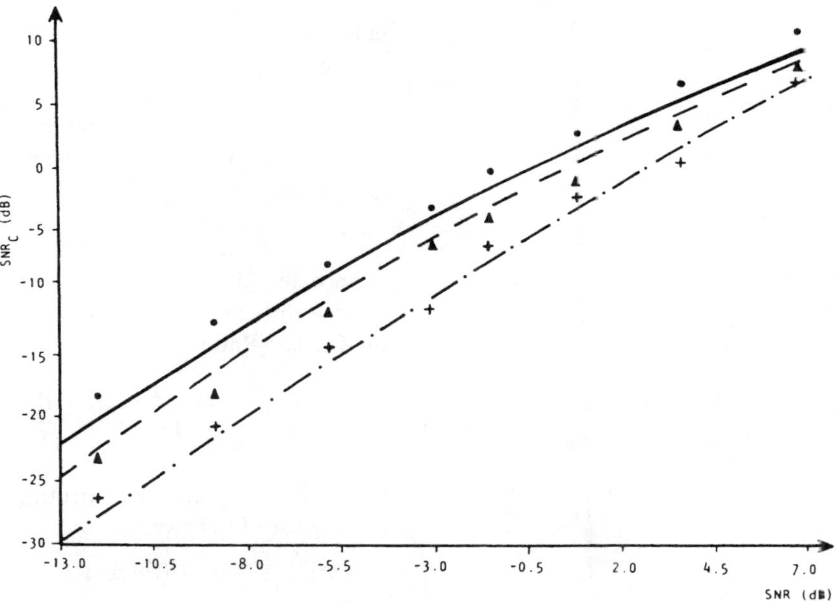

Fig. 6. Theoretical and measured SNR_C for $T = 5.5$ ms; measured: (+) 5 kHz band-limited filter, (▲) 1.5 kHz low-pass filter, (•) matched filter; theoretical: – · 5 kHz band-limited filter, – – 1.5 kHz low-pass filter, — matched filter.

substituting (6) and (7) with $c = 2500$ gives

$$SNR_C = \frac{2.0\ SNR^2}{SNR + 145T}. \tag{18}$$

The theoretical signal-to-noise ratios of (16), (17), and (18) are plotted in Fig. 6 for $T = 5.5$ ms.

Using the experimental setup and procedures described in the previous section, the signal-to-noise ratio SNR_C versus SNR was measured for the three cases and for $T = 5.5$ ms. The delay estimation mean-square error σ_Δ^2, which is a function of SNR_C, was also measured. Fig. 6 shows the experimental results for SNR_C as compared to the theoretical SNR_C versus SNR. Fig. 7(a) and (b) give the results for the measurement of σ_Δ^2 for observation intervals of 4.0 and 1.0 ms, respectively. The threshold effect due to anomalies is clearly evident in the 4.0 ms data. The improvement in σ_Δ^2 offered by the matched filter in the cross correlator is apparent.

The question of the optimum filter for $h(t)$ in the cross correlator now arises. A slightly different form of a generalized cross correlator was studied by Hassab and Boucher [3] in which the two filters of Fig. 5 are replaced by a single filter $g(t)$ following the cross correlator block. They show that the optimum $g(t)$ for the low signal-to-noise case is specified by

$$G(\omega) = \frac{|S(\omega)|^2}{\Phi_{n_1}(\omega)\Phi_{n_2}(\omega)} \tag{19}$$

where $\Phi_{n_1}(\omega)$ and $\Phi_{n_2}(\omega)$ are the power spectral densities of the noise processes $n_1(t)$ and $n_2(t)$, respectively. Now, taking the Fourier transform of $R_{y_1 y_2}(\tau)$ from (13) gives

$$F[R_{y_1 y_2}(\tau)] = R_1(\omega) R_2^*(\omega) |H(\omega)|^2$$

from which it is seen that for the two forms of the generalized correlator to be equivalent we require

$$|H(\omega)|^2 = G(\omega).$$

This requirement is satisfied when $h(t)$ is the matched filter. This follows since in this case

$$H(\omega) = \frac{2S^*(\omega)e^{-j\omega T_0}}{N_0}$$

and

$$|H(\omega)|^2 = 4|S(\omega)|^2/N_0^2. \tag{20}$$

Comparing (20) with (19), it is clear that the matched filter is the optimum filter for the cross correlator estimator. This is true only for the low signal-to-noise ratio case which holds in our application.

VI. Discussion

The two delay estimators considered in Sections IV and V are now compared on the basis of structure and error performance. The estimator of Fig. 3 requires the estimation of two peak positions. When utilizing matched filters before the peak detectors, each channel becomes a maximum likelihood peak position estimator. In effect, the matched filters generate the cross correlations of $s(t)$ with $s_1(t) + n_1(t)$ and with $s_2(t) + n_2(t)$. The signal-to-noise ratio SNR_M is linear in SNR with a 6.8 dB improvement.

The cross correlator estimator of Fig. 5 requires the estimation of a single peak position. The noise performance of this estimator is poorer than the first. This is primarily due to the term $R_{n_1' n_2'}(\tau)$ in (13). The signal-to-noise ratio SNR_C is a nonlinear function of SNR and at low values of SNR the cross correlator degrades the signal-to-noise ratio. The estimator performance is optimized by using the matched filter for $h(t)$ in Fig. 5. Fig. 8(a) and (b) show the delay estimation variance σ_Δ^2 for the two estimators for observation intervals of 4.0 and 1.0 ms, respectively. The estimator of Fig. 3 is superior by approximately 3 dB.

For the matched filter to realize its full SNR improvement, the evoked response must match the form of the filter response. For this reason, it is intended to extend the measurements to

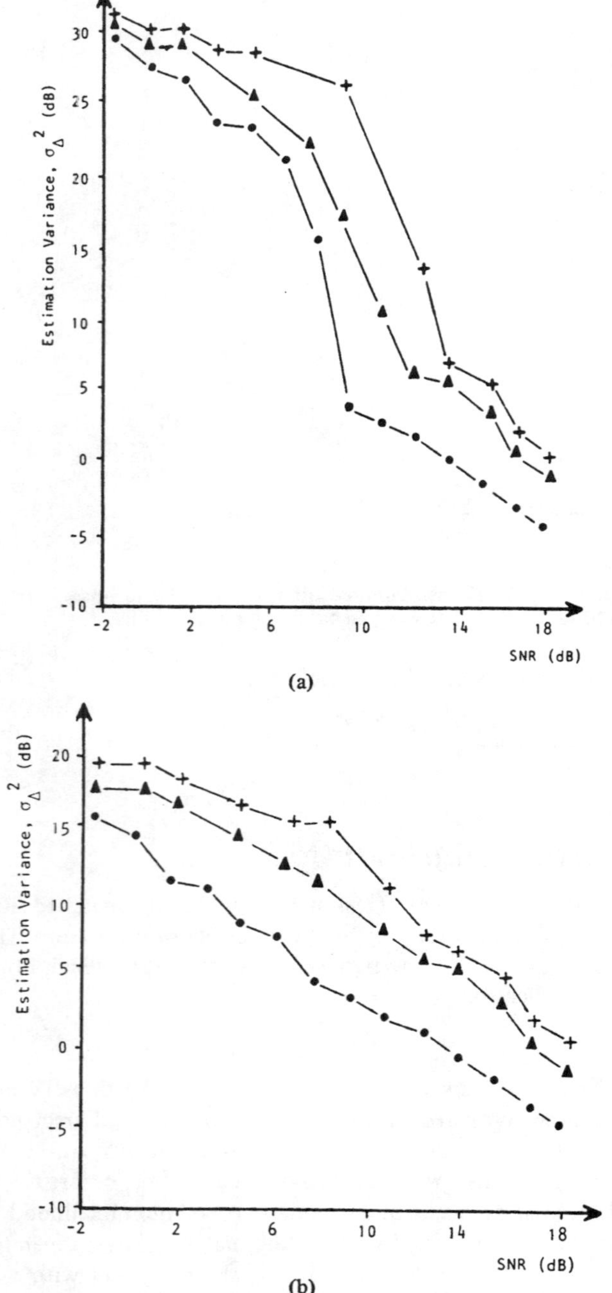

Fig. 7. Cross correlator estimator measured error variance σ_Δ^2 for (a) T_1 = 4.0 ms, (b) T_1 = 1.0 ms, and prefilters (+) 5 kHz band-limited, (▲) 1.5 kHz low-pass, (•) matched (measured in units of sample intervals).

include abnormal and a wider range of conduction velocities. It is worth noting that a mismatch between the filter and an evoked response will not introduce a deterministic bias error, but simply reduce the SNR below that of the matched case.

VII. Conclusions

The noise performance equations have been obtained for the matched filter and cross correlator in the nerve conduction velocity measurement. The matched filter shows a 6.8 dB SNR improvement over the conventional 5 kHz band-limiting filter. The cross correlator, while requiring the estimation of only one peak position, has a significantly poorer noise performance than the estimator of Fig. 3. The matched filter for $h(t)$ gives the optimum cross correlator estimator configuration. The practical effect of the improvement offered by the matched filter is that for a given estimation error, fewer evoked responses are required.

Appendix A

Since the matched filter-peak detector is a maximum likelihood estimator, the estimation error variance is bounded by the Cramér–Rao bound [2] as

$$\mathrm{Var}[\hat{T}_{p_1}] \geqslant \frac{N_0}{2}\int_0^T \left[\frac{\partial s(t - T_{p_1} + T_0)}{\partial T_{p_1}}\right]^2 dt \tag{A1}$$

where $T > T_0 + \alpha_1$. Substituting 4 in (A1), evaluating and making use of (7) gives

$$\mathrm{Var}[\hat{T}_{p_1}] \geqslant 0.14/(Bc\ \mathrm{SNR}).$$

For the high signal-to-noise ratio case which results in the absence of anomalies, the variance is well approximated by this bound, giving

$$\mathrm{Var}[\hat{T}_{p_1}\,|\text{no anomaly}] \simeq 0.14/(Bc\ \mathrm{SNR}).$$

For the low signal-to-noise ratio case which gives rise to anomalies, $\hat{T}_{p_1}$ is equally likely to be any value in the observation range T_1 over which the peak search is carried out. Thus, the variance given that an anomaly has occurred is given by

$$\mathrm{Var}[\hat{T}_{p_1}\,|\text{anomaly}] = E[(T_{p_1} - \hat{T}_{p_1})^2\,|\text{anomaly}] = T_1^2/6$$

since both T_{p_1} and $\hat{T}_{p_1}$ are uniformly distributed over T_1 s and $\hat{T}_{p_1}$ is independent of T_p under anomalous conditions.

The probability of an anomaly P_0 is easily bounded by the union bound [10] as

$$P_0 \leqslant \frac{2T_1W - 1}{(2\pi E/N_0)^{1/2}} \exp\,[-E/2N_0]$$

where W is the equivalent rectangular bandwidth of $s(t)$. The total error variance σ_p^2 may now be written in terms of the high and low signal-to-noise ratio cases as

$$\sigma_p^2 = \mathrm{Var}[\hat{T}_{p_1}\,|\text{no anomaly}]\,[1 - P_0] + \mathrm{Var}[\hat{T}_{p_1}\,|\text{anomaly}]\,P_0 \simeq 0.14/(Bc\ \mathrm{SNR}) + \frac{T_1^2(2T_1W - 1)}{6(2\pi E/N_0)^{1/2}} \exp\,[-E/2N_0]$$

and, using (6) and (7), this becomes

$$\sigma_p^2 \simeq \frac{0.14}{Bc\ \mathrm{SNR}} + \frac{T_1^2(2T_1W - 1)}{6(7.4B\ \mathrm{SNR}/c)^{1/2}} \exp\,[-0.6B\ \mathrm{SNR}/c].$$

Appendix B

Considering the numerator of (14), we have

$$R_{s's'}^2(\tau - \Delta)|_{\tau=\Delta} = R_{s's'}^2(0),$$

but $R_{s's'}(0)$ is the energy of $s'(t)$ and hence,

$$R_{s's'}(0) = \int_{-\infty}^{\infty} |S'(f)|^2\, df = \int_{-\infty}^{\infty} |S(f)H(f)|^2\, df.$$

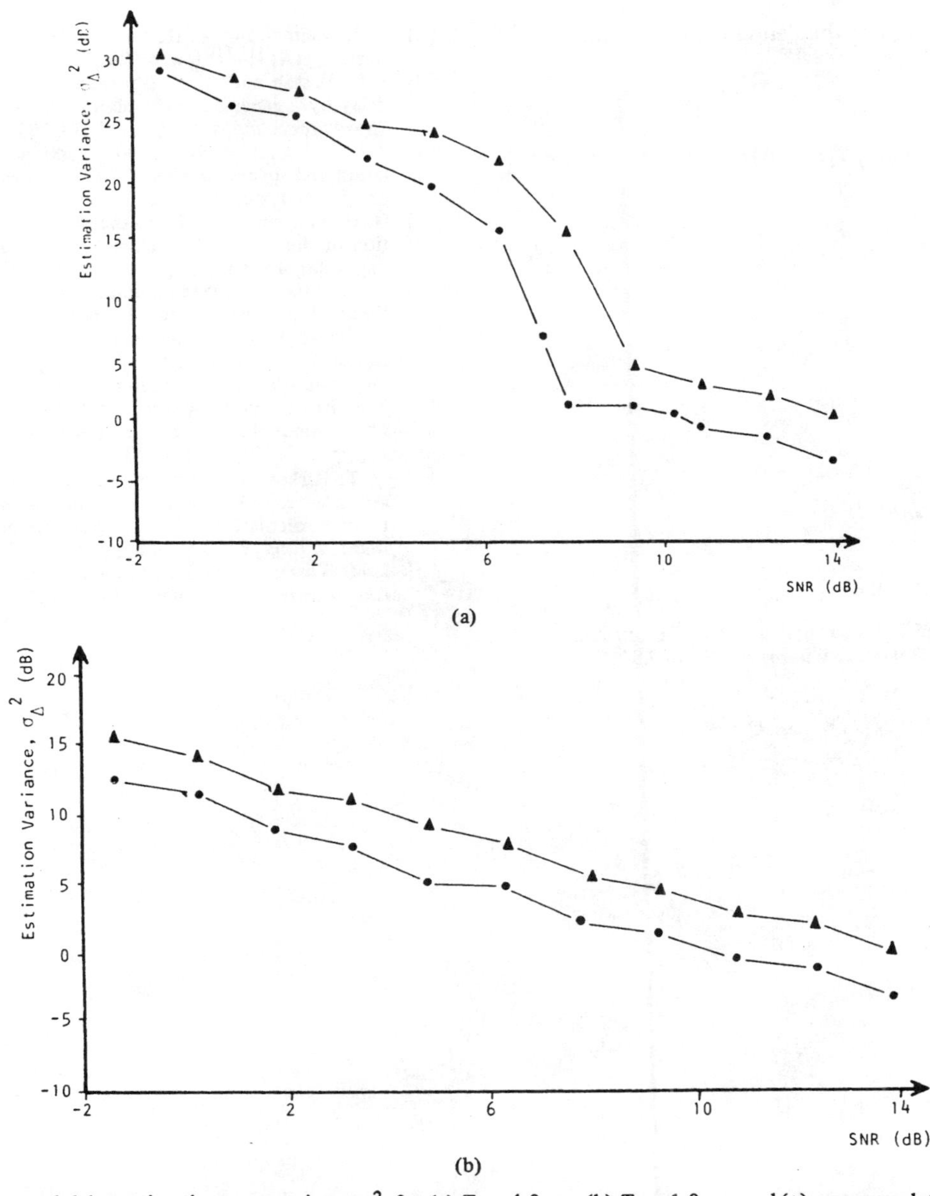

Fig. 8. Measured delay estimation error variance σ_Δ^2 for (a) $T_1 = 4.0$ ms, (b) $T_1 = 1.0$ ms, and (▲) cross correlator estimator, (●) matched filter estimator (measured in units of sample intervals).

Considering the first term of the denominator we have

$$E[\hat{R}^2_{s'n'_2}(\tau + \alpha_1)]\,|_{\tau=\Delta}$$

$$= E\left[\int_0^T \int_0^T s'(t-\alpha_1) n'_2(t+\Delta) s'(u-\alpha_1) \cdot n'_2(u+\Delta)\, du\, dt\right]$$

$$= \int_0^T \int_0^T s'(t-\alpha_1) s'(u-\alpha_1) R_{n'_2 n'_2}(u-t)\, du\, dt$$

and for $T \gg 1/B$

$$\simeq \int_0^T s'(t-\alpha_1)\, dt \int_{-\infty}^{\infty} S'(f) e^{-j2\pi f\alpha_1} \Phi_{n'_2 n'_2}(f) e^{j2\pi f t}\, df$$

$$= \int_{-\infty}^{\infty} S'^*(f) e^{j2\pi f\alpha_1} S'(f) e^{-j2\pi f\alpha_1} \Phi_{n'_2 n'_2}(f)\, df$$

$$= \int_{-\infty}^{\infty} |S(f)|^2 |H(f)|^4 N_0/2\, df.$$

Similarly,

$$E[\hat{R}^2_{n'_1 s'}(\tau - \alpha_2)]\,|_{\tau=\Delta} = \int_{-\infty}^{\infty} |S(f)|^2 |H(f)|^4 N_0/2\, df.$$

Considering the last term of the denominator, we have

$$E[\hat{R}^2_{n_1' n_2'}(\tau)]\,|_{\tau=\Delta}$$

$$= E\left[\int_0^T \int_0^T n_1'(t)\, n_2'(t+\Delta)\, n_1'(u)\, n_2'(u+\Delta)\, dt\, du\right]$$

$$= \int_0^T \int_0^T R^2_{n_1' n_1'}(u-t)\, dt\, du$$

and for $T \gg 1/B$

$$\approx \int_0^T dt \int_{-\infty}^{\infty} \Phi^2_{n_1' n_1'}(f)\, df$$

$$= T \int_{-\infty}^{\infty} |H(f)|^4\, N_0^2/4\, df.$$

References

[1] M. P. Smorto and J. V. Basmajian, *Clinical Electroneurography.* Baltimore, MD: Williams & Wilkins, 1979, pp. 19-57.

[2] R. E. Zeimer and W. H. Tranter, *Prinicples of Communications.* Boston, MA: Houghton Mifflin, 1976, pp. 413-416.

[3] J. C. Hassab and R. E. Boucher, "Optimum estimation of time delay by a generalized correlator," *IEEE Trans. Acoust., Speech, Signal Processing*, vol. ASSP-27, pp. 373-380, Aug. 1979.

[4] P. A. Parker and P. Kelly, "Nerve conduction velocity from correlation and spectral analysis," in *Proc. 9th Ann. Northeast Bioeng. Conf.*, 1981, pp. 381-386.

[5] G. H. Van der Vliet, J. Holsheimer, and D. Bingmann, "Calculation of the conduction velocity of short nerve fibers," *Med. Biol. Eng. Comput.*, vol. 18, pp. 749-757, Nov. 1980.

[6] V. Jasrotia, "Nerve conduction velocity measurements," M.Sc. thesis, Univ. New Brunswick, Fredericton, N.B., Canada, 1981.

[7] E. Shwedyk, R. Balusubramanian, and R. N. Scott, "A nonstationary model for the electromyogram," *IEEE Trans. Biomed. Eng.*, vol. BME-24, pp. 417-424, Sept. 1977.

[8] P. A. Parker and R. N. Scott, "Statistics of the myoelectric signal from monopolar and bipolar electrodes," *Med. Biol. Eng.*, vol. 11, pp. 591-596, Sept. 1973.

[9] A. T. Barker, B. H. Brown, and I. L. Freeston, "Modeling of an active nerve fiber in a finite volume conductor and its application to the calculation of surface action potentials," *IEEE Trans. Biomed. Eng.*, vol. BME-26, pp. 53-56, Jan. 1979.

[10] J. M. Wozencraft and I. M. Jacobs, *Principles of Communication Engineering.* New York: Wiley, 1965.

Local Steam Transit Time Estimation in a Boiling Water Reactor

LJILJANA KOSTIĆ

Abstract—A new correlation method, the so-called "rehocence" [1] or "smoothed coherence transform" (SCOT) [4], for transit time estimation is applied in a boiling water reactor (BWR). The transit time of a propagating perturbation of the coolant density between two points along the propagation path could be measured by noise analysis technique. This is usually performed by directly observing the peak in the cross-correlation function of the stochastic signals measured at two points in the propagation path, or indirectly by determining the slope of the phase. The used rehocence method presents the transit time directly in the same way as in the ordinary cross correlation technique, but with a better resolution, even when the measured signals are contaminated by a narrow-band limited internal noise coming from the global noise of the neutron flux fluctuation.

I. Introduction

Boiling water reactors are more noisy than other power reactors. The major noise source in these reactors, defined as "boiling noise," originates from discrete *steam* bubbles delivered to the steam space. The formation and movement of the steam bubbles in these reactors give rise to a spatial correlation field of the *neutron flux fluctuations*, which may be detected by in-core detectors. The power spectral density function of the neutron flux fluctuations (measured by means of an in-core detector) exhibits two parts—a global part in the lower frequency range, and a local part in the higher frequency range. The global part results from reactivity effects on the whole core, and is limited by a break frequency of 1 Hz. The local part of the spectrum results from local flux disturbances, caused by steam bubbles moving in the vicinity of an in-core detector. This effect, which covers the higher frequency part beyond 1 Hz up to about 30 Hz, makes possible determination of transit times related to the propagation of a perturbation containing high frequencies. The spatial resolution in studying such perturbation is just the sensitivity length of the local component.

In this paper, we investigate the local component of the neutron noise in the core of the BWR.

A new technique of correlating *stochastic signals*, known as "*smoothed coherence transform*" or "*rehocence*" (coherence read backward), is applied and compared with the ordinary cross correlation technique in the time and frequency domain.

Manuscript received July 17, 1980; revised October 20, 1980.
The author is with the Boris Kidrič Institute of Nuclear Sciences, 11001 Belgrade, Yugoslavia.

II. Mathematical Background

In boiling water reactors, the coolant is typically composed of both the liquid and the vapor phases of water. The coolant enters the core with a certain degree of subcooling, i.e., subcooled boiling takes place near the core inlet. At some position up the core, the average temperature of the coolant exceeds saturation temperature, i.e., the region of bulk boiling is reached.

In the lower part of the core, where the steam content is rather low, vapor bubbles are generated at the walls of the fuel elements. The detaching bubbles form a bubbly flow in the liquid. As the steam content increases rapidly along the channel, coalescence of the bubbles becomes frequent and the slug-flow regime (long vapor bubbles separated by liquid slugs) is entered. Further increase of the steam content gives way to the churn-turbulent flow (bubbles break up randomly and vapor flows in a chaotic manner through the liquid) which results ultimately in the annular-flow regime (the liquid phase collects on the wall, the vapor streams into the inner core). In the upper portion of the rod bundle, where the annular-flow regime is dominant, the formation of the vapor at the wall may cease and steam is generated via evaporation at the interface between the liquid film and steam core.

Using a one-dimensional (axial) model [2], the local flux component of the neutron noise field in a BWR can be written approximately as

$$L(t,z) = K\alpha(t,z) \tag{1}$$

where

$L(t,z)$ = function of the flux at a position z
$\alpha(t,z)$ = fluctuation of the void fraction at a position z
K = proportionality constant.

Let us now consider two axial positions z_1 and z_2 ($z_2 \geqslant z_1$) and write the conservation of the fluctuation of the steam as

$$\alpha(t,z_2) = K_\alpha \cdot \alpha(t - D_{12}, z_1) + (\cdots) \tag{2}$$

where D_{12} = *transit time* of the steam between the two positions considered.

The first term on the right-hand side (RHS) of (2) accounts for the propagation between the positions z_1 and z_2. The bracket stands for the contribution coming from steam generation between the two positions considered. Using (1) and (2) we write

$$L(t,z_2) = K_L \cdot L(t - D_{12}, z_1) + (\cdots). \tag{3}$$

Reprinted from *IEEE Trans. Acoust., Speech, Signal Processing*, vol. 29, no. 3, pt. 2, pp. 555–560, June 1981.

If the second term on the RHS of the above equation could be neglected, the neutron signals in the two positions would differ only in a multiplying factor, K_L, and a time shift, D_{12}. Comparing the two signals, the actual value of the time-shift, i.e., the transit time of the steam, could be determined. In the case of BWR the steam generation term can be quite important and mask the propagation effect represented by the first term. To obtain measured values of the transit time we have to evaluate the cross-correlation function or the cross spectrum of the two signals.

Following the usual treatment, let us define the *cross-correlation function* between the flux fluctuations in the lower and upper positions as

$$R_{z_1,z_2}(\tau) = \langle L(t,z_1)\, L(t+\tau, z_2)\rangle \tag{4}$$

where the bracket stands for time averaging, i.e.,

$$\langle L(t,z_1)\, L(t+\tau,z_2)\rangle = \lim_{T\to\infty} \frac{1}{2T}\int_{-T}^{+T} L(t,z_1)\, L(t+\tau,z_2)\, dt. \tag{5}$$

A further quantity of interest is the autocorrelation function of the fluctuation of the flux at position z. Let us write the usual definition as

$$R_z(\tau) = R_{z,z}(\tau) = \langle L(t,z)\, L(t+\tau,z)\rangle. \tag{6}$$

The autocorrelation function describes the correlation between values of flux-fluctuation at a given position separated by a time lag τ. The cross correlation describes the correlation of quantities separated both in time and space. The observation time T can be finite, provided it is large compared to any characteristic time of the system.

As values of the flux-fluctuation at a given position are most strongly correlated if the time lag separating them is zero, the autocorrelation has its maximum at $\tau = 0$ when it represents the variance of the fluctuation of the flux.

$$R_z(0) \geqq |R_z(\tau)|. \tag{7}$$

To evaluate the cross-correlation function, let us substitute (3) into (4) and write

$$R_{z_1,z_2}(\tau) = \langle L(t,z_1)\, L(t+\tau-D_{12}, z_1)\rangle + \langle L(t,z_1)(\cdots)\rangle. \tag{8}$$

As the fluctuation of the flux at the lower position and the steam generated between the two positions are not correlated, the second term on the RHS of the above equation is zero. Using the definition of the autocorrelation function given in (6) we obtain

$$R_{z_1,z_2}(\tau) = K_L \cdot R_{z_1}(\tau - D_{12}). \tag{9}$$

In view of (7) the cross-correlation function of the two signals has a peak at a time lag value

$$\tau = D_{12}. \tag{10}$$

The location of the "propagation peak," i.e., the transit time of the steam, can be evaluated from the measured curve.

The derivation shows the main advantage of the correlation technique. By evaluating the cross-correlation function one eliminates the second term on the RHS of (3). It is this term which jeopardizes the determination of the transit time by direct inspection of the noise signals.

An alternative representation of the noise phenomena can be made via the frequency variable rather than the time variable. To this end, let us define the cross-power spectral density function (CPSD) and the autopower spectral density function (APSD) of the neutron noise as the Fourier transforms of the corresponding correlation functions

$$G_{z_1,z_2}(f) = \int_{-\infty}^{+\infty} e^{-j2\pi f\tau} R_{z_1,z_2}(\tau)\, d\tau \tag{11}$$

$$G_z(f) = \int_{-\infty}^{+\infty} e^{-j2\pi f\tau} R_z(\tau)\, d\tau. \tag{12}$$

The autospectrum is a real valued even function of frequency. The cross spectrum is complex, and can be conveniently expressed in complex polar notation such that

$$G_{z_1,z_2}(f) = |G_{z_1,z_2}(f)|\, e^{j\Phi_{z_1,z_2}(f)} \tag{13}$$

where $|G_{z_1,z_2}|$ is the gain, and the angle Φ_{z_1,z_2} is the phase lag between the two signals.

Using (9), (11), and (12) one obtains that in the actual case the cross spectrum can be written as

$$G_{z_1,z_2}(f) = K_L \cdot G_{z_1}(f)\, e^{-j2\pi f D_{12}}. \tag{14}$$

Measuring the *phase* lag between noise signals of two axially placed in-core neutron detectors, the transit time of the steam can be evaluated from the relation

$$\Phi_{z_1,z_2} = -2\pi f D_{12}. \tag{15}$$

For the autospectrum, (1), (6), and (12) result in the relation

$$G_z(f) = K^2 G_z^{\alpha}(f) \tag{16}$$

where

$$G_z^{\alpha}(f) = \int_{-\infty}^{+\infty} e^{-j2\pi f\tau}\, R_z^{\alpha}(\tau)\, d\tau, \tag{17}$$

i.e., the autospectrum of the noise signal of an in-core detector is proportional to the autospectrum of the local fluctuation of the void fraction.

The complex coherence function $\gamma(f)$ between the two processes is defined as

$$\gamma(f) = \frac{G_{z_1,z_2}(f)}{[G_{z_1}(f)\cdot G_{z_2}(f)]^{1/2}}. \tag{18}$$

If we suppose that the two signals, corresponding to the neutron flux fluctuations in two axial positions in BWR, differ only in a multiplying factor and a time shift, the upper relation becomes

$$\gamma(f) = K_\gamma e^{-j2\pi f D_{12}}. \tag{19}$$

The inverse Fourier transform of $\gamma(f)$ is given by

$$\Gamma(t) \equiv \mathfrak{F}^{-1}[\gamma(f)] = K_\gamma \delta(t - D_{12}). \tag{20}$$

This function is the so-called "rehocence function" or "smoothed coherence transform." In this case it directly depends on the frequency range of the signals and has a peak at a time equal to the delay time, D_{12}.

The difference between the cross-correlation function method $(R_{z_1,z_2}(t) = \mathfrak{F}^{-1}[G_{z_1,z_2}(f)])$ and the rehocence method $(\Gamma(t) = \mathfrak{F}^{-1}[\gamma(f)])$ is that in the complex coherence function the magnitude of the CPSD is weighted for all frequencies by corresponding APSD's. If global noise is present, it appears not only in the CPSD but also in both APSD's. By manipulation via (18) this influence is reduced, allowing, under certain conditions, visualization of a local peak by the rehocence function where it may be obscured in the cross-correlation function.

In order to recapitulate the principle of transit time determination in BWR by cross-correlation methods, we consider a streaming fluid and two detectors "viewing" some kind of fluctuations (density, temperature, etc.).

The signals from two detectors can be written as

$$s_2(t) = l(t - D_{12}) + g(t) + n_2(t) \tag{21}$$

$$s_1(t) = l(t) + g(t) + n_1(t) \tag{22}$$

where $s_1(t)$ and $s_2(t)$ are respectively the ac components of the signals from the lower (z_1) and upper (z_2) detector positions in the core. $l(t)$ is the local signal component of the noise, $g(t)$ is the global signal component of the noise and $n(t)$ is background noise (detection noise + steam generation noise + other noise). D_{12} is the transit time of the propagating perturbation between the two detectors. It appears as a delay time in the local signal component which is the only term containing information about the propagation of the fluid.

Assuming that local and global components are independent, and that $n_1(t)$ and $n_2(t)$ are uncorrelated and independent, both from the other components, the CPSD takes the form

$$G_{z_1,z_2}(f) = e^{-j2\pi f D_{12}} \cdot G^l(f) + G^g(f) \tag{23}$$

where $G^l(f)$ and $G^g(f)$ are the APSD's of the local and global parts of the signal, respectively.

Let us assume the APSD's of the local and global components are flat with the magnitudes L_0 and G_0 and with the break frequencies f_l and f_g, respectively, $(f_l \gg f_g)$. In this case the corresponding cross-correlation function will be

$$R_{z_1,z_2}(\tau) = \int_{-\infty}^{+\infty} e^{j2\pi f\tau} \cdot G_{z_1,z_2}(f)\, df$$

$$= L_0 \frac{1}{\pi} \cdot \frac{\sin\{2\pi f_l(\tau - D_{12})\}}{(\tau - D_{12})} + G_0 \frac{1}{\pi} \cdot \frac{\sin\{2\pi f_g(\tau)\}}{\tau}. \tag{24}$$

Normally, the transit time can be determined quite easily from the local part as appearing in the above equation. Trouble arises if the magnitude of the local peak is small with respect to the global peak and the transit time is also small, so that the small local peak is "hidden" under the big global peak. Formulating this statement in mathematical terms, it follows from the above equation that the cross-correlation function method for transit time determination fails, if

$$L_0 f_l \ll G_0 f_g \tag{25}$$

and

$$D_{12} \ll 1/(2f_g). \tag{26}$$

Let us assume for simplicity that the APSD's of the uncorrelated background noise $n(t)$ are equal for both positions. Thus, for the APSD's of the signals $s_1(t)$ and $s_2(t)$, we obtain the following.

$$G_{z_1}(f) = G_{z_2}(f) = G^l(f) + G^g(f) + G^n(f). \tag{27}$$

By inserting (23) and (27) into (18), the complex coherence function becomes

$$\gamma(f) = \frac{e^{-j2\pi f D_{12}} \cdot G^l(f) + G^g(f)}{G^l(f) + G^g(f) + G^n(f)}. \tag{28}$$

The above equation shows the disadvantage of the coherence method. The cross-correlation function method is not influenced by the independent background noise $n(t)$, which figures in the rehocence method [demoninator of (28)]. If this noise source is large in comparison to the other terms, the rehocence method is not applicable.

Let us assume that the independent background sources are white noise

$$G^n(f) = N_0. \tag{29}$$

In this case, $\gamma(f)$ has the form

$$\gamma(f) = \begin{cases} \dfrac{e^{-j2\pi f D_{12}} \cdot L_0 + G_0}{L_0 + G_0 + N_0}, & 0 \leqslant f \leqslant f_g \\ \dfrac{e^{-j2\pi f D_{12}} \cdot L_0}{L_0 + N_0}, & f_g \leqslant f \leqslant f_l \\ 0, & f > f_l. \end{cases} \tag{30}$$

In the case of a BWR we are interested in considering a dominant global component. Therefore, we can make the simplifying assumptions

$$N_0 \ll L_0, \quad N_0 \ll G_0 \tag{31}$$

and

$$G_0 \gg L_0. \tag{32}$$

Then (30) reduces to the form

$$\gamma(f) = \begin{cases} 1, & 0 \leqslant f \leqslant f_g \\ e^{-j2\pi f D_{12}}, & f_g \leqslant f \leqslant f_l \\ 0, & f > f_l. \end{cases} \tag{33}$$

In order to see better the advantage and limitations of the method, let us calculate the rehocence function resulting from (33) via Fourier transform. Thus, we obtain

$$\Gamma(t) = \frac{1}{\pi} \cdot \frac{\sin [2\pi f_l(\tau - D_{12})]}{(\tau - D_{12})} - \frac{1}{\pi} \cdot \frac{\sin [2\pi f_g(\tau - D_{12})]}{(\tau - D_{12})} + \frac{1}{\pi} \cdot \frac{\sin [2\pi f_g \tau]}{\tau}. \tag{34}$$

The RHS of (34) consists of three terms. The condition for the appearance of a well-defined local peak is that

$$f_l \gg f_g. \tag{35}$$

If f_l becomes close to f_g, then the third term in the rehocence function, which does not contain any transit time information, dominates and the rehocence method breaks down.

The rehocence method is useful for transit time determination if there is a sufficiently large frequency region where the local component dominates over the global component, which actually is the situation in the analysis of neutron noise in BWR.

III. Experimental Procedure and Data Processing

To demonstrate the usefulness of the proposed "rehocence" method, power reactor noise measurements were performed in the core of the Mühleberg BWR. Neutron noise signals from LPRM's (Local Power Range Monitors) were measured along the axial direction. There are four in-core detectors in the Mühleberg BWR, labeled A, B, C, and D in order from the lower part of the core.

The signals for each pair of detectors were amplified by PAR-113 amplifiers (Princeton Applied Research Lab). The mean detector currents were suppressed by ac operation of the amplifiers. The amplified noise signals were analyzed by an Omniferous Dual Channel FFT-400 Analyzer (Nicolet Scientific Corporation). This equipment can simultaneously evaluate the autospectra, the cross spectrum, the transfer function, and the coherence function.

In order to determine the cross-correlation function and the rehocence function for the same time signals, a corresponding algorithm was developed for the CDC-3600 computer.

The noise measurements were performed in a frequency range of 0–50 Hz. The noise data coming from LPRM's were collected for about 30 min.

IV. Results

There are three possibilities to estimate the transit time.

1) The correlation function produces a delta function-like peak when the fluctuation is white and little contaminated by background noise. This, however, is not usually the case, but the signal is band limited and accompanied by internal or external background noise. In the case of a BWR steam bubble transit time measurement, the fluctuating signal is distributed internally by the predominating global noise. Thus, a filtering is needed to eliminate this unwanted component.

2) Another technique is to plot the phase of the cross-power spectrum versus frequency and to calculate the time lag from the slope of the plot. This, however, does not present the result directly. Moreover, when a narrow-band noise contaminates the signal, the phase angle plot becomes discontinuous (also, the phase angle is computed in the principal value of $(-\pi/2, \pi/2)$, making the plot discontinuous) and makes estimation of the straight line difficult.

3) The third technique uses a rehocence method which produces a delta function, irrespective of the frequency band of the signal. It is a certain kind of "automatic" high-pass filtering. It does not remove the global noise, but diminishes it greatly. The method has a disadvantage which limits its usefulness, that it is sensitive to uncorrelated background noise. The presence of uncorrelated background noise, on the other hand, does not influence the cross-correlation function method.

The measured normalized autospectra (LPRM measurements) can be seen in Fig. 1. The results correspond to the finding [2], [3] that the existence of the local component brings about a high-frequency contribution which is rather sensitive to axial position in the core. The magnitude of the spectra at a given frequency increases with increasing axial position because of the increase of coolant void fraction. Systematic change of the spectra with axial position in the core indicates a strong coupling between the local component of the noise and actual thermohydraulic behavior.

The change of the phase between two axial-spaced detectors with the frequency and the distance between the detectors is shown in Fig. 2. In the higher frequency range, the phase between two detectors is expected to have a linear behavior. However, experimental results show characteristic oscillations around this line. From the slope of this line, the mean transit time of bubbles is determined.

Fig. 3 is the result of cross-correlation and rehocence functions based on the same raw data as the phase shifts.

The rehocence functions (or smoothed coherence transforms) retain well-defined peaks, while it is difficult to identify the peaks of the cross-correlation functions, particularly in the lower positions in the core (A-B) where the local component is not as dominant as in the higher positions.

V. Conclusions

The new method for estimating the transit time of a propagating void stochastic information in a BWR was compared with the ordinary cross-correlation technique and was found more effective in producing a well-defined time-delay peak. This method automatically produces the time-delay peak even when the measured signals are contaminated by a narrow-band limited *internal noise*, coming from the global noise of the neutron flux fluctuations.

The relationship between the two methods often used for transit time estimation, i.e., the method of the cross-correlation function in the time domain and that of the CPSD phase angle diagram in the frequency domain, has been explained. The

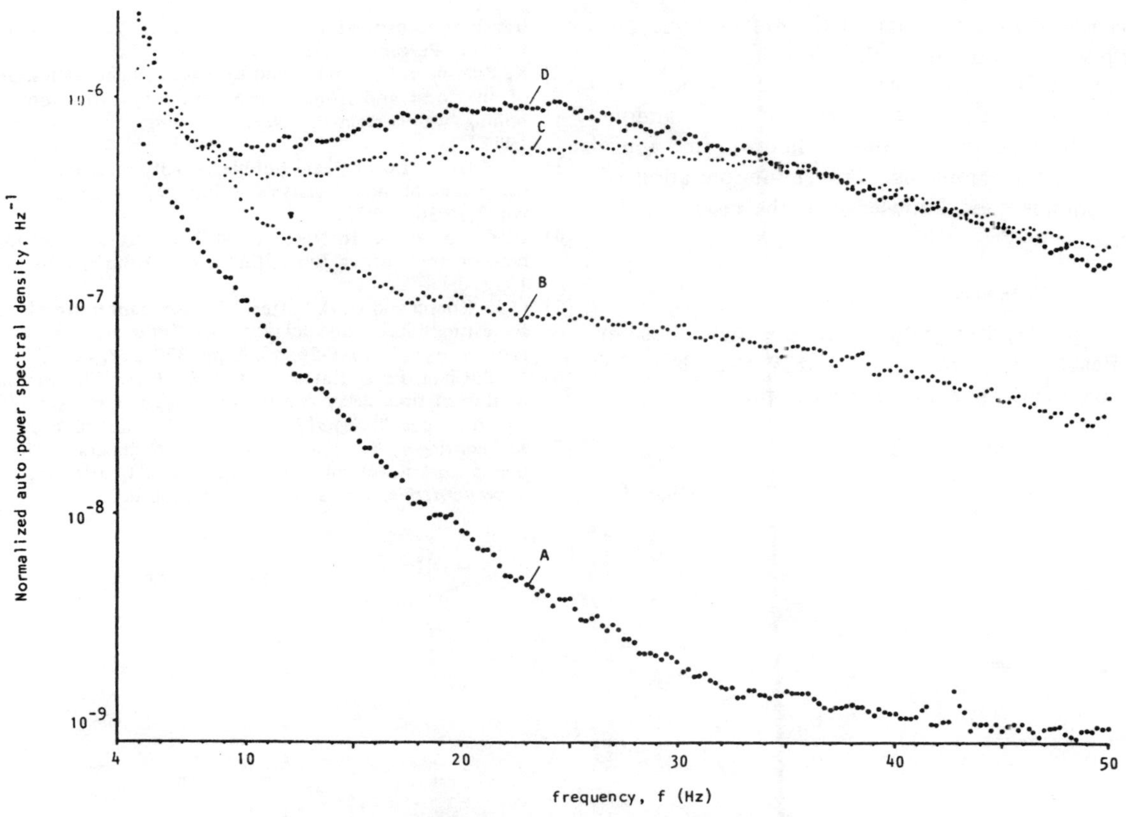

Fig. 1. Autopower spectral densities versus frequency.

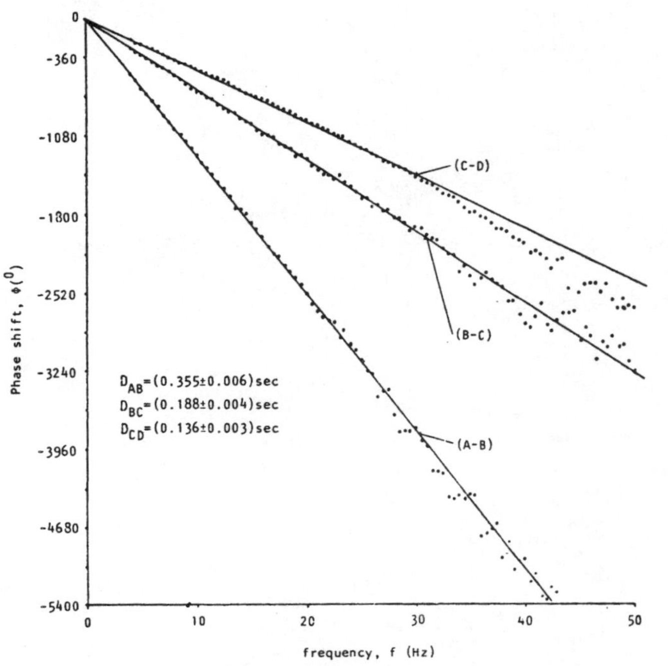

Fig. 2. Phase shift versus frequency.

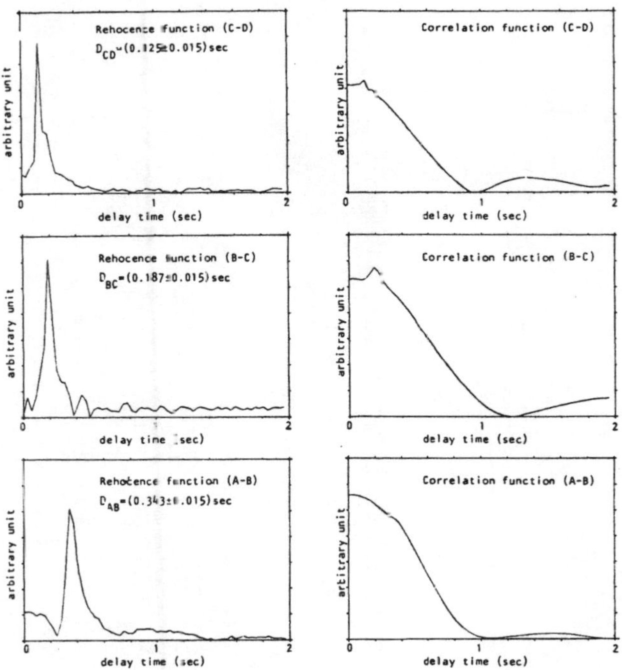

Fig. 3. Rehocence and cross-correlation functions.

new method is essentially a variation of the previous one, but it presents the results directly in the time domain.

The rehocence (or SCOT) is a useful ad hoc technique for analyzing time-delay characteristics between two random processes. It may be expected that this method will find applications in many nuclear problems. Proper interpretation of the rehocence requires good knowledge of the model of the received processes.

Acknowledgment

The author expresses her gratitude to the Swiss Federal Institute for Reactor Research and to BKW Mühleberg for enabling the experimental part of the investigations.

References

[1] H. Nishihara and H. Konishi, "A new correlation method for transit-time estimation," in *Progress in Nuclear Energy*, vol. 1. London: Pergamon Press, 1977, pp. 219–229.

[2] K. Behringer, G. Kosály, and Lj. Kostić, "Theoretical investigation of the local and global components of neutron noise field in a boiling water reactor," *Nucl. Sci. Eng.*, vol. 63, pp. 306–318, 1977.

[3] Lj. Kostić, "Local steam bubble velocity measurements in a BWR using reactor noise analysis technique," *Atomwirt. Atomtechn.*, vol. 7, p. 363, 1976.

[4] G. C. Carter, A. H. Nuttall, and P. G. Cable, "The smoothed coherence transform," *Proc. IEEE*, vol. 61, no. 10, pp. 1497–1498, 1973.

[5] C. H. Knapp and G. C. Carter, "The generalized correlation method for estimation of time delay," *IEEE Trans. Acoust., Speech, Signal Processing*, vol. ASSP-24, no. 4, pp. 320–327, 1976.

[6] Y. T. Chan, R. V. Hattin, and Y. B. Plant, "The least squares estimation of time delay and its use in signal detection," *IEEE Trans. Acoust., Speech, Signal Processing*, vol. ASSP-26, no. 3, 1978.

[7] K. Behringer, G. Kosály, and H. Nishihara, "Theoretical remarks on a novel correlation method of transit time estimation," *Atomkernergie/Kerntechnik*, to be published.

Coherence established between atmospheric carbon dioxide and global temperature

Cynthia Kuo, Craig Lindberg & David J. Thomson

Mathematical Sciences Research Center, AT&T Bell Labs, Murray Hill, New Jersey 07974, USA

The hypothesis that the increase in atmospheric carbon dioxide is related to observable changes in the climate is tested using modern methods of time-series analysis. The results confirm that average global temperature is increasing, and that temperature and atmospheric carbon dioxide are significantly correlated over the past thirty years. Changes in carbon dioxide content lag those in temperature by five months.

DURING the past century (see, for example, refs 1, 2), scientists have studied the possibility that the climate is influenced by changes in the atmospheric concentration of CO_2 caused by industrial and agricultural activities[3-5]. Recently, because of the potentially serious consequences of the greenhouse effect[6,7], the problem has been studied more intensively[8,9] with a view towards observing climatic effects attributable to the increase of atmospheric greenhouse gases. For example, the output of numerical models of the global climate has been compared with measurements[10,11], and the significance of the temperature increase has been tested using various straight-line segment and parametric models (ref. 12 and J. Seater, manuscript in preparation). But present numerical models of the atmosphere are crude, and comparisons between the time series representing the real data and predictions of the atmospheric models are difficult to interpret. Because the available data are short time series, conventional statistical methods are unreliable, and detection of the greenhouse effect remains controversial[13-18].

The longest modern series of precise CO_2 concentration measurements begins in March 1958 and consists of monthly values collected by Keeling at the summit of Mauna Loa in Hawaii[19]. Although these data are from a single station, they are typical of measurements made since 1974 at several sites[20]. Because we are interested in the time-series aspects of the problem, we use the longer Keeling series. The data have an upward trend that is readily visible (the upper curve in Fig. 1), an obvious annual component and irregular fluctuations. Five missing values were interpolated using a stochastic least-squares procedure on the residuals from a quadratic polynomial plus the first five annual harmonics. These interpolated values have an estimated error of 0.35 p.p.m. and are noted in Fig. 1.

Hansen and Lebedeff[21] created a time series of monthly global-average surface-air-temperature changes from January 1880 to December 1988. This series is a weighted average over stations, formed by subtracting the average January temperature during the reference period 1951–80 from all the January data, and repeating this for the other months, eliminating seasonal variations. Displayed as the lower curve in Fig. 1, the temperature series also increases with time but its fluctuations are relatively larger than those in the CO_2 record.

Here we apply multiple-window time-series methods (which are efficient for short series) to estimate the trends and power spectra of the Hansen–Lebedeff average global surface temperature series and Keeling CO_2 concentration measurements as well as the coherence between the two. This analysis shows that from 1880 to 1988, the average global temperature increased by 0.0055 ± 0.00096 °C yr^{-1}, and the probability that this slope is positive exceeds 99.99%. Furthermore, the monthly CO_2 concentration and global temperature series from 1958 to 1988 are coherent over much of the Nyquist frequency band from 0 to 6 cycle yr^{-1}; the probability that the level of coherence observed from 0 to 2 cycle yr^{-1} occurred by chance is $\sim 2 \times 10^{-6}$. Not only do both series have increasing trends that are highly significant, but there are linear relations between many of their oscillatory components. We interpret this as evidence that the changes in atmospheric CO_2 concentration are closely related to changes in global temperature.

Models

Many of the contradictions in the literature about the analysis of climate data can be traced to the use of inappropriate or oversimplified models (for review, see, for example, refs 22, 23). Statistical techniques that are vulnerable to difficulties include those based on parametric models (such as low-order autoregressive moving-average representations) and methods plagued by more insidious problems caused by implicit assumptions of time-series stationarity. The cavalier use of parametric models can lead to misspecification difficulties[24] because

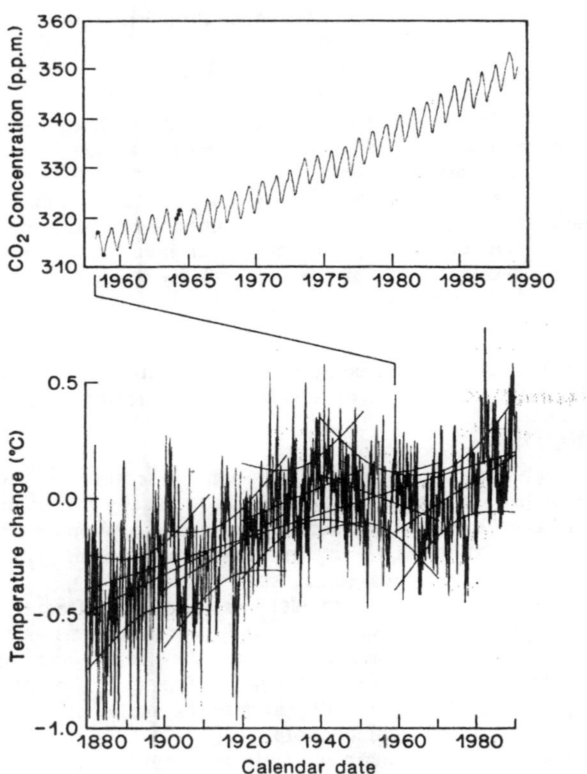

FIG. 1 The upper curve is the monthly Keeling CO_2 concentration data for March 1958 to December 1988. The five interpolated points are marked. The lower curve is the Hansen–Lebedeff average global temperature series for 1880 to 1988. The short line segments and hyperbolic arcs define trends and 95% confidence regions over 30-yr intervals, and the long straight line shows the general trend.

Reprinted with permission from *Nature*, vol. 343, no. 6260, pp. 709–714, February 1990.

unanticipated effects (such as the modulation of the annual cycle discussed below) either are undetected, are attributed incorrectly or are simply lumped in as part of 'residual variance'. As shown below, the CO_2 concentration and global temperature series have complicated spectral density functions, so these series are not likely to be modelled by any simple form in either the time or frequency domains. The fundamental flaw of parametric models, including numerical atmospheric models and time-series models, is that they make no provision for the unexpected.

We represent each time series by the less restrictive decomposition into a parametric trend and a non-parametric residual. Because one cannot directly discriminate between the many possible functional forms for the trends from such short records, we represent the trend in each time series by the simplest non-constant function of time, a straight line of non-zero slope which can be thought of as the first two terms in the Taylor series expansion of the true trend. It may be argued that this semi-parametric representation is subject to the misspecification problems mentioned above, but, because we subject the residuals to greater scrutiny than we do the trends, features not included in the trends will still be present and studied in the residuals. Moreover, we find that including quadratic terms does not change the results of the following analysis significantly.

Therefore, we express each time series

$$\{x(0), x(1), x(2), \ldots, x(N-1)\}$$

as the sum of a constant, a linear trend and a stationary time series specified only by its spectral representation

$$x(t) = a_0 + a_1(t - t_{\mathrm{ref}}) + e(t); \quad t = 0, 1, \ldots, N-1 \tag{1}$$

where t_{ref} is a reference time, a_0 and a_1 are constants and the residual time series $e(t)$ has the spectral representation

$$e(t) = \int_{-1/2}^{1/2} e^{i2\pi ft}\, dX(f) \tag{2}$$

for all t. $dX(f)$ is the differential of a generalized Fourier transform and is known as an 'orthogonal increment process'. (This is an extension of the representation used in ref. 25, where the CO_2 series was decomposed into a trend, an annual component and a residual.) The annual component is given by the first moment of $dX(f)$, and the power spectrum, or power spectral density, $S(f)$, of the residuals is, by definition, the second moment of $dX(f)$:

$$S(f)\, df = \mathbf{E}\{|dX(f)|^2\} \tag{3}$$

where $\mathbf{E}$ is the statistical expected-value operator. In a stationary series, values of $dX(f)$ at distinct frequencies are uncorrelated[26].

Estimation

Although the time- and frequency-domain representations of time series are formally equivalent[27], we usually find it more informative to analyse time series in the frequency domain where the effects of different physical processes can be easier to distinguish. With short time series, such as the CO_2 and global temperature records, it is difficult to resolve different frequencies and simultaneously obtain statistically significant results. Our approach is to use a variant of the multiple-window method of spectrum analysis[28,29] that, although it does not eliminate all the problems associated with short series, makes statistically efficient use of the available data.

Most frequency-domain methods are strictly valid only for the analysis of stationary data, not series with embedded trends such as the CO_2 and global temperature records. Using the multiple window procedure described below, we estimate the trends in each series, subtract off these terms, and estimate the spectrum of the residuals. Finally, we test the residuals for stationarity to see if our assumptions were violated.

Estimating the average a_0 and trend a_1 by ordinary least squares can produce misleading results if the residuals $\{e(t)\}_{t=0}^{N-1}$ have a non-white (non-flat) spectrum and are therefore correlated[30,31]. The residual spectrum need only be roughly flat in a small frequency interval around the origin for multiple-window regression, however, to produce valid estimates of a_0, a_1, the spectrum of the residuals and their errors.

Multiple windows

The multiple-window method uses the orthogonal sequences of N elements that optimally concentrate the spectral energy in a frequency band of width $2B$ centred on a frequency f, that is, between the frequencies $f-B$ and $f+B$. These sequences are the lowest-order $\lfloor 2BT \rfloor$ discrete prolate spheroidal sequences of ref. 32, where T is the duration of the observed series. These 'Slepian sequences' form a basis on which the data in the

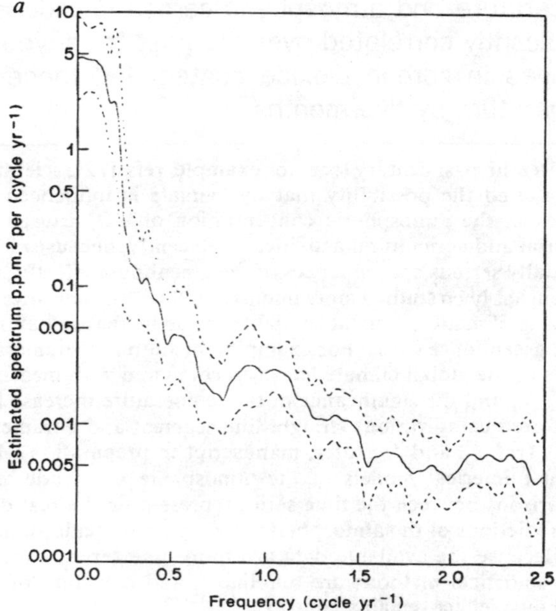

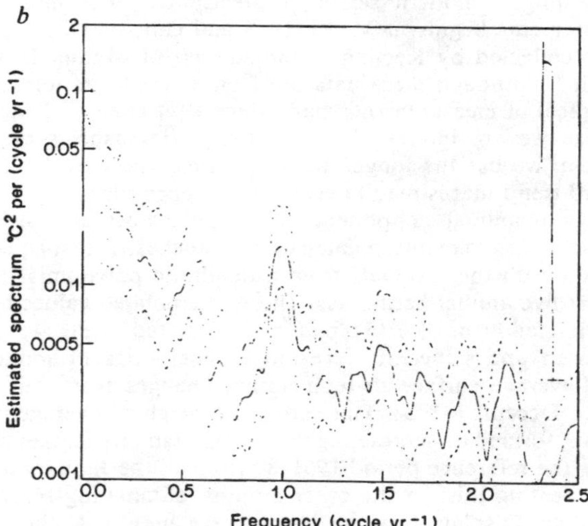

FIG. 2 *a*, Multiple-window spectral estimate of trend-subtracted CO_2 concentration series. The upper and lower dashed curves are 5 and 95% confidence limits obtained by jackknifing on the windows. Harmonics of the annual cycle have been subtracted. *b*, Multiple-window spectral estimate of detrended global temperature series for the interval 1880 to 1988. The dashed lines give the 5 and 95% confidence intervals as in *a*. The insert on the upper right shows the effective spectral window on the same amplitude and frequency scale. Note that the power around 1 cycle yr^{-1} is not a periodic component. (Spectra done at higher resolution show that this feature is asymmetric with more power on the lower sideband than on the upper, indicating combined amplitude and phase modulation of the annual cycle.)

frequency band can be expanded. The coefficients of this expansion depend on frequency and are obtained by taking the discrete Fourier transform of the product of the data with each Slepian sequence. The lowest-order coefficient is similar to a direct spectrum estimate using a conventional data window or taper. A multiple-window estimate of the spectrum is, however, an adaptively weighted average of the $2BT$ frequency-dependent coefficients.

Because of their energy-concentration properties, the Slepian sequences are the data windows that are most resistant to spectral leakage[28]. For example, the estimate of the temperature-series spectrum in Fig. 2*b* is produced using 11 Slepian sequences with $2B = 0.11$ cycle yr^{-1}; sidelobes of the effective spectral window shown in the insert are unobservable on the scale of Fig. 2*b*.

Choosing the parameter B for a multiple-window estimate of a spectrum involves a tradeoff between resolution and variance. The variance of this spectrum estimate is proportional to $1/(2BT)$. Thus, increasing B improves statistical reliability but decreases the resolution of the estimate.

Trend estimates

To estimate the coefficients a_0 and a_1 in equation (1) by conventional time-domain regression, one minimizes the sum of the squares of the residuals $\sum_{t=0}^{N-1} e^2(t)$ with respect to the coefficients. By Parseval's formula,

$$\sum_{t=0}^{N-1} e^2(t) = \int_{-1/2}^{1/2} |\tilde{e}(f)|^2 \, df \qquad (4)$$

where $\tilde{e}(f)$ is the discrete Fourier transform of the residuals $\{e(t)\}_{t=0}^{N-1}$, so minimization in either the time or frequency domain is equivalent. To avoid the energy associated with the harmonics of the annual cycle and other high-frequency processes, however, we minimize only the integral over the low frequencies

$$\int_{-W}^{W} |\tilde{e}(f)|^2 \, df; \quad W = BT/N < \tfrac{1}{2} \qquad (5)$$

and ignore the higher-frequency components of the residuals, resulting in estimates expressible in terms of the Slepian sequences.

The multiple-window approach has several advantages when the residuals are autocorrelated. First, the 'observations' in the multiple-window regression are closer to independent gaussian random variables than the original time-domain data $\{x(t)\}_{t=0}^{N-1}$, and therefore the multiple-window coefficient estimates are closer to maximum-likelihood estimates than are the estimates from ordinary least-squares analysis. Similarly, the number of degrees of freedom and error estimates are easily calculated using the Slepian sequences, and the multiple-window method is often more statistically efficient than ordinary least-squares analysis (C. Lindberg, manuscript in preparation). In conventional regression, the endpoints of the time series (for example, the abnormally high temperatures of the last decade) can have an inordinately large influence on the coefficient estimates[33]. This effect is largely eliminated in this approach. Finally, it is relatively easy to check the assumptions that have been made in representing the data by a particular model.

Trends

For the monthly Keeling CO_2 series (March 1958 to December 1988, $T \approx 29.8$ yr), estimates of the average and trend were obtained using $t_{ref} = 1975.0$, and a bandwidth of $2B = 0.39$ cycle yr^{-1}. This value of B confines the spectral energy associated with the trend components to frequencies f with $|f| \leq 0.195$ cycle yr^{-1}, avoiding energy associated with the annual cycle and its harmonics. This value of B also results in a spectrum that is locally white in the resolution band (at frequencies above 0.2 cycle yr^{-1}, the CO_2 residual spectrum drops rapidly, as shown below). The time-bandwidth product $2BT \approx 11.6$ gives eleven leakage-resistant windows. The multiple-window procedure results in the estimates $a_0 = 331.9 \pm 0.44$ p.p.m. and $a_1 = 1.191 \pm 0.053$ p.p.m. yr^{-1}.

For the monthly Hansen-Lebedeff global temperature series from January 1880 to December 1988, we obtained estimates of the average and trend using $t_{ref} = 1934.5$ and a resolution bandwidth of $2B = 0.128$ cycle yr^{-1}. This value of $2B$ was chosen to avoid the frequency band above 0.07 cycle yr^{-1} where the residual global temperature spectrum decreases rapidly, as shown below. Using 12 Slepian sequences, we obtain $a_0 = -0.106 \pm 0.030$ °C and $a_1 = 0.00554 \pm 0.00096$ °C yr^{-1}.

For comparison, estimates from ordinary least-squares analysis agree with the multiple-window estimates to three significant figures but, because of the lower values of the spectrum at higher frequencies, underestimate their standard deviations by a factor of five.

To assess the significance of this estimated global temperature slope, we note that Milankovitch theory predicts that at present the Earth should be cooling by ~0.0004 °C yr^{-1} (refs 34, 35). Thus, solar variability aside, the null hypothesis is that the temperature trend should be slightly negative. To test this hypothesis we use a t statistic $t = [0.00554 - (-4 \times 10^{-4})]/0.000961 = 6.18$ which, as 12 windows were used and two parameters were estimated, is characterized by approximately 10 degrees of freedom. Therefore, given that the low-frequency spectrum is approximately white (see Fig. 2*b*), the slope is greater than the Milankovitch prediction with probability 99.995% (ref. 36). Using a narrower bandwidth leads to fewer Slepian sequences with resistance to spectral leakage, fewer degrees of freedom and an underestimate of the slope significance. A wider bandwidth results in an invalid t statistic, as the residual spectrum decreases rapidly at frequencies >0.07 cycle yr^{-1} and so violates the 'locally white' assumption. Neither a jackknife variance estimate[37] (a non-parametric statistic sensitive to both non-gaussianity and non-stationarity, determined from the set of spectrum estimates computed with each of the windows deleted in turn), nor the stationarity test of ref. 38 show the residual series to be non-stationary. High-resolution quadratic inverse spectrum estimates provide some evidence for a ripple on the spectrum at frequencies <0.07 cycle yr^{-1}, consistent with a 'recurrence time' in the temperature data of ~28 yr, but the amplitude of this ripple is not enough to change the significance of the slope.

Finally, it has been argued that the negative trend in the temperature record between 1940 and 1970 invalidates the conclusion that the temperature is increasing over the long term. To test this, we repeated the trend calculations for overlapping 30-year subsections of the Hansen-Lebedeff series. The estimated trends for each interval are shown as the short straight lines through the temperature record in Fig. 1, and the 95%-confidence region of each is bounded by the hyperbolic arcs[39]. The line associated with the linear trend of the entire temperature record remains in the corridor collectively outlined by the subsection error bounds.

Spectrum estimates

We estimate the spectrum of the residual series $\{e(t)\}_{t=0}^{N-1}$ by subtracting the periodic components (detected by a statistical F-test[28]) from the detrended data, and by making an adaptively weighted multiple-window estimate of the power spectral density $S(f)$ of the residuals[28]. No frequency components above 2.5 cycle yr^{-1} are shown to avoid artefacts from the unequal lengths of the months.

Figure 2*a* is the low-frequency part of the spectrum estimate of the CO_2 residuals using $2B = 0.39$ cycle yr^{-1} and 11 windows. The spectrum is not white, so estimates using ordinary least-squares analysis of the CO_2 trend error bounds are invalid. The variance of the estimate has been calculated by jackknifing over windows[37,40,41], and the resulting 5% and 95% confidence limits are shown. These error bounds are consistent with the

χ^2-distributed estimate (with 22 degrees of freedom) expected from stationary gaussian data. Because of the shortness of this series, it is unlikely that details in the low-frequency end of the spectrum have been resolved; the plotted spectrum is a compromise between frequency resolution and statistical significance.

Figure 2*b* shows an estimated spectrum of the Hansen-Lebedeff global average temperature residual series. To allow resolution of more details in the spectrum, a bandwidth of $2B = 0.11$ cycle yr^{-1} was used, resulting in 11 Slepian sequences

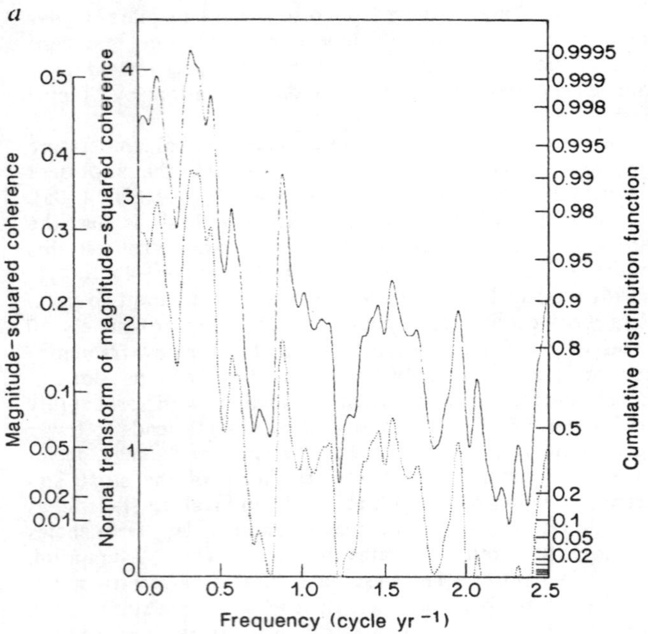

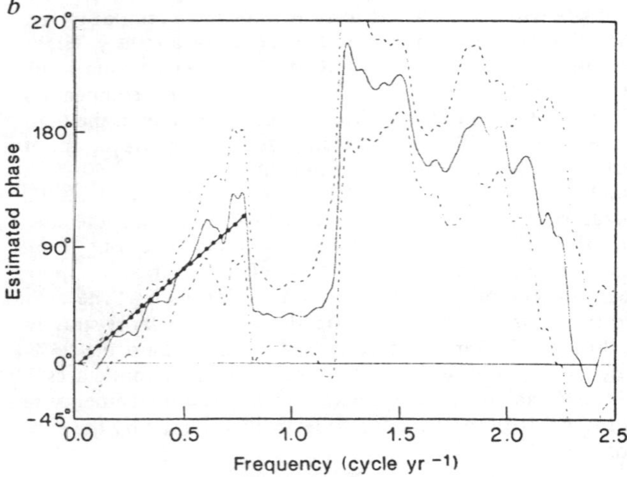

FIG. 3 *a*. Magnitude-squared coherence (MSC), between the Keeling CO_2 series and the global temperature series during 1958-1988. The vertical axis on the far left is the value of the MSC, and the axis on the left of the graph is the value of the coherence magnitude transformed by $\tanh^{-1}$. On this scale the coherence estimates should be roughly gaussian with unit standard deviation[49]. The vertical axis on the right gives the cumulative distribution function for incoherent series. The upper (solid) curve represents the transformed coherence values $\{\tanh^{-1}(|C(f)|)\}$, and the lower (dashed) curve is offset by one standard deviation as determined by jackknifing over windows. *b*, Phase of coherence between the Keeling CO_2 series and the global temperature series during 1958-1988 corresponding to the MSC in *a*. The two dashed lines show the ± 1 s.d. limits obtained by jackknifing. The low-frequency trend in the phase corresponds to a delay of ~5 months, whereas the 'hole' near 1 cycle yr^{-1} reflects the predominance of other effects near this frequency. The poorer limits on the phase at higher frequencies is a consequence of the lower coherences there. These jackknife estimates agree with gaussian theory.

($2BT = 108 \times 0.11 = 11.88$). As noted in ref. 21, this spectrum exhibits substantial power at periods near integer multiples of the annual cycle. Because the *F*-test shows that this power is not a simple periodic signal in the data (recall that the Hansen-Lebedeff referencing procedure has already subtracted the annual cycle) we examine it further below.

Relations between the series

That both the CO_2 and global temperature data have positive slopes does not prove that the two series are related. As the discussion on spurious correlation in ref. 42 makes clear, the presence of the trends in each series will cause simple time-domain correlations between the two series to be high. If it is found, however, that the fluctuations of the two detrended series are coherent over a band of frequencies, then it is more likely that the two series are related.

The frequency-domain analogue of correlation, coherence[43-45], has been applied to meteorological data for many years[46]. Conventional estimates of coherence between single sections of two records are the smoothed (by a moving average) complex product of the discrete Fourier transforms of the two series, and so can be badly biased if the phase changes over the averaging band. Section-averaging methods are inappropriate for short series; not only does dividing these series into subsections result in poor frequency resolution, but the correlation between the subsections produces unreliable coherence estimates. The multiple-window approach described in refs 28 and 41 provides a less biased estimate of coherence $C(f)$ which is suitable for short series, allowing the extraction of more information from the same data.

Figure 3*a* shows the multiple-window magnitude-squared coherence between the Keeling CO_2 and global temperature residuals from 1958 to 1988 (produced using the same bandwidth and number of windows as the spectrum estimate of the CO_2 residuals). It is remarkable that the two series have a coherence above the 90% confidence level at frequencies <0.6 cycle yr^{-1} with coherence exceeding 98% over much of this low-frequency band. Because multiple-window coherence estimates spaced $2B$ apart are essentially independent, the probability of observing such levels across a wide band by chance from independent series is very low, $\sim 2 \times 10^{-6}$. Thus, not only are the trend components (at frequencies ~0 cycle yr^{-1}) of both time series increasing, but the residuals of the two series are also coherent with high confidence in the low-frequency band.

Figure 3*b* is a plot of the phase of the multiple-window coherence between the two residual (trend-subtracted) series. The phase of the coherence at 0 cycle yr^{-1} is zero and, because both trends are positive, is independent of whether trends are included or not. Between 0 and 0.8 cycle yr^{-1} the phase is roughly linear, corresponding to the CO_2 series lagging the temperature series by ~5 months (calculated by taking the slope of this linear section of the phase[44], $\sim 120°/(360° \times 0.8$ cycle yr^{-1})) in agreement with arguments that natural positive feedback mechanisms in the carbon cycle causes carbon dioxide to lag temperature in some frequency bands[19] (R. Marston, personal communication). Current knowledge of these complicated interactions involving solar forcing, the Southern Oscillation and exchange of CO_2 with the oceans on various timescales, is summarized in section 6.6 of ref. 19. Also, in agreement with Keeling's hypothesis that ocean processes are dominant, we find that the coherence between the CO_2 and the Southern Hemisphere average temperature records is slightly higher than that for the Northern Hemisphere and that the observed delay of ~5 months between the global temperature and CO_2 is also seen in the Southern Hemisphere phase. As the Northern Hemisphere average temperature leads CO_2 by ~3 months, part of the ~5-month delay must be due to the transport time from the Southern Hemisphere to Mauna Loa.

The hole in the phase curve near 1 cycle yr^{-1} occurs because the temperature spectrum there is dominated by a different

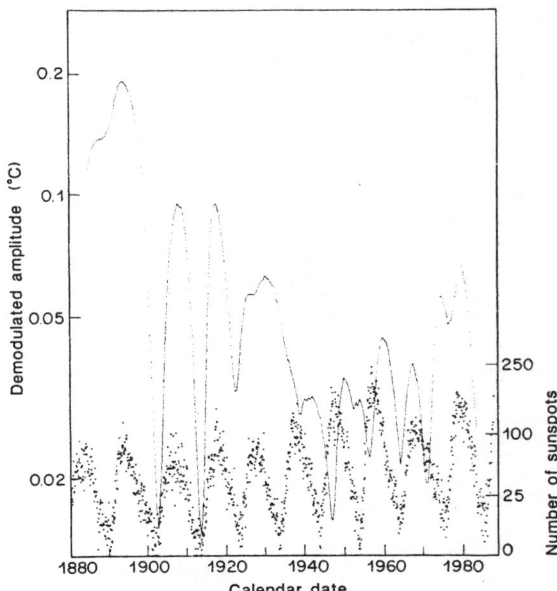

FIG. 4 The solid curve shows the amplitude of the complex demodulate of the Northern Hemisphere temperature record between 0.9 and 1.1 cycle yr^{-1}. A plot of the sunspot numbers is shown below. Detailed examination of the coherence between the demodulate and the solar cycle shows that these two have about a 1/(30 yr) difference in frequency (corresponding to the stronger lower sideband mentioned in Fig. 2*b*).

physical mechanism, and because of their different phase characteristics, we analyse the Northern and Southern Hemisphere temperature records separately in this frequency band. To do this, a multiple-window bandpass-filtered temperature series is formed by expanding the residual temperature series on a set of Slepian sequences that have most of their spectral energy confined between 0.9 and 1.1 cycle yr^{-1}. The resulting 'complex demodulate' for the Hansen–Lebedeff Northern Hemisphere temperatures is shown by the solid line in Fig. 4. The sunspot record (solid dots in Fig. 4) is clearly similar to the demodulated temperature data, suggesting that the annual signal is modulated by fluctuations in solar output. Detailed analysis shows that coherence between the sunspot and filtered temperature series is significant, and similar to the non-stationary structure found in ref. 47. The phase of the sunspot-filtered temperature coherence has a linear drift, corresponding to a frequency difference between these two oscillations of $1/\Delta$, where $\Delta = 27$–30 yr, possibly related to the low-frequency ripple in the temperature spectrum mentioned above. The results for the Southern Hemisphere are similar.

Discussion

The coherence results presented here provide significant evidence that the average global temperature and CO_2 concentration from 1958 to 1988 are linearly related at many frequencies. But caution must be exercised in interpreting this result as suggesting that the variations in atmospheric CO_2 are causing the changes in global temperature, even though there are plausible physical mechanisms linking the two series. Apparent correlations that are used to postulate causality can sometimes be misleading, as in the case involving timing of earthquakes[48]. In addition, one should be particularly cautious in interpreting the coherence when the series analysed are as short as these; climatic and solar variations often are of longer duration than these records.

From atmospheric chemistry, global temperature depends nonlinearly on CO_2 concentration (T. Graedel, personal communication). The procedure used here implicitly uses a linear dependence. Bispectral estimates provide evidence of quadratic terms, although the shortness of these series makes this difficult to quantify. Also, except for the solar modulation of the temperature series near 1 cycle yr^{-1}, we have ignored the cyclostationary properties of these two series (that is, the statistics of these series vary periodically).

A more complete analysis would include estimates of the coherences between the various global average temperature time series, records of atmospheric CO_2 concentration, human CO_2 production, sunspots, volcanic activity and the Southern Oscillation Index, which are all high in various frequency bands and have complicated phase interrelations. For example, the coherence between the detrended Keeling CO_2 series and the Southern Oscillation Index is high near 0.4 and 2.4 cycle yr^{-1}. If we calculate the coherence between the CO_2 and global temperature series from which terms describing their linear dependences on the Southern Oscillation Index and sunspot record have been subtracted (called a partial coherence), the low-frequency magnitude-squared coherence increases to almost 0.7 whereas it decreases near 0.3 cycle yr^{-1}. Several of these series also exhibit an oscillation at an apparent period of ~15 years. Further analysis of their multivariate relations will be described elsewhere in more detail. □

Received 30 October 1989; accepted 8 January 1990.

1. Tyndall, J. *Phil. Mag.* **22,** 160–194; **22,** 273–285 (1863).
2. Arrhenius, S. *Phil. Mag.* **41,** 237–276 (1896).
3. Houghton, R. A. in *The Changing Carbon Cycle: A Global Analysis* (eds Trabalka, J. R. & Reichle, D. E.) 175–193 (Springer, New York, 1986).
4. Schlesinger, W. H. in *The Changing Carbon Cycle: A Global Analysis* (eds Trabalka, J. R. & Reichle, D. E.) 194–220 (Springer, New York, 1986).
5. Rotty, R. M. & Marland, G. *The Changing Carbon Cycle: A Global Analysis* (eds Trabalka, J. R. & Reichle, D. E.) 474–490 (Springer, New York, 1986).
6. Seidel, S. & Keyes, D. *Can We Delay a Greenhouse Warming?* Rep. No. EPA 23010:4001 (US Environmental Protection Agency, Washington, 1983).
7. Budyko, M. I. *Climatic Changes* Ch. 7 (Waverly, Baltimore, 1977).
8. World Meteorological Organization *Rep. World Conf. Changing Atmosphere* (World Meteorological Organization, Geneva, 1989).
9. National Research Council *Changing Climate* (Natn Acad. Press, Washington, DC, 1983).
10. Barnett, T. P. & Schlesinger, M. E. *J. geophys. Res.* **92,** 14772–14780 (1987).
11. Preisendorfer, R. W. & Barnett, T. P. *J. atmos. Sci.* **40,** 1884–1896 (1983).
12. Solow, A. R. *J. Clim. appl. Met.* **26,** 1401–1405 (1987).
13. Maddox, J. *Nature* **334,** 9 (1988).
14. Kerr, R. A. *Science* **244,** 1041–1043 (1989).
15. Broecker, W. S. *Science* **245,** 451 (1989).
16. Schlesinger, M. E. *Science* **245,** 451 (1989).
17. Risby, J. *Science* **245,** 451–452 (1989).
18. Solow, A. R. & Broadus, J. M. *Clim. Change* (in the press).
19. Keeling, C. D., Bacastow, R. B., Carter, A. F., Piper, S. C. & Whorf, T. P. *Am. geophys. Un., Geophys. Monogr.* **55,** 165–236 (1989).
20. Gammon, R. H., Komhyr, W. D. & Peterson, J. T. in *The Changing Carbon Cycle: A Global Analysis* (eds Trabalka, J. R. & Reichle, D. E.) 1–15 (Springer, New York, 1986).
21. Hansen J. & Lebedeff, S. *J. geophys. Res.* **92,** 13345–13372 (1987).
22. Shapiro R. *J. atmos. Sci.* **36,** 1105–1116 (1979).
23. Pittock, A. B. *Rev. Geophys. Space Phys.* **16,** 400–420 (1978).
24. Tukey, J. W. *Scripps Inst. Oceanogr. Ref. Ser.* **84–5,** 100–103 (1984).
25. Cleveland, W. S., Freeny, A. E. & Graedel, T. E. *J. geophys. Res.* **88,** 10934–10946 (1983).
26. Priestley, M. B. *Spectral Analysis and Time Series* (Academic, New York, 1981).
27. Brillinger, D. R. *Time Series, Data Analysis and Theory* (Holt, Rinehart & Winston, New York, 1975).
28. Thomson, D. J. *Proc. IEEE* **70,** 1055–1096 (1982).
29. Park, J., Lindberg, C. R. & Vernon, F. L. III *J. geophys. Res.* **92,** 12675–12684 (1987).
30. Grenander, U. *Ann. math. Statist.* **25,** 252–272 (1954).
31. Brillinger, D. R. *Biometrika* **76,** 23–30 (1989).
32. Slepian, D. *Bell System Tech. J.* **57,** 1371–1429 (1978).
33. Belsley, D. A., Kuh, E. & Welsch, R. E. *Regression Diagnostics: Identifying Influential Data and Sources of Collinearity* (Wiley, New York, 1980).
34. Hays, J. D., Imbrie, J. & Shackleton, N. J. *Science* **194,** 1121–1132 (1976).
35. Imbrie, J. & Imbrie, J. Z. *Science* **207,** 943–953 (1980).
36. Pearson, E. S. & Hartley, H. O. *Biometrika Tables for Statisticians* 3rd edn Vol. 1, (Cambridge University Press, London, 1970).
37. Thomson, D. J. & Chave, A. D. *Advances in Spectrum Estimation* (ed. Haykin, S.) Ch. 2 (Prentice-Hall, Englewood Cliffs, New Jersey, in the press).
38. Thomson D. J. *Bell System Tech. J.* **56,** 1769–1815; **56,** 1983–2005 (1977).
39. Daniel, C. & Wood, F. S. *Fitting Equations to Data* (Wiley, New York, 1980).
40. Lindberg, C. R. & Park, J. *Geophys. J. R. astr. Soc.* **91,** 795–836 (1987).
41. Vernon, F. L. III, thesis, Univ. of California (San Diego) (1989).
42. Yule, G. U. & Kendall, M. G. *An Introduction to the Theory of Statistics* 14th edn (Hafner, New York, 1965).
43. Tukey, J. W. in *Advanced Seminar on Spectral Analysis of Time Series* (ed. Harris, B.) 25–46 (Wiley, New York, 1967).

44. Bloomfield, P. *Fourier Analysis of Time Series: An Introduction* (Wiley, New York, 1976).
45. Freiberger, W. F. in *Time Series Analysis* (ed. Rosenblatt, M.) 244-259 (Wiley, New York, 1963).
46. Panofsky, H. A. in *Advanced Seminar on Spectral Analysis of Time Series* (ed. Harris, B.) 109-132 (Wiley, New York, 1967).
47. Thomson, D. J *Phil. Trans. R. Soc.* (in the press).
48. Knopoff, L *The Moon* Ch. 2 (Reidel, Dordrecht, 1970).
49. Kendall, M. & Stuart, A. *The Advanced Theory of Statistics* 4th edn, Vol 1, Ch. 16 (Macmillan, New York, 1979).

ACKNOWLEDGEMENTS. C. K. was an intern in the 1989 AT&T Bell Labs Summer Research Program. We thank S. Lebedeff, C. D. Keeling and W. Cleveland for data and T. Graedel, C. D. Keeling, J. Kruskal, L. J. Lanzerotti, J. Lagarias, P. Linhart and C. Maclennan for comments on the manuscript.

Smoothed Coherent Transform Method for Measuring Displacement by Means of Planar Holographic Detection

HUA LEE AND GLEN WADE

Abstract—The smoothed coherent transform (SCOT) method has been widely applied to delay estimation for time signals. The development of the method was based upon the shifting property of Fourier transformation. It is shown that SCOT can be employed also to measure three-dimensional displacement in holographic imaging. This is done by observing the spatial frequency spectrum distribution of the wavefield and modifying the estimation process. In this way a three-dimensional displacement can be accurately determined by using a two-dimensional detection array.

Consider a source object $s(x, y, z)$ and a planar holographic detection system at $z = z_0$. The propagating wavefield from the source is detected by the holographic system. Let $r(x, y)$ be the complex wavefield distribution over the receiving plane

$$r(x, y) = H\{s(x, y, z)\} + n_1(x, y), \tag{1}$$

where H represents the wave propagation operator and $n_1(x, y)$ denotes the noise term. Suppose the source is shifted by a displacement vector $\overline{D}$

$$\overline{D} = [d_x, d_y, d_z]^T, \tag{2}$$

where d_x, d_y, and d_z are the displacement's components in the x, y, and z directions respectively. The shifted source distribution can be written as

$$s'(x, y, z) = s(x - d_x, y - d_y, z - d_z), \tag{3}$$

and the received complex wavefield corresponding to the shifted source is $r'(x, y)$:

$$\begin{aligned} r'(x, y) &= H\{s'(x, y, z)\} + n_2(x, y) \\ &= H\{s(x - d_x, y - d_y, z - d_z)\} + n_2(x, y), \end{aligned} \tag{4}$$

where $n_2(x, y)$ is the noise term.

Assume that the noise is additive, uncorrelated, and has zero mean. Then the SCOT method for measuring displacement can be formulated by

$$Q_{rr'}(x, y) = \iint_{-\infty}^{\infty} W(f_x, f_y) P_{rr'}(f_x, f_y) \cdot \exp [j2\pi(f_x x + f_y y] \, df_x \, df_y, \tag{5}$$

where f_x and f_y are the spatial frequencies in the x and y direction respectively, $P_{rr'}$ is the cross spectrum of r and r' and W is a real and positive weighting function [1]. $W(f_x, f_y)$ is given by

$$W(f_x, f_y) = [P_r(f_x, f_y) P_{r'}(f_x, f_y)]^{-1/2}, \tag{6}$$

where P_r and $P_{r'}$ are the power spectra of r and r' respectively. The power spectra, the weighting function, and the Fourier integral are all two-dimensional because the detection system is planar. Due to normalization by the weighting function, the amplitude distribution of the weighted cross-power spectrum in the integral has very little variation.

$$\begin{aligned} W(f_x, f_y) P_{rr'}(f_x, f_y) &\approx \exp [-j2\pi\langle \overline{f}, \overline{D}\rangle] \\ &= \exp [-j2\pi(f_x d_x + f_y d_y + f_z d_z)], \end{aligned} \tag{7}$$

where $\langle \overline{f}, \overline{D}\rangle$ denotes the inner product of the frequency vector $\overline{f}$ and the displacement vector $\overline{D}$, $\overline{f} = [f_x, f_y, f_z]^T$ and f_z is the spatial frequency in z direction.

Assume that the evanescent wave components are attenuated. The wave components that can be detected must satisfy

$$f_x^2 + f_y^2 + f_z^2 = \lambda^{-2}, \tag{8}$$

where λ is the wavelength [2]. Using (8), we can write (7) as a function of f_x and f_y.

$$\begin{aligned} W(f_x, f_y) P_{rr'}(f_x, f_y) &\approx \exp [-j2\pi\langle \overline{f}, \overline{D}\rangle] \\ &= \exp [-j2\pi(f_x d_x + f_y d_y)] \\ &\quad \cdot \exp [-j2\pi(\lambda^{-2} - f_x^2 - f_y^2)^{1/2} d_z]. \end{aligned} \tag{9}$$

Employing the following Fourier transform relation

$$\begin{aligned} &F^{-1} \{\exp [-j2\pi(\lambda^{-2} - f_x^2 - f_y^2)^{1/2} d_z]\}, \quad \text{for } 0 \leqslant f_x^2 + f_y^2 \leqslant \lambda^{-2}, \\ &= \frac{\exp \{j[2\pi(x^2 + y^2 + d_z^2)^{1/2}/\lambda - \pi/2] \operatorname{sgn}(d_z)\}}{\lambda(x^2 + y^2 + d_z^2)^{1/2}}, \end{aligned} \tag{10}$$

where

$$\operatorname{sgn}(d_z) = \begin{cases} 1, & d_z > 0; \\ 0, & d_z = 0; \\ -1, & d_z < 0, \end{cases}$$

we find

Manuscript received April 12, 1982; revised May 30, 1983. This research was supported by the Naval Air Development Center under Grant N62269-81-C-0299.

The authors are with the Department of Electrical and Computer Engineering, University of California, Santa Barbara, CA 93106.

Reprinted from *IEEE Trans. Sonics Ultrason.*, vol. 30, no. 5, pp. 330–331, September 1983.

$$Q_{rr'}(x,y) = \frac{\exp\{j[2\pi((x-d_x)^2+(y-d_y)^2+d_z^2)^{1/2}/\lambda - \pi/2]\,\mathrm{sgn}(d_z)\}}{\lambda[(x-d_x)^2+(y-d_y)^2+d_z^2]^{1/2}}. \tag{11}$$

The amplitude distribution of $Q_{rr'}(x,y)$ has a single peak at (d_x, d_y) with the maximum value $1/(\lambda|d_z|)$:

$$|Q_{rr'}(x,y)| = \frac{1}{\lambda[(x-d_x)^2+(y-d_y)^2+d_z^2]^{1/2}}. \tag{12}$$

The displacements in the x and y directions can be determined by the location of the unique peak value of $|Q_{rr'}(x,y)|$. The value of the maximum indicates the absolute value of the displacement in the z direction:

$$\max\{|Q_{rr'}(x,y)|\} = \frac{1}{\lambda|d_z|}. \tag{13}$$

The ambiguity concerning the sign of d_z can be resolved by observing the phase variation of $Q_{rr'}(x,y)$ about (d_x, d_y). If the displacement is in the positive z direction, the phase variation of $Q_{rr'}(x,y)$ has a local minimum at (d_x, d_y). If the displacement is in the negative z direction, it has a local maximum.

We conclude that a three-dimensional displacement vector can be determined by using a planar holographic detection system. The displacement components parallel to the receiving plane can be indicated by the peak location of the amplitude of the SCOT output. The displacement component perpendicular to the receiving plane can be estimated by the amplitude and phase variations of the SCOT output about the peak. In practice, the integration shown in the theoretical analysis will be replaced by summation and Fourier transform will be performed by using FFT algorithm. The accuracy of the estimation depends largely upon the size of the receiving aperture.

REFERENCES

[1] G. C. Carter, A. H. Nuttall, and P. G. Cable, "The smoothed coherent transform (SCOT)," *Proc. IEEE*, vol. 61, pp. 1497–1498, Oct. 1973.

[2] J. W. Goodman, *Introduction to Fourier Optics.* New York: McGraw-Hill, 1968, p. 49.

Optimum Passive Bearing Estimation in a Spatially Incoherent Noise Environment

VERNE H. MACDONALD AND PETER M. SCHULTHEISS

Department of Engineering and Applied Science, Yale University, New Haven, Connecticut 06520

This paper studies the minimum bearing error attainable with a linear passive array. Signal and noise are stationary Gaussian processes with arbitrary power spectra, and the noise is assumed to be statistically independent from hydrophone to hydrophone. The Cramér–Rao technique is used to set a lower bound on the rms bearing error and the results are compared with the bearing error of a slightly modified split-beam tracker. The latter reaches the lower bound for a two-element array and comes very close to reaching it for a linear array with an arbitrary number of equally spaced hydrophones. Thus, the modified split-beam tracker is very nearly optimal for the uniformly spaced array. Comparisons of split-beam tracker error with the Cramér–Rao lower bound are also obtained for nonuniform hydrophone spacings.

INTRODUCTION

This paper deals with the problem of optimum bearing estimation by means of a passive sonar array. The following basic assumptions are made:

(1) Signal and noise are stationary Gaussian processes.

(2) The signal comes from a source sufficiently remote so that its wavefront may be regarded as planar over the dimensions of the receiving array.

(3) The noise waveshapes received by different hydrophones are statistically independent of each other and of the signal.

(4) The receiving array is linear.

The problem is approached from two points of view. *First*, the Cramér–Rao technique is used to set an absolute lower bound on the rms error attainable with any continuous, unbiased bearing estimator. *Second*, a slightly modified version of the conventional split-beam tracker is analyzed and shown to yield an rms error at or close to the lower bound for most practically interesting array configurations. Taken together, these results imply that the split-beam tracker is very nearly optimal and that the search for significantly better unbiased bearing estimators is bound to be fruitless under the postulated conditions.

The analysis is quite general with regard to the number and distribution of hydrophones as well as spectral properties of signal and noise. However, it is asymptotic in one sense: The observation time is assumed to be long compared with the signal and noise correlation time and with the travel time of the signal wavefront across the array. Since typical observation times tend to be of the order of a few hundred milliseconds or more, while typical signal bandwidths would amount to at least a few thousand radians per second, this analytically very convenient assumption does not appear to be unduly restrictive for arrays of moderate size.

I. CRAMÉR–RAO LOWER BOUND

The general geometry of the receiving array and the incident signal wavefront is sketched in Fig. 1. There are M hydrophones arranged in a linear, but not necessarily regularly spaced array. The objective is to measure the angle θ of the incident signal wavefront relative to the array axis. The available data consist of the M random processes $x_1(t;\theta)$, $x_2(t;\theta)$, $\cdots$, $x_M(t;\theta)$, the outputs of the M hydrophones, each observed for a period of T seconds and each parametrically dependent

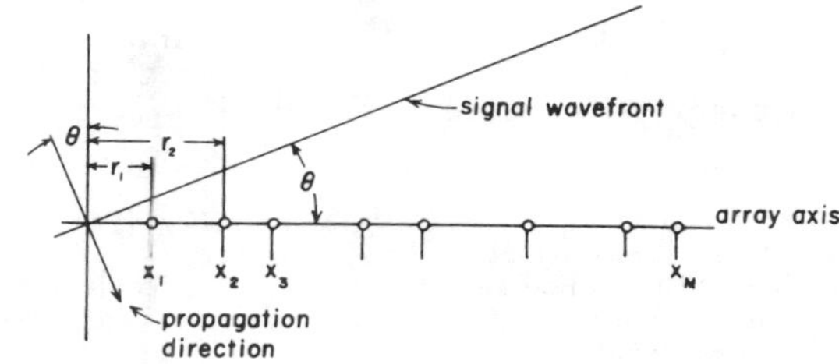

FIG. 1. Receiving array with plane-wave signal.

Reprinted with permission from the *J. of the Acoustical Soc. of America*, vol. 46, no. 1, pt. 1, pp. 37–43, July 1969.

on the angle θ. The problem therefore reduces to the classical one of estimating the parameter θ from observation of $(x_1 \cdots x_M)$.

Cramér[1] and Rao[2] have demonstrated in very general terms that any unbiased estimate $\hat{\theta}$ of θ has a variance $D^2(\hat{\theta})$ satisfying the inequality

$$D^2(\hat{\theta}) \geqq [E\{|\partial \log L(\theta)/\partial\theta|^2\}]^{-1}, \tag{1}$$

where $L(\theta)$ is the likelihood function of the received data and $E\{\ \}$ denotes the mathematical expectation of the bracketed quantity.

Probably the most straightforward approach to the evaluation of Eq. 1 would be to consider the output of each hydrophone sampled at an appropriate rate and then write the likelihood function in terms of these samples. Unfortunately, if the number of hydrophones is at all large or if there is statistical dependence between samples of a given hydrophone output, the likelihood function becomes so complicated that it is no longer practical to carry out the manipulations specified in Eq. 1.[3] If one hopes to solve the general problem, one must therefore find a representation of the hydrophone outputs which drastically simplifies the form of $L(\theta)$.

One such representation is the set of Fourier coefficients of the hydrophone outputs

$$X_i(\omega_l;\theta) = \int_{-T/2}^{T/2} x_i(t;\theta)e^{-j\omega_l t}dt, \tag{2}$$

where $\omega_l = 2\pi l/T$.

In terms of these coefficients, the available data set assumes the form

$$X^T = X_1(\omega_1;\theta)\cdots X_M(\omega_1;\theta), X_1(\omega_2;\theta), \cdots X_M(\omega_2;\theta), X_1(\omega_N;\theta), \cdots X_M(\omega_N;\theta) \tag{3}$$

Here X^T is the transpose of the column vector X and ω_N is the highest processed frequency.

The vector X completely specifies the various hydrophone outputs and can therefore be used without loss of information in place of the actual time functions. The hydrophone outputs are jointly Gaussian by assumption and Eq. 2 is a linear transformation. It follows that $L(\theta)$ has a complex Gaussian probability density in MN dimensions,

$$L(\theta) = 1/[(\pi)^{MN}\det R](e^{-X^T R^{-1} X^*})/[(\pi)^{MN}\det R]. \tag{4}$$

X^* is the complex conjugate of X and det R is the determinant of the moment matrix R defined by

$$R = E[X^* X^T]. \tag{5}$$

According to Eqs. 5 and 2, the elements of R are given by

$$E\{X_p^*(\omega_l)X_q(\omega_n)\} = \int_{-T/2}^{T/2} dt \int_{-T/2}^{T/2} du e^{j(\omega_l t - \omega_n u)} E\{x_p(t;\theta)x_q(u;\theta)\}. \tag{6}$$

If the signal wavefront at the pth hydrophone is delayed Δ_p seconds relative to the origin of coordinates in Fig. 1, then

$$x_p(t;\theta) = \mathrm{s}(t-\Delta_p) + \mathrm{n}_p(t), \tag{7}$$

where $\mathrm{s}(t)$ is the signal waveshape and $n_p(t)$ the noise waveshape at the pth hydrophone. The parameter θ is contained in Δ_p through the relation

$$\Delta_p = (r_p/c)\sin\theta, \tag{8}$$

c being the velocity of sound.

Substituting Eq. 7 in Eq. 6 and using the postulated independence of noise from hydrophone to hydrophone, as well as between signal and noise, one obtains

$$E[X_p^*(\omega_l)X_q(\omega_n)] = \int_{-T/2}^{T/2} dt \int_{T/2}^{T/2} du\ e^{j(\omega_l t - \omega_n u)} \times [R_s(t-u+\Delta_q-\Delta_p) + R_N(t-u)\delta_{pq}], \tag{9}$$

where $R_s(\tau)$ and $R_N(\tau)$ are the signal and noise autocorrelation functions, respectively, and δ_{pq} is the Kroneker delta function:

$$\delta_{pq} = \begin{matrix} 1 & p=q \\ 0 & p\neq q. \end{matrix} \tag{10}$$

At this point, it becomes convenient to express the autocorrelations in terms of the corresponding spectral functions:

$$R_s(\tau) = \frac{1}{2\pi}\int_{-\infty}^{\infty} S(\omega)e^{j\omega\tau}d\omega \tag{11}$$

$$R_N(\tau) = \frac{1}{2\pi}\int_{-\infty}^{\infty} N(\omega)e^{j\omega\tau}d\omega \tag{12}$$

Substituting Eqs. 11 and 12 in Eq. 9 one obtains after some straightforward computation

$$E[X_p^*(\omega_l)X_q(\omega_n)] = \frac{T^2}{2\pi}\int_{-\infty}^{\infty} d\omega[\mathrm{S}(\omega)+\mathrm{N}(\omega)\delta_{pq}]e^{j\omega(\Delta_q-\Delta_p)} \times \left[\frac{\sin(\omega+2\pi l/T)T/2}{(\omega+2\pi l/T)T/2}\right]\cdot\left[\frac{\sin(\omega+2\pi n/T)T/2}{(\omega+2\pi n/T)T/2}\right]. \tag{13}$$

---

[1] H. Cramér, *Mathematical Methods of Statistics* (Princeton University Press, Princeton, N. J., 1951), Sec. 32.3.

[2] C. R. Rao, "Information and the Accuracy Attainable in the Estimation of Statistical Parameters," Bull. Calcutta Math. Soc. **37**, No. 3, 81–91 (1945).

[3] The first effort to apply the Cramér–Rao inequality to the bearing estimation problem known to the authors does indeed take this approach. It is contained in a report by R. A. McDonald ["Processing of Data From Sonar Systems," Vol. 2, Rep. No. 12, General Dynamics/Electric Boat Co. (SUBIC program) (1964)] and deals only with a two-hydrophone array and white signal and noise spectra.

The final two factors in Eq. 13 differ from zero only over an ω interval of the order of $1/T$. Over such an interval the remainder of the integrand is essentially constant if

(A) $S(\omega)$ and $N(\omega)$ do not vary significantly over frequency ranges of the order of $1/T$ rad/sec (i.e., the signal and noise correlation times are short as compared with T).

(B) $|\Delta_q-\Delta_p|\ll T$ for all p and q (i.e., the signal wavefront sweeps across the array in a time short as compared with T).

Under these conditions one has from the orthogonality of the final two factors in Eq. 13

$$E[X_p^*(\omega_l)X_q(\omega_n)] = \begin{cases} 0 & \text{for } l\neq n \\ T[S(\omega_n)+N(\omega_n)\delta_{pq}]e^{j\omega_n(\Delta_q-\Delta_p)} & \text{for } l=n. \end{cases} \quad (14)$$

The crucial simplification of the likelihood function results from the first line of Eq. 14. The asymptotic absence of cross-correlation of Fourier coefficients associated with different frequencies is, of course, well known for a single stationary process.[4] What has been demonstrated above is that the additional assumption (B) extends this useful property to the vector process X. Hence the various frequency components of the quadratic form in the exponent of Eq. 4 separate completely and one is left to deal in effect with an $M\times M$ matrix rather than an $MN\times MN$ matrix.

One can now proceed with the various steps required for the evaluation of Eq. 1. From Eq. 4

$$\log L(\theta)=-MN\log(\pi)-\log\det R-X^TR^{-1}X^*. \quad (15)$$

A tedious but straightforward computation shows[5] that $\det R$ is independent of θ so that all but the last term of Eq. 15 vanishes after differentiation with respect to θ. Next, one differentiates, squares, and averages as demanded by Eq. 1. The procedure is again quite tedious but basically straightforward.[5] The result is

$$D^2(\hat{\theta})\geq\left\{\frac{\cos^2\theta}{c^2}\times\sum_{n=1}^{N}\omega_n^2\frac{[S^2(\omega_n)/N^2(\omega_n)]}{1+M[S(\omega_n)/N(\omega_n)]}\times\sum_{p=1}^{M}\sum_{q=1}^{M}(r_p-r_q)^2\right\}^{-1}. \quad (16)$$

Recognizing that the increment $\Delta\omega$ from frequency to frequency is $(2\pi)/T$ and that the spectral functions do not vary significantly over such a frequency interval (by assumption), one can convert the frequency sum into an integral. Thus, the standard deviation of $\hat{\theta}$ has the lower bound

$$D(\hat{\theta})\geq\frac{(2\pi)^{\frac{1}{2}}c}{(T)^{\frac{1}{2}}\cos\theta}\left\{\sum_{p=1}^{M}\sum_{q=1}^{M}(r_p-r_q)^2\times\int_0^{\omega_N}d\omega\,\omega^2\frac{[S^2(\omega)/N^2(\omega)]}{1+M[S(\omega)/N(\omega)]}\right\}^{-\frac{1}{2}}. \quad (17)$$

Considerable simplification results if the hydrophones are uniformly spaced. In that case,

$$r_p-r_q=(p-q)d, \quad (18)$$

where d is the distance between adjacent hydrophones. The double sum in Eq. 17 can now be evaluated analytically, and one obtains

$$D(\hat{\theta})\geq\frac{2(3\pi)^{\frac{1}{2}}c}{d(T)^{\frac{1}{2}}(\cos\theta)M(M^2-1)^{\frac{1}{2}}}\times\left\{\int_0^{\omega_N}d\omega\,\omega^2\frac{[S^2(\omega)/N^2(\omega)]}{1+M[S(\omega)/N(\omega)]}\right\}^{-\frac{1}{2}}. \quad (19)$$

Uniform hydrophone spacing is actually not the best choice from the point of view of minimizing rms error. If the over-all array length is fixed, it is not difficult to demonstrate that the double sum in Eq. 17 is formally maximized by placing half of the hydrophones at each end of the array. In that case, the lower bound becomes

$$D(\theta)=\frac{2(\pi)^{\frac{1}{2}}c}{L(T)^{\frac{1}{2}}M\cos\theta}\times\left\{\int_0^{\omega_N}d\omega\,\omega^2\frac{[S^2(\omega)/N^2(\omega)]}{1+M[S(\omega)/N(\omega)]}\right\}^{-\frac{1}{2}} \quad (20)$$

where L is the array length. With equally spaced hydrophones the array length can be expressed as $L=d(M-1)$, so that, for M substantially in excess of unity, Eq. 19 exceeds Eq. 20 by a factor of $\sqrt{3}$. This factor represents an upper bound on the improvement in bearing accuracy which might be obtained by deviating from uniform hydrophone spacing. In practice, of course, the assumption of noise independence from hydrophone to hydrophone becomes quite unrealistic as one crowds hydrophones together near both ends of the array, unless the noise consists predominantly of self-noise generated in the hydrophones or associated circuitry.

II. SPLIT-BEAM TRACKER

The Cramér–Rao inequality sets a lower bound on rms error, but it does not guarantee that this bound can actually be reached or even closely approached. For

[4] W. B. Davenport and W. L. Root, *Random Signals and Noise* (McGraw-Hill Book Co., N. Y., 1958), Sec. 6.4.

[5] J. B. Lewis, V. H. MacDonald, and P. M. Schultheiss, "Processing of Data From Sonar Systems," Vol. 6, Rep. No. 37, General Dynamics/Electric Boat Co. (SUBIC Program), 1968.

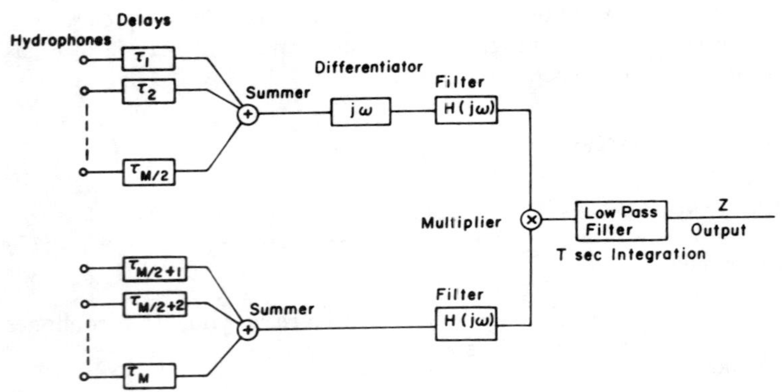

FIG. 2. Idealized split-beam tracker.

that reason, it becomes important to study physically realizable bearing estimators and to compare their performance with the lower bound. If some realizable instrumentation actually reaches the lower bound, then it furnishes an efficient estimate and the associated rms error is the absolute minimum attainable with any unbiased instrumentation.

Figure 2 shows a somewhat idealized version of the conventional split-beam tracker. The 90° phase shift between the two array halves is obtained by a differentiator in one channel. The filters with transfer function $H(j\omega)$ in each channel shape the spectrum in a manner to be discussed shortly.

The average output $\bar{z}$ of a split-beam tracker assumes the general form shown in Fig. 3 when plotted as a function of steering angle α. For the simple symmetrical noise field postulated in the present analysis $\bar{z}$ passes through zero at the true target bearing θ.

The actual output z of Fig. 2 is, of course, only an approximation to $\bar{z}$ because the smoothing time of the low-pass filter is finite. The value of α at which $z(t;\alpha)$ passes through zero gives an estimate of θ. If excursions of $z(t;\alpha)$ are almost entirely confined to the linear segment near θ in Fig. 3, the rms error $D(\theta)$ of the bearing estimate is

$$D(\theta)=\frac{\sigma_z}{|\partial\bar{z}/\partial\alpha|}\bigg|_{\alpha=\theta}, \tag{21}$$

where σ_z stands for the standard deviation of z. In other words, the rms bearing error is the rms fluctuation of the tracker output divided by the slope of the average tracker output near the target bearing. Loss of track would generally occur if instantaneous deviations of the tracker-output exceed the linear range of Fig. 3 at frequent intervals. Hence, Eq. 21 should be a realistic expression for the rms error whenever sustained tracking is possible.

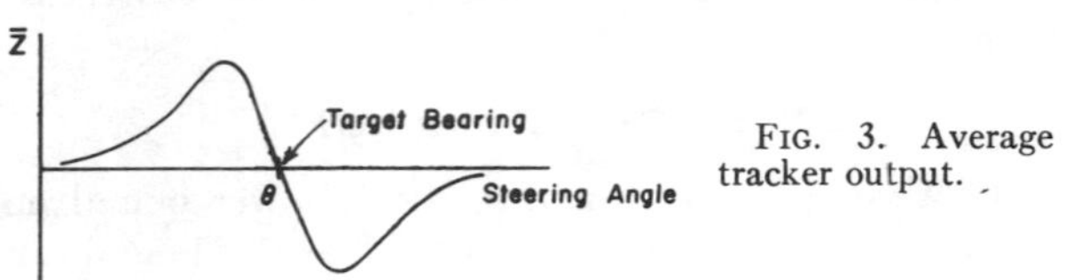

FIG. 3. Average tracker output.

The computation of σ_z and $\partial\bar{z}/\partial\alpha$ is tedious but quite straightforward. For integration times T long as compared with the correlation times of the signal and noise processes, one obtains the general expressions

$$\sigma_z^2=\frac{2\pi}{T}\int_{-\infty}^{\infty}d\omega\ \omega^2|H(j\omega)|^4[G_1(\omega)G_2(\omega)-G_{12}^2(\omega)], \tag{22}$$

$$\frac{\partial\bar{z}}{\partial\theta}=2\int_0^{\infty}d\omega\ \omega|H(j\omega)|^2\frac{\partial}{\partial\alpha}\{\mathrm{Im}[G_{12}(\omega)]\}, \tag{23}$$

where $G_1(\omega)$ and $G_2(\omega)$ are the power spectra of y_1 and y_2 (summer outputs 1 and 2), respectively. $G_{12}(\omega)$ is the cross-spectral density of y_1 and y_2. The symbol Im [ ] stands for the imaginary part of the bracketed quantity.

Another straightforward computation yields the required spectral functions:

$$G_1(\omega)|_{\alpha=\theta}=G_2(\omega)|_{\alpha=\theta}=(M^2/4)S(\omega)+(M/2)N(\omega), \tag{24}$$

$$G_{12}(\omega)|_{\alpha=\theta}=(M^2/4)S(\omega), \tag{25}$$

where, as before, $S(\omega)$ and $N(\omega)$ are the signal and noise spectra at any given hydrophone. Furthermore

$$\left[\frac{\partial}{\partial\alpha}\mathrm{Im}[G_{12}(\omega)]\right]_{\alpha=\theta}=\frac{1}{c}\omega S(\omega)\cos\theta\sum_{p=1}^{M/2}\sum_{q=(M/2)+1}^{M}(r_p-r_q). \tag{26}$$

Now substituting Eqs. 22–26 into Eq. 21 one obtains the desired rms error

$$D(\theta)=\frac{(2\pi)^{\frac{1}{2}}cM}{2T^{\frac{1}{2}}\cos\theta\left|\sum_{p=1}^{M/2}\sum_{q=(M/2)+1}^{M}(r_p-r_q)\right|}\times\left(\int_{-\infty}^{\infty}d\omega\ \omega^2|H(j\omega)|^4[N^2(\omega)+MS(\omega)N(\omega)]\right)^{\frac{1}{2}}\Big/\int_{-\infty}^{\infty}d\omega\ \omega^2|H(j\omega)|^2S(\omega). \tag{27}$$

The final problem is selection of the proper spectrum-shaping filter $H(j\omega)$. A simple variational argument shows that the ratio of integrals in Eq. 27 is minimized by the choice

$$|H(j\omega)|^2 = \frac{[S(\omega)/N^2(\omega)]}{1+M[S(\omega)/N(\omega)]}. \quad (28)$$

Equation 28 is clearly a generalization of the Eckart filter,[6] to which it reduces for low signal-to-noise ratios. With Eq. 28 substituted in Eq. 27, one reaches the final result

$$D(\theta) = \frac{\pi^{\frac{1}{2}} cM}{2T^{\frac{1}{2}} \cos\theta \left| \sum_{p=1}^{M/2} \sum_{q=M/2+1}^{M} (r_p - r_q) \right|} \times \left\{ \int_0^{\omega_N} d\omega\, \omega^2 \frac{[S^2(\omega)/N^2(\omega)]}{1+M[S(\omega)/N(\omega)]} \right\}^{-\frac{1}{2}} \quad (29)$$

where ω_N is the upper limit of the processed frequency range as in Sec. I.

If the hydrophones are uniformly spaced, the sum in Eq. 29 can be evaluated analytically[7] and one finds the following equivalent of Eq. 19

$$D(\theta) = \frac{\pi^{\frac{1}{2}} 4c}{T^{\frac{1}{2}} d\, M^2 \cos\theta} \times \left\{ \int_0^{\omega_N} d\omega\, \omega^2 \frac{[S^2(\omega)/N^2(\omega)]}{1+M[S(\omega)/N(\omega)]} \right\}^{-\frac{1}{2}}. \quad (30)$$

If half of the hydrophones are placed at each end of the array, the configuration which minimizes the lower bound on the rms error, then $D(\theta)$ becomes

$$D(\theta) = \frac{2\pi^{\frac{1}{2}} c}{LT^{\frac{1}{2}} M \cos\theta} \times \left\{ \int_0^{\omega_N} d\omega\, \omega^2 \frac{[S^2(\omega)/N^2(\omega)]}{1+M[S(\omega)/N(\omega)]} \right\}^{-\frac{1}{2}}. \quad (31)$$

Equation 31 is identical with Eq. 20, indicating that the split-beam tracker is strictly optimal for this special array geometry. As in our discussion of Eq. 20 we must, of course, keep in mind that this hydrophone arrangement is probably incompatible with the assumed independence of noise from hydrophone to hydrophone.

[6] C. Eckart, "Optimal Rectifier Systems for the Detection of Steady Signals," Tech. Rep. S10 Ref. 52-11, Scripps Inst. Oceanogr., Berkeley, Calif. (Mar. 1952).

[7] For low-input signal-to-noise ratios Eq. 30 reduces to the form given by T. Usher, J. Acoust. Soc. Amer. **35**, pp. 912–920, Eq. 40, (1965).

III. CONCLUSIONS

We can now compare the rms error of the split-beam tracker with Cramér–Rao lower bound. In general, from Eq. 17 and Eq. 29

$$\frac{D(\theta)}{D(\hat{\theta})} = \frac{M}{2\sqrt{2}} \left[\sum_{p=1}^{M} \sum_{q=1}^{M} (r_p - r_q)^2 \right]^{\frac{1}{2}} \Bigg/ \left| \sum_{p=1}^{M/2} \sum_{q=M/2+1}^{M} (r_p - r_q). \right| \quad (32)$$

The ratio is totally independent of the signal and noise power as well as spectral properties, but it does vary somewhat with array geometry. For a uniformly spaced array one finds

$$D(\theta)/D(\hat{\theta}) = (\tfrac{4}{3})^{\frac{1}{2}} [1-(1/M^2)]^{\frac{1}{2}}. \quad (33)$$

Thus there is only a very small variation with M. The ratio increases monotonically from 1 at $M=2$ to $(\frac{4}{3})^{\frac{1}{2}}$ at $M\rightarrow\infty$. Hence, for a uniformly spaced linear array, the split-beam tracker with modified Eckart filter is very nearly optimal regardless of array size, signal-to-noise ratio or spectral properties of signal and noise.

For the array geometry yielding absolute minimum error (half of the hydrophones at each end), we have already seen in Sec. II that the ratio, Eq. 32, reduces to unity. This may be viewed as a special case of Eq. 33 with $M=2$, for the $M/2$ hydrophones at each end may now be regarded as a single hydrophone with increased signal to noise ratio.

Some additional insight into the general case can be gained by restricting the array geometries to configurations symmetrical about some arbitrary midpoint, which will be chosen as the origin of coordinates. Then Eq. 32 reduces to

$$D(\theta) = (M/2)^{\frac{1}{2}} \left\{ \left[\sum_{p=(M/2)+1}^{M} r_p^2 \right]^{\frac{1}{2}} \Bigg/ \sum_{p=(M/2)+1}^{M} r_p \right\}. \quad (34)$$

The denominator of Eq. 34 is $(2/M)$ times the center of gravity of the right array half. In fact, the denominator of Eq. 32 [hence the double sum in Eq. 27] is simply $2/M$ times the spacing between the centers of gravity of the two array halves. Thus the effect of array geometry on the split-beam tracker is completely described by this spacing of centers of gravity.

By applying the Cauchy–Schwarz inequality to Eq. 34, one can readily establish that $D(\theta)/D(\hat{\theta}) \geqq 1$. A somewhat more laborious calculation yields an upper bound on the ratio of rms errors. Combining the two results one obtains

$$1 \leqq D(\theta)/D(\hat{\theta}) \leqq (M/2)^{\frac{1}{2}}. \quad (35)$$

The upper bound is reached when one hydrophone is located at each end of the array and all others are concentrated near the center. Once again, we have a geometry at odds with the postulated noise independence from hydrophone to hydrophone. Such an arrangement moves the centers of gravity of each array half closer and closer to the origin as M increases, which causes the performance of the split-beam tracker to deteriorate. In practice one would clearly avoid geometries of this type. Thus Eq. 35 asserts the existence of configurations which cause the performance of the split-beam tracker to fall far short of the lower bound, but it does not claim that these unfavorable arrangements are of great practical interest.

Perhaps a better feeling for the practical problem can be gained by starting with a uniformly spaced array and then modifying the hydrophone locations in each array half symmetrically in such a manner as to leave the array length and the centers of gravity fixed. Under these restraints one finds for large M

$$D(\theta)/D(\hat{\theta})=\sqrt{2}. \tag{36}$$

It therefore appears not unreasonable to assert that the split-beam tracker comes quite close to the lower bound for most practically interesting linear array geometries and for all values of other parameters such as spectral shape and signal-to-noise (S/N) ratio.

With this conclusion in mind, it now becomes interesting to examine the dependence of the rms error on some of the "other parameters." If the signal and noise spectra are similar in shape over the entire processed frequency band one has from Eq. 29

$$D(\theta)\propto\begin{cases}(\mathrm{S/N})^{-1} & \text{for } M(\mathrm{S/N})\ll 1\\ (\mathrm{S/N})^{-\frac{1}{2}} & \text{for } M(\mathrm{S/N})\gg 1.\end{cases} \tag{37}$$

Here S/N is the (frequency invariant) signal-to-noise ratio. $(\mathrm{S/N})^{-1}$ dependence is a behavior frequently associated with incoherent systems, whereas $(\mathrm{S/N})^{-\frac{1}{2}}$ dependence is a general characteristic of coherent systems. Note that the dividing line between the two types of behavior occurs where $M(\mathrm{S/N})=1$, i.e., where the postbeamforming S/N ratio reaches unity.

Another interesting feature is the dependence of $D(\theta)$ upon L and M for the uniformly spaced array. From Eq. 30 we have

$$D(\theta)\propto\begin{cases}(dM^2)^{-1}=L^{-1}M^{-1} & \text{for } M(\mathrm{S/N})\ll 1,\\ (dM^{\frac{3}{2}})^{-1}=L^{-1}M^{-\frac{1}{2}} & \text{for } M(\mathrm{S/N})\gg 1.\end{cases} \tag{38}$$

Thus the M dependence becomes weaker when the postbeamforming signal to noise ratio exceeds unity.

Finally, the expressions for $D(\theta)$ and $D(\hat{\theta})$ [Eqs. 29 and 17] exhibit an anomaly which deserves comment. Both vary as $(\cos\theta)^{-1}$, hence approach infinity in the endfire direction. This is particularly disturbing in the case of the Cramér–Rao inequality which should yield an absolute lower bound. To explain the apparent discrepancy we must keep in mind three facts:

(a) We have evaluated the Cramér–Rao lower bound for *unbiased* estimators.

(b) Bearing information is coded into the physical data through the signal delay from hydrophone to hydrophone, which is proportional to $\sin\theta$.

(c) Our results are asymptotic, assuming long observation time and hence small rms error.

When $\theta=\pi/2$, $\sin\theta=1$. The noise perturbed signal will certainly yield some instantaneous estimates of $\sin\theta$ in excess of unity whose interpretation in terms of θ becomes quite arbitrary. A procedure to circumvent the problem, perhaps assigning $\theta=\pi/2$ to all such values, clearly produces a biased estimate. In practice, of course, the long smoothing time condition (*c*) prevents estimates of $\sin\theta$ from exceeding unity with any significant probability, except perhaps in the immediate vicinity of endfire. Then an essentially unbiased estimate of θ can be obtained and our result represents an accurate bound on the variance of such an estimate.

The discussion of the previous paragraph does raise one interesting question: Would it be possible to achieve substantial reductions in rms error by abandoning the requirement of unbiased estimation? With a bias function $b(\theta)$ the total rms error is

$$\text{rms ERROR}=[b^2(\theta)+\text{ERROR VARIANCE}]^{\frac{1}{2}}, \tag{39}$$

where the Cramér–Rao procedure now sets the following lower bound on the variance

$$\text{ERROR VARIANCE}\geq\frac{(1+db/d\theta)^2}{E\{|\partial\log L(\theta)/\partial\theta|^2\}}. \tag{40}$$

This differs from Eq. 1 only by the factor $(1+db/d\theta)^2$ which is independent of the averaging operation and carries through unchanged to Eq. 20. For any given θ, one can clearly choose the bias such that $b(\theta)=0$, $db/d\theta=-1$ and hence obtain a total rms error of zero. In rough physical terms, this implies that the estimator records all values near θ as θ. The price one pays is, of course, larger error at other bearings. In searching for a "good" bias function, one can, therefore, hope at best for a function which is "good" in some average sense, the average being taken over all possible bearings. Such an averaging procedure presupposes a priori knowledge of target bearing statistics and therefore has an element of arbitrariness. In the absence of more precise knowledge one might perhaps assume a uniform distribution of bearing and proceed to calculate the corresponding "best" bias function. Under conditions limiting the rms error to a small fraction of a radian, so that useful bearing estimates can be obtained, the result is a very small bias except for bearings extremely close to endfire. Aside from the endfire direction, where the linear array is a poor bearing estimator in any case, it therefore does not appear that the use of bias would result in practically significant improvements.

ACKNOWLEDGMENT

The work reported in this paper was supported by the Office of Naval Research through a prime contract with General Dynamics/Electric Boat, as part of the SUBIC (Submarine Integrated Control) program.

TIME DELAY ESTIMATOR FOR EEG ANALYSIS BASED ON INFORMATION THEORY

Nicolaas J.I. Mars

University of Leiden
Section Medical Data Processing
Leiden, The Netherlands

ABSTRACT

The estimation of time delay between simultaneously recorded EEG (electroencephalographic) signals is important for the determination of the area in the brain which is responsible for the synchronous activity seen during seizures in epilepsy patients. Methods based on the assumption of linear pathways of propagation in the brain -like the cross-correlation function- have been proposed with limited success. We propose a new analysis method -based on information theory- which does not require this assumption. In our approach, the Average Amount of Mutual Information (AAMI) is computed between pairs of signals, for a range of lag-values, analogous to the cross-correlation function. Estimation of AAMI requires the estimation of the simultaneous probability density function of the signal pairs. For this purpose, an iterative estimation procedure has been developed. Simulations on artificial signals, as well as applications to real EEG signals, show that this procedure is an improvement over the use of the cross-correlation function.

INTRODUCTION

The study of EEGs, recorded during epileptic seizures in humans, is difficult because the pathways of propagation of epileptic activity in the brain are highly non-linear. Conventional techniques for measuring time delays between derivations, based on the cross correlation function, require the assumption of linear channels of propagation.

In this paper we describe a new method for the analysis of non-linear no-memory systems, using the concept of Average Amount of Mutual Information (AAMI). A measure for Mutual Information was introduced in Shannon's landmark paper [1] and extended and generalized by Gel'fand and Yaglom [2]. AAMI can be used to characterize the relation between input and output signals of non-linear systems.

Although the main goal in the development of our method has been the construction of a tool for the estimation of time delay in non-linear systems, we feel that our approach is of more general use. Recently, reports on the application of a comparable technique to studies in psychiatry have appeared [3].

THEORY

Using the mathematical theory of communication of Shannon, Gel'fand and Yaglom extended and generalized Shannon's measure of mutual information [2]. Their measure provides the amount of information about a random vector contained in another such vector.

This measure of mutual information can be defined for stochastic variables with realizations which only assume a finite number of discrete values and for stochastic variables in which the realizations assume continuous values. We will show the definition of Average Amount of Mutual Information (AAMI) only for the continuous case.

We consider two stochastic variables X and Y. The AAMI is defined by

$$\mathrm{AAMI} = \int_{-\infty}^{\infty}\int_{-\infty}^{\infty} f_{XY}(x,y) \log \frac{f_{XY}(x,y)}{f_X(x).f_Y(y)} \, dx \, dy \quad (1)$$

in which $f_X(x)$ is the marginal probability density function of X, $f_Y(y)$ the marginal density function of Y and $f_{XY}(x,y)$ the simultaneous probability density function of X and Y. (When $f_{XY}(x,y) = 0$, the corresponding value in the integration is taken to be zero.)

The unit is bits, nats or hartleys, if the base of the logarithm is respectively 2, e (2.71828...) or 10

AAMI is equal to zero if the variables are independent and greater than zero if the variables are interdependent.

Estimation of AAMI requires knowledge of the joint and marginal densities. Because these densities are normally not known a priori, they have to be estimated. We will first discuss methods for estimating these densities.

Reprinted from *Proc. ICASSP '82*, vol. 2, pp. 733–735, May 1982.

ESTIMATION OF PROBABILITY DENSITY FUNCTIONS

Choice of method

The problem of estimating a probability density function from a number of samples has been discussed extensively in the literature. Wegman [4,5] and Fryer [6] have given reviews of this literature.

We have used a kernel estimator, which appeared to be the most suitable for our application. This estimator has been studied by Parzen [7].

The kernel estimator in general is given by:

$$f_N(y_1,y_2,\ldots y_M) = \frac{1}{N}\sum_{n=1}^{N}\prod_{m=1}^{M} \frac{1}{h_m(N)h} K\left(\frac{y_m - x_{mn}}{h_m(N)}\right)$$

Here M is the number of dimensions of the density, N the number of M- dimensional samples, $h_m(N)$ the kernel width for N samples, y_m the variable in the m'th dimension, x_{mn} the n'th sample value in the m'th dimension and K(.) the kernel function.

Parzen [7] has derived conditions for K(.) which assure asymptotic unbiasedness, asymptotic consistency and uniform consistency. These properties can be achieved with a wide class of kernel functions.

Epanechnikov [8] has used this freedom in the choice of K(.) to derive the non-negative kernel form and kernel width which give the minimum relative global approximation error over all densities, in case the true probability density function has a Taylor expansion in all its arguments everywhere under the restriction that $h_m(N) = h(N)$, i.e., that the kernel width is independent of m.

The optimal kernel function derived by Epanechnikov is given by:

$$K_{opt}(y) = \frac{3}{4\sqrt{5}}\left(1 - \frac{y^2}{5}\right) \qquad \text{if } -\sqrt{5} < y < +\sqrt{5}$$

$$= 0 \text{ elsewhere}$$

Note that the optimal kernel function has a simple quadratic form and is independent of the true probability density and of the sample size. The function has finite support which makes the computation easier.

Within this class of optimal kernel functions, the optimal kernel width which minimizes the relative global approximation error of the density to be estimated is given by [8]:

$$h_{opt}(N) \simeq \left(\frac{ML^M}{ND}\right)^{1/(M+4)}$$

in which:

$$L = \int_{-\infty}^{\infty} K^2_{opt}(y)\, dy = \int_{-\sqrt{5}}^{\sqrt{5}} \left(\frac{3}{4\sqrt{5}}\left(1 - \frac{y^2}{\sqrt{5}}\right)\right)^2 dy = \frac{3}{5\sqrt{5}}$$

and:

$$D = \int\cdots\int \sum_{m=1}^{M} \left(\frac{\partial^2}{\partial x_m} f(x_1,\ldots,x_M)\right)^2 dx_1\ldots dx_M$$

The kernel width which is optimal for the estimation of the density is thus a function of the number of samples and of the density itself.

We have used the Epanechnikov kernel estimator in an iterative way, in which an estimated kernel width is used to provide an estimate of the probability density function, which in turn is used to provide an improved estimate of the optimal kernel width, etc. We have been unable to prove the convergence of this iterative procedure. In our experiments with estimation of densities based on samples from known probability density functions, we have not experienced any problems in this respect.

Computation of AAMI

The algorithm described above is used to compute the density values on a rectangular grid within a range of -3σ to $+3\sigma$. To compute the AAMI value, the integrand of (Eq. 1) is integrated, using Simpson's rule [9, p. 231]. The use of the range -3σ to $+3\sigma$ for the estimation of the probability density function implies that the densities are assumed to be zero outside that range.

We tried two different methods for the the computation of (Eq. 1). In the first method the marginal densities $f_X(x)$ and $f_Y(y)$ were obtained by numerical integration of $f_{XY}(x,y)$. In the second approach the marginal densities were estimated independently, using the one-dimensional Parzen estimator. Extensive simulations showed that the second method always gave superior results (i.e., closer to the theoretically expected values). In the applications to be described the second method has been used.

To assess the feasibility of the computational procedure described and the practical usefulness of AAMI, we performed experiments with three known systems: a linear system, a rectifying system and a squaring system. The input signal for all three systems was recursively low-pass filtered Gaussian noise. The effects of number of samples, correlation between samples, and signal- to-noise ratio have been studied. Very high correlations between samples (> .95) cause our estimator of the probability density functions to degenerate. However, this problem can be lessened by adding known amounts of white noise to the signals. Sequences of 256 to 1024 samples appeared to be usable for the purpose of estimating reliable AAMI-values.

APPLICATION TO TIME DELAY ESTIMATION

To estimate time delays in non-linear systems we computed the value of AAMI between the input and output signal for a range of lag-values, in analogy to the crosscorrelation function. The lag-value where the AAMI reaches a maximum is an estimate of the time delay of the system studied.

Simulations have been performed using the same three systems described above for a wide range of signal-to-noise ratio, number of samples and correlation between samples. These simulations showed that for linear systems the AAMI-function is almost as good an estimator of time delay as the cross correlation function. For non-linear systems the cross correlation function is useless (in the two non-linear systems described, the cross correlation between input and output signals is zero); the performance of the AAMI-function is only marginally less than for linear systems.

APPLICATION TO BIOLOGICAL SYSTEMS

Our motivation for the development of this time delay estimator comes from a medical problem: the localization of an epileptogenic focus in the brain of patients suffering from epileptic seizures. Localization of this focus is be done by computing time delays between EEG-signals recorded during epileptic seizures with electrodes implanted in the brain.

To validate our approach. we analysed seizures recorded with implanted electrodes from dogs with artificially induced epileptogenic foci. The results showed that the (known) focus localization could be found by our method with a sufficient degree of reliability.

Having stood this first test, we applied the method to recordings obtained from patients which have undergone surgical treatment for intractable epilepsy and in whom the localization of the focus is more or less known (post-operatively).

We are currently assessing the outcome of these experiments as a prerequisite to the clinical use of our method for focus localization. More detailed results from this research will be reported elsewhere [10].

REFERENCES

[1] C.E. Shannon and W. Weaver 1949. _The Mathematical Theory of Communication_ University of Illinois Press, Chicago, Ill..

[2] Gel'fand, I.M. and Yaglom, A.M. 1959. Calculation of the amount of information about a random function contained in another such function. _American Mathematical Society Translations_ Vol. 12, pp. 199- 246.

[3] Inouye, T., Yagasaki, A., Takahashi, H. and Shinosaki, K. 1981. The dominant direction of interhemispheric EEG changes in the linguistic process. _Electroencephalography and Clinical Neurophysiology_ Vol. 51, pp. 265- 275.

[4] Wegman, E.J. 1972. Nonparametric probability density estimation: I. A summary of available methods. _Technometrics_ Vol. 14, No. 3, pp. 533- 546.

[5] Wegman, E.J. 1972. Nonparametric probability density estimation: II. A comparison of density estimation methods. _Journal of Statistical Computation and Simulation_ Vol. 1, pp. 225- 245.

[6] Fryer, M.J. 1977. A review of some non-parametric methods of density estimation. _Journal of the Institute of Mathematics and its Applications_ Vol. 20, pp. 335- 354.

[7] Parzen, E. 1962. On estimation of a probability density function and mode. _Annals of Mathematical Statistics_ Vol. 35, pp. 1065- 1076.

[8] Epanechnikov, V.A. 1969. Non-parametric estimation of a multivariate probability density. _Theory of Probability and its Applications_ Vol. 14, pp. 153- 158.

[9] A.H. Stroud 1971 _Approximate calculation of multiple integrals_ Prentice-Hall, Englewood Cliffs, N.J.

[10] N.J.I. Mars and G.W. van Arragon 1982 Submitted for publication in. _Electroencephalography and Clinical Neurophysiology_.

Passive Systems Theory with Narrow-Band and Linear Constraints: Part I–Spatial Diversity

JOSÉ M. F. MOURA AND ARTHUR B. BAGGEROER, MEMBER, IEEE

Abstract—This paper studies passive problems where the receiver extracts from the source radiated signature information concerning the parameters defining the relative source/receiver geometry.

A model encompassing the fundamental global and local characteristics for passive positioning and navigation is presented. It considers narrow-band signals, imposes linear constraints on the geometry, and exhibits explicitly the symmetry between the space and time aspects. The analysis concentrates on questions of global geometry identifiability, emphasizing the passive global range acquisition.

The maximum-likelihood processor is analyzed by studying the ambiguity structure associated with inhomogeneous passive narrow-band tracking. Bounds on the global and local mean-square error performance are studied and tested via Monte Carlo simulations. By considering two limiting geometries, a distant and a close observer, simple approximate expressions for the mean-square errors are presented and compared to the exact bounds.

Herein the study is restricted to stationary geometries where the source is located by an extended array (spatial diversity). Subsequent papers generalize the study to moving sources (temporal diversity) and to coupled geometries.

I. INTRODUCTION

IN MANY tracking problems the observers are of the passive type, capable only of receiving waves without control over them. These passive systems are relevant in a variety of fields: oceanography (locating drifting buoys) [18], [21], meteorology (tracking radiosondes or balloon-borne devices) [9], passive sonar (positioning submersibles, commercial fish finders) [1], nagivation (obtaining position fixes, collision avoidance systems) [10], and so forth.

Fig. 1 shows typical navigational configurations. In Fig. 1 (a) and (c) an omnidirectional receiver R determines its position relative to an ensemble of beacons $B_1, \cdots, B_N$, or relative to a moving satellite. In Fig. 1(b) and (d) a fixed extended platform R_1 to R_N or a moving omnidirectional receiver uses a single beacon as positional reference. Geometries (a) and (b) are stationary; the basic characteristic is the spatial separation exhibited by the source or the receiver. Configurations (c) and (d) are not fixed; the spatial separation observed is synthetically generated by the relative motions. Generally the navigational problem is complex, inducing on the signal a space/time coupled diversity structure.

Manuscript received March 16, 1977; revised June 15, 1977. This work was supported by the J.S.E.P., under Contract DAAB07-71-C-0300.

J. M. F. Moura is with Centro De Análise e Processamento de Sinais, Complexo Interdisciplinar and the Department of Electrical Engineering, Instituto Superior Técnico, Lisboa-1, Portugal.

A. B. Baggeroer is with Research Laboratory of Electronics and the Department of Electrical Engineering and Computer Science at M.I.T., Cambridge, MA 02139.

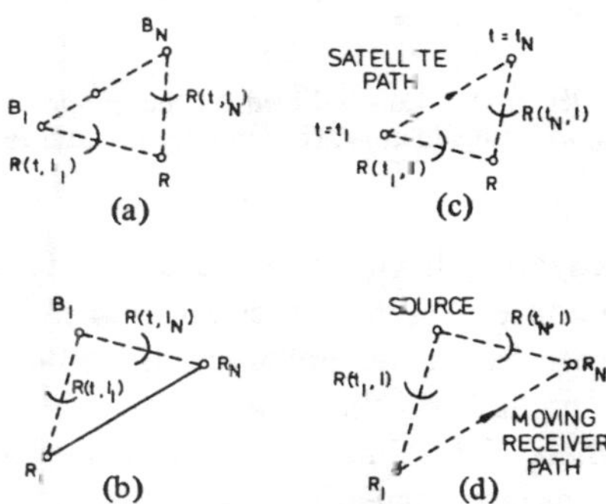

Fig. 1. Typical navigational configurations: (a) Extended source (beacons B_1 to B_N). (b) Extended receiver. (c) Moving source. (d) Moving receiver.

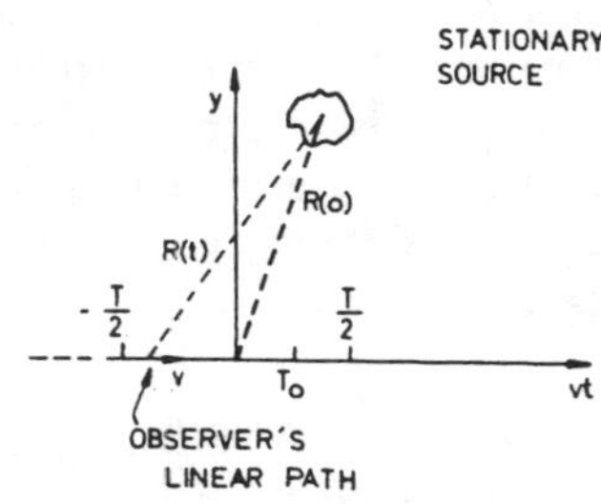

Fig. 2. Passive tracking global geometry.

Reversing the roles in the preceding navigation examples, a positioning application is obtained. In the latter it is the stationary or moving receiver, of known position and dynamics, that tracks a stationary or moving source. Given the dualism, the discussion is restricted to positioning problems.

In Fig. 2, the distance between the source and the linear array's geometric center has the parabolic form of Fig. 3(a), thereby inducing the time variant Doppler modulation indicated in Fig. 3(b). Observation of the change in frequency of the wave emitted by the moving source enables measurement of its speed (Doppler phenomenon). But, at point T_0 of the stationary Doppler modulation (closest point of approach), the line defined by the array's geometric center and the source is normal to the path. Many practical positioning and navigational techniques use this elementary observation.

This paper, and the accompanying Parts II [14] and III [15], are concerned with performance analysis and design of

Reprinted from *IEEE J. Oceanic Eng.*, vol. 3, no. 1, pp. 5–13, January 1978.

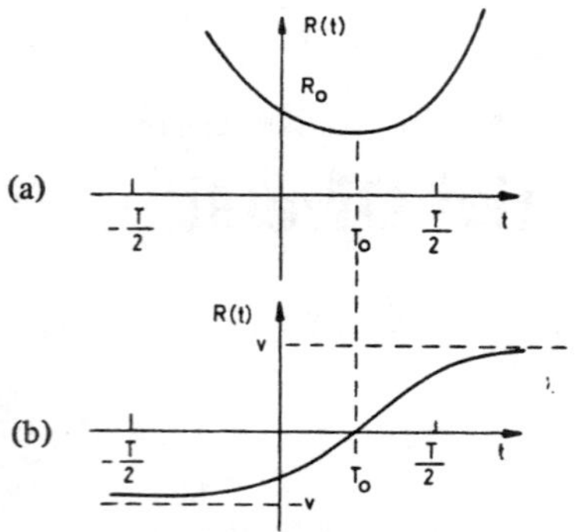

Fig. 3. Doppler modulation induced on the temporal signal structure. (a) Range history. (b) Doppler modulation.

receivers extracting from the signal's structure all available information concerning the source position and dynamics. The study concentrates on the global identifiability of the parameters, with emphasis on the range acquisition. Part I analyzes problems where the geometry is stationary (no relative dynamics), e.g., as in triangulation with an extended source (navigation) or an extended array (positioning). Part II [14] considers those applications where both the source and the observer are pointwise, as, for example, in Doppler location with a single beacon (navigation) or with an omnidirectional array (positioning). Finally, Part III [15] studies the situations where the signal has a spatial and a temporal structure, concentrating on the resulting coupling issues. In [12] this theory is applied to the design of algorithms achieving global acquisition and tracking for passive positioning and navigation problems of practical significance.

II. PRELIMINARY CONSIDERATIONS ON PASSIVE TRACKING

There are two modeling issues. The first involves the signal structure; the second concerns the global geometry and the relative dynamics.

The study is here restricted to the following.

1) A class of random wave forms, in which narrow-band signals multiplied by a Rayleigh–Gaussian random parameter are observed, imbedded in an additive, spatially homogeneous, temporally white Gaussian noise process.

2) A planar geometry with linear constraints. The receiving antenna is either omnidirectional or linear and the source/receiver dynamics are either stationary (with no relative motions) or linear (e.g., source following a deterministic constant-speed linear path). In [12] a more realistic model is discussed where the linear path is disturbed by random accelerations.

Under the preceding assumptions the resulting wave form distribution exhibits in time and space a "narrow-band" modulated structure, and thus is temporally nonstationary and spatially inhomogeneous.

Although the ideal tracking system would use all available information conveyed by the received wave forms the analysis is concentrated on the phase modulations of the narrow-band signals. In particular, the observed changes in the signal strength occurring either across the extended receiving antenna or during the finite time observation interval are ignored. These are of practical significance only, for example, when the total array dimensions are much larger than the source/receiver separation and hence represent higher order corrections that will not be pursued. However the model does take into account transmission losses due to fading, internal wave phenomenon, and medium inhomogeneities, by considering, besides an additive measurement noise, a multiplicative random-type disturbance (Rayleigh channel).

Narrow-band passive tracking has received considerable attention [7], [11], [17]. In most of the studies the following three assumptions are made.

S1. Far-field geometry: The wavefronts are assumed planar; the global and/or local wavefront curvature is neglected.

S2. Decoupling: The spatial and temporal processing aspects are decoupled.

S3. Finite parameter context: The relative source/receiver dynamics are stationary or deterministic. Passive tracking is reduced to a finite parameter estimation problem.

Under S1–S3 the problem simplifies to a "bearings only" situation, wherein the observable source/receiver parameter is the bearing angle and/or the source (radial) velocity. Ranging is accomplished either by an auxiliary active system or by *ad hoc* procedures such as simple triangulation or Doppler counting.

As far as we know, some preliminary analytical work taking into account the (spatial) curvature of the waveforms has only recently been reported [4], [5]; moreover, inhomogeneous waveforms have received scant attention in other applied areas. An exception is in optics [8] where quadratic approximations to the waveform curvature are usual in Fresnel diffraction studies. Also in seismic profiling, new techniques [3], [19] explore the signal's nonlinear spatial structure. Wave theory is still another área which considers a mixture of plane and inhomogeneous waves when finding the distribution of a field scattered by a rough surface [2].

Passive narrow-band tracking when only S1 is assumed has previously been studied [13]. The motions were modeled by a stochastic finite-dimensional dynamic system. A spatial/time integrated approach, with planar wavefront structure, was developed based on first-order approximations to the infinite-dimensional filter. Analysis substantiated by Monte Carlo simulations showed that the filter tracked only the local dynamics and lacked global range observability. This set of papers explores the spatially inhomogeneous and/or the temporally nonstationary character of the waveforms (spatial and temporal curvature). The absence of the hypothesis S1 is an underlying characteristic of the study.

With the point source emitting $S(t)$, the signal at a sensor with spatial coordinate l, and at time t is

$$s(t, l) = S(t - \tau(t, l)) \tag{1}$$

where $\tau(t, l)$ is the travel time delay. For a homogeneous medium

$$\tau(t, l) = \frac{R(t - \tau(t, l))}{c} \cong \frac{R(t, l)}{c} \tag{2}$$

where $R(t, l)$ is the source/sensor separation and c the medium propagation velocity.

In the present paper the geometry is stationary. The signal exhibits a spatial diversity structure

$$\tau(t, l) = \tau(l) \tag{3}$$

i.e., the delay is independent of the time variable. In Part II [14] the signal for processing presents a temporal diversity structure, i.e.,

$$\tau(t, l) = \tau(t). \tag{4}$$

Finally, Part III [15] considers more general geometries, where space and time coupling arise.

III. STATIONARY SOURCE MODEL

Fig. 4(a) illustrates the parametrization for a planar geometry where a stationary point source is being tracked by a linear array oriented along the l axis. Fig. 4(b) shows the problem where a moving omnidirectional sensor (e.g., a short or nonlinear array) with known speed v locates a stationary source. These passive tracking geometries, being space and time dual versions of the same problem, are discussed in the context of the first one. In the sequel they are simply referred to as synthetic observer/stationary source (SOSS) problems.

The distance from the source to the array element at location l is given by the range function

$$R(t, l, A) = \{R_0{}^2 + l^2 - 2lR_0 \sin\theta\}^{1/2},$$

$$l \in \left[-\frac{L}{2}, \frac{L}{2}\right], \qquad t \in \left[-\frac{T}{2}, \frac{T}{2}\right]. \tag{5}$$

In (5)

$$A = [R_0 \sin\theta]^T \tag{6}$$

is the parameter vector defining the relative geometry; it is modeled as unknown and nonrandom. In (6) T denotes matrix transposition.

The source radiates narrow-band signals, which are received across the observing antenna as

$$r(t, l) = \sqrt{2}\, Re\{\tilde{r}(t, l) \exp j\omega_c t\} \tag{7}$$

where

$$\tilde{r}(t, l) = \tilde{s}(t, l, A) + \tilde{w}(t, l) \tag{8}$$

$\tilde{w}(t, l)$ is a zero mean, spatially and temporally "white," Gaussian noise with double spectral height of N_0. The signal complex envelope is

$$\tilde{s}(t, l, A) = \sqrt{E_r}\,\tilde{b}\tilde{s}_n(t, l, A) \tag{9}$$

with $\tilde{s}_n(t, l, A)$ being a normalized signal

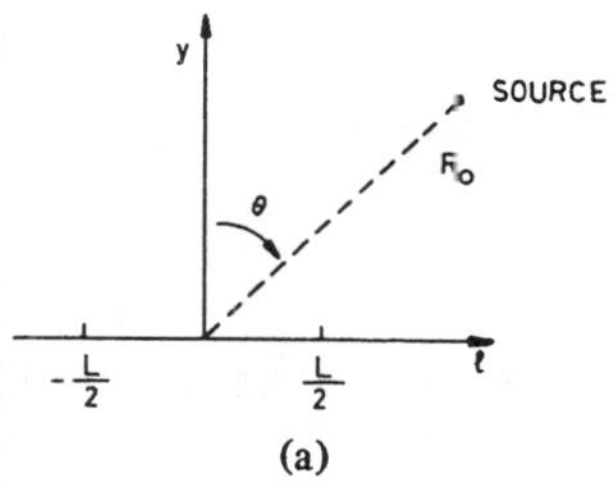

(a)

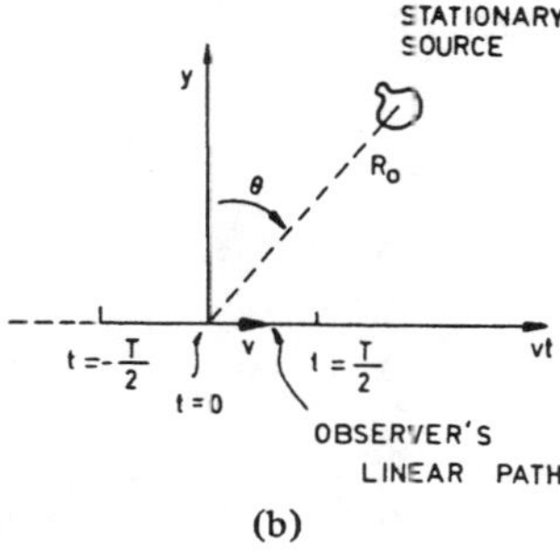

(b)

Fig. 4. Planar geometry for a stationary source: (a) Stationary array. (b) Moving omnidirectional array.

$$\tilde{s}_n(t, l, A) = \frac{1}{\sqrt{LT}} \exp\left[-j\frac{2\pi}{\lambda} R(t, l, A)\right],$$

$$t \in \left[-\frac{T}{2}, \frac{T}{2}\right], \qquad l \in \left[-\frac{L}{2}, \frac{L}{2}\right]. \tag{10}$$

$E_r = PLT$ = total energy received during the observation interval $[-T/2, T/2]$, by an array of geometric dimension L; P = signal power; $\lambda = c/f = 2\pi c/\omega_c$ = wavelength; f = carrier frequency; c = medium waveform speed propagation. $\tilde{b} = b \exp j\psi$, with b Rayleigh and ψ uniformly distributed random variables.

The zero mean Gaussian random variable $\tilde{b}$, with variance $E\,|\,\tilde{b}\,|^2 = 2\sigma_b{}^2$ independent of the measurement noise $\tilde{w}(t, l)$, accounts for model inaccuracies, e.g., radiated signal power variations about some nominal value, fading in the transmission medium, etc. More importantly, the presence of $\tilde{b}$ in (9) represents a structural model constraint: the lack of knowledge of the signal absolute phase (incoherent receiver). It precludes the estimation of the travel time delay, which measures the range in synchronized systems [6]. The paper studies then alternative procedures for range acquisition exploring the signal nonlinear phase structure.

IV. AMBIGUITY STRUCTURE

Under the model assumptions of Section III, the optimum receiver is a maximum-likelihood (ML) processor, e.g. [20], chapter X; it maximizes a monotonic function of the ML function on the parameter space Ω

$$\ln A_1(\overline{A}) = \frac{1}{N_0} \frac{\overline{E}_r/N_0}{1 + \overline{E}_r/N_0} \,|\,\tilde{L}(\overline{A})\,|^2 \tag{11}$$

where

$$\bar{E}_r = \text{average received energy} = (2\sigma_b^2)E_r \tag{12}$$

and

$$\tilde{L}(\bar{A}) = \int_{-T/2}^{T/2} dt \int_{-L/2}^{L/2} dl\, \tilde{r}(t, l)\tilde{s}^*(t, l, A). \tag{13}$$

With an inner product notation

$$\tilde{L}(\bar{A}) = \langle \tilde{r}, \tilde{s}(\bar{A}) \rangle. \tag{14}$$

$L(\bar{A})$ is a Gaussian random variable, with statistics

$$E\tilde{L}(\bar{A}) = \langle \tilde{s}(A), \tilde{s}(\bar{A}) \rangle \tag{15}$$

$$E\{[\tilde{L}(A_1) - E\tilde{L}(A_1)][\tilde{L}(A_2) - E\tilde{L}(A_2)]^*\} = \langle \tilde{s}(A_1), \tilde{s}(A_2) \rangle \tag{16}$$

where A is the actual source vector, and A_1, A_2 are scanning parameter values.

A practical procedure in two steps maximizes (11) over the continuous parameter space Ω. In the first stage Ω is made discrete by a grid; the vertex at which the ML function is maximum is chosen as a coarse estimate. In the second stage a finer search about the previous value returns the approximate ML estimate A_{ml}.

Like in active radar, the structure of the signal autocorrelation

$$\Psi(A, \bar{A}) \triangleq \langle \tilde{s}_n(A), \tilde{s}_n(\bar{A}) \rangle \tag{17}$$

and its squared modulus, the generalized ambiguity function (GAF)

$$\Phi(A, \bar{A}) = |\Psi(A, \bar{A})|^2 \tag{18}$$

play an important role in dimensioning the grid and evaluating the algorithm's mean square performance (see (15) and (16)). In (17) the index n indicates that the signal normalized version (10) is used.

In (11)–(18) the statistical assumptions are reflected in the multiplying gain, while the geometry affects the GAF structure. This factoring is a result of the signal and noise model assumed.

Basically the following two issues have to be pursued.

1) The GAF's main lobe structure, i.e., its local description on Ω about the source location.

2) The GAF's secondary structure, i.e., its relative maxima global distribution, size and rate of falloff. These points are studied next, using first an approximation to GAF and then the general expression (18).

A. Approximate Analysis

The analysis of the GAF is carried out by assuming a polynomial approximation to the phase range difference defined in (19). For a source in the so-called Fresnel zone, a second-order expansion is appropriate leading to

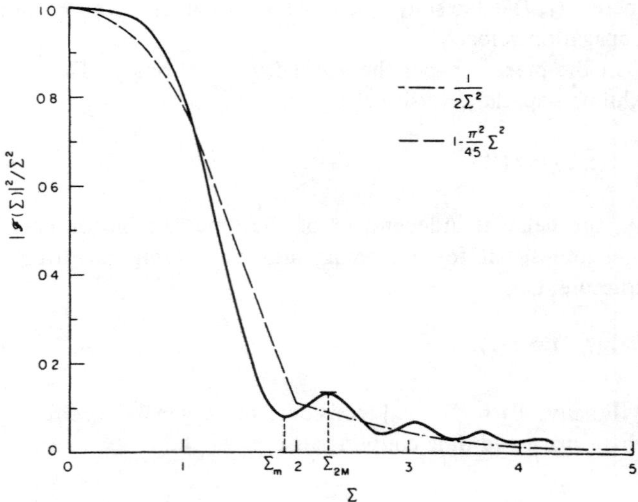

Fig. 5. Ambiguity Fresnel structure.

$$\Delta R(l, A, \bar{A}) \triangleq R(l, A) - R(l, \bar{A}) = \sum_{i=0}^{n} \Delta_i l^i \tag{19}$$

$$= \Delta R_0 - \Delta \sin\theta l + \Delta\left(\frac{\cos^2\theta}{R_0}\right)\frac{l^2}{2} \tag{20}$$

where $\Delta_0 = \Delta R_0 = R_0 - \bar{R}_0$ and likewise for the remaining quantities in (19) and (20). For the quadratic phase (20), the GAF is

$$\Phi(A, \bar{A}) \cong \frac{1}{\Delta\Sigma}[F(\Sigma_+) - F(\Sigma_-)]^2 \tag{21}$$

where

$$\Delta\Sigma = \Sigma_+ - \Sigma_-, \; \Sigma_\pm = \left(\pm\frac{L}{2} + \frac{\Delta_1}{2\Delta_2}\right)$$

$$F(\Sigma) = \int_0^\Sigma \exp jt^2\, dt = \text{Fresnel exponential integral.}$$

Along the $\Delta_1 = \Delta \sin\theta$ axis the GAF has the sinc-squared structure[1]

$$\Phi(A, \bar{A}) = \text{sinc}^2\left[\frac{2\pi}{\lambda}\Delta\sin\theta\frac{L}{2}\right] \tag{22}$$

whose second largest global maximum is reduced to about 4.5 percent of the value at the origin.

Along the radial $\Delta(\cos^2\theta/R_0)$ axis it has the Fresnel structure

$$\Phi(A, \bar{A}) = |F(\Sigma)/\Sigma|^2 \tag{23}$$

[1] sinc $X = \sin X/X$.

illustrated in Fig. 5, with second maximum equal to 0.132 and where $\Sigma = [(\pi/\lambda)\Delta(\cos^2\theta/R_0)]^{1/2}L/2$. Fig. 5 presents also local (about $\Sigma = 0$) and asymptotic (large Σ) quadratic expansions approximating the Fresnel ambiguity structure (23). Equivalent graphical displays may be obtained for (21). However the two-dimensional studies are carried out with the exact expression for the GAF.

B. Graphical Analysis

Fig. 6 presents a three-dimensional and a contour plot of the GAF for the actual source values $A_a = [0.6 \times 10^5 \text{ ft} \; \sin 15°]^T$, and an array $L = R_0/2$. Equivalent diagrams for distant (small array) and close (large array) observers are in [12]. These figures display elliptical patterns for the main-lobe equal-height contours and negligible secondary peaks (less than 20 percent of the GAF's maximum value). An asymptotic analysis, based on the method of stationary phase, e.g. [16], is pursued in [12] leading to bounds on the GAF's rate of falloff similar to the one shown in Fig. 5.

V. MAIN-LOBE QUADRATIC DESCRIPTION

The elliptical patterns of Fig. 6(b) legitimize a quadratic analysis of the GAF's main lobe. Retaining terms up to the second order in the Taylor's series expansion

$$\Phi(A, A_a) \cong 1 - \Delta A^T M \Delta A \tag{24}$$

where $\Delta A = A - A_a$, and

$$M = [M_{ij}] = -\frac{1}{2}\left[\frac{\partial^2\Phi(A, A_a)}{\partial A_i \partial A_j}\bigg|_{A=A_a}\right] \tag{25}$$

is defined as the mean-square spread matrix.

The first minimum of the GAF occurs approximately when

$$Q(\Delta A) \triangleq \Delta A^T M \Delta A \cong 1. \tag{26}$$

Equation (26) measures the extent of the GAF's main lobe.

After algebraic computations

$$M_{ij} = M_{0_{ij}} - M_{1_{ij}} \tag{27}$$

with

$$M_0 = [M_{0_{ij}}] = \left(\frac{2\pi}{\lambda}\right)^2 \frac{1}{2X}\int_{-X}^{X} [\nabla_{A_a} R(l, A)] \cdot [\nabla_{A_a} R(l, A)]^T \, dl \tag{28}$$

and

$$M_1 = [M_{1_{ij}}] = \left(\frac{2\pi}{\lambda}\right)^2 \left[\frac{1}{2X}\int_{-X}^{X} \nabla_{A_a} R(l, A)\, dl\right] \cdot \left[\frac{1}{2X}\int_{-X}^{X} \nabla_{A_a} R(l, A)\, dl\right]^T. \tag{29}$$

(a)

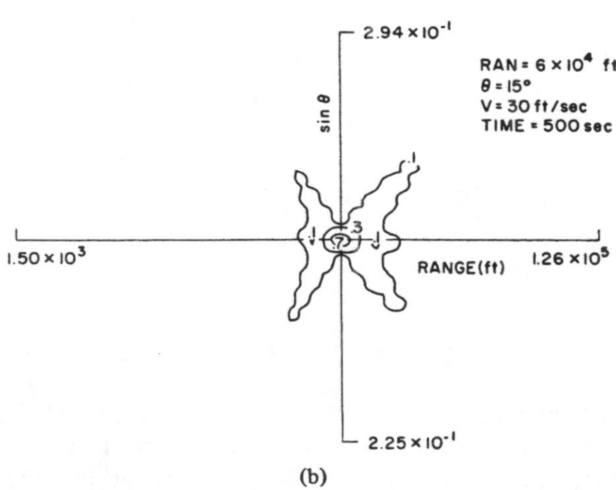

(b)

Fig. 6. SOSS ambiguity structure: (a) Three-dimensional plot. (b) Contour plot.

In (28) and (29) ∇_{A_a} is the gradient column vector operator with respect to the parameter vector A, evaluated at the source location $A = A_a$. Apart from scaling factors, the elements of M depend exclusively on the bearing angle and on the geometric parameter $X = L/2R_0$.

Although (28) and (29) may be integrated [12], due to their analytical complexity, the closed form expressions obtained are of little practical relevance. Rather, local, asymptotic, and graphical studies with respect to X are pursued. However, since in section V the interest lies on the inverse M^{-1}, the det M, and the cross correlation $\rho_{R \cdot \sin\theta}$, the subsequent analysis concentrates on these entities.

A. Distant Observer (Local Analysis) (Fresnel Zone)

For a distant observer geometry, i.e., when $X = L/2R_0 < 1$, a Taylor series study in X is pursued. Truncation after the first nonzero order term leads to

$$M^{-1} \cong \left(\frac{\lambda}{2\pi}\right)^2 \begin{bmatrix} 45/\cos^4\theta X^4 & -33\sin\theta/R_0\cos^2\theta X^2 \\ = & 3/R_0{}^2X^2 \end{bmatrix} \tag{30}$$

$$\det M \cong (2\pi/\lambda)^4 R_0{}^2 \cos^4\theta X^6/3^3 \times 5 \tag{31}$$

$$\rho_{R_0 \sin\theta} = (M^{-1})_{12}/[(M^{-1})_{11}(M^{-1})_{22}]^{1/2} \tag{32}$$

$$\cong -11 \sin\theta X/\sqrt{15}. \tag{33}$$

B. Close Observer (Asymptotic Analysis)

Neglecting the amplitude attenuation effects across the observing array, the asymptotic analysis is restricted to the phase information. As $X \to \infty$

$$M^{-1} \cong \left(\frac{\lambda}{2\pi}\right)^2 \begin{bmatrix} \frac{2}{\pi}\cos\theta X - \cos 2\theta & \sin 2\theta\left(\frac{X}{\pi} - \cos\theta\right)\Big/R_0 \\ = & \left(\frac{2}{\pi}\cos\theta\sin^2\theta X + \cos^2\theta\cos 2\theta\right)\Big/R_0{}^2 \end{bmatrix} \tag{34}$$

$$\det M \cong \left(\frac{2\pi}{\lambda}\right)^4 R_0{}^2\pi\Big/2\cos\theta X \tag{35}$$

and

$$\rho_{R_0 \sin\theta}{}^2 \cong 1 - \pi/2\sin^2\theta\cos\theta X, \qquad \theta \neq 0, \frac{\pi}{2}. \tag{36}$$

For large synthetic arrays the receiver is highly sensitive to the relative geometry. The cross correlation for $\theta \neq 0, \pi/2$ tends asymptotically to 1 and M becomes singular. Intuitively this behavior reflects that, from phase information and for an infinitely large array, the identifiable parameter is the distance $R_0 \cos\theta$ from the source to the observer.

C. Graphical Representations

The exact closed form expressions for the diagonal elements of M^{-1} are presented in Figs. 7 and 8 as functions of X.

The nominal conditions assumed are $R_0 = 6 \times 10^4$ ft, $\theta = 15°$. The figures also display the local and asymptotic tangent equations (30) and (34), respectively.

Observe the quadratic (convex cup) behavior of the range and bearing mean-square spreads. It reflects two different phenomena: the main-lobe flatness and shearing. For small X (distant observer) the main lobe is spread out at the origin (source location) corresponding to large uncorrelated inverses of the second-order derivatives of the GAF. As X increases the main lobe gets sharper, but a shearing effect occurs leading to a cross correlation which decreases monotonically to -1. As a consequence, the spread functions (diagonal elements of M^{-1}) attain a minimum at a certain value of X (dependent on the relative geometry) and then increase monotonically.

VI. MEAN-SQUARE PERFORMANCE

Section IV described a two-stage implementation of the ML receiver, consisting of a crude search followed by a finer one. Global acquisition of the source parameters is performed by the first step of the algorithm. This section studies the errors associated with the range and the bearing estimates, concentrating on a mean-square performance analysis.

A. Cramer-Rao Bounds

For the model described in section III, a lower bound to the mean-square error is given by the Cramer-Rao inequality

$$\Lambda_\epsilon = E[(A_{ml} - A_a)(A_{ml} - A_a)^*] \geqslant J^{-1} \tag{37}$$

where J is the Fisher information matrix

$$J = -E\left[\frac{\partial^2 \ln \Lambda_1(A)}{\partial A_i \partial A_j}\bigg|_{A=A_a}\right]. \tag{38}$$

Under general regularity conditions, satisfied by this parameter estimation problem

$$J = GM \tag{39}$$

with M defined by (25) and

$$G = 2\frac{\bar{E}_r}{N_0}\frac{\bar{E}_r}{N_0 + \bar{E}_r}. \tag{40}$$

B. Total Bounds

Let A_j represent the jth component of the source parameter vector A. The mean-square error on the estimate $A_{j_{ml}}$ is

$$\sigma_{\text{tot}_j}{}^2 = E(A_{j_\epsilon}{}^2) = E(A_j - A_{j_{ml}})^2 = \sigma_{gl_j}{}^2 + \sigma_{\text{loc}_j}{}^2 \tag{41}$$

where the global and local mean-square error components are

$$\sigma_{gl_j}{}^2 = E(A_j{}^2 \mid \xi)\Pr(\xi) \tag{42}$$

$$\sigma_{\text{loc}_j}{}^2 = E(A_{j_\epsilon}{}^2 \mid \xi^c)[1 - \Pr(\xi)]. \tag{43}$$

In (42) and (43) ξ is the event that a decision error or diversion occurs, i.e., that the first step of the algorithm returns the wrong grid vertex; ξ^c is the complement of ξ; and Pr (ξ) is the probability of the event ξ.

The computation of the various quantities in (41) to (43) depends on the design and dimension of the grid discretizing the parameter space. Assuming that the grid cells are the ellipses determined by (26), from the negligible side-lobe structure of the ambiguity function, it follows that the coarse

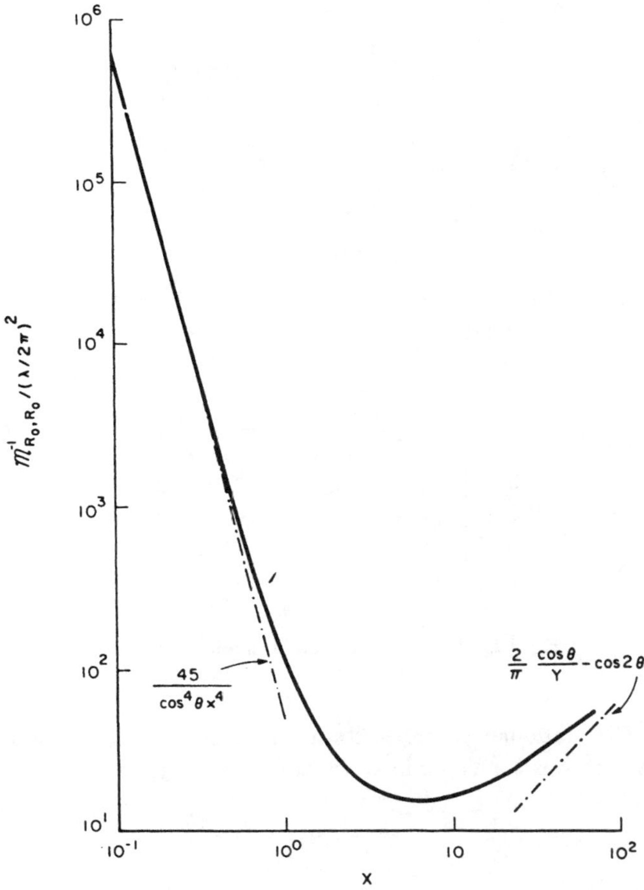

Fig. 7. Inverse range mean-square spread versus $X = L/2\, R_0$.

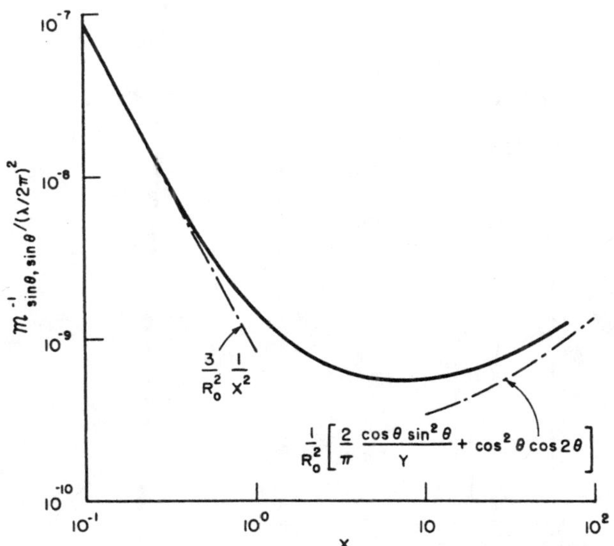

Fig. 8. Inverse bearing mean-square spread versus $X = L/2\, R_0$.

search step is equivalent to a multiple-hypothesis decision-testing problem with M orthogonal signals transmitted over a Rayleigh channel.

The *a priori* unknown source location is restricted to a rectangular region Ω, and

$$\Omega = \mathop{X}_{j=1}^{2} [A_{j_m}, A_{j_M}] = [R_{0_m}, R_{0_M}] \, X \, [\sin\theta_m, \sin\theta_M]. \tag{44}$$

For large-energy signal-to-noise ratio and large number M of grid cells, the probability of error is

$$\Pr(\xi) \cong \left(\ln M - \frac{1}{2M} + \gamma\right) \Big/ (E_r/N_0) \tag{45}$$

and the conditional mean-square error

$$E(A_j{}^2 \mid \xi) = \frac{(\Delta_M A_j)^2}{6} \tag{46}$$

where $\Delta_M A_j = A_{j_M} - A_{j_m}$ and γ is the Euler constant. In (45) M is given by

$$M = \frac{V_\Omega}{\pi} (\det \mathcal{M})^{1/2} \tag{47}$$

where V_Ω is the volume of Ω, computed from (44).

Finally $E(A_j{}^2 \mid \xi^c)$ is approximated by the Cramer-Rao inequality (37).

Substituting (30) and (31), or (34) and (35) in (42)-(47), and these in (41) lead to analytical expressions for the mean-square errors for the distant and close observer geometries, respectively.

C. Graphical Analysis

Figs. 9 and 10 illustrate the range and bearing total mean-square performance as functions of the geometric parameter X. The geometrical and statistical conditions assumed are as follows:

$$R_0 = 6 \times 10^4 \text{ ft}$$

$$\theta = 15^\circ$$

$$\text{SNR} = \text{signal-to-noise ratio} = 0 \text{ dB}$$

$$\Delta_M R_0 = 6 \times 10^5 \text{ ft}$$

$$\Delta_M = 5^\circ$$

$$\sigma_b{}^2 = \frac{1}{2}$$

$$\lambda = 50 \text{ ft}.$$

The *a priori* range uncertainty $\Delta_M R_0$ is ten times the actual range R_0; however, it follows from Fig. 9 that the ML algorithm can focus globally the range parameter, i.e., it achieves $\sigma_{R_0} \ll R_0$.

For $X_1 \cong 10^{-2}$, the standard deviation for the range-estimation error is about 7.5 percent of R_0, while for $X_2 \cong 10^{-1}$ (a ten times larger array), it is reduced to only 5 percent of R_0. This results from the fact that for smaller X, the local errors dominate the global ones, while for larger X, it is the contrary; as X increases the total errors depart from the

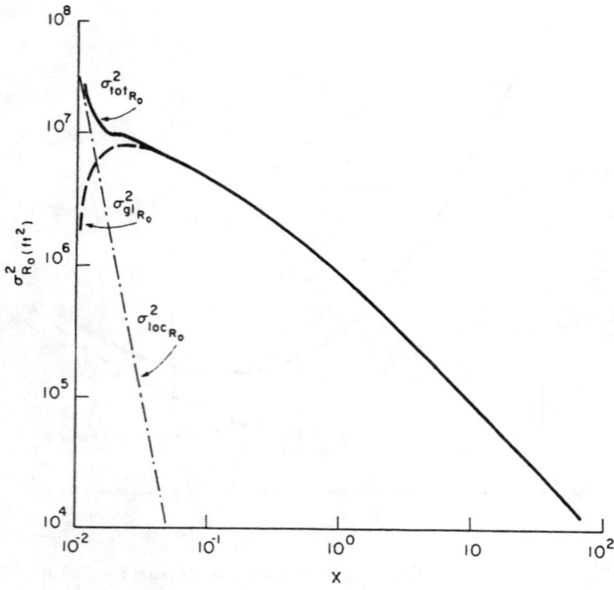
Fig. 9. Total range mean-square error versus $X = L/2\,R_0$.

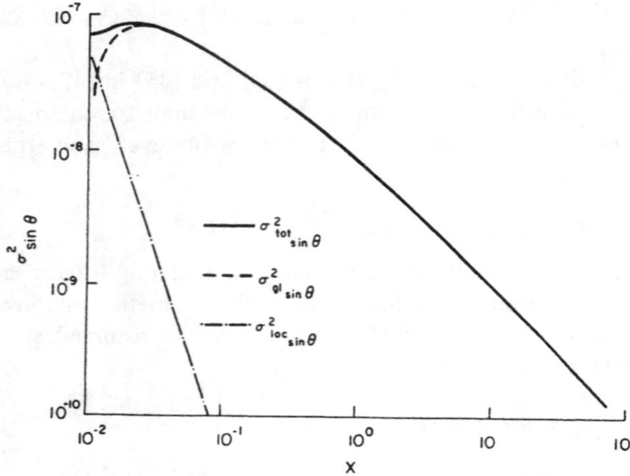
Fig. 10. Total bearing mean-square error versus $X = L/2\,R_0$.

Cramer-Rao bounds and approach the global mean-square error components, with a logarithm-type decaying.

Being independent of the statistical background, the behavior observed is the sole reflexion of the geometry on the performance. In fact the following is true:

1) The increase in X sharpens the main lobe; maintaining the *a priori* uncertainty, it enlarges the total number of grid cells augmenting Pr (ξ).

2) A change in SNR affects only the scaling of the ordinate axis, or, equivalently, just implies a parallel translation of the curves. This pattern of variation is characteristic of the Rayleigh model assumed and of the two-step algorithm. In practice, as it happens with active radar, the problem is circumvented by resorting to multiple independent observations (e.g., finite coherence time); the estimation procedure described is then efficient in the SNR sense.

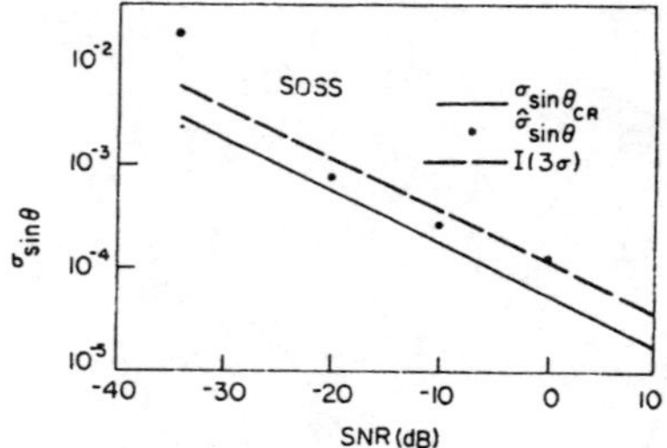
Fig. 11. Range Monte Carlo simulation results.

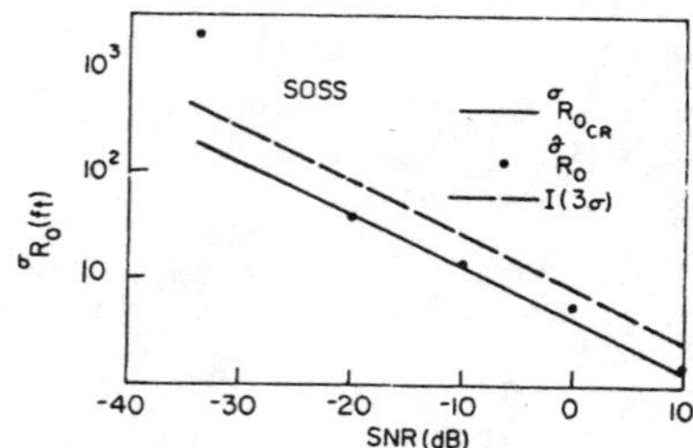
Fig. 12. Bearing Monte Carlo simulation results.

The stationary observer/stationary source passive-tracking problem was simulated in a digital computer, for a configuration where

$$R_0 = 6 \times 10^3$$

$$\theta = 0^\circ$$

$$\lambda = 50 \text{ ft}$$

$$\sigma_b{}^2 = 1$$

$$X = .331$$

$$T = 1 \text{ s}$$

and the inter-spacing between the array elements was $\Delta L = \lambda/2$.

For several SNR values, Figs. 11 and 12 study the convergence between the theoretical performance and the results of Monte Carlo simulations. They exhibit a threshold phenomenon, since for SNR > -30 dB the statistical results are within the three-variance confidence interval $I\,(3\sigma)$ of the Cramer-Rao bounds, while the point SNR $= -34$ dB is about one order of magnitude apart from the Cramer-Rao bound.

VII. CONCLUSION

A model has been presented encompassing the global and local characteristics of the passive-tracking problem. It exhibits global observability for both the range and the bearing. The parameters are demodulated from the wavefront curvature, e.g., as sensed by a spatially extended observer. Underlying all the work presented is the absence of the far-field (planar wavefront) assumption.

The ambiguity structure and the theoretical and practical limitations on the mean-square performance have been analyzed. For a distant observer geometry (Fresnel zone) the estimation errors are practically uncoupled; the statistical mean-square performance follows closely the values predicted by the Cramer–Rao bounds. The bearing estimation depends on the linear effects (e.g., linear delay across the array or observed Doppler), while the range focusing is achieved from the second-order modulations (spherical curvature or chirp modulation). As the geometry changes from broadside ($\theta = 0°$) to endfire ($\theta = \pi/2$), the range performance deteriorates due to the reduction of the effective array length. For larger values of X, the two-step implementation of the ML receiver exhibits threshold effects, with the global errors dominating the mean-square error performance.

The work was not restricted to high SNR's, and remains valid even for large *a priori* uncertainty. However, the Monte Carlo simulations suggested a lower bound on the SNR, around –30 dB, below which the analysis is significantly misleading, i.e., the bounds are not tight.

The model assumed a highly coherent signal disturbed by incoherent noise components. Nevertheless, the results reflecting the geometry on the mean-square performance and the receiver structure are still valid for more general signal models [12]. This is true whenever the statistical and the geometrical effects factor as in (11).

In Part II [14] an omnidirectional observer acquires a source moving along a deterministic path. The range, speed, and bearing defining the source/receiver geometry are measured by exploring the temporal diversity induced on the received signals. In Part III [15] configurations are considered, where a spatially extended observer tracks a moving source. The paper concentrates on the study of the resulting space/time coupled processing.

REFERENCES

[1] R. W. Bass, R. E. Mortensen, V. D. Norum, B. Shawaf, and H. W. Sorenson, "ASW target motion and measurement models," CSA Tech. Rep., 72-024-01, Sept. 1972.

[2] Beckmann and Spizzichino, *Scattering of Electromagnetic Waves from Rough Surfaces.* New York, Pergamon, 1963.

[3] A. B. Baggeroer, "High resolution velocity/depth spectra estimation for seismic profiling," in *IEEE Int. Conf. Eng Ocean Environ.*, vol. II, 1974.

[4] W. J. Bangs, "Array processing with generalized beam-formers," doctoral thesis, Yale Univ., New Haven, CN, Sept. 1971.

[5] W. J. Bangs and P. M. Schultheiss, "Space-time processing for Optimal Parameter Estimation," in *Signal Processing*, J. W. R. Griffiths, P. L. Stocklin, and C. Van Schooneveld, Eds. New York: Academic Press, 1973.

[6] L. S. Cahoon and M. J. Hinich, "A method for locating targets using range only," *IEEE Trans. Inform. Theory*, vol. IT-22, pp. 217–225, Mar. 1976.

[7] M. A. Gallop and L. W. Nolte, 'Bayesian detection of targets of unknown location," *IEEE Trans. Aerosp. Electron. Syst.*, vol. AES-10, pp. 429-435, 1974.

[8] *Proc. IEEE, Special Issue on Rays and Beams*, vol. 62, pp. 1409–1618, Nov. 1974.

[9] *IEEE Trans. Geosci. Electron., Special Issue on Data Collection from Multiple Earth Platforms*, vol. GE-13, Jan. 1975.

[10] K. D. McDonald, "A survey of satellite-based systems of navigation, position surveillance, traffic control and collission avoidance," *Navigation: J. Inst. Navigation*, vol. 20, no. 4, Winter, 1973–74.

[11] T. P. McGarty, "The effect of interfering signals on the performance of angle of arrival estimates," *IEEE Trans. Aerosp. Electron. Syst.*, vol. AES-10, pp. 70–77, 1974.

[12] J. M. F. Moura, "Passive systems theory with applications to positioning and navigation," doctoral thesis, M. I.T., June 1975. —, R.L.E. Rep. no. 490, M.I.T., Cambridge, MA, Apr. 28, 1976.

[13] J. M. F. Moura, H. L. Van Trees, and A. B. Baggeroer, "Space time tracking by a passive observer," 4th Symp. on Nonlinear Estimation Theory and Its Applications, San Diego, CA., Sept. 1973.

[14] J. M. F. Moura, "Passive systems theory with narrow-band and linear constraints. Part II: Temporal diversity," to be submitted for publication.

[15] J. M. F. Moura, "Passive systems theory with narrow-band and linear constraints. Part III: Spatial/temporal diversity," to be submitted for publication.

[16] A. Papoulis, *Systems and Transforms with Applications in Optics.* New York: McGraw-Hill, 1967.

[17] L. P Seidman, "Bearing estimation with a linear array," *IEEE Trans. Audio and Electroacoust.*, vol. AU-19, pp. 147–157, 1971.

[18] R. C. Spindel and R. P. Porter, "Precision tracking systems for sonobuoys," in *IEEE Int. Conf. Eng. Ocean Environ.*, vol. II, 1974.

[19] M. T. Taner and F. Koehler, "Velocity spectra-digital computer derivation and applications of velocity function," *Geophys*, vol. 34, no. 6, pp. 859–881, Dec. 1969.

[20] H. L. Van Trees, *Detection, Estimation and Modulation Theory: Part II.* New York: Wiley, 1971.

[21] E. E. Westerfield, "Determination of position of a drifting buoy by means of the Navy navigation satellite system," in *IEEE Conf. Eng. Ocean Environ.*, pp. 443–446, 1972.

Multisensor Multitarget Time Delay Vector Estimation

LAWRENCE C. NG, SENIOR MEMBER, IEEE, AND YAAKOV BAR-SHALOM, FELLOW, IEEE

***Abstract*—This paper presents the results of an investigation of the optimum and suboptimum signal processors for passive time delay vector estimation in a multisensor, multitarget environment. Target-sensor geometry is assumed stationary over the observation interval and the signal spectra are assumed known. Using the maximum likelihood estimation (MLE) approach, the optimum time delay vector estimator is derived and its performance analyzed in terms of the Cramer–Rao lower bound (CRLB). Analytical closed-form expressions are obtained for the case of two targets with two sensors and one target with *M* sensors. The bias and variance characteristics of a conventional time delay processor in a multitarget environment are discussed; also presented are performance comparisons between the optimum and the conventional generalized cross-correlation (GCC) processor. In addition, a novel, suboptimum multitarget multisensor postcorrelation processor (MMPCP) is proposed. Its performance has been analyzed and verified through extensive simulation.**

The study of the optimum processor realization shows that it requires a coupled, multichannel processor. The required implementation is considerably more complex than the conventional GCC approach. The examination of the post-GCC multitarget processing capability shows that, in general, the MMPCP shows a marked improvement over the GCC estimator and that the performance degradation is acceptable with respect to the optimum processor.

I. Introduction

In passive sonar signal processing, a basic variable being measured is the time difference of arrival (TDOA) or, simply, time delay between two sensor arrays. Since time delay from a target to a group of sensors is related to the target-sensor geometry, then the true parameters of interest, for example, the target location parameters, can be obtained from a set of time delay measurements.

Time delay measurement has been mechanized through the calculation of a GCC function [1]–[4]. The GCC is derived as an MLE operating on a finite data sequence. Another approach is based on a combined ARMA signal modeling and MLE [5], [6].

In the literature, the GCC is formulated under a single target assumption. Therefore, in the presence of multiple targets, there exist multiple correlation peaks. The interference of multiple correlation peaks causes performance degradation to the GCC processor. The extent of this degradation is a function of signal spectral characteristics, signal-to-noise ratio (SNR), signal-to-interference ratio (SIR), and the relative time delay separation.

Our study of the multisensor, multitarget signal processor is motivated by 1) the increased interest in tracking multiple targets [7], and 2) the improved ranging and bearing estimation capability using multiple sensors with long baseline [8]. In addition, the improvement in advanced technology both in sensor and computation also calls for a reexamination of the design of the optimum signal processor to handle a more complex environment such as that involving multiple sensors and targets.

In the literature [9]–[12], interference studies deal primarily with the optimum signal processing in the spatial and frequency domain. This paper, however, discusses the interference problem in the time delay variable. We believe that a detailed study of the general time delay vector estimation process is justified for reasons stated below.

Because the general multisensor, multitarget processor is very complex, this study proposes a two-step approach: namely, first obtain the time delay vector estimates, and second, map the estimated time delays into geometry parameters using the geometry constraint (usually nonlinear). The advantages of this proposed approach are as follows.

1) For a distortionless channel, the correlation (or ambiguity) functions are always symmetrical in the time delay variables. Functional symmetry not only will facilitate the design of automatic detection and tracking algorithms, but also make the estimation process more robust and linear when operating at or near the CRLB. Thus, performing detection and parameter estimation in the time delay variables yields a gain in additional operating range by avoiding nonlinearity losses as is the case for the direct, geometry parameter estimation approach.

2) Because the ambiguity function is symmetrical in the time delay variables, the statistics of the resulting estimates are also symmetrically distributed; whereas, in the direct geometry parameter case, the statistics are skewed. Since these measurements are, in general, further processed for target state estimation, skewed measurement statistics will degrade the state estimation performance.

3) Finally, time delay vector estimation is a more complex problem to analyze because it involves higher di-

Manuscript received May 21, 1984; revised December 20, 1985. This work was performed under the auspices of the U.S. Department of Energy by the Lawrence Livermore National Laboratory under Contract W-7405-ENG-48. The work of Y. Bar-Shalom was supported by NAVELEX under Contract N00039-85-C-0636.

L. C. Ng was with the Naval Underwater Systems Center, New London, CT. He is now with Lawrence Livermore National Laboratory, University of California, Livermore, CA 94550.

Y. Bar-Shalom is with the Department of Electrical Engineering and Computer Science, University of Connecticut, Storrs, CT 06268.

IEEE Log Number 8608138.

Reprinted from *IEEE Trans. Acoust., Speech, Signal Processing,* vol. 34, no. 4, pp. 669–678, August 1986.

mensionality. Therefore, further understanding of the general time delay vector estimation case will yield further insight to the internal structure and the applications of the general multisensor, multitarget processor.

Thus, the following approach is taken. First, the optimum multisensor, multitarget time delay vector processor and the performance bound are presented in Sections II and III. Then various realizations of the optimum processor are considered in Section IV. A number of suboptimum approaches is described in Section V. The performance of the optimum processor is compared to that of the new and other suboptimum processors in Section VI. Finally, we summarize the findings of this study in Section VII.

II. Problem Formulation

Let the M sensor array output waveforms from J acoustic sources be written as

$$y_i(t) = \sum_{j=1}^{J} a_{ij} s_j(t + D_{ij}) + n_i(t); \qquad t \in [0, T], \quad (1)$$

where $i = 1, 2, \cdots, M$; a_{ij} is the known signal attenuation factor for the jth target to the ith sensor; and D_{ij} is the propagation time delay from target j to sensor i; $s_j(t)$ and $n_j(t)$ are band-limited signal and noise processes—zero-mean, Gaussian, mutually uncorrelated, stationary in time, and homogeneous in space. In addition, we assume that the observation time T is large compared to the wavefront transit time across all sensors. Under these assumptions, the Fourier representation of (1) from T seconds of observation can be written as [13]

$$\boldsymbol{\alpha}_k = \sum_{j=1}^{J} \beta_{kj} \boldsymbol{v}_{kj} + \boldsymbol{\eta}_k; \qquad k = 1, 2, \cdots, B \quad (2a)$$

where

$$\boldsymbol{\alpha}_k = (\alpha_{k1}\alpha_{k2} \cdots \alpha_{kM})^T \quad (2b)$$

$$\boldsymbol{v}_{kj} = (\alpha_{1j}\, e^{j\omega_k D_{1j}} a_{2j}\, e^{j\omega_k D_{2j}} \cdots a_{Mj}\, e^{j\omega_k D_{Mj}})^T \quad (2c)$$

$$\boldsymbol{\eta}_k = (\eta_{k1}\eta_{k2} \cdots \eta_{kM})^T, \quad (2d)$$

and B is the highest frequency component of either the signal or the noise processes, $\omega_k = 2\pi k/T$ denotes the kth discrete frequency component and $(\ )^T$ denotes transposition. Note that for a real input waveform, the Fourier coefficients are conjugate symmetric, i.e., $\alpha_{-ki} = \alpha_{ki}^*$. Therefore, for a zero-mean band-limited process, only the positive frequency components need be considered. For a sufficiently long observation time interval, it has been shown that α_k is a complex, zero-mean, Gaussian vector uncorrelated for different discrete frequencies [14].

The probability density function (pdf) of α_k conditioned on the time delay matrix $D = \{D_{ij}\}$; $i = 1, 2, \cdots, M$; $j = 1, 2, \cdots, J$ can be written as [15]

$$p(\boldsymbol{\alpha}_k|D) = \pi^{-M}|R_k|^{-1} \exp\{-\boldsymbol{\alpha}_k^* R_k^{-1} \boldsymbol{\alpha}_k\}. \quad (3)$$

The pdf of the complete observation vector $\boldsymbol{\alpha} = (\boldsymbol{\alpha}_1, \boldsymbol{\alpha}_2, \cdots, \boldsymbol{\alpha}_B)^T$ is

$$p(\boldsymbol{\alpha}|D) = \pi^{-MB} \prod_{k=1}^{B} |R_k|^{-1} \exp\{-\boldsymbol{\alpha}_k^* R_k^{-1} \boldsymbol{\alpha}_k\}, \quad (4)$$

where it can be shown that $E(\boldsymbol{\alpha}_k) = 0$, $E(\boldsymbol{\alpha}_k \boldsymbol{\alpha}_l) = R_k \delta_{kl}$, where δ_{kl} is the Kronecker delta function. The covariance matrix R_k is given by

$$R_k = \sum_{j=1}^{J} S_{kj} \boldsymbol{v}_{kj} \boldsymbol{v}_{kj}^* + N_k Q_k, \quad (5)$$

where S_{kj} is the discrete signal power spectrum, N_k is the noise power spectrum, and Q_k is the normalized noise covariance matrix. Now denote the mn element of the matrix $P_{kj} = \boldsymbol{v}_{kj} \boldsymbol{v}_{kj}^*$ as P_{kj}^{mn}; then from (2c) we obtain

$$P_{kj}^{mn} = a_{mj} a_{nj} \exp\{-j\omega_k(D_{nj} - D_{mj})\}. \quad (6a)$$

But the time delay difference can be written as

$$D_{nj} - D_{mj} = (D_{nj} - D_{n-1,j}) + (D_{n-1,j} - D_{n-2,j}) + \cdots + (D_{m+1,j} - D_{m,j})$$
$$= \tau_{n-1,j} + \tau_{n-2,j} + \cdots + \tau_{mj}, \quad (6b)$$

where $\tau_{ij} = D_{i+1,j} - D_{i,j}$ is defined as the intersensor time delay. Since the time delay between any two sensors can be written as a linear combination of these time delays, then given M sensors and J targets, one could consider that there is a total of $J(M - 1)$ independent time delay pairs from a possibility of $JM(M - 1)/2$. Let $\boldsymbol{\tau} = (\boldsymbol{\tau}_1^T, \boldsymbol{\tau}_2^T, \cdots, \boldsymbol{\tau}_J^T)^T$ where $\boldsymbol{\tau}_j = (\tau_{1j}, \tau_{2j}, \cdots, \tau_{(M-1)j})^T$ be the vector of $P = J(M - 1)$ independent time delays, then from (4) the log likelihood function can be written as

$$\Lambda(\boldsymbol{\tau}) = -MB \log(\pi) - \sum_{k=1}^{B} \Lambda_k(\tau) \quad (7a)$$

where

$$\Lambda_k(\boldsymbol{\tau}) = \log|R_k| + \boldsymbol{\alpha}^* R_k^{-1} \boldsymbol{\alpha}_k. \quad (7b)$$

The MLE of $\boldsymbol{\tau}$ is obtained by finding $\hat{\boldsymbol{\tau}}$ such that $\Lambda(\boldsymbol{\tau})$ is maximum. The necessary condition for the location of the maximum is

$$\nabla\Lambda(\boldsymbol{\tau}) = -\sum_{k=1}^{B} \nabla\Lambda_k(\boldsymbol{\tau}) = \mathbf{0}, \quad (8)$$

where ∇ is the gradient operator.

III. The MLE Solution

In principle, the MLE of $\boldsymbol{\tau}$ is obtained by solving (8) numerically. However, in order to investigate and simplify the structure of the resulting signal processor, (8) must be reduced to its simplest form. This has been done [13], and the resulting necessary condition is given by

$$Z_{ij} = \sum_{k=1}^{B} j\omega_k [|\tilde{h}_{kj}|^2 \boldsymbol{\alpha}_{kj}^* \tilde{Q}_{kj}^{-1} V_{kj}(\boldsymbol{\Phi}_i - \tilde{b}_i^{kj} 1_M) \cdot V_{kj}^* \tilde{Q}_{kj}^{-1} \boldsymbol{\alpha}_k - \tilde{b}_i^{kj}] = 0, \quad (9a)$$

where $Z_{ij} = \partial\Lambda(\boldsymbol{\tau})/\partial\tau_{ij}$, 1_M is an $M \times M$ matrix of 1's, V_{kj} is a diagonal matrix whose elements correspond to the elements of the vector v_{kj}, and

$$\tilde{b}_i^{kj} = \tilde{a}_{kj} \text{ tr } (\tilde{Q}_{kj}^{-1} V_{kj} \Phi_i V_{kj}^*) \tag{9b}$$

is the bias correction term and for $i = 1, 2, \cdots, M - 1$ and $j = 1, 2, \cdots, J$. Note that the $J(M - 1)$ equations from (9a) are coupled and must be solved jointly for the stationary point. Furthermore, it can be shown [13] that the bias correction terms in (9b) reduce to zero only for the single target case. For sufficiently long observation time such that the frequency samples are dense over the frequency bands of the signal and noise, the summation in (9a) can be replaced by integration.

In addition, parameters given in (9a) can be written in continuous form as follows:

$$\Phi_i^{mn} = \begin{cases} 1; & \text{if } n \leq i \leq m - 1 \\ -1; & \text{if } m \leq i \leq n - 1 \\ 0; & \text{otherwise} \end{cases} \tag{10a}$$

$$|\tilde{h}_j(\omega)|^2 = \frac{S_j(\omega)/\tilde{N}_j^2(\omega)}{1 + G_j(\omega)\, S_j(\omega)/\tilde{N}_j(\omega)} \tag{10b}$$

$$\tilde{a}_j(\omega) = \tilde{N}_j(\omega)\, |\tilde{h}_j(\omega)|^2 \tag{10c}$$

$$\tilde{b}_i^j(\omega) = \tilde{a}_j(\omega) \text{ tr } \{\tilde{Q}_j^{-1}(\omega)\, V_j(\omega)\, \Phi_i(\omega)\, V_j^*(\omega)\} \tag{10d}$$

$$\tilde{Q}_j^{-1}(\omega) = \tilde{N}_j(\omega) \left(\sum_{\substack{i=1 \\ i \neq j}}^{J} S_i(\omega)\, P_i(\omega) + N(\omega)\, Q(\omega) \right)^{-1} \tag{10e}$$

where

$$\tilde{N}_j(\omega) = \sum_{\substack{i=1 \\ i \neq j}}^{J} S_i(\omega) + N(\omega) \tag{10f}$$

$$P_j(\omega) = V_j(\omega)\, 1_M\, V_j^*(\omega) \tag{10g}$$

and

$$G_j(\omega) = v_j^*(\omega)\, \tilde{Q}_k^{-1}(\omega)\, v_j(\omega). \tag{10h}$$

Let τ_i be the ith element of the time delay vector $\boldsymbol{\tau}$, then the associated CRLB is given by [13], [16]

$$\text{var } (\hat{\tau}_i) \geq (J^{-1})_{ii}$$

$$\geq \left[\sum_{k=1}^{B} \text{tr} \left(-\frac{\partial R_k^{-1}}{\partial \tau_j} \frac{\partial R_k}{\partial \tau_i} \right)_{\tau = \tau_0} \right]_{ii}^{-1}$$

$$\text{for } i = 1, 2, \cdots, P, \tag{11}$$

where J is the Fisher's information matrix and $\boldsymbol{\tau}_0$ is the true value of $\boldsymbol{\tau}$.

IV. Optimum Processor Realization

From (9a), the optimum multisensor, multitarget time delay signal processor can be realized and is shown in Fig. 1. There are a total of $J(M - 1)$ parallel processing channels for the general case of M sensors and J targets. For simplicity, only one processing channel is shown. Note that the processing channels are tightly coupled. The signal conditioning filters depend on the time delay parameters from other processing channels as well. This is an order of magnitude more complex compared to the single target case. A number of suboptimum realizations can be found whose performance can be evaluated with respect to the optimum processor as will be shown in later sections. However, in order to gain further insight into the specific structure of the optimum time delay vector processor, we shall present results for a number of relatively simple but important cases.

Case 1: One Target and Two Sensors ($J = 1$, $M = 2$)

For convenience, we assume $Q(\omega) = I$; i.e., the noise processes are equal in power and uncorrelated between sensors. Noise processes being uncorrelated between sensors is a reasonable assumption since in practice sensors are separated at least by a half-wavelength spacing.

The steering vector is $v = (1\ e^{j\omega\tau})^T$. The following relations can be verified easily:

$$\tilde{Q}^{-1}(\omega) = I,\ G(\omega) = v^*(\omega)\, \tilde{Q}^{-1}(\omega)\, v(\omega) = 2,$$

$$b(\omega) = \text{tr } \{\tilde{Q}^{-1}(\omega)\, v(\omega)\, \Phi_1(\omega)\, V^*(\omega)\} = 0$$

$$|\tilde{h}(\omega)|^2 = \frac{S(\omega)/N^2(\omega)}{1 + 2S(\omega)/N(\omega)}$$

and

$$\Phi_1 = \begin{bmatrix} 0 & -1 \\ 1 & 0 \end{bmatrix}; \quad V = \begin{bmatrix} 1 & 0 \\ 0 & e^{j\omega\tau} \end{bmatrix}. \tag{12}$$

Using the above relations, (9a) can be written as

$$Z(\tau) = \frac{1}{2\pi} \int_0^\infty j\omega |\tilde{h}(\omega)|^2\, \alpha_1(\omega)\, \alpha_2^*(\omega) \exp\{j\omega\tau\}\, d\omega$$

$$= \frac{\partial}{\partial \tau} R_{12}(\tau) = 0 \tag{13}$$

where

$$R_{12}(\tau) = \int_0^\infty |\tilde{h}(\omega)|^2\, \alpha_1(\omega)\, \alpha_2(\omega) \exp\{j\omega\tau\} \frac{d\omega}{2\pi}, \tag{14}$$

is the GCC function studied extensively in the literature [1], [2]. The optimum estimate is determined by locating the peak of the GCC function or, equivalently, the corresponding null of its derivative. Fig. 2 presents the standard realization of this processor. The CLRB of this estimator can be determined from (11) as follows:

$$\text{var } (\hat{\tau}) \geq \left[\sum_{k=1}^{B} \text{tr} \left(-\frac{\partial R_k^{-1}}{\partial \tau} \frac{\partial R_k}{\partial \tau} \right) \right]^{-1} \tag{15a}$$

where

$$\frac{\partial R_k}{\partial \tau} = j\omega_k S_k V_k \Phi_1 V_k^* \tag{15b}$$

$$\frac{\partial R_k^{-1}}{\partial \tau} = -j\omega_k |\tilde{h}_k|^2\, V_k \Phi_1 V_k^*, \tag{15c}$$

and therefore,

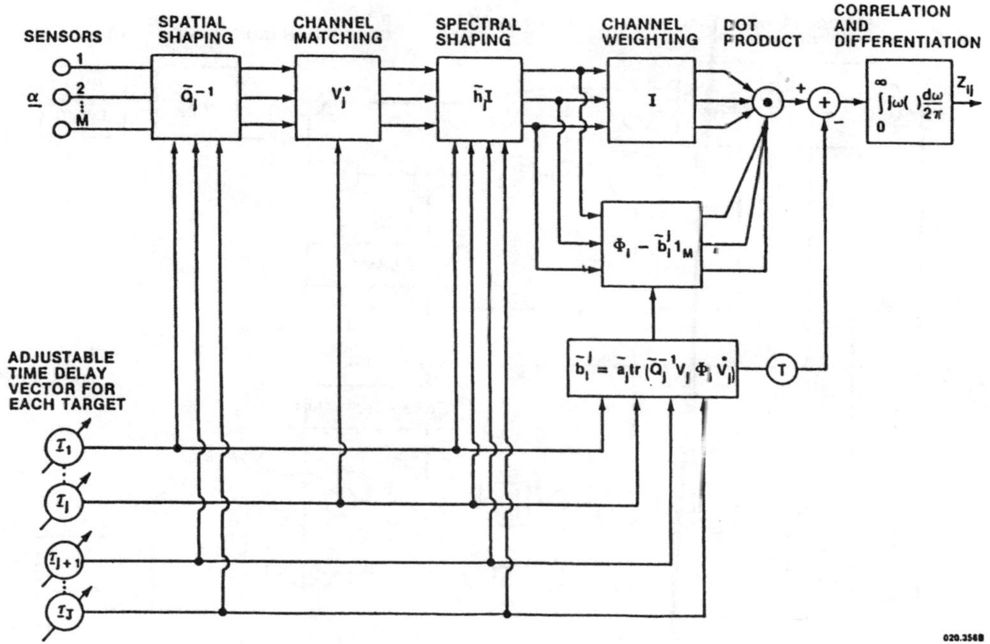

Fig. 1. A multisensor, multitarget time delay processing channel.

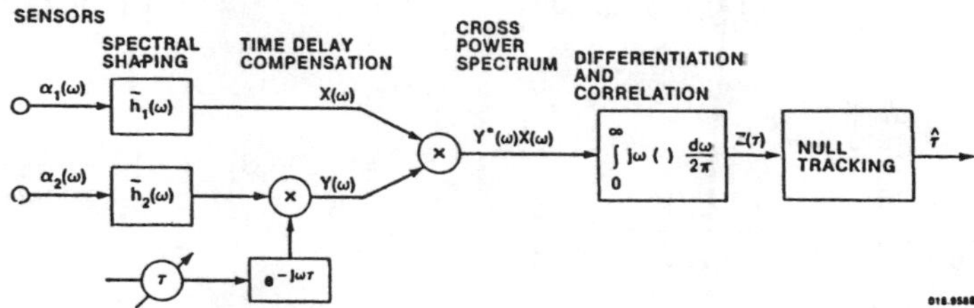

Fig. 2. Optimum two-sensor one-target time delay processor.

$$\operatorname{tr}\left(-\frac{\partial R_k^{-1}}{\partial \tau}\frac{\partial R_k}{\partial \tau}\right) = 2\omega_k^2|\bar{h}_k|^2 S_k. \tag{15d}$$

Thus, the CRLB is

$$\operatorname{var}(\hat{\tau}) \geq \left(2\sum_{k=1}^{B}\omega_k^2\frac{S_k^2/N_k^2}{1+2S_k/N_k}\right)^{-1} \tag{15e}$$

$$\geq 2\pi\left(2T\int_0^\infty \frac{C(\omega)}{1-C(\omega)}\omega^2\,d\omega\right)^{-1}, \tag{15f}$$

which is identical to the expression obtained by Carter [1] and others. Note that T is the observation time and $C(\omega) = S^2(\omega)/(S(\omega)+N(\omega))^2$ is known as the magnitude square coherence (MSC) function.

Case 2: Two Targets and Two Sensors ($J = 2$, $M = 2$)

Again assume $Q(\omega) = I$ for simplicity. The parameter vector consists of two elements; i.e., $\Theta = (\tau_1, \tau_2)^T$, the time delays to target number one and target number two. From (9a) the optimum estimates can be obtained by solving simultaneously the two likelihood equations as follows:

$$\frac{\partial \Lambda(\tau_1, \tau_2)}{\partial \tau_1} = \int_0^\infty j\omega[|\bar{h}_1|^2\,\alpha^*\tilde{Q}_1^{-1}V_1(\Phi_1 - \bar{b}_1 1_M) \cdot V_1^* Q_1^{-1}\alpha - T\bar{b}_1]\frac{d\omega}{2\pi} = 0 \tag{16a}$$

$$\frac{\partial \Lambda(\tau_1, \tau_2)}{\partial \tau_2} = \int_0^\infty j\omega[|\bar{h}_2|^2\,\alpha^*\tilde{Q}_2^{-1}V_2(\Phi_2 - \bar{b}_2 1_M) \cdot V_2^*\tilde{Q}_2^{-1}\alpha - T\bar{b}_2]\frac{d\omega}{2\pi} = 0. \tag{16b}$$

The optimum two-target two-sensor time delay processor derived from (16a) and (16b) is shown in Fig. 3. It can be seen that the optimum processor is considerably more complex than the GCC derived from a single target assumption.

Using (11), and after some straightforward but tedious algebraic manipulations, the CRLB of the time delay estimates for the two-target two-sensor case can be written as follows [13], [17]:

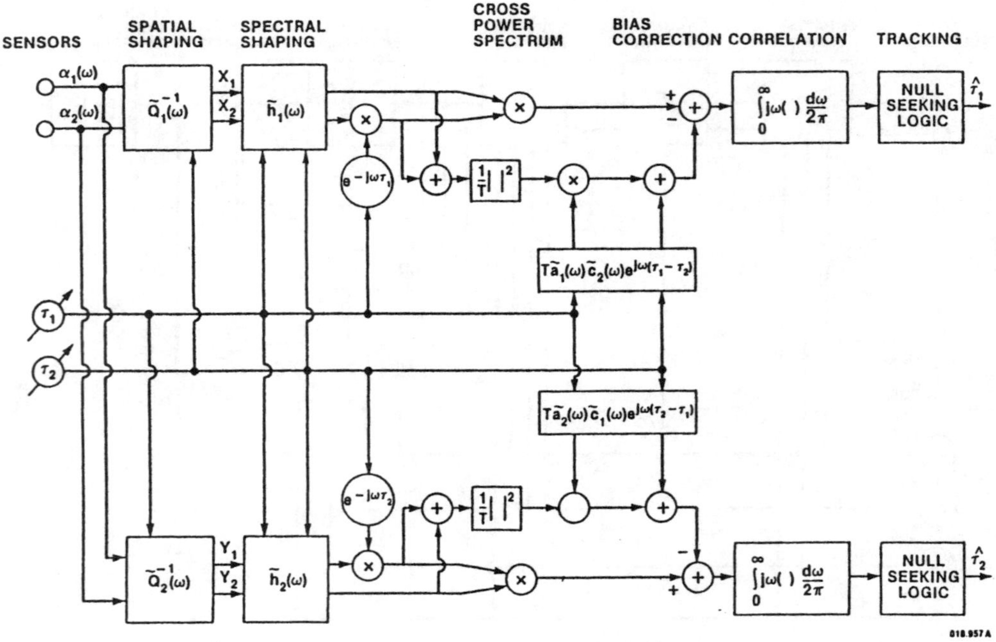

Fig. 3. Optimum two-sensor two-target time delay processor.

$$\mathrm{var}\,(\hat{\tau}_1) \geq \frac{2\pi}{(1 - M_{12}^2)} \left\{ 2T \int_0^\infty \omega^2 \left[\frac{(S_1/N)^2}{1 + G_1 S_1/(S_2 + N)} \right] \cdot \left[\frac{1 + S_2/N}{1 + 2S_2/N} \right]^2 \gamma_1 \, d\omega \right\}^{-1} \tag{17a}$$

$$\mathrm{var}\,(\hat{\tau}_2) \geq \frac{2\pi}{(1 - M_{12}^2)} \left\{ 2T \int_0^\infty \omega^2 \left[\frac{(S_2/N)^2}{1 + G_2 S_2/(S_1 + N)} \right] \cdot \left[\frac{1 + S_1/N}{1 + 2S_1/N} \right]^2 \gamma_2 \, d\omega \right\}^{-1} \tag{17b}$$

where M_{12} is the coefficient of mutual dependence and is related to the elements of the Fisher's information matrix; i.e., $M_{12} = J_{12}/\sqrt{(J_{11}J_{22})}$. Quantities S_1, N, S_2, G_1, G_2, γ_1, and γ_2 are all function of frequency and are given by

$$G_1 = (1 + S_2/N) \left[2 - \left(\frac{S_2/N}{1 + 2S_2/N} \right) |v_1^* v_2|^2 \right] \tag{18a}$$

$$G_2 = (1 + S_1/N) \left[2 - \left(\frac{S_1/N}{1 + 2S_1/N} \right) |v_2^* v_1|^2 \right] \tag{18b}$$

$$\gamma_1 = 1 - \left(\frac{S_2/N}{1 + S_2/N} \right)^2 \sum_{n=0}^{3} A_n \cos^n \omega\Delta_{12} \tag{18c}$$

$$\gamma_2 = 1 - \left(\frac{S_1/N}{1 + S_1/N} \right)^2 \sum_{n=0}^{3} B_n \cos^n \omega\Delta_{12}, \tag{18d}$$

where $\Delta_{12} = \tau_1 - \tau_2$ is the time delay separation between targets 1 and 2. Coefficients A_n and B_n are further defined by the following:

$$A_0 = 1 - A_2 = 4 \cdot \left[\frac{S_1/N}{1 + G_1 S_1/(S_2 + N)} \right] \left[\frac{1 + S_2/N}{1 + 2S_2/N} \right] - 1 \tag{18e}$$

$$A_1 = -A_3 = -4 \left[\frac{S_1/N}{1 + G_1 S_1/(S_2 + N)} \right] \left[\frac{S_2/N}{1 + 2S_2/N} \right] \tag{18f}$$

$$B_0 = 1 - B_2 = 4 \left[\frac{S_2/N}{1 + G_2 S_2/(S_1 + N)} \right] \left[\frac{1 + S_1/N}{1 + 2S_1/N} \right] \tag{18g}$$

$$B_1 = -B_3 = -4 \left[\frac{S_2/N}{1 + G_2 S_2/(S_1 + N)} \right] \left[\frac{S_1/N}{1 + 2S_1/N} \right]. \tag{18h}$$

For the purpose of illustrating the two-sensor two-target CRLB performance behavior, we assume that signal (S_1), interference (S_2), and noise (N) processes have identical power spectral distribution. Fig. 4 shows the frequency characteristics of the multitarget shaping function $\gamma_1(\omega)$. Two effects are apparent: 1) the function $\gamma_1(\omega)$ reduces the low-frequency band contribution to the CRLB for increasing interference-to-signal ratio (ISR); 2) $\gamma_1(\omega)$ is a monotonically increasing function with frequency and approaching unity at upper frequency bands. The effect of reducing the low-frequency band contribution as ISR increases is to increase the CRLB for the signal parameter. This is shown in Fig. 5 where the degradation ratio (signal CRLB normalized by the single target case) is plotted as a function of time delay separation (in units of inverse bandwidth) for a number of ISR values. The degradation ratio is oscillatory as a function of time delay separation. Two major peaks are observed. The first occurs when time delay separation approaches zero and the other occurs at 4 times the inverse bandwidth. Note that the first peak goes to infinity at zero while the second decreases as ISR decreases. Thus, we observe that targets with identical

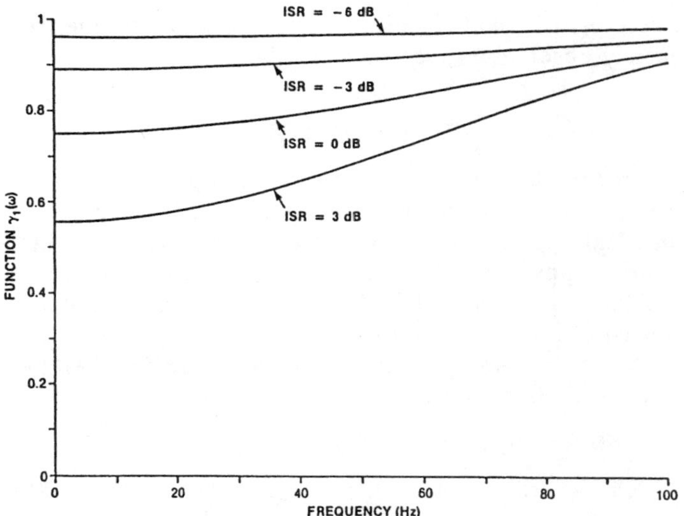

Fig. 4. Frequency response of the function $\gamma_1(\omega)$ at different interference-to-signal ratios (SNR = 0 dB).

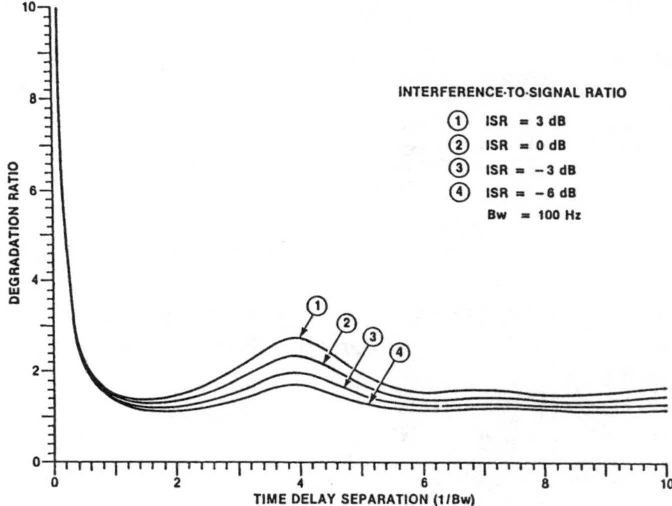

Fig. 5. Two-sensor two-target Cramer–Rao lower bound degradation ratio versus time delay separation (SNR = 0 dB).

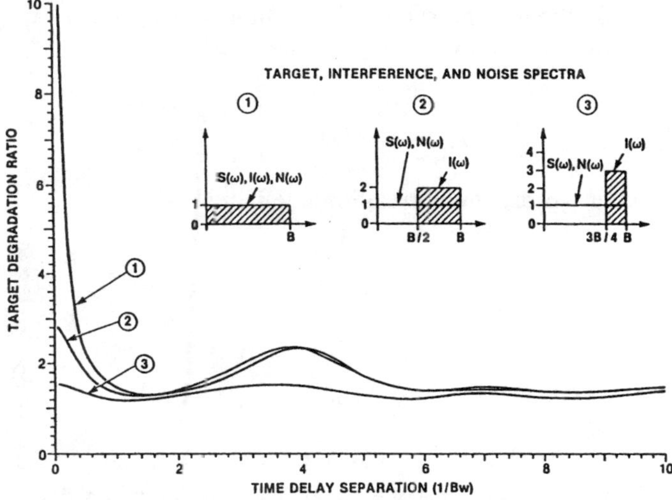

Fig. 6. Two-sensor two-target CRLB degradation ratio versus time delay separation and interference spectrum (SNR = 0 dB, INR = 0 dB).

spectra and spatial location yield the worst estimate since they are not separable in frequency or in time delay space. In order to see the effect of different signal and interference spectra on the resulting estimate, we fix the ISR but vary the interference spectrum. We also choose a noise bandwidth significantly larger than that in the previous example. The results are presented in Fig. 6. Two observations can be made: 1) degradation ratio is no longer infinite at zero time delay separation with nonidentical signal and interference spectra; and 2) comparison between Figs. 5 and 6 shows that the degradation ratio as a function of time delay separation is bandwidth independent.

Case 3: One Target and M Sensors (J = 1, M = M)

For M sensors and with $Q(\omega) = I$, the parameter set consists of $M - 1$ independent time delays. For convenience, we select the intersensor time delays as the parameter set of interest. Denoting this parameter set by $\Theta = (\tau_1, \tau_2, \cdots)^T$, then the optimum estimate of this set of time delay parameters is given by the $M - 1$ likelihood equations

$$\frac{\partial \Lambda(\Theta)}{\partial \tau_i} = \int_0^\infty j\omega\{|h|^2 \,\boldsymbol{\alpha}^* Q^{-1} V[\Phi_i - \tilde{b}_i 1_M] \cdot V^* Q^{-1} \boldsymbol{\alpha} - T\tilde{b}_i\} \frac{d\omega}{2\pi}$$

$$= \int_0^\infty j\omega |h|^2 \,\boldsymbol{\alpha}^* V \Phi_i V^* \boldsymbol{\alpha} \frac{d\omega}{2\pi}$$

$$= 0; \qquad \text{for } i = 1, 2, \cdots, M - 1 \tag{19}$$

where the relations $\tilde{b}_i = 0$, $Q^{-1} = I$ have been used and from (10b) one obtains

$$|h(\omega)|^2 = \frac{S(\omega)/N^2(\omega)}{1 + MS(\omega)/N(\omega)}. \tag{20}$$

The implementation of (19) yields a focused beamformer realization [13].

The CRLB of each element of the time delay vector is identical [13] and is given by

$$\text{var}(\hat{\tau}_i) \geq \left[M \sum_{k=1}^{B} \omega_k^2 \frac{S_k^2/N_k^2}{1 + MS_k/N_k}\right]^{-1}; \qquad \text{for } i = 1, 2, \cdots, M - 1, \tag{21a}$$

or in continuous form

$$\text{var}(\hat{\tau}_i) \geq 2\pi\left[MT \int_0^\infty \frac{S^2(\omega)/N^2(\omega)}{1 + MS(\omega)/N(\omega)} \omega^2 \, d\omega\right]^{-1}. \tag{21b}$$

The time delay CRLB between any two adjacent sensors improves with increased M, the total number of sensors, and T, the observation time. Furthermore, it can be shown [13] that the time delay estimate between any two sensors also has the same CRLB. In addition, the covariance of any two time delay estimates is

$$\operatorname{cov}(\hat{\tau}_i, \hat{\tau}_j) = \begin{cases} -\frac{1}{2} \operatorname{var}(\hat{\tau}_i); & \text{if } |i-j| = 1 \\ 0; & \text{if } |i-j| > 1. \end{cases} \tag{22}$$

In other words, the correlation coefficient is

$$\rho_{ij} = \frac{\operatorname{cov}(\hat{\tau}_i, \hat{\tau}_j)}{(\operatorname{var}(\hat{\tau}_i)\operatorname{var}(\hat{\tau}_j))^{1/2}}$$

$$= \begin{cases} -\frac{1}{2}; & \text{if } |i-j| = 1 \\ 0; & \text{if } |i-j| > 1. \end{cases} \tag{23}$$

V. Suboptimum Processors

In previous sections we have discussed in some detail the optimum time delay vector estimator. The result yields a highly coupled multichannel processor. In addition, we obtained analytical expressions of the estimators' CRLB for a number of important cases. For practical applications, it is desired to seek suboptimum realizations which can substantially simplify the required complexity of implementation. In this section we consider two suboptimum realizations: 1) a single target assumption suboptimum processor; and 2) a multisensor multitarget post-correlation processor (MMPCP). We will briefly describe the approaches and then evaluate their performance with respect to the optimum multisensor, multitarget CRLB derived previously.

A single target assumption processor can be obtained from the multitarget processor by setting all interference power spectra to zero. For the two-sensor case, the resulting processor reduces to the conventional GCC processor as was discussed earlier. The time delay is obtained from the GCC by locating the peak of the GCC function (assuming that SNR is sufficiently high so that a dominant peak can be detected). In the presence of interference, the resulting estimates are known to be biased. One could also obtain an analytical expression for the resulting root-mean-square errors. The performance of this processor in the presence of interference has been studied widely elsewhere in the context of the localization variables (i.e., range and bearing) [9], [19], [20].

The MMPCP is a post-GCC processor. It provides additional multitarget processing capability by correlating the observed GCC outputs with an assumed multiparameter matching function. Thus, with this approach, the conventional GCC processor needs no modification and serve simply as the front-end processor for the MMPCP.

In the following paragraphs, we provide a description of this approach and its performance bound. In addition, all performance predictions are verified via simulations. Comparisons will be made to the optimum multitarget processor (OMP) presented in the previous sections.

From (14), the GCC output can be written as

$$R(\tau) = \sum_{k=-B}^{B} |h_k|^2 \alpha_{1k}\alpha_{2k}^* \, e^{j\omega_k \tau}, \tag{24}$$

where h_k, α_{1k}, and α_{2k} are the shaping filter and the frequency observables, respectively. Now (24) can be written as

$$R(\tau) = \overline{R}(\tau) + W(\tau), \tag{25}$$

where $\overline{R}(\tau)$ denotes the deterministic component and $W(\tau)$ the random component of the noisy GCC output. For practical application, (24) is usually realized via a fast Fourier transform (FFT). An alternate formulation of this problem can be found in [21]. Thus, the GCC outputs consist of a discrete set of time delay observations. Now let $\Delta\tau$ be the sampling time such that $T = N\Delta\tau$, then the discrete GCC output can be written as

$$R(n\Delta\tau) = \overline{R}(n\Delta\tau) + W(n\Delta\tau);$$

$$n = 0, \pm 1, \pm 2, \cdots, \pm N/2. \tag{26}$$

From (24), the deterministic and the random components of $R(n\Delta\tau)$ can be obtained. For example, taking the expected value on both sides of (24), the deterministic component is

$$\overline{R}(n\Delta\tau) = \sum_{k=-B}^{B} |h_k|^2 \, \overline{\alpha_{1k}\alpha_{2k}^*} \, e^{j\omega_k n\Delta\tau}$$

$$= \sum_{j=1}^{J} \sum_{k=-B}^{B} S_{kj}|h_k|^2 \, e^{j\omega_k n\Delta\tau},$$

$$= \sum_{j=1}^{J} S_j \rho_j(n\Delta\tau - \tau_j) \tag{27}$$

where τ_j is the TDOA of target j. Furthermore, the jth target power S_j and the normalized autocorrelation are given by

$$S_j = \sum_{k=-B}^{B} S_{kj} \tag{28a}$$

$$\rho_j(n\Delta\tau - \tau_j) = \sum_{k=-B}^{B} (S_{kj}/S_j)|h_k|^2 \, e^{j\omega_k(n\Delta\tau - \tau_j)}. \tag{28b}$$

The covariance of the random component is given by [13]

$$\Lambda_{nm} = E\{W(n\Delta\tau)W(m\Delta\tau) - \overline{W(n\Delta\tau)W(m\Delta\tau)}\}$$

$$= \sum_{k=-B}^{B} |h_k|^4 [G_{11}(k)\, G_{22}(k)\, e^{-j\omega_k m\Delta\tau}$$

$$+ G_{12}^2(k)\, e^{j\omega_k m\Delta\tau}]\, e^{j\omega_k n\Delta\tau} \tag{29a}$$

where

$$G_{ii}(k) = \overline{\alpha_{ik}\alpha_{ik}^*} = \sum_{j=1}^{J} S_{kj} + N_{ik}; \qquad i = 1, 2 \tag{29b}$$

and

$$G_{12}(k) = \overline{\alpha_{1k}\alpha_{2k}^*} = \sum_{j=1}^{J} S_{kj}\, e^{-j\omega_k \tau_j}. \tag{29c}$$

Now define the $2J$ unknown parameter vector by $\Theta = (S_1, S_2, \cdots, S_J; \tau_1, \tau_2, \cdots, \tau_J)^T$ and the assumed matching function by

$$h_n(\Theta) = \sum_{j=1}^{J} S_j \rho_j (n\Delta\tau - \tau_j). \tag{30}$$

Then the observation (24) can be written as

$$R(n\Delta\tau) = h_n(\Theta) + W(n\Delta\tau);$$
$$n = 0, \pm 1, \pm 2, \cdots, \pm N/2. \tag{31}$$

In matrix notation, this can be written as

$$Z = H(\Theta) + W \tag{32a}$$

where

$$Z = (R(-N\Delta\tau/2), \cdots, R(N\Delta\tau/2))^T \tag{32b}$$

$$H(\Theta) = (h_{-N/2}(\Theta), \cdots, h_{N/2}(\Theta))^T \tag{32c}$$

and

$$W = (W(-N\Delta\tau/2), \cdots, W(N\Delta\tau/2))^T, \tag{32d}$$

are $N + 1$-dimensional vectors.

Note that the noise vector is zero mean with matrix covariance

$$\Lambda = E\{WW^T\}, \tag{33}$$

where the mn element of Λ is given by (29a). Using an LMS criteria, the best estimate of the unknown parameter vector Θ is obtained by minimizing the function

$$J(\Theta) = (Z - H(\Theta))^T (Z - H(\Theta)). \tag{34}$$

This is the essence of MMPCP and the best estimate of Θ is given by

$$\hat{\Theta} = \arg \min_{\Theta} J(\Theta). \tag{35}$$

Let Θ_0 be the true parameter value. Then the covariance of the estimation error is known and given by [13]

$$\text{cov}(\hat{\Theta} - \Theta_0) = A^{-1}(\Theta_0)\, B^T(\Theta_0)\, \Lambda\, B(\Theta_0)\, A^{-1}(\Theta_0) \tag{36a}$$

$$B(\Theta_0) = \left. \frac{\partial H(\Theta)}{\partial \Theta} \right|_{\Theta = \Theta_0} \tag{36b}$$

and

$$A(\Theta_0) = B^T(\Theta_0)\, B(\Theta_0).$$

VI. Simulation Results

Fig. 7 presents the simulation procedure. The multitarget GCC output observation vector (31) was generated by adding observation noise sequence with the prescribed covariance to the noise-free GCC component. The noise vector was generated using the following procedure: 1) calculating the multitarget GCC covariance matrix from (29a), 2) factoring the matrix into lower and upper triangular matrices using the Gramm–Schmidt orthogonalization procedure, and 3) multiplying the lower triangular matrix by a white noise vector to produce the desired correlated observation noise vector. Finally, the GCC observation vector was processed by the MMPCP (or the matched parameter estimator).

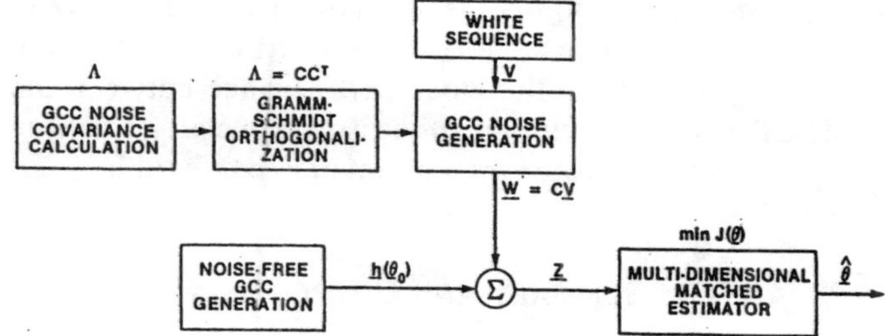

Fig. 7. Post-GCC multitarget processor simulation.

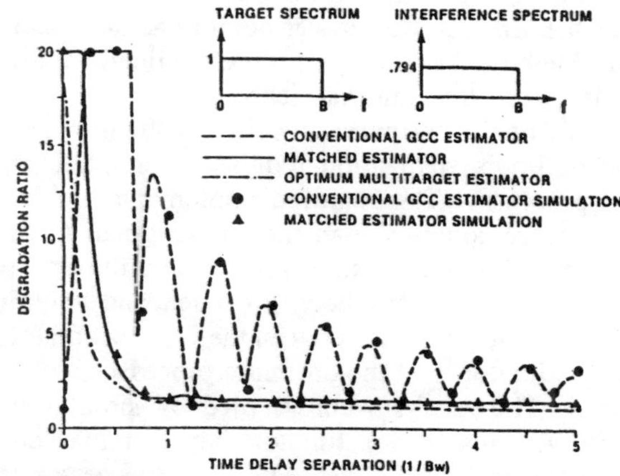

Fig. 8. Degradation ratio versus time delay separation (SNR = 0 dB, INR = −1 dB, $S: O \rightarrow B$, $I: O \rightarrow B$, $N: O \rightarrow B$).

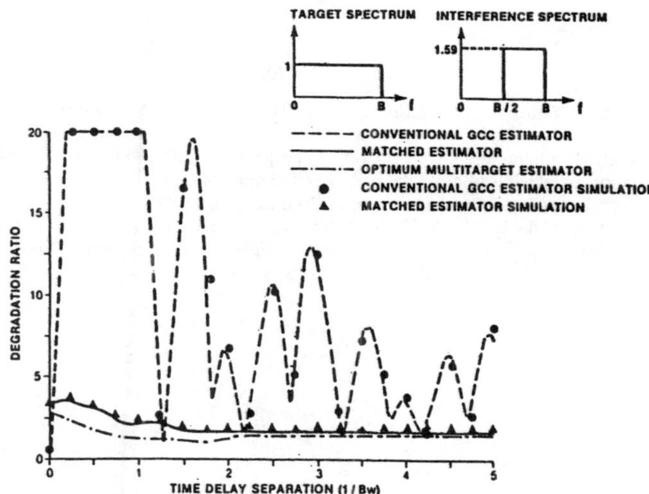

Fig. 9. Degradation ratio versus time delay separation (SNR = 0 dB, INR = −1 dB, $S: O \rightarrow B$, $I: B/2 \rightarrow B$, $N: O \rightarrow B$).

Figs. 8 and 9 show the degradation ratio (normalized variance ratio of multiple targets to single target) performance of a conventional GCC estimator, the post-GCC matched estimator, and the optimum multitarget estimator as a function of time delay separation for the two-target two-sensor case. While Fig. 8 assumed identical broadband signal and interference spectra, Fig. 9 assumed the interference occupied the upper-half of the frequency band but maintained the same interference power. Note that the conventional GCC rms was clipped at a degradation ratio of 20 for time delay separation less than the reciprocal signal bandwidth. Both Figs. 8 and 9 show the marked improvement of the matched estimator (MMPCP) over the GCC estimator. Note the close performance between the MMPCP and the optimum processor especially for time delay separation greater than an inverse bandwidth.

VII. Summary and Conclusions

The optimum multitarget time delay vector processor has been derived using the maximum likelihood approach. The performance of the optimum estimator in terms of the Cramer–Rao lower bound has also been examined. Analytical expressions were obtained for a number of simplified but important cases.

The study on the optimum processor realization showed that the optimum realization required a coupled, multichannel processor. The required implementation is considerably more complex than the conventional GCC approach. A suboptimum multisensor, multitarget postcorrelation processor has been proposed and examined. Performance comparisons between the GCC estimator, the matched estimator, and the optimum processor have been conducted. The results were verified by simulation. In summary, the matched estimator shows a marked improvement over the GCC estimator and the performance degradation is deemed acceptable with respect to the optimum processor.

References

[1] *IEEE Trans. Acoust., Speech, Signal Processing (Special Issue on Time Delay Estimation)*, June 1981.

[2] C. H. Knapp and G. C. Carter, "The generalized correlation method for estimation of time delay," *IEEE Trans. Acoust., Speech, Signal Processing*, vol. ASSP-24, pp. 320–327, Aug. 1976.

[3] W. R. Hahn and S. A. Tretter, "Optimum processing for delay-vector estimation in passive signal arrays," *IEEE Trans. Inform. Theory*, vol. IT-19, pp. 608–614, Sept. 1973.

[4] J. C. Hassab and R. E. Boucher, "Optimum estimation of time-delay by a generalized correlator," *IEEE Trans. Acoust., Speech, Signal Processing*, vol. ASSP-27, pp. 373–380, Aug. 1979.

[5] B. Friedlander, "An ARMA modeling approach to multitarget tracking," in *Proc. 19th IEEE Conf. Dec. Contr.*, Dec. 1980.

[6] G. Su and M. Morf, "The signal subspace approach for multiple wideband emitter location," *IEEE Trans. Acoust., Speech, Signal Processing*, vol. ASSP-31, pp. 1478–1492, Dec. 1983.

[7] Y. Bar-Shalom, "Tracking methods in a multitarget environment," *IEEE Trans. Auto. Contr.*, vol. AC-23, pp. 618–626, Apr. 1978.

[8] W. R. Hahn, "Optimum signal processing for passive range and bearing estimation," *J. Acoust. Soc. Amer.*, vol. 54, no. 3, pp. 201–207, 1973.

[9] H. Cox, "Resolving power and sensitivity to mismatch of optimum array processor," *J. Acoust. Soc. Amer.*, vol. 54, no. 3, pp. 771–785, 1973.

[10] F. Schweppe, "On the angular resolution of multiple targets," *Proc. IEEE*, vol. 2, pp. 294–305, Sept. 1972.

[11] J. Capon, "High resolution frequency wave number spectrum analysis," *Proc. IEEE*, vol. 57, pp. 1408–1418, Aug. 1969.

[12] M. Wax and T. Kailath, "Optimum localization of multiple sources in passive arrays," *IEEE Trans. Acoust., Speech, Signal Processing*, vol. ASSP-31, pp. 1210–1218, Oct. 1984.

[13] L. C. Ng and Y. Bar-Shalom, "Optimum multisensor, multitarget time delay estimation," Naval Underwater Systems Center, NUSC Tech. Rep. TR 6757, Apr. 20, 1983.

[14] W. S. Hodgkiss and L. W. Nolte, "Covariance between Fourier coefficients representing the time waveforms observed from an array of sensors," *J. Acoust. Soc. Amer.*, vol. 59, pp. 582–590, Mar. 1976.

[15] N. R. Goodman, "Statistical analysis based on a certain multivariate complex Gaussian distribution (an introduction)," *Ann. Math. Stat.*, vol. 34, pp. 152–157, 1963.

[16] H. VanTrees, *Detection, Estimation, and Modulation Theory, Part I.* New York: Wiley, 1968.

[17] L. C. Ng, "Optimum multisensor, multitarget localization and tracking," Ph.D. dissertation, Univ. Conn., Storrs, 1983. Also, Naval Underwater Systems Center, NUSC Tech. Rep. 6931, Apr. 20, 1983.

[18] N. Owsley and G. Swope, "Time delay estimation in a sensor array," *IEEE Trans. Acoust., Speech, Signal Processing*, vol. ASSP-29, pp. 519–524, June 1981.

[19] V. C. Anderson and P. Rudnick, "Rejection of a coherent signal arrival at an array," *J. Acoust. Soc. Amer.*, vol. 45, no. 2, pp. 406–410, 1969.

[20] T. P. McGarty, "The effect of interfering signals on the performance of angle of arrival estimates," *IEEE Trans. Aerosp. Electron. Syst.*, vol. AES-10, pp. 70–77, Jan. 1974.

[21] W. K. Fischer and L. C. Ng, "Improved time delay estimation in the presence of interference," in *IEEE Conf. Proc. 1983 ICASSP*, Apr. 1983.

Active High Resolution Time Delay Estimation for Large BT Signals

Marie-Agnès Pallas and Geneviève Jourdain

***Abstract*—This paper is concerned with the active identification of a multipath propagation channel. The channel is described by a parametric model and we estimate the set of unknown parameters. We achieve time delay estimation (TDE) by using high resolution methods (HRM) on signals with a large time-bandwidth product. We describe two different high resolution time delay estimation (HRTDE) methods: a temporal method and a frequency method. The performance gain of these methods is shown to be about four times better in comparison with the "classical" TD resolution methods. The frequency HRTDE method is applied to real data obtained from an ocean acoustic experiment. Although classic methods cannot distinguish close signal components, our method yields estimates of the delay differences and the attenuation associated with each propagation path.**

I. Introduction

DATA transmission and physical medium sounding are both concerned with active identification of the propagation medium. The important class of multipath media is considered here; it occurs in such different fields as underwater acoustics [1], [2], radioastronomy [3], seismics, ionospheric [4] or urban radio communication [5], or nondestructive control.

We suppose in this paper that some prior information on the propagation is available, so that the impulse response of the propagation filter (supposed to be linear) is known, except for a finite set of undetermined parameters. Then multipath medium identification reduces to the estimation of these unknown parameters, basically the time delays and attenuations along each path. Classical parameter estimation is performed using the criterion of maximum likelihood. This step leads an optimal estimator and the Cramer–Rao bound helps to evaluate its performance [6]. Many studies have been devoted to classical time delay estimation, both in the passive [7], [8] and in the active [9]–[11], [24] sense, or to the comparison of both cases [11], [20]. However, parameter estimation complexity increases rapidly as soon as close paths are involved in the propagation [13]. High resolution methods may help in this case. These methods have been widely developed in spectral and spatial analysis, but their use for time delay estimation is recent. The first suggestions for it were presented by Hou and Wu in 1982 [14].

In this paper, we develop two separate methods, specially adapted to large bandwidth duration (BT) product time-resolvant signals. Their novelty is a two-step matched filtering procedure, which leads to a two-step classification of time delays and to a SNR improvement.

Manuscript received November 10, 1988; revised March 27, 1990. This work was supported in part by the Délégation Générale de l'Armement.

M. A. Pallas is with INRETS, 69672 Bron Cedex, France.

G. Jourdain is with CEPHAG, UA 346, INPG/IEG, F-38402 St. Martin d'Hères Cedex, France.

IEEE Log Number 9042260.

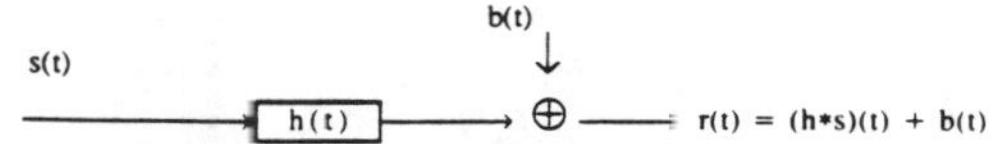

Fig. 1. Model of the propagation channel.

II. Multipath Propagation

We suppose that the propagation channel may be represented by a linear component $h(t)$ describing the multipath propagation with additive random noise (Fig. 1).

Bandpass signals are required for transmission. We take a special interest here in deterministic amplitude-modulated signals, with carrier ν_0:

$$s(t) = c(t) \sin 2\pi\nu_0 t \tag{1}$$

which have a single low-pass component (but it would be also possible to envisage both amplitude and phase modulated signals). If B is the bandwidth of $c(t)$, we impose: $B < \nu_0$. The performance of $s(t)$ is determined by the choice of $c(t)$. $c(t)$ is supposed to be time resolvant with a large BT product.

Let us consider a p-path propagation channel. Each path is supposed to introduce a delay τ, an attenuation α, and a phase ϕ. The received signal is

$$r(t) = \sum_{k=1}^{p} \alpha_k c(t - \tau_k) \sin\left[2\pi\nu_0(t - \tau_k) + \phi_k\right] + b(t). \tag{2}$$

The delays τ_k are supposed to be nonrandom. The attenuations α_k and the phases ϕ_k will be sometimes random and sometimes deterministic. $b(t)$ is a white noise of power spectral density (psd) $\gamma_0/4$.

We will work on the low-pass components of the received signals around the carrier ν_0, that is, on the complex amplitude:

$$A(t) = \sum_{k=1}^{p} \alpha_k e^{j\Phi_k} c(t - \tau_k) + B(t). \tag{3}$$

The phase Φ_k combines the propagation medium phase ϕ_k, a possible demodulation phase due to asynchronous demodulation, and a term depending on the delay τ_k [13]. $c(t)$ is the complex amplitude of the emitted signal. $B(t)$ may be approximately considered as a complex white noise of psd γ_0 [6], [13]. $(\alpha_k, \Phi_k, \tau_k, k = 1 \cdots p)$ form the unknown parameter set we intend to estimate.

Reprinted from *IEEE Trans. Sig. Proc.*, vol. 39, no. 4, pp. 781–788, April 1991.

TABLE I

	Multipath Channel Identification (Time Domain)	Array Processing: Source Location
model	$A(t_i) = \sum_{k=1}^{p} \alpha_k e^{j\Phi_k} c(t_i - \tau_k) + B(t_i)$ $i = 1, \cdots, N$	$X_i(f) = \sum_{k=1}^{p} g_k S_{ik}(f, \theta_k) + b_i(f)$ $i = 1, \cdots, N$
N	number of time samples	number of receivers
p	number of paths	number of sources
unknown data	delay τ_k	directions θ_k
	attenuation $\alpha_k e^{j\Phi_k}$	amplitude g_k
known data	signal $c(t)$	steering vector $S(f, \theta)$

TABLE II

	Multipath Channel Identification (Frequency Domain)	Spectral Analysis: Pure Frequencies Estimation	Source Location (Plane Waves, Equally Spaced Receivers)
model	$H_c(\nu_i) = \sum_{k=1}^{p} \alpha_k e^{j\Phi_k} e^{-2j\pi\nu_i\tau_k}$ $i = 1, \cdots, N$	$y(t_i) = \sum_{k=1}^{p} a_k e^{j(2\pi\nu_k t_i + \phi_k)} + b(t_i)$ $i = 1, \cdots, N$	$X_i(f) = \sum_{k=1}^{p} g_k(f) e^{j\frac{2\pi kd \sin\theta_k}{\lambda}} + b_i(f)$ $i = 1, \cdots, N$
N	number of frequency samples	number of time samples	number of receivers
p	number of paths	number of pure frequencies	number of sources
unknown parameters	delay τ_k	frequencies ν_k	directions θ_k
	attenuation $\alpha_k e^{j\Phi_k}$	amplitudes $a_k e^{j\Phi_k}$	amplitudes c_k

III. Analogy of Time Delay Estimation (TDE) with Spectral and Spatial Analysis

The equation (3) involving the whole set of unknown parameters recalls the formulation of source location problems in array processing without any hypotheses on the steering vector at the frequency f. This point was underlined in [15] and is summed up in Table I.

Let $H_c(\nu)$ be the low-pass-component transfer function of the propagation filter:

$$H_c(\nu) = \sum_{k=1}^{p} \alpha_k e^{j\Phi_k} e^{-2j\pi\nu\tau_k}. \tag{4}$$

$H_c(\nu)$ is composed of p complex sinusoids, whose periods $1/\tau_k$ are directly linked to the unknown delays τ_k.

Equation (4) is quite similar to those arising in spectral analysis for pure frequency estimation, or in array processing for source location when the array is linear with equally spaced receivers and when the propagation involves plane waves. This similarity is detailed in Table II.

Each of these analogies suggests a typical methodology, leading to both cases, which we call, respectively, in the following the "temporal method" and the "frequency method."

IV. High Resolution Time Delay Estimation in the Time Domain

This method operates directly on the temporal signal through the model (3) defined in the previous section. The idea was first proposed by Brückstein *et al.* [15] and is based on the MUSIC algorithm.

A. Principle [15]

Let us consider the model (3). R is the covariance matrix of the sampled observation vector:

$$R = E\{AA^+\} = CDC^+ + \Gamma_B \tag{5}$$

where $C = [c_1 \ c_2 \ \cdots \ c_p]$ is the deterministic signal matrix; each column of C is $c_i^T = [c(t_0 - \tau_i) \ c(t_1 - \tau_i) \ \cdots \ c(t_{N-1} - \tau_i)]$; D is the covariance matrix of the random attenuations $\alpha_k e^{j\Phi_k}$; $\Gamma_B = E\{BB^+\} = \sigma_B^2 I$ is the covariance matrix of the noise, supposed to be white and independent of the other propagation characteristics.

If the attenuations are not completely correlated, the matrix D is full rank and we may obtain the signal subspace from matrix R using its eigenvalues and eigenvectors. The unknown delays τ_k may then be estimated from the function

$$f(\tau) = \frac{c^+(\tau)c(\tau)}{\sum_{k=p+1}^{N} \left|c^+(\tau)v_k\right|^2} \tag{6}$$

where v_k, $k = p + 1, \cdots, N$ are the noise subspace eigenvectors. This function $f(\tau)$ has p maxima at values $\tau = \tau_k$, $k = 1, \cdots, p$.

The covariance matrix of the attenuations may then be estimated from

$$D = (C^+C)^{-1}C^+(R - \sigma_B^2 I)C(CC^+)^{-1}. \tag{7}$$

It is well known that: i) when some signal components happen to be fully correlated, or when the attenuations are nonrandom, this method is not directly applicable. However, a signal component decorrelation technique, generally named "spatial smoothing" in source locating problems [22], can be used as long as the matrix C may be written as a Vandermonde matrix; and ii) when the noise is not white, a whitening step may be introduced which will affect matrix C at the same time. Furthermore, the MUSIC time delay estimation method, as described above, may be practically implemented only for signals $c(t)$ with a low BT product. Here, the number N of samples describing the time spreading of $c(t)$ cannot be large as N describes the observation matrix dimension together with the computational complexity. We describe in the following the new step of time spreading reduction, which nevertheless allows the use of the above principle.

B. Time-Spreading Reduction

As stated in Section II, we are concerned with large BT signals, and, more precisely, biphase-shift keying (BPSK) signals whose modulation code $c(t)$ is a maximal length binary sequence (MLBS). The correlation function of such PSK signals has several interesting properties, one of which is its informative time-spreading of $2/B$, a factor of $BT/2$ smaller than the initial PSK signal.

We introduce *a matched filtering step* on the received signal ($c(t)$ is known):

$$Y(t) = \int_{(T)} A(\theta)c(t-\theta)\,d\theta = \sum_{k=1}^{P} \alpha_k e^{j\phi_k}\Gamma_c(t-\tau_k) + \beta(t) \tag{8}$$

where $A(t)$ is the complex demodulated received signal given in (3), $\Gamma_c(t)$ is the autocorrelation function of $c(t)$; $\beta(t)$ is the noise signal $c(t)$ cross correlation; $\beta(t)$ is a nonwhite random noise of power spectral density $\gamma_\beta(\nu) = \gamma_0 |c(\nu)|^2$, with $c(t) \rightleftharpoons c(\nu)$. We take advantage of two properties: the matched filter provides a *signal-to-noise* ratio gain of 10 log BT and the time resolution of $\Gamma_c(t)$ allows *some isolated signal arrivals to be separated* for focusing the analysis only on very close arrivals.

We then substitute $Y(t)$, given in (8), to $A(t)$ in (5) and apply the MUSIC time delay estimation method by computing the covariance matrix:

$$R_y = C_\Gamma \cdot D \cdot C_\Gamma^+ + \Gamma_\beta$$

$$R_y = E\{YY^+\}$$

$$D = E\{GG^+\}$$

$$C_\Gamma = \begin{bmatrix} \Gamma_c(t_0-\tau_1) & \Gamma_c(t_0-\tau_2) & \cdots & \Gamma_c(t_0-\tau_p) \\ \Gamma_c(t_1-\tau_1) & \cdot & & \\ \cdot & \cdot & & \\ \Gamma_c(t_{N-1}-\tau_1) & \Gamma_c(t_{N-1}-\tau_2) & \cdots & \Gamma_c(t_{N-1}-\tau_p) \end{bmatrix}$$

is the new signal matrix

$$\Gamma_\beta = E\{\beta\beta^+\} \neq \sigma_\beta^2 I. \tag{9}$$

We assume in the following that the signal-to-noise ratio is large enough after the matched filtering, so that we need not introduce any whitening step in the processing. The implementation of the method is diagrammed in Fig. 2.

C. Simulations

We present here some results obtained on simulated data. The parameters of the emitted BPSK signal are BT product = 127, bandwidth B = 15 Hz, duration of one sequence T = 8.45 s, and correlation time of the signal $\theta = T/127$ = 67 ms.

We simulate a two path propagation medium, with respective delays τ_1 and τ_2, for different values of the delay difference $\Delta\tau = |\tau_2 - \tau_1|$ smaller than the classical "time resolution" θ of the emitted signal.[1] We define the complex random attenuations

[1]In fact, it is not the true theoretical time resolution power of the signal; this depends on SNR.

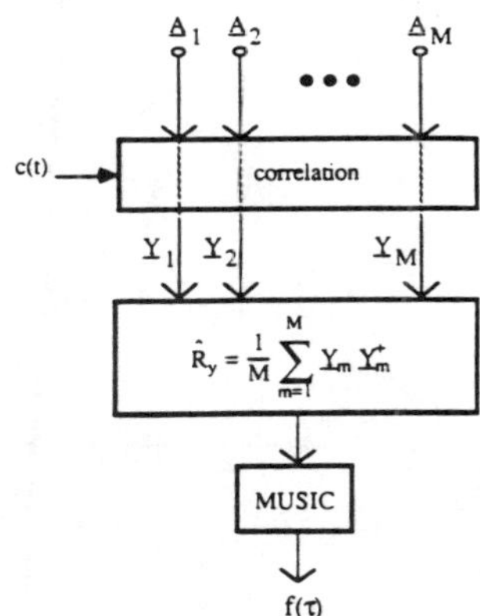

Fig. 2. Implementation of the time delay estimate in the time domain.

such that $\alpha_1 = \alpha_2$ and the phases ϕ_1 and ϕ_2 are independent, uniformly distributed random variables. The signal is corrupted by white noise, the SNR being −10 dB (it becomes 11 dB after the matched filtering). For a given set of parameters, K = 40 independent realizations of the observed signal are thus simulated.

Although the actual number of propagation paths is p = 2, we tested the method in both cases $\hat{p}$ = 2 and $\hat{p}$ = 3 for robustness.

1) $\Delta\tau$ = 54 ms = 0.8 θ (Fig. 3): As shown in Fig. 3, we may guess the presence of two components from the envelope of the signal, but their respective times of arrival are not discernable from the position of the maxima. We draw the function $f(\tau)$ obtained from the MUSIC time delay estimation method. Results are quite similar for $\hat{p}$ = 2 or $\hat{p}$ = 3; $f(\tau)$ exhibits two wide distinct peaks at the true locations τ_1 and τ_2.

2) $\Delta\tau$ = 17 ms = 0.25 θ (Fig. 4): In Fig. 4, both delays are so close that the presence of two signal arrivals cannot be determined from the initial correlation signal $Y(t)$ (Fig. 4(a)). Nevertheless, the HR method still points out the actual values for τ_1 and τ_2, regardless of whether we choose $\hat{p}$ = 2 or $\hat{p}$ = 3 (Figs. 4(b) and (c)).

From all our simulation investigations, we can conclude that:

1) We have been able to estimate τ_1 and τ_2 exactly for delays as close as $\Delta\tau$ = 0.25 θ, beyond which only one maximum appeared in $f(\tau)$ and the method failed;
2) in the range of availability, τ_1 and τ_2 could be exactly estimated, even if we overestimated the number p of propagation paths;
3) one must keep in mind that this method is restricted to random propagation mediums with at least partially decorrelated components; some statistically independent realizations of the signal must be observed in order to estimate properly the covariance matrix R.

V. High Resolution Time Delay Estimation in the Frequency Domain

We present here a second method, based on the frequency-domain model (4) of the propagation medium transfer function.

The first step (Fig. 5) consists in estimating the propagation transfer function. In the case of close paths, this transfer function is the sum of undamped complex exponentials of close periods, which high resolution methods may identify. Several HR methods can be implemented in this case. In this paper, we will

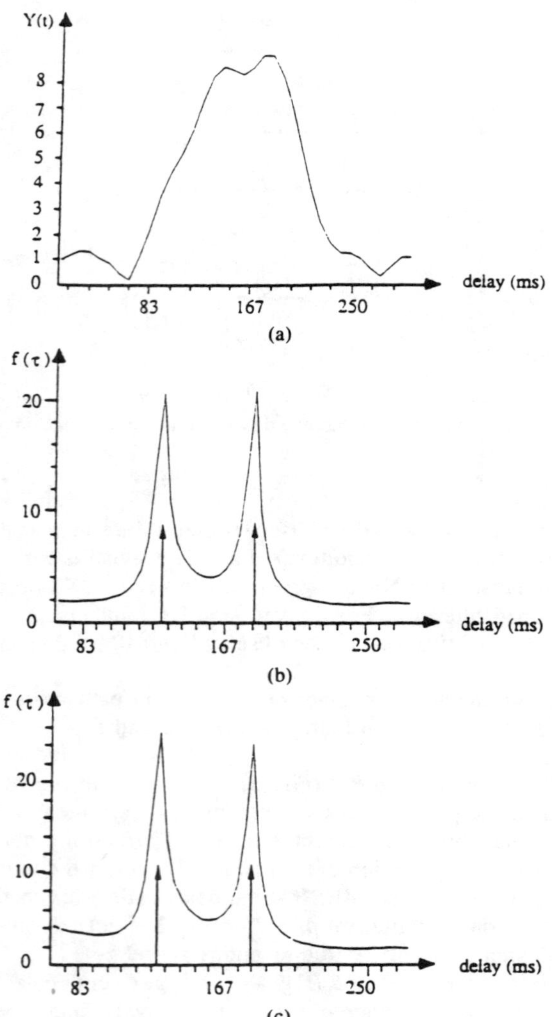

Fig. 3. HRTDE for two close paths with $\Delta\tau$ = 54 ms = 0.8 θ. (a) Envelope of the simulated signal. (b) Result of the temporal method for $\hat{p}$ = 2. (c) Result of the temporal method for $\hat{p}$ = 3.

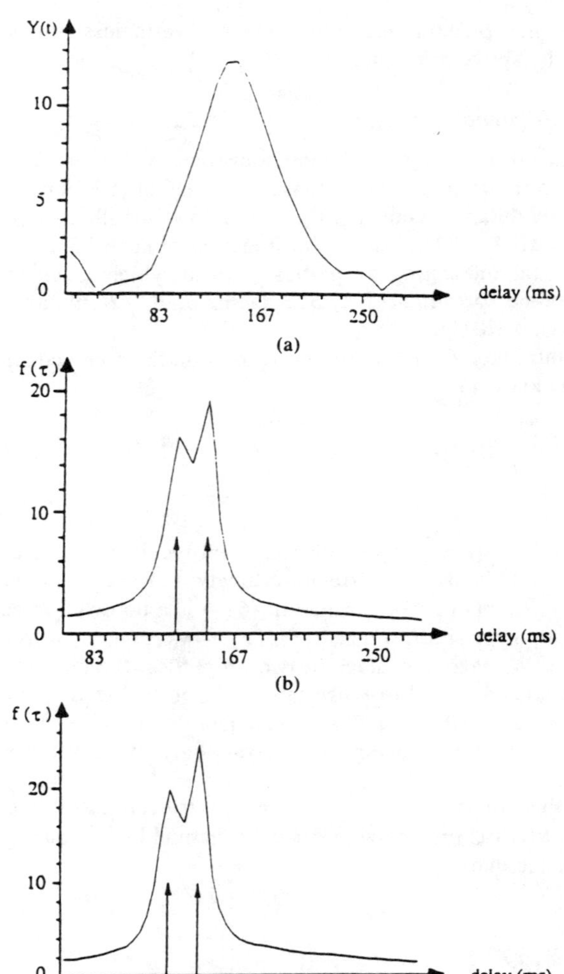

Fig. 4. HRTDE HRTDE for the two close paths with $\Delta\tau$ = 17 ms = 0.25 θ. (a) Envelope of the simulated signal. (b) Result of the temporal method for $\hat{p}$ = 2. (b) Result of the temporal method for $\hat{p}$ = 3.

use the Tufts–Kumaresan [17] method, now widely described in the literature and which has proved to be effective in many cases.

A. Propagation Transfer Function Estimation

The transfer function estimation is simply achieved in different successive steps, as represented in Fig. 6 [23].

As in the time-domain method, the first step consists of a matched filter. At the output of this step, we obtain, as in (8)

$$Y(t) = \Gamma_{Ac}(t) = \int_{(T)} A(\theta)c(t-\theta)\,d\theta$$

$$= \sum_{k=1}^{P} \alpha_k e^{j\Phi_k}\Gamma_c(t-\tau_k) + \beta(t). \qquad (10)$$

Next, after isolating the informative part of $Y(t)$ with an appropriate window (we choose here the rectangular window because the signal is time limited), we perform a Fourier transform on $\Gamma_{Ac}(t)$, $t \in [t_1, t_2]$:

$$\gamma_{Ac}(\nu) = \sum_{k=1}^{P} \alpha_k e^{j\Phi_k}\gamma_c(\nu)e^{-2j\pi\nu\tau_k} + B_1(\nu). \qquad (11)$$

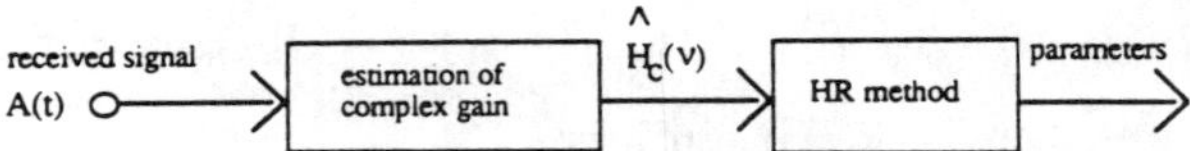

Fig. 5. High resolution time delay estimation in the frequency domain.

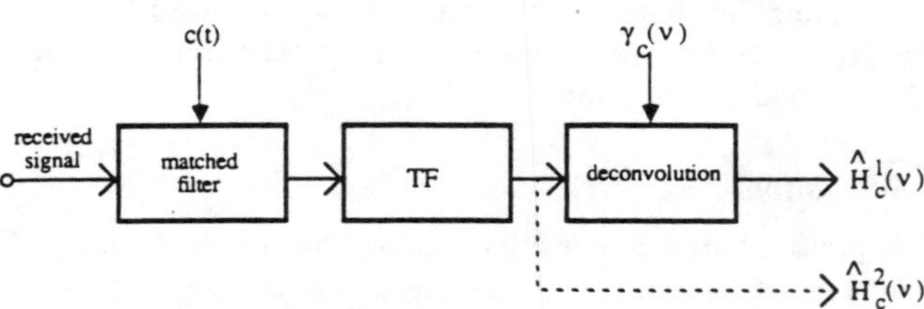

Fig. 6. Estimation of the channel transfer function.

$B_1(\nu)$ is a nonstationary random noise with uncorrelated components [18].

Finally, a simple deconvolution operation is achieved by dividing $\gamma_{Ac}(\nu)$ by the spectral density $\gamma_c(\nu)$ of $c(t)$ (at least over the bandwidth where γ_c is sufficiently >0). This leads to an estimate $\hat{H}_c^1(\nu)$ [23], within the emitted signal bandwidth, of the transfer function $H_c(\nu)$ given in (4):

$$\hat{H}_c^1(\nu) = \frac{\gamma_{Ac}(\nu)}{\gamma_c(\nu)} = \sum_{k=1}^{P} \alpha_k e^{j\phi_k} e^{-2j\pi\nu\tau_k} + B_2(\nu), \qquad \gamma_c(\nu) \neq 0$$
$$= H_c(\nu) + B_2(\nu). \tag{12}$$

$B_2(\nu)$ is a random noise with nonstationary uncorrelated components. Its covariance is

$$E\{B_2(\nu)B_2^*(\nu')\} = \frac{\gamma_0}{\gamma_c(\nu)}\delta(\nu - \nu'), \qquad \gamma_c(\nu) \neq 0. \tag{13}$$

Let us note that if $\gamma_c(\nu)$ may be considered as constant over some bandwidth, the deconvolution is needless and (11) directly leads to $\hat{H}_c^2(\nu)$ (Fig. 6). Moreover, $B_1(\nu)$ is a white noise over that bandwidth.

B. Time Delay Estimation: Tufts–Kumaresan SVD Method

The Tufts–Kumaresan SVD algorithm has been widely described in the literature [17]. We will only mention here some outstanding points useful to our development [18]. Let us consider N sampling points of $\hat{H}_c(\nu)$ ($\hat{H}_c^1(\nu)$ or $\hat{H}_c^2(\nu)$) in the frequency bandwidth of $c(t)$:

$$\hat{H}_c(\nu_1), \hat{H}_c(\nu_1 + \Delta\nu), \cdots, \hat{H}_c(\nu_1 + (N-1)\Delta\nu).$$

From these N points, the method consists of determining an L-order prediction vector describing at best the signal part (nonnoisy) of the observed sequence. All unknown parameters may be deduced from the prediction vector characteristics. If p is the number of signal components (here the number of propagation paths involved in the interval $[t_1, t_2]$), N, L, and p must fulfil the following inequality in order to lead to a solution:

$$p \leq L \leq N - \frac{p}{2}.$$

But the method performance is not uniform all along that interval, for a given couple (N, p), as shown by Tufts and Kumaresan [17] or Ouamri [16]. In addition, in our case of time-delay estimation, if $\gamma_c(\nu)$ is not constant over the observation bandwidth, the precision of the estimate $\hat{H}_c(\nu)$ depends on the value of $\gamma_c(\nu)$ relative to noise and the choice of N dominates the performance of the method (at a given sampling rate).

C. Statistical Performance of the Method [13]

In the following, we consider the particular case of BPSK signals described in Section IV-C. The shape of $\gamma_c(\nu)$ is then well known: it is a sinc^2 (.) versus ν. Consequently, $\hat{H}_c(\nu)$ will be better estimated in the center of the main lobe of the PSK signal than near the bounds. As the frequency sampling is fixed $\Delta\nu = 0.47$ Hz, 64 samples will represent the main lobe. We have performed many numerical simulations of a two-path channel [13]. Our first study was the determination of the best operating conditions, defined by (N, L).

Considering the case of two identical paths ($a_1 = a_2$, $\phi_1 = \phi_2$), with a delay difference $\Delta\tau = 42$ ms $= 0.6\,\theta$ (where θ was introduced in Section IV-C) and a SNR of 0 dB, we tested the influence of N (for a given ratio L/N). The delay estimates proved to be influenced by the value N whereas the amplitudes and phases had little effect. But we concluded that the standard deviation of each estimate decreased when N increased as long as N remains smaller than 64, so that the badly estimated border points of the main lobe are not used. Afterwards, we looked in the same way for the best L/N ratio: ($N = 50$, $L = 30$) proved to be the best operating condition in regard to delay, attenuation, and phase estimates. This result appears rather similar to the optimal choice $L/N = 3/4$ proposed by Tufts and Kumaresan [25].

The optimal conditions being defined, we then studied the method's performance (bias and variance) with respect to the path's closeness $\Delta\tau$ (from $\Delta\tau = \theta$ to $\theta/2$) and to the SNR. Our investigation was first restricted to two-path simulations, with identical attenuations $\alpha_1 = \alpha_2$ and $\phi_1 = \phi_2 = 0$. Figs. 7 to 10 present the bias and variance for each kind of parameter estimates (τ_i, $\Delta\tau$, a_i, ϕ_i) of the two-path channel as they depend on SNR.

The theoretical performance of the optimal estimates (in the maximum likelihood sense) parameters of a two-path channel have been previously studied elsewhere [10], [20], [24], and as a comparison, the Cramer–Rao bound for the delay estimates (respectively, delay difference estimate) have been superimposed in Fig. 7 (respectively, Fig. 8) for $\Delta\tau = \theta$.

Our results are summarized as follows:

1) The general trend of the variance of all estimates is increasing when $\Delta\tau$ or SNR decreases. This is still the case for the biases with respect to the SNR.

2) The delay estimates are precise as compared with the Cramer–Rao bound, but there is often some bias (Fig. 7(a)).

3) The estimate $\hat{\Delta\tau} = \hat{\tau}_2 - \hat{\tau}_1$ is always more precise than the delay estimates themselves and is close to the Cramer–Rao bound (Fig. 8).

4) The amplitude estimates are poor (Fig. 9).

5) The phase estimates appear to be very precise (e.g., standard deviation $<1°$ at SNR 5 dB for the studied $\Delta\tau$) (Fig. 10).

Consequently, this parameter estimation method is satisfactory regarding delays (and especially delay differences) and phase. However, we must notice that delay estimate biases are relatively high, but the low standard deviation may make this method more attractive than other nonbiased methods.

VI. Application to an Underwater Acoustic Experiment

In this section, we apply HRTDE to real data obtained from an acoustic experiment in the ocean. The active transmission involved two 140 km distant boats. The source and the receiver are both close to the sea surface. The transmitted signal is a BPSK signal, whose characteristics have been described in Section IV-B. The SNR is -9 dB at the reception hydrophone. We try here to analyze a set of close arrivals that classical methods cannot easily separate. Fig. 11 shows an example of the received signal envelope at the output of the matched filter: from this, we cannot conclude for the presence of one or more components in the signal. However, in this physical underwater propagation case [13], the ray theory suggests that there may be some close paths.

Previous analysis of the data [13] has shown that the propagation channel could be considered as a quasi-deterministic filter. Therefore, the MUSIC HRTDE we developed in Section IV cannot be applied to this case. However, the frequency domain HRTDE can be. We use the optimal operating conditions previously determined through the simulated data. Of course, the number of paths is *a priori* unknown. As we have not considered here the specific problem of path detection, which requires another complete specific study, we arbitrarily perform

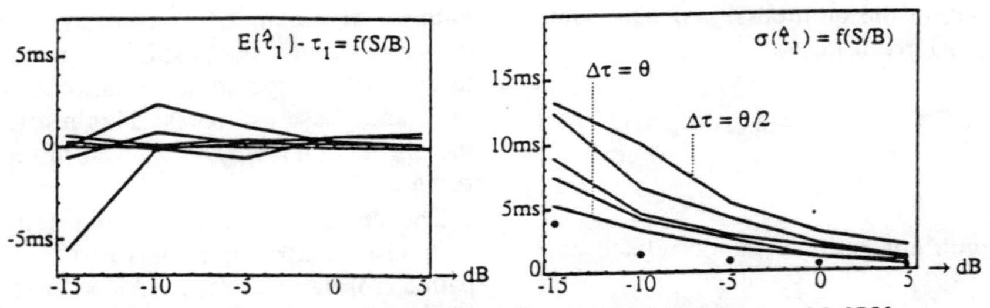

Fig. 7. Performance of the frequency domain HRTDE (for $\Delta\tau = \theta\ 0.875\theta\ 0.50\theta\ 0.625\theta\ 0.5\theta$): delay estimate. (●: Cramer-Rao bound for $\Delta\tau = \theta$.)

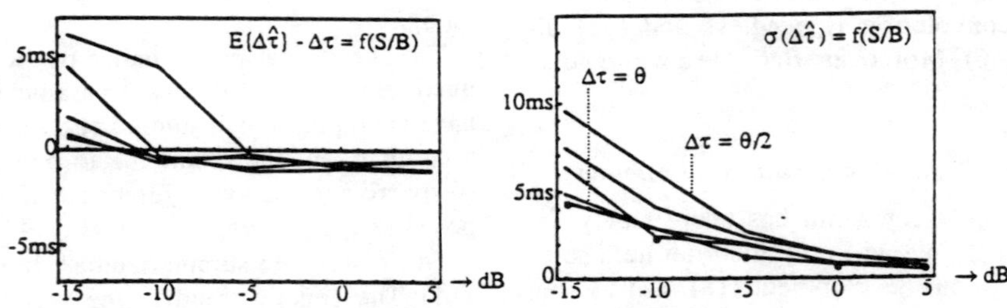

Fig. 8. Performance of the frequency domain HRTDE (for $\Delta\tau = \theta\ 0.875\theta\ 0.75\theta\ 0.625\theta\ 0.5\theta$): delay difference estimate. (●: Cramer-Rao bound for $\Delta\tau = \theta$.)

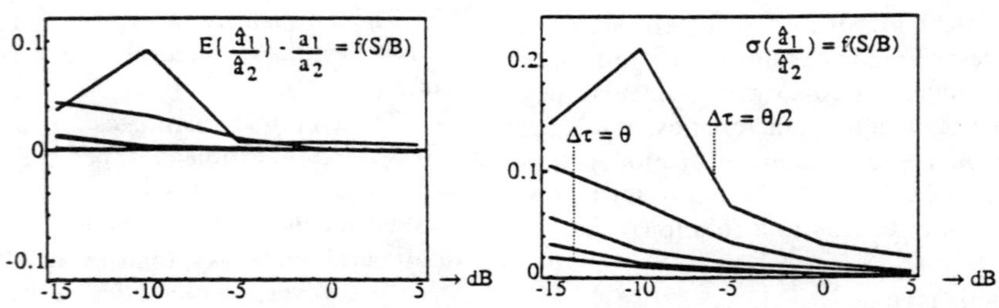

Fig. 9. Performance of the frequency domain HRTDE (for $\Delta\tau = \theta\ 0.875\theta\ 0.75\theta\ 0.625\theta\ 0.5\theta$): amplitude estimate.

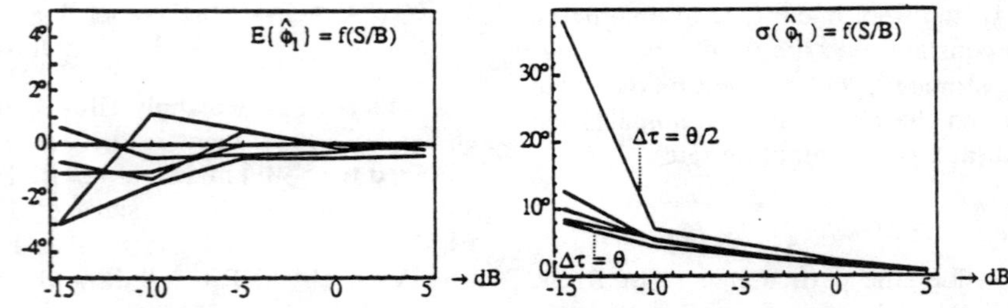

Fig. 10. Performance of the frequency domain HRTDE (for $\Delta\tau = \theta\ 0.875\theta\ 0.75\theta\ 0.625\theta\ 0.5\theta$): phase estimate.

the frequency domain HRDTE for values $\hat{p} = 2$ to 5. The selection of the best solution is then a difficult question, for which we combine different criteria. First we consider the distance of the prediction filter zeros to the unit circle in the Tufts–Kumaresan algorithm. As the final criterion, we choose the least square error between the initial time signal envelope and the envelope simulated from the estimated parameters.

For these data, $\hat{p} = 2$ was the best choice. Fig. 12 draws two superimposed envelopes: envelope 1 is the received signal; envelope 2 is the best envelope, reconstructed from the estimated path characteristics, whose estimated amplitudes and delays are exhibited by the arrows. The measured delay difference between both paths is 51 ms, i.e., 0.76 times the resolution of classical methods. Obviously, the actual value $\Delta\tau$ is unknown. The resolution of the propagation equations may generally provide an idea of $\Delta\tau$. Nevertheless, accurate solution of these equations requires some precise knowledge of the propagation conditions (exact source and receiver locations, physical characteristics of the ocean such as bathymetry and water temperature, etc.), which is seldom available.

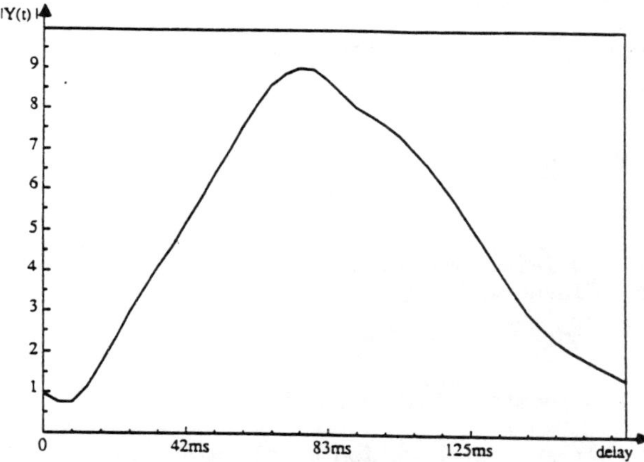

Fig. 11. Real data after matched filtering.

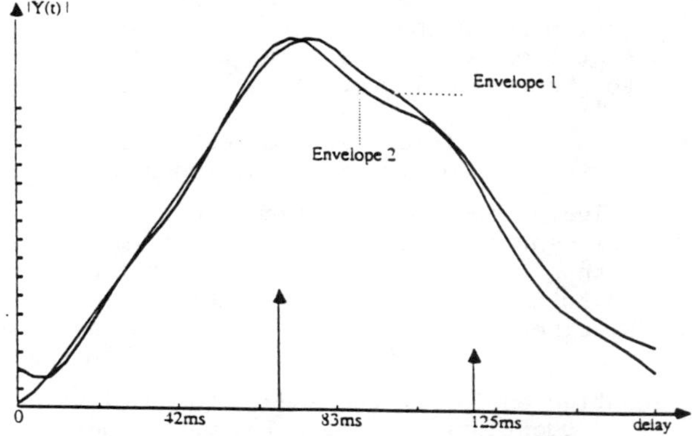

Fig. 12. Real data (1) and estimated received signal (2).

VII. Conclusion

Two high resolution time delay estimation methods have been developed to perform active identification of multipath propagation media in cases where classical methods do not provide an easy optimal procedure. The first one is a time domain method, first reducing the temporal spreading of the signal through matched filtering, and then applying the high resolution MUSIC algorithm. Good resolution and precision, as well as robustness with regards to an overestimated number of signal components, have been exhibited through the first simulated data processing. However, in order to apply this method, the propagation channel must be random, with at least partially decorrelated path characteristics. The frequency domain method reduces, after some steps, to the identification of complex sinusoids, which is a problem widely addressed in the literature. No particular *a priori* model is imposed here as, in case of correlated random components, the smoothing decorrelation technique is available. Some disadvantage is however linked to the channel transfer function estimation, which must be realized cautiously. Its performance has been studied by simulations for two identical close paths; it seems satisfactory, especially concerning delay and phase estimates. The question of path detection becomes now important, which will allow proper data analysis.

References

[1] R. J. Urich, *Principles of Underwater Sound for Engineers*. New York: McGraw-Hill, 1967.

[2] G. Hakizimana, G. Jourdain, and G. Loubet, "Adapted codes for communication through multipath channel," in *Coding Theory and Applications*, G. Cohen and P. Godlewski, Eds. Springer, 1988, pp. 173–182.

[3] R. Price and P. F. Green, "Signal processing in radar astronomy and communication via fluctuating multipath media," M.I.T., Lincoln Lab., Lexington, MA, Rep. 234, 1960.

[4] F. Chavand, Goutelard, C. Desage, and J. P. Van Uffelen, "Système de transmission à codes pseudo orthogonaux adaptés au canal HF," in *Proc. Coll. GRETSI* (Nice, France), 1983, pp. 511–517.

[5] G. L. Turin, "Introduction to spread-spectrum antimultipath techniques to urban digital radio," *Proc. IEEE*, pp. 321–353, Mar. 1980.

[6] H. L. Van Trees, *Detection, Estimation, and Modulation Theory*, vols. 1 and 3. New York: Wiley, 1971.

[7] G. C. Carter, "Time-delay estimation for passive sonar signal processing," *IEEE Trans. Acoust., Speech, Signal Processing*, vol. ASSP-29, no. 3, June 1981.

[8] J. Ianniello, "High resolution multipath time delay estimation for broadband random signals," presented at ICASSP, Dallas, TX, Apr. 1987.

[9] G. Jourdain and M. A. Pallas, "Multiple time delay estimation in underwater acoustic propagation," in *Stochastic Processes in Underwater Acoustics*, C.R. Baker, Ed. Springer, 1986.

[10] M. A. Pallas and G. Jourdain, "Joint estimation of close delays and application to underwater acoustics," presented at EUSIPCO 86, La Haye, The Netherlands, Sept. 1986.

[11] P. Schultheiss and K. Wagner, "Active and passive localization: Similarities and differences," in *Proc. NATO ASI Underwater Acoust. Data Processing* (Kingston, Canada), July 1988.

[12] A. H. Quazi, "An overview on the time delay estimate in active and passive systems for target localization," *IEEE Trans. Acoust., Speech, Signal Processing*, vol. ASSP-29, no. 3, pp. 527–533, June 1981.

[13] M. A. Pallas, "Identification active d'un canal de propagation à trajets multiples," Thèse de Doctorat, INPG, Genoble, France, Feb. 23, 1988.

[14] Z.-Q. Hou and Z. D. Wu, "A new method for high resolution estimation of time delay," presented at ICASSP, 1982.

[15] A. M. Brückstein, T. J. Shan, and T. Kailath, "The resolution of overlapping echoes," *IEEE Trans. Acoust., Speech, Signal Processing*, vol. ASSP-33, no. 6, pp. 1357–1367, Dec. 1985.

[16] A. Ouamri, "Etude des performances des méthodes d'identification à haute résolution et application à l'identification des échos par une antenne linéaire multicapteurs," Thèse de Doctorat d'Etat, Univ. Paris-Sud Orsay, France, June 1986.

[17] R. Kumaresan and D. W. Tufts, "Estimating the angles of arrival of multiple plane waves," *IEEE Trans. Aerosp. Electron. Syst.*, vol. AES-19, no. 1, pp. 134–139, Jan. 1983.

[18] M. A. Pallas and G. Jourdain, "Estimation de retards et méthodes haute résolution," in *Proc. GRETSI* (Nice, France), June 1987, pp. 97–100.

[19] I. Kirsteins and A. H. Quazi, "Resolution of closely spaced multipaths via linear prediction," NUSC Internal Rep., 1987.

[20] I. Kirsteins and A. H. Quazi, "Exact maximum likelihood time delay estimation for deterministic signals," in *Proc. EUSIPCO* (Grenoble, France), Sept. 1988.

[21] G. Jourdain, "Advanced methods for the investigation of the underwater channel," in *Proc. NATO ASI Underwater Acoust. Data Processing* (Kingston), July 1988.

[22] T. H. Shan, M. Wax, and T. Kailath, "On spatial smoothing for direction-of-arrival estimation of coherent signals," *IEEE Trans. Acoust., Speech, Signal Processing*, vol. ASSP-33, no. 4, pp. 806–811, Aug. 1985.

[23] J. S. Bendat and A. G. Piersol, *Random Data: Analysis and Measurement Procedures*. Wiley-Interscience, 1971.

[24] R. Tremblay, G. C. Carter, and D. W. Lyttle, "A practical approach to the estimation of amplitude and time delay parameters of a composite signal," *IEEE J. Ocean. Eng.*, vol. OE-12, no. 1, pp. 273–278, Jan. 1987.

[25] D. W. Tufts and R. Kumaresan, "Estimation of frequencies of multiple sinusoids: Making linear prediction perform like maximum likelihood," *Proc. IEEE*, vol. 70, no. 9, pp. 975–987, Sept. 1982.

MODE AND TIME DELAY ESTIMATION FOR NON-DESTRUCTIVE EVALUATION SYSTEMS

J.Pearson, C J Macleod and T S Durrani

Department of Electronic Science and Telecommunication
University of Strathclyde, Glasgow, U.K.

ABSTRACT

This paper describes the development of a multipath and multimode propagation model for NDE. The mode content of a multipath is estimated using Maximum Likelihood (ML) estimation based on a receiving array delay vector. The model analysis allows a defect to be located by employing both the mode estimates and a ML estimation of the shear and longitudinal propagation angles. Some aspects of the attendant processing hardware are also discussed.

I INTRODUCTION

In recent years the demand for reliable and robust techniques for the non-destructive evaluation (NDE) of various industrial structures and systems has increased considerably. In situations ranging from nuclear reactors to offshore oil exploration and extraction platforms, there are stringent requirements on structural safety which can best be met by using sophisticated NDE systems. Ultrasonic based NDE should supply the required sophistication provided algorithms for comprehensive defect location and sizing are available. Inherent in this is the requirement for signal processing techniques to overcome the effects of multipath signal transmission and mode conversions.

NDE systems have, in general, utilised large crystal transmitting and receiving transducers which do not indicate the path taken by the outward or the return waves nor their mode of propagation. By replacing these large crystals with ultrasonic arrays more detailed information may be extracted concerning:

a) the transmission path to a reflection point,

b) the path taken by any received echo,

c) the mode or modes propagating within a path and the angles of propagation.

Relating this knowledge to the test specimen geometry, reflections from the specimen boundaries may be distinguished from defects and inclusions, thus reducing false detections. Also, detection of multiple reflections from a given point will increase the likelihood of a defect being located at this point.

A multimode propagation model is derived here for identifying both the receiving path and its mode composition. The path may contain multi-shear and multi-longitudinal modes generated for each propagation angle and the model formulation is such that 'a priori' knowledge of the transmission path to the reflecting point is not required.

Subsequent analysis consists of estimating the number of shear and longitudinal modes, α_S and α_L respectively, present in a receiving path. A multi-element receiving array is used in conjunction with a time delay vector which is explicity related to the receiving wavefront curvature. Maximum Likelihood estimation algorithms are then used to derive estimates of α_S and α_L, and the propagation angles of the shear and longitudinal modes in question. The location of the defect may then be obtained using these estimates.

II THE LAYERED MODEL

The problem of parameter estimation for the location of a point source has been widely covered [1-3] and relates particularly to sonar and radar applications where a direct path propagating at a single or variable velocity, is assumed between the source and the receiving array.

The application of ultrasonic NDE to narrow channel specimens results in reflections and mode conversions at the material walls, and from the defect, and gives rise to multipath and multimode propagation. A given multipath may contain velocity variation along its length with an associated change in propagation angle. The model presented here accounts for the multipath and modes experienced in practical NDE by identifying the multipath mode content.

Consider the transmitting (TX) and receiving (RX) arrays shown in figure 1, and a defect located at (x_c, z_c) with reference to an arbitrary origin (0, 0). Each path may be considered to comprise a shear and longitudinal mode with the mode content determined by the number of skips taken by each mode as defined by figure 2. Note that in the layered diagram, the RX and TX paths may not

Reprinted from *Proc. ICASSP '82*, vol. 1, pp. 395–398, May 1982.

coincide at the same point c due to a difference in the integer number of skips contained in each path.

The interaction of ultrasound with defects has been studied by a number of authors and in particular Baborovsky et al[4] have shown Schlieren photographs of the interaction and its resulting reflections. In the context of the present paper, the curvature of the reflected wavefronts reported by them is of specific interest. Baborovsky et al have given examples of the effect of defect size on this curvature, and from this we have concluded that for defects less than 2 λ (λ = wavelength of the ultrasound impinging on the defect), the wavefront has a finite curvature.

The model considers the RX array to operate in the near far-field where it is assumed the wavefront is non-planar and the defect is a point source at the centre of the wavefront curvature. The TX beam impinging on the defect may be approximated by a ray from the TX array to the defect. This ray may also consist of multiples of a shear and longitudinal mode.

Assuming Snell's Law applies at the material walls and each TX or RX path consists of shear and longitudinal modes only, any particular path illustrated in figure 1 may be described by the layered model [5] of figure 2.

Consider each RX array element to be a spatial sampler of the return wavefront as shown in figure 3. The path to a given element may be approximated by a ray of the overall RX beam. Any RX ray will consist of one possible shear mode and one longitudinal mode which are related by Snell's Law. The ray may alternate between these two components due to mode conversions but no shear or longitudinal modes having different propagation angles will be created. The number of shear and longitudinal skips contained in a ray will be equal for each element.

The total measured transit time for the ith receiving path is

$$\tau_i = \alpha_S \; t_{S_i} + \alpha_L \; t_{L_i} + t_o \quad --- \quad (1)$$

where τ_i = time delay between transmission and reception for the ith path,

t_{S_i} and t_{L_i} are the shear and longitudinal mode transit times across one skip distance, respectively,

t_o = total transmission transit time between the TX array and the defect. t_o will be equal for each RX element.

i = 1, 2, 3 - - - N-1 and N is the total number of RX elements detecting the given return.

t_S and t_L are related to the ith ray propagation angles thus:

$$t_{S_i} = D/\mathrm{COS}\,(\theta_{S_i}) \quad --- \quad (2)$$

$$t_{L_i} = D/\mathrm{COS}\,(\theta_{L_i}) \quad --- \quad (3)$$

where θ_{S_i} and θ_{L_i} are the shear and longitudinal propagation angles for the ith ray respectively and D is the specimen thickness.

θ_{S_i} and θ_{L_i} to a first approximation are given by:

$$\theta_{S_i} = \mathrm{SIN}^{-1}\,(v_S(t_{i+1} - t_i)/(x_{i+1} - x_i)) \quad --- \quad (4)$$

$$\theta_{L_i} = \mathrm{SIN}^{-1}\,(v_L(t_{i+1} - t_i)/(x_{i+1} - x_i)) \quad --- \quad (5)$$

where t_i is the measured transit time delay between transmission and reception for the ith RX element, v_S and v_L are the shear and longitudinal propagation velocities respectively and x_i is the location of the ith RX element from the x origin.

The first order angular approximation in equations (4) and (5) for a pair of RX elements and a constant propagation velocity between a far-field source and the receiver, has been shown to yield the angle of the ray bisecting the elements[6]. Similarly, for a multimode path, equations (4) and (5) may be shown to yield the angles of the bisecting ray of a receiving element pair. The transit time difference measured between these elements can be used to obtain θ_{S_i} and θ_{L_i}.

To compensate for the bisecting angle discrepancy, τ_i of equation (1) is approximated to the average measured transit time to the ith and i^{th} + 1 elements.

Errors in estimating θ_{S_i} and θ_{L_i} will invalidate the equality in that equation and thus lead to delay-angle misalignment.

III MAXIMUM LIKELIHOOD ESTIMATION

The receiving channel wave propagation characteristics may be obtained by identifying the optimum values of α_S, α_L, and θ_S and θ_L from time delay estimates, for each return wavefront where θ_S and θ_L represent the overall estimate of propagation angles for a wavefront. The optimum values of these parameters are obtained using Maximum Likelihood estimation.

Expressing equation (1) in vector form, the time delay measurement error may be defined as

$$\mathcal{E} = T - AP \quad --- \quad (6)$$

where $T' = [\tau_i \; \tau_{i+1} \; --- \; \tau_{N-1}]$ = path delay vector

$$A' = \begin{bmatrix} t_{S_1} & t_{S_2} & --- & t_{S_{N-1}} \\ t_{L_1} & t_{L_2} & --- & t_{L_{N-1}} \\ 1 & 1 & --- & 1 \end{bmatrix} = \text{mode transit time skip vector}$$

$P' = [\alpha_S \; \alpha_L \; t_o]$ = parameter vector

' represents matrix transpose

Considering the errors to be normally distributed, the likelihood function of the parameter vector P, may be expressed as:

$$L(T/P) = (2\pi)^{-(N-1)/2} |V|^{-\frac{1}{2}} \exp[-\tfrac{1}{2}(T-AP)'V^{-1}(T-AP)] \quad --- (7)$$

where V = error covariance matrix.

$|V|$ = determinant of V

Taking the logarithm of both sides of equation (7), maximising the expression with respect to P and finally equating to zero gives:

$$\nabla_p[\ln L(T/P)] = -2A'V^{-1}T + 2A'V^{-1}AP = 0 \quad --- (8)$$

where ∇ = differential matrix operator

Hence,

$$P_{OPT} = (A'V^{-1}A)^{-1}A'V^{-1}T. \quad --- (9)$$

The vector P_{OPT} gives the ML estimates,

$\hat{\alpha}_S$, $\hat{\alpha}_L$ and $\hat{t}_o$.

The latter may be used to determine the TX path to the defect.

In practice, it is reasonable to assume that the errors will be uncorrelated and therefore equation (9) reduces to the standard least squares solution:

$$P_{OPT} = (A'A)^{-1}A'T \quad --- (10)$$

The estimates $\hat{\alpha}_S$ and $\hat{\alpha}_L$ may be constrained by considering the values which α_S and α_L may jointly hold. Referring to figure 2, α_S and α_L cannot be jointly real but both may be integer valued. If both estimates are real, it may be concluded that one parameter has been over estimated at the expense of the other. Hence the fractional part of both estimates may be added to or subtracted from, in turn, the remaining variables constrained to an integer value. This results in new estimates of $\hat{\alpha}_S$ and $\hat{\alpha}_L$, $\hat{\alpha}_{1S}$ and $\hat{\alpha}_{1L}$ respectively. Velocity differences may be accounted for by first multiplying this fractional term by the ratio of the propagating velocities.

Applying this constraint to $\hat{\alpha}_S$ and $\hat{\alpha}_L$ results in eight possible integer/real combinations of $\hat{\alpha}_{1S}$ and $\hat{\alpha}_{1L}$. Extraction of the optimum pairing is obtained using a minimum least squares error criteria of the form $e^2 = (T - AP_{OPT})'(T - AP_{OPT})$ for all $\hat{\alpha}_{1S}$ and $\hat{\alpha}_{1L}$ pairings. Improved estimates of α_S and α_L may be obtained using this constraint.

IV TIME DELAY ESTIMATION

The techniques presented in this paper depend critically on the estimation of T. Two methods for time delay estimation have been studied: a) clipped correlation techniques and (b) threshold detection. An array controller has been designed[5] that performs the main processing tasks associated with the defect location. The return signal to the array controller is first hard clipped by the receiving channels and the resulting pulse waveforms are sampled and stored in emitter coupled logic memory for processing.

Software implementation of a) above is achieved by cross correlating this record with a stored reference bit pattern representing the transmitted signal.

An alternative approach is the detection of the received signal against a variable amplitude threshold which translates the problem of time delay estimation to that of the estimation of pulse widths, and more importantly, pulse positions.

The estimates obtained from these methods are used to create the time delay vector T, used for the ML estimates.

V DEFECT LOCATION

The constrained ML estimates $\hat{\alpha}_{1S}$ and $\hat{\alpha}_{1L}$, may be used to estimate the defect coordinates (x_c, z_c):

$$\hat{x}_c = \hat{\alpha}_{1S}.D.TAN(\hat{\theta}_S) + \hat{\alpha}_{1L}.D.TAN(\hat{\theta}_L) + x_R \quad --- (11)$$

$$\hat{z}_c = (1 - \hat{\alpha}_f).D \quad --- (12)$$

$\hat{\theta}_S$ and $\hat{\theta}_L$ represent the ML estimates of equations (4) and (5) giving an overall shear and longitudinal propagating angle with respect to a central reference RX element located at x_R (figure 2).

The estimate $\hat{z}_c$ is based on that fractional part of the real valued $\hat{\alpha}_{1S}$ or $\hat{\alpha}_{1L}$, denoted by $\hat{\alpha}_f$.

For the specific case where $\hat{\alpha}_{1S} = \hat{\alpha}_{1L}$, $\hat{\alpha}_f$ will be zero for an odd number of skips and will be unity for an even number of skips.

The formulation presented does not require knowledge of the TX path, the angle of TX propagation nor the TX beam width thereby relaxing the constraints on the required TX beam profile and hence simplifying TX array design. However, if required, the analysis can be extended to identify the TX path using $\hat{t}_o$.

VI COMMENTS

The model presented in this paper enables estimates of defect locations to be made based on time delay measurement used in conjunction with ML techniques. Multimode and multipath propagation from the defect to the receiver can be catered for in the analysis.

The following sources of error and their effect on time delay/mode estimation are currently under investigation.

(a) cross talk between receiving elements.
(b) non-spherical wavefronts.
(c) finite word length effects.

(d) problems arising from ill conditioning due to small differential time delays between adjacent receiving elements.

VII REFERENCES

1. G.C. Carter, "Time Delay Estimation for Passive Sonar Signal Processing", IEEE ASSP, Vol. ASSP-29, No. 3, June 1981, pp 463-470.

2. W.R. Hahn, "Optimum Signal Processing for Passive Sonar Range and Bearing Estimation" JASA, Vol. 58, No. 1, July 1975, pp 201-207.

3. N.L. Owsley and G.R. Swope, "Time Delay Estimation in a Sensor Array", IEEE ASSP, Vol. ASSP-29, No. 3, June 1981, pp 519-523.

4. V.M. Baborovsky et al, "The Response of Ultrasound to Defects", Ultrasonics Int., 1975 Conference Proc., Session 3, pp 46-53.

5. T.S. Durrani, C.J. MacLeod et al, "Processing Techniques for the Inspection of Offshore Structures", Proc. of IEEE Int. Conference on ASSP, March 30, 1981, pp 564-567.

6. M.J. Jacobson, "Correlation of a Finite Distance Point Source", JASA, Vol. 31, No. 4, April 1959, pp 448-453.

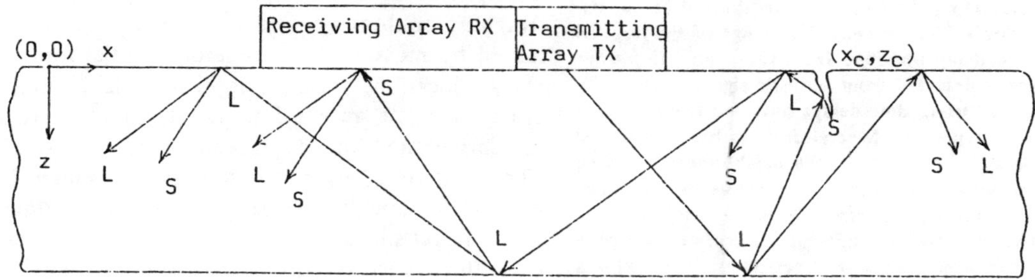

Figure 1: Multipath and Multimode Propagation

L= Longitudinal Mode
S= Shear Mode

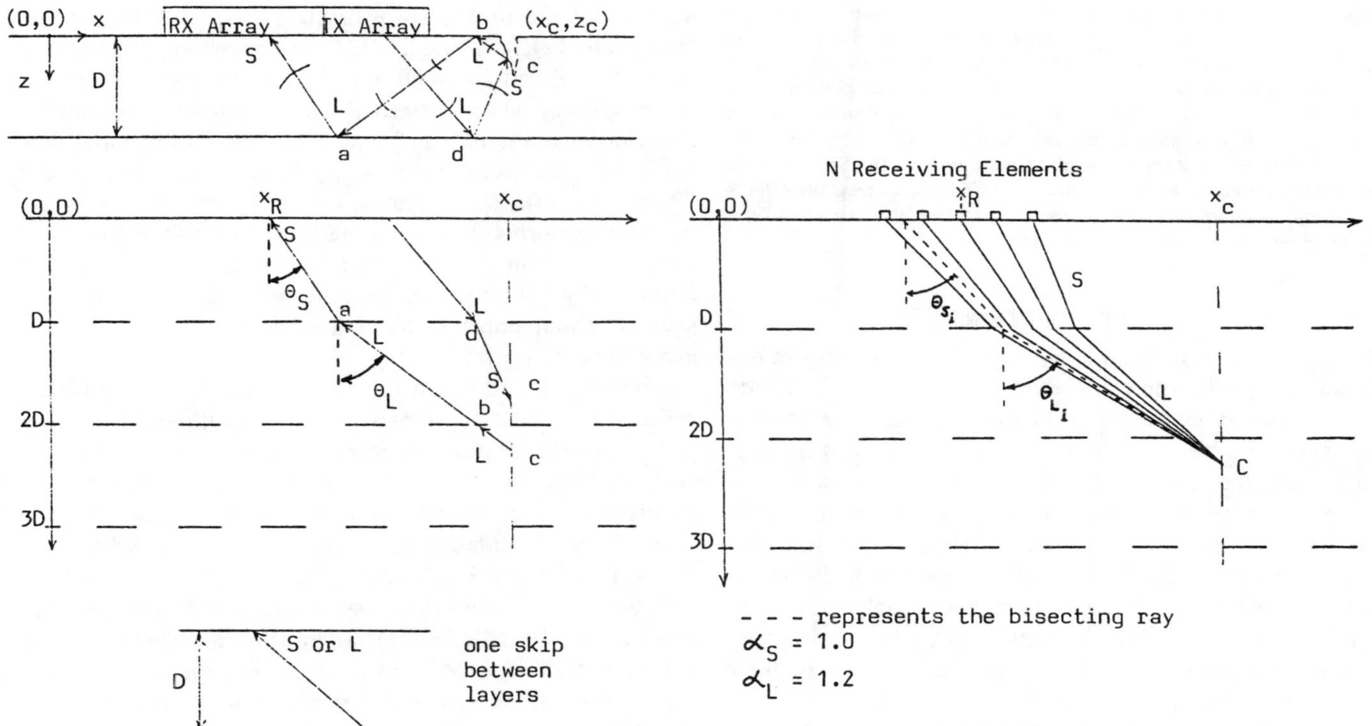

Figure 2: Layered Model

Figure 3: Receiving Array Spatial Sampling

An Overview on the Time Delay Estimate in Active and Passive Systems for Target Localization

AZIZUL H. QUAZI, MEMBER, IEEE

Abstract—Sonar and radar systems not only detect targets but also localize them. The process of localization involves bearing and range estimation. These objectives of bearing and range estimation can be accomplished actively or passively, depending on the situation. In *active sonar* or radar *systems*, a pulsed signal is transmitted to the target and the echo is received at the receiver. The range of the target is determined from the time delay obtained from the echo. In *passive sonar systems*, the target is detected from acoustic signals emitted by the target, and it is localized using time delays obtained from received signals at spacially separated points. Several authors have calculated the variance of the *time delay estimate* in the neighborhood of true time delays and have presented their results in terms of coherence function and *signal and noise autospectra*. Here we analyze these derivations and show that they are the same for the case of low *signal-to-noise ratio* (SNR). We also address a practical problem with a target-generated wide-band signal and present the Cramér–Rao lower bound on the variance of the time delay estimate as a function of commonly understood terms such as SNR, *bandwidth*, *observation time*, and *center frequency* of the band. The analysis shows that in the case of low SNR and when signal and noise autospectra are constants over the band or signal and noise autospectra fall off at the same rate, the minimum *standard deviation* of the time delay estimate varies inversely to the SNR, to the square root of the product of observation time and bandwidth, and to the center frequency (provided $W^2/12 f_0^2 \ll 1$, where W = bandwidth and f_0 = center frequency of the band). The only difference in the case of a high SNR is that the standard deviation varies inversely to the square root of the SNR, and all other parameter relationships are the same. We also address the effects of different signal and noise autospectral slopes on the variance of the time delay estimate in passive localization.

I. Introduction

SONAR and radar systems not only detect the targets but also find the location and velocity of the target. To locate a target using an *active system*, a pulse is transmitted to the target and the echo is received. The range of a target is determined using the *time delay* between the transmission of a pulse and the reception of its echo. To estimate this time delay, the system must determine the instant when the echo arrives. Generally, this is accomplished by matched filter or correlation where the "clean" reference signal, i.e., transmitted signal, is available. The time delay is estimated by measuring the peak of the output processor (matched filter or correlator), but exactly when this peak occurs is uncertain owing to the noise added to the echo signal.

To locate a target using a *passive system*, the sonar system receives a signal generated by the target and noise at spatially separated points. This system provides bearing and range information on a target by comparing its received signal at a multiplicity of widely spaced points along the length of its own ship or along a towed array. The target's bearing is determined by measuring the time differential for received signals at two locations. This time differential is obtained by cross correlating the received signals from these two points and measuring the time displacements of the correlogram peak. The range of a target is determined by measuring the difference of the time differential at two pairs of points.

The target's location may be determined either in the active or in the passive system by measuring the time delays that are obtained from correlation peaks. Therefore, the accuracy of the *time delay estimate* is critical to the accuracy of a target's bearing and range estimation.

Considerable research has been conducted to estimate the variance of the time delay estimate [1]–[7]. Helstrom [1], Woodward [2], and Wahlen [3] have presented the variance of time delay errors about true time delay, especially for an active system, where a clean reference deterministic signal is available in terms of signal energy to noise autospectral density and root-mean-square (rms) bandwidth. Knapp and Carter [4] have shown the variance of a time delay estimate in the neighborhood of true time delay in terms of coherence function. Hahn [5], Schultheiss [6], and Tomlinson and Sorokowsky [7] have presented the variance of the time delay estimate about unbiased mean time delay in terms of *signal and noise autospectra*.

Sometimes it is difficult for a user to determine whether the variance of time delay errors about a true time delay in the passive system obtained by several authors in terms of coherence function and signal and noise autospectra is the same or different [4]–[7]. An attempt to answer this question is made here. The results obtained by Knapp and Carter [4], Schultheiss [6], Tomlinson and Sorokowsky [7], and Hahn [5] are unified and the analysis shows that these results are the same for the case of low SNR (SNR $\ll 1$), where the wide-band signal has a flat spectrum and the noise is white. Furthermore, the variance of time delay errors as a function of commonly understood terms such as SNR, *bandwidth*, *observation time*, *center frequency*, and ratio of the bandwidth to center frequency are calculated and presented. Also, a practical problem with the wide-band signal and noise in the passive system is addressed, and the effect of change of SNR, observation time, bandwidth, center frequency, the ratio of bandwidth to center frequency, and signal and noise autospectral falloff with frequency on

Manuscript received April 4, 1980; revised October 9, 1980. This work was supported by the Naval Underwater Systems Center's Wide Aperture Array Electronics Project.

The author is with the Naval Underwater Systems Center, New London Laboratory, New London, CT 06320.

Reprinted from *IEEE Trans. Acoust., Speech, Signal Processing*, vol. 29, no. 3, pt. 2, pp. 527–533, June 1981.

the variance or *standard deviation* of time delay errors are investigated. In addition, the standard deviation of time delay errors in the passive and active systems are compared.

II. Variance of Time Delay Estimate

A. Active System

To measure the range of a target in an active system, it is necessary to estimate the time delay D at which the echo arrives at the receiver. If the time from the transmission of the pulse is measured, the range of the target is $PD/2$, where P is the speed of sound or electromagnetic wave. The received signal will consist of a deterministic signal that is corrupted with white noise of spectral density $N_0/2$.

It has been shown that the Cramér-Rao lower bound of variance of time delay errors about true time delay is [1]-[3]

$$\sigma_D^2 \geq \frac{1}{d^2\beta^2} \tag{1}$$

where

$$d^2 = \frac{2E}{N_0}, \tag{2}$$

$$\beta^2 = \frac{\int_{-\infty}^{\infty} \omega^2 |F(\omega)|^2 \, d\omega}{\int_{-\infty}^{\infty} |F(\omega)|^2 \, d\omega}, \tag{3}$$

E = energy of the signal $S(t)$, $F(\omega) = \int_{-\infty}^{\infty} S(t) e^{-j\omega t} \, dt$, and β = a measure of bandwidth.

Woodward [2] has shown that if the value of d^2 is about 8 or more, then the variance of the time delay estimate about true time delay can be estimated without ambiguity.

Here we assume that the signal autospectrum is two sided and extends from f_1 to f_2 Hz (and also $-f_1$ to $-f_2$ Hz) with spectral density of $S_0/2$ W/Hz. Then

$$\beta^2 = \frac{2\int_{f_1}^{f_2} (2\pi f)^2 S_0/2 \; 2\pi \, df}{2\int_{f_1}^{f_2} S_0/2 \; 2\pi \, df} = (2\pi)^2(f_2^2 + f_1 f_2 + f_1^2)/3 \tag{4}$$

$$\sigma_D^2 \geq \frac{1}{\frac{2E}{N_0}\frac{4\pi^2}{3}(f_2^2 + f_1 f_2 + f_1^2)} = \frac{1}{\frac{2ST}{N_0(f_2 - f_1)}\frac{4\pi^2}{3}(f_2 - f_1)(f_2^2 + f_1 f_2 + f_1^2)} = \frac{3}{8\pi^2 T}\frac{1}{\text{SNR}}\frac{1}{(f_2^3 - f_1^3)} \tag{5}$$

where signal power $S = S_0(f_2 - f_1)$, noise power $N = N_0(f_2 - f_1)$, observation time = T, and signal-to-noise ratio = S/N = SNR. Therefore, the standard deviation of the time delay estimate about true time delay is

$$\sigma_D \geq \left(\frac{3}{8\pi^2 T}\right)^{1/2} \frac{1}{\sqrt{\text{SNR}}} \frac{1}{\sqrt{f_2^3 - f_1^3}}. \tag{6}$$

Equation (6) is valid for any SNR and may be written in terms of bandwidth W and center frequency f_0 as

$$\sigma_D = \left(\frac{1}{8\pi^2}\right)^{1/2} \frac{1}{\sqrt{\text{SNR}}} \frac{1}{\sqrt{TW}} \frac{1}{f_0} \frac{1}{\sqrt{1 + W^2/12 f_0^2}} \tag{7}$$

where

$$f_1 = f_0 - \frac{W}{2}, \quad \text{and} \quad f_2 = f_0 + \frac{W}{2}.$$

Equation (7) shows that σ_D is inversely proportional to the

1) square root of the SNR;
2) square root of the product of bandwidth and time;
3) center frequency and, also, is a function of the ratio of bandwidth and center frequency.

B. Passive System

No signal is transmitted in the passive system. The received signals are composed of signals generated by target and noise. It is assumed that the target signal and noise are not correlated and are a stationary random process. We calculate the variance of the time delay estimate in the neighborhood of true time delay using derivations provided by several authors as a function of SNR, center frequency, bandwidth, and observation time and then compare these results [4]-[7].

1) Time Delay Estimate at Low SNR, Approach 1: Knapp and Carter [4] have shown that the Cramér-Rao lower bound variance of the time delay estimate about the true value using the coherence function is

$$\sigma_D^2 \geq \left\{2T \int_0^{\infty} (2\pi f)^2 \frac{|\gamma(f)|^2}{[1 - |\gamma(f)|^2]} \, df\right\}^{-1} \tag{8}$$

where $\gamma(f)$ is coherence function and T is observation time, and

$$|\gamma(f)|^2 = \frac{G_{ss}^2(f)}{[G_{ss}(f) + G_{nn}(f)]^2} \tag{9}$$

where $G_{ss}(f)$ is the signal autospectrum and $G_{nn}(f)$ is the noise autospectrum. Let

$$\frac{G_{ss}^2(f)}{[G_{ss}(f) + G_{nn}(f)]^2} \equiv \frac{S^2}{(S+N)^2} \tag{10}$$

and, then,

$$\frac{|\gamma(f)|^2}{[1 - |\gamma(f)|^2]} = \frac{S^2}{(S+N)^2} \Big/ \left[1 - \frac{S^2}{(S+N)^2}\right] \simeq (\text{SNR})^2 \tag{11}$$

when SNR $\ll$ 1.

Substituting the value of

$$|\gamma(f)|^2/[1 - |\gamma(f)|^2]$$

from (11) into (8), we get

$$\sigma_D^2 \geq \left\{2T \int_0^{\infty} (2\pi f)^2 \frac{S^2(f)}{N^2(f)} \, df\right\}^{-1}. \tag{12}$$

Assuming the signal and noise autospectra are constant over the band extending from f_1 to f_2 Hz with S_0 and N_0 W/Hz, respectively, we see that

$$\sigma_D^2 \geqslant \left(\frac{3}{8\pi^2 T}\right) \frac{1}{(S_0/N_0)^2} \frac{1}{(f_2^3 - f_1^3)}. \tag{13}$$

Therefore, for low SNR, the standard deviation of the time delay estimate is

$$\sigma_D \geqslant \left(\frac{3}{8\pi^2 T}\right)^{1/2} \frac{1}{\text{SNR}} \frac{1}{\sqrt{f_2^3 - f_1^3}} \tag{14}$$

where

$$S = S_0(f_2 - f_1)$$
$$N = N_0(f_2 - f_1).$$

2) Time Delay Estimate at Low SNR, Approach 2: Schultheiss [6] has shown that the minimum variance of time delay elements about the true time delay is given by

$$\sigma_D^2 \geqslant \frac{2\pi}{T} \left\{ \int_0^\infty \frac{M\omega^2 S^2(\omega)/N^2(\omega)}{1 + MS(\omega)/N(\omega)} d\omega \right\}^{-1} \tag{15}$$

where M is equal to the number of hydrophones or arrays that are utilized to measure the time delays. In case of a low SNR (SNR $\ll 1$), i.e., $MS(\omega)/N(\omega) \ll 1$ and $M = 2$, (15) reduces to

$$\sigma_D^2 \geqslant \frac{2\pi}{T} \left\{ \int_0^\infty 2\omega^2 S^2(\omega)/N^2(\omega)\, d\omega \right\}^{-1}$$
$$= \frac{3}{8\pi^2 T} \frac{1}{(S_0/N_0)^2} \frac{1}{f_2^3 - f_1^3}. \tag{16}$$

In (16), we have assumed that the signal and noise autospectra are flat, extend from f_1 to f_2 Hz, and yield

$$\sigma_D \geqslant \left(\frac{3}{8\pi^2 T}\right)^{1/2} \frac{1}{\text{SNR}} \frac{1}{\sqrt{f_2^3 - f_1^3}} \tag{17}$$

where

$$S = S_0(f_2 - f_1)$$
$$N = N_0(f_2 - f_1).$$

3) Time Delay Estimate at Low SNR, Approach 3: Hahn [5] and Tomlinson and Sorokowsky [7] have shown that the variance of the time delay estimate about the true time delay is given by

$$\sigma_D^2 = \frac{\dfrac{1}{2\pi T} \displaystyle\int_{-\infty}^{\infty} \omega^2 [N_1(\omega) N_2(\omega) + S(\omega)\{N_1(\omega) + N_2(\omega)\}]\, d\omega}{\left[\dfrac{1}{2\pi} \displaystyle\int_{-\infty}^{\infty} \omega^2 S(\omega)\, d\omega\right]^2} \tag{18}$$

where $S(\omega)$, $N_1(\omega)$, and $N_2(\omega)$ are autospectra of the signal $S(t)$, noise $N_1(t)$, and noise $N_2(t)$, respectively.

At low SNR (SNR $\ll 1$), (18) may be approximated by

$$\sigma_D^2 \cong \frac{\dfrac{1}{2\pi T} \displaystyle\int_{-\infty}^{\infty} \omega^2 N_1(\omega) N_2(\omega)\, d\omega}{\left[\dfrac{1}{2\pi} \displaystyle\int_{-\infty}^{\infty} \omega^2 S(\omega)\, d\omega\right]^2}. \tag{19}$$

Assuming that $N_1(\omega) = N_2(\omega) = N(\omega)$ and that the signal autospectrum $S(\omega)$ and noise autospectrum $N(\omega)$ are constant over the band extending from f_1 to f_2 Hz with densities $S_0/2$ and $N_0/2$, respectively, we can rewrite (19) as follows:

$$\sigma_D^2 \cong \frac{\dfrac{1}{2\pi T}\, 2 \displaystyle\int_{f_1}^{f_2} (2\pi f)^2 \frac{N_0^2}{4}\, 2\pi\, df}{\left[\dfrac{1}{2\pi}\, 2 \displaystyle\int_{f_1}^{f_2} (2\pi f)^2 \frac{S_0}{2}\, 2\pi\, df\right]^2}$$
$$= \left(\frac{3}{8\pi^2 T}\right) \frac{1}{(S_0/N_0)^2} \frac{1}{(f_2^3 - f_1^3)} \tag{20}$$

$$\sigma_D \cong \left(\frac{3}{8\pi^2 T}\right)^{1/2} \frac{1}{\text{SNR}} \frac{1}{\sqrt{f_2^3 - f_1^3}} \tag{21}$$

where

$$S = S_0(f_2 - f_1)$$
$$N = N_0(f_2 - f_1).$$

Notice that (14), (17), and (21) are the same. A comment may be in order. In terms of coherence function and signal and noise autospectra, the basic derivations [see (8) and (15)] for the Cramér–Rao lower bound of the variance of the time delay estimate about true time delay are the same as those shown in (13) and (16), and σ_D^2 is presented in terms of SNR, W, T, f_1, and f_2. Therefore, in the case of low SNR with constant signal and noise autospectra over the band, we can generalize that the standard deviation of the time delay estimate in the neighborhood of true time delay is the same as shown in (14), (17), and (21). The equations may be further simplified as a function of SNR, product of bandwidth and observation time, and center frequency:

$$\sigma_D = \frac{1}{\sqrt{8\pi^2}} \frac{1}{\text{SNR}} \frac{1}{\sqrt{TW}} \frac{1}{f_0} \frac{1}{\sqrt{1 + W^2/12 f_0^2}} \tag{22}$$

or

$$\sigma_D \sim \frac{1}{\text{SNR}} \qquad T, W, f_0 \text{ constant}$$
$$\sim \frac{1}{\sqrt{T}} \qquad \text{SNR}, f_0, W \text{ constant}$$
$$\sim \frac{1}{f_0} \qquad \text{SNR}, T, W \text{ constant, and } W \ll f_0$$
$$\sim \frac{1}{\sqrt{W}} \qquad \text{SNR}, T, f_0 \text{ constant}$$
$$\sim \frac{1}{\sqrt{1 + W^2/12 f_0^2}} \frac{1}{\sqrt{W}} \frac{1}{f_0} \qquad \text{SNR}, T \text{ constant.}$$

Notice that the only difference between an active and a passive system when estimating the variance of the time delay is the term $\sqrt{SNR}$ with the SNR, as shown in (6) and (21); all other terms are the same.

4) Time Delay Estimate at High SNR: So far the analytical results of standard deviation σ_D of the time delay estimate for low SNR have been presented. For completeness, the results for the standard deviation of the time delay estimate when the SNR is high (SNR $\gg$ 1) have been presented. It is now possible to show algebraically and by approximations that (8), (15), and (18) yield

$$\sigma_D \cong \left(\frac{3}{4\pi^2 T}\right)^{1/2} \frac{1}{\sqrt{SNR}} \frac{1}{\sqrt{f_2^3 - f_1^3}}. \tag{23}$$

Observe that the standard deviation of the time delay estimate in the case of high SNR, as shown in (23), varies inversely to the square root of the SNR, whereas it varies inversely to the SNR in the case of low SNR, as shown in (14), (17), and (21). Notice that (23) is similar to (6), which is valid for the active system. The only difference is that (23) is $\sqrt{2}$ times higher than (6). This is intuitively appealing because, in the case of the passive system (23), both the received signals are corrupted by noise; but, in the case of the active system (6), a clean reference or transmitted signal is available for correlation.

5) Effects of Signal and Noise Autospectral Falloff on Time Delay Estimate: Up to now, in the analysis of variance of the time delay estimate, we have assumed that signal and noise autospectra are constants over the band W, which extends from f_1 to f_2 Hz. However, in practice, neither the signal autospectrum nor the noise autospectrum in underwater acoustics is constant over the bandwidth. Therefore, we address a practical problem where signal and noise autospectra fall off and investigate the effects of spectral falloff on the variance of the time delay estimate.

Assuming that the signal and noise autospectra extend from f_1 to f_2 Hz and signal and noise autospectra fall off at the rate $10p$ dB/decade $\{S_0(f_1/f)^p\}$ and $10n$ dB/decade $\{N_0(f_1/f)^n\}$, we can rewrite (12) or (16) as follows:

$$\sigma_D^2 = \left\{2T \int_{f_1}^{f_2} (2\pi f)^2 f_1^{2p-2n} f^{2n-2p} \left(\frac{S_0}{N_0}\right)^2 df\right\}^{-1}$$

$$= \frac{1}{8\pi^2 T} \frac{1}{(S_0/N_0)^2} \frac{1}{f_1^{2(p-n)}} \frac{3+2(n-p)}{f_2^{3+2(n-p)} - f_1^{3+2(n-p)}},$$

$$(p-n) \neq 1.5, \quad SNR \ll 1 \tag{24}$$

$$\therefore \sigma_D = \left(\frac{1}{8\pi^2 T}\right)^{1/2} \frac{1}{S_0/N_0} \frac{1}{f_1^{p-n}} \frac{[3+2(n-p)]^{1/2}}{[f_2^{3+2(n-p)} - f_1^{3+2(n-p)}]^{1/2}},$$

$$(p-n) \neq 1.5, \quad SNR \ll 1 \tag{25}$$

where

$$S_0/N_0 = SNR \frac{1-p}{1-n} (f_1^{n-p}) \frac{f_2^{1-n} - f_1^{1-n}}{f_2^{1-p} - f_1^{1-p}}, \qquad p \neq 1, \quad n \neq 1$$

and

$$\sigma_D = \left(\frac{1}{8\pi^2 T}\right)^{1/2} \frac{1}{S_0/N_0} \frac{1}{f_1^3 \ln f_2/f_1},$$

$$\text{only for} \quad (p-n) = 1.5. \tag{26}$$

Notice that when n and p equal zero, (24) yields to (13), as expected. In other words, when n and p are equal to zero, which implies that the signal and noise autospectra are constants over the band, the variance of the time delay estimated is expected to be equal to (13), and it is.

Also observe that when n and p are equal, signal and noise autospectra are falling off at the same rate, and the variance σ_D^2 is the same as that in (13). So, the variance does not change when signal and noise autospectra fall off at the same rate. On the other hand, if the noise autospectrum falls off faster than the signal autospectrum, then the variance σ_D^2 decreases and the reverse is true if the signal autospectrum falls off faster than the noise autospectrum.

A comment may be in order. In derivations of (8) and (15) a shaping filter was utilized before correlation to obtain the Cramér-Rao lower bound of the variance of the time delay estimate; but in the case of (18), the shaping filter was not utilized to obtain the variance of the time delay estimate. However, if the shaping filter is utilized to obtain the minimun variance, then (18) will yield the same result shown in (13) or (20) [8].

III. Analysis Results

An investigation was made of the effects of SNR, bandwidth, observation time, and center frequency on the standard deviation of the time delay error about mean time delay, which is assumed to be unbiased. In Fig. 1, the standard deviation σ_D is plotted against the SNR in the range of -10 to -20 dB. The signal and noise autospectra are constants over the band $W = 4000$ Hz. The center frequency $f_0 = 4000$ Hz ($f_1 = 2000$ Hz and $f_2 = 6000$ Hz), and the observation time 60 and 120 s. It is seen from Fig. 1 that σ_D varies inversely with the SNR in the passive system, whereas σ_D varies inversely to the square root of the SNR in the active system. By doubling the integration time from 60 to 120 s, σ_D decreases, which is an improvement of 1.5 dB (10 log 2) in both the active and passive systems. It is found in Fig. 1 that σ_D is higher by a factor of $1/(SNR)^{1/2}$ in the passive system compared with the result in the active system, as expected.

In order to see the effect of change in center frequency f_0 on standard deviation σ_D, the latter is plotted in Fig. 2 as a function of SNR with the center frequency as a parameter. Fig. 2 shows that σ_D decreases with increasing center frequency if W and T are held constant. In other words, if the constant bandwidth W is moved along the frequency line with increasing frequency, keeping other parameters such as SNR, T, and W constant, then σ_D will decrease. Most probably this is a result of an increase in oscillations due to increasing frequency. Therefore, it indicates that one can measure the position of the correlation peak more accurately if the center frequency is increased; i.e., the uncertainty in peak position is decreased with increasing frequency. Fig. 3 shows σ_D versus f_0 for SNR's of -10, -15, and -20 dB, with $W = 400$ Hz and $T = 120$ s. By

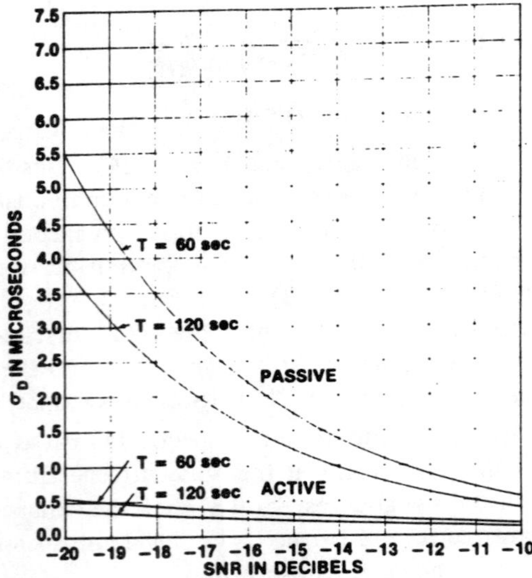

Fig. 1. Standard deviation of time delay estimate as a function of SNR for active and passive systems with different integration times. $W = 4000$ Hz and $f_0 = 4000$ Hz.

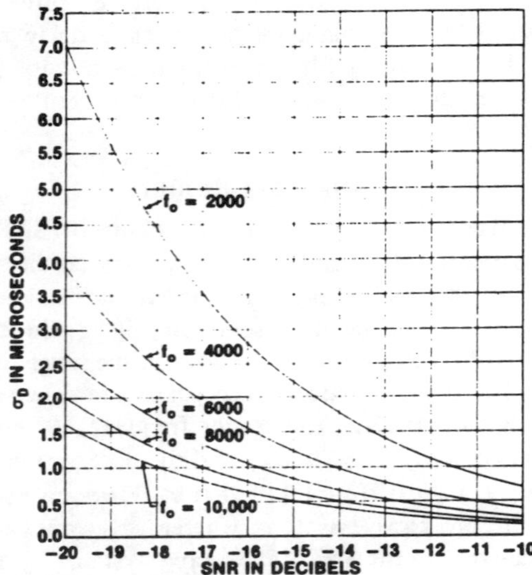

Fig. 2. Standard deviation of time delay estimate as a function of SNR with center frequency f_0 as a parameter. $W = 4000$ Hz and $T = 120$ s.

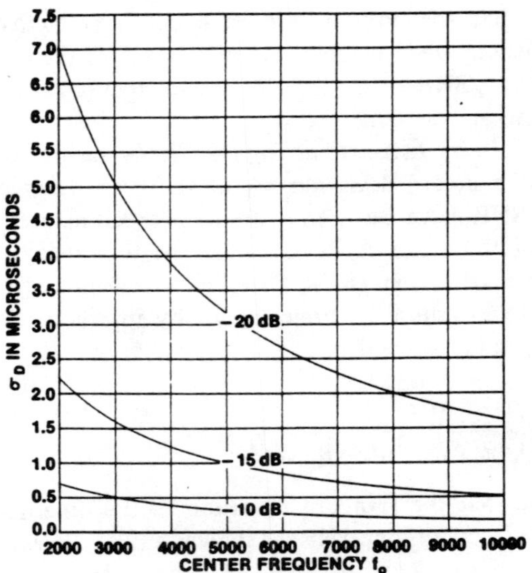

Fig. 3. Standard deviation of time delay estimate as a function of center frequency f_0 at different SNR's. $T = 120$ s and $W = 4000$ Hz.

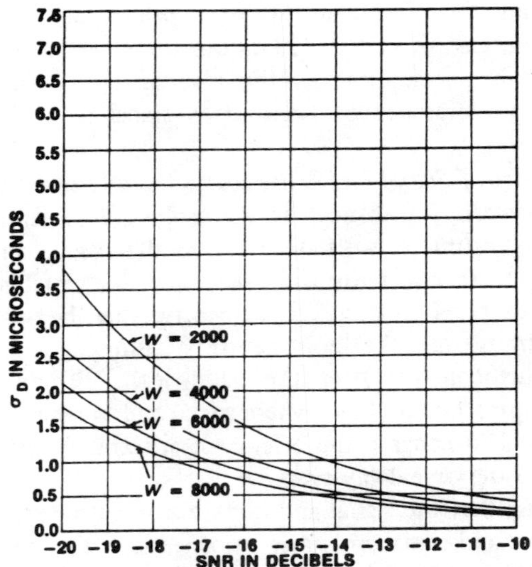

Fig. 4. Standard deviation of time delay estimate as a function of SNR with bandwidth W as a parameter. $T = 120$ s and $f_0 = 6000$ Hz.

doubling the center frequency, an improvement of about 3 dB (10 log 2) in σ_D is possible.

Fig. 4 shows the effect of change in bandwidth W on the standard deviation of the time delay estimate σ_D, which decreases with increasing bandwidth for constant T, f_0, and SNR.

So far, a result of the analysis of the variance of the time delay estimate when the signal and noise autospectra are constants over the band has been presented. Now, a result of the analysis for when the signal and noise autospectra fall off and how it affects the variance or standard deviation of the time delay estimate is shown.

Fig. 5 illustrates the standard deviation of the time delay estimate as a function of SNR in the range of -10 to -20 dB. The integration time is 120 s and the bandwidth is 4000 Hz ($f_1 = 2000$ Hz and $f_2 = 6000$ Hz). Also, the figure shows that the spectral falloff does not affect σ_D when n and p are equal. In other words, as long as the autospectral falloff of signal and noise are equal, σ_D does not change compared with the standard deviation of the time delay estimate when signal and noise autospectra are constants over the band.

Fig. 6 shows the effect of signal autospectral falloff with frequency when the noise autospectrum is constant over the band. It is evident from the figure that σ_D increases with increasing signal autospectral falloff, as expected.

Fig. 7 shows the effect of noise autospectrum falloff, with the frequency keeping the signal autospectrum constant over the band on the standard deviation of the time delay estimate. It is seen from the figure that σ_D decreases with increasing falloff of the noise autospectrum, as expected.

IV. Conclusion and Summary

An attempt has been made here to calculate the Cramér-Rao lower bound of variance for the time delay estimate about

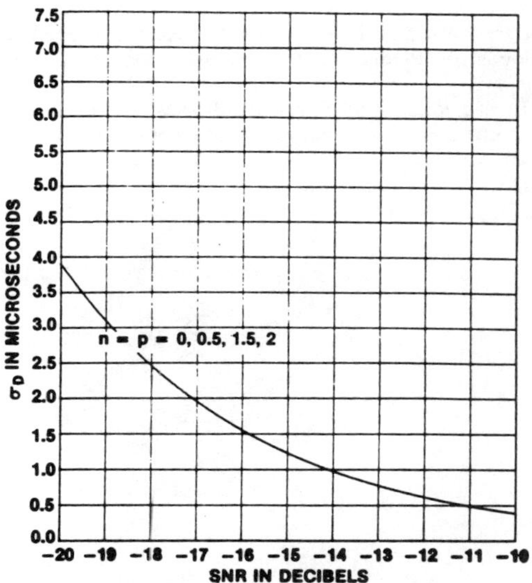

Fig. 5. Standard deviation of time delay estimate as a function of SNR. Signal and noise autospectral fall off at same rate. $n = p$, $T = 120$ s, $W = 4000$ Hz, and $f_0 = 4000$ Hz.

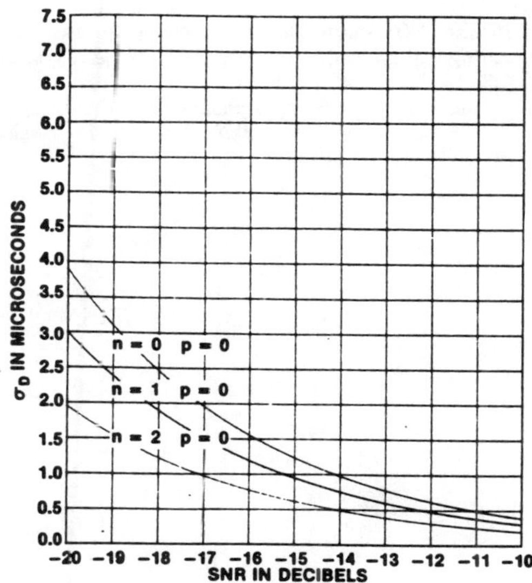

Fig. 7. Standard deviation of time delay estimate as a function of SNR. Signal autospectrum is constant over the band. Noise autospectrum falls off at a rate of 0, −10, and −20 dB/decade. $T = 120$ s, $f_0 = 4000$ Hz, and $W = 4000$ Hz.

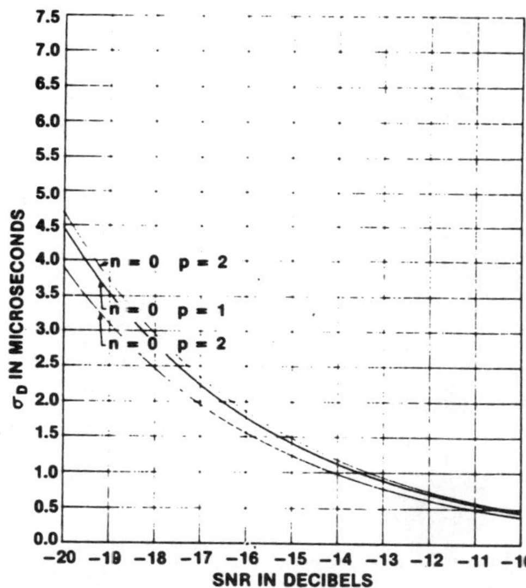

Fig. 6. Standard deviation of time delay estimate as a function of SNR. Noise autospectrum is constant over the band. Signal autospectrum falls off at a rate of 0, −10, and −20 dB/decade over the band. $T = 120$ s, $f_0 = 4000$ Hz, and $W = 4000$ Hz.

the time delay in active and passive systems as used in target localization. The derivations, obtained in various forms, have been analyzed and are presented here: standard deviation of time delay estimate in terms of commonly understood terms such as SNR, bandwidth, observation time, and center frequency. The noteworthy results for the standard deviation of the time delay estimate σ_D in the neighborhood of true time delay in the case of low SNR are:

- The standard deviation σ_D varies inversely to the square root of the SNR in the case of an active system, whereas σ_D varies inversely to the SNR in the case of the passive system provided observation time, bandwidth, and center frequency remain constant.
- σ_D varies inversely to the center frequency for constant W, T, SNR, and $W^2/12f_0^2 \ll 1$.
- σ_D varies inversely to the square root of bandwidth for constants SNR, f_0, T, and $W^2/12f_0^2 \ll 1$.
- σ_D remains constant so long as the signal and noise autospectra fall off at the same rate or are constant over the band.
- σ_D increases if the signal autospectrum falls off faster than the noise autospectrum.
- σ_D decreases if the noise autospectrum falls off faster than the signal autospectrum.

In the case of high SNR (SNR $\gg$ 1), the standard deviation of the time delay estimate varies inversely to the square root of the SNR, whereas σ_D varies inversely to the SNR in case of a low SNR. The effects of other parameters such as T and W remain the same for σ_D in both low and high SNR's.

Future research will include validating the minimum variance of the time delay estimate using simulation or experimental results. Also an investigation of how useful is the Cramér–Rao bound of variance in predicting the performance of bearing and range estimation at low, high, and "in-between" values of SNR is deemed advisable.

Acknowledgment

The author wishes to thank Dr. G. C. Carter and Dr. A. H. Nuttall for their helpful discussions and criticism.

References

[1] C. W. Helstrom, *Statistical Theory of Signal Detection*. New York: Pergamon, 1968, pp. 274–319.

[2] P. M. Woodward, *Probability and Information Theory, With Applications to Radar*. New York: Pergamon, 1953, pp. 81–99.

[3] D. A. Wahlen, *Detection of Signals in Noise*. New York: Academic, 1971, pp. 337–339.

[4] C. H. Knapp and G. C. Carter, "The generalized correlation method for estimation of time delay," *IEEE Trans. Acoust., Speech, Signal Processing*, vol. ASSP-24, pp. 320–327, Aug. 1976.

[5] W. R. Hahn, "Optimum signal processing for passive sonar range and bearing estimation," *J. Acoust. Soc. Amer.*, vol. 58, pp. 201–207, July 1975.

[6] P. M. Schultheiss, "Locating a passive source with array measurements–A summary result," in *Proc. ICASSP '79*, Washington, DC, Apr. 24, 1979, pp. 967–970.

[7] G. Tomlinson and J. Sorokowsky, "Accuracy prediction report, Raploc error model," IBM, Manassas, VA, June 30, 1977, pp. B1–B7.

[8] V. H. MacDonald and P. M. Schultheiss, "Optimum passive bearing estimation in a spatially incoherent noise environment," *J. Acoust. Soc. Amer.*, vol. 46, pp. 37–43, July 1969.

Effect of sound-speed profile on differential time-delay estimation

E. Richard Robinson and Azizul H. Quazi
Naval Underwater Systems Center, New London, Connecticut 06320

(Received 4 April 1984; accepted for publication 4 September 1984)

The sensitivity of correlograms to various sound-speed profiles is being studied in order to estimate the impact of refracted multipath acoustic energy on differential time-delay estimation (TDE). To do this, we utilize the generic sonar model to compute the crosscorrelation function for an array of two separated receivers in realistic ocean environments. The effects of the obtained correlogram peaks that show the differential time delays were compared with those predicted when assuming isovelocity profiles. The influence of a signal source's range and bearing, as well as the vertical separation of the receivers on the correlograms, is demonstrated for each environment considered. Our results indicate that TDE may be significantly influenced by the processing of highly refracted rays of many different types due to the selected sound-speed profile. Also, the Cramer–Rao lower bounds to the standard deviation of time-delay errors at various signal-to-noise ratios are compared with the results that were derived using the generic sonar model and found to be in close agreement.

PACS numbers: 43.60.Gk, 43.30.Cq

INTRODUCTION

Recently time-delay estimation (TDE) has drawn special attention in a variety of acoustic underwater systems. For example, it may be used to obtain the bearing and range of a signal source when a high degree of accuracy is required. Because of this need for precision, there has been considerable research in the area of the TDE.[1]

This article investigates the effects of multipath refraction within the ocean medium on crosscorrelation and, thus on TDE. For a given receiver-target geometry, correlograms are computed for several realistic ocean environments using the generic sonar model.[2] The effects on the correlograms of changes in target range and bearing, as well as the vertical separation of receivers, are investigated. Also, although it is quite well known that the refraction of energy due to temperature gradients in the oceans plays an important role in computing transmission loss, its effect on TDE has not been extensively explored.[3]

Some typical results of a study are presented to determine how different sound-speed profiles and their subsequent effect on refraction and multipath structure impact TDE. In addition, for selected cases, Cramer–Rao predicted lower bound to the standard deviation of time-delay errors and those obtained using the generic sonar model are compared (Table I).

I. THEORETICAL APPROACH AND DEVELOPMENT

Crosscorrelation models simulate the crosscorrelation coefficient for a given target range. Therefore, we first compute the crosscorrelation function of acoustic signals plus noise versus time delay using the inverse Fourier transform of the cross-spectral density. (See Fig. 1 for the geometry that illustrates this model.) Then, the signal's spectral density values for each receiver are generated at ranges R_1 and R_2. Finally, the function is normalized by the mean pressure squared at each hydrophone to produce the crosscorrelation coefficient.

Our objective is to investigate the sensitivity of the crosscorrelation coefficient to the sound-speed profile. For each such profile, we note the effects of changes in target range, bearing, and vertical-receiver separation. Only two realistic profiles were used; thereby, the same target-receiver configuration typified both in- and below-layer cases by merely changing the ocean's sound-speed profile. To demonstrate the significance of refraction on correlograms, certain cases were compared with results obtained assuming isovelocity conditions.

The accuracy of the differential TDE depends on the signal-to-noise ratio (SNR), observation time, effective bandwidth, and center frequency of the signal and noise and their spectral characteristics.[4] Here we discuss the effects of signal and noise spectral characteristics on the Cramer–Rao lower bounds of TDE and, hence, on bearing- and range-estimation capabilities.

The Cramer–Rao lower bound of the variance σ_D^2 of a TDE that is assumed to be unbiased is[4–7]

$$\sigma_D^2 = \frac{1}{8\pi^2}\frac{1}{TW}\frac{1}{f_{\rm rms}^2}\frac{1+\mathrm{SNR}_1+\mathrm{SNR}_2}{\mathrm{SNR}_1\cdot\mathrm{SNR}_2}, \qquad (1)$$

where

SNR_1 and SNR_2 = power SNRs over the band W Hz (extended from f_1–f_2 Hz) at receivers 1 and 2 (Fig. 1),

TABLE I. Environmental and system parameters.

Receiver depth	156 m
Target depth	159 m
Bottom depth	5486 m
Wind speed	13 kn.
Bottom province	2
Bandwidth	500 Hz
Center frequency	450 Hz
SNR	40 dB

Reprinted with permission from the *J. of the Acoustical Soc. of America*, vol. 77, no. 3, pp. 1086–1090, March 1985.

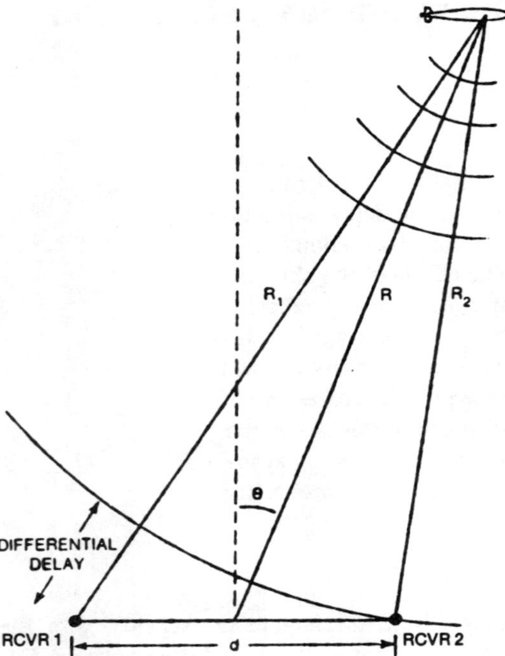

FIG. 1. Target and receiver geometry for differential TDE by crosscorrelation.

$$f_{\rm rms} = f_0\sqrt{1+\frac{W_2}{12f_0^2}}\ \text{Hz},\quad W=f_2-F_1\ \text{Hz},$$

and

$$f_0 = f_1 + W/2 = f_2 - W/2\ \text{Hz},$$

$T=$ observation time in s.

Therefore, the standard deviation of the time errors is given by

$$\sigma_D = \frac{1}{2\pi}\frac{1}{\sqrt{2TW}}\frac{1}{f_{\rm rms}}\frac{\sqrt{1+2\,\text{SNR}}}{\text{SNR}},$$

$$\text{SNR}_1 = \text{SNR}_2 = \text{SNR}, \tag{2a}$$

$$\simeq \frac{1}{2\pi}\frac{1}{\sqrt{2TW}}\frac{1}{f_{\rm rms}}\frac{1}{\text{SNR}},\quad \text{for SNR} \ll 1, \tag{2b}$$

and

$$\simeq \frac{1}{2\pi}\frac{1}{\sqrt{2TW}}\frac{1}{f_{\rm rms}}\frac{1}{\sqrt{\text{SNR}/2}},\quad \text{for SNR} \gg 1. \tag{2c}$$

The consequence of different and unequal signal and noise spectral falloff rates on σ_D are described elsewhere.[8]

Notice that Eq. (2) shows that σ_D is independent of the true differential time delay and depends only on the SNR, the observation time, and the signal and noise characterstics.

II. RESULTS

In our results we show predictions based on the cross-correlation model using the sound-speed profiles in Figs. 2(a) and (b). The profile in Fig. 2(a) is characterized by the rather deep-surface duct, where the layer depth occurs at 251 m. In contrast, the profile in Fig. 2(b) shows a relatively shallow layer (48.8 m) followed by a steep thermocline region. Figures 3 and 4 were generated using the profiles in Figs. 2(a) and 2(b), respectively. Table I lists the remaining environmental and system parameters used in making our predictions. In addition, model generated and analytically predicted time-delay errors at a low SNR are compared.

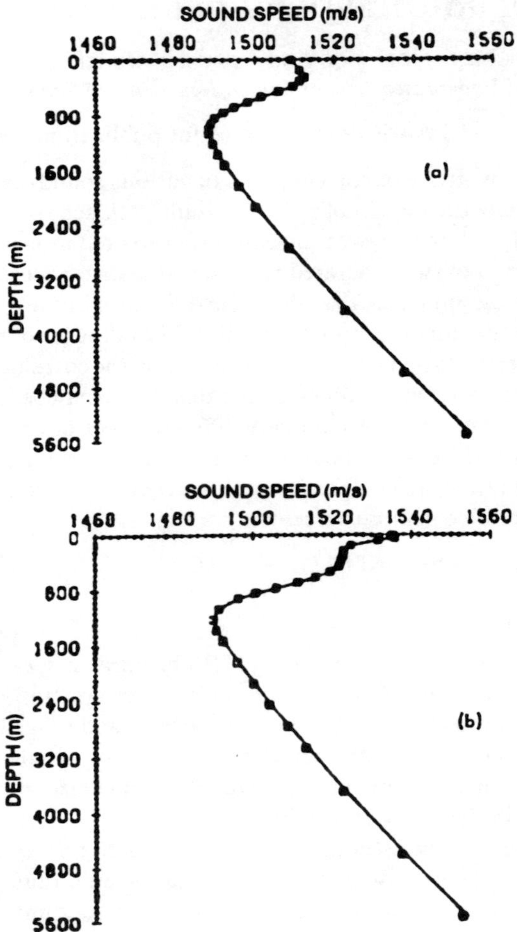

FIG. 2. (a) Ocean sound speed versus depth, layer depth 252 m. (b) Ocean sound speed versus depth, layer depth 49 m.

A. Deep layer

Figure 3(a), a typical correlogram, depicts both receivers at a depth of 156 m with a 300-m separation and a 159-m target that is off broadside at a range of 15 km. The peak positions indicate the differential time delays of signals arriving at the two receivers. Figure 3(b) shows that the effect on the correlogram of increasing the range to 25 km is rather insignificant. At broadside, the peak position is at a differential delay of 0, as expected. Also, because we are considering relatively high SNRs, the peak amplitude is close to 1 at both ranges.

With the target at 25 km, the array is steered 30 degrees from broadside [Fig. 3(c)]. Note that the correlation peak is at 104 ms, i.e., within 1 ms of the theoretically predicted peak position. Close agreement was anticipated because, in the deep layer, the rays are not undergoing rapid changes in speed, the deep-layer channeling produces rays of a similar type with little difference in travel time, and the dominant

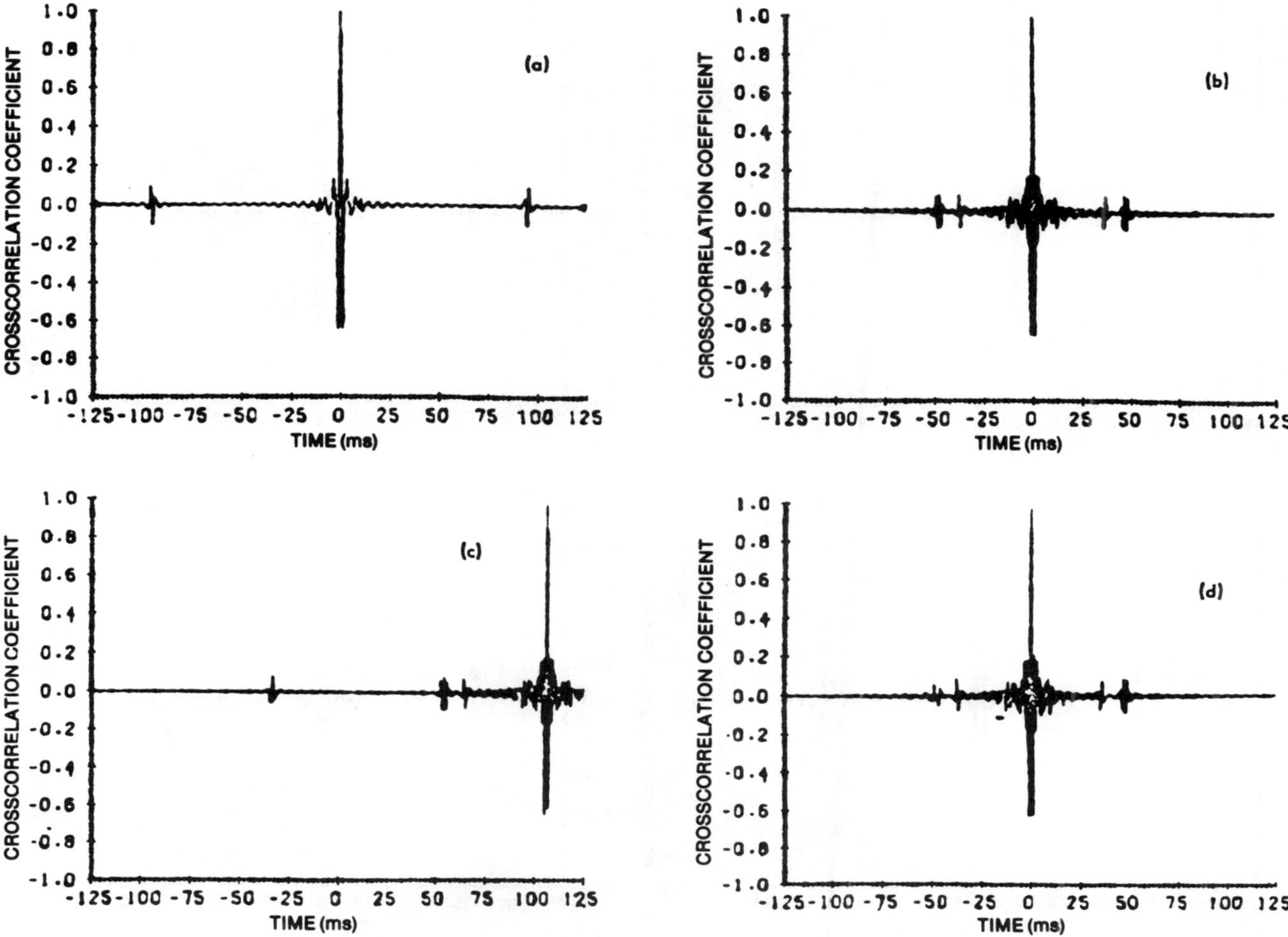

FIG. 3. (a) Crosscorrelation coefficient versus time delay at broadside, range 15 km. (b) Crosscorrelation coefficient versus time delay at broadside, range 25 km. (c) Crosscorrelation coefficient versus time delay at 30° off from broadside, range 25 km. (d) Crosscorrelation coefficient versus time delay at broadside, range 25 km, vertical-receiver separation 5 m.

peak amplitude is not sensitive to a change in target bearing. A comparison of Figs. 3(d) and 3(a) indicates that a change in the vertical separation of the two receivers has little significance, thereby reflecting the closeness in corresponding travel times to the two receivers.

B. Shallow layer

Figure 4(a)–4(d) correspond to Fig. 3(a)–3(d), with only the sound-speed profile changed. These predictions were made using the profiles in Fig. 2(b). It is expected that placing the receivers and target in the steep thermocline of this profile will significantly affect the correlograms. A comparison of Fig. 4(a) and 4(b) shows that increasing the target range from 15–25 km results in more multipath activity at 25 km than in the previously discussed deep-layer case. In the 30-deg case in Fig. 4(c), two significant effects are apparent: (1) The amplitude of the peak is reduced substantially and (2) the peak position at 98 ms is of the order of 6 ms less than the position expected for the corresponding isovelocity case. These effects are due to the rapid changes that occur in the thermocline region. Since the range differential of the two receivers is of the order of 160 m, corresponding travel times have increases approaching 100 ms. Because these increases were not at the same rate for each corresponding ray path, the correlation function was degraded.

Figure 4(d) demonstrates the sensitivity of the correlogram to vertical separation of the receivers in the shallow-layer environment. When comparing Figs. 3(d) and 4(d) note the total collapse of the peak at zero differential delay in Fig. 4(d). This, perhaps, unexpected impact on the correlogram is due to the effect of coherent processing many different types of ray paths at the two depths. Whenever travel-time differences are of the order of one-half wavelength or greater, this degree of degradation has been noted.

A comparison of GSM model-predicted-time-delay errors (due to low SNR) with those obtained from theoretical calculations via Eq. (2b) is given in Table II. There we list time-delay errors for the case considered in Fig. 3(c). Note that the model-predicted errors are the mean of five samples. It can be seen that the simulator model results are in close agreement in this range of the SNR. In fact, it has been shown[9] that the Cramer–Rao lower bound of variance for time delay errors about true time delay is

$$\sigma_D^2 \geqslant 1/d^2\beta^2, \tag{3}$$

where

$$d^2 = 2E/N_0, \tag{4}$$

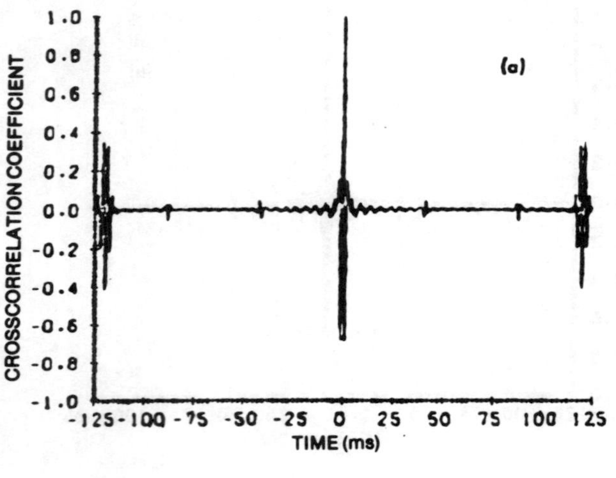

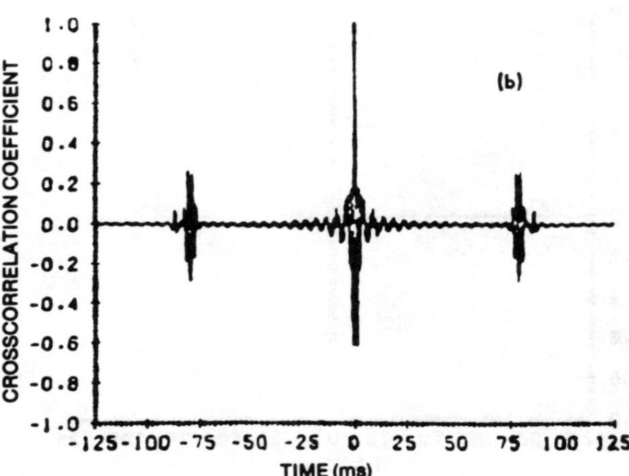

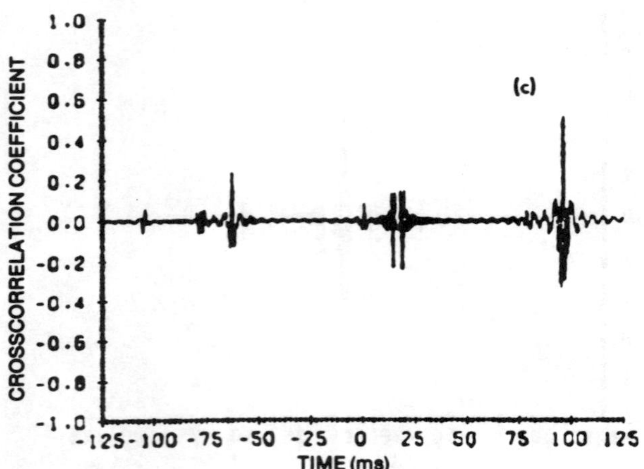

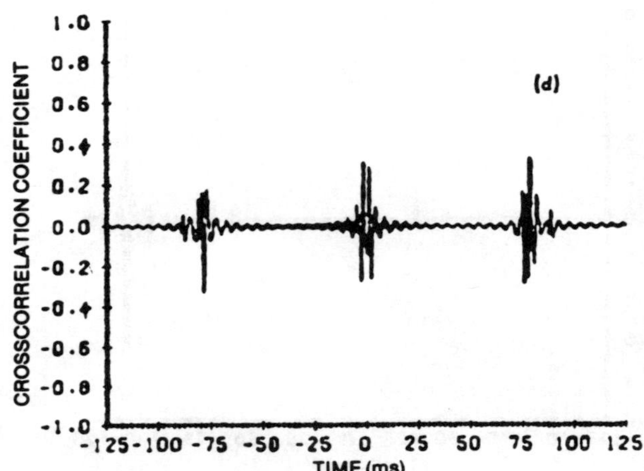

FIG. 4. (a) Crosscorrelation coefficient versus time delay at broadside, range 15 km. (b) Crosscorrelation coefficient versus time delay at broadside, range 25 km. (c) Crosscorrelation coefficient versus time delay at 30° off from broadside, range 25 km. (d) Crosscorrelation coefficient versus time delay at broadside, range 25 km, vertical-receiver separation 5 m.

and

$$\beta^2 = \frac{\int_{-\infty}^{\infty} \omega^2 F(\omega)^2 \, d\omega}{\int_{-\infty}^{\infty} F(\omega)^2 \, d\omega}, \tag{5}$$

where

$E =$ energy of the signal $S(t)$,

$$F(\omega) = \int_{-\infty}^{\infty} S(t)^{-j\omega t} dt,$$

and

$\beta =$ a measure of bandwidth.

Woodward further has shown[9] that if the value of $d^2 \geqslant 9$ dB (which corresponds to the range of values in this case), then the variance of the TDE about true time delay can be estimated without peak ambiguity.

TABLE II. Theoretical versus model-predicted TDE standard deviations.

SNR (dB)	σ (Theoretical) (ms)	σ (Model predicted) (ms)
0	0.04	0.025
−2	0.06	0.050
−4	0.10	0.082
−6	0.16	0.145
−8	0.26	0.200

III. CONCLUSION

Our results have indicated that TDE is very sensitive to different sound-speed profiles. We have found, assuming isovelocity conditions, that differences between these predicted time delays and the delays obtained for realistic ocean environments may greatly exceed the standard deviation of time-delay errors. These errors, which are a function of SNRs, observation time, and system parameters, are used to determine the accuracy of bearing and range estimation. Differences of this magnitude were demonstrated by varying target bearing and range for a given source-receiver geometry. Also demonstrated was the significant effect of increasing the vertical-receiver separation. Therefore, the extreme complexities of a multipath environment combined with marked refraction of the ray paths play a significant role in TDE. Future research will include a more detailed analytic solution of the effect of signal and noise characteristics on the peak position and amplitude variations of correlograms as a function of the sound-velocity profile.

ACKNOWLEDGMENT

The authors wish to thank Dr. Henry Weinberg, of NUSC, for his many helpful discussions.

[1]IEEE Trans. Acoust. Speech Signal Process. **ASSP-29** (3) (June 1981).

[2]H. Weinberg, *Generic Sonar Model*, NUSC Technical Report 5971 C (Naval Underwater Systems Center, New London, CT, December 1981).

[3]E. R. Robinson and A. H. Quazi, "The impact of refraction on time-delay estimation," in *Proc. ICASSP 1983* (GPO, Washington, DC, 1983), pp. 981–984.

[4]A. H. Quazi, "An overview on the time delay estimate in active and passive systems for target localization," IEEE Trans. Acoust. Speech Signal Process. **ASSP-29**, 527–533 (1981).

[5]W. R. Hahn, "Optimum signal processing for passive sonar range and bearing estimation," J. Acoust. Soc. Am. **58**, 201–207 (1975).

[6]P. M. Schultheiss, "Locating a passive source with array measurements—a summary results," in *Proc. ICASSP 1979* GPO, Washington, DC, 1979), pp. 967–970.

[7]C. H. Knapp and G. C. Carter, "The generalized corelation method for estimation of time delay," IEEE Trans. Acoust. Speech Signal Process. **ASSP-24**, 320–327 (1976).

[8]A. H. Quazi, "Effects of signal and noise spectral slopes on the time delay estimates in passive localization," in *Proc. ICASSP 1982* (GPO, Washington, DC, 1981), pp. 1265–1268.

[9]P. M. Woodward, *Probability and Information Theory, with Application to Radar* (Pergamon, New York, 1953), pp. 81–99.

ADAPTIVE COHERENCE ESTIMATION ON BRIEF INTRACARDIAC RECORDINGS

Alan Sahakian, Kristina Ropella, Jeffrey Baerman and Steven Swiryn

Departments of Biomedical and Electrical Engineering, Northwestern University; The Division of Cardiology, Department of Medicine, Evanston Hospital and Northwestern University Medical School; Evanston, IL

ABSTRACT

The magnitude-squared coherence (MSC) spectrum is useful in characterizing the degree of organization of cardiac rhythms and therefore in detecting cardiac arrhythmias such as fibrillation. Adaptive techniques are considered for the estimation of the magnitude-squared coherence spectrum of two simultaneous intracardiac signals. Estimators based on single and multiple-pass algorithms are proposed. Adaptive estimates of the MSC spectrum are shown to effectively separate fibrillatory from several non-fibrillatory rhythms on the basis of five seconds of data. This work may have significance for sophisticated anti-tachyarrhythmia devices.

INTRODUCTION

We have previously reported the utility of the magnitude-squared coherence (MSC) spectrum of two simultaneous intracardiac electrical recordings in characterizing the degree of organization of a cardiac rhythm [1,2,3]. An important issue which arises when estimating the MSC spectrum by a straightforward DFT-based approach involves the signal segmentation. A three-way tradeoff exists between temporal and spectral resolution, as well as the bias and variance of an MSC estimate [4,5]. Adaptive MSC spectrum estimation [6] may offer advantages, especially with time-varying spectra.

THEORETICAL BASIS

For our purposes, the MSC spectrum can be defined as:

$$MCS(f) = \frac{|S_{xy}(f)|^2}{S_{xx}(f)\ S_{yy}(f)} \qquad (1)$$

where signals x and y are two simultaneously-acquired intracardiac recordings. The MSC spectrum ranges in value from 0, if the two signals have no correlated components at a given frequency, to unity if the relationship between the two signals can be entirely explained by a linear model. Thus poorly-organized rhythms (e.g. fibrillation) give rise to lower coherence values than the more-highly organized rhythms (e.g. sinus rhythm or a regular tachycardia). If the cross and auto power spectra expressed in (1) are estimated using the DFT then these must be calculated as experimental averages for meaningful results.

Another method of estimating the MSC spectrum would be to first estimate the transfer functions which could be used to transform signal x into signal y ($H_{xy}(f)$) and signal y into x ($H_{yx}(f)$). The MSC spectrum can then be estimated as the product of these functions

$$MSC(f) = H_{xy}(f)\ H_{yx}(f) \qquad (2)$$

The well known Widrow-Hopf LMS algorithm [7] can be used to adaptively estimate the forward ($h_{xy}(t)$) and backward ($h_{yx}(t)$) transfer functions in the time domain. This adaptive estimation is performed by a pair of adaptive cancellers. Since these transfer functions are estimated in the time domain as impulse responses, they are transformed to the frequency domain and multiplied.

The convergence coefficient (μ) which is used in the weight-updating step of the LMS algorithm must be large for rapid convergence but small for stability. As discussed by Youn [8], we initially used a time-varying convergence coefficient which is based on signal variance. However, we found that this may cause problems with instability and misconvergence in the application at hand, and eventually chose to use a fixed value of μ.

The MSC estimate which results from this simple set of steps may contain values which are complex or out of the range of zero to unity. These errors occur since the two adaptive structures do not, in general, converge on two solutions which are complex conjugates of each other. Nutall [9] proposes estimation constraints to avoid such errors.

RESULTS

We used the adaptive technique to estimate MSC spectra on five-second intracardiac recordings, and compared two approaches. In the first the signals were passed through the adaptive estimator once. In the second approach we circulated the signals through the adaptive estimator 20 times (empirically found to be yield a nearly-static weight set). In both cases the initial weight set was zero.

A sample rate of 120 Hz and filter length of 21 were chosen. Thirty five rhythms (17 non-fibrillatory and 18 fibrillatory) from 20 patients were analyzed using both approaches. The non-fibrillatory rhythms included six sinus rhythm, two PSVT, five monomorphic ventricular tachycardia and four atrial flutter. The fibrillatory rhythms included 12 atrial and six ventricular examples (with all signals taken from the fibrillating chamber).

Reprinted from *Proc. of the 11th International Conference of the IEEE Engineering in Medicine and Biology Society,* vol. 1, pp. 224–225, November 1989.

Typical MSC spectra for fibrillatory and non-fibrillatory rhythms estimated by the two approaches are presented in figure 1. Figure 2 summarizes the results. With the single-pass approach, the two groups are well separated. The mean MSC values (over the entire 60-Hz band) for the non-fibrillatory rhythms were 0.097 to 0.423 (mean of the mean MSC values +/- standard deviation of the mean MSC values: 0.205 +/- 0.087). For the fibrillatory rhythms the mean MSC values were 0.007 to 0.088 (0.048 +/- 0.028), which is significantly lower ($p < 0.05$), and does not overlap the non-fibrillatory range.

Using the 20-pass approach, we found mean MSC values for non-fibrillatory rhythms which ranged from 0.202 to 0.686 (0.488 +/- 0.139). For the fibrillatory rhythms mean MSC values ranged from 0.035 to 0.150 (0.086 +/- 0.034), which is significantly lower ($p < 0.0005$), and again does not overlap.

CONCLUSION

We have considered the adaptive estimation of magnitude-squared coherence spectra on short-duration intracardiac recordings. Non-fibrillatory and fibrillatory rhythms are well separated by mean MSC values. A multiple-pass algorithm improves the inter-group separation.

REFERENCES

[1] K. Ropella, A. Sahakian, J. Baerman and S. Swiryn, "The coherence spectrum: a quantitative discriminator of fibrillatory and non-fibrillatory cardiac rhythms," to appear in Circulation.

[2] A. Sahakian, K. Ropella, J. Baerman and S. Swiryn, "Coherence measures of cardiac arrhythmias from intra cardiac and epicardial leads," Comp. in Cardiol. 1988 (in press).

[3] A. Sahakian, K. Ropella, J. Baerman and S. Swiryn, "Median frequency and coherence measures of atrial and ventricular fibrillation," Proc. 10th IEEE EMBS conf., pp16-17, Nov. 1988.

[4] C. Carter, C. Knapp and A. Nuttall, "Estimation of the magnitude-squared coherence function via overlapped fast Fourier transform processing," IEEE Trans. Audio and Electroacoust. vol. AU-21:4, pp337-344, 1973.

[6] D.H.Youn, A class of adaptive methods for estimating coherence and time delay functions, Ph.D. thesis, Kansas State University, University Microfilms, Ann Arbor, 1982.

[7] B. Widrow and S. Stearns, Adaptive signal processing, Prentice Hall, 1985.

[8] D.H. Youn, N. Ahmed and C. Carter, "Magnitude-squared coherence function estimation: an adaptive approach," IEEE Trans ASSP, Vol ASSP-31, pp137-142, Feb. 1983.

[9] A.H. Nutall, "Direct coherence estimation via a constrained least-squares linear-predictive fast algorithm," Proc. ASSP conf, pp1104-1107, 1982.

Please address correspondence to:
Alan Sahakian, Assistant Professor,
EE/CS Dept. Northwestern University
Evanston, IL 60208 (312)-491-7007

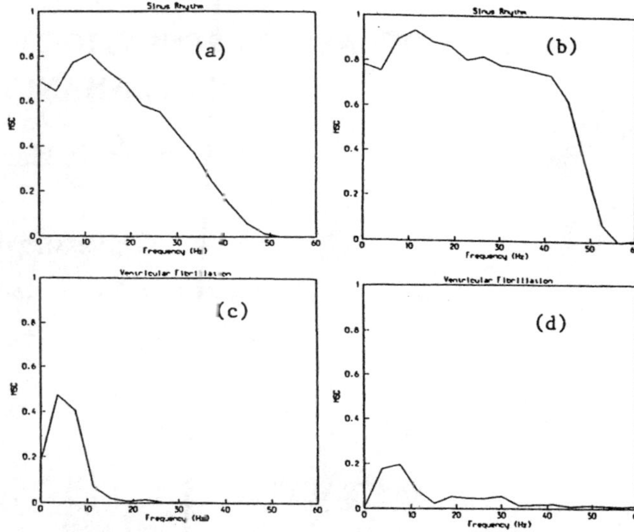

Figure 1. Typical adaptively-estimated MSC spectra for sinus rhythm (a and c) and ventricular fibrillation (b and d). The single-pass (a and b) and multiple-pass (c and d) approaches are also compared.

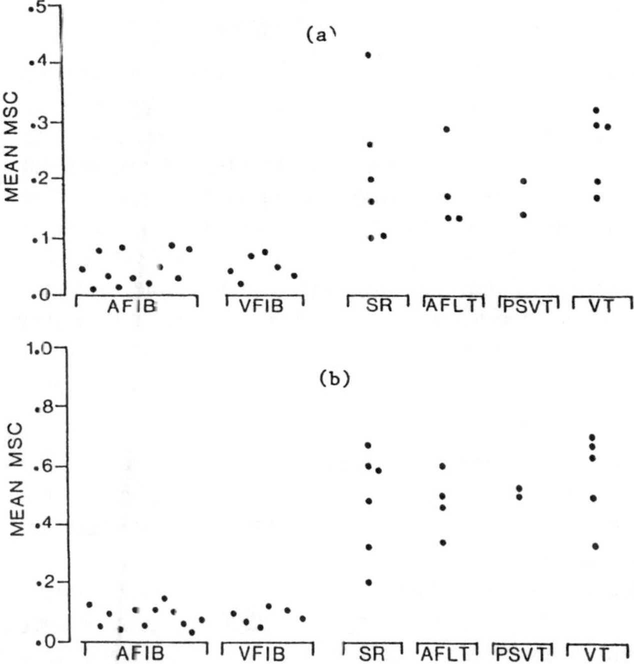

Figure 2. Results for the single-pass (a) and twenty-pass (b) approaches. Although separation is complete with the single-pass approach, the groups move further apart with the multiple-pass approach.

TIME DELAY COMPENSATION FOR ADAPTIVE MULTICHANNEL SPEECH ENHANCEMENT SYSTEMS

K. U. Simmer, P. Kuczynski, A. Wasiljeff

University of Bremen,

Department of Physics and Electrical Engineering,

P.O. Box 330 440, D-2800 Bremen 33, FRG

ABSTRACT

Several algorithms for adaptive multichannel speech enhancement have been tested in an office room and in an anechoic chamber using different noise sources. Temporal synchronisation requires time delay estimation and compensation of the desired speech signal received at different sensors.

1. INTRODUCTION

As most digital speech processing systems have been designed for noise-free environments, the presence of background noise can seriously degrade their performance. In this paper we discuss algorithms for adaptive noise reduction of speech signals using multichannel systems with several acoustic sensors. Non - stationarity of the speech signal and the low spatial coherence of the received noise signals create problems as well as reverberation and echoes that predominate over direct path noise signals in a typical office environment.

2. SYSTEM DESCRIPTION

The receiving system consists of a planar array of four omnidirectional microphones placed at the corners of a square of 60×60 cm. The desired signals are produced by different persons. Either white random noise emitted from a loudspeaker or a hair dryer are used as noise sources. The signals are digitized by a DSP32C signal processor board using a sampling rate of 16 kHz. They pass a beamstearing unit with four delays. Automatic adjustment of these delays is part of the following study.

3. ALGORITHMS FOR ADAPTIVE MULTICHANNEL PROCESSING OF NOISY SPEECH SIGNALS

Algorithms due to Frost [1], Griffiths-Jim [2], Duvall [3] and Compernolle [4] have been investigated in this paper. The noise suppression filter of Kaneda and Tohyama [5] has been generalized to N sensors by the authors.

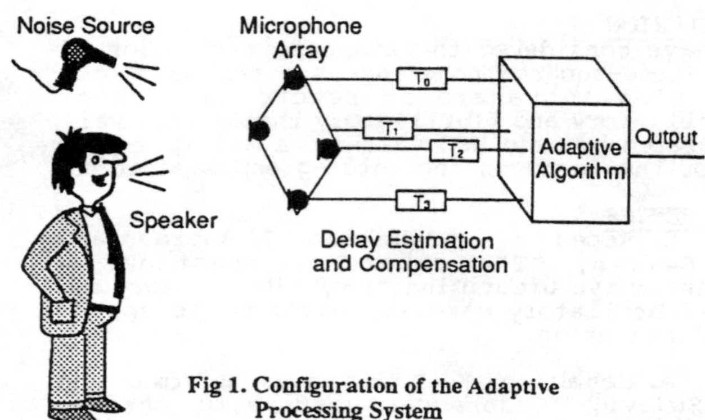

Fig 1. Configuration of the Adaptive Processing System

The algorithm is based on the following assumptions. The speaker has to be close to the sensor array, the distance d between two sensor elements has to obey the spatial sampling theorem $d \leq \lambda_{pitch} / 2$ with regard to the pitch-wavelength $\lambda_{pitch} = c_{sound} / f_{pitch}$, desired speech signal and noise signals are assumed to be uncorrelated, the noise signals are spatially incoherent, i.e. the correlation of the noise signals at two different sensors is close to zero. Cross-spectra $X_i \cdot X_j^*$ of the input signals on sensor pairs of the array are used to estimate the transfer function:

$$W(f) = \frac{\frac{2}{N \cdot (N-1)} \sum_{i=0}^{N-2} \sum_{j=i+1}^{N-1} \mathrm{Re}\{ X_i(f) \cdot X_j^*(f) \}}{\frac{1}{N} \sum_{i=0}^{N-1} |X_i(f)|^2} \quad (1)$$

75% overlapping Hanning windows (256 samples) are used for short-time FFT of the signals. The short-time spectra of the N channels are averaged and multiplied by the transferfunction (1). Inverse transformations and overlapp and add technique yield the filtered output.

Reprinted with permission from *Proc. ISSSE '92*, pp. 660–663, September 1992.

4. TIME DELAY ESTIMATION

An important problem in practical systems is the temporal synchronisation of the desired speech signals. The signals received by N sensors pass a beam steering unit (see fig. 1). The delays T_i of the unit should be automatically adjusted to steer the beam of the array into the direction of the speaker. Time delay estimation of the received signals has to be performed under difficult conditions. The signal source is subject to irregular motions, as the speaker moves his head. In time intervals with low signal to noise ratio as unvoiced speech or speech-pauses the array may look into a wrong direction if no counter-measures are taken. The basic idea proposed in this paper is a switched beam steering unit. The position of the speaker is determined by delay estimation between signal arrivals at different sensors during intervals with maximum signal to noise ratio. The steering direction of the array is hold during intervals with low signal to noise ratio. The input signals can be modelled as

$$\begin{aligned} x_i(t) &= s(t) + n_i(t) \\ x_j(t) &= s(t+D_{ij}) + n_j(t) \end{aligned} \tag{2}$$

where the speech signal s(t) is assumed to be uncorrelated with the spatially incoherent noise sources $n_i(t)$ and $n_j(t)$. To determine the time delay D_{ij} we compute the cross-correlation function:

$$\begin{aligned} R_{x_i x_j}(\tau) &= \mathbf{E}\,[x_i(t)\, x_j(t-\tau)] \\ &= \mathbf{E}\,[(s(t) + n_i(t))\cdot(s(t+D_{ij}-\tau) + n_j(t-\tau))] \\ &= \mathbf{E}\,[s(t)\cdot s(t-(\tau-D_{ij}))] \\ &= R_{SS}(\tau-D_{ij}) \end{aligned} \tag{3}$$

since

$$\begin{aligned} \mathbf{E}\,[n_i(t)\cdot n_j(t-\tau)] = \mathbf{E}\,[n_i(t)\cdot s(t+D_{ij}-\tau)] = \mathbf{E}\,[s(t)\cdot n_j(t-\tau)] \\ = 0 \end{aligned} \tag{4}$$

To estimate the time-delay D_{ij} it is necessary to find the argument τ, that maximizes the value of (3). The maximum value of (3) can also be interpreted as the mean power of the desired speech signal. To distinguish between speech intervals with maximum SNR and speech-pauses with a low SNR the temporal coherence function is used:

$$\gamma_{x_i x_j}(\tau) = \frac{R_{x_i x_j}(\tau)}{\sqrt{R_{x_i x_i}(0)\cdot R_{x_j x_j}(0)}} \tag{5}$$

Same noise power at the different channels gives:

$$R_{x_i x_i}(0) = R_{x_j x_j}(0) = R_{SS}(0) + R_{nn}(0) \tag{6}$$

Therefore it is possible to estimate the signal to noise ratio by computing the coherence function at $\tau = D_{ij}$:

$$\gamma_{x_i x_j}(D_{ij}) = \frac{R_{SS}(0)}{R_{SS}(0) + R_{nn}(0)} = \frac{1}{1 + \dfrac{1}{R_{SS}(0)/R_{nn}(0)}} \tag{7}$$

The temporal coherence function (5) is used to estimate the time delay D_{ij} during the speech-sequence. The maximum magnitude of the coherence function as given by (7) determines whether a speech signal is present or not. Formula (7) shows that a coherence function close to one corresponds to a high signal to noise ratio and yields reliable estimates of the time delays D_{ij}. The proposed coherence detector works as follows:

The time delays D_{ij} are estimated only if the maximum of the coherence function (7) exceeds a preset threshold level (e.g. 0.6). The steering direction of the array is hold during intervals with low SNR.

5. RESULTS

Figure 2a shows a three dimensional plot of a short time coherence function of a segmented speech signal, with 1.4 sec duration. Each segment has a length of 1024 samples, which is equivalent to 1/16 second at a sampling rate of 16kHz. The segments are windowed using 75% overlapping Hanning windows. The speaker is moving from right to left as it is clearly seen in the contour plot of figure 2b. In the case where the speaker does not move and stays at a fixed position, the maximum of the coherence function remains at the same delay (12 samples), as is shown in figure 3a and b. Regions with low coherence maximum occur at time intervalls with unvoiced speech. The delay and hence the direction of the speaker cannot be determined in these coherence valleys. Figure 4 shows estimated delays for a single word of 1 sec duration as a function of time. The upper figure 4a shows delay estimation without coherence detector. Figure 4b shows the excellent performance of the proposed coherence detector for a threshold level of 0.6. Figure 4c gives the corresponding maximum of the coherence function. Figure 5 gives the standard deviation of the delay estimation error of the described coherence detector level for different algorithms. Above a coherence threshold of 0.6 the simple short time coherence function as defined in (5) gives the lowest delay estimation error of all algorithms investigated in this paper. Figure 6 shows the noise reduction as defined in [10] as function of frequency for different microphone distances d for Frost´s algorithm and figure 7 for the algorithm proposed by the authors. A large microphone distance d gives better noise reduction for lower frequencies. However one should keep in mind that violating the spatial sampling condition with regard to the pitch wavelength as mentioned in chapter 3 leads to ambiguities in the measured delays. Although all investigated algorithms work rather well under the artificial conditions of an anechoic chamber (fig. 8) the proposed algorithm (1) seems to give better performance in an office environment (fig. 9).

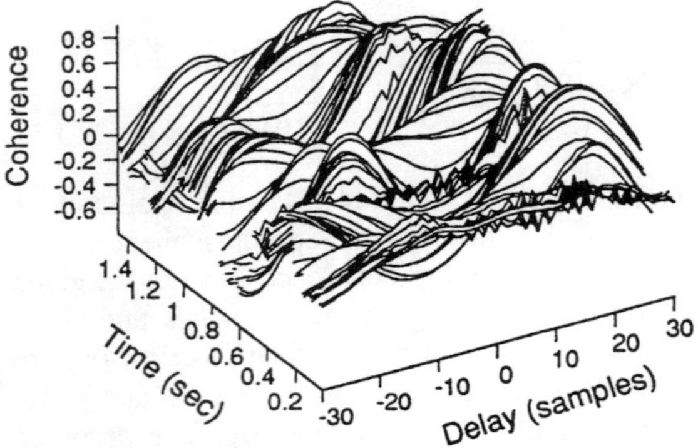

Fig. 2a Short time coherence of segmented speech signal with a moving speaker.

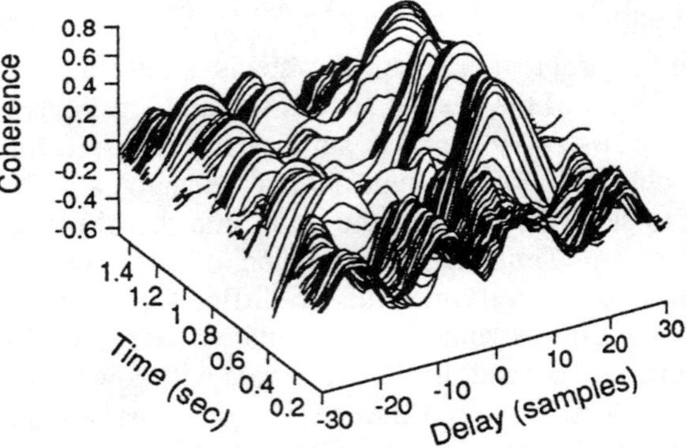

Fig. 3a Short time coherence of a segmented speech signal. Speaker at a fixed position.

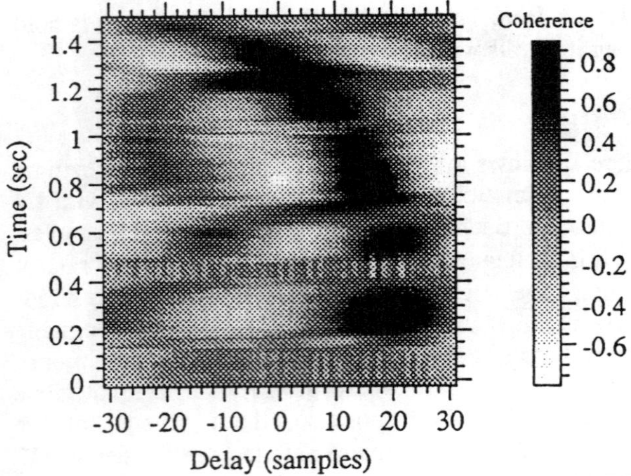

Fig. 2b Grey scale contour plot of **2a**.

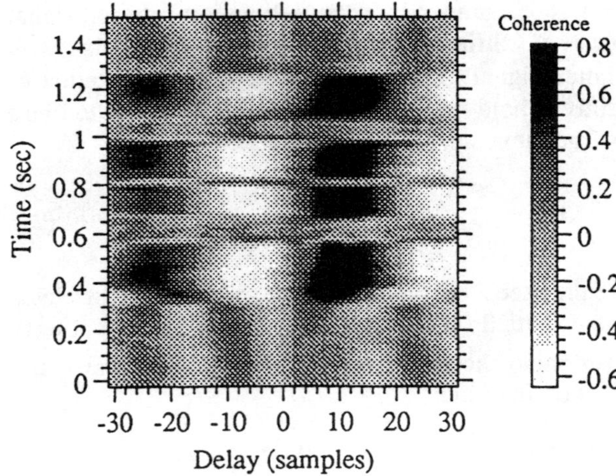

Fig. 3b Grey scale contour plot of **3a**.

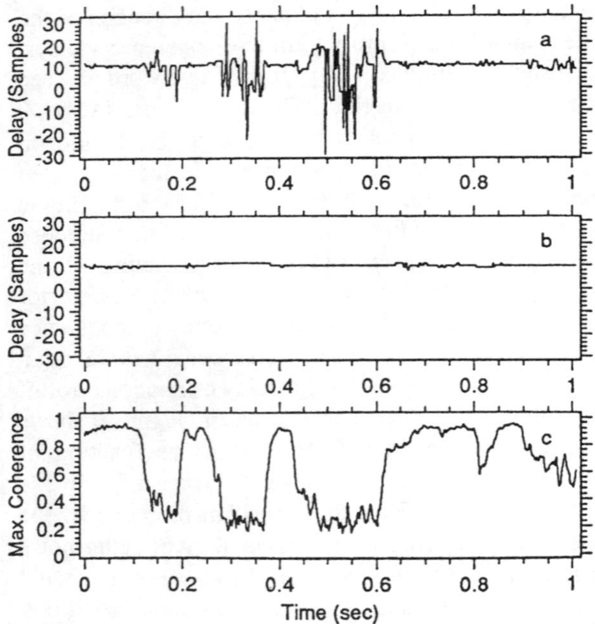

Fig. 4 Delay estimation with corresponding coherence. (True delay D = 10).

a) Delay estimation without coherence detector.

b) Delay estimation with coherence detector.

c) Maximum of the coherence function.

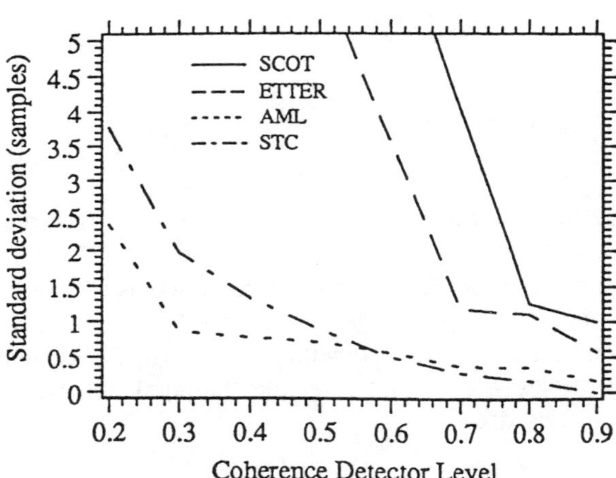

Fig. 5 Standard deviation of time delay estimates as function of the described coherence detector level. (Single spoken word, average SNR 6dB).

SCOT: Smoothed Coherence Transform. Estimated using 4 disjoint data segments of 512 samples [6],[7].

ETTER: Adaptive Estimation of Time Delay [8].

AML: Approximate Maximum Likelihood Estimator [6]. Estimated using 8 data segments of 256 samples

STC: Short Time Coherence as defined in (5) using one Hanning window of 2048 samples.

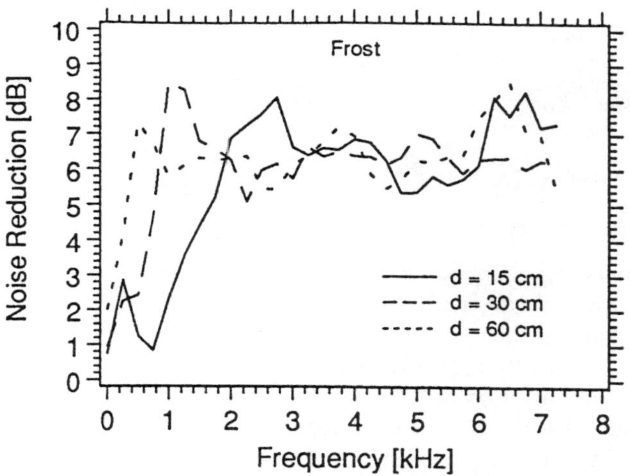

Fig. 6 Noise reduction as function of frequency, different microphone distances d as parameter. (Frost's algorithm).

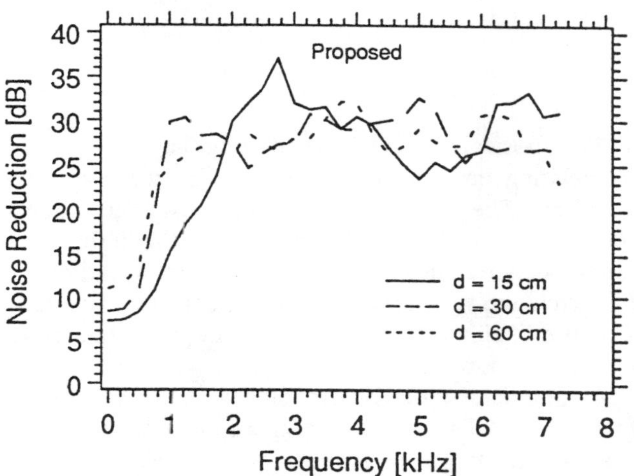

Fig. 7 Noise reduction as function of frequency, different microphone distances d as parameter. (Algorithm proposed by the authors).

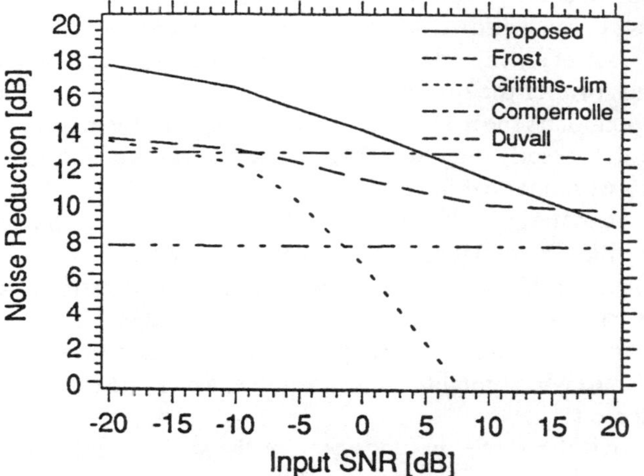

Fig. 8 Noise reduction as function of SNR in an anechoic chamber.

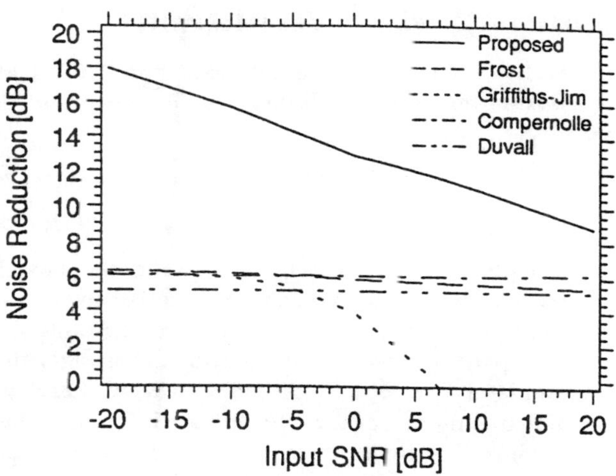

Fig. 9 Noise reduction as function of SNR in an office room.

References

[1] O.L. Frost, III, "An Algorithm for Linearly- Constrained Adaptive Array Processing," Proc. IEEE, vol. 60, no.8, pp. 926-935, Aug. 1972.

[2] L. J. Griffiths, C. W. Jim, "An Alternative Approach to Linearly Constrained Adaptive Beamforming" IEEE Trans. Antennas Propagat. vol. 30, pp. 27-34, Jan. 1982.

[3] B. Widrow et al., "Signal Cancellation Phenomena in Adaptive Antennas: Causes and Cures," IEEE Trans. Antennas Propag., vol AP-30, p. 469, May 1982.

[4] D. Van Compernolle: "Switching Adaptive Filters for Enhancing Noisy and Reverberant Speech from Microphone Array Recordings", Proc. Int. Conf. Acoustics, Speech and Signal Processing, ICASSP-90, Albuquerque, pp. 833-836.

[5] Y. Kaneda, M. Tohyama, "Noise Suppression Signal Processing Using 2-Point Received Signals", Electronics and Communications in Japan, Vol.67-A, pp.19-28, Apr. 1984.

[6] C. H. Knapp, G. C. Carter, "The Generalized Correlation Method for Estimation of Time Delay", IEEE Trans. Acoust., Speech, Signal Processing, vol. ASSP-24, pp. 320-327, Aug. 1976.

[7] K. Scarbrough, N. Ahmed, G. C. Carter, "On the Simulation of a Class of Time Delay Estimation Algorithms", IEEE Trans. Acoust., Speech, Signal Processing, vol. ASSP-29, pp. 534-540, June 1981.

[8] D. M. Etter, S. D. Stearns: "Adaptive Estimation of Time Delay in Sampled Data Systems", IEEE Trans. Acoust., Speech, Signal Processing, vol. 29, pp. 582-587, June 1981.

[9] R. Zelinski: "Adaptive Einstellung der Signallaufzeit für ein Geräuschunterdrückungssystem mit Mikrofongruppe", ITG - Fachbericht 107, pp.307 - 312, 1989.

[10] K. U. Simmer, A. Wasiljeff: "Analysis and Comparison of Systems for Adaptive Array Processing of Speech Signals in a Noisy Environment", Treizième Colloque GRETSI, Juan-Les-Pins, pp. 529 - 532, 1991.

Least-squares time-delay estimation for transient signals in a multipath environment

R. J. Vaccaro, C. S. Ramalingam, and D. W. Tufts
Department of Electrical Engineering, The University of Rhode Island, Kingston, Rhode Island 02881

R. L. Field
Naval Research Laboratory Detachment, Stennis Space Center, Mississippi 39529-5004

(Received 15 June 1991; revised 1 October 1991; accepted 9 February 1992)

The problem of estimating the arrival times of overlapping ocean-acoustic signals from a noisy received waveform that consists of scaled and delayed replicas of a deterministic transient signal is considered. It is assumed that the transmitted signal and the number of paths in the multipath environment are known, and an algorithm is developed that gives least-squares estimates of the amplitude and time delay of each path. A method is given to ensure that the global minimum of the error surface is found in spite of the existence of numerous local minima. The algorithm is then extended to the case in which the transmitted signal is not known precisely, but is assumed to belong to a parametric class of signals. The extended algorithm additionally obtains the parameters that characterize the transmitted signal. The algorithm is demonstrated on the class of signals consisting of gated sinusoids, using both simulated and experimental data.

PACS numbers: 43.60.Gk, 43.30Wi, 43.30.Ma

INTRODUCTION

Time delay estimation is a well-known problem occurring frequently in the fields of sonar, radar, and geophysics. In this problem, the received waveform consists of delayed and scaled replicas of the transmitted signal. This is the result of multiple reflections and attenuation of the signal in the channel.

The received waveform $r(t)$ can be described mathematically as

$$r(t) = \sum_{k=1}^{M} a_k s(t - \tau_k) + n(t), \quad 0 \leqslant t \leqslant T, \tag{1}$$

where $s(t)$ is the transmitted signal, a_k is the attenuation factor for path k, τ_k is the time delay for path k, M is the number of different paths, and $n(t)$ is the inevitable noise component corrupting the received signal. We say that $s(t)$ is a *deterministic* signal, which means that we do not assume any statistical properties for the signal such as its power spectral density. The signal waveform itself is used by the algorithms developed in this paper. For the purposes of this paper, we define a *transient signal* to be a signal whose duration is less than 2 s.

Even though Eq. (1) has been stated in continuous time, any practical implementation using digital processing techniques would deal only with discrete-time samples. Hence, in what follows we will consider only discrete-time signals. Further, we will assume that the noise is white Gaussian. We will also assume that the number of paths M is known. This latter assumption is not as restrictive as it appears to be, since in many cases the number of paths can be determined quite reasonably from the bathymetry of the channel, the approximate locations of the transmitter and receiver, the sound velocity profile, and a propagation model.[1]

The classical method for estimating the times of arrival is correlating the received waveform with the transmitted waveform.[2] The peaks in the correlator output give the estimates of the arrival times. It can be shown that if the signals are separated in time by more than the duration of the signal autocorrelation function, the correlator is equivalent to the maximum-likelihood estimator.[3] Other approaches are given in Refs. 4, 5, 6, and 7.

A completely different approach was recently proposed by Kirsteins.[8,9] The basic idea of this approach is to view the problem in the frequency domain. Since a delay in the time domain is equivalent to multiplication by an exponential in the frequency domain, the corresponding frequency domain problem is one of fitting weighted complex exponentials to the spectrum of the received signal. Utilizing an iterative method of fitting complex exponentials as in Refs. 10 and 11, this approach provides a way of estimating the times of arrival. However, this algorithm does not work if the Fourier transform of the source signal has any nulls (zeros) in the regions of its spectral support. Even if a spectral sample does not fall exactly on a null, a small sample value could cause serious numerical ill-conditioning of the algorithm. Hence we cannot apply this method for a source signal such as a gated sinusoid. More importantly, the delay estimates obtained *via* this method are biased and do not correspond to the true parameter estimates, as demonstrated in Sec. IV A. In addition, contiguous frequency samples have to be considered, which might not be desirable, as will be explained later.

All the above methods require the source signal to be known. In this paper, we develop an algorithm for the case of a known source, and then extend this algorithm to the case in which the source signal is not known precisely, but is as-

Reprinted with permission from the *J. of the Acoustical Soc. of America*, vol. 92, no. 1, pp. 210–218, July 1992.

sumed to belong to a parametric class of signals. That is, we assume that the transmitted signal waveform depends on the values of certain constant parameters (i.e., the parameters are *not* random variables). Our objective, then, will be to obtain good estimates for the delays and at the same time extract the parameters of the source signal. An example of a parametric class of signals is the class of rectangular pulses. A signal in this class is characterized by three parameters: duration, amplitude, and starting point.

In the next section, we set up the least-squares problem to be solved. In Sec. II, we develop an algorithm assuming that the transmitted signal is known. Section III extends this algorithm to the case when the signal is unknown, but can be described by a set of parameters. In Sec. IV we give the results of the known signal and the unknown signal algorithms using simulated as well as experimental data. The transmitted signal is a gated sinusoid in these examples. In Sec. V we give a discussion of the results. Finally, Sec. VI contains the conclusions.

I. THE LEAST-SQUARES ESTIMATOR

Assuming the source signal to belong to a parametric class of signals, the sampled received waveform can be modeled as

$$r[n] = \sum_{k=1}^{M} a_k s[n-\tau_k;\theta] + w[n], \quad 0 \leqslant n \leqslant N-1, \tag{2}$$

where θ is the vector of parameters that characterize the source signal. If the noise $w[n]$ is white Gaussian, then the least-squares estimator is also the maximum likelihood estimator.[12] The squared error function is given by

$$E(\theta,\tau_k,a_k) = \sum_{n=0}^{N-1} \left[r[n] - \sum_{k=1}^{M} a_k s[n-\tau_k;\theta] \right]^2, \tag{3}$$

and the objective is to minimize E with respect to all of its arguments.

In the above time-domain formulation of the problem, we are restricted to values of τ_k that are integer multiples of the basic sampling interval. If a more accurate estimate is required, then one has to resort to interpolation. This inconvenience can be avoided by transforming the problem to the frequency domain, where τ_k can take on a continuum of values.

The data record in (1) exists in the time interval $[0,T]$. If this data record is set to zero for $t<0$ and $t>T$, then the summand in Eq. (3) will be zero outside the range $0 \leqslant n \leqslant N-1$. Thus we can extend the summation range from minus infinity to plus infinity in (3) and invoke Parseval's theorem to get

$$E(\theta,\tau_k,a_k) = \frac{1}{2\pi} \int_{-\pi}^{\pi} \left| R(\omega) - S(\omega;\theta) \sum_{k=1}^{M} a_k e^{-j\tau_k\omega} \right|^2 d\omega, \tag{4}$$

where $R(\omega)$ and $S(\omega;\theta)$ are the discrete-time Fourier transforms of $r[n]$ and $s[n;\theta]$, respectively.

The problem, now, is to approximate $R(\omega)$ by a weighted sum of complex exponentials. For computer implementation, we sample in the frequency domain at intervals of $\Delta\omega$. Hence,

$$e^{-j\tau_k\omega} \rightarrow e^{-j\tau_k n \Delta\omega} = e^{j\lambda_k n},$$

where

$$\lambda_k = -\tau_k \Delta\omega. \tag{5}$$

The integrals are then replaced by sums and the squared error function is given approximately by (after omitting the scale factor $1/2\pi$)

$$E(\theta,\lambda_k,a_k) = \sum_{n=0}^{N-1} \left| R[n] - S[n;\theta] \sum_{k=1}^{M} a_k e^{j\lambda_k n} \right|^2, \tag{6}$$

where $S[n;\theta] = S(n\Delta\omega;\theta)$ and $R[n] = R(n\Delta\omega)$. The above error expression will be used to develop both the known signal and the unknown signal algorithms. For the known signal case, the parameter vector θ in Eq. (6) is fixed and is not included in the minimization, whereas for the unknown signal case Eq. (6) is minimized with respect to all the parameters.

II. THE KNOWN SIGNAL ALGORITHM FOR NARROWBAND SIGNALS

In this section we consider the case when the source signal is known. In the next section, we show how the known-signal algorithm can be used iteratively in the case when the signal is not known.

When the signal is narrow band, most of the energy of the signal is concentrated in the passband. For example, if the signal were a gated sinusoid, most of its energy is concentrated near the main lobe in the frequency domain. Recall that the spectrum of a gated sinusoid is a sinc function centered at the frequency of the sinusoid. In this case, we do not have to include all frequency points in the minimization but consider only those values near the main lobe. Going even further, one need not again include *all* frequency points near the main lobe, but *only those which have significant signal energy*. That is, one can threshold the signal spectrum and consider only those points which are above this value. This has the advantage of working only with those frequency points having good signal-to-noise ratios. Also, since the received and the modeled signal are real, only half the signal spectrum needs to be considered because of its conjugate symmetry. The expression for the error in Eq. (6) is thus rewritten as,

$$E(\lambda,\tilde{a}) = \sum_{n \in \mathcal{N}} \left| R[n] - S[n] \sum_{k=1}^{M} a_k e^{j\lambda_k n} \right|^2, \tag{7}$$

where λ is the vector of scaled delays [see Eq. (5)] and $\tilde{a}$ is a vector of amplitudes. (The tildes are used now to avoid cumbersome notation in subsequent developments.) The notation $n \in \mathcal{N}$ means that the summation is over those values of n which belong to the set $\mathcal{N}$, and not necessarily over contiguous values. The set $\mathcal{N}$ consists of those frequency points at which the magnitude of the signal spectrum exceeds a threshold. A rule-of-thumb is to set this threshold to roughly one-twentieth to one-tenth the peak magnitude of the signal spectrum. Since matrix notation leads to succinct expressions, let us define the following:

$$\boldsymbol{\lambda} = (\lambda_1 \quad \lambda_2 \quad \cdots \quad \lambda_M)^T,$$
$$\tilde{\mathbf{r}} = (R[q_1] \quad R[q_2] \quad \cdots \quad R[q_N])^T,$$
$$\tilde{\mathbf{a}} = (a_1 \quad a_2 \quad \cdots \quad a_M)^T,$$
$$\mathbf{S} = \mathrm{diag}(S[q_1] \quad S[q_2] \quad \cdots \quad S[q_N]),$$
$$\mathbf{A}(\boldsymbol{\lambda}) = \begin{pmatrix} e^{j\lambda_1 q_1} & e^{j\lambda_2 q_1} & \cdots & e^{j\lambda_M q_1} \\ e^{j\lambda_1 q_2} & e^{j\lambda_2 q_2} & \cdots & e^{j\lambda_M q_2} \\ \vdots & \vdots & \vdots & \vdots \\ e^{j\lambda_1 q_N} & e^{j\lambda_2 q_N} & \cdots & e^{j\lambda_M q_N} \end{pmatrix},$$
$$\widetilde{\mathbf{P}}(\boldsymbol{\lambda}) = \mathbf{S}\mathbf{A}(\boldsymbol{\lambda}), \tag{8}$$

where $q_k \in \mathcal{N}$, $k = 1,2,\ldots,N$ denote the frequency samples that are selected. Hence Eq. (7) can be recast as

$$E(\boldsymbol{\lambda},\tilde{\mathbf{a}}) = \|\tilde{\mathbf{r}} - \widetilde{\mathbf{P}}(\boldsymbol{\lambda})\tilde{\mathbf{a}}\|^2. \tag{9}$$

This is the error expression considered in Ref. 9. However, in that work the amplitudes were allowed to be complex. As mentioned in the Introduction, this leads to a biased estimate of the delay parameters if the SNR were not sufficiently high (we show this in Sec. IV A). But an important advantage of allowing complex-valued amplitudes is that the corresponding error surface is reasonably smooth, making it easier to find its global minimum. On the other hand, while the global minimum of the error surface defined by Eq. (9) with the amplitudes constrained to be real yields the true delay parameters, finding this minimum is extremely difficult as the surface is characterized by numerous closely spaced local minima. We now outline a procedure that leads us to this global minimum, starting with the error surface corresponding to the complex amplitudes.

The key step that helps us to achieve the transition from the biased error surface (due to the complex amplitudes) to the true error surface (due to the real amplitudes) is to rewrite Eq. (9) explicitly in terms of real and imaginary parts of each of the terms involved, and add a penalty term to the imaginary part of the complex amplitude. That is, let

$$E_\alpha(\boldsymbol{\lambda},\mathbf{a}) = \left\| \begin{pmatrix} \mathbf{r}_r \\ \mathbf{r}_i \end{pmatrix} - \begin{pmatrix} \mathbf{P}_r & -\mathbf{P}_i \\ \mathbf{P}_i & \mathbf{P}_r \end{pmatrix} \begin{pmatrix} \mathbf{a}_r \\ \mathbf{a}_i \end{pmatrix} \right\|^2 + \alpha\|\mathbf{a}_i\|^2, \tag{10}$$

where

$$\mathbf{r}_r = \mathrm{Re}\{\tilde{\mathbf{r}}\}, \quad \mathbf{r}_i = \mathrm{Im}\{\tilde{\mathbf{r}}\},$$
$$\mathbf{P}_r = \mathrm{Re}\{\widetilde{\mathbf{P}}(\boldsymbol{\lambda})\}, \quad \mathbf{P}_i = \mathrm{Im}\{\widetilde{\mathbf{P}}(\boldsymbol{\lambda})\},$$
$$\mathbf{a}_r = \mathrm{Re}\{\tilde{\mathbf{a}}\}, \quad \mathbf{a}_i = \mathrm{Im}\{\tilde{\mathbf{a}}\}.$$

Rewriting the right-hand side of Eq. (10) as a single norm, we get

$$E_\alpha(\boldsymbol{\lambda},\mathbf{a}) = \left\| \begin{pmatrix} \mathbf{r}_r \\ \mathbf{r}_i \\ 0 \end{pmatrix} - \begin{pmatrix} \mathbf{P}_r & -\mathbf{P}_i \\ \mathbf{P}_i & \mathbf{P}_r \\ 0 & \alpha\mathbf{I} \end{pmatrix} \begin{pmatrix} \mathbf{a}_r \\ \mathbf{a}_i \end{pmatrix} \right\|^2 = \|\mathbf{r} - \mathbf{P}(\boldsymbol{\lambda})\mathbf{a}\|^2, \tag{11}$$

where

$$\mathbf{P}(\boldsymbol{\lambda}) = \begin{pmatrix} \mathbf{P}_r & -\mathbf{P}_i \\ \mathbf{P}_i & \mathbf{P}_r \\ 0 & \alpha\mathbf{I} \end{pmatrix}, \quad \mathbf{r} = \begin{pmatrix} \mathbf{r}_r \\ \mathbf{r}_i \\ 0 \end{pmatrix}, \quad \mathbf{a} = \begin{pmatrix} \mathbf{a}_r \\ \mathbf{a}_i \end{pmatrix}. \tag{12}$$

It is easy to see that for $\alpha = 0$, the error function in Eq. (11) allows the amplitudes to be complex since there is no penalty attached to the imaginary part. As α increases, the penalty on the imaginary part increases, and in the limit as $\alpha \to \infty$, the amplitudes are constrained to be purely real. Note that the model for the channel assumes the amplitudes to be real.

With the above modification of the error expression, the procedure to minimize $E_\alpha(\boldsymbol{\lambda},\mathbf{a})$ is as follows: Initially, α is set equal to zero. Then $E_0(\boldsymbol{\lambda},\mathbf{a})$ is minimized with respect to $\boldsymbol{\lambda}$ and $\mathbf{a}$. The minimum yields a biased estimate of $\boldsymbol{\lambda}$, but one which is reasonably close to the true value. To estimate the true delay values, α is increased gradually and the minimum of $E_\alpha(\boldsymbol{\lambda},\mathbf{a})$ is found for each α. The minimum of $E_\infty(\boldsymbol{\lambda},\mathbf{a})$ yields the final parameter estimates. This method of using a quadratic penalty function to impose a nonlinear constraint is known to converge to a minimum of the true error surface.[13]

For any fixed $\boldsymbol{\lambda}$ and α, the best $\mathbf{a}$ which minimizes $E_\alpha(\boldsymbol{\lambda},\mathbf{a})$ is given by

$$\mathbf{a} = (\mathbf{P}^H\mathbf{P})^{-1}\mathbf{P}^H\mathbf{r}. \tag{13}$$

Substituting this value of $\mathbf{a}$ in Eq. (11) we get

$$E_\alpha(\boldsymbol{\lambda}) = \|[\mathbf{I} - \mathbf{P}(\mathbf{P}^H\mathbf{P})^{-1}\mathbf{P}^H]\mathbf{r}\|^2 = \|\mathbf{P}^\perp\mathbf{r}\|^2, \tag{14}$$

where $\mathbf{P}^\perp = \mathbf{I} - \mathbf{P}(\mathbf{P}^H\mathbf{P})^{-1}\mathbf{P}^H$, and now the error is a function of the delay parameter vector $\boldsymbol{\lambda}$ only. In the next subsection, we present a Gauss–Newton approach to find the minimum of $E_\alpha(\boldsymbol{\lambda})$.

A. A Gauss–Newton approach to minimizing $E_\alpha(\boldsymbol{\lambda})$

A common algorithm for minimizing an objective function, which is expressed as a squared norm, is the Gauss-Newton algorithm.[13,14] This algorithm uses a first-order perturbation expansion to convert the nonlinear minimization problem to a linear one. A sequence of linear problems are then solved until the solutions converge. In this section, we derive the formulas needed to implement the Gauss–Newton algorithm for the error function in Eq. (11).

We begin by considering the value of the error function at an increment $\Delta\boldsymbol{\lambda}$ away from the nominal value of $\boldsymbol{\lambda}$, i.e.,

$$E_\alpha(\boldsymbol{\lambda} + \Delta\boldsymbol{\lambda}) = \|[\mathbf{I} - \mathbf{P}(\boldsymbol{\lambda} + \Delta\boldsymbol{\lambda})(\mathbf{P}^H(\boldsymbol{\lambda} + \Delta\boldsymbol{\lambda}) \times \mathbf{P}(\boldsymbol{\lambda} + \Delta\boldsymbol{\lambda}))^{-1}\mathbf{P}^H(\boldsymbol{\lambda} + \Delta\boldsymbol{\lambda})]\mathbf{r}\|^2. \tag{15}$$

To simplify the above equation, we consider a first-order perturbation expansion of $\mathbf{P}(\boldsymbol{\lambda} + \Delta\boldsymbol{\lambda})$:

$$\mathbf{P}(\boldsymbol{\lambda} + \Delta\boldsymbol{\lambda}) \doteq \mathbf{P} + \Delta\mathbf{P}, \tag{16}$$

where $\doteq$ means "equal to first order," and $\Delta\mathbf{P}$ is a matrix which is a linear function of $\Delta\boldsymbol{\lambda}$. Substituting this expansion of $\mathbf{P}(\boldsymbol{\lambda} + \Delta\boldsymbol{\lambda})$ into Eq. (15) and retaining only first order terms (see Ref. 15 for details) we get

$$E_\alpha(\boldsymbol{\lambda}+\Delta\boldsymbol{\lambda}) \doteq \|\mathbf{P}^\perp\mathbf{r} - \mathbf{P}^\perp(\Delta\mathbf{P})(\mathbf{P}^H\mathbf{P})^{-1}\mathbf{P}^H\mathbf{r} - \mathbf{P}(\mathbf{P}^H\mathbf{P})^{-1}(\Delta\mathbf{P})^H\mathbf{P}^\perp\mathbf{r}\|^2, \qquad (17)$$

where the superscript H denotes complex-conjugate transpose. Using the definition of $\mathbf{P}$ from Eq. (12) it can be verified that

$$\Delta\mathbf{P} = -\underbrace{\begin{pmatrix} \mathbf{QP}_i & \mathbf{QP}_r \\ -\mathbf{QP}_r & \mathbf{QP}_i \\ 0 & 0 \end{pmatrix}}_{\mathbf{S}} \begin{pmatrix} \Delta\Lambda & 0 \\ 0 & \Delta\Lambda \end{pmatrix}, \qquad (18)$$

where

$$\mathbf{Q} = \mathrm{diag}\{q_1 \quad q_2 \quad \cdots \quad q_N\},$$
$$\Delta\boldsymbol{\lambda} = (\Delta\lambda_1 \quad \Delta\lambda_2 \quad \cdots \quad \Delta\lambda_M)^T,$$
$$\Delta\Lambda = \mathrm{diag}\{\Delta\boldsymbol{\lambda}\}.$$

Since

$$\mathrm{diag}\{x_1 \quad x_2 \quad \cdots \quad x_L\}\begin{pmatrix} y_1 \\ y_2 \\ \vdots \\ y_L \end{pmatrix} = \mathrm{diag}\{y_1 \quad y_2 \quad \cdots \quad y_L\}\begin{pmatrix} x_1 \\ x_2 \\ \vdots \\ x_L \end{pmatrix},$$

we can pull $\mathrm{diag}\{\Delta\boldsymbol{\lambda};\Delta\boldsymbol{\lambda}\}$ to the end in the second and third terms on the right-hand side of Eq. (17), and further simplify to get

$$E_\alpha(\boldsymbol{\lambda}+\Delta\boldsymbol{\lambda}) = \|\mathbf{x} - \mathbf{B}\Delta\boldsymbol{\lambda}\|^2, \qquad (19)$$

where

$$\mathbf{x} = \mathbf{P}^\perp\mathbf{r},$$
$$\mathbf{B}' = \mathbf{P}^\perp\mathbf{S}\cdot\mathrm{diag}\{(\mathbf{P}^H\mathbf{P})^{-1}\mathbf{P}^H\mathbf{r}\} + \mathbf{P}(\mathbf{P}^H\mathbf{P})^{-1}\cdot\mathrm{diag}\{\mathbf{S}^T\mathbf{P}^\perp\mathbf{r}\},$$
$$\mathbf{B} = \mathbf{B}'(:,1{:}M) + \mathbf{B}'(:,(M+1){:}2M).$$

(Note that we have used standard MATLAB notation in the definition of $\mathbf{B}$, which is derived from $\mathbf{B}'$ by summing the first M columns with its second M columns.) From Eq. (19) we can solve for the best $\Delta\boldsymbol{\lambda}$ (in the least-squares sense) as

$$\Delta\boldsymbol{\lambda} = \mathbf{B}^\#\mathbf{x} = (\mathbf{B}^H\mathbf{B})^{-1}\mathbf{B}^H\mathbf{x}. \qquad (20)$$

Now, $\boldsymbol{\lambda}$ is replaced by $\boldsymbol{\lambda}+\Delta\boldsymbol{\lambda}$ and the procedure is repeated till $\|\Delta\boldsymbol{\lambda}\|_2 < \epsilon$. In many cases, it might be desirable not to make large changes in $\boldsymbol{\lambda}$ to avoid oscillations. If so, the least-squares solution in Eq. (20) is replaced by a constrained least-squares solution,[16] where $\|\Delta\boldsymbol{\lambda}\|_2$ is restricted to not to exceed a specified value, say δ.

An initial value for the parameter vector $\boldsymbol{\lambda}$ must be estimated before (20) can be used to obtain improved parameter estimates. A standard procedure for generating initial estimates is the *coordinate descent algorithm* which is described in Refs. 17 and 18. We used the coordinate descent algorithm on $E_\alpha(\boldsymbol{\lambda})$ given in (14) with $\alpha = 0$ to obtain initial estimates for the known signal algorithm.

III. THE UNKNOWN SIGNAL ALGORITHM

We now address the problem of time-delay estimation when the source signal is unknown. In this case, the source parameter vector $\boldsymbol{\theta}$ will also enter the minimization in Eq. (3). In the algorithm outlined below, we assume the source to belong to a parametric class of signals, viz., gated sinusoids of unknown frequency and duration.

First, we observe that the source signal can completely be specified (except for a real-valued scale factor) by its frequency, duration, starting point, and phase. If the frequency and duration are such that there are many cycles of the signal present, then we need not precisely determine the phase, which can be set equal to zero. Also, since all time delays are relative to the source pulse, the starting point can be assumed to be at $t=0$, without loss of generality.

With the above assumptions, the overall problem reduces to minimizing

$$\min_{\lambda,a,f,d} \sum_{n\in\mathcal{N}} \left| R[n] - S[n;f,d] \sum_{k=1}^{M} a_k e^{j\lambda_k n} \right|^2,$$

where f and d are the frequency and duration of the transmitted signal, respectively. This expression is not easy to minimize with respect to all the parameters simultaneously. Instead, we resort to an iterative minimization technique outlined below.

(1) Obtain initial estimates of f^0 and d^0, i.e., of the unknown frequency and duration, respectively.

(2) Use f^i and d^i in the known signal algorithm to estimate $\boldsymbol{\lambda}^i$ and $\mathbf{a}^i$.

(3) Using the estimated $\boldsymbol{\lambda}^i$ and $\mathbf{a}^i$, calculate f^{i+1} and d^{i+1}.

(4) Check for convergence to return to step 2.

We now show how to calculate f^{i+1} and d^{i+1} mentioned in step 3. Consider the error function for a given $\boldsymbol{\lambda}$ and $\mathbf{a}$:

$$E(f,d) = \sum_{n\in\mathcal{N}} \left| R[n] - S[n;f,d] \sum_{k=1}^{M} a_k e^{j\lambda_k n} \right|^2.$$

Here, $E(f,d)$ is nonlinear in f and d and an analytical minimization is again very difficult. We simplify the minimization by taking advantage of the fact that d is a discrete variable since the signal is sampled, i.e., only finite number of values for d are possible. Hence, $E(f,d)$ can be minimized as follows:

(1) Carry a search over values of d near the previous estimate.

(2) For each d, find the best value of f using a gradient descent technique.

(3) Repeat step 2 until minimum is found.

The final question is how to obtain f^0 and d^0. In our model we assume that the first path is the least attenuated one and that its amplitude is greater than that of the noise. There may or may not be pulses from other paths overlapping with the first. With these assumptions we get rough initial guesses for f^0 and d^0.

(i) Obtain the envelope of the signal.

(ii) Obtain the high-amplitude portion of the received waveform by finding that part of the signal envelope which exceeds a threshold. Let (t_1, t_2) be the interval in which the envelope exceeds the threshold. Choose $d^0 = (t_2 - t_1)/2$.

(iii) f^0 is obtained using any standard frequency estimator on the signal in the interval (t_1, t_2).

This completes the description of the unknown source algorithm.

IV. ALGORITHM PERFORMANCE

In this section we show the performance of the algorithms described in the previous sections using synthetic signals as well as experimental data. We first present a one-path example to illustrate the nature of the $E_0(\lambda)$ and $E_\infty(\lambda)$ error surfaces. We then demonstrate the effect of increasing α on $E_\alpha(\lambda)$. Finally, results on experimental data using the known- and the unknown-signal algorithms are presented.

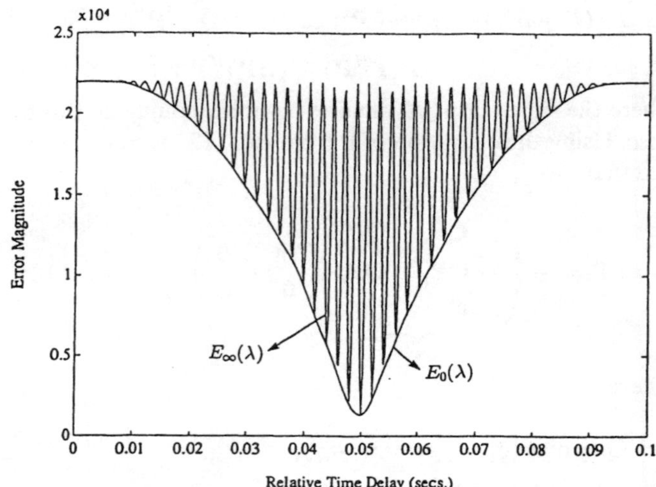

FIG. 2. Error surfaces $E_0(\lambda)$ and $E_\infty(\lambda)$ for the high-SNR received signal of Fig. 1.

A. A one-path example to compare error surfaces

As mentioned in Sec. II, the true parameter estimates are obtained from the global minimum of $E_\infty(\lambda)$. However, its minimum is difficult to find. On the other hand, the error surface $E_0(\lambda)$ is easier to work with, but its global minimum yields an increasingly biased estimate of λ with decreasing SNR. We now illustrate the nature of $E_0(\lambda)$ and $E_\infty(\lambda)$ error surfaces with a one-path example to highlight the difficulties involved.

A synthetic received signal is considered. It was constructed by delaying a 244-Hz, 40-ms duration sinusoid by 50 ms and adding computer generated white Gaussian noise. The error surface (plotted as a function of *relative* delay) should have a minimum at $t = 0.05$ s. We consider two cases, viz., high and low SNRs. (We use the terms "high" and "low" SNR in a qualitative sense.) Figure 1 shows the high SNR signal. In Fig. 2 we show the corresponding error surfaces, $E_0(\lambda)$ and $E_\infty(\lambda)$. The smooth error surface is $E_0(\lambda)$, while the sinusoidal surface is $E_\infty(\lambda)$. Both have the same global minimum, but $E_0(\lambda)$ is easier to minimize, since it is unimodal. However, when the SNR is not high enough, the global minima of these two surfaces can be distinctly different. Figure 3 shows the low SNR received waveform, with the corresponding error surfaces in Fig. 4. In this example, the minimum of $E_0(\lambda)$ is approximately at $t = 0.049$ s, while that of $E_\infty(\lambda)$ is still at $t = 0.05$ s. It can be seen that if the time-delay obtained from the biased error surface is used, the true error can be quite large. This is more so in the M-path case. Hence, one has to minimize $E_\infty(\lambda)$ to obtain unbiased parameter estimates. This fact was first noted in Ref. 19.

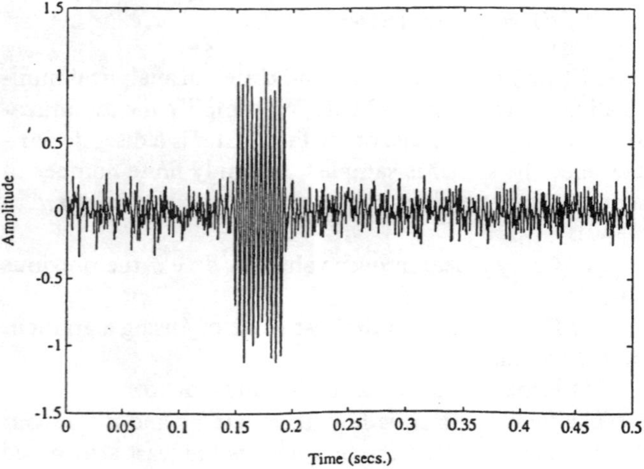

FIG. 1. Received signal at high SNR for a synthetic one-path example.

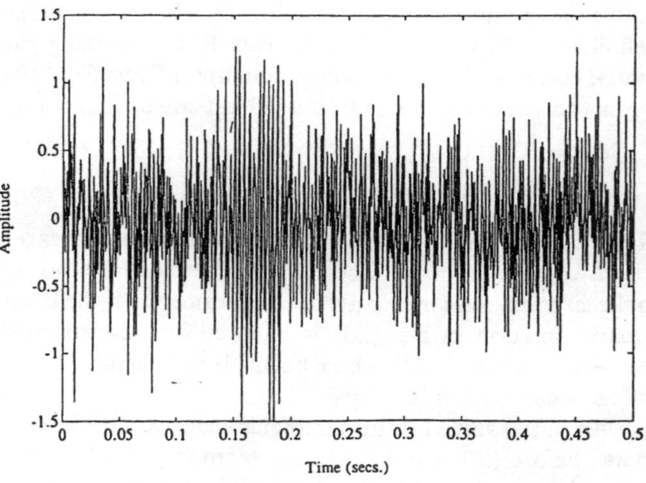

FIG. 3. Received signal at low SNR for a synthetic one-path example.

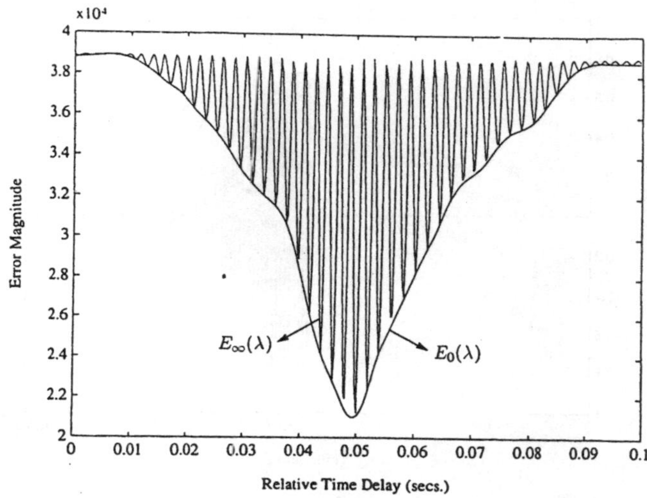

FIG. 4. Error surfaces $E_0(\lambda)$ and $E_\infty(\lambda)$ for the low-SNR received signal of Fig. 3.

B. The effect of increasing α

In order to obtain unbiased estimates, we need to go from the minimum of $E_0(\lambda)$ to the true minimum, i.e., that of $E_\infty(\lambda)$. This is achieved by increasing α. In Fig. 5, the smooth error surface corresponds to $\alpha = 0$, i.e., for which the amplitudes are complex. As α increases, the norm of $\mathbf{a}_i$ decreases and $E_\alpha(\lambda)$ is gradually transformed to the true error surface, as can be seen from Fig. 5.

C. Performance with experimental data

1. The experiment

Transient data were gathered in the Atlantic Ocean on a bottom-mounted receiver in 780 m of water. The experimental geometry is shown in Fig. 6.

The acoustic source was at a depth of 40 m and transmitted a 244-Hz gated sinusoid of 40-ms duration. The source signature was recorded using a hydrophone mounted on the source. The signature is shown in Fig. 7. The pulse was transmitted as the source ship drifted over the bottom receiver shown in Fig. 6. The horizontal range is estimated to be 100 m.

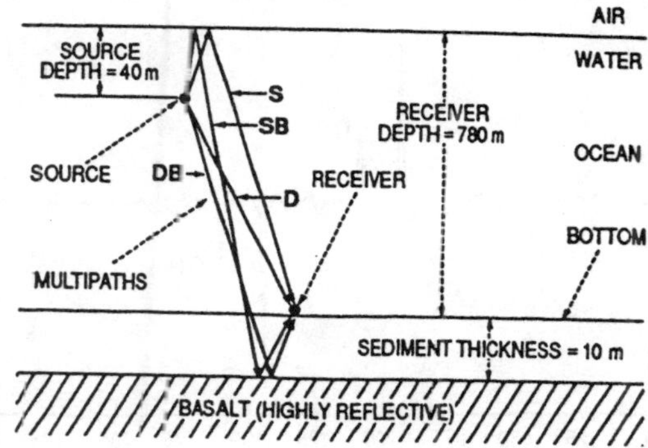

FIG. 6. The geometry of the channel used for the experiments.

The ocean bottom is characterized by a thin sediment layer over a highly reflecting basalt as shown in Fig. 6. The sediment varies in thickness from 0 to 20 m. For this problem, a 10-m sediment thickness was chosen. This environment was modeled with a fast field program, SAFARI.[20,21] A broadband Gaussian pulse is transmitted in the model. The model predicts four paths shown in Fig. 6 (note that Fig. 6 is an artist's sketch of the four paths). Path D is the direct path, path DB is the direct path reflection off the basalt, path S is surface reflected, and path SB is the path reflected from the ocean surface and the basalt. The model ocean impulse response is shown in Fig. 8 with the four paths labeled. Pressure release surfaces such as the air/ocean interface cause a 180° phase shift in the reflected signal, causing the negative peaks seen in Fig. 8. The received signal is shown in Fig. 9.

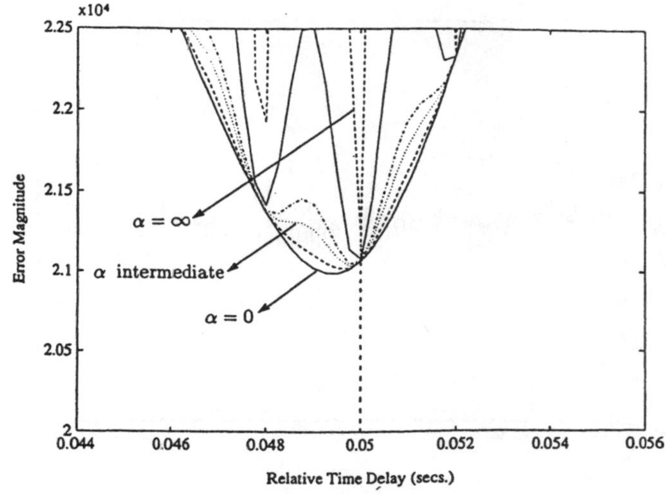

FIG. 5. Effect of increasing α on the constrained error surface $E_\alpha(\lambda)$.

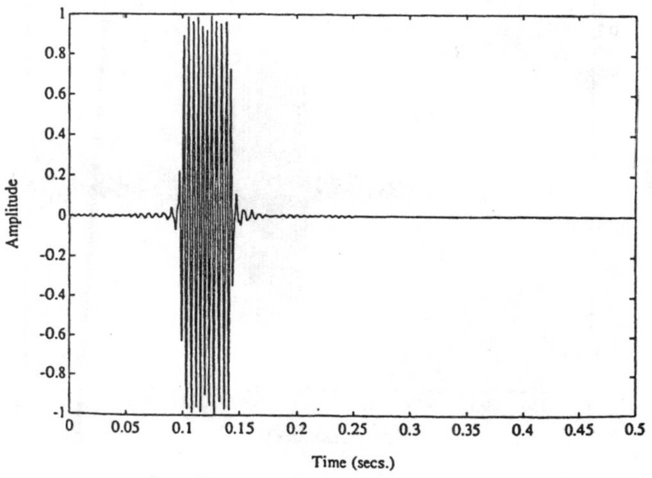

FIG. 7. The transmitted (source) signal: a 244-Hz gated sinusoid.

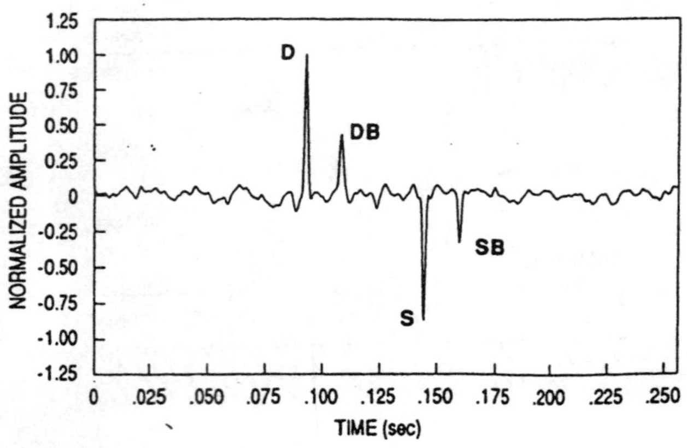

FIG. 8. Broadband Gaussian pulse propagated with SAFARI, source depth = 40 m, range = 0.1 km.

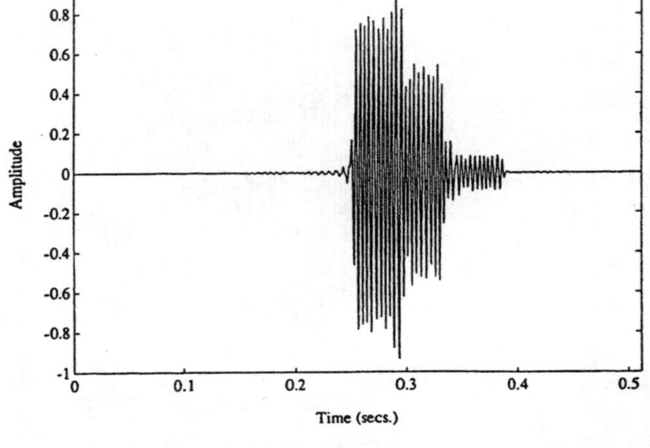

FIG. 10. The reconstructed received signal using channel parameters estimated by the known-signal algorithm.

2. The known signal algorithm

Using the known signal algorithm presented in Sec. II, the time delays and amplitudes were estimated. With these parameters, the received signal was reconstructed (see Fig. 10). The residual error between the received signal and the reconstructed signal is shown in Fig. 11. It is seen that the fit is very good, indicating the accuracy of the estimates. The estimated parameters are shown in Table I in the column labeled "K.S. estimate." Note that the signs of the estimated amplitudes need not necessarily agree with that predicted by the channel model (see Fig. 8), which shows that the first two amplitudes are positive while the second two are negative. The signs of the amplitudes were not constrained in this algorithm; however, such a constraint would be easy to impose using an additional penalty function.

3. The unknown signal algorithm

Next we applied the unknown signal algorithm outlined in the previous section and estimated the parameters of the transmitted signal, as well as the delays and amplitudes in the received signal. These estimates, along with those obtained from the known signal algorithm, are presented in Table I.

The estimated source parameters agree quite closely with the nominal values (244 Hz and 40 ms) of the actual source signal. The channel parameters, too, agree with those obtained by the known signal algorithm, although they are not in the same order. The first three paths identified by the known signal algorithm correspond to paths 1, 3, and 4 found by the unknown signal algorithm. There are many different parameter sets for this problem which all give reasonably good fits to the data. The efficacy of the unknown signal algorithm can be judged by reconstructing a signal using the estimated parameters, and computing the residual error between the received and the reconstructed signals. The reconstructed signal is formed by convolving the modeled source signal (a zero-phase, 245-Hz sinusoid of 43-ms duration) with the estimated channel. Convolution is accomplished by multiplication in the frequency domain fol-

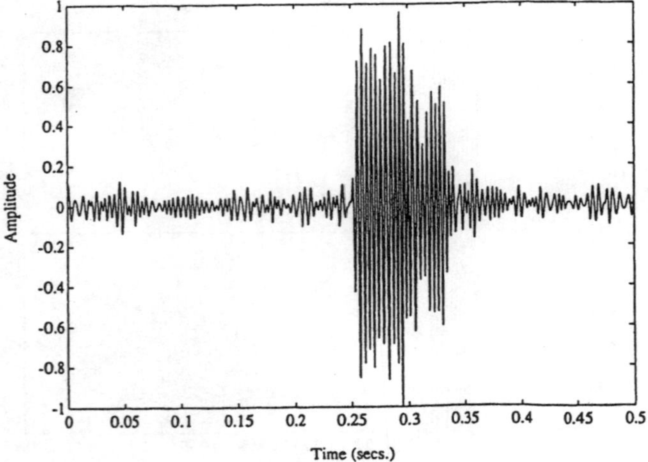

FIG. 9. A record of experimental data containing a received signal with four overlapping paths.

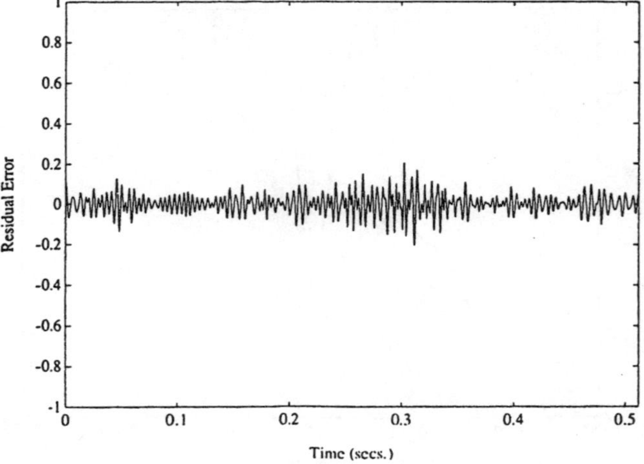

FIG. 11. The residual error (received signal minus reconstructed signal) for the known-signal algorithm.

TABLE I. Parameter estimates using the known signal and the unknown signal algorithms.

	K.S. estimate	U.S. estimate
f (Hz)		245
D (ms)		43
τ_1	0.1544[a]	0.2547[b]
$\tau_2-\tau_1$	0.0360	0.0198
$\tau_3-\tau_1$	0.0453	0.0360
$\tau_4-\tau_1$	0.0898	0.0435
a_1	0.7913	0.8287
a_2	0.4925	−0.1597
a_3	−0.1724	0.4926
a_4	0.0864	0.1810

[a] Relative to the source pulse shown in Fig. 7.
[b] Relative to estimated source pulse starting at $t=0$.

lowed by an inverse DFT. The corresponding residual error is shown in Fig. 12. The fit in this case is poorer compared to the known signal algorithm. The reason for this is explained in the next section.

V. DISCUSSION

The biased error surface corresponding to the complex amplitudes is easier to minimize as it is reasonably smooth. For the one-path example of Sec. IV A, the surface was seen to be unimodal, i.e., has no local minima. In higher dimensions, the surface is still reasonably smooth as opposed to the $\alpha=\infty$ case, *but is no longer unimodal.* Hence, one has to begin any minimizing routine with a reasonable initial estimate for λ to reach the minimum of $E_0(\lambda)$, to avoid getting stuck in a local minimum. We have found that the coordinate descent algorithm provides good initial estimates. The error surface being smooth for $\alpha=0$ is not true for an arbitrary signal. If, for example, the source signal consisted of

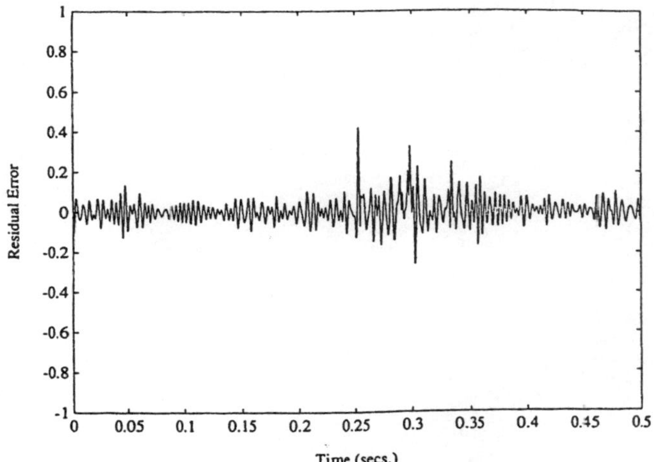

FIG. 12. The residual error (received signal minus reconstructed signal) for the unknown-signal algorithm.

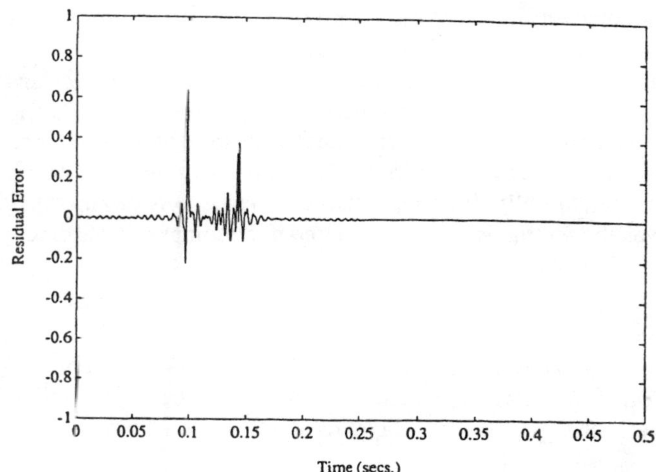

FIG. 13. The residual error between the source signal of Fig. 7. and the optimum zero-phase sinusoid parametric model.

multiple sinusoids, even for $\alpha=0$ the error surface would possess numerous local minima. We are currently developing a technique which will overcome this problem by partitioning the frequency axis into bins. An algorithm similar to that given in this paper is used in each bin.

The rather large residual error obtained from the unknown signal algorithm is now explained. This poor fit is not due to any inadequacy in the unknown signal algorithm *per se*, but due to the fact that the actual source could not accurately be modeled as a perfect sinusoid. To demonstrate this, we ran the unknown signal algorithm by considering the recorded source to be the received signal and tried to model it by a sinusoid with $M=1$. The estimated frequency and duration for the source were found to be $f=243$ Hz and $d=44$ ms, respectively. The residual error is shown in Fig. 13. There is considerable mismatch at the beginning and at the end of the source pulse due to gradual signal build up and decay. This, therefore, is the reason for the spikes in the residual error of Fig. 12, wherein the unknown signal algorithm assumed a perfectly sinusoidal source.

VI. CONCLUSIONS AND FURTHER WORK

The residual error for the experimental data of Sec. IV C using the known signal algorithm was quite small. We tried the same procedure on a different experimental data set which used a chirp signal as the source pulse. The bottom reflecting surface was very rough in this case. The known signal algorithm was unable to improve the fitting error beyond a certain point. There still appeared to be considerable signal structure in the error residual, indicating scope for a better fit. We feel that this might be due to inadequacy of the channel model given in Eq. (1); further work is being carried out to verify this.

To summarize, in this paper we pointed out the importance of constraining the amplitudes to be real. But the error minimization in this case turned out to be difficult. The complex amplitude case is easier to solve but yields biased parameter estimates. We then presented an algorithm which finds the minimum of the true error surface starting with the biased error surface. When initial parameter estimates are obtained using coordinate descent, our algorithm finds the global minimum of the error surface in spite of the existence of numerous local minima. The source was assumed to be known, initially. Finally, even if the source were unknown, it was shown that it can be estimated if it belongs to a parametric class of signals.

ACKNOWLEDGMENTS

This work was supported by the ONR funded Acoustic Transients ARI, program element No. 0601153N, under NRL–SSC program management (Dr. Edward Franchi).

[1] R. Field, "Transient signal distortion in a multipath environment," in *Proceedings of Oceans '90* (IEEE, Piscataway, NJ, 1990), pp. 111–114.

[2] G. Carter, "Time delay estimation for passive sonar signal processing," IEEE Trans. Acoust. Speech Signal Process. **ASSP-29**, 463–470 (June 1981).

[3] J. E. Ehrenberg, T. E. Ewart, and R. D. Morris, "Signal Processing Techniques for Resolving Individual Pulses in a Multipath Signal," J. Acoust. Soc. Am. **63**, 1861–1865 (1978).

[4] S. Senamato and D. G. Childers, "Signal resolution via digital inverse filtering," IEEE Trans. Aerosp. Electron. Syst. **AES-8**, 633–640 (1972).

[5] B. M. Bell and T. E. Ewart, "Separating multipaths by global optimization of a multidimensional matched filter," IEEE Trans. Acoust. Speech Signal Process. **ASSP-34**, 1029–1037 (Oct. 1986).

[6] J. O. Smith and B. Friedlander, "Adaptive multipath delay estimation," IEEE Trans. Acoustics Speech Signal Process. **ASSP-33**, 812–822 (August 1985).

[7] R. Tremblay, G. Carter, and D. Lytle, "A practical approach to the estimation of amplitude and time-delay parameters of a composite signal," IEEE J. Oceanic Eng. **OE-12**, 273–278 (1987).

[8] I. P. Kirsteins, "High Resolution Time Delay Estimation," in *IEEE Proceedings ICASSP 87*, Dallas, TX (IEEE, New York, 1987), pp. 451–454.

[9] I. P. Kirsteins and A. H. Quazi, "Exact Maximum Likelihood Time Delay Estimation for Deterministic Signals," in *Proceedings of EUSIPCO-88*, Grenoble, France, Sept. 1988 (Elsevier, Amsterdam, 1988).

[10] A. G. Evans and R. Fischl, "Optimal least squares time-domain synthesis of recursive digital filters," IEEE Trans. Audio Electroacoust. **AU-21**, 61–65 (February 1973).

[11] R. Kumaresan, L. Scharf, and A. Shaw, "An algorithm for pole-zero modeling and spectral analysis," IEEE Trans. Acoust. Speech Signal Process. **ASSP-34**, 637–640 (June 1986).

[12] C. W. Helstrom, *Statistical Theory of Detection* (Pergamon, New York, 1960).

[13] P. Gill, W. Murray, and M. Wright, *Practical Optimization* (Academic, New York, 1981).

[14] J. Dennis and R. Schnabel, *Numerical Methods for Unconstrained Optimization and Nonlinear Equations* (Prentice-Hall, Englewood Cliffs, NJ, 1983).

[15] E. Maragakis, "Time Delay Estimation in a Multipath Environment," Master's thesis, University of Rhode Island, 1990.

[16] G. Golub and C. V. Loan, *Matrix Computations* (Johns Hopkins U.P., Baltimore, MD, 1984).

[17] D. Luenberger, *Introduction to Linear and Nonlinear Programming.* Reading, (Addison-Wesley, Reading, MA, 1973).

[18] J. Cadzow, "Signal processing via least squares error modeling," IEEE Signal Process. Mag. **7**, 12–31 (October 1990).

[19] R. Vaccaro, E. Maragakis, and R. Field, "Transient Signal Extraction in a Multipath Environment," in Proc. Oceans '90, Washington, DC, pp. 115–118 (September 1990).

[20] H. Schmidt, "SAFARI (Seismo-Acoustic Fast Field Algorithm for Range Independent Environment) User's Guide," Technical report, SACLANT Undersea Research Centre, Rep. SR-113 (1988).

[21] J. Leclere and R. Field, "Comparison of time-domain parabolic equation and measured ocean impulse responses," Proc. Oceans '90, pp. 125–128 (September 1990).

Calculation of the conduction velocity of short nerve fibres

G. H. van der Vliet J. Holsheimer

Bio-information group, Department of Electrical Engineering, Twente University of Technology, Enschede, The Netherlands

D. Bingmann

Physiological Institute I, University of Münster, BRD

Abstract—*The conduction velocity v of a nerve fibre is calculated from the time delay Δ of a propagating action potential between two recording sites along the fibre. However, the conventional method of determining Δ cannot be applied to short nerve fibres. Therefore several linear signal analysis methods for the estimation of Δ have been compared with regard to the reproducibility of their results obtained from pairs of simultaneously recorded action potentials at several small inter-electrode distances. It was found that estimating Δ from the cross-correlogram and as a second one a variant of this method (maximum likelihood time delay estimation) give the most reliable values of v in short nerve fibres.*

Keywords—*Conduction velocity, Time delay, Nerve-fibre, Action potential*

1 Introduction

THE CONDUCTION velocity of a nerve fibre can be calculated from the time delay of a propagating action potential (a.p.) between two recording sites along the fibre. This is a well known method in clinical and experimental neurophysiology. The time delay between the peaks of two extracellularly and referentially recorded a.p.s can be determined directly, for instance by means of a memory oscilloscope.

In order to obtain a reliable estimate of the conduction velocity, the time delay between the a.p.s has to be large in comparison with the duration of the a.p. (1–2 ms). Normally, the distance between the recording electrodes is at least several centimetres. This method cannot be applied to very short nerve fibres, e.g. the carotid sinus nerve. Under these conditions the two recording electrodes have to be situated only a few millimetres apart and the recorded a.p.s will overlap in time (see Fig. 2). The estimation of the time delay is complicated because in most experiments the a.p.s from one fibre have different waveforms at the two electrodes, which might result from asymmetric recording conditions. Hence, in these short nerve fibres one cannot expect the close relationship between some parameters of the a.p. waveform and the conduction velocity, which has been described in vagal fibres by PAINTAL (1966).

In order to determine the conduction velocity of a nerve fibre under these experimental conditions we used different functions in the time- and frequency domain. In order to select the best estimator, the time delays calculated from these functions were compared for a series of experiments.

First received 22nd August 1979 and in revised form 25th January 1980

0140–0118/80/060749 + 09 \$01·50/0

2 Theoretical aspects

2.1 General aspects

A straight unmyelinated active nerve fibre in a volume conductor has been considered. The electrical properties of the axon are supposed not to change with the length co-ordinate x of the fibre. This means that the shape of the membrane action potential (m.a.p.) is constant and that a.p.s are propagated with a constant velocity v.

The relation between the m.a.p. m at x_1 and x_2 at a distance $L = x_2 - x_1$ can be given by

$$m(t, x_2) = m(t, x_1 + L) = m\left(t - \frac{L}{v}, x_1\right) \quad (1)$$

t denotes the time variable. The quotient L/v is the *time delay* Δ of the propagating m.a.p. between x_1 and x_2. So eqn. 1 can be written as

$$m_2(t) = m_1(t - \Delta) \quad (2)$$

Since a.p.s have a limited duration, a value of T can be found, for which the two partly overlapping m.a.p.s $m_1(t)$ and $m_2(t)$ occur between t_0 and $t_0 + T$. So if m denotes the deviation from the resting membrane potential, $m_1(t) = 0$ for $t \leq t_0$ and $t \geq t_0 + T$. This means that m_1 and m_2 can be Fourier-transformed:

$$M_2(j\omega) = M_1(j\omega)e^{-j\omega\Delta} \quad (3)$$

For the determination of the conduction velocity v, however, we are not recording m.a.p., but extracellular potentials close to an isolated fibre or an active fibre in a small bundle, against a reference electrode (earth) situated at some distance in the volume conducting medium.

The extracellular action potential (e.a.p.), represented by $e(t, x)$, is supposed to be related by a given function to the m.a.p. $m(t, x)$, and is dependent on the position of the recording and reference electrodes with respect to the active nerve fibre (ROSENFALCK, 1969). For two simultaneous measurements near the

Reprinted from *Medical and Biological Engineering and Computing*, vol. 18, no. 6, pp. 749–757, November 1980.

fibre with respect to a common reference the relationships between m and e can be written in the frequency domain as follows:

$$E_1(j\omega) = F_1(j\omega)M_1(j\omega) \tag{4}$$

$$E_2(j\omega) = F_2(j\omega)M_2(j\omega) \tag{5}$$

So $F_1(j\omega)$ and $F_2(j\omega)$ are transfer functions of the respective m.a.p. and e.a.p.

The quotient of the two transfer functions is given by the so-called *relation function* (defined for the frequency band for which $E_i(j\omega) \neq 0$ and $M_i(j\omega) \neq 0$):

$$F(j\omega) = \frac{F_2(j\omega)}{F_1(j\omega)} = \alpha(\omega)e^{j\beta(\omega)} \tag{6}$$

From eqn. 6 we see that this relation function can be divided into two parts: the *amplitude ratio function* $\alpha(\omega)$ and the *difference-in-phase-shift function* $\beta(\omega)$.

If the two e.a.p.s have the same shape, then

$$F(j\omega) = \alpha(\omega) = 1 \quad \text{and} \quad \beta(\omega) = 0$$

for all values of ω.

From eqns. 3, 4, 5 and 6, the relationship between the two e.a.p.s in the frequency domain can be written as follows:

$$E_2(j\omega) = E_1(j\omega)\alpha(\omega)e^{-j[\omega\Delta - \beta(\omega)]} \tag{7}$$

Under experimental conditions, however, random noise from the biological preparation and the recording instrumentation, which is assumed to be additive, will influence the measurements. So the recorded signals r_i at the electrodes will be composed of the e.a.p. (e_i) and noise (n_i):

$$\underline{r}_1(t) = e_1(t) + \underline{n}_1(t) \tag{8}$$

$$\underline{r}_2(t) = e_2(t) + \underline{n}_2(t) \tag{9}$$

The e.a.p. e_i is assumed to occur between t_0 and $t_0 + T$, so $e_i(t) = 0$ for $t \leq t_0$ and $t \geq t_0 + T$. $\underline{n}_i(t)$ is the realisation of a stochastic (noise) process, which is supposed to be ergodic.

When using a data window $w(t)$, with $w(t) = 1$ for $t_0 \leq t \leq t_0 + T$ and $w(t) = 0$ otherwise, the Fourier transforms of the recorded signals $\underline{r}_i(t)w(t)$ are as follows:

$$\underline{R}_1(j\omega, T) = E_1(j\omega) + \underline{N}_1(j\omega, T) \tag{10}$$

$$\underline{R}_2(j\omega, T) = E_2(j\omega) + \underline{N}_2(j\omega, T) \tag{11}$$

The generation of the recorded signals as described here is summarised in Fig. 1.

Each realisation $\underline{r}_i(t)w(t)$ has a *noise-signal-ratio function*:

$$\underline{\eta}_i(\omega, T) = \frac{|\underline{N}_i(j\omega, T)|}{|\underline{E}_i(j\omega)|} \tag{12}$$

Now the *bandwidth* of a recorded action potential (r.a.p.) $\underline{r}_i(t)w(t)$ can be defined as the frequency band for which the mean-square value of the noise-signal-ratio function $msv[\underline{\eta}_i(\omega, T)] \leq 1$.

The *conduction velocity* v of the nerve fibre concerned can be calculated from an estimation of the time delay between the two r.a.p.s $\underline{r}_1(t)w(t)$ and $\underline{r}_2(t)w(t)$:

$$\hat{v} = L/\hat{\Delta} \tag{13}$$

We may conclude that a discrepancy between the estimated time delay $\hat{\Delta}$ and the real time delay Δ can be the result of two different processes:

(*a*) $\beta(\omega) \neq 0$ will result in a systematic error or bias

(*b*) $msv[\underline{\eta}_i(\omega, T)] \neq 0$ will result in a random error.

2.2 *Estimation of time delay*

Several methods for calculation of time delay between r.a.p.s are known from the literature (PAINTAL 1966; HUTCHINSON *et al.*, 1970; KRAUSE *et al.*, 1972; KALBFLEISCH *et al.*, 1972; HARDY, 1973). These methods result in several estimators for time delay $\hat{\Delta}$, which have been compared in this investigation in order to find which was the most reliable.

In principle, several linear analysis methods for ergodic signals (cf. BENDAT and PIERSOL, 1971) can be adapted for calculation of the time delay Δ between two simultaneously recorded r.a.p.s which are considered as two partially deterministic signals. As described in the previous part the choice of the data window $w(t)$ is determined by the position in time of the r.a.p. In the time domain we will use some well defined spots on the r.a.p. waveform(*a*) as well as the cross correlogram(*c*); in the frequency domain Δ will be derived from the phase spectrogram(*b*). Other methods, which will be discussed here, are calculation of a 'maximum likelihood time delay estimate' for series of pairs of r.a.p.s, according to CARTER (1976)(*d*) and derivation of Δ from the energy spectrogram of the differential signal of the two recorded signals (bipolarly recorded action potential), according to the method used in electromyography (LINDSTRÖM and MAGNUSSON, 1977)(*e*). The random error as a result of additive noise, related to the estimators of Δ, will be discussed too.

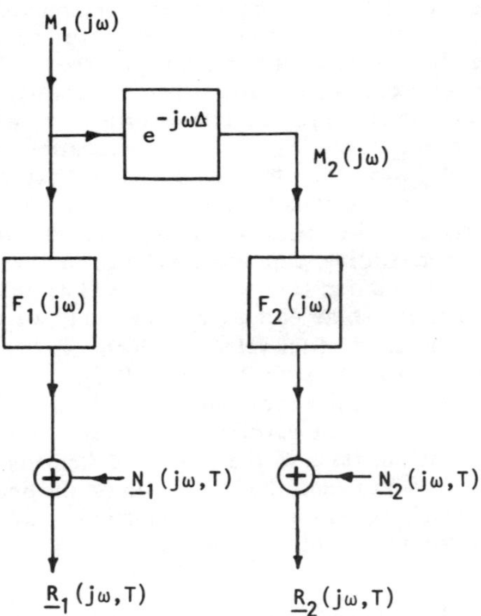

Fig. 1 *Schematic representation (in the frequency domain) of the generation of the recorded signals $\underline{R}_i(j\omega, T)$ from the membrane action potential $M_i(j\omega)$ at two sites of a nerve fibre and an additive random noise component $\underline{N}_i(j\omega, T)$*

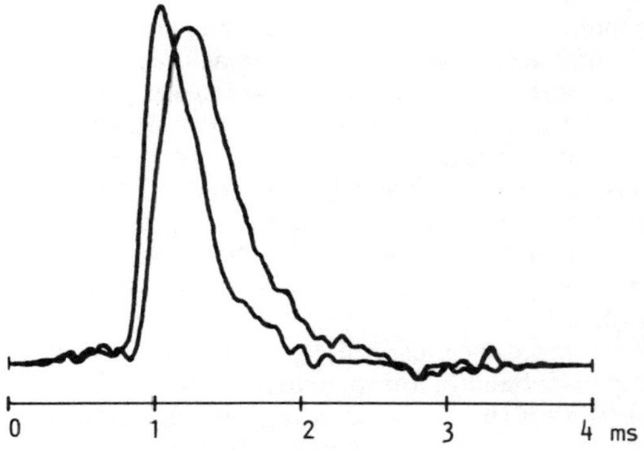

Fig. 2 *Pair of simultaneously recorded action potentials from a sinus nerve preparation at L = 2·2 mm, having different waveforms and a time delay*

(a) Direct time delay

The conventional method for estimation of the time delay of a propagating action potential between two recording sites is measuring the time interval between the corresponding peaks of the r.a.p.s. When the shape of the two r.a.p.s is different, time delays estimated in this way can be quite different from the m.a.p. time delay Δ. The systematic error as well as the noise-induced error depend on the choice of the reference spot (peak, half the peak value, first point of inflection). Thereby the systematic error cannot be determined, while the noise-induced error can be estimated from the variability in the time intervals calculated from a number of pairs of r.a.p.s.

(b) Phase time delay

The *phase spectrogram* of two e.a.p.s $e_1(t)$ and $e_2(t)$ is, according to eqn. 7:

$$\phi_{E12}(\omega) = \arg[E_1(j\omega)/E_2(j\omega)] = \omega\Delta - \beta(\omega) \quad (14)$$

After dividing $\phi_{E12}(\omega, T)$ by the radial frequency ω, we find the *phase time delay spectrogram*:

$$\Delta_{ph}(\omega) = \phi_{E12}(\omega)/\omega = \Delta - \beta(\omega)/\omega \quad (15)$$

When $\Delta_{ph}(\omega)$ is the estimator of Δ, an unknown systematic error $-\beta(\omega)/\omega$ is made. When the e.a.p.s are affected by additive random noise, the phase spectrogram of $e_1(t)$ and $e_2(t)$ is estimated from $\underline{r}_1(t)w(t)$ and $\underline{r}_2(t)w(t)$. Using eqns. 10 and 11, we find:

$$\hat{\phi}_{E12}(\omega, T) = \underline{\phi}_{R12}(\omega, T) = \phi_{E12}(\omega, T) + \arg\left[\frac{1 + \underline{N}_1(j\omega, T)/E_1(j\omega)}{1 + \underline{N}_2(j\omega, T)/E_2(j\omega)}\right] \quad (16)$$

So the calculated phase spectrogram ϕ_{R12} is an approximation of the phase spectrogram ϕ_{E12}.

In order to calculate the variance $\sigma_\Delta^2(\omega)$ of the noise-induced error in the estimated phase time delay $\hat{\Delta}(\omega) = \underline{\phi}_{R12}(\omega, T)/\omega$ the noise components n_1 and n_2 are supposed to have a Gaussian distribution. If in some frequency range $msv[\underline{\eta}_1(\omega, T)]$ and $msv[\underline{\eta}_2(\omega, T)]$ have small values ($< 0{\cdot}02$), we will find the variance of the noise-induced error:

$$\sigma_\Delta^2(\omega) \approx \{msv[\underline{\eta}_1(\omega, T)] + msv[\underline{\eta}_2(\omega, T)]\}/2\omega^2 \quad (17)$$

The main problem of this method is finding the best frequency band for calculation of the phase time delay as an estimate of the real time delay Δ. It is impossible to correct the calculated time delay for the unknown systematic error $-\beta(\omega)/\omega$. So the only criterion for choosing a frequency band can be the mean-square value of the noise-signal-ratio function in that frequency band of the two r.a.p.s The condition is that the variance $\sigma_\Delta^2(\omega)$ of the phase time delay, which can be derived from this function according to eqn. 17, must be small. $\sigma_\Delta^2(\omega)$ can be found experimentally from the phase time delay spectrograms of a series of pairs of r.a.p.s. The noise-induced error can be reduced by averaging.

(c) Cross time delay

Another estimator of the time delay Δ is the position of the maximum of the *cross correlogram* C_{e12} of the two e.a.p.s:

$$C_{e12}(\tau, T) = \int_{t_0}^{t_0+T} e_1(t)e_2(t+\tau)\,dt \quad (18)$$

From eqn. 18 it can be calculated that $C_{e12}(\tau, T) = f(\tau)C_{e11}(\tau - \Delta, T)$. So the cross correlogram is the convolution of the inverse Fourier transform of the relation function $F(j\omega)$ with the autocorrelogram $C_{e11}(\tau, T)$ shifted over Δ. The cross correlogram $C_{e12}(\tau, T)$ has its maximum at $\tau = \Delta_c$. If the two e.a.p.s have the same shape $[F(j\omega) = 1]$ the cross time delay Δ_c is an unbiased estimator of the m.a.p. time delay Δ.

When $F(j\omega) \neq 1$, we have to take into account a systematic error in the estimated time delay. We calculated that Δ_c is equal to a weighed average of the phase time delay spectrogram of the two e.a.p.s, so the systematic error $\Delta_c - \Delta$ is a weighed average of $-\beta(\omega)/\omega$. The weighing function was found to be proportional to the product of the amplitude spectra $|E_1(j\omega)|\,|E_2(j\omega)|$ and the squared frequency. However, since $\beta(\omega)$ is not known we cannot calculate $\Delta_c - \Delta$.

When the e.a.p.s are affected by additive random noise, then $C_{e12}(\tau, T)$ is approximated by the cross correlogram of $\underline{r}_1(t)w(t)$ and $\underline{r}_2(t)w(t)$. So in determining the cross time delay Δ_c a noise-induced error is made, which was found to be the weighed average of $[\underline{\phi}_{R12}(\omega, T) - \phi_{E12}(\omega, T)]/\omega$. It is rather complex to calculate the variance σ_Δ^2 of the noise-induced error from η_1 and η_2, but this value can be found experimentally. The noise-induced error can be reduced by averaging.

(d) Maximum-likelihood time-delay estimate

A variant of the cross-time-delay method is described by CARTER (1976). By this method the maximum-likelihood estimate of the time delay between the linearly related components of two signals is calculated. After adaptation this method can be applied to a series of pairs of r.a.p.s. The time delay is given by the position of the maximum of:

$$C'_{r12}(\tau, T) = \frac{1}{2\pi}\int_{-\infty}^{\infty} \exp[j\hat{\phi}_{E12}(\omega, T)] \times \frac{\gamma_{R12}^2(\omega, T)}{1 - \gamma_{R12}^2(\omega, T)} e^{j\omega\tau}\,d\omega \quad (19)$$

The first factor behind the integration sign is an estimate of the normalised cross energy spectrogram of $e_1(t)$ and $e_2(t)$. The second factor is a weighing function derived from the squared coherence function

$\gamma_{R12}^2(\omega, T)$, which can also be estimated from this series of pairs of recorded signals $\underline{r}_i(t)w(t)$. The time delay to be found by this method is a weighted average of the mean phase time-delay spectrogram. It was calculated that this weighting function is approximately proportional to $1/\sigma_\Delta^2(\omega)$ (see eqn. 17).

(*e*) *Energy spectrograms*

It is expected that action potentials recorded from fast conducting nerve fibres contain more high-frequency components than action potentials from slowly conducting fibres.

So it might be possible to establish a relation between the conduction velocity and some parameter of the energy spectrogram of a single r.a.p.

On the other hand the energy spectrogram of a bipolar recording from a nerve fibre rather contains information about the delay of the propagating action potential between the two recording sites, according to LINDSTRÖM and MAGNUSSON (1977). From eqn. 7 it can be calculated that the energy spectrogram $|E_2(j\omega) - E_1(j\omega)|^2$ of the (noise-free) differential signal from the two electrodes will show minima (dips) for $\omega\Delta - \beta(\omega) = 2k\pi$ ($k = 1, 2, \ldots$). So the time delay can be estimated from the 'first dip-frequency' f_{d1} ($k = 1$):

$$\hat{\Delta} = \frac{1}{f_{d1}} = \Delta - \frac{\beta(\omega_{d1})}{\omega_{d1}} \qquad (20)$$

and/or from $(f_{d2} - f_{d1})$ etc. Apart from an extra slope error in the energy spectrogram (VAN DER VLIET, 1978) a systematic error $-\beta(\omega_{d1})/\omega_{d1}$ is made.

In spite of the advantages of a bipolar recording (little recording artefacts), this method can only be used under certain conditions: f_{d1} has to lie within the frequency band of the r.a.p. as defined before (see eqn. 12 and further).

A comparison of the methods discussed above, leads to the following conclusions:

(i) if the two simultaneously recorded action potentials have exactly the same shape, i.e. there is no systematic error $[\beta(\omega) = 0]$ and no noise-induced error $[\underline{N}_i(j\omega, T) = 0]$, all methods will give exactly the same result: all estimators will give the real time delay Δ.

(ii) if the two r.a.p.s have different shapes, i.e. the relation function $F(j\omega) \neq 1$ and/or they are are disturbed by additive random noise $[\underline{N}_i(j\omega, T) \neq 0]$, the methods may give different estimates $\hat{\Delta}$ as a result of differences in sensitivity for these two influences.

3 Material and methods

3.1 Experimental procedures

In order to test different methods for the estimation of the conduction velocity of short nerve fibres, r.a.p.s were measured from single- and few-fibre preparations of the carotid sinus nerve and the vagal nerve. Experiments were carried out on cats anaesthetised with sodium pentobarbitone (50 mg per kg), paralysed with gallamine triethiodide and ventilated artificially. Body temperature was kept constant at 37°C. The carotid sinus nerve was cut close to the glossopharyngeal nerve in order to obtain fibre preparations as long as possible (4–5 mm). All *in-vivo* fibre preparations were embedded in paraffin oil.

Recordings were made with two platinum wire electrodes (diam. 0·1 mm) fitted in a Perspex holder. The interelectrode distance L could be varied. A third platinum wire in the same holder was used to fix the cut end of the fibre preparation, which was placed on the bent tips of the recording electrodes. Recordings were made simultaneously from the two electrodes against a common reference electrode, which was situated in the neck muscles near the recording site. From the measurements it could be concluded that usually only action potentials from one fibre were recorded.

High-impedance low-noise f.e.t. preamplifiers with a frequency band of approximately 1–10000 Hz were used. Within this frequency range no significant difference in phase shift between the two parallel amplification systems could be detected.

3.2 Signal storage and pre-processing

The two simultaneously recorded signals were sampled and stored on-line by a PDP-12 computer with 10-bit analogue-to-digital convertors. Each signal was sampled at a rate of 18500 per second. This sampling rate is high enough, since the power of the recorded signals beyond 6 kHz can be neglected (see Fig. 5). The interval between sampling of the two signals was 12·3 μs. This interval has to be taken into account when calculating the time delay Δ. Recordings of approximately 10 seconds from several preparations were stored on LINC-tape. Afterwards up to 15 pairs of r.a.p.s from each recording were selected for further data processing on a PDP-11/40 computer.

Each time series holding one r.a.p. had a record length of 4 ms (75 samples). The r.a.p.s were superimposed on low-frequency noise. From each time series the low-frequency noise component, approximated by a linear interpolation between the first and the last (75th) sample, was subtracted. This procedure resulted in new 4 ms time series from which the first and last sample had a value zero. These time series were multiplied with a cosine-taper data window and extended to time series of 128 points by adding sample points with a value zero. The time series obtained this way were used for estimating the time delay Δ by the methods discussed in the previous part (Section 2).

3.3 Data processing

(*a*) *Direct time delay*

Under the experimental conditions mentioned above the time delays Δ were expected to be in the same range as the sample interval ($T_s = 54$ μs). In order to get an accurate estimation of the time delay, a much better resolution in the time domain was required. This resolution was obtained by interpolation using fast Fourier transforms. After transformation of each r.a.p. time series into the frequency domain the complex Fourier series was extended by inserting (equidistant) frequency points—having a value zero—beyond the original Nyquist frequency. After inverse Fourier transformation a new time series having smaller intervals was obtained. In our case the resolution in the time domain was increased 16 times, resulting in 3·4 μs intervals. This interpolation procedure is acceptable since the power of the analogue signal beyond the Nyquist frequency (9250 Hz) can be neglected. From the interpolated time series of two simultaneously recorded r.a.p.s the time delay between well defined spots of the waveforms (peak value, half the peak value, first point of inflection) could easily be calculated.

(*b*) *Phase time delay*

The discrete phase time delay spectrogram was calculated from the two Fourier transforms of the time series $\underline{r}_1(t)w(t)$ and $\underline{r}_2(t)w(t)$, according to eqns. 14 and 15. The resolution in the frequency domain was increased 8 times (to 18 Hz) by extending the time series with additional sample points with value zero. By calculating the deviation $\hat{\sigma}_\Delta(\omega)$ of the phase time-delay spectrograms of a number of successive pairs of r.a.p.s, the frequency band could be chosen for calculating the phase time delay as an estimator of the m.a.p. time delay Δ [see Section 2.2(*b*)].

(*c*) *Cross time delay*

The discrete cross correlogram of the sampled r.a.p. was calculated according to:

$$C_{r12}(nT_s) = \sum_{m=0}^{M-1} \{\underline{r}_1(mT_s)\underline{r}_2[(m-n)T_s]\} \qquad (21)$$

M being the record length of each r.a.p. time series. The cross time delay, with a resolution $T_s = 54$ μs, is approximately given by the value of nT_s corresponding to the maximum value of C_{r12}. In order to get a more accurate estimate of the cross time delay, the resolution was increased 16 times by interpolating the discrete cross correlogram, as has been described in a previous part [Section 3.3(*a*)].

The same result can be obtained by calculation of the discrete cross energy spectrogram of the two time series and inverse Fourier transformation after insertion of frequency points.

(*d*) *Maximum-likelihood time-delay estimate*

The maximum-likelihood time-delay estimates were calculated according to eqn. 19 from series of pairs of r.a.p.s. Therefore the squared coherence functions were estimated in the same way as has been done for ergodic signals (CARTER, 1976). The resolution was increased 16 times by interpolating the resulting time functions. Standard deviations of the maximum-likelihood time-delay estimator were calculated from the results of several groups of some r.a.p. pairs in each series.

(*e*) *Energy spectrogram*

The discrete energy spectrogram of an r.a.p. $\underline{r}(t)w(t)$ was calculated from the Fourier transform of this time series, by squaring the modulus of $\underline{R}(j\omega, T)$. The resolution in the frequency domain was increased 8 times (to 18 Hz) by adding sample points with value zero to the original time series.

In order to get an estimate of the bandwidth (see eqn. 12 and further) of an r.a.p., *msv* $|\underline{N}(j\omega, T)|$ can be estimated by averaging the energy spectrograms of a number of epochs (length T) from the noise signal preceding this r.a.p. The bandwidth was found by comparing this averaged noise spectrogram with the energy spectrogram of the r.a.p. The bandwidth can also be estimated by means of the squared coherence function, if a number of pairs of r.a.p. are available. In this case the mean bandwidth of the two series of r.a.p.s is given by the frequency band for which $\gamma^2_{R12}(\omega, T) \geq 0{\cdot}25$.

4 Results

In Section 2 it was concluded that all presented methods of determining the conduction velocity will give exactly the same result if the two r.a.p.s have identical waveforms. So calculations on a simulated pair of identical action potentials could be used for testing the data-processing programs. The experimentally obtained pairs of r.a.p.s had different shapes (see Fig. 2). Thus the different time-delay estimates had to be compared with respect to their reliability. Results will be presented from three recordings from the same carotid sinus nerve preparation at inter-electrode distances $L = 1{\cdot}1$, $1{\cdot}7$ and $2{\cdot}2$ mm. From each recording a series of over 10 pairs of r.a.p.s were used for estimating the time delay Δ by the different methods.

(*a*) *Direct time delay*

In Fig. 2 the interpolated time series of a pair of r.a.p.s at $L = 2{\cdot}2$ mm are shown. For each pair of r.a.p.s the time delay was calculated from the positions of the peaks (peak time delay), the centres of the rising flanks (flank time delay) and the peaks of the first derivatives (point-of-inflection time delay). From each recording series the means and deviations of the estimated conduction velocities were calculated. The results are shown in Table 1. From these results we see that the three methods give different results at all values of L tested. Thereby the peak propagation velocities are much smaller than the values obtained by the other two methods. This result can be explained by the differences in waveform of the two r.a.p.s, as can be seen in Fig. 2. From Table 1 it can also be seen that $\hat{v}$ obtained from the flank time delays had the smallest deviation at all three values of L.

(*b*) *Phase time delay*

In Fig. 3*a* the average and the deviation curve of $\hat{\Delta}(\omega)$ of 15 pairs of r.a.p.s at $L = 2{\cdot}2$ mm are shown. It can be seen that the mean of $\hat{\Delta}(\omega)$ varies with frequency. Most other recordings also gave varying mean values. So it was impossible to calculate $\hat{v}$ simply from the average curve. The only criterion for choosing a frequency band for calculating $\hat{\Delta}(\omega)$ was the variance [see Section 2.2(*b*)]. In Fig. 3*a* the deviation is relatively small within the frequency range of 300–3800 Hz. In most recordings the frequency range having a relatively small deviation value was restricted to 500–2000 Hz. But also within this frequency range the mean of $\hat{\Delta}(\omega)$ was not constant in most results. In the results presented here the deviation of $\hat{\Delta}(\omega)$ had a minimum at about 1000 Hz. At this frequency—and for comparison at 500 Hz and at 2000 Hz—$\hat{v}$ was calculated from the mean of $\hat{\Delta}(\omega)$ (see Table 1). From this Table it can be seen that the values of $\hat{v}$ calculated from $\hat{\Delta}(\omega)$ at 1000 Hz are only slightly different at the three interelectrode distances. The values of $\hat{v}$ calculated from $\hat{\Delta}(\omega)$ at 500 Hz and at 2000 Hz show larger differences at the three values of L.

(*c*) *Cross time delay*

In Fig. 4*a* the interpolated cross correlogram of a pair of r.a.p.s at $L = 2{\cdot}2$ mm is shown. It can be seen

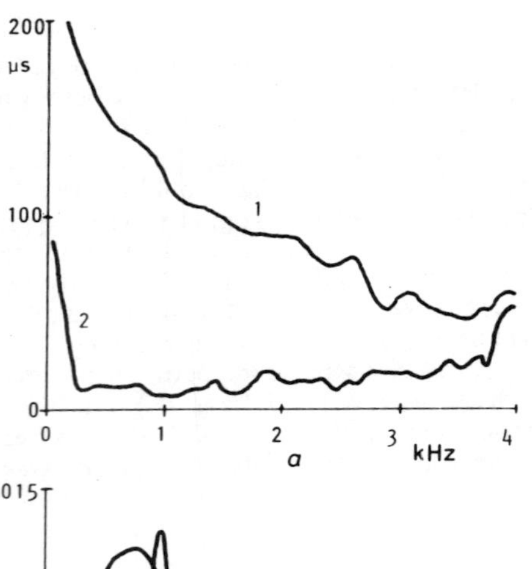

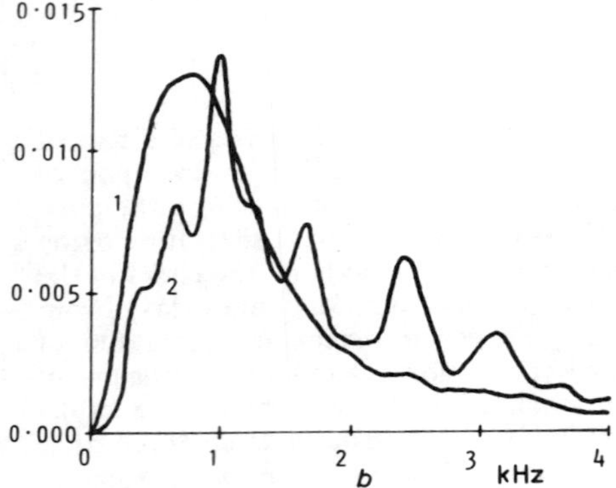

Fig. 3(*a*) *Average phase time delay (1) and deviation (2) as a function of frequency, obtained from 15 pairs of simultaneously recorded action potentials at L = 2·2 mm from a sinus nerve preparation*

(*b*) *Phase-time-delay weighing functions for the cross time delay (1) and for the maximum-likelihood time-delay estimate (2) derived from the same 15 pairs of recorded action potentials*

that the position of the maximum differs from $\tau = 0$. In Table 1 the mean and deviation of $\hat{v}$ calculated from Δ_c at each value of L are presented. The three mean values are slightly different, but the relatively small deviation areas are amply overlapping.

(d) Maximum-likelihood time-delay estimate

In Fig. 4*b* the time function as a result of the application of the method of CARTER (1976) to the 15 pairs of r.a.p.s at $L = 2{\cdot}2$ mm is shown. It can be seen that the position of the maximum differs from $\tau = 0$. In Table 1 the conduction velocities calculated from the resulting time delays are presented. The three values are slightly different. The deviations were estimated as follows: from different combinations of four pairs of r.a.p.s at one value of L, the maximum-likelihood time-delay estimate was calculated. The deviation of the resulting values of $\hat{v}$ was calculated and multiplied by a factor $\sqrt{4}$ in order to allow a comparison with the other deviations from Table 1 (in case $n = 4$).

In Fig. 3*b* the (phase time delay) weighing functions for the cross-time delay (1) and for the maximum-likelihood time-delay estimate (2), as described in Sections 2.2(*c*) and (*d*), are shown. The two curves were estimated from the same 15 pairs of

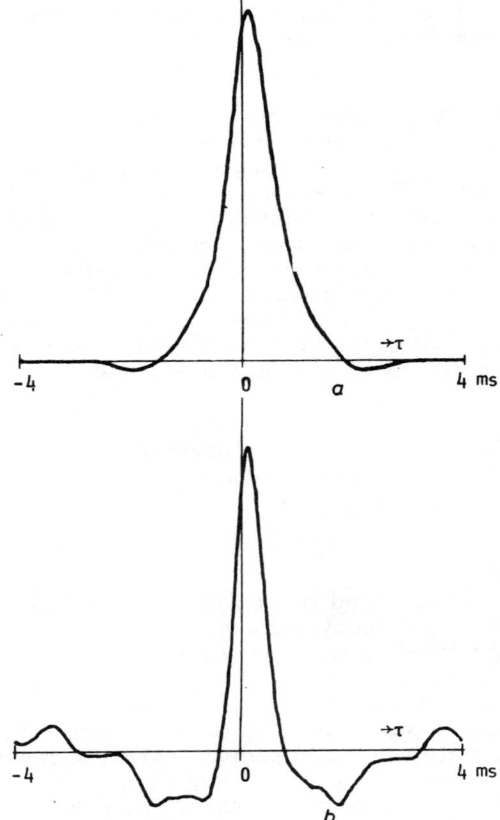

Fig. 4(*a*) *Interpolated cross correlogram of a pair of simultaneously recorded action potentials from a sinus nerve preparation at L = 2·2 mm*

(*b*) *Interpolated time function obtained by application of the maximum-likelihood time-delay estimate method on 15 pairs of simultaneously recorded action potentials at L = 2·2 mm from the same sinus nerve preparation*

Table 1. Average conduction velocities and deviations calculated from different time delay estimates of n pairs of simultaneously recorded action potentials at interelectrode distances L (in m/s). Deviations also in percentage of average values

Time delay estimate	$L = 1{\cdot}1$ mm, $n = 13$		$L = 1{\cdot}7$ mm, $n = 13$		$L = 2{\cdot}2$ mm, $n = 15$	
(*a*) (*i*) peak time delay	15·3 ± 5·5	(36%)	12·0 ± 2·2	(18%)	11·2 ± 1·9	(17%)
(*ii*) flank time delay	23·9 ± 2·6	(11%)	23·6 ± 2·4	(10%)	28·2 ± 1·7	(6%)
(*iii*) point of inflection t.d.	31·4 ± 15·1	(48%)	22·7 ± 5·4	(24%)	39·3 ± 11·8	(30%)
(*b*) (*i*) phase t.d. at 500 Hz	15·1 ± 2·1	(14%)	16·2 ± 1·9	(12%)	14·3 ± 1·1	(8%)
(*ii*) phase t.d. at 1000 Hz	19·3 ± 2·7	(14%)	17·7 ± 1·4	(8%)	17·9 ± 1·1	(6%)
(*iii*) phase t.d. at 2000 Hz	18·6 ± 5·4	(29%)	23·3 ± 5·1	(22%)	24·2 ± 3·6	(15%)
(*c*) cross time delay	18·0 ± 1·4	(8%)	18·5 ± 1·3	(7%)	18·2 ± 0·7	(4%)
(*d*) max. likelihood t.d.e.	21·1 ± 3·3	(16%)	20·6 ± 3·2	(16%)	21·4 ± 2·9	(14%)

r.a.p.s at $L = 2{\cdot}2$ mm. The two time-delay estimates will be found by multiplication of $\hat{\Delta}(\omega)$ [Fig. 3*a*(1)] with the corresponding weighing factor at each discrete frequency and adding the resulting values of the whole frequency range (0–4 kHz).

(*e*) *Energy spectra*

The analysis of monopolarly recorded r.a.p.s did not yield clear relations between v and parameters of the energy spectrogram (VAN DER VLIET, 1978). From Fig. 5 it can be seen that the bandwidth of the r.a.p. was limited to approximately 4000 Hz. This value was found for most recordings. This bandwidth is too small for the estimation of v from bipolar recordings as described in Section 2.2(*e*), when using small values of L. The first dip frequency $f_{d1} \simeq v/L$ to be expected in the energy spectrogram will not be within the frequency band of the r.a.p. At $L = 3$ mm, for example, only values of v smaller than 12 ms^{-1} can be estimated by this method.

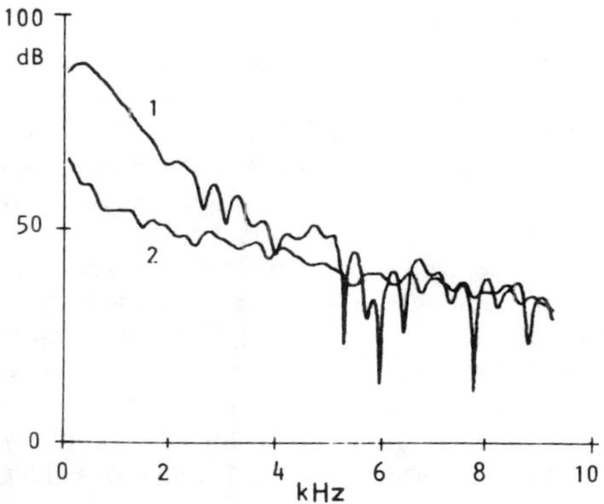

Fig. 5 *Energy spectrogram of a monopolarly recorded action potential (1) and the average energy spectrogram of a number of epochs of the preceding noise signal (2)*

5 Discussion and conclusions

From Table 1 it can be seen that the most reproducible results are obtained when applying the cross correlogram or the maximum-likelihood time-delay estimator. Each of these methods gives only small differences between the mean conduction velocities calculated at different interelectrode distances L. From Table 1 it can also be seen that the cross-time delay had the weakest sensitivity for additive random noise at all three values of L. For all three values of L there is a difference of about 10% between the values of $\hat{v}$ obtained by these two methods. Since the cross-time delay and the maximum-likelihood time-delay estimate can be considered a weighted average of the phase time-delay spectrogram, this difference is due to a difference between the weighting functions concerned. In the results presented in Fig. 3*b* the centre of the weighting function determining the maximum-likelihood time-delay estimate (2) is at a higher frequency than the centre of the weighting function determining the cross-time delay (1). Since $\hat{\Delta}(\omega)$ in this experiment decreases with increasing frequency [see Fig. 3*a*(1)], Δ_c will be the largest one, and thus $\hat{v}$ calculated from Δ_c will be smaller than $\hat{v}$ calculated from the maximum-likelihood time-delay estimate.

The direct time delays calculated for different reference spots, as well as $\hat{\Delta}(\omega)$ calculated at different frequencies, had large deviations and the mean values of $\hat{\Delta}$ obtained at different values of L differed widely. As the energy spectrograms of bipolar r.a.p.s did not yield reliable information about v, only the cross-time delay and the maximum-likelihood time-delay estimate turned out to be useful for the determination of v in short nerve fibres. The suitability of both methods, however, depends on the systematic and random errors which should be minimised by the recording and data processing techniques.

In Section 2.1 the *systematic error* was supposed to occur due to a difference between the transfer functions $F_1(j\omega)$ and $F_2(j\omega)$, as expressed in the relation function $F(j\omega) = \alpha(\omega)e^{j\beta(\omega)}$ eqn. 6. It was found that the effect of an amplitude ratio function $\alpha(\omega) \neq 1$ on $\hat{\Delta}$ is relatively small, in contrast with the effect of a difference-in-phase shift function $\beta(\omega) \neq 0$. From Fig. 3*a* it can be seen that the systematic error in $\hat{\Delta} = \Delta_{ph}(\omega)$ gets a lower value with increasing frequency. In other experiments the slope of the phase time-delay spectrogram could be quite different, even with a positive instead of a negative slope. So the systematic error varies with the experimental conditions. An important factor will be the difference in position of the recording electrodes with respect to earth (STEIN and PEARSON, 1971), and, in the case of myelinated fibres, the position of the recording electrodes with respect to the nodes of Ranvier (HUXLEY and STÄMPFLI, 1949; MARKS and LOEB, 1976).

Another possible systematic error source is a difference in m.a.p. waveform at the two electrodes, when the distal electrode is situated too near to the cut end of the nerve fibre. This cut end will cause a change in shape of the m.a.p. near the end, owing to changes in ion gradients across the membrane and to a decrease in resistance between the intra- and extracellular fluid. This possible difference in m.a.p. will be reflected in a difference in e.a.p., and thus in r.a.p. waveforms.

Different shapes of the two m.a.p. might also reflect a difference in local conduction velocity. So the estimated conduction velocity $\hat{v}$ is an average value over the piece of nerve fibre between the electrodes. A systematic error can also be introduced by the recording systems, owing to differences in phase shift of the two systems and crosstalk between them. However, in this series of experiments this type of error can be neglected. Still another type of systematic error can be introduced by the nerve fibre length between the electrodes. $\hat{v}$ is calculated from $\hat{\Delta}$ and L (eqn. 13). L is a good estimate of the fibre length between the electrodes only if the fibre is stretched up to its normal *in situ* length. So the estimated conduction velocity $\hat{v} = \hat{L}/\hat{\Delta}$.

A possible *random error source* is the discretisation error in the time domain. After interpolation of the time series the resolution $\Delta t = 3{\cdot}4$ μs. Supposing that the difference ε between the real and the discrete values has a uniform distribution between two discrete moments, then the standard deviation of ε is $\sigma_\varepsilon = \sqrt{(\Delta t)^2/12} = 0{\cdot}98$ μs. This discretisation error is small in comparison with other random errors. For instance, the deviation of the cross-time delay was > 4 μs at the three values of L. When calculating the direct time delay, the discretisation error is made twice, resulting in $\sigma_\varepsilon = 1{\cdot}39$ μs.

In Section 2.1 an unmyelinated nerve fibre has been considered. But recordings and calculations were also made of *myelinated fibres* in which the propagation of m.a.p. is not a continuous process:

m.a.p.s are only generated at the nodes of Ranvier. However, estimation of v of myelinated fibres is also possible with the methods discussed here, since the e.a.p. waveform at a location between two nodes is a time function in between the two neighbouring nodal e.a.p.s (HUXLEY and STÄMPFLI, 1949; MARKS and LOEB, 1976). Because this internodal e.a.p. is, owing to internodal membrane current, not the result of linear interpolation of the two nodal e.a.p.s, a small systematic error will be introduced. From the results presented in Table 1—calculated from the cross-time delays and from the maximum-likelihood time-delay estimates—it can be seen that $\hat{v}$ of this probably myelinated fibre is about the same at different values of L ($v \simeq 20$ ms^{-1}).

FIDONE and SATO (1969) estimated v of myelinated carotid chemoreceptor A fibres directly from the oscilloscope display of two monopolarly, simultaneously recorded a.p.s. They found that $\hat{v}$ ranged from 4 to 53 ms^{-1}, with a median of 16 ms^{-1} and a semi-interquartile range from 11 to 21 ms^{-1}. The median value is in the same range as the values shown in Table 1. From a comparison of these results, however, no conclusions can be drawn, since it is unknown to what extent the method and the properties of different fibres contribute to the large variability found by FIDONE and SATO. As a whole, the presented results lead to the conclusion that the conduction velocity of short unmyelinated, as well as myelinated, nerve fibres can be estimated reliably via the determination of the cross-time delay and by the calculation of the maximum-likelihood time-delay estimate. Thereby the cross-time delay has the weakest sensitivity for random noise.

References

BENDAT, J. S. and PIERSOL, A. G. (1971) Random data, analysis and measurement procedures. Wiley & Sons, New York.

CARTER, G. C. (1976) Time delay estimation. Techn. Report 5335, NUSC, New London Laboratory.

FIDONE, S. J. and SATO, A. (1969) A study of chemoreceptor and baroreceptor A and C fibres in the cat carotid nerve. *J. Physiol.*, **205**, 527–548.

HARDY, W. L. (1973) Propagation speed in myelinated nerve. I. Experimental dependence on external Na^+ and on temperature. *Biophysical J.*, **13**, 1054–1070.

HUTCHINSON, N. A., KOLES, Z. J. and SMITH, R. S. (1970) Conduction velocity in myelinated nerve fibres of Xenopus Laevis. *J. Physiol.*, **208**, 279–289.

HUXLEY, A. F. and STÄMPFLI, R. (1949) Evidence for saltatory conduction in peripheral myelinated nerve fibres. *J. Physiol.* **108**, 315–339.

KALBFLEISCH, E. W., OLSON, D. O. and HALAS, E. S. (1972) Bipolar electrode phase shift analysis of multiple unit activity with digital output. *Physiol. & Behav.*, **9**, 873–876.

KRAUSE, D. J., STEADMAN, J. W. and WILLIAMS, Th. W. (1972) Effect of record length on noise-induced error in the cross correlation estimate. *IEEE Trans. on Syst. Man and Cybern.*, **SMC-2**, 255–261.

LINDSTRÖM, L. H. and MAGNUSSON, R. I. (1977) Interpretation of myoelectric power spectra: a model and its applications. *Proc. IEEE*, **65**, 653–662.

MARKS, W. B. and LOEB, G. E. (1976) Action currents, internodal potentials and extracellular records of myelinated mammalian nerve fibers derived from node potentials. *Biophysical J.*, **16**, 655–668.

PAINTAL, A. S. (1966) The influence of diameter of medullated nerve fibres of cats on the rising and falling phases of the spike and its recovery. *J. Physiol.*, **184**, 791–811.

ROSENFALCK, P. (1969) Intra- and extracellular potential fields of active nerve and muscle fibres. Akademisk Forlag, Copenhagen.

STEIN, R. B. and PEARSON, K. G. (1971) Predicted amplitude and form of action potentials recorded from unmyelinated nerve fibres. *J. Theor. Biol.*, **32**, 539–558.

VLIET, G. H. VAN DER (1978) Methoden ter bepaling van de geleidingssnelheid van dunne zenuwvezels. Report no. 182, Bio-information Group, Twente University of Technology, Enschede.

TIME DELAY ESTIMATION USING AN EIGENSTRUCTURE BASED SPECTRAL METHOD

Georges Vezzosi, Philippe Nicolas*

Laboratoire de Traitement du Signal, IRISA, Université de Rennes I
Campus de Beaulieu, 35042 Rennes Cedex, France

Abstract. Under the conditions that the number of sensors is much greater than the number of sources, and that the sensor-to-sensor differential attenuation factor of each coherent wavefront received by the array is negligible, a new method of time delay estimation is presented. The method relies upon a spatial separation of the sources at each frequency. An application to the simultaneous estimation of the relative sensor positions and source directions with a random array is also presented.

I. Introduction

Time delay estimation of stationary signals received on a spatially distributed array of sensors is a classical non linear problem, well known for its threshold effects [1], for which numerous techniques have been designed, such as the generalized cross correlator, and more recently several parametric methods [3-5]. The generalized cross correlator is known to be optimal in the maximum likelihood sense for a single source and two sensors, and to behave poorly in the case of multiple sources with overlapping spectra and near-by time delays. Such situations are handled in principle by parametric methods, at least when the signals can be modelized as AR or ARMA processes with a few number of dominant poles, and the array contains a few number of sensors. However, the behaviour of these methods in the case of large arrays and complicated wideband signals described by high order models is unknown.

This paper takes basically the opposite viewpoint with respect to the parametric methods of [3-5], namely:
(i) the array is large (so that many time delays are to be estimated), and the number N of sensors is much greater than the number P of sources ;
(ii) no parametric dependence for the signal spectra is postulated ; we only require the signals to be wideband, in order to make the time delay estimation meaningful ; (for narrowband signals centered around the frequency f_0 and reasonable observation times, it is not time delays but phase angles, i.e. time delays defined up to $1/f_0$, which can be estimated) ;
(iii) the coherent signals impinging the array are just delayed and scaled versions of the signals emitted by the sources, without differential attenuation factor between two sensors. This allows spatial separation of the sources at each frequency.

Taking the first sensor as reference, the observation $x_n(t)$ at the n-th sensor ($1 \leqslant n \leqslant N$) can thus be written as :

$$x_n(t) = \sum_{p=1}^{P} s_p(t+D_{np}) + b_n(t), \quad 0 \leqslant t \leqslant T , \tag{1}$$

with $D_{11}=D_{12}=\ldots=D_{1P}=0$. The parameter D_{np} defines the time delay of the p-th source between the first and the n-th sensors. The P functions $\{s_1(t),\ldots,s_P(t)\}$ are P stationary uncorrelated wideband processes, uncorrelated with the noise $\{b_n(t)\}$. The noise itself is assumed to be spatially white, i.e. its cross spectral matrix (CSM) is colinear to the identity. It is worth noticing that the (N-1)P time delays $\{D_{np}\}$ are treated throughout as free parameters, as is the case when the antenna geometry is unknown, the chief concern being then to recover, up to an orthogonal transformation, the sensor positions and the source directions from the set $\{D_{np}\}$.

The proposed method is a spectral multistep procedure which hinges upon the following idea : at any frequency f, the array response to a coherent source specified by its set of time delays $\{D_2, \ldots ,D_N\}$ is colinear to the steering vector

$$\underline{d}_\theta = [1 , e^{i\theta_2} , \ldots , e^{i\theta_N}]^T , \tag{2}$$

where

$$\theta_n = 2\pi f\ D_n \tag{3}$$

denotes the (absolute) phase angle between the first and the n-th sensors, and

$$\underline{\theta} = [\theta_2 , \ldots , \theta_N]^T \tag{4}$$

is the corresponding phase vector. If P such sources are present, the CSM of the observation takes the form :

$$R(f) = a(f)\ \underline{I} + \sum_{p=1}^{P} \gamma_p(f)\ \underline{d}_{\underline{\theta}_p}\ \underline{d}^*_{\underline{\theta}_p} \tag{5}$$

where $a(f)$ and $\gamma_p(f)$ denote respectively the noise and the p-th source power. Assume that an estimate of the principal values of each phase vector can be extracted from the CSM at several frequencies. Taking advantage of the wide band property of the sources, and using the linearity of the (absolute) phase angles with the frequency, one can chain the phase vectors produced by the same sources, and recover from the chains the desired time delays by a least square criterion.

II. Estimating the phase vectors

The so-called signal subspace methods for array processing [6-7] are now well established when the steering vectors depend upon a very few number of parameters, as, for instance, steering vectors of the form $[1 , e^{i\theta} , \ldots , e^{i(N-1)\theta}]^T$ which arise for linear equispaced arrays. As far as the number of sensors is much greater than the number of sources, they can be extended to the case where the phase angles are treated as free parameters [8]. Let $\hat{\underline{u}}_1 , \ldots , \hat{\underline{u}}_P$ denote the P highest normalized eigenvectors of the spectral matrix estimate. Let $\underline{v} = [v_n] = z_1\ \hat{\underline{u}}_1 + \ldots + z_P\ \hat{\underline{u}}_P$, where $z_p = x_p + iy_p$ is a complex number, be the generic vector in the estimated signal subspace. We seek in the signal subspace the vectors whose components have their modulus as close to one as possible. This can be done by solving in the least square sense either

$$|v_n| = 1 \quad , \quad 1 \leqslant n \leqslant N \tag{6}$$

or

$$|v_n|^2 = 1 \quad , \quad 1 \leqslant n \leqslant N \tag{7}$$

This work was supported by the Centre Electronique de l'Armement, 35170, Bruz, France.

*presently with the Saclant ASW Research Center, Viale San Bartolomeo 400, I-19026 - San Bartolomeo (La Spezia), Italy.

Reprinted from *Proc. of the 25th IEEE Conference on Decision and Control*, vol. 2, pp. 949–952, December 1986.

Both systems are non linear with respect to their real unknows $\{x_p, y_p\}$. One can show that solving for (6) amounts to optimizing the projection of the steering vector model onto the signal subspace. Both systems are asymptotically equivalent. The second system is slightly simpler, being quadratic with respect to the $\{x_p, y_p\}$. The resolution can be performed by Newton's method with stepsize control. To avoid spurious solutions, a strong degree of overdetermination ($N \gg 2P-1$) is required.

The error covariance of each phase vector is also available [8]. Let $\lambda_1 \geq \lambda_2 \geq \dots \geq \lambda_P$ and $\underline{u}_1, \dots, \underline{u}_P$ denote respectively the P highest eigenvalues and corresponding eigenvectors of the exact CSM, to which we associate the signal subspace projector

$$\underline{\Pi} = \sum_{p=1}^{P} \underline{u}_p \, \underline{u}_p^* \tag{8}$$

and its complement $\underline{\Pi}^{\perp} = \underline{I} - \underline{\Pi}$. Let $(z_1, \dots, z_P)$ denote the components of the steering vector $\underline{d}_\theta$ in the signal subspace :

$$\underline{d}_\theta = z_1 \underline{u}_1 + \dots + z_P \underline{u}_P \quad . \tag{9}$$

Assume that the CSM estimate follows a Wishart's distribution with $K = BT$ degrees of freedom, as is the case when this estimate is obtained by a Fourier method. The error covariance of the phase vector estimate $\hat{\underline{\theta}} = \underline{\theta} + \Delta\underline{\theta}$ can be written as the product :

$$\underline{\Sigma} = E \, \Delta\underline{\theta} \, \Delta\underline{\theta}^T = c \, \underline{A}^{-1} \quad , \tag{10-a}$$

where c is the scalar

$$c = \frac{1}{BT} \sum_{p=1}^{P} |z_p|^2 \, a\lambda_p/(a-\lambda_p)^2, \; a = \text{noise power}, \tag{10-b}$$

and $\underline{A} = [a_{jk}]_{2 \leq j,k \leq N}$ denotes the matrix of order N-1 defined by :

$$a_{jk} = 2 \, \mathrm{Re} \, \{ e^{-i(\theta_j - \theta_k)} \, \underline{e}_j^T \, \underline{\Pi}^{\perp} \, \underline{e}_k \} \; , \tag{10-c}$$

where e_j stands for the j-th unit coordinate vector. In the case of a single source, the result assumes the simple form

$$\underline{\Sigma} = \frac{N}{BT} \frac{a \, \lambda_1}{(a-\lambda_1)^2} \frac{1}{2} (\underline{I} + \underline{1} \, \underline{1}^T) \quad , \tag{11}$$

which is the Cramer-Rao bound [9].

III. Chaining the phase vectors and estimating the time delays

Let D_{max} be the supremum of the absolute values of the time delays. The CSM are available on a set of equispaced frequency bins with frequency spacing Δf

$$\{ f_m = m \, \Delta f \quad , \quad 1 \leq m \leq M \} \; ,$$

with $D_{max} \, \Delta f < 1/2$ to enable the recovery of the time delays from the phase angles. The CSM estimates are assumed statistically independent (this condition is approximately satisfied when the CSM are estimated by a Fourier method). The CSM processing at frequency f provides Q(f) phase vectors whose components lie between $\pm\pi$. To each of them, we associate a covariance matrix by means of (10). Let us assume that the Q(f) phase vectors are numbered from 1 to Q(f). A chain of phase vectors is then a sequence of integers :

$$(a_1 \; , \; a_2 \; , \; 0 \; , \; \dots \; , \; a_m \; , \; \dots \; , \; a_M) \quad ,$$

where the zero means that the chain contains no element at the corresponding frequency. It is required to find the chains which correspond to the same source, namely the sequences of phase vectors whose elements satisfy a linear regression (modulo 2π).

III.1. The elementary triplets [10]

Let $\phi(x)$ be the usual sawtooth function :

$$\phi(x) = x \quad \text{for} \quad -\pi < x \leq \pi \; ,$$
$$\phi(x + 2\pi) = \phi(x) \quad ,$$

which converts any phase angle x into its principal value. If $\underline{\theta} = [\theta_n]$ is a vector, we note $\phi(\underline{\theta})$ the vector $[\phi(\theta_n)]$.

Consider three equispaced frequency bins $f < f' < f''$ with frequency spacing $d \, \Delta f$, where d is an integer called the order of the triplet. Let $Q = Q(f)$, $Q' = Q(f')$, $Q'' = Q(f'')$ be the whole number of phase vectors found at these frequencies. Assume that $Q Q' Q''$ is different from zero, and choose at each of these frequencies the phase vectors :

$$\hat{\underline{\theta}} = [\hat{\theta}_n] \; , \quad \hat{\underline{\theta}}' = [\hat{\theta}'_n] \; , \quad \hat{\underline{\theta}}'' = [\hat{\theta}''_n] \; , \quad 2 \leq n \leq N \; ,$$

which are respectively indexed by a , b , c. Assuming that these vectors are not spurious solutions, they can be written up to 2π as the sum of the exact absolute phase angles and an error :

$$\hat{\underline{\theta}} = \phi(\theta + \Delta\theta) \quad , \quad E \, \Delta\underline{\theta} \, \Delta \, \underline{\theta}^T = \underline{\Sigma} \; ,$$

with similar relations at frequencies f' and f". Consider the vector

$$\underline{\Delta} = \phi(\hat{\underline{\theta}} + \hat{\underline{\theta}}'' - 2\hat{\underline{\theta}}') = \phi(\underline{\theta} + \underline{\theta}'' - 2\underline{\theta}' + \Delta\underline{\theta} + \Delta\underline{\theta}'' - 2\Delta\underline{\theta}') \quad , \tag{12}$$

and neglect the case where $\underline{\theta} + \underline{\theta}'' - 2\underline{\theta}'$ has at least one of its components close to $(2k+1)\pi$. Then $\underline{\Delta}$ is asymptotically gaussian, with mean

$$\underline{\mu} = \phi(\underline{\theta} + \underline{\theta}'' - 2\underline{\theta}') \; ,$$

and since the CSM measured at f, f' and f" are statistically independent, its covariance is given by :

$$\underline{\Gamma} = \underline{\Sigma} + \underline{\Sigma}' + 4\underline{\Sigma}'' \quad . \tag{13}$$

As the absolute phase angles vary linearly with the frequency, the mean of $\underline{\Delta}$ equals zero when the three vectors $(\hat{\underline{\theta}}, \hat{\underline{\theta}}', \hat{\underline{\theta}}'')$ come from the same source. Thus, finding the good triplets amounts to testing the nullity of the mean of a gaussian vector with a known covariance. The optimal test (UMP invariant, minimax for $\underline{\mu}^T \underline{\Gamma}^{-1} \underline{\mu} = C^{te}$) is based on the comparison to a threshold of the statistic

$$T(a\,,\,b\,,\,c) = \underline{\Delta}^T \, \underline{\Gamma}^{-1} \, \underline{\Delta} \; , \tag{14}$$

which under the null hypothesis follows a chi-2 distribution with N-1 degrees of freedom.

This suggests apossible solution : for each of the $Q Q' Q''$ triplets (a,b,c) which can be constructed with the phase vectors found at f, f' and f", compute $\underline{\Delta}$ by (12), $\underline{\Gamma}$ by (13) (replacing every where the unknown quantities by their estimates), T(a,b,c) by (14) and compare it with a χ^2_{N-1}.

At low signal to noise ratio (SNR), the conditions of asymptotic normality of the phase angles are not always met, and the simulations show that some triplets, although correct, are incompatible with a χ^2_{N-1}. Therefore, the procedure is modified as follows.

Let (a_o, b_o, c_o) be a correct triplet. Assume that none of the points a_o, b_o and c_o are common to more than one source. Then, even if the value of $T(a_o, b_o, c_o)$ is inconsistent with a χ^2_{N-1}, it can be inferred that the path (a_o, b_o, c_o) minimizes T(a,b,c) among all paths between f and f" going respectively through a_o, b_o and c_o. This suggests to retain only the triplets which satisfy the condition

$$T(a_o, b_o, c_o) = \inf_{b,c} T(a_o, b, c) = \inf_{a,c} (a, b_o, c) = \inf_{a,b} T(a, b, c_o) \quad . \tag{15}$$

The computation of T(a,b,c) at all frequencies, all orders and for all triplets is numerically too expensive. In practice, the computational load can be kept within reasonable limits if one replaces the complete whitening based on the inverse $\underline{\Gamma}^{-1}$ by an approximate one, defined by the matrix

$$\underline{B} = (\underline{I} + \underline{1}\,\underline{1}^T)^{-1} = \underline{I} - \underline{1}\,\underline{1}^T / N \quad ,$$

derived in the case of a single source (see 11). One then computes

$$X = \underline{\Delta}^T \underline{B}\,\underline{\Delta} = \sum_{n=2}^{N} \Delta_n^2 - (\sum_{n=2}^{N} \Delta_n)^2 / N \quad ,$$

$$EX = \text{trace } (\underline{\Gamma}\ \underline{B})$$

and replaces everywhere T by $T' = X / EX$.

III.2. Chaining the phase vectors

At low SNR, a thirty percent proportion of erroneous triplets can be reached, and therefore a technique based merely on the concatenation of triplets of order one at successive frequency bins would fail. Many chaining procedures can be designed. The selected one relies upon an iterative continuation of elementary validated chains. Throughout, two chains are called compatible if their non vanishing elements located at the same positions are identical.

A. Initialization. Any triplet of order d (a-b-c) defines obviously two links of order d ((a-b) and (b-c)), and a link of order 2d (a-c)). The triplet (a-b-c) is said to be validated if the link (a-c) and one of the two links (a-b) or (b-c) are found in other triplets, and if (a-b-c) is compatible with all triplets going through a, b or c. At the first step we seek for all validated triplets.

B. Concatenation. Whenever two chains (e.g. the validated triplets) are compatible and have at least two non vanishing elements in common, they are concateneted and treated as a single chain.

C. Continuation. Consider a chain such that, for some frequency f, no element has been affected so far. The triplets going through its non vanishing elements allow us to determine, among all phase vectors available at frequency f, those which are connected to the chain. Each of them is indexed by the observed number of links. If some phase vector exhibits a number of links much greater than the others, the chain is extended by this element.

Thus, the chaining procedure is initialized by step A, then steps B ane C are applied until the chains become constant.

III.3. Estimating the time delays

Assume that a chain of phase vectors has been found, which lie between the frequencies $f_1 = m_1\,\Delta f$ and $f_2 = m_2\,\Delta f$. Consider the n-th component of this sequence $(2 \leq n \leq N)$:

$$[\hat{\theta}_n(m_1)\,,\ \ldots\,,\ \hat{\theta}_n(m_2)] \quad , \tag{16}$$

to which corresponds the sequence of absolute exact phase angles

$$[2\pi\ m\ \Delta f\ D_n]_{m_1 \leq m \leq m_2} \quad . \tag{17}$$

Given the sequence of phase angles (16), we want to estimate the time delay D_n. At high SNR, the phase angles are estimated with a good precision, no data is lacking in (16), a fast procedure can be designed by unwrapping the phase angles and fitting to them a linear regression [10]. At low SNR, the phase unwrapping suffers a considerable loss of efficiency, and the time delays are recovered by minimizing the least square criterion

$$\Phi(D) = \sum_{m=m_1}^{m_2} \left| e^{i\,\hat{\theta}_n(m)} - e^{2i\pi\,m\,\Delta f\,D} \right|^2 / \operatorname{var} \hat{\theta}_n(m) \quad , \tag{18}$$

which amounts to maximizing :

$$\Psi(D) = \sum_{m=m_1}^{m_2} \cos(\hat{\theta}_n(m) - 2\pi m\,\Delta f\,D) / \operatorname{var} \hat{\theta}_n(m) \quad .$$

Both procedures are asymptotically equivalent, and lead to a variance given by :

$$\operatorname{var} \hat{D}_n = \frac{1}{(2\pi\Delta f)^2} \left[\sum_{m_1}^{m_2} \frac{m^2}{\operatorname{var} \hat{\theta}_n(m)} \right]^{-1} \quad .$$

Furthermore, if K is the number of non-vanishing elements in the chain, the minimum value of $\Phi(D)$ in (18) follows in principle a chi-2 distribution with K-1 degrees of freedom, which gives a means to detect erroneous chains.

IV. Application to random arrays

Consider a random array of N sensors located at positions $\vec{r}_1 = 0\,,\ \vec{r}_2\,,\ \ldots\,,\ \vec{r}_N$, receiving P wideband stationary planewaves with directions $\vec{k}_1\,,\ \ldots\,,\ \vec{k}_P$ $(\|\vec{k}_p\| = 1)$. For simplicity, we consider here the 2-D case where the sources and the sensors are distributed in a plane, the 3-D case being analogous [10]. Assuming a propagation velocity of unity, the time delays $\{D_{np}\}$ are given by

$$D_{np} = \vec{r}_n \cdot \vec{k}_p \quad , \quad 2 \leq n \leq N \quad , \quad 1 \leq p \leq P \quad .$$

This can be written compactly as

$$\underline{D} = \underline{R}^T\,\underline{K} \quad , \tag{19}$$

where $\underline{D} = [D_{np}]$ is the $(N-1)\times P$ matrix of time delays, $\underline{R}$ is the $2\times(N-1)$ matrix of positions, and $\underline{K}$ is the $2\times P$ matrix of directions. Given an estimate $\hat{\underline{D}}$ of $\underline{D}$, the problem is to solve the non linear system (19) with respect to its unknowns $\underline{R}$ and $\underline{K}$, up to an orthogonal transformation. This problem is meaningful as soon as

(i) N is ≥ 3 and the sensors do not line up ;

(ii) P is ≥ 3 and the source directions are not colinear to each other [10].

Under these assumptions, the matrix $\underline{D}$ has rank two and solving for (19) amounts to finding a decomposition of $\underline{D}$ where the right factor has its columns normalized to one. It turns out a procedure broken into three steps :

1. Reduce the matrix $\hat{\underline{D}}$ to a matrix of rank two $\underline{U}\ \underline{\Delta}\ \underline{V}^T$ by SVD ;

2. Find the upper triangular matrix

$$\underline{T} = \begin{bmatrix} a & c \\ o & b \end{bmatrix}$$

in such a way that the product $\underline{T}\ \underline{V}^T$ has the squared norm of its columns as close as possible to unity (this a linear problem with respect to the unknowns $\{a^2\,,\ b^2\,,\ c^2\,,\ ac\}$;

3. Compute the source directions $\underline{K}$ by means of $\underline{T}\ \underline{V}^T$ and the sensors positions $\underline{R}^T$ by means of $\underline{U}\ \underline{\Delta}\ \underline{T}^{-1}$. Both are defined up to an orthogonal transformation

$$\underline{Q} = \begin{bmatrix} \cos\theta & -\varepsilon\sin\theta \\ \sin & \varepsilon\cos\theta \end{bmatrix} \quad , \quad \varepsilon = \pm 1 \quad . \tag{20}$$

As an example, consider an array of N=16 sensors randomly distributed in a square of side 40 (Fig. 1), where the length unit is set to the half minimum wavelength received by the array. Four sources are present, whose bearings with regard to the x-axis equal 10°, 20°, 40°, 60°. The sources have the same spectrum, which in normalized frequencies is flat in the band [0.15 , 0.35] and vanishes elsewhere. At each frequency where the signal is present, the SNR equals -10 dB. The bandwidth [0 , 0.5] is decomposed into frequency bins with a frequency spacing Δf of 1/128. The CSM are synthetized according to a Wishart distribution with 150 degrees of freedom. For the display of the results, the arbitrary unitary matrix $\underline{Q}$ in (20) is chosen so as to minimize the sum of squares between the matrix of exact positions $\underline{R}^T$ and $\underline{U}\ \underline{\Delta}\ \underline{T}^{-1}\ \underline{Q}$. The results are depicted on Fig. 1 where the crosses represent the true positions and the center of the circles their estimates, and on TAB. I which summarizes the estimated source directions. They illustrate the potentialities of the method.

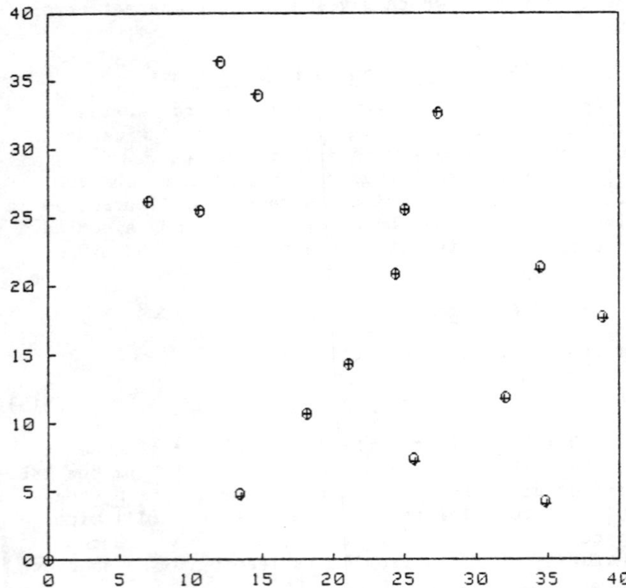

Fig. 1. Comparison between the true sensor positions and their estimates. Number of sensors : $N = 16$; number of sources : $P = 4$; SNR = - 10 dB at each frequency ; BT = 150.

Source Number	True Bearing	Estimated Bearing
1	10°	9°.80
2	20°	19°.83
3	40°	40°.11
4	60°	60°.13

TAB. I. Comparison between the true bearings and their estimates.

V. Conclusion

Under the conditions that the number of sensors is much greater than the number of sources, and that the gradient of amplitude of each coherent wavefront across the array is negligible, an eigenvalue spectral method was designed for the problem of passive time delay estimation in the general case of multiple wideband sources and multiple sensors. Applied to random arrays, the method allows us to reconstruct the sensor geometry and the source directions for values of SNR and precisions of the cross spectral matrices which are current in applications (SNR = - 10 dB, BT = 150). Such a purpose seems to be unrealizable via generalized cross correlation methods because of the closeness of some of the time delays involved. Up to our knowledge, the question whether parametric methods can be tailored to perform as well under the same conditions remains open.

References

[1] S.K. Chow and P.M. Schultheiss, "Delay estimation using narrow-band processes," IEEE Trans. Acoust., Speech, Signal Processing, vol. ASSP-29, pp. 478-484, June 1981.

[2] C.H. Knapp and G.C. Carter, "The generalized correlation method for estimation of time delay," IEEE Trans. Acoust., Speech, Signal Processing, vol. ASSP-24, pp. 320-327, Aug. 1976.

[3] B. Porat, B. Friedlander, "Estimation of spatial and spectral parameters of multiple sources", IEEE Trans. Inform. Theory, vol. IT-29, pp. 412-425, May 1983.

[4] A. Nehorai, G. Su and M. Morf, "Estimation of time difference of arrival by pole decomposition," IEEE Trans. Acoust., Speech, Signal Processing, vol. ASSP-31, pp. 1478-1491, December 1983.

[5] J.J. Fuchs, "State-Space modeling and estimation of time differences of arrival," IEEE Trans. Acoust., Speech, Signal Processing, vol. ASSP-34, pp. 232-244, April 1986.

[6] G. Bienvenu and L. Kopp," Optimality of high resolution array processing using the eigensystem approach," IEEE Trans. Acoust., Speech, Signal Processing, vol. ASSP-31, pp. 1235-1248, Oct. 1983.

[7] R. Schmidt, "Multiple emitter location and signal parameter estimation," in Proc. RADC Spectr. Estimat. Workshop, Rome, NY, 1979, pp. 243-258.

[8] G. Vezzosi, "Estimation of phase angles from the cross spectral matrix," IEEE Trans. Acoust., Speech, Signal Processing, vol. ASSP-34, pp. 405-422, June 1986.

[9] W.R. Hahn and S.A. Tretter, "Optimum processing for delay vector estimation in passive sonar arrays," IEEE Trans. Inf. Theory, vol. IT 19, pp. 608-614, Sept. 1973.

[10] P. Nicolas, "Localisation de sources ponctuelles avec une antenne de géométrie inconnue," thèse de l'Université de Rennes I, France, Sept. 1985.

Bibliography

(Other excellent papers and publications had to be omitted for reasons of space; the interested researcher is encouraged to look up relevant papers for further study.)

Abatzoglou, T. J. 1986. A fast local maximum likelihood estimator for time delay estimation. *IEEE Trans. ASSP* 34 (2): 375–78.

Abel, J. S., and J. O. Smith. 1989. Source range and depth estimation from multipath range difference measurements. *IEEE Trans. ASSP* 37 (8): 1157–65.

Abutaleb, A. S. 1986. Time-varying time delay estimation and nonlinear filtering. *20th Asilomar Conference on Signals, Systems, and Computers* (November): 656–60.

Adams, W. B., J. P. Kuhn, and W. P. Whyland. 1980. Correlator compensation requirements for passive time-delay estimation with moving source or receivers. *IEEE Trans. ASSP* 28 (2): 158–68.

Akaike, H., and Y. Yamanouchi. 1963. On the statistical estimation of frequency response function. *Annals of the Inst. of Stat. Math.* 14:23–56.

Al-Hussaini E. K., and S. A. Kassam. 1984. Robust Eckart filters for time delay estimation. *IEEE Trans. ASSP* 32 (5): 1052–63.

Altes, R. A. 1979. Target position estimation in radar and sonar, and generalized ambiguity analysis for maximum likelihood parameter estimation. *Proc. IEEE* 67 (6): 920–30.

Amos, D. E., and L. H. Koopmans. 1963. Tables of the distribution of the coefficient of coherence for stationary bivariate Gaussian processes. Office of Technical Services, U.S. Department of Commerce. Also, Sandia Corporation, Albuquerque, Monograph SCR-483, 1963.

Andermo, I., and A. Sjölund. 1979. Laser velocity meter with correlation technique. *Acta IMEKO,* 23–29.

Anderson, T. W. 1958. *An introduction to multivariate statistical analysis.* New York: John Wiley.

Avitzour, D. 1991. Time delay estimation at high signal-to-noise ratio. *IEEE Trans. Aerospace and Electronic Systems* 27 (2): 234–37.

Azaria, M., and D. Hertz. 1984. Time delay estimation by generalized cross correlation methods. *IEEE Trans. ASSP* 32 (2): 280–85.

Baggeroer, A. B., W. A. Kuperman, and H. Schmidt. 1988. Matched field processing: Source localization in correlated noise as an optimum parameter estimation problem. *JASA* 83 (2): 571–87.

Bangs, W. J., III. 1971. Array processing with generalized beam-formers. Ph.D. diss., Yale University.

Bangs, W. J., III, and P. M. Schultheiss. 1973. Space-time processing for optimal parameter estimation. Proceedings of the NATO Advanced Study Institute on Signal Processing, Loughborough, England, August 1972. In *Signal Processing,* J. W. R. Griffiths, P. L. Stocklin, and C. van Schooneveld, eds., 577–90. New York: Academic Press.

Barankin, E. W. 1949. Locally best unbiased estimates. *Annals of Mathematical Statistics* 20:477–501.

Bar-David, I., and M. Levy. 1978. Advent of nonregularity in photon-pulse delay estimation. *IEEE Trans. Information Theory* 24 (1): 122–24.

Bar-Shalom, Y. 1978. Tracking methods in a multitarget environment. *IEEE Trans. Automatic Control* 23 (4): 618–626.

Beck, M. S. 1981. Correlation in instruments: cross correlation flowmeters. *J. Physics, E: Scientific Instruments* 14:7–19.

Behringer, K., G. Kosály, and H. Nishihara. 1980. Theoretical remarks on a novel correlation method of transit time estimation, *Atomkernenergie/Kerntechnik* 36 (4): 292–96.

Bendat, J. S. 1978. Statistical errors in measurement of coherence functions and input/output quantities. *J. Sound and Vibration* 59 (3): 405–21.

Bendat, J. S., and A. G. Piersol. 1966. *Measurement and analysis of random data.* New York: John Wiley.

———. 1971. *Random data: Analysis and measurement procedures.* New York: Wiley-Interscience. (Second edition, revised and expanded, 1986.)

———. 1980. *Engineering applications of correlation and spectral analysis.* New York: Wiley-Interscience.

Benignus, V. A. 1969a. Computation of coherence and regression spectra using the FFT. *COMMON,* Proceedings of the Houston Meeting (December 1968).

———. 1969b. Estimation of the coherence spectrum and its confidence interval using the fast Fourier transform. *IEEE Trans. Audio and Electroacoustics* 17 (2): 145–50.

———. 1969c. Estimation of coherence spectrum of non-Gaussian time series populations. *IEEE Trans. Audio and Electroacoustics* 17 (3): 198–201.

———. 1970. Correction to "Estimation of the coherence spectrum and its confidence interval using the fast Fourier transform." *IEEE Trans. Audio and Electroacoustics* 18 (3): 320.

Bethel, R. E., and R. G. Rahikka. 1987. An optimum first-order time delay tracker. *IEEE Trans. Aerospace and Electronic Systems* 23 (6): 718–25.

———. 1990. Optimum time delay detection and tracking. *IEEE Trans. Aerospace and Electronic Systems* 26 (5): 700–12.

Betz, J. W. 1984a. Performance of the deskewed short-time correlator. *ICASSP 84* (1):15.5/1–4.

———. 1984b. Comparison of the deskewed short-time correlator and the maximum likelihood correlator. *IEEE Trans. ASSP* 32 (2): 285–94.

———. 1985. Effects of uncompensated relative time companding on a broad-band cross correlator. *IEEE Trans. ASSP* 33 (3): 505–10.

Bingham, C., M. D. Godfrey, and J. W. Tukey. 1967. Modern techniques of power spectrum estimation. *IEEE Trans. Audio and Electroacoustics* 15 (2): 56–66.

Blackman, R. B., and J. W. Tukey. 1958. *The measurement of power spectra*. New York: Dover Publications.

Blake, W. K., and R. V. Waterhouse. 1977. The use of cross-spectral density measurements in partially reverberant sound fields. *J. Sound and Vibration* 54 (4): 589–99.

Bloomfield, P. 1976. *Fourier analysis of time series: An introduction*. New York: John Wiley.

Bogert, B. P., and J. F. Ossanna. 1966. The heuristics of cepstrum analysis of a stationary complex echoed Gaussian signal in stationary Gaussian noise. *IEEE Trans. Information Theory* 12 (3): 373–80.

Bohmann, J., H. Meyr, R. Peters, and G. Spies. 1984. A signal processor for a noncontact speed measurement system. *IEEE Trans. Vehicular Technology* 33 (1): 14–23.

Bohmann, J., H. Meyr, and G. Spies. 1982. A digital signal processor for high-precision noncontact speed measurement of rail-guided vehicles. *32nd IEEE Vehicular Technology Conference,* San Diego, 454–62.

Böhme, J. F. 1985. Source-parameter estimation by approximate maximum likelihood and nonlinear regression. *IEEE J. Oceanic Engineering* 10 (3): 206–12.

Borys, A. 1989. Solids velocity estimation in two-phase turbulent flow. *Proceedings Mustererkennung* (October): 531–35.

Boucher, R. E., and J. C. Hassab. 1981. Analysis of discrete implementation of generalized cross correlator. *IEEE Trans. ASSP* 29 (3, pt. 2): 609–11.

Boudreau, D., and P. Kabal. 1990. Joint gradient-based time delay estimation and adaptive filtering. *23rd IEEE International Symposium on Circuits and Systems,* New Orleans, 4:3165–69.

Boudreaux-Bartels, G. F. 1985. Time-varying signal processing using the Wigner-distribution time-frequency signal representation. In *Advances in geophysical data processing,* vol. 2. JAI Press.

Bouvet, M., and S. C. Schwartz. 1988. Underwater noises: Statistical modeling, detection, and normalization. *JASA* 83 (3): 1023–33.

Bracewell, R. N. 1986. *The Fourier transform and its applications.* New York: McGraw-Hill.

Bradley, J. N., and R. L. Kirlin 1984. Delay estimation by expected value. *IEEE Trans. ASSP* 32 (1): 19–27.

Brigham, E. O. 1988. *The fast Fourier transform and its applications.* Englewood Cliffs, N.J.: Prentice Hall.

Brillinger, D. R. 1974. Fourier analysis of stationary processes. *Proc. IEEE* 62 (12): 1628–43.

———. 1975. *Time series: Data analysis and theory.* New York: Holt, Rinehart and Winston. Also, San Francisco: Holden-Day, 1981 (expanded edition).

———. 1979. Confidence intervals for the crosscovariance function. In *Selecta Statistica Canadiana,* vol. 5, "Mathematical Statistics." Hamilton, Ontario, Canada: Department of Mathematics, McMaster University, 1–16.

———. 1981. The key role of tapering in spectrum estimation. *IEEE Trans. ASSP* 29 (5): 1075–76.

Bruckstein, A. M., T.-J. Shan, and T. Kailath. 1985. The resolution of overlapping echoes. *IEEE Trans. ASSP* 33 (6): 1357–67.

Bryn, F. 1962. Optimum signal processing of three-dimensional arrays operating on Gaussian signals and noise. *JASA* 34 (3): 289–97.

Bucy, R. S., J. M. F. Moura, and A. J. Mallinckrodt. 1983. A Monte Carlo study of absolute phase determination. *IEEE Trans. Information Theory* 29 (4): 509–20.

Cabot, R. C. 1981. A note on the application of the Hilbert transform to time delay estimation. *IEEE Trans. ASSP* 29 (3, pt. 2): 607–9.

Cadzow, J. A., and O. M. Solomon, Jr. 1987. Linear modeling and the coherence function. *IEEE Trans. ASSP* 35 (1): 19–28.

Carter, G. C. 1972. Estimation of the magnitude-squared coherence function (spectrum). Technical Report of the Naval Underwater Systems Center, New London, Conn., no. TR 4343 (AD 743945).

———. 1976a. Time delay estimation. Ph.D. diss. University of Connecticut, Storrs. Published as a Technical Report of the Naval Underwater Systems Center, New London, Conn., no. 5335 (April 9, 1976).

———. 1976b. The role of coherence in time delay estimation. *Proceedings of the NATO Advanced Study Institute on Signal Processing with Emphasis on Underwater Acoustics,* Portovenere, LaSpezia, Italy. In *Aspects of signal processing with emphasis on underwater acoustics.* G. Tacconi, ed. (NATO ASI Series C, Mathematical and Physical Sciences, vol. 33, pt. 1). Dordrecht/Boston: D. Reidel Publishing Company, 1977, 251–56. Also, Technical Document of the Naval Underwater Systems Center, New London, Conn., no. TD 5507 (August 27, 1976).

———. 1977a. Receiver operating characteristics for a linearly thresholded coherence estimation detector. *IEEE Trans. ASSP* 25 (1): 90–92.

———. 1977b. Variance bounds for passively locating an acoustic source with a symmetric line array. *JASA* 62 (4): 922–26.

———. 1978a. A brief description of the fundamental difficulties in passive ranging. *IEEE J. Oceanic Engineering* 3 (3): 65–66.

———. 1978b. Optimum element placement for passive bearing estimation in unequal signal-to-noise ratio environments. *IEEE Trans. ASSP* 26 (4): 365–66.

———. 1979a. Passive ranging errors due to receiving hydrophone position uncertainty. *JASA* 65 (2): 528–30.

———. 1979b. Comments on "On errors in some papers on array processing." *JASA* 66 (5): 1547–48.

———. 1980. Bias in magnitude-squared coherence estimation due to misalignment. *IEEE Trans. ASSP* 28 (1): 97–99.

———, guest ed. 1981a. Special issue on time delay estimation. *IEEE Trans. ASSP* 29 (3, pt. 2): 461–623.

———. 1981b. Time delay estimation for passive sonar signal processing. *IEEE Trans. ASSP* 29 (3, pt. 2): 463–70.

(Reprinted in *Selected papers in multidimensional digital signal processing*. Multidimensional Signal Processing Committee of the IEEE ASSP Society, ed., 451–58. New York: IEEE Press, 1986.)

———. 1985. Coherence and time delay estimation. Course 9 of the NATO Advanced Study Institute, Session XLV, Les Houches, Haute-Savoie, France. In *Signal processing/traitement du signal,* J.-L. Lacoume, T. S. Durrani, and R. Stora, eds., vol. 2, 515–71. Amsterdam: Elsevier Science Publishers.

———. 1987. Coherence and time delay estimation. *Proc. IEEE* (2): 236–55.

———. 1991. Coherence and time delay estimation. IEEE/EAB Video Tutorial of the presentation at ICASSP 91, product no. HV0230-3. IEEE Educational Activities Department, 445 Hoes Lane, P.O. Box 1331, Piscataway, N.J. 08855-1331, (908)562-5499.

Carter, G. C., and P. B. Abraham. 1980. Estimation of source motion from time delay and time compression measurements. *JASA* 67 (3): 830–32.

Carter, G. C., and C. R. Arnold. 1971. Some practical considerations of coherence estimation. Technical Memorandum of the Naval Underwater Systems Center, New London, Conn., no. TD113-19-71.

Carter, G. C., and J. F. Ferrie. 1979. A coherence and cross spectral estimation program. In *Programs for digital signal processing,* chap. 2.3, 2.3/1–8. Digital Signal Processing Committee of the IEEE ASSP Society, ed. New York: IEEE Press.

Carter, G. C., and C. H. Knapp. 1975. Coherence and its estimation via the partitioned and modified chirp-Z transform. *IEEE Trans. ASSP* 23 (3): 257–64.

———. 1977. Time delay estimation. *ICASSP* 77: 357–60.

Carter, G. C., C. H. Knapp, and A. H. Nuttall. 1973a. Estimation of the magnitude-squared coherence function via overlapped fast Fourier transform processing. *IEEE Trans. Audio and Electroacoustics* 21 (4): 337–44.

———. 1973b. Statistics of the estimate of the magnitude-coherence function. *IEEE Trans. Audio and Electroacoustics* 21 (4): 388–89.

Carter, G. C., and A. H. Nuttall. 1972. Statistics of the estimate of coherence. *Proc. IEEE* 60 (4): 465–66.

———. 1980a. A brief summary of a generalized framework for power spectral estimation. *Signal Processing* 2 (4): 387–90.

———. 1980b. On the weighted overlapped segment averaging method for power spectral estimation. *Proc. IEEE* 68 (10): 1352–54.

———. 1983. Analysis of a generalized framework for spectral estimation—Part I: The technique and its mean value. *Proc. Inst. Electrical Engineers, Part F (Communications, Radar, and Signal Processing)* 130 (3): 239–41. Also in Proceedings of the Conference on "Spectral Analysis and Its Use in Underwater Acoustics," London, England, April 1982. (Reprinted in *Modern spectrum analysis II,* S. B. Kesler, ed., 205–7. New York: IEEE Press, 1986.)

Carter, G. C., A. H. Nuttall, and P. G. Cable. 1972. The smoothed coherence transform (SCOT). *Technical Memorandum of the Naval Underwater Systems Center.* New London, Conn., no. TC-159-72.

———. 1973. The smoothed coherence transform. *Proc. IEEE* 61 (10): 1497–98.

Champagne, B., M. Eizenman, and S. Pasupathy. 1989. Exact maximum likelihood time delay estimation. *ICASSP 89,* vol. 4:2633–36.

Chan, Y. T., ed. 1989. *Underwater acoustic data processing.* Proceedings of the NATO Advanced Study Institute on Underwater Acoustic Data Processing, Kingston, Ontario, Canada, July 1988 (NATO ASI Series E, Applied Sciences, vol. 161). Dordrecht/Boston/London: Kluwer Academic Publishers.

Chan, Y. T., J. G. Bryan, and G. A. Lampropoulos. 1986. Determining time delay via frequency estimation. *ICASSP 86,* 4:2803–6.

Chan, Y. T., R. V. Hattin, and J. B. Plant. 1978. The least squares estimation of time delay and its use in signal detection. *IEEE Trans. ASSP* 26 (3): 217–22.

Chan, Y. T., and R. K. Miskowicz. 1984. Estimation of coherence and time delay with ARMA Models. *IEEE Trans. ASSP* 32 (2): 295–303.

Chan, Y. T., J. M. F. Riley, and J. B. Plant. 1979. A parameter estimation approach to time delay estimation. *ICASSP 79,* 128–131.

———. 1980a. A parameter estimation approach to time-delay estimation and signal detection. *IEEE Trans. ASSP* 28 (1): 8–15.

———. 1980b. A Wiener filter approach to coherence estimation. *ICASSP 80* 2:646–49.

———. 1981. Modeling of a time delay and its application to estimation of nonstationary delays. *IEEE Trans. ASSP* 29 (3, pt. 2): 577–81.

Chandra, N., and C. Knapp. 1990. Optimal hydrophone placements under random perturbations. *IEEE Trans. ASSP* 38 (5): 860–64.

Chazan, D., M. Zakai, and J. Ziv. 1975. Improved lower bounds on signal parameter estimation. *IEEE Trans. Information Theory* 21 (1): 90–93.

Chen, C.-T., and S. C. Schwartz. 1984. Multi-sensor array time-delay estimation with small phase incoherences. Technical Report of the Office of Naval Research, Statistics and Probability Branch, Arlington, Va. Report no. 15.

Chestnut, P. C. 1982. Emitter location accuracy using TDOA and differential Doppler. *IEEE Trans. Aerospace and Electronic Systems* 18 (2): 214–18.

Chiang, H.-H., and C. L. Nikias. 1990. A new method for adaptive time delay estimation for non-Gaussian signals. *IEEE Trans. ASSP* 38 (2): 209–19.

Childers, D. G., ed. 1978. *Modern spectrum analysis.* New York: IEEE Press.

Ching, P. C., and Y. T. Chan. 1988. Adaptive time delay estimation with constraints. *IEEE Trans. ASSP* 36 (4): 599–602.

Ching, P. C., Y. T. Chan, and K. M. Tse. 1987. Constrained adaptive estimation of time delay with multipath. *Proceedings of the 11th GRETSI Symposium on Signal and Image Processing,* Nice, France. June 1987, vol. 1:305–8.

Chow, P. L., and L. Maestrello. 1981. Statistical estimation of correlation for nonstationary aircraft noise. *JASA* 70 (3): 735–39.

Chow, S.-K. 1978. Parameter estimation of narrow-band processes with arrays. Ph.D. diss., Yale University.

Chow, S.-K., and P. M. Schultheiss. 1981. Delay estimation using narrow-band processes. *IEEE Trans. ASSP* 29 (3, pt. 2): 478–84.

Clarkson, P. M., and M. V. Dokic. 1989. Real-time adaptive filters for time delay estimation. *Proceedings of the 32nd Midwest Symposium on Circuits and Systems,* Champaign, Ill. 2:891–94.

Cleveland, W. S., and E. Parzen. 1975. The estimation of coherence, frequency response, and envelope delay. *Technometrics* 17 (2): 167–72.

Cooley, J. W., P. A. W. Lewis, and P. D. Welch. 1970. The application of the fast Fourier transform algorithm to the estimation of spectra and cross-spectra. *Sound and Vibration* 12 (3): 339–52.

Cooley, J. W., and J. W. Tukey. 1965. An algorithm for the machine calculation of complex Fourier series. *Math. of Computation* 19 (90): 297–301.

Cote, G. L., and M. D. Fox. 1988. Comparison of zero-crossing counter to FFT spectrum of ultrasound Doppler. *IEEE Trans. Biomedical Engineering* 35 (6): 498–502.

Cox, H. 1968a. Interrelated problems in estimation and detection I. *NATO Advanced Study Institute on Signal Processing with Emphasis on Underwater Acoustics.* Enschede, the Netherlands.

———. 1968b. Interrelated problems in estimation and detection II. *NATO Advanced Study Institute on Signal Processing with Emphasis on Underwater Acoustics.* Enschede, the Netherlands.

———. 1973. Resolving power and sensitivity to mismatch of optimum array processors. *JASA* 54 (3): 771–85.

Cramér, H. 1946. *Mathematical methods of statistics.* Princeton, N.J.: Princeton University Press.

Cusani, R. 1989. Fast techniques for time delay estimation. *MELECON 89: Mediterranean Electrotechnical Conference Proceedings,* Lisbon, Portugal, 177–80.

Daku, B. L. F. 1990. The accuracy in locating an underwater acoustic source in a multipath environment. Ph.D. diss., Department of Electrical Engineering, University of Saskatchewan, Saskatoon.

Daku, B. L. F., C. M. McIntyre, and J. E. Salt. 1989. An artificial intelligence approach to multipath localization and tracking. *Proceedings of the NATO Advanced Study Institute on Underwater Acoustic Data Processing,* Kingston, Ontario, Canada. In *Underwater acoustic data processing,* Y. T. Chan, ed., 561–65 (NATO ASI Series E, Applied Sciences, vol. 161). Dordrecht/Boston/London: Kluwer Academic Publishers.

d'Assumpcao, H. A., and G. E. Mountford. 1984. An overview of signal processing for arrays of receivers. *J. EEE Australia* 4:6–19.

Davenport, W. B., Jr. 1970. *Probability and random processes: An introduction for applied scientists and engineers.* New York: McGraw-Hill.

Davenport, W. B., Jr., and W. L. Root. 1958. *An introduction to the theory of random signals and noise.* New York: McGraw-Hill.

deFigueiredo, R. J. P., and A. Gerber. 1983. Separation of superimposed signals by a cross-correlation method. *IEEE Trans. ASSP* 31 (5): 1084–89.

deFigueiredo, R. J. P., T. T. Pham, and C. L. Hu. 1988. A constrained frequency-domain Prony's algorithm with application to time delay estimation. *ICASSP 88* 5:2610–13.

Demin, A. A., and E. A. Mikhaylovskiy. 1978. Estimation of length and delay of an intensity pulse of a point-source signal. *Radiotekhnika i Elektronika* 23 (8): 1745–48.

Dianat, S. A., and T. K. Sarkar. 1983. A finite step adaptive technique for time-delay estimation. *ASSP Spectrum Estimation Workshop II,* Tampa, Fla., 67–69.

Dobie, R. A., and M. J. Wilson. 1989. Analysis of auditory evoked potentials by magnitude-squared coherence. *Ear and Hearing* 10 (1): 2–13.

Eckart, C. 1952. Optimal rectifier systems for the detection of steady signals. University of California, Scripps Institute of Oceanography, Marine Physical Laboratory, Report SIO 12692, SIO Reference 52–11.

Ehrenberg, J. E., T. E. Ewart, and R. D. Morris. 1978. Signal-processing techniques for resolving individual pulses in a multipath signal. *JASA* 63 (6): 1861–65.

Enochson, L. D., and N. R. Goodman. 1965. Gaussian approximations to the distribution of sample coherence. Bulletin of the Air Force Flight Dynamics Laboratory, Research and Technology Division, Air Force Systems Command, Wright-Patterson Air Force Base, Ohio, no. AFFDL-TR-65-67.

Etter, D. M. 1979. Digital signal processing with an adaptive delay element. Ph.D. diss., University of New Mexico, Albuquerque.

Etter, D. M., and S. D. Stearns. 1979. Convergence properties of the adaptive delay element in delay-lock loops and frequency tracking. *13th Asilomar Conference on Circuits, Systems, and Computers,* 534–37.

———. 1981. Adaptive estimation of time delays in sampled data systems. *IEEE Trans. ASSP* 29 (3, pt. 2): 582–87.

Favennec, J.-M., B. Georgel, and J. Masson. 1982. Time delay estimation: Application to flow rate measurement of cooling fluid in nuclear power plants. *ICASSP 82* 1:379–82.

Feintuch, P. L., N. J. Bershad, and F. A. Reed. 1981. Time delay estimation using the LMS adaptive filter—Dynamic Behavior. *IEEE Trans. ASSP* 29 (3, pt. 2): 571–76.

Ferguson, B. G. 1989. Improved time-delay estimates of underwater acoustic signals using beamforming and prefiltering techniques. *IEEE J. Oceanic Engineering* 14 (3): 238–44.

Ferretti, G., C. Maffezzoni, and R. Scattolini. 1991. Recursive estimation of time delay in sampled systems. *Automatica* 27 (4): 653–61.

Ferrie, J. F., C. W. Nawrocki, and G. C. Carter. 1975. Applications of the partitioned and modified chirp Z-transform to oceanographic measurements. *IEEE Trans. ASSP* 23 (2): 243–44.

Fertner, A., and A. Sjölund. 1986. Comparison of various time delay estimation methods by computer simulation. *IEEE Trans. ASSP* 34 (5): 1329–30.

Fetzer, G. J., G. K. F. Lee, D. Byers, and K. A. Dallabetta. 1982. Microprocessor-based dynamic programming for time delay estimation of deterministic and stochastic signals. *16th Asilomar Conference on Circuits, Systems, and Computers,* 498–502.

Fischer, W. K., and L. C. Ng. 1983. Improved time delay estimation in the presence of interference. *ICASSP 83* 2:899–902.

Fisher, R. A. 1950. *Contributions to mathematical statistics.* New York: John Wiley. (Chapter 14 was originally published in 1928 as "The general sampling distribution of the multiple correlation coefficient" in *Proceedings of the Royal Society, Series A* (Mathematical and Physical Sciences) 121: 654–73.)

Foster, M. R., and N. J. Guinzy. 1967. The coefficient of coherence: Its estimation and use in geophysical data processing. *Geophysics* 32:602–16.

Fox, M. D. 1978. "Multiple crossed-beam ultrasound Doppler velocimetry," *IEEE Trans. Sonics and Ultrasonics* 25 (5): 281–96.

Fox, M. D., and W. M. Gardiner. 1988. Three-dimensional Doppler velocimetry of flow jets. *IEEE Trans. Biomedical Engineering* 35 (10): 834–41.

Friedlander, B. 1983. A parametric technique for estimation of delay and Doppler. *ICASSP 83* 2:887–90.

———. 1984. On the Cramér-Rao bound for time delay and Doppler estimation. *IEEE Trans. Information Theory* 30 (3): 575–80.

———. 1984. Accuracy of source localization using multipath delays. *IEEE Trans. Aerospace and Electronic Systems* 24 (4): 346–59.

Friedlander, B., and B. Porat. 1982. A parametric technique for time delay estimation. *ICASSP 82* 1:416–19.

———. 1984. A parametric technique for time delay estimation. *IEEE Trans. Aerospace and Electronic Systems* 20 (6): 729–35.

Furgason, E. S. 1982. Optimal operation of ultrasonic correlation systems. *Proceedings of the IEEE Ultrasonic Symposium,* San Diego, 2:919–28.

Gabay, E., and S. J. Merhav. 1976. Identification of linear systems with time-delay operating in a closed loop in the presence of noise. *IEEE Trans. Automatic Control* 21 (5): 711–16.

Gabriel, K. J. 1983. Comparison of three correlation coefficient estimators for Gaussian stationary processes. *IEEE Trans. ASSP* 31 (4): 1023–25.

Gel'fand, I. M., and A. M. Yaglom. 1959. Calculation of the amount of information about a random function contained in another such function. *American Mathematical Society Translations* 12:199–246.

Gerlach, A. A. 1978. Motion-induced coherence degradation in passive systems. *IEEE Trans. ASSP* 26 (1): 1–15.

———. 1983. Performance characteristics of passive coherence estimators. Technical Report of the Naval Research Laboratory, Systems Engineering and Advanced Concepts Branch, Acoustics Division, Washington, D.C., no. 8729.

Gevins, A. S., C. L. Yeager, S. L. Diamond, J. P. Spire, G. M. Zeitlin, and A. H. Gevins. 1975. "Automated analysis of the electrical activity of the human brain (EEG): A progress report," Proc. IEEE; vol. 63, pp. 1382–99, Oct.

Gibson, M. M. 1981. Delay estimation of disturbances on the basilar membrane. *IEEE Trans. ASSP* 29 (3, pt. 2): 621–23.

Gibson, M. M., L. M. Kitzes, J. E. Rose, and J. E. Hind. 1974. Average phase angle of discharges of neurons in the anteroventral cochlear nucleus of the cat. *JASA* 55 (2): 467.

Gish, H., and D. Cochran. 1987. Invariance of the magnitude-squared coherence estimate with respect to second-channel statistics. *IEEE Trans. ASSP* 35 (12): 1774–76.

Glangeaud, F. 1981. Signal processing for magnetic pulsations. *J. Atmospheric and Terrestrial Phys.* 43 (9): 981–98.

Glangeaud, F., and C. Latombe. 1983. Identification of electromagnetic sources. *Annales Geophysicae* 1 (3): 245–51.

Glisson, T. H., C. I. Black, and A. P. Sage. 1969. On digital replica correlation algorithms with applications to active sonar. *IEEE Trans. Audio and Electroacoustics* 17 (3): 190–97.

———. 1970. On sonar signal analysis. *IEEE Trans. Aerospace and Electronic Systems* 6 (1): 37–50.

Goblirsch, D. M., and D. T. Horak. 1985. Error analysis of correlation methods for estimation of time delays in sensor signals. *Proceedings of the American Control Conference,* Boston, 1:526–29.

Godfrey, K. R. 1969. The theory of the correlation method of dynamic analysis and its application to industrial processes and nuclear power plants. *Measurement and Control* 2:T65–T72.

Gold, B., and C. M. Rader. 1969. Digital Processing of Signals. New York: McGraw-Hill.

Goodman, N. R. 1957. On the joint estimation of the spectra, cospectrum, and quadrature spectrum of a two-dimensional stationary Gaussian process. New York University Science Paper 10 (AD 134919).

———. 1963. Statistical analysis based on a certain multivariate complex Gaussian distribution (an introduction). *Annals of Mathematical Statistics* 34 (1): 152–77.

Gosselin, J. J. 1977. Comparative study of two-sensor (magnitude-squared coherence) and single-sensor (square-law) receiver operating characteristics. *ICASSP 77*: 311–14.

Graham, W. J. 1982. The multipath coherence function for uncorrelated underwater channels. *JASA* 72 (3): 908–15.

Griffiths, L. J. 1975. Rapid measurement of digital instantaneous frequency. *IEEE Trans. ASSP* 23 (2): 207–21.

Groves, G. W., and E. J. Hannan. 1968. Time series regression of sea level on weather. *Reviews of Geophysics* 6 (2): 129–74.

Hahn, W. R. 1975. Optimum signal processing for passive sonar range and bearing estimation. *JASA* 58 (1): 201–7.

Hahn, W. R., and S. A. Tretter. 1973. Optimum processing for delay-vector estimation in passive signal arrays. *IEEE Trans. Information Theory* 19 (5): 608–14.

Halvorsen, W. G., and J. S. Bendat. 1975. Noise source identification using coherent output power spectra. *J. Sound and Vibration* 9 (8): 15–24.

Hamilton, M. 1986. Exploitation of multipath arrivals in passive ranging. Ph.D. diss., Yale University.

Hamon, B. V., and E. J. Hannan. 1974. Spectral estimation of time delay for dispersive and nondispersive systems. *J. Royal Stat. Soc., Series C (Applied Statistics)* 23 (2): 134–42.

Hannan, E. J. 1970a. *Multiple time series.* New York: John Wiley.

———. 1970b. Measuring the velocity of a signal. In *Perspectives in Probability and Statistics,* J. Gani, ed., 227–37.

Hannan, E. J., and P. M. Robinson. 1973. Lagged regression with unknown lags. *J. Royal Stat. Soc., Series B (Methodological)* 35 (2): 252–267.

Hannan, E. J., and P. J. Thomson. 1971. The estimation of coherence and group delay. *Biometrika* 58 (3): 469–81.

———. 1973. Estimating group delay. *Biometrika* (2): 241–53.

———. 1981. Delay estimation and the estimation of coherence and phase. *IEEE Trans. ASSP* 29 (3, pt. 2): 485–90.

Harper, R. M., R. J. Sclabassi, and T. Estrin. 1974. Time series analysis and sleep research. *IEEE Trans. Automatic Control* 19 (6): 932–43.

Hass, W. H., and C. S. Lindquist. 1981. A synthesis of frequency domain filters for time delay estimation. *IEEE Trans. ASSP* 29 (3, pt. 2): 540–48.

Hassab, J. C. 1974. Time delay processing near the ocean surface. *J. Sound and Vibration* 35 (4): 489–501.

———. 1989. *Underwater signal and data processing.* Cleveland, Ohio: CRC Press.

Hassab, J. C., and R. E. Boucher. 1976. A probabilistic analysis of time delay extraction by the cepstrum in stationary Gaussian noise. *IEEE Trans. Information Theory* 22 (4): 444–54.

———. 1979a. A quantitative study of optimum and suboptimum filters in the generalized correlator. *ICASSP 79:* 124–27.

———. 1979b. Optimum estimation of time delay by a generalized correlator. *IEEE Trans. ASSP* 27 (4): 373–80.

———. 1981a. An experimental comparison of optimum and suboptimum filters' effectiveness in the generalized correlator. *J. Sound and Vibration* 76 (1): 117–28.

———. 1981b. Performance of the generalized cross correlator in the presence of a strong spectral peak in the signal. *IEEE Trans. ASSP* 29 (3, pt. 2): 549–55.

Hassab, J. C., B. W. Guimond, and S. C. Nardone. 1981. Estimation of location and motion parameters of a moving source observed from a linear array. *JASA* 70 (4): 1054–61.

Hassan, A. F., E. K. Al-Hussaini, and M. Bakry. 1982. Nonparametric detectors for signal detection and time delay estimation. *ICASSP 82* 1:387–90.

Hatamian, M., and D. J. Anderson. 1985. Time delay estimation in presence of jitter. *ICASSP 85* 1:161–64.

Hatsell, C. P., and L. W. Nolte. 1974. Optimal detection of a signal with time-varying carrier phase. *IEEE Trans. Aerospace and Electronic Systems* 10 (6): 788–94.

Haubrich, R. A. 1965. Earth noise, 5 to 500 millicycles per second: 1. Spectral stationarity, normality, and nonlinearity. *J. Geophys. Research* 70 (6): 1415–27.

Heimdal, P., and F. Bryn. 1973. Passive ranging techniques. *Proceedings of the NATO Advanced Study Institute on Signal Processing,* Loughborough, England. In *Signal Processing,* J. W. R. Griffiths, P. L. Stocklin, and C. van Schooneveld, eds., 261–70. New York: Academic Press.

Hero, A. O., and S. C. Schwartz. 1984. Alternatives to the generalized cross correlator for time delay estimation. *ICASSP 84* 1:15.4/1–4.

———. 1985. A new generalized cross correlator. *IEEE Trans. ASSP* 33 (1): 38–45.

———. 1988. Poisson models and mean-squared error for correlator estimators of time delay. *IEEE Trans. Information Theory* 34 (2): 287–303.

Hertz, D. 1986. Time delay estimation by combining efficient algorithms and generalized cross-correlation methods. *IEEE Trans. ASSP* 34 (1): 1–7.

Hertz, D., and M. Azaria. 1985. Time delay estimation between two phase shifted signals via generalized cross-correlation methods. *Signal Processing* 8 (2): 235–57.

Hinich, M. J., and M. C. Bloom. 1981. Statistical approach to passive target tracking. *JASA* 69 (3): 738–43.

Hinich, M. J., and C. S. Clay. 1968. The application of the discrete Fourier transform in the estimation of power spectra, coherence, and bispectra of geophysical data. *Reviews of Geophysics* 6 (3): 347–63.

Hinich, M. J., and G. R. Wilson. 1992. Time delay estimation using the cross bispectrum. *IEEE Trans. SP* 40 (1): 106–13.

Ho, K. C., and P. C. Ching. 1990. A new constrained least mean square time-delay estimation system. *IEEE Trans. Circuits and Systems* 37 (8): 1060–64.

Hodgkiss, W. S., and R. K. Brienzo. 1990. Broadband source detection and range/depth localization via full-wavefield (matched field) processing. *ICASSP 90* 5:2743–46.

Hodgkiss, W. S., and L. W. Nolte. 1976. Covariance between Fourier coefficients representing the time waveforms observed from an array of sensors. *JASA* 59 (3): 582–90.

Hou, Z.-Q., and Z.-D. Wu. 1982. A new method for high resolution estimation of time delay. *ICASSP 82* 1:420–23.

Hurd, H. L., and N. L. Gerr. 1991. Graphical methods for determining the presence of periodic correlation. *J. Time Series Analysis* 12 (4): 337–50.

Hyde, D. W., and A. H. Nuttall. 1969. Linear pre-filtering to enhance correlator performance. *Technical Memorandum of the Naval Underwater Systems Center.* New London, Conn., no. 2020-34-69.

Ianniello, J. P. 1981. Threshold effects in time delay estimation via cross-correlation. *IEEE OCEANS Conference,* Boston, 2:998–1001.

———. 1982a. Threshold effects in time delay estimation using narrowband signals. *ICASSP 82* 1:375–78.

———. 1982b. Time delay estimation via cross-correlation in the presence of large estimation errors. *IEEE Trans. ASSP* (6): 998–1003.

———. 1983. The Cramér-Rao lower bound on multipath time delay estimation. *Technical Memorandum of the Naval Underwater Systems Center.* New London, Conn., no. 831067.

———. 1984. Comparison of the Ziv-Zakai lower bound on multipath time delay estimation with autocorrelator performance. *ICASSP 84* 1:15.3/1–4.

———. 1985a. Lower bounds on the worst-case probability of large error for two channel time delay estimation. *ICASSP 85* 4:1754–57.

———. 1985b. Lower bounds on worst-case probability of large error for two channel time delay estimation. *IEEE Trans. ASSP* 33 (5): 1102–10.

———. 1986. Large and small error performance limits for multipath time delay estimation. *IEEE Trans. ASSP* 34 (2): 245–51.

———. 1987. High-resolution multipath time delay estimation for broadband random signals. *ICASSP 87* 1:447–50.

———. 1988. High-resolution multipath time delay estimation for broadband random signals. *IEEE Trans. ASSP* 36 (3): 320–27.

Ianniello, J. P., E. Weinstein, and A. J. Weiss. 1983. Comparison of the Ziv-Zakai lower bound on time delay estimation with correlator performance. *ICASSP 83* 2:875–78. IEEE ASSP Digital Signal Processing Committee, ed. 1979. *Programs for Digital Signal Processing.* New York: IEEE Press.

Iltis, R. A. 1989. Joint channel, Doppler shift and time-delay estimation for spread-spectrum communications. *23rd Asilomar Conference on Signals, Systems, and Computers* 1:206–10.

———. 1990. Joint estimation of PN code delay and multipath using the extended Kalman filter. *IEEE Trans. Communications* 38 (10): 1677–85.

Jarvis, H. F., Jr., and L. C. Ng. 1983. Comparison of error detectors for time delay tracking. *ICASSP 83* 2:903–6.

Jasrotia, V. S., and P. A. Parker. 1983. Matched filters in nerve conduction velocity estimation. *IEEE Trans. Biomedical Engineering* 30 (1): 1–9.

Jenkins, G. M., and D. G. Watts. 1968. *Spectral analysis and its applications.* San Francisco: Holden-Day.

Johnson, G. W., D. E. Ohlms, and M. L. Hampton. 1983. Broadband correlation processing. *ICASSP 83* 2:583–86.

Jones, R. H. 1969. Phase free estimation of coherence. *Annals of Math. Stat.* 40 (2): 540–48.

Jourdain, G., and M.-A. Pallas. 1986. Multiple time delay estimation in underwater acoustic propagation. In *Stochastic Processes in Underwater Acoustics,* C. R. Baker, ed. New York: Springer.

Kalman, R. E., and R. S. Bucy. 1961. New results in linear filtering and prediction theory. *Trans. ASME (Series D), Journal of Basic Engineering* 83 (1): 95–108.

Kassam, S. A., and H. V. Poor. 1983. Robust signal processing for communication systems. *IEEE Comm. Magazine* 21 (1): 20–28.

Kay, S. M., and S. L. Marple. 1981. Spectrum analysis—A modern perspective. *Proc. IEEE* 69 (11): 1380–1419.

Kelly, E. J., and R. P. Wishner. 1965. Matched-filter theory for high-velocity accelerating targets. *IEEE Trans. Military Electronics* 9 (1): 56–69.

Kemerait, R. C., and D. G. Childers. 1972. Signal detection and extraction by cepstrum techniques. *IEEE Trans. Information Theory* 18 (6): 745–59.

Kenefic, R. J. 1981. A Bayesian approach to time delay estimation. *IEEE Trans. ASSP* 29 (3, pt. 2): 611–14.

Kesler, S. B., ed. 1986. *Modern spectrum analysis, II.* New York: IEEE Press.

Kido, K., M. Ishigame, and M. Abe. 1978. Super directive spectrum analyzer by use of moving microphones, *ICASSP 78,* 824–27.

Kido, K., T. Ito, and Y. Takebayashi. 1978. A method of estimating the transfer function of the system having sharp resonances. *Proceedings of Inter-Noise 78,* San Francisco, 971–74.

Kirlin, R. L. 1978. Augmenting the maximum likelihood delay estimator to give maximum likelihood direction. *IEEE Trans. ASSP* 26 (1): 107–8.

———. 1983. Optimal delay estimation in a multiple sensor array having spatially correlated noise. *ICASSP 83* 2:895–98.

Kirlin, R. L., and J. N. Bradley. 1982. Delay estimation simulations and a normalized comparison of published results. *IEEE Trans. ASSP* 30 (3): 508–11.

Kirlin, R. L., and L. A. Dewey. 1985. Optimal delay estimation in a multiple sensor array having spatially correlated noise. *IEEE Trans. ASSP* 33 (6): 1387–96.

Kirlin, R. L., D. F. Moore, and R. F. Kubichek. 1981. Improvement of delay measurements from sonar arrays via sequential state estimation. *IEEE Trans. ASSP* 29 (3, pt. 2): 514–19.

Kirsteins, I. P. 1987. High resolution time delay estimation. *ICASSP 87* 1:451–54.

Kirsteins, I. P., and A. H. Quazi. 1988. Exact maximum likelihood time delay estimation for deterministic signals. *Proc. EUSIPCO-88,* Grenoble, France (Amsterdam: Elsevier).

Kletter, D., and H. Messer. 1989. The role of third order spectrum in maximum-likelihood time delay estimation of a random multitone signal in noise. *ICASSP 89* 4:2310–13.

Knapp, C. H., and G. C. Carter. 1976. The generalized correlation method for estimation of time delay. *IEEE Trans. ASSP* 24 (4): 320–27.

———. 1977. Estimation of time delay in the presence of source or receiver motion. *JASA* 61 (6): 1545–49.

Knight, W. C., R. G. Pridham, and S. M. Kay. 1981. Digital signal processing for sonar. *Proc. IEEE* 69 (11): 1451–1506.

Koopmans, L. H. 1964. On the coefficient of coherence for weakly stationary stochastic processes. *Annals of Math. Stat.* 35:532–49.

———. 1974. *The spectral analysis of time series.* New York: Academic Press.

Kostic, L. 1981. Local steam transit time estimation in a boiling water reactor. *IEEE Trans. ASSP* 29 (3, pt. 2): 555–60.

Kroenert, J. T. 1982. Some comments on bias/misalignment effects in the magnitude squared coherence estimate. *IEEE Trans. ASSP* 30 (3): 511–13.

Krolik, J. L. 1983. "Constrained time delay estimation via Zer-crossing methods," M.A.Sc. thesis, Department of Electrical Engineering, University of Toronto, Ontario.

Krolik, J. L., M. Eizenman, and S. Pasupathy. 1986. Time delay estimation via generalized correlation with adaptive spatial prefiltering. *ICASSP 86* (3): 1893–96.

———. 1988. Time delay estimation of signals with uncertain spectra. *IEEE Trans. ASSP* 36 (12): 1801–11.

Krolik, J. L., M. Joy, S. Pasupathy, and M. Eizenman. 1984. A comparative study of the LMS adaptive filter versus generalized correlation methods for time delay estimation. *ICASSP 84* 1:15.11/1–4.

Kuhn, J. P. 1978. Detection performance of the smooth coherence transform (SCOT). *ICASSP 78,* 678–83.

Kuhn, J. P., et al. 1981. Time delay detection and estimation techniques for moving sources. *Proceedings of the 24th Midwest Symposium on Circuits and Systems.*

Kumaresan, R., and D. W. Tufts. 1983. Estimating the angles of arrival of multiple plane waves. *IEEE Trans. Aerospace and Electronic Systems* 19 (1): 134–39.

Kuo, C., C. Lindberg, and D. J. Thomson. 1990. Coherence established between atmospheric carbon dioxide and global temperature. *Nature* 343 (6260): 709–14.

Lassahn, G. D., and A. G. Baker. 1982. Errors in cross-correlation peak location. *Trans. ASME, J. Dynamic Systems, Measurement, and Control* 104 (2): 194–99.

Lee, G. K. F., and G. J. Fetzer. 1987. Time delay estimation of deterministic and stochastic signals using dynamic programming. *Int. J. Control and Computers* 15 (3): 81–86.

Lee, H., and G. Wade. 1983. Smoothed coherent transform method for measuring displacement by means of planar holographic detection. *IEEE Trans. Sonics and Ultrasonics* 30 (5): 330–31.

Lee, J. S., and J. K. Hammond. 1989. Time-varying filter modelling of the sound field due to a moving source and time-delay estimation. *ICASSP 89* 4:2105–8.

Leroy-Hebert, S., and A. Plaisant. 1989. The multipath coherence function for correlated random channels and a moving source. *Proceedings of the NATO Advanced Study Institute on Underwater Acoustic Data Processing,* Kingston, Ontario, Canada. In *Underwater acoustic data processing,* Y. T. Chan, ed., 93–97 (NATO ASI Series E, Applied Sciences, vol. 161). Dordrecht/Boston/London: Kluwer Academic Publishers.

Levin, M. J. 1965. Power spectrum parameter estimation. *IEEE Trans. Information Theory* 11 (1): 100–107.

Li, F., R. J. Vaccaro, and D. W. Tufts. 1990. Unified performance analysis of subspace-based estimation algorithms. *ICASSP 90* 5:2575–78.

Li, Y. T., and A. L. Kurkjian. 1981. Arrival time estimation using iterative signal reconstruction from the phase of the cross-spectrum. *Proceedings of the 2nd International Symposium on Computer-Aided Seismic Analysis and Discrimination,* 87–91. North Dartmouth, Mass.

Liggett, W. S., Jr. 1972. Passive sonar processing for noise with unknown covariance structure. *JASA* 51 (1, pt. 1): 24–30.

Lindsey, W. C., and H. Meyr. 1977. Complete statistical description of the phase-error process generated by correlative tracking systems. *IEEE Trans. Information Theory* 23 (2): 194–202.

Lourtie, I. M. G. 1988. Optimal estimation of time delays with nonstationary signals. Ph.D. diss., Instituto Superior Técnico, Lisbon, Portugal.

Lourtie, I. M. G., and G. C. Carter. 1990. Signal detection in the presence of inaccurate multipath time delay modelling. *ICASSP 90* 5:2751–54.

Lourtie, I. M. G., and J. M. F. Moura. 1985a. Delay estimation with nonstationary signals and correlated observation noises. *Proceedings of the NATO Advanced Study Institute on Adaptive Methods in Underwater Acoustics,* Lüneburg, Germany. In *Adaptive methods in underwater acoustics,* H. G. Urban, ed., 235–47 (NATO ASI Series C, Mathematical and Physical Sciences, vol. 151). Dordrecht/Boston/Lancaster: D. Reidel Publishing Company.

———. 1985b. Time delay determination: Maximum likelihood and Kalman-Bucy-type structures. *ICASSP 85* 4:1750–53.

———. 1988. Optimal estimation of time-varying delay. *ICASSP 88* 5:2622–25.

———. 1989. Joint delay and signal determination. *Proceedings of the NATO Advanced Study Institute on Underwater Acoustic Data Processing,* Kingston, Ontario, Canada. In *Underwater acoustic data processing,* Y. T. Chan, ed., 531–36 (NATO ASI Series E, Applied Sciences, vol. 161). Dordrecht/Boston/London: Kluwer Academic Publishers.

———. 1991. Multisource delay estimation: Nonstationary signals. *IEEE Trans. SP* 39 (5): 1033–48.

Ludeman, L. C. 1979a. A procedure for obtaining more accurate timing information for the sound ranging problem. Technical Report of the U.S. Army Electronics Research and Development Command, Atmospheric Sciences Laboratory, White Sands Missile Range, N.M., no. ASL-CR-79-0100-1.

———. 1979b. Bias and variance of a sound ranging estimator. Presented at the 1979 Conference on Time Delay Estimation and Applications, Naval Postgraduate School, Monterey, Calif.

———. 1979c. A software package for estimating time differences for artillery sound ranging applications. Technical

Report of the U.S. Army Electronics Research and Development Command, Atmospheric Sciences Laboratory, White Sands Missile Range, N.M., no. ASL-CR-79-0100-6.

———. 1980. Multisignal time difference estimator with application to the sound ranging problem. *ICASSP 80* 3: 800–803.

Lunde, E. B. 1973. Wave front stability in the ocean. *Proceedings of the NATO Advanced Study Institute on Signal Processing*, Loughborough, England. In *Signal processing*, J. W. R. Griffiths, P. L. Stocklin, and C. van Schooneveld, eds., 271–79. New York: Academic Press.

MacDonald, V. H., and P. M. Schultheiss. 1969. Optimum passive bearing estimation in a spatially incoherent noise environment. *JASA* 46 (1, pt. 1): 37–43.

Maki, B. E. 1986. Interpretation of the coherence function when using pseudorandom inputs to identify nonlinear systems. *IEEE Trans. Biomedical Engineering* 33 (8): 775–79.

Maragakis, E. 1990. Time delay estimation in a multipath environment. Master's thesis, University of Rhode Island, Kingston.

Marple, S. L., Jr. 1987. *Digital spectral analysis with applications.* Englewood Cliffs, N.J.: Prentice Hall.

Mars, N. J. I. 1982. Time delay estimator for EEG analysis based on information theory. *ICASSP 82* 2:733–35.

Mars, N. J. I., and G. W. Van Arragon. 1981. Time delay estimation in nonlinear systems. *IEEE Trans. ASSP* 29 (3, pt. 2): 619–21.

McCannon, T. E. 1985. Application of robust filtering techniques to time delay estimation in the Arctic environment. *ICASSP 85* 1:165–67.

McVicar, G. N., and P. A. Parker. 1988. Spectrum dip estimator of nerve conduction velocity. *IEEE Trans. Biomedical Engineering* 35 (12): 1069–76.

Meyr, H. 1976. Delay-lock tracking of stochastic signals. *IEEE Trans. Communications* 24 (3): 331–39.

Meyr, H., H. Ryser, and C. Zimmer. 1975. Noncontact speed measurement by correlation techniques. *Hasler Review* 8 (2): 50–64.

Meyr, H., and G. Spies. 1984. The structure and performance of estimators for real-time estimation of randomly varying time delay. *IEEE Trans. ASSP* 32 (1): 81–94.

Meyr, H., G. Spies, and J. Bohmann. 1982. Real-time estimation of moving time delay. *ICASSP 82* 1:383–86.

Moddemeijer, R. 1989. Delay-estimation with application to electroencephalograms in epilepsy. Ph.D. diss., University of Twente, Enschede, the Netherlands.

———. 1991. On the determination of the position of extrema of sampled correlators. *IEEE Trans. SP* 39 (1): 216–19.

Mohanty, N. C. 1974. Estimation of delay of M PPM signals in Laguerre communications. *IEEE Trans. Communications* 22 (5): 713–14.

Moose, R. L., and T. E. Dailey. 1985. Adaptive underwater target tracking using passive multipath time-delay measurements. *IEEE Trans. ASSP* 33 (4): 777–87.

Moose, R. L., and P. M. Godiwala. 1985. Passive depth tracking of underwater maneuvering Targets. *IEEE Trans. ASSP* 33 (4): 1040–44.

Morgan, D. R., and T. M. Smith. 1989. Coherence effects on the detection performance of quadratic array processors, with applications to large-array matched-field beamforming. *ICASSP 89* 4:2787–90.

Moura, J. M. F. 1979a. Passive systems theory with narrow-band and linear constraints: Part II—temporal diversity. *IEEE J. Oceanic Eng.* 4 (1): 19–30.

———. 1979b. Passive systems theory with narrow-band and linear constraints: Part III—spatial/temporal diversity. *IEEE J. Oceanic Eng.* 4 (3): 113–19.

———. 1981. The hybrid algorithm: A solution to acquisition and tracking. *JASA* 69 (6): 1663–72.

Moura, J. M. F., and A. B. Baggeroer. 1978. Passive systems theory with narrow-band and linear constraints: Part I—spatial diversity. *IEEE J. Oceanic Eng.* 3 (1): 5–13.

Moura, J. M. F., and M. J. D. Rendas. 1989. Sensitivity of range localization in a multipath environment. *ICASSP 89* 4:2629–32.

Moura, J. M. F., H. L. Van Trees, and A. B. Baggeroer. 1973. Space/time tracking by a passive observer. *Proceedings of the 4th Symposium on Nonlinear Estimation Theory and Its Applications*, San Diego, Calif.

Namazi, M., and J. A. Stuller. 1987. A new approach to signal registration with the emphasis on variable time delay estimation. *IEEE Trans. ASSP* 35 (12): 1649–60.

Nehorai, A. 1983. Algorithms for system identification and source location. Ph.D. diss., Department of Electrical Engineering, Stanford University.

Nehorai, A., and P. Stoica. 1988. Adaptive algorithms for constrained ARMA signals in the presence of noise. *IEEE Trans. ASSP* 36 (8): 1282–91.

Nehorai, A., G. Su, and M. Morf. 1983. Estimation of time differences of arrival by pole decomposition. *IEEE Trans. ASSP* 31 (6): 1478–92.

Nettheim, N. 1966. The estimation of coherence. Technical Report of the Office of Naval Research, Logistics and Mathematical Statistics, Washington, D.C., no. 5.

Ng, L. C. 1983. Optimum multisensor, multitarget localization and tracking. Ph.D. diss., University of Connecticut. Also, Technical Report of the Naval Underwater Systems Center, New London, Conn., no. 6931.

Ng, L. C., and Y. Bar-Shalom. 1986. Multisensor multitarget time delay vector estimation. *IEEE Trans. ASSP* 34 (4): 669–78.

Nikias, C. L., and R. Pan. 1988a. Time delay estimation in unknown Gaussian spatially correlated noise, *ICASSP 88* 5:2638–41.

———. 1988b. Time delay estimation in unknown Gaussian spatially correlated noise. *IEEE Trans. ASSP* 36 (11): 1706–14.

Nikias, C. L., and M. R. Raghuveer. 1983. A new class of high-resolution and robust multi-dimensional spectral estimation algorithms. *ICASSP 83* 2:859–62.

———. 1987. Bispectrum estimation: A digital signal processing framework. *Proc. IEEE* 75 (7): 869–91.

Nishihara, H., and H. Konishi. 1977. A new correlation method for transit-time estimation. In *Progress in Nuclear Energy,* vol. 1, 219–29. London: Pergamon Press.

Nuttall, A. H. 1971. Spectral estimation by means of overlapped FFT processing of windowed data. Report of the Naval Underwater Systems Center, New London, Conn., no. 4169.

———. 1975. Estimation of cross-spectra via overlapped fast Fourier transform. Report of the Naval Underwater Systems Center, New London, Conn., no. 4169-S.

———. 1981. Invariance of distribution of coherence estimate to second-channel statistics. *IEEE Trans. ASSP* 29 (1): 120–22.

———. 1983. Analysis of a generalized framework for spectral estimation—Part II: reshaping and variance results. *Proceedings of the Institution of Electrical Engineers, Part F (Communications, Radar, and Signal Processing)* 130 (3): 242–45. Also in *Proceedings of the Conference on Spectral Analysis and Its Use in Underwater Acoustics,* London, England, April 1982. (Reprinted in *Modern spectrum analysis II,* S. B. Kesler, ed., 208–11. New York: IEEE Press, 1986.)

Nuttall, A. H., and G. C. Carter. 1976. Bias of the estimate of magnitude-squared coherence, *IEEE Trans. ASSP* 24 (6): 582–83.

———. 1980. A generalized framework for power spectral estimation. *IEEE Trans. ASSP* 28 (3): 334–35.

———. 1981. An approximation to the cumulative distribution function of the magnitude-squared coherence estimate. *IEEE Trans. ASSP* 29 (4): 932–34.

———. 1982. Spectral estimation using combined time and lag weighting. *Proc. IEEE* 70 (9): 1115–25.

Nuttall, A. H., and D. W. Hyde. 1969. A unified approach to optimum and suboptimum processing for arrays. Report of the Naval Underwater Systems Center, New London, Conn., no. 992.

Officer, C. B. 1958. *Introduction to the theory of sound transmission.* New York: McGraw-Hill.

Oppenheim, A. V., and J. S. Lim. 1981. The importance of phase in signals. *Proc. IEEE* 69 (5): 529–41.

Oppenheim, A. V., and R. W. Schafer. 1975. *Digital signal processing.* Englewood Cliffs, N.J.: Prentice Hall.

Otnes, R. K., and L. D. Enochson. 1972. *Digital time series analysis.* New York: Wiley-Interscience.

Owsley, N. L., and G. R. Swope. 1981. Time delay estimation in a sensor array. *IEEE Trans. ASSP* 29 (3, pt. 2): 519–23. Also, *Proceedings of the NATO Advanced Study Institute on Underwater Acoustics and Signal Processing,* Kollekolle, Copenhagen, Denmark, August 1980. In *Underwater Acoustics and Signal Processing,* L. Bjørnø, ed., 421–32 (NATO ASI Series C, Mathematical and Physical Sciences, vol. 66). Dordrecht/Boston/London: D. Reidel Publishing Company.

Ozard, J. M., and B. C. Zelt. 1985. Signal coherence model for widely spaced sensors in shallow water with rough boundaries. *Canadian Acoustics* 13 (1): 20–39.

Paik, C. H., G. L. Cote, J. S. DaPonte, and M. D. Fox. 1988. Fast Hartley transforms for spectral analysis of ultrasound Doppler signals,'' *IEEE Trans. Biomedical Eng.* 35 (10): 885–88.

Pallas, M.-A., and G. Jourdain. 1986. Joint estimation of close delays and application to underwater acoustics. *Proceedings of EUSIPCO-86,* La Haye, the Netherlands.

———. 1991. Active high resolution time delay estimation for large BT signals. *IEEE Trans. SP* 39 (4): 781–88.

Pallas, M.-A., N. Martin, and J. Martin. 1987. Time delay estimation by autoregressive modelization. *ICASSP 87* 1: 455–58.

Papoulis, A. 1965. *Probability, random variables, and stochastic processes.* New York: McGraw-Hill.

Parker, P. A., and P. Kelly. 1981. Nerve conduction velocity from correlation and spectral analysis. *Proceedings of the 9th Annual Northeast Bioengineering Conference,* 381–86.

Pasupathy, S. 1978. Optimal and conventional array processing. *Canadian Elec. Eng. J.* 3 (1): 27–31.

Pasupathy, S., and W. J. Alford. 1980. Range and bearing estimation in passive sonar. *IEEE Trans. Aerospace and Electronic Systems* 16 (2): 244–49.

Patzewitsch, J. T., M. D. Srinath, and C. I. Black. 1978. Nearfield performance of passive correlation processing sonars. *JASA* 64 (5): 1412–23.

Pearson, A. E., and C. Y. Wuu. 1980. System identification of pure time delay with finite time data. *Proceedings of the 19th IEEE Conference on Decision and Control,* Albuquerque, 1:739–40.

Pearson, J., C. J. Macleod, and T. S. Durrani. 1982. Mode and time delay estimation for non-destructive evaluation systems. *ICASSP 82* 1:395–98.

Pedersen, P. C., P. A. Lewin, and L. Bjørnø. 1988. Application of time-delay spectrometry for calibration of ultrasonic transducers. *IEEE Trans. Ultrasonics, Ferroelectrics, and Frequency Control* 35 (2): 185–205.

Pedersen, P. C., and C. Pinto. 1982. Technique for ultrasound group velocity measurements in dispersive media with multipath interference. *IEEE Trans. Sonics and Ultrasonics* 29 (6): 352–58.

Petropulu, A., C. L. Nikias, and J. G. Proakis. 1988. Cumulant cepstrum of FM signals and high-resolution time delay estimation. *ICASSP 88* 5:2642–45.

Pflug, L. A., G. E. Ioup, J. W. Ioup, R. L. Field, and J. H. Leclere. 1993. Time delay estimation for deterministic transients using second and higher order correlations. To be published in *JASA*.

Piersol, A. G. 1978. Use of coherence and phase data between two receivers in evaluation of noise environments. *J. Sound and Vibration* 56 (2): 215–28.

———. 1981. Time delay estimation using phase data. *IEEE Trans. ASSP* 29 (3, pt. 2): 471–77.

Poor, H. V., 1980. On robust Wiener filtering. *IEEE Trans. Automatic Control* 25 (3): 531–36.

———. 1983. Robust matched filters. *IEEE Trans. Information Theory* 29 (5): 677–87.

———. 1987. Uncertainty tolerance in underwater acoustic signal processing. *IEEE J. Oceanic Eng.* 12 (1): 48–65.

Porat, B., and B. Friedlander. 1983. Estimation of spacial and spectral parameters of multiple sources. *IEEE Trans. Information Theory* 29 (3): 412–25.

Prasad, S., M. S. Narayanan, and S. R. Desai. 1985. Time-delay estimation performance in a scattering medium. *IEEE Trans. ASSP* 33 (1): 50–60.

Pratt, A. R. 1979. Local estimation of delay parameter following robust detection. *ICASSP 79,* 132–35.

Priestley, M. B. 1981. *Spectral analysis and time series,* vol. 2, *Multivariate series, prediction and control.* New York: Academic Press.

Quazi, A. H. 1981a. Effects of signal and noise spectral slopes on time delay estimates in passive localization. *ICASSP 81* 3:1265–68.

———. 1981b. An overview on the time delay estimate in active and passive systems for target localization. *IEEE Trans. ASSP* 29 (3, pt. 2): 527–33.

———. 1983. Lower bounds on target localization errors for various signal and noise characteristics. Presented at Neuvième Colloque sur le Trait du Signal et Ses Applications, Nice, France.

Quazi, A. H., and D. T. Lerro. 1985. Passive localization using time delay estimates with sensor positional errors. *JASA* 78 (5): 1664–1670.

Raghuveer, M. R., and C. L. Nikias. 1984. A parametric approach to bispectrum estimation. *ICASSP 84* 3:38.1/1–4.

———. 1985a. Bispectrum estimation for short length data. *ICASSP 85* 3:1352–55.

———. 1985b. Bispectrum estimation: A parametric approach. *IEEE Trans. ASSP* 33 (5): 1213–30.

Rao, P., C. Griffin, and F. J. Taylor. 1988. Time-delay estimation using the Wigner distribution. *22nd Asilomar Conference on Signals, Systems, and Computers,* 726–29.

Reddy, S. N., and S. K. Sarna. 1980. The use of coherence function in biological signal processing. *Digest of the 8th Canadian Medical and Biological Engineering Conference,* 150–51.

Reed, F. A., P. L. Feintuch, and N. J. Bershad. 1981. Time delay estimation using the LMS adaptive filter—Static behavior. *IEEE Trans. ASSP* 29 (3, pt. 2): 561–71.

Remley, W. R. 1963. Correlation of signals having a linear delay. *JASA* 35 (1): 65–69.

———. 1966. Doppler dispersion effects in matched filter detection and resolution. *Proc. IEEE* 54 (1): 33–39.

Rendas, M. J. D., and J. M. F. Moura. 1990. Cramér-Rao bounds for passive range and depth in a vertically inhomogeneous medium. *ICASSP 90* 5:2779–82.

Riley, J. M. F. 1978. The least squares estimation of time delay. M.Eng. thesis, Department of Electrical Engineering, Royal Military College of Canada.

Robinson, E. R., and A. H. Quazi. 1983. The impact of refraction on time-delay estimation. *ICASSP 83* (3): 981–84.

———. 1985. Effect of sound-speed profile on differential time-delay estimation. *JASA* 77 (3): 1086–90.

Rosenberger, J. C. 1989. Passive localization. *Proceedings of the NATO Advanced Study Institute on Underwater Acoustic Data Processing,* Kingston, Ontario, Canada. In *Underwater acoustic data processing,* Y. T. Chan, ed., 511–24 (NATO ASI Series E, Applied Sciences, vol. 161). Dordrecht/Boston/London: Kluwer Academic Publishers.

Roth, P. R. 1971. Effective measurements using digital signal analysis. *IEEE Spectrum* 8 (4): 62–70.

Ryan, W. E., and J. M. Schumpert. 1985. Time delay estimation for spread spectrum signals using generalized cross correlation. *ICASSP 85* 1:168–71.

Sackman, G. L., and S. C. Shelef. 1979. The use of time-delay/phase difference trace functions for bearing estimation in arrays. *13th Asilomar Conference on Circuits, Systems, and Computers,* 354–58.

Sahakian, A., K. Ropella, J. Baerman, and S. Swiryn. 1988. Median frequency and coherence measures of atrial and ventricular fibrillation. *Proceedings of the 10th International Conference of the IEEE Engineering in Medicine and Biology Society,* New Orleans, 1:16–17.

———. 1989. Adaptive coherence estimation on brief intracardiac recordings. *Proceedings of the 11th International Conference of the IEEE Engineering in Medicine and Biology Society,* Seattle, 1:224–25.

Salt, J. E. 1987. Mean square registration error for quantized one-dimensional signals—Approximation and bounds. Ph.D. diss., University of Saskatchewan.

Salt, J. E., and A. G. Wacker. 1989. Optimistic and pessimistic approximations to variance of time delay estimators. *IEEE Trans. ASSP* 37 (5): 634–41.

Santopietro, R. F. 1977. The origin and characterization of the primary signal, noise, and interference sources in the high frequency electrocardiogram. *Proc. IEEE* 65 (5): 707–13.

Sasaki, K., T. Sato, and Y. Nakamura. 1977. Holographic passive sonar. *IEEE Trans. Sonics and Ultrasonics* 24 (3): 193–200.

Saunders, K. D., and F. C. Hamrick. 1982. A note on cross spectrum and coherence calculations. *J. Geophys. Research* 87 (C12): 9699–9703.

Sauvet-Carof, M. T. 1992. Experimental results of the differential Doppler compensated cross correlator and its application in target motion analysis. *JASA* 92 (1): 231–37.

Scannell, E. H., Jr., and G. C. Carter. 1978a. Confidence bounds for magnitude-squared coherence estimates (with computer listings). *ICASSP 78,* 670–73. Also, G. C. Carter and E. H. Scannell, Jr., Technical Document of the Naval Underwater Systems Center, New London, Conn., no. 5881.

———. 1978b. Confidence bounds for magnitude-squared coherence estimates. *IEEE Trans ASSP* 26 (5): 475–77.

Scarbrough, K. 1984. Analysis of time delay estimator performance. Ph.D. diss., Kansas State University. Published as a Technical Report of the Naval Underwater Systems Center, New London, Conn., no. TR 7203.

Scarbrough, K., N. Ahmed, and G. C. Carter. 1980. An experimental comparison of the cross correlation and SCOT techniques for time delay estimation. *ICASSP 80* 3:807–10.

———. 1981a. Some considerations of time delay estimation. *ICASSP 81* 3:1261–64.

———. 1981b. On the simulation of a class of time delay estimation algorithms. *IEEE Trans. ASSP* 29 (3, pt. 2): 534–40.

———. 1982. Comparison of two methods for time delay estimation of sinusoids. *Proc. IEEE* 70 (1): 90–92.

Scarbrough, K., N. Ahmed, D. H. Youn, and G. C. Carter. 1982. On the SCOT and Roth algorithms for time delay estimation. *ICASSP 82* 1:371–74.

Scarbrough, K., G. C. Carter, and R. J. Tremblay. 1984. Implications of threshold effects for coherent and incoherent processing techniques of time delay estimation. *ICASSP 84* 1:15.1/1–4.

Scarbrough, K., R. J. Tremblay, and G. C. Carter. 1983. Performance predictions for coherent and incoherent processing techniques of time delay estimation. *IEEE Trans. ASSP* 31 (5): 1191–96.

Schmidt, H., 1985. Resolution bias errors in spectral density, frequency response, and coherence function measurements, I: General theory. *J. Sound and Vibration* 101 (3): 347–62.

Schmidt, R. O. 1972. A new approach to geometry of range difference location. *IEEE Trans. Aerospace and Electronic Systems* 8 (6): 821–35.

Schultheiss, P. M. 1979. Locating a passive source with array measurements: A summary of results. *ICASSP 79,* 967–70.

Schultheiss, P. M., E. Ashok, and J. P. Ianniello. 1983. Optimum and sub-optimum source localization with sensors subject to random motion. *JASA* 74 (1): 131–42.

Schultheiss, P. M., and J. P. Ianniello. 1980. Optimum range and bearing estimation with randomly perturbed arrays. *JASA* 68 (1): 167–73.

Schultheiss, P. M., and K. Wagner. 1989. Active and passive localization: Similarities and differences. *Proceedings of the NATO Advanced Study Institute on Underwater Acoustic Data Processing,* Kingston, Ontario, Canada. In *Underwater acoustic data processing,* Y. T. Chan, ed., 215–32 (NATO ASI Series E, Applied Sciences, vol. 161). Dordrecht/Boston/London: Kluwer Academic Publishers, 1989.

Schultheiss, P. M., and E. Weinstein. 1979. Estimation of differential Doppler shifts. *JASA* 66 (5): 1412–19.

———. 1981. Lower bounds on the localization errors of a moving source observed by a passive array. *IEEE Trans. ASSP* 29 (3, pt. 2): 600–7.

Segal, M., and E. Weinstein. 1989. Spatial and spectral parameter estimation of multiple source signals. *ICASSP 89* 4: 2665–68.

Seidman, L. P. 1971. Bearing estimation error with a linear array. *IEEE Trans. Audio and Electroacoustics* 19 (2): 147–57.

Shah, S. A., E. A. Woodruff, and R. B. Northrop. 1988. An adaptive IPFM/SDC controller for the closed-loop sodium nitroprusside regulation of post-operative blood pressure. *Proceedings of the 10th International Conference of the IEEE Engineering in Medicine and Biology Society,* New Orleans, 2:519–20.

Simaan, M. 1985. A frequency-domain method for time-shift estimation and alignment of seismic signals. *IEEE Trans. Geoscience and Remote Sensing* 23 (2): 132–38.

Simmer, K. U., P. Kuczynski, and A. Wasiljeff. 1992. Time delay for adaptive multichannel speech enhancement systems. *Proceedings of the URSI International Symposium on Signals, Systems, and Electronics,* Paris, France.

Smith, J. O., and B. Friedlander. 1984. Adaptive multipath delay estimation. *ICASSP 84* 1:15.9/1–4.

———. 1985a. Adaptive interpolated time-delay estimation. *IEEE Trans. Aerospace and Electronic Systems* 21 (2): 180–99.

———. 1985b. Adaptive multipath delay estimation. *IEEE Trans. ASSP* 33 (4): 812–22.

Solomon, O. M., Jr. 1985. Linear modeling and the coherence function: An algebraic approach. Ph.D. diss., University of New Mexico.

Spindel, R. C. 1985. Signal processing in ocean tomography. *Proceedings of the NATO Advanced Study Institute on Adaptive Methods in Underwater Acoustics,* Lüneburg, Germany. In *Adaptive methods in underwater acoustics,* H. G. Urban, ed., 687–710 (NATO ASI Series C, Mathematical and Physical Sciences, vol. 151). Dordrecht/Boston/Lancaster: D. Reidel Publishing Company, 1985.

Spindel, R. C., and R. P. Porter. 1974. Precision tracking system for sonobuoys. *IEEE OCEANS Conference,* Halifax, Nova Scotia, Canada, 2:162–65.

Stearns, S. D. 1979. Applications of the coherence function in comparing seismometers. Technical Report of Sandia National Laboratories, Albuquerque, no. SAND79-1633. Available from National Technical Information Service, U.S. Department of Commerce, 5285 Port Royal Road, Springfield, VA 22161.

———. 1981. Tests of coherence unbiasing methods. *IEEE Trans. ASSP* 29 (2): 321–23.

Stein, S. 1981. Algorithms for ambiguity function processing. *IEEE Trans. ASSP* 29 (3, pt. 2): 588–99.

Stuller, J. A. 1987. Maximum-likelihood estimation of time-varying delay—Part I. *IEEE Trans. ASSP* 35 (3): 300–13.

Suzuki, T., N. Sakabe, and Y. Miyashita. 1982. Power spectral analysis of auditory brain stem responses to pure tone stimuli. *Scandanavian Audiology* 11:25–30.

Swerup, C. 1978. On the choice of noise for the analysis of the peripheral auditory system. *Biology and Cybernetics* 29: 97–104.

Takebayashi, Y., and K. Kido. 1978. Effect of the shape of time window in estimating the transfer characteristics of the acoustic system having group delay or reverberation. ASA and ASJ Joint Meeting. *JASA* 64 (Supplement no. 1).

Takeuchi, T., M. Sako, and S. Yoshida. 1990. Multipath delay estimation for indoor wireless communication. *40th IEEE Vehicular Technology Conference,* 401–6, Orlando.

Thomson, D. J. 1982. Spectrum Estimation and Harmonic Analysis. *Proc. IEEE* 70 (9): 1055–96.

———. 1990. Quadratic-inverse spectrum estimates: Applications to palaeoclimatology. *Philosophical Transactions of the Royal Society, Series A (Physical Sciences and Engineering)* 332 (1627): 539–97.

Thomson, D. J., and A. D. Chave. 1991. Jackknifed error estimates for spectra, coherences, and transfer functions. In *Advances in spectrum analysis and array processing,* S. Haykin, ed., 58–113. Englewood Cliffs, N.J.: Prentice Hall.

Tick, L. J. 1967. Estimation of coherency. In *Spectral analysis of time series,* Bernard Harris, ed., 133–52. New York: John Wiley.

Tremblay, R. J., G. C. Carter, and D. W. Lytle. 1987a. A practical approach to the estimation of amplitude and time-delay parameters of a composite signal. *IEEE J. Oceanic Engineering* 12 (1): 273–78.

———. 1987b. A practical approach to the estimation of amplitude and time delay parameters of a composite signal in non-white Gaussian noise. *ICASSP 87* (1): 467–70.

Tribolet, J. M. 1977. A new phase unwrapping algorithm. *IEEE Trans. ASSP* 25 (2): 170–77.

Tugnait, J. K. 1989. Time delay estimation in unknown spatially correlated Gaussian noise using higher-order statistics. *23rd Asilomar Conference on Signals, Systems, and Computers* 1:211–15.

———. 1991. On time delay estimation with unknown spatially correlated Gaussian noise using fourth-order cumulants and cross cumulants. *IEEE Trans. SP* 39 (6): 1258–67.

Tukey, J. W. 1967. In *Spectral analysis of time series,* Bernard Harris, ed., Chapter 2. New York: John Wiley.

Urick, R. J. 1983. *Principles of underwater sound,* 3rd ed. New York: McGraw-Hill. (First edition was in 1967 under the title *Principles of underwater sound for engineers.*)

Vaccaro, R. J., E. Maragakis, and R. L. Field. 1990. Time delay estimation for ocean acoustic transient signal extraction in a multipath environment. *ICASSP 90* 5:2923–26.

Vaccaro, R. J., C. S. Ramalingam, D. W. Tufts, and R. L. Field. 1992. Least-squares time-delay estimation for transient signals in a multipath environment. *JASA* 92 (1): 210–18.

van der Vliet, G. H., J. Holsheimer, and D. Bingmann. 1980. Calculation of the conduction velocity of short nerve fibres. *Medical and Biological Engineering and Computing* 18 (6): 749–57.

van Langevelde, H. J., R. van der Heiden, and C. van Schooneveld. 1990. Phase lags from multiple flux curves of OH/IR stars. *Astronomy and Astrophysics* 239:193–204.

van Schooneveld, C., and D. J. Frijling. 1981. Spectral analysis: On the usefulness of linear tapering for leakage suppression. *IEEE Trans. ASSP* 29 (2): 323–29.

van Schooneveld, C., and M. Sondag. 1983. Spectrum analysis: On the application of data windowing to Kay and Marple's results. *Proc. IEEE* 71 (6): 776–79.

Van Trees, H. L. 1968. *Detection, estimation, and modulation theory, Part I—Detection estimation and linear modulation theory.* New York: John Wiley.

———. 1971a. *Detection, estimation, and modulation theory, Part II—Nonlinear modulation theory.* New York: John Wiley.

———. 1971b. *Detection, estimation, and modulation theory, Part III—Radar/sonar signal processing and Gaussian signals in noise.* New York: John Wiley.

Vezzosi, G., and P. Nicolas. 1986. Time delay estimation using an eigenstructure based spectral method. *Proceedings of the 25th IEEE Conference on Decision and Control,* Athens, Greece, vol. 2:949–52.

Wasiljeff, A. 1975. Spatial horizontal coherence of acoustic signals in shallow water. SACLANT ASW Research Center, LaSpezia, Italy, SACLANTCEN Memorandum SM-68.

Wax, M. 1982. The joint estimation of differential delay, Doppler, and phase. *IEEE Trans. Information Theory* 28 (5, pt. 1): 817–20.

Wax, M., and T. Kailath. 1983. Optimum localization of multiple sources by passive arrays. *IEEE Trans. ASSP* 31 (5): 1210–18.

Weinstein, E. 1982a. Optimal source localization and tracking from passive array measurements. *IEEE Trans. ASSP* 30 (1): 69–76.

———. 1982b. Measurement of the differential Doppler shift. *IEEE Trans. ASSP* 30 (1): 112–17.

———. 1982c. Performance analysis of time delay estimators. WHOI Technical Report.

Weinstein, E., and D. Kletter. 1983. Delay and Doppler estimation by time-space partition of the array data. *IEEE Trans. ASSP* 31 (6): 1523–35.

Weinstein, E., and A. J. Weiss. 1984. Fundamental limitations in passive time-delay estimation—Part II: Wide-band systems. *IEEE Trans. ASSP* 32 (5): 1064–77.

Weiss, A. J., and Z. Stein. 1987. Optimal below threshold delay estimation for radio signals. *IEEE Trans. Aerospace and Electronic Systems* 23 (6): 726–30.

Weiss, A. J., and E. Weinstein. 1982. Composite bound on the attainable mean-square error in passive time-delay estimation from ambiguity prone signals. *IEEE Trans. Information Theory* 28 (6): 977–79.

———. 1983. Fundamental limitations in passive time delay estimation—Part I: Narrow-band systems. *IEEE Trans. ASSP* 31 (2): 472–86.

Welch, P. D. 1967. The use of fast Fourier transform for the estimation of power spectra: A method based on time averaging over short, modified periodograms. *IEEE Trans. Audio and Electroacoustics* 15 (2): 70–73.

Whitsel, H. K., and L. W. Griswold. 1978. Speed sensors for high speed surface craft. Presented at the 5th Ship Control Systems Symposium, Annapolis.

Widrow, B., J. R. Glover, Jr., J. M. McCool, J. Kaunitz, C. S. Williams, R. H. Hearn, J. R. Zeidler, E. Dong, and

R. C. Goodlin. 1975. Adaptive noise cancelling: Principles and applications. *Proc. IEEE* 63 (12): 1692–1716.

Wiener, N. 1930. Generalized harmonic analysis. *Actuarial Math.* 55:117–258.

Wilson, M. J., and R. A. Dobie. 1987. Human short-latency auditory responses obtained by cross-correlation. *Electroencephalography and Clinical Neurophysiology* 66: 529–38.

Wuu, C. Y. 1982. A finite time parameter estimation scheme for system identification and signal processing with unknown delays. Ph.D. diss., Brown University.

Wuu, C. Y., and A. E. Pearson. 1983. On time delay estimation involving received signals. *ICASSP 83* 2:871–74.

Yaglom, A. M. 1962. *An introduction to the theory of stationary random functions.* Englewood Cliffs, N.J.: Prentice Hall.

Youn, D. H. 1982. A class of adaptive methods for estimating coherence and time delay functions. Ph.D. diss., Department of Electrical Engineering, Kansas State University. University Microfilms, Ann Arbor, 1982.

Youn, D. H., N. Ahmed, and G. C. Carter. 1981. A method for generating a class of time-delayed signals. *ICASSP 81* 3:1257–60.

———. 1982a. Estimation of magnitude-squared coherence function: An adaptive approach. *ICASSP 82* 2:1100–3.

———. 1982b. On using the LMS algorithm for time delay estimation. *IEEE Trans. ASSP* 30 (5): 798–801.

———. 1983a. Magnitude-squared coherence function estimation: An adaptive approach. *IEEE Trans. ASSP* 31 (1, pt. 1): 137–42.

———. 1983b. An adaptive approach for time delay estimation of band-limited signals. *IEEE Trans. ASSP* 31 (3): 780–84.

Youn, D. H., S.-N. Chiou, and V. J. Mathews. 1986. Adaptive phase transform processors for time delay estimation. *JASA* 80 (1): 188–94.

Youn, D. H., and V. J. Mathews. 1983. On using the short-time unbiased spectral estimation algorithm for estimating time delays and magnitude squared coherence functions. *ASSP Spectrum Estimation Workshop II,* Tampa, 60–64.

Young, H. J. 1978. Underwater sound arrival angle estimation by multiple cross-correlation measurements. *ICASSP 78,* 659–64.

Yu, T. K., J. H. Seinfeld, and W. H. Ray. 1974. Filtering in nonlinear time delay systems. *IEEE Trans. Automatic Control* 19 (4): 324–33.

Yuen, C. K. 1979. Comments on modern methods for spectrum estimation. *IEEE Trans. ASSP* 27 (3): 298–99.

Zakarevicius, R. A. 1979. Group delay estimation with wideband signals. *IEEE Trans. Communications* 27 (12): 1908–10.

Zhang, W., and M. Raghuveer. 1991. Nonparametric bispectrum-based time-delay estimators for multiple sensor data. *IEEE Trans. SP* 39 (3): 770–74.

Zhao, Z., and Z.-Q. Hou. 1984. The generalized phase spectrum method for time delay estimation. *ICASSP 84* 3:46.2/1–4.

Zheng, W.-X., and C.-B. Feng. 1990. Identification of stochastic time lag systems in the presence of colored noise. *Automatica* 26 (4): 769–79. (Original version presented at the 8th IFAC/IFORS Symposium on Identification and System Parameter Estimation, Beijing, People's Republic of China, August 1988.)

Ziomek, L. J., and C. D. Behrle. 1989. Localization of multiple broadband targets via frequency domain adaptive beamforming for planar arrays. *JASA* 85 (3): 1236–44.

Ziv, J., and M. Zakai. 1969. Some lower bounds on signal parameter estimation. *IEEE Trans. Information Theory* 15 (3): 386–91.

Author Index

Index

A

D

E

F

M

N

O

P

R

S

T

U

V

W

Y

Z

Editor's Biography

G. Clifford Carter was born and raised in Oak Park, Illinois; he is a 1967 graduate of the U.S. Coast Guard Academy and earned the M.S.E.E. degree in 1972 writing a thesis on Coherence Estimation. In 1976 he earned the Ph.D. degree in E.E. Both graduate degrees were from the University of Connecticut; his Ph.D. thesis was on Time Delay Estimation.

Dr. Carter was elected Fellow of the IEEE in 1988 for his contributions to the theory of coherence and time delay estimation. He is an electronics engineer at NUWC, New London, and an adjunct professor at the University of Connecticut. He teaches graduate-level courses in digital signal processing, detection, estimation, and communications theory; he has lectured throughout the United States, Canada, and Europe and is an internationally recognized expert in signal processing.

In addition to being a signal processing expert, Dr. Carter is an experienced technical manager and the recipient of numerous awards for his line and program management accomplishments including organizational and program growth, getting signal processing algorithms to sea for testing and development of numerous cooperative ventures with industry. Dr. Carter served in Washington at the Office of Naval Research, where he was a technical area manager for several of the U.S. Navy's applied research programs.